Objective Electrical Engineering 2e

G.K. Mithal

G K Publications (P) Ltd

CL MEDIA (P) LTD.

Revised Edition : 2017
First Edition : 2002

ISBN : **978-93-5144-977-5**

Typeset by : *CL Media DTP Unit*

Administrative and Production Offices

Published by : **CL Media (P) Ltd.**
KH No. 1027, Ramnagar, Salempur,
Rajputana, Roorkee (U.K.)-247667

Marketed by : **G.K. Publications (P) Ltd.**
H-205, Sector-63
NOIDA - 201307 (U.P.)
☎ (0120) - 4267181, 82, 84
Telefax : 4267183

For product information :
Visit ***www.gkpublications.com*** or email to ***gkp@gkpublications.com***

Preface

In the recent years PSUs like CIL, HAL, BSNL, AAI, BHEL, BEL, SAIL, GAIL, IOCL, HPCL, ONGC, DMRC, DRDO, ISRO, Staff Service Selection and others are preferring to hire *Engineers* for various positions *which has resulted* in a large job opportunities for Engineering graduates. Moreover there are more than 37 PSUs hire through the GATE score card. As these PSUs also offer job security and decent perks, many candidates are attracted towards it, gradually increasing the competition level. The campus recruitment tests for all private companies also include a test on technical questions and a large number of such questions are asked in the interview.

Unlike other competitive examinations, preparing for GATE and vacancy based exams of these PSUs is not an easy task as each exam has its own pattern, syllabus and trend. Most of these exams like to throw surprises based on the vacancies, leaving students amazed during their exam preparation with very less time for preparation. However, you can certainly guarantee yourself a smooth journey if you start early and plan it well. As the technical section of these exams hold the maximum scoring scope; it is very important to pay enough attention to this section.

This series focuses on the technical section of various exams conducted by PSUs such as CIL, HAL, BSNL, AAI, BHEL, BEL, SAIL, GAIL, IOCL, HPCL, ONGC, DMRC, DRDO, ISRO and Railways for the Engineers. These books help to prepare for all the exams at a single go which saves time instead of preparing separately for each PSU's exam. This work is a revision of one of the best selling series in this segment. In order to make students thorough with the variety of questions, the series provides them with questions segregated into two sections. The first section includes a set of conceptual questions under each topic and the second section provides previous year's questions of exams such as GATE, IES and various PSUs exams. Each question in the later section has been tagged with the exam name to make the preparation all the more easier. This combination of conceptual questions and previous year's questions would completely solve the purpose of the students for a quick practice with complete preparation for the exam. The books in this series can also be helpful to clear the technical section of various campus recruitment tests.

GKP has also launched an Android App to provide you an update on all upcoming vacancies in the technical segment and it also has a lot of added content to aid your preparation.

We hope this little effort of ours will be helpful in achieving your dreams. If you have any suggestions on improvement of this book, you can write to us at gkp@gkpublications.com.

All the best!

Contents

1 CHAPTER

Electrical and Electronic Measurements

EXERCISE – I

1. The dimension of force in SI system are
   (*a*) LMT^{-1} (*b*) LMT^{-2}
   (*c*) LMT (*d*) LMT^2
2. The dimension of power is
   (*a*) ML^2T^{-2} (*b*) ML^2T^{-3}
   (*c*) M^2LT^{-3} (*d*) M^2LT^{-2}
3. The dimension of torque in SI unit is
   (*a*) ML^2T^2 (*b*) ML^2T^{-2}
   (*c*) $ML^{-2}T^{-2}$ (*d*) $M^2L^2T^{-2}$
4. For defining standard metre, wave length of which material is considered
   (*a*) Helium (*b*) Neon
   (*c*) Xenon (*d*) Krypton
5. The number of basic SI units is
   (*a*) 4 (*b*) 5
   (*c*) 6 (*d*) 7
6. Supplementary units added to the basic SI units are
   (*a*) 2 (*b*) 3
   (*c*) 4 (*d*) 7
7. Farad is the unit of
   (*a*) inductance
   (*b*) voltage
   (*c*) current
   (*d*) capacitance
8. The unit newton/coulomb is the unit of
   (*a*) electric field intensity
   (*b*) electric flux density
   (*c*) electro-motive force
   (*d*) capacitance
9. The dimension of magnetic flux denisty is
   (*a*) IMT^2 (*b*) $I^{-1}MT^{-2}$
   (*c*) IMT^{-1} (*d*) $I^{-1}MT^{-1}$
10. Electrolyte solution in a standard saturated weston cell is
    (*a*) cadmium sulphite
    (*b*) potassium sulphate
    (*c*) magnesium sulphate
    (*d*) zinc sulphate
11. The emf of a saturated Weston cell drops by increase in temperature at a rate of
    (*a*) 4μVm°C (*b*) 40μVm°C
    (*c*) 4mVm°C (*d*) 40mVm°C
12. The internal resistance of a saturated Weston standard cell is of the order of
    (*a*) a fraction of ohm
    (*b*) a few ohm
    (*c*) practically zero ohm
    (*d*) a few kilo-ohm
13. Purely mechanical instruments can be used for dynamic measurements because they have
    (*a*) high inertia
    (*b*) large time constant
    (*c*) high response time
    (*d*) all of these
14. The span of a zero = centered voltmeter having a scale from 10 V to 10 V is
    (*a*) 0 V (*b*) – 10 V
    (*c*) 10 V (*d*) 20 V
15. The resolution of an indicating instrument is
    (*a*) the smallest change in the output reading due to drifting of pointer
    (*b*) the smallest change in the applied stimulers that will indicate the detectable change in the deflection
    (*c*) the difference between various readings for the same applied stimulus
    (*d*) none of these
16. In which part of the scale does the pointer indicate most accurately
    (*a*) In the first third of the scale
    (*b*) In the first half of the scale
    (*c*) In about middle of the scale
    (*d*) In last third of the scale
17. The desirable static characteristics of a measuring system is/are
    (*a*) accuracy (*b*) sensitivity
    (*c*) reproducibility (*d*) all of these

18. The reliability of an instrument refers to

(*a*) measurement of changes due to temperature variation

(*b*) degree to which repeatability continues to remain within specified limits

(*c*) the life of the instrument

(*d*) the extent to which the characteristics remain linear

19. The efficiency of an instrument is defined as the ratio of the measured quantity at full scale of the power taken by the instrument at

(*a*) one-fourth scale

(*b*) half scale

(*c*) three-fourth scale

(*d*) full scale

20. To measure 2 volts, if one selects 0 – 100 volts range voltmeter which is accurate within ± 1%, the error in his/her measurement may be upto

(*a*) ±0.02% (*b*) ± 1%

(*c*) ±2% (*d*) ± 50%

21. A thermometer is calibrated from 150°C to 200°C. The accuracy specified is ±0.25%., the maximum static error in the measurement is

(*a*) ±0.5°C (*b*) 0.375°C

(*c*) ±0.125°C (*d*) ±0.0125°C

22. The degree of reproductively among several independent measurements of same true value under reference conditions is known as

(*a*) accuracy (*b*) precision

(*c*) linearity (*d*) calibration

23. The errors committed by a person in the measurement are

(*a*) gross errors

(*b*) random errors

(*c*) instrumental errors

(*d*) environmental errors

24. The current I through a resistance R is measured with uncertainties

$$I = 4A \pm 0.5\%, R = 100\Omega \pm 0.2\%$$

The uncertainty in measurement of power is

(*a*) 1600W ± 0.1%

(*b*) 1600W ± 0.02%

(*c*) 1600W ± 0.05%

(*d*) 1600W ± 1.2%

25. A temperature probe having a first order response with a time constant of 1 second is given a step input from 50° to 0°C. The temperature in °C after 0.6 second is

(*a*) 18.4 (*b*) 25

(*c*) 27.4 (*d*) 45

26. Which of the following is absolute instrument?

(*a*) Power factor meter

(*b*) Ammeter

(*c*) Wattmeter

(*d*) Tangent galvanometer

27. An instrument which gives total quantity of energy passed through in a given time is called

(*a*) integrating instrument

(*b*) indicating instrument

(*c*) recording instrument

(*d*) digital instrument

28. Which of the following types of instrument is an integrating instrument?

(*a*) Powor factor meter

(*b*) Energy meter

(*c*) Wattmeter

(*d*) Frequency meter

29. The torque produced in an indicating instrument by utilizing magnetic, electrodynamic, thermal, chemical and electrostatic effect is known as

(*a*) controlling torque

(*b*) deflecting torque

(*c*) damping torque

(*d*) restoring torque

30. The controlling torque in a spring controlled meter is proportional to

(*a*) θ (*b*) θ^2

(*c*) $1/\theta$ (*d*) $1/\theta^2$

31. At a steady deflected position of an indicating instrument, the moving system is subjected to

(*a*) deflecting torque only

(*b*) deflecting and controlling torque

(*c*) deflecting, controlling and damping torque

(*d*) damping and controlling torque

32. An instrument gives maximum deflection for any amount of quantity passed through it. Which of following pair is present?

(*a*) Deflecting and controlling force

(*b*) Deflecting and damping force

(*c*) Damping and controlling force

(*d*) Damping, controlling and deflecting force

33. Which of the following set of torques is provided in deflection galvanometer?

(*a*) Deflecting and controlling

(*b*) Controlling and damping

(*c*) Deflecting and damping

(*d*) Deflecting, controlling and damping

34. A hair spring attached to the moving system is used

(*a*) damping torque

(*b*) controlling torque

(*c*) balancing torque

(*d*) deflecting torque

35. The movement of the moving element of an electrical indicator is dependent on

(*a*) restoring torque

(*b*) number of turns on the coil

(*c*) resistance of the indicator circuit

(*d*) all of these

36. Three force responsible for reduction of oscillations of the pointer in an ammeter is

(*a*) controlling force (*b*) damping force

(*c*) deflecting force (*d*) none of these

37. If the damping torque is more than the critical damping, the instrument is called

(*a*) underdamped

(*b*) overdamped

(*c*) under critically damped

(*d*) over critically damped

38. Air friction damping is used in the instrument which is

(*a*) moving iron

(*b*) moving coil

(*c*) induction

(*d*) hot wire

39. The controlling torque in gravity controlled meter is proportional to

(*a*) cos θ

(*b*) sin θ

(*c*) tan θ

(*d*) θ

40. The purpose of providing a mirror behind the pointer in a measuring instrument is

(*a*) the scale is illuminated through mirror

(*b*) with the help of mirror it can be seen weather the pointer is bend or not

(*c*) The mirror is semi-transparent 50 as the allow the observation of the interior of the instrument

(*d*) Reading errors due to inclined observations are eliminated by removing parallax between the pointer and its image in the mirror

41. The internal resistance of the milli-ammeter must be very low for

(*a*) high sensitivity

(*b*) high accuracy

(*c*) maximum voltage drop across the meter

(*d*) minimum effect on the current in the circuit

42. Consider the following statements.

The causes of error in the measurement of temperature using a thermistor are

1. selfheating
2. poor sensitivity
3. non-linear characteristics

Of these statements

(*a*) 1, 2 and 3 are correct

(*b*) 1 and 2 are correct

(*c*) 2 and 3 are correct

(*d*) 1 and 3 are correct

43. A 35V dc supply is connected across a combined resistance of 600 ohms and an unknown resistance of R ohms in series. A Voltmeter having a resistance of 1.2 μ Ω is connected across 600 ohms resistor and reads 5V. The resistance R will be

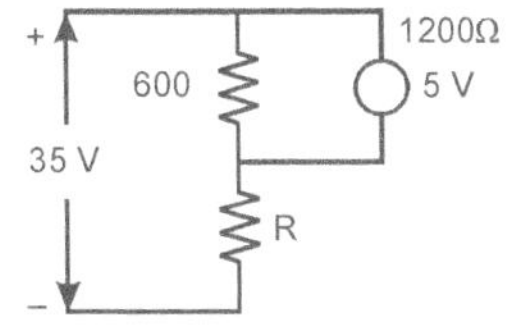

(*a*) 120 ohms (*b*) 500 ohms

(*c*) 1.7 K ohms (*d*) 2.4 K ohms

44. The important characteristics of a frequency counter are given below:

1. Time base accuracy
2. Least significant bit count
3. Gain of the input amplifier.

The more important characteristic(s) responsible for the overall accuracy of frequency measurement using the counter would include

(*a*) 1 and 2 (*b*) 2 and 3

(*c*) 1 and 3 (*d*) 2 alone

45. Which of the following will improve the mutual coupling between primary and secondary circuits?

(*a*) Transformer oil of high break down voltage

(*b*) High reluctance magnetic core

(*c*) Winding material of high resistivity

(*d*) Low reluctance magnetic core

46. In a dual beam oscilloscope

(*a*) there are two separate vertical input and two separate horizontal inputs

(*b*) there are two separate vertical inputs and there is only one set of horizontal deflection plates

(*c*) there is only one vertical input but there are two separate horizontal deflection plates

(*d*) there is only one vertical and one horizontal input

47. A resistance of 105 ohms is specified using significant figures as indicated below:

1. 105 ohms
2. 105. ohms
3. 0.000105 Mega ohms

Among these,

(*a*) 1 represents greater precision than 2 and 3

(*b*) 2 represents greater precision but 1 and 3 represent same precision

(*c*) 2 and 3 represent same precision

(*d*) none of these

48. The given figure shows that the bridge connection for frequency measurement. C and R are variables and ganged together. For balanced condition, the expression for frequency is $f = \dfrac{1}{2\pi \text{ CR}}$ when

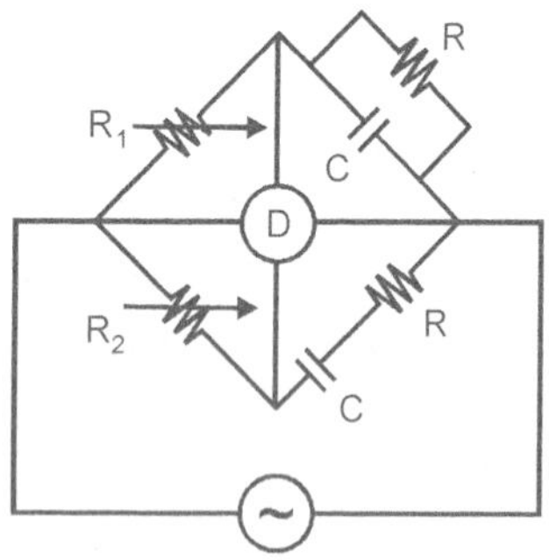

(*a*) $R_1 = R_2$ (*b*) $R_1 = 2R_2$

(*c*) $R_1 = \dfrac{R_2}{2}$ (*d*) $R_1 = 3R_2$

49. The arms of a wheatstone bridge are shown in the given figure. For the balanced conditions, the least tolerance value of R_4 will be

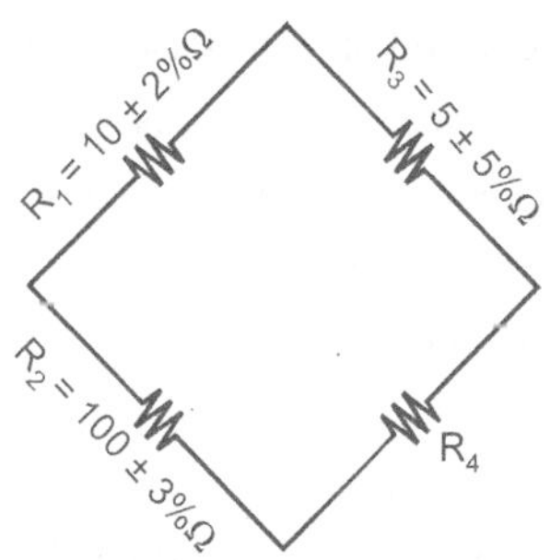

(*a*) $50 \pm 2\%\ \Omega$

(*b*) $50 \pm 3\%\ \Omega$

(*c*) $50 \pm 5\%\ \Omega$

(*d*) $50 \pm 10\%\ \Omega$

50. A strain gauge bridge measures the strain in a cantilever where the gauge is fixed. With strain $\in$, the gauge resistance increases form $110\ \Omega$ to $110.52\ \Omega$. If the gauge factor is 2.03, then the strain in the cantilever will be

(*a*) 2.32×10^{-3}

(*b*) $3\ 15 \times 10^{-3}$

(*c*) 3.81×10^{-3}

(*d*) 433×10^{-3}

51. Match List-I with List-II and select the correct answer using the codes given below the Lists:

	List-I		**List-II**
A	Low value of R	1.	Scherring bridge
B.	High-Q inductor	2.	Maxwell bridge
C.	Low-Q inductor	3.	Kelvin double bridge
D.	High voltage capacitors	4.	Hay bridge

Codes:

	A	B	C	D
(*a*)	1	2	4	3
(*b*)	1	4	2	3
(*c*)	3	2	4	1
(*d*)	3	4	2	1

52. The current through the current coil of a wattmeter is given by,

$$i = (1 + 2 \sin \omega t)\text{ A}$$

and the voltage across the pressure coil is,

$$v = (2 + 3 \sin 2\,\omega t)\text{ V.}$$

The wattmeter will read

(*a*) 8.00 W

(*b*) 5.05 W

(*c*) 2.0 W

(*d*) 1.0 W

53. A dc supply of 35 V is connected across 600 ohm resistance in series with an unknown resistance. A voltmeter having an impedance of 1.2 k Ω is connected across 600 ohm resistance which reads 5V. The unknown resistance is of

(*a*) 500 Ω

(*b*) 1.7 kΩ

(*c*) 2.2 kΩ

(*d*) 2.4 kΩ

54. A circuit draws a current I when a single phase ac voltage V is applied to it. If the power factor is cos ϕ, then the dimensions of VI cos ϕ would be

(*a*) ML^3T^{-2} (*b*) ML^2T^3

(*c*) ML^3T^3 (*d*) ML^2T^{-3}

55. Consider the following statements:

In a measuring instrument,

1. linearity is more important than sensitivity
2. high precision indicates high accuracy
3. accuracy cannot be better than resolution

Of these statements

(*a*) 1,2 and 3 are correct
(*b*) 1 and 2 are correct
(*c*) 2 and 3 are correct
(*d*) 1 and 3 are correct

56. The force between two charged particles is given by $F = \frac{1}{4\pi \in_0} \cdot \frac{Q_1 Q_2}{r^2}$ where the symbols have their usual meanings. The dimensions of $\in_0$ in free space in S. I. System are

(*a*) $M^{-1}L^{-3}T^2A^4$
(*b*) $M^{-1}L^{-3}T^{-4}A^2$
(*c*) $ML^{-3}T^4A^3$
(*d*) $M^{-1}L^{-3}T^2A^2$

57. Match List I with List II and select the correct answer using the codes given below the lists:

List-I

(*Measuring bridge*)

A. Maxwell bridge
B. Anderson bridge
C. Heaviside Campbell bridge

List-II

(*Unknown self- inductance is compared with*)

1. Known value of self-inductance
2. Known value of capacitor
3. Known value of mutual inductance

Codes:

	A	B	C
(*a*)	1	2	3
(*b*)	1	3	2
(*c*)	3	1	2
(*d*)	2	1	3

58. A band pass signal having the frequency contents in the range of 1 MHz to 2 MHz is to be sampled. The original signal can be reconstructed, if the minimum sampling rate is

(*a*) 10^6 samples/s
(*b*) 2×10^6 samples/s
(*c*) 4×10^6 samples/s
(*d*) 6×10^6 samples/s

59. To measure the angular position θ of a shaft which is free to rotate continuously and if the measurement is to be carried out with high precision, the transducer to be selected would be

(*a*) potentiometer (*b*) synchro
(*c*) resolver (*d*) tachometer

60. The sampling rate of a data acquisition system is 1000 samples per second. To avoid antialising, a low pass filter is to be employed. The cut-off frequency for the low pass filter should be

(*a*) ≤ 500 Hz
(*b*) ≥ 1000 Hz
(*c*) $\geq$ the largest signal frequency
(*d*) $\leq$ the largest signal frequency

61. Match List I with List II and select the correct answer using the codes given below the lists

List I

A. SAR type ADC
B. Flash ADC
C. Dual slope ADC
D. Counter-DAC

List II

1. Setting time dependent on the input
2. Works well even in noisy environment
3. $(2n-1)$ comparators required for n-bit
4. Settling time for n-bits type ADC (single RAMP) is $(n+2)$ t clock

Codes:

	A	B	C	D
(*a*)	3	4	1	2
(*b*)	4	3	1	2
(*c*)	3	4	2	1
(*d*)	4	3	2	1

62. Two sinusoidal signals of frequency f and $3f$ are applied as x and y inputs respectively to an oscilloscope. Which one of the following patterns could be observed on the screen?

(*a*) 

(*b*) 

(*c*) 

(*d*) 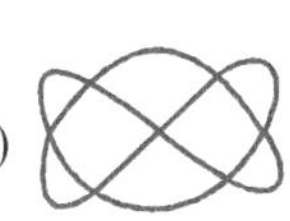

63. Which of the following compensations are provided in the induction type energy-meter for correct measurement of energy?

1. Friction compensation
2. Power factor compensation
3. Frequency compensation
4. Overload compensation

Select the correct answer from the codes given below:

Codes:

(*a*) 1, 2 and 3
(*b*) 1, 2 and 4
(*c*) 1, 3 and 4
(*d*) 2, 3 and 4

64. A figure of the shape appears on a CRT. If the horizontal frequency 60 kHz, then the vertical frequency will be

(*a*) 30 kHz (*b*) 40 kHz
(*c*) 90 kHz (*d*) 120 kHz

65. A rectifier type ammeter indicates a current reading of 7.07 A when a sinusoidal current $i(t) = 10 \sin \omega t$ is passed through it.

A current having a waveform shown in the given figure is passed through the rectifier type instrument.

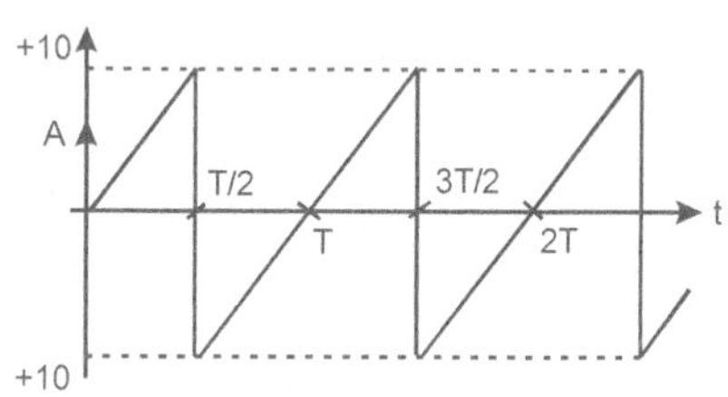

The true RMS current in the circuit is

(*a*) 5.55 A (*b*) 5.77 A
(*c*) 7.07 A (*d*) 5.00 A

66. The figure shows a balanced Maxwell bridge.

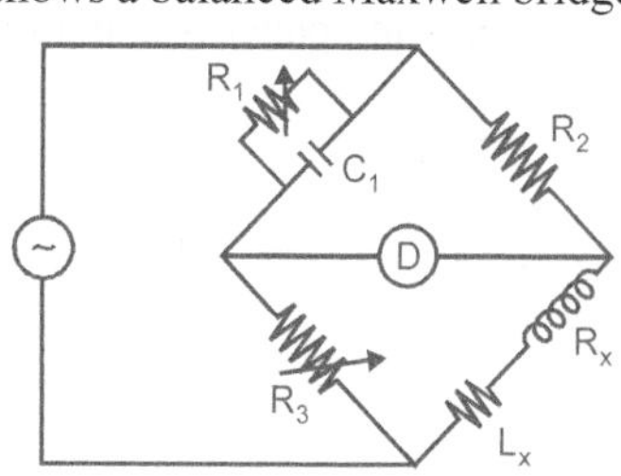

The value of L_x would be

(*a*) $\frac{R_2R_3}{R_1}$ (*b*) $\frac{R_2R_3}{C_1}$

(*c*) $R_1R_2R_3C_1$ (*d*) $R_2R_3C_1$

67. Which one of the following statements about internal resistance of measuring instruments is true?

(*a*) The internal resistance of ammeters should be very small and that of voltmeters very high
(*b*) The internal resistance of ammeters should be very high and that of voltmeters very low
(*c*) The internal resistance of ammeters and voltmeters should be very small
(*d*) The internal resistance of ammeters and voltmeters should be very high

68. Which of the material is used for the swamping resistance that is connected in series with the working coil of a voltmeter?

(*a*) Constantan (*b*) Manganin
(*c*) Eureka (*d*) Nichrome

69. Swamping resistance is used to compensate error due to

(*a*) stray magnetic field
(*b*) temperature variations
(*c*) large supply variations
(*d*) none of these

70. Eddy current damping cannot be used for moving iron instruments because

(*a*) weight to the instrument will increase
(*b*) presence of permanent magnet required for this purpose will affect the deflection and hence the instrument readings
(*c*) size of the instrument will increase
(*d*) eddy current will pass through the iron and thereby course loss

71. The deflecting torque of a moving iron instruments is proportional to

(*a*) I (*b*) I^2
(*c*) $I^{1/2}$ (*d*) $I^{3/2}$

72. In MI instruments, the deflection is proportional to

(*a*) $\frac{\delta L}{\delta \theta}$ (*b*) $\frac{\delta \theta}{\delta L}$

(*c*) $\frac{\delta^2 L}{\delta \theta^2}$ (*d*) $\frac{\delta^2 \theta}{\delta L^2}$

73. The deflecting torque of a moving iron instrument is

(*a*) $I^2 \frac{\delta L}{\delta \theta}$ (*b*) $\frac{1}{2} I^2 \frac{\delta L}{\delta \theta}$

(*c*) $I \frac{\delta L}{\delta \theta}$ (*d*) $\frac{1}{2} I \frac{\delta L}{\delta \theta}$

74. Four voltmeters W, X, Y, Z are specified as follows

Voltmeter W : 100 V, 2 mA

Voltmeter X : 100 V, 100 Ω/V

Voltmeter Y : 10000 Ω, 1 mA

Voltmeter Z : 100 V, 1 mA

The most suitable voltmeter for measurement of 100 V

(*a*) W (*b*) X

(*c*) Y (*d*) Z

75. A spectrum analyser displays

(*a*) different frequency amplitudes w.r.t. time

(*b*) p-p amplitude of modulating signal

(*c*) different signal amplitudes w.r.t frequency

(*d*) Lissajous figures

76. When, C_1, C_2, C_3, C = Capacitance

M_1, M_2, M = Mutual inductances

r, R_1, R_2, R_3, R_4 = resistances

L = Self inductance

ω = angular frequency

which of the following is dimenssionally incorrect?

(*a*) $C = \dfrac{1}{\omega^2 M}$

(*b*) $r = \dfrac{R\,\omega^2 M_1 M_2}{R_2 + \omega^2 L^2}$

(*c*) $R_4 = \dfrac{R_2\left(1+\omega^2 R^2 C_3^2\right)}{\omega^2 R_1 R_2 C_3^2}$

(*d*) $I = \dfrac{V\omega M}{\left[(\omega M + R_2)^2 + \omega^2 L_1 L_2 R_1^{\,2}\right]^{1/2}}$

77. A digital voltmeter has a read-out range from 0 to 9999 counts. When full scale reading is 9.999 V, the resolution of the full scale reading is

(*a*) 0.001

(*b*) 1000

(*c*) 3 digit

(*d*) 1mV

78. The secondary of a current transformer under operating conditions is short-circuited to avoid

(*a*) break in primary winding

(*b*) insulation break-down

(*c*) core saturation and high voltage induction

(*d*) high voltage surge.

79. Hysteresis motors are used for Hi-Q record players because of their

(*a*) constant (synchronous) speed

(*b*) extremely steady torque

(*c*) insensitivity to supply voltage fluctuations

(*d*) non-dependence on centrifugal switch requirement

80. The resistances of two coils of a wattmeter are 0.01 ohm and 1000 ohms respectively and both are non-inductive. The load current is 20 A and the voltage across the load is 30 V. In one of the two ways of connecting the voltage coil, the error in the reading would be

(*a*) 0.1% too high

(*b*) 0.2% too high

(*c*) 0.15% too high

(*d*) zero

81. Consider the following statements regarding the causes of error in current transformers:

1. Some exciting mmf is required by the primary winding to produce of flux.
2. The flux density in the core is not a linear function of the magnetising force
3. There is some magnetic leakage in the secondary winding
4. There is significant power consumption in the metering current

Of these statements

(*a*) 1 and 2 are correct

(*b*) 1, 2 and 3 are correct

(*c*) 2 and 4 are correct

(*d*) 1, 3 and 4 are correct

82. Match List-I with List-II and select the correct answer by using the codes given below the lists:

Electrical Thermal System Analogy

List-I	List-II
A. Resistance	1. Temperature
B. Voltage	2. Heat flow rate
C Charge	3. Resistance
D. Current	4. Heat flow

Codes:

	A	B	C	D
(*a*)	1	2	3	4
(*b*)	2	1	4	3
(*c*)	3	1	4	2
(*d*)	4	1	2	3

83. Holes are drilled on the opposite sides of the disc of an induction type energy meter to

(*a*) avoid creep on no load

(*b*) balance the dise

(*c*) dissipate heat energy due to eddy currents

(*d*) increase the deflecting torque

84. A resistance strain gauge with a gauge factor of 3 is fastened to a steel member subjected to a stress of 100 N/mm^2. The modulus of elasticity of steel is approximately 2×10^3 N/mm^2. The percentage change in resistance is

(*a*) 1.5 (*b*) 1

(*c*) 0.15 (*d*) 0.1

85. In an LVDT, the two secondary voltage

(*a*) are independent of the core position

(*b*) vary unequally depending on the core position

(*c*) vary equally depending on the core position

(*d*) are always in phase quadrature

86. The quantity $\frac{1}{\sqrt{\epsilon_o \mu_o}}$ in SI units has the

(*a*) value 330 m/s (*b*) value 1.73×10^4

(*c*) dimension LT^{-1} (*d*) None of these

87. A photo-cell is illuminated by a small bright source placed one meter away. When the same source of light is placed two metres away, the electrons emitted by the photocathode

(*a*) Each carry on quarter of their previous energy

(*b*) Each carry one quarter of their previous moments

(*c*) Are half as numerous

(*d*) Are one quarter as numerous

88. The principle of operation of an LVDT is based on variation of

(*a*) self-inductance (*b*) mutual inductance

(*c*) reluctance (*d*) permeance

89. Suppose the potential and current coil of a wattmeter are changed accidently and supply is given.

Then,

(*a*) wattmeter will not read

(*b*) current coil will burn out

(*c*) potential coil will burn out

(*d*) nothing will happen

90. The resistance of a thermistor is 5000 Ω at 20°C and its resistance temperature coefficient is 0.04/°C. A measurement with a lead resistance of 10Ω will cause an error of

(*a*) 0.05°C (*b*) 0.1°C

(*c*) 0.4°C (*d*) 0.8°C

91. In a digital voltmeter the oscillator frequency is 400 kHz, the ramp voltage falls from 88V to 0V in 20 msec. The number of pulses counted by the counter is

(*a*) 8000 (*b*) 4000

(*c*) 3200 (*d*) 1600

92. When testing a coil having a resistance of 10 ohms, resonance occurred when the oscillator frequency was 10 MHz and the rotating capacitor was set at 500/2 pF. The effective value of Q of the coil is

(*a*) 200 (*b*) 254

(*c*) 314 (*d*) 542

93. A signal of 10 mV at 75 MHz is to be measured. Which of the following instrument can be used?

(*a*) VTVM

(*b*) Cathode ray oscilloscope

(*c*) Moving iron voltmeter

(*d*) Digital multimeter

94. Which one to the following is the main reason for insertion of two strain gauges in the adjacent arms of a bridge circuit?

(*a*) Achievement of higher sensitivity

(*b*) Elimination of radio frequency interference

(*c*) Elimination of effect of environmental temperature

(*d*) To facilitate quick balancing of the bridge

95. A wheatstone bridge requires change of 6 ohms in the unknown arm of the bridge to produce a change in deflection of 3mm of the galvanometer. The sensitivity of the instrument is

(*a*) 0.5 percent (*b*) 2.0

(*c*) 0.5 mm/ohm (*d*) 2.0 ohm/mm

96. Which of the following instrument is free from hysteresis and eddy current errors?

(*a*) Moving iron instrument

(*b*) Electrostatic instrument

(*c*) Moving coil-permanent magnet type instrument

(*d*) Moving coil dynamometer type instrument

97. Joule/Coulomb is the unit of

(*a*) electric field potential

(*b*) potential

(*c*) charge

(*d*) none of these

98. A voltmeter has a uniform scale with 100 divisions. The full-scale reading is 10 V and could be read upto 1/10 of a scale division with some degree of certainty. It's resolution is

(*a*) 0.1 V (*b*) 0.02 V

(*c*) 0.01 V (*d*) 0.001 V

99. In a semiconductor strain gauge, the change in resistance on application of strain is mainly due to the change in its
   (*a*) length
   (*b*) diameter
   (*c*) resistivity
   (*d*) length and diameter

100. In a 5 × 7 dot matrix format
   (*a*) 64 bits are required to store 64 alphanumeric characters
   (*b*) 560 bits are required to store 64 alphanumeric characters
   (*c*) 1120 bits are required to store 64 alphanumeric characters
   (*d*) 2240 bits are required to store 64 alphanumeric characters

101. Which one of the following causes the disc in an induction type of energy meter to rotate in the opposite direction?
   (*a*) The braking magnet is faulty
   (*b*) Both current coil and voltage coil are wrongly connected
   (*c*) Either current coil or voltage coil is wrongly connected
   (*d*) The load is highly reactive

102. Which one of the following statements is not correct?
   (*a*) It is not possible to have precise measurements which are not accurate
   (*b*) Correctness in measurements requires both accuracy and precision
   (*c*) Reproducibility and consistency are expressions that best describe precision in measurements
   (*d*) An instrument with 2% accuracy is better than another with 5% accuracy

103. Values of VSWR measured by terminating the load end of a transmission line first with horn, then with a short and finally with a matched load, will be respectively
   (*a*) 20, 1.9 and 1.01
   (*b*) 20, 1.01 and 1.9
   (*c*) 1.9, 1.01 and 20
   (*d*) 1.9, 20 and 1.01

104. An advantage which a VTVM has over a nonelectronic voltameter is
   (*a*) lower power consumption
   (*b*) lower input impedance
   (*c*) the ability to measure wider ranges of voltage and resistance
   (*d*) greater portability

105. A 3-phase star-connected symmetrical load consumes P watts of power from a balanced supply. If the same load is connected in delta to the same supply, the power consumption will be
   (*a*) P
   (*b*) $\sqrt{3}$ P
   (*c*) 3 P
   (*d*) not determinable from the given data

106. It can be stated that
   (*a*) C.T. operates at almost the same flux density as a P.T.
   (*b*) C.T. operates with a higher flux density than a P.T.
   (*c*) C.T. operates with considerably lower flux density than a P.T.
   (*d*) no generalisation can be made with regard to the flux densities in C.T. and P.T.

107. A 0-300K voltmeter has a guaranteed accuracy of 2 per cent of full-scale reading. If the voltage measured by this instrument is 180V, then the limiting error will be
   (*a*) less than 2%
   (*b*) less than 3% but greater than 2%
   (*c*) less than 4% but greater than 3%
   (*d*) 4%

108. The controlling torque and the electromagnetic damping of the fluxmeter are
   (*a*) small and heavy respectively
   (*b*) heavy and small respectively
   (*c*) both heavy
   (*d*) both small

109. Which one of the following statements regarding Hay bridge and Maxwell bridge is correct?
   (*a*) Hay bridge is same as Maxwell bridge having resistor R_1 in parallel with C_1 in one arm
   (*b*) Hay bridge has R_1, in series with C_1 with Maxwell bridge has R_1 in parallel with C_1
   (*c*) Hay bridge has R_1 in parallel with C_1 and Maxwell bridge has R_1 in series with C_1
   (*d*) Hay bridge and Maxwell bridge are similar but Hay bridge is used to measure unknown capacitance while Maxwell bridge is used to measure unknown inductance

110. If an induction type energy meter runs fast, it can be slowed down by adjusting the
   (*a*) lag
   (*b*) light load
   (*c*) position of braking magnet and making it move closer to the centre of the disc
   (*d*) position of braking magnet and making it move away from the centre of the disc

111. The aperture time of an A to D converter is given by (symbols have the usual meaning)

(*a*) $\frac{E_m}{2\pi f.\Delta E}$ (*b*) $\frac{2\pi f}{\Delta E\ .E_m}$

(*b*) $\frac{\Delta E}{2\pi f\ .E_m}$ (*d*) $\frac{2\pi f.E_m}{\Delta E}$

112. A thermocouple produces a voltage of 50 mV. Its internal resistance is 50 Ω. The resistance of its leads is 10 Ω. The output is read by a PMMC meter having an internal resistance of 120 Ω. The output voltage indicated will be

(*a*) 50mV (*b*) 40mV

(*c*) 33.3mV (*d*) 25.0mV

113. The following units are used in a data acquisition system:

1. Central Processing Unit
2. Memory
3. Transducer
4. ADC

The sequence of placement of these units from input to output is

(*a*) 4, 3, 1, 2

(*b*) 3, 4, 1, 2

(*c*) 3, 4, 2, 1

(*d*) 4, 3, 2, 1

114. An Ayrton shunt is used to make a D'Arsonval galvanometer into a

(*a*) single range voltmeter

(*b*) single range ammeter

(*c*) multi range voltmeter

(*d*) multi range ammeter

115. The instantaneous power taken by a balanced three-phase load supplied from a balanced three-phase source is

(*a*) zero

(*b*) a constant

(*c*) apulsating function with a non-zero average

(*d*) a lternating with a zero average

116. In the measurement of strain by Wheatsone bridge, the maximum sensitivity is obtained by a combination of

(*a*) an active and a dummy strain gauges

(*b*) two active and two dummy strain gauges

(*c*) three active and one dummy strain gauges

(*d*) four active strain gauges

117. Which one of the following is most suitable for measuring very small capacitances?

(*a*) Transformer ratio-arm bridge

(*b*) Schering bridge

(*c*) Wien'sbridge

(*d*) De-Saute bridge

118. Match List-I (Terms) with List-II (Definitions) and select the correct answer using the codes given below the lists:

List-I

A. Resolution

B. Sensitivity

C. Accuracy

D. Precision

List-II

1. Closeness with which the instrument reading approaches the true value
2. Reproducibility of measurements
3. Smallest change in measured value to which the instrument can respond
4. Ratio of response of the instrument to the input variable.

Codes:

	A	B	C	D
(*a*)	4	3	1	2
(*b*)	3	4	2	1
(*c*)	3	4	1	2
(*d*)	4	3	2	1

119. The A to D converter used in a digital instrument could be

1. successive approximation converter type
2. flash converter type
3. dual slope converter type

The correct sequence of the increasing order of conversion time taken by these type is

(*a*) 1, 2, 3

(*b*) 2, 1, 3

(*c*) 3, 2, 1

(*d*) 3, 1, 2

120. An analog voltage signal whose highest significant frequency is 1 kHz is to be digitally coded with a resolution of 0.01% covering the voltage rage of 0 to 10 volts. To avoid loss of information the MINIMUM number of bits in the digital code should be

(*a*) 4 (*b*) 8

(*c*) 10 (*d*) 14

121. The problem of electro static coupling in a transformer is acute at

(*a*) low frequencies

(*b*) power frequencies

(*c*) high frequencies

(*d*) high load on the transformer

122. If the bridge shown in the given figure is balanced, then the value of the inductance 'Z' will be

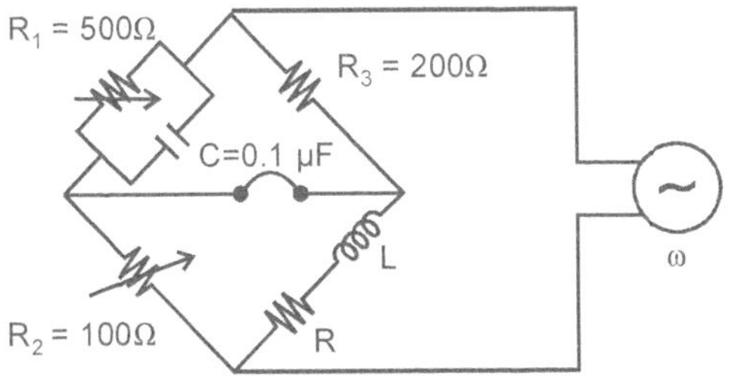

(*a*) 100 mH

(*b*) 20 mH

(*c*) 2 mH

(*d*) dependent on supply frequency ω

123. In measuring the distributed capacitance 'C' of a certain coil using Q-meter, initial resonance at frequency 2 MHz is obtained by turning the capacitance set to 430 pF. If resonance at 4 MHz is obtained by turning the capacitance set to 100 pF, then the value of C_d would be

(*a*) 10 pF

(*b*) 20 pF

(*c*) 100 pF

(*d*) 200 pF

124. The residual voltage at the output of a commercial LVDT is due to

(*a*) improper construction of the core

(*b*) unequal secondaries

(*c*) interwinding capacitances and harmonics generated by the core

(*d*) primary and secondary winding resistances

125. Input impedance of a CRO measured at 10 kHz is Z_1 and at 10 MHz is Z_2. Which one of the following-relations is correct?

(*a*) $|Z_1| > |Z_2|$ (*b*) $|Z_1| = |Z_2|$

(*c*) $|Z_1| < |Z_2|$ (*d*) $|Z_1| = 1000|Z_2|$

126. The principle of bridge duplex working involves

(*a*) potential balance

(*b*) current balance

(*c*) resistance balance

(*d*) impedance balance

127. Match List I with List II and select the correct answer using the codes given beiow the lists:

List-I (*Types of A/D converters*)

A. Dual Slope

B. Counter-Ramp

C. Successive Approximation

D. Simultaneous

List-II (*Properties of A/D converters*)

1. Fixed conversion time; depends on the number of bits
2. High speed operation
3. Hum rejection
4. Conversion time dependent on signal amplitude
5. Large conversion time

Codes:

	A	B	C	D
(*a*)	3	2	5	4
(*b*)	2	3	4	1
(*c*)	3	4	1	2
(*d*)	4	1	2	5

128. How many steps are there in the output of a 10-bit D/A converter?

(*a*) 1023 (*b*) 1024

(*c*) 10 (*d*) 100

129. Rectifier Moving Coil Instruments respond to

(*a*) peak value, irrespective of the nature of the waveform

(*b*) average value, for all waveforms

(*c*) rms value for all waveforms

(*d*) rms value, for symmetrical square waveforms

130. In the bridge given in figure, the reading of the high impedance voltmeters is

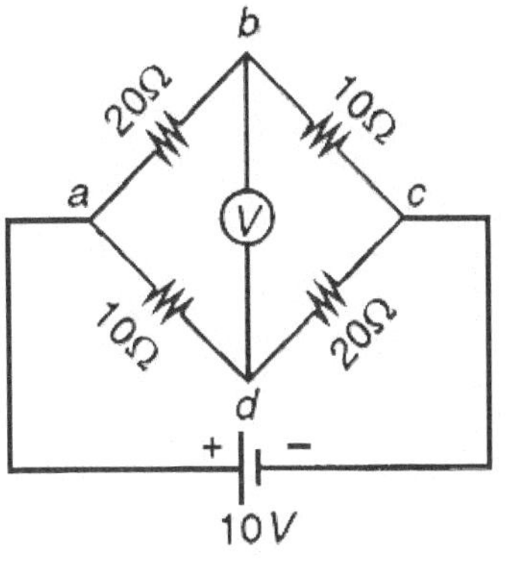

(*a*) zero

(*b*) 6.66 K

(*c*) 4.20 V

(*d*) 3.33 V

131. A resistance strain gauge with a gauge factor of 2 is fastened to a steel member subjected to a stress of 10.5×10^4 kN/m^2. The modulus of elasticity of steel is 2.1×10^3 kN/m^2. The change of resistance due to stress in the strain gauge is

(*a*) 0.1% (*b*) 0.2%

(*c*) 1% (*d*) 10%

132. When a current $i(f) = 5 + 10 \cos 314t$ is measured by an electrodynamic ammeter, the meter will read

(*a*) $\sqrt{5^2 + 10^2}$ (*b*) $\sqrt{5^2 + \frac{10^2}{2}}$

(*c*) $\sqrt{5^2 + \left(\frac{10}{2}\right)^2}$ (*d*) $\sqrt{\frac{5^2 + 10^2}{2}}$

133. In a PMMC instrument, the torque/weight ratio is

(*a*) high (*b*) low

(*c*) zero (*d*) infinity

134. In PMMC instruments, the scale is

(*a*) non-linear

(*b*) logarithmic

(*c*) exponential

(*d*) uniformlly divided

135. The PMMC meter can measure

(*a*) only ac quantities

(*b*) only dc quantities

(*c*) both ac and dc quantities

(*d*) only very high frequency quantities

136. The scale of a dynamometer type instrument marked in terms of rms value would be

(*a*) uniform through out

(*b*) non uniform crowded near full scale

(*c*) non uniform crowded at the beginning

(*d*) non uniform corwded around mid-scale

137. A meter that is capable of measuring dc only is

(*a*) moving coil

(*b*) moving iron

(*c*) thermo-couple

(*d*) none of these

138. The type of instrument to be used for measuring dc voltage is

(*a*) moving coil meter

(*b*) dynamometer type meter

(*c*) inductive type meter

(*d*) rectifier type instrument

139. Which of the following instruments will have poorest overloading capacity?

(*a*) Moving coil instruments

(*b*) Inductive type instruments

(*c*) Permanent magnet instruments

(*d*) Hot wire instruments

140. Which type of the instruments has highest power consumption?

(*a*) Moving iron

(*b*) PMMC

(*c*) Hot wire

(*d*) All of these instruments have identical power consumption

141. Which of the following are used to increase the range of ammeters?

(*a*) Multipliers

(*b*) Shunts

(*c*) Control spring

(*d*) Potential transformers

142. A moving instrument has a resistance of 0.6 Ω and a full scale deflection current of 0.1 A. To convert it into an ammeter of 0-15 A range, the resistance of the shunt should be

(*a*) 0.6 Ω (*b*) 0.06 Ω

(*c*) 0.1 Ω (*d*) 0.004 Ω

143. Measuring range of a voltmeter can be extended by using

(*a*) high shunt resistance

(*b*) high series resistance

(*c*) low shunt resistance

(*d*) low series resistance

144. A current $i = (10 + 10 \sin t)$ amperes is passed through an ideal moving iron type ammeter. Its reading will be

(*a*) zero

(*b*) 10 A

(*c*) $\sqrt{150}$A

(*d*) $10\sqrt{2}$A

145. If two 300 V full-scale voltmeters V_1 and V_2 having sensitivities of 100*k* Ω/V and 150*k* Ω/V are connected in series to measure 500V, then

(*a*) V_1 and V_2 will read 250V each

(*b*) V_1 will read 200 V and V_2 will read 300 V

(*c*) V_1 will read 300V and V_2 will read 200 V

(*d*) V_1 and V_2 will read 0V each

146. A-7 bit successive approximation DVM will convert the analog input into digital output in a period of

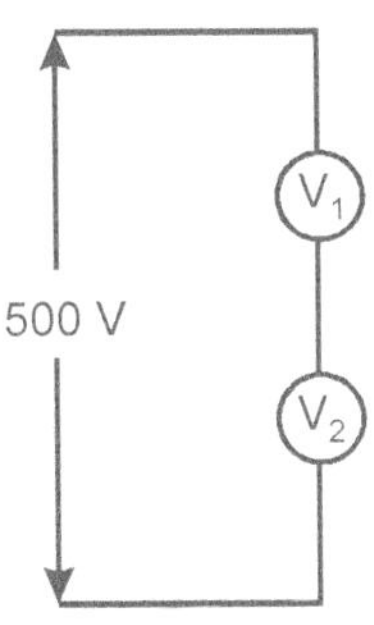

(*a*) 7 clock pulses

(*b*) 8 clock pulses

(*c*) $\frac{\text{input signal amplitude}}{\text{full-scale output amplitude}} \times 7$ clock pulses

(*d*) $\frac{\text{input signal amplitude}}{\text{full-scale output amplitude}} \times 8$ clock pulses

147. For a thermocouple pair (A, B), the extension wires (C, D)

(*a*) should be identical pair elements

(*b*) should have identical temperature-emf relationship

(*c*) can be of any two dissimilar materials

(*d*) should have very smal temperature emf sensitivity

148. A 230V, 10A, single-phase energymeter makes 90 revolutions in 3 minutes at half load rated voltage and unity pF. If the meter constant is 1800 revolutions /kWh, then its error at half load will be

(*a*) 13.04% slow

(*b*) 13.04% fast

(*c*) 15% slow

(*d*) 15% fast

149. For the voltmeter circuit shown in the given figure, the basic D' Arsonval meter used has full-scale current of 1 mA and meter resistance (R_m) of 100 ohms. The values of the series resistance R_1 and R_2 required for 10V range and 50V range will be respectively

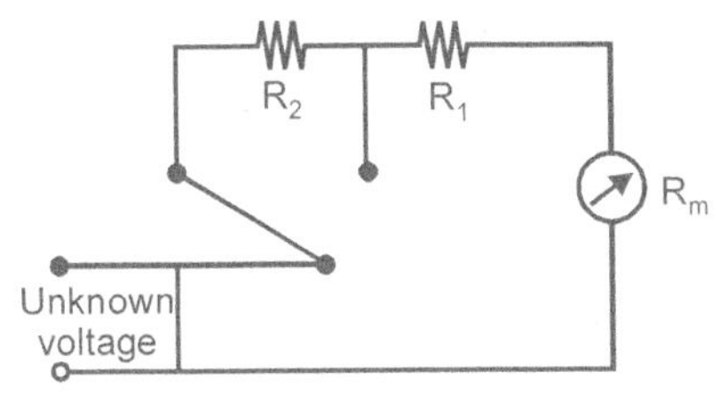

(*a*) 9.9 kΩ and 40 kΩ

(*b*) 10 kΩ and 50 kΩ

(*c*) 20 kΩ and 30 kΩ

(*d*) 200 kΩ and 250 kΩ

150. Which of the following units are present in a spectrum analyser?

1. Mixer
2. Saw-tooth generator
3. Local oscillator

Select the correct answer using the codes given below:

Codes:

(*a*) 1, 2 and 3

(*b*) 1 and 2

(*c*) 1 and 3

(*d*) 2 and 3

151. If the practical units of voltage and current were each made 20 times as large as they are at present, what would be the consequent alteration in the size of the unit of capacitance?

(*a*) 200 times (*b*) 60 times

(*c*) 20 times (*d*) Nil

152. In a flux meter, the controlling torque is

(*a*) produced by weights attached to be moving coil

(*b*) produced by spring

(*c*) not provided at all

(*d*) provided by crossed coil mechanism

153. The X – and Y – inputs of a CRO are respectively –V sin ωt and –V sin ωt. The resulting Lissajous pattern will be

(*a*) a straight line

(*b*) a circle

(*c*) an ellipse

(*d*) a figure of eight

154. An indicating instrument is more sensitive if its torque to weight ratio is

(*a*) much larger than unity

(*b*) of the order unity

(*c*) much less than unity

(*d*) made deflection - dependent

155. In a two-wattmeter method of measuring power, one of the wattmeters is reading zero watts. The power factor of the circuit is

(*a*) zero (*b*) 1

(*c*) 0.5 (*d*) 0.8

156. In distortion factor meter, the filter is used to suppress

(*a*) dc component

(*b*) odd harmonics

(*c*) even harmonics

(*d*) fundamentals

157. Which of the following measurements can be made with the help of a frequency counter?

1. Fundamental frequency of input signal
2. Frequency components of the input signal at least upto third harmonic
3. Time interval between two pulses
4. Pulse width

Select the correct answer using the codes given below:

Codes:

(*a*) 1, 3 and 4
(*b*) 1, 2 and 3
(*c*) 2 and 4
(*d*) 1 and 2

158. The limiting errors of measurement of power consumed by and the current passing through a resistance are ±* 1.5% and ± 1% respectively. The limiting error of measurement of resistance will then be

(*a*) ±0.5%
(*b*) ±1.0%
(*c*) ±2.5%
(*d*) ±3.5%

159. Doppler shift principle is used in the measurement of

(*a*) temperature
(*b*) frequency
(*c*) speed
(*d*) pressure

160. The recording head in a magnetic tape responds to

(*a*) electrical signal and creates a magnetic signal
(*b*) thermal signal and creates a magnetic signal
(*c*) magnetic signal and creates an electrical signal
(*d*) thermal signal and creates an electrical signal

161. The bandwidth requirement of an FM telemetry channel is

(*a*) equal to that of an AM telemetry channel
(*b*) smaller than that an AM telemetry channel
(*c*) about 100 times that of an AM telemetry channel
(*d*) about ten times that of an AM telemetry channel

162. In the figure shown below, the voltmeter has a total resistance of 50 kΩ and reads 50 V, then the

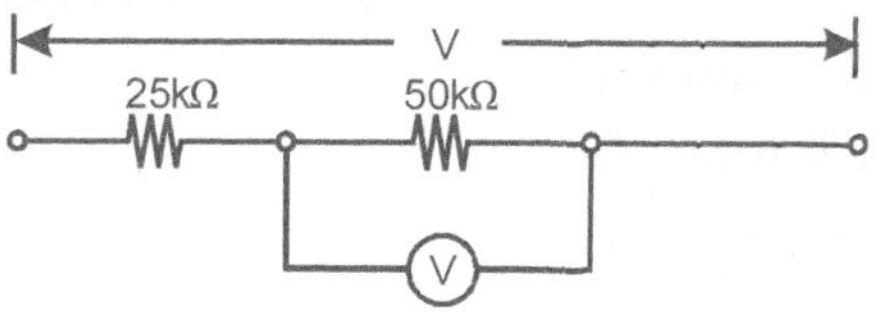

(*a*) 50 V (*b*) 25 V
(*c*) 100 V (*d*) 15 V

163. The voltmeter of sensitivity 1000 Ω/V is connected across the 50 kΩ resistor in the circuit shown in the figure. The voltmeter reading will be

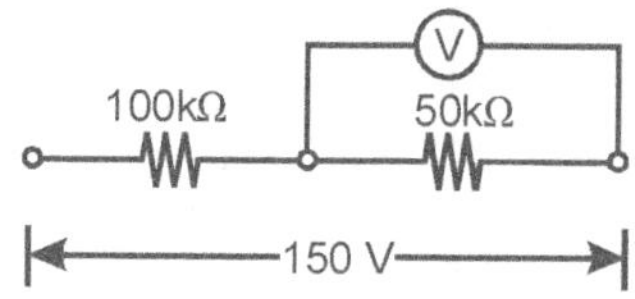

(*a*) 50 V (*b*) 45 V
(*c*) 30 V (*d*) 25 V

164. Which of the following you will prefer to extend the range of an ac voltmeter?

(*a*) Low series resistance
(*b*) High resistance in parallel
(*c*) CT
(*d*) PT

165. High ac voltages are usually measured with

(*a*) magnetic voltmeter
(*b*) inductive voltmeter
(*c*) potential transformers with voltmeters
(*d*) current transformers and voltmeters

166. The power measurement in balanced 3ϕ circuit can be done by

(*a*) one wattmeter method
(*b*) two wattmeter method
(*c*) three wattmeter method
(*d*) any of the above methods

167. In the measurement of 3-phase power by two wattmeter method, if the two wattmeter readings are equal, the power factor of the circuit is

(*a*) 0.8 lagging (*b*) 0.8 leading
(*c*) zero (*d*) unity

168. In a two wattmeter method of measuring power in a 3-phase system one of the wattmeters reads negative implying

(*a*) wattmeter connection is faulty
(*b*) load is unbalanced
(*c*) power flow is in the reverse direction
(*d*) power factor is less than 0.5

169. W_1 and W_2 are the readings of two wattmeters used to measure power of a 3-phase balanced load. The reactive power drawn by the load is

(*a*) $W_1 + W_2$
(*b*) $W_1 - W_2$
(*c*) $\sqrt{3}(W_1 + W_2)$
(*d*) $\sqrt{3}(W_1 - W_2)$

170. Induction wattmeters can be used to measure

(*a*) ac power (*b*) dc power

(*c*) ac or dc power (*d*) none of these

171. Integrating meters are used for the measurement of

(*a*) voltage (*b*) current

(*c*) ac or dc power (*d*) energy

172. Creeping is the phenomenon which occurs in

(*a*) voltmeter (*b*) wattmeter

(*c*) energy meter (*d*) ammeter

173. Match List-I with List-II and select the correct answer using the codes given below the Lists:

List-I	**List-II**
A. Cesium	1. Standard resistance
B. Manganin	2. Atomic clock
C. Standard voltage	3. Instrument interface
D. IEEE4888	4. Josephson junction

Codes:

	A	B	C	D
(*a*)	2	1	3	4
(*b*)	2	1	4	3
(*c*)	1	2	3	4
(*d*)	1	2	4	3

174. If one of the central springs of a Permanent Magnet Moving Coil ammeter is broken then when connected it will read

(*a*) zero

(*b*) half of the correct value

(*c*) twice the correct value

(*d*) an infinite value

175. LCD displays are preferred over LED displays because they

(*a*) are more reliable

(*b*) consume less power

(*c*) respond quickly

(*d*) are cheaper

176. High resistances are provided with a guard terminal in order to

(*a*) protect the resistance against stray electrostatic field

(*b*) bypass the leakage current

(*c*) protect the resistance against overloads

(*d*) protect the resistance against stray electromagnetic field

177. In the case of the RLC circuit shown in the given figure, the voltage across the R, L and C would be respectively

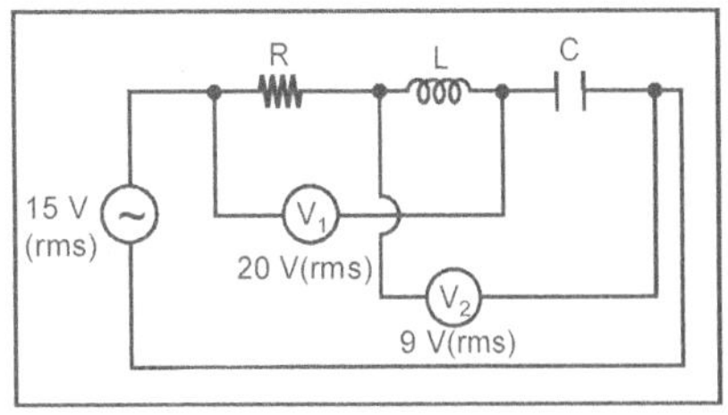

(*a*) 12V, 16V and 7V or 25V

(*b*) 16V, 12V and 7V or 25V

(*c*) 7K, 16V and 12V

(*d*) 16V, 12V and 25V

178. Match List-I with List-II and select the correct answer using the codes given below the lists:

List-I	**List-II**
A. Resistance	1. $[M^{-1}L^{-2}T^{4}I^{2}]$
B. emf	2. $[M^{-1}L^{-2}T^{2}I^{-2}]$
C. Capacitance	3. $[MLT^{-3}I^{-1}]$
	4. $[MLT^{-3}I^{-2}]$

Codes:

	A	B	C
(*a*)	1	2	3
(*b*)	4	2	1
(*c*)	3	4	2
(*d*)	4	3	1

179. In order to obtain fast responsed the meters should have

(*a*) no damping

(*b*) a damping coefficient of > 1

(*c*) a damping coefficient of about 0.7

(*d*) critical damping

180. Which one of the following resistance configurations is best suited for the construction of a low resistance?

(*a*) 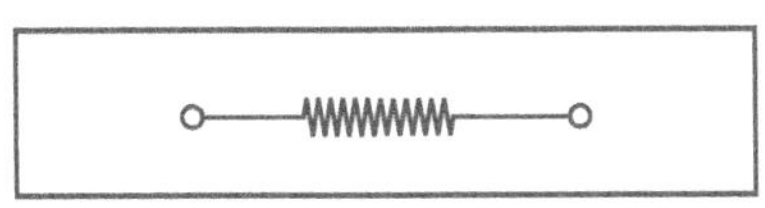

(*b*) 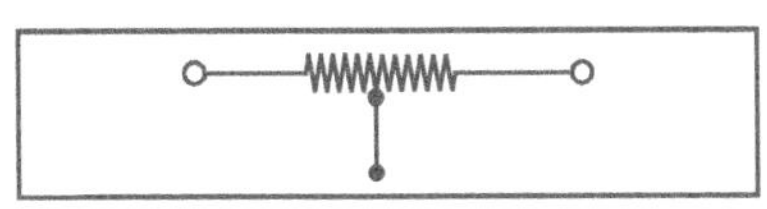

(*c*) 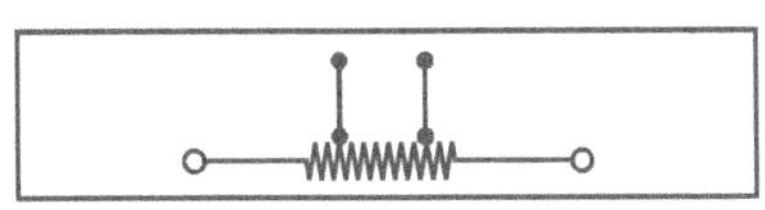

(*d*) 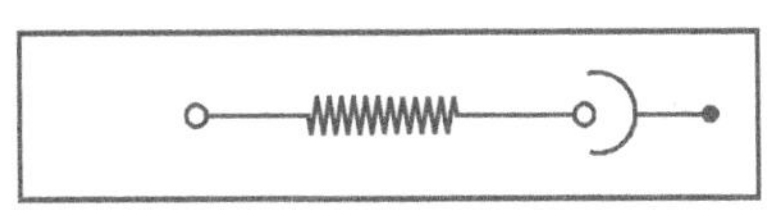

181. Half effect device can be used to
(*a*) multiply two signals
(*b*) divide one signal by another on an instantaneous basis
(*c*) add two signals
(*d*) substract one signal from another

182. Dummy strain gauge is used in conjunction with the main strain gauge to
(*a*) calibrate the system
(*b*) compensate temperature effects
(*c*) improve sensitivity
(*d*) reduce strain on the main gauge

183. Match List-I (Transducer) with List-II (Input/ Output variables) and select the correct answer using the codes given below the Lists:

List-I
A. Electrodynamic generator
B. Venturimeter
C. Pirani gauge
D. Spring balance

List-II
1. Gas pressure to resistance change
2. Force to displacement
3. Motion to voltage
4. Flow rate to pressure

Codes:

	A	B	C	D
(*a*)	2	1	4	3
(*b*)	2	4	1	3
(*c*)	3	4	1	2
(*d*)	3	4	2	1

184. In a Q-meter, a small resistance R is added to the series resonance circuit to inject the oscillatory voltage to the circuit. If R_s is the apparent series resistance of the circuit at resonance, then the value of the actual Q will be equal to

(*a*) observed Q $\dfrac{1}{1+\dfrac{R}{R_s}}$

(*b*) observed Q $\left(1+\dfrac{R}{R_s}\right)$

(*c*) observed Q $\dfrac{1}{1+\dfrac{R_s}{R}}$

(*d*) observed Q $\left(1+\dfrac{R_s}{R}\right)$

185. What voltage would a voltmeter with impedance 20000 Ω and range 0–1 V shown in the circuit given below?

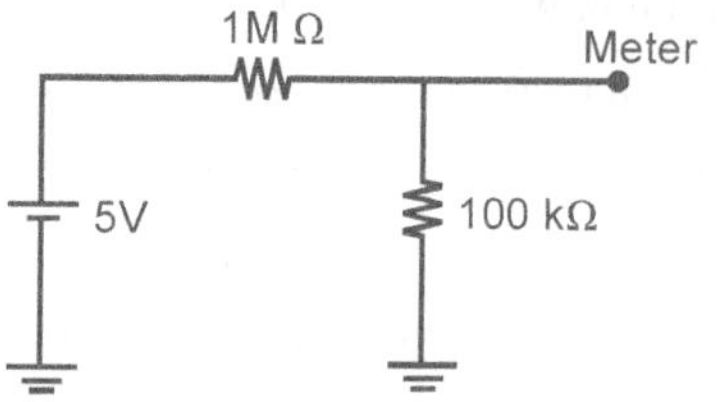

(*a*) 82mV (*b*) 100mV
(*c*) 118mV (*d*) 5V

186. The resistance from 0.1 Ω in to 1 .0 Ω can be measured easily with precision better than 0.5% by
(*a*) differential galvanometer
(*b*) Kelvin double bridge
(*c*) potentiometer
(*d*) Wheatstone bridge

187. Which of the following is used to measure the resistance of an electrical installation?
(*a*) Multimeter
(*b*) Ammeter and voltmeter
(*c*) Maxwell's bridge
(*d*) Megger

188. The insulation resistance of a domestic wiring is to be measured. The instrument preferred is
(*a*) Kelvin's double bridge
(*b*) ohm-meter
(*c*) potentiometer
(*d*) megger

189. The most commonly used null detector in power frequency ac bridge is a
(*a*) vibrational galvanometer
(*b*) D'Arsonval galvanometer
(*c*) Ballistic galvano meter
(*d*) tachometer

190. Anderson bridge is used for the measurement of
(*a*) time period
(*b*) phase difference
(*c*) inductance
(*d*) capacitance

191. Which of the following has low resistance?
(*a*) Armature resistance of large generator
(*b*) 60Ω lamp resistance
(*c*) Field winding resistance of alternator
(*d*) Insulation resistance of machine

192. Which instrument has the lowest resistance?

(*a*) Ammeter (*b*) Voltmeter

(*c*) Frequency meter (*d*) Megger

193. An ohmmeter is basically

(*a*) an ammeter (*b*) a voltmeter

(*c*) a multimeter (*d*) none of these

194. All meters used for measuring current, voltage and resistance are basically

(*a*) voltmeters (*b*) ammeters

(*c*) multimeters (*d*) none of these

195. Loss of charge method is used for measurement of

(*a*) low R (*b*) high R

(*c*) low L (*d*) high L

196. If the transients during switching of a power supply are to be studied, which oscilloscope will be preferred?

(*a*) An ordinary oscilloscope with high frequency sweep generator

(*b*) Dual beam oscilloscope

(*c*) Dual trace oscilloscope

(*d*) Storage oscilloscope

197. The shunt type ohmmeter is NOT suitable for high resistance measurements because

(*a*) very low resistance of the meter would short the high unknown resistance

(*b*) scale is highly cramped for high resistance values

(*c*) full scale value of the meter may be exceeded

(*d*) battery cannot supply the necessary current for proper meter deflection

198. Match List-I (Type of CRO) with List-II (Appropriate use) and select the correct answer using the codes given below the lists:

List-I

A. Storage

B. Electrostatic deflection

C. Magnetic deflection

D. Multitrace

List-II

1. Voltage and current transient studies
2. Comparison of waveforms in time domain
3. Television receiver
4. Accurate voltage measurements

Codes:

	A	B	C	D
(*a*)	4	1	2	3
(*b*)	1	4	2	3
(*c*)	4	1	3	2
(*d*)	1	4	3	2

199. A wattmeter has a range of 1000 W with an error of ± 1% of full scale deflection. If the true power passed through it is 100 W, then the relative error would be

(*a*) ± 10% (*b*) ± 5%

(*c*) ± 1% (*d*) ± 0.5%

200. In a permanent magnet moving coil instrument, if the control spring is replaced by another one having a higher spring constant, then the natural frequency and damping ratio will

(*a*) decrease

(*b*) increase and decrease respectively

(*c*) decrease and increase respectively

(*d*) increase

201. In the bridge shown in the figure

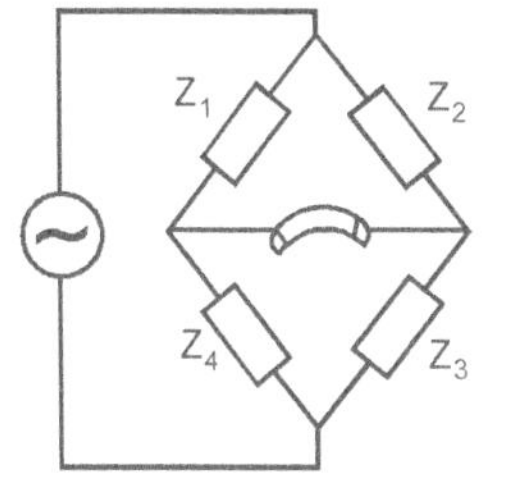

$Z_1 = 450\ \Omega$

$Z_2 = (300 - j\,600)\ \Omega$

$Z_3 = (200 + j\,100)\ \Omega$

The value of unknown impedance Z_4 at balance is

(*a*) $(100 + j\,150)\ \Omega$, (*b*) $(100 - j\,150)\ \Omega$

(*c*) $(300 + j\,100)\ \Omega$ (*d*) $(0 + j\,150)\ \Omega$

202. A 150V moving iron voltmeter of accuracy class 1% reads 75 V when used in a circuit under standard conditions. The maximum possible percentage error in the reading is

(*a*) 0.5 (*b*) 1.0

(*c*) 2.0 (*d*) 4.0

203. Consider the following statements in respect of thermoelectric instruments:

1. They indicate the rms value of current or voltage.
2. They suffer from waveform errors
3. They can be used for frequency ranges of the order MHz
4. They have a low overload capacity,.

Which of these statements are correct?

(*a*) 1, 3 and 4

(*b*) 1, 2 and 4

(*c*) 1, 2 and 3

(*d*) 2, 3 and 4

204. The ac bridge shown in the given figure will remain balanced if impedance Z consists of

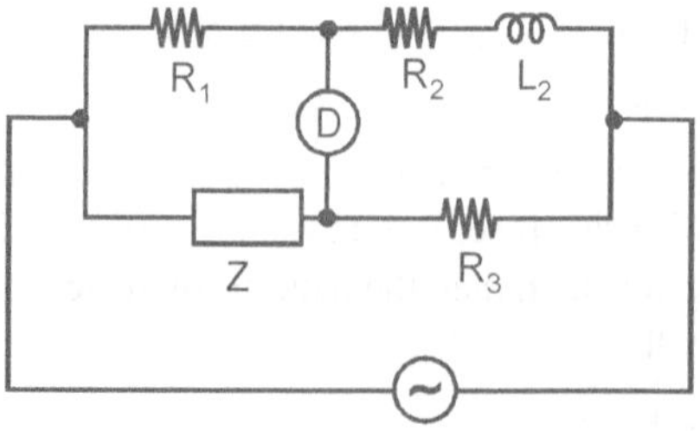

(*a*) resistance and inductance in series
(*b*) resistance and capacitance in parallel
(*c*) capacitance only
(*d*) inductance only

205. In a ballistic galvanometer, the deflecting torque is proportional to

(*a*) the current through coil
(*b*) square of current through coil
(*c*) square-root of current through coil
(*d*) sine of measurand

206. In the ac bridge shown in the given figure, the impedances Z_1, Z_2, Z_3 and Z_4 in ohms at the supply frequency are as indicated. Z_3 is a variable impedance. The bridge

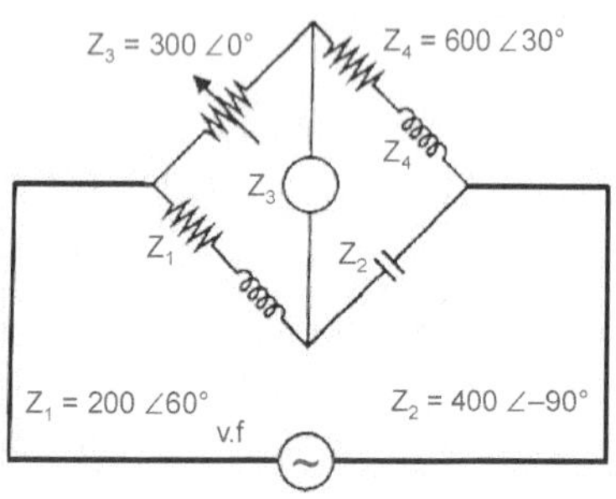

(*a*) is balanced with the indicated values of impedances
(*b*) can be balanced if $Z_3 = 400 \angle 0°$
(*c*) can be balanced if $Z_4 = 600 \angle 60°$
(*d*) cannot be balanced with the given configuration

207. A single-phase energymeter is operating on 230V, 50 Hz supply with a load of 20 A for two hours at upf. The meter makes 1380 revolutions in that period. The meter constant is

(*a*) 695 rev/kWh (*b*) 150 rev/kWh
(*c*) 0.15 rev/kWh (*d*) 1/150 rev/kWh

208. A 100μA ammeter has an internal resistance of 100μA. For extending its range to measure 500μA, the shunt required is of resistance (in Ω)

(*a*) 20.0 (*b*) 22.22
(*c*) 25.0 (*d*) 50.5

209. A 35–V source is connected to a series circuit of 600 ohm and R as shown. If a voltmeter of internal resistance 1.2 kilo ohms is connected across 600 ohm resistor, it reads 5 V. The value R is

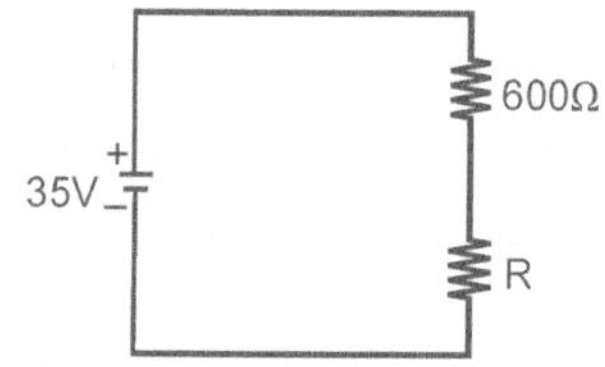

(*a*) 1.2 kΩ (*b*) 2.4 kΩ
(*c*) 3.6 kΩ (*d*) 7.2 kΩ

210. Anderson bridge is used to measure

(*a*) L (*b*) C
(*c*) V (*d*) I

211. Hay's bridge is suitable for the measurement of

(*a*) inductances with Q > 10
(*b*) inductance with Q < 10
(*c*) capacitors with high dissipating factors
(*d*) capacitors with law dissipating factors

212. In Wein bridge, the output frequency is determined by

(*a*) R-C combination
(*b*) R-L combination
(*c*) R-L-C combination
(*d*) L-C combination

213. Dissipation factor of a capacitor can be determined by using a

(*a*) De Saury's bridge
(*b*) Anderson bridge
(*c*) Hay's bridge
(*d*) Kelvin's double arm bridge

214. Most commonly used ac bridge circuit for the measurement of capacitance is

(*a*) Maxwell Wien bridge
(*b*) Kelvin's bridge
(*c*) De Sauty bridge
(*d*) Schering bridge

215. Frequency can be measured by using

(*a*) Max well's bridge
(*b*) Schering bridge
(*c*) Heaviside campbell bridge
(*d*) Wein's bridge

216. Which of the following bridge is frequency sensitive?

(*a*) Wheatstone
(*b*) Maxwell
(*c*) Anderson
(*d*) Wein

217. In audio-frequency oscillators, the frequency can be determined by

(*a*) Owen bridge (*b*) Anderson bridge

(*c*) Wein bridge (*d*) Schering bridge

218. A vibration galvanometer is turned

(*a*) by changing the length and tension of vibrating coil

(*b*) by attaching weight to the vibrating coil

(*c*) by changing its damping constant

(*d*) all of these

219. AC bridges

(*a*) have leakage error and eddy current errors only

(*b*) have residual errors, frequency errors and wave form errors only

(*c*) both (*a*) and (*b*)

(*d*) are free from errors

220. In a ballistic galvanometer, the inertia of the moving system is large so that

(*a*) it is practically stationary during the period the electricity is passing though the coil

(*b*) it is accelerating fast as soon as the coil is energised

(*c*) the amplitude of oscillation is small

(*d*) the frequency of oscillations is large

221. The voltmeter of choice for measuring the emf oi a 100V source would be

(*a*) 100V, 1 mA

(*b*) 100V, 2 mA

(*c*) 100V, 100 kΩ/K

(*d*) 100V, 100 Ω/K

222. If an energy meter disc makes 10 revolutions in 100 seconds when a load of 450W is connected to it, the meter constant (in rev/kWh) is

(*a*) 1000

(*b*) 500

(*c*) 1600

(*d*) 800

223. A Q-meter is supplied with an oscillator having a 500 mV output voltage. While testing an unknown inductor, the voltage across the variable capacitance of the Q-meter, measured by a VTVM, is obtained as 10V. The Q-factor of the inductor is

(*a*) 5 (*b*) 10

(*c*) 20 (*d*) 100

224. An ADC normally used in a 3-digit multimeter is of

(*a*) dual slope integration type

(*b*) voltage to frequency converter type

(*c*) flash (or parallel) type

(*d*) successive approximation type

225. If Q_e is the effective Q of the coil, C is the resonance capacitance and Q_d is the distributed capacitance, then the true Q in a Q-meter will be

(*a*) $Q_e = \left[\left(\frac{C+C_d}{C}\right)\right]$ (*b*) $Q_e = \left(\frac{C}{C+C_d}\right)$

(*c*) $Q_e = \left[\frac{(C_d)}{(C+C_d)}\right]$ (*d*) $Q_e = \left(\frac{C+C_d}{C_d}\right)$

226. Two voltages v_1 and v_2 are connected to the X and Y plates of a CRO as shown in the given figure. To get a circular pattern on the CRO screen, R should be adjusted to a value of

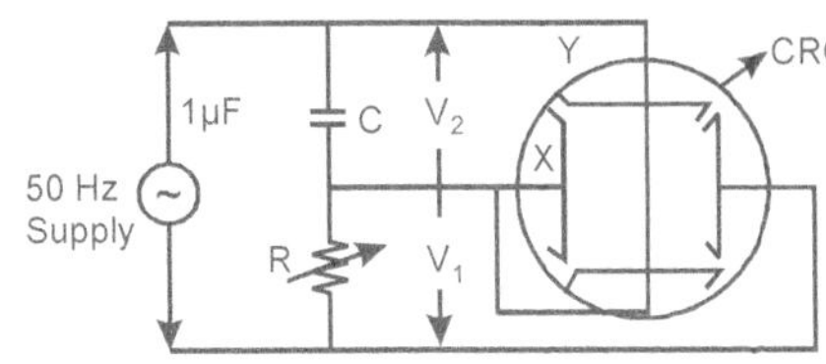

(*a*) 1kΩ (*b*) 16 kΩ

(*c*) 2.35 kΩ (*d*) 3.18 kΩ

227. The natural frequency of vibration of a reed used in a vibrating reed frequency meter, can be increased by

(*a*) increasing the gap between the reed and the electromagnet

(*b*) decreasing the gap between the reed and the electromagnet

(*c*) increasing the length of the reed

(*d*) decreasing the length of the reed

228. Match List-I (Bridges) with List-II (Quantity measured and select the correct answer using the codes given below the lists:

List-I

A. Anderson Bridge

B. Kelvin Double Bridge

C. Schering Bridge

D. Wien Bridge

List-II

1. Frequency
2. Resistance
3. Inductance
4. Capacitance

Codes:

	A	B	C	D
(*a*)	3	2	1	4
(*b*)	3	2	4	1
(*c*)	2	1	3	4
(*d*)	2	3	4	1

229. A shunt type ohmmeter is shown in the figure. With R_x disconnected, the meter reads full scale. 'S' represents the meter current as a fraction of full scale current with R_x connected such that

$$S = \frac{R_x}{R_x + R_p}$$

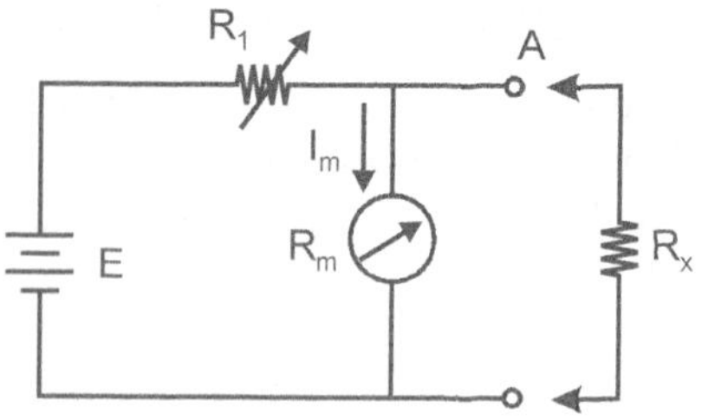

The value of R_p is given by

(*a*) R_m

(*b*) $R_1 + R_m$

(*c*) $\frac{R_1 R_x}{(R_1 + R_m)}$

(*d*) $\frac{R_1 . R_m}{(R_1 + R_m)}$

230. In the potentiometer circuit shown in the given figure, the value of unknown voltage 'E' under balanced condition will be

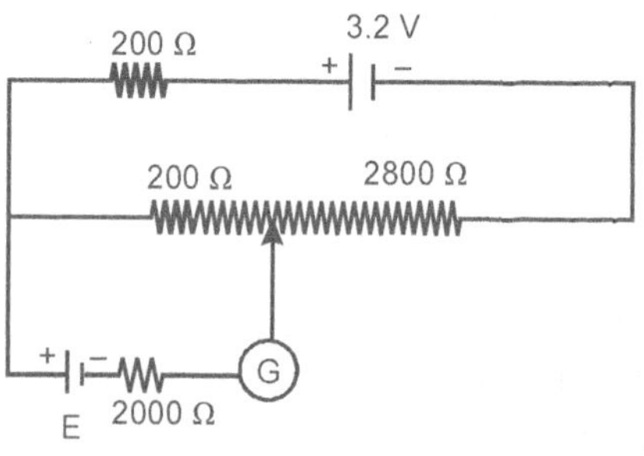

(*a*) 200mV

(*b*) 2.8V

(*c*) 3V

(*d*) 3.2V

231. A PMMC zero to 10A ammeter is not provided with any controlling mechanism and the moving parts are free to rotate. If a current of IA dc is passed through the moving coil, then the instrument

(*a*) will read 1A

(*b*) will read 10A

(*c*) pointer will continually rotate

(*d*) pointer will remain stationary

232. Match List-I (Expression for torque) with List-II (Type of instrument) and select the correct answer using the codes given below the lists:

List-I

A. $\phi_1 \phi_2 \cos\phi$

B. $I_1 I_2 \cos\phi$

C. ki

List-II

1. Ferrodynamic instruments
2. D'Arsonval instruments
3. Electrodynamometer instruments

Codes:

	A	B	C
(*a*)	2	1	3
(*b*)	3	2	1
(*c*)	1	3	2
(*d*)	3	1	2

233. A PMMC instrument uses a coil of resistance 0.2 ohm and produces full-scale deflection with a steady current of 10A. If this instrument is used as an ammeter without the addition of any external resistance, then the range of the instrument will be zero to

(*a*) 0.2A (*b*) 1A

(*c*) 2A (*d*) 10A

234. Which of the following measures the magnetic flux directly?

(*a*) Ballistic galvanometer

(*b*) Grassot fluxmeter

(*c*) Parmeameter

(*d*) All of these

235. Campbell's bridge method is used to measure

(*a*) copper loss

(*b*) iron loss

(*c*) both iron and copper loss

(*d*) none of these

236. Sphere gap voltmeter, when measuring ac voltage measures

(*a*) rms value (*b*) average value

(*c*) peak value (*d*) none of these

237. VTVM can be used to measure

(*a*) dc voltage

(*b*) ac voltage of high frequency

(*c*) dc voltage and ac voltage upto the order of 5 MHz frequency

(*d*) ac voltage of low frequency

238. For measuring voltages in the electric circuit, VTVM is preferred to conventional meter, because conventional meters

(*a*) are costly
(*b*) are less portable
(*c*) are not precise
(*d*) load on the circuit

239. The input impedance of a TVM as compared to that of a VTVM is

(*a*) same (*b*) not comparable
(*c*) high (*d*) low

240. FETs are used in amplifiers to get

(*a*) high input impedance
(*b*) low input impedance
(*c*) high output impedance
(*d*) low output impedance

241. FET input stage is used in an amplifier to increase its

(*a*) output impedance
(*b*) output impedance
(*c*) frequency band width
(*d*) power handling capacity

242. Which of the following exhibits very high input impedance?

(*a*) *p-n-p* transistor (*b*) *n-p-n* transistor
(*c*) FET (*d*) none of these

243. An average reading VTVM uses one diode connected in series with an external resistance. A high value of series resistance is used so that the instrument should have

(*a*) high input impedance
(*b*) a linear V-I characteristic
(*c*) low power consumption
(*d*) all of these

244. Complex waveforms are most accurately measured with

(*a*) rms responding electronic voltmeter
(*b*) average responding electronic voltmeter
(*c*) peak responding electronic voltmeter
(*d*) both average and peak responding electronic voltmeters

245. Two voltmeters have the same range 0 – 400V. The internal impedances are 30,000 ohms and 20,000 ohms. If they are connected in series and 600V be applied across them, the reading are

(*a*) 360 V and 240 V
(*b*) 300 V each
(*c*) 400 V and 200 V
(*d*) one of the meters out of range and other 100 V

246. Two voltmeters one with a full-scale reading of 100V and another with a full-scale reading of 200 V, are connected in series across a 100V supply. The internal resistance of both meters is the same. What are the reading?

(*a*) 100V and 0 (*b*) 75V and 25V
(*c*) 67V and 33V (*d*) 50V and 50V

247. The resistance of a circuit is found by measuring current flowing and the power fed into the circuit. If the limiting errors in the measurement of power and current are ± 1.5% and ± 1.0% respectively, the limiting error in the measurement of resistance will be

(*a*) ± 1% (*b*) ± 1.5%
(*c*) ± 2.5% (*d*) ± 3.5%

248. The term 'artificial aging' in instrument is associated with

(*a*) springs (*b*) permanent magnets
(*c*) controlling torques (*d*) damping

249. In a permanent magnet moving coil instrument, if the control spring is replaced by another one having a higher spring constant, then the natural frequency and damping ratio will

(*a*) decrease
(*b*) increase and decrease respectively
(*c*) decreases and increase respectively
(*d*) increase

250. Which of the following are piezo-electric substances?

1. Barium titanate
2. Lead titanate
3. Lead zirconic
4. Cadmium sulphate

(*a*) 1, 2 and 4
(*b*) 1, 3 and 4
(*c*) 1, 2 and 3
(*d*) 2, 3 and 4

251. Match List-I with List-II and select the correct answer by using the Codes given below the lists:

List-I	**List-II**
A. Resistance	1. $M^{-1}L^{-2}T^{2}I^{-2}$
B. Inductance	2. $ML^{2}T^{-2}I^{2}$
C. Capacitance	3. $M^{-1}L^{-2}T^{-4}I^{2}$
D. Reluctance	4. $ML^{2}T^{-2}I^{-2}$

Codes:

	A	B	C	D
(*a*)	1	2	3	4
(*b*)	2	1	3	4
(*c*)	4	2	3	1
(*d*)	4	3	2	1

252. Which meter is suitable for the measurement of 10 mV at 50 MHz?

(*a*) Moving-iron voltmeter

(*b*) VTVM

(*c*) Moving-coil voltmeter

(*d*) CRO

253. Which voltmeter would you select for measuring 50.000V direct current?

(*a*) Moving coil voltmeter

(*b*) Hot wire voltmeter

(*c*) Electrodynamometer

(*d*) Electrostatic voltmeter

254. Which of the following conditions are to be satisfied so that the common variable shaft of resistance R_1 and R_3 can be graduated in frequency to measure the frequency of E under balanced condition?

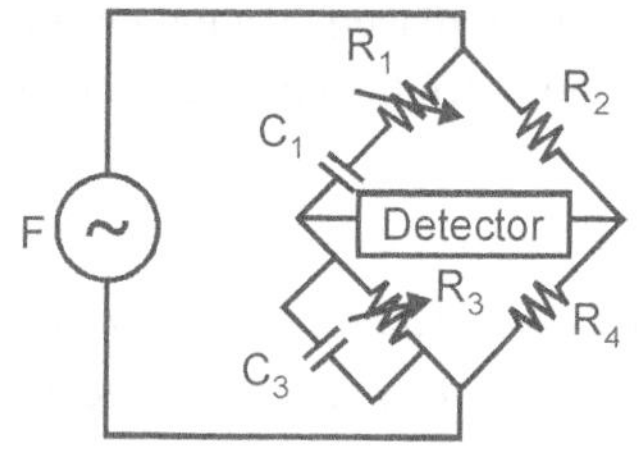

1. $R_1 = R_3$ 2. $C_1 = C_3$

3. $R_2 = 2R_4$ 4. $R_2 = R_4$

Select the correct answer using the codes given below:

(*a*) 1 and 4   (*b*) 1 and 2

(*c*) 2 and 4   (*d*) 1, 2 and 3

255. What are the other methods of measuring power in a three phase circuit without using wattmeter?

(*i*) One voltmeter and one ammeter

(*ii*) Two voltmeters and two ammeters

(*iii*) Three voltmeters

(*iv*) Three ammeters

(*a*) *i* and *ii* only

(*b*) *iii* and *iv* only

(*c*) *i* and *iii* only

(*d*) *iv* only

256. Two voltmeters of 0–300V range are connected in parallel to an ac circuit. One voltmeter is moving iron type and reads 200V. If the other instrument is moving coil type, its reading will be

(*a*) $200\sqrt{3}$ volts

(*b*) 200 × 1.41 volts

(*c*) slightly less than 200V

(*d*) zero

257. Which one of the following terms is NOT associated with a sample-and-hold circuit?

(*a*) Conversion time

(*b*) Acquisition time

(*c*) Aperture time

(*d*) Sample mode

258. Consider the following statements about LVDT as a transducer:

1. The relationship between input displacement and output voltage is almost linear
2. The range of displacement that can be measured is wide
3. It does not form a loading on the mechanical system

Of these statements are correct

(*a*) 1, 2 and 3   (*b*) 2 and 3

(*c*) 1 and 2   (*d*) 1 and 3

259. An average response rectifier type electronic ac voltmeter has a dc voltage of 10V applied to it. The meter reading will be

(*a*) 7.1 V   (*b*) 10.0V

(*c*) 11.1V   (*d*) 22.2V

260. Environmental pollution is not caused by

(*a*) atmospheric air and water pollution

(*b*) high humidity content in atmospheric air

(*c*) noise pollution

(*d*) radioactive fallout

261. The position telemetering system using synchros is a/an

(*a*) pulse telemetering system

(*b*) rf telemetering system

(*c*) dc telemetering system

(*d*) ac telemetering system

262. Which one of the following telemetering system is a digital system?

(*a*) Pulse code modulation (PCM) system

(*b*) Position telemetering system using synchros

(*c*) DC voltage telemetering system

(*d*) Pulse duration modulation (PDM) system

263. An analog transducer has range 0–10V. Calculate bits of an A/D converter if resolution is 5mV

(*a*) 10   (*b*) 9

(*c*) 11   (*d*) none of these

264. The P.M. telemetry as compared with A.M. telemetry requires a channel that is

(*a*) equal to that of A.M. telemetry

(*b*) smaller than what is required for A.M. telemetry

(*c*) 100 times of that for A.M. telemetry

(*d*) 10 times of that required for A.M. telemetry

265. Dummy strain gauges are used for

(*a*) compensation of temperature changes

(*b*) increasing the sensitivity of bridge

(*c*) compensating for different expansion

(*d*) calibration of strain gauges

266. A thermo couple

(*a*) has a low time constant when it is bare

(*b*) has a low time constant if it is provided with a sheath

(*c*) has the same time constant whether it is bare or is provided with sheath

(*d*) none of these

267. Which one of the following detectors is used in bridge measurements of 50 Hz?

(*a*) Vibration galvanometer

(*b*) Ballistic galvanometer

(*c*) D'Arsonval galvanometer

(*d*) Truivector meter

268. In a Wheatstone bridge, the known resistances are correct to within ± 0.1%. The accuracy to which an unknown resistance can be measured is

(*a*) 0.4% (*b*) 0.3%

(*c*) 0.2% (*d*) 0.1%

269. The three impedances of an ac bridge shown below are

$Z_1 = 20\angle 60\Omega$

$Z_2 = 400\angle -90°$

$Z_3 = 300\angle 0°\Omega$

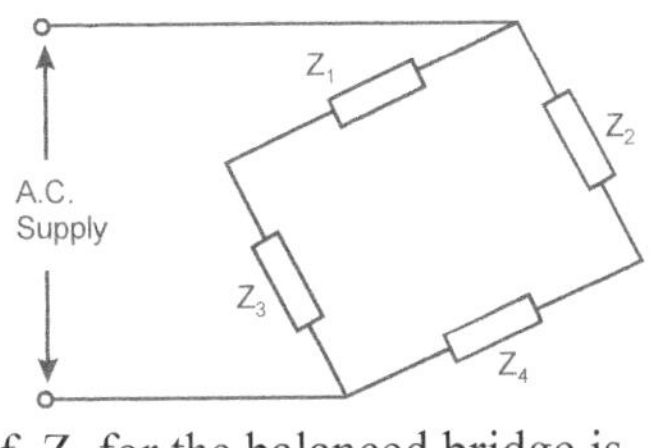

The value of Z_4 for the balanced bridge is

(*a*) $150\angle 150°\Omega$ (*b*) $150\angle -30°\Omega$

(*c*) $600\angle 30°\Omega$ (*d*) $600\angle -150°\Omega$

270. Match List-I (Unit) with List-II (Equivalent unit) and select the correct answer using the codes given below the lists:

List-I	List-II
A. Weber	1. Ampere-second per volt
B. Coulomb	2. Volt-second
C. Farad	3. Ampere-second
	4. Volt per second

Codes:

	A	B	C
(*a*)	1	2	3
(*b*)	2	3	1
(*c*)	2	3	4
(*d*)	3	1	4

271. A Manganin swamp resistance is connected in series with a moving coil ammeter consisting of a milli-ammeter and a suitable shunt in order to

(*a*) minimise the effect of temperature variation

(*b*) obtain large deflecting torque

(*c*) reduce the size of the meter

(*d*) minimise the effect of stray magnetic fields

272. A spectrum analyser can be described as

(*a*) voltage selective frequency meter

(*b*) current selective frequency meter

(*c*) frequency selective voltmeter

(*d*) a filter

273. In noise-level measurement, the output of the microphone pick-up is weighted suitably by electrical filters so as to

(*a*) provide linearised response at different frequencies

(*b*) provide frequency spectra at different frequencies

(*c*) filter the radio frequency interference

(*d*) provide a response similar to that of the human ear

274. Consider the following statements:

A linear variable differential transformer (LVDT) has

1. one primary winding
2. two exactly similar secondary windings
3. a toroidal magnetic core

Of these statements

(*a*) 2 and 3 are correct

(*b*) 1 and 2 are correct

(*c*) 1 and 3 are correct

(*d*) 1, 2 and 3 are correct

275. Consider the following statements:

A 'phantom' load used while testing a high capacity energymeter,

1. consist of inductances and capacitances so that there is no energy loss during testing
2. saves energy during testing because its value changes very rapidly from the highest to the lowest, thus enabling quick measurement
3. involves supplying the voltage circuit with the required voltage and the current circuit from a separate low voltage supply

Of these statements

(*a*) 1, 2 and 3 are correct

(*b*) 1 and 2 are correct

(*c*) 1 and 3 are correct

(*d*) 2 and 3 are correct

276. Consider the following statements:

A capacitive transducer working on the principle of change of capacitance with change of displacement, exhibits non-linear characteristics. The response of these transducers can be made linear by

1. using differential arrangement
2. using an op-amp
3. working them over a small displacement range

Of these statements

(*a*) 1 and 2 are correct

(*b*) 2 and 3 are correct

(*c*) 1 and 3 are correct

(*d*) 1, 2 and 3 are correct

277. How many comparators are there in an ADC with 0–5 V input and 8 bit output assuming a flash converter?

(*a*) 5

(*b*) 8

(*c*) 40

(*d*) 255

278. A capacitive transducer is shown in the given figure. Sensitivity 'S' of the transducer is given

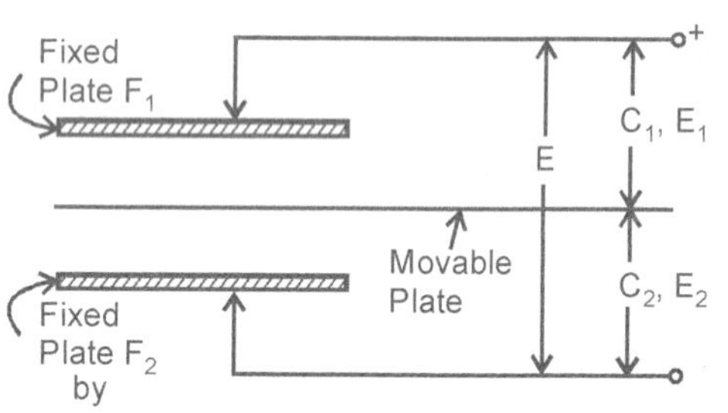

by

(*a*) $\dfrac{E}{2d}$ (*b*) $\dfrac{2E}{d}$

(*c*) $\dfrac{E}{d}$ (*d*) $\dfrac{E}{4d}$

279. Which of the following devices cannot be used to measure pressure?

(*a*) Strain gauge

(*b*) LVDT

(*c*) Piezo-electric crystal

(*d*) Pyrometer

280. Pyrometer is used to measure

(*a*) strain

(*b*) pressure

(*c*) displacement

(*d*) temperature

281. A flow meter that is independent of liquid density is

(*a*) rotameter

(*b*) electromagnetic flow meter

(*c*) venterimeter

(*d*) orifice meter

282. Which of the following transducer is used for transmitting as well as receiving the acoustic energy in an ultrasonic flow meter?

(*a*) LVDT

(*b*) RTD

(*c*) Piazoelectric crystals

(*d*) Strain gauge

283. Instrument transformers are known to introduce magnitude and phase errors in measurements. These are primarly due to

(*a*) improper connection on the primary side

(*b*) measurement errors inherent in the meter connected to the transformer secondary

(*c*) open and short circuit parameters of the instrument transformers

(*d*) none of these

284. A capacitive transformer working on the principle of change of capacitance with change of displacement exhibits non linear characteristics. The response of these transducers can be made linear by using

(*a*) differential arrangements

(*b*) use of an op-amp.

(*c*) working them over a small displacement range

(*d*) all of these

285. Quartz and Rochelle salt belong to

(*a*) natural group of piezo-electric materials

(*b*) synthetic group of piezo-electric materials

(*c*) natural or synthetic group of piezo-electric materials provided they are properly polarized

(*d*) none of these

286. Piezo-electric crystals produce an emf

(*a*) when external mechanical force is applied to it

(*b*) when external magnetic field is applied

(*c*) when radiant energy stimulates the crystal

(*d*) when the junction of two such crystal is heated

287. Piezo-electric crystals are used for the measurement of

(*a*) temperature

(*b*) velocity

(*c*) sound

(*d*) none of these

288. If an ac voltage wave is corrupted with an arbitrary number of harmonics, then the overall voltage waveform differs from its fundamental frequency component in terms of

(*a*) only the peak values

(*b*) only therms values

(*c*) only the average values

(*d*) all the three measures (peak, rms, and average values)

289. A digital voltmeter has $4\frac{1}{2}$ digit display.

The 1 Volt range can read upto

(*a*) 1.0000

(*b*) 1.1111

(*c*) 1.9999

(*d*) 19999

290. A 50 Hz ac voltage is measured with a moving iron voltmeter and a rectifier type ac voltmeter connected in parallel. If the meter readings are V_1 and V_2 respectively and the meters are free from calibration errors, then the form factor of the ac voltage may be estimated as

(*a*) $\frac{V_1}{V_2}$ (*b*) $1.11\frac{V_1}{V_2}$

(*c*) $2\frac{V_1}{V_2}$ (*d*) $\pi\frac{V_1}{2V_2}$

291. A basic *d* Arsonval movement with internal resistance 100 Ω and full scale current 1 mA is to be converted into a multirange dc voltmeter with voltage ranges 0-10V, 0-30V, 0-100V and 0-300V. The circuit arrangement is shown in the given figure. The value of R_1 is

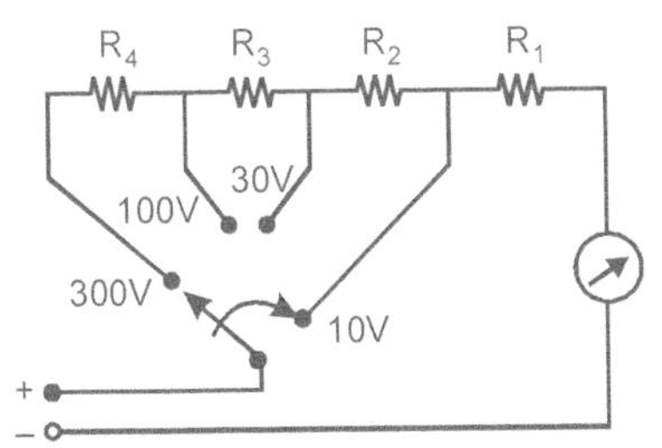

(*a*) 0.0 Ω (*b*) 100 Ω

(*c*) 9.9 kΩ (*d*) 10 kΩ

292. In hygrometers the principle of measurement is

(*a*) change in resistance of salts with humidity

(*b*) change in microwave power using klystron

(*c*) change in thermal conductivity using thermistor

(*d*) none of these

293. The sensitivity of an instrument is the

(*a*) smallest increment in the input that can be detected with certainty

(*b*) largest input change to which the instrument fails to respond

(*c*) ratio of the change in the magnitude of the output to the corresponding change in the magnitude of the input

(*d*) closeness of the output values for repeat applications of constant input

294. In measurements made using a Q-meter, high impedance elements should preferably be connected in

(*a*) star (*b*) delta

(*c*) series (*d*) parallel

295. The ratio of the readings of two wattmeters connected to measure power in a balanced 3-phase load is 5 : 3 and the load is inductive. The power factor of load is

(*a*) 0.917 lead

(*b*) 0.917 lag

(*c*) 0.6 lead

(*d*) 0.6 lag

296. How many time base circuits does a dual trace CRO have?

(*a*) 1 (*b*) 2

(*c*) 3 (*d*) 4

297. A Lissajous pattern on an oscilloscope has 5 horizontal tangencies and 2 vertical tangencies. The frequency of the horizontal input is 1000 Hz. What is the frequency of the vertical input?

(*a*) 400 Hz (*b*) 2500 Hz

(*c*) 4000 Hz (*d*) 5000 Hz

298. In the case of power measurement by two-wattmeter method in a balanced 3-phase system with a pure inductive load

(*a*) both the wattmeters will indicate the same value but of opposite sign

(*b*) both the wattmeters will indicate zero

(*c*) both the wattmeters will indicate the same value and of the same sign

(*d*) one wattmeter will indicate zero and the other some non-zero value

299. Holes are drilled on the opposite sides of the disc of an induction type energy meter to

(*a*) avoid creep on no load

(*b*) balance the dise

(*c*) dissipate heat energy due to eddy currents

(*d*) increase the deflecting torque

300. The bridges suitable for the measurement of an unknown inductance in terms of a known capacitance would include
(*a*) Maxwell and Hay
(*b*) Maxwell and Schering
(*c*) Hay and Schering
(*d*) Maxwell, Hay and Schering

301. In the case of power measurement by two wattmeter method in a balanced 3-phase system with a pure inductive load,
(*a*) both the wattmeters will indicate the same value but of opposite sign
(*b*) both the wattmeters will indicate zero
(*c*) both the wattmeters will indicate the same value and of the same sign
(*d*) one wattmeter will indicate zero and the other will indicate some non-zero value

302. The wire material of strain gauges should have
(*a*) high resistivity and high thermo-emfs
(*b*) low resistivity and high thermo emfs
(*c*) low resistivity and low thermo-emfs
(*d*) high resistivity and low thermo-emfs

303. Which of the following should be incorporated in the RTD to make a temperature sensing bridge most sensitive to temperature.
(*a*) Platinum
(*b*) Nickel
(*c*) Thermistor
(*d*) Copper

304. The thermo-electric effect was first observed by
(*a*) Seeback (*b*) Thermas young
(*c*) Pirani (*d*) Thermus

305. For surface temperature measurement one can use
(*a*) strain gauge (*b*) diaphragm
(*c*) RTD (*d*) thermo-couple

306. LVDT windings are wound on
(*a*) steel sheets (laminated)
(*b*) aluminum
(*c*) ferrite
(*d*) copper

307. An X-Y recorder has X-input and Y-input as 1 sin 314t and 1 $\sin\left(314t+\frac{\pi}{2}\right)$ respectively. The trace generated is a
(*a*) straight line (*b*) parabola
(*c*) circle (*d*) ellipse

308. Bourdon tubes are made of
(*a*) copper (*b*) aluminium
(*c*) metal alloys (*d*) phosphorus

309. Resolution of a potentiometric transducer depends on
(*a*) diameter of wire
(*b*) length of wire
(*c*) material of wire
(*d*) excitation voltage

310. The sensitivity factor of strain gauge is normally of the order of
(*a*) 1 to 1.5
(*b*) 1.5 to 2.0
(*c*) 0.5 to 1
(*d*) 5 to 10

311. A 200 Ω metallic strain gauge with gauge factor = 2.0 when subjected to a strain of 500 micro-strain will have a total change in resistance of
(*a*) 0.2 Ω (*b*) 40 mΩ
(*c*) 2 Ω (*d*) 1.0 Ω

312. Semiconductor of strain gauges are made of
(*a*) silicon (*b*) gold
(*c*) silver (*d*) nickel

313. 'Creep' in energy meters can be prevented by
(*a*) using extra turns of the voltage coil
(*b*) having two holes on opposite sides of the disc
(*c*) using a stronger brake magnet
(*d*) using steel laminations of high permeability

314. In a low power-factor wattmeter, compen-sating coil is employed in order to
(*a*) neutralize the capacitance effect of pressure coil
(*b*) compensate for the inductance of pressure coil
(*c*) compensate for the error caused by power loss in the pressure coil
(*d*) reduce the error caused by the eddy-currents

315. A 10V full-scale voltmeter having 100 $k\Omega$/V sensitivity is used to measure the output of a photovoltaic cell having an internal resistance of 1 MΩ the voltmeter reads 5V. The voltage generated by the photovoltaic cell is
(*a*) 5V (*b*) 1V
(*c*) 10V (*d*) 8V

316. The minimum number of wattmeter(s) required to measure 3-phase, 3-wire balanced or unbalanced power is
(*a*) 1 (*b*) 2
(*c*) 3 (*d*) 4

317. The effect of stray magnetic fields on the actuating torque of a portable instrument is maximum when the operating field of the instrument and the stray field are

(*a*) perpendicular (*b*) parallel
(*c*) X inclined at 60° (*d*) inclined at 30°

318. Match List-I with List-II and select the correct answer using the codes given below the lists:

List-I (*Parameter measured*)

A. Audio frequency
B. High Q inductor
C. Dielectric loss
D. Low Q inductor

List-II (*Bridge*)

1. Maxwell bridge
2. Hay bridge
3. Wien bridge
4. Schering bridge

Codes:

	A	B	C	D
(*a*)	3	4	2	1
(*b*)	3	2	4	1
(*c*)	4	2	1	3
(*d*)	1	4	2	3

319. The two wattmeter method is used to measure active power on a three phase, three wire system. If the phase voltage is unbalanced, then the power reading is

(*a*) affected by both negative sequence and zero sequence voltages
(*b*) affected by negative sequence voltages but not by zero sequence voltages
(*c*) affected by zero sequence voltages but not by negative sequence voltages
(*d*) not affected by negative or zero sequence voltages

320. A current $[1 + \sqrt{2}\sin(314t + 30°) + 2\sin(1570t + 20°)$ amperes is measured with a thermocouple type ammeter. The meter will read

(*a*) 2A
(*b*) $\sqrt{2}$ A
(*c*) $(3 + \sqrt{2})$A
(*d*) 4 A

321. The impedances of a basic ac bridge arms are

$Z_{AB} = 250\ \Omega$
$Z_{BC} = 100\ \Omega\angle 60°$
$Z_{CD} = 400\ \Omega\angle 30°$
Z_{DA} = Unknown

If ω = 1000 radians/s, the components of the unknown arm has the resistance $500\sqrt{3}\ \Omega$, in

(*a*) series with a capacitor of 2 μF
(*b*) parallel with a capacitor of 2 μF
(*c*) series with an inductor of 2 μH
(*d*) parallel with an inductor of 2 μH

322. A dual-slope analog-to-digital converter uses an N-bit counter. When the input signal V_a is being integrated, the counter is allowed to count up to a value

(*a*) equal to 2^N
(*b*) equal to 3^N
(*c*) proportional to V_a
(*d*) inversely proportional to V_a

323. An integrating digital voltmeter measures

(*a*) true average value (*b*) rms value
(*c*) peak value (*d*) peak to peak value

324. Which one of the following pairs is correctly matched?

(*a*) Thermistor...Self-generating transducer
(*b*) Semiconductor strain gauges... Piezoresistivity
(*c*) Magnetostrictive transducer... Change in mutual resistance
(*d*) Magnetoelectric transducer ... Change in self-inductance

325. An integrated circuit is

(*a*) a complicated circuit
(*b*) an integrated device
(*c*) much costlier than a single transistor
(*d*) fabricated on a tiny silicon chip

326. The gain of an "op-amp" is nearly

(*a*) 10^3 (*b*) 10^4
(*c*) 10' (*d*) 10^6

327. The input impedance of the operational amplifier is

(*a*) infinite
(*b*) zero
(*c*) very high but not infinite
(*d*) very small

328. The circuit shown is

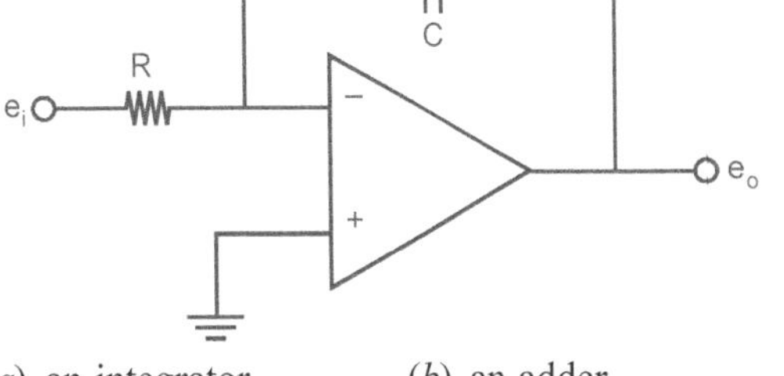

(*a*) an integrator (*b*) an adder
(*c*) a differentiator (*d*) a substractor

329. In a transistor Hartley oscillator

(*a*) inductive feedback is used

(*b*) untapped coil is used

(*c*) entire coil is in the output circuit

(*d*) none of these

330. The oscillator circuit that uses a tapped coil in turned circuit is

(*a*) multi-vibrator

(*b*) Hartley

(*c*) armstrong

(*d*) colpitis

331. For generating a 1 kHz signal, the most suitable circuit is

(*a*) Hartley oscillator

(*b*) Colpitts oscillator

(*c*) Unturned collector oscillator

(*d*) Wien-bridge oscillator

332. To generate a 1 MHz signal, the most suitable oscillator is

(*a*) Colpitt' s oscillator

(*b*) Phase-shift oscillator

(*c*) Wien-bridge oscillator

(*d*) None of these

333. The type of multivibrator used for generation of clock pulses is

(*a*) monostable multivibrator

(*b*) a stable multivibrator

(*c*) bistable multivibrator

(*d*) none of these

334. For measurement and control Instruments should preferably have

(*a*) dead zone and dead time

(*b*) linear output and fast response

(*c*) non-linear output

(*d*) a highly damped response

335. A current measuring instrument can be used to measure voltage by

(*a*) adding resistance in series

(*b*) adding resistance in parallel

(*c*) using larger wire in the coil

(*d*) using a high resistance shunt

336. A differential transformer is a

(*a*) variable inductance transducer

(*b*) variable pressure transducer

(*c*) constant displacement transducer

(*d*) constant pressure transformer

337. A single channel digital storage oscilloscope uses a 10 bit, 10^7 samples per second Analog-to-Digital Converter. For a 100 kHz sine wave input, the number of samples taken per cycle of the input will be

(*a*) 10^7 (*b*) 10^4

(*c*) 10^3 (*d*) 10^2

338. The inductance of a high Q inductor can be measured using a

(*a*) Schering bridge

(*b*) Wien bridge

(*c*) Maxwell bridge

(*d*) Hay bridge

339. A thermocouple ammeter gives full-scale deflection at 10 A. When the reads one fifth of the scale, the current will be

(*a*) 2A (*b*) 4A

(*c*) 4.47A (*d*) 5.78A

340. Electronic voltmeter provides more accurate readings in high resistance circuits as compared to a non-electronic voltmeter because of its

(*a*) high VI ohm ratings

(*b*) high ohm/Kratings

(*c*) low meter resistance

(*d*) high resolution

341. Many modern capacitance meters measure capacitance by making the unknown capacitance an element in a charge amplifier. A schematic of the same is given below. If the maximum permissible output voltage is limited to 1 V_{peak}, due to slow rate constraints, then the maximum C_X will be

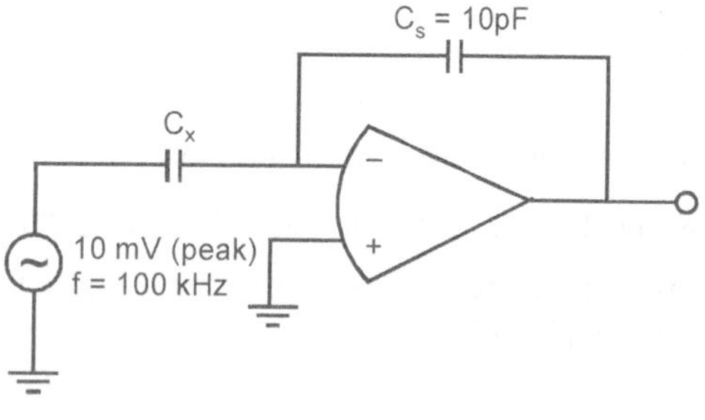

(*a*) 10 pF (*b*) 100 pF

(*c*) 1000 pF (*d*) 10 nF

342. A wattmeter has a range of 1000W with an error of ± 1% full scale deflection. If the true power passed through it is 100 W, then the relative arror would be

(*a*) ±10% (*b*) ±5%

(*c*) ±1% (*d*) ±0.5%

343. Which one of the following has the highest accuracy?
(*a*) Standard resistance
(*b*) Standard inductance
(*c*) Standard capacitance
(*d*) Standard mutual inductance

344. The resistance of a shunt for a precision grade ammeter can be best measured by
(*a*) De Saury bridge
(*b*) Scherring bridge
(*c*) Maxwell bridge
(*d*) Kelvin double bridge

345. A high frequency as signal is applied to a PMMC instrument. If the rms value of the a/c signal is 2 V, then the reading of the instrument will be
(*a*) zero
(*b*) 2V
(*c*) $2\sqrt{V}$
(*d*) $4\sqrt{2}V$

346. Two 100 μA full scale PMMC meters are employed to construct a 10V and a 100V full scale voltmeter. These meters will have figures of merit (sensitivities) as
(*a*) 10 kΩ/V and 10 kΩ/V
(*b*) 100 kΩ/V and 10 kΩ/V
(*c*) 10 kΩ/V and 100 kΩ/V
(*d*) 10 kΩ/V and 1 kΩ/V

347. The capacitance and loss angle of a capacitor can be accurately measured by
(*a*) Kelvin'bridge
(*b*) Andersen's bridge
(*c*) Schering bridge
(*d*) Carey-Foster's bridge

348. In the balanced Wheatstone bridge shown in the figure, if the value of R_6 is increased, the current I_2

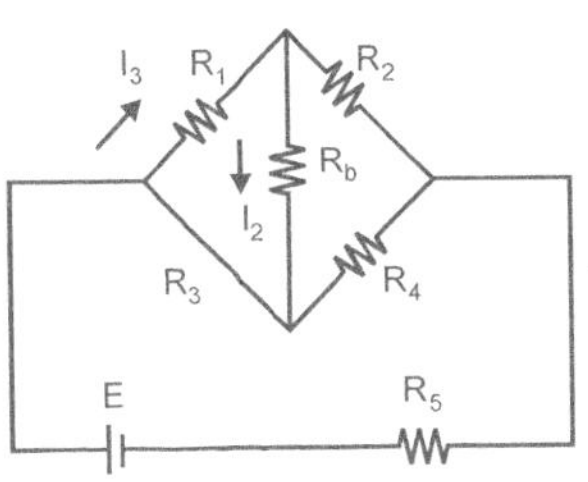

(*a*) will increase
(*b*) will decrease
(*c*) will remain in unchanged
(*d*) all of these

349. In a single-phase induction type energy meter, the lag adjustment is done to ensure that
(*a*) current coil flux lags the applied voltage by 90°
(*b*) pressure coil flux lags the applied voltage by 90°
(*c*) pressure coil flux is in phase with the applied voltage
(*d*) current coil flux lags the pressure coil flux by 90°

350. Swamping resistance is a resistance which is added to the moving coil of a meter to
(*a*) reduce the full-scale current
(*b*) reduce the temperature error
(*c*) increase the sensitivity
(*d*) increase the filed strength

351. A current transformer has a phase error of + 3°. The phase angle between the primary and secondary current is
(*a*) 3° (*b*) 177°
(*c*) 180° (*d*) 183°

352. The ac bridge shown in the given figure will remain balanced if impedance Z consists of

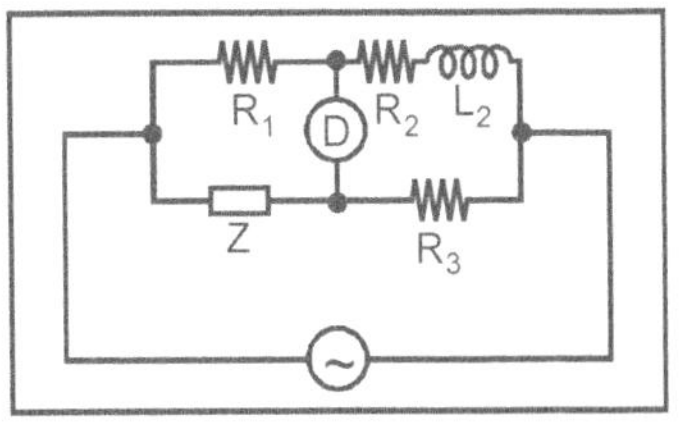

(*a*) resistance and inductance in series
(*b*) resistance and capacitance in parallel
(*c*) capacitance only
(*d*) inductance only

353. A current of $[2+\sqrt{2}\sin(\sin 314t + 30) + 2\sqrt{2}\cos(952t+45)]$ is measured with a thermocouple type, 5A full-scale, class 1 meter. The meter reading would lie in the range
(*a*) 5A ± 1%
(*b*) $(2+3\sqrt{2})$A ± 1%
(*c*) 3A ± 1.7%
(*d*) 2A ± 2.5%

354. Modern electronic multimeters measure resistance by
(*a*) using a bridge circuit
(*b*) using an electronic bridge compensator for nulling
(*c*) forcing a constant current and measuring the voltage across the unknown resistor
(*d*) applying a constant voltage and measuring the current through the unknown resistor

355. A resistance strain gauge is fastened to a beam subjected to a strain of 1×10^{-6}, yielding a resistance change of 240 μΩ. If the original resistance of the strain gauge is 120 Ω, the gauge factor would be

(*a*) 5 (*b*) 2
(*c*) 1 (*d*) 0.2

356. To eliminate 50 Hz pick-up in a dual slope DVM, the minimum period of integration of the input signal is

(*a*) 1 *ms* (*b*) 20 *ms*
(*c*) 1 *s* (*d*) 100 *s*

357. The current T through a resistance R is measured with the following uncertainties

I 4A ± 0.5%,

R = 100Ω ± 0.2%

If power is computed from these two measured quantities, the uncertainty in the power computed will be

(*a*) ± 0.01% (*b*) ± 0.29%
(*c*) ± 0.7% (*d*) ± 1.2%

358. To achieve the optimum transient response, the indicating instruments are so designed as to

(*a*) be critically damped
(*b*) be undamped
(*c*) provide damping which is slightly more than the critical value
(*d*) provide damping which is slightly less than the critical value

359. The dimensional equation of resistance is

(*a*) $L^2 M T^{-2} I^{-4}$ (*b*) $L^2 M T^{-2} I^{-2}$
(*c*) $L^2 M^2 T^{-3} I^{-2}$ (*d*) $L^2 M T^{-3} I^{-2}$

360. Which one of the following sets of building blocks mainly decides the accuracy of a frequency counter?

(*a*) Crystal and ADC
(*b*) ADC and DAC
(*c*) DAC and gate width generator
(*d*) Gate width generator and crystal

361. A voltage of $\{200\sqrt{2}\ \sin 314t + 6\sqrt{2}\ \sin (942 + 30°) + 8\sqrt{2}\ \cos(1570t + 30°)\}$V is given to a harmonic distortion meter. The meter will indicate a total harmonic distortion of approximately

(*a*) 5.0% (*b*) 6.5%
(*c*) 7.5% (*d*) 8.5%

362. The iron loss in a magnetic specimen can be determined by using a

(*a*) frequency meter (*b*) Q-meter
(*c*) CRO (*d*) weston cell

363. An audio-oscillator uses

(*a*) positive feedback
(*b*) negative feedback
(*c*) both positive and negative feedback
(*d*) none of these

364. Wien bridge oscillator is basically a

(*a*) pulse generator
(*b*) sin wave generator
(*c*) square wave generator
(*d*) triangular wave generator

365. In a Wien-bridge oscillator with R = 5k Ω and C = 0.01 μF, the frequency of oscillation will be

(*a*) 7.8 kHz (*b*) 3.2 kHz
(*c*) 2.5 kHz (*d*) 780 Hz

366. The main advantage of a crystal oscillator is that its output is

(*a*) a constant frequency
(*b*) dc
(*c*) 50 Hz to 60 Hz
(*d*) variable frequency

367. In a CRO astigmatism is

(*a*) source of generating fast electrons
(*b*) media for absorbing secondary emission electrons
(*c*) an addition focus control
(*d*) time-delay control in the vertical deflection system

368. The purpose of the synchronising control in a CRO is to

(*a*) focus the spot on the screen
(*b*) lock the display of signal
(*c*) adjust the amplitude of display
(*d*) control the intensity of the spot

369. The patterns used to measure phase and frequency with a cathode ray oscilloscope are called

(*a*) Lissajous pattern
(*b*) Faraday's pattern
(*c*) Ohm's pattern
(*d*) Phillips pattern

370. Time delay and phase can be measured by using a

(*a*) VTVM
(*b*) CRO
(*c*) TVM
(*d*) PMMC

371. The Lissajous pattern with equal voltages of equal frequency and phase shift by 90° is

(*a*) straight line (*b*) circle
(*c*) ellipse (*d*) dot

372. Match List-I with List-II and select the correct answer using the codes given below the Lists:

List-I

A. Meggar
B. Spectrum analyser
C. Schering bridge
D. Digital counter

List-II

1. Measurement of loss angle in a dielectric
2. Measurement of frequency
3. Measurement of insulation resistance
4. Measurement of harmonics

Codes:

	A	B	C	D
(*a*)	1	2	3	4
(*b*)	1	2	4	3
(*c*)	4	3	2	1
(*d*)	3	4	1	2

373. A zero to 300V voltmeter has a guaranteed accuracy of 1% full scale reading. The voltage measured by the instrument is 83V. The per cent limiting error is

(*a*) 0.95 (*b*) 1.81
(*c*) 3.62 (*d*) 4.85

374. In moving-coil meters, damping is provided by

(*a*) eddy current disk
(*b*) the aluminium frame of the coil
(*c*) damping vare in the air-tight chamber
(*d*) the coil spring attached to the moving mechanism

375. In the circuit shown in the figure, if the ammeter indicates 1A, and the voltmeter having an internal resistance of 1 kΩ indicated 100V, then the value of R would be

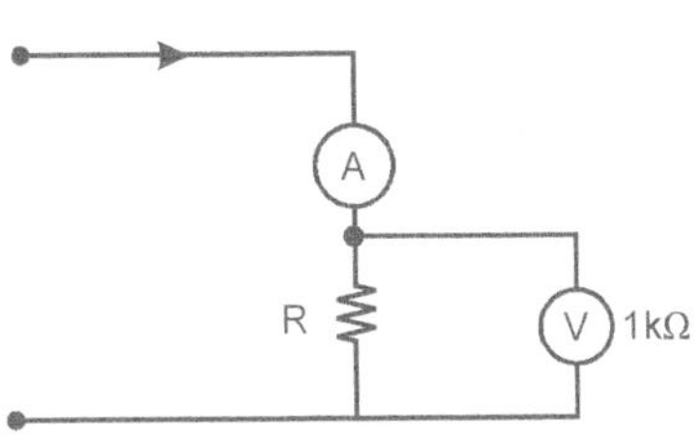

(*a*) 111.11 Ω (*b*) 105.2 Ω
(*c*) 100 Ω (*d*) 90.9 Ω

376. If the largest frequency present in the measured signal and number of bits used in the binary code, are respectively 'f_m' and 'n', then the minimum bandwidth of a pulse code modulation channel used for telemetry would be

(*a*) $\frac{f_m}{2n}$ (*b*) $\frac{f_m}{n}$
(*c*) nf_m (*d*) $2nf_m$

377. Consider the following statements regarding the sources of error in a Q-meter:

1. If a coil with a resistance R is connected in the direct measurement mode and if the residual resistance of the Q-meter is 0.1 R, then the measured Q of the coil world be 1.1 times the actual Q.
2. If the inductance to be measured is less than 0.1 μM, the error due to the presence of residual inductance cannot be neglected.
3. The presence of distributed capacitance in a coil modifies the effective Q of coil.

Of these statements

(*a*) 1, 2 and 3 are correct
(*b*) 1 and 2 are correct
(*c*) 2 and 3 are correct
(*d*) 1 and 3 are correct

378. The maximum percentage quantization error for a 12-bit analog to digital converter is

(*a*) ±0.00076%
(*b*) ±0.012207%
(*c*) ±3.125%
(*d*) ±4.17%

379. Match List-I with List II and select the correct answer using the codes given below the lists:

List-I

A. Display device (lowest power)
B. High speed display device
C. Bright perfect numeral display
D. Perfect light transmitting device

List-II

1. Light emitting diode
2. Nixie Tube
3. Liquid crystal display
4. Optical fibre

Codes:

	A	B	C	D
(*a*)	1	2	3	4
(*b*)	2	4	3	1
(*c*)	3	1	2	4
(*d*)	4	3	2	1

380. In an analog data acquisition unit, the sequence of the blocks starting from the input will be

(*a*) transducer, recorder, signal conditioner
(*b*) transducer, signal conditioner, recorder
(*c*) signal conditioner, transducer, recorder
(*d*) signal conditioner, recorder, transducer

381. Match List-I with List-II and select the correct answer using the codes given below the lists:

List-I

A. Strain

B. Temperature

C. Angular velocity

List-II (*Type of transducers*)

1. Magnetostrictive transducer
2. Magnetoelectric transducer
3. Encoder
4. Thermocouple

Codes:

	A	B	C
(*a*)	1	4	3
(*b*)	2	1	3
(*c*)	3	4	2
(*d*)	1	4	2

382. Siemens is a unit for measuring

(*a*) conductance

(*b*) resistance

(*c*) flux density

(*d*) electric field

383. The time base signal in a CRO is a

(*a*) sinusoidal signal

(*b*) square wave signal

(*c*) sawtooth signal

(*d*) triangular wave signal

384. In a CRO, the time base signal is applied to

(*a*) 7-plates

(*b*) X-plates

(*c*) either X or Y plates

(*d*) both X and Y plates

385. The time base of a CRO is developed by

(*a*) square waveform

(*b*) sawtooth waveform

(*c*) sine waveform

(*d*) output from a build in clock

386. In a CRT the highest positive potential is given to

(*a*) focusing electrodes

(*b*) cathode

(*c*) vertical deflection plates

(*d*) post deflection acceleration anode

387. Thermistor are used as basic sensing elements in a

(*a*) average responding electronic voltmeter

(*b*) peak to peak responding electronic voltmeter

(*c*) true rms responding electronic voltmeter

(*d*) both average and peak to peak responding electronic voltmeter

388. In VTVMs the ac voltage measurements corresponds to

(*a*) saw-tooth wave input

(*b*) square-wave input

(*c*) sine-wave input

(*d*) any waveform at the input

389. Which of the following is used to measure the low impedance components?

(*a*) series connection Q-meter

(*b*) parallel connection Q-meter

(*c*) either (*a*) or (*b*)

(*d*) none of these

390. To prevent the loading of a circuit winder test, the input impedance of the CRO must be

(*a*) very low

(*b*) very high

(*c*) inductive

(*d*) none of these

391. The source of electron beam in a CRO is

(*a*) electron gun

(*b*) grid

(*c*) focussing anode

(*d*) vertical deflection plates

392. In a CRO the quantity to be measured is applied across

(*a*) focusing electrodes

(*b*) cathode

(*c*) Y-plates

(*d*) X-plates

393. Double precision constants in comparision with single precision constants have twice the

(*a*) significant digits in decimal represen-tation

(*b*) accuracy

(*c*) range of magnitude

(*d*) number of binary digits for representation

394. A zero to 300V voltmeter has an error of ± 2% of the full-scale deflection. If the true voltage is 30V, then the range of readings on this voltmeter would be

(*a*) 20V to 40V

(*b*) 24V to 36V

(*c*) 29.4V to 30.6V

(*d*) 29.94V to 30.06V

395. Consider the following statements:

The absolute value of current can be measured by

1. dynamometer type moving coil ammeters
2. tangent galvanometers
3. current balance instruments

Of these statements

(*a*) 1, 2 and 3 are correct (*b*) 1 and 2 are correct

(*c*) 1 and 3 are correct (*d*) 2 and 3 are correct

396. Which of the following are the advantages of a balanced bridge vacuum tube voltameter in comparison with a conventional VTVM?

1. Higher input impedance
2. Meter zero has less tendency to shift
3. Effects of changes due to variation of valve characteristics are minimised
4. Power supply fluctuation have a smaller effect on the measuring circuit

Select the correct answer using the codes given below:

(*a*) 3 and 4 (*b*) 1 and 2

(*c*) 1,2 and 3 (*d*) 1, 2, 3 and 4

397. An electrodynamometer type wattmeter is connected (as shown in the given figure) in a 3-phase supply and having a 3-phase balanced load. E and I are the values of phase voltage and current and ϕ is the phase angle between them. The wattmeter reading will be

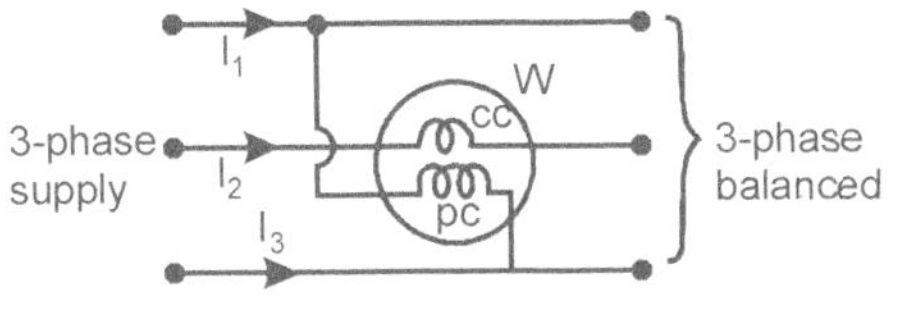

(*a*) proportional to EI sin ϕ

(*b*) proportional to EI cos ϕ

(*c*) proportional to EI tan ϕ

(*d*) zero

398. A compensated wattmeter has its reading corrected for error due to

(*a*) the frequency

(*b*) friction

(*c*) power consumed in current coil

(*d*) power consumed in potential coil

399. While measuring power in a three-phase load by two-wattmeter method, the reading of the two wattmeter will be equal and opposite when

(*a*) power factor is unity

(*b*) load is balanced

(*c*) phase angle is between 60° and 90°

(*d*) the load is purely inductive

400. The Wheatstone bridge method of resistance measurement is ideally suitable for the measurement of resistance values in the range of

(*a*) 0.001 Ω to 1 Ω (*b*) 0.1 Ω to 100 Ω

(*c*) 100 Ω to 10 kΩ (*d*) 100 kΩ to 10 MΩ

401. The coils having self-inductances of 10 mH and 15 mH have an effective inductance of 40 mH, when connected in series-aiding. What will be the equivalent inductance if we connect them in series-opposing?

(*a*) 20 mH (*b*) 10 mH

(*c*) 5 mH (*d*) zero

402. The major cause of creeping in an energy meter is

(*a*) over-compensation for friction

(*b*) mechanical-vibrations

(*c*) excessive voltage across the potential coil

(*d*) stray magnetic fields

403. Pulses of a frequency of 1 MHz are applied to the time base selector of a digital frequency meter which consists of 6 frequency dividers, each dividing the incoming frequency by a factor of 10. The time-base setting at the output of 4th frequency divider starting from the input is

(*a*) 1 ms (*b*) 10 ms

(*c*) 100 ms (*d*) none of these

404. In the Maxwell bridge as shown in given figure, the values of resistance R_x and inductance L_x of a coil are to be calculated after balancing the bridge. The component values are shown in the figure at balance. The values of R_x and L_x will respectively be

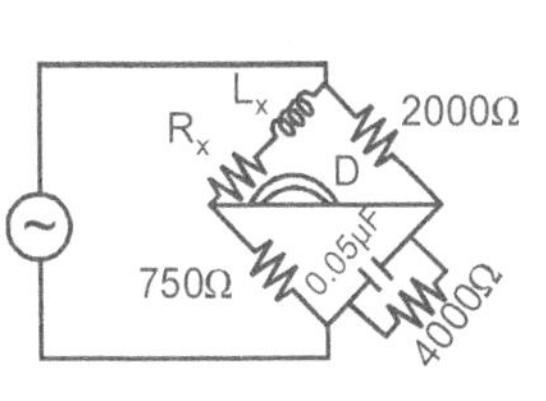

(*a*) 370 Ω, 75 mH (*b*) 75 Ω, 150 mH

(*c*) 37.5 Ω, 75 mH (*d*) 75 Ω, 75 mH

405. Consider the following statements:

DC potentiometer is the best means available for measurement of dc voltages because

1. The precision in measurement is independent of the type of detector used
2. It is based on null-balance technique
3. It is possible to standardize before a measurement is undertaken
4. It is possible to measure dc voltages ranging in value from millivolts to hundreds of volts.

Of these statements

(*a*) 2 and 3 are correct (*b*) 1 and 4 are correct

(*c*) 2 and 4 are correct (*d*) 3 and 4 are correct

406. Galvanometer type recorders use
(*a*) vibration galvanometer
(*b*) ballistic galvanometer
(*c*) D'Arsonval galvanometer
(*d*) tangent galvanometer

407. In a telemetry system used for measuring inaccessible non-electrical quantity, the following sub-systems are used at the measuring end:
1. Transmitter
2. Transducer
3. Amplifier
4. Modulator

(*a*) 2,3,1,4 (*b*) 2, 3, 4, 1
(*c*) 3, 2, 1, 4 (*d*) 3, 2, 4, 1

408. The power delivered to a three-phase load can be measured by the use of 2 wattmeter only when the
(*a*) load is balanced
(*b*) load is unbalanced
(*c*) 3-phase load is connected to the source through 3 wires
(*d*) 3-phase load is connected to the source through 4 wires

409. Which of the following are the features of a charge amplifier?
1. Its output voltage is proportional to charge
2. Its sensitivity is independent of the capacitance of the crystal and connecting cable
3. Its signal to noise ratio tends to be small
4. Its time constant is low

Select the correct answer using the codes given below:
Codes:
(*a*) 1, 2 and 3 (*b*) 2, 3 and 4
(*c*) 1, 3 and 4 (*d*) 1, 2 and 4

410. Match List-I (Measurements) with List-II (Transduction principles) and select the correct answer using the codes given below the lists:

List-I	List-II
A. Rpm	1. Variable resistance
B. Temperature	2. Hall effect
C. Static pressure	3. Piezoelectric effect
D. Current	4. Variable reluctance
	5. Variable capacitance

Codes:

	A	B	C	D
(*a*)	4	1	5	2
(*b*)	1	4	5	2
(*c*)	2	1	3	5
(*d*)	4	1	2	3

411. Which of the following features determine the precision of an integrating type digital voltmeter'
1. Time constant of the integrator
2. Input impedance of the integrator
3. Reference voltage of the comparator
4. Opening time of the gate before counter

Select the correct answer using the codes given below:
Codes:
(*a*) 1, 2 and 3 (*b*) 1, 2 and 4
(*c*) 1, 3 and 4 (*d*) 2, 3 and 4

412. In the circuit given in the figure, the limiting error in the power dissipation 'I^2 R' in the resistor R is

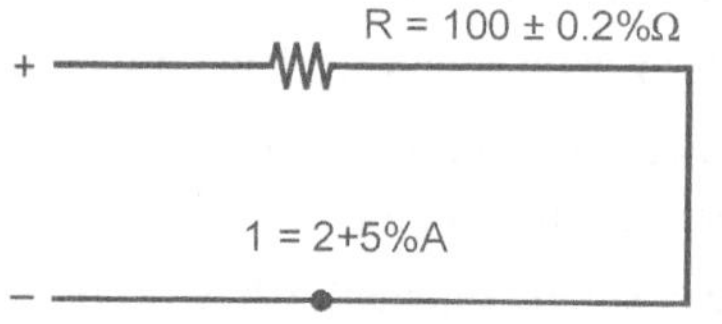

(*a*) 1.2% (*b*) 5.2%
(*c*) 10.2% (*d*) 25.2%

413. In a digital voltmeter, 'over-ranging' implies that
(*a*) the next four digits are switched on
(*b*) 1/2 digit is switched off
(*c*) 1/2 digit is switched on
(*d*) an over-range indicator starts glowing

414. Two voltmeters A and B are connected in series across dc line. A reads 100 volts and has a resistance 100 ohms/volt. B has a total resistance of 15,000 ohms. What is the line voltage?
(*a*) 220 V (*b*) 400 V
(*c*) 250 V (*d*) 230 V

415. The rotor of a synchro-transmitter
(*a*) uses a salient pole type of construction
(*b*) uses a cylindrical pole type of construction
(*c*) may use either salient pole of cylindrical type of construction
(*d*) none of these

416. It is found that 10A current passes through a resistor of 10 ohms value. The resistor has 5% tolerance. The error in measurement of current can be as high as 5%. What is the maximum error in measurement of power, calculated from the measured value of I and the nominal value of resistor?
(*a*) 100 W (*b*) 150 W
(*c*) 50 W (*d*) 200 W

417. Different damping systems are mentioned. Choose the best for PMMC meter.
(*a*) oil (*b*) air
(*c*) eddy current (*d*) electrodynamic

EXERCISE – II

1. The deflection depends on the average value in
   (*a*) Moving-iron meter
   (*b*) Rectifier meter
   (*c*) Hot-wire meter
   (*d*) Moving-coil meter **[NTPC]**

2. Some wire-wound resistors have bifilar winding. This type of winding is used to
   (*a*) increase the thermal stability
   (*b*) reduce the tolerance
   (*c*) reduce the inductance of winding
   (*d*) double the power rating of the resistors **[NTPC]**

3. The practical unit of energy is kWh. The unit of energy in the SI is the joule. The number of joules in kWh is
   (*a*) 3.6
   (*b*) 3.6×10^6
   (*c*) 3.6×10^{-6}
   (*d*) none of these **[BHEL]**

4. The secondary winding of a current transformer is open when current is flowing in the primary. Then,
   (*a*) there will be high current in primary
   (*b*) there will be very high secondary voltage
   (*c*) the transformer will burn out immediately
   (*d*) the meter will burn out **[NTPC]**

5. Ionic wind voltmeter is used for measuring
   (*a*) low voltage
   (*b*) interwinding capacitance
   (*c*) leakage inductance
   (*d*) high voltage **[BHEL]**

6. Consider the following statements regarding an electromagnetic flowmeter;
   1. A.C. excitation is used to avoid polarisation of the fluid.
   2. The meter calibration is changed as the viscosity of the fluid changes.
   3. Stainless steel pipes can be used for measuring the flow of fluids of high conductivtiy.

   Of these statements,
   (*a*) 1, 2 and 3 are correct
   (*b*) 1 and 2 are correct
   (*c*) 2 and 3 are correct
   (*d*) 1 and 3 are correct **[NTPC]**

7. A Hall-effect transducer with Hall coefficient $K_H = -1 \times 10^{-8}$ is required to measure a magnetic field of 10,000 gauss. A 2 mm bismuth slab is used as the transducer with a current of 3 A. The output voltage of the transducer will be
   (*a*) -15×10^{-6} V (*b*) -7.5×10^{-6} V
   (*c*) -20×10^{-6} V (*d*) -22.5×10^{-6} V **[BHEL]**

8. Schering bridge is used to
   (*a*) determine dielectric loss
   (*b*) determine the inductance
   (*c*) measure low resistance
   (*d*) measure mutual inductance **[NTPC]**

9. Which of the following instruments can be used for measurement of alternating current only?
   (*a*) Permanent magnet type ammeter
   (*b*) Induction type ammeter
   (*c*) Moving-iron voltmeter
   (*d*) Moving-iron ammeter **[NTPC]**

10. Electrodynamometer can be used also to measure
    (*a*) Time, frequency, flux
    (*b*) Flux, power, voltage
    (*c*) Power, power factor, reactive power
    (*d*) Power, power factor, reactive power, frequency **[BHEL]**

11. A single pulse has a
    (*a*) single frequency component
    (*b*) continuous frequency spectrum
    (*c*) spectrum of even harmonics
    (*d*) spectrum of odd harmonics **[BHEL]**

12. V_{RN}, V_{YN} and V_{BN} are the instantaneous line to neutral voltages and i_R, i_Y and i_B are instantaneous line currents in a balanced three-phase circuit the computation
    $$V_{RN}(i_Y - i_B) - (V_{YN} - V_{BN})\, i_R$$
    will yield a quantity proportional to
    (*a*) active power
    (*b*) power factor
    (*c*) reactive power
    (*d*) complex power **[GATE]**

13. A milliammeter can be used as
    (*a*) voltmeter and ammeter
    (*b*) wattmeter
    (*c*) ohmmeter
    (*d*) frequency meter **[NTPC]**

14. A Galvanometer has
(*a*) air-friction damping
(*b*) fluid-friction damping
(*c*) spring coil damping
(*d*) eddy current damping **[IOCL]**

15. X and Y plates of a CRO are connected to unequal voltages of equal frequency with phase shift of 90°. The Lissajous figure on the screen will be
(*a*) circle (*b*) straight line
(*c*) ellipse (*d*) figure of eight **[IOCL]**

16. PMMC type instruments normally use
(*a*) air-friction damping
(*b*) fluid-friction damping
(*c*) eddy current damping
(*d*) none of these **[NTPC]**

17. No eddy current and hysteresis losses occur in
(*a*) Electro-static instruments
(*b*) PMMC type instruments
(*c*) Moving-iron instruments
(*d*) Electrodynamometer instruments **[NHPC]**

18. For power measurement of three-phase circuit by two wattmeter method, when the value of power factor is less than 0.5 lagging
(*a*) one of the wattmeters will read zero
(*b*) both give the same readings
(*c*) one of the wattmeter connections will have to be reversed
(*d*) pressure coil of the wattmeter will become ineffective **[IOCL]**

19. A 1000 ohms/V meter is used to measure a resistance on 150 V scale. The meter resistance is
(*a*) 150 kΩ (*b*) 1 k
(*c*) 6.67 ohms (*d*) 0.001 ohm **[BHEL]**

20. Manganin is
(*a*) an insulator material
(*b*) a semiconductor material
(*c*) used for making inductor
(*d*) an alloy used make standard resistors **[BHEL]**

21. The preferred damping condition for indicating instruments is
(*a*) critically damped
(*b*) a damping coefficient of 0.8 to 1
(*c*) over damped
(*d*) underdamped **[NTPC]**

22. The standard secondary voltage for a P.T. is
(*a*) 1 V (*b*) 5 V
(*c*) 50 V (*d*) 110 V **[NTPC]**

23. The shunt used in a milliammeter will
(*a*) extend the range and reduce the meter resistance
(*b*) extend the range and increase the resistance
(*c*) decrease the range and meter resistance
(*d*) not affect the meter **[NTPC]**

24. In two wattmeter method of measuring 3-phase power, power factor is 0.5, then one of the wattmeter will read
(*a*) $\frac{W}{2}$ (*b*) Zero
(*b*) $\sqrt{2}W$ (*d*) $\frac{W}{\sqrt{2}}$ **[IOCL]**

25. With a variable reluctance tachometer having 60 rotor teeth, the counter records 3600 counts per minute. The device speed, in rpm, would be
(*a*) 36 (*b*) 60
(*c*) 360 (*d*) 3600 **[BHEL]**

26. A P.T. is a device which is
(a) electro-statically coupled
(b) electrically coupled
(c) electromagnetically coupled
(d) conductivity coupled **[PGCIL]**

27. Electrostatic instruments are normally used for
(*a*) low current measurements
(*b*) high current measurements
(*c*) low voltage measurements
(*d*) high voltage measurements **[PGCIL]**

28. In three-phase power measurement the power factor of load will be
(*a*) $\cos \tan^{-1} \sqrt{3}\frac{(W_1 - W_2)}{W_1 + W_2}$
(*b*) $\frac{(W_1 + W_2)}{W_1 - W_2}$
(*c*) $\sqrt{3}\frac{W_1 - W_2}{W_1 + W_2}$
(*d*) $\tan^{-1} \frac{W_1 - W_2}{\sqrt{W_1 + W_2}}$ **[PGCIL]]**

29. In a Q-meter, in inductor tunes to 2 MHz with 450 pF and to 4 MHz with 90 pF. The distributed capacitance of the inductor is
(*a*) 30 pF (*b*) 45 pF
(*c*) 90 pF (*d*) 360 pF **[IOCL]**

30. Among the measuring instruments for current and voltage the instrument that can be used as a secondary standard is

(*a*) Thermocouple (*b*) Electrodynamometer

(*c*) Moving iron (*d*) Electrostatic **[NTPC]**

31. The full-scale deflection current of an ammeter is 1 mA and its internal resistance is 100 ohms. If this meter is to have full deflection at 5 A, what is the value of the shunt resistance to be used?

(*a*) 49.99 ohms (*b*) 1/49.99 ohm

(*c*) 1 ohm (*d*) 2 ohms **[BHEL]**

32. In a moving-iron meter, the deflecting torque is proportional to

(*a*) square of the current through the coil

(*b*) current through the coil

(*c*) sine the measurand

(*d*) square-root of the measurand **[BHEL]**

33. An oscilloscope indicates

(*a*) peak to peak value of voltage

(*b*) d.c. value of voltage

(*c*) r.m.s value

(*d*) average value **[BHEL]**

34. The resistance of a 125 ohm strain gauge changes by one ohm for 4000 micro strain. The gauge factor of the strain gauge is

(*a*) 1.5 (*b*) 2.0

(*c*) 2.5 (*d*) 3 **[BEL]**

35. In a dynamometer, wattmeter used for polyphase measurement, the two power readings are equal and opposite,

(*a*) The load is reactive

(*b*) The load is resistive

(*c*) The load is inductive

(*d*) The load is capacitive **[BEL]**

36. When using ohmmeter, applied voltage is to be disconnected from the circuit because

(*a*) voltage source will increase resistance

(*b*) current will decrease resistance

(*c*) the ohmmeter has its own internal battery

(*d*) none of these **[BEL]**

37. When ac voltage is connected to a* PMMC meter, then

(*a*) meter will get damaged

(*b*) reading is zero

(*c*) pointer will oscillate to and fro

(*d*) pointer will not move at all **[NTPC]**

38. The advantage of Hay's bridge over Maxwell's inductance-capacitance bridge is because

(*a*) its equations for balance do not contain any frequency term

(*b*) it can be used for measurement of inductance of high Q coils

(*c*) it can be used for measurement of inductance of low Q coils

(*d*) none of these **[NTPC]**

39. Sensitivity inaccuracy of a recording instrument means

(*a*) the amount of input required to produce unit pen deflection

(*b*) the smallest signal required to produce detectable output

(*c*) the maximum error in sensitivity displayed by a pen

(*d*) degree to which the instrument is not sensitive enough to repeat readings **[PGCIL]**

40. The value of resistance of an earthing electrode depends upon

(*a*) shape and material of electrode

(*b*) depth to which electrode is driven into earth

(*c*) specific resistance of soil

(*d*) all of these **[PGCIL]**

41. A Merz Price Maximum Demand Inducto Indicates

(*a*) maximum demand

(*b*) average maximum demand over a specified period of time

(*c*) maximum energy consumption

(*d*) all of these **[PGCIL]**

42. The power in a 3 phase four wire circuit can be measured by using

(*a*) 2 wattmeters (*b*) 4 wattmeters

(*c*) 3 wattmeters (*d*) 1 wattmeter **[IOCL]**

43. Thermocouple instruments can be used for a frequency range

(*a*) upto 100 Hz

(*b*) upto 5000 Hz

(*c*) upto 1 MHz

(*d*) 50 MHz and above **[IOCL]**

44. A metal strain gauge factor of two, its nominal resistance is 120 ohms. If it undergoes a strain of 10^{-5}, the value of change of resistance in response to the strain is

(*a*) 240 ohms

(*b*) 2×10^{-5}

(*c*) 2.4×10^{-3} ohms

(*d*) 1.2×10^{-3} ohms **[GATE]**

45. The sensitivity of a voltmeter with a full scale indication of 500μA having an internal resistance of 250 ohms is
(*a*) 2000 ohm/V (*b*) 50000 ohm/V
(*c*) 1000 ohm/V (*d*) 2000 ohm/V **[PGCIL]**

46. The disadvantage of a Maxwell bridge is that
(*a*) both inductance and Q of a coil connot be measured
(*b*) measurement is not independent of frequency
(*c*) induction cannot be measured over wide range of Q
(*d*) number of components is large **[IOCL]**

47. The instrument used for the measurement of power at 500 MHz is
(*a*) bolometer
(*b*) compensated wattmeter
(*c*) electrodynamometer wattmeter
(*d*) none of these **[NTPC]**

48. The total power delivered to a three-phase load is equal to
(*a*) algebraic difference of two-wattmeter readings
(*b*) algebraic sum of two-wattmeter readings
(*c*) vectorial difference of two-wattmeter readings
(*d*) vectorial sum of the two-wattmeter readings **[NTPC]**

49. A very accurate voltmeter gives inaccurate readings when used to measure voltage across a law resistance because
(*a*) the sensitivity of the meter is high
(*b*) the sensitivity of the meter is too low
(*c*) the higher scale has been selected
(*d*) the voltmeter is drawing too low current **[BHEL]**

50. Two meters X and Y require 40 mA and 60 mA, respective to give full-scale deflection, then
(*a*) X is more sensitive
(*b*) Y is more sensitive
(*c*) both are equally sensitive
(*d*) sensitivity cannot be judged with the given information **[BEL]**

51. For an instrument to have a wide range, the instrument should have
(*a*) linear scale
(*b*) square-law scale
(*c*) exponential scale
(*d*) logarithmic scale **[BEL]**

52. When selecting a measuring instrument, be sure its range is
(*a*) higher than the value being measured
(*b*) below that of the circuit being tested
(*c*) mid-way on the scale
(*d*) on the lower third of the scale **[BEL]**

53. The principle of pirani gauge is based on
(*a*) humidity of the medium
(*b*) thermal conductivity of the medium
(*c*) computability of the medium
(*d*) none of these **[BEL]**

54. Creeping in energy-meters implies that
(*a*) fast rotation of the rotor with current coil excited
(*b*) slow rotation of the rotor with current coil excited
(*c*) slow rotation of the rotor with only voltage coil excited
(*d*) none of these **[IOCL]**

55. The potentiometer wire should have
(*a*) high specific resistance and high temperature coefficient
(*b*) high specific resistance and low temperature coefficient
(*c*) low specific resistance and low temperature coefficient
(*d*) low specific resistance and low temperature coefficient **[IOCL]**

56. A multimeter has different shunts, which increase the
(*a*) current range
(*b*) voltage ranges
(*c*) resistance range
(*d*) impedance range **[BHEL]**

57. The potential coil of a wattmeter is designed for minimum inductance. Therefore the current through it is
(*a*) inversely proportional to the voltage and out of phase with it
(*b*) proportional to the voltage and out of phase with it
(*c*) proportional to the voltage and in phase with it
(*d*) inversely proportional to the voltage and in phase with it **[BHEL]**

58. Two wattmeter method can be used for measurement of Power
(*a*) for balanced power
(*b*) for unbalanced power
(*c*) for reactive power
(*d*) for all three purposes mentioned above **[BHEL]**

59. The material of wires used for making resistance standards is usually
(*a*) Manganin
(*b*) Nichrome
(*c*) Copper
(*d*) Phosphore Bronze **[NTPC]**

60. Air friction damping should not be used where the deflecting torque in the instrument is produced due to

(*a*) magnetic field

(*b*) electrostatic field

(*c*) therm-electric emf

(*d*) none of these **[NTPC]**

61. The frequency range of moving iron instrument is

(*a*) audio frequency band 20 Hz to 20 kHz

(*b*) very low frequency band 10 Hz to 30 kHz

(*c*) low frequency band 30 Hz to 300 kHz

(*d*) power frequencies 0 to 125 Hz **[NHPC]**

62. The controlling torque in a meagar is provided by

(*a*) spring

(*b*) weights attached to the moving system

(*c*) it does not need any controlling torque

(*d*) none of these **[NHPC]**

63. A rectifier type instrument is connected to 100 V d.c. and is operated in the d.c. measuring mode. It reads

(*a*) 111 V (*b*) 90 V

(*c*) 50 V (*d*) 100 V **[PGCIL]**

64. For high accuracy the meter has

(*a*) supporting mirror

(*b*) hand drawn scale

(*c*) printed scale

(*d*) scale with sharp pointer **[PGCIL]**

65. The instrument used normally to check the insulation resistance is

(*a*) multimeter (*b*) ohm-meter

(*c*) Tong-tester (*d*) Megger **[NHPC]**

66. Why is a MISC meter not recommended for d.c. measurement?

(*a*) The meter is calibrated for a.c. and its error I for d.c. would be high

(*b*) The meter does not respond to dc. signals

(*c*) The error is high due to hysteresis effect

(*d*) The error is high due to eddy-current effect **[NTPC]**

67. A thermocouple ammeter is used to measure a 10 MHz sine wave and it indicates a current of 2A in a pure resistance. The peak current in this waveform is

(*a*) 2 A

(*b*) 2. 83 A

(*c*) 0. 2 A

(*d*) 20 A **[BHEL]**

68. In dc potentiometer measurements, a second reading is often taken after reversing the polarities of the dc supply and unknown voltage and the average of the two readings is taken. This is with a view to eliminate the effects of

(*a*) ripples in the dc supply

(*b*) stray magnetic field

(*c*) stray thermal emf

(*d*) erroneous standardisation **[BHEL]**

69. A 0-10 amps ammeter is guaranteed for 2% accuracy full-scale reading. What will be % limiting error when meter is measuring 5 amps?

(*a*) 2% (*b*) 4%

(*c*) 5% (*d*) 10% **[BEL]**

70. The basic principle required to be satisfied for ac/dc measurement by a meter is

(*a*) deflection $\propto$ current

(*b*) deflection $\propto$ (current)^2

(*c*) deflection = controlling torque

(*d*) electrodynamic damping **[BEL]**

71. Electrodynamic type wattmeters have large errors while measuring power in ac circuits at low power factor conditions, since the voltage across and the current through the

(*a*) current coil are not in phase

(*b*) current coil are not in quadrature

(*c*) pressure coil are not in phase

(*d*) pressure coil are not in quadrature **[NTPC]**

72. A PMMC meter, as a voltmeter, it is more sensitive if the coil resistance is (1) and as an ammeter it is more sensitive if the coil resistance is (2)

(*a*) (1) low (2) high

(*b*) (1) high (2) low

(*c*) (1) low (2) low

(*d*) (1) high (2) high **[BEL]**

73. The output of a piezoelectric crystal has

(*a*) low amplitude and low impedance

(*b*) high amplitude and high impedance

(*c*) low amplitude and high impedance

(*d*) high amplitude and low impedance **[IOCL]**

74. A photo conductive cell is useful for

(*a*) high frequency applications

(*b*) medium frequency applications

(*c*) low frequency applications

(*d*) all of these **[NHPC]**

75. The resolution of a digital voltmeter with 5-digit display is

(*a*) $\frac{1}{5}$ (*b*) $\frac{1}{1000}$

(*c*) $\frac{1}{100000}$ (*d*) 0.2% **[NHPC]**

76. A low voltage is measured on a higher scale of a voltmeter. The measurement would have

(*a*) low accuracy (*b*) low resolution

(*c*) low precision (*d*) all of these **[NHPC]**

77. The dynamic characteristic of capacitive transducers are similar to those of

(*a*) low pass filters (*b*) high pass filters

(*c*) notch filter (*d*) band stop filters **[IOCL]**

78. Sphere gaps are used for measurement of

(*a*) instantaneous values of voltage

(*b*) rms values of voltage

(*c*) peak value of voltage

(*d*) average value of voltage **[IOCL]**

79. In a Q-meter, distributed capacitance of coil is measured by changing the capacitance of the turning capacitor. The value of turning capacitor are C_1 and C_2 for resonant frequencies F_1 and $2F_1$ respectively. The value of distributed capacitance is

(*a*) $\frac{C_1 - C_2}{2}$ (*b*) $\frac{C_1 - 2C_2}{3}$

(*c*) $\frac{C_1 - 4C_2}{3}$ (*d*) $\frac{C_1 - 3C_2}{2}$ **[IOCL]**

80. The dissipation factor of capacitor can be measured by using a

(*a*) potentiometer (*b*) campbell bridge

(*c*) Schering bridge (*d*) galvanometer

[BHEL]

81. The best method of measuring a resistance of the order of 100 kilo ohms is

(*a*) ammeter-voltmeter method

(*b*) potentiometer method

(*c*) Kelvin-double bridge method

(*d*) ohm meter **[BHEL]**

82. Magneto elastic transducers work on the principle of

(*a*) change of dimension with change of applied stress

(*b*) change of permeability with change in stress

(*c*) change of coercive force with change of stress

(*d*) none of these **[BEL]**

83. A $3\frac{1}{2}$ digit, 2V full scale slope ADC has its integration time set to 300 ms. If the input to the ADC is $(1 + 1 \sin 314\,t)$V, then the ADC output will be

(*a*) 1.000 (*b*) 1.999

(*c*) 1.414 (*d*) 1.500 **[GATE]**

84. The moving iron voltmeter indicate

(*a*) the same value for dc and ac voltages

(*b*) lower values for ac voltages than for corresponding dc voltages

(*c*) higher values for ac voltage than for corresponding dc voltages

(*d*) none of these **[PGCIL]**

85. The horizontal amplifier in the C.R.O. should be designed for

(*a*) high frequency signals with a fast rise time

(*b*) high amplitude signals with a slow rise time

(*c*) high amplitude signals with a fast rise time

(*d*) low amplitude signals with a fast rise time **[PGCIL]**

86. Electronic voltmeters which use rectifiers employ negative feedback. This is done

(*a*) to increase the overall gain

(*b*) to improve stability

(*c*) to overcome non-linearity of diodes

(*d*) none of these **[PGCIL]**

87. The standardization of ac potentiometers is done by

(*a*) directly using ac standard voltage sources

(*b*) using dc standard sources and transfer instruments

(*c*) using dc standard sources and d' Arsonval galvanometer

(*d*) using ac standard sources and transfer instruments

[BEL]

88. Temperature inside a boiler furnace can be measured by

(*a*) mercury thermometer

(*b*) biometallic thermometer

(*c*) optical pyrometer

(*d*) thermistor **[NHPC]**

89. The main advantage of a strip-chart recorder is that it provides a

(*a*) record you can see at a glance

(*b*) record for long unattended periods

(*c*) daily record on a single sheet

(*d*) record of voltage and current, using the same instrument **[BHEL]**

90. The minimum number of wattmeters required to measure the real power in an, N-phase system with unbalanced load is

(*a*) N – 1 (*b*) N

(*c*) N + 1 (*d*) none of these **[BHEL]**

91. A rectifier is used in an instrument for the purpose of

(*a*) measuring high voltage values

(*b*) measuring high current values

(*c*) converting ac to dc

(*d*) making the instrument more stable **[BEL]**

92. Which of the following is electro mechanical device?

(*a*) Direct current motor

(*b*) LVDT

(*c*) Induction relay

(*d*) None of these **[BHEL]**

93. Hydrometer is used to determine

(*a*) specific gravity of liquids

(*b*) relative humidity

(*c*) specific gravity of solids

(*d*) specific gravity of gases **[BHEL]**

94. The name plate of a single phase Energy meter generally provides

(*a*) meter constant (*b*) voltage and frequency

(*c*) current (*d*) all of these **[ONGC]**

95. Electrodynamic effect is used for operation of

(*a*) wattmeter (*b*) voltmeter

(*c*) ammeter (*d*) all of these **[ONGC]**

96. Very small resistances can be measured more accurately by

(*a*) Resistance box (*b*) Kelvin-double bridge

(*c*) Anderson bridge (*d*) Schering bridge **[BHEL]**

97. Electrostatic instruments work on the principle

(*a*) repulsion of unlike poles

(*b*) attraction of like poles

(*c*) attraction between positive and negative charge

(*d*) repulsion between negative charges **[NTPC]**

98. A phase sequence indicator is used to determine the phase sequence of any energized

(*a*) 3-phase line (*b*) single phase line

(*c*) dcline (*d*) telephone line **[NTPC]**

99. A 0-10 mA PMMC ammeter reads 4 mA in a circuit. Its bottom control spring snaps suddenly. The meter will now read nearly

(*a*) 10mA (*b*) 8 mA

(*c*) 2mA (*d*) zero **[GATE]**

100. For a given frequency, the deflecting torque of an induction ammeter is directly proportional to

(*a*) current^2 (*b*) current^3

(*c*) $\sqrt{\text{current}}$ (*d*) current **[GATE]**

101. If the fault current is 2000A, the relay setting is 50% and CT ratio is 400/5, the plug setting multiplier will be

(*a*) 25A (*b*) 15A

(*c*) 50A (*d*) 10A **[GATE]**

102. The scale of a voltmeter is uniform. Its type is

(*a*) moving iron

(*b*) induction

(*c*) moving coil permanent magnet

(*d*) moving coil dynamometer **[GATE]**

103. The moving coil in a dynamometer wattmeter is connected

(*a*) in series with the fixed coil

(*b*) across the supply

(*c*) in series with the load

(*d*) across the load **[GATE]**

104. An advantage of a permanent magnet moving coil instrument is that it is

(*a*) free from friction error

(*b*) has high (torque/weight of the moving parts) ratio

(*c*) has low (torque/weight of the moving parts) ratio

(*d*) can be used on both ac and dc **[GATE]**

105. Kelvin double bridge is best suited for the measurement of

(*a*) inductance (*b*) capacitance

(*c*) low resistance (*d*) high resistance **[GATE]**

106. The primary current in a current transformer is dictated by

(*a*) the secondary burden

(*b*) the core of the transformer

(*c*) the load current

(*d*) none of these **[GATE]**

107. A dynamometer type wattmeter responds to the

(*a*) average value of active power

(*b*) average value if reactive power

(*c*) peak value of active power

(*d*) peak value of reactive power **[GATE]**

108. A certain oscilloscope with 4 cm by 4 cm screen has its own sweep output fed to its input. If the *x* and *y* sensitivities are same, the oscilloscope will display a

(*a*) triangular wave (*b*) diagonal line

(*c*) sine wave (*d*) circle **[GATE]**

109. A transfer instrument employed in the standardization of a polar type ac potentiometer is a/an

(*a*) electrostatic instrument

(*b*) thermal instrument

(*c*) dynamometer instrument

(*d*) moving coil instrument **[GATE]**

110. Two helical springs are used in a D' Arsonval meter movement because

(*a*) it improves damping

(*b*) it compensates for temperature changes

(*c*) it controls the defecting torque effectively

(*d*) it improves torque-weight ratio **[BHEL]**

111. The material most preferred for control spring is

(*a*) German silver (*b*) Platinum silver

(*c*) Silicon bronze (*d*) Phosphor bronze **[NTPC]**

112. Usually a C.T. has

(*a*) poorer overload capacity than a P.T.

(*b*) the same overload capacity as a P.T.

(*c*) a higher overload capacity than a P. T.

(*d*) no overload capacity **[BEL]**

113. In the circuit shown, through oversight, the meters have been interchanged. This is likely to result in

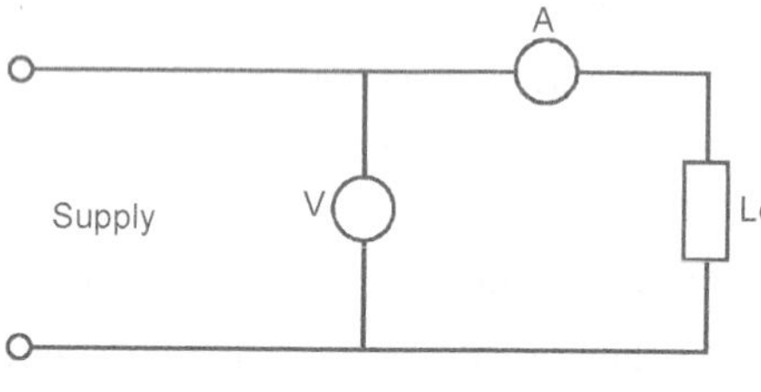

(*a*) damage to ammeter (*b*) damage to voltmeter

(*c*) damage to both (*d*) no damage **[IOCL]**

114. The following remarks pertain to the sensitivity of a PMMC meter and electro-dynamometer (ED). Choose the correct one.

(*a*) ED is less than a PMMC

(*b*) They are the same

(*c*) PMMC is less than ED

(*d*) It depends upon the purpose of measurement **[BEL]**

115. The advantage of the LVDT over the inductive bridge-type transducer is that

(*a*) it produces high output voltage for small changes in core position

(*b*) it can be used for operation at temperatures as low as -265°C and as high as 500°C

(*c*) it can measure displacement of a moving object

(*d*) the error due to stray magnetic field is less **[BEL]**

116. The current coil of a wattmeter is connected to the C.T. of R-phase. The potential coil is connected across Y and B phases. The wattmeter measures

(*a*) active power in R phase

(*b*) active power of Y phase

(*c*) reactive power in R phase

(*d*) power proportional to 3-phase power if the load is balanced **[BHEL]**

117. In modern electronic multimeter, a FET or MOSFET is preferred over BJT because

(*a*) its input resistance is high

(*b*) its input resistance is high and does not change with range switch

(*c*) its input resistance is low

(*d*) it is cheaper **[NTPC]**

118. Which one of the following statements is NOT correct?

(*a*) If everything else is equal, then a 10 bit digital ramp ADC will have a better resolution but a longer conversion time than an 8 bit ADC

(*b*) The conversion time for a successive approximation increase with the increase in input voltage

(*c*) A flash ADC does not contain a DAC

(*d*) VCO is the main component of a voltage to frequency ADC **[IAS]**

119. Given figure shows the input attenuation of a multimeter. If the meter reads fullscale with 10V at 'M' with the range switch at position 'p' then the required voltage at '*r*' to obtain full-scale deflection with the range switch at position '*r*' will be

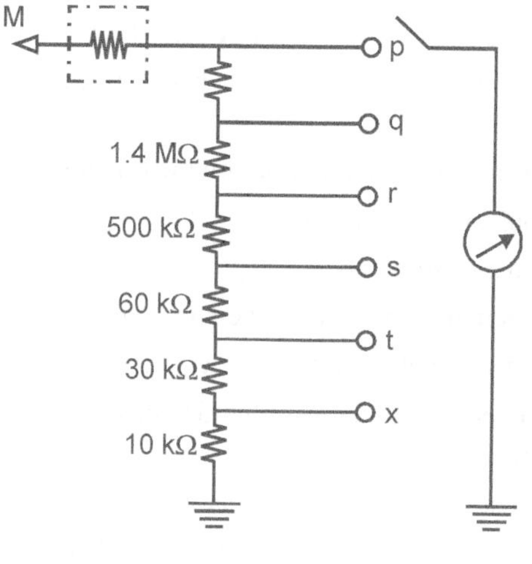

(*a*) 10V

(*b*) 200V

(*c*) 100V

(*d*) none of these **[IAS]**

120. Which of the following units are needed for designing a data acquisition system using 8050 microprocessor?

1. An A/D converter
2. A D/A converter
3. A programmable peripheral device
4. A timer

Select the correct answer by using the codes given below:

Codes:

(*a*) 1 and 4 (*b*) 3 and 4

(*c*) 1 and 3 (*d*) 2 and 4 **[IAS]**

121. Match List-I (Type of instrument) with List-II (Suitability for measurement) and select the correct answer using the codes given below the lists:

List-I

A. Moving iron

B. Electrodynamic

C. Induction

D. Electrostatic

List-II

1. ac/dc voltage only
2. ac/dc voltage/current only
3. ac voltage/current/power only
4. ac/dc voltage/current/power
5. dc voltage/current

Codes:

	A	B	C	D
(*a*)	2	4	3	1
(*b*)	2	4	1	3
(*c*)	1	2	3	5
(*d*)	1	2	5	3

[IAS]

122. Horizontal deflection in a CRO is due to E sin ωt while vertical deflection is due to E sin $(\omega t + \theta)$ with a positive θ. Consider the following patterns obtained in the CRO.

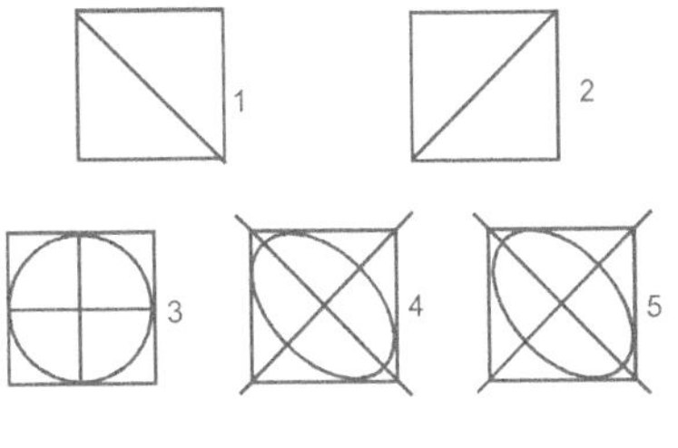

The correct sequence of these patterns in increasing order of the values of θ is:

(*a*) 3, 2, 5, 1, 4

(*b*) 3, 2, 4, 5, 1

(*c*) 2, 3, 5, 1, 4

(*d*) 2, 3, 5, 4, 1 **[ES]**

123. In the circuit shown in the given figure, L_2 is a known self-inductance of resistance r_x If R_3 and R_4 are non-inductive resistors and D is a detector, 4 then ,4 is

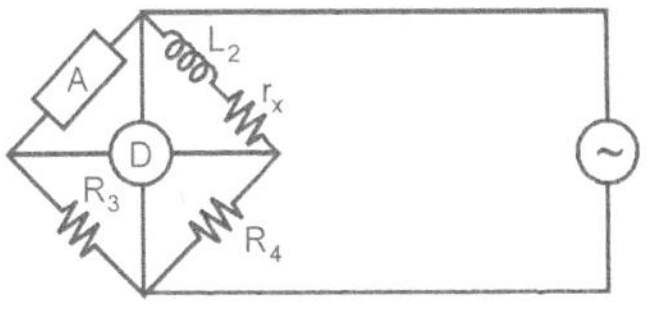

(*a*) an unknown self-inductance having resistance R_1

(*b*) an unknown capacitor

(*c*) an unknown non-inductive resistance *r*

(*d*) a pair of inductance and capacitance **[IAS]**

124. Let $\frac{dM}{d\theta}$ be the rate of change of mutualt inductance between the fixed and moving coils with respect to the angular deflection θ in an electrodynamic spring controlled ammeter. Which one of the following is the correct condition for obtaining a linear scale in the ammeter?

(*a*) $\theta \frac{dM}{d\theta}$ is constant

(*b*) $\frac{1}{\theta}\frac{dM}{d\theta}$ is constant

(*c*) $\theta^2 \frac{dM}{d\theta}$ is constant

(*d*) $\frac{1}{\theta^2}\frac{dM}{d\theta}$ is constant **[IAS]**

125. Which of the following are the requirements of a CRO (cathode ray oscilloscope) for display of high frequency signals (mega hertz)?

1. Input capacitance should be low
2. Cathode ray tube should be provided with post-acceleration
3. Short persistence phosphors should be used in the screen
4. Electromagnetic focussing and deflection system should be used

Select the correct answer using the codes given below:

Codes:

(*a*) 1, 2 and 4 (*b*) 1, 3 and 4

(*c*) 1, 2 and 3 (*d*) 2, 3 and 4 **[IAS]**

126. Magneto-electric transducers work on the principle of

(*a*) change of dimensions with change of applied stress

(*b*) change of permeability with change in stress

(*c*) change of co-ercive force with change of stress

(*d*) none of these **[BHEL]**

127. In a gravity controlled instrument, the deflection angle is proportional to

(*a*) measurand

(*b*) square of the measurand

(*c*) sine-inverse of the measurand

(*d*) sine of the measurand **[IOCL]**

128. To measure flux, devices used are based on

(*a*) voltaic effect

(*b*) piezo-electric effect

(*c*) hall effect

(*d*) photo-voltaic effect **[IOCL]**

129. To the *y*-input of a CRO we feed a signal defined by 10 sin 100*t*. To the *x* input, we fred the signal 10 cos 100*t*. The gain for both *x* channel and *y*-channel is the same. The screen shows

(*a*) sinusoidal signal

(*b*) a straight line

(*c*) an ellipse

(*d*) a circle **[PGCIL]**

130. In a digital frequency meter, the Schmidt trigger is used for

(*a*) sinusoidal waveforms into rectangular pulses

(*b*) scaling of sinusoidal waveforms

(*c*) providing time base

(*d*) none of these **[NTPC]**

131. Low resistance is measured by

(*a*) De sauty's bridge

(*b*) Maxwell's bridge

(*c*) Kelvin's double bridge

(*d*) Wien bridge **[PGCIL]**

132. The dielectric loss can be measured by,

(*a*) Wien bridge

(*b*) Moving-iron meter

(*c*) Moving-coil meter

(*d*) Electro-static meter **[NTPC]**

133. A permanent magnet moving coil ammeter has a coil resistance of 99Ω and F.S.D. current of 0.1 mA. Shunt resistance is 1Ω current through the meter at 0.5 F.S.D. is

(*a*) 0.07 mA (*b*) 0.05 mA

(*c*) 0.025 mA (*d*) 0.1 mA **[IOCL]**

134. The maximum deflection that is possible with a PMMC movement is

(*a*) 360° (*b*) 90°

(*c*) 120° (*d*) 100° **[BHEL]**

135. The full-scale deflection current of a meter is 1 mA and its internal resistance is 100 ohms. This meter is to have full deflection when 100V is measured. What is the value of series resistor to be used?

(*a*) 99.90 k ohms (*b*) 100 k ohms

(*c*) 99.99 ohms (*d*) 100 ohms **[IOCL]**

136. The absolute measurement of resistance is done by

(*a*) Ohm's law method

(*b*) Bridge method

(*c*) Rayleigh method

(*d*) Lorenz method **[BHEL]**

137. The internal resistance of an ammeter should be

(*a*) very small (*b*) medium

(*c*) high (*d*) infinity **[BHEL]**

138. The resistance can be measured most accurately by

(*a*) voltmeter-ammeter method

(*b*) bridge method

(*c*) multimeter

(*d*) megger **[IOCL]**

139. The only voltmeter that reads directly voltage and not by the indirect method of current passing through a resistor is

(*a*) Moving iron (*b*) Koving coil

(*c*) Electrostatic (*d*) Thermocouple

[NTPC]

140. The purpose of applying post deflection potential is to

(*a*) increase deflection sensitivity

(*b*) speed up the electrons

(*c*) increase the brightness of the spot

(*d*) keep deflection angle unchanged **[IOCL]**

141. Standardization of potentiometers is done in order that, they become

(*a*) accurate

(*b*) precise

(*c*) accurate and direct reading

(*d*) accurate and precise **[NTPC]**

142. The internal resistance of an unsaturated Weston cell used as a secondary standard is

(*a*) < 1 Ω (*b*) 100 Ω

(*c*) 600 Ω (*d*) 1500 Ω **[NTPC]**

143. The percentage limiting error, in the case of an instrument reading of 8.3V with a 0 to 150V voltmeter having a guaranteed accuracy of 1% full-scale reading is

(*a*) 1.810% (*b*) 0.181%

(*c*) 18.10% (*d*) 0.0018%

[NTPC]

144. The reading of an electrodynamometer ammeter when a current $I = 8 - 6\sqrt{2}\sin\left(\omega t + \frac{\pi}{2}\right)$ passed through will be

(*a*) 8 A (*b*) $6\sqrt{2}$ A

(*c*) $8 - 6\sqrt{2}$ A (*d*) 10 A **[IOCL]**

145. A milli ammeter has resistance in the range of

(*a*) Ohms (*b*) kΩ

(*c*) MΩ (*d*) hundredths of ohm **[NHPC]**

146. An *n*-bit ADC using V_R as reference voltage has a resolution (in volts) of

(*a*) $\frac{V_R}{2^n}$ (*b*) $\frac{V_R}{n}$

(*c*) $\frac{V_R}{2^{n-1}}$ (*d*) $\frac{V_R}{2n}$ **[IOCL]**

147. The current in a circuit consists of a DC component of 10A superposed on a 50 Hz sinusoidal component of rms value of 10A. The reading shown by a PMMC ammeter will be

(*a*) $10\sqrt{2}$ (*b*) $\frac{20}{\pi}$

(*c*) 10 A (*d*) $10 + \frac{10}{\pi}$ **[BEL]**

148. High ac voltages are measured with

(*a*) current transformers and voltmeter

(*b*) voltage transformers and voltmeter

(*c*) electrostatic voltmeter

(*d*) induction voltmeter **[BEL]**

149. An ammeter has resistance in the range of

(*a*) Ohms

(*b*) kΩ

(*c*) MΩ

(*d*) hundredths of ohm **[BEL]**

150. A CRT has a deflection factor of 18 V/cm. The amount of deflection seen on the screen for deflection voltages of 54 V is

(*a*) 3 cm (*b*) 18 cm

(*c*) 54 cm (*d*) 0.056 cm **[IOCL]**

151. The precision of an instrument indicates its ability to reproduce a certain reading with a

(*a*) drift

(*b*) resolution

(*c*) shift

(*d*) consistency **[IOCL]**

152. A 20,000 Ω/V meter is used to measure voltage in the 250 V range (*a*) across a 20,000 Ω resistor (*b*) a 2 MΩ resistor. The meter will cause loading across

(*a*) both

(*b*) 20,000 Ω resistor only

(*c*) 2 MΩ resistor

(*d*) none **[NTPC]**

153. A dc electronic voltmeter using chopper stabilization is free from errors due to

(*a*) low CMRR

(*b*) amplifier drift

(*c*) source output impedance

(*d*) interference **[NTPC]**

154. The principle of operation of an LVDT is based on variation of

(*a*) self-inductance

(*b*) mutual inductance

(*c*) reluctance

(*d*) permeance **[ES]**

155. A digital voltmeter has $4\frac{1}{2}$ digit display. The 1 volt range can read upto

(*a*) 1.0000 (*b*) 1.1111

(*c*) 1.9999 (*d*) 19999 **[ES]**

156. In the statement "the wattmeter commonly used for power measurement at commercial frequencies is of the X-type. This meter consists of two coil systems, the fixed system being the Y-coil and moving system being the Z-coil." X, Y and Z stand respectively for

(*a*) dynamometer, voltage and current

(*b*) dynamometer, current and voltage

(*c*) induction, voltage and current

(*d*) induction, current and voltage **[ES]**

157. A temperature sensitive transducer is subjected to a sudden temperature change. It takes 10 seconds for the transducer to reach equilibrium condition (five time constants).

The time taken by the transducer to read half of the temperature difference will be nearly

(*a*) 1.38 second (*b*) 5.00 second

(*c*) 8.62 second (*d*) second **[ES]**

158. In CRT aquadag carries

(*a*) aqueous solution of graphite

(*b*) sweep voltage

(*c*) secondary emission electrons

(*d*) none of these **[ES]**

159. Match List-I with List-II and select the correct answer using the codes given below the lists:

List-I	List-II
A. LED	1. Optical scattering
B. LCD	2. Glow discharge
C. NIXIE	3. Electromagnetic radiation emitter

Codes:

	A	B	C
(*a*)	1	3	2
(*b*)	2	1	3
(*c*)	3	1	2
(*d*)	3	2	1

[ES]

160. The resistance of a thermistor is 5000Ω at 20°C and its resistance temperature coefficient is 0.04/°C. A measurement with a lead resistance of 10Ω will cause an error of

(*a*) 0.05°C (*b*) 0.1°C
(*c*) 0.4°C (*d*) 0.8°C **[ES]**

161. The meter constant of a single-phase 240V induction watt-hour meter is 400 revolutions per kWh. The speed of the meter disc for a current of 10 amperes at 0.8 p.f. lagging will be

(*a*) 12.8 rpm (*b*) 16.02 rpm
(*c*) 18.2 rpm (*d*) 21.1 rpm **[ES]**

162. In a particular form of frequency meter, a 1 μF capacitor is connected across a symmetrical square wave signal of 1 volt peak value. If the average value of the current taken by the capacitor, after full wave rectification is measured as 2 mA, then the frequency of the signal will be

(*a*) 1000/μHz (*b*) 500 Hz
(*c*) 1000 Hz (*d*) 1000 μHz **[ES]**

163. The measurement of very low and very high frequencies is invariably done using a frequency counter/timer in the

(*a*) frequency measurement mode only
(*b*) period measurement mode only
(*c*) frequency and period measurement modes respectively
(*d*) period and frequency measurement modes respectively **[ES]**

Directions: *The following items consist of two statements, one labelled as 'Assertion A' and the other labelled as 'Reason R'. You are to examine these two statements carefully and decide if the Assertion A and the Reason A are individually true and if so, whether the Reason is a correct explanation of the Assertion. Select your answers to these item using the codes given below and mark your answer sheet accordingly.*

Codes:

(*a*) *Both A and R are true and R is the correct explanation of A*
(*b*) *Both A and R are true but R is not a correct explanation of A*
(*c*) *A is true but R is false*
(*d*) *A is false but R is true*

164. **Assertion (A):** For High-Q coils, inductance measurement is better done with Hay's Bridge rather than with Maxwell's Bridge.

Reason (B): Hay's Bridge uses the standard capacitor in series with a variable resistor while Maxwell's Bridge uses it in parallel. **[IAS]**

165. **Assertion (A) :** A permanent magnet moving coil instrument does not work satisfactorily when the frequency of the signal is high.

Reason (R) : The power loss in the moving coil becomes excessive at high frequency. **[IAS]**

166. **Assertion (A):** Low resistances are not measured by Wheatstone bridge arrangements.

Reason (R) : The effects of contact and lead resistances detrimental to low resistance measurements, are present in large amounts in Wheatstone bridge method. **[ES]**

167. **Assertion (A):** In order to produce a light spot at the point at which the electron beam strikes the CRO screen, the inner surface of the screen is coated with a phosphor.

Reason (R): The phosphor absorbs the kinetic energy of the bombarding electrons and remits energy at a higher frequency. **[IAS]**

168. A voltmeter with a resistance of 10 is connected with two other 10 resistors across a 200V3-phase mains, as shown in the figure.

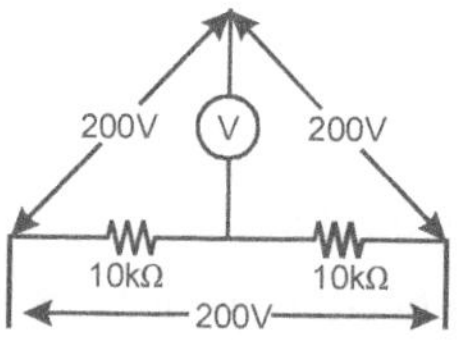

Assertion (A) : The voltmeter reading will be 115K

Reason (R): The connections are similar to set connection **[ES]**

169. **Assertion (A):** Data acquisition systems are widely used in a variety of industrial and scientific areas.

Reason (R): Data acquisition systems often use magnetic tape recorders. **[IAS]**

170. **Assertion (A):** The vertical deflecting plates of a CRT are kept farther away from the screen as compared to the horizontal deflecting plates.

Reason (R) : This improves accuracy in measurements. **[ES]**

171. Assertion (A) : The resistance scale of the electronic multimeter reads increasing resistance from left to right.

Reason (R) : Electronic multimeter draws lower current for high measured resistance. **[ES]**

172. Assertion (A) : In a good quality cassette audio tape recorder, ac bias is provided for recording purposes to improve the quality.

Reason (R) : The ac bias provided for recording in an audio tape recorder takes care of the non-linearity of the recording head. **[ES]**

173. Assertion (A) : Q-meter cannot be used with a capacitive element.

Reason (R) : The Q value of a coil is the ratio of ωL to its resistance R. **[ES]**

174. Assertion (A) : Fast events can be captured and displayed by a real-time spectrum analyzer.

Reason (R): All components of a signal are presented simultaneously in the frequency domain. In swept type spectrum analyzer, frequencies are supplied one at a time at a particular rate. **[ES]**

175. Assertion (A): Vertical deflection sensitivity of a CRO is higher than that of the horizontal system.

Reason (R): Sweep voltage is applied to the horizontal deflecting plates. **[IAS]**

176. Assertion (A): The bridge shown in the figure is balanced by first adjusting R_1 for inductive balance and then adjusting R_3 for resistive balance; and this is repeated till balance is achieved.

Reason (R): For medium-Q coils, the resistance effect is not pronounced by balance is reached after a few adjustments. **[ES]**

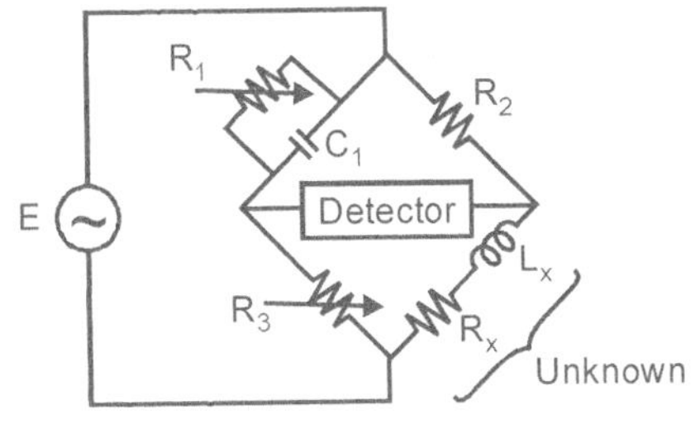

177. Assertion (A): The VTVM is a balanced bridge voltmeter.

Reason (R): VTAM is the transistorised version of VTVM. **[ES]**

178. Assertion (A): General purpose dynamometer type wattmeter cannot indicate the correct value of power at low power factors.

Reason (R): The presence of self-inductance in the pressure coil circuit introduces an error in the indicated value which increases appreciably with decreases in power factor of the load. **[ES]**

179. Assertion (A): While using an ohmmeter for the measurement of resistance, the meter is to be disconnected from the external voltage source, if already connected.

Reason (R): No current is needed for meter movement. **[ES]**

180. Assertion (A): A variance is dependent on the mean value of the set of data.

Reason (R): Variance is the difference of the mean squared values and the square of the mean value of the set cf data. **[ES]**

181. Assertion (A): The opening of the secondary circuit of a current transformer results in a very high voltage in the secondary winding with consequent severe strain on the insulation and danger to the operator.

Reason (R): The high magnetic forces acting on the core, if suddenly removed, may leave, behind considerable residual magnetism in the core. **[ES]**

182. Assertion (A): Eddy current damping is not suitable in a moving iron type instrument.

Reason (R): Deflecting torque of a moving iron type instrument is proportional to the square of the current. **[ES]**

183. Assertion (A): To achieve good resolution in a spectrum analyser, the bandwidth should be narrow.

Reason (R): Resolution in any instrument is proportional to its accuracy. **[ES]**

184. Assertion (A): The basic principle of operation of a Q-meter is based on the property of a series-resonant circuit.

Reason (R): If a fixed voltage is applied to a series resonant circuit, the voltage developed across its capacitor is Q times the applied voltage. **[ES]**

185. Assertion (A): The screen of the CRT is coated with phosphor on the inside.

Reason (R): Phosphor absorbs the kinetic energy of the bombarding electrons and remits energy at a frequency in the visible region. **[IAS]**

186. Assertion (A): Dual-slope A/D converter provides high accuracy in A/D conversion approach in digital multimeter.

Reason (R): Dual-slope A/D converter provides high accuracy in A/D conversion while at the same time suppressing the hum effect on the input signal. **[ES]**

187. Assertion (A): The frequency response shown in the given figure is that of a vibration galvanometer.

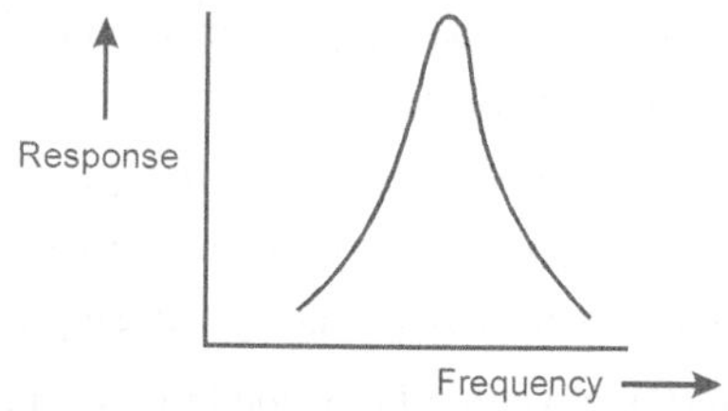

Reason (R): A vibration galvanometer is the most suitable detector in ac potentiometric method of measurement. **[ES]**

188. Assertion (A): The resolution of a digit voltmeter is 0.001.

Reason (R): Addition of $3\frac{1}{2}$ digit to a digital voltmeter increases the range of the meter. **[ES]**

189. Assertion (A): spectrum analyzer is an instrument which displays the spectrum of the input waveform with respect to the frequency.

Reason (R): A spectrum analyzer consists of an internal sweep voltage generator connected to the X-deflecting plate of the CRT. **[ES]**

190. Assertion (A): Electrometers are used at the output stage of electrode pH meters.

Reason (R): In electrode type pH meters, the potential is measured using reference and measuring electrodes. **[ES]**

191. Assertion (A): Precision is a necessary prerequisite to accuracy.

Reason (R): Precision guarantees accuracy. **[ES]**

ANSWERS

EXERCISE – I

1. (*b*)	**2.** (*b*)	**3.** (*b*)	**4.** (*d*)	**5.** (*d*)	**6.** (*a*)	**7.** (*d*)	**8.** (*a*)	**9.** (*b*)	**10.** (*a*)
11. (*b*)	**12.** (*b*)	**13.** (*c*)	**14.** (*d*)	**15.** (*b*)	**16.** (*c*)	**17.** (*d*)	**18.** (*b*)	**19.** (*d*)	**20.** (*d*)
21. (*a*)	**22.** (*b*)	**23.** (*a*)	**24.** (*d*)	**25.** (*c*)	**26.** (*d*)	**27.** (*a*)	**28.** (*b*)	**29.** (*b*)	**30.** (*a*)
31. (*b*)	**32.** (*b*)	**33.** (*d*)	**34.** (*b*)	**35.** (*d*)	**36.** (*b*)	**37.** (*b*)	**38.** (*a*)	**39.** (*b*)	**40.** (*d*)
41. (*d*)	**42.** (*d*)	**43.** (*d*)	**44.** (*a*)	**45.** (*d*)	**46.** (*b*)	**47.** (*b*)	**48.** (*c*)	**49.** (*d*)	**50.** (*a*)
51. (*d*)	**52.** (*c*)	**53.** (*d*)	**54.** (*d*)	**55.** (*a*)	**56.** (*b*)	**57.** (*a*)	**58.** (*c*)	**59.** (*b*)	**60.** (*a*)
61. (*b*)	**62.** (*c*)	**63.** (*b*)	**64.** (*d*)	**65.** (*b*)	**66.** (*d*)	**67.** (*a*)	**68.** (*b*)	**69.** (*b*)	**70.** (*b*)
71. (*b*)	**72.** (*a*)	**73.** (*b*)	**74.** (*d*)	**75.** (*a*)	**76.** (*d*)	**77.** (*d*)	**78.** (*c*)	**79.** (*a*)	**80.** (*c*)
81. (*b*)	**82.** (*c*)	**83.** (*a*)	**84.** (*c*)	**85.** (*b*)	**86.** (*c*)	**87.** (*d*)	**88.** (*b*)	**89.** (*b*)	**90.** (*a*)
91. (*a*)	**92.** (*c*)	**93.** (*b*)	**94.** (*)	**95.** (*c*)	**96.** (*b*)	**97.** (*b*)	**98.** (*c*)	**99.** (*d*)	**100.** (*d*)
101. (*a*)	**102.** (*a*)	**103.** (*d*)	**104.** (*a*)	**105.** (*c*)	**106.** (*c*)	**107.** (*c*)	**108.** (*a*)	**109.** (*b*)	**110.** (*b*)
111. (*b*)	**112.** (*c*)	**113.** (*b*)	**114.** (*d*)	**115.** (*b*)	**116.** (*c*)	**117.** (*b*)	**118.** (*c*)	**119.** (*b*)	**120.** (*d*)
121. (*c*)	**122.** (*c*)	**123.** (*a*)	**124.** (*c*)	**125.** (*b*)	**126.** (*b*)	**127.** (*a*)	**128.** (*a*)	**129.** (*b*)	**130.** (*d*)
131. (*c*)	**132.** (*b*)	**133.** (*a*)	**134.** (*d*)	**135.** (*b*)	**136.** (*c*)	**137.** (*a*)	**138.** (*a*)	**139.** (*d*)	**140.** (*c*)
141. (*b*)	**142.** (*d*)	**143.** (*b*)	**144.** (*c*)	**145.** (*b*)	**146.** (*b*)	**147.** (*a*)	**148.** (*c*)	**149.** (*a*)	**150.** (*a*)
151. (*c*)	**152.** (*c*)	**153.** (*a*)	**154.** (*a*)	**155.** (*c*)	**156.** (*d*)	**157.** (*a*)	**158.** (*d*)	**159.** (*c*)	**160.** (*a*)
161. (*a*)	**162.** (*)	**163.** (*c*)	**164.** (*d*)	**165.** (*c*)	**166.** (*d*)	**167.** (*d*)	**168.** (*d*)	**169.** (*d*)	**170.** (*a*)
171. (*d*)	**172.** (*c*)	**173.** (*b*)	**174.** (*c*)	**175.** (*b*)	**176.** (*b*)	**177.** (*a*)	**178.** (*d*)	**179.** (*c*)	**180.** (*c*)
181. (*a*)	**182.** (*b*)	**183.** (*c*)	**184.** (*d*)	**185.** (*a*)	**186.** (*b*)	**187.** (*d*)	**188.** (*d*)	**189.** (*a*)	**190.** (*c*)
191. (*a*)	**192.** (*a*)	**193.** (*a*)	**194.** (*b*)	**195.** (*b*)	**196.** (*d*)	**197.** (*b*)	**198.** (*d*)	**199.** (*a*)	**200.** (*b*)
201. (*d*)	**202.** (*c*)	**203.** (*a*)	**204.** (*b*)	**205.** (*a*)	**206.** (*d*)	**207.** (*b*)	**208.** (*c*)	**209.** (*b*)	**210.** (*a*)
211. (*a*)	**212.** (*a*)	**213.** (*a*)	**214.** (*d*)	**215.** (*d*)	**216.** (*d*)	**217.** (*c*)	**218.** (*a*)	**219.** (*c*)	**220.** (*a*)
221. (*c*)	**222.** (*d*)	**223.** (*c*)	**224.** (*c*)	**225.** (*a*)	**226.** (*d*)	**227.** (*d*)	**228.** (*b*)	**229.** (*d*)	**230.** (*a*)

231. (*b*) 232. (*c*) 233. (*d*) 234. (*b*) 235. (*b*) 236. (*c*) 237. (*c*) 238. (*d*) 239. (*d*) 240. (*a*)
241. (*b*) 242. (*c*) 243. (*d*) 244. (*a*) 245. (*a*) 246. (*d*) 247. (*d*) 248. (*b*) 249. (*b*) 250. (*c*)
251. (*d*) 252. (*d*) 253. (*d*) 254. (*d*) 255. (*a*) 256. (*b*) 257. (*a*) 258. (*d*) 259. (*c*) 260. (*b*)
261. (*c*) 262. (*a*) 263. (*c*) 264. (*d*) 265. (*a*) 266. (*a*) 267. (*a*) 268. (*b*) 269. (*d*) 270. (*b*)
271. (*a*) 272. (*c*) 273. (*d*) 274. (*b*) 275. (*c*) 276. (*d*) 277. (*d*) 278. (*c*) 279. (*d*) 280. (*d*)
281. (*b*) 282. (*c*) 283. (*c*) 284. (*d*) 285. (*a*) 286. (*a*) 287. (*a*) 288. (*d*) 289. (*c*) 290. (*b*)
291. (*c*) 292. (*a*) 293. (*c*) 294. (*d*) 295. (*b*) 296. (*c*) 297. (*b*) 298. (*a*) 299. (*a*) 300. (*a*)
301. (*a*) 302. (*d*) 303. (*c*) 304. (*a*) 305. (*d*) 306. (*c*) 307. (*c*) 308. (*c*) 309. (*a*) 310. (*b*)
311. (*a*) 312. (*a*) 313. (*b*) 314. (*b*) 315. (*c*) 316. (*b*) 317. (*b*) 318. (*a*) 319. (*d*) 320. (*a*)
321. (*a*) 322. (*b*) 323. (*a*) 324. (*b*) 325. (*d*) 326. (*c*) 327. (*c*) 328. (*a*) 329. (*a*) 330. (*b*)
331. (*b*) 332. (*a*) 333. (*a*) 334. (*b*) 335. (*a*) 336. (*a*) 337. (*d*) 338. (*d*) 339. (*c*) 340. (*b*)
341. (*c*) 342. (*a*) 343. (*a*) 344. (*d*) 345. (*a*) 346. (*b*) 347. (*c*) 348. (*b*) 349. (*b*) 350. (*b*)
351. (*b*) 352. (*b*) 353. (*c*) 354. (*c*) 355. (*b*) 356. (*c*) 357. (*d*) 358. (*d*) 359. (*d*) 360. (*d*)
361. (*a*) 362. (*c*) 363. (*a*) 364. (*b*) 365. (*b*) 366. (*a*) 367. (*c*) 368. (*b*) 369. (*a*) 370. (*b*)
371. (*b*) 372. (*d*) 373. (*c*) 374. (*d*) 375. (*a*) 376. (*c*) 377. (*c*) 378. (*d*) 379. (*c*) 380. (*b*)
381. (*d*) 382. (*a*) 383. (*c*) 384. (*b*) 385. (*b*) 386. (*d*) 387. (*c*) 388. (*c*) 389. (*a*) 390. (*b*)
391. (*a*) 392. (*c*) 393. (*d*) 394. (*b*) 395. (*d*) 396. (*d*) 397. (*a*) 398. (*b*) 399. (*d*) 400. (*b*)
401. (*b*) 402. (*a*) 403. (*b*) 404. (*a*) 405. (*a*) 406. (*a*) 407. (*b*) 408. (*c*) 409. (*c*) 410. (*a*)
411. (*b*) 412. (*c*) 413. (*b*) 414. (*c*) 415. (*c*) 416. (*b*) 417. (*d*)

EXERCISE – II

1. (*b*) 2. (*c*) 3. (*b*) 4. (*b*) 5. (*d*) 6. (*d*) 7. (*a*) 8. (*a*) 9. (*b*) 10. (*d*)
11. (*b*) 12. (*c*) 13. (*a*) 14. (*d*) 15. (*a*) 16. (*c*) 17. (*a*) 18. (*c*) 19. (*a*) 20. (*d*)
21. (*b*) 22. (*d*) 23. (*a*) 24. (*b*) 25. (*b*) 26. (*a*) 27. (*d*) 28. (*a*) 29. (*a*) 30. (*b*)
31. (*b*) 32. (*a*) 33. (*a*) 34. (*b*) 35. (*a*) 36. (*c*) 37. (*b*) 38. (*b*) 39. (*c*) 40. (*d*)
41. (*b*) 42. (*c*) 43. (*b*) 44. (*c*) 45. (*d*) 46. (*c*) 47. (*a*) 48. (*b*) 49. (*b*) 50. (*a*)
51. (*d*) 52. (*a*) 53. (*b*) 54. (*c*) 55. (*b*) 56. (*a*) 57. (*c*) 58. (*d*) 59. (*a*) 60. (*d*)
61. (*d*) 62. (*c*) 63. (*a*) 64. (*b*) 65. (*d*) 66. (*c*) 67. (*b*) 68. (*c*) 69. (*b*) 70. (*b*)
71. (*c*) 72. (*a*) 73. (*c*) 74. (*c*) 75. (*c*) 76. (*d*) 77. (*b*) 78. (*c*) 79. (*c*) 80. (*c*)
81. (*d*) 82. (*b*) 83. (*b*) 84. (*b*) 85. (*b*) 86. (*c*) 87. (*b*) 88. (*c*) 89. (*b*) 90. (*a*)
91. (*c*) 92. (*b*) 93. (*a*) 94. (*d*) 95. (*d*) 96. (*b*) 97. (*c*) 98. (*a*) 99. (*c*) 100. (*a*)
101. (*d*) 102. (*b*) 103. (*b*) 104. (*b*) 105. (*c*) 106. (*c*) 107. (*a*) 108. (*b*) 109. (*b*) 110. (*b*)
111. (*d*) 112. (*c*) 113. (*d*) 114. (*a*) 115. (*a*) 116. (*c*) 117. (*b*) 118. (*b*) 119. (*c*) 120. (*c*)
121. (*a*) 122. (*d*) 123. (*a*) 124. (*a*) 125. (*d*) 126. (*b*) 127. (*c*) 128. (*c*) 129. (*d*) 130. (*a*)
131. (*c*) 132. (*d*) 133. (*b*) 134. (*a*) 135. (*a*) 136. (*d*) 137. (*a*) 138. (*b*) 139. (*c*) 140. (*c*)
141. (*c*) 142. (*c*) 143. (*c*) 144. (*d*) 145. (*a*) 146. (*a*) 147. (*c*) 148. (*b*) 149. (*b*) 150. (*a*)
151. (*d*) 152. (*c*) 153. (*c*) 154. (*b*) 155. (*c*) 156. (*b*) 157. (*a*) 158. (*a*) 159. (*b*) 160. (*a*)
161. (*a*) 162. (*a*) 163. (*d*) 164. (*a*) 165. (*c*) 166. (*a*) 167. (*c*) 168. (*c*) 169. (*a*) 170. (*b*)
171. (*a*) 172. (*b*) 173. (*d*) 174. (*a*) 175. (*d*) 176. (*d*) 177. (*d*) 178. (*a*) 179. (*d*) 180. (*c*)
181. (*b*) 182. (*b*) 183. (*c*) 184. (*a*) 185. (*a*) 186. (*a*) 187. (*b*) 188. (*b*) 189. (*a*) 190. (*d*)
191. (*c*)

EXPLANATIONS

EXERCISE - I

1. $$\text{Force} = \text{mass} \times \text{acceleration} = MLT^{-2}$$

2. $$\text{Power} = \frac{\text{Work}}{\text{Time}} = \frac{ML^2T^{-2}}{T} = ML^2T^{-3}$$

3. $$\text{Torque} = \text{force} \times \text{distance} = MLT^{-2} \times L = ML^2T^{-2}$$

9. We have charge, $Q = IT$

$$\text{emf}, E = \frac{\text{Work done}}{\text{Charge}} = ML^2T^{-3}I^{-1}$$

Flux is given by, $$E = N\frac{d\phi}{dt}$$

$\therefore$ $$[\phi] = ML^2T^{-3}I^{-1}$$

Now, flux density, $$B = \frac{\text{Flux}}{\text{Area}} = [MT^{-2}I^{-1}]$$

20. $$\text{Error} = 100 \times \frac{1}{100} = \pm 1V$$

When measuring 2V,

$$\text{Relative error} = \pm\frac{1}{2} \times 100 = \pm 50\%$$

21. Maximum static error $= \pm\dfrac{0.25 \times 200}{100} = \pm 0.5°C$

24. $$P = I^2R$$

$\therefore$ $$\frac{dP}{P} = 2\frac{dI}{I} + \frac{dR}{R} = \left(2\times\frac{0.5}{100} + \frac{0.2}{100}\right)\times 100$$

$$\frac{dP}{P} = \pm 1.2\%$$

And $$P = 4 \times 4 \times 100 = 1600W$$

25. $$\theta = \theta_o + (\theta_e - \theta_o)\exp(^{-t}/_T) = 0 + (50 - 0)\,e^{-(0.6/1)} = 27.4$$

43. Equivalent resistance of 600 ohm and 1200 ohm

$$= \frac{600 \times 1200}{1800} = 400 \text{ ohm}$$

We have $$\frac{R}{35-5} = \frac{400}{5}$$

$\Rightarrow$ $$R = 2400 \text{ ohms} = 2.4\,k\Omega$$

48. Under balanced condition, we have

$$R_1\left(R + \frac{1}{j\omega c}\right) = R_2\frac{\frac{1}{j\omega c}}{R + \frac{1}{j\omega c}}$$

$$\Rightarrow \frac{R_1(j\omega c R + 1)}{j\omega c} = \frac{R_2R}{j\omega cR + 1}$$

$$\Rightarrow R_1(-c^2\omega^2R^2 + 1 + j\omega cR) = jR_2c\omega R$$

$$\therefore \quad 2c\omega RR_1 = C\omega RR_2$$

$$\Rightarrow R_1 = \frac{R_2}{2}$$

and $$c^2\omega^2R^2 + 1 = 0$$

$$\Rightarrow \omega = \frac{1}{Rc}$$

$$\therefore \phi = \frac{1}{2\pi Rc}$$

49. Under balanced condition

$$R_1R_4 = R_2R_3$$

$$\therefore R_4 = \frac{R_2R_3}{R_1} = \frac{5 \times 100}{10} = 50\Omega$$

Relative limiting error in unknown resistance

$$\frac{dR_4}{R_4} = \pm\left(\frac{dR_2}{R_2} + \frac{dR_1}{R_1} + \frac{dR_3}{R_3}\right) = \pm(3 + 5 + 1) = \pm 10\%$$

50. We have,

Gauge factor, $$G = \frac{\Delta R/R}{\text{Strain}}$$

$\therefore$ $$\text{Strain} = \frac{0.52}{110 \times 2.03} = 2.32 \times 10^{-3}$$

52. Here, $i = 1 + 2\sin\omega t$,

$$v = 2 + 3\sin 2\omega t$$

$$P = Vi$$

$$\therefore = (1 + 2\sin\omega t)(2 + 3\sin 2\omega t) = 2 + 4\sin\omega t + 3\sin 2\omega t + 6\sin\omega t\sin 2\omega t$$

Wattmeter gives average values of power, hence

$$P = 2W$$

53.

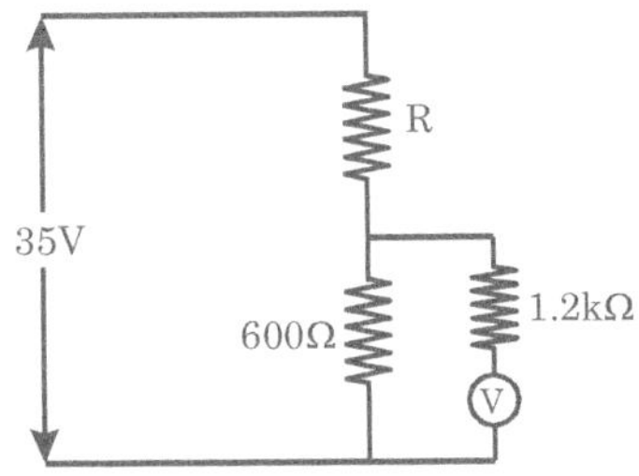

Combined resistance of 1.2 μΩ and 600Ω

$$= \frac{1200 \times 600}{1800} = 400\ \Omega$$

Voltage across combined resistance = 5V

∴ Voltage across unknown resistance

$$= 35 - 5 = 30\ \text{V}$$

∴ Unknown resistance,

$$R = \frac{30}{5} \times 400 = 2.4\ \text{k}\Omega$$

54. We have,

$$\text{Power} = \text{VI}\cos\phi$$

Now, $$\text{Power} = \frac{\text{Work done}}{\text{Time}} = \frac{\text{ML}^2\text{T}^{-2}}{\text{T}} = \text{ML}^2\text{T}^{-3}$$

Hence dimension of VI cos $\phi = \text{ML}^2\text{T}^{-3}$

56. $$F = \frac{1}{4\pi\varepsilon 0}\frac{Q_1Q_2}{r^2}$$

∵ $q = it$

∴ $[Q] = [AT]$

$[F] = MLT^{-2}$

$[v] = L$

∴ $$[\epsilon] = \frac{A^2T^2}{MLT^{-2}L^2} = M^{-1}L^{-3}T^{-4}A^2$$

58. To reconstruct original signal, minimum sampling frequency should be atleast twice the original signal frequency.

∴ Sampling frequency

$$= 2 \times 2 \times 10^6$$
$$= 4 \times 10^6 \text{ samples/s}$$

60. To avoid autioliasing and to limit band width of the analog signal to less than half sampling frequency in order to eliminate frequency folding.

∴ Cut-off frequency of the low pass filter should be

less than $\frac{1000}{2} = 500$ Hz

62. $$\frac{f_y}{f_x} = \frac{3f}{f} = 3 = \frac{\text{Number of vertical tangency}}{\text{Number of horizontal tangency}}$$

65. Rms value of current $= \sqrt{\frac{1}{T}\int_0^T I^2(t)\,dt}$

$$= \left[\frac{1}{T}\int_o^{T/2}\left(\frac{20}{T}t\right)^2 dt + \int_{T/2}^{T}\left(\frac{20t}{T} - 20\right)^2 dt\right]^{1/2}$$

$$= \left[\frac{400}{3} + \frac{400}{2} - 300\right]^{1/2} = 5.77$$

66. At balance we have

$$\frac{z_1}{z_2} = \frac{z_3}{z_x}$$

$$\Rightarrow \quad z_x = \frac{z_2 z_3}{z_1}$$

$$\Rightarrow \quad z_x = \frac{R_2R_3}{R_1}(R_1C_1j\omega + 1)$$

$$\Rightarrow \quad R_x + j\omega L_x = \frac{R_2R_3}{R_1} + j\omega R_2R_3C_1$$

$$\therefore \quad L_x = R_2R_3C_1$$

74. To measure 100V, the voltmeter choosen should be such that its rated voltage should be equal to the voltage to be measured and this should have highest possible resistance.

77. Resolution $= \frac{9.999}{9999} = 1 \times 10^{-3}\text{V} = 1\ \text{mV}$

80. We have, $r_{cc} = 0.01\Omega$

$r_{vc} = 1000\Omega$

We can connect current coil and voltage coil can be connected. The wattmeter reads high by an amount equal to power loss in current coil in one case and in voltage coil in the other. Hence power loss in two connections are

(*a*) $I_c^2\, r_{cc} = 20^2 \times 0.01 = 4\text{W}$

(*b*) $\frac{V^2}{r_{vc}} = \frac{30^2}{1000} = 0.9\text{W}$

Percentage errors are

(*i*) $\frac{4}{600} \times 100 = 0.66$

(*ii*) $\frac{0.9}{600} \times 100 = 0.15$

84. Gauge factor $= \frac{\Delta R/R}{\Delta L/L}$

But we have $\frac{\Delta L}{L} = \frac{100}{2 \times 10^{-5}}$

$$\therefore \quad \frac{\Delta R}{R} = \frac{3\times100}{2\times10-5} = 1.5\times10^{-3}$$

Hence percentage change in resistance

$$= \frac{\Delta R}{R}\times100$$
$$= 1.5\times10^{-3}\times100$$
$$= 0.15$$

90. A change of 1°C in temperature is reflected as a change of $5000\times0.04 = 200$ ohms of resistance in the thermistor resistance

∴ A resistance change of 10Ω is equivalent to

$$\frac{10}{200} = 0.05°C$$

91. Number of pulse $= 400\times10^{3}\times20\times10^{-3}$
$= 8000$

92.
$$Q = \frac{L\omega_o}{R} = \frac{1}{C\omega_o R}$$
$$= \frac{1}{\frac{500}{20}\times10^{-2}\times2\pi\times10^{7}\times10} = 314$$

95.
$$\text{Sensitivity} = \frac{\text{Output response}}{\text{Input}}$$
$$= \frac{3}{6} = 0.5 \text{ mm/ohm}$$

98. Resolution $= \frac{10}{100}\times\frac{1}{10} = 0.01$ V

107. Accuracy $= 300\times\frac{2}{100} = 6$V

when measuring 180 V, limiting error

$$= \frac{6}{180}\times100 = 3.33\%$$

112. All the three resistance will be in series

∴ Voltage indicated by meter

$$= \frac{120}{120+60}\times50$$
$$= 33.3\text{mV}$$

120. Resolution $= \frac{0.01}{100} = 10^{-4}$

Now let there be "N" bits needed to have this resolution

$$\therefore \quad \frac{1}{2^N} = 10^{-4}$$

$$\Rightarrow \quad 2^N = 10000$$

Hence minimum numbers of bit,

$$N = 14$$

122. To have balanced bridge, we have

$$\frac{R_1}{\frac{jR_1\omega c_1+1}{R_3}} = \frac{R_2}{R+j\omega L}$$

$$\Rightarrow \quad \frac{R_1}{1+j\omega R_1c_1} = \frac{R_2R_3}{R+j\omega L}$$

$$\Rightarrow R_1R + jR_1\omega L = R_3R_2 + j\omega R_1c_1R_3R_2$$

Equating imaginary part

$$L = c_1R_3R_2$$
$$= 0.1\times10^{-6}\times200\times100$$
$$= 2\text{ mH}$$

123.
$$f_1 = \frac{1}{2\pi\sqrt{(c_2+c_d)^L}} = 2\text{ MHz}$$

$$f_2 = \frac{1}{2\pi\sqrt{(c_2+c_d)^L}} = 4\text{ MHz}$$

$$\therefore \quad \frac{f_2}{f_1} = 2$$

$$\therefore \quad 2 = \sqrt{\frac{c_1+c_d}{c_2+d_d}}$$

Squaring, $$4 = \frac{430+c_d}{100+c_d}$$

$$\therefore \quad c_d = 10\text{ pF}$$

128. Total number of steps $= 2^{10} = 1023$

130. Current in each branch abc and adc,

$$I = \frac{10}{30} = \frac{1}{3}\text{A}$$

$$\therefore \quad V_{bd} = -\left(20\times\frac{1}{3}\right)+10\times\frac{1}{3}$$

$$= -\frac{10}{3}\text{V} = -3.33\text{ V}$$

Hence reading of voltmeter = 3.33 V

131. We have, gauge factor,

$$G = \frac{\Delta R/R}{\text{Strain}}$$

$$\therefore \quad \frac{\Delta R}{R} = G\times\frac{F}{A\in}$$

$$= \frac{2\times10.5\times10^{4}\times10^{3}}{2.1\times10^{8}\times10^{-3}} = 10^{-3}$$

$$\therefore \quad \%\frac{\Delta R}{R} = 0.1\%$$

132. Electrodyamic ammeter reading

$$= \sqrt{5^2 + \left(\frac{10}{\sqrt{2}}\right)^2} = \sqrt{5^2 + \frac{10^2}{2}}$$

142. Voltage across the shunt

$$= 0.6 \times 0.1 = 0.06 \text{ V}$$

Current in the shunt

$$= 15 - 0.1 = 14.9$$

$\therefore$ Resistance of the shunt

$$= \frac{0.06}{14.9} = 0.004\Omega$$

144. Moving iron meter gives rms value of input current

$$\therefore \text{ Reading} = \sqrt{10^2 + \left(\frac{10}{\sqrt{2}}\right)^2} = \sqrt{150}$$

145. We have voltmeters V_1 and V_2 are connected in series, hence

$$\frac{V_1}{V_2} = \frac{100 \times 10^3 \times 300}{150 \times 10^3 \times 300} = \frac{1}{1.5}$$

Now $\quad V_1 + V_2 = 500$

$\therefore \quad V_1 = 200 \text{ V},$

$V_2 = 300 \text{ V}$

148. Energy consumed as indicated by energy meter

$$= \frac{90}{1800} = 0.05 \text{ kwh}$$

Actual energy consumed

$$= 230 \times \frac{10}{2} \times 1 \times \frac{3}{60} \times 10^{-3}$$

$$= 0.0575 \text{ kwh}$$

$\therefore \quad \text{Error} = 0.05 - 0.0575 = -0.0075$

Hence meter is slow by

$$\frac{0.0075}{0.05} \times 100 = 15\%$$

149. For 10V range switch is thrown to pass current only through "R_1"

Now, full scale current = 1 mA

For 10V range: $\quad 10^{-3} = \dfrac{10}{100 + R_1}$

$\Rightarrow \quad R_1 = 9.9 \times 10^{-3}\Omega$

For 50V rang : $\quad 10^{-3} = \dfrac{50}{(100 + R_1 + R_2)}$

$\Rightarrow \quad R_1 + R_2 = 49.9 \times 10^3$

$\Rightarrow \quad R_2 = (49.9 - 9.9) \times 10^3 = 40\text{k}\Omega$

151. We have, $\quad q = cV$

$$\therefore \quad C = \frac{q}{V} = \frac{it}{V}$$

Since both voltage and current units are made 20 times,

hence $\quad$ new value of $C = \dfrac{20it}{20V} = \dfrac{it}{V}$

Hence size of unit of capacitance will be same

156.

$$P = I^2R$$

$$R = \frac{P}{I^2}$$

$\therefore$ Limiting error in the measurement of resistance

$$\frac{dR}{R} = \pm\left(\frac{dP}{P} + 2\frac{dI}{I}\right) + 100$$

$$= \pm\left(\frac{1.5}{100} + \frac{2 \times 1}{100}\right) \times 100$$

$$= \pm\left(\frac{3.5}{100}\right) \times 100 = \pm 3.5\%$$

162.

$$\text{Current} = \frac{50}{50} = 1 \text{ mA}$$

$\therefore$ Total voltage across the current

$$= (25 + 50) \times 1 = 75\text{V}$$

163. Voltage across 50 kΩ $= \dfrac{50}{150} \times 150 = 50\text{V}$

$\therefore$ Voltmeter resistance

$$= 1000 \times 50 = 50 \text{ k}\Omega$$

Hence voltmeter ready

$$= \frac{\frac{50 \times 50}{2}}{100 + \frac{50 \times 50}{2}} \times 150$$

$$= \frac{25}{125} \times 150 = 30\text{V}$$

167.

$$\tan\theta = \sqrt{3}\,\frac{\omega - \omega}{\omega + \omega} = 0$$

$\Rightarrow \quad \theta = 0$

$\therefore \quad$ Power factor $= \cos 0 = 1.0$ pF.

168. When power factor is below 0.5, one of the wattmeter will give negatives indication. Under these condition in order to read the wattmeter, we must either reverse the current or pressure coil connections.

177.

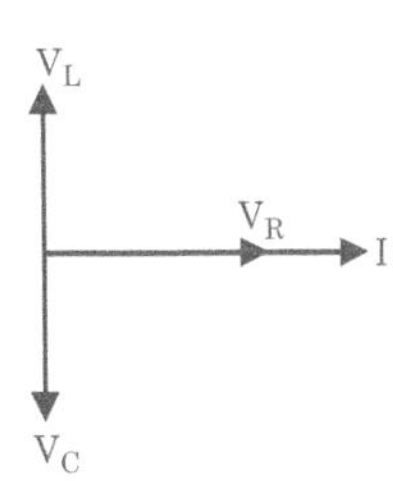

$$V_R^2 + V_L^2 = (20)^2 = 400 \quad ...(i)$$
$$V_L \pm V_C = 9 \quad ...(ii)$$
$$V_R^2 + (V_L - V_c)^2 = 15^2 \quad ...(iii)$$
$$\therefore \quad V_R^2 = 144$$
$$\Rightarrow \quad V_R = 12$$

From equation (*i*), $144 + V_L^2 = 400$

$$\Rightarrow \quad V_L^2 = 256$$
$$\Rightarrow \quad V_R = 16$$

From equation (*ii*), $V_L \pm V_c = 9$

$$\therefore \quad V_c = 16 \pm 9 = 7, 25 \text{ V}$$

178. $$\text{Emf} = \frac{\text{work done}}{\text{charge}}$$

Charge, $Q = I.t$

$= [T. I]$

$\therefore \quad emf = [ML^2T^{-3}I^{-1}]$

Resistance, $[R] = \dfrac{E}{I} = [ML^2T^{-3}I^{-2}]$

Capacitance, $[C] = \dfrac{Q}{E} = M^{-1}L^{-2}T^4I^2$

179. Setting of the potentiometer at 1.0138 instead of 1.0183. Therefore, working current in the slide wire is more than what it would have been if setting was at 1.0183, i.e. correct value for applied voltage of 1.0183V

$$\therefore \quad \frac{I}{I_c} = \frac{1.0183}{1.0138}$$

If setting is a 0.0183 and then applied voltage is also 1.0183, the standardisation is correct.

185. We can draw circuit as follow

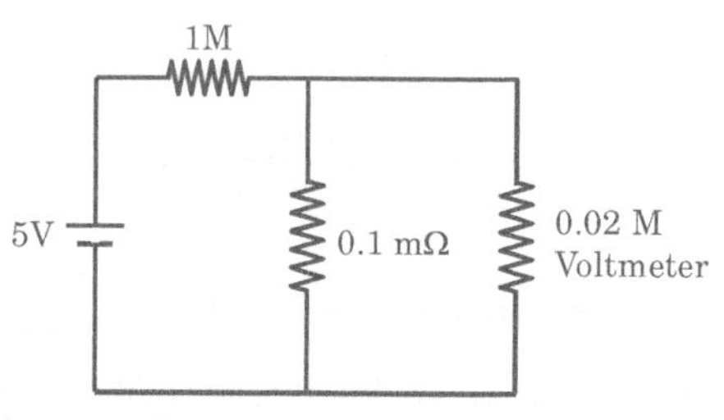

Resistance of parallel connection

$$= \frac{0.1 \times 0.02}{0.1 \times 0.02}$$

$$= \frac{0.002}{01/2} = \frac{1}{50} \text{M}\Omega$$

Total resistance of the circuit

$$= 1 + \frac{1}{60} = \frac{61}{60}\Omega$$

$\therefore$ Voltmeter reading

$$= \frac{5 \times \frac{1}{60}}{\frac{61}{60}} \text{volt} = 82 \text{ mV}$$

199. $$\text{Error} = 100 \times \frac{1}{100} = 10 \text{ W}$$

When measured power is 100W

$$\text{Relative error} = \pm \frac{10}{100} \times 100 = \pm 10\%$$

201. Here, $z_1 = 450\Omega$

$z_2 = (300 - j600)\Omega$

$z_3 = (200 - j100)\Omega$

Now, $$\frac{z_2}{z_3} = \frac{z_1}{z_4}$$

$$\therefore \quad z_4 = \frac{z_1 z_3}{z_2} = \frac{450(200 + j100)}{300 - j600}$$

$$= j150\,\Omega$$

202. $$\text{Error} = \frac{1}{100} \times 150 = 1.5\text{V}$$

Now when measuring 75 V,

$$\text{Percentage error} = \frac{1.5}{75} \times 100 = 2.0$$

205. At balance, $$\frac{Z_3}{Z_4} = \frac{Z_1}{Z_2}$$

$$\therefore \quad \frac{Z_3}{Z_4} = \frac{300\angle 0}{600\angle 30} = \frac{1}{2}\angle -30°$$

$$\frac{Z_1}{Z_2} = \frac{200\angle 60}{400\angle -90} = \frac{1}{2}\angle 150$$

Since phase angles are not matching, hence bridge cannot be balanced with the given configuration

207. $$\text{Energy consumed} = \frac{230 \times 20 \times 1 \times 2}{1000} \text{kWh}$$

$$= 9.2 \text{ kWh}$$

Revolution made = 1380

$$\text{Meter constant} = \frac{\text{Revolution made}}{\text{Energy consumed}}$$

$$= \frac{1390}{9.2} = 150 \; revs/kWh.$$

209. Equivalent resistance of 600Ω and 1200Ω

$$= \frac{600 \times 1200}{600 + 1200} = 400\Omega$$

$$\therefore \quad \text{Current in the circuit} = \frac{5}{400} = 0.0125$$

$$\text{Voltage across R} = 35 - 5 = 30\text{ V}$$

$$\therefore \quad R = \frac{30}{0.0125} = 2400\Omega = 2.4\text{ k}\Omega$$

221. Voltmeter used should have internal resistance of as high of possible, IIIrd voltmeter is of highest internal resistance hence it should be used.

223. $$Q = \frac{10}{0.5} = \frac{100}{5} = 20$$

226. To get circular pattern, we should have

$$V_1 = V_2$$

$$\therefore \quad R_i = \frac{1}{2\pi fc}.i$$

$$\Rightarrow \quad R = \frac{1}{2\pi fc}$$

$$= \frac{1}{2\times 3.14\times 50\times 1\times 10^{-6}}$$

$$= 3184.7\Omega = 3.18\text{k}\Omega$$

229. When R_x is not connected,

$$I_{fs} = \frac{E}{R_1 + R_m} i$$

$$\text{Now,} \quad I_m = \frac{R_x}{R_x + R_m} \cdot \frac{E}{R_1 + \frac{R_m R_x}{R_m + R_x}}$$

$$= -\frac{E}{R_x + R_m} \cdot \frac{R_x}{R_1 + \frac{R_m R_x}{R_m + R_x}}$$

$$= \frac{ER_x}{R_1(R_m + R_x) + R_m R_x}$$

$$= \frac{ER_x}{R_x(R_1 + R_m) + R_1 R_m}$$

$$= \frac{E}{(R_1 + R_m)} \cdot \frac{R_x}{R_x + \frac{R_1 R_m}{R_1 + R_m}}$$

$$= I_{fs} \frac{R_x}{R_m + R_p}$$

$$\text{where,} \quad R_p = \frac{R_1 R_m}{R_1 + R_m}$$

230. Under balanced condition, there is no current flowing through the galvanometer, hence "E" will equal voltage across 200Ω resistor.

$$\therefore \quad E = \frac{3.2\times 200}{2800 + 200 + 200}$$

$$= \frac{3.2\times 200}{3200} = 200\text{ mV}$$

231. Since there is no controlling force, hence pointer will move to full scale and hence it will read 10A

245. Reading of the first voltmeter,

$$V_1 = \frac{30000}{30000 + 20000}\times 600$$

$$= 360$$

Reading of the second voltmeter

$$= \frac{20000}{30000 + 20000}\times 600 = 240\text{ V}$$

246. Reading of voltmetrs depend on their internal resistance when connected in series, since both the voltmeters have same resistance, hence their reading will be same and therefore both voltmeter will read $\frac{100}{2} = 50\text{V}$

247. $$P = I^2 R$$

$$\therefore \quad \frac{dP}{P} = \pm\left(2\frac{dI}{I} + \frac{dR}{R}\right)$$

$$= \pm\left(\frac{1.5}{100} + \frac{2}{100}\right)\times 100$$

$$= \pm 3.5\%$$

249. We have $J\frac{d^2\theta}{dt^2} + f\frac{d\theta}{dt} + k\theta = 0$

$$\Rightarrow \quad Js^2\theta(s) + fs\theta(s) + K\theta(s) = 0$$

$$\text{Natural frequency,} \quad \omega_n = \sqrt{\frac{k}{J}}$$

$$\therefore \quad 2\zeta\omega_n = \frac{f}{J}$$

$$\Rightarrow \quad \zeta = \frac{f}{2\omega_n J} = \frac{f}{2\sqrt{kJ}}$$

On increasing spring constant, natural frequency ω_n will increase, and damping factor "ζ" will decrease

254. Under balanced condition

$$\left(R_1 + \frac{1}{jc_1 w}\right)R_4 = R_2\left(\frac{R_1\frac{1}{jc_2 w}}{R_3\frac{1}{jc_3 w}}\right)$$

$$= \frac{R_2 R_3}{1 + jR_3 c_s w}$$

$$\Rightarrow \quad \left(R_1 + \frac{1}{jc_1 w}\right)(1 + jR_3 c_3 w) = \frac{R_2 R_3}{R_4}$$

$$\text{It gives} \quad R_1 + R_3\frac{c_3}{c_1} = \frac{R_2 R_3}{R_4} \qquad ...(i)$$

and $R_1R_3c_3w = \frac{1}{c_1w}$ ...(*ii*)

$\therefore \quad w = \frac{1}{\sqrt{R_1R_3c_1c_3}}$

If $c_1 = c_3$, then from equation (*i*)

$$R_1 + R_3 = \frac{R_2R_3}{R_4}$$

Further if $R_2 = 2R_4$, the condition to be maintained to satisfy the balance equation is $R_1 = R_3$ and common variable shaft of resistance R_1 and R_3 can be graduated in frequency to measure frequency

$$\omega \propto \frac{1}{R_1}$$

Thus conditions to be satisfied are

$$c_1 = c_3,$$
$$R_2 = 2R_4$$

259. Rectifier type instrument calibrated to rms value, hence reading $= 10 \times 1.11 = 11.1$ V

263. $\text{Resolution} = \frac{5\times10^{-3}}{10} = 0.0005\text{V}$

We should have number of bits such that

$$\frac{1}{2^N} \le 0.0005$$

$\therefore$ Minimum number of bits = 11

264. In wheatstone bridge,

$$R_xR_4 = R_2R_3$$

$\therefore \quad R_x = \frac{R_2R_3}{R_4}$

$\therefore \quad \frac{dR_x}{R_x} = \pm\left(\frac{dR_2}{R_2}+\frac{dR_3}{R_3}+\frac{dR_4}{R_4}\right)$

$$= \pm(0.1+0.1+0.1)$$
$$= \pm 0.3\%$$

269. For balance, $\frac{Z_2}{Z_4} = \frac{Z_1}{Z_3}$

$\therefore \quad Z_4 = \frac{Z_2Z_3}{Z_1}$

$$= \frac{400\angle{-90°}\times 300\angle{0}}{20\angle{60°}}$$

$$= 600\angle{-150°}$$

277. Number of comparators

$$= 2^N - 1 = 2^8 - 1 = 255$$

289. Total number of digits = 5

And it is a $4\frac{1}{2}$ digit display

$\therefore$ 1 volt range can be read upto 1.9999V

290. $\text{Form factor} = \frac{\text{RMS values of the wave}}{\text{Mean value of the wave}}$

Moving iron instrument will show RMS value. Rectifier voltmeter is calibrated to read RMS value of sinusoidal voltage is with form factor of 1.11.

$\therefore$ Mean value of the applied voltage $= \frac{V_2}{1.11}$

Hence form factor $= \frac{V_1}{V_2/1.11}$

$$= 1.11\frac{V_1}{V_2}$$

291. $\frac{10}{R_1+100} = 10^{-3}$

$\therefore \quad R_1 + 100 = 1000$ ohm

$\Rightarrow \quad R_1 = 9900$ ohm

295. $\tan\theta = \sqrt{3}\frac{5w-3w}{5w+3w} = \sqrt{3}\frac{2w}{8w} = 0.433$

$\Rightarrow \quad \theta = 23.413$

$\therefore \quad p.f. = \cos\theta = 0.917$ lagging

297. $\frac{f_y}{f_x} = \frac{5}{2}$

$\therefore \quad f_y = \frac{5}{2}\times f_x = \frac{5}{2}\times 1000$

$$= 2500 \text{ Hz}$$

307. X – Y recorder acts as CRO to print on the paper. Then X-input is "1 sin 314t" and *y*-input is 1 sin $\left(314+\frac{\pi}{2}\right)$

We see that, there is a phase difference of 90° between them. Hence trace generated will be circle.

311. Gauge factor, $G = \frac{\Delta R/R}{\Delta l/l}$

$\Rightarrow \quad \frac{\Delta R}{R} = \frac{\Delta l}{l} = 2\times 500\times 10^{-6}$

$\therefore \quad \Delta R = 200\times 2\times 500\times 10^{-6}$

$$= 0.2\Omega$$

315. Resistance of the voltmeter,

$$R = 10\times 100 = 1000\text{ k}\Omega = 1\text{ M}\Omega$$

Since voltmeter is connected across the cell and the voltage across it is 5V,

Now resistance of cell and voltmeter resistance are of same value, hence voltage drop across them will be same, hence voltage generated by the cell = 10 V

320. Meter reading will be given by

$$I_m = \sqrt{I_{dc}^2 + \left(\frac{I_{m1}}{\sqrt{2}}\right)^2 + \left(\frac{I_{m2}}{\sqrt{2}}\right)^2}$$

$$= \sqrt{(1)^2 + \left(\frac{\sqrt{2}}{\sqrt{2}}\right)^2 + \left(\frac{2}{\sqrt{2}}\right)^2}$$

$$= \sqrt{1+1+2} = 2\text{A}$$

321. $Z_{DA} \cdot Z_{BC} = Z_{AB} \cdot Z_{CD}$

$\therefore$ $Z_{DA} = 500\sqrt{3} - 500j$

Here $R = 500\sqrt{3}$

and $X = 500j$

We have $X = \frac{1}{Cjw}$

$\therefore$ $C = \frac{1}{jw.X}$

$$= \frac{1}{j \times 1000 \times C - 500j} = 2\mu F$$

Hence capacitor of 2μF should be in series.

322. Rectifier type instrument measures rms value of voltage

Now $V_{avg} = \frac{V_{rms}}{\text{Form factor}}$

We have $V(p-p) = 2V$

$\therefore$ $V_m = 1V$

324. Thermocouple ammeter is calibrated to square value of current

$\therefore$ $\frac{\theta_1}{\theta_2} = \frac{I_1^2}{I_2^2}$

$\Rightarrow$ $\frac{1}{0.2} = \frac{10^2}{I_2^2}$

$\Rightarrow$ $I_2 = 4.47$

341. $\frac{V_0}{V_1} = \frac{1/c_s w}{1/c_x w}$

$\therefore$ $\frac{1}{10 \times 10^{-3}} = \frac{c_x}{c_s}$

$\Rightarrow$ $c_x = c_s \times 10^{+2}$

$= 10 \times 10^{-12} \times 10^{+2} = 1000$ pF

342. $\text{Error} = 1000 \times \frac{1}{100} = 10\text{W}$

True power = 100 W

$\therefore$ Relative error $= \frac{10}{100} \times 100 = \pm 10\%$

345. PMMC instrument only functions for dc voltage. when high frequency signal is applied, then reading will be zero.

351. We can draw phasor diagram as shown.

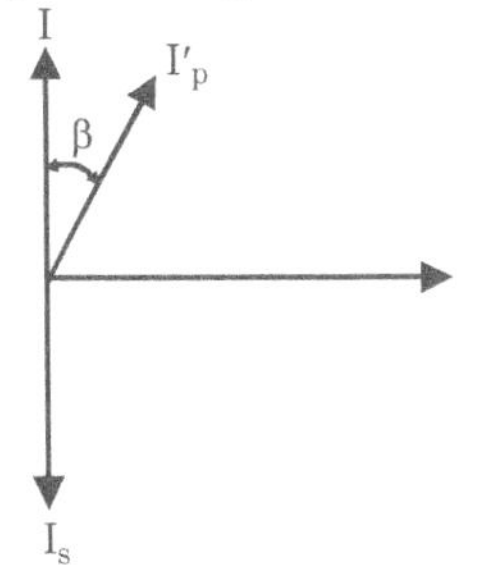

Phase error, $\beta = 3°$

$\therefore$ Phase angle between the primary and secondary currents

$= 180 - 3$

$= 177°$

355. Gauge factor, $G = \frac{\Delta R/R}{\Delta l/l}$

$$= \frac{240 \times 10^{-6}}{1 \times 10^{-6} \times 120} = 2$$

357. We know $P = I^2R$

$\therefore$ $\frac{dP}{P} = 2\frac{dI}{I} + \frac{dR}{R}$

$= \pm(2 \times 0.5 + 0.2)$

$= \pm 1.2\%$

361. $\text{T.H.P} = \sqrt{\frac{E_2^2 + E_3^2 + E_4^2 + \ldots}{E_1}}$

$$= \sqrt{\frac{(6\sqrt{2})^2 + (8\sqrt{2})^2}{200\sqrt{2}}} \times 100 = 5.0\%$$

373. $\text{Error} = 300 \times \frac{1}{100} = 3\text{V}$

When reading is 83V, then

$\%\text{limiting error} = \frac{3}{83} \times 100 = 3.62\%$

375. $\frac{R \times 1000}{R + 1000} \times 1 = 100$

$\Rightarrow$ $1000R = 100R + 100000$

$\Rightarrow$ $R = 111.11\Omega$

394. $\text{Maximum error} = \frac{3\cancel{00} \times 2}{1\cancel{00}} = \pm 6\text{V}$

$\therefore$ Range of reading = 30 – 6 to 30 + 6

= 24 V to 36 V

395. Let full scale output be 1V

∴ Quantization interval,

$$Q = \frac{1}{2^n} = \frac{1}{2^{12}} = 2.44 \times 10^{-4}$$

∴ Quantization error

$$= \frac{Q}{2\sqrt{3}} = \frac{2.44\times10^{-4}}{2\sqrt{3}}$$

$$= 0.7044 \times 10^{-4}$$

∴ % quantization error

$$= \frac{0.7044\times10^{-4}}{1}\times 100$$

$$= 0.007044\%$$

397. Phasor diagram in the given condition can be drawn as shown.

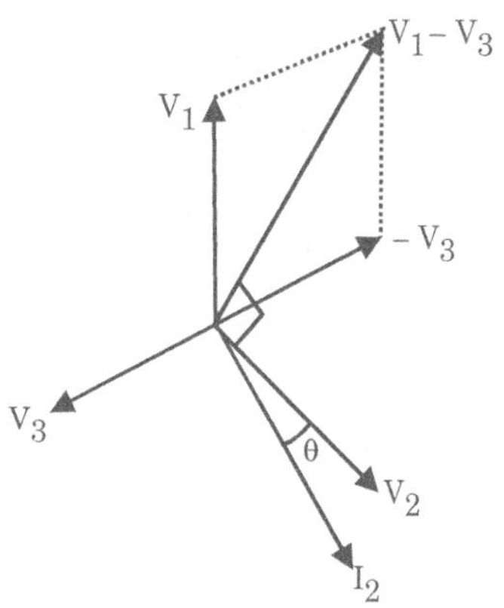

∴ Wattmeter rading ∝ EI cos (θ + 90)

⇒ Wattmeter rading ∝ EI sin ϕ

399. $\tan\theta = \sqrt{3}\,\frac{w_1 - w_2}{w_1 + w_2}$

When reading of two wattmeters are equal and opposite, then

$$\tan\theta = \sqrt{3}\,\frac{w-w}{w+w} = \infty$$

⇒ $\theta = 90°$

Hence the lead is purely inductive

401. $L_{eq} = L_1 + L_2 + K\sqrt{L_1L_2}$

In first case when series adding, then

$$40 = 10 + 15 + K\sqrt{10\times15}$$

⇒ $K = 1.2247$

Now when they are connected in series opposing, then

$$L_{eq} = 10 + 15 - K\sqrt{10\times15}$$

$$= 25 - 1.2247 \times \sqrt{150}$$

$$= 10.00\,\text{mH}$$

403. $f = \frac{1\times10^6}{1\times10^4} = 100$

∴ $T = \frac{1}{100} = 10\text{ ms}$

404. $z_1z_4 = z_2z_3$

∴ $(R_1 + jwL_1).\frac{R_4.\frac{1}{jwc_4}}{R_4 + \frac{1}{jwc_4}} = R_2R_3$

Equating real and imaginary parts, we have

$$R_x = \frac{R_2R_3}{R_4};$$

and $L_x = R_2R_3c_4$

∴ $R_x = \frac{2000\times750}{4000} = 370\Omega$

$$L_x = 2000 \times 750 \times 0.05 \times 10^{-6}$$

$$= 75000 \times 10^{-6}\,\text{H} = 75\,\text{mH}$$

414. Reading of voltmeter

$$B = \frac{15000}{100} = 150\text{V}$$

Reading of voltmeter

$$A = 100\text{ V}$$

∴ Total line voltage = 100 + 150 = 250V

EXERCISE – II

7. Output voltage, $E_H = K_H\,IB/t$

$$= -\frac{1\times10^{-8}\times3\times10000\times10^{-5}}{2\times10^{-3}}$$

$$= -15 \times 10^{-6}\text{ V}$$

Note : 1 gauss = 10^{-5} Wb/m²

12.

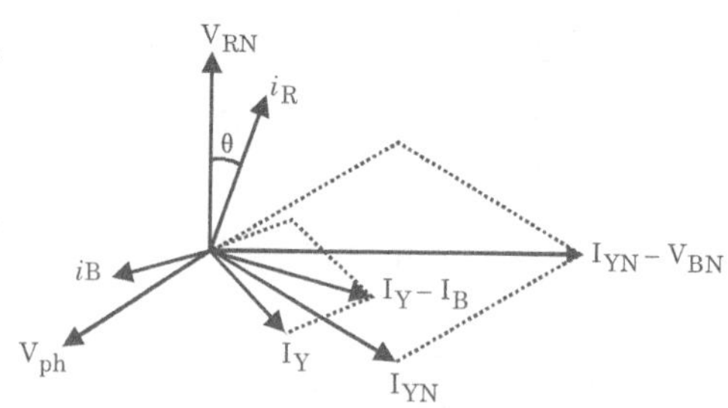

From above phasor diagram

$$V_{RN}(i_Y - i_B) = V_{Ph}I_L\cos(90-\theta)$$

$$(V_{YN} - V_{BN})i_R = V_L\,I_{Ph}\cos(90+\theta)$$

∴ $V_{RN}(i_Y - i_B) - (V_{YN} - V_{BN})i_R$

$$= \sqrt{3}V_{Pn}\,I_{Ph}\cos(90-\theta)$$

$$- \sqrt{3}V_{Ph}\,I_{Ph}\cos(90+\theta)$$

$= \sqrt{3}V_{Ph}\, I_{Ph}\, [\sin\theta + \sin 90 \sin\theta]$

$= 2\sqrt{3}V_{Ph}\, I_{Ph} \sin\theta$

= Reactive power

18. When power factor is below 0.5, one of the wattmeters will give negative indication. Under these conditions in order to read the wattmeter, we must either reverse the current coil or the pressure coil connections.

19. Meter resistance = 150 × 1000Ω = 150kΩ

24. To have 0.5 power factor,

$$\cos\theta = 0.5$$

$$\therefore \quad \theta = 60$$

We have $\tan\theta = \sqrt{3}\dfrac{w_1 - w_2}{w_1 + w_2}$

Hence, one of the wattmeter "w_2" should be zero to have power factor of 0.5

25. $$\text{Speed} = \frac{\text{pulses per second}}{\text{number of teeth}} = \frac{3600}{60} = 60 \text{ rpm}$$

29. Distributed capacitance,

$$c_d = \frac{c_1 - 4c_2}{3} = \frac{450 - 4\times 90}{3} = 30 \text{ pF}$$

34. Gauge factor,

$$G_f = \frac{\Delta R/R}{\text{Strain}} = \frac{\frac{1}{125}}{4000\times 10^{-6}} = 2.0$$

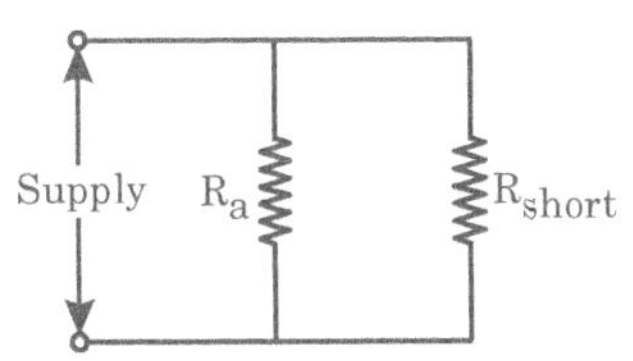

35. $$\tan\theta = \sqrt{3}\frac{w-(-w)}{w+(-w)} = \infty$$

$$\therefore \quad \theta = 90$$

i.e., power factor angle is 90°.

thus the load is reactive.

44 Gauge factor, $G_f = \dfrac{\Delta R/R}{\Delta l/l}$

$$\therefore \quad 2 = \frac{\Delta R/120}{10^{-5}}$$

$$\Rightarrow \quad \Delta R = 2\times 120\times 10^{-5} = 2$$

45. Full scale voltage $= 500\times 10^{-6}\times 250$

$= 125000\times 10^{-6}$

$= 0.125$ V

Internal resistance = 250 ohms

∴ Sensitivity of the meter

$$= \frac{250}{0.125} = 2000 \text{ ohm/V}$$

63. For a rectifier type instrument with sinusoidal input and with full wave rectifier, the a.c. senstivity is

$$s_{ac} = 0.9\, sdc$$

Hence when connected to 100 *vdc*,

$$\text{Reading} = \frac{100}{0.9} = 111\text{V}$$

67. Thermocouple ammeter reads rms value, its reading is 24.

We have $I_{rms} = \dfrac{I_m}{\sqrt{2}}$

$$\therefore \quad I_m = \sqrt{2}\times I_{rms} = \sqrt{2}\times 2 = 2.83 \text{ A}$$

69. Minimum error $= 10\times\dfrac{2}{100} = 0.2$ A

Hence when measuring 5 ampere, then limiting error $= \dfrac{2.0}{5}\times 100 = 4\%$

83. For 4 bits $\left(3\frac{1}{2}\text{digit}\right)$ and 2V ref, each bits is $\dfrac{2}{15} = 0.133$V.

For an integration time of 300 ms input signal time of 20 ms, the number of cycles in

$$15 = 15\times 0.133 = 1.999$$

99. The spring gives controlling torque and is connected in series with the coil. If spring is snapped, there will be no deflection

101. $$\text{P.S.M.} = \frac{2000}{\frac{400}{5}\times 2.5} = 10$$

108. Given that *x* and *y* sensitivities are same. Since its own sweep is fed to the *y* input and *x* has already same input, hence display will be line.

We know $P = I^2R$

$$\frac{dP}{P} = 2\frac{dI}{I} + \frac{dR}{R} = 2\times\frac{5}{100} + \frac{0.2}{100} = \frac{10.2}{100} = 10.2\%$$

113. Ammeter has low resistance hence when directly connected acroos line, then it will burn because of high current in it. Voltmeter already has high resistance, hence when connected in series with load lesser current will flow through it, hence it will not be damaged.

119. At position P,

$$\text{Reading of meter} = \frac{10}{R_{probe}+6}\times 6$$

At r, $\quad \text{reading} = \frac{V_{ref}}{R_{probe}+6}\times 0.6$

If both reading at P and r are full-scale, then

$$\frac{10\times 6}{R_{probe}+6} = \frac{V_{ref}\times 6}{R_{probe}+6}$$

$\therefore \quad V_{ref} = 100\text{V}$

129. Since phase difference between signals at x and y-input is 90°, hence the screen shows a circle.

133. Shunt resistance value of 1Ω is quite low compared to the coil resistance value of 99Ω, hence

Meter current at 0.5,

$$\text{F.S.D.} = 0.1\times 0.5 = 0.05\text{ mA}$$

134. $\quad 550\text{ P} = I^2 R$

$$\therefore \quad \frac{dP}{P} = 2\frac{dI}{I}+\frac{dR}{R}$$

$$= 2\times\frac{5}{100}+\frac{1}{100} = \frac{15}{100}$$

$$\therefore \quad dP = \frac{15}{100}\times I^2 R$$

$$= \frac{15}{100}\times 100\times 10 = 150\Omega$$

135. Total value of resistance

$$= \frac{100}{1\times 10^{-3}} = 100000 = 100\text{ k}\Omega$$

Meter resistance $= 0.1\text{ k}\Omega$

∴ Series resistance required

$$= 100 - 0.1$$

$$= 99.9\text{ k ohms}$$

143. $\quad \text{Error} = \frac{150\times 1}{100} = 1.5\text{V}$

∴ % limiting error when measuring 8.3V

$$= \frac{1.5}{8.3}\times 100 = 18.1\%$$

144. We have, $\quad i = 8 - 6\sqrt{2}\sin\left(wt+\frac{\pi}{2}\right)$

$$\therefore \text{Reading of the meter} = \sqrt{8^2+\left(\frac{6\sqrt{2}}{\sqrt{2}}\right)^2}$$

$$= \sqrt{8^2+6^2} = 10\text{A}$$

147. PMMC meter responds only to dc component, hence reading of the meter will be 10A.

150. $\text{Deflection} = \frac{54}{18} = 3\text{ cm.}$

152. Meter resistance $= 20000\times 250$

$$= 5000000$$

$$= 5\times 10^6$$

$$= 5\text{ M}\Omega$$

2 MΩ resister is equivalent to meter resistance, hence meter will cause loading across 2 MΩ resister only.

155. In $4\frac{1}{2}$ digit display, one extra bit is their, hence meter reading will be of 5 digits hence on 1V range it can read up to 1.9999V

157. $\quad 5\tau = 10\text{ sec}$

$\Rightarrow \quad \tau = 2\text{ sec}$

Now, $\quad \theta = \theta_o(1-e^{-t/T})$

$\Rightarrow \quad \frac{\theta}{\theta_o} = 1 - e^{-t/T}$

$\Rightarrow \quad \frac{1}{2} = -e^{-t/T}$

$\Rightarrow \quad e^{-t/z} = 1/2$

$\therefore \quad t = 1.38\text{ sec.}$

160. A change of 1°C in temperature is reflected as a change of $500\times 0.04 = 200$ ohms of resistance in thermistor resistance

Hence resistance change of 10 ohm is equivalent to

$$= \frac{10}{200} = 0.05°\text{C}$$

161. 400 revolutions → 1 kWh

When power is 1 kW, then speed

$$= \frac{400}{600} = \frac{20}{3}\text{ rpm}$$

$$\text{Power} = \frac{240\times 10\times 0.8}{1000}$$

$$= 1.92\text{ kW}$$

$$\therefore \quad \text{Speed} = \frac{20}{3}\times 1.92$$

$$= 12.8\text{ rpm.}$$

$$\text{Frequency} = \frac{2}{2\pi\sqrt{10^{-6}}}$$

$$= \frac{1000}{\pi}$$

■■

2
CHAPTER

Control Systems

EXERCISE – I

1. With the feedback system, the transient response
   (*a*) decays slowly (*b*) decays rapidly
   (*c*) rises slowly (*d*) rises quickly
2. Potentiometers are used in control system
   (*a*) to improve frequency response
   (*b*) to improve time response
   (*c*) to improve stability
   (*d*) as error sensing transducer
3. A pair of synchros is used in ac position control system.
   (*a*) true (*b*) false
   (*c*) perhaps true (*d*) may be false
4. A free gyro with no restraining forces is used in space to measure
   (*a*) velocity of a vehicle
   (*b*) acceleration
   (*c*) angular momentum
   (*d*) angular position of vehicle with respect to gyro axis
5. Damping due to back emf is observed in
   (*a*) armature controlled dc motor
   (*b*) field controlled dc motor
   (*c*) ac servo-motor
   (*d*) none of these
6. Which one of the following statements is not true for an ac servomotor?
   (*a*) Has low inertia rotor
   (*b*) Is a single phase motor
   (*c*) Has slip torque characteristic as straight line with negative slope
   (*d*) Reference voltage frequency is high compared to signal frequency
7. For stability of an ac servo motor
   (*a*) a negative slope on the torque speed curve is necessary
   (*b*) linearized torque-speed curve is essential
   (*c*) the ratio of the rotor reactance to rotor resistance should be high
   (*d*) none of these
8. For a pair of potentiometers acting as an error device of a dc servo system
   (*a*) input is voltage, output is position
   (*b*) input is position, output is voltage
   (*c*) input is position, output is position
   (*d*) input is voltage, output is voltage
9. The transfer function of a system is the Laplace transform of its
   (*a*) square wave response
   (*b*) step response
   (*c*) ramp response
   (*d*) impulse response
10. The transfer function of a system is defined as the
    (*a*) Laplace transform of the impulse response
    (*b*) step response
    (*c*) response due to an exponentially varying input
    (*d*) all of these
11. Transfer function of a control system depends on
    (*a*) initial conditions of input and output
    (*b*) system parameters alone
    (*c*) nature of input
    (*d*) nature of output
12. The damping ratio of the characteristic equation $s^2 + 2s + 8 = 0$ is
    (*a*) 0.353 (*b*) 0.350
    (*c*) 0.30 (*d*) 0.333
13. The overshoot of the system having the transfer function $\frac{16K}{s\left(s^2+2s+16\right)}$ for a unit step input applied would be
    (*a*) 60% (*b*) 44%
    (*c*) 20% (*d*) 10%
14. Root locus diagram can be used to determine
    (*a*) absolute stability
    (*b*) relative stability
    (*c*) conditional stability
    (*d*) none of these

15. Root locus diagram
   (*a*) is always symmetrical about horizontal axis
   (*b*) may be symmetric about horizontal axis
   (*c*) cannot be symmetric about horizontal axis
   (*d*) is symmetric about the vertical axis

16. A root locus is symmetrical about
   (*a*) real axis
   (*b*) imaginary axis
   (*c*) both real and imaginary axis
   (*d*) none of these

17. When gain K of the open loop transfer function is varied from zero to infinity the closed loop system
   (*a*) may become unstable
   (*b*) stability is not affected
   (*c*) always become unstable
   (*d*) stability is improved

18. A system with gain margin close to unity or a phase margin close to zero is
   (*a*) highly stable
   (*b*) oscillatory
   (*c*) relatively stable
   (*d*) none of these

19. The total response of a system is denoted by

$$y(t)=\frac{1}{2}\left(2-e^{-3t}\right)$$

The steady state and transient responses respectively are
   (*a*) $2, 3t$
   (*b*) $1, \frac{1}{2}e^{-3t}$
   (*c*) $\frac{1}{2}, e^{-3t}$
   (*d*) $-\frac{1}{2}e^{-3t}, 1$

20. The PM of a stable system when the gain is 0 dB is
   1. a positive number in degrees
   2. a negative number in degrees
   3. about – 40° to – 60°
   4. less than 90°

   Of these the incorrect statement are
   (*a*) all
   (*b*) 4
   (*c*) 2, 3
   (*d*) 1, 4

21. The ac motor used in servo applications is a
   (*a*) single-phase induction motor
   (*b*) two-phase induction motor
   (*c*) three-phase induction motor
   (*d*) synchronous motor

22. A plant has the following transfer function:

$$G(s)=\frac{1}{s^2+0.2s+1}$$

For a step input it is required that the response settles to within 2% of its final value. The plant settling time is
   (*a*) 20 sec
   (*b*) 40 sec
   (*c*) 35 sec
   (*d*) 45 sec

23. $G(s)=\frac{1-s}{s(s+2)}$. The system with this transfer function is operated in closed-loop with unity feedback. The closed loop system is
   (*a*) stable
   (*b*) unstable
   (*c*) marginally stable
   (*d*) conditionally stable

24. A closed-loop nonlinear system is shown in the figure.

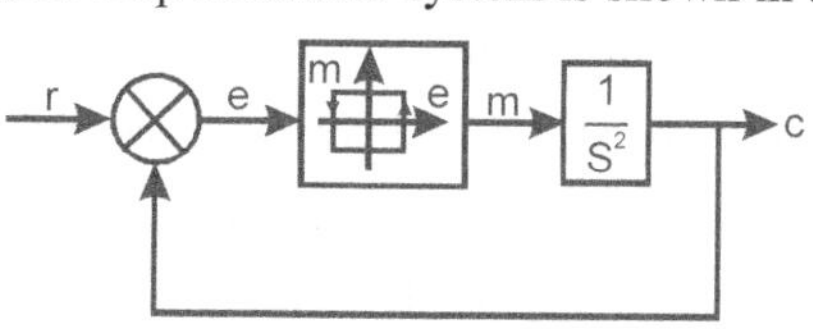

The phase-plane plot in e-e plane is :

(*a*)
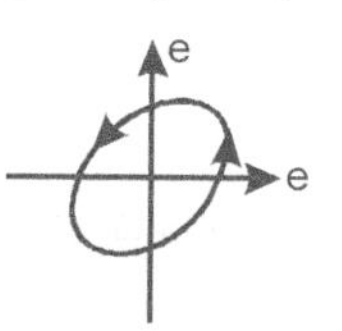

(*b*)
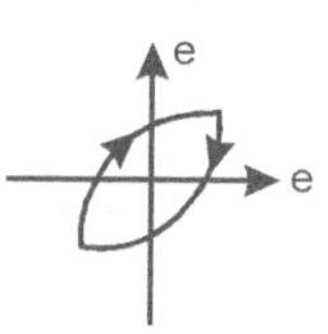

(*c*)
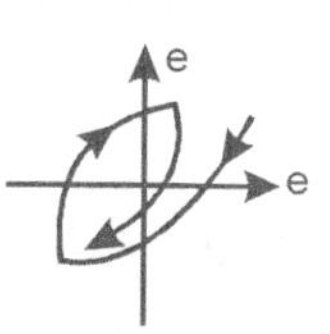

(*d*)
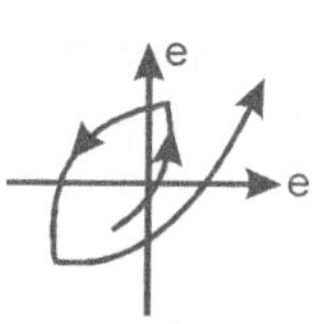

25. Consider the following polynomials:

1. $s^4 + 7s^3 + 17s^2 + 17s + 6$
2. $s^4 + 11s^3 + 41s^2 + 61s + 30$
3. $s^4 + s^3 + 2s^2 + 3s + 2$

Among these polynomials, those which are Hurwitz would include

(*a*) 1 and 2 (*b*) 2 and 3

(*c*) 1 and 3 (*d*) 1, 2 and 3

26. Consider the following sampled data systems:

1.
2.
3.

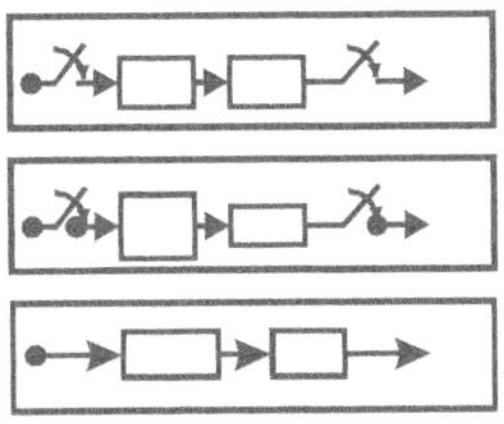

Which of these systems are similar?

(*a*) 1 and 2 (*b*) 1 and 3

(*c*) 2 and 3 (*d*) 1, 2 and 3

27. A closed-loop system is shown in the given figure. The noise transfer function

$\frac{C_n(s)}{N(s)}$ [$C_n(s)$ = output corresponding to noise input N(*s*)] is approximately

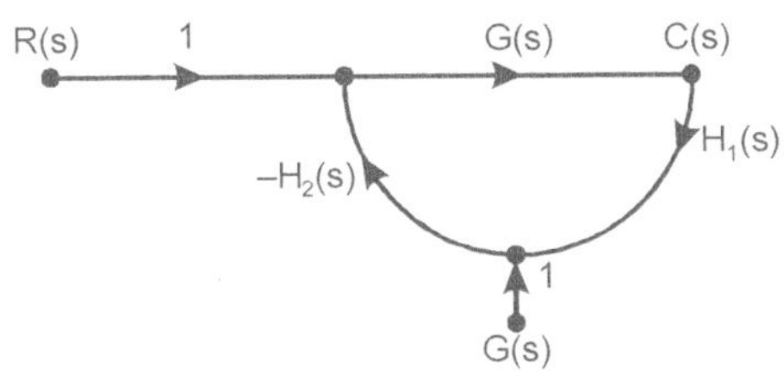

(*a*) $\frac{1}{G(s)H_1(s)}$ for $|G(s)H_1(s)H_2(s)| >> 1$

(*b*) $-\frac{1}{H_1(s)}$ for $|G(s)H_1(s)H_2(s)| >> 1$

(*c*) $-\frac{1}{H_1(s)H_2(s)}$ for $|G(s)H_1(s)H_2(s)| >> 1$

(*d*) $\frac{1}{G(s)H_1(s)H_2(s)}$ for $|G(s)H_1(s)H_2(s)| >> 1$

28. Consider the following statements regarding a linear system $y = f(x_2)$

1. $f(x_1 + x_2) = f(x_1) + (x_2)$
2. $f[x(t+T) = f[x(t)] + f[x(T)]$
3. $f(Kx) = Kf(x)$

(*a*) 1, 2 and 3 are correct (*b*) 1 and 2 are correct

(*c*) 3 alone is correct (*d*) 1 and 3 are correct

29. The characteristic equation of a linear control system is $s^2 + 5Ks + 10 = 0$

The root-loci of the system for $0 < K < \infty$ is

(*a*)

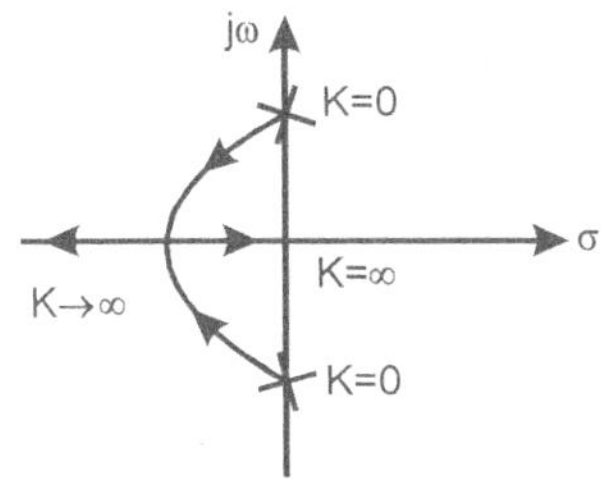

(*b*)

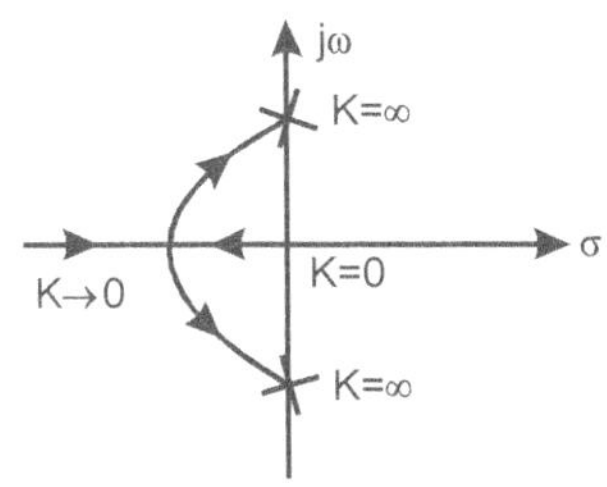

(*c*)

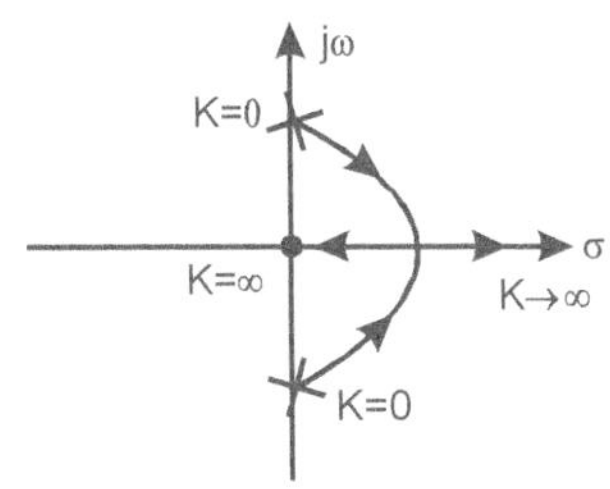

(*d*)

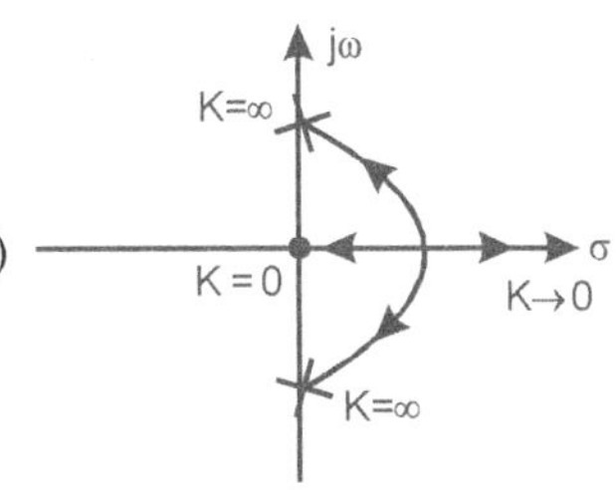

30. A feedback control system is shown in the given figure. The system is stable for all positive values of K, if

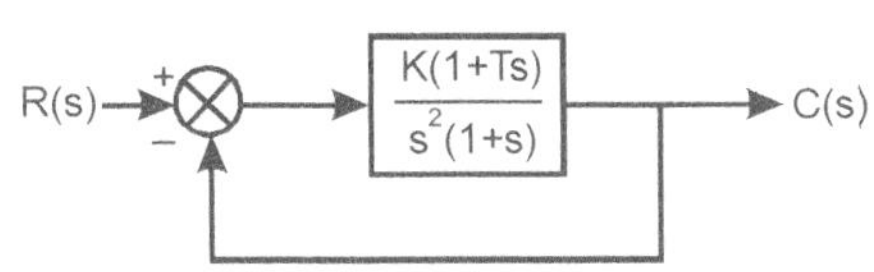

(*a*) T = 0 (*b*) T < 0

(*c*) T > 1 (*d*) 0 < T < 1

31. The minimum of states necessary to describe the network shown in the figure in a state variable form is

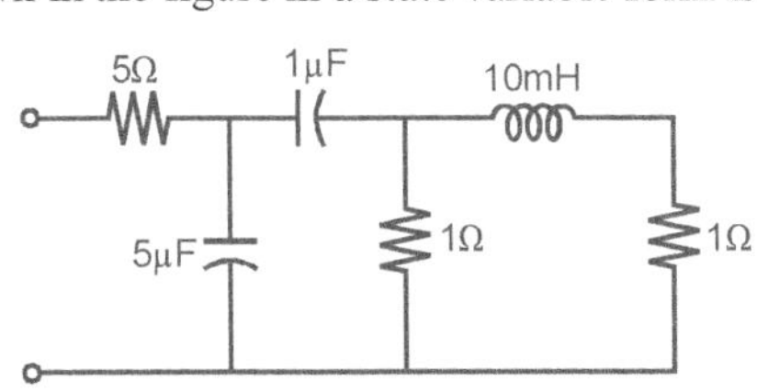

(*a*) 2 (*b*) 3

(*c*) 4 (*d*) 6

32. The state transition matrix for the system $\dot{X} = AX$ with initial state X(0) is

(a) $(sI - A)^{-1}$

(b) $e^{At}X(0)$

(c) Laplace inverse of $[(sI - A)^{-1}]$

(d) Laplace inverse of $[(sI - A)^{-1}X(0)]$

33. The signal flow graph of a system is shown in the given figure, in this graph, the number of three non-touching loops is

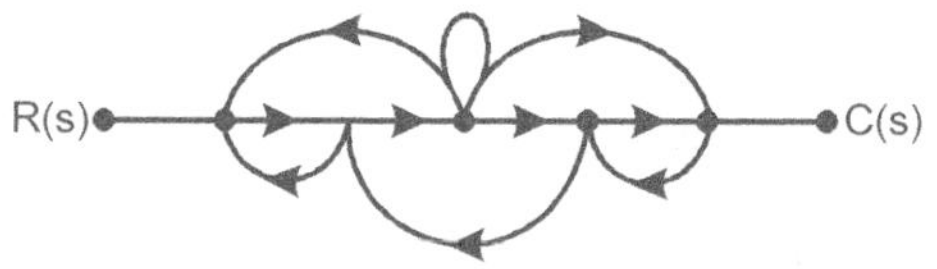

(a) zero (b) 1

(c) 2 (d) 3

34. In the following block diagram

$G_1 = \frac{10}{s}$,

$G_2 = \frac{10}{(s+1)}$, $H_1 = s + 3$

and $H_2 = I$

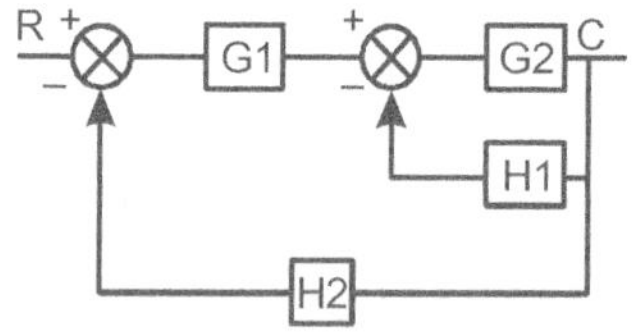

The overall transfer function C/R is given by

(a) $\frac{10}{11s^2 + 31s + 10}$

(b) $\frac{100}{11s^2 + 31s + 100}$

(c) $\frac{100}{11s^2 + 31s + 10}$

(d) $\frac{100}{11s^2 + 31s}$

35. The advantages of Nyquist stability test are

(a) it guides in stabilising an unstable system

(b) it enables to predict closed loop stability from open loop results

(c) it is applicable to experimental results of frequency response to open loop system

(d) all of these

36. Which of the following system is unstable?

(a) $G(s)H(s) = \frac{K}{(T_1 s + 1)(T_1 2 + 1)}$

(b) $G(s)H(s) = \frac{K(s+1)}{s^2(s+4)(s+5)} K > 09$

(c) $G(s)H(s) = \frac{K(s+2)}{(s+1)(s-3)} K > 2$

(d) $G(s)H(s) = \frac{K}{(T.\ s+1)} -1 < 1 < 8$

37. The root locus of a system has four separate loci. The system can have

(a) four poles and four zeros

(b) four poles or four zeros

(c) six poles and two zeros

(d) two poles and two zeros

38. In a closed loop control system

(a) control action is independent of the output

(b) output is independent of input

(c) there is no feedback

(d) control action is dependent of the output

39. A system is described by the state equation

$$\begin{bmatrix} \dot{x}_1 \\ \dot{x}_2 \end{bmatrix} = \begin{bmatrix} 2 & 0 \\ 0 & 2 \end{bmatrix} \begin{bmatrix} x_1 \\ x_2 \end{bmatrix} + \begin{bmatrix} 1 \\ 1 \end{bmatrix} u$$

The state transition matrix of the system is

(a) $\begin{bmatrix} e^{2t} & 0 \\ 0 & e^{2t} \end{bmatrix}$ (b) $\begin{bmatrix} e^{-2t} & 0 \\ 0 & e^{t} \end{bmatrix}$

(c) $\begin{bmatrix} e^{2t} & 1 \\ 1 & e^{2t} \end{bmatrix}$ (d) $\begin{bmatrix} e^{-2t} & 0 \\ 0 & e^{-2t} \end{bmatrix}$

40. Roots of characteristic equations are

(a) poles of closed loop

(b) poles of open loop

(c) zero of closed loop

(d) zero of open loop

41. Laplace transform of $\frac{e^{-at} - e^{-bt}}{b - a}$ is

(a) $\frac{1}{(s+a)(s-b)}$ (b) $\frac{1}{(s-a)(s+b)}$

(c) $\frac{1}{(s-a)(s-b)}$ (d) $\frac{1}{(s+a)(s+b)}$

42. The output of a feedback control system must be a function of

(*a*) reference and input

(*b*) reference and output

(*c*) output and feedback signal

(*d*) input and feed back signal

43. The Nyquist plot for a control system is shown in figure I. The Bode plot for the same system will be as in

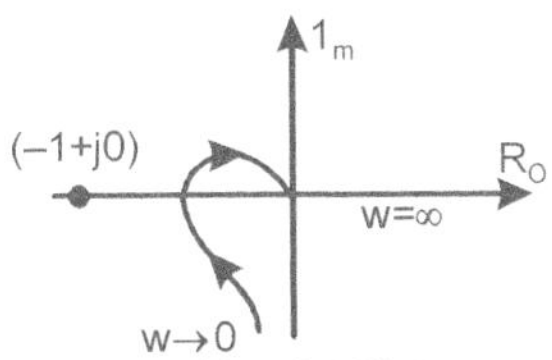

Nyquist Plot

(*a*)

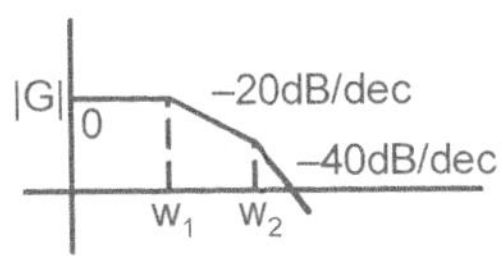

(*b*)

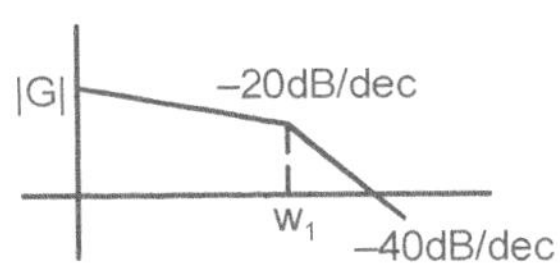

(*c*)

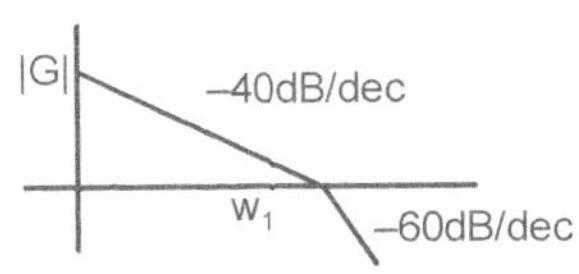

(*d*)

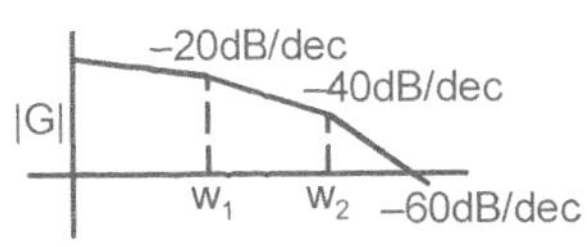

44. When deriving the transfer function of a linear element

(*a*) both initial conditions and loading are taken into account

(*b*) initial conditions are taken into account but the element is assumed to be not loaded

(*c*) initial conditions are assumed to be zero but loading is taken into account

(*d*) initial conditions are assumed to be zero and the element assumed to be not loaded

45. The two-port network shown in the figure can be described as

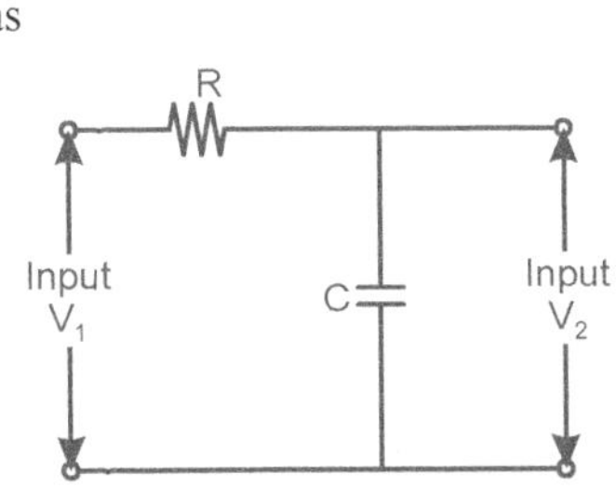

(*a*) phase lead network, differentiating circuit or low-pass filter

(*b*) phase lag network, integrating circuit or low-pass filter

(*c*) phase lag network, differentiating circuit or high-pass filter

(*d*) phase lead network, integrating circuit or high-pass filter

46. Which one of the four signal flow graphs shown in (*a*), (*b*), (*c*) and (*d*) represents the block diagram shown in the figure?

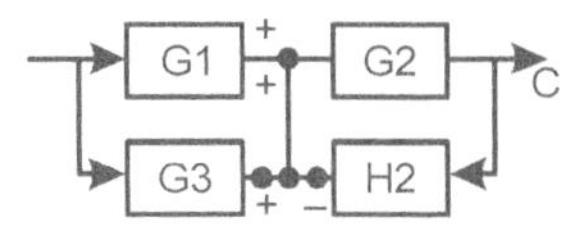

(*a*)

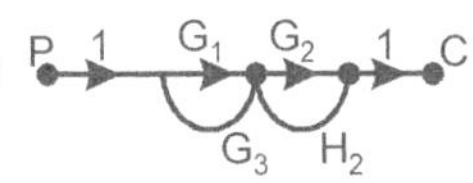

(*b*)

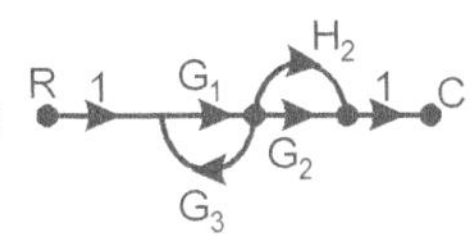

(*c*)

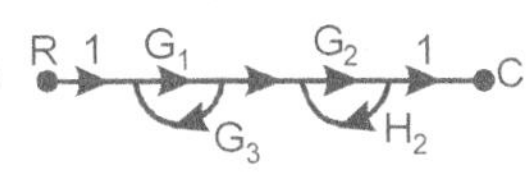

(*d*)

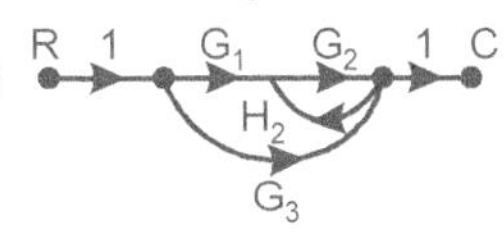

47. The transfer function C(*s*)/R(*s*) of the system, whose block diagram is given below

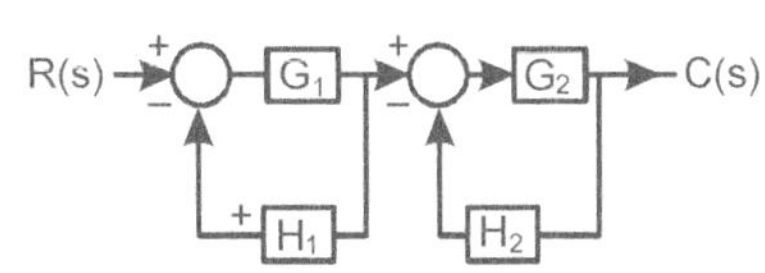

(*a*) $\dfrac{G_1G_2}{1+G_1H_1+G_2H_2-G_1G_2H_1H_2}$

(*b*) $\dfrac{G_1G_2}{1+G_1H_1+G_2H_2+G_1G_2H_1H_2}$

(*c*) $\dfrac{G_1G_2}{1+G_1H_1+G_2H_2}$

(*d*) $\dfrac{G_1(1+G_2H_2)G_2(1+G_1H_1)}{1+G_1H_1+G_2H_2+G_1G_2H_1H_2}$

48. Consider the following properties attributed to state model of a system:

1. State model is unique.
2. State model can be derived from the system transfer function.
3. State model can be derived for time variant systems.

(*a*) 1, 2 and 3 are correct (*b*) 1 and 2 are correct

(*c*) 2 and 3 are correct (*d*) 1 and 3 are correct

49. The state representation for a second-order system is given by

$$\dot{x}_1 = x_1 + u$$

$$\dot{x}_2 = x_1 - 2x_2 + u$$

Consider the following statements regarding the above systems:

1. The system is completely state-controllable.

2. If x_1 is the output, then the system is output controllable.

3. If x_2, is the output, then the system is output controllable.

Of these statements:

(*a*) 1, 2 and 3 are correct

(*b*) 1 and 2 are correct

(*c*) 2 and 3 are correct

(*d*) 1 and 3 are correct

50. The polar plot of a type-1, 3-pole, open-loop system is shown in the given figure. The closed-loop system is

(*a*) always stable

(*b*) marginally stable

(*c*) unstable with one pole on the right half *s*-plane

(*d*) unstable with two poles on the right half *s*-plane

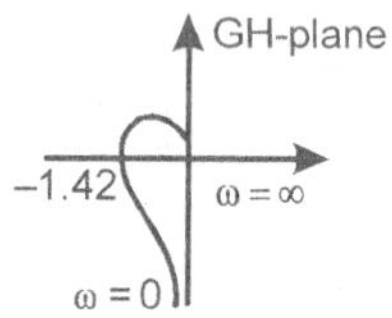

51. The closed-loop gain of the system shown in the given-figure is

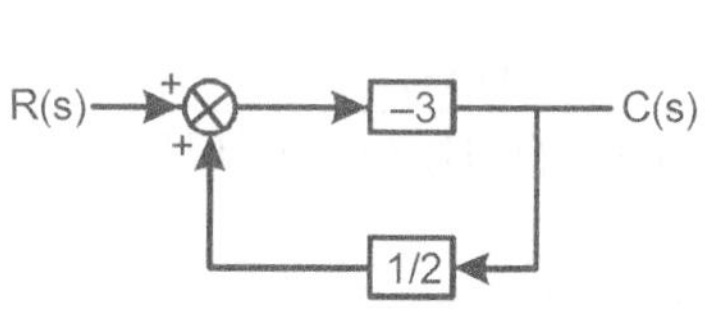

(*a*) $-\frac{9}{5}$ (*b*) $-\frac{6}{5}$

(*c*) $\frac{6}{5}$ (*d*) $\frac{9}{5}$

52. The first two rows of Routh's tabulation of a fourth-order system are:

S^5	1	10	5
S^4	1	20	

The number of roots of the system lying on the right half of *s*-plane is

(*a*) zero (*b*) 2

(*c*) 3 (*d*) 4

53. A lag compensation network

(*a*) increases the gain of the original network without affecting the stability

(*b*) does not affect stability on increased gain

(*c*) reduces the steady-state error

(*d*) reduces the speed of response

54. Maximum phase-lead of the compensator

$$D(s) = \frac{(0.5s+1)}{(0.05s+1)} \text{ is}$$

(*a*) 52 deg at 4 rad/sec

(*b*) 52 deg at 10 rad/sec

(*c*) 55 deg at 12 rad/sec

(*d*) None of the answers is correct

55. The TF of a simple RC network functioning as a controller is $GC(s) = \frac{s+z_1}{s+P_1}$. The condition for RC network to set as a phase lead controller is

(*a*) $p_1 < z_1$

(*b*) $p_1 = 0$

(*c*) $p_1 = z_1$

(*d*) $p_1 > z_1$

56. The final value of function is equal to

$$F(s) = \frac{5}{s(s^2+s+2)}$$

(*a*) zero (*b*) $\frac{2}{5}$

(*c*) $\frac{5}{2}$ (*d*) 5

57. For the block diagram given in the figure,

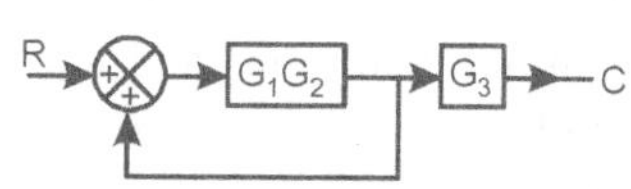

the expression of $\frac{C}{R}$ is

(*a*) $\frac{G_1G_2G_3}{1-G_1G_2}$

(*b*) $\frac{G_1G_2}{1-G_1G_2G_3}$

(*c*) $\frac{G_1G_2\,G_3}{1\text{-}G_1G_2G_3}$

(*d*) $\frac{G_1G_2}{G_3(1-G_1G_2)}$

58. Match List I with List II and select the correct answer using the codes given below the lists:

List I (*Signal flow graphs*) | **List II** (*Simplification*)

A. B. C. D. | 1. 2. 3. 4. 5.

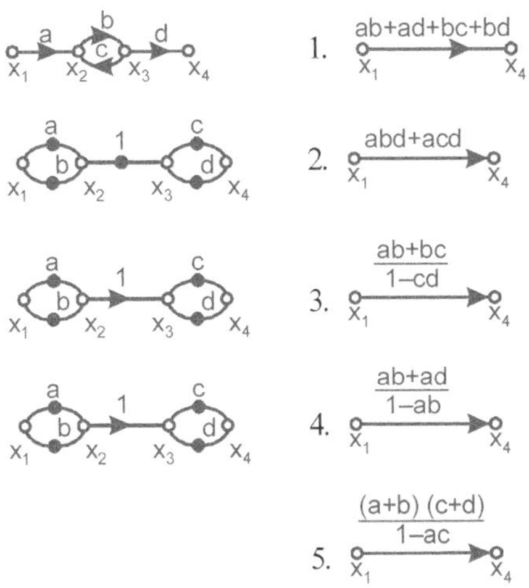

	A	B	C	D
(*a*)	2	1	5	4
(*b*)	3	4	2	1
(*c*)	2	1	4	3
(*d*)	1	2	2	5

59. The open-loop transfer function of the control system shown in the given figure is

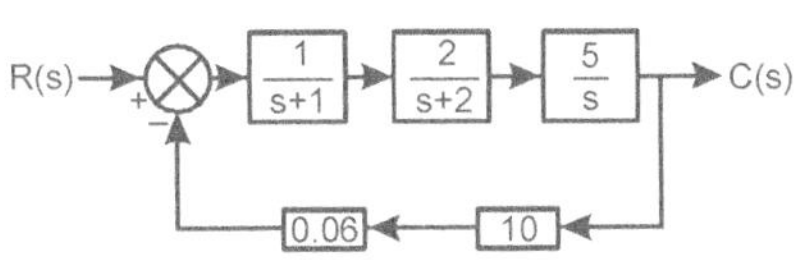

(*a*) $\frac{10}{s(s+1)(s+2)}$ (*b*) 0.6

(*c*) $\frac{6}{s(s+1)(s+2)}$ (*d*) $\frac{10}{s^3+3s^2+2s+6}$

60. Given a system represented by equations,

$$X = \begin{bmatrix} 0 & 1 \\ -2 & -3 \end{bmatrix} x + \begin{bmatrix} 0 \\ 1 \end{bmatrix} u,$$

$$Y = [1\ 0]x$$

The equivalent transfer function representation G(*s*) of the system is

(*a*) $G(s) = \frac{1}{s^2+5s+2}$ (*b*) $G(s) = \frac{1}{s^2+3s+2}$

(*c*) $G(s) = \frac{3}{s^2+3s+2}$ (*c*) $G(s) = \frac{2}{s^2+2s+2}$

61. Consider the following statements: Addition of resistance to the rotor of a two-phase ac servomotor will

1. increase the overall torque
2. increase the stability of operation
3. linearize its torque-speed characteristics
4. lead to attainment of maximum stalling torque

Of these statements:

(*a*) 1, 2 and 4 are correct
(*b*) 2, 3 and 4 are correct
(*c*) 2 and 3 are correct
(*d*) 1, 3 and 4 are correct

62. A servomotor is usually designed to have

(*a*) high rotor inertia and high bearing friction
(*b*) high rotor inertia and low bearing friction
(*c*) low rotor inertia and high bearing friction
(*d*) low rotor inertia and low bearing friction

63. The response C(*t*) of a system to an input *r*(*t*) is given by the following differential equation:

$$\frac{d^2C(t)}{dt^2} + 3\frac{dc(t)}{dt} + 5C(t) = 5r(t)$$

The transfer function of the system is given by

(*a*) $G(s) = \frac{5}{s^2+3s+5}$ (*b*) $G(s) = \frac{1}{s^2+3s+5}$

(*c*) $G(s) = \frac{3s}{s^2+3s+5}$ (*d*) $G(s) = \frac{s+3}{s^2+3s+5}$

64. The Nyquist plot of a closed-loop system is shown in the given figure. The plot indicates that the

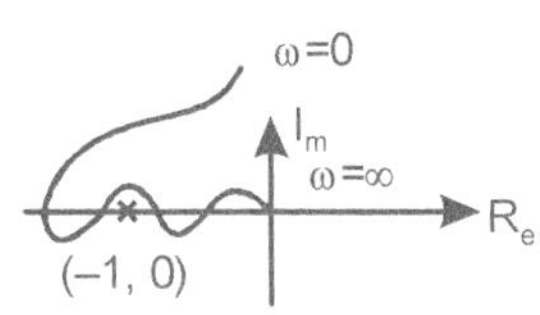

(*a*) system is unstable
(*b*) system is stable
(*c*) system is critically stable
(*d*) stability of the system cannot be determined

65. Consider the following statements:

In a feedback control system, lead compensator

1. increases the margin of stability
2. speeds up transient response
3. does not affect the system error constant

Of these statements

(*a*) 2 and 3 are correct (*b*) 1 and 2 are correct
(*c*) 1 and 3 are correct (*d*) 1, 2 and 3 are correct

66. An electromechanical closed-loop control system has the following characteristic equation

$$s^3 + 6Ks^2 + (K+2)s + 8 = 0,$$

where K is the forward gain of the system. The condition for closed-loop stability is

(*a*) K = 0.528

(*b*) K = 2

(*c*) K = 0

(*d*) K = −2.528

67. The poles of a continuous time oscillator are

(*a*) on the negative *s*-plane

(*b*) on the positive *s*-plane

(*c*) on both the negative and positive *s*-planes

(*d*) conjugates on the imaginary axis

68. The system shown in the given figure, is subjected to a unit ramp input.

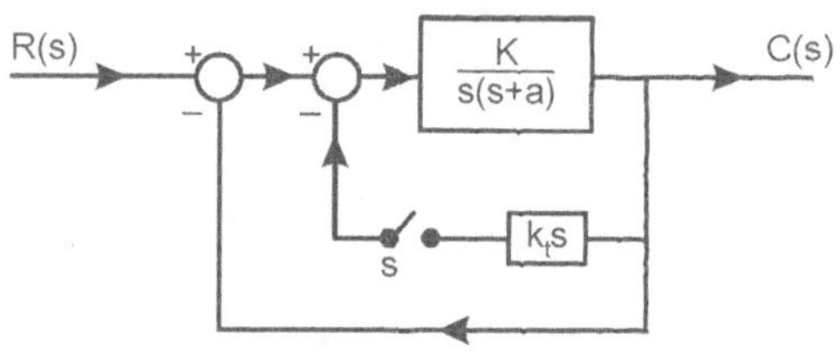

On close the switch '*s*'

(*a*) steady state error will increase and damping coefficient ξ, will decrease

(*b*) both steady state error and damping coefficient ξ will increase

(*c*) both steady state error and damping coefficient ξ will decrease

(*d*) steady state error will decrease and damping coefficient ξ will increase

69. 1. Transfer function can be obtained from the signal flow graph of the system
2. Transfer function typically characterizes linear time variant system
3. Block diagram of the system can be obtained from its transfer function given the ratio of output to input in frequency domain of the system.
4. Transfer function gives the ratio of output toinput in frequency domain of the system.

Which of the following is the correct combination about the four statements stated above?

(*a*) only (1) and (2) are correct

(*b*) only (2), (3) and (4) are correct

(*c*) only (3) and (4) are correct

(*d*) only (1), (2) and (4) are correct

70. If the signal flow graph shown in the given figure has M number of forward paths and P number of individual loops, then

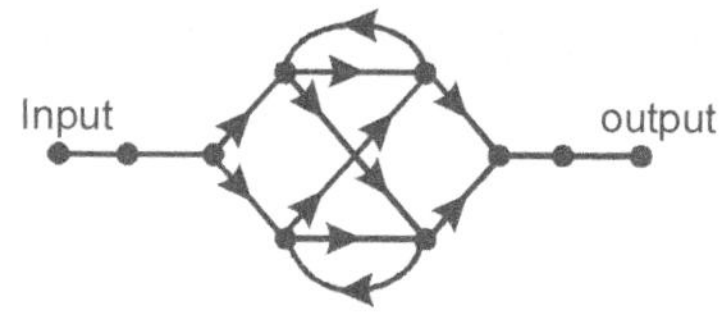

(*a*) M = 6, P = 6 (*b*) M = 6, P = 4

(*c*) M = 4, P = 6 (*d*) M = 4, P = 4

71. The open -loop transfer function of a unity-feedback control system is:

$$G(s) = \frac{K(s+10)(s+20)}{s^2(s+2)}$$

The closed-loop system will be stable if the value of K is

(*a*) 2 (*b*) 3

(*c*) 4 (*d*) 5

72. Consider the single-input system described by the vector-matrix state equation

$$\dot{\underline{X}} = A\,\underline{X}(t) + Bu(t)$$

and the output equation

$$y(t) = C\,\underline{X}(t), \text{ where } \underline{X}(t)$$

is the state vector, $u(t)$ is the control input and

$$A = \begin{bmatrix} 0 & 1 \\ -1 & -2 \end{bmatrix} \quad B = \begin{bmatrix} 0 \\ 1 \end{bmatrix} \quad C = [1\ 1]$$

The system is

(*a*) controllable and observable

(*b*) controllable but unobservable

(*c*) uncontrollable but observable

(*d*) uncontrollable and unobservable

73. The characteristic equation of a unity feedback control system is given by

$$s^3 + K_1 s^2 + s + K_2 = 0$$

Consider the following statements in this regard:

1. For a given value of K_1 all the root-locus branches will terminate at infinity for K_2 in the positive direction.
2. For a given value of K_2 all the root-locus branches will terminated at infinity for variable K_1 in the positive direction.
3. For a given value of K_2 only one root-locus branch will terminate at infinity for variable K_1 in the positive direction.

Of these statements

(*a*) 1 and 2 are correct (*b*) 3 alone is correct

(*c*) 2 alone is correct (*d*) 1 and 3 are correct

74. A system is described by

$$x = \begin{bmatrix} 0 & 1 \\ -2 & -3 \end{bmatrix} x + \begin{bmatrix} 0 \\ 1 \end{bmatrix} u$$

$$y = [1\ 0]x$$

The transfer function is

(*a*) $\dfrac{1}{s^2+2s+3}$ (*b*) $\dfrac{1}{2s^2+3s+1}$

(*c*) $\dfrac{1}{s^2+3s+2}$ (*d*) $\dfrac{1}{3s^2+2s+1}$

75. Consider the following statements about the z-transform, X (z) of the sequence

$x(n) = 2^n$ for $n \geq 0$

$= 0$ for $n < 0$

(ROC denotes region of convergence in the z-plane)

1. $X(z) = \dfrac{1}{1-2z^{-1}}$ ROC : $|z| > 2$

2. $X(z) = 1 + \dfrac{2z^{-1}}{1-2z^{-1}}$ ROC : $|z| > 2$

3. $X(z) = \dfrac{1}{1-2z^{-1}}$ ROC : $|z| > 2$

(*a*) 1 and 3 (*b*) 1 and 2

(*c*) 2 and 3 (*d*) 1, 2 and 3

76. The first element of each of the rows of a Routh Hurwitz stability test showed the sign as follows:

Rows	I	II	III	IV	V	VI	VII
Signs	+	–	+	+	+	–	+

The number of roots of the system lying the right half of s-plane is

(*a*) 2 (*b*) 3

(*c*) 4 (*d*) 5

77. The open-loop transfer function of a unity feedback control system is given by

$$G(s) = \frac{K}{s(s+1)}$$

If the gain K is increased to infinity, then the damping ratio will tend to become

(*a*) $\dfrac{1}{\sqrt{2}}$ (*b*) 1

(*c*) 0 (*d*) ∞

78. The characteristic equation of a feedback control system is

$$2s^4 + s^3 + 3s^2 + 5s + 10 = 0$$

The number of roots in the right half of s-plane are

(*a*) zero (*b*) 1

(*c*) 2 (*d*) 3

79. The asymptotic approximation of the log-magnitude versus frequency plot of a minimum phase system with real poles and one zero is shown in the figure. Its transfer functions is

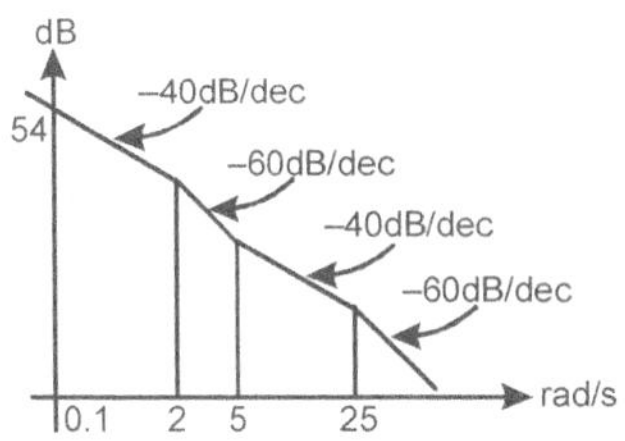

(*a*) $\dfrac{20(s+5)}{s(s+2)(s+25)}$

(*b*) $\dfrac{10(s+5)}{s(s+2)^2(s+25)}$

(*c*) $\dfrac{20(s+5)}{s^2(s+2)(s+25)}$

(*d*) $\dfrac{50(s+5)}{s^2(s+2)(s+25)}$

80. A linear second-order system with the transfer function

$$G(s) = \frac{49}{s^2+16s+49}$$

is initially at rest and is subject to a step input signal. The response of the system will exhibit a peak overshoot of

(*a*) 16% (*b*) 9%

(*c*) 2% (*d*) zero

81. The performance specifications for a unity feedback control system having an open-loop transfer function

$$G(s) = \frac{K}{s(s+1)(s+2)} \text{ are}$$

1. Velocity error coefficient $K_v > 10\text{sec}^{-1}$
2. Stable closed-loop operation

The value of K, satisfying the above specifications, is

(*a*) $K < 6$

(*b*) $6 < K < 10$

(*c*) $K > 10$

(*d*) None of these

82. If R and X are the resistance and inductive reactance respectively, D and L are the diameter and length respectively of the rotor of a two-phase *ac* servomotor, then the ratios $\frac{D}{L}$ and $\frac{X}{R}$ will be respectively

(*a*) largo and large (*b*) large and small
(*c*) small and large (*d*) small and small

83. The magnitude - frequency response of a control system is shown in the figure. The value of ω_1 and ω_2 are respectively

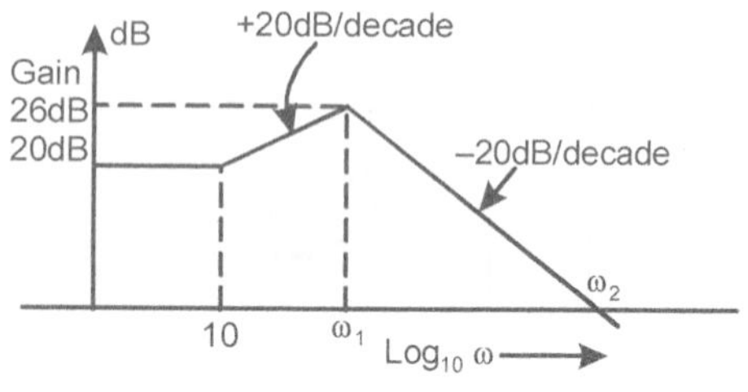

(*a*) 10 and 200 (*b*) 20 and 200
(*a*) 20 and 400 (*d*) 100 and 400

84. A linear 2nd order single input continuous time system is described by the following set of differential equations:

$$\dot{x}_1 = -2x + 4x_2$$

$$\dot{x}_2 = 2x_1 - x_2 - u(t)$$

Where x_1 and x_2 are state variables and $u(t)$ is the input. The system is

(*a*) controllable and stable
(*b*) controllable but unstable
(*c*) uncontrollable and unstable
(*d*) uncontrollable but stable[IAS]

85. For $\phi(s) = \begin{bmatrix} \frac{s+6}{s^2+6s+5} & \frac{1}{s^2+6s+5} \\ \frac{-5}{s^2+6s+5} & \frac{s}{s^2+6s+5} \end{bmatrix}$

the coefficient matrix A is

(*a*) $\begin{bmatrix} 6 & -5 \\ -6 & 0 \end{bmatrix}$

(*b*) $\begin{bmatrix} 5 & -5 \\ 0 & -6 \end{bmatrix}$

(*c*) $\begin{bmatrix} 6 & 0 \\ -5 & -6 \end{bmatrix}$

(*d*) $\begin{bmatrix} 0 & 1 \\ -5 & -6 \end{bmatrix}$

86. The characteristic equation of a closed-loop system is given by:

$$s^4 + 6s^3 + 11s^2 + 6s + K = 0$$

Stable closed-loop behaviour can be ensured when gain K is such that

(*a*) $0 < K < 10$ (*b*) $K > 10$
(*c*) $-\infty \le K < \infty$ (*d*) $0 < K \le 20$

87. Which of the following quantities are analogous in a Force-Current analogy?

1. Displacement and Inductance
2. Velocity and Voltage
3. Mass and Capacitance

Select the correct answer using the codes given below:

(*a*) 1, 2 and 3 (*b*) 1 and 2
(*c*) 2 and 3 (*d*) 1 and 3

88. Consider the following features:

1. Squirrel-cage or drag-up type rotor.
2. Nearly linear torque-speed charact-eristics.
3. High X/R ratio.

Which of these features are possessed by a two- phase ac servomotor?

(*a*) 1, 2 and 3 (*b*) 1 and 2
(*c*) 2 and 3 (*d*) 1 and 3

89. The open-loop transfer functions with unity feedback are given below for different systems:

(*a*) $G(s) = \frac{2}{s+2}$ (*b*) $G(s) = \frac{2}{s(s+2)}$

(*c*) $G(s) = \frac{2}{s^2(s+2)}$ (*d*) $G(s) = \frac{2(s+1)}{s(s+2)}$

Among these system the unstable system is

(*a*) 1 (*b*) 2
(*c*) 3 (*d*) 4

90. The number of roots in the left-half of *s*-plane for the equation $s^3 - 4s^2 + s + 6 = 0$, would be

(*a*) 1 (*b*) 2
(*c*) 3 (*d*) 4

91. The transfer function for a system is

$$G(s) = \frac{K(s+2)}{(s+1)(s+3)(s+4)}, K > 0$$

Two branches (loci) of the plot directed along asymptotes are centered at point

(*a*) -3 (*b*) -4
(*c*) -2 (*d*) -1

92. Match List I (Signal flow graph) with List II (Transfer function) and select the correct answer using the codes given below the Lists:

List I

A. 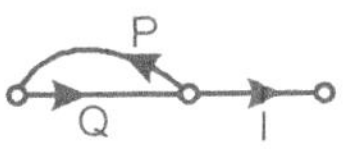

B. 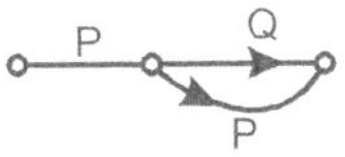

C. 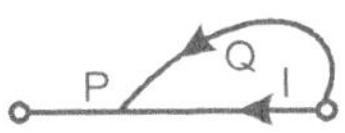

D. 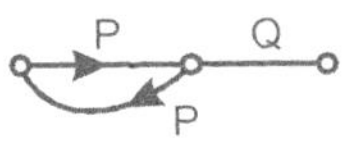

List II

1. $\frac{P}{1-Q}$
2. $\frac{Q}{1-PQ}$
3. $\frac{PQ}{1-PQ}$
4. $\frac{PQ}{1-P^2}$

Codes:

	A	B	C	D
(*a*)	2	3	4	1
(*b*)	2	3	1	4
(*c*)	3	2	1	4
(*d*)	3	2	4	1

93. The given figure shows the root-locus of open-loop transfer function of a control system. Index:

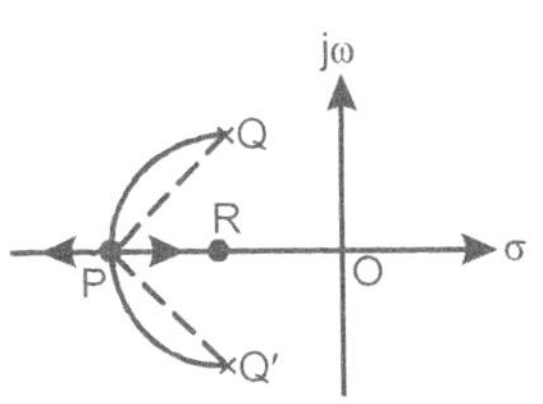

☆ — pole
● — Zero
○ — Origin
PQ = 2.6 = PQ
PR = 1.4
OR = 2.0
OQ = 1.4 = Q

The value of the forward path gain K at the point P is

(*a*) 0.2 (*b*) 1.4
(*c*) 3.4 (*d*) 4.8

94. The system shown in the given figure has second order response with a damping ration of 0.6 and a frequency of damped oscillations of 10 rad/sec. The values of K_1, and K_2 are respectively

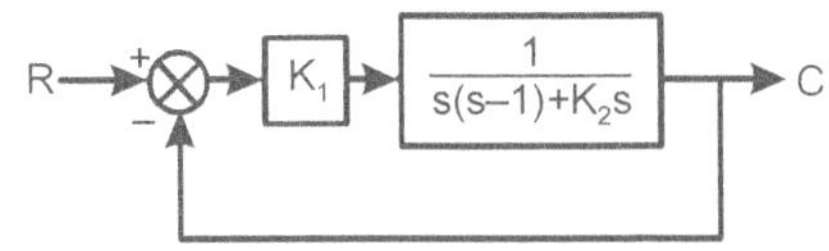

(*a*) 12.5 and 15 (*b*) 156.25 and 15
(*c*) 156.25 and 14 (*d*) 12.5 and 14

95. If the gain of the open-loop control system is halved the gain margin will

(*a*) remain the same (*b*) become half
(*c*) become double (*d*) become four times

96. In the feedback system shown in the given figure, the noise component of output is given by (assume high loop gain at frequency of interest)

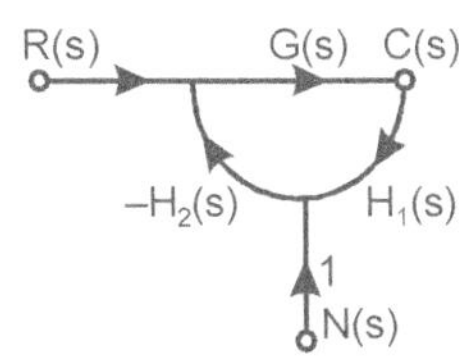

(*a*) $\frac{-N(s)}{H_1(s)}$

(*b*) $\frac{N(s)}{H_1(s)}$

(*c*) $\frac{N(s)}{H_1(s)H_2(s)}$

(*d*) $\frac{-N(s)}{H_1(s)H_2(s)}$

97. The system shown in the given figure has a unit step input. In order that the steady-state error is 0.1, the value of AT required is

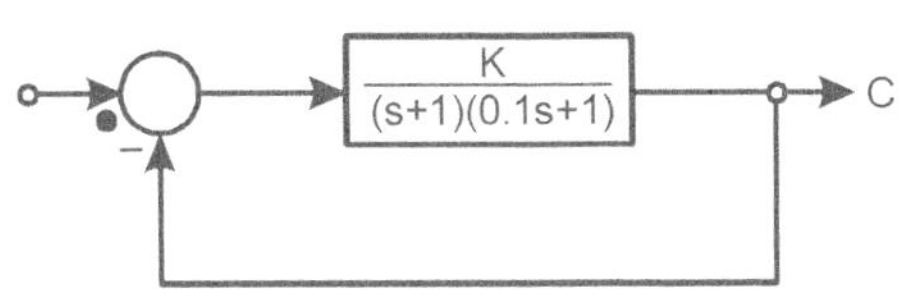

(*a*) 0.1 (*b*) 0.9
(*c*) 1.0 (*d*) 9.0

98. In the derivative error compensation

(*a*) damping decreases and settling time decreases
(*b*) damping increases and settling time increases
(*c*) damping decreases and settling time increases
(*d*) damping increases and settling time decreases

99. Which of the following is not valid in case of signal flow graph?

(*a*) In signal flow graph signals travel along branches only in the marked direction

(*b*) Nodes are arranged from right to left in a sequence

(*c*) Signal flow graph is applicable to linear systems only

(*d*) For signal flow graph, the algebraic equations must be in the form of cause and effect relationship.

100. Damping factor and undamped natural frequency for the position control system is given by

(*a*) $2\sqrt{KJ}, \sqrt{KJ}$ respectively

(*b*) $\frac{K}{2fJ}, \sqrt{K/j}$ respectively

(*c*) $\frac{f}{2\sqrt{KJ}}, \sqrt{KJ}$ respectively

(*d*) $\frac{J}{2\sqrt{Kf}}, \sqrt{KJ}$ respectively

where K = tortional stiffness;

J = moment of inertia,

f = coefficient of viscous function. [ES]

101. The figures shows two equivalent block diagrams

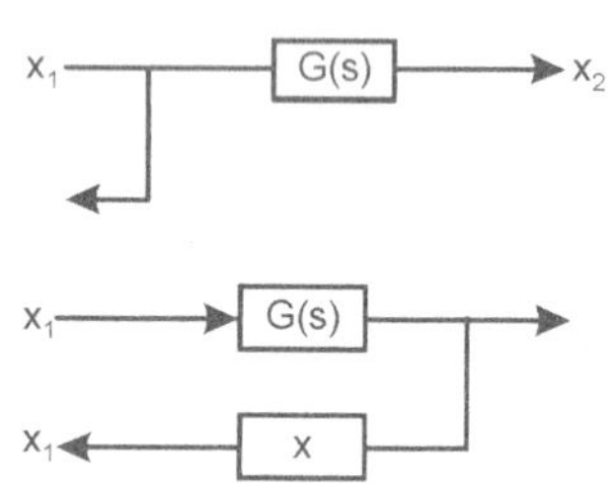

The transfer function of the block marked 'X' is given by

(*a*) $G_1(s)$

(*b*) $\frac{1}{G_1(s)}$

(*c*) 1

(*d*) none of these

102. Which one of the following statements regarding the state transition matrix is correct?

(*a*) $\phi=(0)=0$

(*b*) $\phi(0)=\phi\left(\frac{1}{t}\right)$

(*c*) $\phi(t_1+t_2)=\phi(t_1)+(t_2)$

(*d*) $\phi(t_2+t_1)\,\phi(t_1-t_0)=\phi(t_2-t_0)$

103. The Nyquist polar plot of a system with $G(s)\,H(s)=\frac{s-z}{s(s+p)}$; $z, p>0$, is represented as

(*a*)

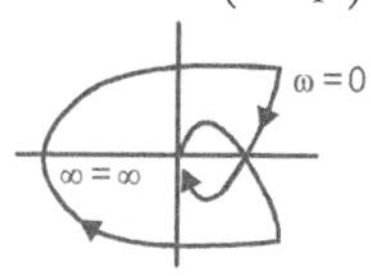

(*b*)

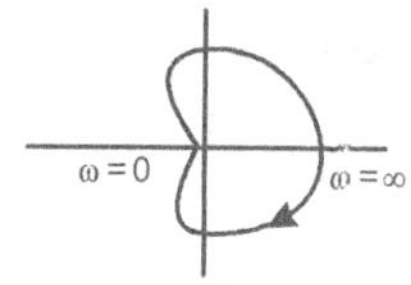

(*c*)

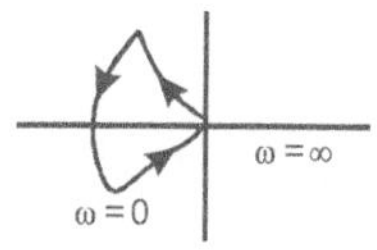

(*d*)

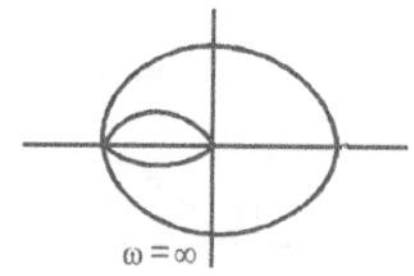

104. The transient response of a system is mainly due to

(*a*) internal forces (*b*) inertia forces

(*c*) friction (*d*) stored energy

105. Well designed first order control systems have a

(*a*) small bandwidth

(*b*) negative time constant

(*c*) a large negative transfer function pole

(*d*) none of these

106. A synchro is used to

(*a*) accelerate a rotating shaft

(*b*) convert an angular position of a shaft into an electrical signal

(*c*) convert linear motion into angular position

(*d*) amplify low frequency signals

107. A position-control system is a

(*a*) automatic regulating system

(*b*) process control system

(*c*) servomechanism

(*d*) stochastic control system

108. The pole-zero plot of a network function H(*s*) is shown in the given figure. Which one of the following will be the correct nature of its frequency response plots?

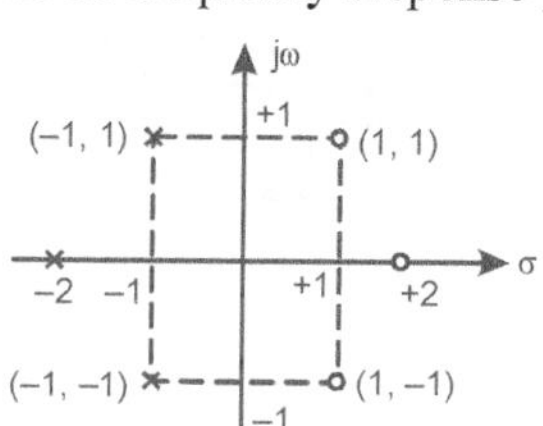

(*a*) The amplitude plot will decay with frequency but the phase plot will be constant at all frequencies

(*b*) The amplitude plot will be constant but the phase plot will vary from + 180° to + 90° as frequency varies from zero to infinity

(*c*) Both the amplitude and phase plots will be constant at all frequencies

(*d*) The amplitude plot will be constant but the phase plot will vary from + 180° to 0° as the frequency varies from zero to infinity

109. A control system having unit damping factor will give

(*a*) critically damped response
(*b*) oscillatory response
(*c*) undamped response
(*d*) no response

110. The unit step response of a second order linear system with zero initial states is given by

$c(t) = 1 + 1.25 \exp(-6t)$

$\sin(8t - \sin^{-1} 1.333), 1 \geq 0$

(*a*) 0.6 and 10 rad/s (*b*) 0.6 and 12.5 rad/s
(*c*) 0.8 and 10 rad/s (*d*) 0.8 and 12.5rad/s

111. In a system with rotating mass, the time constant of the system can be decreased by

(*a*) increasing the inertia of the system
(*b*) output rate feedback
(*c*) increasing input to system
(*d*) reducing friction

112. The overall transfer function of the system in the given figure is

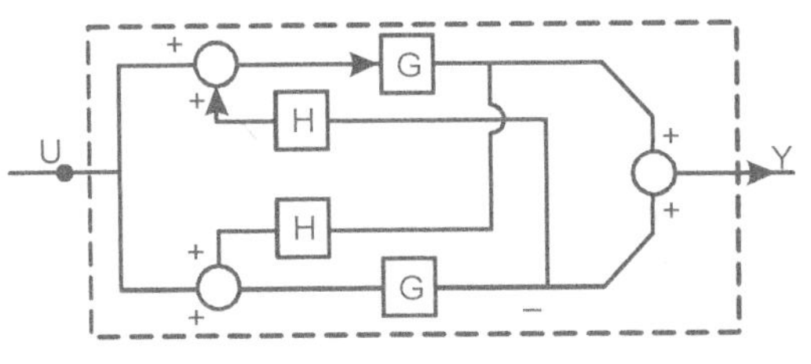

(*a*) $\dfrac{G}{1-GH}$ (*b*) $\dfrac{2G}{1-GH}$

(*c*) $\dfrac{GH}{1-GH}$ (*d*) $\dfrac{2G}{1-H}$

113. The system response can be tested better with

(*a*) sinusoidal input signal
(*b*) unit impulse input signal
(*c*) ramp input signal
(*d*) exponentially decaying signal

114. Consider the following statements regarding root loci:

1. All root loci start from the respective poles of G(*s*) H(*s*).
2. All root loci end at the respective zeros of G(*s*) H(*s*) or go to infinity
3. The root loci are symmetrical about the imaginary axis of the s-plane

Of these statements:

(*a*) 1, 2 and 3 are correct (*b*) 1 and 2 are correct
(*c*) 2 and 3 are correct (*d*) 1 and 3 are correct

115. The output of the controller in a control system is given to

(*a*) sensor
(*b*) comparator
(*c*) amplifier
(*d*) field control element

116. A compared to a closed-loop system, an open-loop system is

(*a*) more stable as well as more accurate
(*b*) less stable as well as less accurate
(*c*) more stable but less accurate
(*d*) less stable but more accurate

117. In control system compensation methods, lead compensation network

(*a*) achieves the desired results through the merits of its attenuation properties at high frequencies
(*b*) in the frequency domain decreases the bandwidth but increase phase margin
(*c*) in the frequency domain increases the bandwidth and phase margin
(*d*) improves steady-state accuracy but reduces bandwidth

118. Which of the following motors would be suitable for use as servomotors?

1. Two-phase induction motor
2. DC separately excited motor
3. A Single-phase capacitance start induction motor
4. Universal motor
5. Single-phase synchronous motors

Select the correct answer using the codes given below:

Codes:

(*a*) 1, 2, 3, 4 and 5
(*b*) 2, 3 and 4
(*c*) 1, 2 and 5
(*d*) 1 and 5

119. A system with a proportional controller shows steady-state error to a step input. It can be reduced by

(*a*) rate feedback
(*b*) increasing the controller gain
(*c*) adding integral action to the controller
(*d*) positive feedback

120. Saturation in a stable control can cause

(*a*) conditional stability
(*b*) overdamping
(*c*) low-level oscillations
(*d*) high-level oscillation

121. The presence of negative feedback in a control system

(*a*) increases accuracy and reduces bandwidth

(*b*) reduces bandwidth and increases distortion

(*c*) increases the effect of nonlinearities and reduces distortion

(*d*) reduces distortion and increase bandwidth

122. A step input as shown ir Fig. II is given to the circuit shown in Fig. I. The output is as shown in the Fig. III. The rise time of the circuit response is equal to

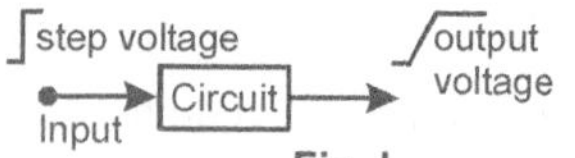

Fig. I

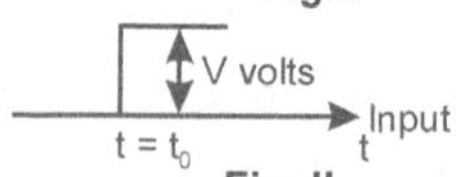

Fig. II

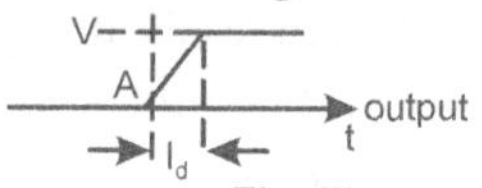

Fig. III

(*a*) t_d

(*b*) the times taken to reach 100% of V from the instant at which the output is 5% of V

(*c*) the time taken to reach is 5% of V

(*d*) the time taken to reach 90% of V from the instant at which the output is 10% of V

123. Consider the following statements regarding constructional features of synchros:

1. Synchro repeater has a three-phase stator and three-phase rotor.
2. Disc shaped rotor of the synchro control transformer offers a low reluctance to the magnetic path.
3. Synchro transmiter has a frictionless bearing

Of these statements

(*a*) 1, 2 and 3 are correct

(*b*) 1 alone is correct

(*c*) 2 alone is correct

(*d*) 2 and 3 are correct

124. A Nyquist plot of a system is shown in the given figure. What is the type of this system?

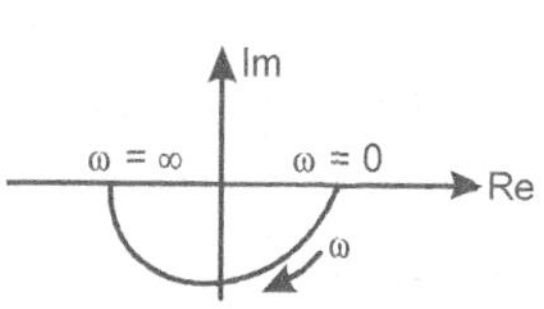

(*a*) 0

(*b*) 1

(*c*) 2

(*d*) 3

125. Match List-1 with List-II and select the correct answer by using the codes given below the lists:

List-I (*Function*)	**List-II** (*Laplace Transform*)
A. $\sin \omega t$	1. $\omega/(s^2+\omega^2)$
B. $\cos \omega t$	2. $s/(s^2+\omega^2)$
C. $\sin \omega t$	3. $\omega/(s^2-\omega^2)$
D. $\cos \omega t$	4. $s/(s^2-\omega^2)$

Codes:

	A	B	C	D
(*a*)	1	2	3	4
(*b*)	2	1	4	3
(*c*)	3	1	4	2
(*d*)	4	1	2	3

126. The impluse response of a system is given by

$$c(t) = \frac{1}{2} e^{-\frac{1}{2}}$$

Which one of the following is its unit step response?

(*a*) $1 - e^{-1/2t}$ (*b*) $1 - e^{-t}$

(*c*) $2 - e^{-t}$ (*d*) $1 - e^{2t}$

127. A synchro transmitter-control transfor-mer pair is excited at 400 Hz ac voltage. The output voltage

(*a*) has constant magnitude but its phase angle is proportional to the angular differences between their rotor angles

(*b*) has a magnitude proportional to the sine of the angular difference between their rotor with its phase angle changing by 180° whenever the sign of the angular difference changes

(*c*) has both magnitude and phase angle proportional to the angular difference between their rotors

(*d*) has constant magnitude but its phase angle changes by 180° whenever the sign of the angular difference changes

128. The intersection of root locus branches with the imaginary axis can be determined by use of

(*a*) Nyquist criterion (*b*) Routh's criterion

(*c*) Polar plot (*d*) None of these

129. A system has a complex conjugate root pair of multiplicity two or more in its characteristic equation. The impulse response of the system will be

(*a*) a sinusoidal oscillation which decays exponentially, the system is therefore stable

(*b*) a sinusoidal oscillation with time multiplier, the system is therefore unstable

(*c*) a sinusoidal oscillation which rises exponentially with time; the system is therefore unstable

(*d*) a dc term and harmonic oscillation

130. Which one of the following transfer functions represents the Bode plot shown in the figure?

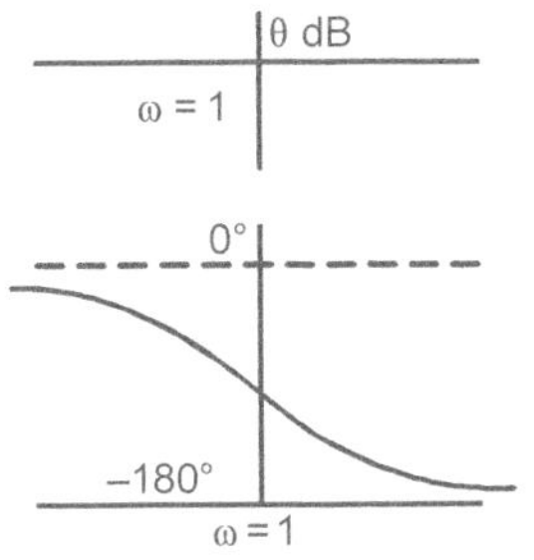

(*a*) $G = \frac{1-s}{1+s}$ (*b*) $G = \frac{1}{(1+s)^2}$

(*c*) $G = \frac{1}{s^2}$ (*d*) $G = \frac{1}{s(1+s)}$

131. A system G(*s*) is expressed in the state variable form as

$X = JX + Bu$

$y = CX + Du$

Consider the following statements with regard to the properties of Jordan canonical matrix J:

1. The diagonal elements of J are poles of G(*s*).
2. All the elements below the principal diagonal are zeros.
3. All the elements above the principal diagonal are zeros.

Among these statements:

(*a*) 1 and 3 are correct (*b*) 1 and 2 are correct

(*c*) 2 alone is correct (*d*) 3 alone is correct

132. On eliminating the feedback loop in the system shown in the figure, it would lead to a simplification with a single edge of gain

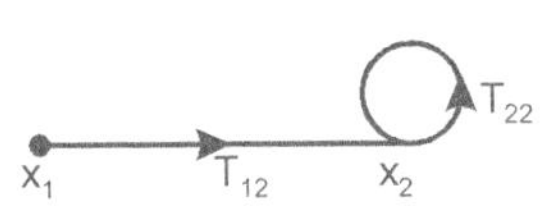

(*a*) $\frac{T_{12}}{1+T_{22}}$ (*b*) $\frac{T_{22}}{1-T_{12}}$

(*c*) T_{12} (*d*) $\frac{T_{12}}{1-T_{22}}$

133. If the unit-step response of a system is a unit-impulse function, then the transfer function of such a system will be

(*a*) 1 (*b*) $1/s$

(*c*) s (*d*) $1/s^2$

134. Consider the following transfer function of a phase-lead network:

$$\frac{1+0.2s}{1+0.05s}$$

The angular frequency for maximum phase shift of the above network is

(*a*) 40 (*b*) 30

(*c*) 20 (*d*) 10

135. The servomotor differs from other motors in the sense that it has

(*a*) high inertia and high torque

(*b*) low inertia and low torque

(*c*) low inertia and high torque

(*d*) entirely different constructions

136. The transfer function of the block diagram in terms of R and U is

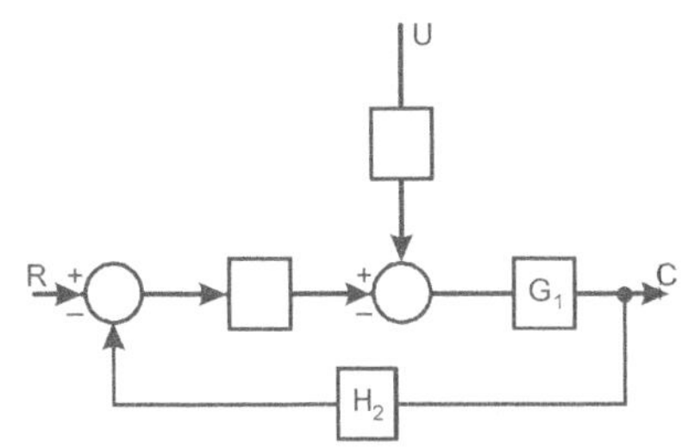

(*a*) $\frac{G_1G_3}{1+G_1G_3H_1}.R + \frac{G_2G_3}{1+G_1G_2G_3H_1}.u$

(*b*) $\frac{G_1G_3}{1+G_1G_3H_1}(R+u)$

(*c*) $\frac{G_1G_3+G_2G_3}{1+G_1G_2H_1}$

(*d*) $\frac{G_1G_3+G_2G_3}{1+G_1H_1+G_2H_1+G_3H_1}$

137. A rotational system is described by the first order differential equation

$2\frac{d\omega}{dt} + \omega = 5$, where the right-hand side

gives the constant torque acting on the system and represents the angular velocity. The solution for ω is given by

(*a*) $5(1-e^{-t/2})$

(*c*) $5e^{-t/2}$

(*c*) $5(1+e^{-t/2})$

(*d*) $5(1-e^{+t/2})$

138. The transfer function C/R of the block diagram given below is

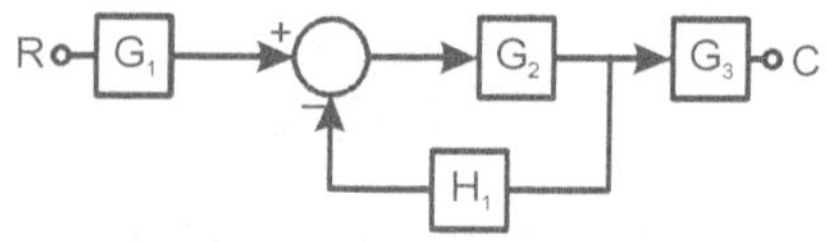

(*a*) $\dfrac{G_1G_2G_3}{1+G_1G_2G_3H_1}$

(*b*) $G_1G_3+\dfrac{G_1}{1+G_2H_1}$

(*c*) $\dfrac{G_1G_2}{1+G_1G_2H_1}+G_3$

(*d*) $\dfrac{G_1G_2G_3}{1+G_2H_1}$

139. In a non-linear control system limit cycle is self sustained oscillations of

(*a*) variable amplitude

(*b*) variable frequency

(*c*) fixed frequency

(*d*) fixed frequency and amplitude

140. Bandwidth is used as a means for specifying the performance of a control system related to the

(*a*) speed of response

(*b*) constant gain

(*c*) relative stability

(*d*) all of these

141. The Nyquist plot of an open-stable -system is shown in the given figure. If (–1,0) point is as shown in the figure, then the closed-loop system will be

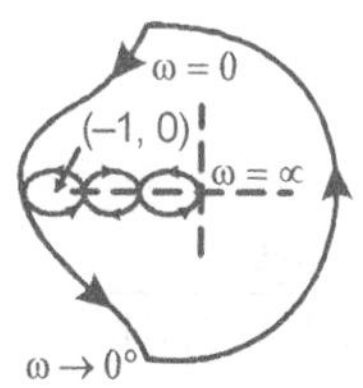

(*a*) unstable with two RHP poles

(*b*) stable

(*c*) unstable with one pole in RHP

(*d*) unstable with three poles in RHP

142. Which of the following is not a desirable feature of modern control system?

(*a*) Quick response

(*b*) Accuracy

(*c*) Correct power level

(*d*) No oscillation

143. When the network function is realised, it will give the network

$$Z(s)=\frac{(s+1)(s+3)}{(s+2)(s+4)}$$

is realised, it will give the network

(*a*) 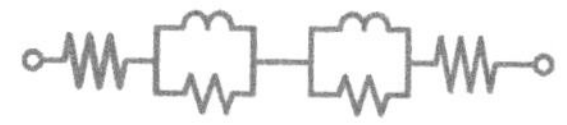

(*b*) 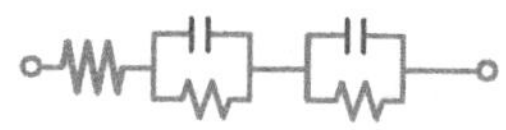

(*c*) 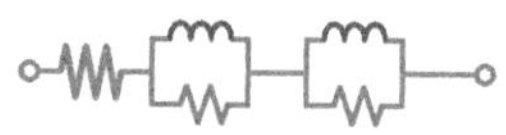

(*d*) 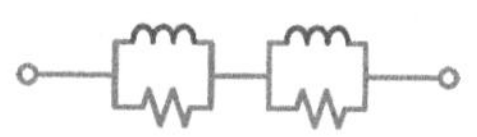

144. The sum of the gains of the feedback paths in the signal flow graph below is

(*a*) *af* + *be* + *cd* +*cbef* + *bcde* + *abcdef*

(*b*) *af* + *be* + *cd* + *abef* + *bcde*

(*c*) *af* + *be* + *cd* + *abef* + *abcdef*

(*d*) *af* +*be* + *cd*

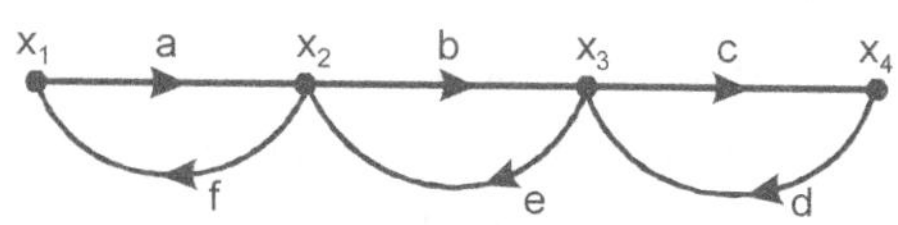

145. Consider the following statements regarding negative feedback in a closed-loop system:

1. It increases sensitivity.
2. It minimises the effect of disturbance.
3. There is possibility of instability.
4. It improves the transient response.

Of these statements:

(*a*) 1, 3 and 4 are correct

(*b*) 1, 2 and 4 are correct

(*c*) 1, 2 and 3 are correct

(*d*) 2, 3 and 4 are correct

146. To detect the error between a reference shaft position and an output shaft position which of the following error detectors can be used?

1. A pair of tachogenerators
2. A pair of potentiometers
3. A synchro pair

Codes:

Select the correct answer using the codes given below:

(*a*) 1, 2 and 3

(*b*) 1 and 2

(*c*) 2 and 3

(*d*) 1 and 3

147. If the unilateral Laplace transform X(*s*) of a signal X(*t*) is, $\frac{7s+10}{s(s+2)}$, then the initial and final values of the signal would be respectively

(*a*) 3.5 and 5

(*b*) zero and 7

(*c*) 5 and zero

(*d*) 7 and 5

148. The error response of a second order system to a step input is obtained as

$E(t) = 1.66\, e^{-8t} \sin(6t + 37°)$

The damping ratio is

(*a*) 0.4 (*b*) 0.5

(*c*) 0.8 (*d*) 1.0

149. The transfer function is $\frac{1+0.5s}{1+s}$:

It represents a

(*a*) lead network

(*b*) lag network

(*c*) lag-lead network

(*d*) proportional controller

150. The signal flow graph shown in the figure has

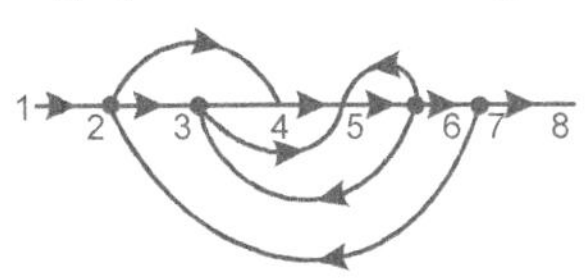

(*a*) two forward paths, four loops and no non-touching loops

(*b*) three forward paths, four loops and no non-touching loops

(*c*) three forward paths, three loops and no non-touching loops

(*d*) tow forward paths, four loops and two non-touching loops

151. Gain margin is the reciprocal of gain at the frequency at which the phase angle becomes

(*a*) 0° (*b*) 90°

(*c*) 180° (*d*) 270°

152. The open-loop transfer function of a system is given by

$$G(s)H(s) = \frac{K(s+1)(s+3)}{s^2+4s+8}.$$

Its root locus diagram is

(*a*) 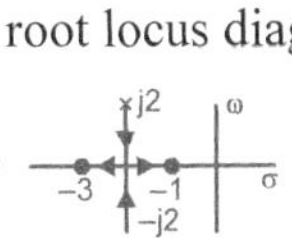

(*b*) 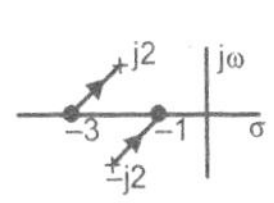

(*c*) 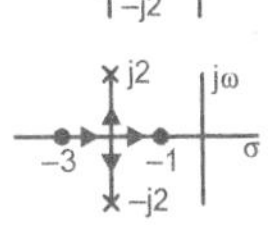

(*d*) 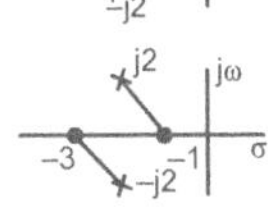

153. Match List I (location of roots) with List II (step response) and select the correct answer using the codes given below the lists :

List I

A. 

B. 

C. 

D. 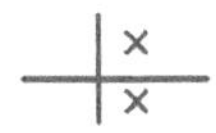

List II

1. 

2. 

3. 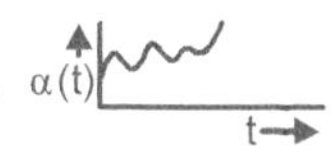

4. 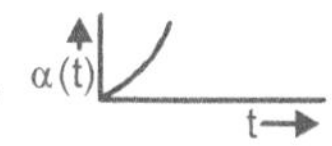

5. 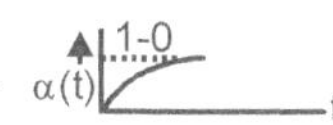

Codes :

	A	B	C	D
(*a*)	4	1	2	3
(*b*)	5	1	2	3
(*c*)	4	2	1	3
(*d*)	5	2	3	4

154. For a system to be stable

(*a*) all poles and zeros of the transfer function must be in the right half of *s*-plane

(*b*) all poles and zeros of the transfer function must be on the imaginary axis

(*c*) all the poles of the transfer function must be in the left half of the s-plane

(*d*) all the zeros of the transfer function must be in the left half of the s-plane

155. The Nyquist plot of an open-loop system with the transfer function

$$G(s)H(s) = \frac{s+2}{(s+1)(s-1)}$$

will be of the form

(*a*) 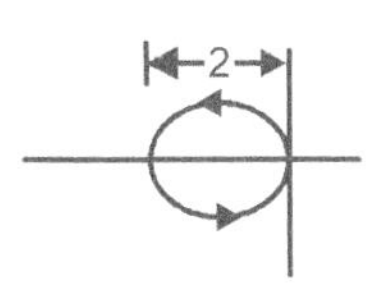

(*b*) 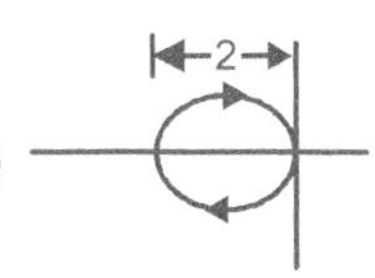

(*c*) 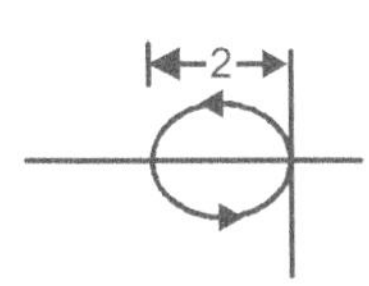

(*d*) 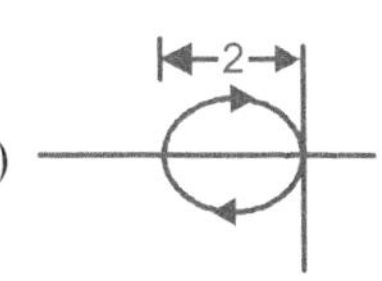

156. Which of the following statements are true of a type I system having unity gain in the forward path and a unity feedback?

1. Positional error constant K_p is equal to infinity
2. Acceleration error constant K_a is equal to zero
3. Steady-state error e_{ss} to a unit step displacement input is equal to one

Select the correct answer using the codes given below

Codes:

(*a*) 1, 2 and 3 (*b*) 1 and 2
(*c*) 2 and 3 (*d*) 1 and 3

157. For the asymptotic Bode plot (dB versus log) of a transfer function with

a factor of the form $\frac{1}{\left(s^2+2\xi_{\omega}s+\omega_n^2\right)}$ the corner frequency (I), slope in dB/ decade (II) and error at the corner frequency (III) will be

	I	II	III
(*a*)	$\xi\omega_n$	+40	Error is ξ dependent
(*b*)	ω_n	−40	Error ξ is dependent
(*c*)	$\xi\omega_n$	+20	− 3dB
(*d*)	ω_n	−20	+3dB

158. With a derivative error feedback control

(*a*) a second order system is converted into a first order system
(*b*) a second order system is converted into a third order system
(*c*) natural frequency of oscillation
(*d*) damping ratio is increased

159. For the feedback system shown in the given figure, the forward path does not affect the system output when KG is

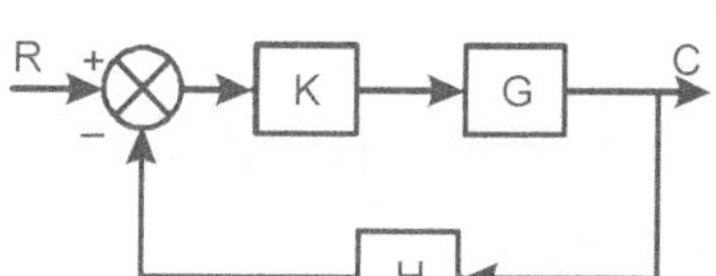

(*a*) small (*b*) negative
(*c*) one (*d*) very large

160. Given $F(s)=\frac{(s+2)}{s(s+1)}$, the initial and final value of $f(t)$ will be respectively

(*a*) 1, 2
(*b*) 2, 1
(*c*) 1, 1
(*d*) 2, 2

161. For a standard feedback control loop, the sensitivity of the closed-loop transfer function T to forward path gain G and feedback path gain H are

	S^T	S_H^T
(*a*)	1/(1 + GH)	− 1/(1 + GH)
(*b*)	−GH(1 + GH)	−H/(1 + GH)
(*c*)	G/(1 + GH)	1/(1 + GH)
(*d*)	1/(1 + GH)	−GH/(1 + GH)

162. Consider the following types of machines:

1. 2-Phase A.C. servomotor
2 Conventional 2-phase induction motor
3. Stepper motor
4. Tachogenerator

The rotors are generally of drag cup construction in the case of

(*a*) 1, 2 and 4 (*b*) 1 and 3
(*c*) 3 and 4 (*d*) 1 and 4

163. The Bode plot of the transfer function G(*s*) = *s* is

(*a*) Zero magnitude and phase shift
(*b*) Constant magnitude and constant phase shaft angle
(*c*) 20 *d*B/decade and phase shift of $\pi/2$
(*d*) − 20 *d*B/decade and constant phase shift angle.

164. Which of the following is not necessarily valid for root locus pattern?

(*a*) The n finite zeros and the m-poles are plotted on the s-plane. Then (m-n) indicates the number of non-finite zeros
(*b*) The number of poles gives the number of loci
(*c*) A value of s on the real axis is a point on the root locus if the total number of poles and zeros on the real axis to the right of the point is even
(*d*) There are as many asymptotes as non-finite zeros

165. A system block diagram is shown in the given figure. The overall transfer function of the system is

$$\frac{C}{R}=\frac{G_1G_2G_3}{1+G_1G_2G_3H_1+G_3G_2-G_3G_2H_3}$$

The value 'X' would be equal to

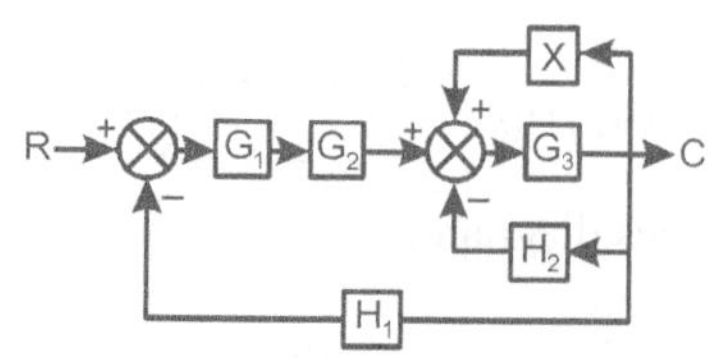

(*a*) H_3 (*b*) G_1H_3
(*c*) G_2H_3 (*d*) G_3H_3

166. If a system has an open loop transfer function $\frac{1-s}{1+s}$, then the gain of the system at frequency of 1 rad/s will be

(*a*) 1 (*b*) $\frac{1}{2}$

(*c*) zero (*d*) – 1

167. Non-linear system cannot be analysed by Laplace transform because

(*a*) it has no zero initial conditions

(*b*) superposition law cannot be applied

(*c*) non-linearity is generally not well defined

(*d*) all of these

168. For the system shown in the given figure, the damping factor 'ξ' and undamped natural frequency 'ω_n' are respectively

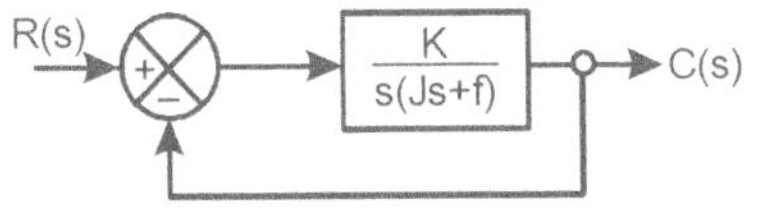

(*a*) $\frac{2\sqrt{KJ}}{f}$ and $\sqrt{\frac{J}{K}}$

(*b*) $\sqrt{\frac{K}{J}}$ and $\frac{f}{2\sqrt{KJ}}$

(*c*) $\frac{f}{2\sqrt{KJ}}$ and $\sqrt{\frac{K}{J}}$

(*d*) $\frac{2f}{\sqrt{KJ}}$ and $\sqrt{\frac{K}{J}}$

169. For the signal-flow graph shown in the given figure the value of x_2 is

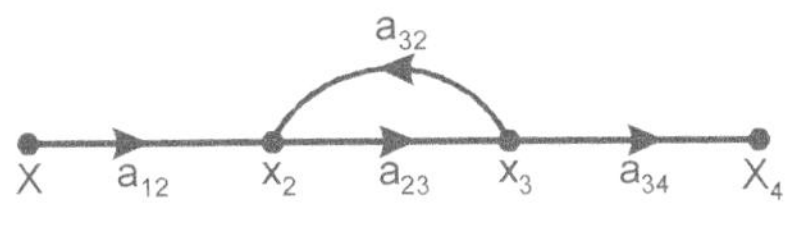

(*a*) $a_{12}x_1$

(*b*) $a_{12}x_1 + a_{32}x_3$

(*c*) $a_{12}x_1 - a_{23}x_3 + a_{32}x_3$

(*d*) $-a_{23}x_3 - a_{34}x_4$

170. Error constant of system are a measure of

(*a*) steady-state response

(*b*) transient-state response

(*c*) relative stability

(*d*) steady-state as well as transient response

171. A control system with excessive noise is likely to sulfer from

(*a*) vibrations

(*b*) oscillations

(*c*) saturation in amplifying stages

(*d*) loss in gain

172. Which of the following is used for Nyquist plot?

(*a*) Pole zero plot

(*b*) Closed loop function

(*c*) Open loop function

(*d*) Characteristic equation

173. Consider the following statements regarding stability analysis of Routh-Hurwitz criterion:

1. For a system to be stable, all the coefficients of the characteristic equation must be present and be of the same sign.
2. If a system is to be stable, there should not be any sign change in the first column of the Routh's array.
3. The order of the auxiliary equation obtained from the elements of the Rouths' table is always odd.

Of these statements:

(*a*) 1 and 2 are correct

(*b*) 2 and 3 are correct

(*c*) 1 and 3 are correct

(*d*) 1, 2 and 3 are correct

174. A system has the following transfer function

$$G(s) = \frac{100(s+5)(s+50)}{s^4+(s+10)(s^2+3s+10)}.$$

The type and order of the system are respectively

(*a*) 4 and 9 (*b*) 4 and 7

(*c*) 5 and 7 (*d*) 7 and 5

175. Which one of the following system is completely state controllable?

(*a*) $\begin{bmatrix} \dot{x}_1 \\ \dot{x}_2 \end{bmatrix} = \begin{bmatrix} -1 & 0 \\ 0 & -2 \end{bmatrix}\begin{bmatrix} x_1 \\ x_2 \end{bmatrix} + \begin{bmatrix} 2 \\ 1 \end{bmatrix}u$

(*b*) $\begin{bmatrix} \dot{x}_1 \\ \dot{x}_2 \end{bmatrix} = \begin{bmatrix} 1 & 0 \\ 0 & -2 \end{bmatrix}\begin{bmatrix} x_1 \\ x_2 \end{bmatrix} + \begin{bmatrix} 2 \\ 0 \end{bmatrix}u$

(*c*) $\begin{bmatrix} \dot{x}_1 \\ \dot{x}_2 \end{bmatrix} = \begin{bmatrix} -1 & 0 \\ 0 & -1 \end{bmatrix}\begin{bmatrix} x_1 \\ x_2 \end{bmatrix} + \begin{bmatrix} 1 \\ 0 \end{bmatrix}u$

(*d*) $\begin{bmatrix} \dot{x}_1 \\ \dot{x}_2 \end{bmatrix} = \begin{bmatrix} -1 & 0 \\ 2 & -2 \end{bmatrix}\begin{bmatrix} x_1 \\ x_2 \end{bmatrix} + \begin{bmatrix} 2 \\ 2 \end{bmatrix}u$

176. Consider the following Nqyuist plots of different control systems:

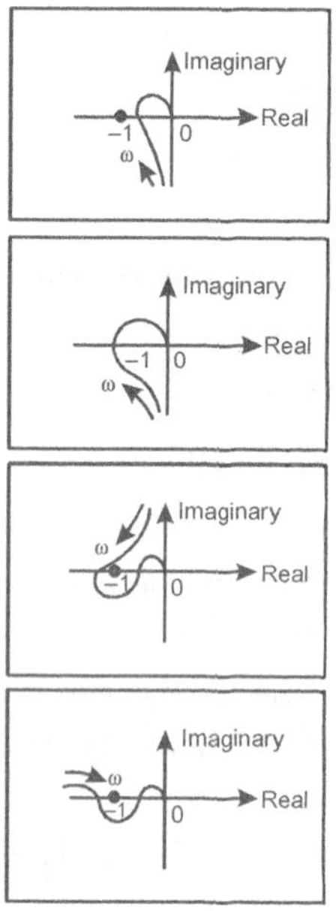

Which of these plot(s) represent(s) a stable system?

(a) 1 alone (b) 2, 3 and 4

(c) 1, 3 and 4 (d) 1,2 and 4

177. Match List I with List II and select the correct answer using the codes given below in the lists:

List I (*Loop transfer function*)

A. $\frac{K(s+1)}{s^2(s+10)}$

B. $\frac{K}{s(s+2)(s^2+2s+2)}$

C. $\frac{K}{s(s+2)(s^2+2s+5)}$

D. $\frac{K}{s(s+4)(s^2+2s+5)}$

List II (*Point* (s) *of root-locus plot*)

1. One real breakaway point
2. Two real breakaway points
3. Tliree real breakaway points
4. One real and one pair of complex conjugate break-away points

Codes:

	A	B	C	D
(a)	1	2	4	3
(b)	1	2	3	4
(c)	2	1	4	3
(d)	2	1	3	4

178. The Laplace transform of the function $f(t)$ is $F(s)$. $u(t)$ represents the unit step functions. The inverse Laplace transform of $e^{-s}F(s)$

(a) $f(t)\,u(t-1)$

(b) $f(t-1)\,u(t)$

(c) $f(t-1)\,u(t-1)$

(d) $\frac{f(t)}{(t-1)}$

179. Consider the system shown in the block diagram in the figure. The signal flow diagram of the system is best represented as

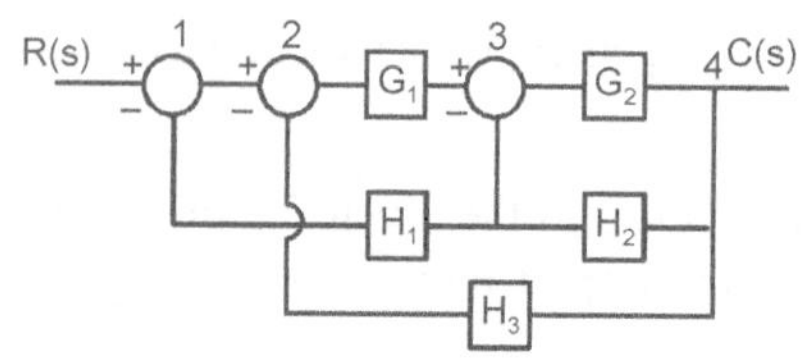

(a)
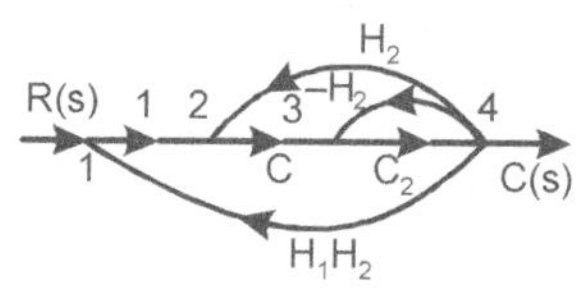

(b)
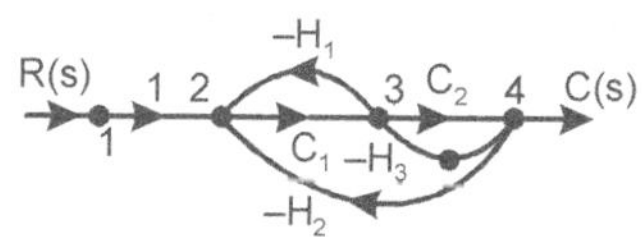

(c)
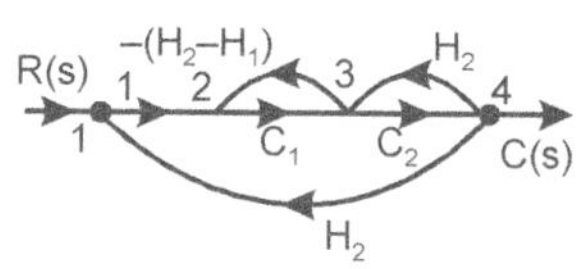

(d)
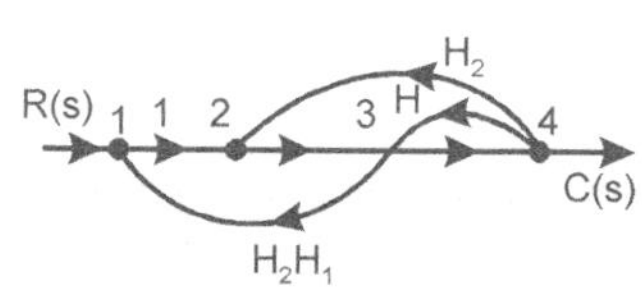

180. A synchro-transmitter and a control transformer have

(a) similar rotor and stators

(b) both rotors and stators of different types

(c) similar stators but different rotors

(d) similar rotor but deferent stators

181. In the figure alongside, spring constant is K, viscous friction coefficient is B, mass is M and the system output motion is $x(t)$ corresponding to input force $F(t)$. Which of the following parameters relate to the above system?

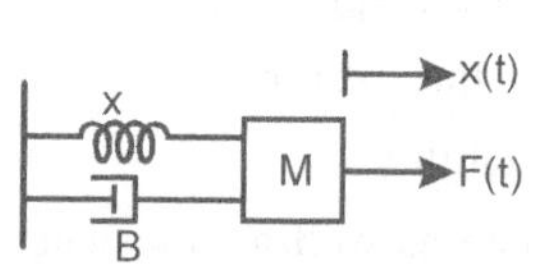

1. The constant $= \dfrac{1}{M}$
2. Damping coefficient $= \dfrac{B}{2\sqrt{KM}}$
3. Natural frequency of oscillation $= \sqrt{\dfrac{K}{M}}$

Select the correct answer using the codes given below :

Codes:

(*a*) 1, 2 and 3 (*b*) 1 and 2

(*c*) 2 and 3 (*d*) 1 and 2

182. Gain margin is the factor by which the system gain can be increased to drive it to the verge of

(*a*) instability (*b*) stability

(*c*) critical damping (*d*) oscillation

183. In force-current Analogy, mass is analogous to

(*a*) current (*b*) resistance

(*c*) capacitance (*d*) voltage

184. Match List I with List II and select the correct answer using the codes given below the lists:

List I (*Signal flow graphs*)

A. 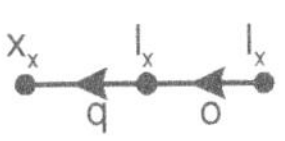

B. 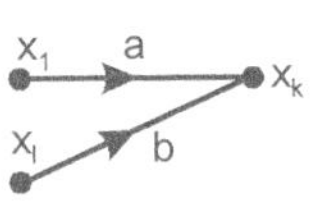

C. 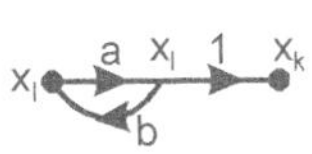

D. 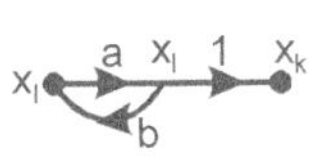

List II (*Expression for output*)

1. $xk = (a + b)x_i$
2. $xk = (ab)x_i$
3. $xk = \dfrac{a}{1-b}x_i$
4. $xk = \dfrac{a}{1-ab}x_i$

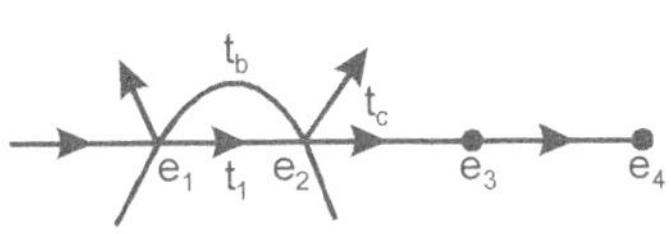

Codes:

	A	B	C	D
(*a*)	1	2	3	4
(*b*)	1	4	3	2
(*c*)	2	1	3	4
(*d*)	2	1	4	3

185.

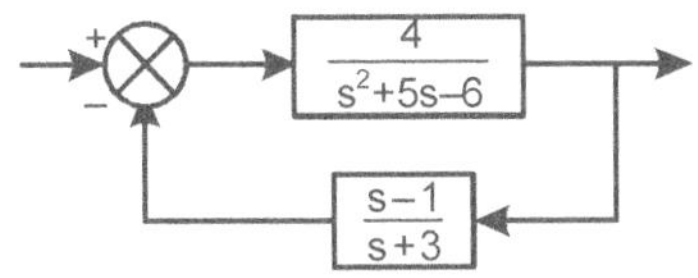

Which one of the following statements is true for the system shown in the above figure

(*a*) Open-loop system is stable but closed-loop system is unstable

(*b*) Both open-loop and closed-loop systems are stable

(*c*) Both open-loop and closed-loop systems are unstable

(*d*) None of these

186. Match List I with List II and select the correct answer using the codes given below the lists:

List I (*Transfer Function*)

A. $\dfrac{1}{(s+\alpha)(s+\beta)}$

B. $\dfrac{1}{(s-\alpha)(s-\beta)}$

C. $\dfrac{1}{(s-\alpha+j\beta)(s-\alpha-j\beta)}$

D. $\dfrac{1}{(s-\alpha+j\beta)(s+\alpha-j\beta)}$

List II (*Transient Response*)

1. 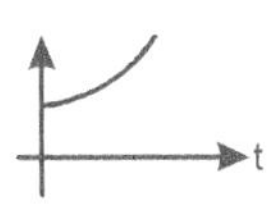

2. 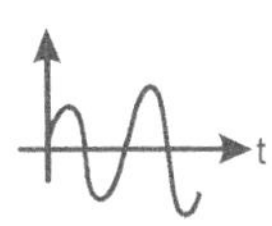

3. 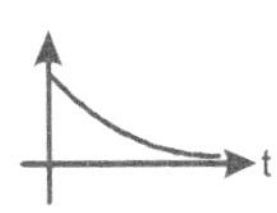

4. 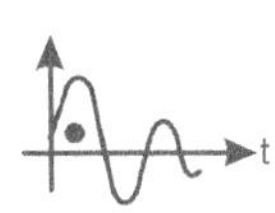

Assume $\alpha > 0$ and $\beta > 0$

Codes:

	A	B	C	D
(*a*)	1	3	2	4
(*b*)	3	1	2	4
(*c*)	3	4	2	1
(*d*)	4	2	1	3

187. Match List I with List II and select the correct answer using the codes given below the lists:

List I (*Nyquist plot*)

A. 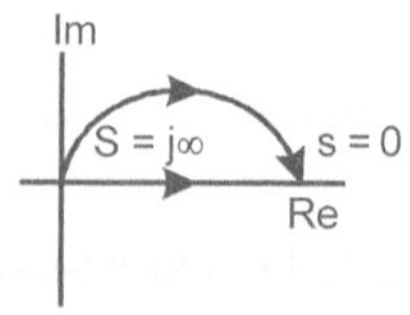

B. 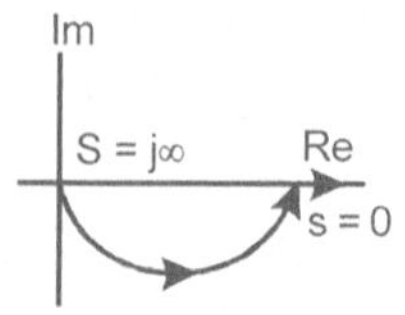

C. 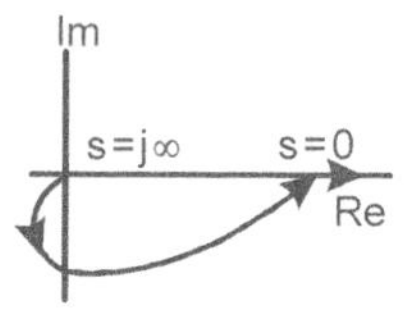

D. 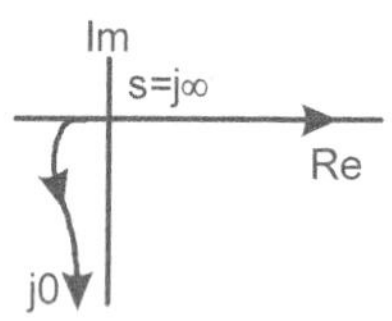

List II (*Transfer function*)

1. $\dfrac{K}{(1+sT_1)(1+sT_2)}$

2. $\dfrac{K}{(1+sT_1)(1+sT_2)(1+sT_3)}$

3. $\dfrac{K}{(1+sT_1)}$

4. $\dfrac{K}{(1-sT_1)}$

5. $\dfrac{K}{s(1+sT_1)}$

Codes:

	A	B	C	D
(*a*)	4	3	1	5
(*b*)	3	1	2	5
(*c*)	4	1	2	5
(*d*)	1	2	3	4

188. $G_A(j\omega)$, $G_B(j\omega)$ and $G_C(j\omega)$, are three transfer functions whose phase variations are shown in the given figure. The magnitude variation with respect to frequency is the same for both $G_A(j\omega)$ and $G_C(j\omega)$ but their phase variation is as indicated by the curves labelled A and C in the given figure. The transfer function $G(j\omega)$ is such that

$\phi_C(j\omega) = \phi_A(j\omega) + \phi_B(j\omega)$

G_A, $G_B G_C$ are respectively

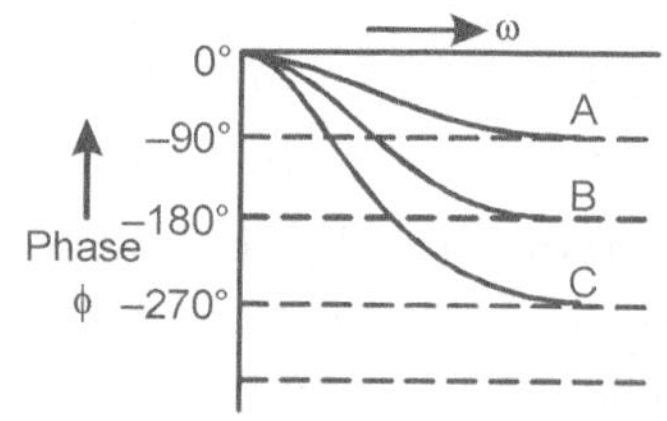

(*a*) minimum phase, all pass and non-minimum phase-functions

(*b*) minimum phase, non-minimum phase and all pass functions

(*c*) all pass, minimum phase and non-minimum phase functions

(*d*) all pass, non-minimum phase and minimum phase functions

189. The gain from source node x_1 output node x_3 for the signal flow graph given below is

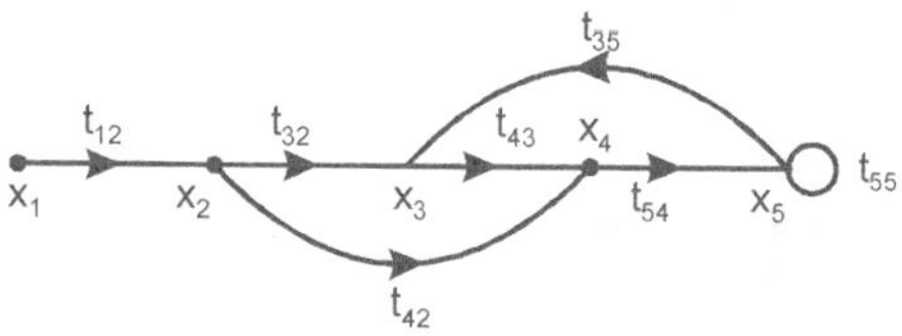

(*a*) $\dfrac{t_{21}t_{32}(t_{43}t_{54}+t_{42})}{(1-t_{55}-t_{43}t_{35}t_{54})}$

(*b*) $\dfrac{t_{21}t_{54}(t_{32}t_{42}-t_{43})}{(1+t_{55}-t_{43}t_{35}t_{54})}$

(*c*) $\dfrac{t_{21}t_{54}(t_{32}t_{43}-t_{42})}{(1-t_{55}+t_{43}t_{35}t_{54})}$

(*d*) $\dfrac{t_{21}t_{54}(t_{32}t_{43}+t_{42})}{(1-t_{55}-t_{43}t_{35}t_{54})}$

190. A plot of open-loop poles zeros of a unity feedbcak control system is shown in the given figure:

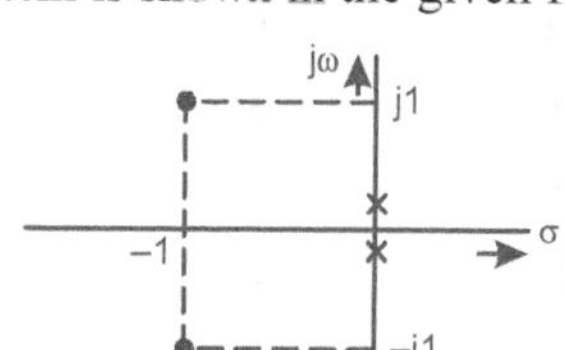

Its root-loucs will be

(a) 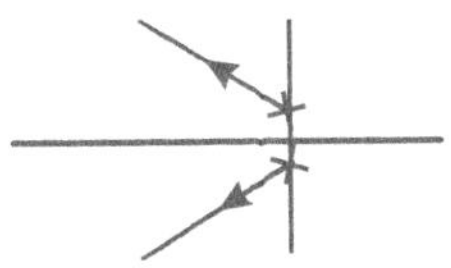

(b) 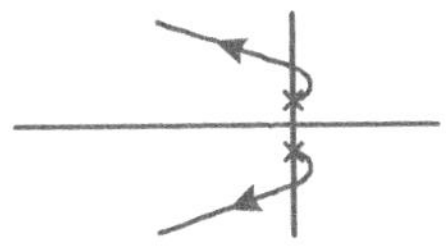

(c) 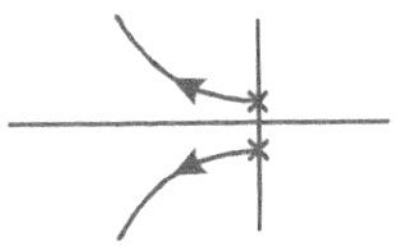

(d) 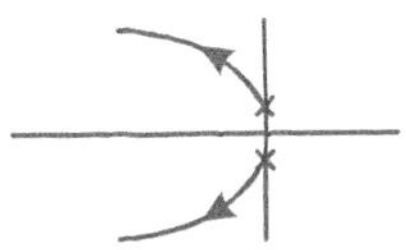

191. Consider the following features:

1. Its poles is located to the left of its zero.
2. It decreases the margin of stability of the system
3. It speeds up the system transient response

Those associated with a iead compensator would include

(a) 1, 2 and 3 (b) 1 and 2

(c) 2 and 3 (d) 1 and 3

192. A control system whose step response is $-0.5\,(1 + e^{-2t})$ is cascaded to another control block whose impulse response is e^{-1}. The transfer functions of the cascaded combination is

(a) $\frac{1}{(s+1)(s+2)}$ (b) $\frac{1}{(s+1)}$

(c) $\frac{1}{(s+2)}$ (d) $\frac{0.5}{(s+1)(s+2)}$

193. The root-locus plot of the open-loop transfer function

$$GH(s) = \frac{K}{s(s+2)(s+4)}$$

(a) 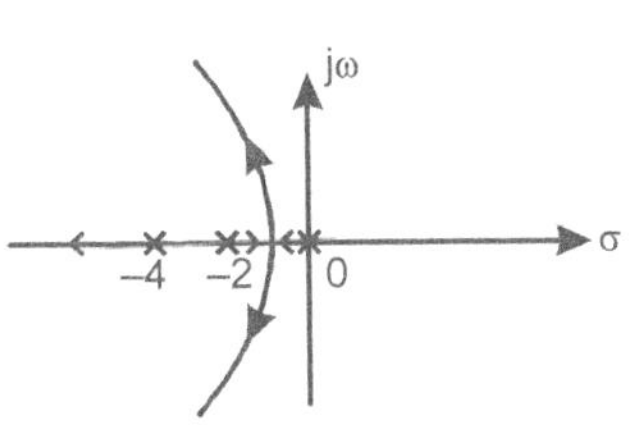

(b) 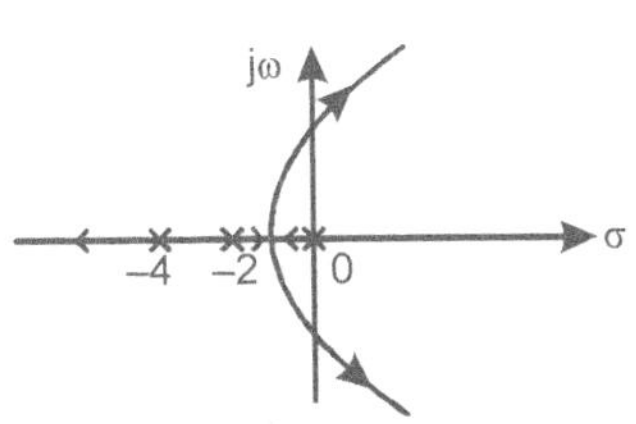

(c) 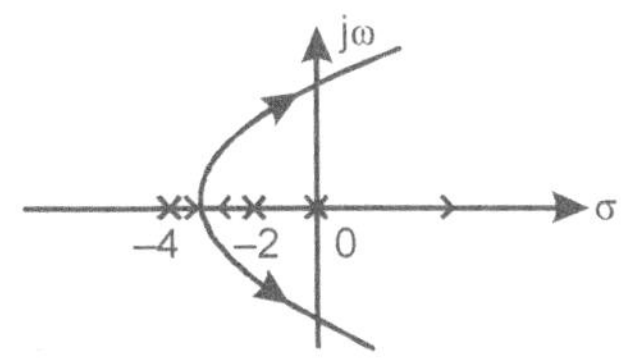

(d) 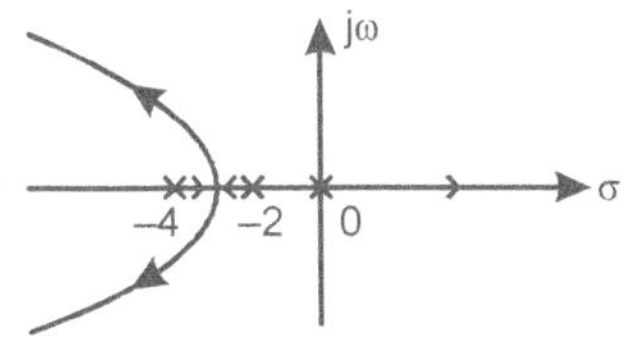

194. For the minimum phase system to be stable

(a) phase margin should be negative and gain margin positive

(b) phase margin should be positive and gain margin negative

(c) both gain margin and phase margin should be positive

(d) both gain margin and phase margin should be negative

195. A step functions is applied to the input of a system and output is of the form $y = t$, the system is

(a) stable

(b) unstable

(c) not necessarily stable

(d) conditionally stable

196. Match List I with List II and select the correct answer using the codes given below the lists:

List I

(*Pools in the s-plane*)

A. 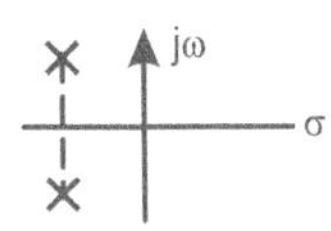

B. 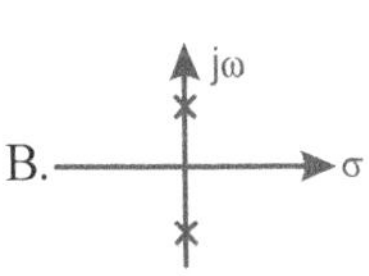

C. 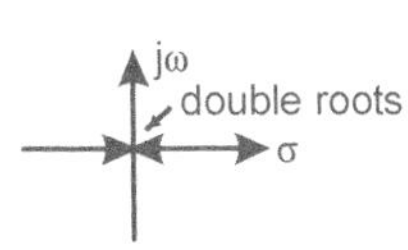

D. 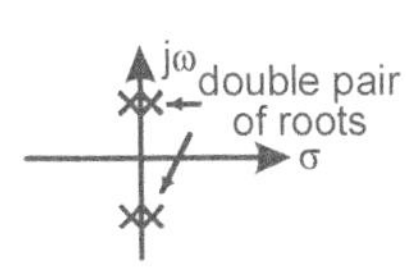

List II

(Corresponding impulse response)

1. 

2. 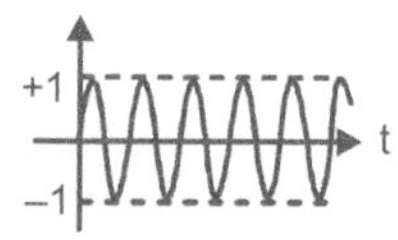

3. 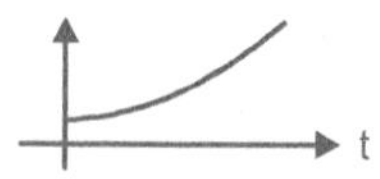

4. 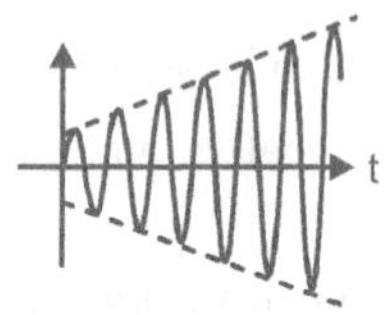

5. 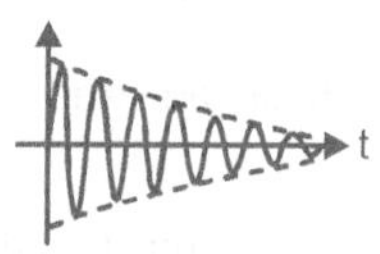

Codes:

	A	B	C	D
(*a*)	5	1	2	4
(*b*)	3	4	2	1
(*c*)	5	2	1	4
(*d*)	4	3	1	5

197. What will be the closed-loop transfer function of a unity feedback control system whose step response is given by

$$c(t) = K[1 - 1.66\, e^{-8t} \sin(5t + 37°)]$$

(*a*) $\dfrac{100K}{s^2+16s+100}$ (*b*) $\dfrac{10}{s^2+16s+100}$

(*c*) $\dfrac{K}{s^2+16s+100}$ (*d*) $\dfrac{10K}{s^2+8s+100}$

198. Bode plots of an open-loop transfer function of a control system are shown in the given figure:

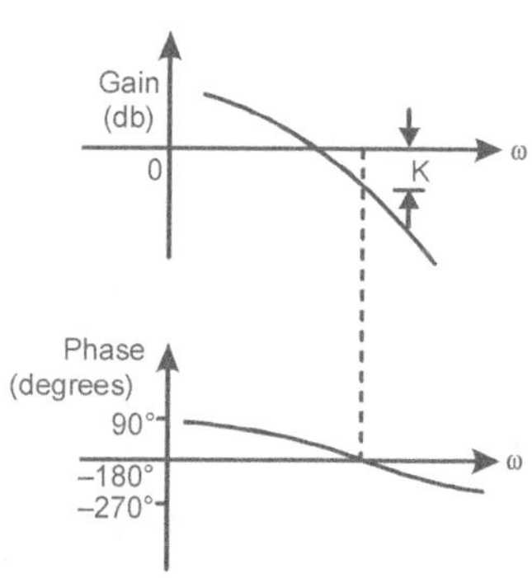

The gain margin of the system is

(*a*) K (*b*) –K

(*c*) $\dfrac{1}{K}$ (*d*) $-\dfrac{1}{K}$

199. Match List I (Matrix) with List-II (Dimensions) for the state equations:

$$X(t) = PX(t) + QU(t)$$

and

$$Y(t) = RX(t) + SU(t)$$

Select the correct answer using the codes given below the lists:

List–I	List–II
A. P	1. $(n \times p)$
B. Q	2. $(q \times n)$
C. R	3. $(n \times n)$
S. S	4. $(q \times p)$

Codes:

	A	B	C	D
(*a*)	1	3	4	2
(*b*)	1	3	2	4
(*c*)	3	1	4	2
(*d*)	3	1	2	4

200. The settling time of a second order system on 2% basis is given by

(*a*) $t_s = \dfrac{4}{\xi\omega_n}$ (*b*) $t_s = \dfrac{\xi\omega_n}{4}$

(*c*) $t_s = \dfrac{4\xi}{\omega_n}$ (*d*) $t_s = 4\xi\omega_n$

201. Unit impulse response of a system in Laplace transform gives

(*a*) transfer function (*b*) system gain

(*c*) unit step (*d*) unit ramp function

202. The given figure shows root-loci of open-loop transfer function G(*s*) H(*s*) of a system. Consider the following inference drawn from this figure:

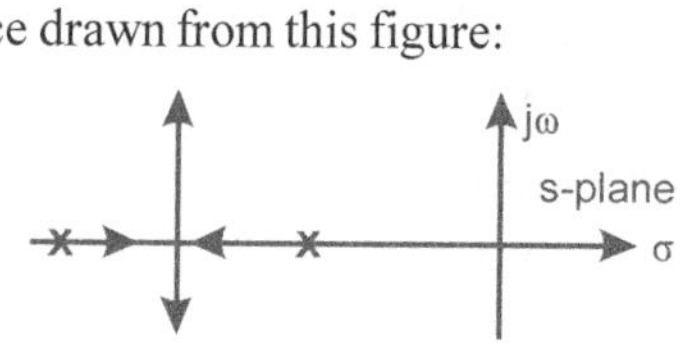

1. It has no zero.
2. It is a stable system.
3. It is a second order system.

Which of these inference are correct?

(*a*) 1, 2 and 3 (*b*) 1 and 2

(*c*) 2 and 3 (*d*) 1 and 3

203. The phase lag produced by transportation delays

(*a*) decreases linearly with frequency

(*b*) increases linearly with frequency

(*c*) inversly proportional to frequency

(*d*) independent of frequency

204. Consider the following statements regarding the transfer function G(s) plotted in the given figure:

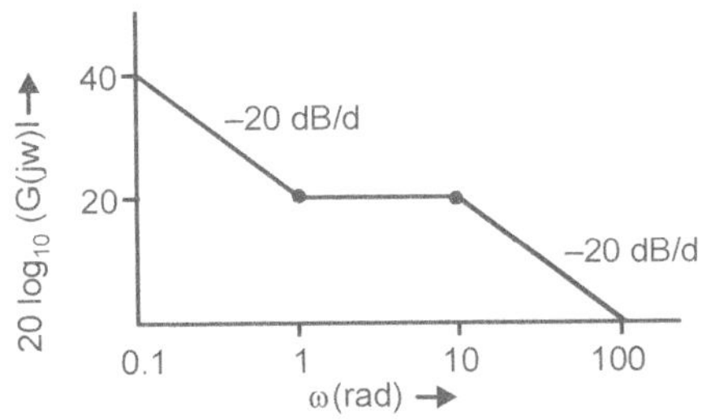

1. G(s) has corner frequencies at $\omega = 0.1$, 1 and 10

2. $G(s) = \dfrac{100(s+1)}{s(s+10)}$

3. The magnitude of 20 $\log_{10}$ $|G(j\omega)|$ at $\omega = 1000$ is -20 dB

Which of these statements are correct?

(*a*) 1, 2 and 3

(*b*) 1 and 2

(*c*) 2 and 3

(*d*) 1 and 3

205. Consider the control system shown in the following figure and the statements given below the figure

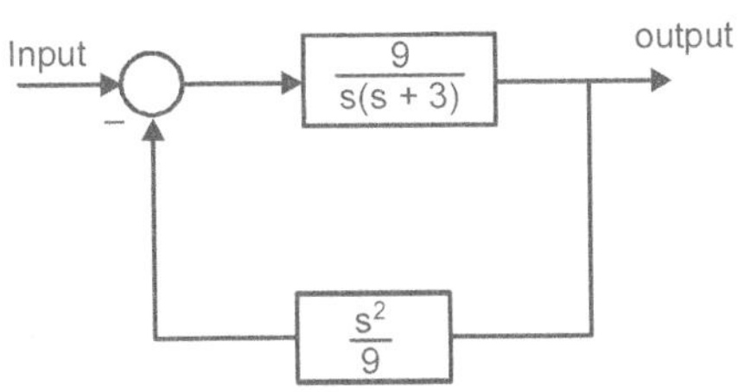

1. The system is of second order

2. Basically, the system is having positive feedback

3. The system is of type 1.

4. The dimension of the output is not the same as that of the input.

Of these statements:

(*a*) 1 and 2 are correct

(*b*) 2 and 4 are correct

(*c*) 2, 3 and 4 are correct

(*d*) 1, 2 and 3 are correct

206. The open-loop transfer function of a unity feedback control system is

$$G(s)\ H(s) = \frac{30}{s(s+1)+(s+1)}$$

where T is a variable parameter. The closed loop system will be stable for all values of

(*a*) $T > 0$

(*b*) $0 < T < 3$

(*c*) $T > 5$

(*d*) $3 < T < 5$

207. The main application of transfer function is in the study of

(*a*) steady state behaviour of systems

(*b*) steady state as well as transient behaviour of systems

(*c*) only transient behaviour of system

(*d*) none of these

208. Match List-I (Plot/diagram/chart) with List-II (Characteristic) and select the correct answer using the codes given below the Lists:

List-I

A. Constant M loci

B. Constant N loci

C. Nichol's chart

D. Nyquist plot

List-II

1. Constant gain and phase shift loci of the closed-loop system

2. Plot of loop gain with variation of w

3. Circle of constant gain for closed-loop transfer function

4. Circles of constant phase shift of closed-loop transfer function

Codes:

	A	B	C	D
(*a*)	3	4	2	1
(*b*)	3	4	1	2
(*c*)	4	3	2	1
(*d*)	4	3	1	2

209. A system is shown in the given figure. The value of K which gives a steady-state error of 20% to a unit step-input is given

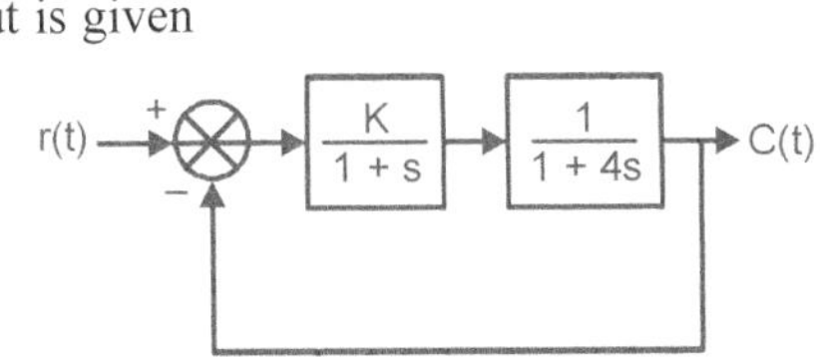

(*a*) 500 (*b*) 100

(*c*) 20 (*d*) 4

210. The response of a system to a unit ramp input is

$$\frac{1}{2}t-\frac{1}{3}u(t)+\frac{1}{8}e^{-4t}$$

Which one of the following is the unit impulse response of the system?

(*a*) $1-e^{-4t}$

(*b*) $2(1-e^{-4t})$

(*c*) e^{-4t}

(*d*) $2e^{-4t}$

211. By a suitable choice of the scalar parameter K, the system shown in the given figure can be made to oscillate continuously at a frequency of

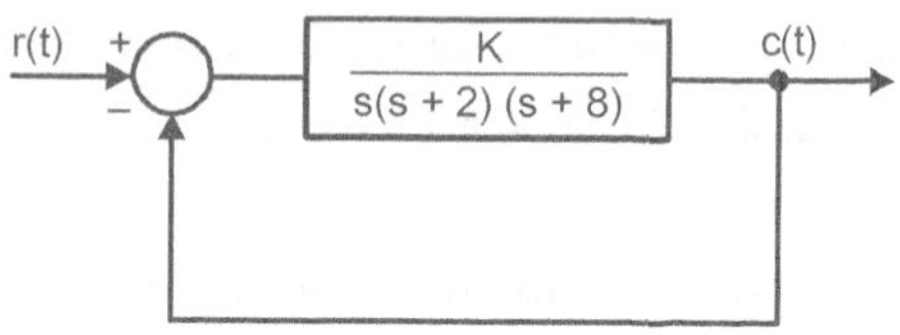

(*a*) 1 rad/s (*b*) 2 rad/s

(*c*) 4 rad/s (*d*) 8 rad/s

212. The impulse response of a linear system is given by

$$g(t) = Ke^{-\frac{1}{\tau}}$$

A tangent is drawn to the impulse response curve at time $t = 0$. The tangent will interest the time axis at

(*a*) $\frac{1}{2}\tau$ (*b*) τ

(*c*) 2τ (*d*) $K\tau$

213. For the function $X(s) = \frac{1}{(s+1)^2(s+2)}$, the residues associated with the simple poles at $s = 0$ and $s = -2$ are respectively

(*a*) $\frac{1}{2}$ and $\frac{1}{2}$ (*b*) 1 and 1

(*c*) -1 and -1 (*d*) $-\frac{1}{2}$ and $\frac{1}{2}$

214. The gain margin of a unity negative feedback system having forward transfer function $\frac{K}{s(sT+1)}$ is

(*a*) infinity

(*b*) kT

(*c*) 1

(*d*) zero

215. The signal flow graph of the system is shown in the given figure. The transfer function $\frac{C(s)}{D(s)}$ of the system is

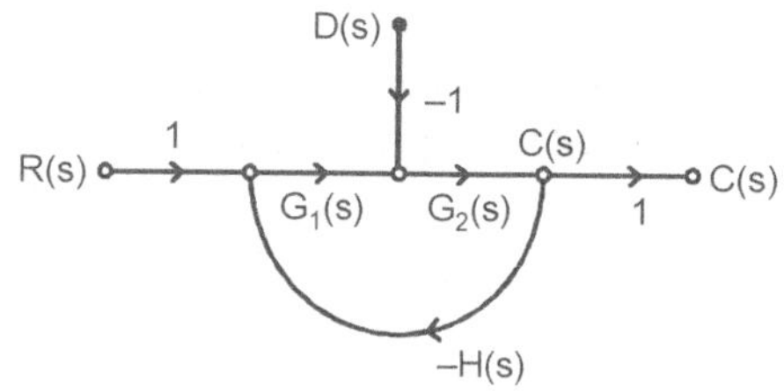

(*a*) $\frac{G_1(s)G_2(s)}{1+G_1(s)H(s)}$

(*b*) $\frac{G_1(s)G_2(s)}{1-G_1(s)G_2(s)H(s)}$

(*c*) $\frac{-G_2(s)}{1+G_1(s)G_2(s)H(s)}$

(*d*) $\frac{-G_2(s)}{1-G_1(s)G_2(s)H(s)}$

216. The state diagram of a system is shown in the given figure:

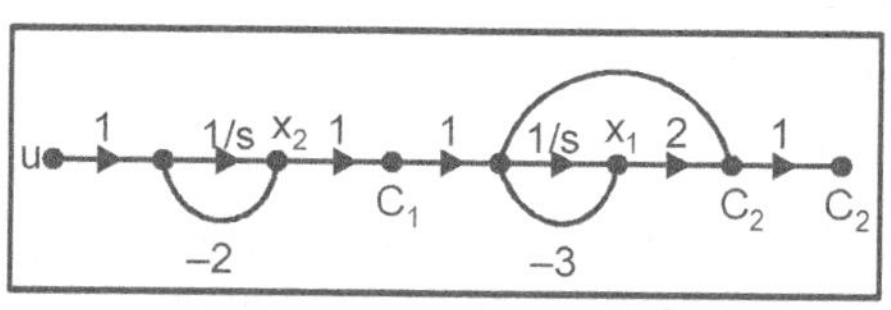

(*a*) controllable and observable

(*b*) controllable but not observable

(*c*) observable but not controllable

(*d*) neither controllable nor observable

217. If a second order system has poles at $-1 \pm j$, then step response of the system will exhibit a peak at

(*a*) 4.5 *s* (*b*) 3.5 *s*

(*c*) 3.14 *s* (*d*) 1 *s*

218. The transfer function of a multi-input multi-output system, with the state-space representation of

$$X = AX + BU \qquad Y = CX + DU$$

where X represents the state, Y the output and U the input vector, will be given by

(*a*) $C(sI-A)^{-1}B$

(*b*) $C(sI-A)^{-1}B+D$

(*c*) $(sI-A)^{-1}B+D$

(*d*) $C(sI-A)^{-1}B$

219. Consider system shown in the given figure with

$$G(s) = \frac{K(s+1)}{s^3 + as^1 + 2s + 1}$$

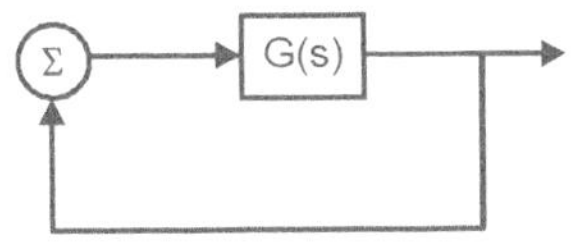

What values of 'K' and 'a' should be chosen so that the system oscillates?

(*a*) $K = 2, a = 2$

(*b*) $K = 2, a = 0.75$

(*c*) $K = 4, a = 1$

(*d*) $K = 4, a = 0.75$

220. Consider the following signal flow graphs:

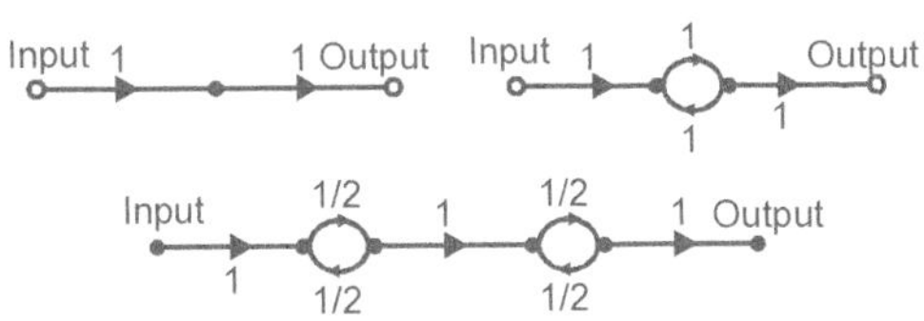

The value of grains is two for

(*a*) 1

(*b*) 2

(*c*) 2 and 3

(*d*) 1, 2 and 3

221. The Nyquist plot of the open-loop transfer function G(s) H(s) is shown in the given figure. It indicates that

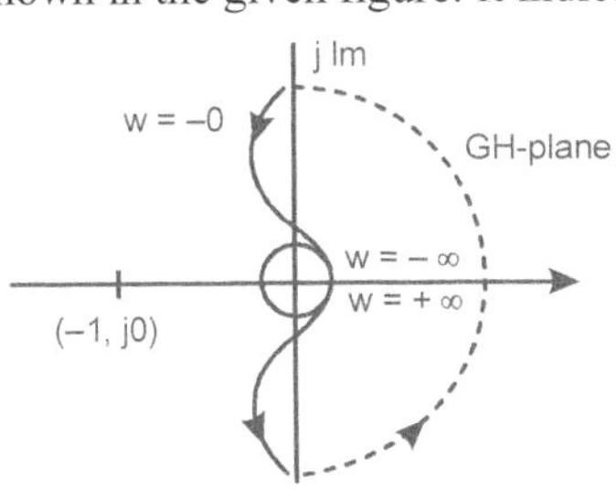

(*a*) the open-loop system is unstable but the close-loop system is stable

(*b*) both open-loop and closed-loop systems are unstable

(*c*) open-loop system is stable but closed loop system is unstable

(*d*) both open-loop and closed-loop systems are stable

222. Consider a system shown in the given figure:

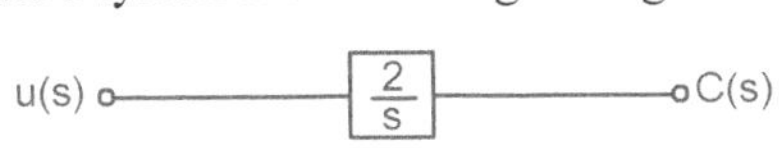

If the system is distributed so that $c(0) = I$, then $c(t)$ for a unit step input will be

(*a*) $1 + t$ (*b*) $1 - t$

(*c*) $1 + 2t$ (*d*) $1 - 2t$

223. For a unit step input, a system with forward path transfer function $G(s) = \frac{20}{s^2}$ and feedback paths transfer function $H(s) = (s + 5)$, has a steady state output of

(*a*) 20 (*b*) 5

(*c*) 0.2 (*d*) zero

224. Given a unity feedback system with open loop transfer function $G(s) = \frac{K(s+2)}{(s+1)^2}$

The correct root-locus plot of the system is

(*a*)

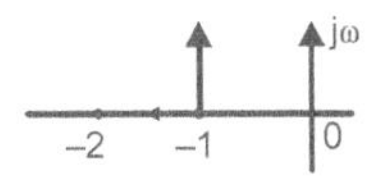

(*b*)

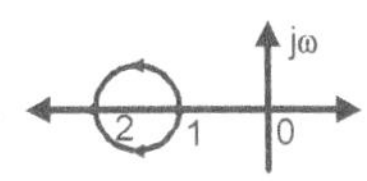

(*c*)

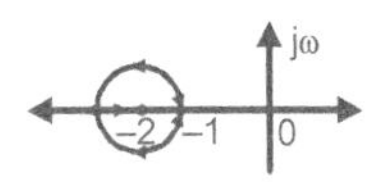

(*d*)

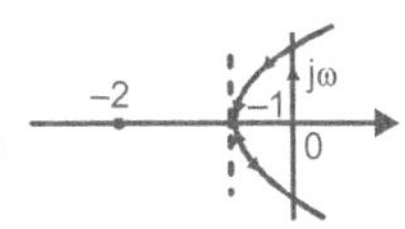

225. Which one of the following statements is true for gain margin and phase margin of two closed-loop systems having loop transfer functions G(s) and exp (–s) G(s) H(s)?

(*a*) Both gain and phase margins of the two systems will be identical

(*b*) Both gain and phase margins of G(s) H(s) will be more

(*c*) Gain margins of the two systems are the same but phase margin of G(s) H(s) will be more

(*d*) Phase margins of the two systems are the same but gain margin of G(s) H(s) will be less

226. Consider the system

$$X(t) = \begin{bmatrix} 1 & 1 \\ 0 & 1 \end{bmatrix} X(t) \begin{bmatrix} b_1 \\ b_2 \end{bmatrix} U(t)$$

$$c(t) = [d_1\ d_2]\, x(t)$$

The conditions for complete state controllability and complete observability is

(*a*) $d_1 > 0$, $b_2 > 0$, b_1 and d_2 can be anything

(*b*) $d_1 0$, $d_2 0$, b_1 and d_2 can be anything

(*c*) $b_1 > 0$, $b_2 > 0$, d_1 and d_2 can be anything

(*d*) $b_1 > 0$, $b_2 > 0$, b_2 and d_1 can be anything

227. A single-phase transmission line of impedance j 0.8 ohm supplies a resistive load of 500 A at 300 V. The sending-end power factor is

(*a*) unity (*b*) 0.8 lagging

(*c*) 0.8 leading (*d*) 0.6 lagging

228. When the input to a system was withdrawn at $t = 0$, its output was found to decrease exponen-tially from 1000 units to 500 units in1.386 seconds. The time constant of the system is

(*a*) 0.500

(*b*) 0.693

(*c*) 1.386

(*d*) 2.000

229. An effect of phase-lag compensation on servo system performance is that

(*a*) for a given relative stability, the velocity constant is increased

(*b*) for a given relative stability, the velocity constant is decreased

(*c*) the bandwidth of the system is increased

(*d*) the time response is made faster

230. In the control system shown in the given figure, the controller which can give zero steady-state error to a ramp input, with K = 9 is

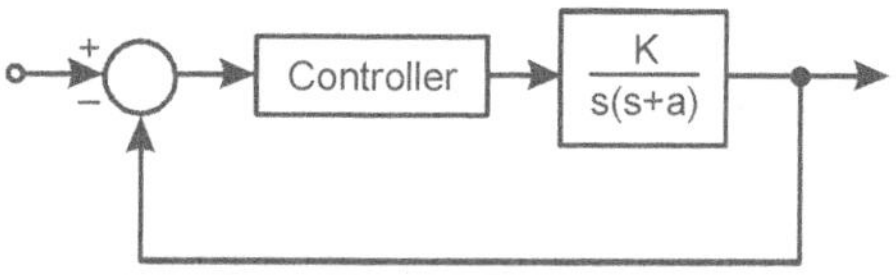

(*a*) proportional type

(*b*) integral type

(*c*) derivative type

(*d*) proportional plus derivative type

231. The transfer function of a control system is given as $T(s) = \frac{K}{s^2 + 4s + k}$

where K is the gain of the system in radians /Amp. For this system to be critically damped, the value of K should be

(*a*) 1

(*b*) 2

(*c*) 3

(*d*) 4

232. If the time response of a system is given by the following equation

$$y(t) = 5 + 3\sin(\omega t + \delta_1) + e^{-3t}\sin(\omega t + \delta_2) + e^{-5t}$$

Then the steady-state part of the above response is given by

(*a*) $5 + 3\sin(\omega t + \delta_1)$

(*b*) $5 + 3\sin(\omega t + \delta_1)\, e^{-3t}\sin(\omega t + \delta_2)$

(*c*) $5 + e^{-5t}$

(*d*) 5

233. The transfer function of a system is $\frac{10}{1+s}$.

When operated as a unity feedback system, the steady-state error to a unit step input will be

(*a*) zero

(*b*) $\frac{1}{11}$

(*c*) 10

(*d*) infinity

234. The impulse response of a system is $5e^{-10t}$ its step response is equal to

(*a*) $0.5\, e^{-10t}$

(*b*) $5(1 - e^{-10t})$

(*c*) $0.5(1 - e^{-10t})$

(*d*) $10(1 - e^{-10t})$

235. A system is represented by

$$\frac{dy}{dt} + 2y = 4t\, u(t)$$

The ramp component in the forced response will be

(*a*) $t\, u(t)$

(*b*) $2t\, u(t)$

(*c*) $3t\, u(t)$

(*d*) $4t\, u(t)$

EXERCISE – II

1. AnON-OFF controller is a
   (*a*) P Controller
   (*b*) integral controller
   (*c*) non-linear controller
   (*d*) PID controller **[NTPC]**

2. The pole-zero plot given below is that of a/an

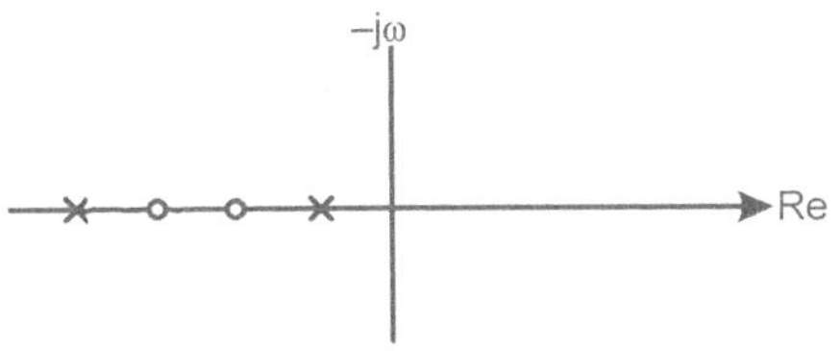

   (*a*) integrator
   (*b*) PD controller
   (*c*) PID controller
   (*d*) Lag-lead compensating network **[NTPC]**

3. 20 *d*B/decade corresponds to
   (*a*) 3 *d*B/octave (*b*) 6 *d*B/octave
   (*c*) 9 *d*B/octave (*d*) 20 *d*B/octave **[NTPC]**

4. The transfer function is $\frac{1+0.5s}{1+s}$. It represents a
   (*a*) lead network (*b*) lag network
   (*c*) lag-lead network (*d*) proportional controller **[NTPC]**

5. If the gain of the open-loop system is doubled, the gain margin
   (*a*) is not affected (*b*) gets doubled
   (*c*) becomes half (*d*) becomes one-fourth **[BHEL]**

6. A lag network for compensation normally consists of
   (*a*) R only (*b*) R and C elements
   (*c*) R and L elements (*d*) R, L and C elements **[BHEL]**

7. The figure below gives two equivalent block diagrams

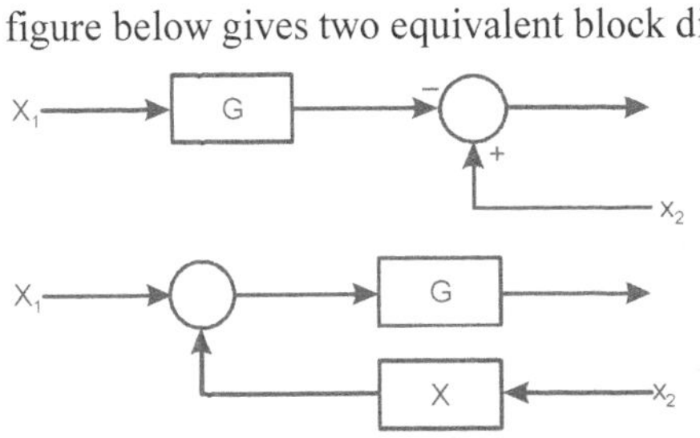

   The value of transfer function of block marked 'X' is given by
   (*a*) G(*s*) (*b*) $\frac{1}{G(s)}$
   (*c*) 1 (*d*) none of these **[BHEL]**

8. The system with transfer function $\frac{K}{s^2(1+sT)}$ is operated in closed-loop with unity feedback. The closed-loop system is
   (*a*) stable
   (*b*) unstable
   (*c*) marginally stable
   (*d*) conditionally stable **[BHEL]**

9. The gain of a system is 10 at some frequency. In terms of dB, it is equal to
   (*a*) 0 (*b*) 1 *d*B
   (*c*) 20 *db* (*d*) 100 *d*B **[NTPC]**

10. The terms in the first column of Routh's array of characteristic equation of certain systems are
$$5, 2, -4, 6, 3.$$
   The number of roots of characteristic in the right half of the *s*-plane is equal to
   (*a*) 2 (*b*) 1
   (*c*) none (*d*) 3 **[NTPC]**

11. If stability error for step input and speed of response be the criteria for design, what controller would you recommend?
   (*a*) P controller
   (*b*) PD controller
   (*c*) PI controller
   (*d*) PID controller **[NTPC]**

12. Let Y(*s*) be the Laplace transformation of the function *y*(*t*), then the final value of the function is
   (*a*) $\lim_{s\to 0} Y(s)$
   (*b*) $\lim_{s\to \infty} Y(s)$
   (*c*) $\lim_{s\to 0} sY(s)$
   (*d*) $\lim_{s\to \infty} sY(s)$ **[BHEL]**

13. A control system is defined by the following mathematical relationship
$$\frac{d^2x}{dt^2}+6\frac{dx}{dt}+5x = 12(1-e^{-2t})$$
   The response of the system as $t \to \infty$ is
   (*a*) $x = 6$
   (*b*) $x = 2$
   (*c*) $x = 2.4$
   (*d*) $x = -2$ **[BHEL]**

14. It is given that

$$G(s) = \frac{K}{s^2(1+sT)}$$

This system is operated in a closed-loop with unity feedback. What is the order and type of the closed-loop system?

(*a*) 3 and 2

(*b*) 2 and 3

(*c*) 3 and 1

(*d*) 3 and 0 **[BHEL]**

15. The sum of combinations of two non-touching loops in the following signal flow graph is

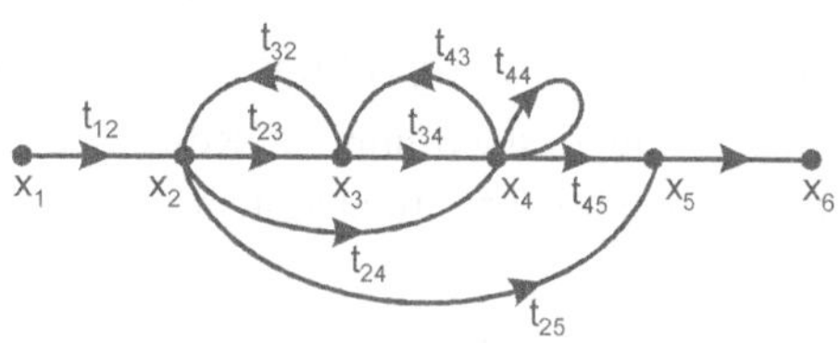

(*a*) $t_{23}\,t_{32} + t_{44}$

(*b*) $t_{23}\,t_{32} + t_{32} + t_{43}$

(*c*) $t_{23}\,t_{32} + t_{34} + t_{44}$

(*d*) $t_{24}\,t_{43}\,t_{32} + t_{44}$ **[ES]**

16. In a control system, the use of negative feedback

(*a*) eliminates the chances of instability

(*b*) increases the reliability

(*c*) reduces the effects of disturbance and noise signals on the forward path

(*d*) increases the influence of variations of component parameters on the system performance **[NTPC]**

17. The open-loop transfer function of a unity feedback control system is

$$G(s) = \frac{1}{(s+2)^2}$$

The closed-loop transfer function will have poles at

(*a*) $-2, -2$

(*b*) $-2, -1$

(*c*) $-2 \pm j$

(*d*) $-2, 2$ **[BHEL]**

18. A current impulse, $5\delta(t)$, is forced through a capacitor C. The voltage, $V_c(t)$ across the capacitor is given by

(*a*) $5t$

(*b*) $5u(t) - C$

(*c*) $\frac{5}{C}t$

(*d*) $\frac{5u(t)}{C}$ **[BHEL]**

19. A unity feedback system has an open-loop transfer function $G(s) = \frac{K}{s(s^2+14s+13)}$

The angles of asymptotes are given by

(*a*) 45°, 135°, 225°

(*b*) 60°, 180°, 300°

(*c*) 90°, 80°, 270°

(*d*) none of these **[BHEL]**

20. Consider a characteristic equation

$$s^4 + 3s^2 + 5s^2 + 6s + K + 10 = 0.$$

The conditions for stability is

(*a*) $K > 5$

(*b*) $-10 \leq K$

(*c*) $K > -4$

(*d*) $-10 < K < -4$ **[BHEL]**

21. State variable approach converts an *n*th order system into

(*a*) *n* second order integro differential equations

(*b*) two differential equations

(*c*) *n* first order differential equations

(*d*) a lower order system **[IOCL]**

22. The frequency at which the Nyquist diagram crosses the negative real axis is known as

(*a*) gain crossover frequency

(*b*) phase crossover frequency

(*c*) damping frequency

(*d*) natural frequency **[IOCL]**

23. For stability poles should be in LH of S-plane. LH includes

(*a*) real and imaginary axes

(*b*) real axis but not *j* axis

(*c*) real axis but not –*j* axis

(*d*) negative real axis but not *j* axis **[IOCL]**

24. A minimum-phase system with no zeros has a phase angle of – 270° at gain crossover frequency. The system is

(*a*) stable

(*b*) unstable

(*c*) marginally stable

(*d*) conditionally stable **[IOCL]**

25. A system is stable

(*a*) if bounded inputs produce bounded outputs

(*b*) if unbounded inputs produce bounded outputs

(*c*) if bounded inputs produce unbounded outputs

(*d*) if all bounded inputs produce bounded outputs **[IOCL]**

26. A temperature control system is

(*a*) a servomechanism

(*b*) a digital system

(*c*) a cascade control system

(*d*) an analogue process control system **[IOCL]**

27. If the open loop transfer function G(*jw*) of a unity feedback system has a phase angle of –150° at the gain cross-over frequency, the phase margin of the system is equal to

(*a*) – 150°

(*b*) +210°

(*c*) 30°

(*d*) none of these **[IOCL]**

28. A linear discrete-time system has the characteristic equation $z^3 - 0.81\,z = 0$. The system

(*a*) is stable

(*b*) is marginally stable

(*c*) is unstable

(*d*) stability cannot be assessed from the given information **[IOCL]**

29. $G(s) = \dfrac{1}{s(1+6s)}$. The system is

(*a*) stable

(*b*) unstable

(*c*) marginally stable

(*d*) conditionally stable **[IOCL]**

30. The transfer function of a closed-loop is

$$F_k(s) = \frac{K(s+3)}{1+s-s^2+s^3}$$

This system is

(*a*) stable

(*b*) unstable

(*c*) marginally stable

(*d*) conditionally stable **[IOCL]**

31. The open-loop transfer function of an unity feedback system is given by

$$G(s) = \frac{K}{s(s+1)(s+2)}$$

The breakaway point of the root locus plot is given by

(*a*) –0.423

(*b*) –0.523

(*c*) –0.700

(*d*) none of these **[ONGC]**

32. The open-loop transfer function of a feedback control system is

$$G(s)\,.\,H(s) = \frac{1}{(s+1)^3}.$$

The gain margin of the system is

(*a*) 2 (*b*) 4

(*c*) 8 (*d*) 16 **[BHEL]**

33. The pole-zero configuration of a phase-lead compensator is given by

(*a*) 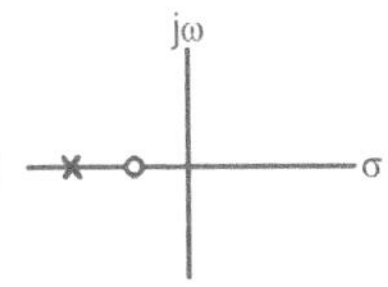

(*b*) 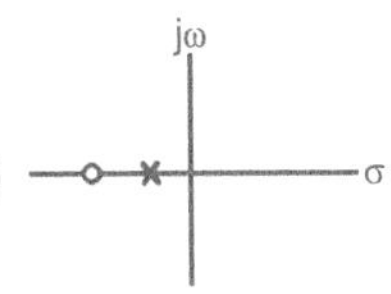

(*c*) 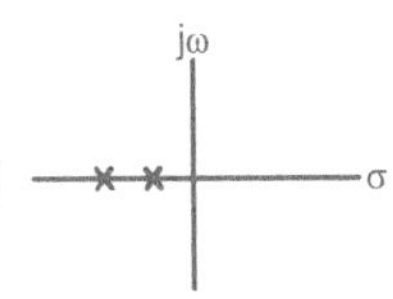

(*d*) 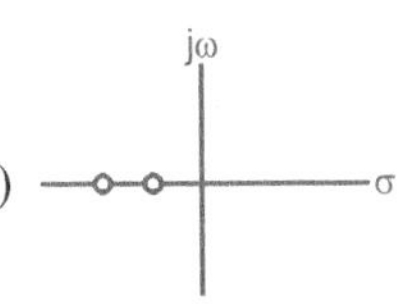

[GATE]

34. In position control systems, the device used for providing rate-feedback voltage is called

(*a*) potentiometer

(*b*) synchro transmitter

(*c*) synchro transformer

(*d*) tachogenerator **[NTPC]**

35. A second order system exhibits 100% overshoot. Its damping coefficient is

(*a*) equal to 0

(*b*) equal to 1

(*c*) <1

(*d*) >1 **[NTPC]**

36. The open-loop transfer function of a feedback system is

$$G(s)\,H(s) = \frac{K}{s(s+4)(s^2+4s+20)}$$

The four branches of root locus originate at

(*a*) $-2, -3, -1+j4 -1-j4$

(*b*) $-1, -2, -3+j4 -3-j4$

(*c*) $0, -4, -3+j4 -2-j4$

(*d*) none of these **[PGCIL]**

37. Which one of the following is an example of an open-loop system?
(*a*) Household refrigerator
(*b*) Respiratory system of an animal
(*c*) Stabilisation of air pressure entering into a mask
(*d*) Execution of a program by a computer **[PGCIL]**

38. A position control system is
(*a*) a process control system
(*b*) an automatic regulatory system
(*c*) a servomechanism
(*d*) a stochastic control system **[PGCIL]**

39. If the system has multiple poles and the $j\omega$-axis, the system is
(*a*) stable
(*b*) unstable
(*c*) marginally stable
(*d*) conditionally stable **[PGCIL]**

40. Microprocessor based control systems can be classified as
(*a*) continuous data based systems
(*b*) learning control systems
(*c*) sampled data control systems
(*d*) stochastic control systems **[PGCIL]**

41. The frequency at which the Nyquist diagram cuts (–1, 0) circle is known as
(*a*) gain crossover frequency
(*b*) phase crossover frequency
(*c*) damping frequency
(*d*) natural frequency **[PGCIL]**

42. Signal flow graph is used to obtain the
(*a*) stability of a system
(*b*) transfer function of a system
(*c*) controllability of a system
(*d*) absorbability of a system **[PGCIL]**

43. A second order system with no zeros has its located at – 3 +j4 and – 3 – j4 in the *s*-plane. The undamped natural frequency and the damping factor of the system are respectively.
(*a*) 4 rad/sec and 0.75
(*b*) 3 rad/sec and 0.60
(*c*) 5 rad/sec and 0.80
(*d*) 5 rad/sec and 0.60 **[PGCIL]**

44. Given a unity feedback system with open-loop transfer function

$G(s) = \dfrac{K}{s(s+1)(s+2)}$, the root locus plot

of the system is of the form

(*a*)
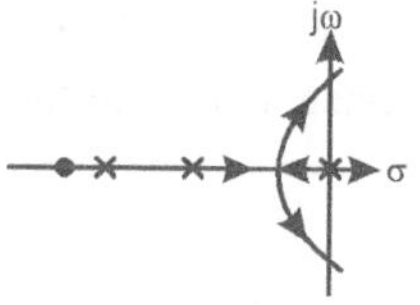

(*b*)
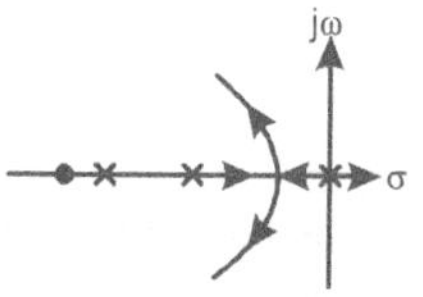

(*c*)
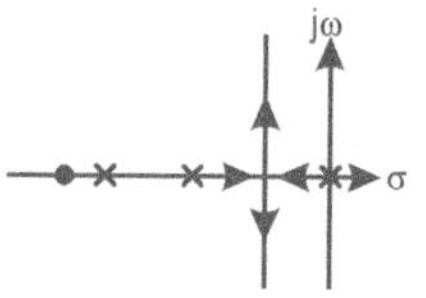

(*d*)
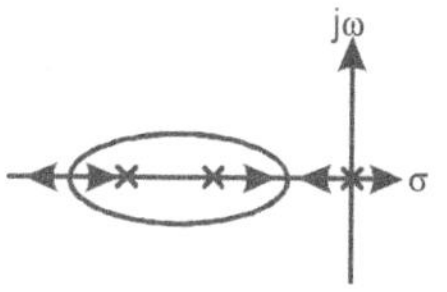

[PGCIL]

45. A unity feedback system has OLTF.

$$G(s) = \frac{9}{s(s+3)}$$

(*a*) Natural frequency = 9
(*b*) Natural frequency = 3
(*c*) Damping ratio = $\dfrac{1}{2}$
(*d*) Damping ratio = $\dfrac{1}{6}$ **[BHEL]**

46. Sinusoidal oscillators are
(*a*) stable
(*b*) marginally stable
(*c*) unstable
(*d*) conditionally stable **[BHEL]**

47. $G(s) = \dfrac{K}{s(s+sT)}$ This system is operated in closed- loop with unity feedback. The closed-loop system is
(*a*) stable (*b*) unstable
(*c*) marginally stable (*d*) conditionally stable
[BHEL]

48. Traffic light on roads are examples of

(*a*) closed-loop system

(*b*) open-loop system

(*c*) both open-loop and closed-loop

(*d*) open-loop but can be made closed-loop **[BHEL]**

49. If $f(t)$ is the step response of a linear time invariant system, then its impulse response is

(*a*) $\frac{d}{dt}f(t)$

(*b*) $\int f(t)\,dt$

(*c*) $f(t)$

(*d*) $\frac{1}{t(t)}$ **[BHEL]**

50. A second order differential equation is given by

$$\frac{d^2x}{dt^2}+5\frac{dx}{dt}+7x=7y.$$

The undamped natural frequency and damping ratio are

(*a*) 1, 5 (*b*) 5, 7

(*c*) $1, \sqrt{7}$ (*d*) $\sqrt{7}$, 0.94 **[BHEL]**

51. An electromechanical device which actuates a train of step angular movements in response to a train of input pulses on one to one basis is

(*a*) synchro control transformer

(*b*) LVDT

(*c*) stepper motor

(*d*) tachogenerator **[NTPC]**

52. If the open loop transfer function is a ratio of a numerator polynomial of degree '*m*' and denominator polynomial of degree '*n*', then the integer n-m represents the number of

(*a*) breakaway points

(*b*) unstable poles

(*c*) separate root loci

(*d*) asymptotes **[NTPC]**

53. In order to stablise the system shown in Figure, T_i should satisfy

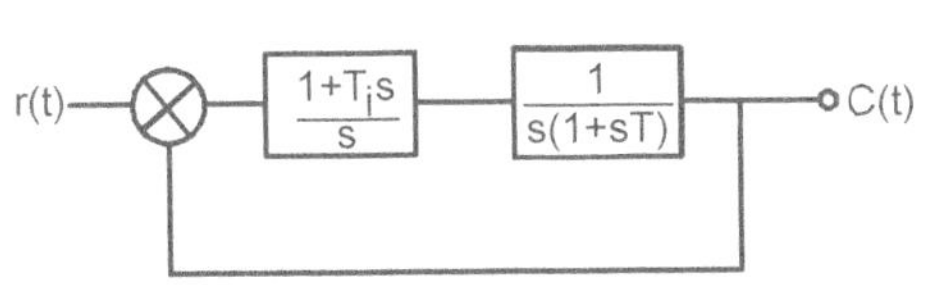

(*a*) $T_i = -T$

(*b*) $T_i = T$

(*c*) $T_i < T$

(*d*) $T_i > T$ **[NTPC]**

54. The purpose of the series quardrature windings in an amplidyne is to

(*a*) neutralise the effect of armature reaction

(*b*) reduce commutation difficulties

(*c*) increase the gain

(*d*) increases the response time **[NTPC]**

55. In frequency domain the resonant frequency of a second order system's transient becomes zero if the damping factor is

(*a*) 0.1 (*b*) 0.5

(*c*) 0.707 (*d*) 1 **[NTPC]**

56. On the C/R for the signal flow graph shown in the figure is

(*a*) $\frac{G_1G_2G_3G_4}{(1+G_1G_2)(1+G_3+G_4)}$

(*b*) $\frac{G_1G_2G_3G_4}{(1+G_1+G_2+G_1G_2)(1+G_3+G_4+G_3G_4)}$

(*c*) $\frac{G_1G_2G_3G_4}{(1+G_1+G_2)(1+G_3+G_4)}$

(*d*) $\frac{G_1G_2G_3G_4}{1+G_1+G_2+G_3+G_4}$ **[NTPC]**

57. A unity feedback system has open-loop transfer function G(*s*), the steady-state error is zero for

(*a*) step input and type-1 G(*s*)

(*b*) ramp input and type-1 G(*s*)

(*c*) step input and type- G(*s*)

(*d*) ramp input and type-0 G(*s*) **[BHEL]**

58. A second order system is given as

$$G(s)=\frac{K}{(1+r_1s)(1+r_2s)}.$$

The gain K is increased. Then the phase margin

(*a*) gets increased

(*b*) gets decreased

(*c*) is unaffected

(*d*) gets decreased and becomes stable **[PGCIL]**

59. The steady state error of a stable, type 0 unity feedback system for a unit step function is

(*a*) 0 (*b*) $\frac{1}{1+K_p}$

(*c*) ∞ (*d*) $\frac{1}{K_p}$ **[HAL]**

60. A linear time-invariant system initially at rest, when subjected to a unit-step input, gives a response

$$y(t) = te^{-t}, t > 0$$

The transfer function of the system is.

(*a*) $\frac{1}{(s+1)^2}$

(*b*) $\frac{1}{s(s+1)^2}$

(*c*) $\frac{s}{(s+1)^2}$

(*d*) $\frac{1}{s(s+1)}$ **[HAL]**

61. Feedback control systems are

(*a*) insensitive to both forward-and feedback-path parameter changes.

(*b*) less sensitive to forward-path parameter changes than to feedback-path parameter changes.

(*c*) less sensitive to forward-path parameter changes than to feedback-path parameter changes.

(*d*) equally sensitive to forward-and feedback-path parameter changes. **[HAL]**

62. The gain margin of a system is –10*d*B. It is increased by 5*d*B. Then the system is

(*a*) stable

(*b*) unstable

(*c*) marginally stable

(*d*) unstable for – 10*d*B itself **[HAL]**

63. Which one of the following is an example of open-loop system?

(*a*) A windscreen wiper

(*b*) Aqualung

(*c*) Respiratory system of an animal

(*d*) A system for controlling anti-rocket missiles **[HAL]**

64. A unity-feedback control system has the open-loop transfer function

$G(s) = \frac{4(1+2s)}{s^2(s+1)}$. If the input to the system

to a unit ramp, the steady state error will be

(*a*) 0

(*b*) 0.5

(*c*) 2

(*d*) infinity **[HAL]**

65. A unity feedback system has an open-loop transfer function

$$G(j\omega) = \frac{K}{j\omega(j0.2\omega+1)(j0.05\omega+1)}.$$

The phase cross-over frequency of the Nyquist plot is given by

(*a*) 1 rad/sec (*b*) 100 rad/sec

(*c*) 10 rad/sec (*d*) none of these **[PGCIL]**

66. The characteristic equation of a feedback control system is given by

$$s^3 + 5s^2 + (K+6)s + K = 0$$

There K > 0 is a scalar variable parameter. In the root locus diagram of the system the asymptotes of the root loci for large values of K meet at a point in the *s*-plane, whose coordinates are

(*a*) (– 3, 0) (*b*) (–2, 0)

(*c*) (–1, 0) (*d*) (2, 0) **[PGCIL]**

67. Tachometer feedback in a *dc* position control system enhances

(*a*) stability (*b*) instability

(*c*) transient response (*d*) gain margin **[PGCIL]**

68. If the gain margin of a certain feedback system is given as 20 *db*, the Nyquist plot will cross the negative real axis at the point

(*a*) $s = -0.05$ (*b*) $s = -0.2$

(*c*) $s = -0.1$ (*d*) none of these **[HAL]**

69. A unity feedback system has open-loop transfer function

$$G(s) = \{25/[s(s+6)]\}$$

The peak overshoot in the step-input response of the system is approximately equal to

(*a*) 5% (*b*) 10%

(*c*) 15% (*d*) 20% **[HAL]**

70. A synchro transmitter-receiver unit is a

(*a*) two-phase *ac* device

(*b*) three-phase *ac* device

(*c*) *dc* device

(*d*) single-phase *ac* device **[HAL]**

71. A second order system has a transfer function given by

$$G(s) = \frac{25}{s^2+8s+25}$$

If the system, initially at rest, in subjected to a unit input at $t = 0$, the second peak in a response will occur at

(*a*) π sec (*b*) $\frac{\pi}{3}$ sec

(*c*) $\frac{2\pi}{3}$ sec (*d*) $\frac{\pi}{2}$ sec **[HAL]**

72. Given the homogeneous state-space equation

$$\dot{x} = \begin{bmatrix} -3 & 1 \\ 0 & -2 \end{bmatrix} x$$

the steady state value

$$x_{ss} = \lim_{t \to \infty} x(t),$$

given the initial state value of

$x(0) = [\,10 -10]^T$, is

(*a*) $x_{ss} = \begin{bmatrix} 0 \\ 0 \end{bmatrix}$ (*b*) $x_{ss} = \begin{bmatrix} -3 \\ -2 \end{bmatrix}$

(*c*) $x_{ss} = \begin{bmatrix} -10 \\ 10 \end{bmatrix}$ (*d*) $x_{ss} = \begin{bmatrix} \infty \\ \infty \end{bmatrix}$ **[NTPC]**

73. A cascade control system has

(*a*) single loop

(*b*) loops within loop

(*c*) a single controller

(*d*) overlapping loops **[NTPC]**

74. A control system is as shown in the given figure. The maximum value of gain K. for which the system is stable is

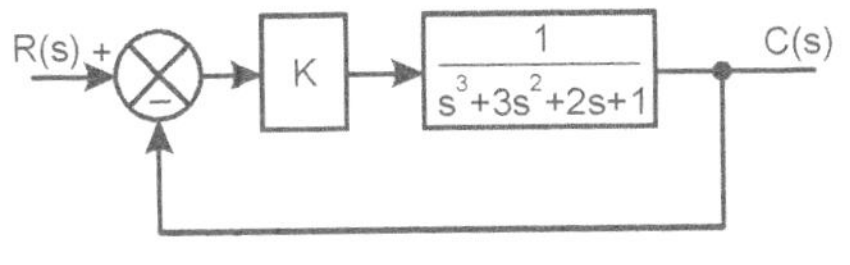

(*a*) $\sqrt{3}$ (*b*) 3

(*c*) 4 (*d*) 5 **[NTPC]**

75. The root locus of UFB is drawn in the figure. The OLTF is given by

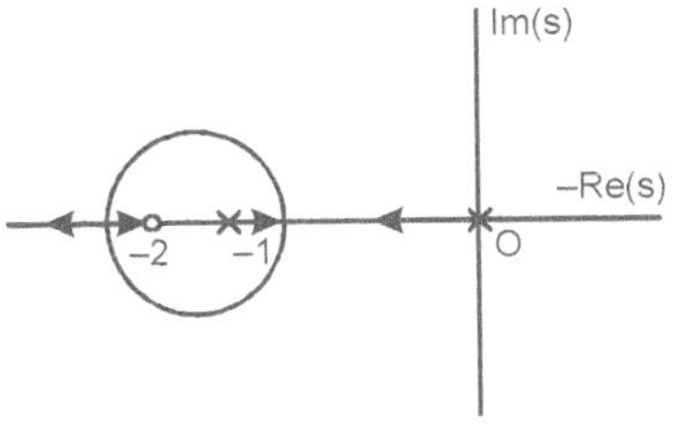

(*a*) $\dfrac{K}{s(s+1)(s+2)}$

(*b*) $\dfrac{K(s+1)}{s(s+2)}$

(*c*) $\dfrac{K(s+2)}{s(s+1)}$

(*d*) $\dfrac{Ks}{(s+1)(s+2)}$ **[NTPC]**

76. A system with a large gain margin or a large phase margin is usually a/an

(*a*) unstable system

(*b*) stable system

(*c*) very stable but very sluggish system

(*d*) highly oscillatory system **[BHEL]**

77. A lag compensation network

A. increases the gain of the original network without affecting stability

B. reduces the steady state error

C. reduces the speed of response

D. permits the increase of gain if phase margin is acceptable.

In the above, correct statements are

(*a*) All

(*b*) A, B

(*c*) B, C, D

(*d*) C, B **[BHEL]**

78. If the Nyquist plot of a certain feedback system crosses the negative real axis at – 0.1 point, the gain margin of the system is given by

(*a*) 0.1

(*b*) 10

(*c*) 100

(*d*) none of these **[BHEL]**

79. The state representation of a second order system is

$$\dot{x}_1 = x_1 + u,$$

$$\dot{x}_2 = x_1 - 2x_2 + u$$

Consider the following statements regarding the above system;

1. The sysetm is completely state contrallable
2. If x_1 is the output, then the system is completely output controllable.
3. If x_2 is the output, then the system is completely output controllable.

Of these statements

(*a*) 1, 2 and 3 are correct

(*b*) 1 and 2 are correct

(*c*) 2 and 3 are correct

(*d*) 1 and 3 are correct **[BHEL]**

80. Straight-lines asymptote Bode magnitude plot for a certain system is shown in the Figure. What will be its transfer function?

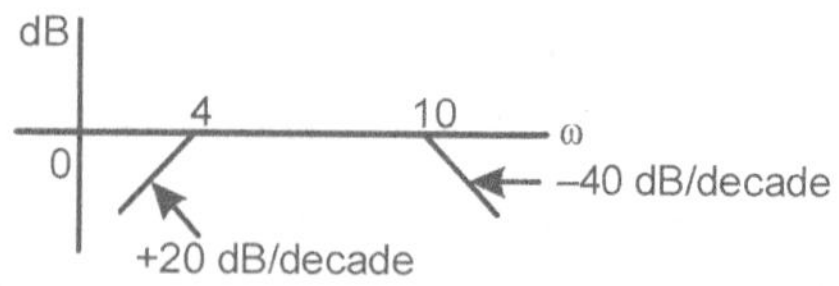

(*a*) $\frac{0.25s}{\left(1+\frac{s}{4}\right)\left(1+\frac{s}{10}\right)^2}$ (*b*) $\frac{4\left(1+\frac{5}{4}\right)}{\left(1+\frac{s}{10}\right)^2}$

(*c*) $\frac{0.25}{\left(1+\frac{s}{4}\right)\left(1+\frac{s}{10}\right)}$ (*d*) $\frac{4s}{\left(1+\frac{s}{4}\right)\left(1+\frac{s}{10}\right)^2}$ **[BHEL]**

81. A lead compensating network

A. improves response time

B. increases resonant frequency

C. stabilises the system with low phase margin

D. enables moderate increase in gain without affecting stability

(*a*) A, B, C, D (*b*) A, D

(*c*) B, C (*d*) B, D **[BHEL]**

82. The open loop control system is one in which

(*a*) output is dependent on control input

(*b*) output is independent on control input

(*c*) only system parameters have effect on the control output

(*d*) none of these **[BHEL]**

83. $s(t)$ is step response and $h(f)$ is impulse of a system. Its response $y(t)$ for any input $u(t)$ is given by

(*a*) $\frac{d}{dt}\int_0^1 s(t-\tau)u(\tau)d\tau$

(*b*) $\int_0^1 s(t-\tau)u(\tau)d\tau$

(*c*) $\int_0^1\int_0^r s(t-\tau_1)u(\tau_1)d\tau_1 d\tau$

(*d*) $\frac{d}{dt}\int^r h(t-\tau)u(\tau)d\tau$ **[GATE]**

84. The impulse response of an initially relaxed linear system is $e^{-2t}U(t)$. To produce a response of $te^{-2t}U(t)$, the input must be equal to

(*a*) $2e^{-t}U(t)$ (*b*) $1/2e^{-2t}U(t)$

(*c*) $e^{-2t}U(t)$ (*d*) $e^{-t}U(t)$ **[GATE]**

85. A network has a pole at $s = -1$ and a zero at $s = -2$. If this network is excited by sinusoidal input, the output

(*a*) leads the input

(*b*) leg the input

(*c*) is in phase with input

(*d*) decays exponentially to zero **[NTPC]**

86. The unit circle of the Nyquist plot transforms into 0 *dB* line of the amplitude plot of the Bode diagram at

(*a*) 0 frequency (*b*) low frequency

(*c*) high frequency (*d*) any frequency **[PGCIL]**

87. If in a linear system an input of 5 sin ωt produces an output of 10 cos ωt, the output corresponding to–15 sin ωt will be equal to

(*a*) $\frac{10}{3}\cos\omega t$ (*b*) $30\sin\omega t$

(*c*) $30\cos\omega t$ (*d*) $30\tan\omega t$ **[PGCIL]**

88. In the following pick out the time-invariant systems

(*i*) $\frac{dy(t)}{dt}+5yt=u(t)$

(*ii*) $t\frac{d^2y(t)}{dt^2}+\frac{dy(t)}{dt}+2yt=v(t)$

(*iii*) $\frac{d^2y(t)}{dt^2}+5\frac{dy(t)}{dt}+6y(t)=w(t)$

(*a*) (*i*) and (*iii*)

(*b*) (*i*) and (*ii*)

(*c*) (*ii*) and (*iii*)

(*d*) (*i*) only **[PGCIL]**

89. The Laplace Transform of $f(t) = K\sin\omega t$ is given by

(*a*) $\frac{K}{s^2-\omega^2}$ (*b*) $\frac{K}{s+\omega}$

(*c*) $\frac{K}{s^2+\omega^2}$ (*d*) $\frac{K\omega}{s^2+\omega^2}$ **[PGCIL]**

90. The Laplace Transform of $\frac{df}{dt}$ is given by

(*a*) $sF(s)-f(0)$ (*b*) $sF(s)+f(0)$

(*c*) $F(s)-sf(0)$ (*d*) $F(s)+sf(0)$ **[PGCIL]**

91. Type of a system depends on the

(*a*) number of its poles

(*b*) difference between the number of poles and zeros

(*c*) number of its real poles only

(*d*) number of poles it has at the origin **[PGCIL]**

92. The transfer function for the state variable representation

$X = AX + Bu,$

$y = CX + Du$, is given by

(*a*) $D + C(sI - A)^{-1}B$

(*b*) $B(sI - A)^{-1}C + D$

(*c*) $D(sI - A)^{-1}B + C$

(*d*) $C(sI - A)^{-1}D + B$ **[GATE]**

93. Frequency ω_n of 4 rad/sec, the values of K and *a* are

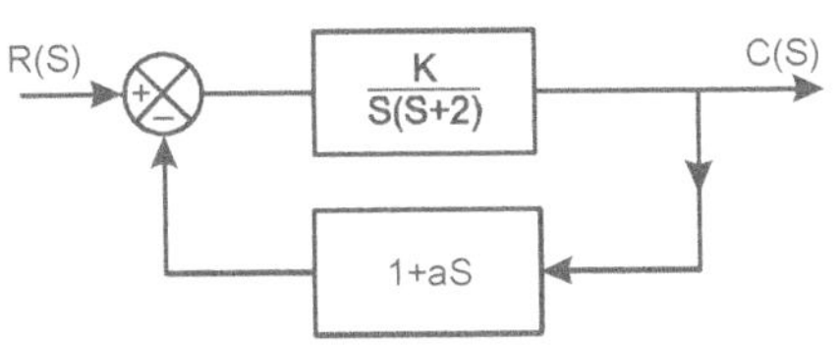

(*a*) $K = 4, a = 0.35$

(*b*) $K = 8, a = 0.455$

(*c*) $K = 16, a = 0.225$

(*d*) $K = 64, a = 0.9$ **[GATE]**

94. Given that the transfer function $G(s)$ is $\frac{K}{s(1 + sT)}$.

State the type and order of the system.

(*a*) 2 and 3 (*b*) 3 and 2

(*c*) 3 and 8 (*d*) 2 and 2 **[BEL]**

95. The root locus plot is shown below. The open-loop system has

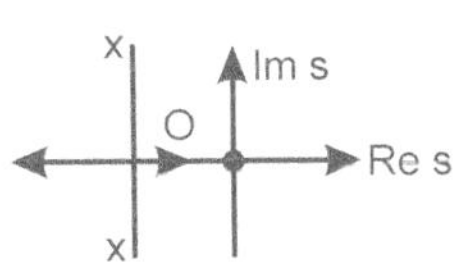

(*a*) two real poles

(*b*) two complex poles and a zero

(*c*) two complex zeros and a pole

(*d*) two complex poles **[BEL]**

96. Routh's array for a system is given below

s^4	1	3	5
s^3	1	2	0
s^2	1	5	
s^1	−3		
s^0	5		

The system is

(*a*) stable

(*b*) unstable

(*c*) marginally stable

(*d*) conditionally stable **[BEL]**

97. The Laplace Transform of the fanction $f(t) = Ke^{-\alpha t}$ is given by

(*a*) $\frac{K}{s + \alpha}$ (*b*) $\frac{K}{s - \alpha}$

(*c*) $\frac{K}{s\alpha}$ (*d*) none of these **[BEL]**

98. The phase angle versus frequency plot is shown below.

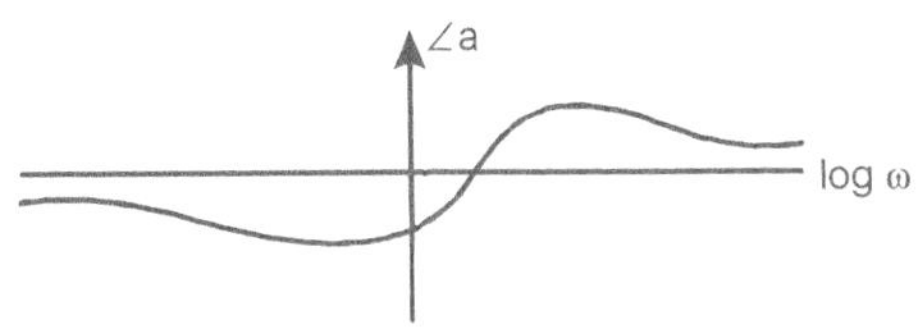

The network is a

(*a*) PID controller

(*b*) Lag-lead network

(*c*) PD controller

(*d*) PI controller **[BEL]**

99. The audible range of 15 kHz is passed through a filter with cut off a 7.5 kHz. The maximum sampling frequency should be

(*a*) 30 kHZ (*b*) 15 kHZ

(*c*) 7.5 kHZ (*d*) 5 kHz **[PGCIL]**

100. The transfer function C/R of the block diagram given below is

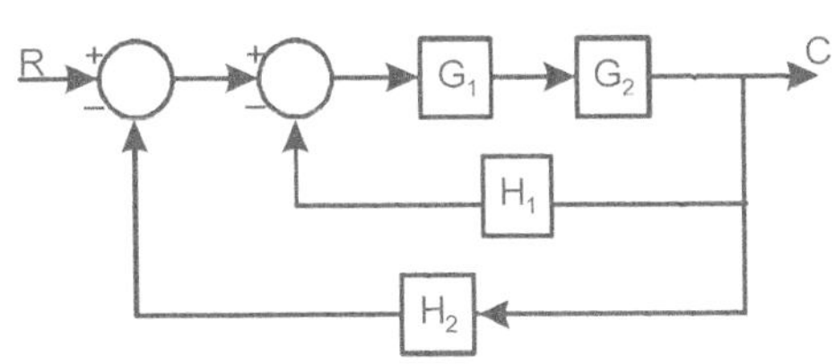

(*a*) $\frac{G_1G_2}{1 + H_1H_2}$

(*b*) $\frac{G_1G_2}{G_1G_2(H_1 + H_2) + 1}$

(*c*) $\frac{G_1G_2}{1 + G_1G_2H_1 + H_2}$

(*d*) $\frac{G_1G_2}{1 + G_2H_1 - G_1H_2}$ **[GATE]**

101. The unit impulse response of a system is given as

$c(t) = -4e^{-t} + 6\,e^{-2t}$.

The step response of the same system for $t \geq 0$ is equal to

(*a*) $-3e^{-2t} + 4e^{-t} + 1$

(*b*) $-3e^{-2t} + 4e^{-t} - 1$

(*c*) $-3e^{-2t} - 4e^{-t} - 1$

(*d*) $3e^{-2t} + 4e^{-t} - 1$ **[GATE]**

102. The step response of a system is

$c(t) = 1 - 5e^{-t} + 10e^{-t} - 6e^{-3t}$

The impulse response of the system is

(*a*) $5e^{-t} - 20e^{-2t} + 18e^{-3t}$

(*b*) $5e^{-t} + 20e^{-2t} + 18e^{-3t}$

(*c*) $5e^{-t} - 20e^{-2t} + 18e^{-3t}$

(*d*) $5e^{-t} - 20e^{-2t} - 18e^{-3t}$

103. The Laplace transform of $e^{-at} \cos \alpha t$ is

(*a*) $\frac{s+\alpha}{(s+\alpha)^2+\alpha^2}$ (*b*) $\frac{s-\alpha}{(s-\alpha)^2+\alpha^2}$

(*c*) $\frac{s+\alpha}{(s-\alpha)^2+\alpha^2}$ (*d*) $\frac{s-\alpha}{(s+\alpha)^2+\alpha^2}$ **[BHEL]**

104. Given a unity feedback system with

$$G(s) = \frac{K}{s(s+4)}$$

The value of K for damping ratio of 0.5 is

(*a*) 1 (*b*) 16

(*c*) 4 (*d*) 2 **[BHEL]**

105. The root locus plot of a system having open loop transferfunction

$$G(s)\,H(s) = \frac{K(s+10)(s+70)}{s^3(s+100)(s+200)}$$

will have angle of asymptotes as

(*a*) 60°, 100° (*b*) 60°, 180°, 300°

(*c*) 60°, 120°, 180° (*d*) 60°, 90°, 1200° **[BHEL]**

106. The functions corresponding to the Bode plot of figure is

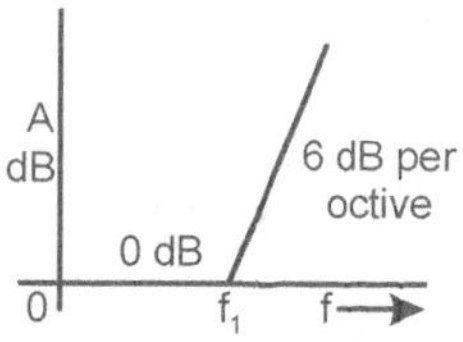

(*a*) $A = jf/f_1$

(*b*) $A = 1/(1 - jf_1/f)$

(*c*) $A = 1/(1 - jf_1/f)$

(*d*) $A = 1 + jf/f_1$ **[BHEL]**

107. The error detectors elements in a control system gives

(*a*) the sum of the reference signal and feedback signal

(*b*) the sum of the reference signal and error signal

(*c*) the difference of the reference signal and feedback signal

(*d*) the difference of the reference signal and output signal **[BHEL]**

108. The phase angle for the transfer function

$$G(l\omega) = \frac{1}{(1+l\omega T)^2}$$

at corner frequency is,

(*a*) 45° (*b*) −45°

(*c*) −90° (*d*) 90° **[NTPC]**

109. If the root locus lies only on the negative real axis then the time response is

(*a*) over damped

(*b*) under damped

(*c*) oscillatory

(*d*) sustains oscillations **[NTPC]**

110. The closed-loop transfer function of a control system is given by

$$\frac{C(s)}{R(s)} = \frac{2(s-1)}{(s+2)(s+1)}$$

For a unit step input, the output is

(*a*) $-3e^{-2t} + 4e^{-t} - 1$

(*b*) $-3e^{-2t} - 4e^{-t} + 1$

(*c*) zero

(*d*) infinity **[GATE]**

111. The natural frequency of oscillations due to dominant roots of the system transfer

functions $\frac{C(s)}{R(s)} = \frac{3}{(s+5)(s^2+2s+3)}$

will be

(*a*) 1.73 rad/sec

(*b*) 2 rad/sec

(*c*) 1.4 rad/sec

(*d*) 3 rad/sec **[PGCIL]**

112. As K appoaches zero the root loci approach towards

(*a*) open loop poles

(*b*) open loop zeros

(*c*) break away points

(*d*) break in points **[PGCIL]**

113. The lag-compensation has a

(*a*) zero nearer to the origin

(*b*) pole nearer to the origin

(*c*) pole at the origin

(*d*) zero at the origin **[PGCIL]**

114. For a stable system

(*a*) the gain crossover occurs earlier than phase crossover

(*b*) the phase crossover occurs earlier than gain crossover

(*c*) the gain crossover and phase crossover frequencies are very nearer to each other

(*d*) the gain crossover and phase crossover frequencies are coincident **[PGCIL]**

115. In exponential series from the state transition matrix is

(*a*) $e^{-[A]t}$ (*b*) $e^{[A]t}$

(*c*) $e^{[A]}$ (*d*) $e^{-[A]}$ **[IOCL]**

116. Which one of the following transfer functions will have greatest maximum overshoot?

(*a*) $\dfrac{2}{s^2+2s+9}$ (*b*) $\dfrac{16}{s^2+2s+16}$

(*c*) $\dfrac{25}{s^2+2s+25}$ (*d*) $\dfrac{36}{s^2+2s+36}$ **[IOCL]**

117. A square matrix is called singular if its

(*a*) determinant is unity

(*b*) determinant is zero

(*c*) determinant is infinity

(*d*) rank is unity **[GATE]**

118. M(*s*) + N(*s*) is a Hurwitz polynomial, where M(*s*) is the even and N(*s*) is the odd part. All the zeros of M(*s*) – N(*s*) are located

(*a*) in the left half of the *s*-plane

(*b*) in the right half of the *s*-plane

(*c*) in the entire *s*-plane

(*d*) on the *j*ω-axis only **[ES]**

119. The matrix of any state-space equations for the transfer functions C(*s*)/R(*s*) of the system, shown in the figure below, is

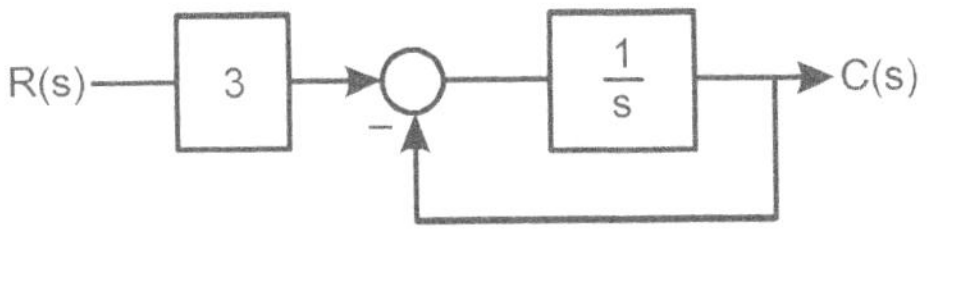

(*a*) $\begin{bmatrix} -1 & 0 \\ 0 & -1 \end{bmatrix}$ (*b*) $\begin{bmatrix} 0 & 1 \\ 0 & -1 \end{bmatrix}$

(*c*) [–1] (*d*) [3] **[GATE]**

120. Match List I with List II and select the correct answer using the codes given below the lists:

List I

A. Transfer function of a lag network

B. State-transition matrix for X = A_x

C. DC amplifier transfer function

D. Steady state error or type-1 system with step-input

List II

1. $L^{-1}(sI - A)^{-1}$

2. $\dfrac{(s+z)}{(s+P)}, \dfrac{z}{P} > 1$

3. $\dfrac{K}{1+sT}$

4. $L^{-1}(sA - I)^{-1}$

5. $\dfrac{(s+z)}{(s+P)}, \dfrac{z}{p} < 1$

6. K(1 + *s*T)

7. zero

8. ∞

	A	B	C	D
(*a*)	1	3	4	5
(*b*)	2	3	1	8
(*c*)	5	4	6	8
(*d*)	2	1	3	7

[ES]

121. Match List I with List II and select the correct answer using codes given below the lists:

List I (*Circuit diagram*)

A.

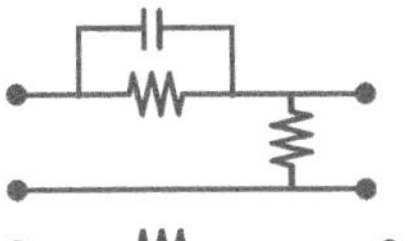

B.

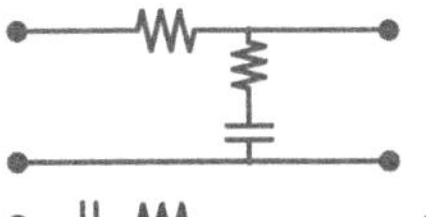

C.

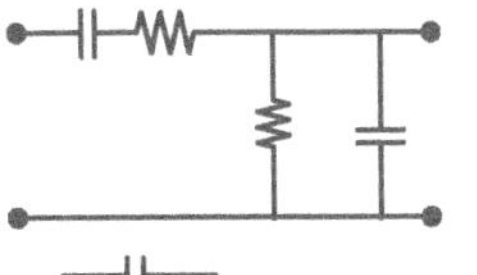

D.

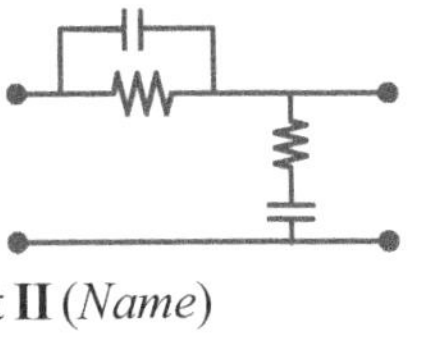

List II (*Name*)

1. Lag network
2. Lead network
3. Lag-lead network
4. Lead-lag network

Codes:

	A	B	C	D
(*a*)	1	2	3	4
(*b*)	2	1	3	4
(*c*)	1	2	4	3
(*d*)	2	1	4	3

[ES]

122. Consider the following analytical methods:

1. Modified z - transform
2. Shanon's theorem
3. State transition matrix
4. Nyquist plot

The methods applicable to sampled data system will include

(a) 1, 2, 3 and 4
(b) 3 and 4
(c) 1 and 2
(d) 1, 2 and 3 [ES]

123. Which of the following are the characteristics of a hydraulic actuator?

1. Sluggish
2. Having capacity to handle large power
3. Having linear operation over a wide range

Select the correct answer using the codes given below

(a) 1, 2 and 3 (b) 1 and 2
(c) 2 and 3 (d) 1 and 3 [ES]

124. A phase-lead comnensator has the transfer function

$$G_c(s) = \frac{10(1+.40s)}{(1+.01s)}.$$

The maximum phase-angle lead provided by this compensator will occur at a frequency ω_m equal to

(a) 50 rad/sec
(b) 25 rad/sec
(c) 10 rad/sec
(d) 4 rad/sec [ES]

125. The open-loop transfer function of a system is given by

$G(s) = \frac{K}{s(s+2)(s+4)}$. The value of K which will cause sustained oscillations in the closed-loop unity feedback system is

(a) 16 (b) 32
(c) 48 (d) 64 [ES]

126. The gain cross-over frequency and bandwidth of a control system are ω_{cu} and ω_{bn} respectively. A phase-lag network is employed for compensating the system. If the gain cross-over frequency and bandwith of the compensated system are ω_{cc} and ω_{bc} respectively, then

(a) $\omega_{cc} < \omega_{cu}$; $\omega_{bc} < \omega_{bu}$
(b) $\omega_{cc} > \omega_{cu}$; $\omega_{bc} < \omega_{bu}$
(c) $\omega_{cc} < \omega_{cu}$; $\omega_{bc} > \omega_{bu}$
(d) $\omega_{cc} > \omega_{cu}$; $\omega_{bc} > \omega_{bu}$ [ES]

127. For a feedback control system of type 2, the steady state error for a ramp input is

(a) infinite
(b) constant
(c) zero
(d) indeterminate [GATE]

128. In most systems an increase in gain leads to

(a) larger damping ratio
(b) smaller damping ratio
(c) constant damping ratio
(d) none of these [GATE]

129. The characteristic equation 1 4 G(s) I(s) = 0 of a system is given by

$$s^4 + 6s^2 + 11s^2 + 6s + K = 0$$

For the system to remain stable, the value of gain K should be

(a) zero
(b) greater than zero but less than 1 0
(c) greater than 10 but less than 20
(d) greater than 20 but less than 30 [ES]

130. A linear system, initially at rest, is subject to an input signal

$r(t) = 1 - e^{-t}(t \geq 0)$
$c(t) = 1 - e^{-2t}$

The response of the system for $t > 0$ is given by The transfer function of the system is
$c(t) = 1 - e^{-2t}$

(a) $\frac{(s+2)}{(s+1)}$ (b) $\frac{(s-1)}{(s-2)}$
(c) $\frac{2(s+2)}{(s+2)}$ (d) $\frac{1(s-2)}{2(s-2)}$ [ES]

131. A system is represented by the block diagram given in the figure.

Which one of the following represents the input-output relationship of the figure?

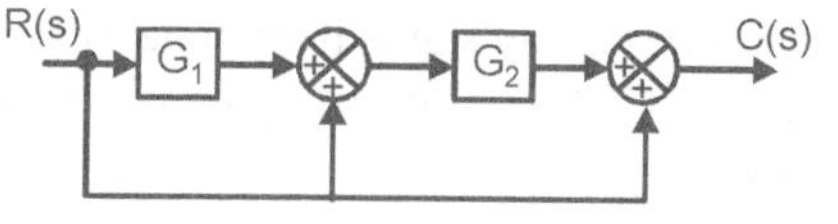

(a) $R(s) \rightarrow \boxed{G_1G_2} \rightarrow C(s)$
(b) $R(s) \rightarrow \boxed{G_1 + G_2} \rightarrow C(s)$
(c) $R(s) \rightarrow 1 + \boxed{G_1 + G_1\ G_2} \rightarrow C(s)$
(d) $R(s) \rightarrow 1 + \boxed{G_2 + G_1\ G_2} \rightarrow C(s)$ [ES]

132. The loop transfer function GH of a control system is given by

$$GH = \frac{K}{s(s+1)(s+2)(s+3)}$$

1. There will be four asymptotes
2. There will be three sperate root loci
3. Asymptotes will intersect at real axis at $\sigma A = -2/3$

Select the correct answer using the codes given below:

Codes:

(*a*) 1 alone

(*b*) 2 alone

(*c*) 3 alone

(*d*) 1, 2 and 3 **[ES]**

133. The open-loop transfer function of a feedback control system is given by

$$G(s)H(s) = \frac{K(s+2)}{s(s+4)(s^2+4s+8)}$$

In the root-locus diagram of the system, the asymptotes of the root loci for large values of K meet a point in the *s*-plane. Which one of the following is the set of co-ordinates of that point?

(*a*) $(-1, 0)$ (*b*) $(-2, 0)$

(*c*) $\left(-\frac{10}{3}, 0\right)$ (*d*) $(2, 0)$ **[ES]**

134. When the signal flow graph is as shown in the figure, the overall transfer functions of the systems, will be

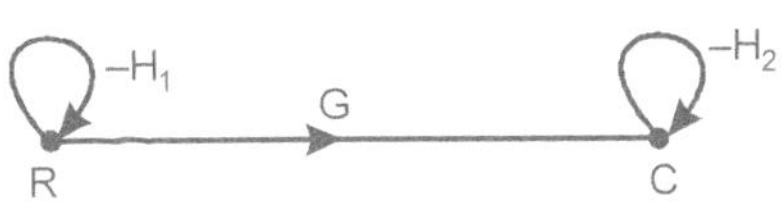

(*a*) $\frac{C}{R} = G$

(*b*) $\frac{C}{R} = \frac{C}{1+H_2}$

(*c*) $\frac{C}{R} = \frac{G}{(1+H_1)(1+H_2)}$

(*d*) $\frac{C}{R} = \frac{G}{1+H_1+H_2}$ **[GATE]**

135. The transfer function of system is given by

$$G(j\omega) = \frac{K}{(j\omega)(j\omega T+1)}, K < \frac{1}{T}$$

Which one of the following is the Bode plot of this function?

(*a*)

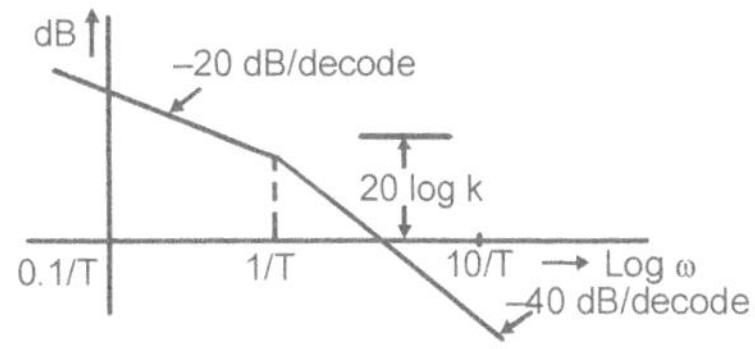

(*b*)

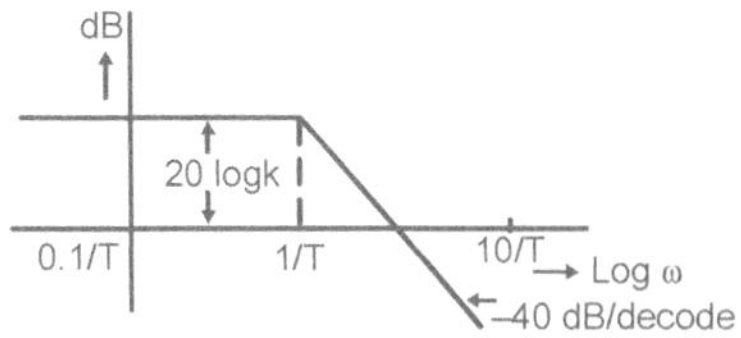

(*c*)

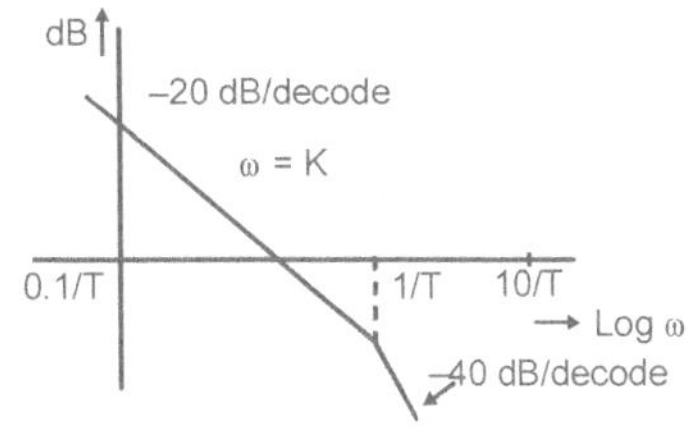

(*d*)

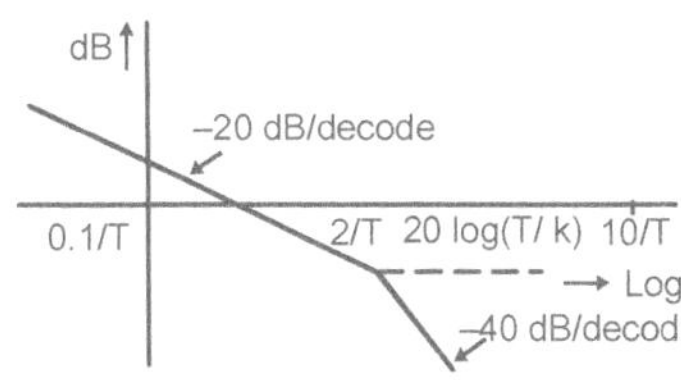

[ES]

136. The second order system defined by $\frac{25}{s^2+5s+25}$ is given a step input. The time taken for the output to settle with in ± 2% of input is

(*a*) 1.2*s* (*b*) 1.6*s*

(*c*) 2*s* (*d*) 0.4*s* **[GATE]**

137. Which input yields natural response?

(*a*) ac input

(*b*) dc input

(*c*) impulse input

(*d*) a ramp input **[IOCL]**

138. The transfer function is $\frac{K}{(s+1)(s+2)(s+3)}$.

The break point will lie between

(*a*) 0 and – 1

(*b*) –1 and – 2

(*c*) – 2 and – 3

(*d*) beyond – 3 **[GATE]**

139. Consider the following statements regarding the advantages of closed-loop negative feedback control-systems over open-loop systems:

1. The overall reliability of the closed-loop systems is more than that of open-loop system.
2. The transient response in the closed-loop system decays more quickly than in the open-loop system.
3. In an open-loop system, closing of loop increases the overall gain of the system.
4. In the closed-loop system, the effect of variation of component parameters on its performance is reduced.

Of these statements

(*a*) 1 and 3 are correct

(*b*) 1 and 2 are correct

(*c*) 2 and 4 are correct

(*d*) 3 and 4 are correct **[ES]**

140. The second order system

$$X = AX \text{ has } A = \begin{bmatrix} -1 & -1 \\ 1 & 0 \end{bmatrix}$$

The values of its damping factor ζ, and natural frequency ω_n are respectively

(*a*) 1 and 1

(*b*) 0.5 and 1

(*c*) 0.707 and 2

(*d*) 1 and 2 **[GATE]**

141. The root loci of a system has four seperate loci. The system can have

(*a*) two poles and two zeros

(*b*) five poles an four zeros

(*c*) four poles and four zeros

(*d*) four poles or four zeros **[GATE]**

142. Consider the following statements regarding time domain analysis of control systems:

1. derivative control improves systems transient performance
2. integral control does not improve system steady state performance
3. integral control can convert a second order system into a third order system

Of these statements

(*a*) 1 and 2 are correct

(*b*) 1 and 3 are correct

(*c*) 2 and 3 are correct

(*d*) 1, 2 and 3 are correct **[ES]**

143. A unity feedback control system has a forward path transfer function

$$G(s) = \frac{10(1+4s)}{s^2(1+s)}.$$

If the system is subjected to an input $r(t) = 1 + t + \frac{t^2}{2} + (t \geq 0)$, the steady-state error of the system will be

(*a*) zero

(*b*) 0.1

(*c*) 10

(*d*) infinity **[ES]**

144. The closed-loop transfer function of a control system is given $\frac{C(s)}{R(s)} = \frac{1}{1+s}$.

For the input $r(t) = \sin t$, the steady state value of $c(t)$ is equal to

(*a*) $\frac{1}{\sqrt{2}}\cos t$

(*b*) 1

(*c*) $\frac{1}{\sqrt{2}}\sin t$

(*d*) $\frac{1}{\sqrt{2}}\sin\left(t - \frac{\pi}{4}\right)$ **[GATE]**

145. The state equation of a dynamic system is given by, $\dot{\underline{x}} = A\,\underline{x}(t)$

$$A = \begin{bmatrix} -1 & 1 & 0 & 0 & 0 \\ 0 & -1 & 1 & 0 & 0 \\ 0 & 0 & -1 & 0 & 0 \\ 0 & 0 & 0 & -3 & 4 \\ 0 & 0 & 0 & -4 & -3 \end{bmatrix}$$

The eigen values of the system would be

(*a*) real non-repeated only

(*b*) real non-repeated and complex

(*c*) real repeated

(*d*) real repeated arid complex **[ES]**

146. The transfer function of a certain system is

$$\frac{Y(s)}{U(s)} = \frac{1}{s^4 + 5s^3 + 7s^2 + 6s + 3}$$

The $\underline{A}$, $\underline{B}$ matrix pair of the equivalent state-space model will be

(*a*) $\begin{bmatrix} 0 & 1 & 0 & 0 \\ 0 & 0 & 1 & 0 \\ 0 & 0 & 0 & 1 \\ -3 & -6 & -7 & 5 \end{bmatrix}, \begin{bmatrix} 0 \\ 0 \\ 0 \\ 1 \end{bmatrix}$

(*b*) $\begin{bmatrix} 0 & 1 & 0 & 0 \\ 0 & 0 & 1 & 0 \\ 0 & 0 & 0 & 1 \\ -3 & -5 & -6 & -7 \end{bmatrix}, \begin{bmatrix} 0 \\ 0 \\ 0 \\ 1 \end{bmatrix}$

(*c*) $\begin{bmatrix} 0 & 1 & 0 & 0 \\ 0 & 0 & 1 & 0 \\ 0 & 0 & 0 & 1 \\ -5 & -7 & -6 & -3 \end{bmatrix}, \begin{bmatrix} 1 \\ 0 \\ 0 \\ 0 \end{bmatrix}$

(*d*) $\begin{bmatrix} 1 & 0 & 0 & 0 \\ 0 & 1 & 0 & 0 \\ 0 & 0 & 1 & 0 \\ -3 & -6 & -7 & -5 \end{bmatrix}, \begin{bmatrix} 0 \\ 0 \\ 0 \\ 1 \end{bmatrix}$ **[ES]**

147. A second order system is given by

$$\frac{d^2y}{dt^2} + 12\frac{dy}{dt} + 100y = 0$$

The damped natural frequency in rad/sec is

(*a*) 100 (*b*) 10

(*c*) $\sqrt{44}$ (*d*) 8 **[IOCL]**

148. The phase portrait of non-linear system is shown in the figure. Here the origin is a

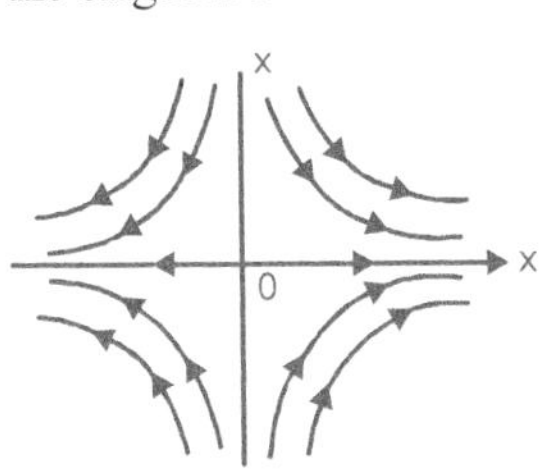

(*a*) stable focus (*b*) vortex

(*c*) stable node (*d*) saddle point **[ES]**

149. Consider the unit-step response of a unity-feedback control system whose open-loop transfer function is

$$G(s) = \frac{1}{s(s+1)}.$$

The maximum overshot is equal to

(*a*) 0.143 (*b*) 0.153

(*c*) 0.163 (*d*) 0.173 **[GATE]**

150. Match List-I with List-II and select the correct answer using the codes given below the List:

List-1 (*Function*)	**List- II** (*Laplace transform*)
A. Unit ramp	1. s
B. Unit step	2. 1
C. Unit impulse	3. $\frac{1}{s}$
D. Unit double	4. $\frac{1}{s^2}$

Codes:

	A	B	C	D
(*a*)	4	3	2	1
(*b*)	3	4	1	2
(*c*)	4	3	1	2
(*d*)	3	4	2	1

[ES]

151. Which ope of the following pairs of phase-plane diagram and types of singularities is NOT correctly matched?

Phase-plane diagram **Singularity**

(*a*) 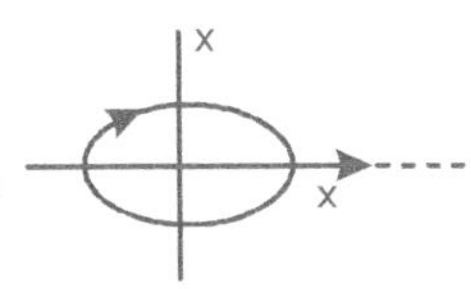

Centre

(*b*) 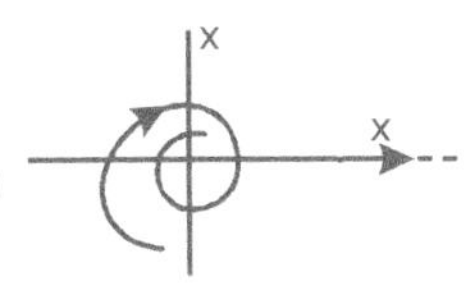

Focus

(*c*) 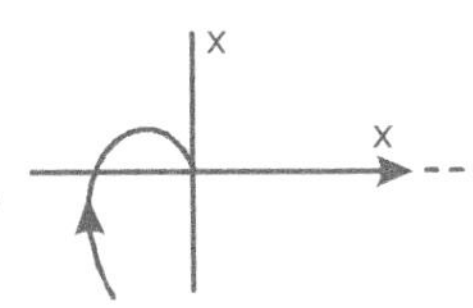

Node

(*d*) 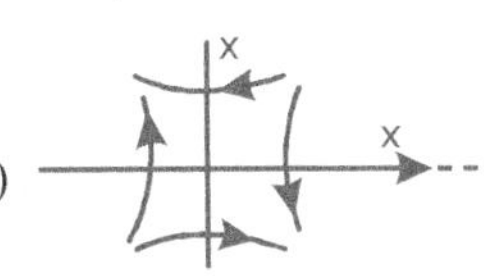

Saddle **[ES]**

152. The transfer function C/R of the system shown in the figure is:

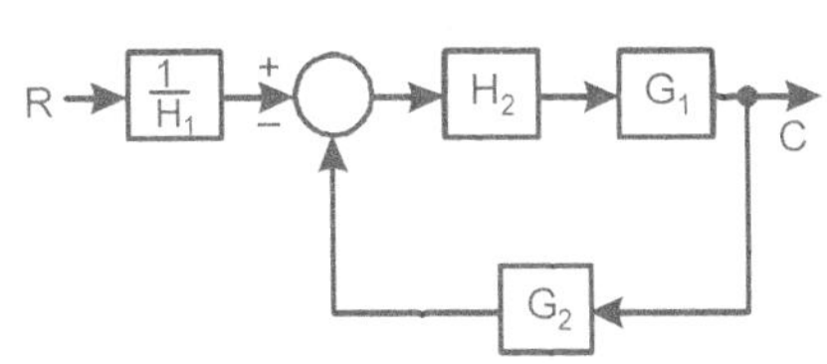

(*a*) $\frac{G_1G_2H_2}{H_1(1+G_1G_2H_2)}$ (*b*) $\frac{G_1H_2}{H_1(1+G_1G_2H_2)}$

(*c*) $\frac{G_2H_2}{H_1(1+G_1G_2H_1)}$ (*d*) $\frac{G_2H_1}{H_2(1+G_1G_2H_2)}$ **[ES]**

153. Which one of the following is open loop?

(*a*) The respiratory system of man

(*b*) A system for controlling the movement of the slide of a copying milling machine

(*c*) A thermostatic control

(*d*) Traffic light control **[ES]**

154. The value of A matrix is $\dot{X} = AX$ for the system described by the differential equation $\ddot{y} + 2\dot{y} + 3y = 0$ is

(*a*) $\begin{bmatrix} 1 & 0 \\ -2 & -1 \end{bmatrix}$ (*b*) $\begin{bmatrix} 1 & 0 \\ -1 & -2 \end{bmatrix}$

(*c*) $\begin{bmatrix} 0 & 1 \\ -2 & -1 \end{bmatrix}$ (*d*) $\begin{bmatrix} 0 & 1 \\ -3 & -2 \end{bmatrix}$ **[ES]**

155. A transfer function has its zero in the right of the s-plane. The function

(*a*) is positive real

(*b*) will give stable impulse response

(*c*) is minimum phase

(*d*) is non-minimum phase **[ES]**

156. If $i(t) = \frac{1}{4}(1 - e^{-2t})\, u(t)$, where $u(t)$ is a unit step voltage, then the complex frequencies associated with $i(t)$ would include

(*a*) $s = 0$ and $s = j2$

(*b*) $s = j2$ and $s = -j2$

(*c*) $s = -j2$ and $s = -2$

(*d*) $s = 0$ and $s = -2$ **[ES]**

157. Which of the following statements regarding an ac servo employing a synchro error detector are correct?

1. Sensitivity of the error detector is measured in volts/degree.
2. The transfer function of the synchro error detector is $\frac{E}{\theta_r - \theta_c}$, where, E = error voltage, r and c are reference and controlled shaft positions.
3. Sensitivity of the error detector has the same sign at the two null positions.

Select the correct answer using the codes given below:

Codes:

(*a*) 1, 2 and 3

(*b*) 1 and 2

(*c*) 2 and 3

(*d*) 1 and 3 **[ES]**

158. Matcn List-I with List-II and select the correct answer using the codes given below the Lists:

List-I (*Properties of frictional force*)

A. 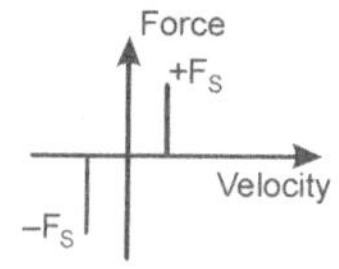

B. 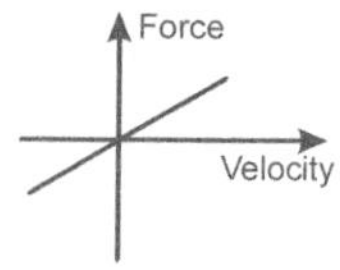

C. 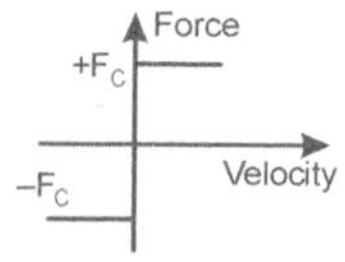

List-II (*Frictional force*)

1. Coulomb friction y
2. Viscous friction
3. Static friction

Codes:

	A	B	C
(*a*)	1	2	3
(*b*)	2	1	3
(*c*)	3	2	1
(*d*)	2	3	1

[AMIE]

159. Which of the following pairs are correctly matched?

Function Laplace Transform

1. te^{at} : $\frac{1}{(s-a)^2}$
2. $1 - e^{at}$: $\frac{a}{s(s-a)}$
3. $e^{-at}\sin\omega t$: $\frac{\omega}{(s+a)^2 + \omega^2}$
4. $1 - \cos\omega t$: $\frac{\omega^2}{(s^2 + \omega^2)}$

Select the correct answer using the codes given below

(*a*) 1, 2 and 3 (*b*) 1, 3 and 4

(*c*) 1, 2 and 4 (*d*) 2, 3 and 4 **[ES]**

160. Type I system, a constant output velocity at steady state will be possible when

(*a*) there is no error

(*b*) there is constant steady state error

(*c*) there is variable steady.state error

(*d*) there is fluctuating error. **[ES]**

161. In regeneration feedback the transfer function is given by

(*a*) $\frac{G(s)}{R(s)} = \frac{G(s)}{1+G(s)H(s)}$

(*b*) $\frac{G(s)}{R(s)} = \frac{G(s)H(s)}{1-G(s)H(s)}$

(*c*) $\frac{G(s)}{R(s)} = \frac{G(s)H(s)}{1+G(s)H(s)}$

(*d*) $\frac{G(s)}{R(s)} = \frac{G(s)}{1-G(s)H(s)}$ **[ES]**

162. If a system is represented by the differential equation, $\frac{d^2y}{dt^2} + \frac{6dy}{dt} + 9y = 0$, then the solution y will be of the form

(*a*) $k_1e^{-t} + k_2e^{-9t}$ (*b*) $(k_1 + k_2t)e^{-3t}$

(*c*) $ke^{-3t}\sin(t+\phi)$ (*d*) $(k_1 + k_2t)e^{3t}$ **[ES]**

163. The unit-impulse response of a unity-feedback control system is given by

$c(t) = -te^{-t} + 2e^{-t}\ (t \geq 0)$

The open-loop transfer function is equal to

(*a*) $\frac{s+1}{(s+2)^2}$

(*b*) $\frac{2s+1}{s^2}$

(*c*) $\frac{s+1}{(s+1)^2}$

(*d*) $\frac{s+1}{s^2}$ **[GATE]**

164. The state transition matrix for the system $\dot{x} = AX + Bu$ is given by

(*a*) e^{AT} (*b*) e^{-AT}

(*c*) $\frac{1}{2}e^{AT}$ (*d*) $\frac{1}{2}e^{-AT}$ **[ES]**

165. If $F(s) = \frac{(s+1)}{s(s+k)}$ and $f(t)$ as $t \to \infty$ is $\frac{1}{2}$ then the value of k is

(*a*) $\frac{1}{2}$ (*b*) 1

(*c*) 2 (*d*) ∞ **[ES]**

166. With symbols having the usual meanings, the Laplace transform of $U(t-a)$ is

(*a*) $\frac{1}{s}$ (*b*) $\frac{1}{s-a}$

(*c*) $e^{-as}\left(\frac{1}{s}\right)$ (*d*) $e^{as}\left(\frac{1}{s}\right)$ **[ES]**

167. The polar plot of a transfer function with ω as the parameter is known as the

(*a*) Nyquist plot

(*b*) Bode plot

(*c*) Root-locus

(*d*) Signal flow graph **[ES]**

168. A 5 × 7 matrix has all its entries equal to –1. The rank of the matrix is

(*a*) 7

(*b*) 5

(*c*) 1

(*d*) Zero **[GATE]**

169. Match List I with List II and select the correct answer using the codes given below the Lists:

List-I (*Response to a unit step input*)

A. 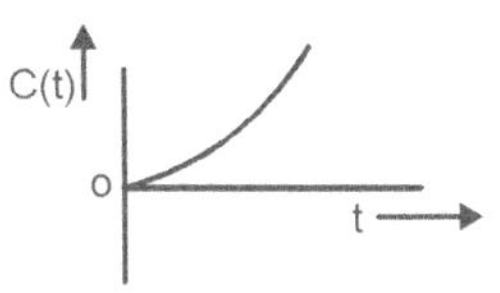

B. 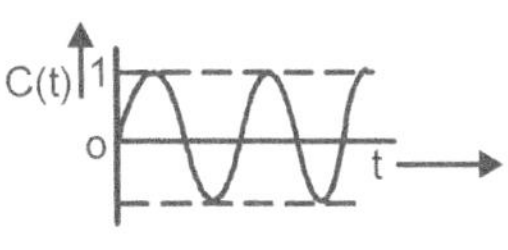

C. 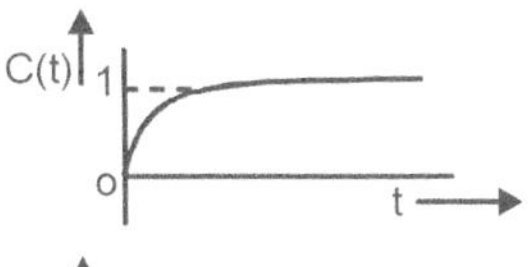

D. 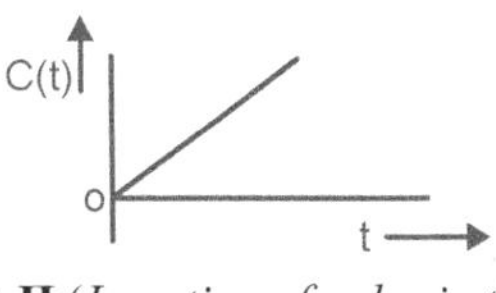

List-II (*Location of poles in the s plane*)

1. One at the origin
2. Two indentical roots on the negative real axis
3. Two on the imaginary axis
4. One on the positive real axis

Codes:

	A	B	C	D
(*a*)	4	3	2	1
(*b*)	3	4	1	2
(*c*)	3	4	2	1
(*d*)	4	3	1	2

[ES]

170. Which of the following is used for Nyquist plot?

(*a*) Characteristic equation
(*b*) Open-loop function
(*c*) Close-loop function
(*d*) None of these **[ES]**

171. Match List I with List II and select the correct answer using the codes given below the lists:

List-I (*Unit*)

A. Synchro transmitter
B. Control Transformer
C. A.C. Servo-motor
D. Stepper motor

List-II (*Type of rotor*)

1. Dumb-bell rotor
2. Drag-cup rotor
3. Cylindrical rotor
4. Toothed rotor
5. Phase wound rotor

Codes:

	A	B	C	D
(*a*)	1	3	2	4
(*b*)	2	4	3	1
(*c*)	1	5	3	2
(*d*)	3	2	1	5

[ES]

172. On an infinitely long transmission line, the attenuation co-efficient α is expressed in terms of the ratio K of the two voltages one kilometer apart on the line, in bels/ km. The value of the co-efficient is given by

(*a*) $a = \log_e K$
(*b*) $a = K^2$
(*c*) $a = 2\log_{10} K$
(*d*) none of these **[ES]**

173. The root-locus of a unity feedback system is shown in the figure. The open loop transient function is given by

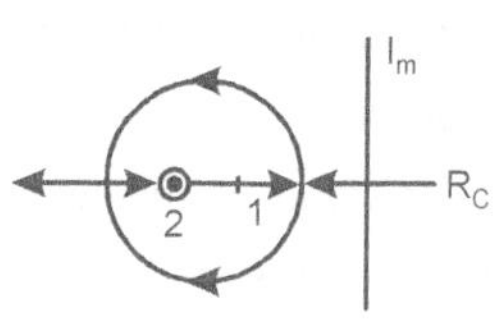

(*a*) $\frac{K}{s(s+1)(s+2)}$ (*b*) $\frac{K(s+1)}{s(s+2)}$

(*c*) $\frac{K(s+2)}{s(s+1)}$ (*d*) $\frac{Ks}{(s+1)(s+2)}$ **[ES]**

174. Laplace transform of the output response of a linear system is the system transfer functions when the input is

(*a*) a step signal
(*b*) a ramp signal
(*c*) an impulse signal
(*d*) a sinusoidal signal **[ES]**

175. The eigen-values of the matrix $\begin{bmatrix} a & 1 \\ a & 1 \end{bmatrix}$ are

(*a*) $(a+1), 0$
(*b*) $a, 0$
(*c*) $(a-1), 0$
(*d*) $0, 0$ **[GATE]**

176. For making an unstable system stable

(*a*) gain of the system should be increased
(*b*) gain of the system should be decreased
(*c*) the number of zeros to the loop transfer function should be increased
(*d*) the number of poles to the loop transfer funtion should be increased. **[ES]**

177. Match List I with List II and select the correct answer using the codes given below the lists:

List-I	List-II
A. Hydraulic actuator	1. Linear device
B. Flapper valve	2. A.C. servo system C.
Potentiometer error	3. Large power to weight ratio
D. Dumb-bell rotor	4. Pneumatic systems

Codes:

	A	B	C	D
(*a*)	4	3	2	1
(*b*)	3	4	2	1
(*c*)	3	4	1	2
(*d*)	4	3	1	2

[ES]

178. A linear system is described by the state equations.

$$\begin{bmatrix} \dot{x}_1 \\ \dot{x}_2 \end{bmatrix} = \begin{bmatrix} 1 & 0 \\ 1 & 1 \end{bmatrix}\begin{bmatrix} x_1 \\ x_2 \end{bmatrix} \begin{bmatrix} 0 \\ 1 \end{bmatrix} r$$

Where r and c are the input and output respectively. The transfer function is

(*a*) $\frac{1}{(s+1)}$ (*b*) $\frac{1}{(s+1)^2}$

(*c*) $\frac{1}{(s-1)}$ (*d*) $\frac{1}{(s-1)^2}$ **[ES]**

179. A simple electric water heater is shown in the given figure. The system can be modelled by

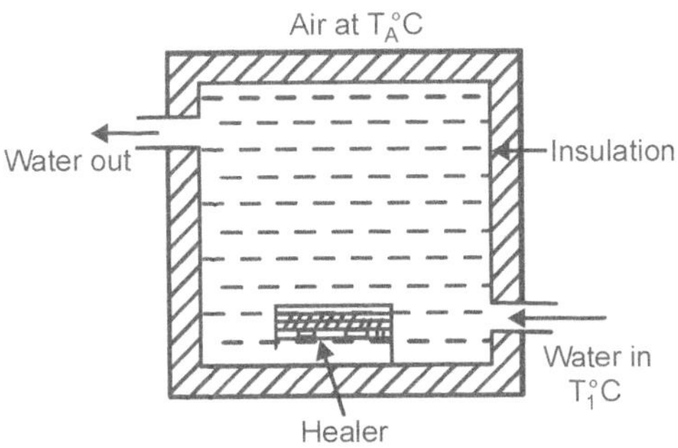

(*a*) a first order differential equation

(*b*) a second order differential equation

(*c*) a third order differential equation

(*d*) an algebraic equation **[ES]**

180. Match List I with List II and select the correct answer using the codes given below the lists:

List-I (*G*(*s*) *H*(*s*))

A. $\dfrac{K}{1+sT_1}$

B. $\dfrac{K}{s(1+sT_1)}$

C. $\dfrac{K}{(1+T_1s)(1+T_2s)}$

D. $\dfrac{K}{s^2(1+sT_1)}$

List-II (*Nyquist plot*)

1. 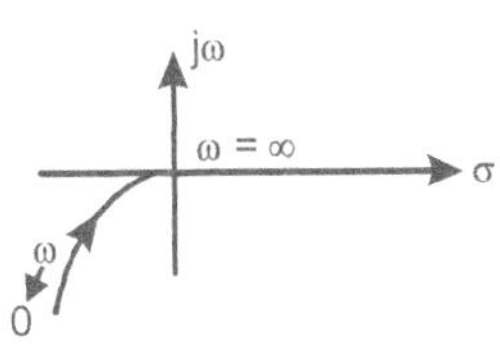

2. 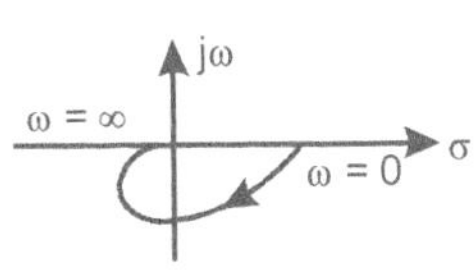

3. 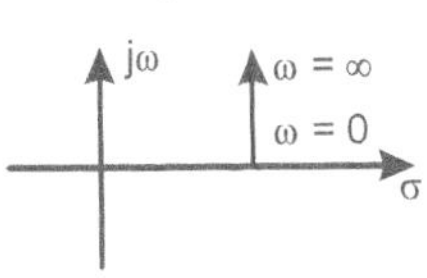

4. 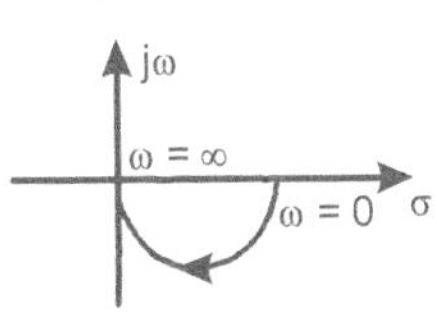

5. 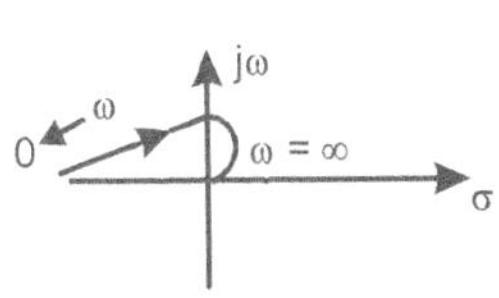

Codes:

	A	B	C	D
(*a*)	1	4	5	3
(*b*)	1	4	3	5
(*c*)	4	1	2	3
(*d*)	4	1	2	5

[ES]

181. For a given gain constant K, the phase-lead compensator

(*a*) reduces the slope of the magnitude curve in the entire range of frequency domain

(*b*) decreases the gain cross-over frequency

(*c*) reduces the phase margin

(*d*) reduces the resonance peak M_p **[ES]**

182. Nyquist stability criterion is employed for determination of

(*a*) absolute stability

(*b*) relative stability

(*c*) both (*a*) and (*b*)

(*d*) none of these **[ES]**

183. For the RC circuit shown in the given figure, V_1 and V_0 are the input and output of the system respectively. The block diagram of the system is represented by

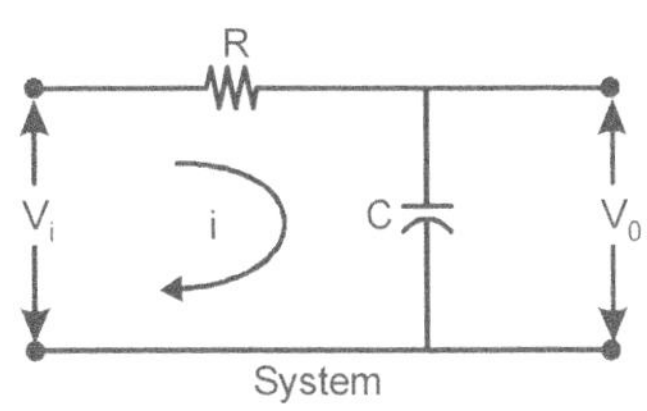

(*a*) 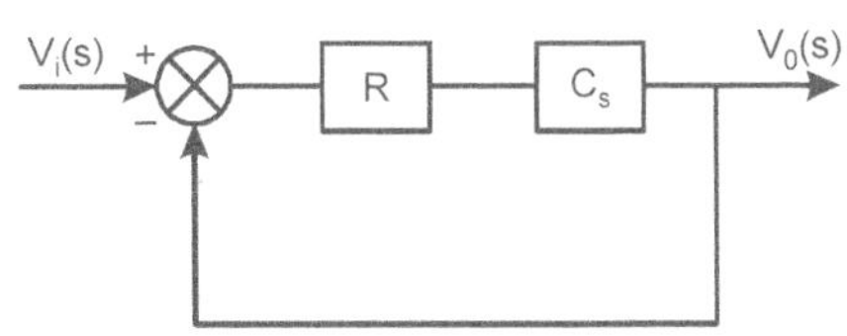

(*b*) 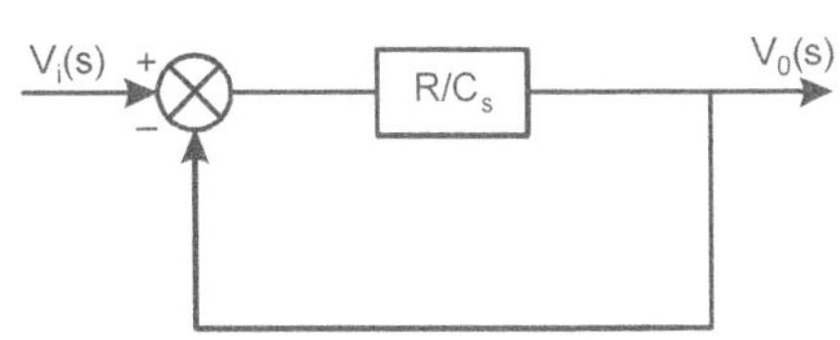

(*c*) 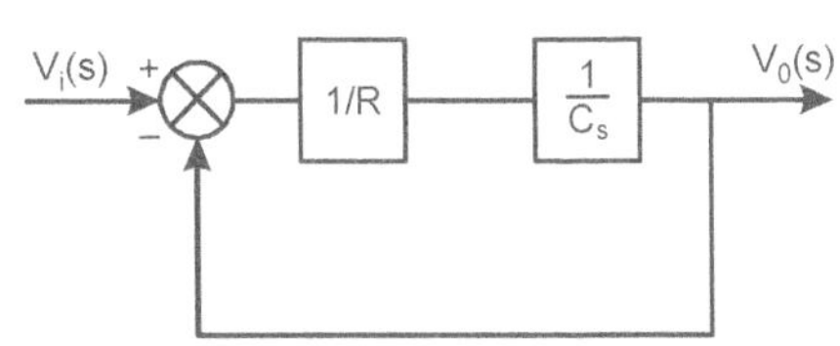

(*d*) 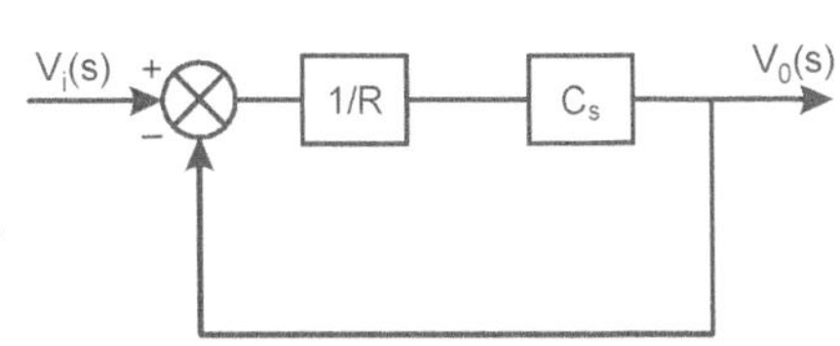

[ES]

184. A network has a pole at $s = -1$ and a zero at $s = -2$. If this network is excited by sinusoidal input, the output

(a) leads the input

(b) lags the input

(c) is in phase with input

(d) decays exponentially to zero **[ES]**

185. For the block diagram shown in the given figure, the limiting values of K for stability of inner loop is found to be X < K < Y, The overall system will be stable if and only if

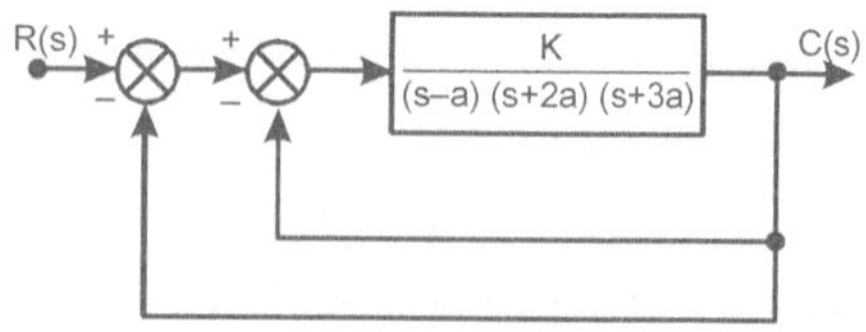

(a) 4X<K<4Y (b) 2X<K<2Y

(c) X<K<Y (d) $\frac{X}{2} < K < \frac{Y}{2}$ **[ES]**

186. A unity feedback second order control system is characterised by

$$G(s) = \frac{K}{s(Js - B)}$$ where

J = moment of inertia

K = system gain

B = viscous damping coefficient

The transient response specification which is NOT affected by variation of system gain is the

(a) peak overshoot

(b) rise time

(c) settling time

(d) damped frequency of oscillations **[ES]**

187. For the determination of stability of a non-linear control system by the describing Function method, the plots of

$G(j\omega)$ and $-\frac{1}{N(x)}$ are drawn as shown in the given figure for an input $x = X \sin \omega t$.

The types of limit cycles at the points P and Q will be respectively

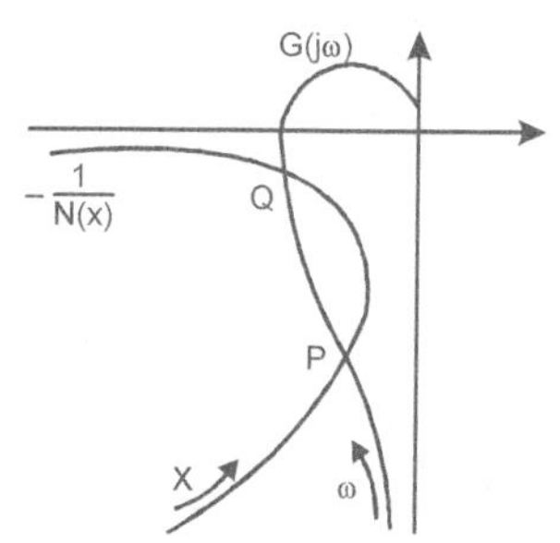

(a) stable and stable (b) stable and unstable

(c) unstable and stable (d) unstable and unstable

[ES]

188. For the system $x + axx + bx = 0$, the equation for the phase-plane isocline of slope 'm' is

(a) $m^2 + am + b = 0$

(b) $m = -ax - \frac{bx}{\dot{x}}$

(c) $m = ax + \frac{bx}{\dot{x}}$

(d) $m^2 + axm + b = 0$ **[ES]**

189. The Laplace transform of the function

$$i(t) = \frac{10s+4}{s(s+1)(s^2+4s+5)}$$ its final value will be

(a) $\frac{4}{5}$ (b) $\frac{5}{4}$

(c) 4 (d) 5 **[ES]**

190. If the unit step response of a network is $(1 - e^{-at})$ then its unit impulse response will be

(a) ae^{-at} (b) $\frac{1}{\alpha} e^{-t/\alpha}$

(c) $\frac{1}{\alpha e^{-t/\alpha}}$ (d) $(1-\alpha)^{-at}$ **[ES]**

191. A system is described by the state equation

$$\dot{x} = AX + BU.$$

The output is given by $Y = CX$

where $A = \begin{bmatrix} -4 & -1 \\ 3 & -1 \end{bmatrix}, B = \begin{bmatrix} 1 \\ 1 \end{bmatrix}, C = [1, 0]$

Transfer function G(s) of the system is

(a) $\frac{s}{s^2+5s+7}$

(b) $\frac{1}{s^2+5s+7}$

(c) $\frac{1}{s^2+3s+2}$

(d) $\frac{1}{s^2+3s+2}$ **[GATE]**

192. To achieve the optimum transient response, the indicating instruments are so designed as to

(a) be critically damped

(b) be undamped

(c) provide damping which is slightly more than the critical value

(d) provide damping which is slightly less than the critical value **[ES]**

193. The Laplace transform of $f(t)$ is $F(s) = \frac{\omega}{s^2 + \omega^2}$, the final value of $f(t)$ is

(a) infinity (b) zero

(c) one (d) none of these **[GATE]**

194. Consider the following statements

1. Nyquist criterion is in frequency domain
2. Bode Plot is in frequency domain
3. Root locus plot is in time domain
4. Routh hermitz's criterion is in time domain

(a) 1, 2 and 3 are correct

(b) 2, 3 and 4 are correct

(c) 1 and 2 are correct

(d) all four are correct **[ES]**

195. Either phenomenon can be used to overcome the

(a) large steady state error

(b) coulombic friction

(c) instability of a system

(d) viscous friction **[ES]**

196. A decade Frequency range is specified by

(a) $\frac{\omega_2}{\omega_1} = 2$ (b) $\frac{\omega_2}{\omega_1} = 10$

(c) $\frac{\omega_2}{\omega_1} = 8$ (d) none of these **[GATE]**

Directions : *The following items consist of two statements, on labelled the 'Assertion A' and the other labelled the 'Reason R'. You are to examine these two statements carefully and decide if the Assertion A and the Reason R are individually true and if so, whether the Reason is a correct explanation of the Assertion. Select your answer to these items using the codes given below and mark your answer sheet accordingly.*

Codes:

(a) Both A and R are true and R is correct explanation of A

(b) Both A and R are true but R is not a correct explanation of A

(c) A is true but R is false

(d) A is false but R is true

197. **Assertion (A) :** The steady-state error of a type 1 system due to acceleration input is zero.

Reason (R) : Type 1 system has a zero positional error constant. **IAS]**

198. **Assertion (A) :** A discrete data system having the characteristic equation $Z^2 - 1.5Z - 1 = 0$ is unstable.

Reason (R) : The negative coeafficients of the characteristics equation in the Z-plane represent an unstable discrete system **[ES]**

199. **Assertion (A) :** The steady-state error in control system is not a function of time.

Reason (R) : The steady state error is defined as $e_{ss} = \lim e(t) t \to \infty$ **[ES]**

200. **Assertion (A) :** The Laplace transform of

$e^{-at} \sin wt$ is $\frac{w}{(s+a)^2 + w^2}$

Reason (R) : If the Laplace transform $f(t) = F$ then Laplace transform of

$e^{-at} f(t)s = F(s + a)$ **[ES]**

201. **Assertion (A) :** The step response of the system having a transfer function

$$G(s) = \frac{10}{s^2 + 12s + 144}$$

will have an overshoot.

Reason (R) : The system has a natural frequency of oscillation equal to 12 rad/s. **[ES]**

202. **Assertion (A) :** A phase lead compensator increases the margin of stability of a control system.

Reason (R) : Phase-lead compensator provides a compensating pole which is located to the right of the compensating zero. **[ES]**

203. **Assertion (A) :** If the open-loop transfer function of a system has unstable roots then it will remain closed-loop unstable.

Reason (R) : The roots of the characteristic equation of a closed loop system are given by

$1 + G(s)H(s) = 0$ **[ES]**

204. **Assertion (A) :** Feedback control systems are often compensated by phase-lead/phase-lag-networks.

Reason (R) : Phase-lead/phase-lag networks are easily realizable and obtainable. **[ES]**

205. **Assertion (A) :** The system with transfer function $G(s) = \frac{10(s-12)}{s(s+2)(s+3)}$ is a minimum phase function.

Reason (R) : It has a zero on the right half of the s-plane **[IAS]**

206. **Assertion (A) :** The steady-state error due to stop input to a system represented by

$$G(s) = \frac{10(1+0.5s)}{s(1+s)(1+2s)}$$ is zero.

Reason (R) : For a type 1 system, the steady-state error due to step input is zero **[IAS]**

207. **Assertion (A) :** In an error detector using a synchro transmitter and a synchro-control transformer, the later is connected to the servo amplifier.

Reason (R) : The synchro transmitter a salient pole device and the control transformer had a low reluctance rotor. **[IAS]**

208. **Assertion (A) :** Phase-lead compensation is generally less effective if the uncom-pensated system is highly unstable.

Reason (R) : Phase-lead compensator adds a zero to the right of the pole of closed-loop system. **[ES]**

209. **Assertion (A) :** The compensating network shown in the given figure is used for reduction of steady-state error in the system response.

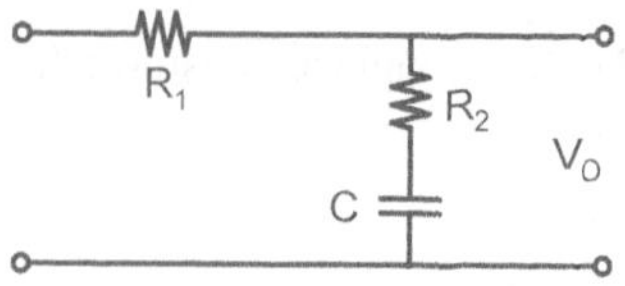

Reason (R) : A lead compensating network reduces system steady-state error. **[ES]**

210. **Assertion (A) :** When plant parameter uncertainties are present, the open-loop structure is potentially superior to the closed-loop structure.

Reason (R) : When feedback is employed, there is a possibility of additional noise or uncertainty in the measurement of the plant signals. **[ES]**

211. **Assertion (A) :** If the transfer function of a system is known, then its impulse response can be calculated.

Reason (R) : The impulse response is the inverse Laplace transform of the transfer function. **[IAS]**

212. **Assertion (A) :** The polar plot is never used to determine the absolute stability of a closed-loop system.

Reason (R) : The Nyquist stability criterion uses this plot. **[IAS]**

213. **Assertion (A) :** For the system having the Nyquist plot as shown in the figure, the gain margin is positive.

Reason (R) : The gain margin of a system is positive of the intercept

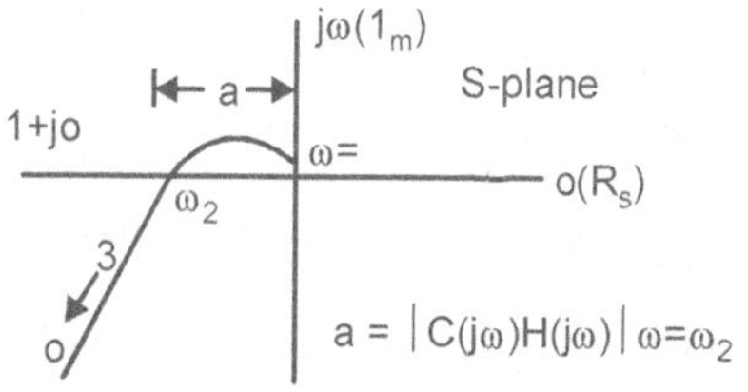

$a = |G|(j\omega)\ H(j\omega)\ \omega = \omega_2$ on the negative real axis. **[ES]**

214. **Assertion (A) :** In a position control system, tachometric feedback is employed to improve the performance of the system.

Reason (R) : Velocity feedback has the effect of increasing the damping ratio without affecting the undamped natural frequency of oscillation of the system. **[ES]**

215. **Assertion (A) :** Whenever a synchro pair is used as an error detector, a two-phase ac motor may be used as an actuator.

Reason (R) : The synchro pair produces an error voltage which is suppressed carrier modulated and a two-phase ac motor could demodulate this. **[IAS]**

216. **Assertion (A) :** On supplying 2-phase voltage to balanced two-phase windings (two indentical windings in space quadrature), a rotating field is developed.

Reason (R) : The amplitude of rotating field is 1.5 times that of each phase. **[ES]**

217. **Assertion (A) :** If the impulse response of a network is available, the response to any arbitrary input can be derived.

Reason (R) : The Laplace transform of unit impulse is unity. **[ES]**

218. **Assertion (A) :** By superimoposing the log magnitude versus phase angle plot of open-loop transfer function of Nichols chart, it is possible to determine the closed-loop frequency response of the system.

Reason (R) : Nichols chart gives constant M and N loci in the gain-phase plane. **[IAS]**

219. **Assertion (A) :** Linearity is essential for the servo amplifiers.

Reason (R) : The servo amplifiers are used to amplify the output of the null detector. **[IAS]**

220. **Assertion (A) :** The sampled data system with its characteristic equation $z^2 - 4z + 4 = 0$ is unstable.

Reason (R) : Sampled data systems with positive roots of the characteristic equation are unstable. **[ES]**

221. **Assertion (A) :** The state steady response of a linear network is termed 'forced response'.

Reason (R) : The forcing function does not have any effect on the transient response of a linear system. **[ES]**

222. **Assertion (A) :** The closed-loop system represented in the given figure is a type-1 system.

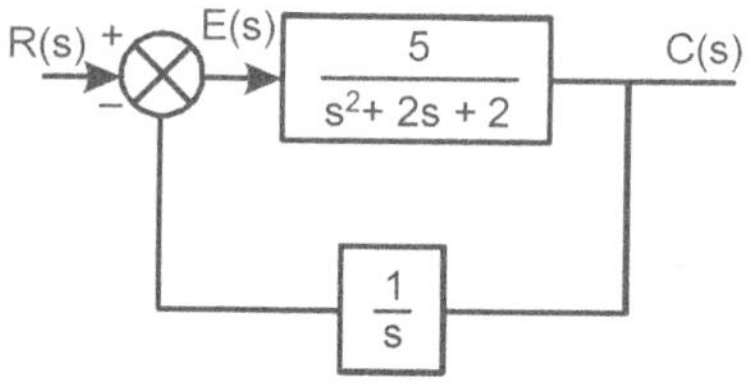

Reason (R) : Number of poles at the origin possessed by the feedback path decides the type of the closed-loop system. **[IAS]**

223. **Assertion (A) :** The state transition matrix $\phi(t)$ related to the solution of the state equation $x(t) = Ax(t) + BU(t)$ has the property that $\phi^{-1}(t) = \phi(-t)$

Reason (R) : It is assumed that the matrix $(sI - A)$ is non-singular so that its inverse exists. **[ES]**

224. **Assertion (A) :** The stability of a closed-loop system can be obtained from the open-loop transfer function $G(j\omega)$ $H(j\omega)$ plot with respect to the critical point $(-1, j0)$ in $G(j\omega)H(j\omega)$ plane.

Reason (R) : The origin of $1 + G(j\omega)\ H(j\omega)$ corresponds to $(-1, j0)$ point in $G(j\omega)\ H(j\omega)$ plane. **[ES]**

225. **Assertion (A) :** Stability is a major problem in closed loop control systems.

Reason (R) : Introduction of feedback affects the location of poles of open-loop system. **[IAS]**

ANSWERS

EXERCISE – I

1. (*b*) 2. (*d*) 3. (*a*) 4. (*d*) 5. (*a*) 6. (*d*) 7. (*a*) 8. (*b*) 9. (*d*) 10. (*a*)
11. (*b*) 12. (*a*) 13. (*b*) 14. (*b*) 15. (*a*) 16. (*a*) 17. (*a*) 18. (*b*) 19. (*b*) 20. (*c*)
21. (*b*) 22. (*b*) 23. (*a*) 24. (*c*) 25. (*a*) 26. (*a*) 27. (*b*) 28. (*d*) 29. (*a*) 30. (*c*)
31. (*b*) 32. (*c*) 33. (*b*) 34. (*b*) 35. (*d*) 36. (*c*) 37. (*a*) 38. (*d*) 39. (*a*) 40. (*a*)
41. (*d*) 42. (*b*) 43. (*d*) 44. (*d*) 45. (*b*) 46. (*a*) 47. (*b*) 48. (*c*) 49. (*c*) 50. (*c*)
51. (*b*) 52. (*b*) 53. (*c*) 54. (*d*) 55. (*d*) 56. (*c*) 57. (*a*) 58. (*b*) 59. (*d*) 60. (*b*)
61. (*c*) 62. (*d*) 63. (*a*) 64. (*b*) 65. (*b*) 66. (*a*) 67. (*d*) 68. (*b*) 69. (*d*) 70. (*b*)
71. (*d*) 72. (*b*) 73. (*a*) 74. (*c*) 75. (*b*) 76. (*c*) 77. (*c*) 78. (*c*) 79. (*d*) 80. (*d*)
81. (*d*) 82. (*d*) 83. (*c*) 84. (*b*) 85. (*d*) 86. (*d*) 87. (*c*) 88. (*b*) 89. (*c*) 90. (*a*)
91. (*a*) 92. (*b*) 93. (*d*) 94. (*c*) 95. (*c*) 96. (*a*) 97. (*d*) 98. (*d*) 99. (*b*) 100. (*c*)
101. (*b*) 102. (*d*) 103. (*d*) 104. (*d*) 105. (*c*) 106. (*b*) 107. (*c*) 108. (*c*) 109. (*a*) 110. (*a*)
111. (*b*) 112. (*b*) 113. (*b*) 114. (*b*) 115. (*d*) 116. (*c*) 117. (*c*) 118. (*c*) 119. (*b*) 120. (*a*)
121. (*d*) 122. (*a*) 123. (*c*) 124. (*a*) 125. (*a*) 126. (*a*) 127. (*b*) 128. (*b*) 129. (*b*) 130. (*a*)
131. (*b*) 132. (*d*) 133. (*c*) 134. (*a*) 135. (*c*) 136. (*a*) 137. (*a*) 138. (*d*) 139. (*d*) 140. (*a*)
141. (*a*) 142. (*d*) 143. (*c*) 144. (*d*) 145. (*d*) 146. (*d*) 147. (*d*) 148. (*c*) 149. (*b*) 150. (*c*)
151. (*c*) 152. (*a*) 153. (*b*) 154. (*c*) 155. (*a*) 156. (*b*) 157. (*b*) 158. (*d*) 159. (*d*) 160. (*a*)
161. (*d*) 162. (*d*) 163. (*c*) 164. (*c*) 165. (*d*) 166. (*a*) 167. (*a*) 168. (*c*) 169. (*b*) 170. (*a*)
171. (*c*) 172. (*c*) 173. (*a*) 174. (*b*) 175. (*c*) 176. (*a*) 177. (*c*) 178. (*c*) 179. (*a*) 180. (*c*)
181. (*c*) 182. (*a*) 183. (*c*) 184. (*c*) 185. (*d*) 186. (*b*) 187. (*a*) 188. (*a*) 189. (*d*) 190. (*d*)
191. (*d*) 192. (*c*) 193. (*b*) 194. (*c*) 195. (*b*) 196. (*c*) 197. (*a*) 198. (*c*) 199. (*d*) 200. (*a*)
201. (*d*) 202. (*a*) 203. (*b*) 204. (*c*) 205. (c) 206. (*c*) 207. (*b*) 208. (*b*) 209. (*d*) 210. (*d*)
211. (*c*) 212. (*b*) 213. (a) 214. (*a*) 215. (*c*) 216. (*a*) 217. (*c*) 218. (*d*) 219. (*b*) 220. (*b*)
221. (d) 222. (*c*) 223. (*d*) 224. (*a*) 225. (*c*) 226. (*a*) 227. (*d*) 228. (*d*) 229. (a) 230. (*b*)
231. (*d*) 232. (*a*) 233. (*b*) 234. (*c*) 235. (*b*)

EXERCISE – II

1. (*c*) 2. (*d*) 3. (*b*) 4. (*b*) 5. (*c*) 6. (*b*) 7. (*b*) 8. (*b*) 9. (*c*) 10. (*a*)
11. (*d*) 12. (*c*) 13. (*c*) 14. (*a*) 15. (*a*) 16. (*c*) 17. (*c*) 18. (*d*) 19. (*b*) 20. (*c*)
21. (*c*) 22. (*b*) 23. (*d*) 24. (*b*) 25. (*d*) 26. (*d*) 27. (*c*) 28. (*a*) 29. (*c*) 30. (*b*)
31. (*a*) 32. (*c*) 33. (*a*) 34. (*d*) 35. (*a*) 36. (*c*) 37. (*c*) 38. (*c*) 39. (*b*) 40. (*c*)
41. (*a*) 42. (*b*) 43. (*d*) 44. (*a*) 45. (*c*) 46. (*b*) 47. (*a*) 48. (*d*) 49. (*a*) 50. (*d*)
51. (*c*) 52. (*d*) 53. (*d*) 54. (*b*) 55. (*c*) 56. (*b*) 57. (*a*) 58. (*b*) 59. (*b*) 60. (*a*)
61. (*c*) 62. (*a*) 63. (*a*) 64. (*a*) 65. (*c*) 66. (*b*) 67. (*a*) 68. (*c*) 69. (*b*) 70. (*a*)
71. (*a*) 72. (*a*) 73. (*b*) 74. (*d*) 75. (*c*) 76. (*c*) 77. (*a*) 78. (*b*) 79. (*c*) 80. (*a*)
81. (*a*) 82. (*b*) 83. (*a*) 84. (*c*) 85. (*b*) 86. (*d*) 87. (*c*) 88. (*a*) 89. (*d*) 90. (*a*)
91. (*d*) 92. (*a*) 93. (*c*) 94. (*a*) 95. (*b*) 96. (*b*) 97. (*a*) 98. (*b*) 99. (*b*) 100. (*b*)
101. (*b*) 102. (*a*) 103. (*b*) 104. (*b*) 105. (*b*) 106. (*d*) 107. (*c*) 108. (*c*) 109. (*a*) 110. (*a*)
111. (*a*) 112. (*a*) 113. (*b*) 114. (*a*) 115. (*a*) 116. (*d*) 117. (*b*) 118. (*c*) 119. (*c*) 120. (*d*)
121. (*d*) 122. (*c*) 123. (*d*) 124. (*a*) 125. (*c*) 126. (*a*) 127. (*c*) 128. (*b*) 129. (*b*) 130. (*c*)
131. (*d*) 132. (*a*) 133. (*b*) 134. (*c*) 135. (*a*) 136. (*b*) 137. (*c*) 138. (*b*) 139. (*c*) 140. (*b*)
141. (*d*) 142. (*b*) 143. (*d*) 144. (*d*) 145. (*d*) 146. (*a*) 147. (*d*) 148. (*d*) 149. (*c*) 150. (*a*)
151. (*b*) 152. (*b*) 153. (*d*) 154. (*a*) 155. (*d*) 156. (*d*) 157. (*a*) 158. (*a*) 159. (*b*) 160. (*b*)
161. (*d*) 162. (*b*) 163. (*b*) 164. (*a*) 165. (*c*) 166. (*c*) 167. (*a*) 168. (*c*) 169. (*a*) 170. (*b*)
171. (*a*) 172. (*d*) 173. (*c*) 174. (*c*) 175. (*a*) 176. (*a*) 177. (*d*) 178. (*d*) 179. (*d*) 180. (*d*)
181. (*a*) 182. (*c*) 183. (*c*) 184. (*b*) 185. (*d*) 186. (*d*) 187. (*c*) 188. (*b*) 189. (*a*) 190. (*a*)
191. (*a*) 192. (*d*) 193. (*d*) 194. (*d*) 195. (*a*) 196. (*b*) 197. (*c*) 198. (*c*) 199. (*a*) 200. (*a*)
201. (*b*) 202. (*c*) 203. (*d*) 204. (*a*) 205. (*d*) 206. (*a*) 207. (*a*) 208. (*a*) 209. (*c*) 210. (*c*)
211. (*a*) 212. (*b*) 213. (*c*) 214. (*a*) 215. (*c*) 216. (*c*) 217. (*a*) 218. (*a*) 219. (*b*) 220. (*c*)
221. (*c*) 222. (*d*) 223. (*b*) 224. (*a*) 225. (*a*)

EXPLANATIONS

EXERCISE – I

12. $s^2 + 2s + 8 = 0$

$\Rightarrow \quad \omega_n = \sqrt{8} = 2.828$

Now $\quad 2\omega_n \xi = 2$

$\therefore \quad \xi = \dfrac{2}{2 \times 2.828} = 0.353$

13. $\omega_n^{\,2} = 16$

$\therefore \quad \omega_n = 4$

Again $\quad 2\xi\omega_n = 2$

$\therefore \quad \xi = 0.25$

$\therefore$ Peak overshoot,

$$M_p = e - \frac{\pi\xi}{\sqrt{1-\xi^2}}$$

$$= e - \frac{(\pi \times 0.25)}{\sqrt{1-(0.25)^2}}$$

$$= e^{-0.81} = 0.44 = 44\%$$

22. $G(s) = \dfrac{1}{s^2 + 0.2s + 1}$

$\omega_n = 1$

$2\xi\omega_n = 0.2$

$\therefore \quad \xi = 0.1$

$\therefore$ Setting time, $\quad t = \dfrac{4}{\xi\omega_n} = \dfrac{4}{0.1 \times 1} = 40$ sec

23. Characteristics equation is given by

$$1 + G(s) = 0$$

$\therefore \quad 1 + \dfrac{1-s}{s(s+2)} = 0$

$\Rightarrow \quad s^2 + 2s + 1 - s = 0$

$$s^2 + s + 1 = 0$$

From this equation, it is clear that both the roots lie on the left half of s-plane, hence the system is stable.

25. In all the polynomials, the coefficients are positive; hence we will construct Routh array for further check.

Take first polynomial

$s^4 + 7s^3 + 17s^2 + 17s + 8 + 6$

s^4	1	17	6
s^3	17	17	
s^2	$\frac{102}{7}$	6	
s^1	$\frac{720}{51}$		
s^0	6		

since all the terms in the first column have positve sign, hence it is hurwitz polynomial.

Take second polynomial

$s^4 + 11s^3 + 41s^2 + 61s + 30$

s^4	1	141	6
s^3	11	61	–
s^2	$\frac{390}{11}$	30	–
s^1	$\frac{672}{1113}$	–	–
s^0	30	–	–

$\therefore$ It is also a Hurwitz Polynomial

Take third polynomial

s^4	1	2	2
s^3	1	3	–
s^2	–1	2	–
s^1	5	–	–
s^0	2	–	–

$\therefore$ It is not a Hurwitz Polynomial.

26. Transfer function of Zero Order Hold (ZOH) is given by

$$G_{no}(s) = \frac{1 - e^{-ST}}{s}$$

We have, $\quad (1 - z^{-1}) = \dfrac{z-1}{z}$

Laplace transform of this "s".

Hence (1) and (2) are equivalent.

27. $\dfrac{C_n(s)}{N(s)}$ is determined by making $R(s) = 0$

From given signal flow diagram

$$\frac{C_n(s)}{N(s)} = \frac{-G(s)H_2(s)}{1 + G(s)H_1(s)H_2(s)}$$

For $G(s)$, $\quad H_1(s)\,H_2(s) >> 1$

$$\frac{C_n(s)}{N(s)} = \frac{-G(s)H_2(s)}{G(s)H_1(s)H_2(s)}$$

$$= \frac{1}{H_1(s)}$$

28. Mathematical model of system is linear if it obeys principle of superposition and homogeneity.

29. $s^2 + 5ks + 10 = 0$

System poles will be for $K = 0$

$\therefore$ Poles $\quad s^2 + 10 = 0$

$\therefore \quad s = \pm\, j\sqrt{10}$

Now we know that the root locus starts from open loop poles and terminates on open loop zeros.

Hence choice is from answers (*a*) and (*c*). Now for all values of K between 0 and ∞ the system is stable. Thus the root locus will be on the left half of *s*-plane, hence the correct answer is (*a*).

30. The characteristic equation for close loop system is given by

$$1+\frac{K(1+Ts)}{s^2(1+s)}=0$$

$$\Rightarrow s^2(1+s)+K+TKs=0$$

$$\Rightarrow s^3+s^2+TKs+K=0$$

Routh Array is

s^3	1	TK
s^2	1	K
s	TK − K	
s^0	K	

For stable system $K>0$

and $TK-K>0$

$\Rightarrow K>1$

32.

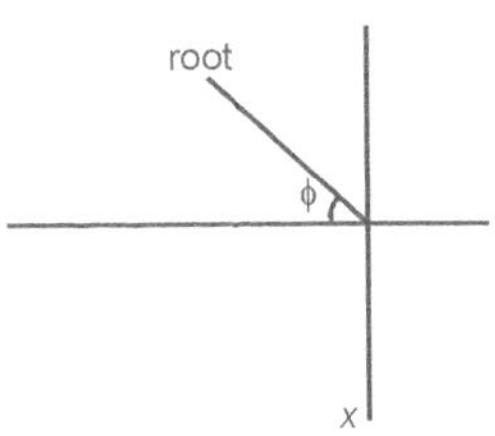

Characteristic equation of the system is given by

$$1+\frac{1}{sT_1(1+sT_2)}=0$$

$$\Rightarrow T_1T_2s^2+sT_1+1=0$$

$$\Rightarrow s^2+\frac{1}{T_2}s+\frac{1}{T_1T_2}=0$$

Now the roots of the characteristic equation are

$$s=\frac{-T_1\pm\sqrt{T_1^2-4T_1T_2}}{2}$$

Now, $\xi=\cos\theta$

For very much under damping, the ξ should be very low; hence for this angle ϕ would be very large hence ϕ will be large only when second term of the root will beimaginary and large, and hence

$\therefore\ T_1^2<<4T_1T^2$

$\Rightarrow T_2>>T_1$

34. Transfer function of the inner loop is

$$TF_1=\frac{G_2}{1+G_2H_1}$$

$$=\frac{\frac{10}{s+1}}{1+\frac{10}{(s+1)}(s+3)}$$

$$=\frac{10}{s+1+10s+30}$$

$$=\frac{10}{11s+31}$$

Overall transfer function is given by

$$T.F.=\frac{G_1.TF_1}{1+G_1.TF_1.H_2}$$

$$=\frac{\frac{10}{5}.\frac{10}{(115+31)}}{1+\frac{10}{5}.\frac{10}{(115+31)}.1}$$

$$=\frac{100}{11s^2+31s+100}$$

39. $A=\begin{bmatrix}2&0\\0&2\end{bmatrix}$

The trasition matrix of the system

$$\phi(t)=L^{-1}(sI-A)^{-1}$$

$$[sI-A]^{-1}=\begin{bmatrix}s&0\\0&s\end{bmatrix}-\begin{bmatrix}2&0\\0&2\end{bmatrix}$$

$$=\begin{bmatrix}(s-2)&0\\0&(s-2)\end{bmatrix}$$

$$[sI-A]^{-1}=\frac{1}{(s-2)^2}\begin{bmatrix}s-2&0\\0&s-2\end{bmatrix}$$

$$L^{-1}[sI-A]^{-1}=-1\begin{bmatrix}\frac{1}{s-2}&0\\0&\frac{1}{s-2}\end{bmatrix}$$

$$=\begin{bmatrix}e^{2t}&0\\0&e^{2t}\end{bmatrix}$$

41. $L\left[\frac{e^{-at}-e^{-bt}}{b-a}\right]=\frac{1}{(b-a)}\left[\frac{1}{(s+a)}-\frac{1}{(s+b)}\right]$

$$=\frac{1}{b-a}\left[\frac{s+b-s-a}{(s+a)(s+b)}\right]$$

$$=\frac{1}{b-a}\left[\frac{b-a}{(s+a)(s+b)}\right]$$

$$=\frac{1}{(s+a)(s+a)}$$

43. From the Nyquist plot, the transfer function will be of the type

$$G(s)H(s)=\frac{K}{s(1+sT_1)(1+sT_2)}$$

The bode plot of this transfer function should be (*d*).

47. $$\frac{C(s)}{R(s)} = \left(\frac{G_1}{1+H_1G_1}\right)\cdot\left(\frac{G_2}{1+H_2G_2}\right)$$

$$= \frac{G_1G_2}{1+G_1H_1+G_2H_2+G_1G_2H_1H_2}$$

51. Close loop gain $= \dfrac{(-3)}{1-(-3)\times\frac{1}{2}}$

$$= \frac{-3}{1+3/2} = -\frac{6}{5}$$

52.

s^5	1	10	5
s^4	2	20	
s^3	$0 = \in$	5	
s^2	$\dfrac{20\in-10}{\in}$	0	
s^1	5		
s^0	0		

When $\in$ tends to zero, then $\dfrac{20\in-10}{\in}$ becomes negative.

Hence number of roots lying on the right half = Number of sign change = 2

56. $$F(s) = \frac{5}{s(s^2+s+2)}$$

$\therefore$ Final value

$$= \lim_{s\to 0} SF_s$$

$$= \lim_{s\to 0}\frac{5}{s^2+s+2}$$

$$= \frac{5}{2}$$

59. Overall transfer function

$$G(s) = \frac{\frac{1}{(s+1)}\times\frac{2}{(s+2)}\times\frac{5}{(s)}}{1+0.6\times\left(\frac{1}{(s+1)}\right)\times\left(\frac{2}{(s+2)}\right)\times\left(\frac{5}{(s)}\right)}$$

$$= \frac{\dfrac{10}{(s+1)(s+2)(s)}}{\dfrac{s(s+1)(s+2)+6}{(s+1)(s+2)(s)}}$$

$$= \frac{10}{s^3+3s^2+2s+6}$$

60. $$\dot{x} = \begin{bmatrix} 0 & 1 \\ -2 & -3 \end{bmatrix} x + \begin{bmatrix} 0 \\ 1 \end{bmatrix} u$$

$$y = [1 \;\; 0]x$$

$$x = \begin{bmatrix} x_1 \\ x_2 \end{bmatrix};\; y = \begin{bmatrix} y_1 \\ y_2 \end{bmatrix}$$

However from the output equation it is seen that y_1 exits, if $y_1 = x_2$

Also $\dot{x}_1 = x_2$

$$\dot{x}_2 = 2x_1 - 3x_2 + u$$

$$x_1 = -2x_1 - 3\,\dot{x}_1 + u$$

Taking Laplace transforming

$$s^2 x_1(s) = 2x_1 - 3.s.x_1(s) + u(s)$$

Since $x_1(s)$ is output function and $u(s)$ the input function, therefore

$$G(s) = \frac{x_1(s)}{u(s)} = \frac{1}{s^2+3x+2}$$

63. $$\frac{d^2c(t)}{dt^2} + 3\frac{dc(t)}{dt}\, 5c(t) = 5r(t)$$

Taking Laplace transform

$$s^2C(s) + 3sc(s) + 5.C(s) = 5r(s)$$

$$\therefore \quad \frac{c(s)}{r(s)} = \frac{5}{s^2+3s+5}$$

64. Since Nyquist plot does not encircle the (–1, 0), hence the system is stable.

66. $$s^3 + 6Ks^2 + (K+2)s + 8 = 0$$

s^3	1	K + 2
s^2	6K	8
s^1	$\dfrac{6K^2+12K-8}{6K}$	
s^0	8	

hence K is given by

$$6K^2 + 12K - 8 = 0$$

$$\Rightarrow 3K^2 + 6K - 4 = 0$$

$$K = \frac{-6 \pm\sqrt{36+48}}{2\times 3} = 0.528, -2.528$$

but K should be positive, hence K = 0.528

68. $$\frac{c(s)}{r(s)} = \frac{K}{s^2+as+K}$$

Damping factor, $\xi = \dfrac{a}{2\sqrt{K}}$

Steady state error,

$$e_{ss} = \lim_{s\to 0} sE(s)$$

$$= \lim_{s\to 0} s.\frac{1}{s^2}.\frac{s^2+as}{s^2+as+K}$$

$$= \frac{a}{K}$$

After closing the switch,

$$\frac{c(s)}{r(s)} = \frac{K}{s^2+s(a+KK_t)+K}$$

which gives $\xi = \dfrac{a + KK_t}{2\sqrt{K}}$

$$e_{ss} = \lim_{s\to 0} s.E(s)$$

$$= \frac{(a + KKt)}{K}$$

Hence both ξ and e_{ss} will increase

71. $$G(s) = \frac{K(s+10)(s+20)}{s^2(s+2)}$$

Characteristic equation is given by

$$1 + \frac{K(s+10)(s+20)}{s^2(s+2)} = 0$$

$\Rightarrow s^3 + 2s^2 + Ks^2 + 30\,K.s + 200K = 0$

Routh array is

s^3	1	30K
s^2	K + 2	200K
s^1	$\dfrac{30K^2 - 140K}{K+2}$	
s^0	200K	

Hence, for system to stable, $k > 0$

or $30\,K^2 - 140K > 0$

or $K > 4.66$

$\therefore$ $K = 5$

72. For controllability,

$$Q_c = [B : AB] = \begin{bmatrix} 0 & 2 \\ 1 & -2 \end{bmatrix}$$

Since $|Q_c| \neq 0$, hence system is controllable

For observability,

$$Q_o = [C^T : A^T C^T]$$

$$= \begin{bmatrix} 1 & -1 \\ 1 & -1 \end{bmatrix}$$

Since $|Q_c| = 0$, hence system is not observable.

73. We have, $s^3 + K_1 s^2 + s + K_2 = 0$

$$\Rightarrow \quad 1 + \frac{K_2}{s(s^2 + K_1 s + 1)} = 0$$

$$1 + \frac{K_2}{s\left(s + \frac{K_1}{2} + \sqrt{\frac{K_1^2}{4}}\right)\left(s + \frac{K_1}{2} - \sqrt{\frac{K_1^2}{4}} - 1\right)}$$

Hence statement (*i*) and (*ii*) are correct.

74. $T(s) = C[sI - A]^{-1}B$

$$= [1\ 0]\begin{bmatrix} \dfrac{s+3}{s^2+3s+2} & \dfrac{1}{s^2+3s+2} \\ \dfrac{-2}{s^2+3s+2} & \dfrac{s}{s^2+3s+2} \end{bmatrix} = \frac{1}{s^2+3s+2}$$

77. $$G(s) = \frac{K}{s(s+1)}$$

Characteristic equation for unity feedback system is

$$1 + \frac{K}{s(s+1)} = 0$$

$\Rightarrow$ $s^2 + s + K = 0$

$\therefore$ $w_n = \sqrt{K}$, $\xi = \dfrac{1}{2\sqrt{K}}$

Thus is $k \to \infty$, ξ will tend to zero.

80. $$G(s) = \frac{49}{s^2 + 16s + 49}$$

$\therefore$ $\omega_n = \sqrt{49} = 7$

and $2\xi\omega_n = 16$

$\Rightarrow$ $\xi = \dfrac{16}{2 \times 7} = 1.1428$

Since the system is overdamped, hence no peak overshoot.

81. $$K_v = \lim_{s\to 0} sG(s)$$

$$= \lim_{s\to 0} \frac{K}{(s+1)(s+2)} = \frac{K}{2}$$

Given $Kv > 10\text{sec}^{-1}$

$\Rightarrow$ $\dfrac{K}{2} > 10\text{sec}^{-1}$

$\Rightarrow$ $K > 20\text{sec}^{-1}$

Hence, correct answer is (*d*)

82. In two-phase servomotor, diameter of the rotor is kept small in order to reduce inertia to obtain good accelerating characteristices. Also $\dfrac{X}{R}$ ratio is kept small to obtain nearly linear torque-speed characteristics.

83. $$26 - 20 = 20 \times \log_{10} \frac{\omega_1}{10}$$

$\Rightarrow$ $\dfrac{\omega_1}{10} = 1.9952$

$\Rightarrow$ $\omega_1 = 19.95 \approx 20$

$\therefore$ $26 = 20 \log \dfrac{\omega_2}{\omega_1}$

or $\omega_2 = 400$

84. $\dot{x}_1 = -2x_1 + 4x_2$

$\dot{x}_2 = 2x_1 - x_2 - u(t)$

$\therefore$ $A = \begin{bmatrix} -2 & 4 \\ 2 & -1 \end{bmatrix}$

$B = \begin{bmatrix} 0 \\ -1 \end{bmatrix}$

$$Q_c = [B : AB]$$

$$[AB] = \begin{bmatrix} -4 \\ 1 \end{bmatrix}$$

$$Q_c = \begin{bmatrix} 0 & -4 \\ -1 & -1 \end{bmatrix}$$

Since $|Q_c| \neq 0$; hence the system is controllable

Now the transfer function,

$$T(s) = C[sI - A]^{-1}B$$

$$= [1\ 0]\begin{bmatrix} s-2 & -4 \\ -2 & s+1 \end{bmatrix}\begin{bmatrix} 0 \\ -1 \end{bmatrix}$$

$$= \frac{-4}{s^2 - s + 10}$$

As one pole lies on the R.H.S., the system is unstable.

85. $$\phi(s) = [sI - A]^{-1}$$

$$\Rightarrow \quad [sI - A] = [\phi(s)]^{-1}$$

$$\Rightarrow \quad \begin{bmatrix} s & 0 \\ 0 & s \end{bmatrix} - [A] = \begin{bmatrix} s & -1 \\ 5s & s+6 \end{bmatrix}$$

$$\Rightarrow \quad [A] = \begin{bmatrix} 0 & 1 \\ -5 & -6 \end{bmatrix}$$

89. If $$G(s) = \frac{N(s)}{D(s)}$$

with unity feedback loop characteristic equation is

$$N(s) + D(s) = 0$$

For the system to be stable, the roots of the characteristic equation should neither lie in RHP, nor should have multiple poles on the imaginary axis.

It can be checked that, real parts of root of equations (1), (2) and (4) are negative. To find the root of $s^2(s+2)+2 = 0$ is slightly difficult. Howver one concludes without actually checking that correct answer is (c).

90. $s^3 - 4s^2 + s + 6 = 0$

s^3	1	1
s^2	–4	1
s^1	$+\frac{10}{4}$	6
s^0	6	

Number of right half roots = 2

$\therefore$ Roots on the left half = 3 – 2 = 1

91. $$\sigma_A = \frac{-1-3-4-(-2)}{3-1}$$

$$= -\frac{6}{2}$$

$$= -3$$

93. $$K = \frac{\text{Product of phasor length from so to open loop poles}}{\text{Product of phasor length from so to open-loop zeros}}$$

$$= \frac{PQ \times PQ^1}{PR}$$

$$= \frac{2.6 \times 2.6}{1.4} = 4.8$$

94. Characteristic equation is

$$1 + GH = 0$$

$$\Rightarrow \quad 1 + \frac{K_1}{s(s+1) + K_2 s} = 0$$

$$\Rightarrow \quad s(s+1) + K_2 s + K_1 = 0$$

$$\Rightarrow \quad s^2 + s(1 + K_2) + K_1 = 0$$

$$\therefore \quad \omega_n^2 = K_1 \text{ and } 2\xi\omega_n = 1 + K_2$$

we have, $\omega_d = \omega_n\sqrt{-\xi^2}$

$$\therefore \quad 10 = \sqrt{K_1}\,.\sqrt{1-(0.6)^2}$$

$$\Rightarrow \quad 10 = 0.8\sqrt{K_1}$$

$$\Rightarrow \quad \sqrt{K_1} = \frac{10}{0.8}$$

$$\Rightarrow \quad K_1 = 156.25$$

Also $\quad 2 \times 0.6 \times 12.5 - 1 + K_2$

Hence $\quad K_2 = 14$

96. $$\frac{C(s)}{N(s)} = \frac{-H_2(s)G(s)}{1 + H_2(s)H_1(s)G(s)}$$

Since loop gain $H_2(s)\,H_1(s)G(s)$ very high, hence

$$\frac{C(s)}{N(s)} = \frac{-H_2(s)G(s)}{H_2(s)H_1(s)G(s)}$$

$$\therefore \quad C(s) = -\frac{N(s)}{H_1(s)}$$

97. For a stop input,

$$\text{steady state error} = \frac{1}{1+K_p}$$

where k_p is proportional error constant

$$K_p = \lim_{s\to 0} G(s)H(s)$$

$$= \lim_{s\to 0}\frac{1}{(s+1)(0.1s+1)}$$

$$\frac{1}{1+K_p} = \frac{1}{1+K} = 0.1 = K$$

$$\therefore \quad K = 9.0$$

110. $c(t) = 1 + 1.25 \exp(-6t) \sin(8t - \tan^{-1} 1.333);\ t \geq 0$

Unit step response of a second order linear system with zero initial state is given by

$$c(t) = 1 - \frac{e - \xi\omega_n t}{\sqrt{1-\xi^2}} \sin \left[\omega_n \sqrt{(1-\xi^2)}.t + \tan^{-1}\frac{\sqrt{1-\xi^2}}{\xi}\right]$$

$\therefore \quad \xi\omega_n = 6$

and $\sqrt{1-\xi^2} = \dfrac{1}{1.25}$

$\Rightarrow \quad \xi = 0.6$

$\therefore \quad \omega_n = 10$ rad/sec.

119. Steady state error, $e_{ss} = \dfrac{1}{1+G(0)}$,

thus increasing the gain will decrease steady-state error.

124. $\angle G(j\omega)H(j\omega)|_{\omega=0} = 0°$, hence its a type-0 system.

126. Step response $= \int$ impulse response (dt)

$$= \int_0^t \frac{1}{2} e^{-t/2}\, dt$$

$$= 1 - e^{-t/2}$$

127. Impulse response,

$$f(t) = L^{-1} \frac{1}{(s^2 + a^2)^2}$$

$$= \frac{1}{a} t \sin at$$

130. Gain of the system is always constant at all the frequencies, hence first transfer function will represent Bode plot.

134. $$\phi = \tan^{-1} \frac{0.2\omega - 0.05\omega}{1 + 0.01\omega^2}$$

$$= \tan^{-1}\left[\frac{0.15\omega}{1+0.01\omega^2}\right]$$

For maximum phase difference,

$$\frac{d\phi}{d\omega} = 0$$

which gives $\quad \omega = 10$ rad/sec.

141. Encirclement of the point $(-1 + j0)$ is $N = 2$

We have $\quad N = P - Z$

$\Rightarrow \quad 2 = P - Z$

Since open loop system is stable, hence

$P = 0$

$\therefore \quad Z = 2$

$\Rightarrow$ there are two poles of closed loop function in the right half of s-phane, hence the close-loop system is unstable.

147. $$X(s) = \frac{7s+10}{s(s+2)}$$

$\therefore \quad$ Initial value $= \lim\limits_{s\to 0} s \times (s)$

$$= \lim_{s\to 0} \frac{7s+10}{s(s+2)}$$

$$= 7$$

Final value $= \lim\limits_{s\to 0} s \times (s)$

$$= \lim_{s\to 0} \frac{7s+10}{s(s+2)}$$

$$= 5$$

148. $E(t) = r(t) - c(t)$

$$= 1 - \left[1 - \frac{e^{-\xi\omega_n}}{\sqrt{1-\xi^2}} \sin(\omega dt + \phi)\right]$$

$$E(t) = \frac{e^{-\xi\omega_n}}{\sqrt{1-\xi^2}} \sin(\omega dt + \phi)$$

Comparing this with given equation

$$\frac{1}{\sqrt{1-\xi^2}} = 1.6$$

$\Rightarrow \quad \xi = 0.8$

155. Since open loop transfer function has one pole at the right hand side, hence Nyquist plot will encircle point (–1, 0) once in counter-clockwise direction.

156. $K_p = G(0)$

$$K_p = \lim_{s\to 0} \frac{1}{sG(s)}$$

Hence K_p for type–1 system is infinity. And K_a would be equal to zero.

And steady-state error e_{ss} to unit step input will be zero.

159. Transfer Function, $T(s) = \dfrac{KG}{1+KGH}$

or $\quad C(s) = R(s)\dfrac{KG}{1+KGH}$

Now if $KG >> 1$, then $C(s) \cong \dfrac{R(s)}{H}$

160. Initial value $= \lim\limits_{s\to 0} s.F(s)$

$$= \lim_{s\to 0} \frac{s+2}{s+1} = 2$$

161. $$S_G^T = \frac{\partial T/T}{\partial T/G} = \frac{\partial T}{\partial G} \times \frac{G}{T}$$

we have $\quad T = \dfrac{G}{1+GH}$

$$\therefore \quad S_G^T = \frac{1}{1+GH}$$

$$\therefore \quad S_G^T = \frac{\partial T/T}{\partial H/H} = \frac{-GH}{1+GH}$$

165. Over all transfer function of the system

$$T(s) = \frac{G_1G_2G_3}{1+G_1G_2G_3H_1+G_3H_2-G_3x}$$

Comparing this transfer function from given transfer function, we have

$$X = G_3H_3$$

168. Closed loop transfer function is given by

$$\frac{C(s)}{R(s)} = \frac{\dfrac{K}{s(Js+f)}}{1+\dfrac{K}{s(Js+f)}}$$

$$= \frac{K}{Js^2+fs+K}$$

$$= \frac{K/J}{s^2+\dfrac{fs}{J}+\dfrac{K}{J}}$$

$$\therefore \quad \omega_n = \sqrt{\frac{k}{J}}$$

$$\xi = \frac{f}{2\sqrt{KJ}}$$

175. Take first $[A] = \begin{bmatrix} -1 & 0 \\ 0 & -2 \end{bmatrix}$; $[B] = \begin{bmatrix} 2 \\ 1 \end{bmatrix}$

Now composite matrix,

$$Q_c = [B:AB] = \begin{bmatrix} 2 & -2 \\ 1 & -2 \end{bmatrix}$$

Now $\begin{bmatrix} 2 & -2 \\ 1 & -2 \end{bmatrix} \neq 0$

Hence rank of the system is 2, and the system is completely controllable.

185. $$G(s) = \frac{4}{s^2+5s-6}$$

Since one coefficient in denominator is negative, hence open loop system is unstable

Now characteristic equation for closed loop is

$$1+G(s)H(s) = 0$$

$$\Rightarrow \quad 1+\left(\frac{4}{5^2+5.s-6}\right)\left(\frac{s-1}{s+3}\right) = 0$$

$$\Rightarrow \quad s^3+8s^2+9s-18+4s-4 = 0$$

$$\Rightarrow \quad s^3+8s^2+13s-22 = 0$$

Since one coefficient is negative, hence closed loop system is also unstable.

192. Step response, $C(s) = +0.5\left[\frac{1}{s}+\frac{1}{s+2}\right]$

$$= +0.5\left[\frac{2s+2}{s(s+2)}\right]$$

$$= \left[\frac{s+1}{s(s+2)}\right]$$

Impulse response,

$$C_2(s) = \frac{1}{s+1}$$

$\therefore$ Transfer function of first block,

$$(G_1) = \frac{s+1}{s+2}$$

Transfer function of second blcok,

$$(G_2) = \frac{1}{s+1}$$

$\therefore$ Overall transfer function,

$$G(s) = G_1G_2$$

$$= \frac{s+1}{s+2}\cdot\frac{1}{s+1} = \frac{1}{s+2}$$

194. The state equation can be written as

$$\begin{bmatrix} \dot{x}_1 \\ \dot{x}_2 \end{bmatrix} = \begin{bmatrix} a & b \\ c & d \end{bmatrix}\begin{bmatrix} x_1 \\ x_2 \end{bmatrix} + \begin{bmatrix} e \\ e \end{bmatrix}u(t)$$

Since $x_1 = x_2$, therefore $a+b = c+d$

Controllability matrix is

$$Q_c = [B:AB]\,[B]$$

$$\text{and } AB = \begin{bmatrix} a & b \\ c & d \end{bmatrix}\begin{bmatrix} e \\ e \end{bmatrix}$$

$$= \begin{bmatrix} (a+b) & e \\ (c+d) & e \end{bmatrix}$$

$$\therefore \quad Q_c = \begin{bmatrix} e & (a+b)e \\ e & (c+d)e \end{bmatrix}$$

$\therefore \quad |Q_c| = 0$ since $a+b = c+d$

Hence the system is uncontrollable.

197. $C(t) = K[1-1.66\,e^{-8t}\sin(5t+37°)]$

We have $\dfrac{1}{\sqrt{1-\xi^2}} = 1.66$

$$\therefore \quad \xi\omega_n = 8$$

$$\Rightarrow \quad \omega_n = \frac{8}{0.8} = 10$$

Hence, closed loop transfer function will be

$$G(s) = \frac{k\omega_n^2}{s^2 + 2\xi\omega_n s + \omega_n^1}$$

$$= \frac{100k}{s^2 + 16s + 100}$$

198. Gain margin = inverse of the gain at 180° phase.

$$= \frac{1}{K}$$

204. Corner frequencies are 1 and 10

At $\omega = 1000$, magnitude

$$= -20 \log \frac{1000}{100}$$

$$= -20 \log_{10}^{10}$$

$$= -20 \text{ dB}$$

$$\text{Transfer function} = \frac{K(s+1)}{s(1+0.15)}$$

Hence, K = Frequencies at which low frequency plot cut the 0dB live

K = 10

∴ Transfer functions

$$G(s) = \frac{10(s+1)}{s(1+0.15)}$$

$$= \frac{100(s+1)}{s(s+10)}$$

205. Transfer function of the system,

$$T(s) = \frac{\dfrac{9}{s(s+3)}}{1 + \dfrac{9}{s(s+3)} \cdot \dfrac{s^2}{9}}$$

Hence statements 2, 3 and 4 are correct.

206.
$$G(s)H(s) = \frac{30}{s(s+1)(s+T)}$$

Hence, characteristic equation will be

$s(s+1)(s+T) + 30 = 0$

or $s^3 + s^2(1+T) + sT + 30 = 0$

Now

s^3	1	T
s^2	$1+T$	30
s^1	$T - \dfrac{30}{(1+T)}$	
s^0	30	

Hence for stable system

$$1 + T > 0 \text{ and } T - \frac{30}{1+T} > 0$$

$T > -6$ or $T > 5$

Thus, correct statements is $T > 5$

207. Total number of root locus branches = Number of poles.

But root locus branches terminating at infinity is equal to number of poles minus number of zeros.

And the angle criteria to be satisfied is $(2I + 1)\pi$, where $I = 0, \pm 1, \pm 2$

209.
$$G(s) = \frac{K}{(1+s)(1+4s)}$$

For step-input steady state error,

$$e_{ss} = \lim_{s \to \infty} \frac{1}{1 + G(s)}$$

$$\therefore \quad 0.20 = \lim_{s \to \infty} \frac{1}{1 + \dfrac{K}{(s+1)(1+4s)}}$$

$$\Rightarrow \quad 0.20 = \frac{1}{1+K}$$

$$\Rightarrow \quad K = 4$$

210.
$$c(t) = \frac{1}{2}t - \frac{1}{8}u(t) + \frac{1}{8}e^{-4t}$$

$$\therefore \quad \frac{dc(t)}{dt} = \frac{1}{2} - \frac{1}{8}\delta(t) - \frac{1}{2}e^{-4t}$$

$$\text{Impulse response} = \frac{d^2c(t)}{dt^2} = 0 + 0 + 2e^{-4t}$$

211. Characteristic equation will be

$$1 + \frac{K}{s(s+2)(s+8)} = 0$$

For a system to oscillate, $G(s)H(s) = -1$

$$\Rightarrow \quad \frac{K}{s(s+2)(s+8)} \times 1 = -1$$

$$\Rightarrow \quad s(s^2 + 10s + 16) + K = 0$$

$$\Rightarrow \quad j\omega(-\omega^2 + 10j\omega + 16) + K = 0$$

For systems to oscillate imaginary part should be zero

or $\quad -j\omega^2 + 16j\omega = 0$

or $\quad \omega = 4$ rad/sec.

Then $\quad -10\omega^2 + K = 0$

or $\quad K = 160$

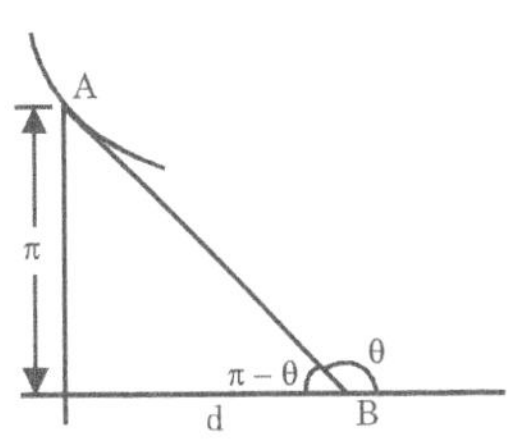

212.

$$g(t) = Ke^{-t/\tau}$$

$$\tan\theta = \left.\frac{dg(t)}{dt}\right|_{t=0}$$

$$= \frac{-K}{\tau}$$

Now, $\tan(\pi-\theta) = -\frac{K}{t}$ ...(*i*)

From the figure, $\tan(\pi-\theta) = -\frac{K}{t}$ ...(*ii*)

From equations (*i*) and (*ii*), we get

$$t = \tau$$

213. Residue associated with $s = 0$

$$K_1 = \lim_{s\to\infty} s\times(s)$$

$$= \lim_{s\to\infty} \frac{1}{(s+1)^3(s+2)}$$

$$= \frac{1}{2}$$

Residue associated with s = –2 is

$$K_2 = \lim_{s\to(-2)} \frac{(s+2)}{s(s+1)^3(s+2)} = \frac{1}{2}$$

214. Nyquist contour of the given transfer function, $G(s) = \frac{1}{s(sT+1)}$ can be drawn as shown below.

From the Nyquist contour gain at phase crossover, frequency = 0

$$\therefore \quad \text{Gain margin} = \frac{1}{0} = \infty$$

216.

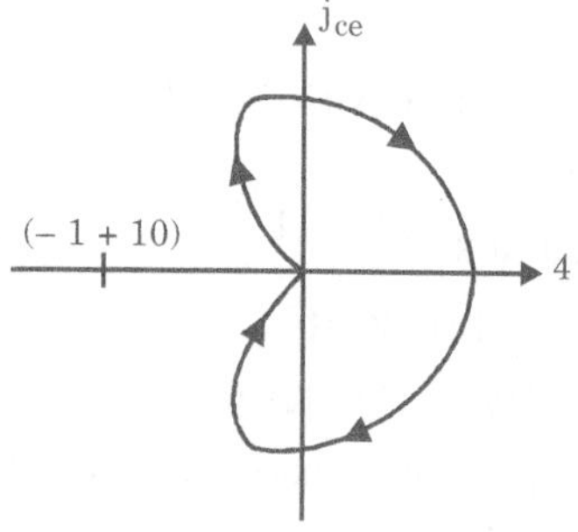

The controllability is in relation to transfer function of a system from one state to another by appropriate input control in a finite time.

From the signal flow graph, we have

$$\dot{x}_1 = -3x_1 + x_2$$

$$\dot{x}_2 = -2x_2 + u$$

and output $C_2 = x_1 + x_2$

$$= [1\ 1]\begin{bmatrix} x_1 \\ x_2 \end{bmatrix}$$

$$\therefore \quad [A] = \begin{bmatrix} -3 & 1 \\ 0 & -2 \end{bmatrix}, [B] = \begin{bmatrix} 0 \\ 1 \end{bmatrix}$$

$$[A][B] = \begin{bmatrix} 1 \\ -2 \end{bmatrix}$$

$$\therefore \quad [B:AB] = \begin{bmatrix} 0 & 1 \\ 1 & -2 \end{bmatrix} = 1$$

Hence system is controllable.

$$C^T = \begin{bmatrix} 1 \\ 1 \end{bmatrix}$$

$$A^T = \begin{bmatrix} -3 & 0 \\ 1 & -2 \end{bmatrix}$$

$$A^TC^T = \begin{bmatrix} -3 \\ -1 \end{bmatrix}$$

$$\therefore \quad [C^T : A^TC^T] = \begin{bmatrix} 1 & -3 \\ 1 & -1 \end{bmatrix} \neq 0$$

Hence system is observable also.

217.

$$\text{Angle, } \theta = \tan^{-1}\frac{1}{1}$$

$$= 45°$$

∴ Damping factor,

$$\xi = \cos 45 = 0.707$$

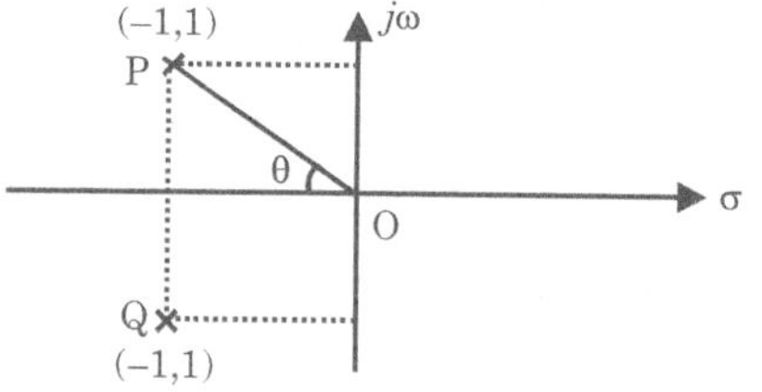

Time for peak overshoot,

$$t_p = \frac{\pi}{\omega_n\sqrt{1-\xi^2}}$$

$$\omega_n = |OP| = \sqrt{1^2+1^2} = \sqrt{2} \text{ rad/sec}$$

$$\therefore \quad t_p = \frac{3.14}{\sqrt{2}\sqrt{1-(0.707)^2}} = 3.14 \text{ seconds.}$$

222.

$$C(s) = \frac{2}{5}.R(s)$$

$$= \frac{2}{5}.\frac{1}{s} = \frac{2}{s^2}$$

$$\therefore \quad c(t) = L^{-1}[C(s)]$$

$$= L^{-1}\left[\frac{2}{s^2}\right] = 1 + 2t \text{ for } t \geq 0$$

223. $G(s)H(s) = \frac{20}{s^2}.(s+5)$

$\therefore$ Steady state error $= \lim_{s\to\infty} \frac{1}{1+G(s)H(s)}$

$= 0$

224. Asymptote's angle,

$$\phi_A = \frac{(2q+1)180}{2-1}$$

For $q = 0 = 180$, we have

$$1 + \frac{K(s+2)}{(s+1)^2} = 0$$

$$K = \frac{-(s+1)^2}{(s+1)} = -\left[\frac{s^2+1+25}{s+2}\right]$$

$$\therefore \quad \frac{dK}{ds} = -\left[\frac{(s+2)(2s+2)-(s^2-2s+1)(1)}{(s+2)^2}\right]$$

$$= -\left[\frac{2s^2+2s+4s+4-s^2-2s-1}{(s+2)^2}\right]$$

$$= -\left[\frac{s^2+4s+3}{(s+2)^2}\right] = -\left[\frac{(s+1)(s+3)}{(s+2)}\right]$$

$\frac{dK}{ds} = 0$ gives s = 1, – 3

$\therefore$ Break away points are (–1, 0) and (–3, 0)

Hence correct plot is as shows in (c).

228. $500 = 1000e^{-t/T}$

$\therefore \quad 0.5 = e^{-t/T}$

Take log of each side,

$$-0.6931 = \frac{-t}{T}$$

$$\Rightarrow \quad T = \frac{1.386}{0.6931} = 2.0$$

230. The transfer function G(*s*) should be to give zero steady-state error for a ramp input, hence type of the G(*s*) should be 2 or G(*s*) should have a term like s^2 as one of the factor in its denominator polynomial. Therefore only an integral type of controller can make overall G(*s*)

of tyep i.e., $\frac{K_1}{s^2(s+a)}$

231. $T(s) = \frac{K}{s^2+4s+K}$

$\omega_n = \sqrt{K}$

Now $2\xi\omega_n = 4$

For critically damped circuit,

$\xi = 1$

$\Rightarrow \quad \omega_n = 2$

$\Rightarrow \quad \sqrt{K} = 2$

$\Rightarrow \quad K = 4$

233. $e_{ss} = \lim_{s\to 0} \frac{1}{1+G(0)}$

$= \frac{1}{1+10}$

$= \frac{1}{11}$

234. Step response $= \int_0^t 5e^{-10t}\,dt$

$= 0.5(1-e^{-10t})$

EXERCISE – II

4. $T.F. = \frac{1+.5s}{1+s}$

Its pole will be given by $s = -1$

and zero is $s = \frac{1}{0.5} = 2$

Hence zero is greater than pole hence its a lag network.

8. $G = \frac{K}{s^2+(1+sT)}$

Characteristic equation is,

$1 + G(s) = 0$

$\Rightarrow \quad 1 + \frac{K}{s^2+(1+sT)} = 0$

$\Rightarrow \quad Ts^3 + s^2 + K = 0$

Constructing Routh array

s^3	T	0
s^2	1	K
s^1	0	
s^0	K	

Since first element in each row is positive and one element is zero; hence the system is unstable.

9. Gain = 20 log 10

= 20 dB

10. Number of roots on the positive half of the s-phane is equal to the number of sign changes.

First sign change is from 2 to –4 and second sign change is from –4 to 6.

17. Closed loop transfer function have poles which are given by the roots of its characteristics equation

$$1 + \frac{1}{(s+2)^2} = 0$$

$$\Rightarrow \quad (s+2)^2 + 1 = 0$$

$$\Rightarrow \quad s^2+4+4s+1=0$$
$$\Rightarrow \quad s^2+4s+5=0$$
$$\Rightarrow \quad s=-2\pm j$$

19. Angle of asymptotes will be given by

$$\phi_A=\frac{(2q+1)180}{n-m}=\frac{(2q+1)180}{3-0}=\frac{(2q+1)180}{3}$$

For $q=0, 1, 2$ $\quad \phi_A=60, 180°$ and $300°$

20. Characteristic equation is

$$s^4+3s^3+5s^2+6s+K+10=r$$

Routh array is

$$\begin{bmatrix} s^4 & 1 & 5 & K+10 \\ s^3 & 3 & 6 & \\ s^2 & 3 & K+10 & \\ s^1 & K+4 & & \\ s^0 & K+10 & & \end{bmatrix}$$

To have stable system

$$K+4>0$$
$$\Rightarrow \quad K>-4$$
$$\text{and} K+10> \quad 0$$
$$\Rightarrow \quad K>-10$$
$$\text{Hence} \quad K>-4$$

24. Phase margin $=180°+\phi$

$$=180°+(-270°)$$
$$=-90°$$

For a stable system phase margin should be positive. Hence the system is unstable.

27. Phase margin $=180+\phi=180-150=30°$

28. Put $z=\dfrac{1+\omega}{1-\omega}$

the characteristic equation is

$$\left(\frac{1+\omega}{1-\omega}\right)^3-0.81\left(\frac{1+\omega}{1-\omega}\right)=0$$

$$\Rightarrow \quad \omega^2+3\omega^2+3\omega+1-0.81-0.81\omega=0$$

$$\omega^3+3\omega^2+2.19\omega+0.19=0$$

ω^3	1	2.19
ω^2	3	0.19
ω^1	1.396	
ω^0	0.19	

Since all the coefficient of the first is positive, hence the system is stable.

29. System transfer function is

$$G(s)=\frac{1}{s(1+6s)}$$

If has a pole at imaginary axis, hence the system is marginally stable.

30. $$F(s)=\frac{K(s+3)}{1+s-s^2+s^3}$$

Characteristic equation is

$$1+s-s^2+s^3=0$$

It has one negative coefficient, hence the system is unstable.

31. $$G(s)=\frac{K}{s(s+1)(s+2)}$$

Its characteristic equations is,

$$1+\frac{K}{s(s+1)(s+2)}=0$$

$$K=-s(s+1)(s+2)$$
$$=-s(s^2+3s+2)$$
$$=-[s^3+3s^2+2s]$$

Now $$\frac{dk}{ds}=0$$

$$\Rightarrow \quad 3s^3+6s+2=0$$

$\therefore$ -1.577 can not be breakaway point, the breakaway point is -0.423

32. $$G(s)H(s)=\frac{1}{(s+1)^3}=\frac{1}{(s+j\omega)^3}$$

Angle $\quad -3\tan^{-1}\omega_1=-180$

or $\quad \tan^{-1}\omega_1=+60$

$$\omega_1=1.732$$

$$|G(s)H(s)| \text{at } \omega_1=\frac{1}{8}$$

$\therefore$ Gain margin $=8$

35. 100% overshoot is only possible when damping factor is zero.

36. All the branches of root locus originate from poles

43. $[s-(-3+j4)][s-(-3-j4)]$

$$=(s+3-j4)(s+3+j4)$$
$$=(s+3)^2-(-j4)^2$$
$$=s^2+9+6s+16$$
$$=s^2+6s+25$$

$\therefore \quad \omega_n^2=25$

$\Rightarrow \quad \omega_n=5$

and $\quad 2\xi\omega_n=6$

$\Rightarrow \quad \xi=\dfrac{6}{2\times 5}=0.6$

45. $$G(s) = \frac{9}{s(s+3)}$$

Characteristic equation is

$$1 + \frac{9}{s(s+3)} = 0$$

$\Rightarrow$ $s^2 + 3s + 9 = 0$

$\therefore$ $\omega_n^2 = 9$

$\Rightarrow$ $\omega_n = 3$

and $2\xi\omega_n = 3$

$\Rightarrow$ $\xi = 0.5$

50. $$\frac{d^2x}{dt^2} + 5\frac{dx}{dt} + 7x = 7y$$

Taking Laplace transform

$$s^2x(s) + 5.s\,x(s) + 7x(s) = 7y(s)$$

$\therefore$ $$\frac{y(s)}{x(s)} = \frac{7}{s^2 + 5s + 7}$$

$\therefore$ $\omega_n = \sqrt{7}$

and $2\xi\omega_n = 5$

$\therefore$ $\xi = 0.94$

53. Characteristic equation of the system is

$$1 + \frac{1 + T_i s}{s} \cdot \frac{1}{s(1 + sT)} = 0$$

or $Ts^3 + s^2 + T_i s + 1 = 0$

Constructing Routh array

s^3	T	T_i
s^2	1	1
s^1	$T_i - T$	
s^0	1	

For stable system all the coefficient in the first column should be positive

56. $$\frac{C}{R} = \left(\frac{G_1}{1+G_1}\right)\left(\frac{G_2}{1+G_2}\right)\left(\frac{G_3}{1+G_3}\right)\left(\frac{G_4}{1+G_4}\right)$$

$$= \frac{G_1G_2G_3G_4}{(1 + G_1 + G_2 + G_1G_2)(1 + G_3 + G_4 + G_3G_4)}$$

57. $$L[r(t-a)] = \frac{e^{-as}}{s^2}$$

62. Gain margin $= -10 + 5$

$= -5$ dB

We have $-5 = 20 \log(\text{G.M})$

$\therefore$ G.M. $= 0.5623$

Since gain margin is positive, hence the system is stable.

64. $$G(s) = \frac{4(1+2s)}{s^2(s+2)}$$

$\therefore$ Steady state error

$$= \lim_{s\to 0} \frac{1}{sG(s)}$$

$$= \lim_{s\to 0} \frac{1}{s.\dfrac{4(1+2s)}{s^2(s+2)}}$$

$$= \lim_{s\to 0} \frac{s(s+2)}{4(1+2s)} = 0$$

66. $s^3 + 5s^2 + (K+6)s + K = 0$

$$\Rightarrow \quad 1 + \frac{K(s+1)}{s^3 + 5s^2 + 6s} = 0$$

$$\Rightarrow \quad 1 + \frac{K(s+1)}{s(s^2 + 5s + 6s)} = 0$$

$$\Rightarrow \quad 1 + \frac{K(s+1)}{s(s+2)(s+3)} = 0$$

$$\Rightarrow \quad \sigma_A = \frac{(-2) + (-3) + (-1)}{3 - 1}$$

$$= \frac{-4}{2} = -2$$

Hence, point is $(-2, 0)$

68. $$20 = 20 \log_{10}\left(\frac{1}{a}\right)$$

$$\Rightarrow \quad \left(\frac{1}{a}\right) = (10)^1$$

$a = 0.1$

71. $$G(s) = \frac{25}{s^2 + 8s + 25}$$

$\therefore$ $\omega_n^2 = 25$

$\Rightarrow$ $\omega_n = 5$

and $2\xi\omega_n = 8$

$$\Rightarrow \quad \xi = \frac{8}{2 \times 5} = 0.8$$

$$\therefore \quad t_p = \frac{\pi}{\omega_n\sqrt{1 - (0.8)^2}} = \frac{\pi}{5\sqrt{0.36}}$$

$$= \frac{\pi}{5 \times 0.6} = \frac{\pi}{3}$$

Hence, time for second peak $= 3t_p = \pi$

75. System poles are at $s = 0, -1$

and zeros are at $s = -2$

$$\therefore \text{ system transfer function } = \frac{K(s+2)}{s(s+1)}$$

80. Corner frequency are 4 and 10.

Value of gain "K" can be calculated as gain at

$$\omega = 1 = -20 \log \frac{4}{1}$$

$$= -12 \text{ dB}$$

$\therefore$ $-12 = 20 \log K$

$\Rightarrow$ $K = 0.2512$

Now frequency slope is 20 dB/decade, hence term $(j\omega)$ will be present.

83. Response for any input $u(t)$ is,

$$y(t) = \int_0^t h(t-z)u(\tau)d\tau$$

where impulse response,

$$u(t) = \frac{d}{dt}[s(t)]$$

where $s(t)$ is the response.

Hence $$y(t) = \frac{d}{dt}\int_0^t s(t-\tau)u(\tau)d\tau$$

84. Transfer function, $G(s) = \dfrac{1}{s+2}$

Laplace transform of output $-\dfrac{1}{(s+2)^2}$

Taking inverse Laplace transform,

Input $= e^{-2t}u(t)$

85.

$$G(s) = \frac{s+2}{s+1}$$

$$G(f\omega) = \frac{2+j\omega}{1+j\omega}$$

This is the transfer function of lag network. Hence the output will lag the input.

93.

$$\frac{C(s)}{(s)} = \frac{\dfrac{K}{s(s+2)}}{1+\dfrac{K}{s(s+2)}(1+as)}$$

$$= \frac{K}{s(s+2)+K(1+as)}$$

$$= \frac{K}{s^2+s(2+Ka)+K}$$

$$K = \omega_n^2 = 16$$

$$2\xi\omega_n = 2 + Ka$$

$$= 2 \times 0.7 \times 4 = 5.6 \quad \Rightarrow = 2 + Ka$$

$\therefore$ $$a = \frac{5.6-2}{16} 0.225$$

101. Step response $= \int$ impulse response

$$= \int(-4e^{-t} + 6e^{-2t})dt$$

$$= (-3e^{-2t} + 4e^{-t}) - 1$$

102. Impulse Response

$$= \frac{d}{dt}(\text{step response})$$

$$= \frac{d}{dt}(1 - 5e^{-t} + 10e^{-2t} - 6e^{-3t})$$

$$= 5e^{-t} + 20e^{-2t} + 18e^{-3t}$$

104.

$$G(s) = \frac{K}{s(s+4)}$$

Characteristic equation is

$$1 + \frac{K}{s(s+4)} = 0$$

$\Rightarrow$ $s^2 + 4 + K = 0$

$\therefore$ $\omega_n = \sqrt{K}$

Also $2\xi\omega_n = 4$

$\Rightarrow$ $\xi = \dfrac{4}{\sqrt{k}} = 0.5$

$\Rightarrow$ $\sqrt{K} = 4$

$\Rightarrow$ $K = 16$

105.

$$\phi_A = \frac{(2q+1)}{m-n} 180°$$

$$= \frac{180}{5-2}(2q+1)$$

For $q = 0, 1, 2,\ \phi_A = 60°, 180°, 300°$

108. Phase angle, $\phi = -2 \tan^{-1} \omega t$

At corner frequency, $\omega = 1/T$

$\therefore$ $\phi = 2 \tan^{-1} \dfrac{t}{t}$

$\Rightarrow$ $\phi = -90°$

111.

$$\omega_n = \sqrt{3}$$

$$= 1.73 \text{ rad/sec.}$$

116. Maximum over shoot,

$$M_P = e^{-\frac{\pi\xi}{\sqrt{1-\xi^2}}}$$

Hence for maximum over shoot damping factor "ξ" should be minimum

We have in all the transfer function

$$2\xi\omega_n = 2$$

$\therefore$ $$\xi = \frac{1}{\omega_n}$$

Hence ω_n should be maximum and maximum value of $\omega_n = \sqrt{36} = 6$

124. $\omega_n = \sqrt{\omega_1\omega_2} = \sqrt{\frac{1}{0.04}\times\frac{1}{0.01}}$

$= 50$ rad/sec

132. $\sigma_A = \frac{0-1-2-3-0}{4-0}$

$= -\frac{6}{4}$

$= -\frac{3}{2}$

Number of asymptote will be 4, and number of separate part will be 4.

Hence 1st statement will be correct only.

143. $G(s) = \frac{10(1+4s)}{s^2(1+s)}$

$\therefore \quad E(s) = \frac{(1+s)(s^2+s+1)}{s^2[s^2(1+s)+10(1+4s)]}$

$\therefore$ Steady state error,

$e_{ss} = \lim_{s\to 0} sE(s)$

$= \infty$

144. $\frac{C(s)}{R(s)} = \frac{1}{1+s}$

$r(t) = \sin t$

$\therefore \quad \omega = 1$ rad/sec

$\therefore$ phase angle of $\frac{C(s)}{R(s)} = \tan^{-1}\omega$

$= -\tan^{-1}(1)$

$= -45$

$\therefore$ and at $\omega = 1$,

magnitude $= \frac{1}{\sqrt{2}}$

$\therefore$ Steady state value of $c(t)$

$= \frac{1}{\sqrt{2}}\sin\left(t-\frac{\pi}{4}\right)$

146. $\frac{Y(s)}{U(s)} = \frac{1}{s^4+5s^3+7s^2+6s+3}$

or $\frac{d^4}{dt^4}y(t)+5\frac{d^3y(t)}{dt^3}+7\frac{d^2}{dt^2}y(t)$

$+ 6\frac{d}{dt}y(t)+3y(t) = u(t)$

Let $x_1(t) = y(t)$

$x_2(t) = \frac{dy(t)}{dt}$

$x_3(t) = \frac{d^2}{dt^2}y(t)$

$x_4(t) = \frac{d^3}{dt^3}y(t)$

Now $\frac{d^4}{dt^4}y(t) = -3y(t)-6\frac{d}{dt}y(t)-7\frac{d^2}{dt^2}y(t)$

$-5\frac{d^3}{dt^3}y(t)+u(t)$

$= -3x_1(t)-6x_2(t)-7x_3(t)$

$-5x_4(t)+u(t)$

$$\begin{bmatrix}\dot{x}_1\\ \dot{x}_2\\ \dot{x}_3\\ \dot{x}_4\end{bmatrix} = \begin{bmatrix}0 & 1 & 0 & 0\\ 0 & 0 & 1 & 0\\ 0 & 0 & 0 & 1\\ -3 & -6 & -7 & -5\end{bmatrix}\begin{bmatrix}x_1\\ x_2\\ x_3\\ x_4\end{bmatrix} + \begin{bmatrix}0\\ 0\\ 0\\ 1\end{bmatrix}u$$

147. $\frac{d^2y}{dt^2}+12\frac{dy}{dt}+100y = 0$

$s^2y(s)+12.s.y(s)+100y(s) = 0$

$\therefore \quad \omega_n^2 = 100$

$\Rightarrow \quad \omega_n = 10$

and $\quad \xi\omega_n = 10$

$\Rightarrow \quad \xi = \frac{12}{20} = 0.6$

$\therefore$ Damped natural frequency,

$\omega_d = \omega_n\sqrt{1-\xi^2}$

$= 10\sqrt{1-(0.6)^2}$

$= 8$

149. $2\xi\omega_n = 1$

and $\quad \omega_n = 1$

$\therefore \quad \xi = 0.5$

$\therefore \quad M_P = e^{\frac{-\pi\xi}{\sqrt{1-\xi^2}}}$

$= 0.163$

154. $\frac{d^2y}{dt^2}+2\frac{dy}{dt}+3y = 0$

Let $x_1 = y$;

$x_2 = \frac{dy}{dt} = \dot{x}_1$

$\therefore \quad \frac{d^2y}{dt^2} = -3y-2\frac{dy}{dt}$

$= -3x_1-2x_2$

$$\begin{bmatrix}\dot{x}_1\\ \dot{x}_2\end{bmatrix} = \begin{bmatrix}0 & 1\\ -3 & -2\end{bmatrix}\begin{bmatrix}x_1\\ x_2\end{bmatrix}$$

156. $$i(t) = \frac{1}{4}(1 - e^{-2t})u(t)$$

Taking Laplace transfer

$$I(s) = \frac{1}{4}\left[\frac{1}{s} - \frac{1}{s+2}\right]$$

$$= \frac{1}{4}\left[\frac{2}{s(s+2)}\right]$$

$$= \frac{1}{2s(s+2)}$$

Hence complex frequencies, $s = 0, s - 2$

162. Characteristic equation of the system is

$$s^2 + 6s + 9 = 0$$

or $$(s+3)^2 = 0$$

$\therefore$ $$s = -3$$

Hence solution will be

$$y = (k_1 + k_2)e^{-3t}$$

163. $$C(t) = -te^{-t} + 2e^{-t}$$

$$C(s) = -\frac{1}{(s+1)^2} + \frac{2}{(s+1)}$$

$$= \frac{-1+2s+2}{(s+1)^2} = \frac{2s+1}{(s+1)^2}$$

Now let open loop gain be G.

$\therefore$ $$\frac{G}{1+G} = \frac{2s+1}{(s+1)^2}$$

$\Rightarrow$ $$(s+1)^2 G = (2s+1)(G+1)$$

$\Rightarrow$ $$G[s^2 + 2s + 1 - 2s - 1] = (2s+1)$$

$$G = \frac{2s+1}{s^2}$$

165. $$F(s) = \frac{(s+1)}{s(s+K)}$$

$$\lim_{s\to\infty} f(t) = \lim_{s\to 0} sF(s) = \lim_{s\to 0} \frac{s+1}{s+K} = \frac{1}{K} = \frac{1}{2}$$

$\therefore$ $$K = 2$$

175. $$|AI - A| = \begin{bmatrix} \lambda - a & -1 \\ -a & \lambda - 1 \end{bmatrix}$$

$$= (\lambda - a) - (\lambda - 1) - a$$

$$= \lambda^2 - a\lambda - \lambda + a - a$$

Put = $\lambda^2 - a\lambda - \lambda = 0$

$\therefore$ $$\lambda = 0, a+1$$

178. Denominatior of the transfer function

$$= |SI - A|$$

$$= \begin{bmatrix} s-1 & 0 \\ -1 & s-1 \end{bmatrix}$$

$$= (s-1)^2$$

Hence the correct answer is (*d*).

179. Let P = wattage of heater

q = water flow rate (kg/s)

Mean temperature of the walls

$$T_m = \frac{T_1 + T_2}{2}$$

$\therefore$ Heat balance equation is

$$\frac{P}{4.2} = q \times 1000 \times (T2 - T1) + sk(T_m - T_A)$$

Hence, under steady state, the system can be modelled by an algebraic equation.

189. $$\text{Final value} = \lim_{s\to 0} s.I.(s)$$

$$= \lim_{s\to 0} \frac{10.s+4}{(s+1)(s^2+4s+5)}$$

$$= \frac{4}{5}$$

193. $$F(s) = \frac{\omega}{s^2 + \omega^2}$$

Final value of this function cannot be determined because it is laplace transform of sinusoidal function, which always oscillating.

■■

3

CHAPTER

Network Theory

EXERCISE – I

1. In the circuit shown in the given figure, the switch s is closed at $t = 0$. The induced voltage v_2 will have a maximum value of

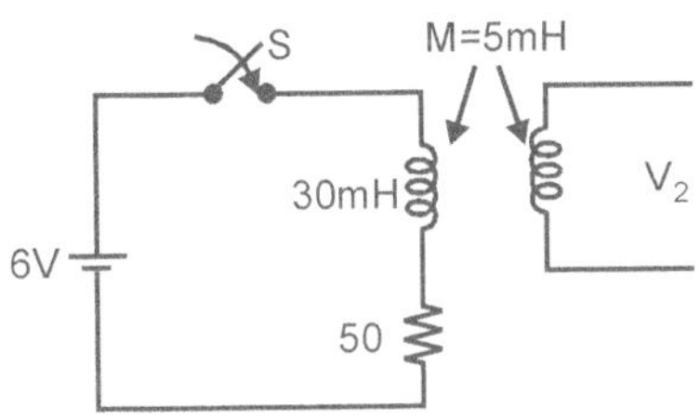

(*a*) 0.6 V (*b*) 1 V

(*c*) 3.78 V (*d*) 6V

2. A coil having a resistance of 5 ohms and inductance of 0.1 H connected in series with a condenser of capacitance 50 μF. A constant alternating voltage of 200 volts is applied to the circuit. The voltage across the coil at resonance is

(*a*) 200 volts

(*b*) 1788 volts

(*c*) 1800 volts

(*d*) 2000 volts

3. A series RLC circuit, consisting of R = 10 ohms, $X_L = 20$ ohms and $X_C = 20$ ohms is connected across an ac supply of 100V(rms). The magnitude and phase angle (with reference to supply voltage) of the voltage across the inductive coil are respectively

(*a*) 100V; 90° (*b*) 100V; –90°

(*c*) 200V; –90° (*d*) 200V; 90°

4. Two coupled coils connected in series have an equivalent inductance of 16 mH or 8 mH depending on the inter-connection. Then the mutual inductance M between the coils is

(*a*) 12 mH (*b*) $8\sqrt{2}$ mH

(*c*) 4 mH (*d*) 2 mH

5. A unit step voltage is applied at $t = 0$ to a series RL circuit with zero initial conditions.

(*a*) It is possible for the current to be oscillatory

(*b*) The voltage across the resistor at $t = 0^+$ is zero

(*c*) The energy stored in the inductor in the steady state is zero

(*d*) The resistor current eventually falls to zero

6. A two-tank system is shown in figure I. The electrical equivalent of this system is

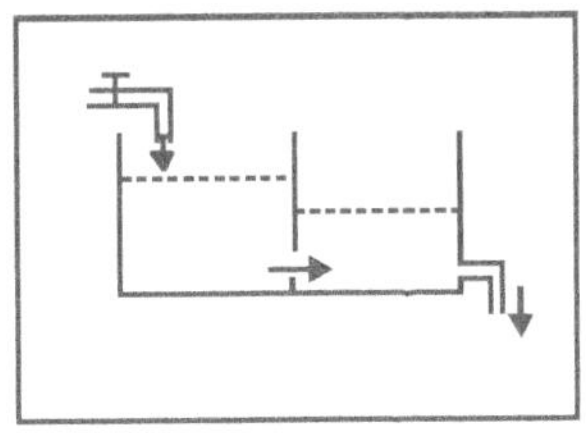

(*a*)

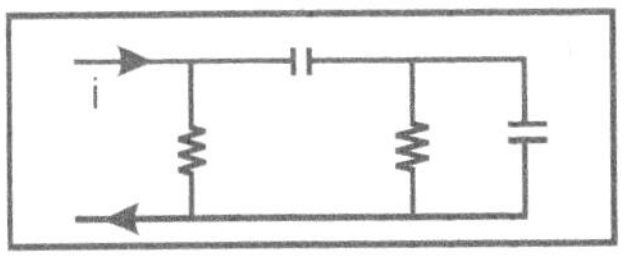

(*b*)

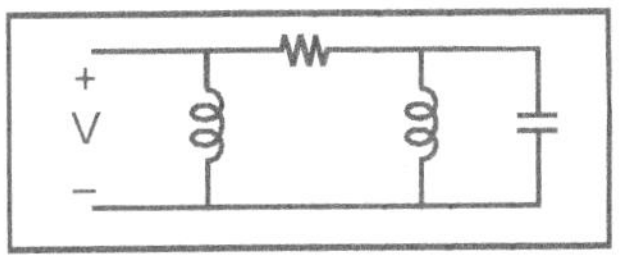

(*c*)

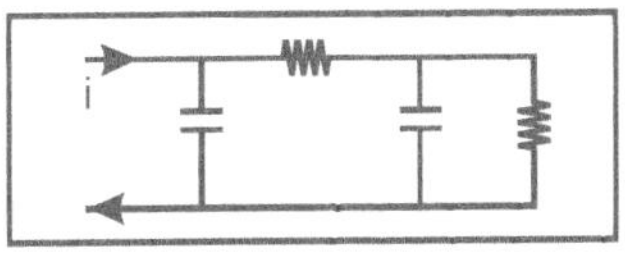

(*d*)

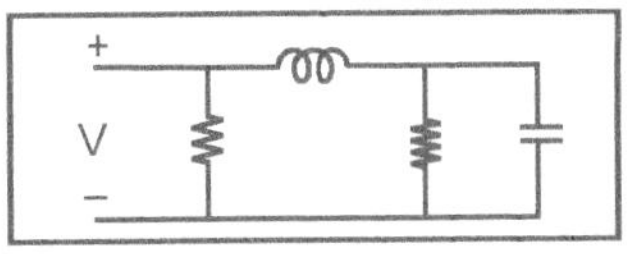

7. The unit impulse response of a system is given by $c(t) = 0.5e^{-t/2}$. Its transfer function is

(*a*) $1/(s+2)$

(*b*) $1/(1+2s)$

(*c*) $2/(1+2s)$

(*d*) $2/(s+2)$

8. In an LCR circuit, supplied from an ac source, the reactive power is proportional to the

(*a*) average energy stored in the electric field

(*b*) average energy stored in the magnetic field

(*c*) sum of the average energy stored in the electric field and that stored in the magnetic field

(*d*) difference between the average energy stored in the electric field and that stored in the magnetic field

9. Consider the following data:

1. Input applied for $t < t_o$
2. Input applied for $\geq t_o$
3. State of the network at $t = t_o$
4. State of the network at $t < t_o$

Among these, those needed for determining the response of a linear network for $t > t_o$ would include

(*a*) 1, 3 and 4 (*b*) 2, 3 and 4
(*c*) 2 and 3 (*d*) 2 and 4

10. The voltage across the 1KΩ resistor between the nodes A and B of the network shown in the given figure is

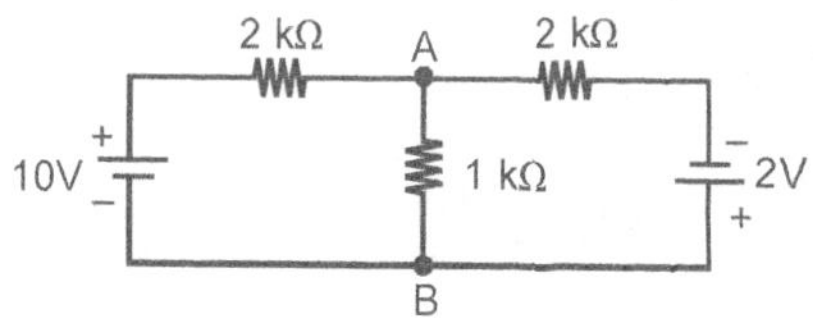

(*a*) 2V (*b*) 3V
(*c*) 4V (*d*) 8V

11. The open-circuit voltage ratio $\frac{V_2(s)}{V_1(s)}$ of the network shown in the given figure is

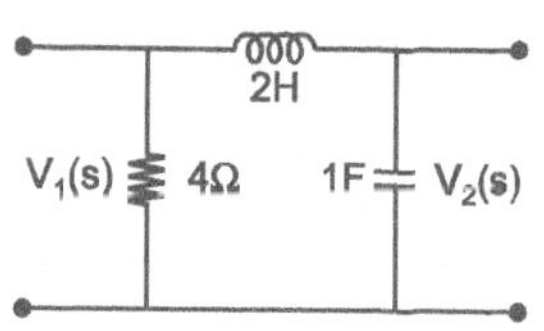

(*a*) $1 + 2s^2$ (*b*) $\frac{1}{1+2s^2}$

(*c*) $1 + 2s$ (*d*) $\frac{1}{1+2s}$

12. Initially, the circuit shown in the given figure was relaxed. If the switch is closed at $t = 0$, the values of $i(0^+) \frac{di}{dt}(0^+)$ and $\frac{d^2i}{dt^2}(0^+)$ will respectively be

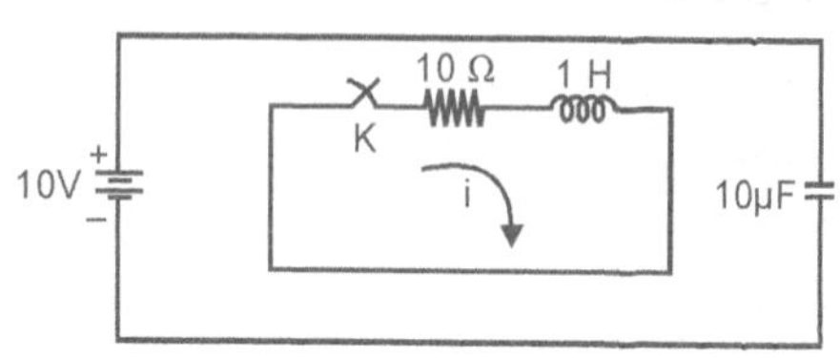

(*a*) 0, 10 and –100 (*b*) 0, 10 and 100
(*c*) 10, 100 and 0 (*d*) 100, 0 and 10

13. A rectangular voltage pulse of magnitude V and duration T is applied to a series combination of resistance R and capacitance C. The maximum voltage developed across the capacitor is

(*a*) V[1 – exp (–T/RC)] (*b*) VT/RC
(*c*) V (*d*) V exp (–T/RC)

14. Which one of the following is a realistic representation of an equivalent circuit of a containing a lossy dielectric?

(*a*) 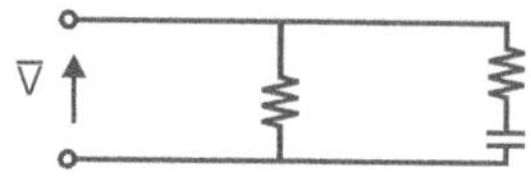

(*b*) 

(*c*) 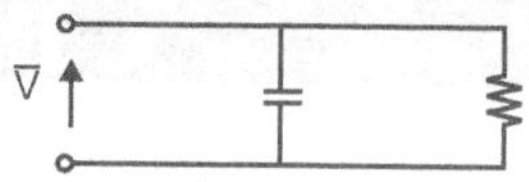

(*d*) 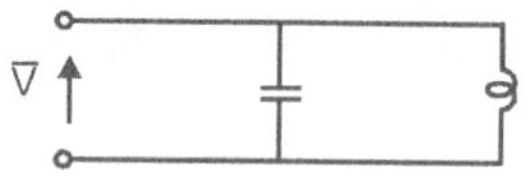

15. Laplace transform of a function $f(t)\,u(t)$ where $f(t)$ is periodic with period T is A(s) times theL.T. of its first period, then

(*a*) $A(s) = s$ (*b*) $A(s) = \frac{1}{1-e^{-Ts}}$

(*c*) $A(s) = \frac{1}{1+e^{-Ts}}$ (*d*) $A(s) = e^{Ts}$

16. Consider the following statements

1. The two-port network shown below does NOT have an impedance matrix representation.

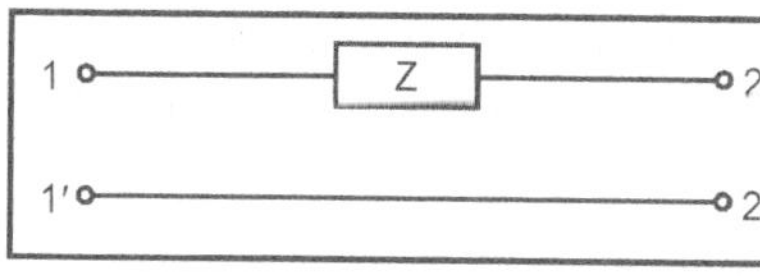

2. The following two-port network does NOT have an admittance matrix representation

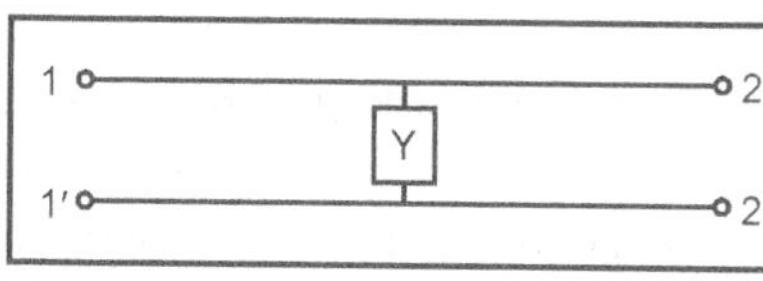

3. A two-port network is said to be reciprocal if it satisfies $z_{12} = z_{21}$ or an equivalent relationship.

Of these statements

(*a*) 1 and 2 are correct (*b*) 1, 2 and 3 are correct
(*c*) 1 and 3 are correct (*d*) none is correct

17. In the circuit shown in figure below, X is an element which always absorbs power. During a particular operation, it sets up a current of 1 amp in the direction shown and absorbs a power P_x. It is possible that X can absorb the same power P_x for another current i, the value of this currents is

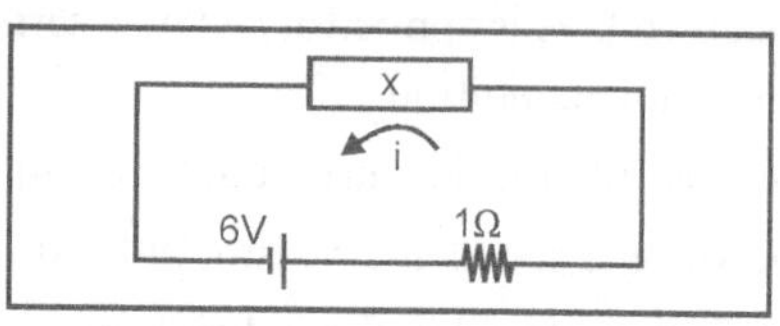

(*a*) $(3 - \sqrt{14})$ amps (*b*) $(3 + \sqrt{14})$ amps
(*c*) 5 amps (*d*) none of these

18. An initially relaxed RC- series network with $R = 2\ M\Omega$, and $C = 1\ \mu F$ is switched on to a 10V step input. The voltage across the capacitor after 2 seconds will be

(*a*) zero (*b*) 3.68 V

(*c*) 6.32 V (*d*) 10 V

19. In the circuit shown in the given figure, with voltameter indicates 30K. The reading of the ammeter will be

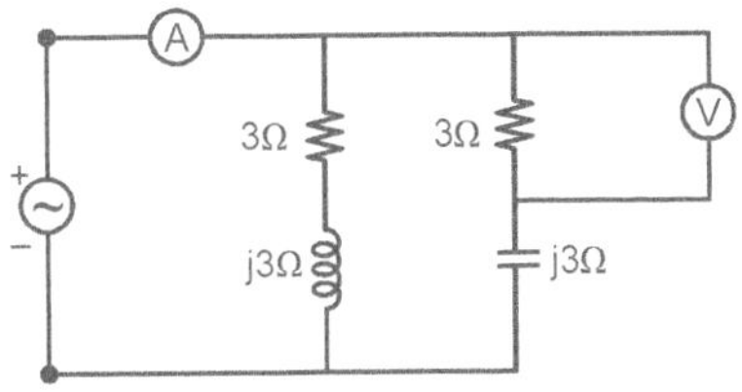

(*a*) 20 A (*b*) $10\sqrt{2}$ A

(*c*) 10 A (*d*) zero

20. Kerr cell is used

(*a*) to produce chemical energy using radiant energy

(*b*) as a master switch operated by finger pressure

(*c*) to modulate a beam of light

(*d*) for interaction of heat and magnetism

21. If an R-L circuit having impedance angle ϕ is switched in when the applied sinusoidal voltage wave is passing through an angle θ, there will be no switching transient if

(*a*) $\theta - \phi = 0°$

(*b*) $\theta + \phi = 0°$

(*c*) $\theta - \phi = 90°$

(*d*) $\theta + \phi = 90°$

22. The pole-zero configuration of an impedance function is given the figure. The network is

(*a*) R-L realizable

(*b*) R-C realizable

(*c*) L-C realizable

(*d*) None of these

23. At a certain current, the energy stored in an ironcored coil is 1000J and its copper loss is 2000W. The time constant (in seconds) of the coil is

(*a*) 0.25

(*b*) 0.5

(*c*) 1.0

(*d*) 2.0

24. The response of an initially relaxed linear circuit to a signal vs is $e^{-2t}\,u(t)$. If the signal is changed to $\left(V_s + 2\frac{dv_s}{dt}\right)$ the response would be

(*a*) $-4e^{-2t}\,u(t)$

(*b*) $-3e^{-2t}\,u(t)$

(*c*) $4e^{-2t}\,u(t)$

(*d*) $5e^{-2t}\,u(t)$

25. Consider the following statements:

The transfer impedances and admittances of a network remain constant when the position of excitation and response are interchanged if the network

1. is linear
2. consists of bilateral elements
3. has high impedance or admittance as the case may be
4. is resonant

Of these statements

(*a*) 1 and 2 are correct

(*b*) 1, 3 and 4 are correct

(*c*) 2 and 4 are correct

(*d*) 1, 2, 3, and 4 are correct

26. A unit impulse input to a linear network has a response R(t) and a unit step input to the same network has response S(t). The response R(t)

(*a*) equals $\frac{dS}{dt}(t)$

(*c*) equals the integral of S(t)

(*c*) is the reciprocal of S(t)

(*d*) has no relation with S(t)[ES]

27. A 3-phase, 3-wire supply feeds a load consisting of three equal resistors connected star. If one of the resistors is open circuited, the percentage reduction in the load will be

(*a*) 75 (*b*) 50

(*c*) 66.66 (*d*) 3.33

28. The value of the current I flowing in the 1 ohm resistor in the circuit shown in the given figure will be

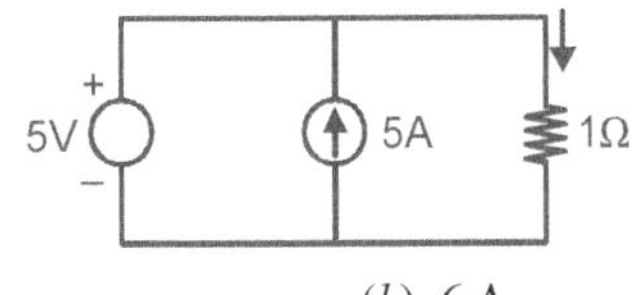

(*a*) 10 A (*b*) 6 A

(*c*) 5 A (*d*) zero

29. A connected planar network has 4 nodes and 5 elements. The number of meshes in its dual network is

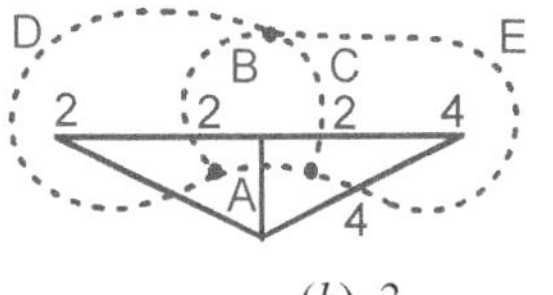

(*a*) 4 (*b*) 3

(*c*) 2 (*d*) 1

30. A rectangular current pulse of duration T and magnitude I has the Laplace transform

(*a*) I/s (*b*) (I/s) exp (−Ts)

(*c*) (I/s) exp (Ts) (*d*) (I/s) [1 − exp(−Ts)]

31. A dc supply of 35 V is connected across 600Ω resistance in series with an unknown resistance. The voltmeter having a resistance of 1.2 kΩ is connected across 600 Ω, resistance which reads 5V. Then the unknown resistance should be

(a) 50 Ω (b) 1.7 kΩ
(c) 7.2 Ω (d) 2.4 kΩ

32. In the circuit shown, the $v(t)$ and $i(t)$ will be

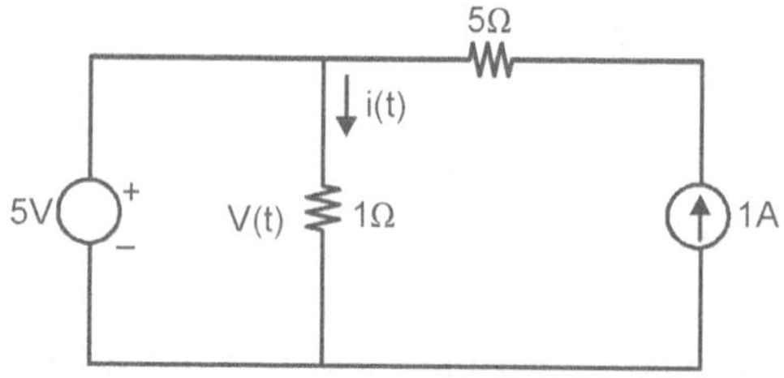

(a) 1 V, 1 A (b) 1 V, 6 A
(c) 5 V, 5 A (d) none of these

33. The source in the circuit shown in a sinusoidal source. The voltage across various elements are marked in the figure. The input voltage is

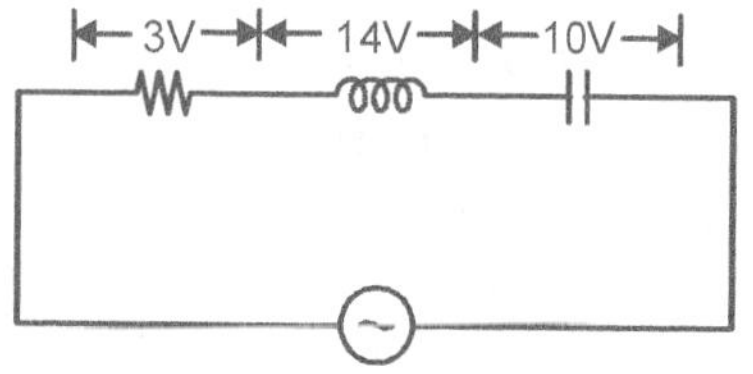

(a) 10 F (b) 5 V
(c) 27 V (d) 24 V

34. In the circuit shown in the given figure, the current 1 through R_l is

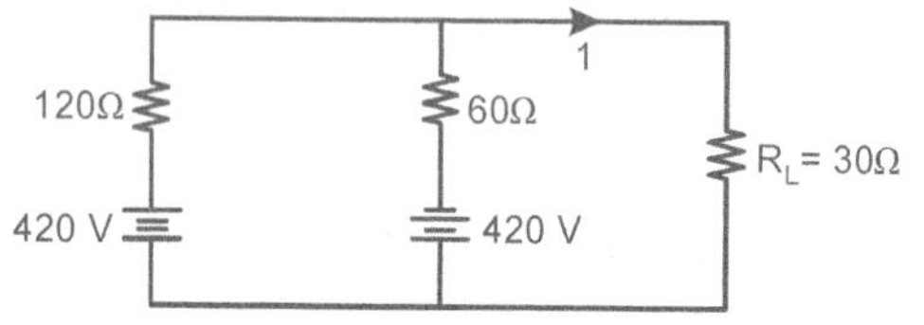

(a) 2 A (b) zero
(c) − 2 A (d) − 6 A

35. The Thevenin impedance across the terminals AB of the given network is

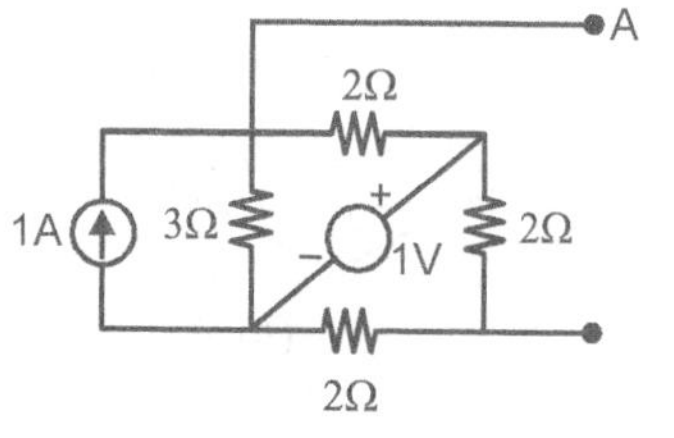

(a) $\frac{10}{3}\Omega$ (b) $\frac{20}{9}\Omega$

(c) $\frac{13}{4}\Omega$ (d) $\frac{11}{5}\Omega$

36. For a two-port symmetrical bilateral network, if A = 3 and B = 1Ω, the value of parameter C will be

(a) 4 s (b) 6 s
(c) 8 s (d) 16 s

37. Two identical coils of negligible resistance, when connected in series across a 50 Hz fixed voltage source, draw a current of 10A. When the terminals of one of the coils are reversed, the current drawn is 8A. The coefficient of coupling between the two coils is

(a) 1/100 (b) 1/9
(c) 4/10 (d) 8/10

38. The definiting equations of a two-port network in terms of the Laplace transform in matrix form are

$$\begin{bmatrix} V_1 \\ I_2 \end{bmatrix} = \begin{bmatrix} 0 & n \\ -n & 0 \end{bmatrix} \begin{bmatrix} I_1 \\ V_2 \end{bmatrix}$$

The network so defined is called

(a) a gyrator
(b) an ideal transformer
(c) loss-less transmission line
(d) T-network with resistive elements

39. The graph of an electrical network has N nodes and B branches. The number of links, L, with respect to the choice of a tree, is given by

(a) B − N + 1 (b) B + N
(c) N − B + 1 (d) N − 2B − 1

40. The network shown in the figure is described by the equation:

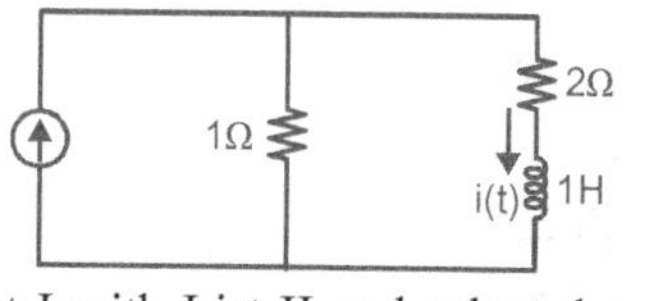

Match List-I with List-II and select the correct answer using the codes given below the lists:

List-I [$i_s(t)$]

A. $u(f)$
B. $e^{2t}\,u(t)$
C. $e^{it}\,u(t)$
D. $\cos tu(t)$

List-II [$i(t)$]

1. $\frac{1}{\sqrt{10}}[\cos(t-\phi)-e^{-3t}\cos\phi]$
2. $\frac{1}{j+3}[e^{it}-e^{-3t}]$
3. $\frac{1}{5}(e^{2t}-e^{-3t})$
4. $\frac{1}{3}(1-e^{-3t})$

Codes:

	A	B	C	D
(*a*)	4	3	1	2
(*b*)	3	4	2	1
(*c*)	3	4	1	2
(*d*)	4	3	2	1

41. Which one of the following pairs of poles and responses is correctly matched?

(*a*) 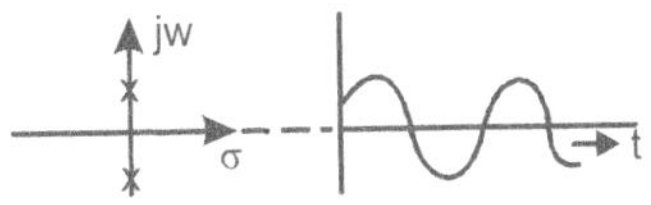

(*b*) 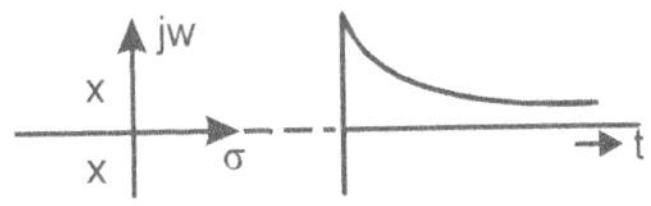

(*c*) 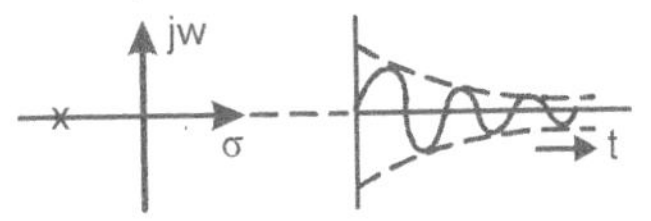

(*d*) 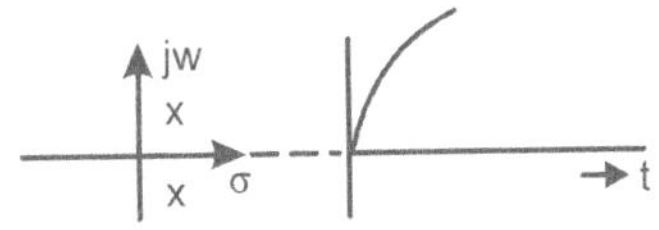

42. The equivalent capacitance of the following circuit is

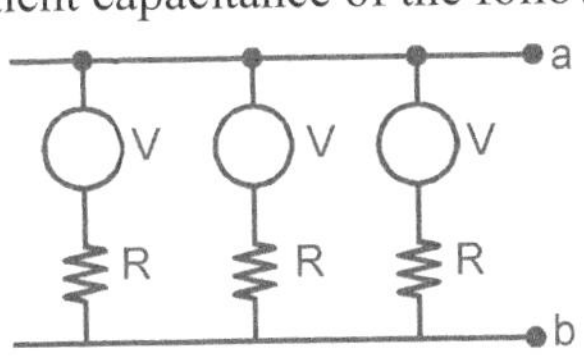

(*a*) 

(*b*) 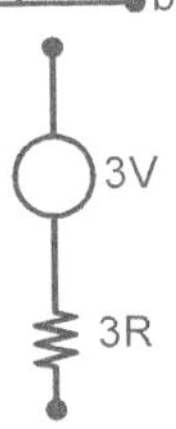

(*c*) 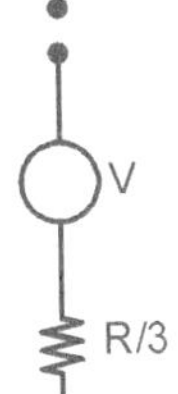

(*d*) 

43. An RLC resonant circuit has a resonance frequency of 1.5 MHz and a bandwidth of 10 kHz. If C= 150 pF, then the effective resistance of the circuit will be

(*a*) 29.5 Ω

(*b*) 14.75 Ω

(*c*) 9.4 Ω

(*d*) 4.7 Ω

44. In addition to the condition that Y(*s*) is real when '*s*' is real, for an admittance function Y(*s*) to be positive real, which of the following conditions are to be satisfied?

1. $\text{Re}\,Y(s) \le 0$, for $\text{Re}\,s \ge 0$
2. $\text{Re}\,Y(s) \ge 0$, for $\text{Re}\,s \ge 0$
3. $|\text{Arg}\,Y(s)| \le |\text{Arg}\,s$ for $|\text{Arg}\,s| \le \pi/2$
4. $|\text{Arg}\,Y(s)| \ge |\text{Arg}\,s|$ for $|\text{Arg}\,s| \le \pi/2$

Select the correct answer using the codes given below:

Codes:

(*a*) 1 and 3 are correct (*b*) 1 and 4 are correct

(*c*) 2 and 3 are correct (*d*) 2 and 4 are correct

45. Current having wave form shown in following is a resistance of 10 ohms. The average power is

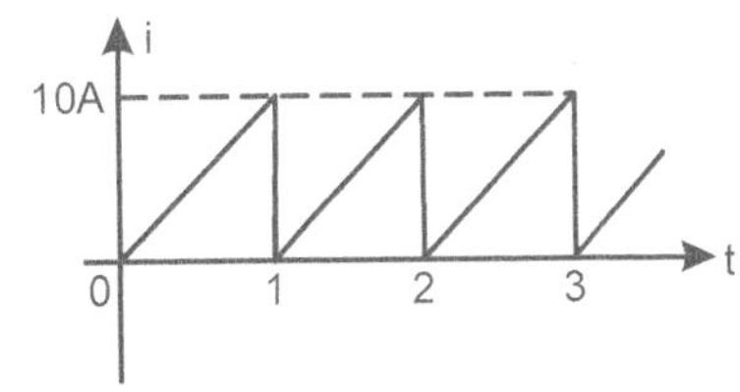

(*a*) $\frac{1000}{1}$ W (*b*) $\frac{1000}{2}$ W

(*c*) $\frac{1000}{3}$ W (*d*) $\frac{1000}{4}$ W

46. An alternating current source having voltage

$$E = 110 \sin(\omega t + \pi/3)$$

is connected in an a.c. circuit. If the current drawn from the circuit varies as $1 = 5 \sin(\omega t - \pi/3)$, then impedance of the circuit will be

(*a*) 22 ohm

(*b*) 16 ohms

(*c*) 30.8 ohm

(*d*) none of these

47. The impulse response of an R-L circuit is a

(*a*) rising exponential function

(*b*) decaying exponential function

(*c*) step function

(*d*) parabolic function

48. The circuit shown in the given figure is in steady-state with switch 'S' open. The switch is closed at $t = 0$. The values of $V_c(0^+)$ and $V_c(\infty)$ will be respectively

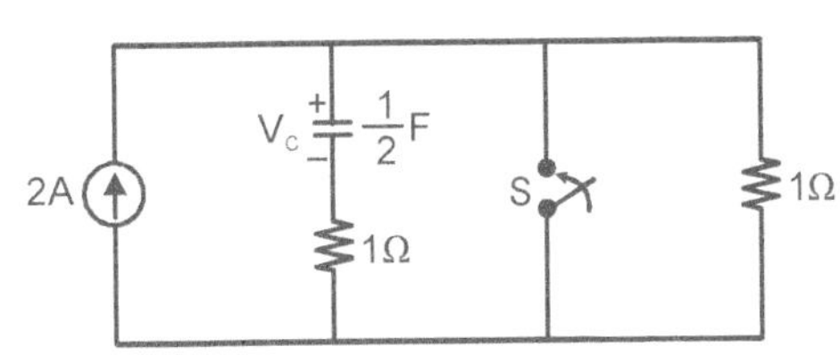

(*a*) 2 V, 0 V (*b*) 0 V, 2 V

(*c*) 2 V, 2 V (*d*) 0 V, 0 V

49. An initially relaxed 100 mH inductor is switched 'ON' at $t = 1$ sec. to an ideal 2A dc current source. The voltage across the inductor would be

(*a*) zero (*b*) $0.25\,\delta(t)$V

(*c*) $0.25\,\delta(t-1)$K (*d*) $0.2t\,u(t-1)$V

50. The correct value of the current $i(t)$ at any instant when K is switched on at $t = 0$ in the network shown in the given figure is

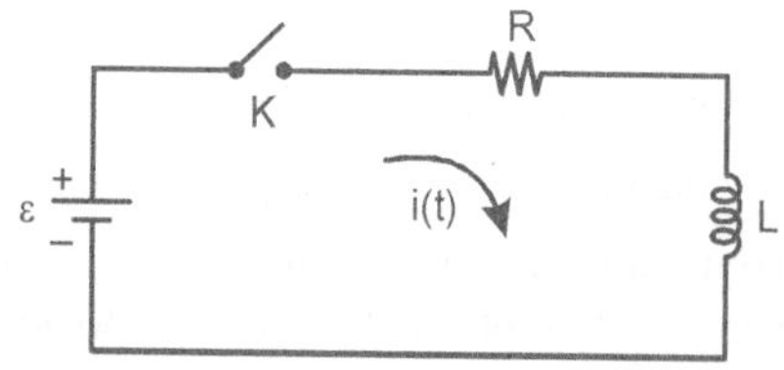

(a) $\frac{E}{R}+\frac{E}{R}e^{(R/L)t}$ (b) $\frac{E}{R}-\frac{E}{R}e^{(R/L)t}$

(c) $\frac{E}{R}+\frac{E}{R}e^{-(R/L)t}$ (d) $\frac{E}{R}-\frac{E}{R}e^{-(R/L)t}$

51. Consider the star network shown in Fig. P 71. The resistance between terminals A and B with C open is 6Ω, between terminals B and C with A open is 11Ω, and between terminals C and A with B open is 9Ω. Then

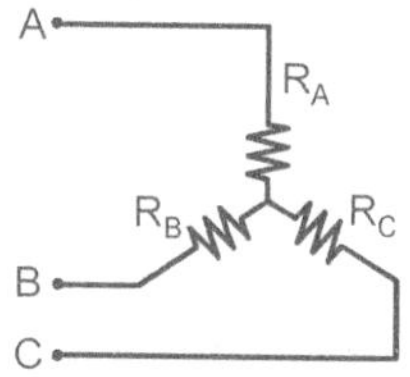

(a) $R_A = 4Ω, R_B = 2Ω, R_C = 5Ω$
(b) $R_A = 2Ω, R_B = 4Ω, R_C = 7Ω$
(c) $R_A = 3Ω, R_B = 3Ω, R_C = 4Ω$
(d) $R_A = 5Ω, R_B = 1Ω, R_C = 10Ω$

52. The Laplace transform of the response of a circuit to some excitation is

$$I(s) = \frac{s+2}{(s+1)^2}$$

The time-domain current $i(t)$ is given by

(a) $e^{-t} + 2te^{-t}$ (b) $2e^{-t} + te^{-t}$
(c) $e^{-t} + te^{-t}$ (d) $e^{-t} + 3te^{-t}$

53. For a two-port network Y_{22} is given by
(a) $Z_{11}/\Delta Y$
(b) $Z_{11}/\Delta Z$
(c) $Z_{22}/\Delta Y$
(d) $Z_{22}/\Delta Z$

54. When all the resistances in the circuit are of one ohm each, the equivalent resistance across the points A and B will be

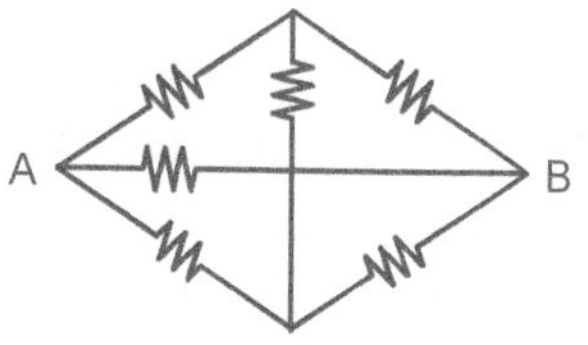

(a) 1Ω (b) 0.5Ω
(c) 2Ω (d) 1.5Ω

55. The normalised energy of the signal $f(t) = \exp(-3t)\,u(t)$ volts is
(a) infinite (b) 1/6 volt^2
(c) 1/9 volt^2 (d) zero

56. A battery of 66 cells has to be charged at 200 amperes, each cell has emf of 1.8 volts. When discharged, internal resistance is 0.0005 ohm and the resistance of the connections between dynamo and battery is 0.02 ohm. To what value must be terminal voltage of the dynamo be adjusted at the commencement of the charge?
(a) 66 volts
(b) 142.6 volts
(c) 129.6 volts
(d) 155 volts

57. The quality factor of R-L-C circuit will increase
(a) R decreases
(b) R increases
(c) voltage increases
(d) Q always increases

58. Resonant frequency f_r of a series RLC circuit is related to half power frequencies f_1 and f_2 as

(a) $f_r = \frac{f_1 + f_2}{2}$
(b) $f_r = \sqrt{f_1 f_2}$
(c) $f_r = f_2 - f_1$
(d) $f_r = \sqrt{f_1} + \sqrt{f_2}$

59. When resonant frequency for an R-L-C parallel circuit is given by $\frac{1}{2\pi\sqrt{LC}}$, the essential condition is that
(a) inductance must be non-resistive
(b) capacitance must be non-leaky
(c) current is minimum at resonance
(d) all of these

60. A balanced 3-phase circuit is energised by a symmetrical 3-phase sinusoidal voltage source. Which one of the following statements best describes the steady-state total 3-phase instantaneous power?
(a) It has a double-frequency sinusoidal variation with respect to time
(b) It varies sinusoidally at fundamental frequency with respect to time
(c) It varies at twice the fundamental frequency superimposed upon a constant
(d) T is independent of time and remains constant

61. The autocorrelation function of a rectangular pulse of duration T, not centred at the origin, is a triangular pulse of duration

(*a*) T and centred at the origin

(*b*) 2T and centred at the origin

(*c*) 2T and not centred at the origin

(*d*) T and not centred at the origin

62. A signal $\cos^2 200$ nt is multiplied by the signal $\sin 200\,\pi t$. The frequencies which will be generated in this process include

(*a*) 50 Hz, 100 Hz, 300 Hz

(*b*) 50 Hz, 200 Hz

(*c*) 100 Hz, 150 Hz, 250 Hz

(*d*) 100 Hz, 200 Hz, 300 Hz[ES]

63. A major advantage of active filters is that they can by realized without using

(*a*) op-amps

(*b*) inductors

(*c*) resistors

(*d*) capacitors

64. Consider the following statements:

If the energy sources are connected to the unenergised system as shown in figures A and b, at $t = 0$, then

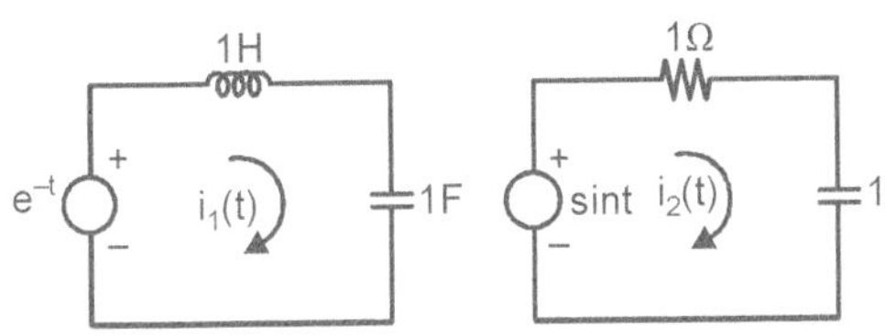

1. i_1(t) and $i_2(t)$ are equal

2. the current circuit in fig. A has one free frequency and two forced frequencies

3. the circuit in fig. B has one free frequency and two forced frequencies

4. the circuit in fig. A has two free frequencies and one forced frequency

Of these statements

(*a*) 1,2 and 3 are correct

(*b*) 1, 3 and 4 are correct

(*c*) 1 and 3 are correct

(*d*) 4 alone is correct

65. The driving-point impedance of an RC network is given by

$$Z(s) = \frac{2s^2 + 7s + 2}{s^2 + 3s + 1}$$

Its canonical realisation will have

(*a*) 6 elements (*b*) 5 elements

(*c*) 4 elements (*d*) 3 elements

66.

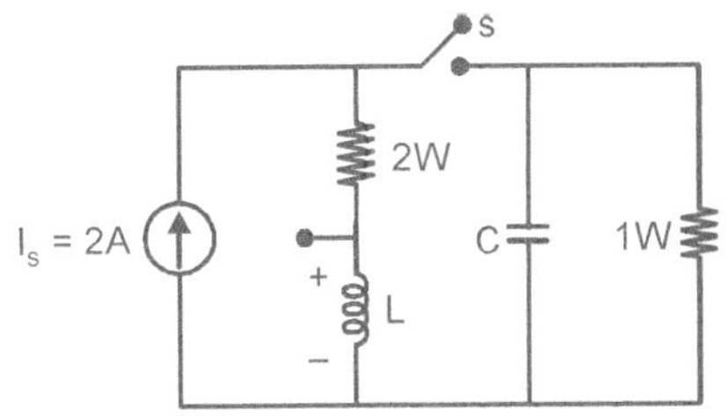

In the circuit shown in the above figure, steady-state is reached with the switch S open .V is closed at $t = 0$. The voltage V_L across L at $/ = 0^+$ will be

(*a*) zero (*b*) 2V

(*c*) 4V (*d*) −4V

67. A connected network of N > 2 nodes has at most one branch directly connecting any pair of nodes.

The graph of the network

(*a*) must have at least N branches for one of more closed paths to exist

(*b*) can have an unlimited number of branches

(*c*) can only have at most N branches

(*d*) can have a minimum number of branches not decided by N

68. The system described by the equation

$$F(s) = s^4 + 2s^3 + 3s^2 + 6s + K,$$

according to Routh-Hurwitz's criteria, is

(*a*) unstable for all values of K

(*b*) stable if $K \geq 0$

(*c*) stable if $K < 0$

(*d*) stable for all values of K

69. The condensers of 20μF and 40μF capacitance are connected in series across a 90Vsupply. After charging, they are removed from the supply and are connected in parallel with positive terminals connected together and similarly the negative terminals. Then the voltage across them will be

(*a*) 90 V (*b*) 60 V

(*c*) 40 V (*d*) 20V

70. If Rg in the circuit shown in the given figure is variable between 20Ω and 30Ω then the maximum power transferred to load R_L will be

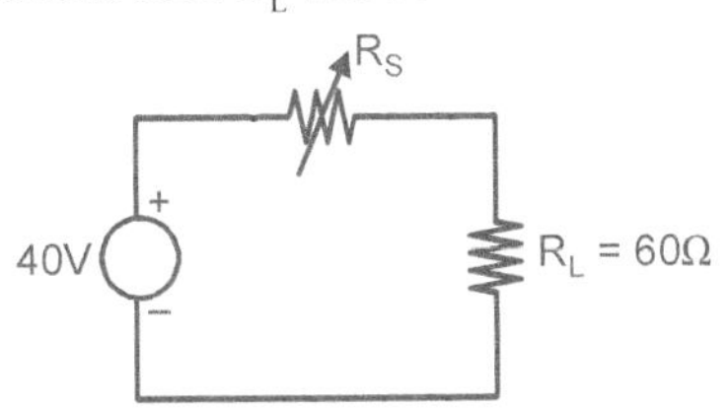

(*a*) 15 W (*b*) 13.33W

(*c*) 6.67 W (*d*) 2.4 W

71. Under conditions of maximum power transfer from an ac source to a variable load

(*a*) the load impedance must also be inductive, if the generator impedance is inductive

(*b*) the sum of the source and load impedances is zero

(*c*) the sum of the source reactance and load reactance is zero

(*d*) the load impedance has the same phase angle as the generator impedance

72. In the network as shown in the figure, the input signal Vs is a 6-volt step, that is, $6u(t)$. The time constant of the network is

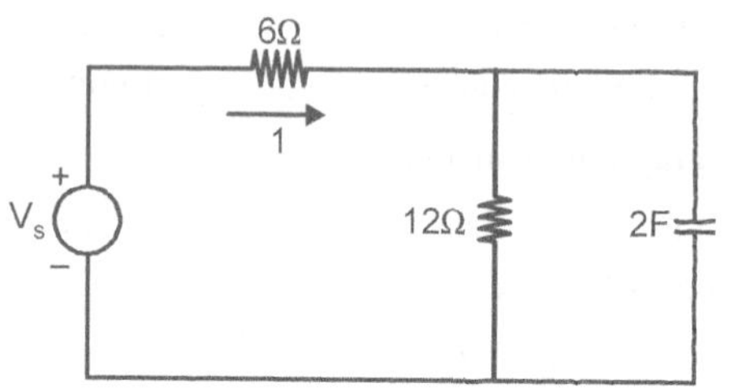

(*a*) 8*s* (*b*) 12*s*

(*c*) 24*s* (*d*) 36*s*

73. In a circuit, the current i(t) has the Laplace transform I(s) $= \frac{3(s+10)}{s(s+12)}$. The final value of $i(t)$ will be

(*a*) 0.25 (*b*) 2.5

(*c*) 3 (*d*) infinity

74. With the usual notations, a two-port resistive network satisfies the condition,

$$A = D\frac{3}{2}B, = \frac{4}{3}C$$

The Z_{11} of the network is

(*a*) 5/3 (*b*) 4/3

(*c*) 2/3 (*d*) 1/3

75. The driving point impedance of the infinite ladder network shown in the figure is

(Given : $R_1 = 2\Omega$ and $R_2 = 1.5\Omega$)

(*a*) 3Ω (*b*) 3.5Ω

(*c*) $\frac{3}{3.5}\Omega$ (*d*) $\left(1-\frac{3}{3.5}\right)\Omega$

76. The resonant frequency of the series circuit shown in the given figure is

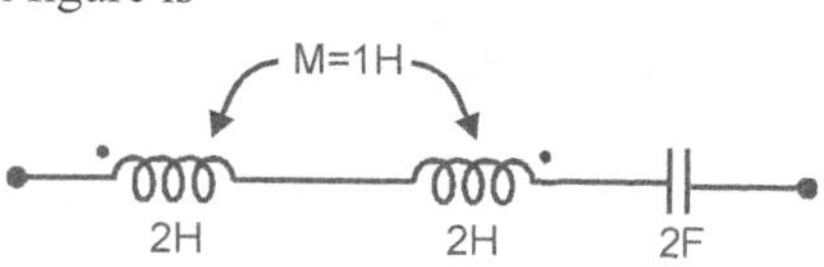

(*a*) $\frac{1}{4\pi}$ Hz (*b*) $\frac{1}{4\pi\sqrt{2}}$ Hz

(*c*) $\frac{1}{2\pi\sqrt{2}}$ Hz (*d*) $\frac{1}{4\pi\sqrt{3}}$ Hz

77. The reciprocal of a network function is

(*a*) an immittance function, if the original function is an omittance function

(*b*) a transfer function, if the original function is a transfer function

(*c*) never an immittance function

(*d*) never a transfer function

78. A set of 3 equal resistors, each of value R, connected in star across RYB in place of load as shown in the given figure consumes the same power as the unbalanced delta connecied load. The value of R_x is

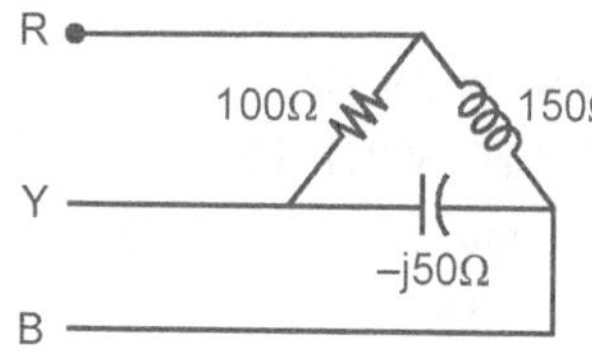

(*a*) 33.33Ω (*b*) 100Ω

(*c*) 173.2Ω (*d*) 300Ω

79. Match List-I with List II and select the correct answer using the codes given below the lists:

List-I (*Time function*)	**List-II** (*Laplace transforms*)
A. $[af_1(t)+bf_2(t)]$	1. $aF1(s)+bF2(s)$
B. $[e^{-at}f(t)]$	2. $sF(s)+f(0^-)$
C. $\left[\frac{df(t)}{dt}\right]$	3. $\frac{1}{s}F(s)$
D. $\left[\int_0^1 f(x)dx\right]$	4. $sF(s)-f(0^-)$
	5. $aF_1(s)+a_2F_2(s)$
	6. $F(s+a)$

Codes:

	A	B	C	D
(*a*)	6	2	3	4
(*b*)	1	6	4	3
(*c*)	2	5	3	4
(*d*)	1	6	3	4

80. Which of the following are the necessary conditions for an entire three-phase system to be balanced?

1. The line voltages are equal in magnitude
2. The phase differences between successive line voltages are equal
3. The impedance in each of the phases are identical

Select the correct answer using the codes given below :

Codes:

(*a*) 1, 2 and 3 (*b*) 1 and 3

(*c*) 1 and 2 (*d*) 2 and 3

81. 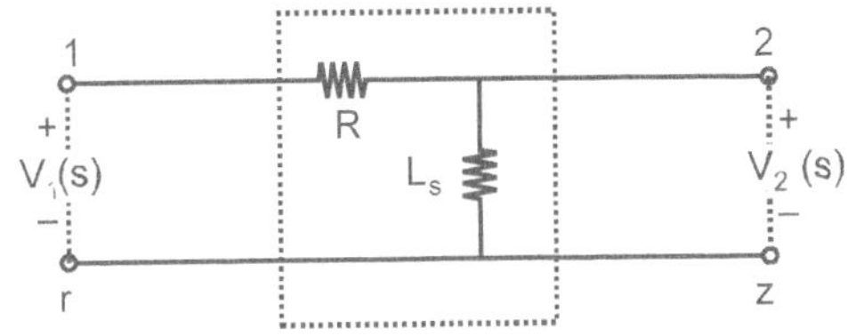

For the 2-port network as shown in the above figure, the one-circuit transferfunction $\frac{V_2(s)}{V_1(s)}$ is

(a) $\frac{s}{s+R/L}$ (b) $\frac{s+R/L}{s}$

(c) $\frac{s}{s+L/R}$ (d) $\frac{s+L/R}{s}$

82. Two ac sources feed a common variable resistive load as shown in the figure. Under the maximum power transfer condition, the power absorbed by the load resistance R_L is

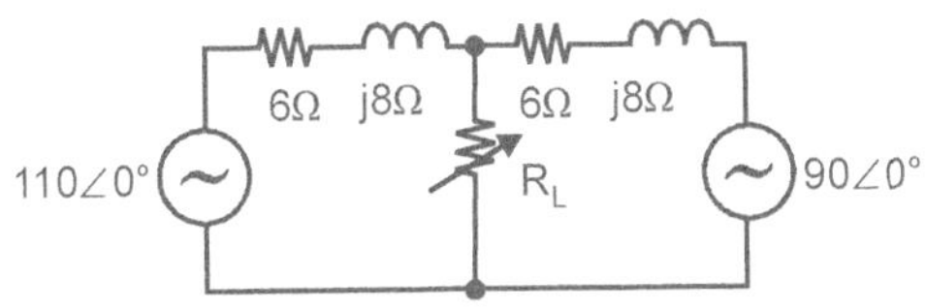

(a) 2200W (b) 1250W

(c) 1000W (d) 625 W

83. Two inductive coils with self-inductances L_1 and L_2 are magnetically coupled in series opposing and in parallel aiding respectively. The mutual inductance between the coils is M The equivalent inductances in the two cases are respectively

(a) $L_1+L_2+2M, \frac{L_1L_2-M^2}{L_1+L_2-2M^2}$

(b) $L_1+L_2+2M, \frac{L_1L_2-M^2}{L_1+L_2+2M}$

(c) $L_1+L_2+2M, \frac{L_1L_2-M^2}{L_1+L_2-2M}$

(d) $L_1+L_2+2M, \frac{L_1L_2-M^2}{L_1+L_2+2M}$

84. Consider the following statements regarding driving-point admittance function having two complex conjugate poles:

1. Closer the poles to $j\omega$-axis, higher the Q of the circuit
2. Value of Q varies inversely as the damping ratio
3. A circuit with low R has low Q

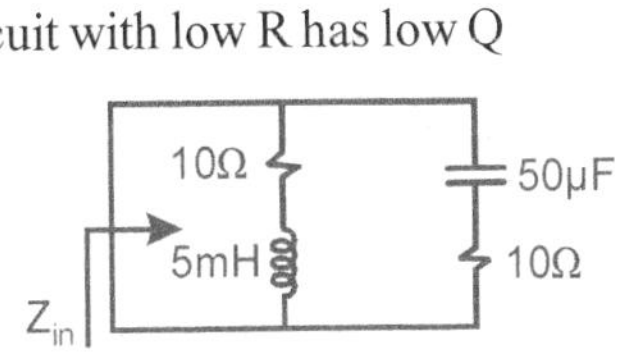

Of these statements

(a) 1 and 3 are correct (b) 2 and 3 are correct

(c) 1 and 2 are correct (d) 1,2 and 3 are correct

85. A network is composed of two subnetworks N_1 and N_2 as shown in the given figure.

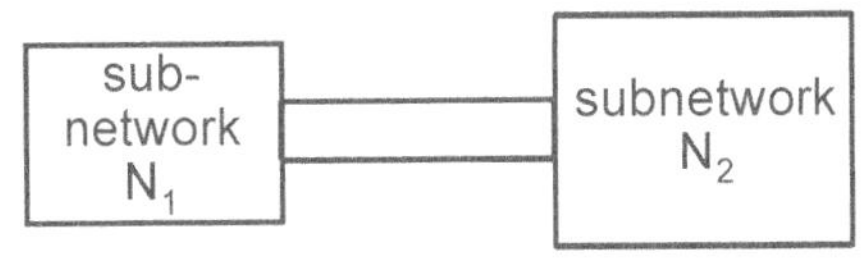

If the subnetwork N_1 contains only linear, bilateral, time invariant elements, then it can be replaced by its Thevenin equivalent even if the subnetwork N_2 contains

(a) a two-terminal element which is non-linear

(b) a non-linear inductance mutually coupled to an element in N_1

(c) an element which is linear, but mutually coupled to some element in N_1

(d) a dependent source the value of which depends upon the voltage or current in some element in N_1

86. A series circuit containing passive elements has the following current and applied voltage:

$v = 200 \sin(2000t + 50°)$,

$i = 4 \cos(2000t + 13.2°)$

The circuit elements

(a) must be resistance and capacitance

(b) must be resistance and inductance

(c) must be inductance, capacitance and resistance

(d) could be either resistance and capacitance or resistance, inductance and capacitance.

87. Which of the following functions represents parabolic characteristic?

(a) $f(t) = t$ (b) $f(t) = t^2$

(c) $f(t) = e^{-at}$ (d) $f(t) = te^{-at}$

88. Which of the following conditions are necessary for the validity of initial value theorem

$$\lim_{s\to\infty} sF(s) = \lim_{s\to\infty} F(t)$$

(a) $f(t)$ and its derivative $f(t)$ must have Laplace transform

(b) If the Laplace transform of $f(t)$ is $F(s)$, the $\lim_{s\to\infty} sF(s)$ must exist

(c) only $f(t)$ must have Laplace transform

(d) both (a) and (b) above

89. For which of the following function the Laplace transform will be $\frac{2}{s^3}$?

(a) $f(t) = t$ (b) $f(t) = 2t$

(c) $f(t) = t^2$ (d) $f(t) = t^3$

90. The Laplace transform of $f(t)$ is F(s). Given

$F(s) = \frac{\omega}{s^2 + \omega^2}$, the final value of $f(t)$ is

(*a*) ∞ (*b*) zero

(*c*) one (*d*) none of these

91. Inverse Laplace transform of $\frac{1}{s-a}$ is

(*a*) sin *at*

(*b*) cos *at*

(*c*) e^{at}

(*d*) e^{-at}

92. Spectrum analyzer is used across the frequency spectrum, of a given signal to study the

(*a*) current distribution

(*b*) voltage distribution

(*c*) energy distribution

(*d*) power distribution

93. For a dc voltage, an inductor

(*a*) is virtually a short-circuit

(*b*) is an open circuit

(*c*) depends on polarity

(*d*) depends on voltage value

94. The power in a 3-phase system is given by $\sqrt{3}\, V_L I_L \cos\phi$, where ϕ is the phase angle between

(*a*) line voltage and line current

(*b*) phase voltage and phase current

(*c*) line voltage and phase current

(*d*) phase voltage and line current

95. When phase sequence at the 3-phase load is reversed

(*a*) phase currents are changed in magnitude

(*b*) phase currents change in phase angle but not in magnitude

(*c*) total power consumed is changed

(*d*) phase powers are changed

96. Match List-I with List-II and select the answer using the codes given below the lists:

List-I (*Location of poles on s-plane*)

A. 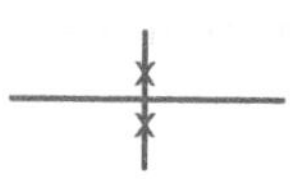

B. 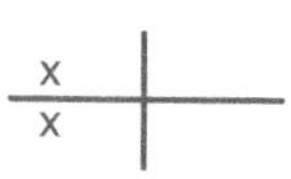

C. 

D. 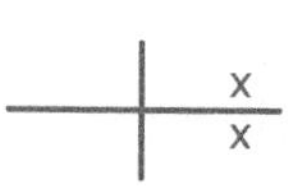

List-II (*Type of response*)

1. 

2. 

3. 

4. 

5. 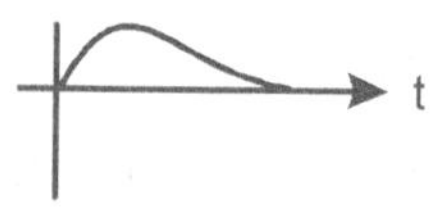

Codes:

	A	B	C	D
(*a*)	4	1	2	3
(*b*)	5	1	4	3
(*c*)	2	3	5	4
(*d*)	2	1	3	4

97. The current read by the ammeter A in the ac circuit shown in the given figure is

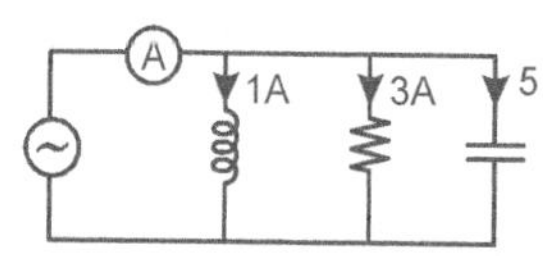

(*a*) 9A (*b*) 5A

(*c*) 3A (*d*) 1A

98. In the given figure, $Z_1 = 10\angle{-60°}$, $Z_2 = 10\angle 60°$, $Z_3 = 50\angle 53.13°$. The Equivalent impedance seen from X–Y is

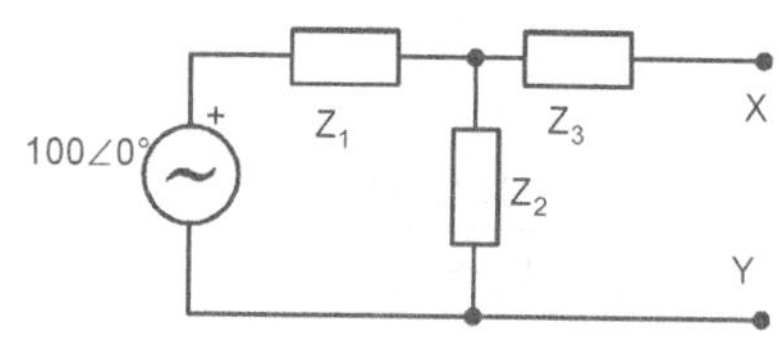

(*a*) 56.6∠45° (*b*) 60∠30°

(*c*) 70∠30° (*d*) 34.4∠65°

99. In the network shown in the given figure, there is no initial current through L_2 and no initial voltage across the C. The switch 'S' is closed at $t = 0$. The current i_{Li} in inductor L_1 and the voltage V_c across C at $= 0^+$ and $t = \infty$ will be

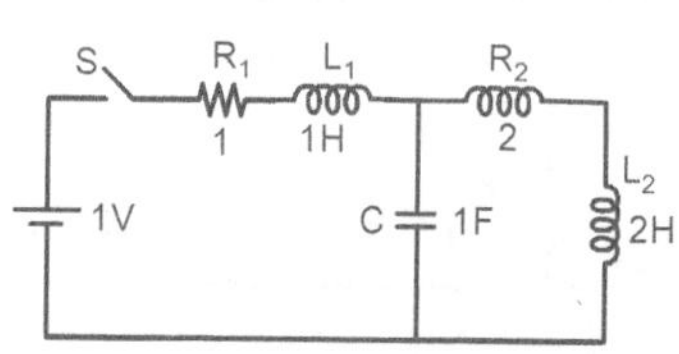

	$i_{Li}(0^+)$	$i_{Li}(\infty)$	$V_c(0^+)$	$V_c(\infty)$
(*a*)	1/3/1	1/3A	2/3 F	2/3 V
(*b*)	0	1/3A	0	1V
(*c*)	1/3A	0	2/3	0
(*d*)	0	1/3A	0	2/3 V

100. In the circuit shown in the given figure, S is open for a long time and steady state is reached. S is closed at $t = 0$. The current I at $t = 0^+$ is

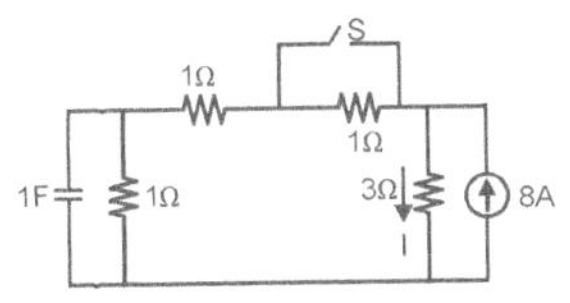

(*a*) 44 (*b*) 3A
(*c*) 2A (*d*) 1A

101. For the circuit given in the figure $V_o = 2V$ and the inductor is initially relaxed. The switch S is closed at $t = 0$. The value of v at $t = 0^+$ is

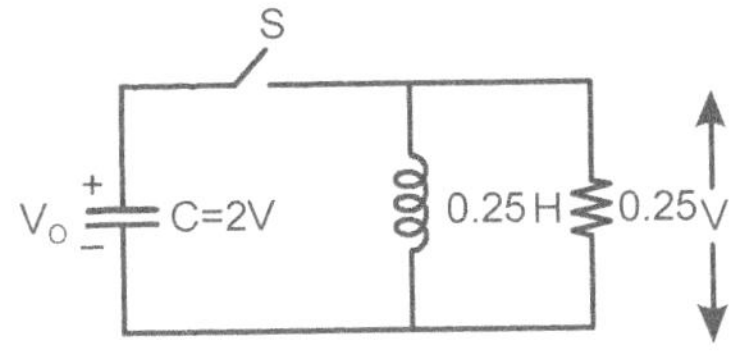

(*a*) 3V
(*b*) 2V
(*c*) 0.5V
(*d*) 0.25V

102. In the given circuit, the value of R that will give critical damping is

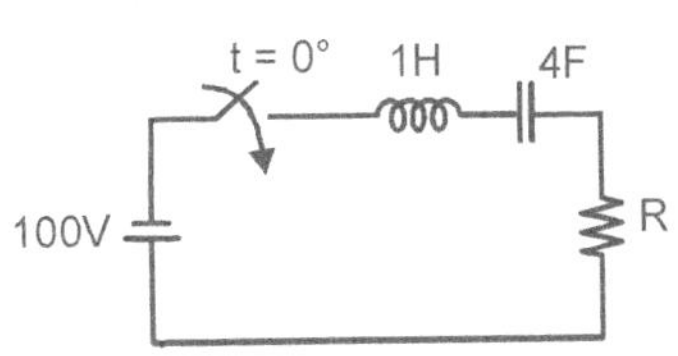

(*a*) 1Ω
(*b*) 2Ω
(*c*) 4Ω
(*d*) 10Ω

103. A transient current in network is:

$i(t) = 2e^{-t} - e^{-5t} \geq 0.$

The pole-zero configuration of I(*s*) is

	Poles	**Zeros**
(*a*)	1, 5	9
(*b*)	−1, −5	−9
(*c*)	2, −1	−1, −5
(*d*)	2, −1	1, 5

104. The y_{21} parameter of the network in the given figure will be

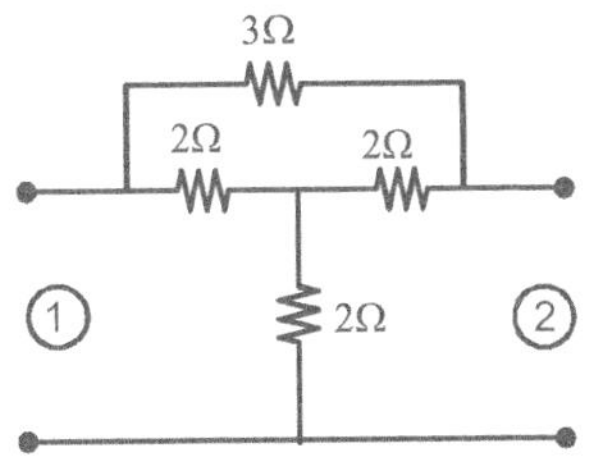

(*a*) $\frac{1}{6}\Omega$ (*b*) $-\frac{1}{6}\Omega$

(*c*) $\frac{1}{3}\Omega$ (*d*) $-\frac{1}{3}\Omega$

105. In two-wattmeter method of power measurement, one of the wattmeters will show negative reading when the load *pf* angle is strictly
(*a*) less than 30°
(*b*) less than 60°
(*c*) greater than 30°
(*d*) greater than 60°

106. A first order linear system is initially relaxed. For a unit step signal *u*(*t*), the response is

$v_1(t) = (1 - e^{-3t})$ for $t > 0$.

If a single 3 *u*(*t*) + 5(*t*) is applied to the same initially relaxed system, the response will be
(*a*) $(3 - 6e^{-3t})\,u(t)$
(*b*) $(3 - 3e^{-3t})\,u(t)$
(*c*) $3\,u(t)$
(*d*) $(3 + 3e^{-3t})\,u(t)$

107. Consider the voltage waveform shown in the given figure. The equation for v(*t*) is

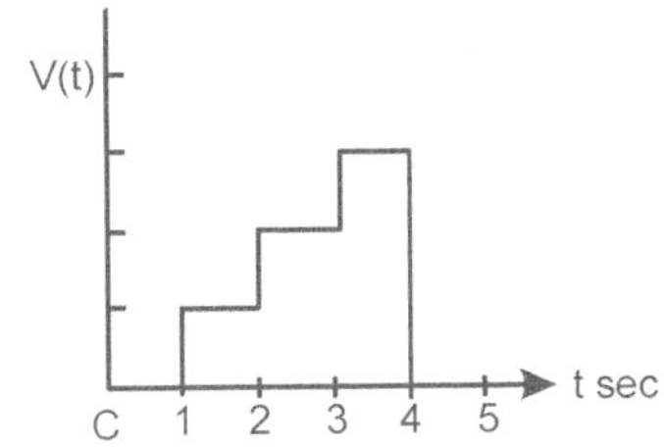

(*a*) $u(t-1) + u(t-2) + u(t-3)$
(*b*) $u(t-1) + 2u(t-2) + 3u(t-3)$
(*c*) $u(t) + u(t-1) + u(t-2) + u(t-4)$
(*d*) $u(t-1) + u(t-2) + u(t-3) - 3u(t-4)$

108. An LC-driving point impedance function is

(*a*) $\frac{s^3 + s^2 + s + 1}{s^2 + 2s + 5}$ (*b*) $\frac{s^4 + 2s^2 + 1}{s^2 + 5}$

(*c*) $\frac{s^4 + 1}{s^3 + 2}$ (*d*) $\frac{2(s^2 + 1)}{2}$

109. When a number of 2-port networks are connected in cascade, the individual

(*a*) Z_{oc}-matrices are added

(*b*) Y_{sc}-matrices are added

(*c*) chain matrices are multiplied

(*d*) H-matrices are multiplied

110. Constant voltage source is

(*a*) active and bilateral (*b*) passive and bilateral

(*c*) active and unilateral (*d*) passive and unilateral

111. A series R-L-C circuit when excited by a 10K sinusoidal voltage source of variable frequency, exhibits resonance at 100 Hz and has a 3 dB bandwidth of 5 Hz. The voltage across the inductor L at resonance is

(*a*) 10V (*b*) $10\sqrt{2}$V

(*c*) $\frac{10}{\sqrt{2}}$V (*d*) 200V

112. If the step response of an initially relaxed circuit is known, then the ramp response can be obtained by

(*a*) integrating the step response

(*b*) differentiating the step response

(*c*) integrating the step response twice

(*d*) differentiating the step response twice

113. The principles of homogeneity and superposition are applied to

(*a*) linear time variant systems

(*b*) non-linear-time variant systems

(*c*) linear time invariant systems

(*d*) non-linear time invariant systems

114. Four networks are shown below in figure (1), (2), (3) and (4)

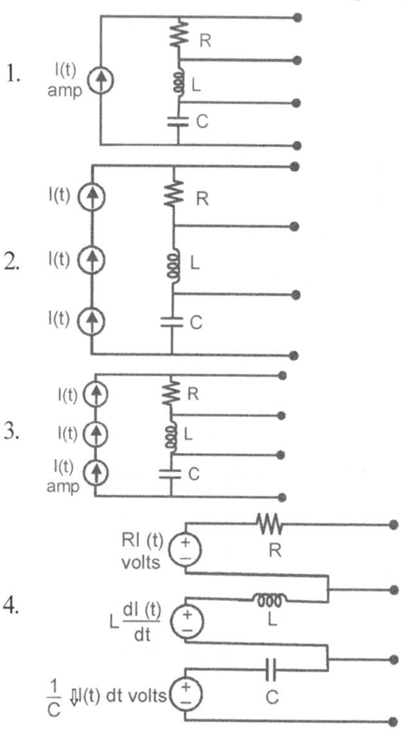

Of these networks,

(*a*) all the four networks are equivalent

(*b*) no two networks are equivalent

(*c*) networks shown in figures (2), (3) and (4) are equivalent

(*d*) networks shown in figures (3) and (4) are equvalent

115. The input-output equation of a system is given by

$$y(n) = x(n^2)$$

where $y(n)$ $x(n)$ represent output and input sequences respectively. The system is

(*a*) linear and causal

(*b*) linear and noncausal

(*c*) nonlinear and causal

(*d*) nonlinear and noncausal

116. In the circuit shown in the given figure, the switch S is closed at $t=0$. The induced voltage V_2 will have a maximum value of

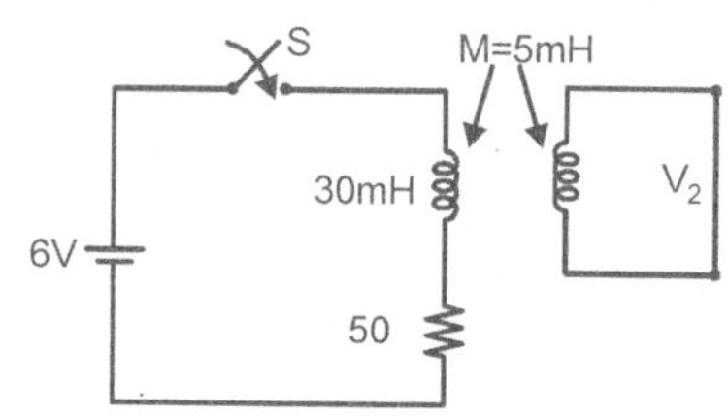

(*a*) 0.6V (*b*) 1V

(*c*) 3.78V (*d*) 6V

117. The transfer function, $T(s) = \frac{s^2}{s^2 + as + b}$ belongs to an active

(*a*) Low-pass filter

(*b*) High-pass filter

(*c*) Band-pass filter

(*d*) Band-reject filter

118. The *z* and *h* parameters of the network shown in the given figure will be

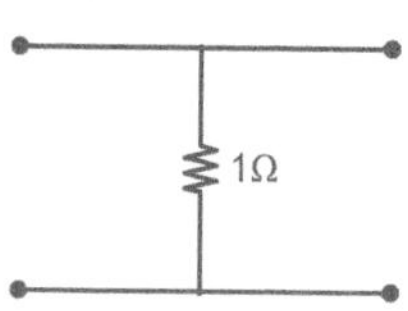

(*a*) $z = \begin{bmatrix} 0 & 1 \\ -1 & 1 \end{bmatrix}; h = \begin{bmatrix} 1 & 1 \\ 1 & 1 \end{bmatrix}$

(*b*) $z = \begin{bmatrix} 1 & 1 \\ 1 & 1 \end{bmatrix}; h = \begin{bmatrix} 0 & 1 \\ -1 & 1 \end{bmatrix}$

(*c*) $z = \begin{bmatrix} 1 & 0 \\ -1 & 1 \end{bmatrix}; h = \begin{bmatrix} -1 & 1 \\ 1 & -1 \end{bmatrix}$

(*d*) $z = \begin{bmatrix} 1 & 0 \\ -1 & 1 \end{bmatrix}; h = \begin{bmatrix} -1 & 1 \\ 1 & -1 \end{bmatrix}$

119. The current read by the ammeter A in the ac circuit shown in the given figure is

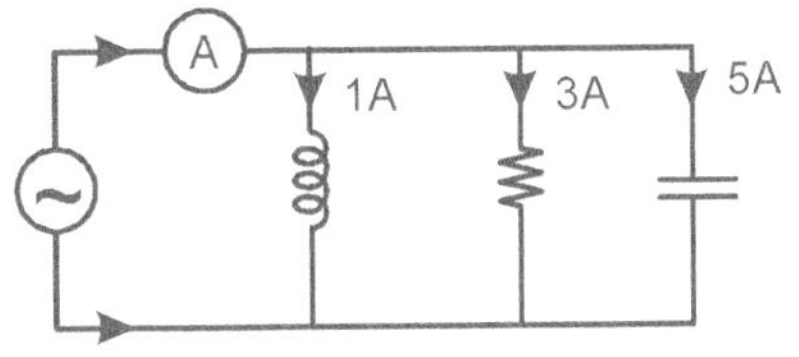

(*a*) 9 A (*b*) 5 A
(*c*) 3 A (*d*) 1 A

120. Consider the following statements regarding a first order system with a proportional (P) controller which exhibits an offset to a step input:

1. increase the gain of the P controller
2. add derivative mode
3. add integral mode Of these statements

(*a*) 1,2 and 3 are correct
(*b*) 1 and 2 are correct
(*c*) 2 and 3 are correct
(*d*) 1 and 3 are correct

121. If the charge on each of the capacitors in the given figure is 4500 μC, what is the total capacitance (in μF), assuming that the voltage distribution across C_1, C_2, C_3 is in the ratio of 2 : 3 : 4?

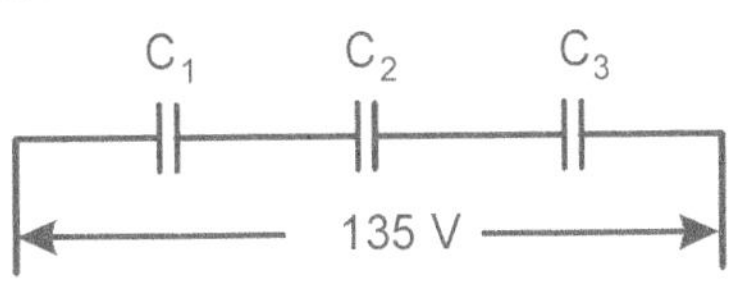

(*a*) 325
(*b*) 11.1
(*c*) 22.2
(*d*) 33.3

122. A phase-lag compensation will
(*a*) improve relative stability
(*b*) increase the speed of response
(*c*) increase bandwidth
(*d*) increase overshoot

123. Which one of the following correctly depicts the output of low-pass filter to a pulse input?

(*a*)
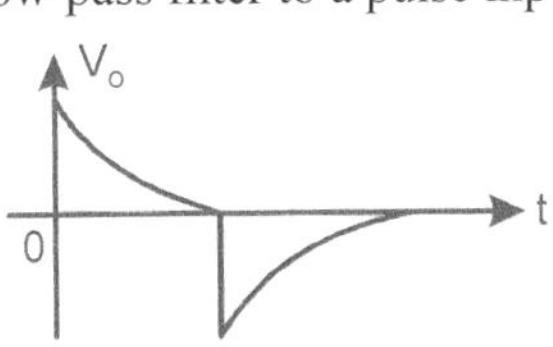

(*b*)
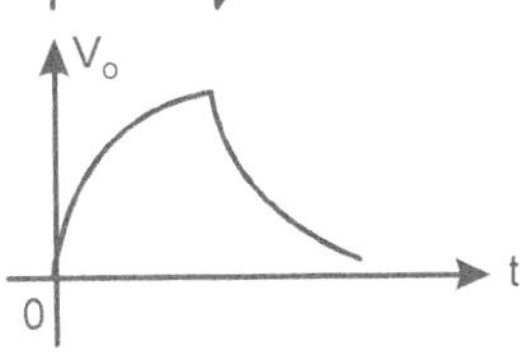

(*c*)
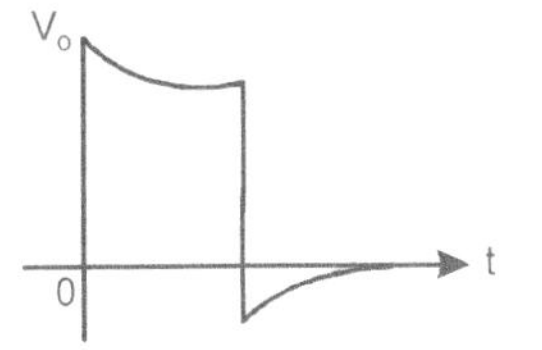

(*d*)
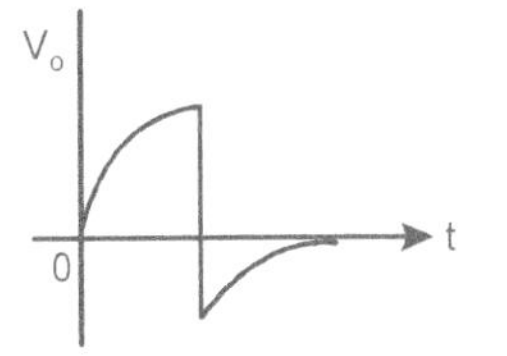

124. The current in the circuit shown in the given figure is

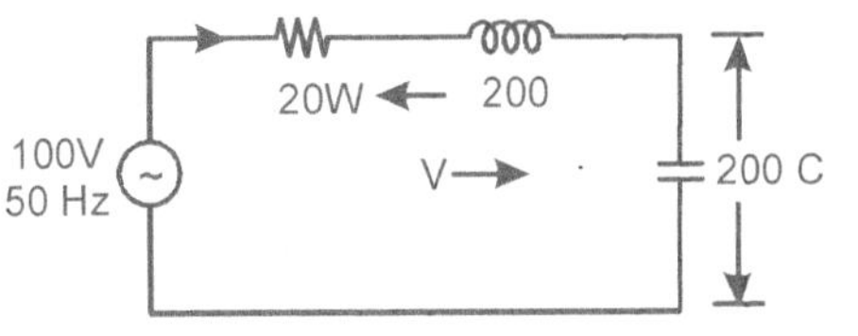

(*a*) 5A (*b*) 10A
(*c*) 15A (*d*) 25A

125. W_1, and W_2 are the readings of two wattmeters used to measure power of a 3-phase balanced load. The reactive power drawn by the load is

(*a*) $W_1 + W_2$ (*b*) $W_1 - W_2$
(*c*) $\sqrt{3}\,(W_1 + W_2)$ (*d*) $\sqrt{3}\,(W_1 - W_2)$

126. In the measurement of 3-phase power by two wattmeter method, if the two wattmeter readings are equal, the power factor of the circuit is

(*a*) 0.8 lagging (*b*) 0.8 leading
(*c*) zero (*d*) unity

127. Power loss in an electrical circuit can take place in
(*a*) inductor only
(*b*) capacitance only
(*c*) inductance and resistance
(*d*) resistance only

128. The *pf* of a practical inductor is
(*a*) unity
(*b*) zero
(*c*) lagging
(*d*) leading

129. As the Q-factor of an R-L-C series resonant circuit increases, its selectivity becomes

(*a*) better (*b*) worse
(*c*) narrower (*d*) none of these

130. A parallel resonant circuit can be employed
(*a*) as a high impedance
(*b*) to reject a small band of frequencies
(*c*) to amplify certain frequencies
(*d*) both (*a*) and (*b*)

131. A resistance of 'R' Ω and inductance of 'L' are connected across 240V, 50 Hz supply. Power dissipated in the circuit is 300 W and the voltage across R is 100V. In order to improve the power factor to unity, the capacitor that is to be connected in series should have a value of

(*a*) 43.7 μF (*b*) 4.37 μF
(*c*) 437 μF (*d*) 4.37 mF

132. Which of the following statement(*s*) is/are true of the circuit shown in the given figure?

1. It is first order circuit with steady-state value of $V=\frac{5}{3}V,\ I=\frac{5}{3}A,$
2. It is second order circuit with steady-state values of V = 1 V; I = 1A.
3. The network function $\frac{V(s)}{I(s)}$ has one pole.
4. The network function $\frac{V(s)}{I(s)}$ has two poles

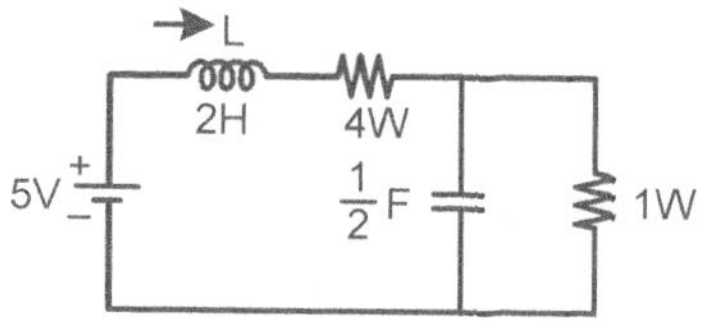

Select the correct answer using the codes given below:

(*a*) 2 and 3 (*b*) 2 and 4
(*c*) 2 alone (*d*) 1 alone

133. For the network shown in the figure,
if $V_s = V_1$ and V = 0, then I = – 5A

and if $V_s = 0$, and $V_1 = 1$, then $I=\frac{1}{2}A$.

The values of I_{sc} and R_i of the Norton' equivalent across AB would be respectively

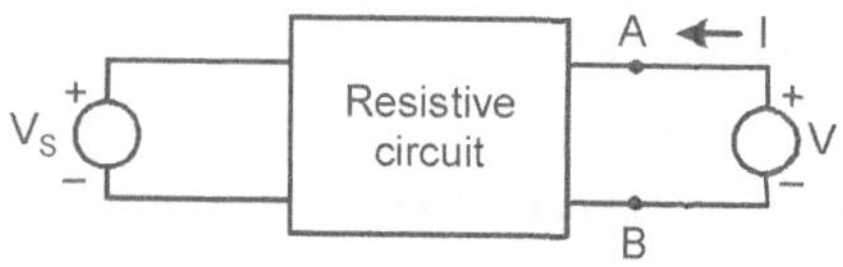

(*a*) – 5A and 2Ω (*b*) 10A and 0.5Ω
(*c*) 5A and 5Ω (*d*) 2.5A and 5Ω

134. Consider the following statements with respect to a series R-L-C circuit under resonance condition:

1. All the applied voltage appears across R.
2. There is no voltage across either L or C.
3. The voltage across L and C equal and equal to their maximum values.

Of these statements

(*a*) 1 alone is correct
(*b*) 2 alone is correct
(*c*) 1 and 3 are correct
(*d*) 1 and 2 are correct

135. A network whose impedance function is $\frac{4(s^2+1)(s^2+9)}{s(s^2+4)}$ is synthesised. It consists of '*n*' LC tank circuits in series with an inductance and/or capacitance. The value of '*n*' is

(*a*) zero (*b*) 1
(*c*) 2 (*d*) 3

136. The realization of the reactance function $Z(s)=\frac{4(s^2+1)(s^2+9)}{s(s^2+4)}$ requires a minimum of

(*a*) 4 inductors and 4 capacitors
(*b*) 3 inductors and 1 capacitor
(*c*) 1 inductor, 1 capacitor and 1 resistor
(*d*) 2 inductors and 2 capacitors

137. The V-1 relation for the network shown in the given figure

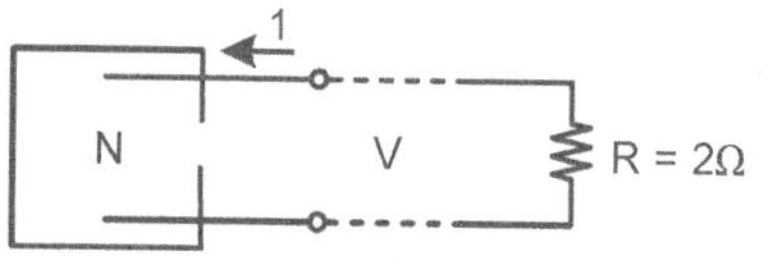

V = 4I – 9.

If now a resistor R = 2Ω is connected across it, then the value of I will be

(*a*) – 4.5 A
(*b*) – 1.5 A
(*c*) 1.5 A
(*d*) 4.5 A

138. In the network shown in the figure, the effective resistance faced by the voltage source is

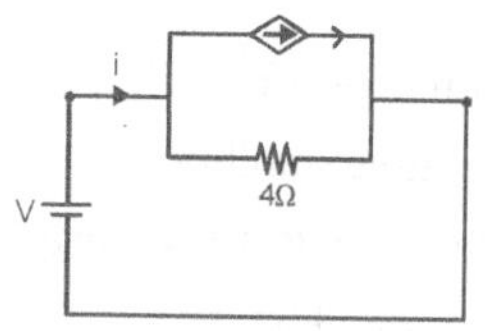

(*a*) 4Ω (*b*) 3Ω
(*c*) 2Ω (*d*) 1Ω

139. For the circuit shown in the given figure, the voltage of the terminal A with respect to earth is

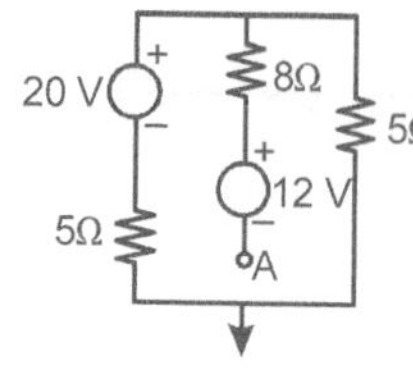

(*a*) –2.0V (*b*) 2.0V
(*c*) –4.0V (*d*) 4.0V

140. 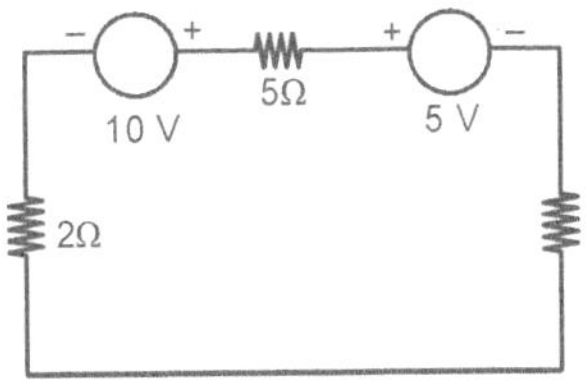

Consider the above circuit.

If the voltage of each source in the given network is doubled, then which of the following statements would be true?

1. Current flowing in the network will be doubled.
2. Voltage across each resistor will be doubled.
3. Power absorbed by each resistor will be doubled.
4. Power delivered by each source will be doubled.

Select the correct answer using the codes given below:

Codes:

(*a*) 1,2, 3 and 4
(*b*) 1 and 2
(*c*) 2 and 3
(*d*) 1,3 and 4

141. The color code of a 1 KΩ resistance is

(*a*) black, brown, red
(*b*) red, brown, brown
(*c*) brown, black, red
(*d*) black, black, red

142. A square wave is fed to an RC circuit. Then

(*a*) voltage across R is square and across C is not square
(*b*) voltage across C is square and across R is not square
(*c*) voltage across both R and C is square
(*d*) voltage across both R and C is not square

143. The current waveform in a pure resistor of 10Ω is shown in the given figure. Power dissipated in the resistor is

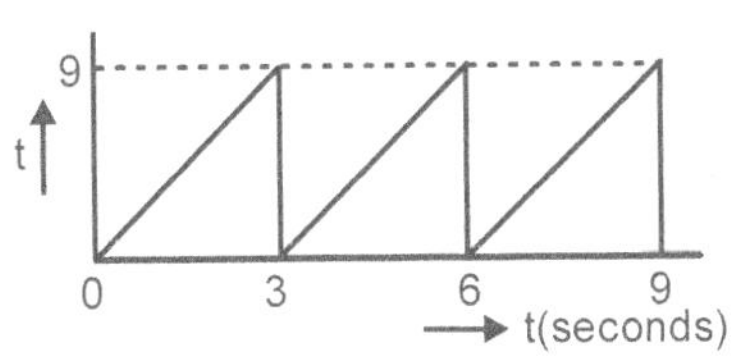

(*a*) 7.29W (*b*) 52.4W
(*c*) 135W (*d*) 270W

144. E_{oa}, E_{ob} and E_{oc} are three phase voltages while E_{ab}, E_{bc} and E_{ca} are the line voltages of a balance three-phase system having a-b-c phase sequence. In relation to E_{oc}, E_{bc} would

(*a*) lag by 30°
(*b*) lead by 30°
(*c*) have the same phase
(*d*) have no definite phase relationship

145. The *v-i* characteristics as seen from the terminal-pair (A, B) of the network of figure is shown in

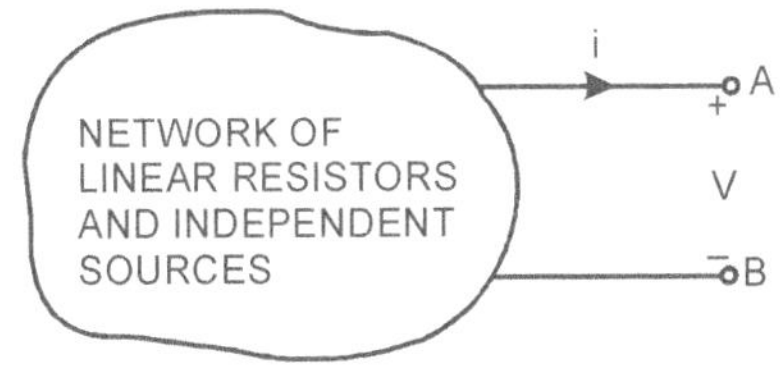

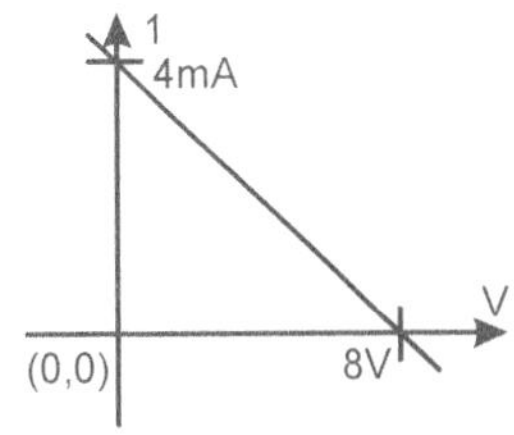

Fig. (*b*).

If an inductance of value 6 mH is connected across the terminal - pair (A, B), the time constant of the system will be

(*a*) 3 μsec
(*b*) 12 sec
(*c*) 32 sec
(*d*) unknown, unless the actual network is specified

146. A coil have a resistance of 5 ohms and inductance of 0.1H connected in series with a condenser of capacitance 50 μF. A constant alternating voltage of 200 volts is applied to the circuit. The voltage across the coil at resonance is

(*a*) 200 volts (*b*) 1788 volts
(*c*) 1800 volts (*d*) 2000 volts

147. The voltage phasor of a circuit is $10\angle 15°$V and the current phasor is $2\angle -45°$A. The active and the reactive powers in the circuit are

(*a*) 10W and 17.32 VAr
(*b*) 5W and 8.66 VAr
(*c*) 20W and 60 VAr
(*d*) $20\sqrt{2}$ W and $10\sqrt{2}$ VAr

148. The driving-point impedance of a reactive network is given by

(*a*) $\frac{(s^2+1)(s^2+2)}{s(s^2+2)(s^2+4)}$ (*b*) $\frac{(s^2+1)(s^2+3)}{s(s^2+2)(s^2+4)}$

(*c*) $\frac{(s^2+1)}{(s^2+2)(s^2+3)}$ (*d*) $\frac{1}{s+1}$

149. If the number of branches in a network is 'B', the number of nodes is 'N' and number of dependent loops is 'L', then the number of independent node equations will be

(*a*) N + L − 1 (*b*) B − 1
(*c*) B − N (*d*) N − 1

150. The open-loop transfer function of a unity negative feedback control system is given by

$$G(s) = \frac{k(s+2)}{(s+1)(s-7)}$$

For K > 6, the stability characteristics of the open-loop and closed-loop configurations of system are respectively

(*a*) stable and stable

(*b*) unstable and stable

(*c*) stable and unstable

(*d*) unstable and unstable

151. Two, two-port networks with transmission parameters A_1, B_1, C_1, D_1 and A_2 B_2, C_2, D_2 respectively are cascaded. The transmission parameter matrix of the cascaded network will be

(*a*) $\begin{bmatrix} A_1 & B_1 \\ C_1 & D_1 \end{bmatrix} + \begin{bmatrix} A_2 & B_2 \\ C_2 & D_2 \end{bmatrix}$

(*b*) $\begin{bmatrix} A_1 & B_1 \\ C_1 & D_1 \end{bmatrix}\begin{bmatrix} A_2 & B_2 \\ C_2 & D_2 \end{bmatrix}$

(*c*) $\begin{bmatrix} A_1A_2 & B_1B_2 \\ C_1C_2 & D_1D_2 \end{bmatrix}$

(*d*) $\begin{bmatrix} (A_1A_2 + C_1C_2) & (A_1A_2 - B_1D_2) \\ C_1A_2 - D_1C_2) & (C_1C_2 + D_1D_2) \end{bmatrix}$

152. The frequency response of two single turned networks have the following characteristics:

Network A has a single half-power frequency;

Network B has two half-power frequencies.

Which of the following statements are true for these two networks?

1. The maximum amplitude response of network A will occur at zero frequency
2. The maximum amplitude response of network B will occur between the two half-power frequencies
3. The phase response of network A at very low frequencies will be very nearly zero
4. The phase response of network B at very low frequencies will be nearly – 180°

Select the correct answer using the codes given below

Codes:

(*a*) 1, 2 and 4

(*b*) 2, 3 and 4

(*c*) 1 and 3

(*d*) 1, 2, 3 and 4

153. The ratio of resistances of a 100W, 220V lamp to that of a 100V, 110V lamp will be at respective voltages

(*a*) 4 (*b*) 2

(*c*) 1/2 (*d*) 1.4

154. Kirchoffs laws are valid for

(*a*) linear circuits only

(*b*) passive time invariant circuits

(*c*) non-linear circuits only

(*d*) both linear and non linear circuits.

155. In the Wheatstone bridge in figure given below if the resistance in each arm is increased by 0.5% then the value of V_{out}, will be

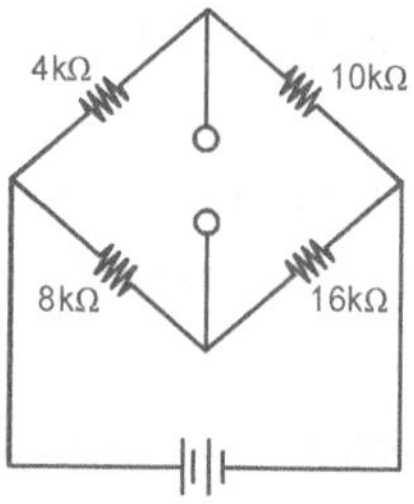

(*a*) 50mV (*b*) 5mV

(*c*) 0.1mV (*d*) zero

156. Which of the following theorems is applicable to both linear and non linear circuits?

(*a*) Superposition

(*b*) Thevenin's

(*c*) Notions

(*d*) None of these

157. The resistance between the opposite faces of 1 m cube is found to be 1Ω. If its length is increased to 2m, with its volume remaining the same, then its resistance between the opposite faces along its length is

(*a*) 2Ω (*b*) 4Ω

(*c*) 1Ω (*d*) 8Ω

158. A wire of length *l* and of circular cross section of radius *r* has a resistance of R ohms. Another wire of same material and of sectional radius 2*r* will have the same resistance R, if the length is

(*a*) 2*l*

(*b*) *l*/2

(*c*) 4*l*

(*d*) l^2

159. Two resistors R_1 and R_2 gives combined resistance of 4.5Ω when in series and 1Ω, when in parallel, the resistances are

(*a*) 2Ω and 2.5Ω

(*b*) 1Ω and 3.5Ω

(*c*) 1.5Ω and 3Ω

(*d*) 4Ω and 0.5Ω

160. Two heaters, rated at 1000W, 250V each are connected in series across a 250V, 50Hz ac mains. The total power drawn from the supply would be

(*a*) 1000W (*b*) 500W

(*c*) 250W (*d*) 2000W

161. If a pulse voltage v(t) of 4V magnitude and 2secs duration is applied to a pure inductor of 1 H, with zero initial current, the current (in amps) drawn at t = 3secs will be

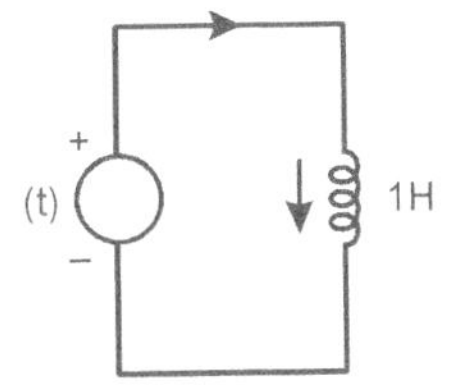

(a) zero

(b) 2

(c) 4

(d) 8

162. Consider the following statements:

I. Any function in 's' which can be expressed as a ratio of any two arbitrary polynomials in 's' can be released as a driving-point function of a passive network.

II. Any function in 's' which can be expressed as a ratio of two arbitrary polynomials in 's', is positive real function.

Of these statements

(a) both I and II are true

(b) I is true but II is false

(c) both I and II are false

(d) 1 is false but II is true

163. A signal is described by

$$s(t) = r(t-a) - r(t-b),\ a < b,$$

where $r(t)$ is a unit ramp function starting at $t = 0$. The signal $s(t)$ is represented as

(a)
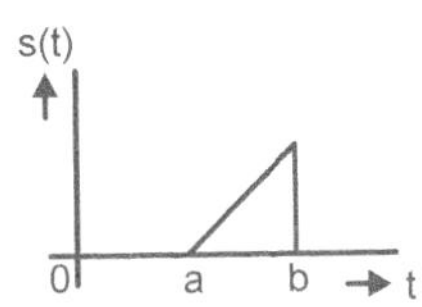

(b)
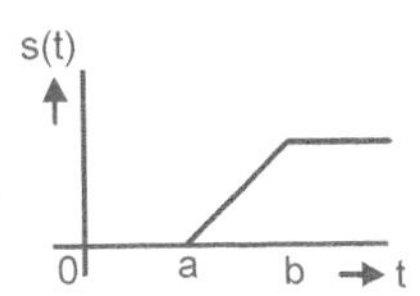

(c)
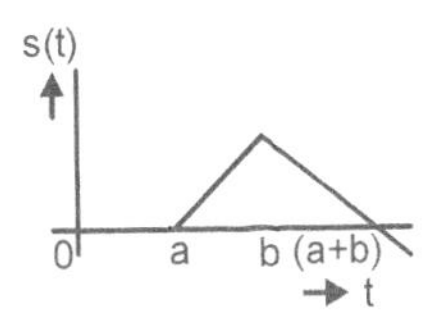

(d)
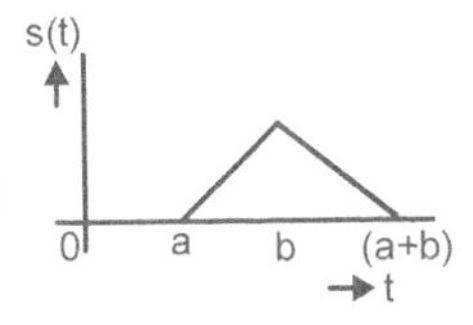

164. The voltage ratio transfer function for the network shown in the given figure under sinusoidal steady state conditions is G($j\omega$).

Match List-I with List-II and select the correct answer using the codes given below the lists:

List-I (*frequency*)	**List-II** ($G(j\omega)$)
1. 0	A. $1\angle 0°$
2. $\dfrac{1}{RC}$	B. $0\angle -90°$
3. α	C. $0.707\angle -45°$

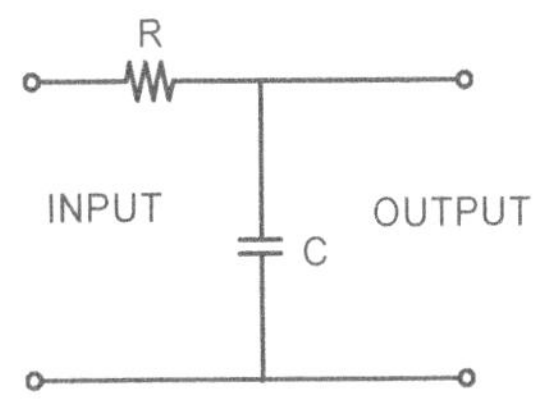

Codes:

	A	B	C
(a)	1	3	2
(b)	2	1	3
(c)	1	2	3
(d)	3	2	1

165. The h-parameters for a two-port network are defined by

$$\begin{bmatrix} E_1 \\ I_2 \end{bmatrix} = \begin{bmatrix} h_{11} & h_{12} \\ h_{21} & h_{22} \end{bmatrix} \begin{bmatrix} I_1 \\ E_2 \end{bmatrix}$$

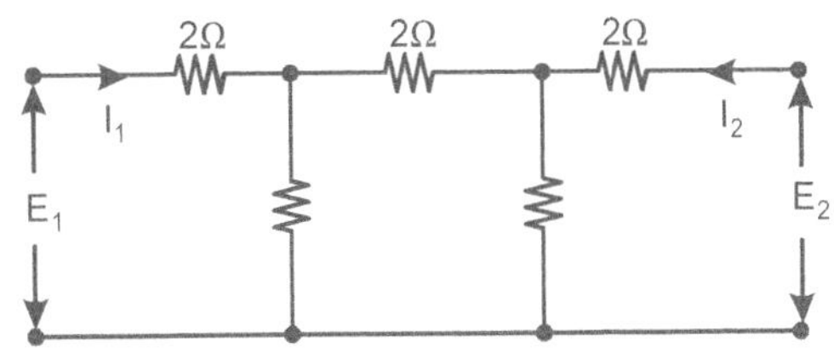

For the two-port network shown in the figure, the value of h_{12} is given by

(a) 0.125 (b) 0.167

(c) 0.625 (d) 0.25

166. The response of a series RLC circuit fed from a fixed rms voltage and variable frequency source is represented graphically in the given figure. Match List-I with List-II and select the correct answer using the codes given below the lists:

List-I (*curve*)	**List-II** (*quantity*)
A. AA	1. Current
B. BB	2. Impedance
C. CC	3. Captive
D. DD	4. Net reactance
	5. Inductile reactance

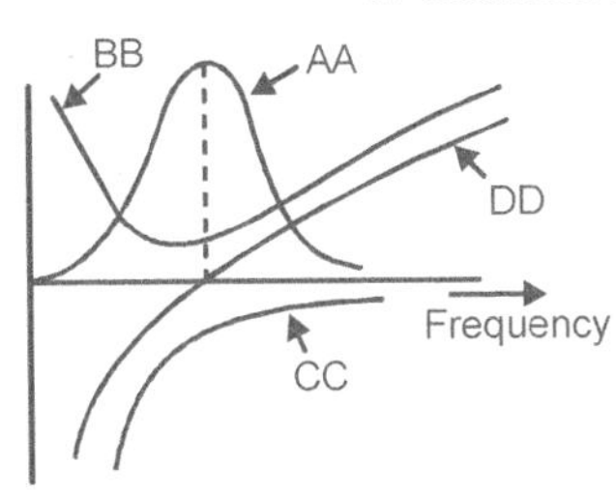

Codes:

	A	B	C	D
(a)	2	1	3	5
(b)	1	2	3	5
(c)	1	2	3	4
(d)	1	2	4	3

167. For the given network, match List-I with List-II and select the correct answer using the codes given below the lists:

List-I	List-II
A. Z_{11}	1. $\frac{1}{3}$
B. Z_{21}	2. $\frac{9}{8}$
C. y_{12}	3. $-\frac{3}{8}$
D. y_{22}	4. 1

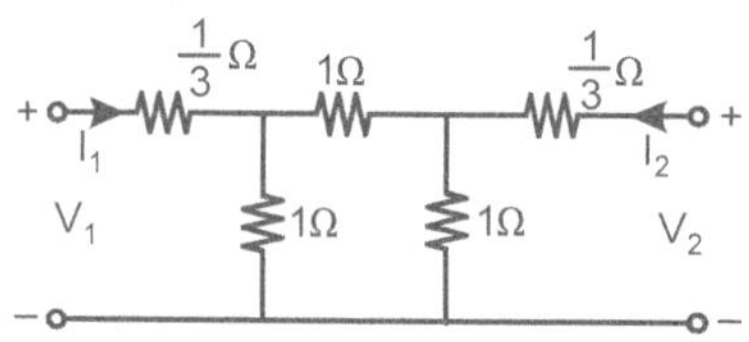

Codes:

	A	B	C	D
(*a*)	4	1	2	3
(*b*)	1	4	3	2
(*c*)	1	4	2	3
(*d*)	4	1	3	2

168. A voltage source with an internal resistance Rs, supplies power to a load R_L. The power delivered to the load varies the R_L as

(*a*) 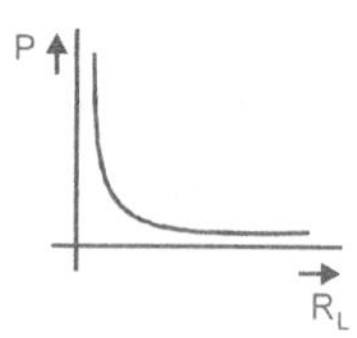
(*b*) 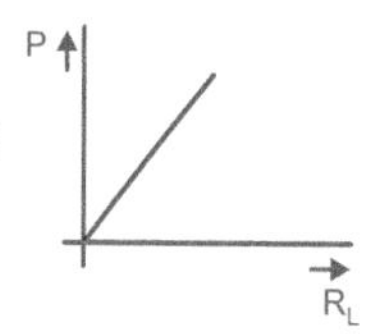
(*c*) 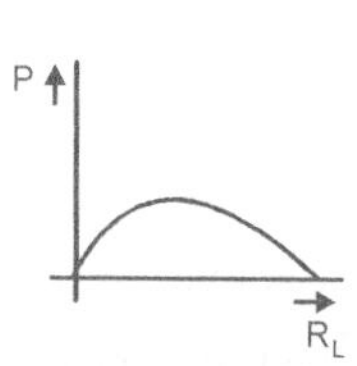
(*d*) 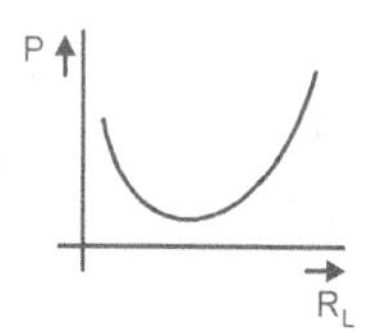

169. A pulse of unit amplitude and width '*a*' is applied to a series RL circuit as shown in the figure. The current *i*(*t*) as '*t*' tends to infinity will be

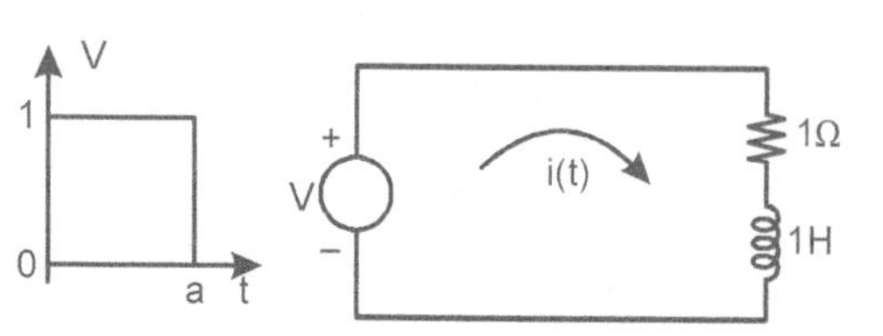

(*a*) zero
(*b*) 1A
(*c*) a value between zero and one depending upon the width of the pulse
(*d*) infinite

170. An inductor with inductance L and initial current I is shown as

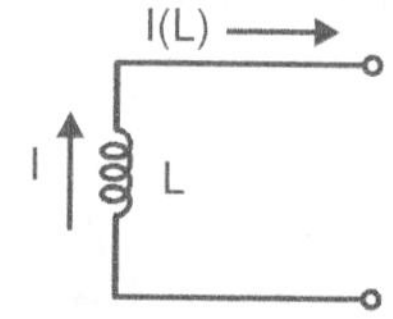

The correct admittance diagram for it is

(*a*) 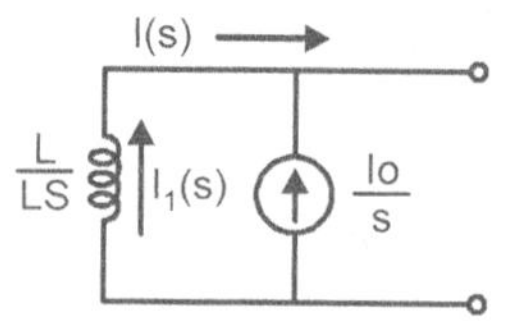

(*b*) 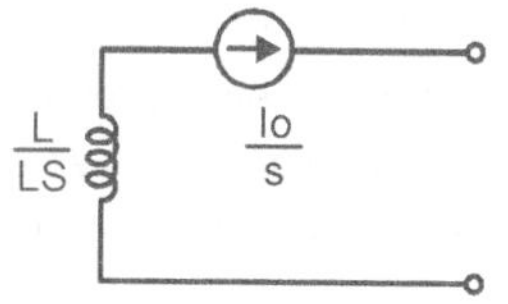

(*c*) 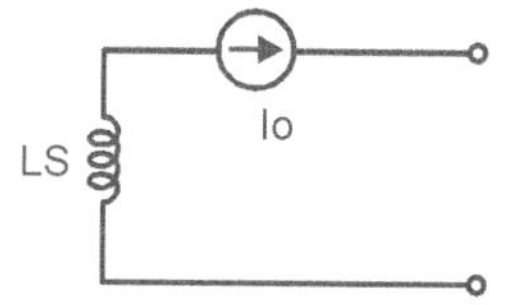

(*d*) 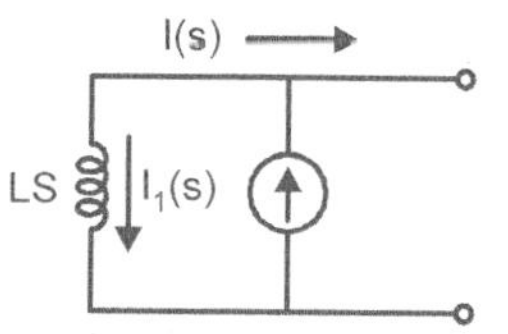

171. The copper armature winding of a motor is subject to an operating temperature of 80°C. The room temperature is 20°C. The percentage change in resistance of the armature winding from cold starting condition to normal running will be nearly

(*a*) 10% (*b*) 16%
(*c*) 20% (*d*) 24%

172. Let *s*(*t*) be the step response of a linear system with zero initial conditions;then response of this system to an input *u*(*t*) is

(*a*) $\int_0^t s(t-\tau)u(\tau)d\tau$

(*b*) $\frac{d}{dt}\left[\int_0^t s(t-\tau)u(\tau)\,d\tau\right]$

(*c*) $\int_0^t s(t-\tau)\left[\int_0^t s(\tau_1)d\tau_1\right]d\tau$

(*d*) $\int_0^t s(t-\tau)^2 u(\tau)d\tau$

173. In the network shown in figure, the moving-iron ammeters A_1 and A_2 read 5A and 12A respectively. The impedance Z_1 can be a pure resistor, or an inductor, or a capacitor, and so is Z_2. The possible readings of the ammeter A_3 are

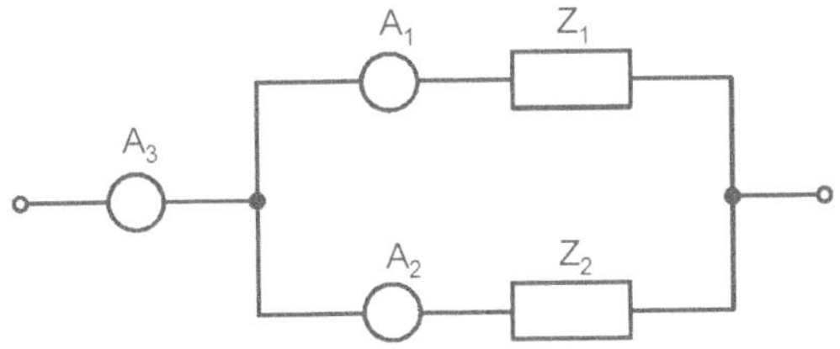

(*a*) $5\sqrt{2}$ (*b*) 13

(*c*) $12\sqrt{2}$ (*d*) 7

174. For the series R-L-C circuit of figure (*a*), below, the partial phasor diagram at a certain of frequency is shown in figure (*b*). The operating frequency of the circuit is

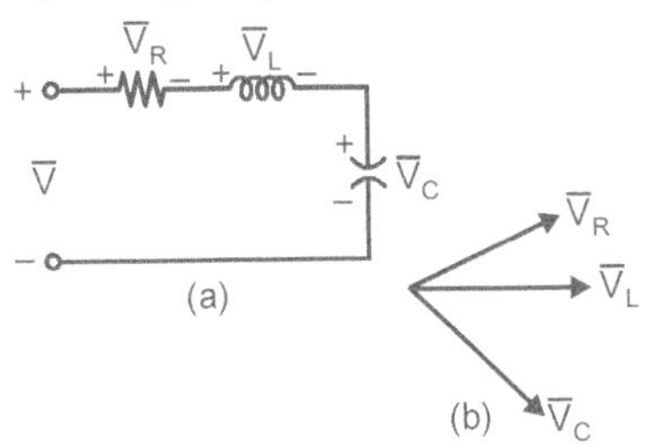

(*a*) equal to the resonance frequency

(*b*) less than the resonance frequency

(*c*) greater than the resonance frequency

(*d*) not zero

175. The impulse response of anR-L circuit is a

(*a*) rising exponential function

(*b*) decaying exponential function

(*c*) step function

(*d*) parabolic function

176. In addition to the condition that Y(*s*) is real when '*s*' is real, for admittance function Y(*s*) to be positive real, which of the following conditions are to be satisfied?

1. Re $Y(s) \leq 0$ for Re $s \geq 0$
2. Re $Y(s) \geq 0$ for Re $s \geq 0$
3. $|\text{Arg } Y(s)| \leq 0$ Arg s | for | Arg s | $\leq /\pi 2$.
4. $|\text{Arg } Y(s)| \geq 0$ Arg s | for | Arg s | $\leq /\pi 2$.

Select the correct answer using the codes given below

(*a*) 1 and 3 are correct (*b*) 1 and 4 are correct

(*c*) 2 and 3 are correct (*d*) 2 and 4 are correct

177. A circuit with a resistor, inductor and a capacitor in series is resonant of f_e Hz. If all the component values are now doubled, the new resonant frequency is

(*a*) $2f_0$ (*b*) Still f_0

(*c*) $f_0/4$ (*d*) $\frac{f_0}{2}$

178. In a three-phase circuit, the power-factor of the load is known to be 0.5 (leading). The power consumed by the balanced load is measured by the two-wattmeter method. Under this condition,

(*a*) both the wattmeter readings are equal

(*b*) one wattmeter reads zero

(*c*) the wattmeter readings are in the ratio 2 :1

(*d*) both the wattmeters read zero

179. In the circuit shown in the figure, the switch S is closed at time $t = 0$. The voltage across the inductance at $t = 0^+$, is

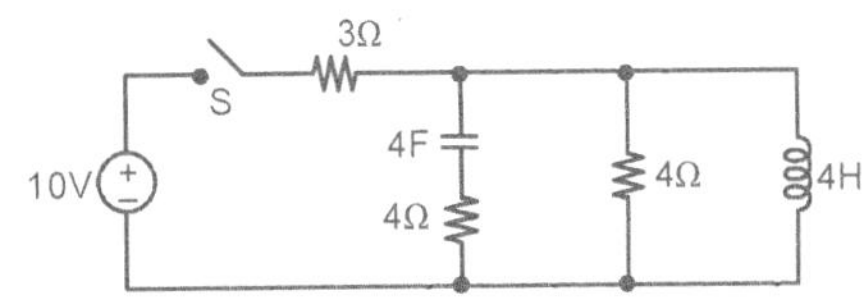

(*a*) 2V (*b*) 4V

(*c*) –6V (*d*) 8V

180. The figures show two different sets of input and output variables for the same two-port resistive network N.I_x is

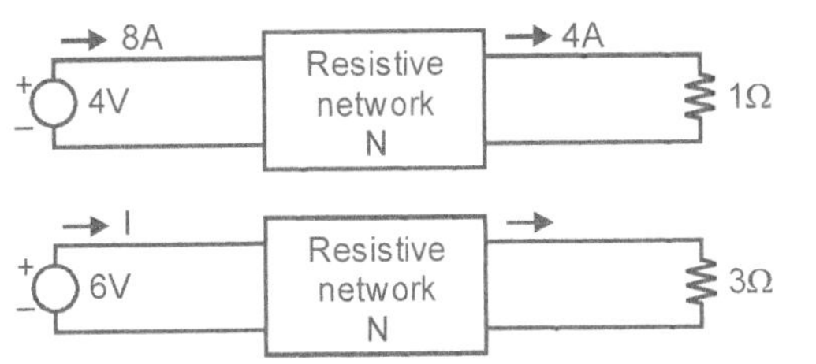

(*a*) 12A

(*b*) 8A

(*c*) 4A

(*d*) $\frac{3}{2}$A

181. Two coils having equal resistances but different inductances are connected in series. The time constant of the series combination is the

(*a*) sum of the time constants of the individual coils

(*b*) average of the time constants of individual coils

(*c*) geometric mean of the time constants of the individual coils

(*d*) product of the time constants of the individual coils

182. The circuit shown in figure, is a

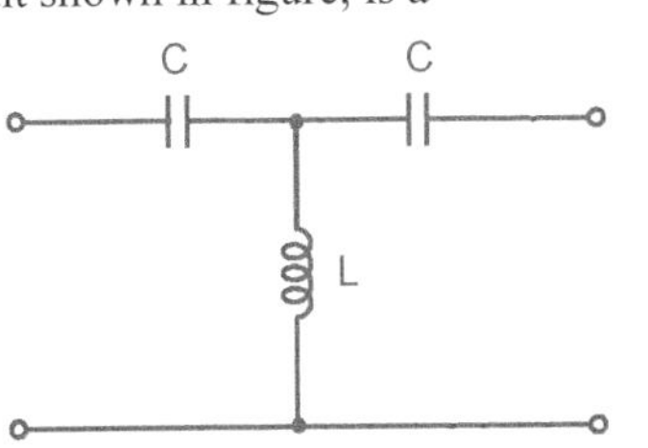

(*a*) Low-pass filter

(*b*) High-pass filter

(*c*) Band-pass filter

(*d*) Differentiator

183. The Laplace transform of v(*t*) shown in the figure is

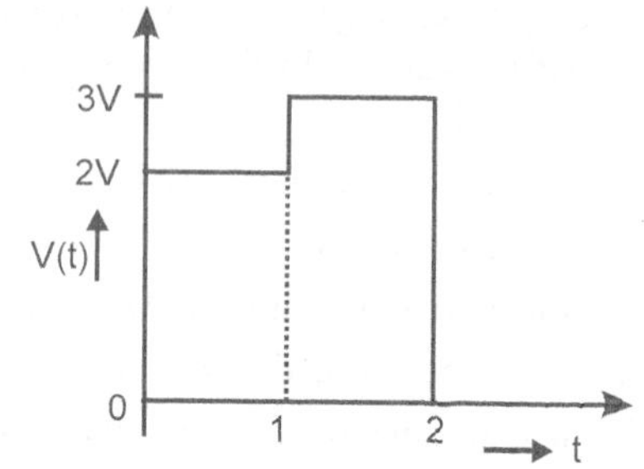

(*a*) $\frac{v}{s}e^{-s} - \frac{3v}{s}e^{-2s}$

(*b*) $\frac{2v}{s} - \frac{3v}{s}e^{-2s}$

(*c*) $\frac{2v}{s} + \frac{v}{s}e^{-s}$

(*d*) $\frac{2v}{s} + \frac{v}{s}e^{-s} - \frac{3v}{s}e^{-2s}$

184. The number of 2μF, 400V capacitors needed to obtain a capacitance value of 1.5 μF rated for 1600V is

(*a*) 12 (*b*) 8
(*c*) 6 (*d*) 4

185. In the circuit shown in the figure, the switch S has been open for a long time. It is closed at $t = 0$. For $t > 0$ the current flowing through the inductor will be given by

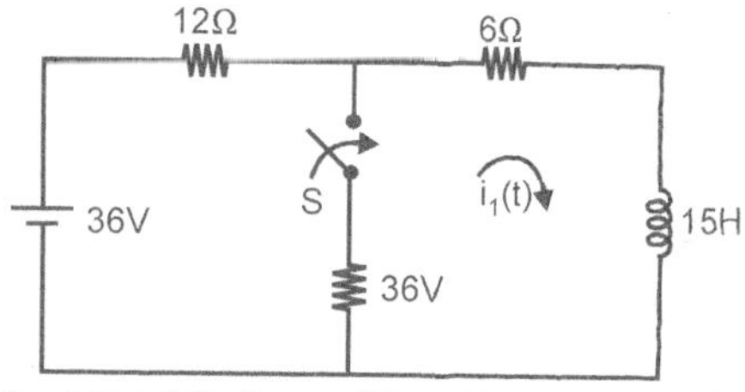

(*a*) $i_L(t) = 1.2 + 0.8e^{-2t}$ (*b*) $i_L(t) = 0.8 + 1.2e^{-2t}$
(*c*) $i_L(t) = 1.2 - 0.8e^{-2t}$ (*d*) $i_L(t) = 0.8 - 1.2e^{-2t}$

186. Like a series R-L-C resonant circuit, a parallel resonant circuit also

(*a*) magnifies current .
(*b*) offer minimum impedance
(*c*) draws maximum current
(*d*) has unity power factor

187. The transient response of the initially relaxed network shown in the figure is

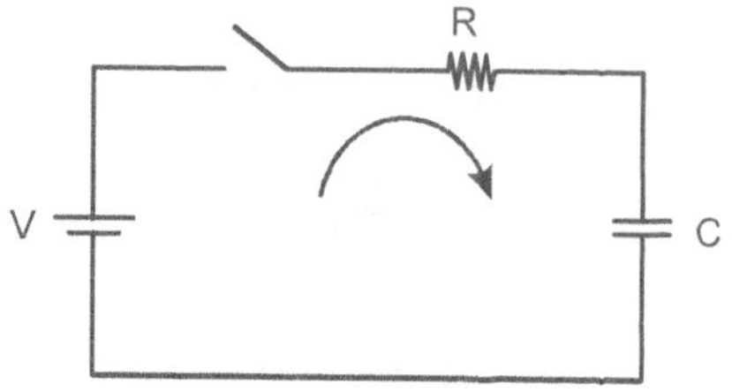

(*a*) $i = \frac{V}{R}e^{-t/RC}$ (*b*) $i = \frac{V}{R}e^{t/RC}$

(*c*) $i = \frac{V}{R}(1 - e^{-t/RC})$ (*d*) $i = \frac{V}{R}(1 + e^{-t/RC})$

188. When a resistor R is connected to a current source, it consumes a power of 18 W. When the same R is connected to a voltage source having the same magnitude as the current source, the power absorbed by R is 4.5W. The magnitude of the current source and the value of R are

(*a*) $\sqrt{18}$A and 1Ω (*b*) 3A and 2Ω
(*c*) 1.4A and 18Ω (*d*) 6A and 0.5Ω

189. The network function $F(s) = \frac{(s+2)}{(s+1)(s+3)}$ represents an

(*a*) RC impedance
(*b*) RL impedance
(*c*) RC impedance and an RL admittance
(*d*) RC admittance and an RL impedance

190. For $V(s) = \frac{(s+2)}{s(s+1)}$, the initial and final value of v(*t*) will be respectively

(*a*) 1 and 1 (*b*) 2 and 2
(*c*) 2 and 1 (*d*) 1 and 2

191. The Fourier transform of a signal g(*t*) is G(*t*). The Fourier transform of the signal *g*(*kf*), where k is a negative constant will be

(*a*) $-kG(f)$ (*b*) $-\frac{1}{k}G(f)$

(*c*) $\frac{1}{k}G\left(\frac{f}{k}\right)$ (*b*) $\frac{1}{|k|}G(-kf)$

192. In a linear network, the ratio of voltage excitation to current response is unaltered when the position of excitation and response are interchanged. This assertion stems from the

(*a*) principle of duality
(*b*) reciprocity theorem
(*c*) principle of superposition
(*d*) equivalence theorem

193. An ac source of voltage E_s and an internal impedance of $Z_s = (R_s + jX_s)$ is connected to a load of impedance $Z_L = (R_L + jX_L)$. Consider the following conditions in this regard:

1. $X_L = X_s$, if only X_L is varied
2. $X_L = X_s$, if only X_L is varied
3. $R_L = \sqrt{R_s^2 + (X_s + X_L)^2}$, if only RL is varied
4. $|Z_L| = |Z_s|$, if the magnitude alone of Z_L is varied, keeping the phase angle fixed. Among these conditions, those which are to be satisfied for maximum power transfer from the source to the load would include

(*a*) 2 and 3
(*b*) 1 and 3
(*c*) 1, 2 and 4
(*d*) 2, 3 and 4

EXERCISE – II

1. If a capacitor is charged by a square wave current source, the voltage across the capacitor is
   (*a*) a square wave
   (*b*) triangular wave
   (*c*) step function
   (*d*) zero **[BEL]**
2. A system defined by $\frac{dy}{dt} + y = x$ is initially at rest for input $x = u \sin 2t$. The forced response is K sin $(2t - \theta)$. The value of θ is given by
   (*a*) $\tan\theta = 1$ (*b*) $\tan\theta = 2$
   (*c*) $\tan\theta = 1/2$ (*d*) $\theta = 0$ **[BHEL]**
3. The standing wave ratio of a 75-Ω transmission line used to feed a 300-Ω resistance load will be
   (*a*) 1 (*b*) 2
   (*c*) 3 (*d*) 4 **[BHEL]**
4. In case of delta connected circuit, when one resistor is open, the power will be
   (*a*) zero (*b*) reduced by 2/3
   (*c*) reduced by 1/3 (*d*) none of these **[BHEL]**
5. System function $H(s) = \frac{1}{s+2}$ for a signal cos $2t$. The steady state response is
   (*a*) $\frac{1}{2}\cos 2t$
   (*b*) $\frac{1}{2\sqrt{2}}$ $\cos(2t - 45°)$
   (*c*) $\frac{1}{2\sqrt{2}}$ $\cos(2t + 45°)$
   (*d*) $\frac{1}{2\sqrt{2}}\cos 2t$ **[BHEL]**
6. Kirchoff's voltage law states that
   (*a*) the total voltage drop in a series circuit is always finite
   (*b*) the sum of emf and voltage drops in a closed mesh is zero
   (*c*) the sum of emf in a series circuit is zero
   (*d*) none of these **[BHEL]**
7. When a source is delivering maximum power to a load, the efficiency of the circuit
   (*a*) is always 50%
   (*b*) depends on the circuit parameters
   (*c*) is always 75%
   (*d*) none of these **[BHEL]**
8. The expression for current $i(t)$ and voltage $v(t)$ for an inductor is given as $v(t) = L\frac{di}{dt}$, where t represents the time. The above inductor is
   (*a*) linear-time invariant
   (*b*) linear-time varying
   (*c*) none-linear-time invariant
   (*d*) non-linear-time varying **[NTPC]**
9. In series LCR circuit, at resonance
   (*a*) current is maximum, power factor is zero
   (*b*) current is maximum, power factor is unity
   (*c*) current is maximum, power factor is unity
   (*d*) none of these **[NTPC]**
10. A network is said to be non-linear if it does not satisfy
    (*a*) super position condition
    (*b*) homogeneity condition
    (*c*) both (*a*) and (*b*)
    (*d*) associative condition **[NTPC]**
11. Which of the following is an active element of a circuit?
    (*a*) Resistance
    (*b*) Inductance
    (*c*) Ideal current source
    (*d*) Capacitance **[NTPC]**
12. Which of the following theorems enables a number of voltage (or current) source to be combined directly into a single voltage (or current) source?
    (*a*) Compensation theorem
    (*b*) Reciprocity theorem
    (*c*) Superposition theorem
    (*d*) Millman's theorem **[NTPC]**
13. The current in an RLC series circuit at resonance is
    (*a*) maximum
    (*b*) minimum
    (*c*) infinity
    (*d*) zero **[BHEL]**
14. Two coupled coils of $L_1 = 0.8$H and $L_2 = 0.2$H have a coupling coefficient of $K = 0.9$. The mutual inductance M is
    (*a*) 0.144 H (*b*) 0.23 H
    (*c*) 0.36 H (*d*) 0.43 H **[BHEL]**
15. For a 2-port netowrk, the output short circuit current was measured with a 1 V source at the input. The value of the current gives
    (*a*) h_{12} (*b*) y_{12}
    (*c*) h_{21} (*d*) y_{21} **[BHEL]**

16. $H(s) = \dfrac{V(s)}{I(s)} = \dfrac{4s+6}{s+2}$ when $i(t)$ is a unit step function, as t tends to infinity is given by the value of $v(t)$

(*a*) 0 (*b*) 3

(*c*) c (*d*) ∞ **[BHEL]**

17. A voltage source $Vs = 200 \cos \omega t$ has impedance across it. The reactive power is 200 VARs and the power factor is 0.866. The power (P_{av}) drawn from the source is

(*a*) 430 watts (*b*) 400 watts

(*c*) 344 watts (*d*) 200 watts **[PGCIL]**

18. A system function is $H(s) = \dfrac{V(s)}{I(s)} = \dfrac{s}{s+3}$

The system is at rest for $t < 0$, $i(t) = 2u(t)$ where $u(t)$ is a unit step u(t) for $t > 0$ is given by

(*a*) $2(1 - e^{-3t})$ (*b*) $\dfrac{2}{3}(1 - e^{-3t})$

(*c*) $2\,e^{-3t}$ (*d*) $\dfrac{2}{3}e^{-3t}$ **[PGCIL]**

19. Which of the following does not have dual relationship?

(*a*) Current-voltage

(*b*) Inductance-capacitance

(*c*) Tie set-cut-set

(*d*) Resistance-reactance **[PGCIL]**

20. A coil with large distributed capacitance has a

(*a*) low resistance (*b*) low θ

(*c*) low resonant frequency (*d*) high resonant frequency **[PGCIL]**

21. A coil (which can be modeled as a series-RL, circuit) has been designed for high-Q performance at a rated voltage and a specified frequency. If the frequency of operation is doubled, and the coil is operated at the same rated voltage, then the Q-factor and the active power P consumed by the coil will be affected as follows:

(*a*) P is doubled, Q is halved

(*b*) P is halved, Q is doubled

(*c*) P remains constant, Q is doubled

(*d*) P- decreases 4 times, Q is doubled **[GATE]**

22. The transfer function H(*s*) of a system is given by

$$H(s) = \frac{Y(s)}{X(s)} = \frac{s+2}{s^2+s+4}$$

Given that under steady-state conditions, the sinusoidal input and output are respectively

$x(t) = \cos 2t$

and $y(t) = \cos(2t + \phi)$,

then the angle ϕ will be

(*a*) 45° (*b*) zero°

(*c*) −45° (*d*) −90° **[IAS]**

23. The rms value of the periodic waveform $e(t)$, shown in the figure is

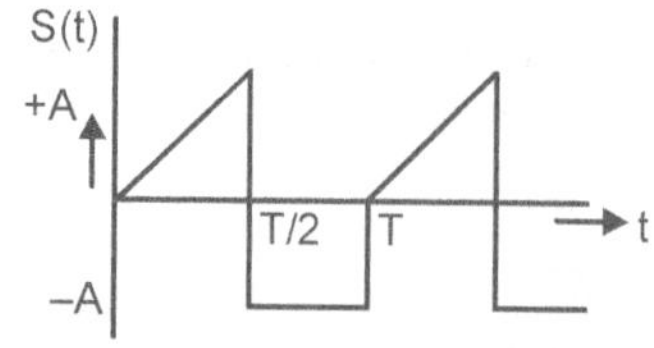

(*a*) $\sqrt{3/2}$A (*b*) $\sqrt{2/3}$A

(*c*) $\sqrt{1/3}$A (*d*) $\sqrt{2}$A **[GATE]**

24. In a series RLC circuit at resonance, the magnitude of the voltage developed across the capacitor

(*a*) is always zero

(*b*) can never be greater than the input voltage

(*c*) can be greater than the input voltage, however, it is 90° out of phase with input voltage

(*d*) can be greater than the input voltage, and is in phase with the input voltage **[GATE]**

25. Energy stored in a capacitor over a cycle, when excited by an a.c. source is

(*a*) the same as that due to a d.c. source of equivalent magnitude

(*b*) half of that due to a d.c. source of equivalent magnitude

(*c*) zero

(*d*) none of these **[GATE]**

26. A practical current source is usually represented by

(*a*) a resistance in series with an ideal current source

(*b*) a resistance in parallel with an ideal current source

(*c*) a resistance in parallel with an ideal voltage source

(*d*) none of these **[GATE]**

27. An ideal voltage source will charge an ideal capacitor

(*a*) in infinite time (*b*) exponentially

(*c*) instantaneously (*d*) none of these **[GATE]**

28. The function $\delta(2n)$ is equal to

(*a*) $\delta(n)$

(*b*) $\dfrac{1}{2}\delta(n)$

(*c*) $2\delta(n)$

(*d*) $2\delta\left(\dfrac{n}{2}\right)$ **[IAS]**

29. Two coupled coils connected in series have an equivalent inductance of 16 mH or 8 mH depending on the interconnection then the mutual inductance m between the coils is

(*a*) 12mH (*b*) $8\sqrt{2}$ mH

(*c*) 4mH (*d*) 2mH **[IOCL]**

30. An ac current is given as

$$i = 10 + 10 \sin 314t$$

the average and rms values of the current are

(*a*) 16.36A, 17.07A (*b*) 10A, 17.07A

(*c*) 10A, 12.25A (*d*) 16.36A, 12.2A **[IOCL]**

31. The rms value of a half wave rectified current is 50A, its rms value for full wave rectification would be

(*a*) 100A (*b*) 70.7A

(*c*) $\frac{50}{\pi}$A (*d*) $\frac{100}{\pi}$A **[IOCL]**

32. In an ac circuit if voltage V = $(a + jb)$ and current I = $(c + jd)$, then the power is given by

(*a*) $ac + ad$ (*b*) $ac + bd$

(*c*) $bc - ad$ (*d*) $bc + ad$ **[AMIE]**

33. In a series RLC circuit, the voltage across inductance will be maximum

(*a*) at resonant frequency

(*b*) just after resonant frequency

(*c*) just before resonant frequency

(*d*) just before and after resonant frequency **[IOCL]**

34. Tellegen's theorem is applicable to

(*a*) linear networks only

(*b*) non-linear networks only

(*c*) both linear and nonlinear networks

(*d*) none of these **[BHEL]**

35. The realization of the reactance function

$Z(s) \frac{4(s^2+1)(s^2+9)}{s(s^2+4)}$ requires a minimum of

(*a*) 4 inductors and 4 capacitors

(*b*) 3 inductors and 1 capacitor

(*c*) 1 inductor, 1 capacitor, and 1 resistor

(*d*) 2 inductors and 2 capacitors **[BHEL]**

36. If $F(s) = \frac{1(s+1)}{s(s+K)}$ and $\underset{t \to \infty}{f(t)} = \frac{1}{2}$, then the value of K is

(*a*) 1/2 (*b*) 1

(*c*) 2 (*d*) ∞ **[BHEL]**

37. Laplace transform of a delayed unit impulse function $\delta(t-1)$is

(*a*) unity (*b*) zero

(*c*) e^{-s} (*d*) s **[AM1E]**

38. In a two wattmeter method of measuring power in a 3-phase system one of the wattmeters reads negative employing

(*a*) wattmeter connection is faulty

(*b*) load is unbalanced

(*c*) power flow is in the reverse direction

(*d*) power factor is less than 0.5 **[AMIE]**

39. Lamps of 40W, 100W and 200W ratings are connected in different phases of a 3-phase, 4-wire supply. If the neutral wire breaks, then the lamp likely to fuse first will be

(*a*) 200 W (*b*) 100 W

(*c*) 40 W (*d*) all lamps **[BHEL]**

40. A circuit component that opposes the change in circuit current is

(*a*) resistance (*b*) capacitance

(*c*) inductance (*d*) conductance **[AMIE]**

41. In R-C series circuit excited by a dc source, the circuit current is maximum at

(*a*) $t = 0^-$ (*b*) $t = 0^+$

(*c*) $t = \infty$ (*d*) $t =$ CR **[BHEL]**

42. Phase reversal in a 4 wire unbalanced load supplied from a balanced 3-phase supply causes change in

(*a*) the power consumed

(*b*) magnitude of phase currents

(*c*) only the magnitude of the neutral current

(*d*) magnitude as well as phase angle of the neutral current. **[BHEL]**

43. Lamp load is being supplied from a 3-phase 4 wire, 230/400V ac supply system. If a 3-phase motor is now switched on across the same supply, then neutral current will

(*a*) increase (*b*) decrease

(*c*) remain unchanged (*d*) unpredictable **[BHEL]**

44. In two wattmeter method of power measurement next, one of the wattmeters will show negative reading when the load power factor angle is strictly

(*a*) less than 30° (*b*) less than 60°

(*c*) greater than 30° (*d*) greater than 60° **[IOCL]**

45. In a series resonant circuit the impedance below resonant frequency is

(*a*) capacitive

(*b*) inductive

(*c*) resistive

(*d*) depends upon the values of circuit components **[IOCL]**

46. A periodic rectangular signal, $x(t)$ has the wave form shown in the figure. Frequency of the fifth harmonic of its spectrum is

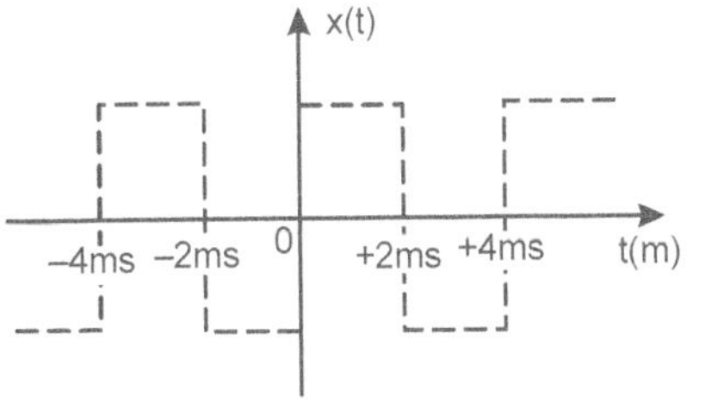

(*a*) 40 Hz (*b*) 200 Hz

(*c*) 250 Hz (*d*) 1250 Hz **[GATE]**

47. With the decrease in supply frequency (capacitive), reactance of the circuit will
   (*a*) decrease
   (*b*) increase
   (*c*) remain the same
   (*d*) unpredictable **[NTPC]**

48. A boiler at home is switched on to the ac mains supplying powers at 230V 50 Hz. The frequency of the instantaneous power consumed is
   (*a*) 0 Hz (*b*) 50 Hz
   (*c*) 100 Hz (*d*) 150 Hz **[GATE]**

49. A circuit component that opposes the change in circuit voltage is
   (*a*) resistance (*b*) capacitance
   (*c*) inductance (*d*) all of these **[AMIE]**

50. An instantaneous change in voltage is not possible in
   (*a*) resistor (*b*) an inductor
   (*c*) a capacitor (*d*) a current source **[AMIE]**

51. The power factor of an ac circuit lies between
   (*a*) 0 and 1 (*b*) –1 and 1
   (*c*) 0 and – 1 (*d*) none of these

52. The maximum efficiency of power transformer to the load is
   (*a*) 25% (*b*) 50%
   (*c*) 75% (*d*) 100% **[NTPC]**

53. Cells are connected in series in order to increase the
   (*a*) current capacity
   (*b*) life of the cells
   (*c*) voltage rating
   (*d*) terminal voltage **[BHEL]**

54. A constant current of 2.8A exists in a resistor. The rms value of current is
   (*a*) 2.8A
   (*b*) about 2A
   (*c*) 1.4A
   (*d*) undifined **[AMIE]**

55. An alternating voltage $e = 200 \sin 314t$ is applied to a device which offers an ohmic resistance of 20Ω to the flow of current in open direction, while entirely preventing the flow in opposite direction. The average value of current will be
   (*a*) 5A (*b*) 3.18A
   (*c*) 1.57A (*d*) 1.10A **[AMIE]**

56. The form factor for dc supply voltage is always
   (*a*) zero
   (*b*) unity
   (*c*) infinity
   (*d*) any value between 0 and 1 **[BHEL]**

57. 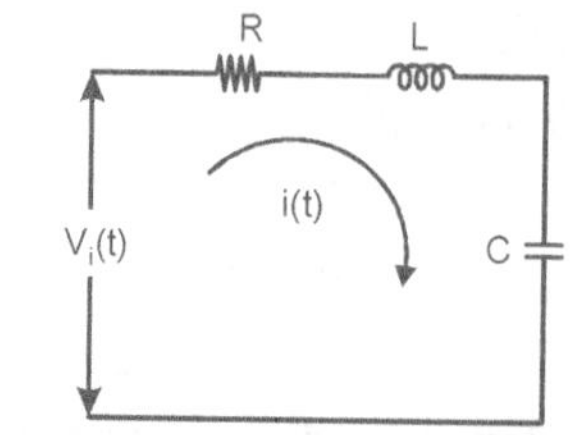

For the network shown in the above figure, $v_i(t)$ is the input and $i(t)$ is the output. The transfer function $I(s)/V_i(s)$ of the network is
   (*a*) $\dfrac{Cs}{LCs^2 + RCs + 1}$ (*b*) $\dfrac{C}{LCs^2 + RCs + 1}$
   (*c*) $\dfrac{Cs}{RCs^2 + LCs + 1}$ (*d*) $\dfrac{C}{RCs^2 + LCs + 1}$ **[IAS]**

58. In the series RC circuit shown in the figure, the voltage across C starts increasing when the d.c. source is switched on. The rate of increase of voltage across C at the instant just after the switch is closed (i.e., at $t = 0^+$) will be

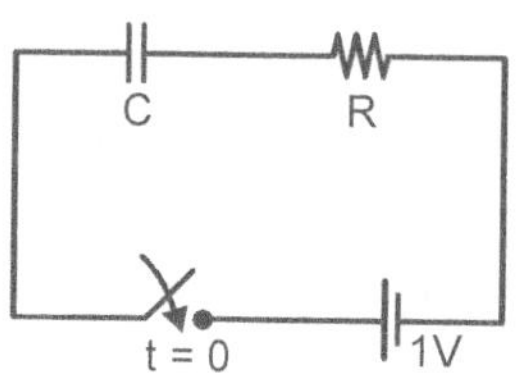

   (*a*) zero (*b*) infinity
   (*c*) RC (*d*) 1/RC **[GATE]**

59. The average value of the square waveform $2 \cos 2t + 4 \cos (2t + \pi/4)$ is nearly
   (*a*) 205
   (*b*) 20.5
   (*c*) 2.05
   (*d*) 0.205 **[PGCIL]**

60. The driving point impedance Z_1 for the circuit below is given by

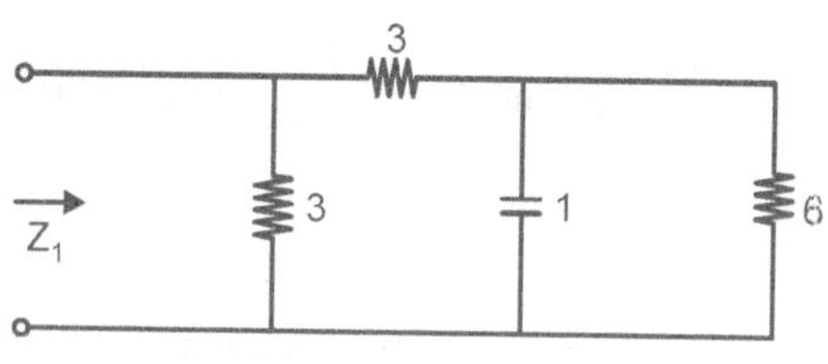

   (*a*) $\dfrac{3\,(s + 1/2)}{2\,(s + 1/3)}$
   (*b*) $\dfrac{3\,(s + 1/3)}{2\,(s + 1/2)}$
   (*c*) $3\left(\dfrac{s + 1/2}{s + 1/3}\right)$
   (*d*) $3\left(\dfrac{s + 1/3}{s + 1/2}\right)$ **[PGCIL]**

61. Tick the correct statement

(*a*) delta connection draws same current as star connection

(*b*) delta connection draws 3 times as much current as star connection

(*c*) delta connection draws $\frac{1}{\sqrt{3}}$ times as much current as star connection

(*d*) delta connection draws $\frac{1}{3}$ times as much current as star connection **[PGCIL]**

62. In a series RC circuit, the voltage across the capacitor and the resistor are 60V and 80V respectively. The input voltage should be

(*a*) $70\angle 45°$V (*b*) $100\angle -37°$V

(*c*) $105\angle -42°$V (*d*) $108\angle -60°$V **[NTPC]**

63. The storage battery generally used in electric power stations is

(*a*) lead acid battery

(*b*) zinc carbon battery

(*c*) nickel cadmium battery

(*d*) none of these **[NTPC]**

64. During discharge of a battery

(*a*) the voltage of cell decreases

(*b*) the voltage of cell increases

(*c*) voltage does not change

(*d*) none of these **[PGCIL]**

65. Two electric bulbs of 50 and 100 watts are given. Which one of the two bulbs will be brighter when they are connected in series to the mains

(*a*) 50W (*b*) 100W

(*c*) 50W and 100W (*d*) none of these **[ES]**

66. For a periodic function, the spectral density and autocorrelation functions would be a

(*a*) Fourier transform pair

(*b*) Laplace pair

(*c*) Z-transform pair

(*d*) Hilbert's transform pair **[ES]**

67. In the figure shown, if we connect a source of 2V, with internal resistance of 1Ω at A′ A, with positive terminal at A′, then the current through R is

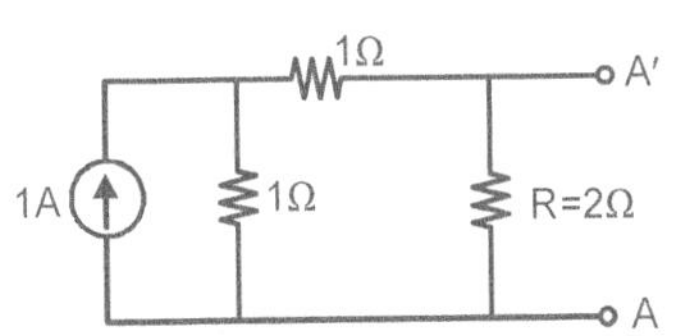

(*a*) 2.4 (*b*) 1.664

(*c*) 1 (*d*) 0.625A **[ES]**

68. A galvanometer is tested in the circuit, where $E = 1.5V$, $R_1 = 1.0\Omega$, $R_2 = 2500\Omega$ and R_3 is variable. With R_3 set at 450Ω, the galvanometer deflection is 140 mm and with R_3 set at 950Ω, the galvanometer deflection is 70 mm. The resistance of the galvanometer is

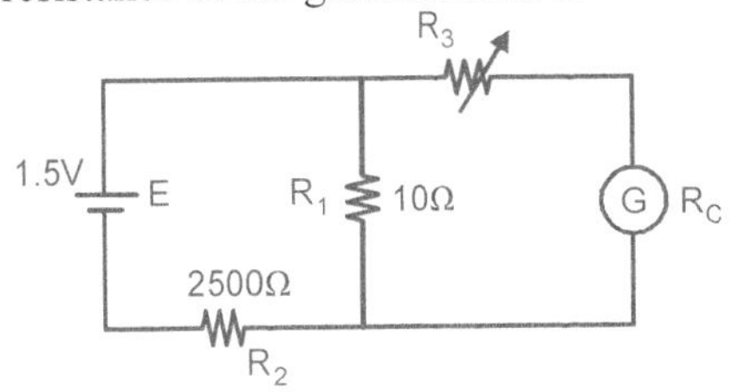

(*a*) 99Ω (*b*) 49Ω

(*c*) 25Ω (*d*) 10Ω **[ES]**

69. In the circuit shown in the figure, X is an element which always absorbs power. During a particular operation, it sets up a current of 1 amp in the direction shown and absorbs a power P_x. It is possible that A'can absorb the same power P_x for another current *i*; the value of this current is

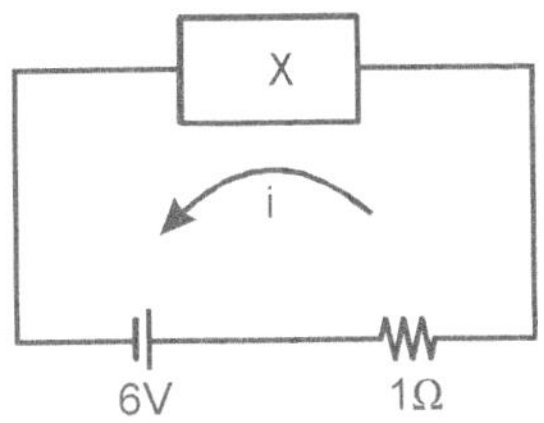

(*a*) $(3-\sqrt{14})$amps (*b*) $(3+\sqrt{14})$amps

(*c*) 5 amps (*d*) none of these **[GATE]**

70. The inverse of the matrix $S = \begin{bmatrix} 1 & -1 & 0 \\ 1 & 1 & 1 \\ 0 & 0 & 1 \end{bmatrix}$ is:

(*a*) $\begin{bmatrix} 1 & 0 & 1 \\ 0 & 0 & 0 \\ 0 & 1 & 1 \end{bmatrix}$ (*b*) $\begin{bmatrix} 0 & 1 & 1 \\ -1 & -1 & 1 \\ 1 & 0 & 1 \end{bmatrix}$

(*c*) $\begin{bmatrix} 2 & 2 & -2 \\ -2 & 2 & -2 \\ 0 & 2 & 2 \end{bmatrix}$ (*d*) $\begin{bmatrix} 1/2 & 1/2 & -1/2 \\ -1/2 & 1/2 & -1/2 \\ 0 & 0 & 1 \end{bmatrix}$

[GATE]

71. Consider the driving-point impedance function of the parallel RC circuit shown in the figure. When the pole of this function is made to come closer to the *j*Ω-axis, the memory of the circuit will

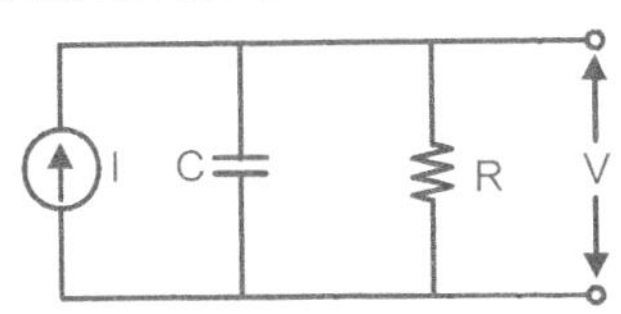

(*a*) increase

(*b*) decrease but not become zero

(*c*) remain unaffected

(*d*) become zero **[ES]**

72. The pole-zero configuration of a network transfer function is shown in the given figure. The magnitude of the transfer function will

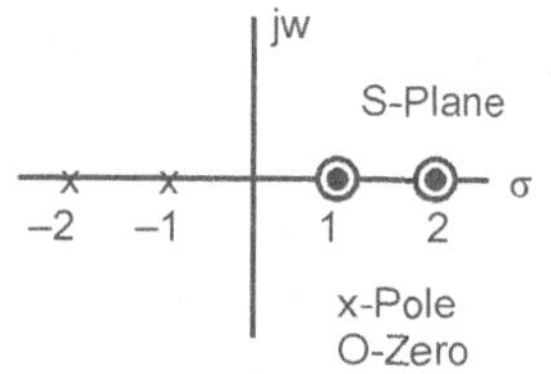

(*a*) decrease with frequency
(*b*) increase with frequency
(*c*) initially increase and then decrease with frequency
(*d*) be independent of frequency **[ES]**

73. At resonance, the parallel circuit of shown in the figure (below) constituted by an iron-cored coil and a capacitor behaves like:

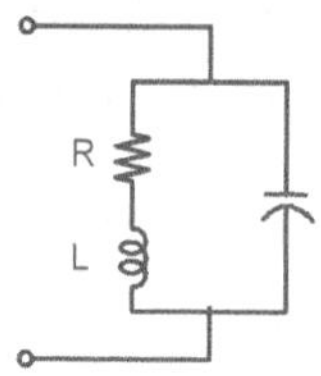

(*a*) an open-circuit
(*b*) a short-circuit
(*c*) a pure resistor of value R
(*d*) a pure resistor of value much higher than R **[GATE]**

74. Consider the following statements regarding the T-network shown in the figure: The zeros of the short-circuit input impedance Z_{is} across 11 are at

1. ω = zero
2. ω = infinity
3. $\omega = \dfrac{2}{\sqrt{LC}}$

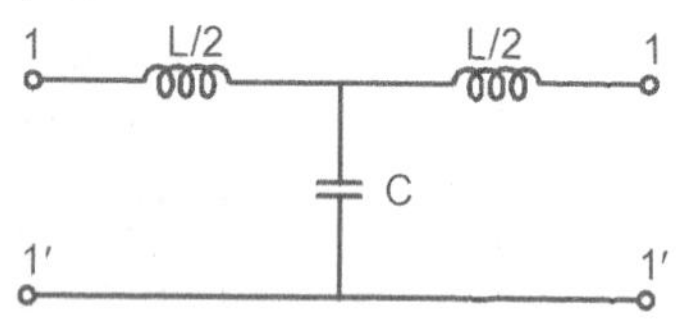

Of these statements
(*a*) 1 alone is correct
(*b*) 1 and 2 are correct
(*c*) 2 and 3 are correct
(*d*) 1 and 3 are correct **[ES]**

75. If all the elements in a particular network are linear, then the superposition theorem would hold, when the excitation is

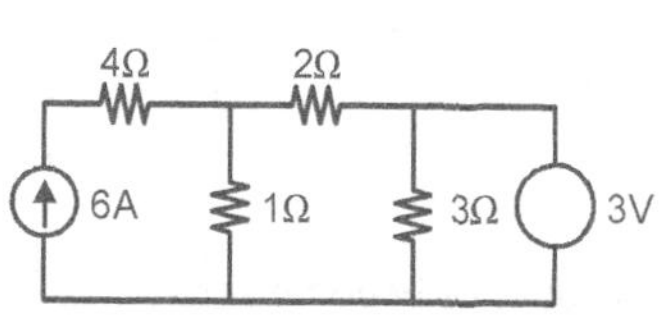

(*a*) dc only (*b*) ac only
(*c*) either ac or dc (*d*) an impulse **[ES]**

76. In the circuit shown in the given figure, the switch S is open for a long time and closed at $t = 0$.
The value of I at $t = 0$ is

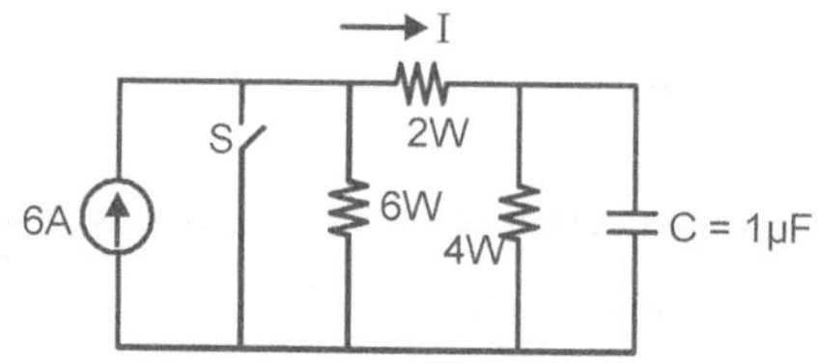

(*a*) –6A (*b*) $-\dfrac{3}{2}$A
(*c*) 3A (*d*) $\dfrac{9}{2}$A **[IAS]**

77. The current in resistor R shown in figure will be

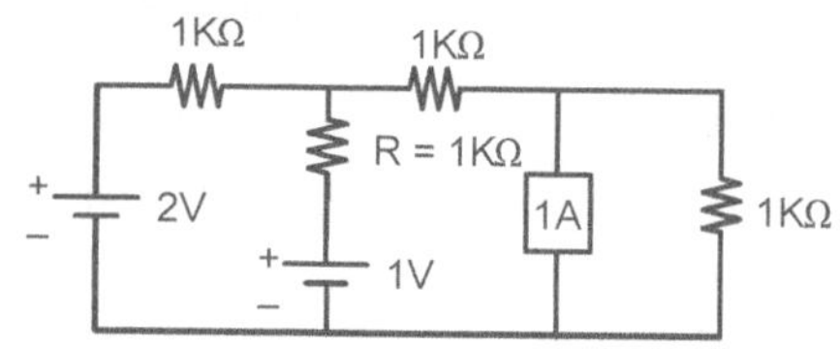

(*a*) 0.2A
(*b*) 0.4A
(*c*) 0.6A
(*d*) 0.8A **[ES]**

78. In the network shown in the figure, the circuit was initially in the steady-state condition with the switch K closed. At the instant when the switch is opened, the rate of decay of current through the inductance will be

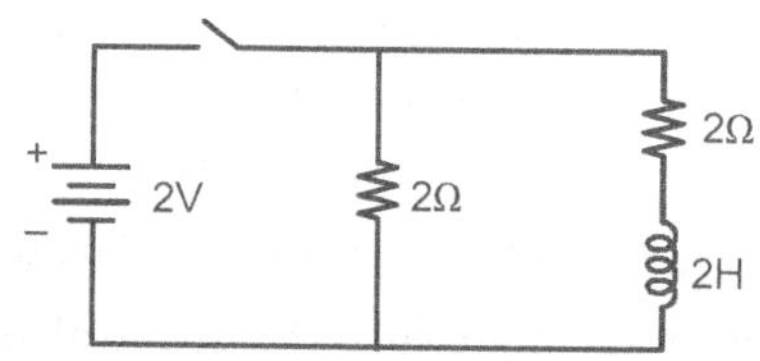

(*a*) zero (*b*) 0.5A/s
(*c*) 1A/s (*d*) 2A/s **[ES]**

79. In the network shown in the given figure, if, the voltage v at the time considered is 20V, then *dv*/*dt* at that time will be

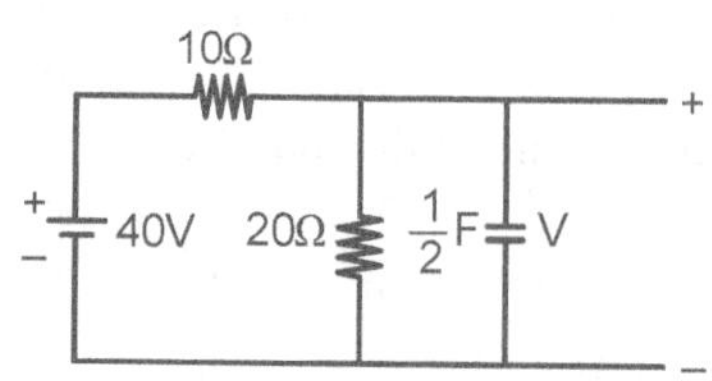

(*a*) $1\dfrac{V}{s}$ (*b*) $2\dfrac{V}{s}$
(*c*) $-2\dfrac{V}{s}$ (*d*) zero **[ES]**

80. In the circuit shown in the figure given below, RC is very small and L/R is very large when compared to the time period of the various input signals. Also $R << R_\xi$.

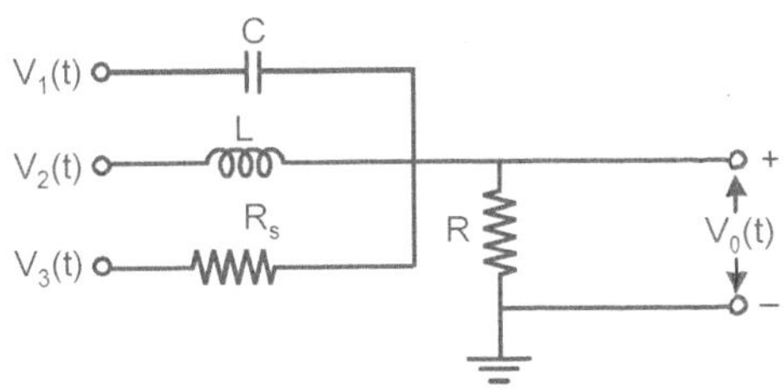

The output $V_o(t)$ is given by

(*a*) $\frac{dv_1(t)}{dt} + \int v_z(t)\,dt + \frac{R}{R_3} v_3(t)$

(*b*) $\int v_1(t)\,dt + \frac{dv_2(t)}{dt} + \frac{R}{R_3}\int v_3(t)\,dt$

(*c*) $RC\frac{dv_1(t)}{dt} - \frac{1}{L/R}\int v_2(t)\,dt + \frac{R}{R_3} v_3(t)$

(*d*) $RC\int V_1(t)\,dt - \frac{1}{L/R}\int v_2(t)\,dt + \frac{R}{R_3} v_3(t)$ **[ES]**

81. The pole-zero configuration of a network transfer function is shown in the given figure The magnitude of the transfer function will

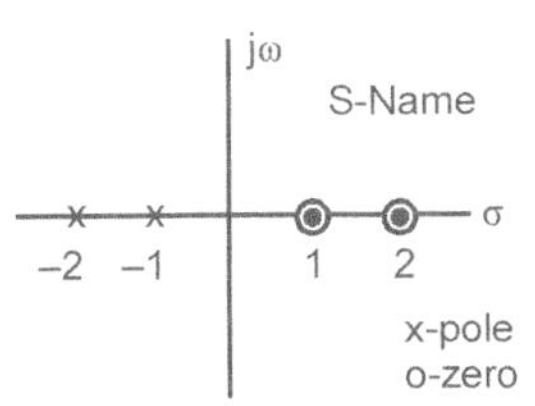

(*a*) decrease with frequency
(*b*) increase with frequency
(*c*) initially increase and then decrease with frequency
(*d*) be independent of frequency **[ES]**

82. For the circuit shown in the figure, the capacitance measured between terminals B and Y will be

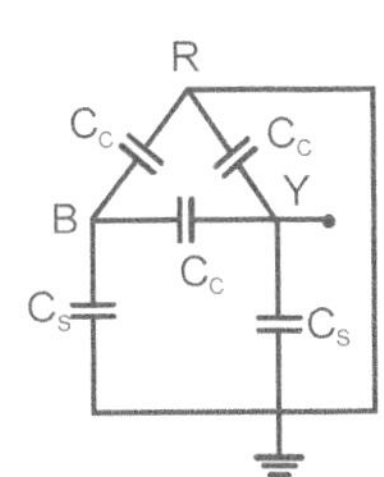

(*a*) $C_c + (C_s/2)$
(*b*) $C_s + (C_c/2)$
(*c*) $(C_s + 3C_c)/2$
(*d*) $3C_c + 2C_s$ **[ES]**

83. Match List-I with List-II and select the correct answer using the codes given below the lists

List-I (*Pole-zero configuration*) — **List-II** (*Number of elements in Foster's first form network*)

A. 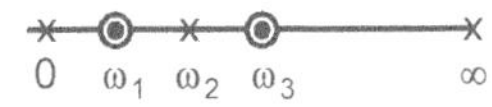 1. Two

B. 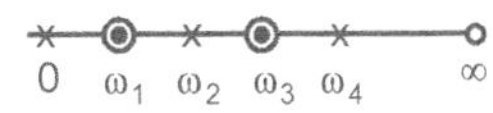 2. Three

C. 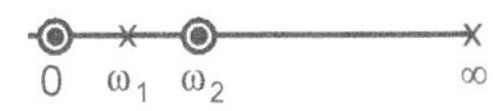 3. Four

D.  4. Five

Codes:

	A	B	C	D
(*a*)	3	2	1	4
(*b*)	4	3	1	2
(*c*)	3	4	2	1
(*d*)	2	1	4	3

[ES]

84. The sinusoidal steady-state voltage gain of the network shown in the given figure Which has magnitude equal to 0.707 at an angular frequency of

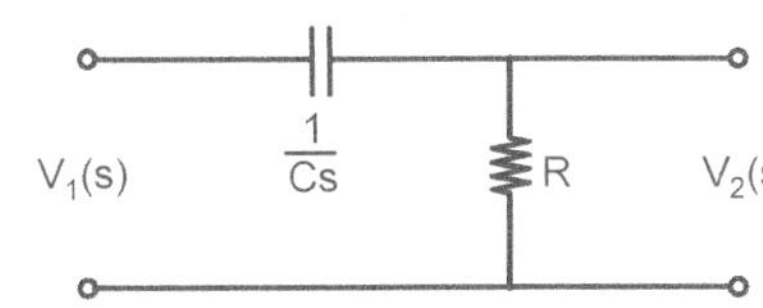

(*a*) zero (*b*) RC rad/s
(*c*) 1/RC rad/s (*d*) 1 rad/s **[ES]**

85. Consider the following statements

A circuit is said to be in resonance when the

1. current delivered by the source is in phase with the source voltage
2. load current is in phase with the load voltage
3. power factor of the circuit is unity

Of these statements:

(*a*) 1 and 3 are correct
(*b*) 1 and 2 are correct
(*c*) 2 and 3 are correct
(*d*) 1, 2 and 3 are correct **[IAS]**

86. In the given figure, the value of R is

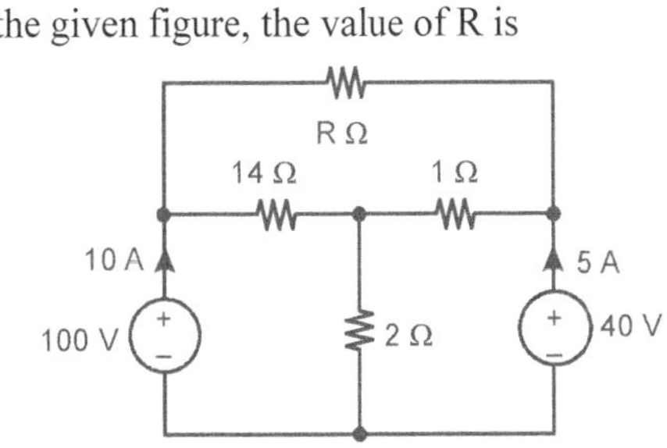

(*a*) 10Ω (*b*) 18Ω
(*c*) 24Ω (*d*) 12Ω **[ES]**

87. If a capacitor is energised by a symmetrical square wave current source, then the steady-state voltage across the capacitor will be a

(*a*) square wave (*b*) triangular wave
(*c*) step function (*d*) impulse function **[ES]**

88. The dc component of the function $\pi|\sin t|$

(*a*) 0 (*b*) 1
(*c*) 2 (*d*) $\frac{2}{\pi}$ **[IAS]**

89. In the circuit shown in the given figure, maximum power will be delivered to the load if the value of '*n*' is

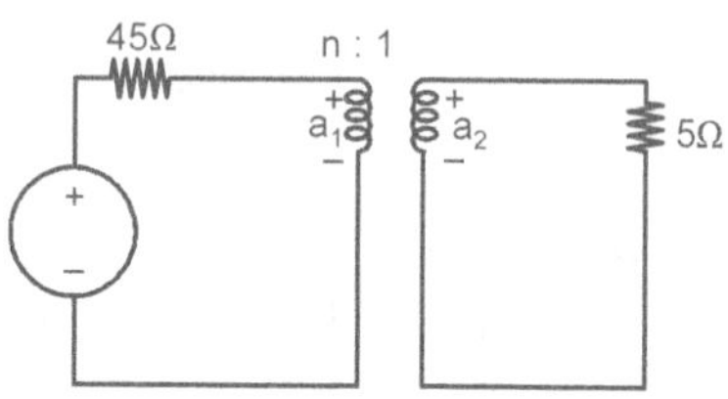

(*a*) 2 (*b*) 3
(*c*) 4 (*d*) 5 **[IAS]**

90. The voltage v_0 for the circuit shown in the given figure will be

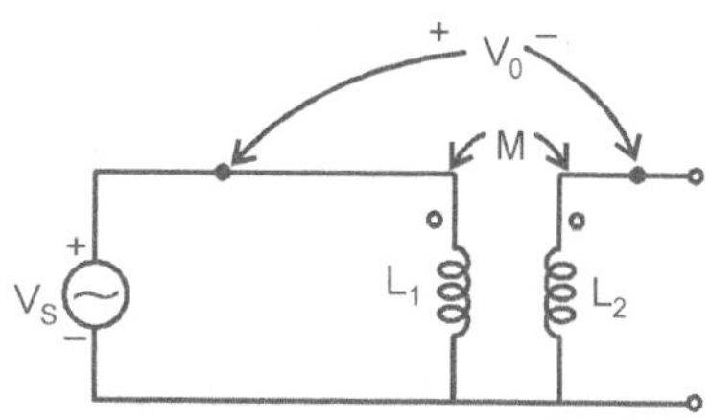

(*a*) $v_s\left(1+\frac{M}{L_1}\right)$ (*b*) $v_s\left(1-\frac{M}{L_1}\right)$

(*c*) $v_s\left(\frac{M}{L_1}-1\right)$ (*d*) $v_s\left(1-\frac{M}{L_2}\right)$ **[IAS]**

91. The dynamic equation of a second-order system is

$$2\frac{d^2y}{dt^2}+4\frac{dy}{dt}+8y=8$$

The damping coefficient is

(*a*) 0.1 (*b*) 0.25
(*c*) 0.5 (*d*) 1.0 **[IAS]**

92. In the circuit shown in the given figure, if the power consumed by the 5 ohm resistor is 10W, then the power factor of the circuit is

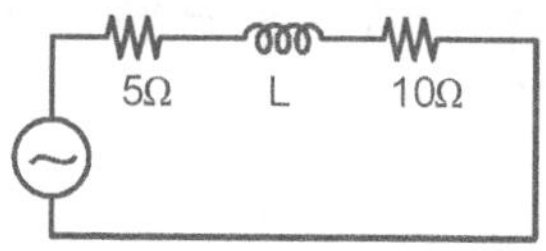

(*a*) 0.8 (*b*) 0.6
(*c*) 0.5 (*d*) zero **[ES]**

93. For the circuit shown in the given figure, the switch 'K' was closed for a long time till steady state conditions reached. At time $t = 0$, the switch 'K' is opened, then the current through inductor will be

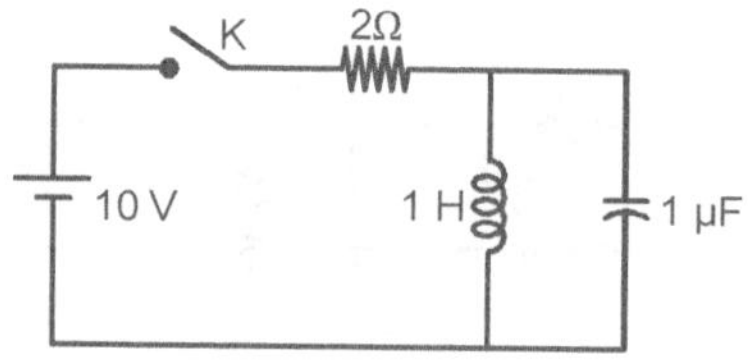

(*a*) 5 cos 10*t*
(*b*) 5 cos 100*t*
(*c*) 5 cos 1000*t*
(*d*) 5 cos 10000*t* **[IAS]**

94. A lamp rated at 10 watts, 50 volts is proposed to be used in 110 volts, system. The wattage and resistance of the resistor to be connected in series with the lamp should be

(*a*) 15 watts, 350 ohms
(*b*) 10 watts, 250 ohms
(*c*) 12 watts, 300 ohms
(*d*) 15 watts, 250 ohms **[IAS]**

95. The RMS value of a half-wave rectified symmetrical square wave current of 2A is

(*a*) $\sqrt{2}$A
(*b*) 1A
(*c*) $1/\sqrt{2}$A
(*d*) $\sqrt{3}$A **[ES]**

96. In the circuit shown in figure -I the switch 'K' was initially at position '1' and a current 'I' was flowing through the inductor 'L' and a voltage 'V_0' existed across the capacitor 'C'.

If at $t = 0$, the switch 'K' is put on the position '2' in the circuit shown in figure-I, which one of the following transformed circuits will give $i(t)$ for $i > 0$?

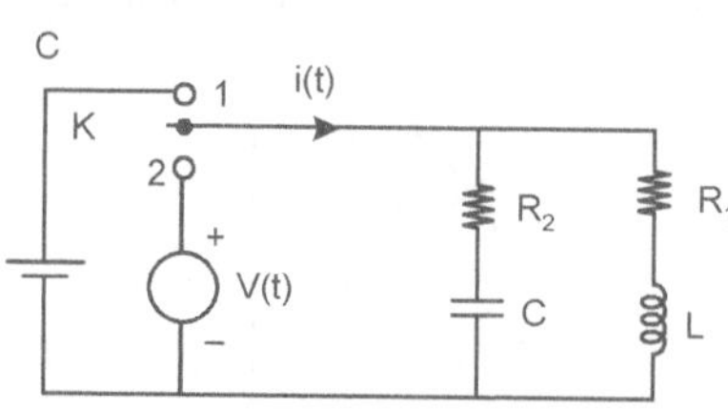

(*a*)

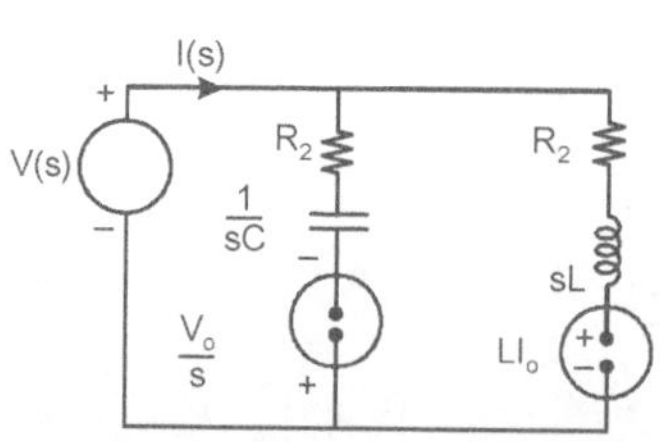

(*b*) 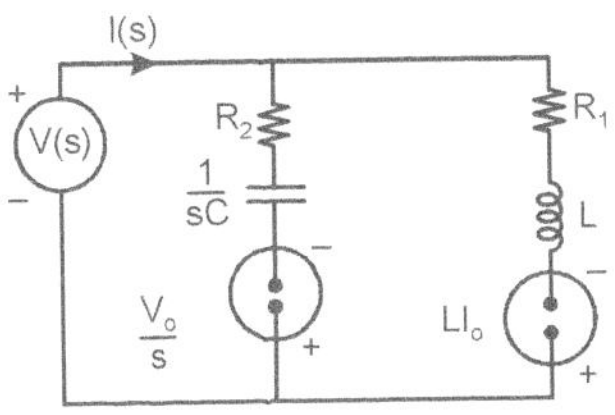

(*c*) 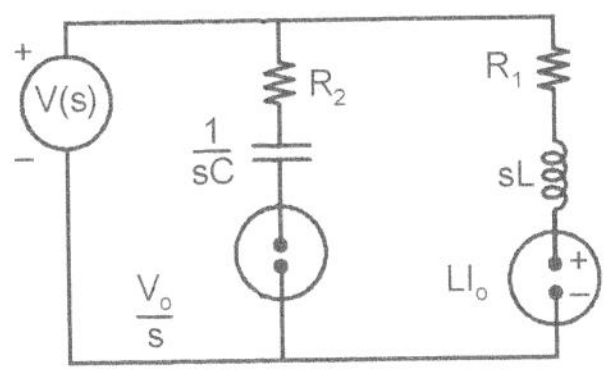

(*d*) 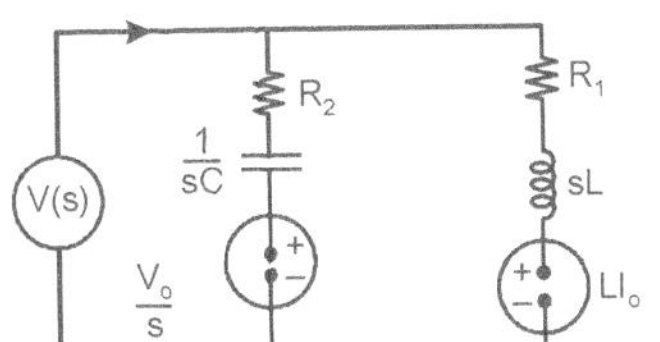 **[ES]**

97. A two-port device is defined by the following pair of equations:

$i_1 = 2v_1 + v_2$ and $i_2 = v_1 + v_2$

Its impedance parameters $(z_{11}, z_{12}, z_{21}, z_{22})$ are given by

(*a*) (2, 1, 1, 1) (*b*) (1, –1, – 1, 2)
(*c*) (1, 1, 1, 2) (*d*) (2, – 1, – 1, 1) **[ES]**

98. In the RL circuit shown in the given figure, the inductor is carrying current I_0, A time $t = 0$ a unit step input is applied. Then the current $i(t)$ at t is given by

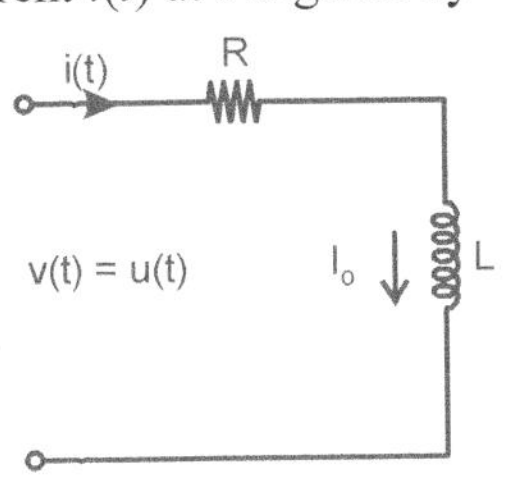

(*a*) $\frac{0.632}{R}$ (*b*) $0.632I_0$

(*c*) $\frac{0.632}{R} + 0.368I_0$ (*d*) $0.632\left(I_0 \frac{1}{R}\right)$ **[IAS]**

99. If the resonant frequency of the circuit shown in figure-I is 1kHz, the resonant frequency of the circuit shown in figure-II will be

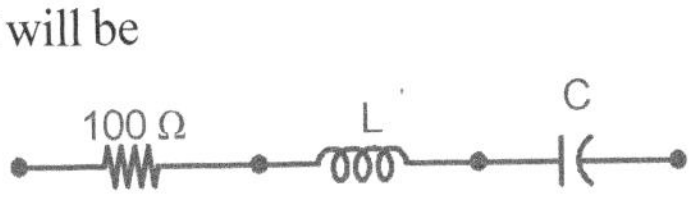

Figure-I

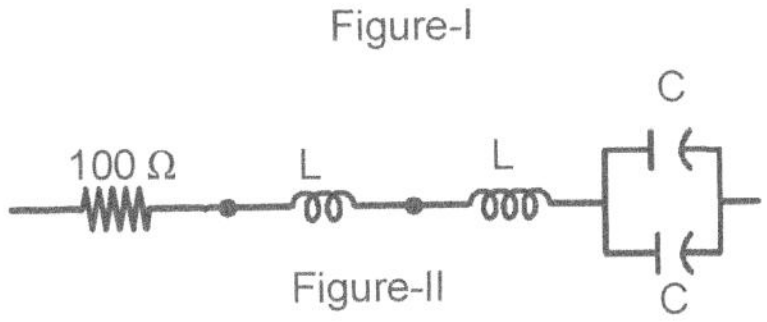

Figure-II

(*a*) 4 kHz (*b*) 2 kHz
(*c*) 0.5 kHz (*d*) 0.25 kHz **[ES]**

100. An impedance match is desired at the 1-1 port of the two-port netowrk shown in the given figure. The match will be obtained when Z_e equals

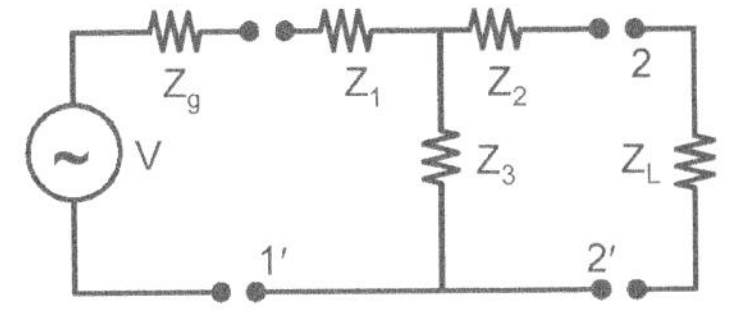

(*a*) $z_1 + z_3$ (*b*) $z_1 + \frac{z_2(z_3 + z_L)}{z_2 + z_3 + z_L}$

(*c*) $\frac{z_1(z_2 + z_3)}{z_1 + z_2 + z_3}$ (*d*) z_1 **[ES]**

101. Consider the following functions:

1. $\frac{s^3 + 3s}{2(s^2 + 1)}$

2. $\frac{s^2 + 3s + 1}{5(s^2 + 1)}$

3. $\frac{s^2 + 1}{3s}$

4. $\frac{3\left(s^3 + \frac{7s}{2}\right)}{s^4 + 14s^2 + 30}$

Which of these are valid driving-point impedance functions?

(*a*) 1, 2 and 3 (*b*) 1, 3 and 4
(*c*) 1, 2 and 4 (*d*) 2, 3 and 4 **[ES]**

102. Match List-I [Input voltage v(*t*)] with List-II [I(*s*), the Laplace transform of *i*(*t*) for the given circuit and select the correct answer using the codes given below the lists:]

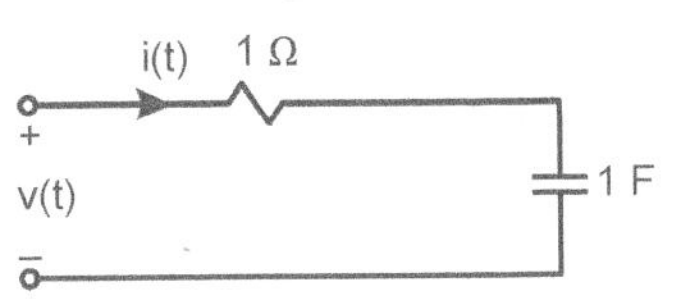

List-I	**List-II**
A. Unit step	1. $\frac{s}{s+1}$
B. Unit ramp	2. $\frac{s}{s+1}$
C. Unit impulse	3. $\frac{s}{(s+1)^2}$
D. sin *t*	4. $\frac{s}{s(s+1)}$

Codes:

	A	B	C	D
(*a*)	2	4	1	3
(*b*)	2	1	4	3
(*c*)	3	1	4	2
(*d*)	3	4	1	2

[ES]

103. Viewed from the terminals AB, the following-circuit shown in the figure can be reduced to an equivalent circuit of single voltage source in series with a single resistor with the following parameters

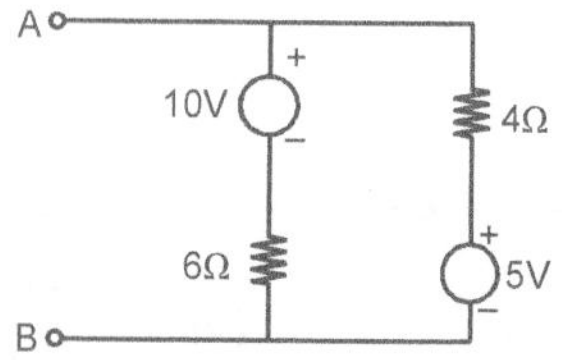

(*a*) 5 volt source in series with 10Ω resistor
(*b*) 7 volt source in series with 2.4Ω resistor
(*c*) 15 volt source in series with 2.4Ω. resistor
(*d*) 1 volt source in series with 10Ω resistor **[ES]**

104. The effective inductance of the circuit across the terminals AB in the figure shown below is

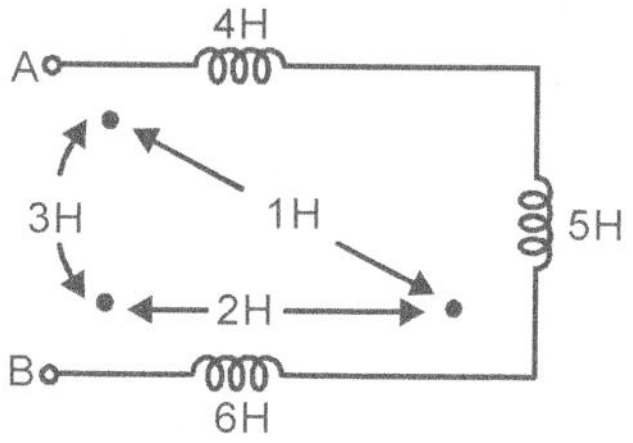

(*a*) 9 H (*b*) 21 H
(*c*) 11 H (*d*) 6 H **[GATE]**

105. Given two coupled inductors L_1 and L_2, their mutual inductance M satisfies

(*a*) $M = \sqrt{L_1^2 + L_2^2}$ (*b*) $M > \frac{(L_1 + L_2)}{2}$
(*c*) $M > \sqrt{L_1 L_2}$ (*d*) $M \le \sqrt{L_1 L_2}$ **[GATE]**

106. In the network shown, the switch is opened at $t = 0$. Prior to that, the network was in the steady-state $V_s(t)$ at $t = 0^+$ is

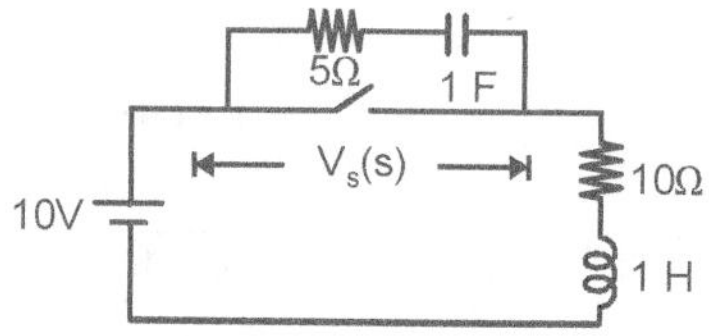

(*a*) 0 (*b*) 5V
(*c*) 10V (*d*) 15V **[ES]**

107. The current through the galvanometer and the direction of current is

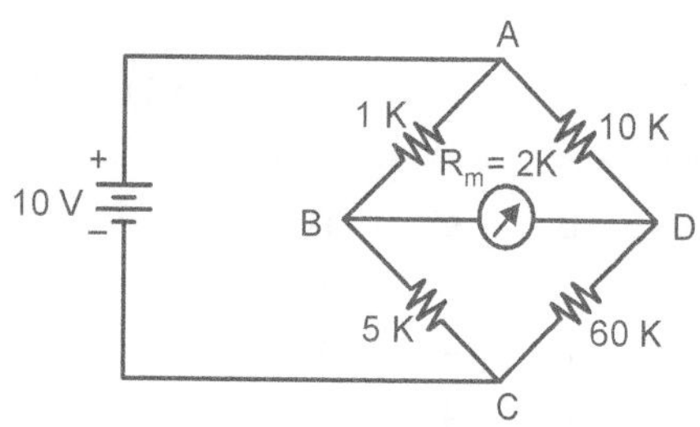

(*a*) 1 μA, BD (*b*) 21 μA, DB
(*c*) 1 μA, DB (*d*) 1.1 μA, DB **[ES]**

108. Match List-I (Parameters) with List-II (Units) and select the correct answer using the codes given below the lists

List-I	List-II
A. h_{11}	1. Dimensionless
B. h_{12}	2. Ohms
C. h_{22}	3. Siemens

Codes:

	A	B	C
(*a*)	1	2	3
(*b*)	1	3	2
(*c*)	2	1	3
(*d*)	3	2	1

[ES]

109. In the figure below, the current of 1 ampere flows through the resistance of

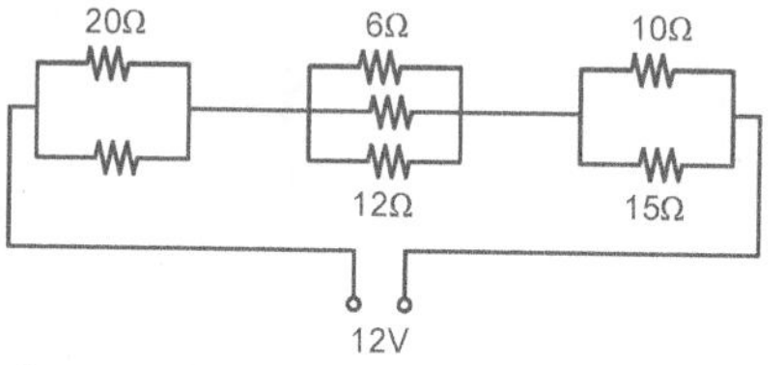

(*a*) 4 ohms (*b*) 20 ohms
(*c*) 30 ohms (*d*) 12 ohms

110. Match List-I with List-II and select the correct answer using the codes given below the lists

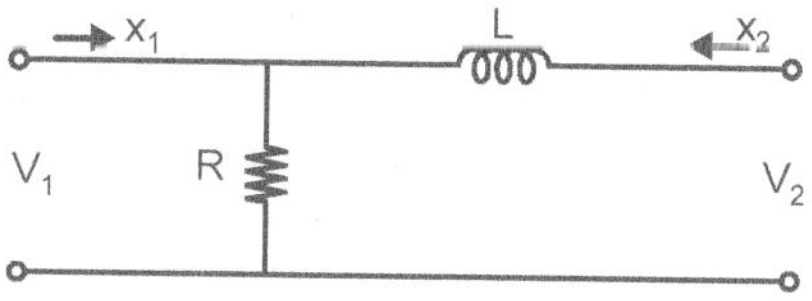

List-I	List-II
A. Z_{11}	1. R
B. Z_{12}	2. R + L
C. Z_{21}	3. $R - L_s$
D. Z_{22}	4. $R + L_s$

Codes

	A	B	C	D
(*a*)	1	2	1	4
(*b*)	2	1	1	3
(*c*)	1	1	1	4
(*d*)	2	1	3	4

111. Two coils in differential connection have self-inductance of 2 mH and 4 mH and a mutual inductance of 0.15 mH. The equivalent inductance of the combination is

(*a*) 5.7 mH
(*b*) 5.85 mH
(*c*) 6 mH
(*d*) 6.15 mH **[IAS]**

112. For the network shown in the above figure below, if the value of C is such that I is in phase with V_s, then I_1 leads I_2 by (where $\omega = 2$ rad/s)

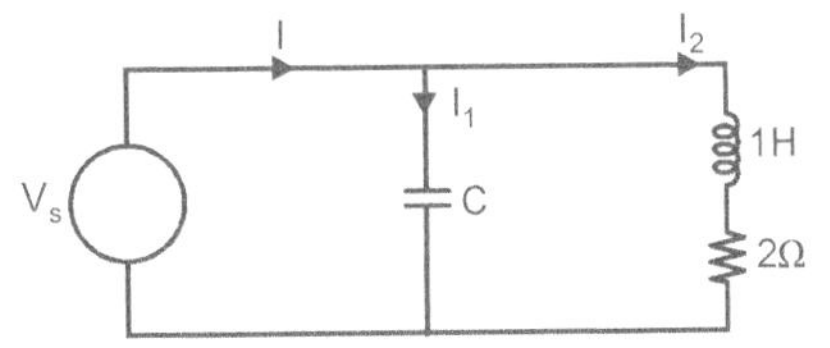

(*a*) zero

(*b*) 45°

(*c*) 90°

(*d*) 135°

113. For the two-port-network shown in the given figure, if the driving-point impedance Z_{11} at port 1 is equal to $\frac{k(s+2)}{(s+5)}$, then the driving-point impedance Z_{22} at port 2 will be

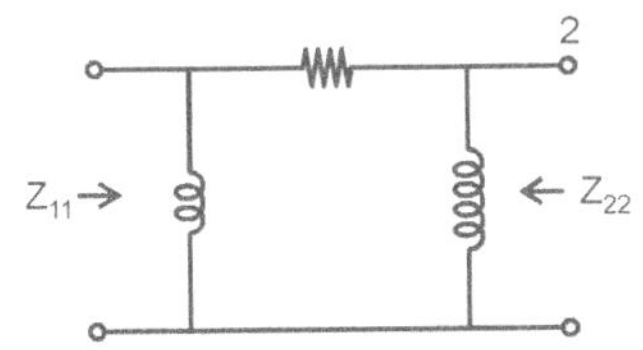

(*a*) $\frac{k_1(s+3)}{(s+5)}$

(*b*) $\frac{k_1(s+2)}{(s+3)}$

(*c*) $\frac{k_1 s}{(s+5)}$

(*d*) $\frac{k_1 s}{(s+2)}$ **[IAS]**

114. In the network shown in the given figure, for the current to be zero in R_L, the value of R should be adjusted to

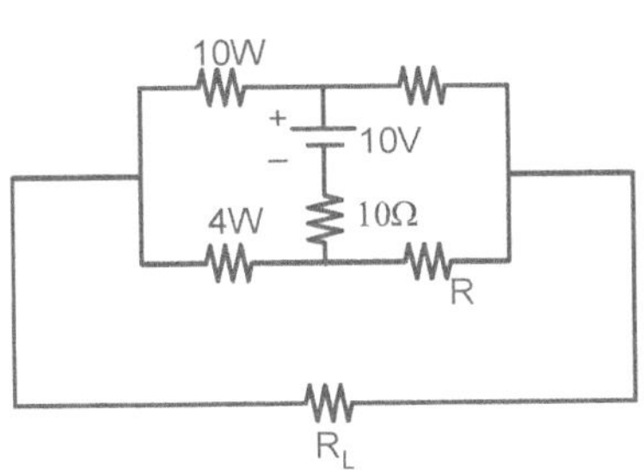

(*a*) 1Ω (*b*) 2Ω

(*c*) 5Ω (*d*) 10Ω **[IAS]**

115. A series RLC circuit draws current at leading power factor at

(*a*) the resonant frequency

(*b*) frequencies less than the resonant frequency

(*c*) frequencies more than the resonant frequency

(*d*) frequencies both less and more than the resonant frequency **[IAS]**

116. In the given network, if $i_R = (4e^{-3t} + 3e^{-4t})$ A and $i_L(0) = 1$A, then ϕ would be equal to

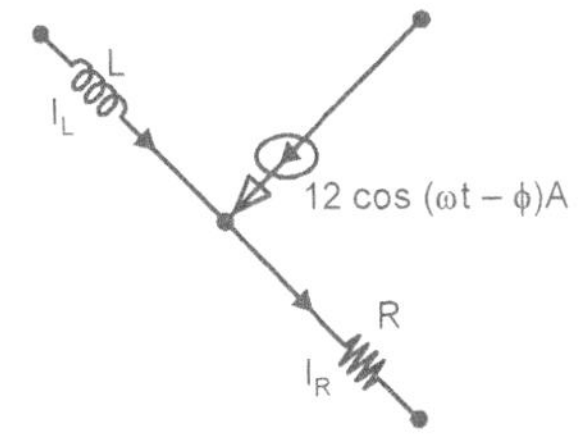

(*a*) π/2 (*b*) π/3

(*c*) π/4 (*d*) π/6 **[ES]**

117. If the driving-point admittance function of a 1-port network is $Y(s) = \frac{Ks}{s+\alpha}$, it can be realized using

(*a*) series combination of R, L

(*b*) parallel combination of R, L

(*c*) series combination of R, C

(*d*) parallel combination of R, C **[IAS]**

118. Match List-I (Definition) with List-II (Parameter) and select the correct answer using the codes giving below the lists:

	List-I	List-II
A.	$\left.\frac{V_2(s)}{I_1(s)}\right\vert_{I_2=0}$	1. z_{21}
B.	$\left.\frac{I_2(s)}{I_1(s)}\right\vert_{V_2=0}$	2. g_{12}
C.	$\left.\frac{I_2(s)}{I_1(s)}\right\vert_{V_1=0}$	3. h_{21}
D	$\left.\frac{I_2(s)}{I_2(s)}\right\vert_{V_1=0}$	4. y_{22}
		5. h_{12}

Codes:

	A	B	C	D
(*a*)	1	5	2	4
(*b*)	1	3	2	4
(*c*)	3	1	4	5
(*d*)	5	3	4	2

[IAS]

119. Which one of the following statements regarding the function whose plot is shown in the figure, is NOT correct?

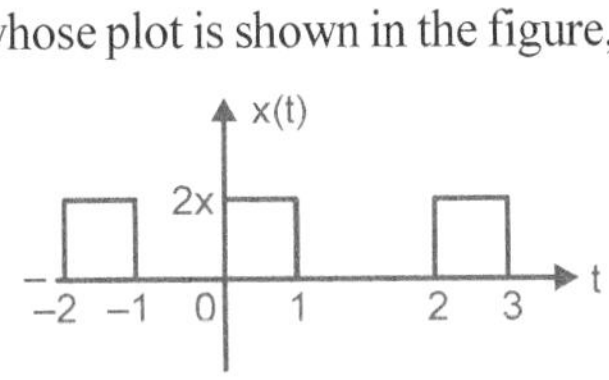

(*a*) It has a d value of π

(*b*) it has a period of 2*s*

(*c*) Its fundamental component has the magnitude 2

(*d*) It contains only even harmonics **[IAS]**

120. The bandwidth of a control system is the range of frequency over which $\left|\frac{C}{R}(j\omega)\right|$ is greater than or R equal to

(*a*) $1\sqrt{2}$ and it depends only on the undamped natural frequency of system

(*b*) 1 and it depends only on the damping factor of the system

(*c*) 1 and it depends only the undamped natural frequency of the system

(*d*) $1/\sqrt{2}$ and it depends on both undamped natural frequency and the damping factor of the system **[IAS]**

121. For the circuit shown in the given figure, the Thevenin equivalent impedance across terminal CD is given by

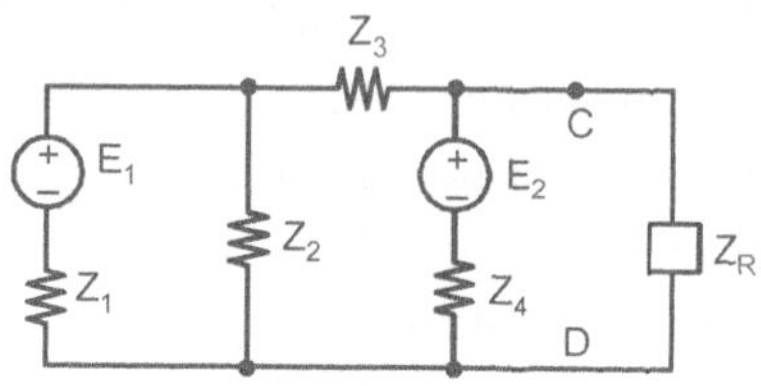

(*a*) $Z_{TH} = \dfrac{Z_4\left[\dfrac{Z_1Z_2}{Z_1+Z_2}+Z_3\right]}{Z_4+Z_3+\dfrac{Z_1Z_2}{Z_1+Z_2}}$

(*b*) $Z_{TH} = \dfrac{Z_4\left[\dfrac{Z_3Z_2}{Z_3+Z_2}+Z_1\right]}{Z_1+Z_4+\dfrac{Z_3Z_2}{Z_3+Z_2}}$

(*c*) $Z_{TH} = \dfrac{Z_1+Z_2+Z_3Z_2}{Z_1+Z_2+Z_4}$

(*d*) $Z_{TH} = \dfrac{Z_3+(Z_1+Z_2Z_4)}{Z_1+Z_2+Z_3+Z_4}$ **[IES]**

122. A sinusoidal voltage source $50\angle 0°$ with an internal impedance $10+j20$ is connected to a load (z_1) which is variable in both resistance and reactance. Match List-I with List-II for maximum power transfer and select the correct answer using the codes given below the lists

List-I	List-II
A. Load impedance	1. 62.5
B. Total impedance	2. $10+j20$
C. Current	3. 2.5
D. Maximum power	4. $10-j20$
	5. $2.5-j2.5$

Codes:

	A	B	C	D
(*a*)	5	2	3	4
(*b*)	5	4	3	1
(*c*)	2	4	6	1
(*d*)	2	5	6	4

[IAS]

123. Hurwitz polynomial has

(*a*) zeros only in the left half of the *s*-plane

(*b*) poles only in the left half of the *s*-plane

(*c*) zeros anywhere in the *s*-plane

(*d*) poles on the $j\omega$ axis only **[ES]**

124. For an all-pass function zeros are in the

(*a*) right half plane (RHP) and poles in the left half plane (LHP)

(*b*) LHP and poles in LHP

(*c*) LHP and poles in RHP

(*d*) RHP and poles in RHP **[ES]**

125.

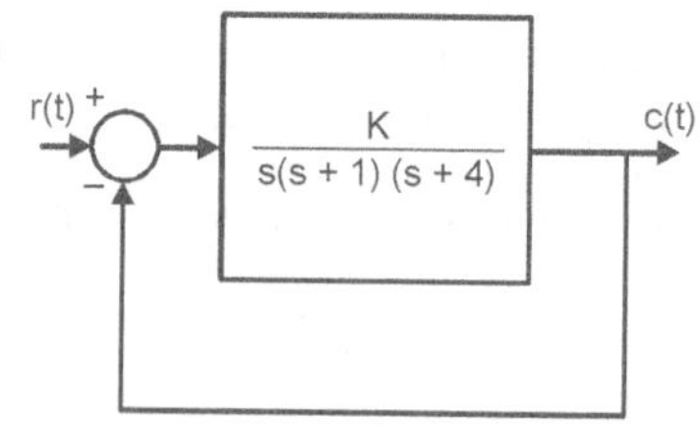

By properly choosing the value of KT, the output C(*t*) of the system as shown in the above figure can be made to socialite sinusoidally at a frequency (in rad/sec) of

(*a*) 1 (*b*) 2

(*c*) 3 (*d*) 4 **[IAS]**

126. Consider the following functions for the rectangular voltage pulse shown in the given figure:

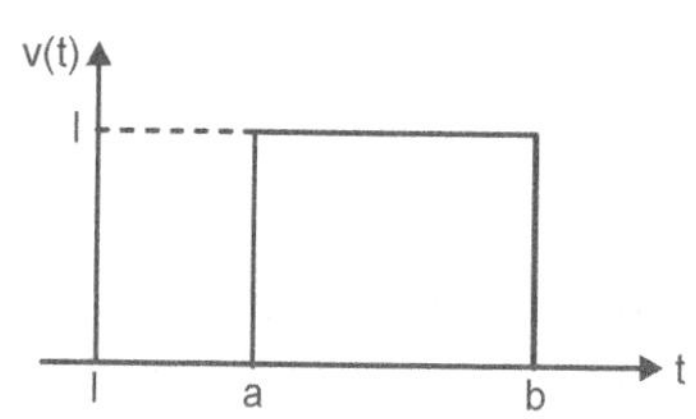

1. $v(t) = u(t-a) - u(t-b)$
2. $v(t) = u(b-t) - u(a-t)$
3. $v(t) = u(b-t) - u(t-a)$
4. $v(t) = u(a-t) - u(t-b)$

Which of these functions describe the pulse shown in the given figure?

(*a*) 1, 2 and 3 (*b*) 1, 2 and 4

(*c*) 2, 3 and 4 (*d*) 1, 3 and 4 **[ES]**

127. Match List-I (Loop concept) with List-II (Junction concept) and select the correct answer using the codes given below the lists:

List-I	List-II
A. Mesh	1. Number of nodes
B. Outside mesh	2. node voltage
C. Mesh current	3. Reference node
D. Number of meshes	4. node

Codes:

	A	B	C	D
(a)	3	4	1	2
(b)	3	4	2	1
(c)	4	3	2	1
(d)	4	3	1	2

[ES]

128. A certain network N feeds a load resistance R as shown in Figure-I. It consumes a power of 'P' W. If an identical network is added as shown in Figure-II, the power consumed by R will be

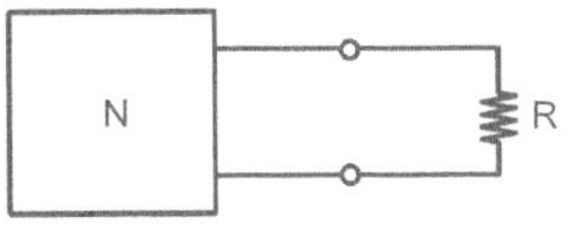

Figure-I

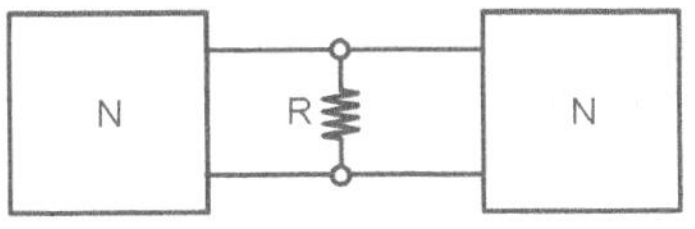

Figure-II

(a) less than P
(b) equal to P
(c) between P and 4P
(d) more than 4P **[ES]**

129. For the network shown in the figure, the voltage V_B will be

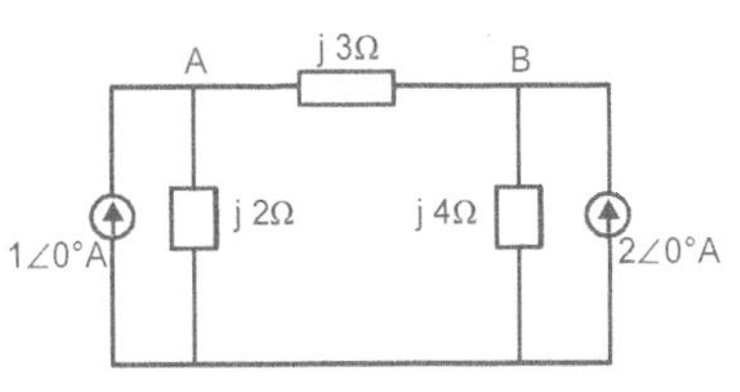

(a) (j5.33)V (b) (5.33)V
(c) (–j5.33)V (d) (j3.33)V **[ES]**

130. In the two-port netowrk shown in the given figure, if $G_{21}(s)$ is $\frac{1}{s_2+1}$, then Z(s) will be

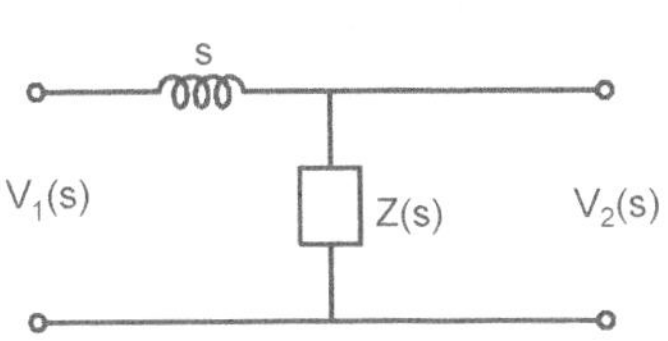

(a) s (b) $s+1$
(c) $\frac{1}{s}$ (b) $s+\frac{1}{s}$ **[ES]**

131. If a two-port network is passive, then we have, with the usual notation, which of the following relationship

(a) $h_{12} = h_{21}$
(b) $h_{12} = -h_{21}$
(c) $h_{11} = h_{22}$
(d) $h_{11} . h_{22} - h_{12} . h_{21} = 1$ **[GATE]**

132.

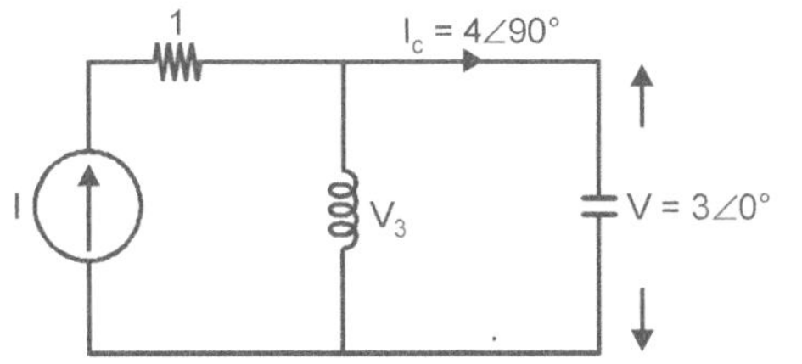

For the above circuit ω = 3 rad/s. If V is taken as reference, then phasor of I is given by

(a) $1\underline{|90°}$ (b) $3\underline{|90°}$
(c) $5\underline{|60°}$ (d) $\sqrt{2}\underline{|-45°}$ **[IAS]**

133. The Laplace transform of the voltage across the capacitor of 0.5F is

$$V(s) = \frac{s+1}{s^3+s^2+s+1}$$

Then the value of the current through the capacitor at $t = 0^+$ given by

(a) 0A
(b) 0.5A
(c) 1.0A
(d) 1.5A **[IAS]**

134. If $u(t)$ and $\delta(t)$ are the step function and the impulse function respectively at $t = 0$, then the Laplace transform of the function

$f(t) = u(t-1)\,\delta(t)$ is equal to

(a) 1 (b) 1/s
(c) 0 (d) $\frac{1}{s+1}$

135. A balanced delta connected load has an impedance of $9\underline{|30°}\Omega$ ohms per phase. What is the impedance per phase of its equivalent star?

(a) $27\underline{|30°}\Omega$
(b) $27\underline{|90°}\Omega$
(c) $3\underline{|30°}\Omega$
(d) $3\underline{|20°}\Omega$ **[ES]**

136. In an ac series RLC circuit, the voltage across R and L is 20V, voltage across L and C is 9V and voltage across RLC is 15V. What is the voltage across C?

(a) 7V (b) 12V
(c) 16F (d) 21V **[ES]**

137. Two resistors of equal value are connected in series across the lines B and C of a symmetrical 3-phase 400 -V system. What is the magnitude of the voltage between the line A and the function of the resistors

(a) $400/\sqrt{3}$ (b) $200/\sqrt{3}$
(c) 400/3 (d) $100\sqrt{3}$ **[ES]**

138. In the network shown in the figure, if the 1F capacitor had an initial voltage of 2V, then which of the following would represent the s-domain equivalent circuits?

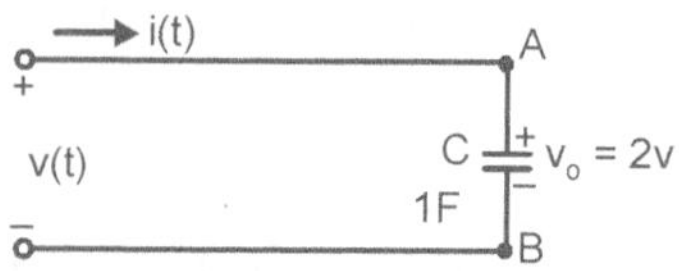

Fig. 1

1. 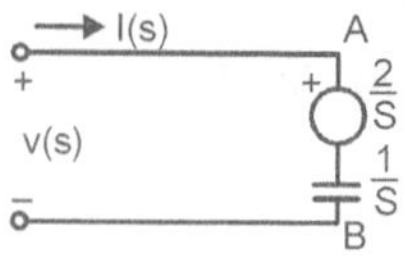

2. 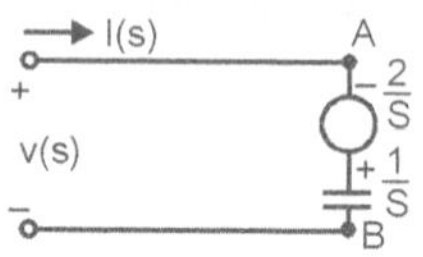

3. 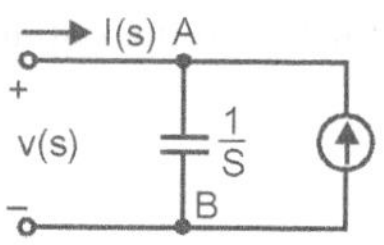

4. 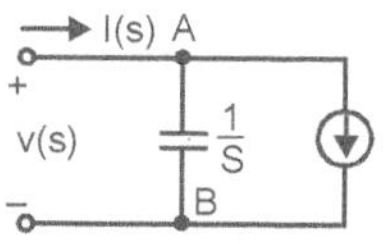

Select the correct answer using the codes given below:

Codes:

(*a*) 1 and 3 (*b*) 1 and 4

(*c*) 2 and 3 (*d*) 2 and 4 **[ES]**

139. For the circuit shown in the given figure, if $V_s = 250 \sin 300t$ and peak of $V_2 = 150$, then the value of L for R = 100 is

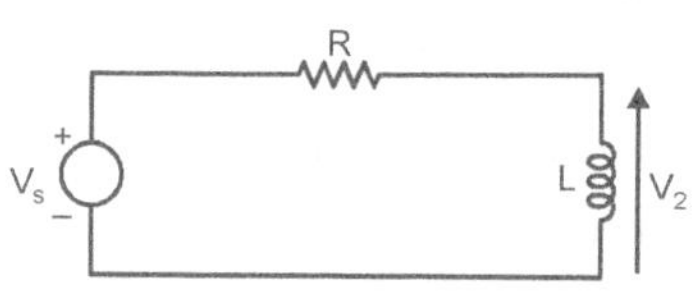

(*a*) 0.50 (*b*) 0.25

(*c*) 0.33 (*d*) 0.41 **[ES]**

140. After closing the switch 'S' at $t = 0$, the current $i(t)$ at any instant 't' in the network shown in the given figure will be

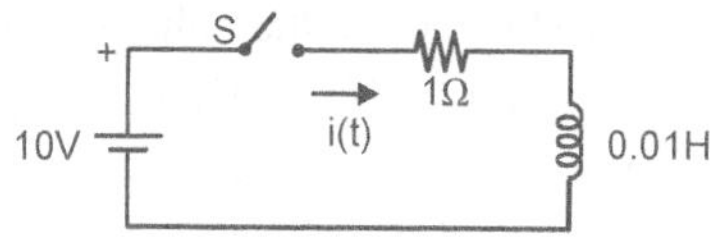

(*a*) $10 + 10e^{100t}$ (*b*) $10 - 10e^{100t}$

(*c*) $10 + 100e^{-100t}$ (*d*) $10 - 10e^{-100t}$ **[ES]**

141. Like a series R-L-C resonant circuit, a parallel resonant circuit also

(*a*) magnifies current.

(*b*) offer minimum impedance

(*c*) draws maximum current

(*d*) has unity power factor

142. For the circuit shown in the given figure, if the input impedance Z_1 at part 0 is given by

$$Z_1 = \frac{K_1(s+2)}{(s+5)}$$

then the input impedance Z_2 at port 2 will be

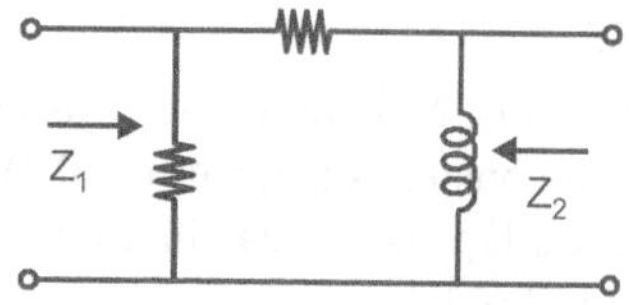

(*a*) $\frac{K_2(s+3)}{(s+5)}$ (*b*) $\frac{K_2(s+2)}{(s+3)}$

(*c*) $\frac{K_2 s}{(s+5)}$ (*d*) $\frac{K_2 s}{(s+2)}$ **[ES]**

143. A T-network is shown in the given figure. Its Y_{sc} matrix will be (units in Siemens)

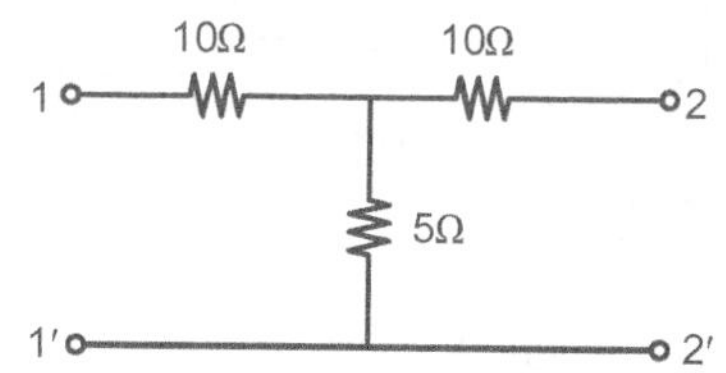

(*a*) $\begin{bmatrix} \frac{10}{100} & \frac{5}{200} \\ \frac{5}{200} & \frac{10}{200} \end{bmatrix}$ (*b*) $\begin{bmatrix} \frac{10}{100} & \frac{-5}{200} \\ \frac{-5}{200} & \frac{10}{200} \end{bmatrix}$

(*c*) $\begin{bmatrix} \frac{15}{100} & \frac{-5}{200} \\ \frac{-5}{200} & \frac{15}{200} \end{bmatrix}$ (*d*) $\begin{bmatrix} \frac{15}{200} & \frac{5}{200} \\ \frac{5}{200} & \frac{15}{200} \end{bmatrix}$ **[ES]**

144. A 10V battery with an internal resistance of 1Ω is connected across a non-linear load whose v-*i* characteristic is given by

$$7i = v^2 + 2v.$$

The current delivered by the battery is

(*a*) 2.5.A (*b*) 5A

(*c*) 6A (*d*) 1A **[ES]**

145. Figure I shows a network. Its pole-zero diagram for the network function $\frac{V_2(s)}{I_t(s)} = Z(s)$ is given in Figure II. If $\lim_{s\to 0} = 1\Omega$, then the values of R, L and C will be

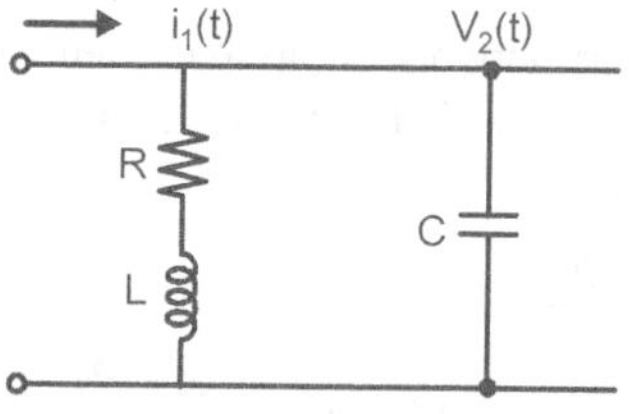

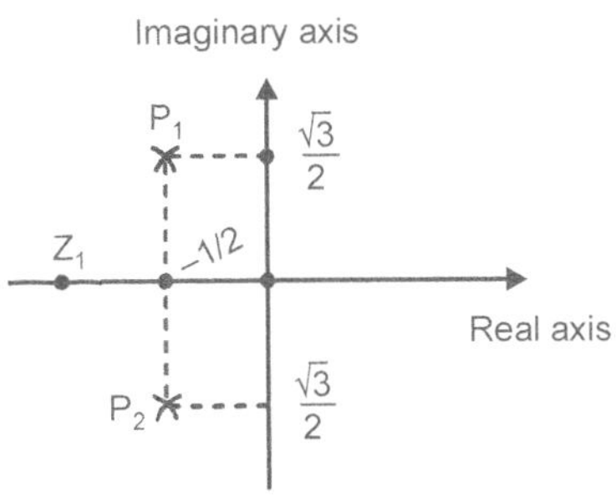

	R (in ohms)	L (inH)	C (inF)
(a)	1	1	1
(b)	1	0.5	0.5
(c)	0.5	1	1
(d)	1	1	0.5

[ES]

146. 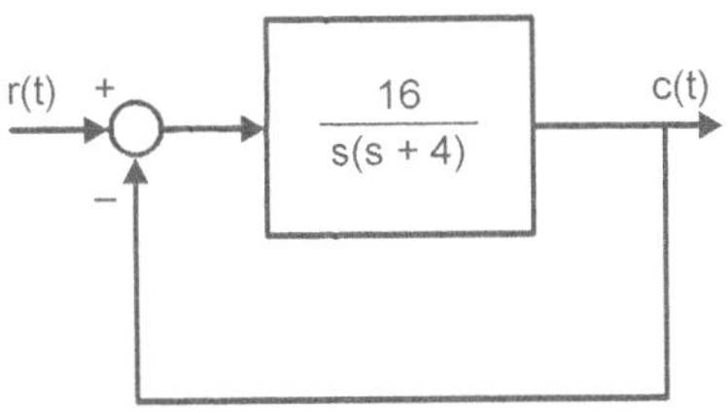

In the system shown in the above figure $r(t) = \sin \omega t$. The steady-state response $C(t)$ will exhibit a resonance peak at a frequency of

(*a*) 4 rad/sec

(*b*) $2\sqrt{2}$ rad/sec

(*c*) 2 rad/sec

(*d*) $\sqrt{2}$ rad/sec **[ES]**

147. The open-loop transfer function of a unity feedback control system is given by

$$G(s) = \frac{K(s+2)}{s(s^2+2s+2)}$$

The centroid and angles of root locus asymptotes are respectively

(*a*) zero and +90°, –90°

(*b*) –2/3 and +60°, –60°

(*c*) zero and +120°, –120°

(*d*) –2/3 and –90°, +90° **[IAS]**

148. A system with the characteristic equation $s^4 + 2s^3 + 11s^2 + 18s + 18 = 0$,

will have closed-loop poles such that

(*a*) all poles lies in the left half of the *s*-plane

(*b*) all poles lies in the right half of the *s*-plane

(*c*) two poles lies symmetrically on the imaginary axis of the *s*-plane

(*d*) no pole lies on the imaginary axis of the *s*-plane **[IAS]**

149. A coil X of 1000 turns and another coil Y of 2000 turns are placed such that 60% of the flux produced by X links Y. A current of 1A in X produces a flux of 0.1 mWb in it. The mutual inductance between the two coils is

(*a*) 0.12 H (*b*) 0.08H

(*c*) 0.06 H (*d*) 0.04H **[IAS]**

150. For the circuit shown in the given figure, the time constant of the basic circuit is

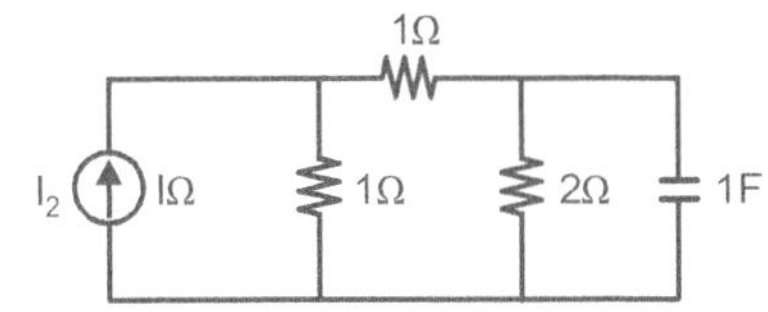

(*a*) 2/3 sec (*b*) 1 sec

(*c*) 2 sec (*d*) 3 sec **[IAS]**

151. Match List-I with List-II and select the correct answer using the codes given below the Lists

List-I

(*Pole-zero configuration of the impedance function of a 1-port LC network*)

A. 

B. 

C. 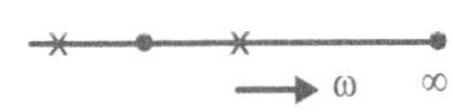

D. 

List-II

(*Cononical network realisation of the LC network*)

1. 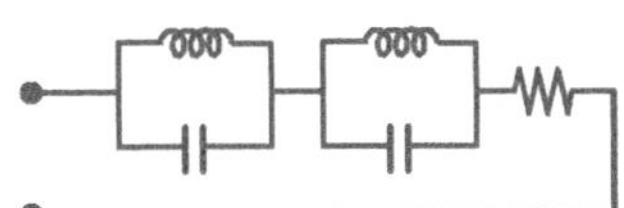

2. 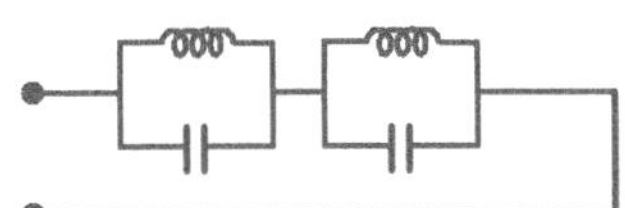

3. 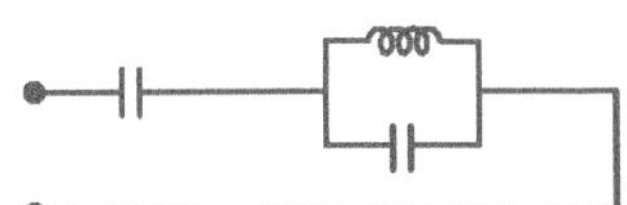

4. 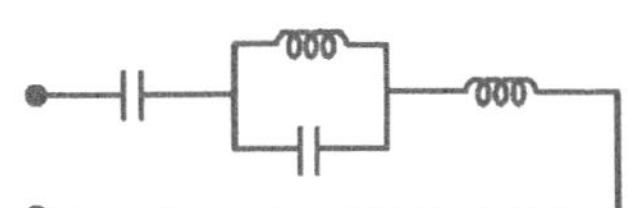

Codes:

	A	B	C	D
(*a*)	1	2	3	4
(*b*)	1	3	2	4
(*c*)	2	1	4	3
(*d*)	2	4	3	1

[ES]

152. An RC driving point impedance function has zeros at $s = -2$ and $s = -5$. The admissible poles for the function are

(*a*) $s = 0$ and $s = -6$

(*b*) $s = -1$ and $s = -3$

(*c*) $s = 0$ and $s = -1$

(*d*) $s = -3$ and $s = -4$ **[IAS]**

153. In a two-port network, the output short-circuit current was measured while the source voltage at the input was 1V; the value of the output current would provide the parameter

(*a*) h_{12} (*b*) y_{12}

(*c*) h_{21} (*d*) y_{21} **[IAS]**

154. For an ac circuit, if $e(t) = 160 \sin(\omega t + 10°)$ and $i(t) = 5 \sin(\omega t - 20°)$, then the reactive power absorbed by the circuit will be

(*a*) 400Vars (*b*) 346.4 Vars

(*c*) 200 Vars (*d*) 136.8 Vars **[IAS]**

155. The current shown is a linear time invariant one and the sources are ideal. Choose from the answers given below, the values of voltage across and current through 1Ω resistor

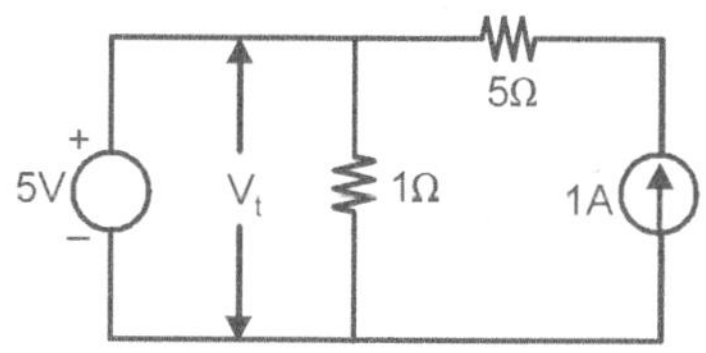

(*a*) –5V, –5A (*b*) 1V, 6A

(*c*) 1V, 1A (*d*) 5V, 5A **[ES]**

156. For the given parallel resonant circuit, match List-I with List-II and select the correct answer using the codes given below the lists

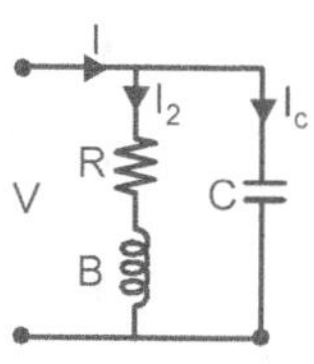

List-I

A. I at resonance

B. I_L

C. Dynamic impedance

List-II

1. $\frac{\omega l}{R}$

2. In phase with voltage

3. $\frac{L}{CR}$

4. Lags the applied voltage

Codes:

	A	B	C
(*a*)	4	2	3
(*b*)	2	4	3
(*c*)	4	2	1
(*d*)	2	4	1

[IAS]

157. A passive 2-port network is in a steady-state. Compared to its input, the steady state output can never offer

(*a*) higher voltage

(*b*) lower impedance

(*c*) greater power

(*d*) pattern regulation

158. Currents I_1, I_2 and I_3 meet at a junction (node) in a circuit. All currents are marked as entering the node. If $I_1 = 6 \sin(\omega t)$ mA and $I_2 = 8 \cos(\omega t)$ mA, then I_3 will be

(*a*) $10 \cos(\omega t + 36.87)$mA

(*b*) $14 \cos(\omega t + 36.87)$mA

(*c*) $-14 \sin(\omega t + 336.87)$mA

(*d*) $-10 \cos(\omega t + 36.87)$mA

159. Match List-I with List-II and select the correct answer using the codes given below the list:

	List-I (*Hybrid parameters*)		List-II (*Relation with immitance parameters*)
A.	h_{11}	1.	$\frac{1}{Y_{11}}$
B.	h_{22}	2.	$\frac{1}{Z_{22}}$
C.	g_{11}	3.	$\frac{1}{Z_{11}}$
D.	g_{22}	3.	$\frac{1}{Y_{22}}$

Codes:

	A	B	C	D
(*a*)	1	2	3	4
(*b*)	3	2	1	4
(*c*)	1	4	3	2
(*d*)	3	4	1	2

[IAS]

160. The transfer function of the system shown in the given figure is

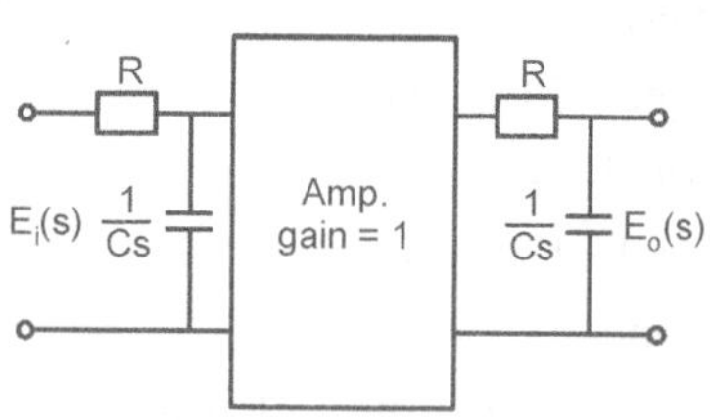

τ = RC

(*a*) $\frac{1}{1 + \tau s}$

(*b*) $\frac{1}{(1 + \tau s)^2}$

(*c*) $\frac{\tau s}{1 + \tau s}$

(*d*) $\frac{\tau s}{(1 + \tau s)^2}$ **[IAS]**

161. The y-parameter 'y_{21}' of the given network

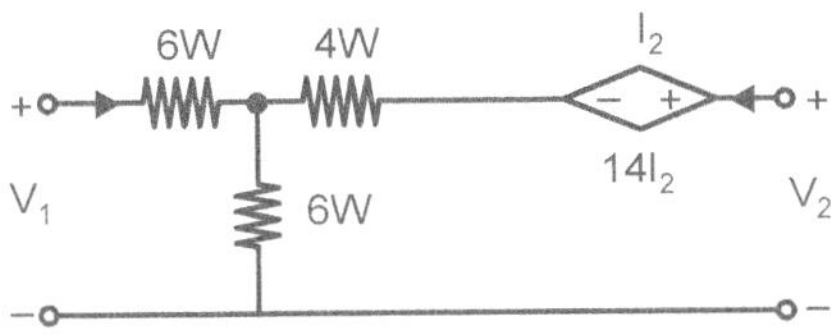

(a) is 12 mho

(b) is 6 mho

(c) is 3 mho

(d) does not exit **[ES]**

162. The Fourier transform of unit impulse function (t) would be

(a) $\dfrac{1}{(2+\omega^2)}$

(b) $(2+\omega^2)$

(c) 1

(d) $\dfrac{1}{2}$ **[IAS]**

163. The Laplace transform of current in an RLC series circuit with $R = 2\Omega$, $L = 1$H and $C = \dfrac{1}{2}$ F is $I(s) = \dfrac{1}{s^2+2s+2}$

The voltage across the inductor 'L' will be

(a) $e^{-t}\sin t\, u(t)$

(b) $e^{-t}\cos t\, u(t)$

(c) $e^{-t}(\sin t + \cos t\, u(t)$

(d) $e^{-t}(\cos t + \sin t\, u(t)$ **[IAS]**

164. The phasor current through the inductance in the given circuit shown in

(a) $\left(\dfrac{10}{\sqrt{2}}\right)\angle -45°$

(b) $\left(\dfrac{10}{\sqrt{2}}\right)\angle 45°$

(c) $5\angle 45°$

(d) $5\angle -45°$ **[IAS]**

165. The maximum value of mutual inductance of two inductively coupled coils with self inductance

$L_1 = 49$ mH and $L_2 = 81$ mH is

(a) 130mH

(b) 63mH

(c) 32mH

(d) 3969mH **[IAS]**

166. The circuit shown in the figure is equivalent to a load of

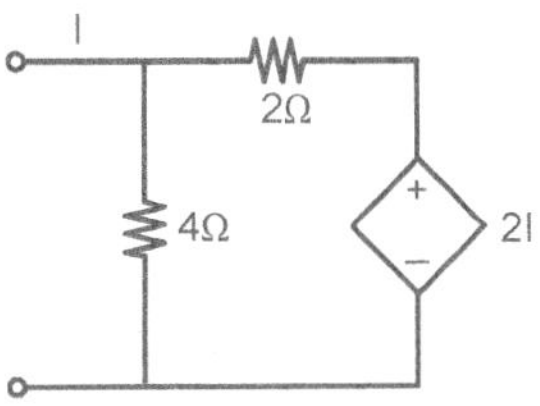

(a) $\dfrac{3}{4}$ ohms

(b) $\dfrac{8}{3}$ ohms

(c) 4 ohms

(d) 2 ohms **[IAS]**

167. The resonant frequency of a series RLC circuit is 1.5 MHz with the resonating capacitor set at 150 pF. If the bandwidth is 10 kHz, then the effective resistance of the circuit would be approximately

(a) 29.5Ω (b) 14.75Ω

(c) 9.4Ω (d) 4.7Ω **[IAS]**

168. The impedance Z(s) of the one-port network shown in the figure is given by

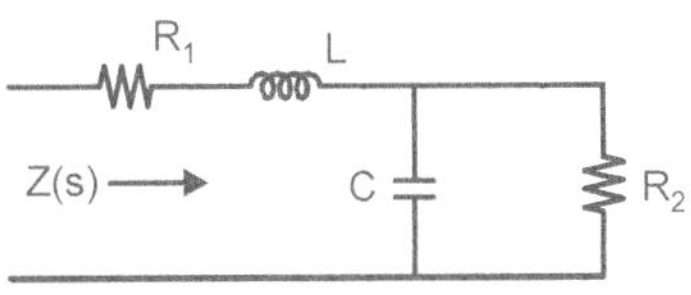

(a) $\dfrac{s^2+s\dfrac{L}{C}+\dfrac{R_1}{CL}}{s+\dfrac{1}{LC}}$

(b) $\dfrac{s^2+s\dfrac{L+R_1R_2C}{R_2LC}+\dfrac{R_1+R_2}{R_2LC}}{S+\dfrac{1}{R_2C}}$

(c) $\dfrac{s^2+s\dfrac{L+C}{R_1R_2}+\dfrac{R_1}{LC}}{s+\dfrac{1}{R_1C}}$

(d) $\dfrac{s^2+s\dfrac{L+R_2C}{R_2LC}+\dfrac{R_1+R_2}{R_2LC}}{s+\dfrac{1}{R_1C}}$ **[IAS]**

169. Consider the following driving-point functions:

$$F_1 = \frac{(s^2+1)(s^2+3)}{s(s^2+4)},$$

$$F_2 - \frac{(s^2+1)(s^2+3)}{(s^2+2)(s^2+4)}$$

$$F_3 = \frac{s(s^2+3)}{(s^2+2)(s^2+4)}$$

$$F_4 = \frac{(s^2+2)(s^2+5)}{s(s^2+4)}$$

Among these, the LC functions would include

(*a*) F_1 and F_2
(*b*) F_4 and F_3
(*c*) F_3 and F_4
(*d*) F_2 and F_4 **[ES]**

170. A 240V single-phase ac source is connected to a load with an impedance of 10 ∠60°Ω A capacitor is connected in parallel with the load. If the capacitor supplies 1250 VAR, the real power supplied by the source is

(*a*) 3600 W (*b*) 2880 W
(*c*) 2400 W (*d*) 1200 W **[ES]**

171. Lists-I and II given below show four elementary circuits and their current response due to unit voltage impulse (*t*). Match List I with List II and select the correct answer using the codes given below the lists:

List-I	**List-II**
(*Elementary circuit*)	(*Current response*)

A. 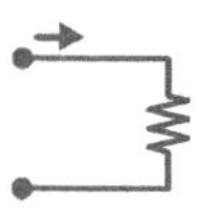 1. 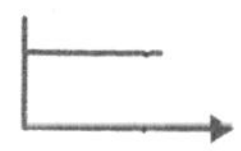

B. 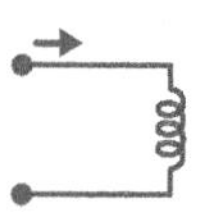 2. 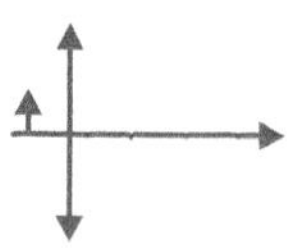

C. 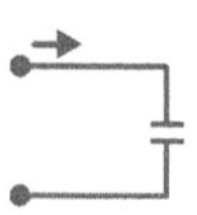 3. 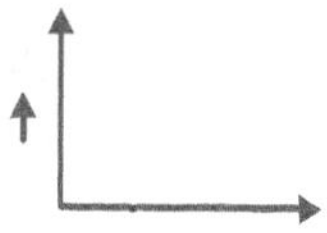

D. 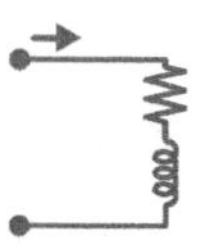 4. 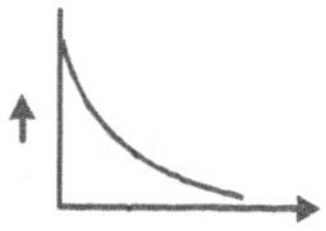

Codes:

	A	B	C	D
(*a*)	1	3	2	4
(*b*)	3	1	2	4
(*c*)	1	3	4	2
(*d*)	3	1	4	2

[IAS]

172. The equivalent inductance of two coils A and B connected as in the given figure is given by

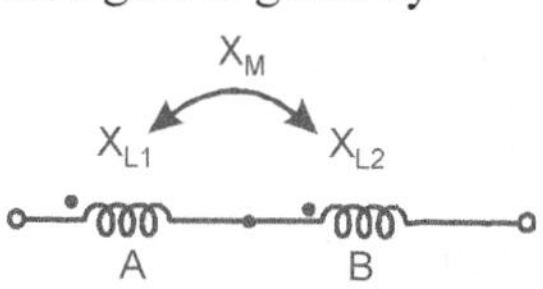

(*a*) $X_{L1} + X_{L2} - 2X_M$ (*b*) $X_{L1} + X_{L2} + X_M$
(*c*) $X_{L1} + X_{L2} - X_M$ (*d*) $X_{L1} + X_{L2} + 2X_M$ **[ES]**

173. If I(*s*) is found to be $\frac{1}{s(s+2)(s+5)}$ then the final value of I (i.e. I at $t = \infty$) will be

(*a*) 10 (*b*) 1.0
(*c*) 0.667 (*d*) 0.1 **[IAS]**

174. The short-circuit admittance matrix of the network shown in the given figure is

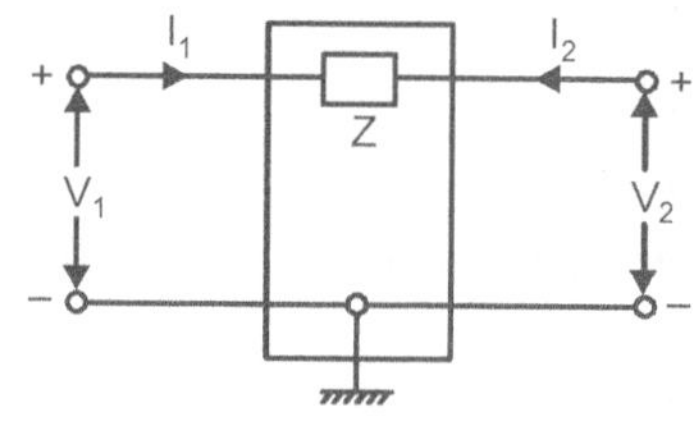

(*a*) $\begin{pmatrix} \frac{1}{Z} & -\frac{1}{Z} \\ -\frac{1}{Z} & \frac{1}{Z} \end{pmatrix}$ (*b*) $\begin{pmatrix} \frac{1}{Z} & -\frac{1}{Z} \\ \frac{1}{Z} & \frac{1}{Z} \end{pmatrix}$

(*c*) $\begin{pmatrix} \frac{1}{Z} & \frac{1}{Z} \\ -1 & 1 \end{pmatrix}$ (*d*) $\begin{pmatrix} \frac{1}{Z} & 1 \\ 1 & \frac{1}{Z} \end{pmatrix}$ **[ES]**

175. The hot resistance of the filament of a bulb is higher than the cold resistance because the temperature coefficient of the filament is

(*a*) negative (*b*) infinite
(*c*) zero (*d*) positive **[ES]**

176. Two incandescent light bulbs of 40W and 60W rating are connected in series across the mains. Then

(*a*) the bulbs together consume 100W
(*b*) the bulbs together consume 50W
(*c*) the 60W bulb glows brighter
(*d*) the 40W bulb glows brighter **[IAS]**

177. For the circuit shown in the given figure, if R = 10Ω, X_L = 0.1Ω and X_C = 0.1Ω, then the current through R is

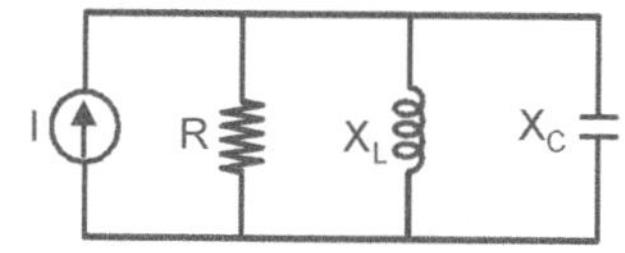

(*a*) 0 (*b*) 101
(*c*) 1 (*d*) 0.11 **[IAS]**

178. In a two-port network, the condition for reciprocity in terms of '*h*'-parameters is

(*a*) $h_{12} = h_{21}$
(*b*) $h_{11} = h_{22}$
(*c*) $h_{11} = -h_{22}$
(*d*) $h_{12} = -h_{21}$ **[ES]**

179. In the circuit shown in the given figure, three-phase supply is connected to a star-connected load. If ammeters A_1, A_2, A_3 read 12.5A, what will be the reading of ammeter A_4?

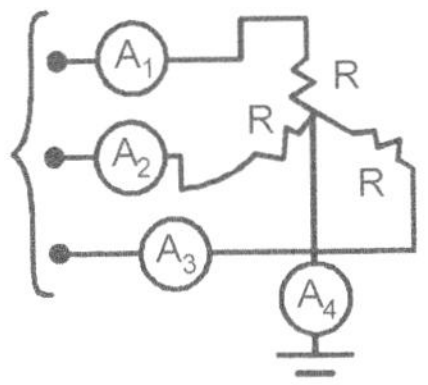

(*a*) 37.5A (*b*) $\frac{37.5}{\sqrt{3}}$

(*c*) $37.5\sqrt{3}$ (*d*) zero **[IAS]**

180. Match List-I (Readings obtained while measuring 3-phase power by two-wattmeter method) with List-II (Power factors for the load) and select the correct answer using the codes given below the lists:

List-I

A. Both the wattmeters read equal values of power but of opposite sign

B. Both the wattmeters read equal values of power and both are of positive sign

C. One wattmeter reads zero and the other reads the complete power

List-II

1. Unity
2. Zero
3. 0.5
4. 0.866

Codes:

	A	B	C
(*a*)	2	3	4
(*b*)	3	1	2
(*c*)	1	4	3
(*d*)	2	1	3

[IAS]

181. In the circuit shown in the given figure, if I_1 = 1.5A, then I_2 will be

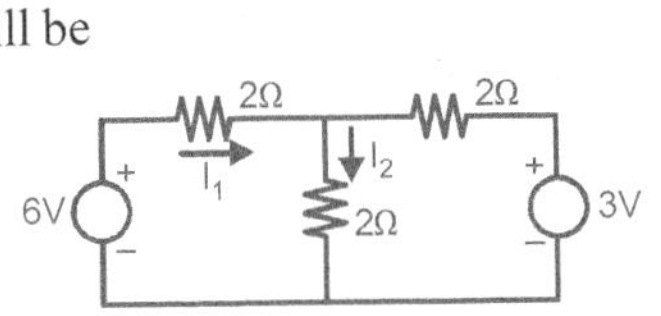

(*a*) 0.5A (*b*) 1.0A

(*c*) 1.5A (*d*) 3.04 **[ES]**

182. When a constant voltage source V is connected to a series R-L circuit with zero initial stored energy in the inductor, the instantaneous value of power supplied to the inductor I is given by

(*a*) $\frac{V^2}{R}(e^{-Rt/L} - e^{-2Rt/L})$ (*b*) $\frac{V^2}{R}(1 - e^{-Rt/L})$

(*c*) $\frac{V^2}{R}e^{-Rt/L}$ (*d*) $\frac{V^2}{R}(1 + e^{-Rt/L})$ **[IAS]**

183. In a linear system, an input of 5 sin *ωt* produces an output of 10 cos *ωt*. The output corresponding to input 10 cost *ωt* will be equal to

(*a*) + 5 sin *ωt* (*b*) – 5 sin *ωt*

(*c*) – 20 sin *ωt* (*d*) 20 sin *ωt* **[IAS]**

184. In a balanced delta connected resistive load when one resistor is open-circulated, then the power drawn by the load

(*a*) is reduced by 1/3 (*b*) is increased by 1/3

(*c*) remains same (*d*) is reduced by 1/2 **[IAS]**

185. The power in a series R-L-C circuit will be half of that a resonance when the magnitude of the current is equal to

(*a*) $\frac{V}{2R}$ (*b*) $\frac{V}{\sqrt{3}R}$

(*c*) $\frac{V}{\sqrt{2}R}$ (*d*) $\frac{\sqrt{2}V}{R}$ **[IAS]**

186. A battery is connected to a resistance causing a current of 0.5 in the circuit. The current drops to 0.4A when additional resistance of 5Ω is connected in series. The current will drop to 0.2A when the resistance is further increased by

(*a*) 10Ω (*b*) 15Ω

(*c*) 25Ω (*d*) 40Ω **[ES]**

187. In a series RLC circuit, the voltage across the inductance as a function of frequency will be maximum

(*a*) just before resonant frequency

(*b*) at resonant frequency

(*c*) just after resonant frequency

(*d*) both just before and just after resonant frequency **[IAS]**

188. The driving-point impedance of the circuit shown in the figure is

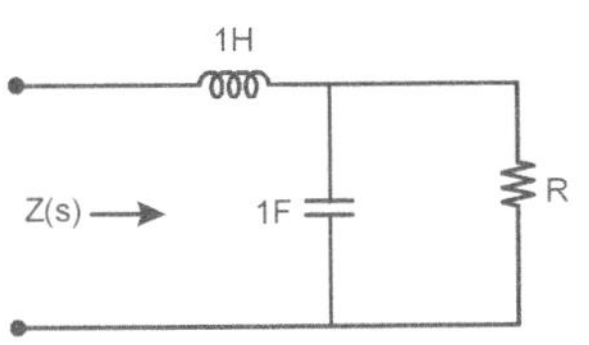

(*a*) $\frac{s^2 + Rs}{1 + Rs}$ (*b*) $\frac{Rs^2 + s + R}{Rs + 1}$

(*c*) $\frac{s^2(R+1)+3}{1+Rs}$ (*d*) $\frac{1}{(1+Rs)^2}$ **[ES]**

189. Consider the following properties:

1. Z(*s*) is real when s is real
2. $\frac{1}{Z(s)}$ is not necessarily positive real
3. All coefficients of Z(*s*) are real and non-negative
4. No poles or zeros of Z(*s*) are in the right half of the *s*-plane

Those properties which are satisfied by a positive real network function $Z(s)$ would include

(*a*) 1,2 and 3 (*b*) 2, 3 and 4

(*c*) 1,2 and 4 (*d*) 1,3 and 4 **[IAS]**

190. A balanced star-connected load with impedances of $20 \angle -30°\Omega$ is supplied from a 3-phase, 4-wire, 173 volt system, the voltages to neutral being $100\angle -90°$, $100\angle 30°$ and $100\angle 150°$V. The current in the neutral wire is

(*a*) 5A (*b*) 8.65A

(*c*) zero (*d*) 50A **[IAS]**

191. Consider the following statements:

For a bilateral network,

1. $A = D$
2. $Z_{12} = Z_{21}$
3. $h_{12} = -h_{21}$

(*a*) 1 , 2 and 3 are correct (*b*) 1 and 2 are correct

(*c*) 1 and 3 are correct (*d*) 2 and 3 are correct **[ES]**

192. The Thevenin equivalent of a network is as shown in the given figure. For maximum power transfer to the variable and purely resistive load R_L, its resistance should be

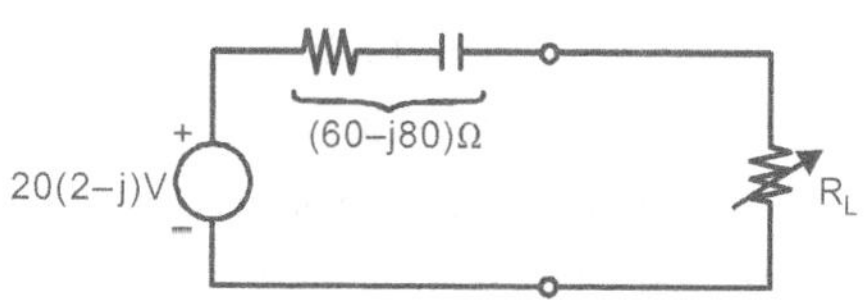

(*a*) 60Ω (*b*) 80Ω

(*c*) 100Ω (*d*) infinitely **[ES]**

193. Two two-port network α and β having ABCD parameters as

$A_\alpha = 4 = D_\alpha$ $A_\beta = 3 = D_\beta$

and

$B_\alpha = 5, C_\alpha = 3$ $B_\beta = 4$ and $C_\beta = 2$

are connected in cascade in the order of α, β. The equivalent 'A' parameter of the combination is

(*a*) 17 (*b*) 22

(*c*) 24 (*d*) 31 **[ES]**

194. A linear transformer and its T-equivalent circuit are shown in Figure-I and Figure-II respectively. The values of inductance L_a, L_b and L_c are respectively

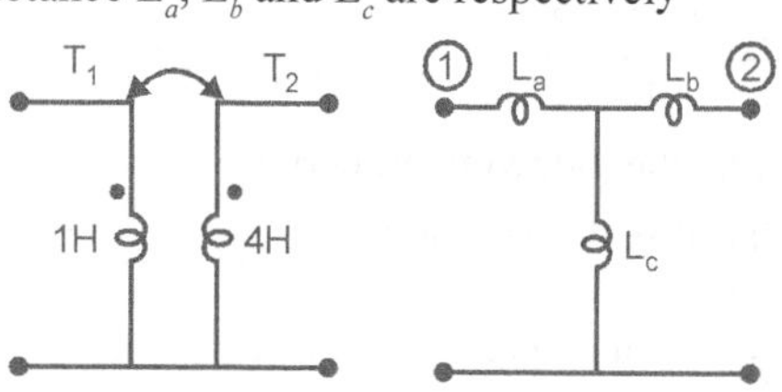

Fig.I *Fig.II*

(*a*) 1H, –2H and 2H (*b*) –1H, 2H and 2H

(*c*) 3H, 6H and –2H (*d*) 3H, 6H and 2H **[ES]**

195. One of the two wattmeters has read zero in the two-wattmeter method of power measurement. This indicated that the load phase angle is

(*a*) 0° (*b*) 30°

(*c*) 61° (*d*) 90° **[IAS]**

196. In case of a 'lossy capacitor', its series equivalent resistance value will be

(*a*) negligible

(*b*) very small

(*c*) small

(*d*) large **[PGCIL]**

197. Two capacitors each of breakdown voltage 500V are connected in parallel. The breakdown rating of the combination will be

(*a*) 500V (*b*) 650V

(*c*) 1000V (*d*) 2500V **[PGCIL]**

198. For a capacitor of 10μF, the potential difference is increased uniformally from to 600V in 2 seconds. The charging current will be

(*a*) 1 mA (*b*) 3 mA

(*c*) 5mA (*d*) 10mA **[PGCIL]**

199. Which of the following statements about electric voltage is true?

(*a*) voltage is the directed movement of electrons

(*b*) voltage causes current to flow

(*c*) voltage is the irregular movement of electrons

(*d*) voltage is not always needed to course the flow of current **[PGCIL]**

200. The self-induced voltage of a coil depends upon

(*a*) the voltage applied to the coil

(*b*) the shape of the iron core

(*c*) the shape of the bobbin

(*d*) the number of turns of the coil **[BEL]**

201. A voltage source having an internal impedance of $8 + j6$ ohms supplies power to a resistive load. What should be the load resistance of maximum power transferred to it?

(*a*) 8 ohms

(*b*) 6 ohms

(*c*) 10 ohms

(*d*) $\sqrt{10}$ ohms **[BEL]**

202. The transfer function of a system

$Z(s) = \dfrac{V(s)}{I(s)} = \dfrac{s}{s+3}$ the system is at rest for $t < 0$. What will be the value of $V(f)$ for $t \geq 0$ if $i(t) = 3u(t)$, where $u(t)$ is a unit step?

(*a*) e^{-t} (*b*) $2e^{-3t}$

(*c*) $3e^{-3t}$ (*d*) $4e^{-3t}$ **[BEL]**

203. If Thevenin's equivalent resistance of the given circuit seen from the open terminals is 2Ω, then the value of 'R' will be

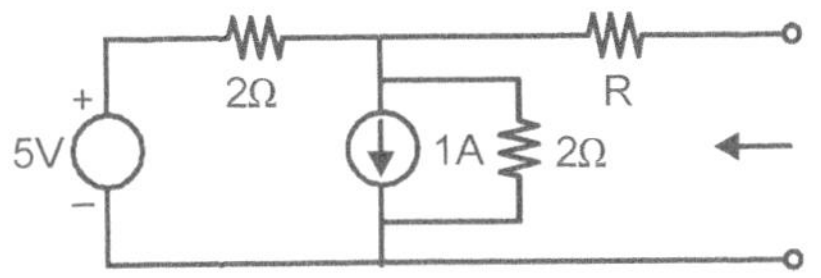

(*a*) 4Ω (*b*) 2Ω
(*c*) 1Ω (*d*) zero **[IAS]**

204. The derivative of a step function is
(*a*) zero
(*b*) an impulse function
(*c*) a ramp function
(*d*) a constant **[IAS]**

205. Match List-I with List-II for the two-port network shown in the given figure and select the correct answer using the codes given below the lists:

List-I	**List-II**
A. Z_{11}	1. R
B. Z_{12}	2. R+L
C. Z_{21}	3. R−Ls
D. Z_{22}	4. R+Ls

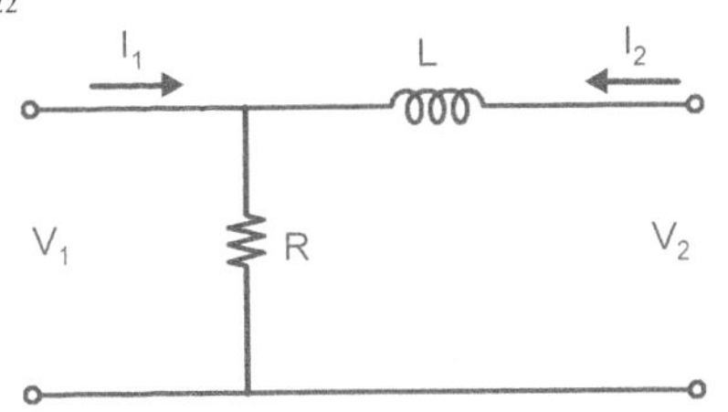

Codes:

	A	B	C	D
(*a*)	1	2	1	4
(*b*)	2	1	1	3
(*c*)	1	1	1	4
(*d*)	2	1	3	4

[IAS]

206. What is the rms value of the voltage waveform shown in the figure?

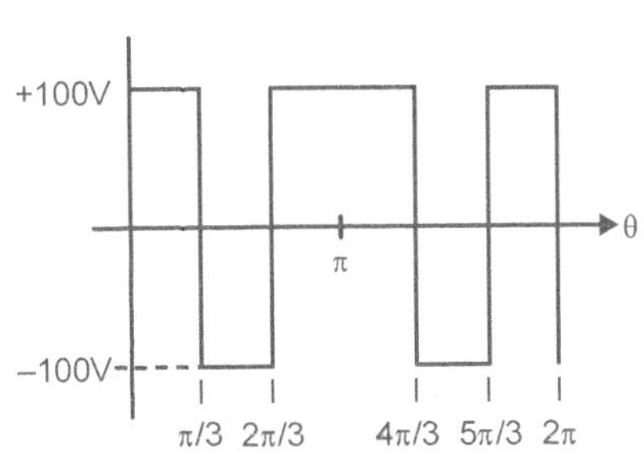

(*a*) 200/πV
(*b*) 100/πV
(*c*) 200V
(*d*) 100V **[IAS]**

207. The time constant of the network shown in the figure is

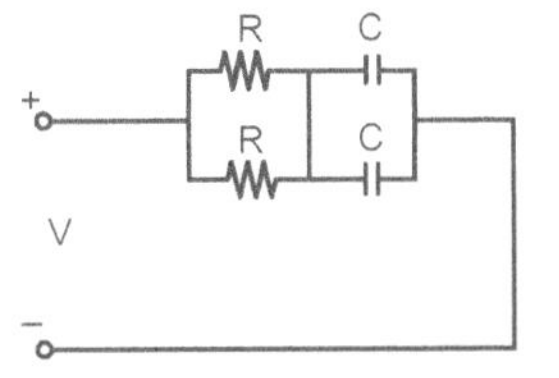

(*a*) CR (*b*) 2CR
(*c*) $\frac{CR}{4}$ (*d*) $\frac{CR}{2}$ **[ES]**

208. In the network shown in the given figure, if the voltage v at the time considered is 20V, then *dv*/*dt* at that time will be

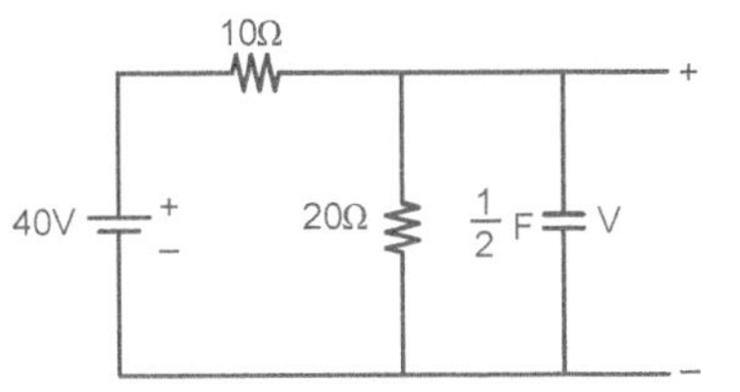

(*a*) 1V/s (*b*) 2V/s
(*c*) −2V/s (*d*) zero **[ES]**

209. Consider the following statements regarding the circuit shown in the given figure:

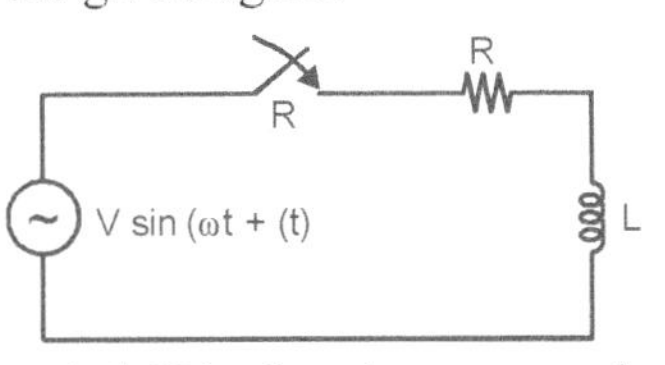

1. If the switch K is closed at a proper instant there will be no transient.
2. The instant at which K is closed such that the transient is zero depends on the frequency of the supply.
3. The instant at which K is closed such that the transient is zero depends on the circuit elements.
4. There will always be a non-zero transient after the switch K is closed.

Of the statements
(*a*) 1 alone is correct
(*b*) 1 and 2 are correct
(*c*) 1 and 3 are correct
(*d*) 4 alone is correct **[IAS]**

210. Two perfectly coupled coils each of one Henry self-inductance are connected in parallel so as to aid each other. The overall inductance is
(*a*) 2H
(*b*) 1H
(*c*) 1/2H
(*d*) zero **[IAS]**

211. In the circuit of the figure, the magnitudes of V_L and V_C are twice that of V_R. The inductance of the coil is

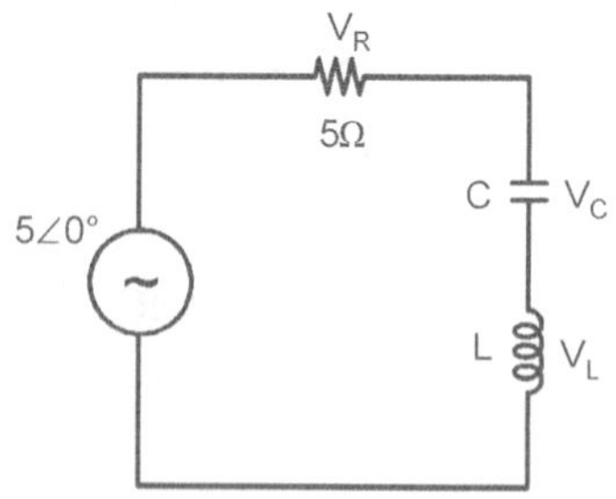

(a) 2.14mH (b) 5.30H
(c) 3.18mH (d) 1.32H **[IAS]**

212. A choke coil having resistance R Ω and inductance L henry is shunted by capacitor of C farads as shown in the given figure. The impedance at resonance is given by

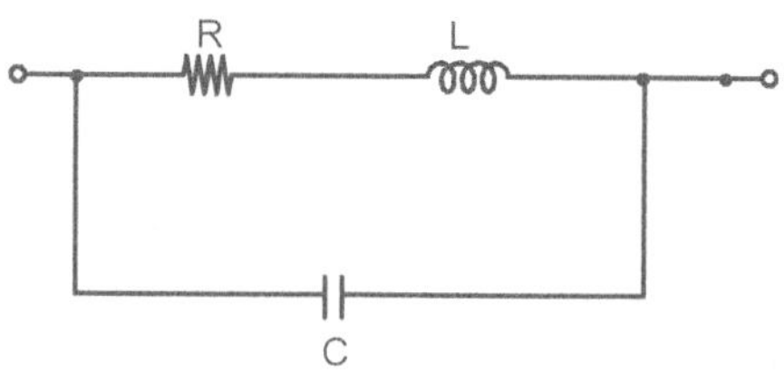

(a) $\frac{R}{LC}$ (b) $\frac{C}{RL}$
(c) $\frac{L}{RC}$ (d) $\frac{1}{RLC}$ **[IAS]**

213. The equivalent capacitance across ab will be:

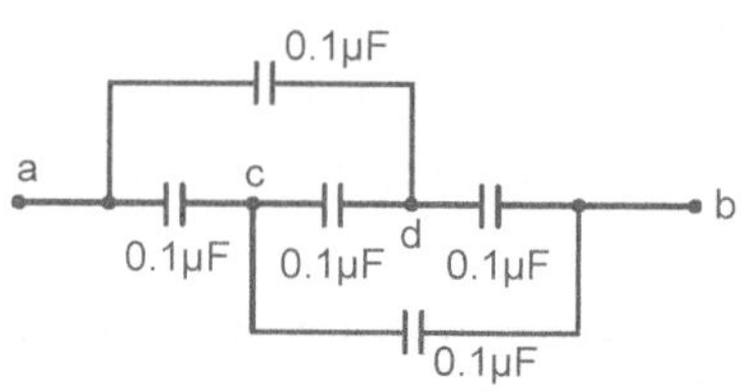

(a) 0.2 μF
(b) 0.1 μF
(c) 0.5 μF
(d) 0 **[ES]**

214. In the circuit shown in the figure,
$v = \cos 2t$, $Z_2 = 1 + j$: C_1 is so chosen that $i = i \cos 2t$. The value of C_1 is

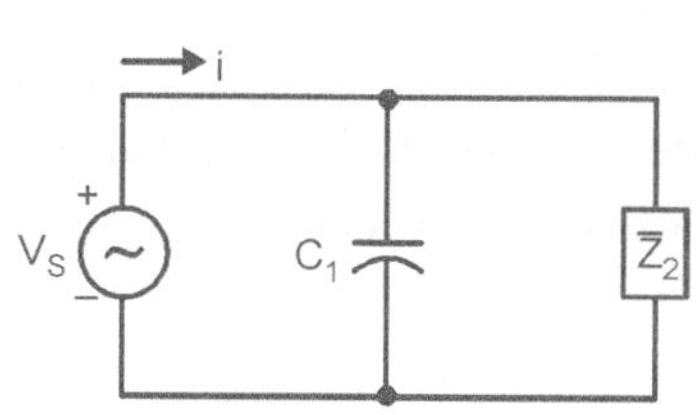

(a) 2F (b) 1F
(c) 0.5F (d) 0.25F **[ES]**

215. In the circuit shown in the given figure, if g_m is 0.5×10^{-3}, then the current I will be

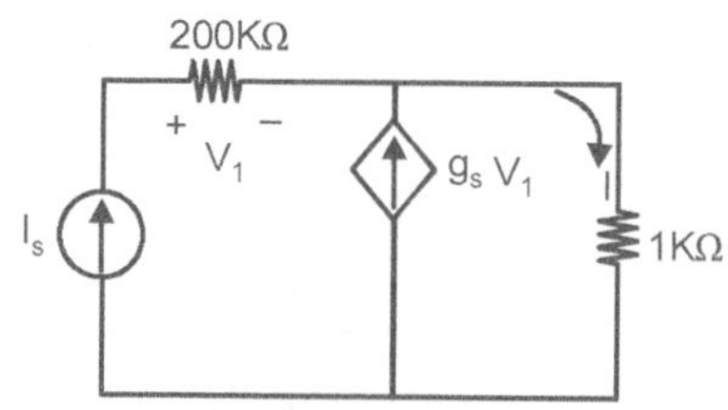

(a) I_s (b) $100I_s$
(c) $101I_s$ (d) $200I_s$ **[ES]**

216. For the network shown in the given figure, the hybrid parameter h_{21} is

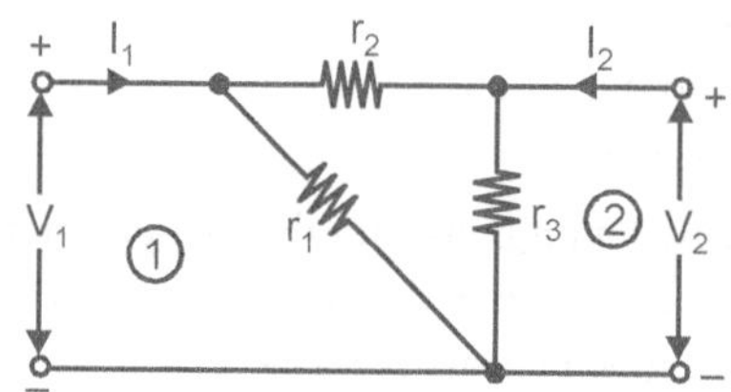

(a) $\frac{r_2}{r_1 + r_2}$ (b) $\frac{r_1 + r_2}{r_1 + r_2 + r_3}$
(c) $-\frac{r_2}{r_1 + r_2}$ (d) $-\frac{r_1}{r_1 + r_2}$ **[ES]**

217. If the steady state current in the circuit given in the figure is $i(t) = I \cos(\omega t + \phi)$, then

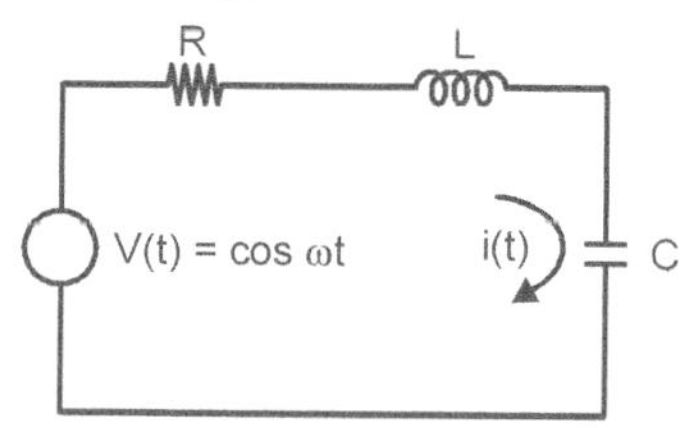

(a) $\tan^{-1}\left[\frac{\omega L - \frac{1}{\omega C}}{R}\right]$ (b) $\tan^{-1}\left[\frac{R}{\omega L - \frac{1}{\omega C}}\right]$
(c) $\tan^{-1}\sqrt{\frac{L}{C}}$ (d) $\tan^{-1}\sqrt{\frac{R}{LC}}$ **[IAS]**

218. Consider the following statements:

1. The two-port network shown below does NOT have an impedance matrix representation.

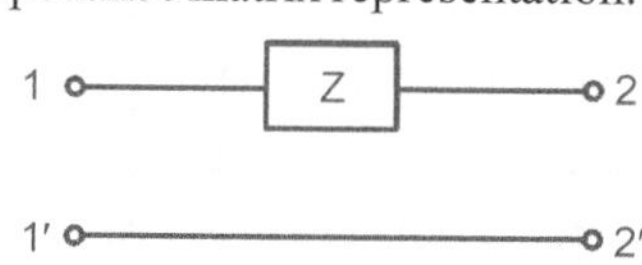

2. The following two-port network does NOT have an admittance matrix representation.

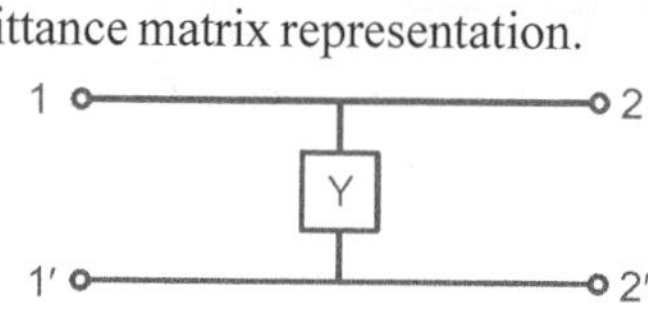

3. A two-port network is said to be reciprocal if it satisfied $z_{12} = r_{21}$ or an equivalent relationship.

Of these statements

(*a*) 1 and 2 are correct

(*b*) 1, 2 and 3 are correct

(*c*) 1 and 3 are correct

(*d*) none is correct **[ES]**

219. The output voltage V_{out} of the given circuit, for a given input signal, as shown in the figure, will be of the form

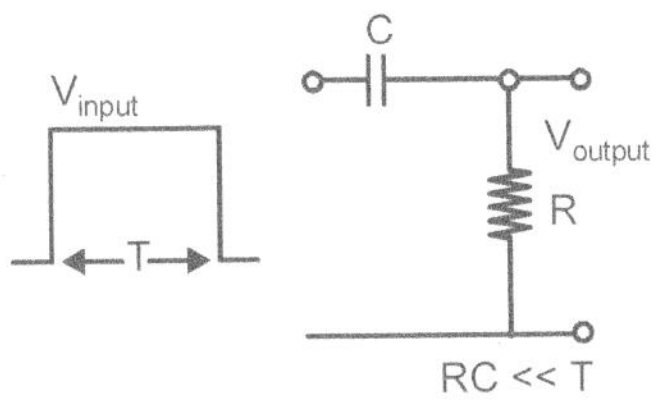

(*a*)

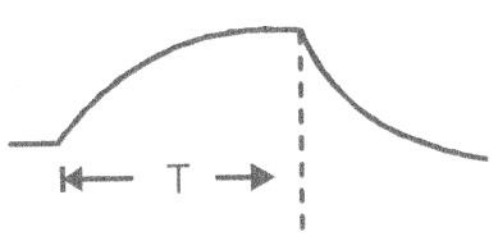

(*b*)

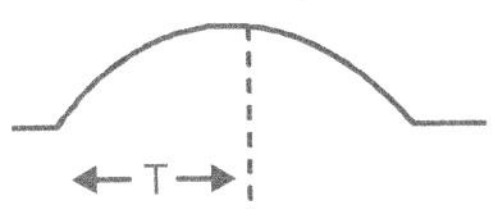

(*c*)

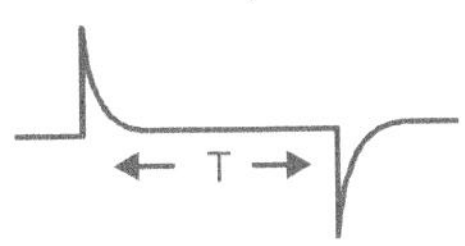

(*d*)

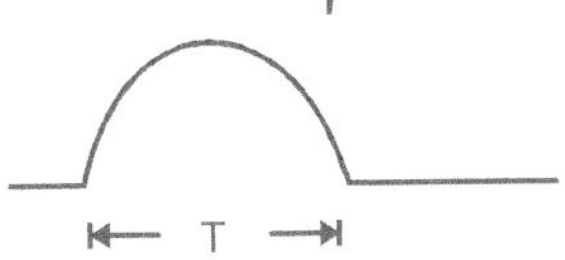

The following items consist of two statements, one labelled the 'Assertion A' and the other labelled the 'Reason R'. You are to examine these two statements carefully and decide if the Assertion A and the Reason R are individually true and if so, whether the Reason is a correct explanation of the Assertion. Select your answers to these eleven items using the codes given below and mark your answer sheet accordingly:

Codes:

(*a*) Both A and R are true and R is the correct explanation of A

(*b*) Both A and R are true but R is NOT a correct explanation of A

(*c*) A is true but R is false

(*d*) A is false but R is true **[IAS]**

220. **Assertion (A):** In the circuit shown in the figure, five current equations are required to calculate the independent branch currents and four voltage equations are required to determine the independent branch voltages.

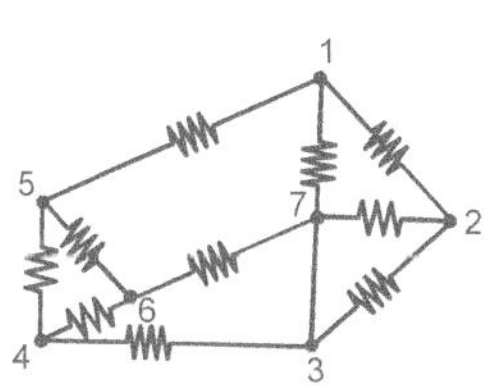

Reason (R): The number of current equations is $b - n + 1$ and the number of voltage equation $n - 1$ where b is the number of branches and n is the number of nodes for a circuit. **[ES]**

221. **Assertion (A):** The impulse response of a network can be obtained from its step response.

Reason (R): The Laplace transform of the impulse response is the transfer function of the network. **[ES]**

222. **Assertion (A):** In a two-port reciprocal network, the open-circuit impedance matrix and short-circuit admittance matrix are symmetrical.

Reason (R): The Y and Z matrices of a two-port network are inverses of each other **[ES]**

223. **Assertion (A):** In the series R-L-C circuit the voltages across elements are 300, 100 and 500 volts resistively.

Reason (R): The resultant voltage is the one which is maximum across any one element. **[ES]**

224. **Assertion (A):** The driving-point impedance and the driving-point admittance are reciprocals of each-other.

Reason (R): The driving-point network functions are the ratios of Laplace transforms of appropriate quantities (voltage and current) at the same pair of terminals. **[IAS]**

225. **Assertion (A):** The poles and zeros of a reactance function alternate on the $j\omega$-axis.

Reason (R): The slope of the reactance curve is always positive. **[ES]**

226. **Assertion (A):** The generated voltage in three phase supply contains harmonics.

Reason (R): The angular variation of the radial component of the magnetic flux in the air-gap is not a pure sinusoid. **[IAS]**

227.

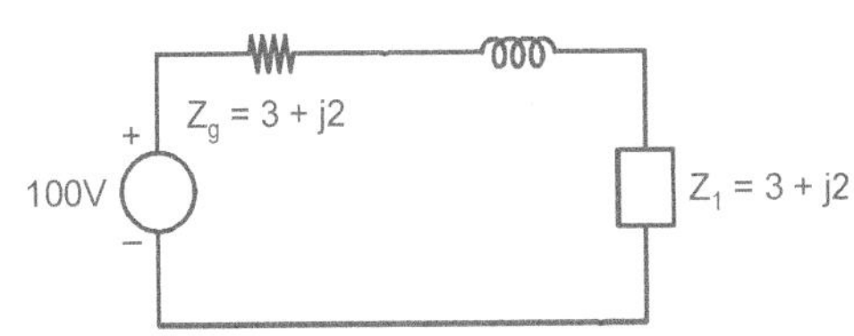

Assertion (A): In the given network, the source is delivering maximum power.

Reason (R): Z_L is the complex conjugate of Z_g **[IAS]**

228. **Assertion (A):** A driving-point function should have its poles and zeros to the left half of A-plane.

Reason (R): Only a positive real function can be realized in network form. **[ES]**

229. **Assertion (A):** The Norton's equivalent of a given networks is the dual of its Thevenin's equivalent.

Reason (R): The Thevenin's and Norton's equivalents of a given network are derivable only when the given network is linear. **[ES]**

230. Assertion (A): All networks can be reduced to Thevenin's equivalent, but only a few selected ones are capable of being converted into Norton's equivalent.

Reason (R): Thevenin's and Norton's equivalents are duals of each other. **[ES]**

231. Assertion (A): For reciprocal two-point networks det

$$\begin{bmatrix} A & B \\ C & D \end{bmatrix} = 1$$

Reason (R): In reciprocal two-point networks, the forward transfer impedance equals the reverse transfer impedance. **[ES]**

232. Assertion (A): The coefficients of the numerator and denominator polynomials of a positive real function $Y(s)$ are real and positive.

Reason (R): A positive real function $Y(s)$ is real when s is real. **[ES]**

233. Assertion (A): The life of a secondary cell is more than that of a primary cell.

Reason (R): The chemical reaction in a secondary cell is reversible. **[IAS]**

234. Assertion (A): If the voltages across R, L and C elements connected in series in an ac circuit are 300V, 100V and 500V respectively,then the total voltage applied will be 500V

Reason (R): The applied voltage in LCR series circuit is equal to the maximum voltage existing across any of these elements. **[IAS]**

235. Assertion (A): In a two-point network (see figure), the Y matrix is the inverse of Z matrix and vice versa.

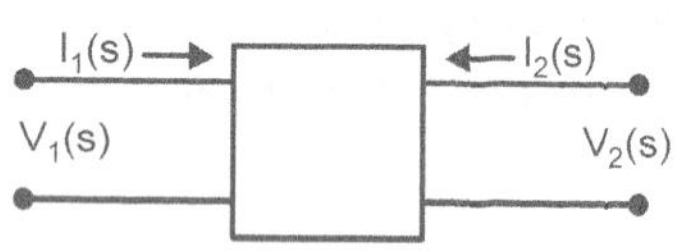

Reason (R): $\begin{bmatrix} I_1(s) \\ I_2(s) \end{bmatrix} = Y \begin{bmatrix} V_1(s) \\ V_2(s) \end{bmatrix}$

$$\begin{bmatrix} V_1(s) \\ V_2(s) \end{bmatrix} = Z \begin{bmatrix} I_1(s) \\ I_2(s) \end{bmatrix}$$ **[ES]**

236. Assertion (A): Resonance occurs in an RLC circuit (simple series on parallel) at

a frequency of $\omega_o = \dfrac{1}{\sqrt{LC}}$

Reason (R): The impedance of the circuit at this frequency is maximum. **[ES]**

237. Assertion (A): The driving point admittance function of a passive network cannot have multiple poles on they $j\omega$-axis.

Reason (R): Multiple poles on the $j\omega$-axis give rise to time domain functions of the type $t \cos \omega t$ and $t \sin \omega t$. **[IAS]**

238. Assertion (A): In a balanced three-phase star connected system, the line voltage is equal to $\sqrt{3}$ times the phase voltage.

Reason (R): The line voltage is the vector sum of the two phase voltages. **[IAS]**

239. Assertion (A): In three-phase star connected network, the neutral point can assume any potential upto phase voltage.

Reason (R): The three-phase voltage system is unbalanced. The three voltages are not equal in magnitude but they are displaced by 120°. **[IAS]**

240. Assertion (A): The Fourier series expansion of the square wave contains only cosine terms.

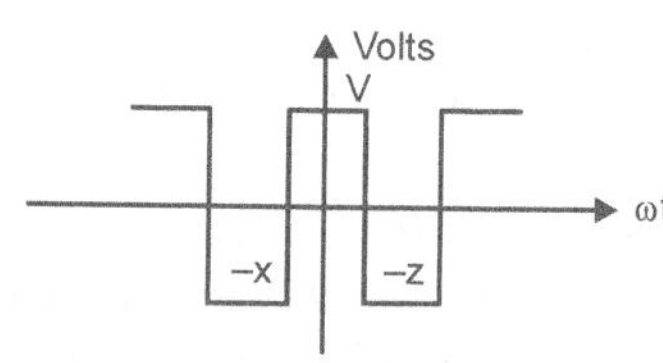

Reason (R): The wave is an even function. **[ES]**

241. Assertion (A): In delta connections Phase Voltage = Line Voltage

Line current = $\sqrt{3}$ phase current

Reason (R): In balanced three phase circuits voltages are equal in magnitude and displaced by 120°. **[ES]**

ANSWERS

EXERCISE – I

1. (*b*) **2.** (*c*) **3.** (*d*) **4.** (*d*) **5.** (*b*) **6.** (*c*) **7.** (*b*) **8.** (*d*) **9.** (*c*) **10.** (*a*)
11. (*b*) **12.** (*a*) **13.** (*a*) **14.** (*c*) **15.** (*b*) **16.** (*b*) **17.** (*c*) **18.** (*c*) **19.** (*b*) **20.** (*c*)
21. (*a*) **22.** (*b*) **23.** (*c*) **24.** (*b*) **25.** (*a*) **26.** (*a*) **27.** (*b*) **28.** (*c*) **29.** (*b*) **30.** (*d*)
31. (*d*) **32.** (*c*) **33.** (*b*) **34.** (*c*) **35.** (*d*) **36.** (*c*) **37.** (*b*) **38.** (*b*) **39.** (*a*) **40.** (*d*)
41. (*a*) **42.** (*c*) **43.** (*d*) **44.** (*c*) **45.** (*c*) **46.** (*a*) **47.** (*b*) **48.** (*a*) **49.** (*c*) **50.** (*d*)
51. (*b*) **52.** (*a*) **53.** (*b*) **54.** (*b*) **55.** (*a*) **56.** (*c*) **57.** (*a*) **58.** (*b*) **59.** (*d*) **60.** (*c*)
61. (*b*) **62.** (*d*) **63.** (*b*) **64.** (*b*) **65.** (*b*) **66.** (*a*) **67.** (*a*) **68.** (*c*) **69.** (*c*) **70.** (*c*)
71. (*c*) **72.** (*a*) **73.** (*b*) **74.** (*b*) **75.** (*a*) **76.** (*a*) **77.** (*a*) **78.** (*b*) **79.** (*b*) **80.** (*a*)
81. (*a*) **82.** (*d*) **83.** (*c*) **84.** (*c*) **85.** (*a*) **86.** (*d*) **87.** (*b*) **88.** (*d*) **89.** (*c*) **90.** (*d*)
91. (*c*) **92.** (*b*) **93.** (*a*) **94.** (*b*) **95.** (*b*) **96.** (*d*) **97.** (*b*) **98.** (*c*) **99.** (*d*) **100.** (*b*)
101. (*b*) **102.** (*a*) **103.** (*b*) **104.** (*d*) **105.** (*d*) **106.** (*c*) **107.** (*d*) **108.** (*d*) **109.** (*c*) **110.** (*c*)
111. (*d*) **112.** (*a*) **113.** (*c*) **114.** (*a*) **115.** (*d*) **116.** (*b*) **117.** (*b*) **118.** (*b*) **119.** (*b*) **120.** (*b*)
121. (*d*) **122.** (*d*) **123.** (*b*) **124.** (*a*) **125.** (*d*) **126.** (*d*) **127.** (*d*) **128.** (*c*) **129.** (*a*) **130.** (*d*)
131. (*a*) **132.** (*a*) **133.** (*c*) **134.** (*a*) **135.** (*b*) **136.** (*d*) **137.** (*c*) **138.** (*a*) **139.** (*a*) **140.** (*b*)
141. (*c*) **142.** (*d*) **143.** (*d*) **144.** (*b*) **145.** (*a*) **146.** (*c*) **147.** (*c*) **148.** (*b*) **149.** (*d*) **150.** (*b*)
151. (*b*) **152.** (*b*) **153.** (*a*) **154.** (*d*) **155.** (*d*) **156.** (*d*) **157.** (*b*) **158.** (*c*) **159.** (*c*) **160.** (*b*)
161. (*d*) **162.** (*c*) **163.** (*b*) **164.** (*a*) **165.** (*b*) **166.** (*c*) **167.** (*d*) **168.** (*c*) **169.** (*a*) **170.** (*a*)
171. (*c*) **172.** (*b*) **173.** (*b,d*) **174.** (*c*) **175.** (*b*) **176.** (*c*) **177.** (*d*) **178.** (*b*) **179.** (*b*) **180.** (*b*)
181. (*b*) **182.** (*b*) **183.** (*d*) **184.** (*a*) **185.** (*a*) **186.** (*d*) **187.** (*a*) **188.** (*d*) **189.** (*c*) **190.** (*d*)
191. (*c*) **192.** (*b*) **193.** (*d*)

EXERCISE – II

1. (*b*) **2.** (*b*) **3.** (*d*) **4.** (*c*) **5.** (*b*) **6.** (*b*) **7.** (*a*) **8.** (*a*) **9.** (*b*) **10.** (*c*)
11. (*c*) **12.** (*d*) **13.** (*a*) **14.** (*c*) **15.** (*d*) **16.** (*b*) **17.** (*c*) **18.** (*c*) **19.** (*d*) **20.** (*d*)
21. (*d*) **22.** (*c*) **23.** (*b*) **24.** (*c*) **25.** (*c*) **26.** (*b*) **27.** (*c*) **28.** (*b*) **29.** (*d*) **30.** (*c*)
31. (*b*) **32.** (*b*) **33.** (*b*) **34.** (*c*) **35.** (*d*) **36.** (*c*) **37.** (*c*) **38.** (*d*) **39.** (*c*) **40.** (*c*)
41. (*b*) **42.** (*d*) **43.** (*c*) **44.** (*d*) **45.** (*a*) **46.** (*d*) **47.** (*b*) **48.** (*c*) **49.** (*b*) **50.** (*c*)
51. (*a*) **52.** (*b*) **53.** (*c*) **54.** (*a*) **55.** (*b*) **56.** (*b*) **57.** (*a*) **58.** (*d*) **59.** (*b*) **60.** (*a*)
61. (*b*) **62.** (*b*) **63.** (*a*) **64.** (*a*) **65.** (*a*) **66.** (*d*) **67.** (*d*) **68.** (*b*) **69.** (*d*) **70.** (*d*)
71. (*a*) **72.** (*d*) **73.** (*d*) **74.** (*d*) **75.** (*c*) **76.** (*c*) **77.** (*a*) **78.** (*d*) **79.** (*b*) **80.** (*c*)
81. (*d*) **82.** (*c*) **83.** (*c*) **84.** (*c*) **85.** (*d*) **86.** (*d*) **87.** (*b*) **88.** (*d*) **89.** (*b*) **90.** (*b*)
91. (*c*) **92.** (*b*) **93.** (*c*) **94.** (*c*) **95.** (*a*) **96.** (*d*) **97.** (*b*) **98.** (*c*) **99.** (*c*) **100.** (*b*)
101. (*a*) **102.** (*a*) **103.** (*b*) **104.** (*c*) **105.** (*d*) **106.** (*a*) **107.** (*b*) **108.** (*c*) **109.** (*d*) **110.** (*c*)
111. (*a*) **112.** (*d*) **113.** (*a*) **114.** (*b*) **115.** (*b*) **116.** (*b*) **117.** (*c*) **118.** (*b*) **119.** (*d*) **120.** (*d*)
121. (*a*) **122.** (*b*) **123.** (*a*) **124.** (*a*) **125.** (*b*) **126.** (*a*) **127.** (*c*) **128.** (*c*) **129.** (*a*) **130.** (*c*)
131. (*d*) **132.** (*a*) **133.** (*b*) **134.** (*c*) **135.** (*c*) **136.** (*a*) **137.** (*b*) **138.** (*a*) **139.** (*b*) **140.** (*d*)
141. (*d*) **142.** (*c*) **143.** (*c*) **144.** (*b*) **145.** (*a*) **146.** (*b*) **147.** (*d*) **148.** (*c*) **149.** (*a*) **150.** (*b*)
151. (*a*) **152.** (*b*) **153.** (*d*) **154.** (*c*) **155.** (*d*) **156.** (*b*) **157.** (*c*) **158.** (*d*) **159.** (*a*) **160.** (*b*)
161. (*d*) **162.** (*c*) **163.** (*b*) **164.** (*a*) **165.** (*b*) **166.** (*b*) **167.** (*d*) **168.** (*b*) **169.** (*c*) **170.** (*b*)
171. (*b*) **172.** (*d*) **173.** (*d*) **174.** (*a*) **175.** (*d*) **176.** (*d*) **177.** (*a*) **178.** (*d*) **179.** (*d*) **180.** (*d*)
181. (*c*) **182.** (*b*) **183.** (*b*) **184.** (*a*) **185.** (*c*) **186.** (*c*) **187.** (*c*) **188.** (*b*) **189.** (*d*) **190.** (*c*)
191. (*d*) **192.** (*c*) **193.** (*b*) **194.** (*b*) **195.** (*c*) **196.** (*d*) **197.** (*a*) **198.** (*b*) **199.** (*b*) **200.** (*d*)
201. (*c*) **202.** (*c*) **203.** (*c*) **204.** (*b*) **205.** (*c*) **206.** (*d*) **207.** (*a*) **208.** (*b*) **209.** (*c*) **210.** (*c*)
211. (*c*) **212.** (*c*) **213.** (*b*) **214.** (*c*) **215.** (*c*) **216.** (*d*) **217.** (*a*) **218.** (*b*) **219.** (*a*) **220.** (*d*)
221. (*b*) **222.** (*d*) **223.** (*c*) **224.** (*a*) **225.** (*a*) **226.** (*a*) **227.** (*a*) **228.** (*b*) **229.** (*b*) **230.** (*d*)
231. (*b*) **232.** (*b*) **233.** (*a*) **234.** (*c*) **235.** (*a*) **236.** (*c*) **237.** (*a*) **238.** (*c*) **239.** (*a*) **240.** (*a*)
241. (*b*)

EXPLANATIONS

EXERCISE – I

1. $$i = \frac{6}{50}(1 - e^{-t/T})$$

where $$T = \frac{30 \times 10^{-3}}{50}$$

$$\frac{di(t)}{dt} = -\frac{6}{50}\left(-\frac{1}{T}e^{-t/T}\right)$$

$$\left.\frac{di(t)}{dt}\right|_{\substack{Max\\ t=0}} = \frac{6}{51T}$$

$$\therefore \quad V_2 = M\frac{di(t)}{dt} = \frac{5 \times 10^{-3} \times 6 \times 50}{50} = 1V$$

2. Current at resonance $= \frac{200}{5} = 40A$

$$\omega^2 = \frac{1}{LC} = \frac{1}{0.1 \times 50 \times 10^{-6}}$$
$$= 20 \times 10^4$$
$$Z_{coil} = \sqrt{R^2 + L^2\omega^2}$$
$$= \sqrt{5^2 + (0.1)^2 \times 20 \times 10^4}$$
$$= \sqrt{2025} = 45 \text{ ohms}$$
$$V_{coil} = 45 \times 40 = 1800 \text{ volts.}$$

3. Since reactances are cancelling each other, the circuit is purely resistive and the current is in phase with the applied voltage. Voltage across the inductor leads the current through it by 90°.

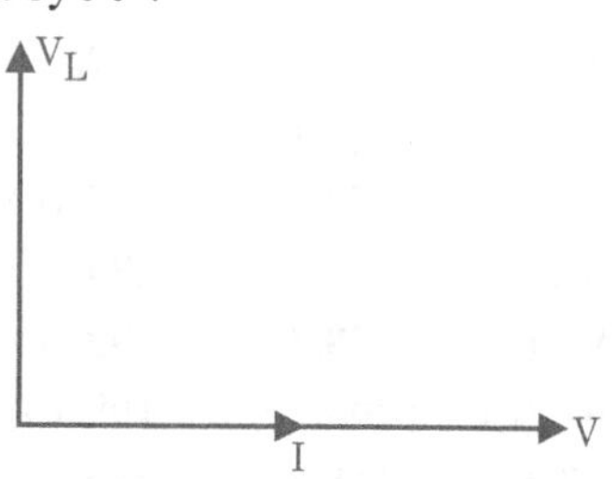

Since $$\vec{I} = \frac{100\angle 0}{\{10 + j(20-20)\}} = 10\angle 0°$$

$$\therefore \quad V_L = j\omega LI$$
$$= 20 \times 10\angle 90$$
$$= 200\angle 90$$

4. When two coils are connected in series, then effective inductance is

$$L = L_1 + L_2 \pm 2M$$

Thus $$L_1 + L_2 + 2M = 16$$
$$L_1 + L_2 \pm 2M = 8$$
$$\therefore \quad M = 2\,mH$$

7. $$C(t) = 0.5e^{-t/2}$$

Transfer function is the impuluse response of the system

$$\therefore \quad T.F. = \frac{0.5}{s + \frac{1}{2}} = \frac{1}{2s+1}$$
$$= \frac{1}{(1+2s)}$$

10. We will determine Thevenin voltage across AB first

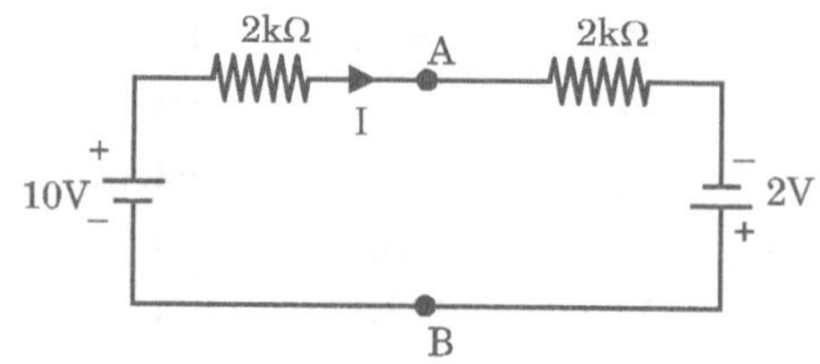

Current I in the circuit,

$$I = \frac{10-2}{2+2} = 2$$

$$\therefore \quad V_{TH} = 4V, R_{TH} = 1k\Omega$$

∴ Voltage across 1kΩ resistor when connected

$$= \frac{4 \times 1}{1+1} = 2V$$

11. $$\frac{V_2(s)}{V_1(s)} = \frac{\frac{1}{Cs}}{\frac{L}{s} \times \frac{1}{Cs}} = \frac{1}{1 + LCs^2} = \frac{1}{1+2s^2}$$

12. From the circuit, $i(0^+) = 0$

Hence from the options $\frac{di}{dt}(0_t) = 10$

Now, $$10\frac{di}{dt} + \frac{d^2i}{dt^2} + \frac{1}{10}i(0^+) = 0$$

$$\therefore \quad \frac{d^2i}{dt^2}(0^+) = -10 \times 10 = -100$$

14. Lossy capacitor is represented by a resistance in parallel in it.

18. Here, $$T = RC = 2 \times 10^6 \times 1 \times 10^{-6}$$
$$= 2\text{sec.}$$

Voltage across the capacitor,

$$v = 10(1\ e^{-2/2}) = 6.32V$$

∴ After 2 sec,

$$v = 10(1 - e^{-2/2})$$
$$= (1 - e^{-1})$$
$$= 6.32\,V$$

19. $V_R = 30\,V$

$$I_2 = \frac{30}{3} = 10A$$

$$V_C = 10 \times 3 = 30V$$

$$V = \sqrt{V_R^2 + V_C^2} = 30\sqrt{2}$$

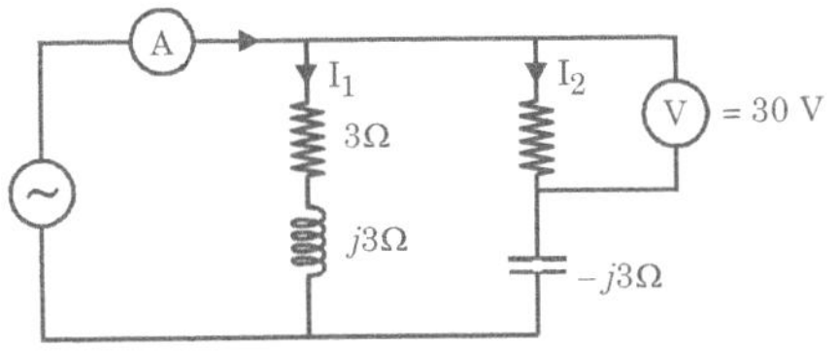

Current I_2 leads the branch voltage by 45°

Current I_1 is also 10A but lags the voltage by 45°

Current I_1 and I_2 have a phase difference of 90°

Thus $\sqrt{I_1^2 + I_2^2} = 10\sqrt{2}$

21. If $v = V_m \sin \omega t$,

$$Z = \sqrt{R^2 + L^2\omega^2} \text{ and } \phi = \tan^{-1}\frac{\omega L}{R}$$

Then $i(t) = \frac{V_m}{2}[\sin(\omega_t - \phi) - \sin(\omega t_0 - \phi)e^{-(t-t_0/T)}]$

where t_0 is the instant of switching

For no transients,

$$\omega t_o = \phi, \text{ or } \theta = \phi, \text{ or } \theta - \phi = 0$$

23. Time constant,

$$T = \frac{L}{R} = \frac{I^2L}{I^2R} = \frac{2\times\frac{1}{2}I^2L}{P}$$

$$= \frac{2\times 1000}{2000} = 1 \text{ sec.}$$

24. When excitation to a linear system is differential, output also is the differentiation of previous output.

$$\therefore \text{ Response} = e^{-2t}u(t) + \frac{d}{dt}[e^{-2t}u(t)]$$

$$= -3e^{-2t}u(t)$$

27. $P_1 = 3\frac{(V\sqrt{3})^2}{R} = 3\frac{V^2}{3\times R} = \frac{V^2}{R}$

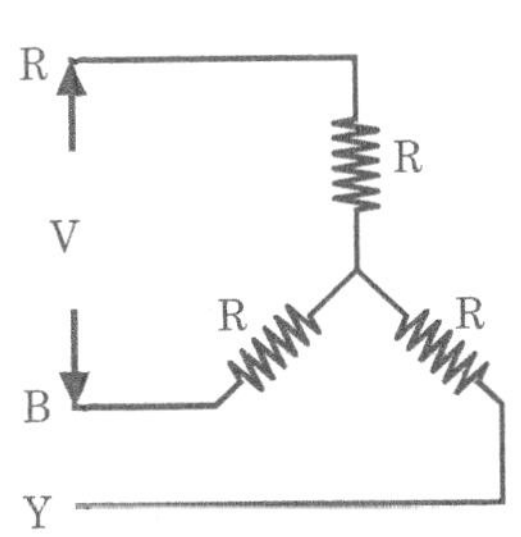

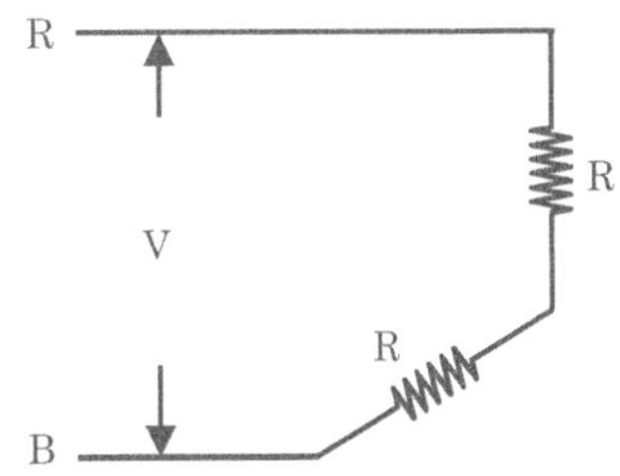

When one of the resistor is opened out,

$$P_2 = \frac{V^2}{2R}$$

$\therefore$ %reduction in load

$$= \frac{V^2/R - V^2/2R}{V^2/R} \times 100 = 50\%$$

28. Using superposition theorem when only voltage is considered and current source is opened

$$I_1 = \frac{5}{1} = 5A$$

When voltage source is short circuited and only current source is considered, then all the current flow throw short circuited path,

i.e. Current through resistor, $I_2 = 0$

$\therefore$ Total current, $I = I_1 + I_2$

$= 5 + 0$

$= 5A$

31. Since combination of voltmeter resistance and 600Ω is 400Ω and voltage across it is 5V, the unknown resistance across while the voltage is 35 – 5 = 30V must be

$$\frac{30}{5} \times 400 = 2.4\,k\Omega$$

32. Voltage across an current through the resistor of 1Ω is constrained by the source connected directly across it, i.e. 5V and 5A.

33. Net voltage across two reactive elements taken together is

$$14 - 10 = 4V$$

$$\therefore \text{ Supply voltage} = \sqrt{3^2 + 4^2} = 5V$$

34. Thevenin resistane $= 420 - \frac{840}{180} \times 120$

$= 420 - 560 = -140V$

Thevenin resistance $= \frac{120 \times 160}{180}$

$= 40$ ohms

$$\therefore \quad I = \frac{-140}{40 + 30} = -2A$$

35. $R_{TH} = \frac{2\times3}{2+3} + \frac{2\times2}{2+2} = \frac{6}{5} + 1 = \frac{11}{5}$

36. Here, $AD - BC = 1$

$A = D = 3$

$\therefore \quad 3\times3 - 1\times C = 1$

$\Rightarrow \quad C = 8$

37.

$$V = ZI$$

$$Z = \omega L + \omega L \pm 2kwL$$

$$\frac{I_1}{I_2} = \frac{Z_2}{Z_1}$$

$$\frac{10}{8} = \frac{2\omega L + 2k\omega L}{2\omega L - 2k\omega L} = \frac{1+k}{1-k}$$

$$\Rightarrow \quad 10 - 10K = 8 + 8K$$

$$\Rightarrow \quad K = \frac{1}{9}$$

40.

$$\frac{di(t)}{dt} + 3i(t) = I_s(t)$$

$$I(o) = 0$$

$$sI(s) + 3I(s) = I_s(s)$$

$$\therefore \quad I(s) = \frac{I_s(s)}{s+3}$$

For $I_s(t) = u(t)$

$$I_s(t) = \frac{1}{s}$$

$$\therefore \quad I(s) = \frac{1}{s(s+3)}$$

$$= \frac{1/3}{s} = \frac{1/2}{(s+3)}$$

$$I(t) = \frac{1}{3}u(t) - \frac{1}{3}e^{-3t}$$

$$= \frac{1}{3}(1 - e^{-3t})$$

For $\quad I_s(t) = e^{2t}u(t)$

$$I_s(t) = \frac{1}{s-2}$$

or $\quad I(s) = \frac{1}{(s+2)(s+3)}$

$$= \frac{1/5}{s-2} - \frac{1/5}{s+3}$$

$$\therefore \quad I(t) = \frac{1}{5}(e^{2t} - e^{-3t})$$

43. Effective resistance,

$$R = \frac{f_2 - f_1}{2\pi f^2 C}$$

$$= \frac{10\times10^3}{2\times3.14(1.5\times10^6)^2\times150\times10^{-12}}$$

$$= 4.7\,\Omega$$

45. Rms value of current $= \sqrt{\frac{1}{1}\int\left[\left(\frac{10}{1}\right)t\right]^2 dt}$

$$= \frac{10}{\sqrt{3}}\text{A}$$

$\therefore \quad$ Power $= I^2R$

$$= \left(\frac{10}{\sqrt{3}}\right)^2 \times 10$$

$$= \frac{1000}{3}\text{W}$$

46. Impedance $= \frac{E_{rms}}{I_{rms}} = \frac{E_{peak}}{I_{peak}}$

$$= \frac{110}{5} = 22 \text{ ohm}$$

47.

$$I(s) = \frac{V(s)}{Z(s)}$$

$$Z(s) = R + Ls$$

$$I(s) = \frac{V(s)}{R + Ls}$$

Also $V(s) = 1$,

therefore $\quad I(s) = \frac{1}{R + Ls}$

$\therefore \quad i(t) = \left(\frac{1}{L}\right)e^{-Rt/L}$:

a decaying exponential function

48. When the switch had been open, the current of 2A flows through R_2 alone. Thus the voltage is 2V, and since there is no current in branch 1, the voltage across the capacitor is also 2V.

$\therefore \quad V_c(0) = 2$

Since after the switching action, voltage across the capacitor cannot change instantaneously.

$$V_c(0^+) = V_c(0^-) = 2V$$

After the switch is closed and steady-state condition reached, the entire current of 2A passes through the closed switch and the voltage across the capacitor becomes zero

49. $i(t) = 2u(t);\ t \geq 1$

$$V_c = L\frac{di}{dt}$$
$$= 100 \times 10^{-3} \times 2\delta(t-1)$$
$$= 0.2\,\delta(t-1)\,V$$

52. $$I(s) = \frac{s+2}{(s+1)^2}$$

$$\therefore \quad I(t) = L^{-1}\left[\frac{s+2}{(s+1)^2}\right]$$
$$= L^{-1}\left[\frac{1}{s+1} + \frac{2}{(s+1)^2}\right]$$
$$= e^{-t} + 2t\,e^{-t}$$

54. It is a balanced bridge network with the addition of a resistor across AB. Thus, total resistance as measured between A and B shall be 0.5 ohm.

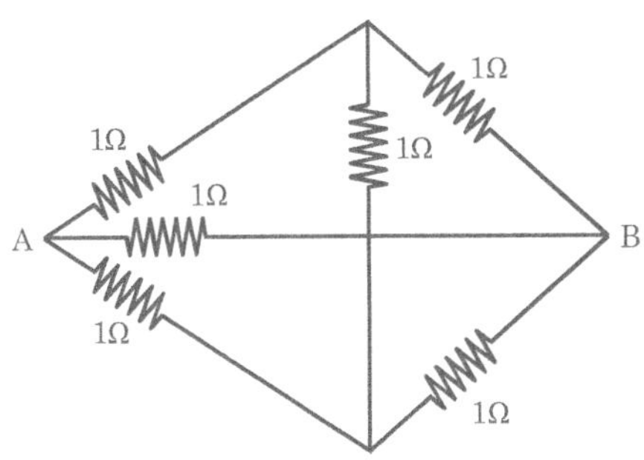

66. Voltage across the inductor can not change instantaneously, hence (voltage) across inductor remains same as before switch opened, and it will be zero

68. $F(s) = s^4 + 2s^3 + 3s^2 + 6s + K$

Constructing Routh array

s^4	1	3 K
s^3	2	6
s^2	0	K
s^1	$\frac{6\in - 2K}{\in}$	
s^0	K	

When $\in \to 0$, becomes negative, hence system is unstable for all values of K.

69. $$\frac{V_1}{V_2} = \frac{C_2}{C_1} = 2$$

But $V_1 + V_2 =$ 90

When capacitors are in parallel, charge on each

$$= C_1V_1 = C_2V_2 = 120\mu C$$

$$\therefore \quad V(C_1 + C_2) = 1200 + 1200 = 2400\mu C$$

$$\Rightarrow \quad V = \frac{2400}{60} = 40V$$

70. For maximum power transfer,

$$Rs = 60\Omega$$

$\therefore$ Maximum power transferred

$$= \frac{V^2}{4R} = \frac{40 \times 40}{4 \times 60} = 6.67V$$

72. Time constant, T = RC

where, R = equivalent resistance across capacitor terminal

73. $$I(s) = \frac{3(s+10)}{s(s+12)}$$

$$\text{Final value of } i(t) = \lim_{s \to 0} sI(s)$$
$$= \lim_{s \to 0} \frac{3.s.(s+10)}{s(s+12)}$$
$$= \lim_{s \to 0} \frac{3(s+10)}{(s+12)} = 2.5A$$

74. We have $V_1 = AV_2 + BI_2$

$$I_1 = CV_2 + DI_2$$

$$\therefore \quad Z_{11} = \left.\frac{V_1}{I_1}\right|_{I_2=0} = \frac{AV_2}{CV_2}$$
$$= \frac{A}{C} = \frac{4}{3}$$

75.

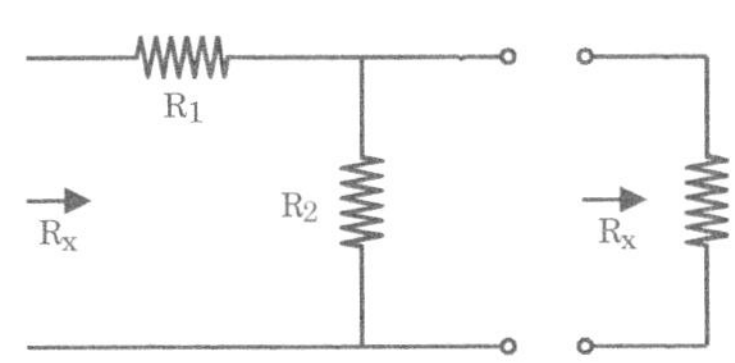

Let equivalent resistance be R_x

$$\therefore \quad R_x = R_1 + \frac{R_2R_x}{R_x + R_2}$$
$$\Rightarrow \quad (R_x + R_2)(R_x - R_1) = R_2R_x$$
$$\Rightarrow \quad R_x^2 - R_xR_1 - R_1R_2 = 0$$
$$\Rightarrow \quad R_x = \frac{R_1 \pm \sqrt{R_1^2 + 4R_1R_2}}{2}$$
$$= \frac{2 \pm \sqrt{4 + 4 \times 2 \times 1.5}}{2}$$
$$= \frac{2 \pm 4}{2} = 3\Omega$$

76. $L_{eq} = 2 + 2 - 2 \times 1 = 2H$

$$\therefore \quad f = \frac{1}{2\pi}\sqrt{\frac{1}{LC}} = \frac{1}{2\pi}\sqrt{\frac{1}{2 \times 2}} = \frac{1}{4\pi}$$

78. We have, $\frac{3(v/\sqrt{3})^2}{R} = \frac{v^2}{100}$

$$\Rightarrow \quad \frac{3v^2}{3R} = \frac{v^2}{100}$$

$$\Rightarrow \quad R = 100\Omega$$

82. Neutral current,

$$I_N = I_1\angle 0^\circ + I_2\angle 120^\circ + I_3\angle -120^\circ$$

86. $v = 200\sin(200t + 50°)$

$i = 4\cos(2000t + 13.2°)$

$= 4\sin(2000t + 103.2)$

Thus current leads the voltage by 53.2 and therefore net reactance of the circuit is capacitive

97.

$$I = \sqrt{3^2 + (5-1)^2}$$

$$= \sqrt{3^2 + 4^2}$$

$$= 5$$

101. Voltage across capacitor discharge through inductor and resistance

At $t = 0^+$, voltage across resistor = 2V

102. For critical damping,

$$R = \sqrt{\frac{4L}{C}} = \sqrt{\frac{4\times 1}{4}} = 1\Omega$$

104.

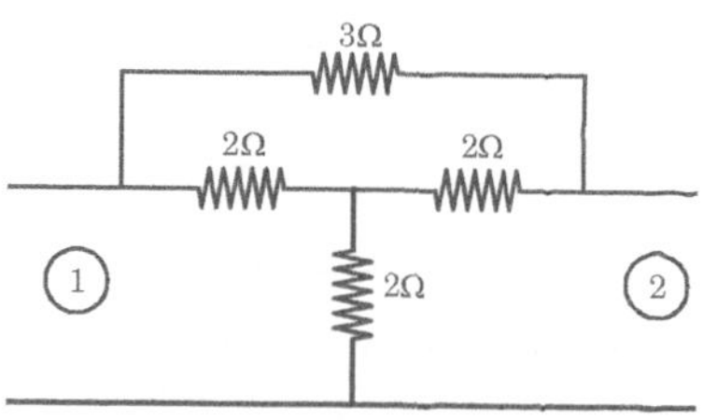

$$I_2 = Y_{21}V_1 + Y_{22}V_2$$

$$Y_{21} = \left.\frac{I_2}{V_1}\right|_{V_2=0} = -\frac{1}{3}\Omega$$

106. Impulse response due to

$$\delta(t) = \frac{d}{dt}(\text{Step response})$$

$$= \frac{d}{dt}(1 - e^{-3t})$$

$$= 3e^{-3t}$$

Response due to

$$3u(t) + \delta(t) = 3(1 - e^{-3t}) + 3e^{-3t}$$

$$= 3u(t)$$

116.

$$i(t) = \frac{6}{50}(1 - e^{-t/T})$$

where $\frac{di(t)}{dt} = -\frac{6}{50}\left(-\frac{1}{T}e^{-t/T}\right)$

$$\left.\frac{di(t)}{dt}\right|_{\substack{max \\ t\to 0}} = \frac{6}{50T}$$

$$\therefore \quad V_2 = M\frac{di(t)}{dt}$$

$$= \frac{5\times 10^{-3}\times 6\times 50}{50}$$

$$= 1V$$

117.

$$T(s) = \frac{s^2}{s^2 + as + b}$$

$$\lim_{s\to 0} T(s) = 0$$

$$\lim_{s\to\infty} T(s) = \lim_{s\to\infty}\frac{s^2}{s^2 + as + b} = 1$$

As frequency approaches to infinity, gain of the transfer function increases from zero to 1. Hence it should be High-pass filter.

119.

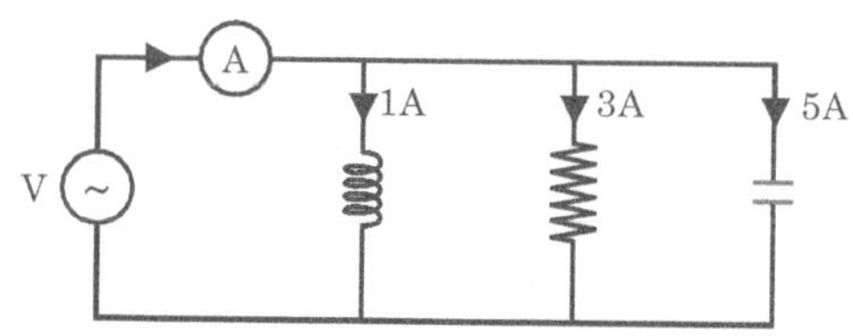

Ammeter reading

$$= \sqrt{3^2 + (5-1)^2}$$

$$= \sqrt{3^2 + 4^2}$$

$$= 5A$$

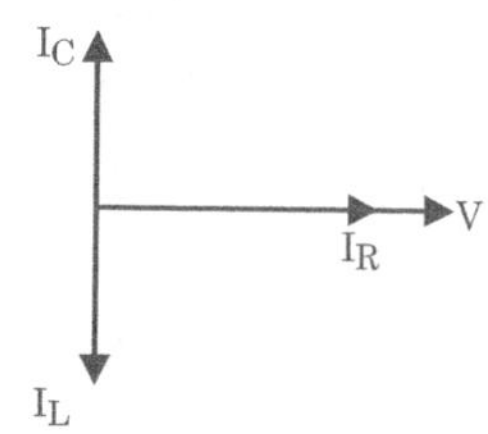

121. $V_1 + V_2 + V_3 = 135$

$$\therefore \quad \frac{Q}{C}\left[\frac{1}{2} + \frac{1}{3} + \frac{1}{4}\right] = 135$$

$$\Rightarrow \quad C = \frac{4500\times 10^{-6}}{135}\times\frac{13}{12} = \frac{325}{9}\times 10^{-6}$$

Total capacitance C_T is given by

$$\frac{1}{C_T} = \frac{1}{C}\left[\frac{1}{2} + \frac{1}{3} + \frac{1}{4}\right] = \frac{13}{120}$$

$$\therefore \quad C_T = \frac{12}{13}\times\frac{325}{9}\times 10^{-6} = 33.3\times 10^{-6}C$$

131. $i^2R = 300, \text{ and } iR = 100$

$\therefore$ Current, $i = \dfrac{300}{100} = 3\text{A}$

$$R = \frac{100}{3} = 33.33$$

Now $\omega L = \dfrac{240}{3} = 72.72$

But $\omega L = \dfrac{1}{\omega C}$

$\therefore$ $\omega C = \dfrac{1}{72.72}$

$\Rightarrow$ $C = \dfrac{1}{2 \times 3.14 \times 50 \times 72.72}$

$$= 43.7\ \mu\text{F}$$

132. Under steady state current,

$$I = \frac{5}{4+1} = 1\text{A}$$

$$V = 1\text{V}$$

Network function,

$$\frac{V(s)}{I(s)} = 2\frac{(s^2+4s+5)}{s+2} \times \frac{1}{(s^2+4s+5)}$$

$$= \frac{2}{s+2}$$

where $V(s)$ is L.T. of the voltage across the capacitor.

Pole $= -2$

133. Short circuit current at terminal AB of the network = 5A, hence

Norton's equivalent current

$$I_{sc} = 5\text{A}$$

$$Ri = \frac{1}{1/2} = 2\Omega$$

137. $V = 4I - 9$

Voltage across $R = -2 \times I$

$\therefore$ $-2I = 4I - 9$

$\therefore$ $I = 1.5\text{A}$

139. No current will be in 8Ω and hence no voltage drop across it.

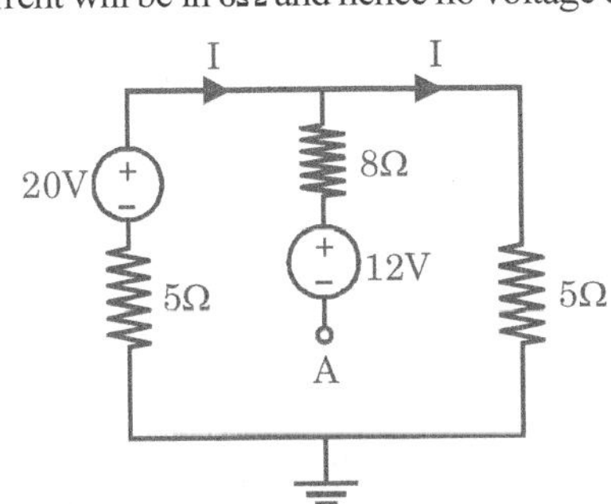

$$I = \frac{20}{5 \times 5} = 2\text{A}$$

$\therefore$ $V_A = -12 + 5 \times 2$

$$= -2\text{V}$$

143. $I_{rms}^2 = \dfrac{1}{3}\displaystyle\int_0^3 \left(\frac{9}{3}t\right)^2 dt$

$$= \frac{1}{3} \times 9 \times \frac{t^3}{3}\bigg|_0^3 = 27\text{A}$$

$\therefore$ Power, $P = I^2R$

$$= (27) \times 10 = 270\ \text{W}$$

145. $R = \dfrac{8}{4 \times 10^{-3}} = 2000\ \Omega$

Time constant $= \dfrac{L}{R} = \dfrac{6 \times 10^{-3}}{2 \times 10^3}$

$$= 3\mu\text{sec.}$$

146. Current at resonance,

$$I = \frac{200}{5} = 40\text{A}$$

Now $\omega^2 = \dfrac{1}{LC} = \dfrac{1}{0.1 \times 50 \times 10^{-6}}$

$$= 200 \times 10^4$$

$$Z_{coil} = \sqrt{R^2 + L^2\omega^2}$$

$$= \sqrt{5^2 + (0.1)^2 \times 20 \times 10^4}$$

$$= 45 \text{ ohms}$$

$\therefore$ $V_{coil} = 45 \times 40 = 1800\ \text{V}$

161. We have

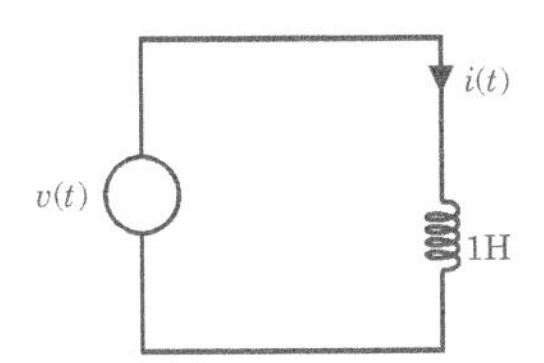

$v(t)$ is given as

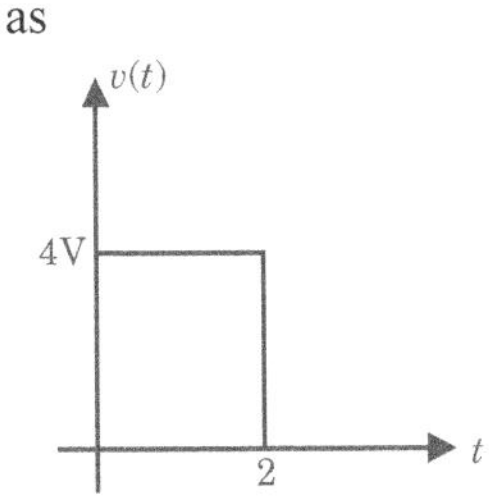

Current can be shown as

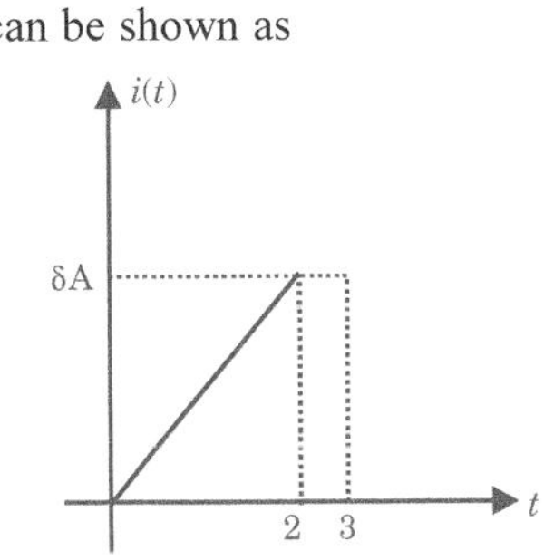

164. $$G(j\omega) = \frac{V_2(j\omega)}{V_1(j\omega)} = \frac{1/j C\omega}{R + \frac{1}{jC\omega}}$$

$$= \frac{1}{Cj\omega R + 1}$$

$$= \frac{1}{\sqrt{C^2\omega^2R^2 + 1}} \angle -\tan^{-1} C\omega R$$

At $\omega = 0, G = 1\angle 0$

At $\omega = \frac{1}{RC}, G = \sqrt{\frac{1}{2}} \angle -45$

At $\omega = \infty, G = 0 \angle -90$

168. $$P = \frac{V^2 R_L}{(R_s + R_L)^2}$$

Power will be maximum at $R_L = R_s$, and decreases again. Hence correct plot will be as shown in (*c*).

169. The circuit has resistance also, eventually the current will die down, hence $i(\infty) = 0$

177. $$f_o = \frac{1}{2\pi\sqrt{LC}}$$

180. Since output port voltage is 4 × 1 = 4V and input voltage is also 4V, the network N consists of merely a shunt element.

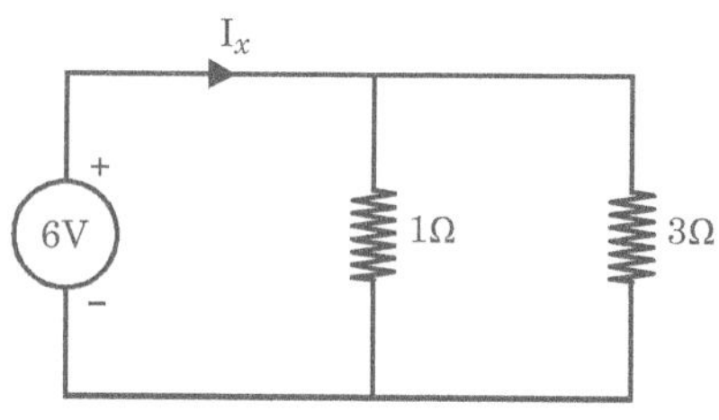

Also since the source current is 8A and load current is 4A, the shunt element should have same resistance as that of the load i.e. 1Ω

$$\therefore \quad I_x = \frac{6}{\frac{3\times 1}{3+1}} = \frac{6\times 4}{3} = 8A$$

183. $$v(t) = 2u(t) + u(t-1) - 3u(t-2)$$

$$\therefore \quad V(s) = \left[\frac{2}{s} + \frac{1}{s}e^{-s} - \frac{3}{s}e^{-2s}\right] v$$

184. Number of capacitors required in series

$$= \frac{1600}{400} = 4$$

Capacitance of one series branch

$$= \frac{2}{4} = 0.5\ \mu F$$

Number of such series chains required to be connected in parallel

$$= \frac{1.5}{0.5} = 3$$

∴ Total numbers of capacitors required

$$= 4 \times 3 = 12$$

185. Switch current at the instant of closing

$$= \frac{36}{12+6} = 2A$$

Under steady-state conditions after closing the switch,

$$\text{Current} = \frac{36}{15} \times \frac{6}{12} = \frac{6}{5} A$$

$$\therefore \quad i_L(t) = 1.2 + 0.8\, e^{-2t}$$

EXERCISE – II

2. $$\frac{dy}{dt} + y = x$$

Taking Laplace Transform

$$sy(s) + y(s) = x(s)$$

$$\therefore \quad \frac{y(s)}{x(s)} = \frac{1}{s+1}$$

Put $s = j\omega$, $$\frac{y(s)}{x(s)} = \frac{1}{1 + j\omega}$$

$$\therefore \quad x = u(t) \sin 2t$$

∴ Input, ω = 2 rad/sec

Now, $$\text{angle } \theta = \tan^{-1}\frac{\omega}{1} = \tan^{-1}\frac{2}{1} = \tan^{-1} 2$$

$$\therefore \quad \tan\theta = 2$$

3. $$SWR = \frac{300}{75} = 4$$

4. Delta connected load can be represented as power in first case is given by

$$P_1 = 3\frac{V^2}{R} = P$$

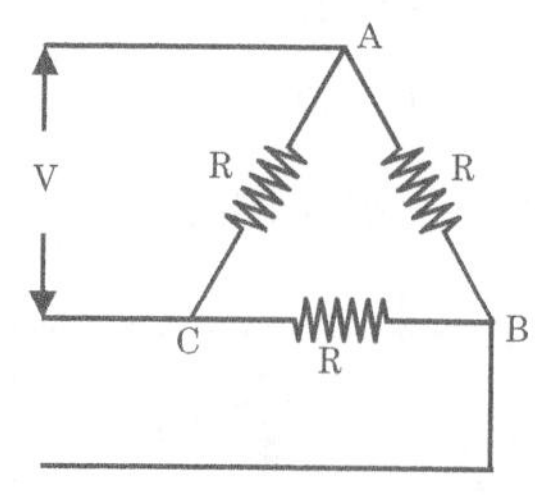

Now when one resister is opened out

$$P_2 = 2\frac{V^2}{R} = \frac{2}{3}P$$

$$\therefore \quad P_1 - P_2 = P - \frac{2}{3}P = \frac{1}{3}P$$

5. $$H(s) = \frac{1}{s+2}$$

Input = cos 2t

$\omega = 2$

At $\omega = 2$, $$|H(s)| = \frac{1}{\sqrt{2^2+2^2}} = \frac{1}{2\sqrt{2}}$$

And angle $\theta = \tan^{-1}\frac{2}{2} = \tan^{-1}(1)$

$\therefore$ Steady state response

$$= \frac{1}{2\sqrt{2}}\cos(2t - 45°)$$

14. $$M = K\sqrt{L_1L_2}$$
$$= 0.9\sqrt{0.8 \times 0.2}$$

16. $$H(s) = \frac{4s+6}{s+2}$$
$$I(s) = \frac{1}{s}$$
$$\therefore \quad V(s) = \frac{4s+6}{s(s+2)}$$
$$\therefore \quad \lim_{t\to\infty} V(t) = \lim_{s\to 0} sV(s)$$
$$= \lim_{s\to 0}\frac{4s+6}{s+2}$$
$$= \frac{6}{2}$$
$$= 3$$

17.

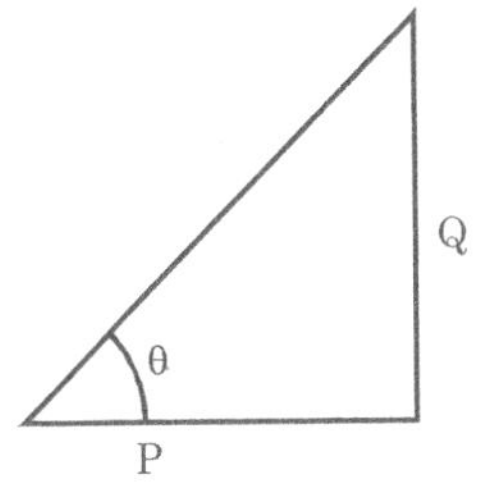

$$\theta = \cos^{-1} 0.866 = 30$$
$$\tan\theta = \frac{Q}{P} = \tan 30$$
$$\therefore \quad P = \frac{Q}{\tan 30°} = \frac{200}{0.5774}$$
$$= 344 \text{ watts}$$

18. $$i(t) = 2u(t)$$
$$\therefore \quad I(s) = \frac{2}{5}$$
$$\therefore \quad V(s) = \frac{s}{s+3}\cdot I(s) = \frac{S}{s+3}\times\frac{2}{s}$$
$$= \frac{2}{s+3}$$
$$\therefore \quad V(t) = 2e^{-3t}$$

21. We have $\omega_2 L = 2\omega_1 L$

R will remain constant

$$\therefore \quad Q_2 = \frac{2\omega_1 L}{R} = 2Q_1, \text{ i.e., Q is doubled}$$
$$I_1 = \frac{V}{\sqrt{\omega_1^2 L^2 + R^2}}$$
$$= \frac{V}{\omega_1 L} \text{ for a high Q coil,}$$

$\omega L >> R$

$$I_2 = \frac{V}{2\omega_1 L} = \frac{I_1}{2}$$
$$\therefore \quad P_2 = R\left(\frac{I_1}{2}\right)^2 = \frac{P_1}{4}$$

22. $$H(j\omega) = \frac{j\omega + 2}{(j\omega)^2 + j\omega + 4}$$
$$= \frac{2 + j\omega}{4 - \omega^2 + j\omega}$$

At $\omega = 2$, angle of $H(j\omega)$

$$= \tan^{-1}\frac{\omega}{2} - \tan^{-1}\frac{\omega}{4-\omega^2}$$
$$= \tan^{-1}\frac{2}{2} - \tan^{-1}\frac{2}{4-4}$$
$$= 45 - 90$$
$$= -45°$$

23. $$e_{rms} = \sqrt{\frac{1}{T}\int_0^T e^2(t)dt}$$
$$= \sqrt{\frac{1}{T}\left[\int_0^{T/2}\left(\frac{2At}{T}\right)^2 dt + \int_{T/2}^{T} A^2 dt\right]}$$
$$= \sqrt{\frac{1}{T}\left[\frac{4A^2}{T^2}\cdot\frac{1}{3}\frac{T^3}{8} + A^2\frac{T}{2}\right]}$$
$$= \sqrt{\frac{A^2}{6} + \frac{A^2}{2}}$$
$$= \sqrt{\frac{2A^2}{3}} = \sqrt{\frac{2}{3}}.A$$

29. When coil are connected adding each other, then

$$L_1 = L + L + 2M$$
$$\Rightarrow \quad 16 = 2L + 2M \quad ...(i)$$

When connected opposing each other

$$L_2 = L + L - 2M$$

$$\Rightarrow \qquad 8 = 2L + 2M \qquad ...(ii)$$

From equations (*i*) and (*ii*), we get

$$M = 2mH$$

30. $I_{av} = 10$

As average of sinusoidal signal is zero

$$I_{rms} = \sqrt{10^2 + \left(\frac{10}{\sqrt{L}}\right)^2}$$

$$= \sqrt{150}$$

$$= 12.12 A$$

32. $P = Re\{VI\}$

$$= Re\{(a+jb)(c-jd)\}$$

$$= ac + bd$$

36. $F(s) = \frac{1}{s} \cdot \frac{(s+1)}{(s+K)}$

$$\underset{t\to\infty}{F(t)} = \lim_{s\to 0} sF(s)$$

$$= \lim_{s\to 0} \frac{s(s+1)}{s.(s+K)} = \frac{1}{K} = \frac{1}{2}$$

$$\therefore \qquad K = 2$$

46. Frequency of the fundamental wave

$$= \frac{1}{4\times 10^{-3}}$$

$$= 250 \text{ Hz}$$

$\therefore$ Frequency of the fifth harmonics

$$= 5 \times 250$$

$$= 1250 \text{ Hz}$$

55. The circuit behaves like a half wave rectifier

$$\therefore \qquad I = \frac{V_m}{\pi R} = \frac{200}{3.14\times 20} = 3.18 A$$

62.

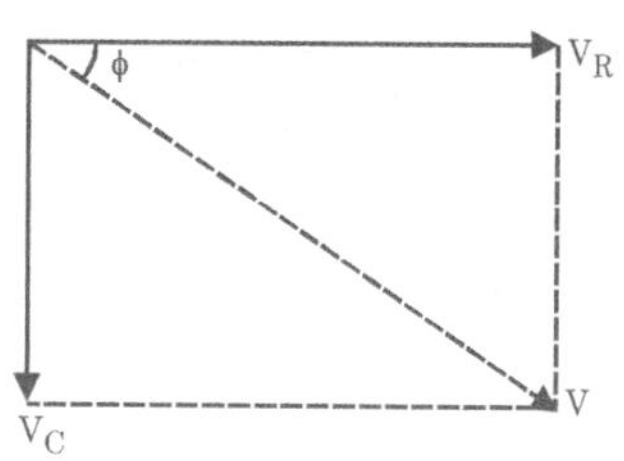

$$V = \sqrt{(60)^2 + (80)^2} = 100$$

$$\theta = \tan^{-1}\frac{60}{80}$$

$$= 36.86$$

$$\approx 37°$$

$$\therefore \qquad V = 100\underline{|-37°}$$

68. $I_C = \frac{1}{1+R_s+R_G} - \frac{E}{2500 + \frac{1.(R_3+R_G)}{1+R_3+R_G}}$

From given data,

$$I_C = \frac{1}{1+450+R_G} \cdot \frac{E}{2500 + \frac{450+R_G}{1+450+R_G}}$$

$$= \frac{1}{1+950+R_G} \cdot \frac{2E}{2500 + \frac{950+R_G}{(1+950R_G)}}$$

$$\Rightarrow 2[2500(451+RG)+450+R_G]$$

$$= 2500(951+R_G)+50$$

$$\Rightarrow \qquad 2501 R_G = 2500(951-902)+50$$

$$\Rightarrow \qquad R_G = 49 + \frac{1}{2500} \approx 49\Omega$$

69. As the element absorbs power, let it be R_x

$$i = 1 = \frac{6}{R_x+1}$$

$$\therefore \qquad R_x = 5$$

$$\Rightarrow \qquad P = \left(\frac{6}{6}\right)^2 \times 5 = 5W$$

For 5A, $\qquad P = 5^2 \times 5 = 125$ W

Take $i = 3 + \sqrt{14}$, $\Rightarrow P = 227$ W

And for $i = 3 - \sqrt{14}$, $\quad \Rightarrow \quad P = 2.75$ W

Hence none of the answer is correct.

73. Input admittance is given by

$$Y_{in}(\omega) = j\omega = \frac{1}{R + j\omega L}$$

$$= \frac{1}{R^2+(\omega L)^2} + j\left[\omega C - \frac{\omega L}{R^2+(\omega L)^2}\right]$$

For resonance Im $[Y_{in}(\omega)]$ is zero

$$\therefore \qquad Y_{in}(\omega) = \frac{R}{R^2+(\omega L)^2}$$

$$\therefore \qquad Z_{in}(\omega) = \frac{R^2+(\omega L)^2}{R}$$

$$= R + \frac{(\omega L)^2}{R} >> R$$

Hence the circuit behaves like a pur resistor of value much higher than R.

76. Using Thevenin theorem, the circuit can be drawn as shown.

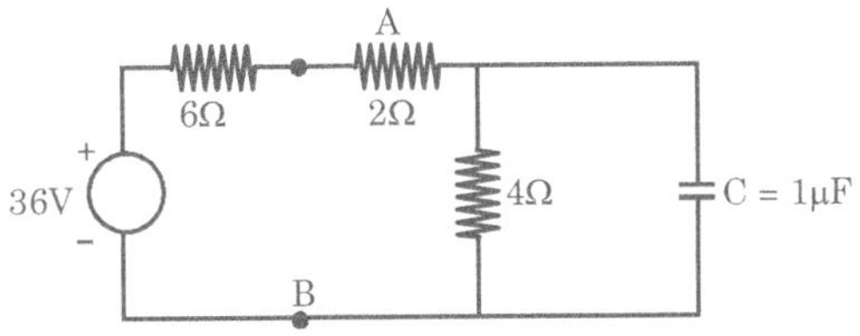

At $t = 0^+$; capacitance is open circuited

than $$I = \frac{36}{12} = 3A$$

77. Converting current source into voltage source

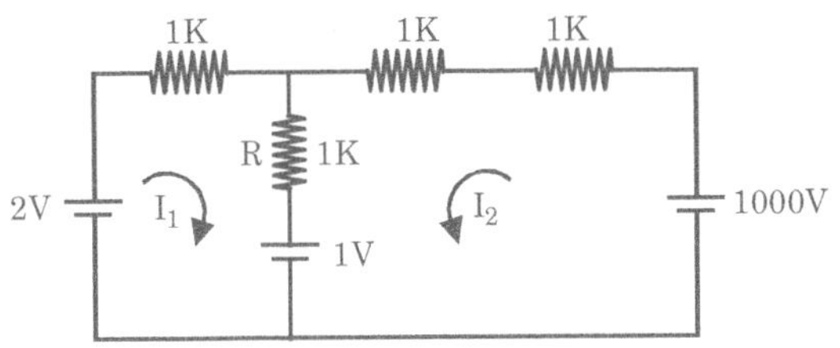

$$1 = (2I_1 + I_2) \times 10^3 \quad ...(1)$$

and $$999 = (I_1 + 3I_2) \times 10^3$$

$\Rightarrow$ $$1 = 0.999 - 3I_2 \quad ...(2)$$

From equation (1),

$$1 = \{2(0.999 - 3I_2) + I_2\}10^3$$

$\Rightarrow$ $$I_2 = \frac{1.997}{5} = 0.3994A$$

$\therefore$ $$I_1 = 0.999 - 3 \times 0.3994$$
$$= -0.1992\text{ A}$$

$\therefore$ Current through $R = I_1 + I_2 = 0.2002$ A

78. Current in the inductor at the instant of opening the switch is 1A. Therefore

$$i(t) = 1.e^{-\frac{1}{L/R}} = e^{-2t}$$

$$\frac{di(t)}{dt} = -2e^{-2t}$$

$$\left.\frac{di(t)}{dt}\right|_{t=0} = 2\text{ A/s}$$

Thus initial rate of current decay is 2A/s

89. For maximum power transfer,

$$Z_1 = \left(\frac{n_1}{n_2}\right)^2 Z_L$$

$\therefore$ $$45 = \left(\frac{n_1}{n_2}\right)^2 \times 5$$

$\Rightarrow$ $$\frac{n_1}{n_2} = 3$$

90. From the circuit,

$$V_s = L_1\frac{dI_1}{dt} \quad ...(i)$$

and $$V_0 = V_s - M\frac{dI_1}{dt} \quad ...(ii)$$

From equation (*i*) and (*ii*), we get

$$V_0 = V_s\left(1 - \frac{M}{L_1}\right)$$

91. $2\frac{d^2y}{dt^2} + 4\frac{dy}{dt} + 8y = 8$

Taking Laplace Transform

$$2s^2Y(s) + 4sY(s) + 8Y(s) = \frac{8}{s}$$

$\therefore$ $$Y(s) = \frac{8}{s(2s^2 + 4s + 8)}$$
$$= \frac{2}{s(s^2 + 2s + 4)}$$

$\therefore$ $$\omega_n^2 = 4$$

$\Rightarrow$ $$\omega_n = 2$$

Now $$2\xi\omega_n = 2$$

$\Rightarrow$ $$\xi = \frac{2}{2 \times 2} = 0.5$$

93. $$\omega = \frac{1}{\sqrt{LC}} = 10^3\text{ rad/sec.}$$

$$I = \frac{V_0}{R}\cos\omega t$$
$$= \frac{10}{2}\cos 1000t$$
$$= 5\cos 1000t$$

94. $$R = \frac{(50)^2}{10} = 250\Omega$$

Since supply voltage is 110V, hence voltage across unknown resistance

$$= 110 - 50 = 60\text{ V}$$

Current in both resistance will be same.

Hence voltage across them will be divided proportional to their resistance.

$\therefore$ Unknown resistance $= \frac{60}{50} \times 250$

$$R_x = 300\text{ ohms}$$

$$\text{Power Rating} = \frac{V^2}{R_x} = \frac{(60)^2}{300} = 12V$$

98. Using KVL, $V - iR = L\frac{di}{dt}$

Since voltage is a unit step function,

$V(t) = u(t)$

$$\therefore \quad \int_{I_0}^{I} \frac{di}{V(t) - iR} = \frac{1}{L}\int_0^t dt$$

$$\Rightarrow \quad i(t) = \frac{V(t)}{R}\left(1 - e^{-\frac{R}{L}t}\right) + I_0 e^{-R/Lt}$$

$$t = \frac{L}{R} \text{ and } i(t) = \frac{0.631}{R} + 0.368 I_0$$

99. $$f = \frac{1}{2\pi\sqrt{LC}}$$

In second figure, $L_{eq} = 2L$; $C_{eq} = 2C$

$$\therefore \quad f_2 = \frac{f_1}{2} = 0.5\,\text{kHz}$$

111. $$L_{eq} = 2 + 4 - 2 \times 0.15 = 5.7\,\text{mH}$$

112. We have,

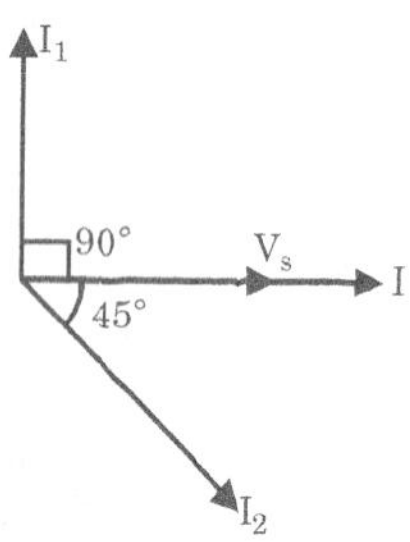

$$I = I_1 + I_2$$

$$= V_s j\omega C + \frac{V_s}{2}$$

$$= V_s j2C + \frac{V_s}{2 + j^2}$$

$$I = 2V_s Cj + \frac{V_s}{4}(1 - j)$$

For I to be in phase with V_s,

$$\text{Img}(I) = 0$$

$$\Rightarrow \quad 2CV_s - \frac{V_s}{4} = 0$$

$$\therefore \quad C = \frac{1}{8}$$

For this value of C, I_1 leads I_2 by 135°

116. From the network

$$i_R = i_L(t) + 12\cos(\omega t - \phi)$$

$$\Rightarrow \quad 4e^{-3t} + 3e^{-4t} = i_L(t) + 12\cos(2t - \phi)$$

At $t = 0$, $i_L(0) = 1 = 7 - 12\cos\phi$

$$\Rightarrow \quad \cos\phi = \frac{1}{2} = \cos\frac{\pi}{3}$$

$$\Rightarrow \quad \phi = \frac{\pi}{3}$$

128. When identical network is connected, then current would be twice hence power will become 4P.

129. $$V_B = j8 - j4 \times \frac{2}{3} = j5.33\text{V}$$

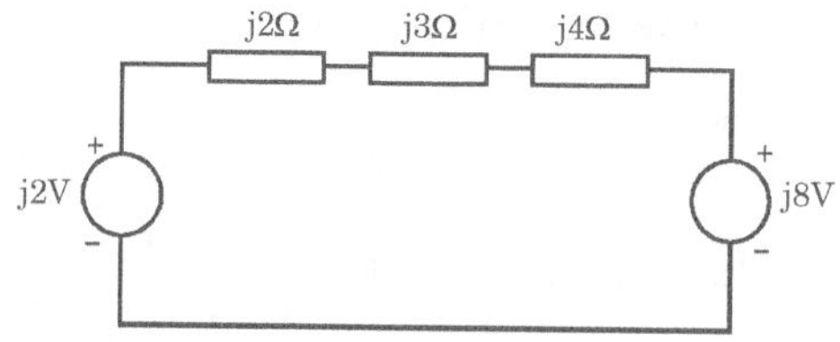

130. $$\frac{V_2(s)}{V_1(s)} = G_{21}(s) = \frac{1}{1 + s^2}$$

$$V_2(s) = Z(s)\,I(s)$$

$$V_1(s) = [s + Z(s)]\,I(s)$$

$$\therefore \quad \frac{V_2(s)}{V_1(s)} = \frac{Z(s)}{s + Z(s)} = \frac{1}{1 + s^2}$$

$$\therefore \quad Z(s) = \frac{1}{s}$$

133. $$V(s) = \frac{s + 1}{s^3 + s^2 + s + 1}$$

$$Z(s) = \frac{1}{j\omega C} = \frac{1}{s \times 0.5} = \frac{2}{s}$$

$$\therefore \quad I(s) = \frac{V(s)}{Z(s)} = \frac{(s + 1)s}{2(s^3 + s^2 + s + 1)}$$

$$I(o) = \lim_{s \to \infty} sI(s)$$

$$= \lim_{s \to \infty} \frac{s^2 + s}{2C\left(s^3 + s^2 + s + 1\right)} = 0.5\text{A}$$

136. $$V_R^2 + V_L^2 = -(20)^2 \quad ...(i)$$

$$V_L^2 - V_C^2 = 9\text{V} \quad ...(ii)$$

$$V_R^2 + (V_L - V_C)^2 = 15^2 \quad ...(iii)$$

From equations (*ii*) and (*iii*), we have

$$V_R^2 = (15)^2 - (9)^2 = 144$$

$$\Rightarrow \quad V_R = 12$$

From equation (*i*),

$$V_L^2 = (20)^2 - (12)^2 = 256$$

$$\Rightarrow \quad V_L = 16$$

$$\therefore \quad V_C = 7\text{V}$$

144. $$7i = V^2 + 2V \quad ...(i)$$

Voltage across resistor,

$$V = 10 - iR = 10 - i \times 1$$

$$V = 10 - i$$

Put this value in equation (*i*),

$$7i = (10 - i)^2 + 2(10 - i)$$

$$\Rightarrow \quad 7i = 100 + i^2 - 20i + 20 - 2i$$

$\Rightarrow \quad i^2 - 29i + 120 = 0$

$\therefore \quad i = \dfrac{29 \pm \sqrt{29^2 - 4\times 120}}{2}$

$= \dfrac{29 \pm 19}{2} = 5\text{A}, 24\text{A}$

Hence correct answer is (*b*).

145. $\dfrac{V_2(s)}{I_2(s)} = \dfrac{V_1(s)}{I_1(s)}$

$= \dfrac{(R + Ls)1/Cs}{R + Ls + 1/Cs}$

$= \dfrac{R + Ls}{LCs^2 + RCs + 1}$

$= \dfrac{1}{C} \cdot \dfrac{s + R/L}{s^2 + \dfrac{R}{L}s + \dfrac{1}{LC}}$

Putting $s^2 + \dfrac{R}{L}s + \dfrac{1}{LC} = 0$

$s = -\dfrac{R}{2L} \pm \sqrt{\dfrac{R^2}{4L^2} - \dfrac{1}{LC}}$

$= -\dfrac{R}{2L} \pm i\sqrt{\dfrac{1}{LC} - \left(\dfrac{R}{2L}\right)^2}$

Thus poles are at $s = -\dfrac{1}{2} \pm j\dfrac{\sqrt{3}}{2}$

$\therefore \quad R = 1\Omega, L = 1H, C = 1F$

150. R_{eq} across the capacitor $= \dfrac{2\times 2}{2+2} = 1\Omega$

$\therefore$ Time constant $= RC$

$= 1 \times 1 = 1$ sec

152. Poles and zeros shoud be alternate, hence correct anwer should be (*b*)

154. $Q = V_{rms} I_{rms} \sin\phi$

$= \dfrac{160}{\sqrt{2}} \times \dfrac{5}{\sqrt{2}} \times \sin 30$

$= 200$ Vars

160. Given circuit can be represented as

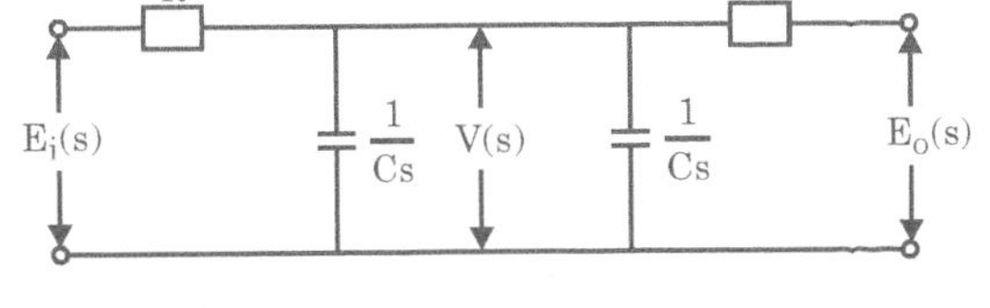

Transfer function,

$\dfrac{E_o(s)}{E_i(s)} = \dfrac{E_o(s)}{V(s)} \cdot \dfrac{V(s)}{E_i(s)}$

$= \dfrac{\dfrac{1}{Cs}}{R + \dfrac{1}{Cs}} \cdot \dfrac{1/Cs}{R + \dfrac{1}{Cs}}$

$= \dfrac{1}{(1 + Zs)^2}$

163. $V(t) = L\dfrac{di}{dt}$

$V(s) = L.sI(s) = \dfrac{s}{s^2 + 2s + 2}$

$= \dfrac{s}{(s+1)^2 + 1}$

$\therefore \quad V(t) = e^{-t}\cos t.u(t)$

164. $i_L(t) = \dfrac{10\cos 2t}{2 + j\omega L} . Z$

$= \dfrac{20\cos 2t}{2 + j2} = \dfrac{10\cos 2t}{1 + j}$

$= \dfrac{10}{\sqrt{2}}\cos(2t - 45°)$

$= \dfrac{10}{\sqrt{2}}\angle -45°$

165. $M = K\sqrt{L_1 L_2}$

For maximum value of mutual inductance K = 1

$\therefore \quad M = 1 \times \sqrt{49 \times 81} = 63$ mH

167. $Q_o = \dfrac{1}{\omega CR} = \dfrac{\omega_0}{\Delta\omega} = \dfrac{f_0}{\Delta f}$

$\therefore \quad R = \dfrac{1 \times \Delta f}{2\pi f_o C \times f_o}$

$= \dfrac{\Delta f}{2\pi f_o^2 C} = 4.7\,\Omega$

177. Since current I is dc, hence capacitor will be open circuited and inductor will be short circuited. Hence all the current will flow through inductor, hence current thought the resistor will be zero.

179. Current in all the phases are same, hence the sytem is balanced, so sum of all the three current will be zero.

181. Using KVL, $\quad 6 = 2I_1 + 2I_2$

$\Rightarrow \quad 2I_2 = 6 - 2 \times 1.5 = 3$

$\Rightarrow \quad I_2 = 1.5\text{A}$

190. System is balanced one and hence current in neutral $= I_{aN} + I_{bN} + I_{cN}$ and for the balanced system, this will be zero.

192. For maximum power transfer,

$$R_L = \sqrt{(60)^2 + (80)^2} = 100\Omega$$

193. Combined ABCD parameter will be given by

$$\begin{bmatrix} A_\alpha & \beta_\alpha \\ C_\alpha & D_\alpha \end{bmatrix}\begin{bmatrix} A_\beta & B_\beta \\ C_\beta & D_\beta \end{bmatrix} = \begin{bmatrix} 4 & 5 \\ 3 & 4 \end{bmatrix}\begin{bmatrix} 3 & 4 \\ 2 & 3 \end{bmatrix}$$

$$= \begin{bmatrix} 22 & 31 \\ 17 & 24 \end{bmatrix}$$

∴ A parameter of the combination = 22

194. For figure-I,

$$V_1 = 1.\frac{di_1}{dt} + 2\frac{di_2}{dt} \quad ...(i)$$

$$V_2 = 2.\frac{di_1}{dt} + 4\frac{di_2}{dt} \quad ...(ii)$$

For figure-II,

$$V_1 = (La + LC)\frac{di_1}{dt} + LC\frac{di_2}{dt} \quad ...(iii)$$

$$V_2 = (Lb + LC)\frac{di_2}{dt} + LC\frac{di_1}{dt} \quad ...(iv)$$

Comparing equations (*i*) and (*ii*), we get

$$Lc = 2\,H$$

and $La + Lc = 1H$

∴ $La = -1H$

Comparing equations (*ii*) and (*iv*), we get

$$Lb + Lc = 4$$

∴ $Lb = 2\,H$

198. $q = cV$

∴ $$i = \frac{dq}{dt} = c\frac{dV}{dt}$$

$$= 10 \times 10^{-6} \times \frac{(600 - 0)}{2} = 3mA$$

201. For maximum power transfer,

resistance load, $R_L = \sqrt{8^2 + 6^2} = \sqrt{100} = 10\,\Omega$

203. We can draw Thevenin Equivalent circuit as shown.

Given: $R + 1 = 2$

⇒ $R = 2 - 1 = \Omega$

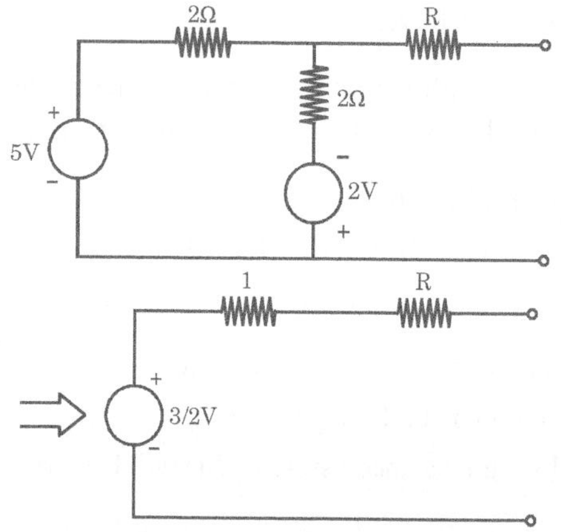

207. Equivalent capacitance across resistor

$$= 2C$$

Equivalent resistance across capacitor

$$= \frac{R}{2}$$

∴ Time constant $= \frac{R}{2}.2C = RC$

208. We have $V = 20V$

∴ Current through 20 ohm resistance

$$= \frac{20}{20} = 1 \text{ amp}$$

Voltage across 10 ohm resistor

$$= 40 - 20 = 20V$$

Then current through the capacitor

$$= 2 - 1 = 1 \text{ amp}$$

∴ $$C\frac{dV}{dt} = \frac{1}{2}\frac{dV}{dt}$$

⇒ $$\frac{dV}{dt} = 2V/s$$

213. Given circuit can be considered as. Wheatstone bridge. The bridge is balanced one hence no current will be flow through *cd*.

∴ $$C = \frac{C}{2} + \frac{C}{2} = C = 0.1\,\mu F$$

214. Voltage and current have to be in same phase

Equivalent impedance will be given by

$$\frac{1}{Z} = \frac{1}{1/j\omega C_1} + \frac{1}{1+j}$$

$$= j\omega C_1 + 1 - j$$

$$= j2C_1 + 1 - j$$

$$= 1 + j(2C_1 - 1)$$

Imaginary part of this must be zero

∴ $2C_1 - 1 = 0$

⇒ $C_1 = 0.5F$

215. We have

I_s will be given by

$$I = I_s \times 200 \times 10^3 \times 0.5 \times 10^{-3} I_s = 101\,I_s$$

216. $$h_{21} = \left.\frac{I_2}{I_1}\right|_{V_2 = 0}$$

Now, $$I_2 = -I_1 \times \frac{r_1}{r_1 + r_2}$$

⇒ $$\frac{I_2}{I_1} = \frac{r_1}{r_1 + r_2}$$

217. $$Z = R + j\left(\omega L - \frac{1}{\omega C}\right)$$

∴ $$\phi = \tan^{-1}\frac{\left(\omega L - \frac{1}{\omega C}\right)}{R}$$

■■

4 CHAPTER Materials and Components

EXERCISE – I

1. A material with unequal, antiparallel atomic magnetic moments is termed as
   (*a*) ferrite (*b*) ferrimagnetic
   (*c*) antiferromagnet (*d*) none of these
2. Magnetic recording tape is most commonly made from
   (*a*) small particles of iron
   (*b*) silicon-iron
   (*c*) ferric-oxide
   (*d*) none of these
3. Magnetic susceptibility of medium is defined as the ratio of
   (*a*) intensity of magnetisation of magnetic flux
   (*b*) magnetic field strength of mmf
   (*c*) intensity of magnetisation to field strength
   (*d*) none of these
4. Which of the following material does not have covalent bonds?
   (*a*) metal (*b*) organic polymers
   (*c*) silicon (*d*) all of these
5. Fermi level is
   (*a*) the highest occupied level at 0°k
   (*b*) the highest available level at 0°k
   (*c*) the highest velocity of electron of 0°k
   (*d*) none of these
6. Match List-I with List-II and select the correct answer using the codes given below the lists:

   List-I (*Application*)
   A. Power transformer
   B. Audio transformer
   C. Pulse transformer
   D. Welding transformer

   List-II (*Core material*)
   1. Si-Fe alloy
   2. Permalloy
   3. Ferrite
   4. Supermalloy
   5. Hypersil

   Codes :

	A	B	C	D
(*a*)	1	2	4	5
(*b*)	2	3	4	5
(*c*)	1	2	3	4
(*d*)	2	3	5	4

7. A material for a spring application must have
   (*a*) high tensile strength
   (*b*) high resistance
   (*c*) high ductility
   (*d*) none of these
8. Which one of the following statements is correct? Super conductors are popularly used for
   (*a*) generating very strong magnetic field
   (*b*) reducing I^2R losses
   (*c*) generating electrostatic field
   (*d*) generating regions free from magnetic field
9. On which of the following factors does the value of critical current density in a superconductor depend?
   (*a*) temperature
   (*b*) applied magnetic field
   (*c*) temperature and applied magnetic field
   (*d*) silsbee's rule
10. A dielectric material has the real part of the dielectric constant (ε_r^1) as 4 and its loss tangent is 0.004. What is the complex dielectric constant (ε_r^*) represented by?
    (*a*) $4+j0.016$
    (*b*) $4-j0.016$
    (*c*) $4+j0.001$
    (*d*) $4-j0.001$
11. Polarization is a measure of
    (*a*) dipole moments per unit volume
    (*b*) voltage gradient to produce electrical breakdown
    (*c*) product of charge and distance
    (*d*) excess charge density

12. Consider the following properties:

1. Perfect diamagnetism.
2. Ionic conductivity.
3. Magnetic field causes an increase in transition temperature.
4. Loss of superconductivity in long circular wire by currents.

Which of these properties are exhibited by a superconductor?

(*a*) 1, 2 and 3
(*b*) 2, 3 and 4
(*c*) 1, 3 and 4
(*d*) 1, 2 and 4

13. Electrons in a completely filled band

(*a*) take part in the process of conduction
(*b*) take part in the process of conduction at low temperature only
(*c*) do not take part in the process of conduction
(*d*) take part in the process of conduction at high temperatures only

14. When the time period of the applied voltage is much shorter than the relaxation time of a polarization process, the loss angles 8 is

(*a*) zero (*b*) between 0 and 90°
(*c*) 90° (*d*) greater than 90°

15. Consider the following statements:

If the temperature is increased, the resistivity of a metal increases because of

1. decrease in carrier concentration
2. an increase in the extent of scattering of carriers
3. increases in density of impurity

Of these statements

(*a*) 1, 2 and 3 are correct
(*b*) 2 alone is correct
(*c*) 1 alone is correct
(*d*) 2 and 3 are correct

16. Consider the following semiconductor diodes

1. Germanium diode
2. Silicon diode
3. Tunnel diode
4. Schottky diode

The correct increasing order of forward voltage drop of these diodes is:

(*a*) 1, 3, 2, 4
(*b*) 1, 2, 3, 4
(*c*) 3, 4, 2, 1
(*d*) 3, 1, 4, 2

17. Magnetic susceptibility of an ideal Type I superconductor in the superconductivity state is

(*a*) $-\infty$
(*b*) -1
(*c*) between 0 and -1
(*d*) zero

18. Two closed-packed crystal structures are

(*a*) bcc and fcc
(*b*) fcc and hcp
(*c*) bcc and hcp
(*d*) fee and sc

19. Which of the following will serve as a donor impurity in silicon?

(*a*) Boron
(*b*) Indium
(*c*) Germanium
(*d*) Antimony

20. A polarized electrolytic capacitor can be used

(*a*) only in sinusoidal circuits
(*b*) in dc and sinusoidal ac circuits
(*c*) only in dc circuits
(*d*) in circuits having any arbitrary waveform of current

21. Lithium Niobat is used in

(*a*) SAW devices
(*b*) LED's
(*c*) the manufacture of optical fibres
(*d*) laser diodes

22. The spins in a ferrimagnetic material are

(*a*) all aligned parallel
(*b*) partially aligned antiparallel without exactly cancelling out sub-lattice magnetism
(*c*) randomly oriented
(*d*) all aligned antiparallel such that the sub-lattice magnetism cancels out exactly

23. If two semiconductor specimen have carrier concentration 'n_1' and 'n_2' and their Hall coefficients are 'R_1' and 'R_2' respectively, then

(*a*) $\dfrac{R_1}{R_2} = \left(\dfrac{n_1}{n_2}\right)^{\frac{1}{2}}$

(*b*) $\dfrac{R_1}{R_2} = \dfrac{n_1}{n_2}$

(*c*) $\dfrac{R_1}{R_2} = \dfrac{n_2}{n_1}$

(*d*) $\dfrac{R_1}{R_2} = \left(\dfrac{n_2}{n_1}\right)^{\frac{1}{2}}$

24. Silicon contains 5×10^{28} atoms per cubic metre. If it is doped with two parts per million of arsenic, then the electron density at room temperature will be approximately

(*a*) 4×10^{23} m^{-3} (*b*) 10^{23} m^{-3}

(*c*) 2×10^{6} m^{-3} (*d*) 10^{6} m^{-3}

25. Which of the following thermocouple is capable of measuring highest temperature?

(*a*) Iridium-rhodium

(*b*) Plantium-rhodium

(*c*) Chromel-alumel

(*d*) Iron-constant

26. In a certain temperature range, the electrical conductivity of a semiconductor increases with increase in temperature. This is because

(*a*) both carrier concentration and mobility of carriers increase with increasing temperature.

(*b*) both carrier concentration and mobility of carriers decrease with increasing temperature.

(*c*) the carrier concentration increases substantially but the mobility of carriers decreases with increase in temperature.

(*d*) the carrier concentration remains constant but the mobility of carriers increases in temperature.

27. Which of the following contains silicon?

(*a*) Brass

(*b*) Solder

(*c*) Glass

(*d*) polythene

28. If the lattice temperature is increased, then the Hall coefficient of a semiconductor will

(*a*) decrease

(*b*) increase

(*c*) first increase to a peak and then decrease

(*d*) remain constant

29. The decreasing order of the electrical resistivities of nichrome, silicon and diamond is

(*a*) nichrome, silicon, diamond

(*b*) silicon, diamond, nichrome

(*c*) diamond, nichrome, silicon

(*d*) diamond, silicon, nichrome

30. Match List-I with List-II and select the correct answer the codes given below the lists:

List-I (*Material*)	**List-II** (*Magnetic*)
A. Paramagnetic	1. 10^{-5}
B. Diamagnetic	2. $10^{3} - 10^{5}$
C. Ferromagnetic	3. 10^{-3}
D. Ferrimagnetic	4. $10 - 10^{2}$

Codes:

	A	B	C	D
(*a*)	3	1	2	4
(*b*)	3	4	2	1
(*c*)	1	3	2	4
(*d*)	4	1	2	3

31. The mean free path for electron drift increases with

(*a*) purity

(*b*) strain hardening

(*c*) elastic modulus

(*d*) none of these

32. Match List-I (Application) with List-II (Semiconductors) and select the correct answer using the codes given below the Lists:

List-I (*Material*)	**List-II** (*Magnetic*)
A. Light emitting diode	1. Si
B. Gunn diode	2. Gap
C. Thyristor	3. InSb
D. Infra-red detector	4. GaAs

Codes:

	A	B	C	D
(*a*)	2	4	3	1
(*b*)	4	2	3	1
(*c*)	4	2	1	3
(*d*)	2	4	1	3

33. Consider the following in relation to the orbital motion of an electron:

1. State of energy momentum.
2. Orbital angular level.
3. Angle between the applied magnetic field and angular momentum.

The quantum numbers l, m and n of an electron in orbit represent respectively

(*a*) 1, 2 and 3 (*b*) 2, 3 and 1

(*c*) 3, 2 and 1 (*d*) 3, 1 and 2

34. The Curie point of Rochelle salt is

(*a*) 1000°C (*b*) 250°C

(*c*) 100°C (*d*) 10°C

35. Which of the following are associated with soft superconductors?

1. Silsbee's rule
2. Meissner effect
3. Faraday rotation
4. Curie-Weiss law

Select the correct answer using the codes given below:

(*a*) 2, 3 and 4 (*b*) 1 and 3

(*c*) 1 and 2 (*d*) 2 and 4

36. Assuming carrier mobility to be temperature independent, it can be shown that pure Si (Eg = 1.1eV) (Eg = 0.7eV) have the same conductivity at a temperature of
(*a*) 191 K
(*b*) 300 K
(*c*) 471 K
(*d*) 1470 K

37. The room temperature resistivity (in-ohm-metre) of pure silicon is
(*a*) 3000
(*b*) 300
(*c*) 30
(*d*) 3

38. Temperature (T) dependence of reciprocal of susceptibilities of three different magnetic materials, I, II and III is shown in the given figure. Among these materials, spontaneous magnetisation will be shown by

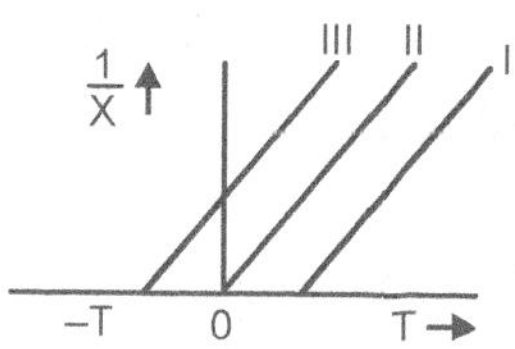

(*a*) II alone
(*b*) II and III
(*c*) I and III
(*d*) I, II and III

39. Hall coefficient of an intrinsic semiconductor is
(*a*) positive under all conditions
(*b*) negative under all conditions
(*c*) zero under all conditions
(*d*) zero at 0 K

40. Some magnetic materials may be classified on the basis of
1. susceptibility
2. saturation
3. spin arrangement
4. nature of hysteresis loop
5. domain structure
6. Critical temperature above which it behaves as a paramagnetic material

Out of these, those which can be used to distinguish between ferri and ferromagnetic materials would include.
(*a*) 1, 3 and 4
(*b*) 2, 3 and 6
(*c*) 3, 4 and 5
(*d*) 3, 5 and 6

41. In ceramic insulators, glazes are used to improve
(*a*) mechanical properties
(*b*) chemical properties
(*c*) electrical properties
(*d*) none of these

42. Dislocation is a
(*a*) lattice defect (*b*) point defect
(*c*) planer defect (*d*) none of these

43. Retentivity of a magnetic material is
(*a*) the ability of retaining residual magnetism
(*b*) the demagnetisation force when field is reversed
(*c*) the ability to cover up lag in flux when field is increasing
(*d*) none of these

44. Which one of the following is NOT true for Sulphur Hexafluoride gas?
(*a*) It is electronegative in nature
(*b*) It is high dielectric strength
(*c*) It is not-toxic
(*d*) It is highly inflammable

45. The losses in a dielectric subject to an alternating electric field are determined by
(*a*) Real part of the complex dielectric constant
(*b*) Imaginary part of the complex dielectric constant
(*c*) Absolute value of the complex dielectric constant
(*d*) Ratio of the magnitudes of the real and imaginary parts of the complex dielectric constant.

46. Match List-I (Materials) with List-II (Equation/ rule) and select the correct answer using codes given below the lists:

List-I (*Materials*)
A. Dielectric material
B. Ferromagnetic material
C. Conductors
D. Superconductor

List-II (*Equation/rule*)
1. Debye equation
2. CurrieWeiss law
3. Matthiessen's rule
4. Meissner effect

Codes:

	A	B	C	D
(*a*)	3	4	1	2
(*b*)	1	4	3	2
(*c*)	3	2	1	4
(*d*)	1	2	3	4

47. Which one of the following is the temperature below which certain materials are antiferro-magnetic and above which they are paramagnetic?

(*a*) Curie temperature

(*b*) Neel temperature

(*c*) Transition temperature

(*d*) Weiss temperature

48. In which one of the following ways can the Hall voltage across an impurity semiconductor crystal be increased ?

(*a*) By increasing the thickness of the crystal

(*b*) By increasing the concentration of impurity atoms in the crystal

(*c*) By increasing the width of the crystal

(*d*) By increasing the current flowing through the crystal

49. A small saw cut is made in permanent magnet of toroidal shape. Which of the following are true for H_0, if H_0 is the magnetic field in the air-gap and H_i is that inside the magnet?

1. $H_0 < |H_i|$
2. $H_0 > |H_i|$
3. H_0 is in the same direction as H_i
4. H_0 and H_i are in opposite directions

Select the correct answer using the codes given below :

Codes:

(*a*) 1 and 3

(*b*) 1 and 4

(*c*) 2 and 3

(*d*) 2 and 4

50. Consider the following statements:

Magnetic field induced orbital dipole moment in an atom is

1. in a directions opposite to that of the applied field
2. larger for larger atoms
3. independent of effective electron mass
4. 1.3 times the zero-field orbital dipole moment

Of these statements

(*a*) 1 and 2 are correct

(*b*) 2 and 3 are correct

(c) 3 and 4 are correct

(*d*) 1 and 4 are correct

51. Which one of the following materials can be used to produce temperatures as low as 250°C?

(*a*) KH_2PO_4

(*b*) $BaTiO_2$

(*c*) CeF_2

(*d*) InGaAs

52. The resistivity and electron density in a metal are 1.54×10^{-8} ohm-m and $5.8 \times 10^{28}/m^3$ respectively. Electron mobility in the metal (in units of $M^2V^{-1}S^{-1}$) is

(*a*) 0.007 (*b*) 0.07

(*c*) 0.7 (*d*) 1.05

53. At 300 K, the density of electron density in a metal are 6×10^{28}/m and its relaxation time is 10^{14} s. The thermal conductivity of the metal is

(*a*) 70.63 W/m/K

(*b*) 123.93 W/m/K

(*c*) 508 W/m/K

(*d*) 1359.63 W/m/K

54. A paramagnetic material is used in an energy conversion process in which the following events occur:

1. Thermal energy is generated.
2. Dipoles are turned.
3. Magnetic field is generated.
4. Material reacts to oppose it.

The correct sequence of these events is

(*a*) 3, 4, 2, 1 (*b*) 1, 4, 2, 3

(*c*) 1, 2, 4, 3 (*d*) 3, 2, 4, 1

55. When a dielectric is subjected to an alternating field, the real part of the complex relative dielectric constant is found to be 2.5. The loss tangent is 0.004. The imaginary part of the relative dielectric constant is

(*a*) 0.01

(*b*) 2.496

(*c*) 25.04

(*d*) 625.63

56. Consider a single crystal of an intrinsic semiconductor. The number of free carriers at the Fermi level at room temperature is

(*a*) half the total number of electrons in the crystal

(*b*) half the number of free electrons in the crystal

(*c*) half the number of atoms in the crystal

(*d*) zero

57. Which of the following statements are true of piezoelectric crystals?

1. Their crystal structure possesses a centre of symmetry.
2. They always exhibit ferroelectricity.
3. They are necessarily poor electrical conductors.
4. They exhibit a relative permittivity less than unity.

Select correct answer using the codes given below:

(*a*) 1, 2 and 4

(*b*) 2, 3 and 4

(*c*) 1 and 3

(*d*) 1, 2 and 3

58. The Hall coefficient of a specimen of doped semiconductor is 3.06×10^{-4} M^3C^{-1} and the resistivity of the specimen is 8.93×10^{-3} ohm meter. The majority carrier mobility (in units of $M^2V^{-1}S^{-1}$) will be

(*a*) 0.0078 (*b*) 0.035

(*c*) 0.151 (*d*) 0.350

59. The energy band gap of germanium is 0.72 eV at 300 K. The fraction of the total number of electrons (conduction band as well as valance bond) in the conduction band at 300 K is

(*a*) 10^{-6} (*b*) 10^{-4}

(*c*) 10^{-2} (*d*) 0.1

60. Which material is generally used for LED?

(*a*) compounds of silica

(*b*) compounds of gallium

(*c*) compounds of phosphorus

(*d*) compounds of sulphur

61. The Hall coefficient of an intrinsic semiconductor is

(*a*) positive under all conditions

(*b*) negative under all conditions

(*c*) zero under all conditions

(*d*) zero at 0 K

62. Which of the following statement(s) is/are correct?

1. Ferromagnetic substances have very large negative values of susceptibility.
2. Diamagnetic substances have very small negative values of susceptibility.
3. Paramagnetic substances have very small positive values of susceptibility.
4. The permeability of diamagnetic substances is slightly greater than unity.

Select the correct answer using the codes given below:

(*a*) 1 and 4

(*b*) 2 and 3

(*c*) 3 and 4

(*d*) 4 alone

63. The nature of crystal inding in germanium is

(*a*) ionic

(*b*) metallic

(*c*) covalent

(*d*) Vanderd Wall type

64. Ultrasonic amplification through electron-phonon interaction is generally achieved using a

(*a*) paramagnetic material

(*b*) piezoelectric material

(*c*) superconductor

(*d*) elemental semiconductor

65. Consider the following statements:

1. The insulation resistance of cable will increase if the length of the cable is increased.
2. For the same overall diameter of cable, the grading of cable will increase the safe working voltage.
3. The normal operating temperature of PVC cable is 70°C
4. The thermal resistance of soil increases as the moisture content of soil increases.

Of these statements

(*a*) 1 and 2 are correct

(*b*) 2 and 3 are correct

(c) 3 and 4 are correct

(*d*) 1 and 4 are correct

66. Consider the following types of semiconductors:

1. *n*-type
2. *p*-type
3. Intrinsic
4. Extrinsic

Which of these types of semiconductors are formed by doping germanium with gallium?

(*a*) 1 and 3 (*b*) 2 and 4

(*c*) 1 and 4 (*d*) 2 and 3

67. Consider the following materials for line conductors:

1. Hard drawn copper
2. Cadmium copper
3. Aluminium
4. Galvanised steel

The correct sequence of the descending order of their electrical conductivities is

(*a*) 1, 2, 4, 3 (*b*) 2, 1, 4, 3

(*c*) 2, 1, 3, 4 (*d*) 1, 2, 3, 4

68. Match List-I with List-11 select the correct answer using the codes given below the lists:

List-I (*Class of material*)	**List-II** (*Material*)
A. Ferromagnetic	1. $YBa_2Cu_2O_7$
B. Superconductor	2. $LiNbO_3$
C. Semiconductor	3. CO
D. Piezoelectric	4. InGaAsP

Codes:

	A	B	C	D
(*a*)	3	1	2	4
(*b*)	3	1	4	2
(*c*)	1	3	2	4
(*d*)	1	3	4	2

69. In a periodic linear chain of atoms, the interatomic chain is 3.14 A. The first Brillouin zone exists between the wave numbers (in units of m^{-1})

(*a*) zero and $+ 2 \times 10^{13}$

(*b*) -3.14×10^{12} and $+3.14 \times 10^{12}$

(*c*) -3.14×10^{12} and zero

(*d*) -10^{10} and $+ 10^{10}$

70. The actual field seen by an atom in dielectric is

(*a*) lower than the applied field

(*b*) higher than the applied field

(*c*) equal to the applied field

(*d*) higher or lower depending upon the particular solid

71. The Hall effect voltage in intrinsic silicon

(*a*) is positive

(*b*) is zero

(*c*) is negative

(*d*) changes sign on application of magnetic field

72. For a soft magnetic material there will be

(*a*) steeply rising magnetization curve

(*b*) relatively narrow and small hysteresis loop

(*c*) less energy loss per cycle of magnetization

(*d*) all of these

73. Silicon content in iron lamination is kept within 5% as it

(*a*) increases hysteresis loss

(*b*) increases cost

(*c*) makes the material brittle

(*d*) reduces the curie point

74. Unbalanced magnetic pull is proportional

(*a*) directly to *d*

(*b*) inversely with *d*

(*c*) directly with d^2

(*d*) inversely with d^2

75. What is suitable shape of plunger magnet to give almost flat force-stroke characteristics

(*a*) flat faced

(*b*) conical

(*c*) stepped

(*d*) tapered

76. Maximum temperature limit for class B insulation is

(*a*) 105°C

(*b*) 120°C

(*c*) 130°C

(*d*) 155°C

77. As the operating voltage and consequently the electric stress on the dielectric of solid type cable is increased from a low value, the dielectric power factors cos ϕ remains almost unchanged up to a certain value of the stress beyond which cos ϕ increase very rapidly. This is due to increase in the

(*a*) resistivity of the electric material

(*b*) ionisation in the voids present in the dielectric

(*c*) core-to-core capacitance of the cable

(*d*) core-to-core capacitance of the cable

78. Which one of the following is the best electrical conductor?

(*a*) Copper (*b*) Aluminium

(*c*) Silver (*d*) Cadmium copper

79. In Hall's effect the output voltage produced across the crystal is due to

(*a*) movement of charge carriers towards one end

(*b*) induced voltage by the applied magnetic filed

(*c*) voltage drop across the crystal due to flow of current through it

(*d*) none of these

80. With fall in temperature the resistivity of all normal metals

(*a*) remains unaffected

(b) increases

(*c*) tends to be zero

(*d*) decreases first and then increases

81. The resistivity of a metal is function of temperature because

(*a*) the electron density varies with temperature

(*b*) the electron gas density varies with temperature

(*c*) the amplitude of vibration of atoms does not vary with temperature

(*d*) none of these

82. Which of the following statements is correct?

(*a*) Copper has partially filled conduction band

(*b*) Diamond has a completely filled conduction band but an empty valance band

(*c*) Silicon has a partially filled conduction band and an empty valence band

(*d*) Energy gap between conduction and valance band in diamond is smaller than in silicon.

83. When a transition occurs to the superconducting state, the magnetic flux is excluded from the material. This is known as

(*a*) Magnetophobic effect

(*b*) Silsbee effect

(*c*) Meissner effect

(*d*) Cooper effect

84. The superconducting transition temperature of lead is 7.26 K. If the initial field at 0 K is 64×10^3 Amp/m, then the critical field at 5 K is

(*a*) 0.3364×10^3 Amp/m

(*b*) 3.364×10^3 Amp/m

(*c*) 33.640×10^3 Amp/m

(*d*) 336.400×10^3 Amp/m

85. What type of magnetic behaviour is observed in a type I superconductor?

(*a*) Perfect diamagnetism

(*b*) Perfect paramagnetism

(*c*) Perfect ferromagnetism

(*d*) Perfect ferrimagnetism

86. Some of the characterizing parameters of a material are

1. magnetic permeability
2. electron relaxation time
3. electron effective mass
4. energy band gap

In case of metals, increase in one of the above parameters decreases its conductivity, while increase in another increases the conductivity. These two parameters are respectively

(*a*) 1 and 3

(*b*) 3 and 2

(*c*) 4 and 3

(*d*) 1 and 2

87. The conductivity of a metal at ultraviolet frequency (10^{14} Hz) approximately equals

(*a*) infinity

(*b*) zero

(*c*) dc conductivity

(*d*) half of dc conductivity

88. Constantan is an alloy composed of

(*a*) 86% copper, 12% manganese and 2% nickel

(*b*) 76% nickel, 21% chromium, 2% manganese and 1 % iron

(*c*) 60% copper and 40% nickel

(*d*) 40% copper, 40% nickel and 20% carbon

89. The number of stress coefficients or constants which characterize a piezoelectric material is

(*a*) 24

(*b*) 18

(*c*) 9

(*d*) 3

90. Which of the following pair (s) is/are correctly matched?

Material	**Application**
1. Rochelle salt	— Phonograph
2. Barium titanate	— Amplifier
3. Quartz	— Oscillator

Select the correct answer using the code given below:

(*a*) Only 2 (*b*) 1 and 2

(*c*) 1 and 3 (*d*) 2 and 3

91. Match list-I with List-II and select the correct answer using the codes given below the lists:

List-I	**List-II**
A. LED	1. Optical scattering
B. LCD	2. Glow discharge
C. NIXIE	3. Electromagnetic radiation emitter

Codes:

	A	B	C
(*a*)	1	3	2
(*b*)	2	1	3
(*c*)	3	1	2
(*d*)	3	2	1

92. Electrical conductivity of copper reduces with addition of a small amount of nickel because

(*a*) the mobility decreases due to creation of new scattering centres for electrons

(*b*) nickel is ferromagnetic

(*c*) less free electrons are available for conduction

(*d*) the electrons of copper become bound to nickel atoms and are not free to move

93. When a solid crystal of sodium is formed, the 3*s*-band will have

(*a*) half as many states as the number of atoms

(*b*) same number of states as there are atoms

(*c*) twice as many states as there are atoms

(*d*) no states at all

94. The relaxation time (τ) in a perfect dielectric is

(*a*) 0 (zero) (*b*) 1

(*c*) $1 < \tau < \infty$ (*d*) ∞

95. Forbidden band is largest in

(*a*) conductor

(*b*) semiconductor

(*c*) insulator

(*d*) reactor

96. Packing fraction of simple cube is

(*a*) 0.48 (*b*) 0.52

(*c*) 0.65 (*d*) 0.89

97. A super-conductor material is

(*a*) copper (*b*) silver
(*c*) mercury (*d*) gold

98. Thermo-couple is based on

(*a*) Seeback effect
(*b*) Thomson effect
(*c*) Peltier effect
(*d*) None of these

99. N-type impurity is

(*a*) Fe (*b*) Ga
(*c*) Sb (*d*) B

100. Piezo-electricity has been observed in

(*a*) nickel (*b*) glass
(*c*) quartz (*d*) mica

101. Iron is a

(*a*) ferro-magnetic
(*b*) ferrimagnetic
(*c*) paramagnetic
(*d*) antiferromagnetic material [AMIE]

102. Magnetic material in VHFapplications is

(*a*) silicon steel (*b*) Alnico
(*c*) cobalt salt (*d*) ferrite

103. Air exhibits

(*a*) ferromagnetism
(*b*) paramagnetism
(c) antiferromagnetism
(*d*) ferrimagnetism

104. Soft magnetic material is

(*a*) tungsten steel
(*b*) alcomax
(*c*) bismuth
(*d*) iron

105. Above curie temperature, ferromagnetic material exhibits

(*a*) paramagnetism
(*b*) antiferromagnetism
(*c*) ferrimagnetism
(*d*) none of these

106. Area of hysteresis loop represents

(*a*) copper loss (*b*) eddy current loss
(*c*) hysteresis loss (*d*) none of these

107. Silicon context in steel reduces

(*a*) hysteresis loss
(*b*) eddy current loss
(*c*) copper loss
(*d*) none of these

108. Maximum flux density in coiled rolled grain oriented steel is

(*a*) 0.5 (*b*) 1.1
(*c*) 1.5 (*d*) 2.1

109. Hysteresis loss is proportional to

(*a*) f (*b*) $f^{1.5}$
(*c*) f^2 (*d*) f^3

110. Low current round conductor used in electric machines is covered by

(*a*) PVC (*b*) enemel
(*c*) polythene (*d*) rubber

111. Line insulator are made of

(*a*) porcelain (*b*) mica
(*c*) marble (*d*) PVC

112. Insulation used in commutator is

(*a*) wood (*b*) PVC
(*c*) mica (*d*) glass

113. High frequency transformer cores are generally made of

(*a*) cast iron
(*b*) mu-metal
(*c*) ferrite
(*d*) graphite

114. The conductivity of material 'A' is half of material 'B'. The ratio of relaxation of 'A' to that of 'B' is

(*a*) 0.5 (*b*) 1
(*c*) 2 (*d*) 4.1

115. Magnetostriction is a phenomenon whereby the magnetisation of a ferromagnetic material leads to a change in

(*a*) relative permeability
(*b*) physical dimensions
(*c*) spontaneous magnetisation
(*d*) magnetic susceptibility

116. Which one of the following classes of materials can be categorised as ferrites?

(*a*) Plastics
(*b*) Metals
(*c*) Alloys
(*d*) Ceramics

117. The materials in which valance electrons are bounded very tightly to their parent atoms are called the

(*a*) ferrites
(*b*) super conductors
(*c*) insulators
(*d*) none of these

118. Match List-I (Magnetic parameters) with List-II (Electric parameters) on the basis of analogy in definition and select the correct answer using the codes given below the lists:

List-I

A. Magnetic field

B. Flux density

C. Magnetization

List-II

1. Electric dipole moment per unit volume
2. Electric force per unit positive charge
3. Dielectric density

Codes :

	A	B	C
(*a*)	2	3	1
(*b*)	3	1	2
(*c*)	1	3	2
(*d*)	1	2	3

119. Match List-I (Materials) with List - II (Applications) and select the correct answer using the codes given below the lists:

List-I	List-II
A. Tantalum	1. Schottky gate
B. Graphite	2. Cryotron
C. Refractory metal	3. IC-technology
D. Polysilicon	4. Furnace

Codes:

	A	B	C	D
(*a*)	1	3	1	4
(*b*)	1	4	2	3
(*c*)	1	3	2	4
(*d*)	2	4	1	3

120. Consider the following statements:

There will be no spontaneous polarization, if

1. there is no hysteresis loop in a plot of polarization against field.
2. there is a hysteresis loop in a plot of polarization against field
3. the material is below the Curie temperature
4. the material is above the Curie temperature

(*a*) 1 and 3

(*b*) 1 and 4

(*c*) 2 and 3

(*d*) 2 and 4

121. Consider the following statements:

In the case of paramagnetic materials there

1. is no permanent dipole moment.
2. are permanent dipole moments.
3. is no alignment of dipoles in the absence of magnetic field.
4. is no interaction among the dipoles.

Which of these statement (s) is are correct?

(*a*) 1 alone (*b*) 3 and 4

(*c*) 2 and 4 (*d*) 2, 3 and 4

122. Consider the following statements:

The energy band gap GaAs

1. is indirect
2. is direct
3. is 1.43 ev
4. corresponds to wavelength of 0.89 μm

(Given: $h = 6.63 \times 10^{34}$ Joule-sec)

Of these statements

(*a*) 1 alone is correct (*b*) 2, 3 and 4 are correct

(*c*) 2 and 3 are correct (*d*) 1 and 4 are correct

123. Magnetostriction is a phenomenon whereby the magnetisation of ferromagnetic material leads to a change in

(*a*) relative permeability

(*b*) physical dimensions

(*c*) spontaneous magnetisation

(*d*) magnetic susceptibility

124. German silver contains

(*a*) 12.5 % silver (*b*) 5 % silver

(*c*) 1 % silver (*d*) no silver

125. When a solid crystal of sodium is formed, the 3*s*-band will have

(*a*) half as many states as the number of atoms

(*b*) same number of states as there are atoms

(*c*) twice as many states as there are atoms

(*d*) no states at all

126. In a superconductor, the electron and hole concentrations are comparable. Which one of the following conditions must be satisfied for the Hall coefficient to be positive?

(*a*) (Hole concentration) × (hole mobility)^2 should be less than (electron concentration) × (electron mobilty)^2

(*b*) (Hole concentration) × (hole mobility)^2 should be more than (electron concentration) × (electron mobility)^2

(*c*) (Hole concentration) × (electron mobility)^2 should be greater than (electron concentration) × (electron mobilty)^2

(*d*) (Hole concentration) × (electron mobility)^2 should be less than (electron concentration) × (electron mobilty)^3

EXERCISE – II

1. Piezoeletric is
   (*a*) material which becomes polarised when stressed
   (*b*) a material which changes dimensions due to applied field
   (*c*) a material that has properties midway between those of magnetic and non magnetic materials
   (*d*) a material in which magnetising force reduces when current flowing is increased **[L&T]**

2. The Wieldemann-Franz law states that the ratio of thermal conductivity to electrical conductivity for a material is approximately constant, independent of the material for a fixed temperature. Moreover, as temperature changes, the ratio of thermal conductivity to electrical conductivity for a material is directly proportional to the temperature of the material. Which of the following is true of the Wiedemann-Franz law?
   I. The Wiedemann-Franz law applies only to ceramics and plastics
   II. The Wiedemann-Franz law applies best to elements metals
   III. The Wiedemann-Franz law is based on the free electron theory of metals in which both thermal energy and electrical charge are transported primarily by free electrons.
   IV. The Wiedemann-Franz law is based on the Pauling's rules for electronegativity, in which the electrical conductivity and thermal conductivity and thermal conductivity of ceramic and plastic materials are related to the percentage of ionicity of the atomic bonds in the structure.
   (*a*) I only (*b*) II only
   (*c*) II and III (*d*) I, II, III and IV **[BHEL]**

3. Which of the following is copper free alloy?
   (*a*) Brass
   (*b*) Phosphor bronze
   (*c*) Invar
   (*d*) Muntz metal **[NTPC]**

4. Which of the following chemical is added to give the blue colour to the glass?
   (*a*) Cobalt oxide
   (*b*) Manganese dioxide
   (*c*) Tin oxide
   (*d*) Calcium phosphate **[BHEL]**

5. For molybdenum, which has a body-centered cubic lattic structure, the number of atoms per unit cell is
   (*a*) 1 (*b*) 2
   (*c*) 4 (*d*) 6 **[ISTE]**

6. Voltage dependent resistors are usually made from
   (*a*) graphite
   (*b*) charcoal
   (*c*) silicon carbide
   (*d*) nichrome **[BHEL]**

7. Silicon diodes are preferred to germanium diodes for high temperature operation because
   (*a*) doping of silicon is a simple processes
   (*b*) rate of increase of reverse saturation current with temperature is more in the case of silicon
   (*c*) the reverse saturation current of silicon diodes is smaller than that of germanium
   (*d*) silicon diodes can be used to rectify even very small voltage **[BEL]**

8. A high, T superconductor has the transaction temperature around
   (*a*) 400 K
   (*b*) 90 K
   (*c*) 20 K
   (*d*) 4 K **[SAIL]**

9. The thermal resistance between the body of a power semiconductor device and the ambient is expressed as
   (*a*) voltage across the device divided by current through the device
   (*b*) average power dissipated in the device divided by the square of the RMS current in the device
   (*c*) average power dissipated in the device divided by the temperature difference from body to ambient
   (*d*) temperature difference from body to ambient divided by average power dissipated in the device **[GATE]**

10. The depletion region or space charge region or transition region in a semiconductor *p-n* junction diode has
    (*a*) electrons and holes
    (*b*) positive ions and electrons
    (*c*) positive ions and negative ions
    (*d*) negative ions and holes
    (*e*) no ions, electrons or holes **[GATE]**

11. Metal with hexagonal close packed structure is
    (*a*) Silver
    (*b*) Iron
    (*c*) Magnetism
    (*d*) Aluminum **[L&T]**

12. A bimetallic strip on cooling bends as shown in figure. It can be concluded that

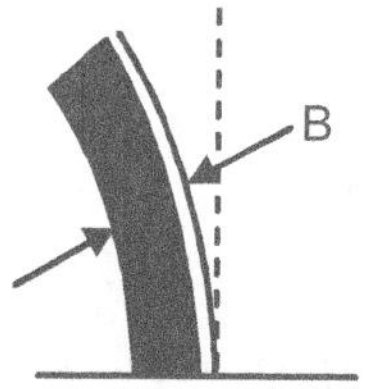

(*a*) material A is denser than material B

(*b*) material A is good conductor of heat than material B

(*c*) material A has higher coefficient of linear expansion than that for material B

(*d*) material A is under compressive stress and material B is under tensile stress **[BHEL]**

13. Above the Curie temperature, the hysteresis loop of a ferroelectric material can be described by a

(*a*) parabola

(*b*) cycloid

(*c*) straight line

(*d*) point

14. Bakelite is

(*a*) acrylate

(*b*) silicon varnish

(*c*) thermoplastic synthetic resin

(*d*) phenol formaldehyde resin **[L&T]**

15. Ferromagnetic materials when heated to temperature above Curie temperature

(*a*) behave as paramagnetic materials

(*b*) become ferrites

(*c*) tend towards superconductivity

(*d*) becomes insulators for heat and electricity **[L&T]**

16. Scintillation is

(*a*) variation of physical dimensions under the influence of magnetic field

(*b*) deionisation of gases under low pressures

(*c*) production of flashes of light by some substances when struck by alpha, beta or gamma rays

(*d*) change in dielectric constant of substances with change in humanity **[L&T]**

17. PVC refers

(*a*) Polymer Chloride

(*b*) Phosphor Vinyl Chloride

(*c*) Polyvinyl Chloride

(*d*) None of these **[L&T]**

18. When a *p*-type semiconductor material is bounded to an *n*-type semiconductor material and terminals are added, the resulting structure does not have the property that

(*a*) ideally, the forward bias resistance is zero.

(*b*) as temperature is increased above room temperature, the material resistance decreases, because the doping material charges are "freed".

(*c*) both holes and electrons contribute to the current flows.

(*d*) the resistance increases with increasing doping **[BHEL]**

19. The coercive force in a ferromagnetic material is related to the

(*a*) area enclosed by the B-H loop

(*b*) intercept on the negative H-axis of the B-H loop

(*c*) intercept on the positive B-axis of the B-H loop

(*d*) maximum value of B on the B-H loop **[BHEL]**

20. Match List-with the List-II and select the correct answer using the codes given below the lists

List-I (*Property*)	**List-II** (*Material*)
A. Ferroelectric	1. TiO_2
B. Dielectric	2. Tourmaline
C. Pyroelectric	3. $LiNbO_3$

Codes:

	A	B	C
(*a*)	3	1	2
(*b*)	3	2	1
(*c*)	2	3	1
(*d*)	2	1	3

[BHEL]

21. Match List-I with List-II and select the correct answer using the codes given below the lists :

List-I

A. Paramagnetic

B. Ferromagnetic

C. Ferrimagnetic

D. Antiferromagnetic

List-II

1. Parallel orientation of equal moments
2. Parallel orientation of unequal moments
3. Antiparallel orientation of unequal moments
4. Antiparallel orientation of equal moments
5. Neglible orientation

Codes:

	A	B	C	D
(*a*)	1	2	4	5
(*b*)	5	1	3	4
(*c*)	4	1	2	3
(*d*)	5	3	1	4

[ES]

22. Which glass is resistant to heat and chemicals?
(*a*) Soda lime glass
(*b*) Silica glass
(*c*) Boro silicate glass
(*d*) Ceramic glass **[Coal India Ltd.]**

23. Which of the following alloy has equal percentage of constituents?
(*a*) Constantan (*b*) Duralumin
(*c*) Gunmetal (*d*) White metal **[ESCORTS]**

24. Which of the following is not a polymer?
(*a*) Polyvinyl acetate
(*b*) Dichlorodifluoromethane
(*c*) Polychroprene
(*d*) PVC **[BHEL]**

25. Above the Curie temperature, the hysteresis loop of a ferroelectric material can be described by a
(*a*) parabola
(*b*) cycloid
(*c*) straight line
(*d*) point **[ES]**

26. Match List-I with List-II and select the correct answer using the codes given below the lists:

List-I
A. Electrostriction
B. Piezoelectric effect
C. Ferroelectricity
D. Dielectric

List-II
1. Polarisation is a linear function of the applied field
2. Polarisation in a non- linear function of the applied field
3. Mechanical distortion produces polarisation
4. Mechanical distortion does not produce polarisation

Codes:

	A	B	C	D
(*a*)	4	3	1	2
(*b*)	3	4	2	1
(*c*)	4	3	2	1
(*d*)	3	4	1	2

[ES]

27. Two group IV elements have the same diamond cubic structure. The one withe larger....... expected to have the smaller energy gap.
(*a*) packing factor
(*b*) coordination number
(*c*) number of valance electrons
(*d*) atomic weight **[ES]**

28. Orientational polarization is
(*a*) inversely proportion to temperature and proportional to the square of the permanent dipole moment
(*b*) proportional to temperature as well as to square of the permanent dipole moment
(*c*) proportional to temperature and inversely proportional to the square of the permanent dipole moment
(*d*) inversely proportional to temperature as well as to the square of dipole moment

29. The mobility of electrons in a material is expressed in units of
(*a*) V/*s* (*b*) $m^2/V^{-1}s^{-1}$
(*c*) m^2/s (*d*) J/K **[ES]**

30. The processes that can be used to make the steel magnetically softer are
1. Annealing
2. Grain growth
3. Decarburization
4. Quenching

(*a*) 1, 2 and 3 (*b*) 2, 3 and 4
(*c*) 2 and 3 only (*d*) 3 and 4 only **[ES]**

31. The electrical conductivity of metal is typically of the order of (in $ohm^{-1}\,m^{-1}$)
(*a*) 10^7 (*b*) 10^5
(*c*) 10^{-4} (*d*) 10^{-6} **[ES]**

32. For a particular material the Hall coefficient was found to be zero. The material is
(*a*) intrinsic semi-conductor
(*b*) extrinsic semi-conductor
(*c*) metal
(*d*) insulator

33. Match List-I with List-II and select the correct answer using the codes given below the lists:

List-I (*Type of magnetic material*)	**List-II** (*Orientation of individual dipole moments*)
A. Ferromagnetic	1. ↓ ↑ ↓ ↑
B. Anti-ferromagnetic	2. ↑ ↓ ↑ ↓
C. Ferrimagnetic	3. ↑↘↖↙
	4. ↑ ↑ ↑ ↑

Codes:

	A	B	C
(*a*)	4	2	1
(*b*)	1	2	4
(*c*)	2	1	3
(*d*)	4	2	3

[ES]

34. The relation between electric polarization and susceptibility indicates that electric polarization is

(*a*) independent of susceptibility

(*b*) inversely proportional to susceptibility

(*c*) proportional to square root of susceptibility

(*d*) proportional to susceptibility **[ES]**

35. Consider the following statements :

"In the case of diamagnetic materials, the magnetic susceptibility is"

1. positive
2. negative
3. independent to temperature
4. inversely proportional to temperature

Of these statements

(*a*) 1 and 3 are correct

(*b*) 2 and 3 are correct

(c) 1 and 4 are correct

(*d*) 2 and 4 are correct **[ES]**

36. Cobalt is

(*a*) ferromagnetic (*b*) ferroelectric

(*c*) dielectric (*d*) paramagnetic

[Coal India Ltd]

37. Ferrites are

(*a*) ferro-magnetic (*b*) ferri-magnetic

(*c*) para-magnetic (*d*) anti-magnetic **[ESCORTS]**

38. Wood is a

(*a*) granular material

(*b*) cellular material

(*c*) homogeneous material

(*d*) amorphous material **[L&T]**

39. The fracture surface energy of a brittle glassy polymer can be several orders of magnitude greater than that for glassy silica. The reason for the very high fracture surface energy observed for brittle polymers is that

(*a*) covalent bonding in the polymer chains is quite high, thus it takes considerable energy to break the polymer chains during fracture.

(*b*) the polymers crosslink extensively as stress is applied, producing three-dimensional network structure that is very resistant to fracture.

(*c*) the applied stress tends to fragment the long polymer chains into shorter molecular segments which fracture less readily than the long polymer chain molecules.

(*d*) during fracture small fibre-like bundles of aligned polymer chains (called a craze) bridge the crack face. The energy required to form, stretch, and break the craze fibrils contributes to the high fracture surface energy that is observed. **[BHEL]**

40. Which of the following crystallographic defects contribute to electronic defects?

1. Thermal vibration
2. Points defects
3. Line defects
4. Surface defects

Selects the correct answer using the codes given below :

Codes:

(*a*) 1, 2 and 3 (*b*) 2, 3 and 4

(*c*) 1, 3 and 4 (*d*) 1, 2 and 4 **[BHEL]**

41. Sodium alumino gels are also known as

(*a*) borosilicate glass

(*b*) zeolites

(*c*) extrinsic semiconductors

(*d*) elastomers **[BHEL]**

42. The magnetization of any magnetic material takes place by

(*a*) electron spin

(*b*) electron orbital motion

(*c*) spin of nucleus about its axis

(*d*) all of these **[NTPC]**

43. Magnetisation is non-linearly related to the applied field in case of

(*a*) diamagnetic material

(*b*) paramagnetic material

(*c*) ferromagnetic material

(*d*) all of these **[NTPC]**

44. The energy required to change the speed of an electron from rest to 0.6sec is nearly

(*a*) 0.085 MeV (*b*) 0.13 MeV

(*c*) 0.29 MeV (*d*) 0.37 MeV **[ES]**

45. If the molecular structure and density of the material are known, then which of the following material parameters may be determined from the

ε_r verses $\frac{1}{T}$ plot?

(*a*) sum of electric and ionic polarizabilities

(*b*) lattice constant

(*c*) breakdown voltage

(*d*) loss tangent **[ES]**

46. The magnetic susceptibility of a paramagnetic material is

(*a*) less than zero

(*b*) less than one but positive

(*c*) greater than one

(*d*) equal to zero **[ES]**

47. The coordination number is twelve for a
   (*a*) H.C.P. structure
   (*b*) F.C.C. structure
   (*c*) S.C. and F.C.C. structure
   (*d*) H.C.P. and F.C.C. structure **[ES]**

48. A type I superconductor maintained at a temperature $T < T_c$ is subjected to a magnetic field $H < H_c$. The superconductor will exhibit
   (*a*) perfect diamagnetism
   (*b*) partial diamagnetism
   (*c*) perfect paramagnetism
   (*d*) partial paramagnetism **[ES]**

49. Piezoelectric effect is generally observed in
   (*a*) insulators
   (*b*) insulators and semiconductors
   (*c*) conductors and superconductors
   (*d*) conductors and semiconductors **[ES]**

50. Given that
   α = Polarisability
   N = Number of dipoles/m^3
   $\in_r$ = Relative dielectric constant of free space
   the Clausius-Mossoti relation is
   (*a*) $\dfrac{\in_r - 1}{\in_r + 1} = \dfrac{N\alpha}{3 \in_0}$
   (*b*) $\dfrac{\in_r - 1}{\in_r + 2} = \dfrac{N\alpha}{3 \in_0}$
   (*c*) $\dfrac{\in_r - 1}{\in_r + 2} = \dfrac{3 \in_0}{N\alpha}$
   (*d*) $\dfrac{\in_r - 1}{\in_r + 2} = \dfrac{3 \in_0}{N\alpha}$ **[IAS]**

51. The electron relaxation time of metal A is 2.7×10^{-4} sec, that of B is 1.35×10^{-4} sec. The ratio of resistivity of B to resistivity of A will be
   (*a*) 4.0 (*b*) 2.0
   (*c*) 0.5 (*d*) 0.25 **[ES]**

52. Which of the following crystallographic defects contributes to electron defects?
   1. Thermal vibration
   2. Point defects
   3. Line defects
   4. Surface defects
   Select the correct answer using the codes given below:
   Codes:
   (*a*) 1, 2 and 3 (*b*) 2, 3 and 4
   (*c*) 1, 3 and 4 (*d*) 1, 2 and 4 **[ES]**

53. A constant current source supplies a current of 300 mA to a load of 1 kΩ. When the load is changed to 100 Ω, the load current will be
   (*a*) 3 A
   (*b*) 300 mA
   (*c*) 30 mA
   (*d*) 100 mA **[BTEUP]**

54. For thermionic emission
   (*a*) a material with high work function is preferable
   (*b*) a material with low work function is preferable
   (*c*) the work function of the material has no importance
   (*d*) none of these **[SLET & TE WB]**

55. Grid current in a triode is of the order of
   (*a*) 1 A
   (*b*) 100 A
   (*c*) 10^{-6} A
   (*d*) 10^{-100} A **[AMIE]**

56. An electron in the conduction band
   (*a*) is bound to its parent atom
   (*b*) has no change
   (*c*) is located near the top of the crystal
   (*d*) has a higher energy than an electron in the valance band **[AMIE]**

57. The resistivity of a semiconductor is of the order of
   (*a*) $10^{-6}\,\Omega^{-m}$
   (*b*) 10^{-6} to $100\,\Omega^{-m}$
   (*c*) $10^{-4}\,\Omega$ to $10^{4}\,\Omega^{-m}$
   (*d*) above $10^{4}\,\Omega^{-m}$ **[AMIE]**

58. The number of electrons in the valance shell of silicon are
   (*a*) 2 (*b*) 4
   (*c*) 8 (*d*) none of these **[AMIE]**

59. In an intrinsic semiconductor
   (*a*) there are no holes in the material
   (*b*) the number of holes is too small
   (*c*) electrons in the material are neutralized by holes
   (*d*) there are no electrons in the material **[AMIE]**

60. In a conductor the valance band and the conduction band
   (*a*) are separated by a large gap
   (*b*) are separated by small gap
   (*c*) are overlapping
   (*d*) none of these **[SCE & TEWB]**

61. The forbidden energy gap for germanium is
   (*a*) 0.3 ev (*b*) 3.5 ev
   (*c*) 0.72 ev (*d*) 1.1 ev **[AMIE]**

62. The temperature coefficient of an intrinsic semiconductor is
(*a*) zero
(*b*) positive
(*c*) negative
(*d*) like that of metals **[AMIE]**

63. In a semiconductor material doping is done to
(*a*) decrease the number of charge carriers
(*b*) change the chemical properties of semiconductors
(*c*) make semiconductor less than 100% pure
(*d*) all the crystal structure of pure semiconductor **[BTEMP]**

64. GaAs LEDs emit radiation in the
(*a*) ultra-violet region
(*b*) violet-blue green range of the visible region
(*c*) visible region
(*d*) infra-red region **[AMIE]**

65. The band-gap of a semiconductor is 1.43ev. Its cut-off wavelength is
(*a*) 1 μm (*b*) 0.81 μm
(*c*) 0.56 μm (*d*) 0.27 μm **[ES]**

66. Match List-I with List-II and select the correct answer by using the codes given below the Lists:

List-I (*Instrument*)
A. Electro Encephalograph
B. Electro Cardiograph
C. Sphygmomanommeter
D. Stethoscope

List-II (*Use*)
1. Diagnostic tool for heart ailment
2. Diagnostic tool for brain element
3. Instrument used for measuring B.P.
4. Instrument used to hear pulse/heart beat

Codes:

	A	B	C	D
(*a*)	2	1	3	4
(*b*)	3	1	2	4
(*c*)	2	1	4	3
(*d*)	1	2	3	4

[ES]

67. The Hall angle of θ of a metal sample is
(*a*) independent of the magnetic flux density B
(*b*) independent of the carrier mobility
(*c*) independent of the density of free carriers
(*d*) dependent on magnetic flux density, carrier mobility and density of free carriers **[ES]**

68. The permeability and permeability of a medium are
(*a*) independent of each other
(*b*) related by the velocity of electromagnetic waves
(*c*) related by the boitzmann's constant
(*d*) all of these **[NTPC]**

69. In a certain temperature range, the electrical conductivity of a semiconductor increases with increase in temperature. This is because
(*a*) both carrier concentration and mobility of carriers increases with increasing temperature
(*b*) both carrier concentration and mobility of carriers decreases with increasing temperature
(*c*) the carrier concentration increases substantially, but the mobility of carrier decreases with increase in temperature
(*d*) the carrier concentration remains constant but the mobility of carrier increases in temperature **[ES]**

70. When n_e and n_h are the electron and hole densities, and μ_e and μ_h are the carrier mobilities, the Hall coefficient is positive when,
(*a*) $n_h\mu_h > n_e\mu_e$
(*b*) $n_h\mu^2_h > n_e > \mu^2 e$
(*c*) $n_h\mu h > \mu_c$
(*d*) $n_h\mu^2_h > ne\mu^2 e$ **[ES]**

71. The ratio of the electrical resistance of a material at 300 K to its electrical resistance at 4 K is 5. This material could be
(*a*) an intrinsic semiconductor
(*b*) an extrinsic semiconductor
(*c*) an insulator
(*d*) a conductor **[IAS]**

72. Match List-I with List-II and select answer by using the codes given below the lists.

List-I	List-II
A. Isotopes	1. Same A, same Z
B. Isotones	2. Same A, Different Z
C. Isobar	3. Different A, Different Z
	4. Same(A – Z), Different Z
	5. Different N, same Z

Codes:

	A	B	C
(*a*)	5	4	2
(*b*)	5	3	2
(*c*)	5	2	3
(*d*)	4	5	3

[ES]

73. U_{92}^{235} nucleus absorbs a slow neutron and undergoes fission into 56^{x140} and 38^{Sr94} nuclei. The other particles produced in this fission process are

(*a*) 1 β and 1 neutron

(*b*) 2 β and 1 neutron

(*c*) 2 neutrons

(*d*) 3 neutrons **[ES]**

74. For N type semiconductor the doping material is

(*a*) a tetravalent material

(*b*) a pentavalent material

(*c*) a trivalent material

(*d*) germanium **[AMIE]**

75. In a *p-n* function without biasing, the *p*-side is

(*a*) negatively charged

(*b*) uncharged

(c) positively charged

(*d*) none of these **[SCE & TE WB]**

76. When a reverse bias is applied to a *p-n* junction, the width of the depletion layer

(*a*) remains the same

(*b*) is increased

(*c*) is decreased

(*d*) may increase or decrease **[AMIE]**

77. In an unbiased junction the thickness of charge depletion region is of the order of

(*a*) 0.005 μm (*b*) 0.5 μm

(*c*) 5 μm (*d*) 10^{-10}m **[ES]**

78. The components of current across a forward biased *p-n* junction are

(*a*) drift current

(*b*) diffusion current

(*c*) drift and diffusion current

(*d*) none of these **[AMIE]**

79. The *p-n* junction diode is a

(*a*) passive device

(*b*) vacuum tube

(*c*) unilateral device

(*d*) bilateral device **[AMIE]**

80. A certain amount of diode current still flows when diode is under reverse bias condition. What is this current called?

(*a*) reverse bias current

(*b*) reverse saturation current

(*c*) reverse diode current

(*d*) diode off current **[AMIE]**

81. The number of minority carriers crossing the function of a *pn*? junction diode depends primarily on

(*a*) concentration of doping impurities

(*b*) magnitude of potential barrier

(*c*) magnitude of forward bias voltage

(*d*) rate of thermal generation of electron hole pairs **[AMIE]**

82. The peak inverse voltage is the maximum voltage that can be applied to a diode without

(*a*) burning (*b*) destruction

(*c*) overheating (*d*) charging **[BTEMP]**

83. Consider the following statements:

In the case of diamagnetic materials, the magnetic susceptibility is

1. positive
2. negative
3. independent of temperature
4. inversely proportional to temperature

Of these statements

(*a*) 1 and 3 are correct

(*b*) 2 and 3 are correct

(*c*) 1 and 4 are correct

(*d*) 2 and 4 are correct

84. Three B-H curves are shown in the figure. The curves X, Y and Z refer respectively to

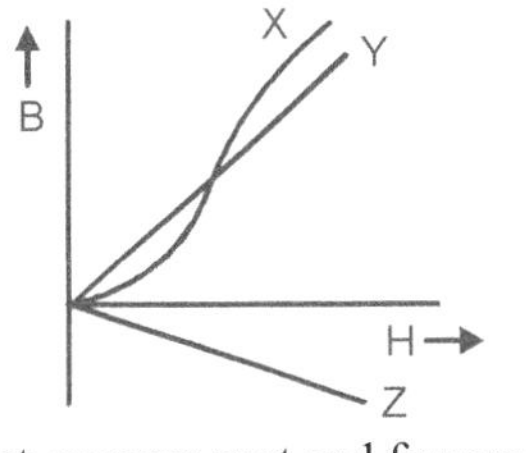

(*a*) diamagnet, paramagnet and ferromagnet

(*b*) paramagnet, ferromagnet and diamagnet

(*c*) ferromagnet, diamagnet and paramagnet

(*d*) ferromagnet, paramagnet and diamagnet **[ES]**

85. The desirable properties of transformer core material are

(*a*) low permeability and low hysteresis loss

(*b*) high permeability and high hysteresis loss

(*c*) high permeability and low hysteresis loss

(*d*) low permeability and high hysteresis loss **[ES]**

86. In an intrinsic semiconductor, the mobility of electrons in the conduction band is

(*a*) zero

(*b*) less than the mobility of holes in the valance band

(*c*) equal to the mobility of holes in the valance band

(*d*) greater than the mobility of holes in the valance band **[ES]**

87. The minority carrier concentration is largely a function of
(*a*) temperature
(*b*) amount of doping
(*c*) forward biasing voltage
(*d*) reverse biasing voltage [NTPC]

88. *n*-type semi conductors
(*a*) are negatively charged
(*b*) are produced when Indium is added as an impurity to Germanium
(*c*) are produced when phosphorous is added as an impurity to silicon
(*d*) none of these [ES]

89. As the temperature of a '*p*' type semiconductor is gradually and continuously increased, the Fermi level will move
(*a*) into the valance band
(*b*) into the conduction band
(*c*) towards the middle of the forbidden gap
(*d*) into the region between the acceptor level and the valance band [ES]

90. Consider the following materials:
1. Pure silica
2. Bakelite
3. Hard rubber
4. Paraffin

The correct sequence of the decreasing order of loss angles (60 Hz) of these dielectric materials is
(*a*) 1, 2, 3, 4
(*b*) 1, 4, 3, 2
(*c*) 4, 1, 2, 3
(*d*) 4, 3, 1, 2 [ES]

91. Silicon carbide reinforced aluminium metal matrix composites find application in
(*a*) the manufacture of transformer cores
(*b*) the manufacture of cutting tools
(*c*) the manufacture of standard resistors
(*d*) aerospace industry [ES]

92. Which one of the following polarisation mechanisms/sets of mechanisms contributes(s) to the static dielectric constant of a homopolar neutral dielectric?
(*a*) electronic
(*b*) electronic and ionic
(*c*) electronic, ionic and dipolar
(*d*) electronic and dipolar [ES]

93. Which one of the following materials is piezoelectric?
(*a*) Pb_2Au (*b*) $BaTiO_3$
(*c*) $MgAl_2O_4$ (*d*) $NiFe_2O_4$ [ES]

94. One of the way to destroy a diode is
(*a*) to exceed the current limit in forward direction
(*b*) to apply small voltage in the reverse direction
(*c*) to dip it in water
(*d*) to drop on floor [AMIE]

95. The resistance of a reverse biased ideal diode is
(*a*) zero (*b*) low
(*c*) negative (*d*) infinity [BTEMP]

96. LEDs are fabricated from
(*a*) silicon (*b*) germanium
(*c*) Si or Ge (*d*) gallium arsenide [ES]

97. For constructing a full wave rectifier
(*a*) at least two diodes are needed
(*b*) more than two diodes are needed
(*c*) atleast four diodes are needed
(*d*) none of these [SCE & TE WB]

98. If the output voltage of a bridge rectifier is 100V the PIV of diode will be
(*a*) $\sqrt{2} \times 100$ V (*b*) $\dfrac{2 \times 100}{\pi}$
(*c*) $\pi \times 100$ V (*d*) $\dfrac{\pi}{2} \times 100$ V [BTEMP]

99. Ripple frequency of the output waveform of a bridge rectifier when fed with a 50Hz sine wave is
(*a*) 100 Hz (*b*) 25 Hz
(*c*) 50 Hz (*d*) 200 Hz

100. In a 3-phase half wave rectifier if the input phase voltage is 200 V, the PIV required for each diode will be
(*a*) 400 V (*b*) 346 V
(*c*) 370 V (*d*) 200 V [AMIE]

101. The primary function of a filter is to
(*a*) minimise ac input variations
(*b*) supress odd harmonics in the rectifier output
(*c*) stabilise dc level of the output voltage
(*d*) remove ripples from the rectified output [BTEMP]

102. The coercive force in ferromagnetic material is related to the
(*a*) area enclosed by the B-H loop
(*b*) intercept on the negative H-axis of the B-H loop
(*c*) intercept on the positive B-axis of the B-H loop
(*d*) maximum value of B on the B-H loop [ES]

103. Cu_2MnAl is
(*a*) piezoelectric (*b*) paramagnetic
(*c*) ferrimagnetic (*d*) ferromagnetic [ES]

104. The effect of doping intrinsic semiconductor is to

(*a*) move the Fermi level away from the centre of the forbidden band

(*b*) move the Fermi level towards the centre the forbidden band

(*c*) change the crystal structure of the semiconductor

(*d*) keep the Fermi level at the middle of the forbidden band **[ES]**

105. The Hall Coefficient of sample (A) of a semiconductor is measured at room temperature. The Hall Coefficient of A at room temperature is 4×10^{-4} m^3 Columb^{-1}. The carriers concentration in sample A at room temperature is

(*a*) $\sim 10^{-4}$ m^{-3}

(*b*) $\sim 10^{-24}$ m^{-3}

(*c*) $\sim 10^{4}$ m^{-3}

(*d*) $\sim 10^{-22}$ m^{-3} **[ES]**

106. Some of electrons are injected into the interior of a conductor surrounded by an insulator. The injected electrons will

(*a*) distribute themselves uniformly

(*b*) distribute themselves randomly

(*c*) be confined at the point of injection

(*d*) travel to the surface of the conductor **[ES]**

107. Match List-I with List-II and select the correct answer by using the Codes given below the lists:

List-I (*Crystal System*)

A. Hexagonal

B. Rhombohedral

C. Triclinic

D. Monoclinic

List-II (*Axial length and interaxial angles*)

1. $a = b = c; y = \alpha = \beta \neq 90°$
2. $a = b \neq c; \alpha = \beta = 90°\ Y = 120°$
3. $a \neq b \neq c; \alpha \neq \beta \neq Y \neq 90°$
4. $a \neq b \neq c; \alpha \neq Y \neq 90° \neq \beta$

Codes:

	A	B	C	D
(*a*)	1	2	3	4
(*b*)	2	3	4	1
(*c*)	1	4	3	2
(*d*)	2	1	3	4

[ES]

108. A piece of copper and another of germanium are cooled from room temperature to 80 K. The resistance of

(*a*) each of them increases

(*b*) each of them decreases

(*c*) copper increases and germanium decreases

(*d*) copper decreases and germanium increases **[ES]**

109. A cylindrical rod of magnetic material fits tightly when. it is inserted into a long solenoid, halfway through. The nature of the force on the rod and the type of material of the rod are listed below. Match List I with List II and select the correct answer using the codes given below the lists:

List-I (*Material*)	**List-II** (*Force*)
A. Diamagnetic	1. Nil
B. Paramagnetic	2. Repulsive
C. Ferromagnetic	3. Attractive

Codes:

	A	B	C
(*a*)	2	2	3
(*b*)	1	3	3
(*c*)	2	3	3
(*d*)	3	2	2

[ES]

110. Selenium is

(*a*) intrinsic semi-conductor

(*b*) extrinsic semi-conductor

(*c*) *p*-type semi-conductor

(*d*) *n*-type semi-conductor

111. An initially long, straight copper bar is deformed into the shape of the letter "S" at room temperature. The bar is then annealed (heated) in a furnace.

Select the correct answer :

(*a*) The number density of dislocations in the bar will be reduced, but the shape of the bar remains unchanged.

(*b*) The bar will straighten out again, since dislocations are eliminated by the annealing.

(*c*) The bar will straighten out again, since the number of dislocations is increased by annealing.

(*d*) Nothing changes on an atomic, microstructural, or macroscopic scale in the bar. The bar just gets hot. **[BILT]**

112. Insulating material used in spark plugs is

(*a*) Rubber (*b*) Porcelain

(*c*) Mica (*d*) Polysterene **[BILT]**

113. Material having highest electrical conductivity is

(*a*) Steel

(*b*) Silver

(*c*) Aluminium

(*d*) Lead **[BILT]**

114. The electronic configuration of an iron atom is

(*a*) $1s^2\,2s^2\,2p^6\,3s^2\,3p^6\,3d^6\,4s^2$

(*b*) $1s^2\,2s^2\,2p^6 3s^2\,3p^6\,3d^3\,4s^2$

(*c*) $1s^2\,2s^2\,2p^6\,3s^2\,3p^6\,3d^8\,4s^2$

(*d*) none of these **[IETE]**

115. The fatigue failure of a material can be avoided by
(*a*) coating the surface
(*b*) shot peening
(*c*) nitriding
(*d*) any of these **[ISTE]**

116. The connecting leads of IC are made of
(*a*) Silicon (*b*) Germanium
(*c*) Aluminium (*d*) Carbon **[BHEL]**

117. Dislocations in materials are
(*a*) point defect
(*b*) line defect
(*c*) planer defect
(*d*) surface defects **[IETE]**

118. A rare and precious metal is
(*a*) Cadmium (*b*) Sodium
(*c*) Calcium (*d*) Potassium

119. Match List I with List II and select the correct answer using the codes given below the lists:

List-I (*Type of cables*)	**List II** (*Dielectric constant of insulation*)
A. Polyvinyl chloride	1. 2.35
B. Paper impregnated	2. 3.8
C. Cross-linked polyethylene	3. 6.0

Codes:

	A	B	C
(*a*)	1	2	3
(*b*)	2	3	1
(*c*)	3	1	2
(*d*)	3	2	1

[IAS]

120. Match List I [Coercive force in (A/m) with List II (Material) and select the correct answer using the codes given below the lists:

List-I (*Application*)	**List-II** (*Core material*)
A. 44000	1. Silicon steel
B. 40	2. Mu alloy
C. 4	3. Supermalloy
D. 0.16	4. Alnico

Codes:

	A	B	C	D
(*a*)	1	4	2	3
(*b*)	4	1	3	2
(*c*)	1	4	3	2
(*d*)	4	1	2	3

[ES]

121. The materials in which valance electrons are bounded very tightly to their parent atoms are called the
(*a*) ferrites
(*b*) super conductors
(*c*) insulators
(*d*) none of these **[NTPC]**

122. With respect to a completely filled band, which one of the following statements is correct?
(*a*) the total wave vector and the total velocity of all electrons in the band is always zero
(*b*) the total wave vector and the total velocity of all the electrons in the band is not necessarily zero
(*c*) both (*a*) and (*b*)
(*d*) none of these **[ES]**

123. The miller indices of the diagonal plane of a cube are
(*a*) 110 (*b*) 111
(*c*) 100 (*d*) 000 **[IETE]**

124. The temperature below which certain materials are antiferromagnetic and above which they are paramagnetic is called
(*a*) weiss temperature
(*b*) curie temperature
(*c*) neel temperature
(*d*) double point temperature **[ES]**

125. The residual resistivity of metals
(*a*) is essentially independent of temperature
(*b*) increases linearly with increasing temperature
(*c*) decreases linearly with increasing temperature
(*d*) initially increases linearly with temperature and then remains constant **[ES]**

126. Consider the following metals:
1. zinc
2. gold
3. silver
4. copper

The correct sequence of the increasing order of their resistivities is
(*a*) 4, 3, 1, 2 (*b*) 3, 4, 2, 1
(*c*) 4, 3, 2, 1 (*d*) 3, 4, 1, 2 **[ES]**

127. If the maximum wavelength of the Balmer line emitted by hydrogen atom—, the wavelength of the next lower Balmer line is
(*a*) $\frac{\lambda}{1.5}$ (*b*) $\frac{20}{27}\lambda$
(*c*) 0.67λ (*d*) 1.5λ **[ES]**

128. In a piezoelectric crystal oscillator, the oscillation or turning frequency is linearly porportional to the

(*a*) mass of the crystal

(*b*) square root of the mass of the crystal

(*c*) square of the mass of the crystal

(*d*) inverse of the square root of the mass of crystal **[ES]**

129. The activity of a radioactive sample goes down to 12.5% in a time of x days. The half-life of this sample is nearly

(*a*) $12x$

(*b*) $8x$

(*c*) $\frac{x}{3}$

(*d*) $3x$ **[ES]**

130. Which of the following is not electromagnetic in nature?

1. alpha rays
2. x-rays
3. gamma rays
4. cathod rays

(*a*) 1 and 2 only

(*b*) 2 and 3 only

(*c*) 3 and 4 only

(*d*) 1 and 4 only **[ES]**

131. Match List-I with List-II and select the correct answer using the codes given below the Lists:

List-I (*Property*)	**List-II** (*Material*)
A. Ferroelectric	1. TiO_2
B. Dielectric	2. Tourmaline
C. Pyroelectric	3. $LiNbO_3$

Codes:

	A	B	C
(*a*)	3	1	2
(*b*)	3	2	1
(*c*)	2	3	1
(*d*)	2	1	3

[ES]

132. In a photostate machine, the drum is made of

(*a*) selenium coating

(*b*) silicon coating

(*c*) cadmium coating

(*d*) zircon coating **[ES]**

133. Consider the following functions:

1. To mask against diffusion or ion implant.
2. To act as a component in MOS devices.
3. To provide low resistivity paths.
4. To facilitate the entry of dopants.

The function of an oxide layer on a silicon wafer would include

(*a*) 1 and 2

(*b*) 2 and 3

(*c*) 3 and 4

(*d*) 1 and 4 **[ES]**

134. Manganese ferrite is a 1 : 1 mixture of

(*a*) MnO and CdO

(*b*) MnO and Fe_2O_3

(c) $MnCl_2$ and Fe_2O_3

(*d*) $MnCl_2$ and CdO **[IAS]**

135. Semiconductor A has a higher band gap than semiconductor B. If both A and B have the same dimensions, the same number of electrons at a given temperature and the same electron and hole mobilities, then

(*a*) A has the same number of holes as B

(*b*) A has a larger number of holes than B

(*c*) A has less number of holes than B

(*d*) any of the above statements at (*a*), (*b*) or (*c*) could be true **[ES]**

136. Match List-I with List-II and select correct answer by using the codes given below the lists:

List-I (*Material*)	**List-II** (*Electrical Conductivity*)
A. Gold	1. 10^{-14} $(\Omega\text{-m})^{-1}$
B. Aluminium	2. 10^5
C. Graphite	3. 3.5×107
D. Diamond	4. 6.3×107

Codes:

	A	B	C	D
(*a*)	1	2	3	4
(*b*)	2	3	4	1
(*c*)	3	4	1	2
(*d*)	4	3	2	1

[ES]

137. In the optical fibre used for communication, the core and cladding material used are respectively

(*a*) pure silica and Ge-doped silica

(*b*) P-doped silica and Ge-doped silica

(*c*) Ge-doped silica and P-doped silica

(*d*) Ge-doped silica and pure silica **[ES]**

138. Magnetostriction

(*a*) induced magnetism

(*b*) volume changes accompanying magnetization

(*c*) rate of magnetisation

(*d*) rate of loss of magnetism **[ES]**

139. Pure metals generally have

(*a*) high conductivity and low temperature coefficient

(*b*) high conductivity and large temperature coefficient

(*c*) low conductivity and zero temperature coefficient

(*d*) low conductivity and high temperature coefficient **[ES]**

140. Which one the following statements is correct?

(*a*) The absence of a hysteresis loop in a plot of polarization against field is proof of the absence of spontaneous polarization.

(*b*) The Curie temperature of a ferroelectric is the temperature above which its spontaneous polarization disappears.

(*c*) The Curie temperature of a ferroelectric is the temperature below which is spontaneous polarization disappears.

(*d*) Barium titanate is ferroelectric because its lattice strains spontaneously above the Curie temperature. **[ES]**

141. Fermi levels is

(*a*) highest occupied energy level at zero Kelvin

(*b*) highest occupied level at 0°C

(*c*) energy level at which electron emission occurs

(*d*) minimum energy level in the conduction band

142. The cooking of the food in a microwave oven is based on the principle of

(*a*) magnetic hysteresis loss

(*b*) dielectric loss

(*c*) both magnetic hysteresis loss and dielectric loss

(*d*) evaporation of water **[IAS]**

143. A ferromagnetic material exhibits different characteristics above and below the

(*a*) Joule's temperature

(*b*) Faraday temperature

(*c*) Curie temperature

(*d*) Neel temperature **[ES]**

144. The correct sequence of increasing order of electrical resistivity of the given materials is

(*a*) diamond doped germanium, silicon, gold

(*b*) gold, silicon, doped germanium, diamond

(*c*) gold, doped germanium, silicon diamond

(*d*) gold, diamond, silicon, doped germanium **[NTPC]**

145. Consider the following statements:

Piezoelectric materials are useful for converting

1. mechanical energy into electrical energy
2. electrical energy into mechanical energy
3. mechanical energy into chemical energy
4. chemical energy into mechanical energy

(*a*) 1 and 2 are correct (*b*) 1, 2, 3 and 4 are correct

(*c*) 1 alone is correct (*d*) 2, 3 and 4 are correct **[ES]**

146. Superconductivity is observed for

(*a*) dc and low frequency

(*b*) ac and high frequency

(*c*) infrared frequencies

(*d*) frequencies having no effect **[NTPC]**

147. Consider the following statements:

Pure germanium and pure silicon are example of

1. direct band-gap semiconductors
2. indirect band-gap semiconductors
3. degenerate semiconductors

Of these statements

(*a*) 1 alone is correct

(*b*) 2 alone is correct

(c) 3 alone is correct

(*d*) 1 and 3 are correct **[ES]**

148. A sample of iron carrying a current is subjected to a magnetic field. Match List-I with List-II to indicate the influence of angle between the magnetic field and current on the resistivity of the sample and select the correct answer using the codes given below the lists:

List-I (*Direction of current*)	**List-II** (*Electrical resistivity*)
A. Parallel to magnetization	1. Increase
B. Perpendicular to magnetization	2. Decrease
	3. Unchanged

Codes:

	A	B
(*a*)	1	2
(*b*)	2	1
(*c*)	2	3
(*d*)	3	1

[ES]

149. If an intrinsic semiconductor is doped with a very small amount of boron, then in the extrinsic semiconductor so formed, the number of electrons and holes will

(*a*) decrease

(*b*) increase and decrease respectively

(*c*) increase

(*d*) decrease and increase respectively **[ES]**

Direction (Q. 150-172) : *The following items consist of two statements, one labelled the 'Assertion A' and the other labelled the 'Reason R'. You are to examine these two statements carefully and decide if the Assertion A and the Reason R are individually true and if so, whether the Reason is correct explanation of the Assertion. Select your answers to these items using the codes given below and mark your answser sheet accordingly.*

Codes:

(*a*) Both A and R are true and R is the correct explanation of A

(*b*) Both A and R are true but R is NOT a correct explanation of A

(*c*) A is true but R is false

(*d*) A is false but R is true

150. Assertion (A): The mobility of the conduction electrons is greater than that of holes.

Reason (R): Recombination of electrons holes obeys a logarithmic decay pattern. **[ES]**

151. Assertion (A): Resistivity of materials is based on deviations from periodicities of the perfect *c* crystal structure.

Reason (R): The mean free path of the electron between two successive collisions of particles is infinite in a perfect crystal with all the nuclei at rest. **[ES]**

152. Assertion (A): In some semiconductors, the Hall coefficient is positive.

Reason (R): In these semiconductors, the mobility of holes is greater than that of electrons. **[ES]**

153. Assertion (A): dc resistance of ferrites is much higher than that of dia-, para-or ferro-magnetic materials.

Reason (R): Ferrites are obtained by replacing divalent ferros ion in ferrimagnetic magnite by another divalent metal such as Mg, Zn etc., but the exchange interactions are antiferromagnetic type. **[ES]**

154. Assertion (A): Intrinsic germanium does not show *n* type behaviour.

Reason (R): In intrinsic germanium, the number of electrons is less than the number of holes. **[ES]**

155. Assertion (A): In intrinsic elemental semi conductors, electrons and holes are created in pairs.

Reason (R): In intrinsic silicon, the electron mobility is 2.81 times greater than the hole mobility. **[ES]**

156. Assertion (A): Transverse waves occur in two independent states of polarization.

Reason (R): Transverse electromagnetic wave does not propagate in either a rectangular or a circular guide. **[ES]**

157. Assertion (A): In an alternating of 10^{14}Hz, the electrical conductivity of a metal is zero.

Reason (R): As the electrons fail to respond to the fact changing electric field, there is no drift. **[ES]**

158. Assertion (A): The BH product is a measure of the energy required for demagnetization.

Reason (R): Energy if required for domain boundary movements. **[ES]**

159. Assertion (A): Microstructural factors such as high dislocation density and fine precipitates contribute to mechanical hardness as well as magnetic hardness.

Reason (R): The domain wall motion in magnetically hard materials is obstructed by the same microstructural features. **[ES]**

160. Assertion (A): Iron, cobalt and nickel are ferromagnetic.

Reason (R): Spin of electrons in third band is responsible for ferromagnetism. **[ES]**

161. Assertion (A): A piezoelectric crystal is used for measurement of varying force only.

Reason (R): The equivalent circuit piezoelectric transducer has a shunt capacitor at its output. **[IAS]**

162. Assertion (A): In a type II superconductor, the magnetization gradually decreases between two critical values of the applied field Hc_1 and Hc_2.

Reason (R): In a type II superconductor, magnetic flux line are pinned by macrostructural imperfection. **[ES]**

163. Assertion (A): Addition of boron to intrinsic or pure silicon in the proportion of one boron atom to 10^{15} silicon increases the conductivity of pure silicon by a factor of 10^3 at room temperature.

Reason (R): Addition of impurities to intrinsic semiconductors increases carrier mobility. **[ES]**

164. Assertion (A): The energy gap in diamond is less than that is silicon.

Reason (R): The potential energy of electrons as a function of location in diamond varies at a higher rate than it does in silicon. **[ES]**

165. Assertion (A): Copper behaves like a dielectric for very high frequency field say of the order of 10^{20} Hz.

Reason (R): For such frequencies relaxation time is about the same as the period of the wave and its propagation is not possible. **[ES]**

166. Assertion (A): Storage of energy in a dielectric is due to shift in relative position of internal charges against normal atomic and molecular forces.

Reason (R): All dielectrics are characterised by the presence of molecules having permanent displacement between the centres of gravity of the positive and negative charges. **[ES]**

167. Assertion (A): Copper normally considered a good conductor, behaves as an imperfect dielectric at a high frequency of 10^{20} Hz corresponding to short X-rays.

Reason (R): At a frequency at 10^{20} Hz, the displacement current is much greater than the conduction current. **[ES]**

168. Assertion (A): Ceramic are non-ductile.

Reason (R): They are sensitive to notches and flows during tensile loading. **[ES]**

169. **Assertion (A):** On application of an electric field, a piezoelectric crystal exhibits a spontaneous mechanical strain.

Reason (R): On application of an electric field, the ferroelectric domains tend to align themselves in the direction of the field. **[ES]**

170. **Assertion (A):** Quartz crystal is preferred over Rochelle salt crystal for commercial piezoelectric applications.

Reason (R): Rochelle salt is mechanically weaker. **[ES]**

171. **Assertion (A):** Quartz crystals are piezoelectric.

Reason (R): Fine watch circuits use quartz crystals. **[ES]**

172. **Assertion (A):** Divalent metals like magnesium are conductors.

Reason (R): In divalent metals, the full valance band overlaps with a empty conduction bane. **[ES]**

ANSWERS

EXERCISE – I

1. (*b*) **2.** (*c*) **3.** (*c*) **4.** (*a*) **5.** (*b*) **6.** (*a*) **7.** (*b*) **8.** (*b*) **9.** (*b*) **10.** (*b*)
11. (*a*) **12.** (*c*) **13.** (*c*) **14.** (*a*) **15.** (*b*) **16.** (*c*) **17.** (*b*) **18.** (*b*) **19.** (*d*) **20.** (*b*)
21. (*c*) **22.** (*b*) **23.** (*c*) **24.** (*b*) **25.** (*a*) **26.** (*c*) **27.** (*c*) **28.** (*a*) **29.** (*d*) **30.** (*a*)
31. (*a*) **32.** (*c*) **33.** (*b*) **34.** (*b*) **35.** (*c*) **36.** (*d*) **37.** (*a*) **38.** (*c*) **39.** (*b*) **40.** (*d*)
41. (*a*) **42.** (*b*) **43.** (*a*) **44.** (*d*) **45.** (*b*) **46.** (*c*) **47.** (*b*) **48.** (*d*) **49.** (*c*) **50.** (*b*)
51. (*c*) **52.** (*a*) **53.** (*b*) **54.** (*d*) **55.** (*a*) **56.** (*b*) **57.** (*c*) **58.** (*b*) **59.** (*d*) **60.** (*b*)
61. (*b*) **62.** (*b*) **63.** (*c*) **64.** (*c*) **65.** (*b*) **66.** (*b*) **67.** (*d*) **68.** (*b*) **69.** (*b*) **70.** (*c*)
71. (*c*) **72.** (*d*) **73.** (*c*) **74.** (*a*) **75.** (*d*) **76.** (*c*) **77.** (*b*) **78.** (*c*) **79.** (*a*) **80.** (*c*)
81. (*d*) **82.** (*a*) **83.** (*c*) **84.** (*c*) **85.** (*a*) **86.** (*b*) **87.** (*a*) **88.** (*b*) **89.** (*a*) **90.** (*d*)
91. (*b*) **92.** (*a*) **93.** (*a*) **94.** (*a*) **95.** (*c*) **96.** (*b*) **97.** (*c*) **98.** (*a*) **99.** (*c*) **100.** (*c*)
101. (*a*) **102.** (*d*) **103.** (*b*) **104.** (*d*) **105.** (*a*) **106.** (*c*) **107.** (*a*) **108.** (*c*) **109.** (*a*) **110.** (*b*)
111. (*a*) **112.** (*c*) **113.** (*c*) **114.** (*a*) **115.** (*b*) **116.** (*d*) **117.** (*c*) **118.** (*a*) **119.** (*b*) **120.** (*b*)
121. (*b*) **122.** (*c*) **123.** (*b*) **124.** (*d*) **125.** (*a*) **126.** (*b*)

EXERCISE – II

1. (*a*) **2.** (*c*) **3.** (*c*) **4.** (*a*) **5.** (*b*) **6.** (*b*) **7.** (*c*) **8.** (*d*) **9.** (*d*) **10.** (*c*)
11. (*c*) **12.** (*c*) **13.** (*b*) **14.** (*d*) **15.** (*a*) **16.** (*c*) **17.** (*c*) **18.** (*d*) **19.** (*b*) **20.** (*a*)
21. (*b*) **22.** (*c*) **23.** (*a*) **24.** (*b*) **25.** (*b*) **26.** (*c*) **27.** (*d*) **28.** (*a*) **29.** (*b*) **30.** (*a*)
31. (*a*) **32.** (*a*) **33.** (*a*) **34.** (*d*) **35.** (*d*) **36.** (*a*) **37.** (*b*) **38.** (*b*) **39.** (*d*) **40.** (*a*)
41. (*b*) **42.** (*d*) **43.** (*c*) **44.** (*b*) **45.** (*a*) **46.** (*b*) **47.** (*d*) **48.** (*a*) **49.** (*a*) **50.** (*b*)
51. (*b*) **52.** (*a*) **53.** (*b*) **54.** (*b*) **55.** (*c*) **56.** (*d*) **57.** (*c*) **58.** (*b*) **59.** (*c*) **60.** (*c*)
61. (*c*) **62.** (*c*) **63.** (*b*) **64.** (*d*) **65.** (*b*) **66.** (*a*) **67.** (*d*) **68.** (*b*) **69.** (*c*) **70.** (*b*)
71. (*d*) **72.** (*a*) **73.** (*b*) **74.** (*b*) **75.** (*b*) **76.** (*b*) **77.** (*b*) **78.** (*c*) **79.** (*c*) **80.** (*b*)
81. (*d*) **82.** (*c*) **83.** (*b*) **84.** (*d*) **85.** (*c*) **86.** (*d*) **87.** (*a*) **88.** (*c*) **89.** (*b*) **90.** (*c*)
91. (*d*) **92.** (*c*) **93.** (*b*) **94.** (*a*) **95.** (*d*) **96.** (*d*) **97.** (*a*) **98.** (*d*) **99.** (*a*) **100.** (*b*)
101. (*d*) **102.** (*b*) **103.** (*d*) **104.** (*a*) **105.** (*d*) **106.** (*d*) **107.** (*d*) **108.** (*d*) **109.** (*a*) **110.** (*a*)
111. (*a*) **112.** (*b*) **113.** (*b*) **114.** (*a*) **115.** (*d*) **116.** (*c*) **117.** (*b*) **118.** (*a*) **119.** (*d*) **120.** (*d*)
121. (*c*) **122.** (*b*) **123.** (*a*) **124.** (*c*) **125.** (*a*) **126.** (*b*) **127.** (*b*) **128.** (*d*) **129.** (*c*) **130.** (*d*)
131. (*a*) **132.** (*a*) **133.** (*a*) **134.** (*b*) **135.** (*b*) **136.** (*d*) **137.** (*d*) **138.** (*d*) **139.** (*b*) **140.** (*b*)
141. (*a*) **142.** (*b*) **143.** (*c*) **144.** (*c*) **145.** (*a*) **146.** (*a*) **147.** (*a*) **148.** (*d*) **149.** (*d*) **150.** (*c*)
151. (*c*) **152.** (*c*) **153.** (*b*) **154.** (*c*) **155.** (*b*) **156.** (*b*) **157.** (*a*) **158.** (*a*) **159.** (*a*) **160.** (*c*)
161. (*d*) **162.** (*a*) **163.** (*c*) **164.** (*c*) **165.** (*a*) **166.** (*a*) **167.** (*a*) **168.** (*a*) **169.** (*a*) **170.** (*c*)
171. (*a*) **172.** (*a*)

EXPLANATIONS

EXERCISE – I

1. Material with antiparallel atomic magnetic moments are called *ferrimagnetic material*, and ferrites are a class of ferrimagnetic material.

3. Susceptibility is given by $x = \frac{M}{H}$

where, M = intensity of magnetisation
H = magnetic field strength

5. Free electrons in a metal can accept a series of discrete energy levels. At absolute zero, all energy levels below a certain value we are filled and all those above W_F (Fermi level) being empty.

10. ϵ_r^* is given by, $\frac{\epsilon_r^{11}}{\epsilon_r^1} = 0.004$

$$\therefore \quad \epsilon_r^{11} = 0.004 \times 4 = 0.016$$

$$\epsilon^* = \epsilon_r^1 - j\,\epsilon_r^{11}$$
$$= 4 - j0.0016$$

17. In superconducting state,

$$B = 0$$
$$M = B - H$$
$$\therefore \quad M = -H$$

$$\text{Susceptibility, } x = \frac{M}{H}$$
$$= -\frac{H}{H}$$
$$= -1$$

24. Na = 5×10^{28} atoms/m^3
Doping = 2 parts per million
i.e., 2×10^{-6}
Electron density = $2 \times 10^{-6} \times 5 \times 10^{28}$
= 10^{23} electrons/m^3

36. At higher temperature, conductivity of both Si & Ge will be nearly same.

52. Thermal conductivity is given by

$$K = \frac{1}{3}\frac{n\pi^2 k^2 Tc}{m}$$

55. Loss tangent, $\tan\delta = \frac{\epsilon_r^{11}}{\epsilon_r^1}$

$$\epsilon_r^{11} = 0.004 \times 2.5 = 0.01$$

58.
$$R_H = \frac{1}{ne}$$
$$j = ne\mu$$
$$\therefore \quad \rho = \frac{1}{ne\mu} = \frac{R_H}{\mu}$$
$$\therefore \quad \mu = \frac{R_H}{p}$$
$$= \frac{3.06 \times 10^{-4}}{8.93 \times 10^{-3}}$$
$$= 0.035$$

84.
$$H_c(T) = H_o\left(1 - \frac{T^2}{T_c^2}\right)$$
$$= 64 \times 10^3 \left[1 - \frac{5^2}{(7.26)^2}\right]$$
$$= 33.64 \times 10^3 \text{ Amp/m}$$

114. Conductivity is given by,

$$\sigma = \frac{ne^2\tau}{m}$$
$$\therefore \quad \frac{\tau_A}{\tau_B} = \frac{\sigma_A}{\sigma_B}$$
$$= \frac{1/2\sigma_B}{\sigma_B}$$
$$= 0.5$$

EXERCISE – II

44.
$$\Delta E = mc^2 - m_o c^2$$
$$= \frac{m_o c^2}{\sqrt{1 - v^2/c^2}} - m_o c^2$$
$$= m_o c^2\left[\frac{1}{\sqrt{1 - v^2/c^2}} - 1\right]$$
$$= 0.51\left[\frac{1}{\sqrt{1 - 0.36}}\right]$$
$$= \frac{0.51}{4}\text{MeV} = 0.13 \text{ MeV}$$

65. $$\in = \frac{hc}{\lambda} = \frac{1.24}{1.43}$$
$$= 0.81\ \mu m$$

98. $$V_o = \frac{2}{\pi} V_s$$

$$\therefore \quad V_s = \frac{\pi}{2} V_o$$
$$= \frac{\pi}{2} \times 100V$$

$$\therefore \text{ PIV} \quad V_s = \frac{\pi}{2} \times 100V$$

99. In full wave rectifier,

frequency of ripples in the output
$$= 2Fs$$
$$= 2 \times 50$$
$$= 100 \text{ Hz}$$

100. PIV of each diode $= 200 \times \sqrt{3}$
$$= 346 \text{ V}$$

129. We have, $12.5\% = 0.125 = \frac{1}{8}$

Now, $$x = x_0 \left(\frac{1}{2}\right)^n$$

$$\Rightarrow \quad \frac{x}{x_0} = \left(\frac{1}{2}\right)^n$$

$$\Rightarrow \quad \frac{1}{8} = \left(\frac{1}{2}\right)^n$$

$$\Rightarrow \quad n = 3$$

$\therefore$ Half life time $= x/3$

■■

5 CHAPTER

POWER ELECTRONICS AND CONTROL OF ELECTRIC DRIVES

EXERCISE - I

1. Conside the following statements:

A free wheeling diode across the load circuit of a two-pulse converter

1. improves the operating power factor.
2. makes the distortion factor $\left(\frac{I_1}{I}\right)$ vary as a function of firing angle.
3. is effective in reducing the smoothing inductance.
4. distorts the line current.

Of these statements:

(*a*) 1, 3 and 4 are correct

(*b*) 1, 2 and 3 are correct

(*c*) 1, 2 and 4 are correct

(*d*) 2, 3 and 4 are correct

2. A single-phase circulating current dual converter is supplying a load. If the triggering angle of one converter is 30°, then the other converter operates at a triggering angle of

(*a*) 30° (*b*) 60°

(*c*) 90° (*d*) 150°

3. Which of the following are the specific features that apply to a current source inverter ?

1. Capacity to feed several motors simultaneously.
2. Capacity to provide straightforward regeneration.
3. Inherent protection in case there is a load short-circuit.

Select the correct answer using the codes given below:

Codes:

(*a*) 1, 2 and 3 (*b*) 1 and 2

(*c*) 1 and 3 (*d*) 2 and 3

4. In a thyristorised converter, a free-wheeling diode is used to

(*a*) add to the conduction current of thyristors

(*b*) oppose the thyristor conduction

(*c*) conduct current during the off period of the thyristor

(*d*) protect the thyristor by providing a shunt path

5. The torque produced by a single-phase induction motor fed through an a.c. voltage controller for speed control is due to

(*a*) fundamental component of current as well as harmonics, both odd and even

(*b*) fundamental component and even harmonic of current

(*c*) fundamental component and odd harmonics of current

(*d*) fundamental component of current alone

6. Use of a reverse conducting thyristor in place of antiparallel combination of thyristor and feedback diode in an inverter

(*a*) effectively minimises the peak commutating current

(*b*) decreases the operating frequency of operation

(*c*) minimises the effects of lead inductances on the commutation performance

(*d*) causes deterioration in the commutation performance

7. Analysis of voltage waveform of a single phase bridge converter shows that it contains x% of 6th harmonic. The 6th harmonic content of the voltage wave form of a 3-phase bridge converter would be

(*a*) less than x% due to an increase in the number of pulses

(*b*) equal to x%, the same as that of the single-phase converter

(*c*) greater than x% due to changes in the input and output voltages of the converter

(*d*) difficult to predict as the analysis of converters is not governed by any generalized theory

8. When compared to those of a symmetrical thyristor, the turn-off time and reverse blocking voltage of an asymmetrical thyristor are respectively

(*a*) large and large

(*b*) large and small

(*c*) small and large

(*d*) small and small

9. Consider the following statements:

The overlap angle of a phase-controlled converter would increase on increasing the

1. Supply voltage
2. Supply frequency
3. Load current
4. Source inductance

Of these statements

(*a*) 1, 2 and 3 are correct

(*b*) 2, 3 and 4 are correct

(*c*) 1, 2 and 4 are correct

(*d*) 1, 3 and 4 are correct

10. In a conventional reverse blocking thyristor

(*a*) external layers are lightly doped and internal layers are heavily doped

(*b*) external layers are heavily doped and internal layers are lightly doped

(*c*) the *p*-layers are heavily doped and the *n*-layers are lightly doped

(*d*) the *p*-layers are lightly doped and the *n*-layers are heavily doped

11. A single-phase bridge rectifier is preferred to the full-wave centre-tap rectifier because its

(*a*) ripple percentage is lower

(*b*) power factor is higher

(*c*) ratio of ripple frequency to line frequency is higher

(*d*) peak inverse voltage across the diode is lower

12. A separately excited dc motor is fed from a phase-controlled converter. The motor exhibits poor speed regulation. This is due to the

(*a*) inductive voltage drop of the converter and possible discontinuous conduction

(*b*) harmonics in the armature current

(*c*) large inductance of the load current

(*d*) poor power factor of the converter

13. The purpose of feedback transformer and diode in complementary commutated inverters is to

(*a*) fundamental

(*a*) improve the power factor of the load

(*b*) return the trapped energy to source

(*c*) improve the input power factor

(*d*) absorb more reactive power from the source

14. The peak input-voltage to a full-wave bridge rectifier is 1000V at 50 Hz. The dc output voltage and ripple respectively are

(*a*) 318 V and 50Hz

(*b*) 318 V and100 Hz

(*c*) 636 V and 50 Hz

(*d*) 636.9 V and 100 Hz

15. In a single-phase semiconductor with resistive load and for a firing angle, each SCR conduction and free wheeling action take place respectively, for

(*a*) $\alpha, 0°$ (*b*) $\pi - \alpha, \alpha$

(*c*) $\pi + \alpha, \alpha$ (*d*) $\pi - \alpha, 0°$

16. Consider the following statements:

Phase-controlled coverters

1. Do not provide smooth variation of output voltage.
2. Inject harmonics into the power system.
3. Draw non-unity power factor (*pf*) current for finite triggering angle.

Which of these statements are correct?

(*a*) 1, 2 and 3 (*b*) 1 and 2

(*c*) 2 and 3 (*d*) 1 and 3

17. Opto couples are used in transistor base drive circuit for

(*a*) amplification of base drive signal

(*b*) isolation between control and power circuits

(*c*) fast switching of power transistor

(*d*) protection against overcurrent

18. Consider the following statements in respect of important features of IGBT:

1. It has a high impedance gate as that of a MOSFET.
2. It has a low ON state voltage as that of a BJT.
3. It can block negative voltages as in the case of GTO.

Which of these statements are correct?

(*a*) 1 and 2 (*b*) 1 and 3

(*c*) 2 and 3 (*d*) 1, 2 and 3

19. The average value of dc voltage of a single-phase semiconverter under continuous conduction is

(*a*) $\dfrac{\sqrt{2}V(1+\cos\alpha)}{\pi}$

(*b*) $\dfrac{\sqrt{2}V(1-\cos\alpha)}{\pi}$

(*c*) $\dfrac{\sqrt{2}V(1+\cos\alpha)}{2\pi}$

(*d*) $\dfrac{\sqrt{2}V(1-\cos\alpha)}{2\pi}$

20. Which one of the following is necessary for a triggering system for thyristors in a line commutator converter?

(*a*) It must use separate power supply

(*b*) It should provide a train of pulses

(*c*) It should be synchronised with the mains supply providing a single pulse of suitable value

(*d*) It should be synchronised with mains providing a train of pulses

21. A single-phase ac voltage regulator is fed from a 50 Hz supply system. If it supplies a load comprising a resistance of 2Ω connected in series with an inductance of 6.36 mH, then the range of firing angle 'α' providing controlled voltage would be

(*a*) $0° < \alpha < 180°$

(*b*) $45° < \alpha < 180°$

(*c*) $90° < \alpha < 180°$

(*d*) $0° < \alpha < 45°$

22. In a 2-pulse bridge converter with freewheeling diode, the width of the diode current pulse is (α is the firing angle)

(*a*) $\pi + \alpha$ (*b*) $\pi - \alpha$

(*c*) π (*d*) 2α

23. In order to obtain static voltage equalization in series-connected SCRs. connections are made of

(*a*) one resistor across the string

(*b*) resistors of different values across each SCR

(*c*) resistors of the same value across each SCR

(*d*) one resistor in series with the string

24. In the chopper circuit given below, E = 220V, L = 1mH, R = 0.25Ω, E_L = 22V, T = 2500 μ sec, T = 1000 μ sec. The value of T_{on} for which the current becomes discontinuous is

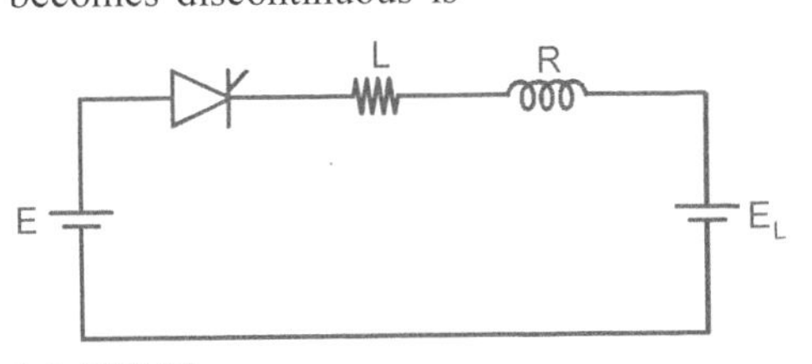

(*a*) 328.25 μ sec

(*b*) 33.25 μ sec

(*c*) 665 μ sec

(*d*) 66.5 μ sec

25. If $\frac{dv}{dt}$ occurring across a thyristor exceeds its permissible value, then the thyristor will

(*a*) remain in the forward blocking state without any damage

(*b*) turn-on only to get damaged

(*c*) turn-on but whether it gets damaged or not will depend on the final follow up current

(*d*) not turn-on but get damaged

26. A single-phase thristorised voltage controller with triggering angle 'α' is supplying an RZ, load of pf angle 'ϕ'. The output voltage can be controlled if

(*a*) $\alpha > \phi$

(*b*) $\alpha = \phi$

(*c*) $\alpha < \phi$

(*d*) a ranges from 0° to 180°

27. The average load current of a dc chopper feeding a pure resistive load is 1 amp. If a variable inductance connected in the load circuit is progressively increased from zero value, keeping the duty ratio unchanged, then the average load current will

(*a*) increase starting from 1

(*b*) decrease with a starting value of 1

(*c*) remain the same at 1

(*d*) increase to some highest value of current and then decrease again to 1

28. Consider the following situations/conditions for a phase-controlled converter:

1. Firing angle is varied in the range of 90° to 180°
2. Firing angle is varied in the range of zero to 90°
3. Feeds a load comprising R_L and additional back emf load.
4. Feeds a load comprising R_L and additional dc voltage source.

A phase-controlled converter will operate as an inverter only when

(*a*) 1 and 3 are simultaneously satisfied

(*b*) 2 and 4 simultaneously satisfied

(*c*) 2 and 3 are simultaneously satisfied

(*d*) 1 and 4 are simultaneously satisfied.

29. A six-pulse non-reversible thyritorised 415V ac converter is controlling a 440Vdc motor. If total drop in the circuit is limited to 10%, then the converter must be locked at an angle of

(*a*) 30° (*b*) 60°

(*c*) 45° (*d*) 90°

30. In sinusoidal PWM, there are m cycles of the triangular carrier wave, in the half cycle of reference sinusoidal signal. If the zero of the reference sinusoid coincides with the zero of the triangular carrier wave, then the number of pulses generated in each half cycles is

(*a*) $\frac{m}{m-1}$ (*b*) $\frac{m+1}{m-1}$

(*c*) 1 (*d*) $\frac{m-1}{m}$

31. A thyristor converter of 415 V, 100 A is operating at rated load. Details of the thyristor used are as follows:

'ON' state power loss = 150W

Thermal resistance:

Junction of case = 0.01°C/W

Case to sink = 0.08°C/W

Sink to atmosphere = 0.09°C/W

Assume ambient temperature as 35°C. The junction temperature for 100% load is

(*a*) 48.5°C (*b*) 54.5°C

(*c*) 60°C (*d*) 62°C

32. If the communication angle of a diode rectifier (due to source inductance effect) is μ, then inductive voltage regulation will be

(*a*) $\frac{1+\cos\mu}{2}$ (*b*) $1+\frac{\cos\mu}{2}$

(*c*) $1-\frac{\cos\mu}{2}$ (*d*) $1-\frac{\cos\mu}{2}$

33. Match List-I with List-II and select the correct answer using the codes given below the lists:

List-I (*Thyristor devices*)

A. Diac

B. Triac

C. SCR

D. GTO

List-II (*Characteristics*)

1. Conducts when positively biased and triggered
2. Conducts in both the directions when triggered
3. Conducts in both the directions without the need for triggering signal
4. Conducts when positively biased and turns off when triggered

Codes:

	A	B	C	D
(*a*)	1	4	3	2
(*b*)	2	1	4	3
(*c*)	3	2	1	4
(*d*)	4	3	2	1

34. A dc chopper is fed from a constant voltage mains the duty ratio of chopper is progressively increased while the chopper feeds aRL load. The per unit current ripple would

(*a*) increase progressively

(*b*) decrease progressively

(*c*) decrease to a minimum value at a = 0.5 and then increase

(*d*) increase to a maximum value at a = 0.5 and then decrease

35. In a single face cycloconverter arrangement for single phase changing the frequency of the supply voltage to an ac supply load (as shown in the figure) the load terminal 'L' should be connected to

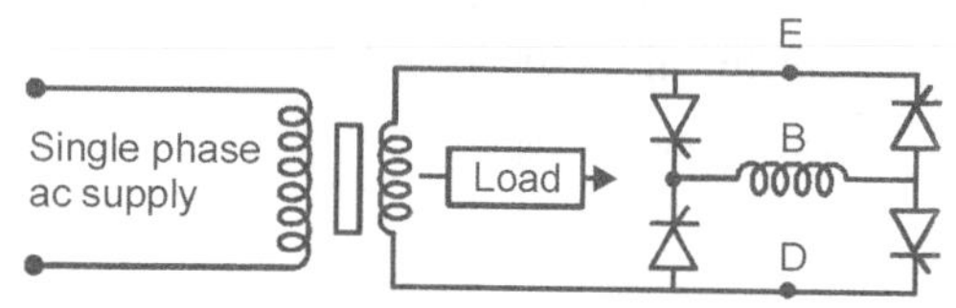

(*a*) A (*b*) B

(*c*) C (*d*) D or E

36. Match List-I with List-II and select the correct answer using the codes given below the lists:

List-I

A. Voltage across R-C parallel load connected through full-wave rectifier bridge

B. Instantaneous power consumed by a resistor

C. Output voltage of a positive clamped circuit

D. Instantaneous power consumed by an R-L circuit

List-II

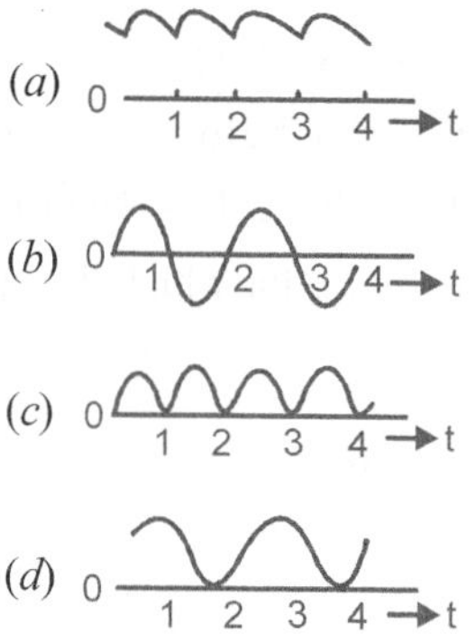

Codes:

	A	B	C	D
(*a*)	1	2	3	4
(*b*)	2	3	4	5
(*c*)	1	3	4	5
(*d*)	2	4	5	1

37. In a two-quadrant dc to dc chopper, the load voltage is varied from positive maximum to negative maximum by varying the time ratio of the chopper from

(*a*) zero to unity

(*b*) unity to zero

(*c*) zero to 0.5

(*d*) 0.5 to zero

38. Given that V_m = maximum value of the phase voltage. The average value of the dc voltage for a 3-phase 3-pulse, uncontrolled rectifier is given by

(*a*) $V_d = \dfrac{3\sqrt{3}}{2\pi} V_m$

(*b*) $V_d = \dfrac{3\sqrt{3}}{\pi} V_m$

(*c*) $V_d = \dfrac{3}{\pi} V_m$

(*d*) $V_d = \dfrac{\sqrt{3}}{\pi} V_m$

39. The main difference between PN-junction diode and Schottky diode is that the recovered charge of

(*a*) PN-junction diode is less than that of Schotty diode and the leakage current is higher than that in Schottky diode

(*b*) Schotty diode is much less than that of the PN-junction diode and the leakage current is higher than that in PN-junction diode

(*c*) Schottky diode is much less than that of PN-junction diode and the leakage current is less than that in PN-junction diode

(*d*) PN-junction diode is less than that of Schottky diode and the leakage current is less than that in Schottky diode

40. A series capacitor commuted inverter can operate satisfactorily

(*a*) if $\dfrac{1}{LC} > \dfrac{R^2}{4L^2}$

(*b*) if $\dfrac{1}{LC} = \dfrac{R^2}{4L^2}$

(*c*) $\dfrac{1}{LC} < \dfrac{R^2}{4L^2}$

(*d*) irrespective of the values of R, L and C

41. A 3-pulse converter feeds an RLE load. The source has a definite inductance causing overlap. The thyristors are ideal. It has an overlap angle 'μ' of 20° at the minimum firing angle 'α'. The current remains constant in the complete range of firing angles. The range of firing angles of the converter would be

(*a*) $0° < \alpha < 180°$ (*b*) $20° < \alpha < 180°$

(*c*) $20° < \alpha < 160°$ (*d*) $0° < \alpha < 160°$

42. Match List-I (Power Electronic Devices) with List-II (Symbols) and select the correct answer using the codes given below the lists:

List-I	List-II
A. GTO Thyristor	1. 
B. Triac	2. 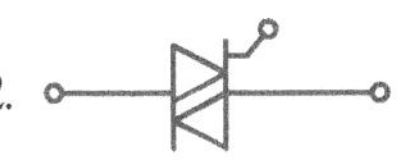
C. IGBT	3. 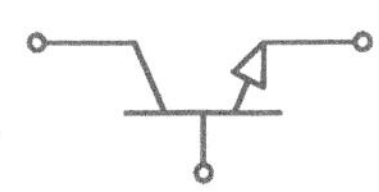
D. BJT	4. 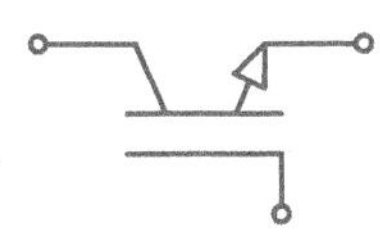

Cedes:

	A	B	C	D
(*a*)	1	2	3	4
(*b*)	1	2	4	3
(*c*)	2	1	3	4
(*d*)	2	1	4	3

43. Consider the following statements:

The source inductance of an ac to dc line commutated phase controlled converter

1. Limits $\dfrac{dv}{dt}$ capability of thyristors.
2. Causes a voltage drop in the dc terminal voltage.
3. Improves the line side power factor.
4. Limits the range of firing angles.
5. Reduces the line side power factor.

Of these statements:

(*a*) 1, 3 and 4 are correct

(*b*) 1, 2 and 4 are correct

(*c*) 2 and 5 are correct

(*d*) 2, 4 and 5 are correct

44. In a 3-phase semi-converter, if firing angle is less than or equal to 60°, then the duration of conduction of each thyristor and diode would be respectively

(*a*) 60° and 60° (*b*) 90° and 30°

(*c*) 120° and 120° (*d*) 180° and 180°

45. Match List-I (Drive) with List-II (Speed control range) and select the correct answer using the codes given below the lists:

List-I

A. Square wave inverter-fed induction motor

B. PWM inverter-fed induction motor

C. Slip energy recovery scheme with cycloconverter in the rotor circuit

D. Cycloconverter-fed induction motor

List-II

1. 50% of synchronous speed to synchronous speed
2. 10% of base speed to beyond base speed (1 : 20)
3. 0 to 1/3 base speed
4. 67% of synchronous speed to 133% of synchronous speed
5. Zero to 150% of base speed (1 : infinity)

Codes:

	A	B	C	D
(*a*)	2	5	4	3
(*b*)	2	5	3	4
(*c*)	3	2	4	1
(*d*)	3	2	1	4

46. A 12-phase converter consists of two sets of bridge rectifiers, one set fed from a delta-connected secondary and the other from a star-connected one. The two sets are connected to the dc motor armature through an inter-phase reactor. The purpose of this inter-phase reactor is to

(*a*) permit each 6-phase converter set to operate independently in the normal manner

(*b*) increase the current output

(*c*) match the voltages to two sets of converters

(*d*) counteract the effect of inductance of the armature

47. A single-phase full bridge inverter can operate in load-commutated mode in case load consists?

(*a*) RLC over damped

(*b*) RLC under damped

(*c*) RLC critically damped

(*d*) None of these

48. In a high voltage dc transmission scheme, reactive power is needed both for the rectifier at the sending-end and, for the inverter at the receiving-end. During the operation of such a dc link the rectifier receives

(*a*) lagging reactive power and the inverter supplies leading reactive power

(*b*) leading reactive power and the inverter supplies lagging reactive power

(*c*) lagging reactive power and the inverter supplies lagging reactive power

(*d*) leading reactive power and the inverter supplies leading reactive power

49. The number of thyristors required for single-phase to single-phase cyclo-converter of the mid-point and for three-phase to three-phase 3-pulse type of cycloconverter are respectively

(*a*) 4 and 6 (*b*) 4 and 4

(*c*) 4 and 18 (*d*) None of these

50. Consider the following statements: Switched mode power supplies are preferred over the continuous types, because they are

1. Suitable for use in both ac and dc.
2. More efficient.
3. Suitable for low-power circuits.
4. Suitable for high-power circuits.

Of these statements

(*a*) 1 and 2 are correct

(*b*) 1 and 3 are correct

(*c*) 2 and 3 are correct

(*d*) 2 and 4 are correct

51. Consider the following statements:

1 BJT has lower power losses than MOSFETs.

2. MOSFETs has lower power losses than IGBTs.

3. SCRCs have lower power losses than MOSFETs

Which of these statements are correct?

(*a*) 1, 2 and 3 (*b*) 1 and 2

(*c*) 2 and 3 (*d*) 1 and 3

52. During the turn-off process of a thyristor, the current flow does not stop at the instant when the current reaches zero but continues to flow to a peak value in the reverse direction. This is due to

(*a*) commutation failure

(*b*) hole storage effect

(*c*) presence of reverse voltage across the thyristor

(*d*) protective inductance in series with the thyristor

53. Consider a 1-pulse converter supplying the armature of a dc motor from a supply transformer. Corresponding wave-forms of the output voltages and currents,

(*i*) during starting and

(*ii*) during motor running conditions are shown below:

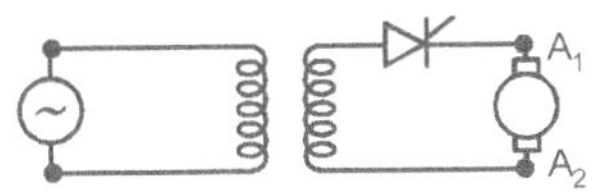

The portion of the output voltage MN during the shut-off period of the thyristor, as shown in the figure, represents

(*a*) the rectified voltage of ac supply harmonics, which is not blocked

(*b*) the voltage induced in the inductances of the armature and transformer because of stoppage of current

(*c*) a voltage produced by the 3rd harmonic magnetising current, as there is no current in the transformer secondary

(*d*) the back emf of the armature

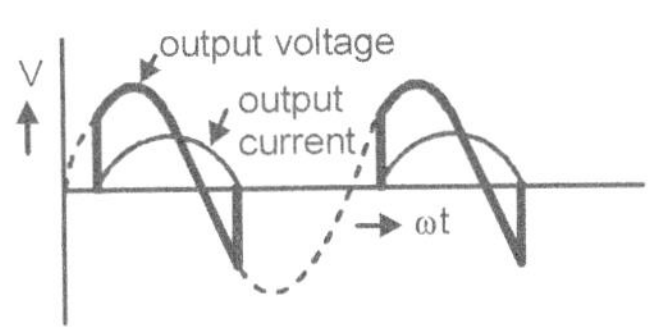

(I) at start

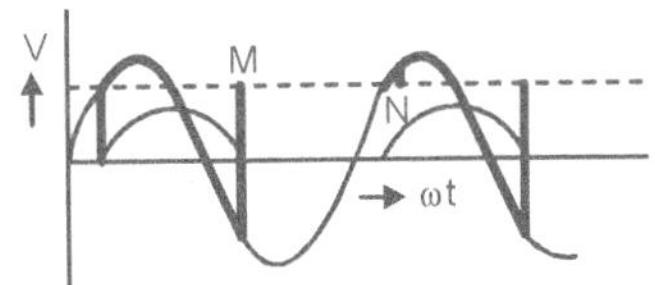

(II) during running

54. If the amplitude of the gate pulse to thyristor is increased, then

(*a*) both delay time and rise time would increase

(*b*) the delay time would increase but the time would decrease

(*c*) the delay time would decrease but the rise time would increase

(*d*) the delay time would decrease while the rise time remains unaffected

55. A forward voltage can be applied to an SCR after its

(*a*) anode current reduces to zero

(*b*) gate recovery time

(*c*) reverse recovery time

(*d*) anode voltage reduces to zero

56. Match List-I with List-II and select the correct answer using the codes given below the lists:

List-I

A. Freewheeling diode

B. Feedback diode

C. Current source inverter

D. Voltage source inverter

List-II

1. Voltage spikes in the output voltage
2. Peaks in the inverter current
3. Inductive loads of phase-controlled conve rter
4. Inductive loads of dc to ac inverters

Codes:

	A	B	C	D
(*a*)	4	3	1	2
(*b*)	3	4	1	2
(*c*)	3	4	2	1
(*d*)	4	3	2	1

57. Consider the following statements :

While operating on a phase controlled converter, the commutation capability of a dc motor deteriorates because of

1. Increase in the rms value of armature current
2. The presence of peaks in the armature current
3. Harmonic content of armature current
4. Discontinuous conduction, if any

Of these statements:

(*a*) 1, 3 and 4 are correct

(*b*) 1, 2 and 3 are correct

(*c*) 1, 2 and 4 are correct

(*d*) 2, 3 and 4 are correct

58. A 3-phase converter feeds a pure resistive load at a firing angle of $\alpha = 60°$. The average value of current flowing in the load is 10A. If a very large inductance is connected in the load circuit, then the

(*a*) average value of current will remain as 10A

(*b*) average value of current will become greater than 10A

(*c*) average value of current will become less than 10A

(*d*) trend of variation of current cannot be predicted unless the exact value of the inductance connected is known

59. In a 3-phase rectifier circuit, thyristor number 1, 2 and 3 are connected respectively to R, Y and B phases of the star-connected transformer secondary. When the current is being commutated from thyristor No. 1 to No. 2, the effect of the transformer leakage and the ac system inductance will be such that it will

(*a*) prolong the conduction in No. 1 and delay the turn-on of No. 2 correspondingly

(*b*) stop the conduction in No. 1 at the scheduled time, but delay the turn-on of No. 2

(*c*) produce conduction in both No. 1 and No. 2 in parallel for an overlapping period through a transient

(*d*) double the voltage output through a commutation transient

60. A single-phase half bridge inverter has a resistive load of 2.4 ohms and a dc input voltage of 48V with a centre tap. The rms output voltage of fundamental frequency is

(*a*) 21.6V (*b*) 43.2V

(*c*) 20.6V (*d*) 40V

61. For speed control of dc motors using controlled rectifiers, armature voltage control gives speed

(*a*) above rated speed

(*b*) below rated spëed

(*c*) above as well as below rated speed

(*d*) none of these

62. Free wheeling diode at the output of a rectifier allows

(*a*) SCR to trigger properly

(*b*) turn-off SCR at the end of half cycle

(*c*) protection of SCR against over voltage

(*d*) none of these

63. A free-wheeling diode is connected across R-L load, because

(*a*) it prevents infinite voltage across switch which breaks current

(*b*) it rectifier current

(*c*) it prevents current in opposite direction

(*d*) none of these

64. A power chopper converts

(*a*) ac to dc (*b*) dc to dc

(*c*) dc to ac (*d*) none of these

65. Dc to dc converter to chopper can

(*a*) only step down dc voltage

(*b*) only step up dc voltage

(*c*) step up or step down dc voltage

(*d*) none of these

66. If a chopper is operating with 50% duty cycle, then the output voltage in terms of input voltage is

(*a*) 25% (*b*) 50%

(*c*) 75% (*d*) 100%

67. In a dc motor the windage loss is proportional to

(*a*) supply voltage

(*b*) square of the supply voltage

(*c*) square of the flux density

(*d*) square of the armature speed

68. The efficiency of a dc shunt generator is maximum when the

(*a*) magnetic losses are equal to mechanical losses

(*b*) field copper losses are equal to copper losses

(*c*) stray losses equal constant losses

(*d*) armature copper losses equal constant losses

69. Efficiency of a dc machine is maximum when the variable losses are equal to

(*a*) constant losses

(*b*) square of the constant losses

(*c*) square root of the constant losses

(*d*) none of these

70. Match List-I (Devices) with the List-II (Properties) and select the correct answer using the codes give below the lists:

List-I

A. Triac

B. Reverse conducting thyristor (RCT)

C. Gate turn-off thyristor (GTO)

D. Amplifying gate thyristor

List-II

1. Good $\frac{di}{dt}$ behavior even at low gate currents.
2. Normally provided with a small continuous negative gate pulse during off state.
3. Negative gate pulse of reverse conduction.
4. No gate pulse for reverse conduction.

Codes:

	A	B	C	D
(*a*)	4	3	1	2
(*b*)	3	4	2	1
(*c*)	3	4	1	2
(*d*)	4	3	2	1

71. If V_m is peak value of an applied voltage in a half-wave rectifier with a large capacitor across the load, then the peak inverse voltage will be

(*a*) $V_m/2$ (*b*) V_m

(*c*) $\sqrt{2}V_m$ (*d*) $2V_m$

72. A line commutated phase-controlled inverter is operating at its inverter limit. There can be a commutation failure if

(*a*) the frequency decreases

(*b*) the voltage increases

(*c*) the frequency increases

(*d*) both voltage and frequency change such that V/*f* is constant

73. When a thyristor is negatively biased,

(*a*) all the three junctions are negatively biased

(*b*) outer junctions are positively biased and the inner junction is negatively biased

(*c*) outer junctions are negatively biased and the inner junction is positively biased

(*d*) the junction near the anode is negatively biased and the one near the cathode is positively biased

74. Consider the following semiconductor devices:

1. Triac
2. Thyristor
3. Amplifying gate thyristor

The correct sequence of these devices in increasing order of their $\frac{dv}{dt}$ capabilities is

(*a*) 1, 3, 2 (*b*) 1,2,3

(*c*) 3, 1, 2 (*d*) 3, 2, 1

75. In a 3-phase bridge rectifier fed from the star connected secondary winding of a transformer, let the voltage to the neutral of the A-phase (phase sequence A, B, C) be $V_m \sin \omega t$. At the instant when the voltage of A-phase is maximum, the output voltage at the rectifier terminals will be

(*a*) $\frac{V_m}{\sqrt{2}}$ (*b*) V_m

(*c*) $1.5\,V_m$ (*d*) $\sqrt{3}V_m$

76. A voltage source inverter is normally employed when

(*a*) source inductance is large and load inductance is small

(*b*) source inductance is small and load inductance is large

(*c*) both source and load inductances are small

(*d*) both source and load inductances are large

77. Consider the following features:

1. Inherent short-circuited protection.
2. Regeneration capability.
3. Need for inverter grade thyristors.
4. Voltage spikes across the load

Which of these are the features associated with current source inverter?

(*a*) 1, 2 and 3

(*b*) 2, 3 and 4

(*c*) 1, 3 and 4

(*d*) 1, 2 and 4

78. A single-phase full-bridge voltage source inverter operating in square wave mode supplies a purely inductive load. If the inverter time period is T, then the time duration for which the feedback diodes conduct in a cycle is

(*a*) T (*b*) $\frac{T}{2}$

(*c*) $\frac{T}{4}$ (*d*) $\frac{T}{8}$

79. Match the 3 curves labelled A, B, C and D with the characteristics given under List-II and select the correct answer using the codes given below the figures:

List-II Control characteristics of

1. 3-pulse converter feeding pure resistance
2. full controlled converter feeding R-L load with perfect smoothing.
3. single-phase ac voltage controller feeding pure inductive load.
4. 6-pulse half-controlled converter.

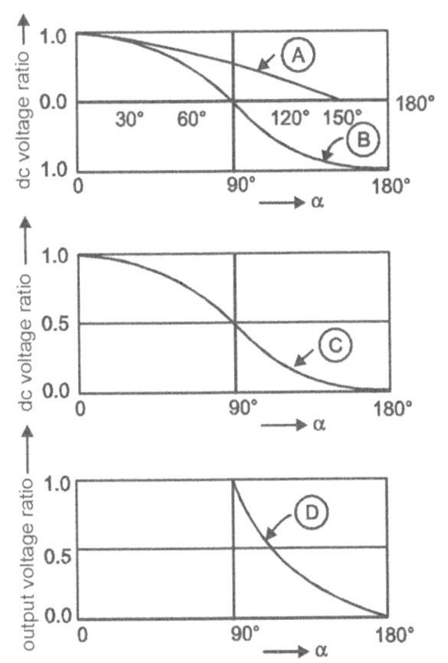

Codes:

	A	B	C	D
(*a*)	1	2	3	4
(*b*)	1	2	4	3
(*c*)	2	1	3	4
(*d*)	2	1	4	3

80. Consider the following statements:

The diodes in a voltage source inverter (McMurray inverter) should be able to

1. Withstand a large voltage in the reverse direction.
2. Carry the commutating current excess of load current.
3. Provide the required reverse bias of the outgoing thyristor.
4. Feedback the reactive current to the source

Of these statements:

(*a*) 1, 2 and 3 are correct
(*b*) 1, 3 and 4 are correct
(*c*) 2, 3 and 4 are correct
(*d*) 1, 2 and 4 are correct

81. A voltage source inverter will have better performance if its

(*a*) load inductance is small and source, inductance is large
(*b*) both load and source inductances are small
(*c*) both load and source inductances are large
(*d*) load inductance is large and source inductance is small

82. A step-up chopper is required to deliver a load voltage of 660V from 220V dc source. If the non-conduction time of SCR is 100 μs, the required pulse-width will be

(*a*) 300 μs (*b*) 250 μs
(*c*) 200 μs (*d*) 150 μs

83. Consider the following statements :

The overlap angle of a phase-controlled converter will increase

1. as the firing angle increases.
2. as the frequency of supply increases.
3. as the supply voltage decreases.

Of these statements:

(*a*) 1, 2 and 3 are correct
(*b*) 2 and 3 are correct
(*c*) 1 and 3 are correct
(*d*) 1 and 2 are correct

84. Match List-I with List-II and select the correct answer using the codes given below the lists:

List-I

A. Semiconverters.
B. Source inductance of the converter.
C. Sequence control of converters.
D. Current limit control of converters.

List-II

1. Additional reactive power loading.
2. Two quadrant operation with improved displacement factor on the ac side
3. Unity displacement factor
4. One quadrant operation with better displacement factor

Codes:

	A	B	C	D
(*a*)	2	1	4	3
(*b*)	2	3	4	1
(*c*)	4	1	2	3
(*d*)	4	3	2	1

85. Which of following PNPN devices has a terminal for synchronising purpose?

(*a*) SCS (*b*) Triac
(*c*) Diac (*d*) SUS

86. Consider the following steps:

1. Reversing connections to the terminals of the capacitor.
2. Changing the position of the capacitor from auxiliary winding circuit to main winding circuit.
3. Reversing supply connection to the main winding
4. Reversing supply connection to the auxiliary circuit.

While installing a new ceiling fan, if the fan motor is found to be rotating in the wrong direction, then the direction of rotation of the motor can be corrected by

(*a*) 1, 2 and 3 (*b*) 1, 2 and 4
(*c*) 1, 3 and 4 (*d*) 2, 3 and 4

87. The ac-to-dc voltage ratio of a converter in the state of discontinuous conduction is

(*a*) dependent upon the firing angle but independent of load time constant
(*b*) dependent upon load time constant but independent of firing angle
(*c*) dependent upon both firing angle and load time constant
(*d*) independent of both firing angle and load time constant

88. A dc chopper has a resistive load of 10 ohms and an input voltage of 220 V. When the chopper switch is 'ON', its voltage drop is 2V and the chopping frequency is 1 kHz. If the duty cycle is 50%, then the average output voltage will be

(*a*) 100V (*b*) 103V
(*c*) 106V (*d*) 109V

89. Consider the following statements :

When gate triggering is employed, a thyristor can withstand higher values of *di/dt* (rate of change of forward current), if

1. the gate current is increased.
2. the rate of rise of gate current is increased.
3. the gate current is decreased.
4. the rate of rise of gate current is decreased.

Of these statements:

(*a*) 3 and 4 are correct
(*b*) 1 and 4 are correct
(*c*) 2 and 3 are correct
(*d*) 1 and 2 are correct

90. A single-phase voltage controller is employed for controlling the power flow from 260V, 50 Hz source into a load consisting of R = 5Ω, and ωL = 12Ω. The value of maximum rms load current and the firing angle are respectively

(*a*) $\frac{260}{10.9}$ A and 0° (*b*) 20 A and 0°

(*c*) 20A and 90° (*d*) $\frac{260}{10.9}$ and 90°

91. Consider the following statements related to diodes, transistors and thyristors:

1. Diode is a 2- terminal device, transistor is a 3-terminal device and thyristor is 4-terminal device.
2. Transistor cannot be used as a converter value while diode and thyristor can be used for this purpose.
3. The main (collector) current is always under the influence of the control current in a transistor while the main current in the conducting state cannot be controlled through the gate in a thyristor.
4. An inverter can be built using transistors or thyristors and not using diodes.

Of these statements:

(*a*) 1, 2, 3 and 4 are correct
(*b*) 2, 3 and 4 are correct
(*c*) 2 and 3 are correct
(*d*) 2 and 4 are correct

92. In dc choppers feeding highly inductive loads, the waveforms for input and output currents are

(*a*) discontinuous and continuous respectively
(*b*) both continuous
(*c*) both discontinuous
(*d*) continuous and discontinuous respectively

93. A single-phase voltage source square-wave inverter feeds pure inductive load. The waveform of the load current will be

(*a*) sinusoidal (*b*) rectangular
(*c*) trapezoidal (*d*) triangular

94. A single-face two-pulse converter feeds an R-L load with insufficient smoothing but the conduction is continuous. If the resistance of the load circuit is increased

(*a*) the ripple content of the load current will remain the same
(*b*) the ripple content of the load current will increase
(*c*) the ripple content of the load current will decrease
(*d*) there is possibility of discontinuous conduction consequent to an increase in the ripple content

EXERCISE - II

1. As the number of phases in a multiphase rectifier are increased the output will **[NTPC]**

(*a*) increase
(*b*) become more smooth
(*c*) decrease
(*d*) the diodes will require high peak inverse voltage

2. A single-phase full bridge VSI has inductor L as the load. For a constant source voltage, the current through the inductor is

(*a*) square wave
(*b*) triangular wave
(*c*) sine wave
(*d*) none of these **[NTPC]**

3. Turn-on and turn-off times of transistor depend on

(*a*) static characteristic
(*b*) junction capacitances
(*c*) current gain
(*d*) none of these **[NTPC]**

4. If the rms source voltage is volts, the minimum and maximum values of firing angles for a single phase, half-wave controlled rectifier, supplying a load with a back emf of 40 volts are

(*a*) 0° and 180°

(*b*) $\alpha = \sin^{-1}\frac{40}{\sqrt{2}V}$ and 180°

(*c*) $\alpha = \sin^{-1}\frac{40}{\sqrt{2}V}$ and $\left(\pi - \sin^{-1}\frac{40}{\sqrt{2}V}\right)$

(*d*) 0° and $\left(\pi - \sin^{-1}\frac{40}{\sqrt{2}V}\right)$ **[IOCL]**

5. In a GTO, anode current begins to fall when gate current
   (*a*) is negative peak at time $t = 0$
   (*b*) is negative peak at $t = t_5$ storage period *ts*
   (*c*) just begins to become negative at $t = 0$
   (*d*) is negative peak at $t = t_5$ + fall time **[BHEL]**
6. When a thyristor gets turned on, the gate drive
   (*a*) should hot be removed as it will turn-off the SCR
   (*b*) may be may not be removed
   (*c*) should be removed
   (*d*) should be removed in order to avoid increased losses and higher junction temperature **[BHEL]**
7. An IGBT has three terminals called
   (*a*) collector, emitter and base
   (*b*) drain, source and base
   (*c*) collector, emitter and gate
   (*d*) none of these **[NTPC]**
8. A register connected across the gate and cathode of an SCR increases its
   (*a*) $\frac{dv}{dt}$ rating (*b*) holding current
   (*c*) noise immunity (*d*) all of these **[NTPC]**
9. For a single phase phase-controlled rectifier with a free wheeling diode across the load
   (*a*) the instantaneous output voltage v_0 is always positive
   (*b*) v_0 may be positive zero or negative
   (*c*) v_0 may be positive or zero
   (*d*) none of these **[BHEL]**
10. Which semiconductor power device out of the following is not a current triggered device?
   (*a*) Thyristor (*b*) G.T.O.
   (*c*) Triac (*d*) MOSFET **[GATE]**
11. In a 3-phase controlled bridge rectifier, with an increase of overlap angle, the output dc voltage
   (*a*) decreases
   (*b*) increases
   (*c*) does not change
   (*d*) depends upon load inductance **[GATE]**
12. Which of the following does not cause permanent damage of an SCR?
   (*a*) High current
   (*b*) High rate of rise of current
   (*c*) High temperature rise
   (*d*) High rate of rise of voltage **[GATE]**
13. In dc choppers, for chopping period T, the output voltage can be controlled by FM by varying
   (*a*) T keeping T on constant
   (*b*) T keeping T off constant
   (*c*) T on keeping T constant
   (*d*) T off keeping T constant **[BHEL]**
14. Which of the following devices does not have negative resistance characteristics?
   (*a*) UJT (*b*) Tunnel diode
   (*c*) SCR (*d*) FET **[BHEL]**
15. Thermal resistance in SCRs has the units of °C W and heat sinks are made from
   (*a*) copper (*b*) steel
   (*c*) aluminium (*d*) none of these **[IOCL]**
16. A single-phase full-wave mid-point thyristor converter uses a 230/200V transformer with centre tap on the secondary side. The P.I.V. per thyristor is
   (*a*) 100 V (*b*) 141.4 V
   (*c*) 200 V (*d*) 282.8 V **[IOCL]**
17. The function of connecting a zener diode in an UJT circuit used for the triggering of SCRs, is to
   (*a*) expedite the generation of triggering pulses
   (*b*) delay the generation of triggering pulses
   (*c*) provide a constant voltage to UJT to prevent erratic firing
   (*d*) provide a variable voltage to UJT as source voltage changes **[IOCL]**
18. The average on-state current for an SCR is 20A for a conduction angle of 120°. Its average on-state current for 60° conduction angle would be
   (*a*) 20 A (*b*) 10 A
   (*c*) less then 20 A (*d*) 40 A **[NTPC]**
19. A 3 phase to 3-phase cycloconverter requires
   (*a*) 18 SCRs for 3-phase device
   (*b*) 18 SCRs for 6-pulse device
   (*c*) 36 SCRs for 3-phase device
   (*d*) 36 SCRs for 6-pulse device **[NTPC]**
20. A single-phase voltage controller feeds an induction motor (A) and a heater (B)
   (*a*) In both the loads, fundamental and harmonics are useful
   (*b*) In A only fundamental and in B only harmonics are useful
   (*c*) In A only fundamental and in B harmonics as well as fundamental are useful
   (*d*) In A only harmonics and in B only fundamentals are useful **[NTPC]**

21. In an inverter with fundamental output frequency of 50 Hz, if third harmonic is eliminated, then frequencies of other components in the output voltage wave, in Hz would be

(*a*) 250, 350, 450, high frequencies

(*b*) 50, 250, 350, 450

(*c*) 50, 250, 350, 550

(*d*) 50, 100, 200, 250 **[BHEL]**

22. The Triac can be used only in

(*a*) inverter

(*b*) rectifier

(*c*) multiquadrant chopper

(*d*) cycloconverter **[GATE]**

23. If a diode is connected in anti-parallel with a thyristor, then

(*a*) both turn-off power loss and turn-off time decrease

(*b*) turn-off power loss decreases but turn-off time increases

(*c*) turn-off power loss increases, but turn-off time decreases

(*d*) Both turn-off power loss and turn-off time increases **[GATE]**

24. The peak inverse voltage in ac to dc converter systems is highest in

(*a*) single-phase full-wave mid-point converter

(*b*) single-phase full converter

(*c*) 3-phase bridge converter .

(*d*) 3-phase half-wave converter **[NTPC]**

25. An ideal chopper operating at a frequency of 500 Hz, supplies a load having resistance of 3 ohms and inductance of gm H from a 60 V battery. The mean value of the load voltage P or on/off ratio of 4/1, (assuming that load is shunted by a perfect-commutating diode and batter is loss-less) is

(*a*) 240 V (*b*) 48 V

(*c*) 15 V (*d*) 4 V **[NHPC]**

26. In a self-controlled synchronous motor fed from a variable frequency inverter

(*a*) the rotor poles invariably have damper winding

(*b*) there are stability problems

(*c*) the speed of the rotor decides stator frequency

(*d*) the frequency of the stator decides the rotor speed **[NHPC]**

27. An inter-group reactor is used in a single-phase cycloconverter circuit to

(*a*) Reduce current-ripples

(*b*) Reduce voltage-ripples

(*c*) Limit circulating current

(*d*) Limit $\frac{di}{dt}$ in the semiconductor switch **[BHEL]**

28. If the gate current of an SCR is increased, then forward breakdown voltage V_{BR} will

(*a*) increase (*b*) decrease

(*c*) not be affected (*d*) infinite **[BHEL]**

29. For normal SCRs, turn-on time is

(*a*) less than turn-off time t_q

(*b*) more than turn-off time t_q

(*c*) equal to turn-off time t_q

(*d*) half of turn-off time t_q **[PGCIL]**

30. Surge current rating of an SCR specifies the maximum

(*a*) repetitive current with sine wave

(*b*) non-repetitive current with rectangular wave

(*c*) non-repetitive current with sine wave

(*d*) repetitive current with rectangular wave **[PGCIL]**

31. A d.c. chopper has an average value of load current of 10 A while feeding a pure resistive load at a given time ratio and frequency. If a large inductance is placed in series with the load resistance and operated at the same switching frequency and the same time ratio, then the average load current **[PGCIL]**

(*a*) remains constant at 10A

(*b*) becomes less than 10A

(*c*) becomes greater than 10A

(*d*) variation cannot be estimated without knowing the value of inductance

32. A single phase diode bridge rectifier supplies a highly inductive load. The load current can be assumed to be ripple free. The ac supply side current waveform will be

(*a*) sinusoidal (*b*) constant dc

(*c*) square (*d*) triangular **[GATE]**

33. The on-state voltage of a GTO is

(*a*) 0.7 V

(*b*) 1.2 V

(*c*) 2.3 V

(*d*) >3 V **[NTPC]**

34. Consider the following statements: Specific features of BJT are

1. Very small on-state resistance
2. Presence of second breakdown
3. Infinite input resistance
4. Good performance in parallel operation

Which of these statements are correct?

(*a*) 1, 2, 3 and 4 (*b*) 1, 2 and 3
(*c*) 3 and 4 (*d*) 1 and 2 **[NTPC]**

35. The number of *p-n* junction in a thyristor is/are

(*a*) 1 (*b*) 2
(*c*) 3 (*d*) 4 **[NTPC]**

36. The three-phase ac to dc converter which requires neutral point connection is

(*a*) 3-phase semiconverter
(*b*) 3-phase full converter
(*c*) 3-phase half-wave converter
(*d*) 3-phase full converter with diodes **[NTPC]**

37. Secondary breakdown occurs in

(*a*) MOSFET but not in BJT
(*b*) both MOSFET and BJT
(*c*) BJT but not in MOSFET
(*d*) none of these **[NHPC]**

38. A cycloconverter is a

(*a*) frequency changer (*fc*) from higher to lower frequency with one state conversion
(*b*) *fc* from higher to lower frequency with two-stage conversion
(*c*) *fc* from lower to high frequency with one-state conversion
(*d*) either (*a*) or (*c*) **[NHPC]**

39. The frequency of the ripple in the output voltage of a 3-phase semiconverter depends upon

(*a*) firing angle and load resistance
(*b*) firing angle and load inductance
(*c*) the load circuit parameters
(*d*) firing angle and the supply frequency **[NHPC]**

40. In single-pulse modulation of PWM inverter, third harmonic can be eliminated if pulse width is equal to

(*a*) 30° (*b*) 60°
(*c*) 120° (*d*) 150° **[BHEL]**

41. In a 3-phase semiconverter the three SCRs are triggered at an interval of

(*a*) 60° (*b*) 90°
(*c*) 120° (*d*) 180° **[BHEL]**

42. Turn-on time of an SCR can be reduced by using a

(*a*) rectangular pulse of high amplitude and narrow width
(*b*) rectangular pulse of low amplitude and wide-width
(*c*) triangular pulse
(*d*) trape zoidal pulse **[BHEL]**

43. Each diode of a 3-phase half-wave diode rectifier conducts for

(*a*) 60° (*b*) 120°
(*c*) 180° (*d*) 90° **[BHEL]**

44. A dc to dc transistor chopper supplied from a fixed voltage dc source feeds a fixed resistive-inductive load and a free-wheeling diode. The chopper operates at 1 kHz and 50% duty cycle. Without changing the value of the average dc current through the load, if it is desired to reduce the ripple content of load current, the control action needed will be

(*a*) increase the chopper frequency keeping its duty cycle constant
(*b*) increase the chopper frequency and duty cycle in equal ratio
(*c*) decrease only the chopper frequency
(*d*) decrease only the duty cycle **[GATE]**

45. Figure I shows a 3-phase inverter fed by a constant voltage source V_{DC} and connected to a balanced resistive load at the output. Each switching device may conduct for 120° or for 180°. The wave form shown in Figure II is figures I, II the

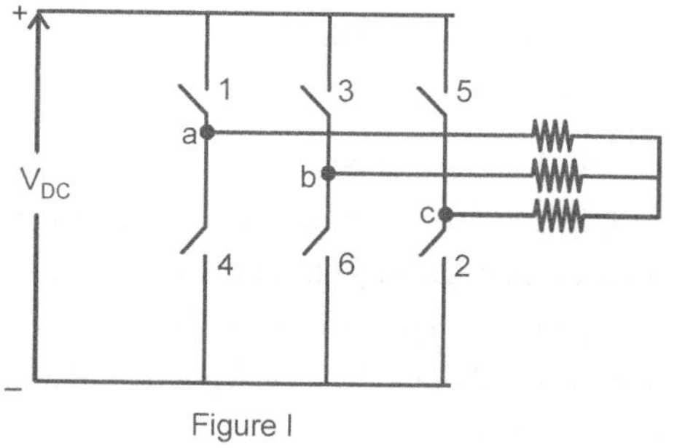

Figure I

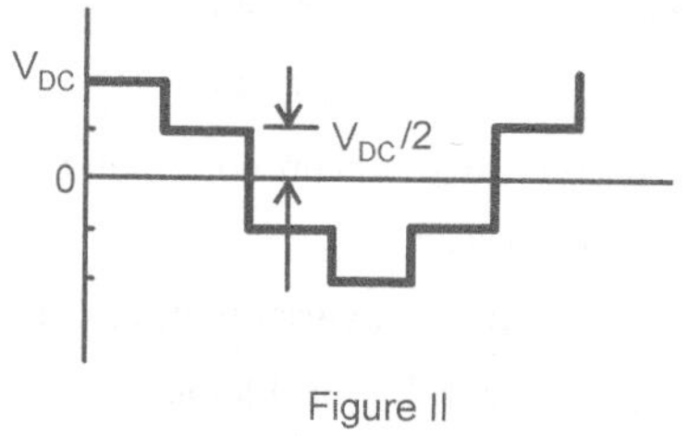

Figure II

(*a*) line voltage with 120° firing
(*b*) load phase voltage with 120° firing
(*c*) line voltage with 180° firing
(*d*) load phase voltaee with 180° firine **[NTPC]**

46. An inverter capable of supplying a balanced three-phase variable voltage variable frequency output is feeding a three-phase induction motor rated for 50Hz and 440V. The stator winding resistances of the motor are negligibly small. During starting, the current inrush can be avoided without sacrificing the starting torque by suitably applying

(*a*) low voltage at rated frequency
(*b*) low voltage keeping the V/*f* ratio constant
(*c*) rated voltage at low frequency
(*d*) rated voltage at rated frequency **[GATE]**

47. The cycle converters requires natural or forced communtation as under

(*a*) natural commutation in both step-up and step-down cycloconverter
(*b*) forced commutation in both step-up and step-down cycloconverters
(*c*) forced commutation in step-up cyclo converters. forced commutation in step-down cycioconverters
(*d*) forced communtation in step-down cycloconverters **[NTPC]**

48. Which one of the following statements is correct?
In a transistor, the reverse saturation current Ico

(*a*) doubles for every 10°C rise in temperature
(*b*) doubles for every 1°C rise in temperature
(*c*) increases linearly with temperature
(*d*) decreases linearly with temperature **[NTPC]**

49. A compared to oscillators, inverters provides

(*a*) distortionless output
(*b*) highly stabilised output
(*c*) high frequency output
(*d*) low frequency output **[NTPC]**

50. The *di/dt* rating of an SCR is specified for its

(*a*) decaying anode current
(*b*) decaying gate current
(*c*) rising gate current
(*d*) rising anode current **[BHEL]**

51. A forward voltage can be applied to an SCR after its

(*a*) anode current reduces to zero
(*b*) gate recovery time
(*c*) reverse recovery time
(*d*) anode voltage reduces to zero **[PGCIL]**

52. A single-phase full bridge inverter can operate in load-communication mode in case load consist of

(*a*) RL
(*b*) RLC under damped
(*c*) RLC over damped
(*d*) RLC critically damped **[PGCIL]**

53. A dc chopper is fed from 100 Vdc. Its load voltage consists of rectangular pulses of duration in sec in an overall cycle time of 3 m sec. The average output voltage and ripple factor for this chopper are respectively

(*a*) 25V, 1 (*b*) 50V, 1
(*c*) 33.33 V, $\sqrt{2}$ (*d*) 33.33V, 1 **[PGCIL]**

54. Which one of the following statements is correct?

For an SCR, $\frac{dv}{dt}$ protection is achieved through at the use of

(*a*) RL in series with SCR
(*b*) RC across SCR
(*c*) L in series with SCR
(*d*) RC in series with SCR **[NPTC]**

55. In order to get best results per unit cost heat sinks on which the thyristors mounted, are made of

(*a*) aluminium (*b*) copper
(*c*) nickel (*d*) stainless steel **[NTPC]**

56. In SCR, the turn-OFF time

(*a*) increases with increase of anode current I_A
(*b*) is independent of I_A
(*c*) varies $1/I_A$
(*d*) varies a I/A_A^2 **[NTPC]**

57. The advantages of SCR over SCS are

(*a*) larger switching time and smaller V_H
(*b*) larger switching time and large V_H
(*c*) smaller switching time and smaller V_H
(*d*) smaller switching time and larger V_H **[NTPC]**

58. SCR is a

(*a*) 2 layer device
(*b*) 3 layer device
(*c*) 4 layer device with one gate
(*d*) 4 layer device with two gates **[BHEL]**

59. In a three phase half-wave rectifier, th.e peak inverse voltage equals

(*a*) $\sqrt{6}$E (*b*) $\sqrt{3}$E
(*c*) 2E (*d*) $\sqrt{3/2}$E

where E is the R.M.S. voltage of secondary **[BHEL]**

60. Holding current in a pnpn diode is the

(*a*) maximum operating current
(*b*) normal operating current
(*c*) current corresponding to breakover voltage
(*d*) minimum current to keep the device ON **[NTPC]**

61. Thermistor is used for measurement of power at
(a) audio frequencies
(b) high frequencies
(c) very high frequencies
(d) microwave frequencies [IOCL]

62. The transformer utilisation factor of a single phase full-wave rectifier is
(a) 0.750 (b) 0.550
(c) 0.330 (d) 0.693 [BEL]

63. Triac is c
(a) 2 terminal bidirectional switch
(b) 3 terminal bidirectional switch
(c) 2 terminal unilateral switch
(d) 3 terminal unilateral switch [BHEL]

64. A bilateral pnpn diode switch
(a) consists of two pnpn diodes in parallel but in opposite order
(b) is a 3 layer semiconductor device
(c) is a unilateral device
(d) is a bilateral device [NTPC]

65. A SCR's, breakover voltages is 12 V and holding current I_H = 5mA. What should be the input voltage V in the circuit for the SCR to drop out if the voltage at drop out point is 0.7V?

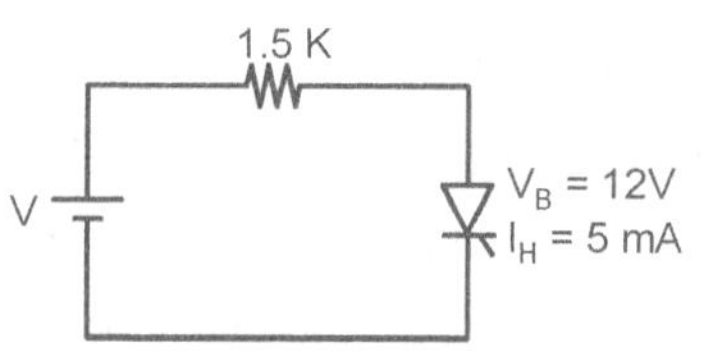

(a) 7.5V (b) 11.3V
(c) 18.8V (d) 8.2V [BEL]

66. SCR uses
(a) no gate
(b) one gate on the p-layer next to cathode
(c) one gate on the n-layer next to anode
(d) two gates [NHPC]

67. Diac is a
(a) 2 terminal bidirectional switch
(b) 2 terminal unilateral switch
(c) 3 terminal bidirectional switch
(d) 3 terminal unilateral switch [NHPC]

68. The ratio of gate current to continuous anode current in an SCR to switch on is
(a) 1 : 10 (b) 1 : 100
(c) 1 : 1000 (d) >1 : 10000 [BEL]

69. A half wave rectified sinusoidal waveform has a peak voltage of 10V. Its average value and rms value of the fundamental component are respectively given by
(a) $\frac{20}{\pi}V, \frac{10}{\sqrt{2}}V$
(b) $\frac{16}{\pi}V, \frac{10}{\sqrt{2}}V$
(c) $\frac{10}{\pi}V, 5V$
(d) $\frac{30}{\pi}V, 5V$ [BEL]

70. In SCR, the turn-ON time
(a) is independent of V_g
(b) decreases with increase of V_g
(c) varies as V_g
(d) varies as V_g
where V_g is the trigger voltage amplitude [NHPC]

71. A voltage source 200 sin 314t supplies resistive load of 50 ohms through a controlled thyristor which performs half-wave rectification. If the firing angle is 30° with respect to supply voltage waveform, the average power in the load is
(a) 90.3 watts
(b) 80.3 watts
(c) 65.8 watts
(d) 70.6 watts [BHEL]

72. Diac is a silicon device with
(a) 3 layers and one gate
(b) 3 layers and no gate
(c) 4 layers and one gate
(d) 4 layers and no gate [BHEL]

73. In an SCR, the breakover voltage V_{BO}
(a) in independent of gate current
(b) increases with the increase of positive gate current
(c) decreases with the increase of positive gate current
(d) may increase or decrease with increase of gate current depending on temperature [BHEL]

74. In the circuit shown, the average value of current

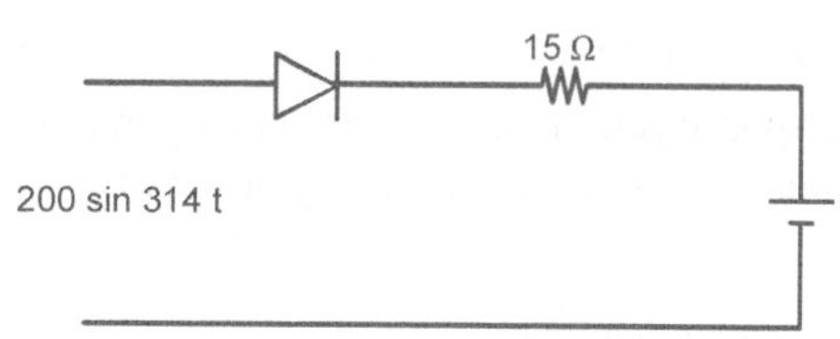

(a) 4.2A (b) 3.68A
(c) 2.7A (d) 5.1A [BEL]

75. A regulated DC supply of 12V is obtained from an unregulated DC supply having the peak value of 28V and peak to peak ripple of 6V. The power dissipation in the regulator for a output current of I_{DC} = 2A is

(*a*) 26W

(*b*) 24W

(*c*) 20W

(*d*) 32W **[BHEL]**

76. In a pnpn diode, the phenomenon of rate effect depends on

(*a*) rate of change of temperature

(*b*) rate of change of applied voltage

(*c*) avalanche breakdown

(*d*) negative resistance in the forward voltage condition **[NTPC]**

77. In SCK, the turn-OFF time

(*a*) increases with increase of T

(*b*) is independent of T

(*c*) varies as 1/T

(*d*) varies as $1/T^2$

where T is the temperature is deg K **[NTPC]**

78. In phase controlled rectifiers the power factor

(*a*) is unaffected by phase control

(*b*) worsens if the delay angle is small

(*c*) worsens for increasing delay angle

(*d*) worsens for large output voltage **[BEL]**

79. In a pnpn diode, breakover takes place when

(*a*) $(\alpha_1 + \alpha_2) = 0.5$

(*b*) $(\alpha_1 + \alpha_2) = 0.9$

(*c*) $(\alpha_1 + \alpha_2) = 1.0$

(*d*) $\alpha_1 = \alpha_2$

where α_1 and α_2 refer to the constituent transistors **[BEL]**

80. The peak current through the resistor in the circuit of Figure assuming the diode to be ideal is

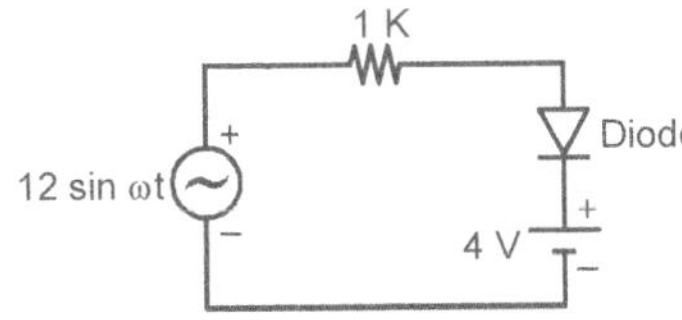

(*a*) 12 mA

(*b*) 4 mA

(*c*) 16 mA

(*d*) 8 mA **[NTPC]**

81. Match List-I (device of thyristor family) with List-II (characteristic) and select the correct answer using the codes given below the lists:

List-I	List-II
A. Asymmetrical thyristor	1. Poor turn-off gain
B. Amplifying gate thyristor	2. Good - capability
C. Triac	3. Poor reverse blocking capability
D. GTO	4. Good turn-off

Codes:

	A	B	C	D
(*a*)	2	4	1	3
(*b*)	2	4	3	1
(*c*)	4	2	3	1
(*d*)	4	2	1	3

[IAS]

82. Consider a dc to dc converter (chopper) having a load resistance of 10Ω and an input voltage of 232V. If the voltage drop across the converter during ON time is 2V and the frequency of operation is 1 kHz at a duty cycle of 0.4, then the average output voltage of the converter will be

(*a*) 94.0V (*b*) 92.8V

(*c*) 92.0V (*d*) 90.8V **[IAS]**

83. Which one of the following statements is correct in connection with inverters?

(*a*) Both voltage source inverter and current source inverter require feedback diodes

(*b*) The current source inverter requires feedback diodes

(*c*) GTO's can be used only in the current source inverters

(*d*) The voltage source inverter requires feedback diodes **[IAS]**

84. A drive suitable for low-speed high-power reversible application is the

(*a*) synchro converter-fed ac drive

(*b*) current source inverter-fed ac drive

(*c*) voltage source (square wave) inverter-fed drive

(*d*) induction motor fed from ac voltage supply **[IAS]**

85. The advantage of dual converter in the circulatory current mode is

(*a*) the low loss in the converter

(*b*) that it permits dynamic braking of connected motor

(*c*) the lower cost of the equipment

(*d*) faster response **[IAS]**

86. A thyristor equivalent of a thytration tube is a

(*a*) SCR (*b*) UJT

(*c*) Diac (*d*) Triac **[ES]**

87. For a step-down chopper, V is the source voltage, 'R' is the load resistance and 'α' is the duty cycle. RMS and average values of thyristor currents for this chopper are, respectively

(*a*) $\frac{\alpha V_s}{R}$ and $\frac{\sqrt{\alpha} V_s}{R}$

(*b*) $\frac{\sqrt{\alpha} V_s}{R}$ and $\frac{\sqrt{\alpha} V_s}{R}$

(*c*) $\frac{\alpha V_s}{R}$ and $\frac{\alpha V_s}{R}$

(*d*) $\frac{\sqrt{\alpha} V_s}{R}$ and $\frac{\alpha V_s}{R}$ **[IAS]**

88. The SCR circuit is shown in the given figure. Which one of the following wave forms represent V_T, when input voltage *v* is sinusoidal?

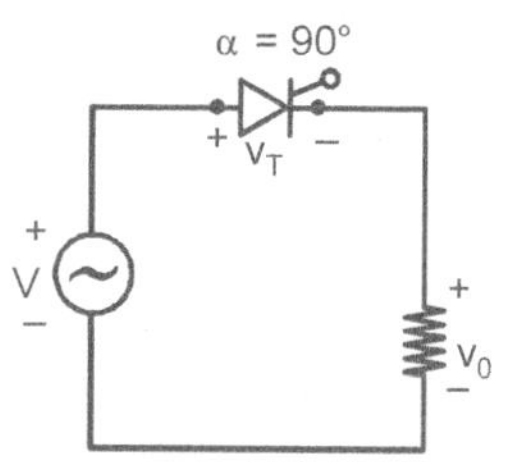

(*a*)
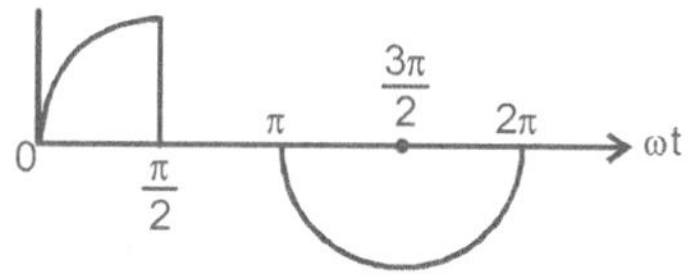

(*b*)
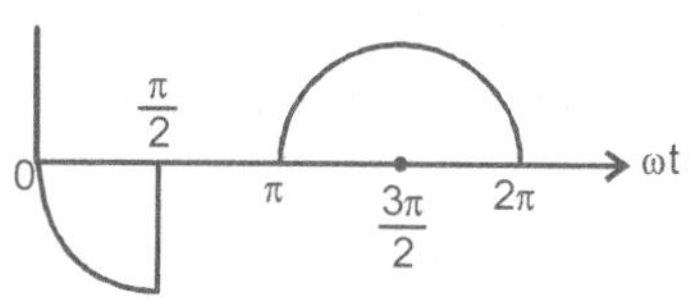

(*c*)
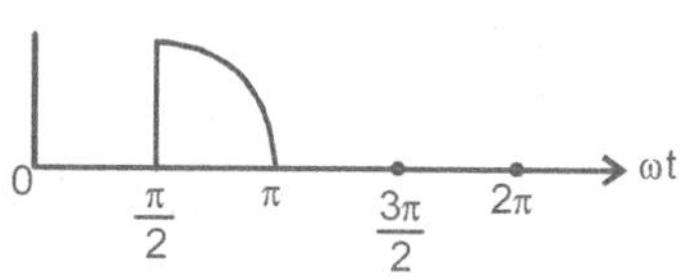

(*d*)
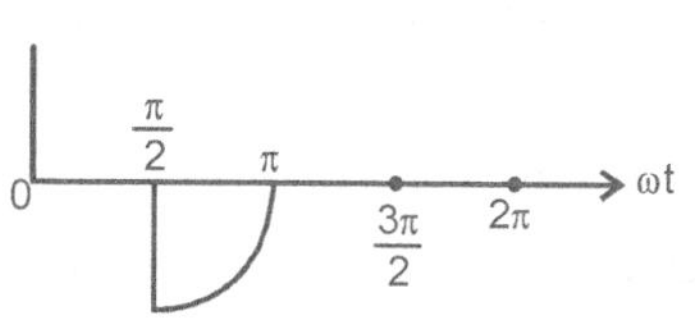
[IAS]

89. Chopper circuits are used to control the speed of *dc* motors which are fed from fixed dc voltage sources. Variables voltage output can be obtained by which of the following methods?

1. Varying the "on" to "off" time ratio
2. Adjusting the wave shape
3. Adjusting the frequency of switching
4. Using a saturable reactor in series with the motor

Select the correct answer using the codes given below:

Codes:

(*a*) 1 and 2 (*b*) 1 and 3

(*c*) 2 and 3 (*d*) 2 and 4 **[IAS]**

90. A H W rectifier feeds into a load which is a series combination of R and L. The conduction period is

(*a*) 0° (*b*) $<180°$

(*c*) $>180°$ (*d*) $=180°$ **[NTPC]**

91. A half-wave SCR controlled circuit with a load resistor $R_L = 10\Omega$ has an applied voltage of 300V rms for a conduction angle of 60°. The reading of a true rms reading ammeter is

(*a*) 9.37A (*b*) 11.97A

(*c*) 30A (*d*) 0A **[NTPC]**

92. A pnpn diode is a

(*a*) negative resistance device

(*b*) voltage controllable device

(*c*) controlled rectifier

(*d*) current controlled negative resistance device **[NTPC]**

93. The pnpn diode

(*a*) is unilateral device

(*b*) is a bilateral device

(*c*) may function either as a unilateral or as a bilateral device

(*d*) functions as a bilateral device depending on the ambient temp. **[NTPC]**

94. In SCR, the turn-ON time

(*a*) is independent of ambient temperature T

(*b*) increases with increase of T

(*c*) varies 1/T

(*d*) varies is $1/T^2$ **[BHEL]**

95. In a SCR,

(*a*) gate current is directly proportional to forward breakover voltage

(*b*) as gate current is raised, forward breakover voltage reduces

(*c*) gate current has to be kept on continuously for conduction

(*d*) forward breakover voltage is low in the forward blocking state **[BEL]**

96. A dc source of 100V supplies a purely inductive load 0.1H The controller is a thyristor in series with the source and load. The latching current specified is 100 mA. The minimum width of the gating pulse to ensure thyristor on is

(*a*) 100μs (*b*) 100s
(*c*) 1 ms (*d*) 50μs **[NTPC]**

97. In SCR, the turn-ON time

(*a*) is independent of anode current I_A
(*b*) increases with increase of I_A
(*c*) varires as $1/I_A$
(*d*) varires as $1/\sqrt{I_A}$ **[PGCIL]**

98. A thyristor is triggered by a pulse train of 5 kHz. The duty ratio is 0.4. If the allowable average power is 100W the maximum allowable gate drive power is

(*a*) $100\sqrt{2}$W (*b*) 50 W
(*c*) 150 W (*d*) 250 W **[BEL]**

99. In a pnpn diode, breakover condition is marked by

(*a*) a sudden decrease in current
(*b*) a sudden increase in current
(*c*) diode getting burnt off
(*d*) a sudden glow taking place **[BEL]**

100. A pnpn diode

(*a*) is always made of silicon
(*b*) is always made of germanium
(*c*) may be made of either silicon or germanium
(*d*) may be made of any semiconductor

101. A half-wave SCR controlled circuit with $R_L = 50\Omega$ conducts for 90° for an applied voltage of 300V sinusoidal rms. If the SCR voltage drop is negligible, the power dissipated by the load is

(*a*) 1800 W (*b*) 91 W
(*c*) 52.36 W (*d*) 0 W **[NTPC]**

102. After firing an SCR, if the gate pulse is removed, the SCR current

(*a*) remains the same
(*b*) reduces to zero
(*c*) rises up
(*d*) rises a little and then falls to zero **[BHEL]**

103. An SCS is a

(*a*) pnpn diode with three terminals
(*b*) pnpn diode with one gate
(*c*) pnpn diode with two gates
(*d*) pnpn diode made of germanium **[BHEL]**

104. A gate-turn-off (GTO) thyristor

(*a*) requires a special turn-off circuit like a thyristor
(*b*) can be turned off by removing the gate pulse
(*c*) can be turned off by a negative current pulse at the gate
(*d*) can be turned off by a positive current pulse at the gate **[IAS]**

105. In a thyristor dc chopper, which type of commutation results in best performance?

(*a*) voltage commutation
(*b*) current commutation
(*c*) load commutation
(*d*) supply commutation **[GATE]**

106. In a 3-phase half-wave diode rectifier, the ratio of average output voltage to per-phase maximum ac voltage is

(*a*) 0.955
(*b*) 0.827
(*c*) 1.654
(*d*) 1.169 **[IAS]**

107. Severe over-voltages are produced during arcing faults in a power system with the neutral

(*a*) isolated
(*b*) solidly earthed
(*c*) earthed through a low resistance
(*d*) earthed through an inductive coil **[IAS]**

108. In dc choppers, per unit ripple is maximum when then duty cycle is

(*a*) 0.2 (*b*) 0.5
(*c*) 0.7 (*d*) 0.9 **[ES]**

109. The $\frac{dv}{dt}$ protection of an SCR is achieved by using

(*a*) RL circuit in series with SCR
(*b*) RC circuit across SCR
(*c*) L in series with SCR
(*d*) RC circuit in series with SCR **[IAS]**

110. Natural commutation of a thyristor takes place when

(*a*) anode current becomes zero
(*b*) gate current becomes zero
(*c*) voltage across the device becomes zero
(*d*) voltage across the device becomes negative **[BHEL]**

Directions (Q. 111-118) : *The following items consists of two statements one labelled the "Assertion A" and the other labelled the "Reason R". You are to examine these two statements, and if so, whether the Reason is a correct explanation of the Assertion. Select your answers to these items using the codes given below and mark your answer sheet accordingly.*

Codes:

(*a*) Both A and R are true and R is the correct explanation of A

(*b*) Both A and R are true but R is not a correct explanation of A

(*c*) A is true but R is false

(*d*) A is false but R is true

111. Assertion (A) : The static magnetic field has zero divergence and non-zero curl.

Reason (R) : The static magnetic field satisfies the flux low. **[ES]**

112. Assertion (A) : The inductive voltage drop caused at the terminals of an *ac* to *dc* converter at given value of local current does not depend upon the firing angle.

Reason (R) : The overlap of conduction of incoming and outgoing thyristors is responsible for the inductive voltage drop. **[IAS]**

113. Assertion (A) : It is strongly recom-mended that the GTO (Gate Turn-Off thyristor) should not be turned 'on' until it has been 'off' for a specified time.

Reason (R) : There may be poor current sharing between the various cathode islands **[ES]**

114. Assertion (A) : Variable frequency square wave inverter providing variable voltage cannot be operated below 10% of base frequency.

Reason (R) : The thyristors get damaged due to high $\frac{di}{dt}$. **[ES]**

115. Assertion (A) : Discontinuous conduction of a power converter is characterised by larger value and average load voltage at a given firing angle compared to the case of continuous conduction.

Reason (R) : The ripple content of the voltage waveform a is larger in the discontinuous conduction mode of the converter. **[ES]**

116. Assertion (A) : A phase controlled line commutated converter can be operated both as rectifier and inverter.

Reason (R) : S.C.R.'s have the property of blocking positive as well as negative voltages and they can be made conducting whenever required during the positive half cycle. **[IAS]**

117. Assertion (A) : Thyristorised current source inverters do no operate at high switching frequency.

Reason (R) : Thrystorised current source inverter is capable of four-quadrant operation **[IAS]**

118. Assertion (A): In *dc* to *dc* step down chopper, free-wheeling diode is mandatory for RL loads.

Reason (R) : A part of energy fed to the RL load is dissipated in the resistance. **[IAS]**

ANSWERS

EXERCISE - I

1. (*a*)	**2.** (*d*)	**3.** (*d*)	**4.** (*c*)	**5.** (*d*)	**6.** (*a*)	**7.** (*a*)	**8.** (*d*)	**9.** (*b*)	**10.** (*b*)
11. (*d*)	**12.** (*a*)	**13.** (*b*)	**14.** (*d*)	**15.** (*b*)	**16.** (*d*)	**17.** (*c*)	**18.** (*d*)	**19.** (*a*)	**20.** (*d*)
21. (*b*)	**22.** (*d*)	**23.** (*c*)	**24.** (*a*)	**25.** (*c*)	**26.** (*a*)	**27.** (*d*)	**28.** (*d*)	**29.** (*d*)	**30.** (*d*)
31. (*d*)	**32.** (*d*)	**33.** (*c*)	**34.** (*d*)	**35.** (*b*)	**36.** (*c*)	**37.** (*a*)	**38.** (*b*)	**39.** (*d*)	**40.** (*a*)
41. (*c*)	**42.** (*b*)	**43.** (*b*)	**44.** (*c*)	**45.** (*d*)	**46.** (*d*)	**47.** (*b*)	**48.** (*a*)	**49.** (*c*)	**50.** (*c*)
51. (*b*)	**52.** (*b*)	**53.** (*b*)	**54.** (*d*)	**55.** (*b*)	**56.** (*b*)	**57.** (*b*)	**58.** (*a*)	**59.** (*c*)	**60.** (*a*)
61. (*b*)	**62.** (*c*)	**63.** (*a*)	**64.** (*b*)	**65.** (*c*)	**66.** (*b*)	**67.** (*d*)	**68.** (*d*)	**69.** (*a*)	**70.** (*b*)
71. (*d*)	**72.** (*c*)	**73.** (*c*)	**74.** (*a*)	**75.** (*b*)	**76.** (*b*)	**77.** (*d*)	**78.** (*b*)	**79.** (*d*)	**80.** (*c*)
81. (*b*)	**82.** (*a*)	**83.** (*b*)	**84.** (*c*)	**85.** (*d*)	**86.** (*c*)	**87.** (*d*)	**88.** (*d*)	**89.** (*b*)	**90.** (*c*)
91. (*a*)	**92.** (*b*)	**93.** (*d*)	**94.** (*d*)						

EXERCISE - II

1. (b)	2. (b)	3. (b)	4. (c)	5. (b)	6. (d)	7. (d)	8. (d)	9. (b)	10. (d)
11. (a)	12. (a)	13. (a, b)	14. (d)	15. (a)	16. (d)	17. (c)	18. (c)	19. (d)	20. (c)
21. (c)	22. (c)	23. (b)	24. (a)	25. (b)	26. (d)	27. (d)	28. (b)	29. (a)	30. (c)
31. (b)	32. (c)	33. (d)	34. (d)	35. (c)	36. (c)	37. (c)	38. (d)	39. (d)	40. (c)
41. (c)	42. (a)	43. (b)	44. (a)	45. (a)	46. (b)	47. (c)	48. (a)	49. (d)	50. (d)
51. (b)	52. (b)	53. (c)	54. (b)	55. (a)	56. (a)	57. (d)	58. (c)	59. (a)	60. (d)
61. (d)	62. (d)	63. (b)	64. (a)	65. (d)	66. (b)	67. (a)	68. (d)	69. (c)	70. (b)
71. (d)	72. (b)	73. (c)	74. (b)	75. (a)	76. (b)	77. (a)	78. (c)	79. (c)	80. (d)
81. (b)	82. (c)	83. (a)	84. (c)	85. (d)	86. (a)	87. (b)	88. (a)	89. (d)	90. (c)
91. (a)	92. (d)	93. (a)	94. (b)	95. (b)	96. (a)	97. (b)	98. (d)	99. (b)	100. (a)
101. (b)	102. (a)	103. (c)	104. (c)	105. (a)	106. (b)	107. (d)	108. (b)	109. (b)	110. (d)
111. (b)	112. (c)	113. (a)	114. (b)	115. (a)	116. (c)	117. (b)	118. (b)		

EXPLANATIONS

EXERCISE - I

1. The power delivered to the load for the same firing angle is more when F_0 is used.
As volt-ampere input is almost same, hence input power factor given by

$$\text{input } pf = \frac{\text{Power delivered to load}}{\text{Input volt ampere}}$$

Also using free wheeling, the load current waveform is improved.

2. In circulating current dual-converter, firing angle of the two converters are so adjusted that

$$\alpha_1 + \alpha_2 = 180°$$

∴ Triggering angle of IInd converter

$$= 180 - 30 = 150°$$

5. When ac voltage controller is used for speed control of a single phase induction motor, only fundamental component is useful in producing the torque. Harmonics in the motor current merely increases the losses and therefore heating of the induction motor.

14.
$$V_o = \frac{2V_M}{\pi}\cos\alpha$$

For $\alpha = 0$,

$$V_o = \frac{2\times 1000}{3.14}\times\cos 0 = 636.9\text{ V}$$

And frequency of ripples will be 100 Hz

21. To have controlled voltage,
firing angle $(\alpha) > \phi$ (load phase angle)
and conduction angle = 180°

$$\phi = \tan^{-1}\frac{\omega L}{R}$$

$$= \tan^{-1}\frac{2\times 3.14\times 50\times 6.36}{2}$$

$$= 90°$$

Hence α should be greater than 90° and should be less than 180°

22. Width of free wheeling diode pulse = π

24. For limit of continuous conduction,

$$\alpha^1 = \frac{T_{oN}}{T}$$

$$= \frac{T_a}{T}\ \text{In}\left[1 + m(e^{T/Ta} - 1)\right]$$

$$T_a = \frac{L}{R} = \frac{6.36\times 10^{-3}}{2}$$

$$= 3.18\times 10^{-3}$$

$$T = 2500\times 10^{-6} = 2.5\times 10^{-3}$$

$$m = \frac{E_L}{E} = \frac{22}{220}0.091$$

$$\alpha^1 = \frac{318\times 10^{-3}}{2.5\times 10^{-3}}$$

$$\ln\left[1+0.091\left(e^{\frac{2.5}{3.18}}-1\right)\right]$$

$$=0.1313$$

$$\therefore \quad T_{on}=0.1313\times 2500$$

$$=328.25\ \mu\text{sec.}$$

26. For controlling load, the minimum value of firing angle, α = load phase angle ϕ; and maximum possible value of a is 180°.

30. In case zero of the triangular wave coincides with the zero the reference sinusoid, there are (N – 1) pulses per her half cycle,

where, $N=\frac{f_c}{2f}$

f_c = frequency of carrier or triangular wave, and

f = frequency of reference wave,

31. Total thermal resistance between junction and ambient

$$\theta_{jA}=\theta_{jc}+\theta_{cs}+\theta_{sA}$$

$$=0.01+0.08+0.09=0.18$$

$$T_j=T_A=P_{av}\times 0.18$$

$$T_j-35=150\times 0.18$$

$$T_j=62$$

$$V_d=\frac{V_0}{2}\left[\cos\alpha+\cos(\alpha+\gamma)\right]$$

where, V_o = rectifier output voltage

when $\alpha=0$

and $\gamma=0$

α = delay angle

γ = commutation angle

Assuming α to be zero,

$$V_d=\frac{V_0}{2}[1+\cos\gamma]$$

Due to source inductance effect,

$$\text{change in voltage}=V_o-\frac{V_0}{2}-\frac{V_0}{2}\cos\gamma$$

$$=\frac{V_0}{2}(1-\cos\gamma)$$

$$\therefore \text{Fractional regulation}=\frac{\frac{V_0}{2}(1-\cos\gamma)}{V_o}$$

$$=\frac{1-\cos\gamma}{2}$$

38. For 3 phase to 3 pulse converter,

$$\alpha=0$$

$$\therefore \quad V_o=\frac{3V_{ml}}{\pi}=\frac{3\sqrt{3}V_m}{\pi}$$

EXERCISE - II

4. Minimum firing angle,

$$\alpha=\sin^{-1}\frac{E_b}{V_{max}}=\sin^{-1}\frac{40}{\sqrt{2}V}$$

Maximum value of firing angle

$$=\pi-\alpha=\pi-\sin^{-1}\frac{40}{\sqrt{2}V}$$

16.

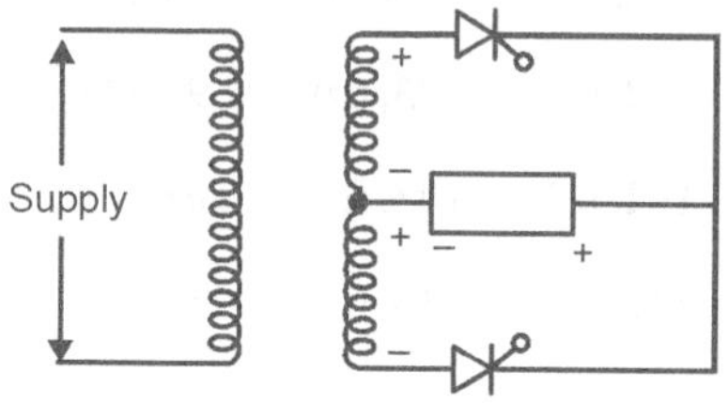

Pick inverse voltage or each Thyristor

$$=2\times V_m$$

$$=2\times 100\times\sqrt{2}$$

$$=282.8\text{V}$$

25. $V_o=\alpha V_{in}=\frac{4}{4+1}\times 60$

$$=\frac{4}{5}\times 60=48\text{ V}$$

40. For single-pulse modulation of PWM inverter

$$V_o=\sum_{n=1,3,5}^{\infty}\frac{4V_s}{n\pi}\sin\frac{n\pi}{2}\sin nd\sin n\omega t$$

Hence to eliminate 3rd harmonics, sin nd = $3d$ = 180

$$\Rightarrow \quad d=60°$$

$$\therefore \quad \text{Pulse width}=2d=120°$$

53. $V_o=\frac{1}{3}\times 100=33.33\text{V}$

$$\text{Rippled factor}=\sqrt{\frac{1}{\alpha}-1}=\sqrt{\frac{1}{1/3}-1}$$

$$=\sqrt{3-1}=\sqrt{2}$$

6

CHAPTER

Electromagnetic Theory

EXERCISE – I

1. The potential difference between the faces A and B of a uniformly polarized infinite slab shown in the figure is :

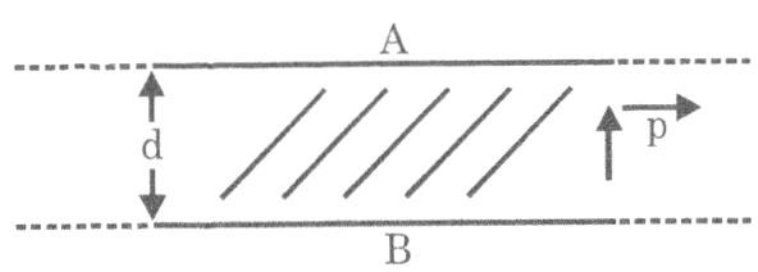

(*a*) $\dfrac{Pd}{\epsilon_0(\epsilon_R - 1)}$ (*b*) $\dfrac{Pd}{\epsilon_0 \epsilon}$

(*c*) $\dfrac{Pd}{\epsilon_0}$ (*d*) $\dfrac{Pd(\epsilon + 1)}{\epsilon_0}$

2. Match List-I with List-II and select the correct answer using the codes given below the lists: (Symbols have the usual meanings)

List-I	List-II
A. Poisson' equation	1. $\nabla^2 \phi = 0$
B. Laplace's equation	2. $\nabla^2 E + K_0^2 E = 0$ where $K_0 = \omega\sqrt{\mu_0 \epsilon_0}$
C. Joule's equation	3. $\nabla^2 \phi = \dfrac{-P}{\varepsilon_0}$
D. Helmholtz's equation	4. $\dfrac{dP}{dV} = U_j \overline{E}.\overline{J}$

Codes:

	A	B	C	D
(*a*)	2	1	4	3
(*b*)	3	4	1	2
(*c*)	3	1	4	2
(*d*)	2	4	1	3

3. Which of the following equations as derived from Ampere's, Faraday's and Gauss's laws are valid for free space?

1. $\overline{\nabla} \times \overline{H} = \dfrac{\partial \overline{D}}{\partial t}$ 2. $\overline{\nabla} \times \overline{H} = \overline{J} + \dfrac{\partial \overline{D}}{\partial t}$

3. $\overline{\nabla} \times \overline{E} = -\dfrac{\partial \overline{B}}{\partial t}$ 4. $\overline{\nabla} \times \overline{D} = 0$

5. $\overline{\nabla} \times \overline{D} = P$ 6. $\overline{\nabla} \times \overline{D} = O$

Select the correct answer from the codes given below:

Codes:

(*a*) 1, 4 and 5 (*b*) 2, 4 and 6

(*c*) 2, 3 and 5 (*d*) 1, 3 and 6

4. A rectangular loop in the *x-z* plane bounded by the lines $x = 0, x = a, z = 0$ and $z = b$, is in a time-varying magnetic field given by

$$\overline{B} = B_0 \cos \omega \overline{a}y$$

where B_0 = constant

ω = angular frequency

$\overline{a}y$ = unit vector in the y-direction

The emf induced in the loop is given by

(*a*) $ab\,B_0 \cos \omega t$ (*b*) $ab\,B_0\,\omega \sin \omega t$

(*c*) $B_0\,\omega \sin \omega t$ (*d*) zero

5. If a charge of one coulomb is to be brought to a distance of one meter from a charge of 2 coulombs, the work required is

(*a*) $\dfrac{-1}{2\pi\epsilon_0}$ N-m

(*b*) $\dfrac{1}{4\pi\epsilon_0}$ N-m

(*c*) $2\pi\epsilon_0$ N-m

(*d*) $4\pi\epsilon_0$ N-m

6. Two intersecting infinite straight wires (shown in the given figure) carry equal, uniform but opposite linear charge densities. The electric field on the bisector of the angle of intersection is

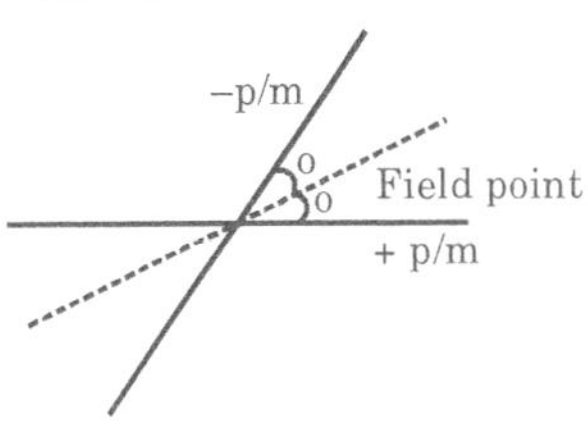

(*a*) constant

(*b*) directed towards the point of intersection

(*c*) directed away from the point of intersection

(*d*) normal to the bisector

7. The electric field $\vec{E}$ (in volts/metre) at the point (1, 1, 0) due to a point charge of + 1μC located at (–1, 1, 1) (coordinates in metres) is

(a) $\frac{10^{-6}}{20\sqrt{5}\pi\varepsilon_0}(2i-k)$

(b) $\frac{10^{-6}}{20\pi\varepsilon_0}(2i-k)$

(c) $\frac{-10^{-6}}{20\sqrt{5}\pi\varepsilon_0}(2i-k)$

(d) $\frac{-10^{-6}}{20\pi\varepsilon_0}(2i-k)$

8. Following equations hold for time-varying fields

A. $\nabla\times\vec{E} = -\frac{\partial\vec{B}}{\partial t}$

B. $E = -\nabla V \frac{\partial\vec{A}}{\partial t}$

C. $\nabla^2 V + -\frac{\partial}{\partial t}(\nabla.\vec{A}) = -\frac{\rho v}{\epsilon}$

D. $\vec{B} = \nabla\times\vec{A}$

E. $\nabla\times\vec{H} = \vec{J} + \epsilon\frac{\partial\vec{E}}{\partial t}$

In the above equations

(a) both F and $\vec{A}$ are completely defined and thus can be evaluated

(b) V is completely defined but not $\vec{A}$

(c) $\vec{A}$ is completely defined but not V

(d) both $\vec{A}$ and V are not completely defined

9. The expression for $\overline{B}$, given that in the free space $\overline{E}$ = 15 cos $(6\pi\times10^8 t - 2\pi z)i_x$ V/m, is

(a) $5\times10^{-8}\cos(6\pi\times10^8 t-2\pi z)i_y$

(b) $-90\pi\times10^8\sin(6\pi\times10^8 t-2\pi z)i_x$

(c) $-45\times10^8\sin(6\pi\times10^8 t-2\pi z)i_y$

(d) $5\times10^{-8}[\cos(6\pi\times10^8 t-2\pi z)i_y - \sin(6\pi\times10^8 t-2\pi z)i_z]$

10. Two electrons moving parallel to each other of the same velocity will

(a) attract each other if they move in the same direction

(b) repel each other more strongly when moving in same direction than when moving in opposite direction

(c) repel each other less strongly when moving in same direction than when moving in opposite direction

(d) repel each other with the same force wnen moving in the same or opposite direction.

11. The region $z < 0$ has $\mu_r = 6$ and the region $z > 0$ has $\mu_r = 4$. If the magnetic flux density flux density in region $z > 0$ is $5\hat{a}_x + 8\hat{a}_z$ m Wb/m², the magnetic field intensity in region $z < 0$ would be

(a) $\frac{5\hat{a}_x+8\hat{a}_z}{4\mu_0}$ mA/m (b) $\frac{5\hat{a}_x+8\hat{a}_z}{6\mu_0}$ mA/m

(c) $\frac{5\hat{a}_x}{4\mu_0}+\frac{8\hat{a}_z}{6\mu_0}$ mA/m (d) $\frac{5\hat{a}_x}{6\mu_0}+\frac{8\hat{a}_z}{4\mu_0}$ mA/m

12. Consider the following statements regarding Maxwell's equations in differential form (symbols have the usual meanings)

1. For free space $\nabla\times\vec{H} = (\sigma + j\omega\epsilon)\vec{E}$

2. For free space $\nabla\times\vec{D} = \rho$

3. For steady current $\nabla\times\vec{H} = J$

4. For static electric field $\nabla.\vec{D} = \rho$

Of these statements:

(a) 1 and 2 are correct

(b) 2 and 3 are correct

(c) 3 and 4 are correct

(d) 1 and 4 are correct

13. An infinite plane sheet of uniform surface charge density σC/m² is lying in the x-y plane. The electric field $\overline{E}$ at a point P (0, 0, z) above the sheet (z > 0) would be (ε_o is the free space permittivity)

(a) $\overline{E} = \frac{\sigma}{2\varepsilon_0}\hat{a}_z$ (b) $\overline{E} = \frac{\sigma}{\varepsilon_0}\hat{a}_z$

(c) $\overline{E} = \frac{\hat{a}_z\sigma}{4\pi\varepsilon_0 z^2}$ (d) $\overline{E} = \frac{\sigma^2}{4\pi\varepsilon_0 z^2}\hat{a}_z$

14. "In all cases of electromagnetic induction, an induced voltage will cause a current to flow in a closed circuit in such a direction that the magnetic field which is caused by that current will oppose the change that produces the current" is the original statement of

(a) Lenz's law

(b) Faraday's law of magnetic induction

(c) Fleming's law of induction

(d) Ampere's law

15. Which law is synonymous to the occurrence of diamagnetism?

(a) Ampere's law (b) Maxwell's law

(c) Coulomb's law (d) Lenze's law

16. In a uniform electric field, field lines and equipotential
(*a*) are parallel to one another
(*b*) intersect at 45°
(*c*) intersect at 30°
(*d*) are orthogonal

17. The electric field lines and equipotential lines
(*a*) are parallel to each other
(*b*) are one and the same
(*c*) cut each other orthogonally
(*d*) can be inclined to each other at any angle

18. If the 36° arc AB of a circle shown in the given figure is uniformly charged leaving the arc BCA uncharged, the electric field at the centre 'O' is $\overline{E}$. If the arc BCA is given the same linear charge density and the arc AB left uncharged, then the electric field at the centre 'O' will be

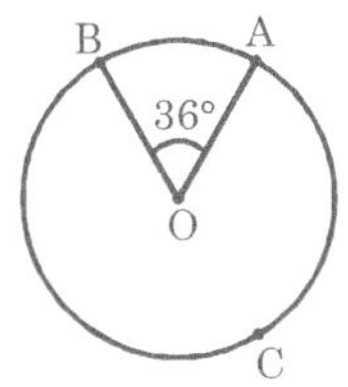

(*a*) $\overline{E}$
(*b*) $-\overline{E}$
(*c*) $9\overline{E}$
(*d*) $-9\overline{E}$

19. When a soap bubble is charged,
(*a*) it contracts
(*b*) it expands
(*c*) it does not undergo any change in size
(*d*) none of these

20. If $\overline{B} = B_0 z \cos \omega t\, \hat{a}_y$ and $\vec{E} = E_x \vec{a}_x$, then
(*a*) $E_x = 0$
(*b*) $E_x = +B_0 z\omega \sin \omega t$
(*c*) $E_t = +B_0 \cos \omega t$
(*d*) $E_x = \frac{1}{2} B_0 z^2 \omega \sin \omega t$

21. Consider the following statements associate with the basic electronic properties of ideal conductors:
1. The resultant field inside is zero.
2. The net charge density in the interior is zero.
3. Any net charge resides on the surface.
4. The surface is always equipotential.
5. The field just outside is zero.

Of these statements :
(*a*) 1, 2, 3 and 4 are correct
(*b*) 3, 4, and 5 are correct
(*c*) 1, 2 and 3 are correct
(*d*) 2 and 3 are correct

22. Two vectors $\vec{A}$ and $\vec{B}$ are such that $\vec{A}+\vec{B} = \overrightarrow{nA}$ where n is a positive scalar. The angle between $\vec{A}$ and $\vec{B}$ is
(*a*) $\pi\sqrt{2}$ (*b*) $\frac{3\pi}{4}$
(*c*) π (*d*) 2π

23. Consider the following statements regarding field boundary conditions:
1. The tangential component of electric field is continuous across the boundary between two dielectrics.
2. The tangential component of electric field at a dielectric conductor boundary is non-zero.
3. The discontinuity in the normal component of the flux density at a dielectric-conductor boundary is equal to the surface charge density on the conductor.
4. The normal component of the flux-density is continuous across the charge-free boundary between two dielectrics.

Of these statements:
(*a*) 1, 2 and 3 are correct
(*b*) 2, 3 and 4 are correct
(*c*) 1, 2 and 4 are correct
(*d*) 1, 3 and 4 are correct

24. In the given figure, the force acting on the conductor PQ is in the direction of

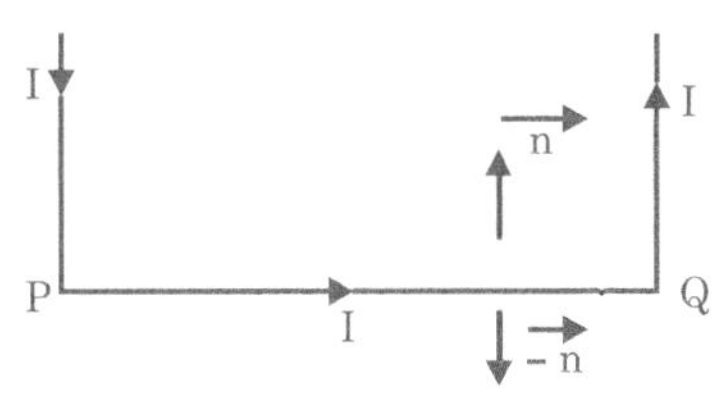

(*a*) PQ (*b*) QP
(*c*) $-\vec{n}$ (*d*) $\vec{n}$

25. The ratio of velocity of propagation of em waves in an overhead transmission line and in a cable with a dielectric of permittivity 4, is
(*a*) 0.25 (*b*) 0.5
(*c*) 2.0 (*d*) 4.0

26. If an em wave travelling in a dissipative medium in the *x*-direction has peak electric field E_1, E_2, E_3 at points x, $x + 1$, $x = 2$ respectively, then
(*a*) $E_2 = \sqrt{E_1 E_3}$ (*b*) $E_1 = \sqrt{E_2 E_3}$
(*c*) $E_3 = \sqrt{E_1 E_2}$ (*d*) $E_2 = \frac{E_1 + E_3}{2}$

27. The relation between the wave impedance (Z_ω) and the intrinsic impedance (Z_i) for a TEM wave is given by

(*a*) $Z_\omega = 0.5\,Z_i$ (*b*) $Z_\omega = 1/Z_i$

(*c*) $Z_\omega = 2\,Z_i$ (*d*) $Z_\omega = Z_i$

28. A circular current-carrying loop and a field point are shown in the above figure. The vector magnetic potential at P is in the direction of

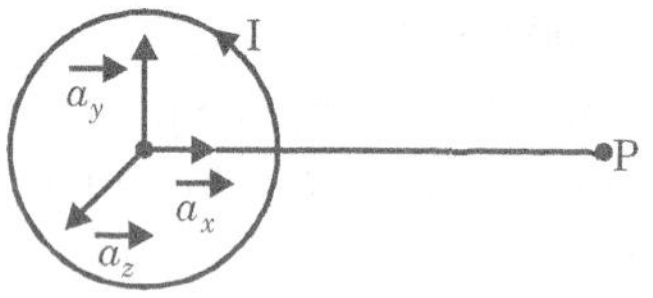

(*a*) $\vec{a}_x$ (*b*) $\vec{a}_y$

(*c*) $\vec{a}_z$ (*d*) $\dfrac{(\vec{a}_x + \vec{a}_y)}{\sqrt{2}}$

29. A current I flows along a circular coil placed in the *x-y* plane with the centre at the origin. The line integral $\int \vec{B}, \vec{dz}$ from minus infinity to plus infinity is

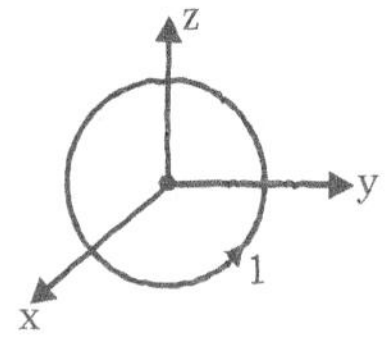

(*a*) zero

(*b*) $\dfrac{\mu I}{2}$

(*c*) μI

(*d*) $2\mu I$

30. The electric field lines and equipotential lines

(*a*) are parallel to each other

(*b*) are one and the same

(*c*) cut each other orthogonally

(*d*) can be inclined to each other at any angle

31. A circular ring carrying a uniformly distributed charge Q and a point charges–Q on the axis of the ring are shown in the figure. The magnitude of the dipole moment of the charge system is

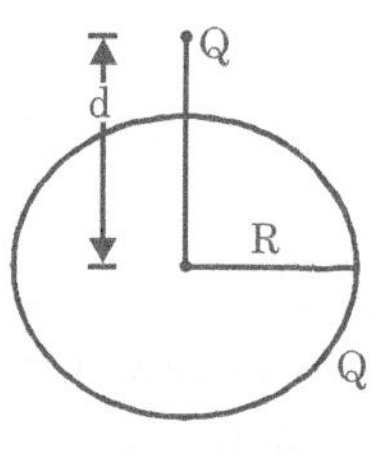

(*a*) Q*d* (*b*) $\dfrac{QR}{d^2}$

(*c*) $Q\sqrt{R^2 + d^2}$ (*d*) QR

32. The equation of a plane wave may be written as (with usual symbols)

(*a*) $\dfrac{\partial^2 E_y}{\partial t^2} = \dfrac{1}{\mu\varepsilon}\dfrac{\partial^2 E_y}{\partial x^2}$ (*b*) $\dfrac{\partial^2 E_y}{\partial x^2} = \dfrac{1}{\mu\varepsilon}\dfrac{\partial^2 E}{\partial t^2}$

(*c*) $\dfrac{\partial^2 E_y}{\partial t^2} = \dfrac{\mu}{\varepsilon}\dfrac{\partial^2 E}{\partial x^2}$ (*d*) $\dfrac{\partial^2 E_y}{\partial t^2} = \dfrac{\varepsilon}{\mu}\dfrac{\partial^2 E}{\partial x^2}$

33. The velocity of a travelling electromagnetic wave in free space is given by

(*a*) $\mu_0\varepsilon_0$

(*b*) $\sqrt{\mu_0\varepsilon_0}$

(*c*) $\dfrac{1}{\sqrt{\mu_0\varepsilon_0}}$

(*d*) $\dfrac{1}{\mu_0\varepsilon_0}$

34. Consider the two fields,

$E = 120\pi \cos(10^6\pi t - \beta x)a_y$ V/m and

$H = A \cos(10^6 \pi t - \beta x)a_z$ A/m

The values of A and β which will satisfy the Maxwell's equation in a linear isotropic, homogenous, loss-less medium with $\varepsilon_r = 8$ and $\mu_r = 2$ will be

	A (in A/m)	**P (in rad/m)**
(*a*)	1	0.0105
(*b*)	1	0.042
(*c*)	2	0.0105
(*d*)	2	0.042

35. Match List-I (Antenna) with List-II (Application/ Characteristics) and select the correct answer using the codes given below the lists:

List-I

A. A quarter wave monopole over ground plane

B. A vertical half-wave dipole

C. Folded dipole

D. Hertzian dipole (length < λ/100)

List-II

1. Radiation resistance << 1Ω
2. Radio broadcast
3. Omnidirectional pattern
4. TV reception

Codes:

	A	B	C	D
(*a*)	2	3	4	1
(*b*)	2	3	1	4
(*c*)	3	4	1	2
(*d*)	3	4	2	1

36. A point charge Q is located on the surface of a sphere of radius R as shown in the figure. The average electric field on the surface of the sphere will be

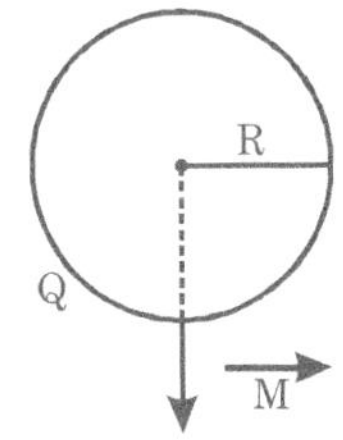

(*a*) infinite

(*b*) $\frac{Q}{4\pi\epsilon_o}\frac{1}{R^2}(-\vec{n})$

(*c*) $\frac{Q}{8\pi\epsilon_o}\frac{1}{R^2}(-\vec{n})$

(*d*) zero

37. A parallel plate capacitor with air as dielectric is charged and then disconnected from the supply. The force between the plates is F. If the capacitor is now immersed in a liquid dielectric of dielectric constant ε, then the force between the plates will be

(*a*) $\varepsilon^2 F$ (*b*) εF

(*c*) F (*d*) F/ε

38. Two point charges Q and –Q are located on two opposite corners of a square as shown in the figure. If the potential at the corner A is taken as 1V, then the potential at B, the centre of the square will be

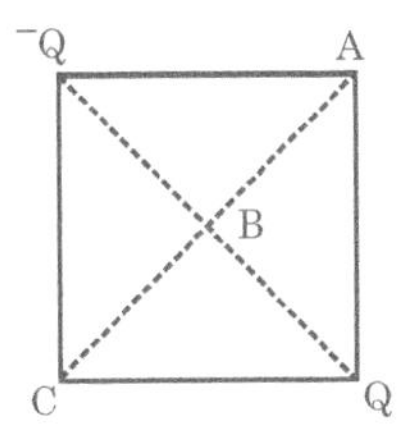

(*a*) zero (*b*) $1/\sqrt{2}$V

(*c*) 1V (*d*) $\sqrt{2}$V

39. The velocity of the plane wave $\sin^2(\omega t - \beta x)$ is

(*a*) $\frac{2\omega}{\beta}$ (*b*) $\frac{\omega}{2\beta}$

(*c*) $\frac{\omega^2}{\beta^2}$ (*d*) $\frac{\omega}{\beta}$

40. A filamentary current of 10π A flows in the negative *z*-direction. The magnetic field at (0, 5, 0)is

(*a*) $\hat{a}_x$ A/m

(*b*) $\hat{a}_y$ A/m

(*c*) $\hat{a}_z$ A/m

(*d*) $(\hat{a}_x + \hat{a}_y)$A/m

41. A straight wire of circular cross-section carries a direct current I (see given figure). If R is the resistance per unit length of the wire, then the Poynting vector at the surface of the wire will be

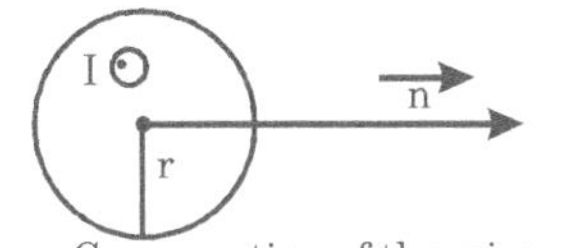

Cross section of the wire

(*a*) $\frac{RI^2}{2\pi r}\vec{n}$ (*b*) $\frac{RI^2}{2\pi r}(-\vec{n})$

(*c*) $\frac{RI^2}{2\pi}\vec{n}$ (*d*) $\frac{RI^2}{2\pi}(-\vec{n})$

42. Two electric dipoles aligned parallel to each other and having the same axis exert a force F on each other, when a distance *d* apart. If the dipoles are a distance 2*d* apart, then the mutual force between them would be

(*a*) F/2 (*b*) F/4

(*c*) F/8 (*d*) F/16

43. If the vectors $\vec{A}$ and $\vec{B}$ are conservative, then

(*a*) $\vec{A}\times\vec{B}$ is solenoidal

(*b*) $\vec{A}\times\vec{B}$ is conservative

(*c*) $\vec{A}+\vec{B}$ is solenoidal

(*d*) $\vec{A}-\vec{B}$ is solenoidal

44. Conservation of charge implies that

(*a*) $\oiint \vec{J}.\vec{ds} = -\frac{d}{dt}\iiint \rho\, dv$

(*b*) $\oiint \vec{J}.\vec{ds} = 0$

(*c*) $\oiint \vec{J}.\vec{ds} = \iiint \rho\, dv$

(*d*) $\oiint \vec{J}.\vec{ds}\ 0$

45. EE 1.5 given a vector field $\vec{F}$,the divergence theorem states that

(*a*) $\oint_s \vec{F}.d\vec{S} = \int_v \vec{\nabla}.\vec{F}.dV$

(*b*) $\oint_s \vec{F}.d\vec{S} = \int_v \vec{\nabla}\times\vec{F}.dV$

(*c*) $\oint_s \vec{F}\times d\vec{S} = \int_v \vec{\nabla}.\vec{F}.dV$

(*d*) $\oint_s \vec{F}\times d\vec{S} = \int_v \vec{\nabla}\times\vec{F}.dV$

46. Given that

$\vec{E}$ = time and space dependent electric intensity vector

μ = permeability of the medium

σ = conductivity of the medium

$\in$ = permittivity of the medium

The electromagnetic wave equation is given by

(*a*) $\nabla^2 \vec{E} - \mu\sigma \frac{\partial \vec{E}}{\partial t} - \propto \in \frac{\partial \vec{E}}{\partial t} = 0$

(*b*) $\nabla^2 \vec{E} - \mu \in \frac{\partial \vec{E}}{\partial t} - \propto \sigma \frac{\partial \vec{E}}{\partial t} = 0$

(*c*) $\nabla^2 \vec{E} - \sigma \in \frac{\partial \vec{E}}{\partial t} - \propto \sigma \frac{\partial \vec{E}}{\partial t} = 0$

(*d*) $\nabla^2 \vec{E} - \frac{\sigma}{\mu}\frac{\partial \vec{E}}{\partial t} - \frac{\mu}{\in}\frac{\partial \vec{E}}{\partial t} = 0$

47. A uniform plane electromagnetic wave with electric component $E = E_m \cos(\omega t - \beta x)$ propagates in vacuum along the positive *x*-direction. The mean Poynting Vector is given by

(*a*) $\frac{1}{2}\frac{\beta \in_0 c^2 E_m}{\omega}$ (*b*) $\frac{1}{2}\frac{\beta \mu_0 c^2 E_m^2}{\omega}$

(*c*) $\frac{1}{2}\frac{\beta \mu_0 c^2 E_m}{\omega}$ (*d*) $\frac{1}{2}\frac{\beta \in_0 c^2 E_m}{\omega}$

48. A positive charge Q is uniformly distributed over a spherical surface of radius '*a*'. Which of the following figures correctly illustrate the variation of electric field intensity E and potential V with the radial distance '*r*'?

1. 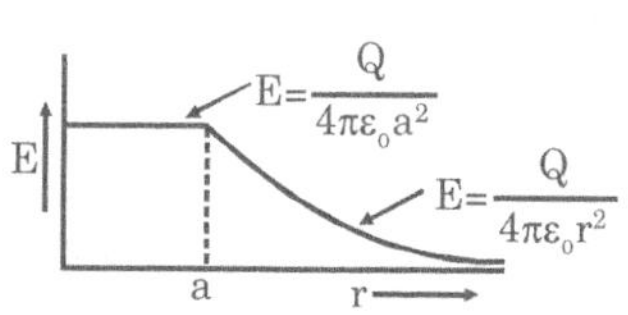

2. 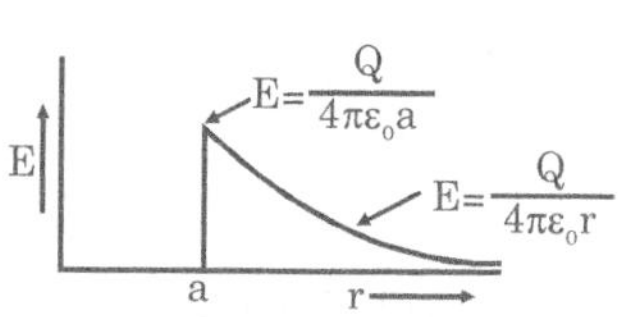

3. 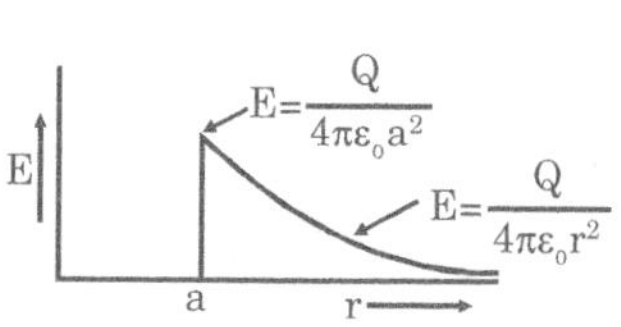

4. 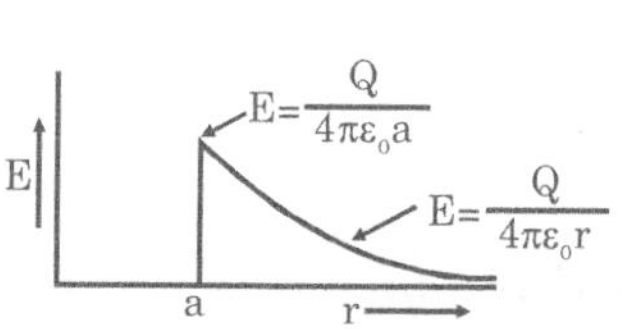

Select the correct answer using the codes given below:

Codes:

(*a*) 1 and 4

(*b*) 2 and 3

(*c*) 2 and 4

(*d*) 1 and 3

49. A slab of uniform magnetic field deflects a moving charged particle by 45° as shown in the figure. The kinetic energy of the charged particle at the entry and exit points in the magnetic field will change in the ratio of

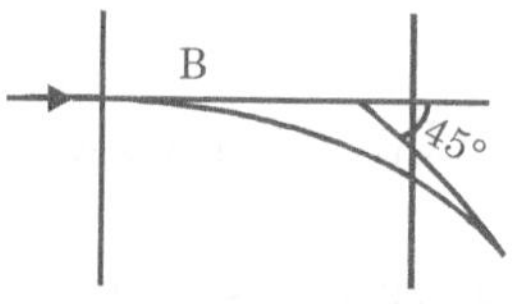

(*a*) $1:\sqrt{2}$

(*b*) $\sqrt{2}:1$

(*c*) 1 : 1

(*d*) 1 : 2

50. The region $z \le 0$ is a perfect conductor. On its surface at the origin, the surface current density $\vec{K}$ is $(5\vec{a}_x - 6\vec{a}_y)$ A/m. If the region $z > 0$ were free space, then the magnetic field intensity $\vec{H}$ in A/m, at the origin would be

(*a*) $\vec{H} = 0$

(*b*) $\vec{H} = 5\vec{a}_x - 6\vec{a}_y$

(*c*) $\vec{H} = 6\vec{a}_x - 5\vec{a}_y$

(*d*) $\vec{H} = -(6\vec{a}_x + 5\vec{a}_y)$

51. Three equal positive charges are located at the corners A , B and D of square as shown in the given figure. The electric field at C due to the charge at B alone is 800 ∠0°V/m. The result field at C due to a combination of all the other three charges is

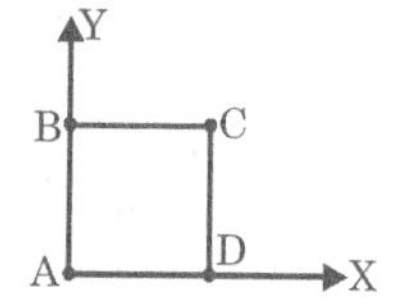

(*a*) $800\sqrt{2}\angle 45°$

(*b*) $400(2\sqrt{2}+1)\angle 45°$

(*c*) $\frac{400}{\sqrt{2}}(2\sqrt{2}+1)\angle 45°$

(*d*) $\frac{400}{\sqrt{2}}(2\sqrt{2}-1)\angle 90°$

52. Match List-I with List-II and select the correct answer using the codes given below the lists:

List-I (*Medium*)	**List-II** (*Expression for intrinsic impedance for plane wave propagation*)
A. Loss-less dielectric	1. $\sqrt{\dfrac{j\omega\mu}{\sigma + j\omega\varepsilon}}$
B. Good conductor	2. $\sqrt{\dfrac{\mu}{\varepsilon}}\left(1 + j\dfrac{\sigma}{2\omega\varepsilon}\right)$
C. Poor conductor	3. $\sqrt{\dfrac{\mu}{\varepsilon}}$
D. Lossy	4. $(1+j)\sqrt{\dfrac{\omega\mu}{2\sigma}}$

Codes:

	A	B	C	D
(*a*)	4	3	2	1
(*b*)	4	3	1	2
(*c*)	3	4	2	1
(*d*)	3	4	1	2

53. Match List-I with List-II and select the correct answer using the codes given below the lists:

List-I	**List-II**
A. $\oint_s \left(\vec{J} + \dfrac{\partial \vec{D}}{\partial t}\right) \hat{n}\, ds$	1. Zero
B. $-\oint_s \dfrac{\partial \vec{B}}{\partial t} . \hat{n}\, ds$	2. $\oint_v \rho\, dv$
C. $-\oint_s \vec{D}.\hat{n}\, ds$	3. $\oint_c \overline{C}.\overline{dl}$
D. $-\oint_s \vec{B}.\hat{n}\, ds$	4. $\oint_c \overline{H}.\overline{dl}$
	5. $\oint_v \overline{B}\, dv$

Codes:

	A	B	C	D
(*a*)	4	3	2	1
(*b*)	3	4	2	5
(*c*)	2	5	4	1
(*d*)	4	2	3	1

54. A small permanent magnet makes small oscillations in a uniform magnetic field as shown in the given figure. The period of oscillations is T. If the strength of the magnetic filed is doubled, the period of oscillations will be

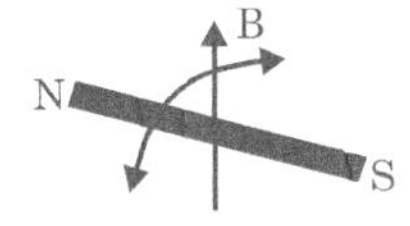

(*a*) 2T

(*b*) $T\sqrt{2}$

(*c*) $T/\sqrt{2}$

(*d*) T/2

55. A circular current carrying loop is shown in the given figure. The vector potential at the point P on the axis of the loop will be

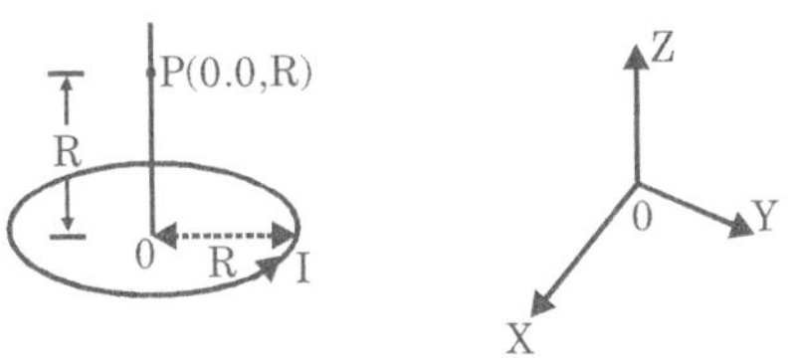

(*a*) zero

(*b*) $\mu_0 \dfrac{(2\pi R)I}{R} \vec{a}_z$

(*c*) $\dfrac{\mu_0 (2\pi R) I}{R\sqrt{2}} \vec{a}_z$

(*d*) $\dfrac{\mu_0 (\pi R)^2 I}{R\sqrt{2}} \vec{a}_z$

56. Tangential component of the electric field on a perfect conductor will be

(*a*) infinite

(*b*) zero

(*c*) same as the normals field component and 90° out of phase

(*d*) same as the normal component but 180° out of phase

57. What is the work done by a constant force $F = (3\hat{x} + 4\hat{y})$ N in moving an object over a straight line path defined by means of the vector $R = (-\hat{x} + 6\hat{y} + 3\hat{z})m$? ($\hat{x}$, $\hat{y}$ and $\hat{z}$ are unit vectors parallel to *x*-*y* and z-axis)

(*a*) zero (*b*) 12 J

(*c*) 15 J (*d*) 21 J

58. The force $\vec{f}$ per unit area on the surface of a conductor, with surface charge density in the presence of an electric field is ($\vec{a}_n$ is unit outward normal to the conductor surface)

(*a*) $\vec{f} = \dfrac{\sigma^2}{2 \epsilon_0} \vec{a}_n$ (*b*) $\vec{f} = \dfrac{\sigma^2}{\epsilon_0} \vec{a}_n$

(*c*) $\vec{f} = \epsilon_0 \sigma^2 \vec{a}_n$ (*d*) $\vec{f} = 0$

59. In the given figure A, B and C are conducting plates and the region X and Y are free space. The electric field inside X will be

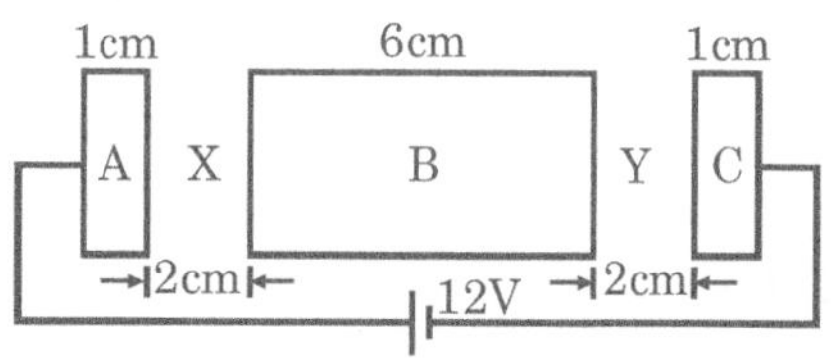

(a) zero
(b) 1 V/cm
(c) 2 V/cm
(d) 3 V/cm

60. Two conducting shells of raddi r_1 and r_2 ($r_1 > r_2$) each with a charge 'Q' are placed far apart in uniform external electric field. The space surrounding each shell is filled with the same dielectric material. In this situation, the
(a) shells will not experience any force
(b) shells will experience identical forces
(c) shell with radius r_1 will experience a greater force
(d) shell with radius r_2 will experience a greater force

61. Consider the following statements relating to a circular disc rotating in a transverse magnetic field B Wb/m^2 as shown in the given figure.

The *emf* generated across outer rim A and centre is proportional to
1. Angular velocity.
2. Flux density.
3. Scuare of the radius of the disc.

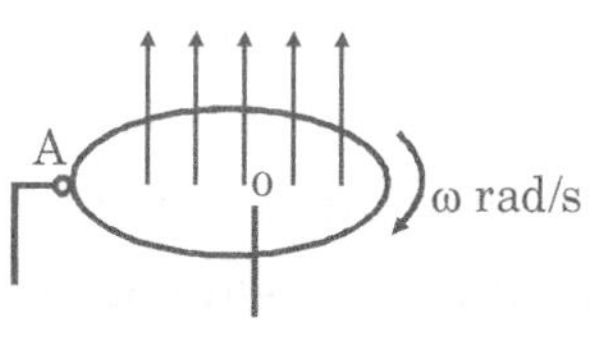

Of these statements:
(a) 1, 2 and 3 are correct
(b) 2 and 3 are correct
(c) 1 and 3 are correct
(d) 1 and 2 are correct

62. The electric potential due to an electric dipole of length L at a point distance r away from it will be doubled if the
(a) length L of the dipole is doubled
(b) r is doubled
(c) r is halved
(d) L is halved

63. For the current element Idl, situated at an arbitrary point, the magnetic vector potential A equals (R is the distance of the observation point from the centre of the current element)
(a) $\mu IDl/(4\pi R)$
(b) $\mu IDl/(4\pi R^2)$
(c) $\mu IDl/(4\pi R)$
(d) $\mu IDl/(2\pi R)$

64. The number of turns of a coil having a time constant T are doubled. Then the new time constant will be
(a) T (b) 2T
(c) 4T (d) T/2

65. An infinite numbers of concentric circular loops carry a current 'I' each but alternately in opposite directions. The radii of the loops are R, 2R, 4R... in geometric progression. The magnetic flux density at the centre of the loops will be
(a) zero (b) $\frac{\mu_0 I}{3R}$
(a) $\frac{\mu_0 I}{4R}$ (b) $\frac{\mu_0 I}{6R}$

66. Two concentric metallic spheres of radii 'a' and 'b' carry charges + Q and – Q respectively as shown in the given figure. Potential at the centre 'P' will be

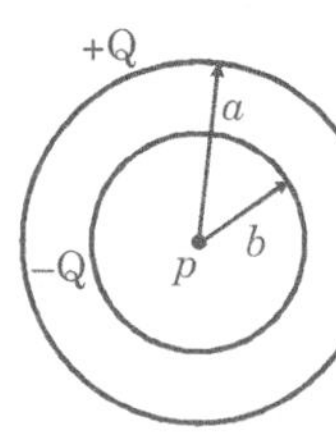

(a) zero (b) $\frac{Q}{4\pi\varepsilon_0 b}$
(c) $\frac{-Q}{4\pi\varepsilon_0 b}$ (d) $\frac{Q}{4\pi\varepsilon_0}\left(\frac{1}{a}-\frac{1}{b}\right)$

67. The term(s) responsible for attenuation in a source-free wave equation

$$\nabla^2\vec{E} - \mu\mu_0 \in\in_0 \frac{\partial^2 \vec{E}}{\partial^2 t} - \mu\mu_0\sigma\frac{\partial \vec{E}}{\partial t} = 0 \text{ is/are given by}$$

(a) $\mu\mu_0\sigma\frac{\partial \vec{E}}{\partial t}$
(b) $\mu\mu_0 \in\in_0 \frac{\partial^2 \vec{E}}{\partial^2 t}$
(c) $\nabla^2\vec{E}$ and $\mu\mu_0\sigma\frac{\partial \vec{E}}{\partial t}$
(d) $\mu\mu_0\sigma\frac{\partial \overline{E}}{\partial t}$ and $\mu\mu_0 \in\in_0 \frac{\partial^2 \vec{E}}{\partial^2 t}$

68. A parallel plate capacitor with air as the dielectric is charged and disconnected from the supply. The force between its plates is P. If the capacitor is now immersed in a liquid of dielectric constant $\in_r$, the force between the plates would be

(*a*) $\in_r^2 P$ (*b*) $\in_r P$

(*c*) P (*d*) $P/\in_r$

69. If the frequency of a plane electromagnetic wave increases four times, the depth of penetration, when the wave is incident normally on a good conductor will

(*a*) increase by factor of two

(*b*) decrease by a factor of four

(*c*) remain same

(*d*) decrease by a factor of two

70. A plane electromagnetic wave in free space is specified by the electric field

$\vec{a}_x\,[20\cos(\omega t - \beta z) + 5\cos(\omega t + \beta z)]$ V/m. The associated magnetic field is

(*a*) $\dfrac{\vec{a}_y}{120\pi}[20\cos(\omega t - \beta z) + 5\cos(\omega t + \beta z)]$ A/m

(*b*) $\dfrac{\vec{a}_y}{120\pi}[20\cos(\omega t - \beta z) - 5\cos(\omega t + \beta z)]$ A/m

(*c*) $\dfrac{\vec{a}_x}{120\pi}[20\cos(\omega t - \beta z) + 5\cos(\omega t + \beta z)]$ A/m

(*d*) $\dfrac{\vec{a}_x}{120\pi}[20\cos(\omega t - \beta z) - 5\cos(\omega t + \beta z)]$ A/m

71. The units of J_s and J_v, the surface and volume current densities are respectively

(*a*) A/m^2, A/m^3 (*b*) A/m, A/m^2

(*c*) A/m, A/m^3 (*d*) A, A/m

72. If the vector field $\vec{A} = \vec{x}a_x + \vec{y}a_y + kz\vec{a}_z$ represents a magnetic field, then the value of K must be

(*a*) 1 (*b*) 2

(*c*) -1 (*d*) -2

73. There is a constant homogeneous magnetic field pointing in vertical direction. A metallic wire in the form of a square is rotated about a horizontal axis passing through the middle points of its opposite arms. The e.m.f. generated in the square is

(*a*) zero

(*b*) finite and constant

(*c*) oscillating

(*d*) varying with time *t* as *t*2

74. Three concentric conducting spherical surfaces of radii R_1, R_2 and R_3 ($R_1 < R_2 < R_3$) carry charges of -1, -2 and 4 coulombs respectively. The charges on the inner and outer surfaces of the outermost sphere will be respectively (in coulombs)

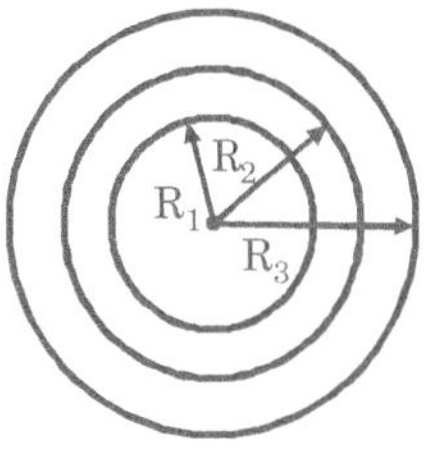

(*a*) 0, 4

(*b*) 3, 1

(*c*) -3, 7

(*d*) -2, 6

75. The concentric rings 1 and 2 carrying equal and uniform charge densities revolve at the same angular speed ω about their common axis passing through O as shown. The ratio of flux densities due to the two rings at O will be

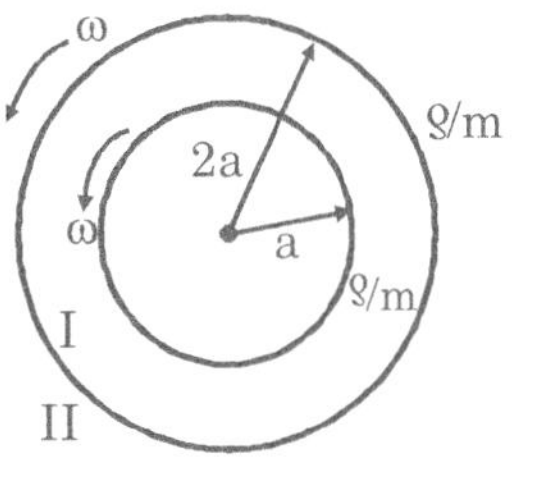

(*a*) 1 : 1

(*b*) 1 : 2

(*c*) 1 : 4

(*d*) 2 : 1

76. Two infinite straight wires crossing each other at right angles carry equal currents. Four square loops of equal size are placed symmetrically as shown in the given figure. The flux will be zero through the loops.

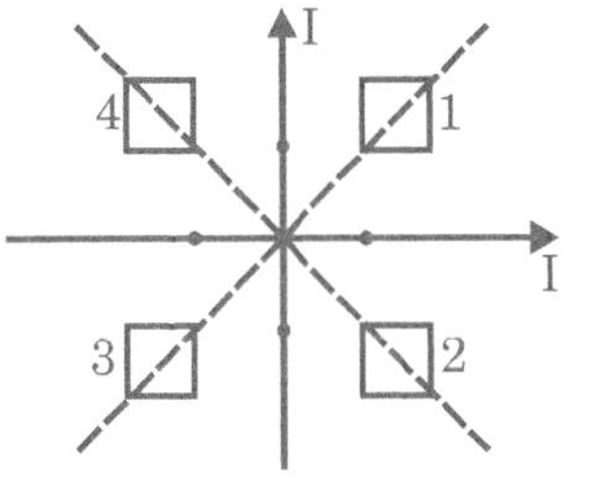

(*a*) 1 and 3

(*b*) 1 and 2

(*c*) 2 and 3

(*d*) 2 and 4

77. A magnetic field $\vec{B} = (\vec{a}_x + \vec{a}_y - 4\vec{a}_z)$ exists at a point. If a test charge moving with a velocity $\vec{v} = v_0(3\vec{a}_x - a_y + 2\vec{a}_z)$ experiences no force at a certain point, the electric field at that point will be

(*a*) $\vec{E} = -v_0(3\vec{a}_x - 2\vec{a}_y - 4\vec{a}_z)$

(*b*) $\vec{E} = -v_0(14\vec{a}_y + 7\vec{a}_z)$

(*c*) $\vec{E} = -\dfrac{v_0}{\mu}(14\vec{a}_y + 7\vec{a}_z)$

(*d*) $\vec{E} = +\dfrac{v_0}{\mu}(14\vec{a}_y + 7\vec{a}_z)$ [ES]

78. Match List-I with List-11 and select the correct answer using the codes given below the lists:

List-I	List-II
A. $\oint_s \vec{B}.\vec{ds}$	1. $\int_s \rho.\, dv$
B. $\oint_s \vec{B}.\vec{ds}$	2. zero
C. $\oint_s \vec{D}.\vec{ds}$	3. $(\sigma + j\omega\varepsilon)\int_s \vec{E}.\, \vec{ds}$
D. $\oint_s \vec{H}.\vec{dl}$	4. $-j\omega\mu\int_s \vec{H}.\, \vec{ds}$

Codes:

	A	B	C	D
(*a*)	2	4	3	1
(*b*)	2	4	1	3
(*c*)	4	2	1	3
(*d*)	4	2	3	1

79. Electric field of a uniform plane wave is given by

$\vec{E} = 15\sin(\omega t - \pi z)\,\vec{a}_x + 15\cos(\omega t - \pi z)\,\vec{a}_y$ V/m.

What is the nature of the polarization of the wave?

(*a*) Linear

(*b*) Elliptical

(*c*) Right-circular

(*d*) Left-circular

80. When the plate area of a parallel plate capacitor is increased keeping the capacitor voltage constant, the force between the plates

(*a*) increases

(*b*) decreases

(*c*) remains constant

(*d*) may increase of decrease depending on the metal making up the plates

81. Match List-I with List-II and select the correct answer using the codes given below the lists:

List-I (*Expression*)	List-II (*Law/theorem/equation*)
A. $\nabla^2 V = \dfrac{P v}{\varepsilon}$	1. Gauss's law
B. $\nabla^2 V = 0$	2. Divergence theorem
C. $\int_s \vec{D}.\, \vec{ds} = Q$	3. Laplaces's equation
D. $\int_s \vec{\nabla}.\vec{D}.\, dv = \int_s \vec{D}.\, \vec{ds}$	4. Poisson's equation
	5. Coulomb's law

Codes:

	A	B	C	D
(*a*)	4	2	5	1
(*b*)	3	1	2	4
(*c*)	2	3	4	5
(*d*)	4	3	1	2

82. The incident wave on a loss-less line carries an average power of 1.0W. The load end reflection coefficient is 1/3. The average power absorbed by the load is

(*a*) 1/3 W (*b*) 2/3 W

(*c*) 4/9 W (*d*) 8/9 W

83. Two positive charges, Q coulombs each, are placed at points (0, 0, 0) and (2, 2, 0) while two negative charges, Q coulombs each in magnitude, are placed at points (0, 2, 0) and (2, 0, 0). The electric field intensity at the point (1, 1, 0) is

(*a*) zero (*b*) $\dfrac{Q}{8\pi\varepsilon_0}$

(*c*) $\dfrac{Q}{4\pi\varepsilon_0}$ (*d*) $\dfrac{Q}{16\pi\varepsilon_0}$

84. Match List-I with List-II and select the correct answer using the codes given below the lists:

List-I (*Laws of magnetostatics*)	List-II (*Equations with which the given laws are associated*)
A. Ampere's Law	1. $\vec{F} = Q[\vec{E} + (\vec{v} + \vec{B})]$
B. Biot-Savart Law	2. $\vec{B} = \dfrac{\mu_0 I}{4\pi}\int \dfrac{\vec{dl} \times \vec{R}}{R^2}$
C. Lorentz Force Law	3. $\vec{\nabla} \times \vec{B} = \mu_0 \vec{j}$
D. Magnetic flux Law	4. $\nabla.\vec{B} = 0$

Codes :

	A	B	C	D
(*a*)	3	2	1	4
(*b*)	2	3	4	1
(*c*)	3	2	4	1
(*d*)	2	3	1	4

85. A quarterwave line will behave as

(*a*) a rectifier

(*b*) an amplifier

(*c*) a transformer

(*d*) a variable capacitance

86. The infinite parallel metal plates are charged with equal surface charge density of the same polarity. The electric field in the gap between the plates is

(*a*) the same as that produced by one plate

(*b*) double of the field produced by one plate

(*c*) dependent on coordinates of field point

(*d*) zero

87. A parallel plate capacitor has an electrode area of 100 mm^2, with a spacing of 0.1 mm between the electrodes. The dielectric between the plates is air with a permittivity of 8.85×10^{-12} F/m. The charge on the capacitor is 100V. The stored energy in the capacitor is

(*a*) 8.85 *p*J (*b*) 440 *p*J

(*c*) 22.1 *n*J (*d*) 44.3 *n*J

88. A capacitor is made up of two concentric spherical shells. The radii of the inner and outer shells are R_1 and R_2 respectively and $\in$ is the permittivity of the medium between the shells. The capacitance of the capacitor is given by

(*a*) $\frac{1}{4\pi \in}\left(\frac{1}{R_1}-\frac{1}{R_2}\right)$

(*b*) $\frac{1}{4\pi \in}\left(\frac{1}{R_1}+\frac{1}{R_2}\right)$

(*c*) $4\pi \in \frac{R_1 R_2}{(R_1+R_2)}$

(*d*) $4\pi \in \frac{R_1 R_2}{(R_1-R_2)}$

89. A plane monochromatic electro-magnetic wave travels in a perfect conducting medium, which is charge-free and external current free. Then,

(*a*) $\vec{E}$ field lags $\vec{B}$ field by $\pi/4$

(*b*) $\vec{E}$ field leads $\vec{B}$ field by $\pi/4$

(*c*) $\vec{E}$ and $\vec{B}$ fields are co-phasal

(*d*) $\vec{E}$ and $\vec{B}$ field differ in phase by $\frac{\pi}{2}$

90. When a lossy capacitor with a dielectric of permittivity ε and conductivity σ operates at a frequency ω, the loss tangent for the capacitor is given by

(*a*) $\omega\sigma/\varepsilon$ (*b*) $\omega\varepsilon/\sigma$

(*c*) $\sigma/\omega\varepsilon$ (*d*) $\sigma\omega\varepsilon$

91. The flux density $\vec{B}$ in a ferromagnetic medium of large extent and permeability μ is as shown in the given figure. Two cavities, cavity I shaped like a thin disc and cavity II long and needle like are cut. The ratio of H_1 to H_2 for air filled cavities is

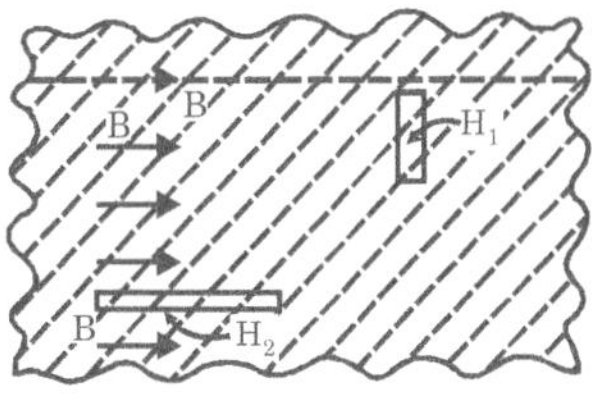

(*a*) zero

(*b*) 1

(*c*) ω

(*d*) μ/μ_0

92. The force with which the plates of a parallel plate capacitor having charge Q and area of each plate A, attract each other is

1. Directly proportional to Q
2. Directly proportional to Q^2
3. Inversely proportional to A

(*a*) 1 and 2 only (*b*) 2 and 3 only

(*c*) 1 and 3 only (*d*) 1, 2 and 3

93. Maxwell's divergence equation for the magnetic field is given by

(*a*) $\overline{\nabla} \times B = O$

(*b*) $\overline{\nabla} . B = O$

(*c*) $\overline{\nabla} \times B = \rho$

(*d*) $\overline{\nabla} . B = \rho$

94. Which of the following statements would be true if one compared the hysteresis loops of hard and soft magnetic materials?

1. Hysteresis loss of hard magnetic material will be less than that of soft material.
2. Coercivity of hard material will be greater than that of soft material.
3. Retentivity of the two materials will always be equal.

Select the correct answer using the codes given below:

Codes:

(*a*) 1 alone (*b*) 2 alone

(*c*) 2 and 3 (*d*) 1, 2 and 3

95. An infinite charged conducting plate is placed between and parallel to two infinite conducting grounded planes as shown. The ratio of charge densities ρ_1 and ρ_2 on the two sides of the plate will be

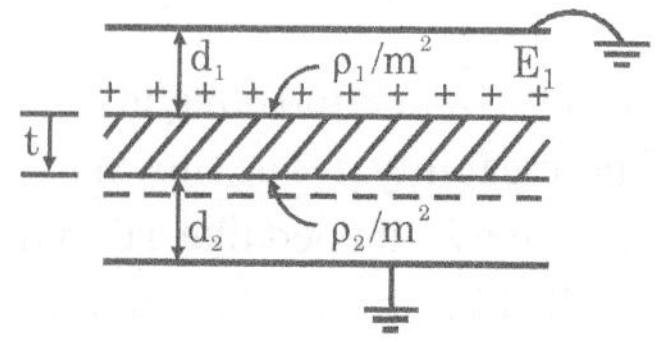

(a) $\frac{d+t}{d_2+t}$ (b) $\frac{d_2+t}{d_2+t}$

(c) $\frac{d_2}{d_1}$ (d) $\frac{d_1}{d_2}$

96. When a closed conducting loop 'C' is moving with a constant velocity 'V' through a non-uniform time-varying magnetic field 'B', the voltage induced in the loop is given by

(a) $e=-\int_S \frac{\partial \overline{B}}{\partial t}\cdot d\bar{s}$

(b) $e=-\int_C (\overline{V}\times\overline{B})\cdot d\bar{l}$

(c) $e=-\int_C (\overline{B}\times\overline{V})\cdot d\bar{l}$

(d) $e=-\int_S \frac{\partial \overline{B}}{\partial t}\cdot d\bar{s}+\int_C (\overline{V}\times\overline{B})\cdot d\bar{l}$

97. An infinite dielectric slab is uniformly polarized as shown in the figure. The electric field inside the slab is given by

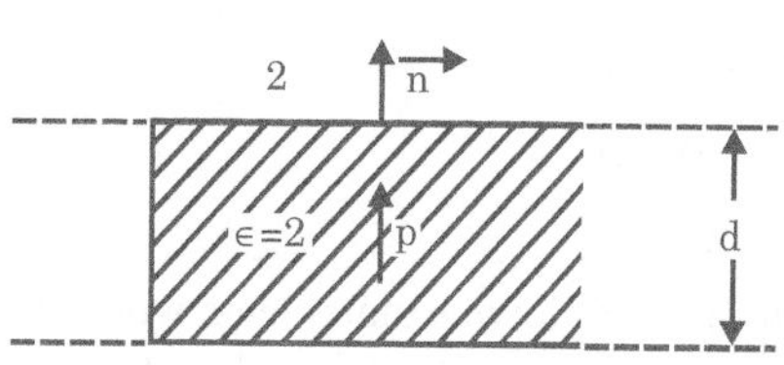

(a) $\frac{2P}{\epsilon_0}\vec{n}$

(b) $\frac{2P}{\epsilon_0}(-\vec{n})$

(c) $\frac{P}{\epsilon_0}(\vec{n})$

(d) $\frac{P}{\epsilon_0}(-\vec{n})$

98. Electric field of a uniform plane wave is given by $\vec{E}=10\cos(37\pi\times10^8 t-\pi z)\vec{a}_x$ Match List-I with List-II pertaining to the above wave and select the correct answer using the codes given below the lists:

List-I (*Parameters*)	**List-II** (*Values in MKS units*)
A. Phase velocity	1. 2
B. Wavelength	2. 3.14
C. Frequency	3. 377
D. Phase constant	4. 1.5×10^8
	5. 3.8×10^8

Codes:

	A	B	C	D
(a)	5	4	3	2
(b)	3	4	2	1
(c)	4	3	1	2
(d)	5	1	4	2

99. The intrinsic impedance of a good conducting medium is given by (symbols have the usual meaning)

(a) $\sqrt{\frac{\mu\omega}{\sigma}}\angle-45°$ (b) $\sqrt{\frac{\omega\sigma}{\mu}}\angle45°$

(c) $\sqrt{\frac{\mu\omega}{\sigma}}\angle45°$ (d) $\sqrt{\mu\sigma\omega}\angle0°$

100. A transmission lihe with a characteristic impedance Z_1 is connected to a transmission line with a characteristic impedance Z_2. If the system is being driven by a generator connected to the first line, then the overall transmission co-efficient will be

(a) $\frac{2Z_1}{Z_1+Z_2}$

(b) $\frac{Z_2}{Z_1+Z_2}$

(c) $\frac{2Z_2}{Z_1+Z_2}$

(d) $\frac{Z_2}{Z_1+Z_2}$

101. If the electric field $\overline{E}=0.1-e^{-t}\,\bar{a}_x$, and $\epsilon=4\epsilon_0$, then the displacement current crossing an area of 0.1 m^2 at $t=0$ will be

(a) zero (b) $0.04\epsilon_0$

(c) $0.4\epsilon_0$ (d) $4\epsilon_0$

102. Which of the following pairs of parameters and expressions is/are correctly matched?

1. Characteristic impedance ... $\frac{\bar{E}}{\bar{H}}\sqrt{\epsilon_r}$
2. Power flow density ... $\bar{V} \times \bar{H}$
3. Displacement current in non-conducting medium ... $\bar{E} \times \bar{H}$

Select the correct answer using the codes given below:

Codes:

(*a*) 1 alone (*b*) 2 and 3
(*c*) 1 and 3 (*d*) 1 and 2

103. Which of the following properties pertain to a circularly polarized wave having no component of electric field in the *x*-direction but having components E*y* and E*z*?

1. E*y* and E*z* are equal in magnitude.
2. Direction of resultant electric vector varies with polarity.
3. Direction of resultant electric vector varies with time.
4. E*y* and E*z* have a time phase difference of 90°.

Select the correct answer from the codes given below:

(*a*) 2 and 4 (*b*) 1 and 3
(*c*) 1, 3 and 4 (*d*) 1 and 2

104. A uniform plane wave travelling is a perfect dielectric is incident normally on the surface of a perfect conductor. Then

(*a*) the wave is transmitted into the conductor without attenuation
(*b*) 50% of the incident wave is transmitted and 50% is reflected
(*c*) a standing wave is set up in the conducting medium
(*d*) a standing wave is set up in the dielectric

105. Two concentric spherical shells carry equal and opposite uniformly distributed charges over their surfaces as shown in the figure. Electric field on the surface of the inner shell will be

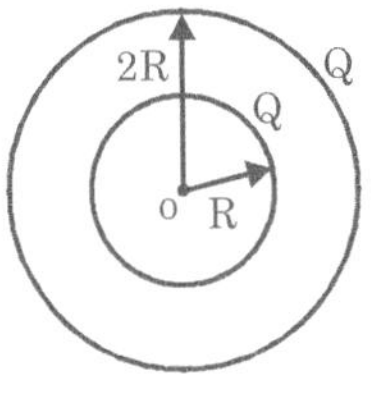

(*a*) zero
(*b*) $\frac{Q}{4\pi \epsilon_0 R^2}$
(*c*) $\frac{Q}{8\pi \epsilon_0 R^2}$
(*d*) $\frac{Q}{16\pi \epsilon_0 R^2}$

106. An infinite number of concentric rings carry a charge Q each alternately positive and negative. Their radii are 1, 2, 4, 8 — meters in geometric progression as shown. The potential at the centre of the rings will be

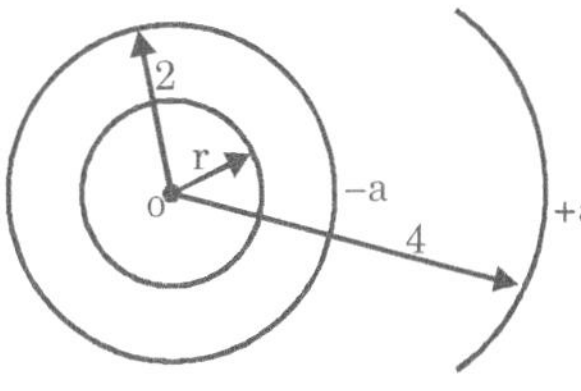

(*a*) zero
(*b*) $\frac{Q}{12\pi \epsilon_0}$
(*c*) $\frac{Q}{8\pi \epsilon_0}$
(*d*) $\frac{Q}{6\pi \epsilon_0}$

107. A circular loop has its radius increasing at a rate of 2 m/s. The loop is placed perpendicular to a constant magnetic field of 0.1 Wb/m². When the radius of the loop is m, the emf induced in the loop will be

(*a*) 0.8πV (*b*) 0.4πV
(*c*) 0.27πV (*d*) zero

108. A sample of *p*-type semiconductor is subjected to the combined effect of mutually perpendicular field E_x and magnetic field B_z as shown in the figure. This would result in the following events:

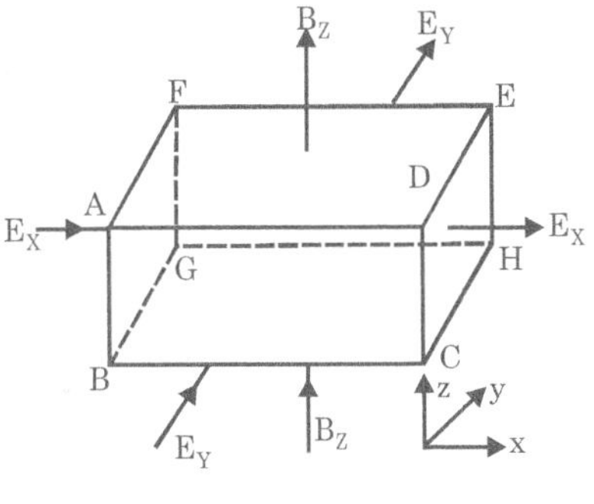

1. Lorentz force drives the majority carriers to the face ABCD.
2. Electric field Ey results and grows.
3. The faces ABCD and EFGH are oppositely charged.
4. Lorentz force on holes is balanced.

The correct sequence of these events is

(*a*) 1, 3, 4, 2 (*b*) 3, 1, 2, 4
(*c*) 1, 3, 2, 4 (*d*) 3, 1, 4, 2

109. The net flux of electric field emanating from the surface 'S' with location of point charges as shown in the given figure is

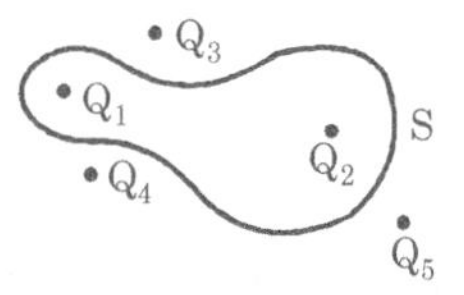

(a) $\dfrac{Q_1+Q_2}{\varepsilon_0}$

(b) $\dfrac{Q_1+Q_2+Q_3+Q_4+Q_5}{\varepsilon_0}$

(c) $\dfrac{Q_3+Q_4+Q_5}{\varepsilon_0}$

(d) $\varepsilon_0(Q_1+Q_2)$

110. If a uniform plane wave propagating in the direction $(4\hat{a}_x-2\hat{a}_y+\hat{a}_z)$ has its E field in the direction $(4\hat{a}_x-2\text{K}\hat{a}_y+4\text{K}\hat{a}_z)$, the the value of K must be

(a) 2 (b) −2
(c) 1 (d) zero

111. The projection of the vector $6\vec{a}_x+2\vec{a}_y-3\vec{a}_z$ along $3\hat{a}_x-4\hat{a}_y$ is

(a) 2 (b) $18\hat{a}_x-8\hat{a}_y$
(c) 10 (d) $21\hat{a}_x-28\hat{a}_y$

112. The region $z \le 0$ is a dielectric of relative permittivity 2 while the region $z \le 0$ is a dielectric of relative permittivity 5. If the electric field intensity in region $z \le 0$ is $\overline{\text{E}}=10\,\hat{a}_z$ kVIm, the potential difference between $(0, 0-10^{-3})$ m and $(0, 0+10^{-3})m$ will be

(a) zero (b) 70 V
(c) 20V (d) 14V

113. Two parallel wires separated by a distance d are carrying a current I in the same direction. The magnetic field along a line running parallel to these wires and midway between them

(a) depends upon I
(b) is zero
(c) depends upon d
(d) depends upon the permeability of medium between the wires

114. If the vector $\overline{\text{B}}=x^2\hat{a}_x-xy\hat{a}_y-\text{K}xz\hat{a}_z$ represents a magnetic field, the value of the constant K must be

(a) 0 (b) 1
(c) 2 (d) 3

115. The electrostatic field on the surface of a conductor at a certain point is $0.3\,\hat{a}_x+0.4\,\hat{a}_y$. If the normal to the surface of the conductor at that point makes an angle θ with respect to x-axis, then value of cos θ will be

(a) 0.8 (b) 0.75
(c) 0.6 (d) 0.3

116. The electric energy density is 'W' at a point in free space where the electric field is $\hat{a}_x\text{E}$. If the electric field is $\hat{a}_x\text{E}-\hat{a}_y\text{E}$, in a dielectric of relative permittivity 4, then the electric energy density at that point will be

(a) zero (b) 2W
(c) 4W (d) W

117. The plane $x = 3$ has a layer of charge density $2n$ C/m². A line charge of density $20n$ C/m is located at $x = 1, z = 4$. The force acting on unit metre length of the line charge is

(a) $\dfrac{2\times10^{-9}\times20\times10^{-9}}{4\pi\varepsilon_0}\hat{a}_x$ N

(b) $\dfrac{2\times10^{-9}\times20\times10^{-9}}{2\pi\varepsilon_0}\hat{a}_z$ N

(c) $\dfrac{2\times10^{-9}\times20\times10^{-9}}{2\varepsilon_0}\hat{a}_y$ N

(d) $\dfrac{2\times10^{-9}\times20\times10^{-9}}{2\varepsilon_0}\hat{a}_x$ N

118. Given that

$\vec{\text{E}}$ = time and space dependent electric intensity vector
μ = permeability of the medium
σ = conductivity of the medium
∈ = permittivity of the medium

The electromagnetic wave equation is given by

(a) $\nabla^2\vec{\text{E}}-\mu\sigma\dfrac{\partial\vec{\text{E}}}{\partial t}-\mu\in\dfrac{\partial\vec{\text{E}}}{\partial t}=0$

(b) $\nabla^2\vec{\text{E}}-\mu\in\dfrac{\partial\vec{\text{E}}}{\partial t}-\mu\sigma\dfrac{\partial\vec{\text{E}}}{\partial t}=0$

(c) $\nabla^2\vec{\text{E}}-\sigma\in\dfrac{\partial\vec{\text{E}}}{\partial t}-\mu\sigma\dfrac{\partial\vec{\text{E}}}{\partial t}=0$

(d) $\nabla^2\vec{\text{E}}-\dfrac{\sigma}{\mu}\dfrac{\partial\vec{\text{E}}}{\partial t}-\dfrac{\mu}{\in}\dfrac{\partial\vec{\text{E}}}{\partial t}=0$

119. Match List-I with List-II and select the correct answer using the codes given below the lists:

List-I	**List-II**
(*Circuit relation for transmission line*)	(*Field relations for space wave*)
A. $Z = R + j\omega L$	1. $\sigma + j\omega\in$
B. $Y = G + j\omega C$	2. μ
C. Series inductance L	3. $j\omega\mu$
D. Shunt capacitance C	4. $\in$

Codes:

	A	B	C	D
(*a*)	1	2	3	4
(*b*)	3	2	4	1
(*c*)	3	1	2	4
(*d*)	4	1	2	3

120. The polarization of a dielectric material is given by

(*a*) $\vec{P} = \varepsilon_r \vec{E}$

(*b*) $\vec{P} = (\varepsilon_r - 1)\vec{E}$

(*c*) $\vec{E} = \vec{E}\,\varepsilon_0(\varepsilon_r - 1)$

(*d*) $\vec{P} = (\varepsilon_r - 1)\in_0$

121. The Lorentz force law is given by

(*a*) $\vec{f} = Q\vec{E}$

(*b*) $\vec{f} = \vec{V} \times \vec{B}$

(*c*) $\vec{f} = Q(\vec{E} + \vec{V} \times \vec{B})$

(*d*) $\vec{f} = Q(\vec{V} \times \vec{B})$

EXERCISE – II

1. Voltage drop in a conductor depends upon

(*a*) the resistance of the conductor

(*b*) the length and the specific resistance of the conductor

(*c*) the cross-section area and the conductivity of the conductor

(*d*) the resistance of the conductor and the current flowing though it **[NTPC]**

2. An air gap is usually inserted in magnetic circuit to

(*a*) increase the flux (*b*) prevent saturation

(*c*) increase mmf (*d*) none of these **[BHEL]**

3. Capacitor of 40μF is charged to a potential difference of 500V. The energy stored in the electrical field between the plates of this condenser is

(*a*) 4.5 joules (*b*) 4 joules

(*c*) 5 joules (*d*) 40 joules **[BHEL]**

4. Which of the following values of alternating voltage should in insulation absolutely with stand?

(*a*) The effective value

(*b*) The arithmetical average value

(*c*) Half the effective value

(*d*) The peak value **[BHEL]**

5. Silicon steel is used in electrical machines because it has

(*a*) low hysteresis loss

(*b*) low retentivity

(*c*) low coercivity

(*d*) none of these **[BHEL]**

6. A magnet is able to attract

(*a*) iron, alluminium and brass

(*b*) iron, cobalt and zinc

(*c*) iron, copper and nickel

(*d*) nickel, cobalt and steel **[NTPC]**

7. What is the force experienced per unit length by a conductor carrying 5A current in positive Z direction and placed in a magnetic field $\overline{B} = (3\vec{a}_x + 4\vec{a}_y)$?

(*a*) $15\vec{a}_x + 20\vec{a}_y$ N/m

(*b*) $-20\vec{a}_x + 15\vec{a}_y$ N/m

(*c*) $20\vec{a}_x - 15\vec{a}_y$ N/m

(*d*) $-20\vec{a}_x - 20\vec{a}_y$ N/m **[NTPC]**

8. What is the value of total electric flux coming out of a closed surface?

(*a*) Zero

(*b*) Equal to volume charge density

(*c*) Equal to the total charge enclosed by the surface

(*d*) Equal to the surface charge density **[NTPC]**

9. A flat slab of dielectric, $\in_r = 5$ is placed normal to a uniform field with a flux density D = 1 coulomb/m^2. The slab is uniformly polarized. What is the polarization P of the slab in coulomb/m^2?

(*a*) 0.8 (*b*) 1.2

(*c*) 4 (*d*) 6 **[NTPC]**

10. Maxwell's divergence equation for the magnetic field is given by

(*a*) $\overline{\nabla} \times \overline{B} = 0$ (*b*) $\overline{\nabla} . \overline{B} = 0$

(*c*) $\overline{\nabla} \times \overline{B} = p$ (*d*) $\overline{\nabla} . \overline{B} = p$ **[IAS]**

11. For a conducting medium with conductivity σ, permeability μ and permittivity ∈, the skin depth for an electromagnetic signal at an angular frequency ω is proportional to

(*a*) σ (*b*) $\frac{1}{\omega}$

(*c*) $\frac{1}{\sqrt{\sigma}}$ (*d*) $\frac{1}{M}$ **[BHEL]**

12. A loss-less transmission line of characteristic impedance Z and $l < \frac{\lambda}{4}$ is terminated at the load end by a short circuit. Its input impedance Z_s is

(*a*) $Z_s = -jZ_0 \tan \beta l$ (*b*) $Z_s = jZ_0 \cot \beta l$

(*c*) $Z_s = -jZ_0 \tan \beta l$ (*d*) $Z_s = -jZ_0 \cot \beta l$

[BHEL]

13. A loss-less transmission line with characteristic impedance of 600Ω is terminated in a purely resistive load of 900Ω. The reflection coefficient is

(*a*) 0.2

(*b*) 0.5

(*c*) 0.667

(*d*) 1.5 **[BHEL]**

14. In a loss less medium the intrinsic impedance $n = 60\pi$ and $\propto_r = 1$. What is the value of the dielectric constant $\in_r$?

(*a*) 2 (*b*) 1

(*c*) 4 (*d*) 8 **[BHEL]**

15. A long cylindrical conductor carries a uniformly distributed current, as shown in the given figure. The ratio of the Poynting's vector at Q and P would be

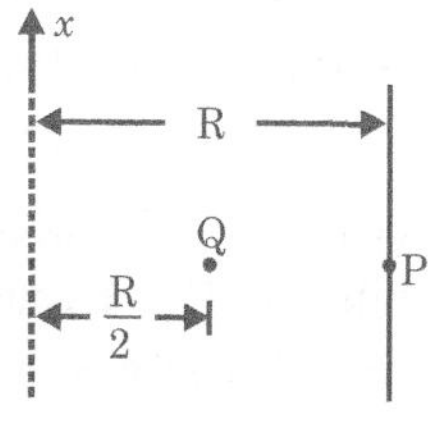

(*a*) 4 : 1 (*b*) 2 : 1

(*c*) 1 : 4 (*d*) 1 : 2 **[IAS]**

16. A current carrying conductor of length *l* is under the influence of a magnetic field having magnetic flux density B. If I is the current flowing through the conductor, which of the following formula is correct for calculating the force exerted on it?

(*a*) $F = B \times l \times I^2$ (*b*) $F = \frac{B \times I}{l}$

(*c*) $F = \frac{I \times l}{B}$ (*d*) $F = B \times I \times l$ **[NTPC]**

17. If the area of hysteresis loop of a material is large, the hysteresis loss in this material will be

(*a*) small (*b*) large

(*c*) zero (*d*) none of these

[NTPC]

18. A plane slab of dielectric having dielectric constant 5, placed normal to a uniform field with a flux density of 2 c/m² is uniformally polarized. The polarisation of the slab is

(*a*) 0.4c/m² (*b*) 1.6c/m²

(*c*) 2.0c/m² (*d*) 6.4c/m² **[NTPC]**

19. Which one of the following formulae is NOT correct for the boundary between two magnetic materials?

(*a*) $B_{N1} = B_{N2}$

(*b*) $B_2 = \sqrt{B_{N2} + B_{t2}}$

(*c*) $H_1 = H_{n1} + H_{t1}$

(*d*) $\hat{a}_{n21} \times (\vec{H}_1 - \vec{H}_2) = \overline{k}$

where $\hat{a}_{n21}$ is a unit vector normal to the interface and directed from region 2 to region 1. **[NTPC]**

20. Which one of the following pairs is not correctly matched?

(*a*) Gauss's Theorem $\oint_s \vec{D}.\vec{ds} \oint_v \nabla.\overline{D}dv$

(*b*) Gauss's Law: $\oint_s \vec{D}.\vec{ds} \oint_v \rho dv$

(*c*) Coulomb's Law: $V = \frac{d\phi_m}{dt}$

(*d*) Stokes'sTheorem: $\oint \vec{\xi}\vec{dl} = \int_s (\nabla \times \vec{\xi})\vec{ds}$ **[BHEL]**

21. The magnitude of emf induced in a wire does not depend on

(*a*) length of wire

(*b*) speed of wire

(*c*) diameter of wire

(*d*) none of these **[PGCIL]**

22. The direction of the magnetic lines of forces is

(*a*) from + to – charges
(*b*) from south to north poles
(*c*) from one end of the magnet to the other
(*d*) from north to south poles **[PGCIL]**

23. Which one of the following gives the approximate value of capacitance between two spheres, whose separation is very much larger than their radii R?

(*a*) $\frac{2\pi}{\epsilon_o R}$ (*b*) $2\pi\epsilon_o R$

(*c*) $\frac{2\pi\epsilon_o}{R}$ (*d*) $\frac{4\pi\epsilon_o}{R}$ **[IOCL]**

24. Which one of the following statements is correct? For a loss less dielectric medium, the phase constant for a travelling wave, β is proportional to

(*a*) ϵ_r (*b*) $\sqrt{\epsilon_r}$

(*c*) $\frac{1}{\epsilon_r}$ (*d*) $\frac{1}{\sqrt{\epsilon_r}}$ **[IOCL]**

25. With the increase in temperature, the insulating property of insulator

(*a*) weakens (*b*) gains
(*c*) remains same (*d*) none of these **[IOCL]**

26. The unit normal to the equipotential surface, ϕ = constant *c* is, in general, given by

(*a*) ϕ (*b*) $-\frac{\nabla\phi}{|\nabla\phi|}$

(*c*) $\frac{\nabla\phi}{|\nabla\phi|}$ (*d*) $\pm\frac{\nabla\phi}{|\nabla\phi|}$ **[IOCL]**

27. When the load impedance is equal to the characteristic impedance of the transmission lines then the reflection coefficient and standing wave ratio are respectively

(*a*) 0 and 0 (*b*) 1 and 0
(*c*) 0 and 1 (*d*) 1 and 1 **[PGCIL]**

28. Three current carrying conductors are shown in the figure. The value of $|\oint B.dl|$ around the closed curve is

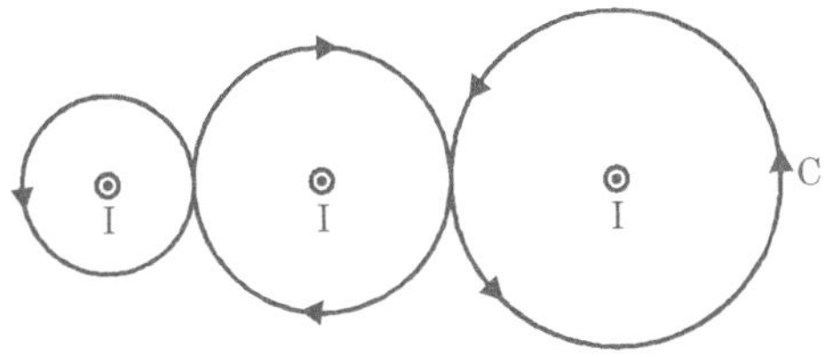

(*a*) $4\mu_0 I$ (*b*) $3\mu_0 I$
(*c*) $2\mu_0 I$ (*d*) $\mu_0 I$ **[PGCIL]**

29. The internal heating of a capacitor is usually attributed to

(*a*) leakage resistance
(*b*) dielectric charge
(*c*) electron movement
(*d*) plate vibration **[NHPC]**

30. The force with which the plates of a parallel plate capacitor having charge Q and area of each plate, A attract each other is

1. directly proportional to Q
2. directly proportional to Q^2
3. inversely proportional to A

(*a*) 1 and 2 only (*b*) 2 and 3 only
(*c*) 1 and 3 only (*d*) 1, 2 and 3 **[NHPC]**

31. Magnetic flux density at a point distance R due to an infinitely long linear conductor carrying a current I is given by

(*a*) $B = \frac{I}{2\pi R}$ (*b*) $B = \frac{\mu I}{2R}$

(*c*) $B = \frac{\mu I}{2\pi R}$ (*d*) $B = \frac{\mu I}{2\pi R^2}$ **[NTPC]**

32. Maxwell's divergence equation for the magnetic field is given by

(*a*) $\vec{\nabla} \times B = 0$ (*b*) $\vec{\nabla}.B = 0$

(*c*) $\vec{\nabla} \times B = \rho$ (*d*) $\vec{\nabla}.B = \rho$ **[NTPC]**

33. Two infinite radial planes with interior angle 30° are held at potentials 0V and 100V as shown in the given figure. If the potential V is a function of ϕ alone, then the equation of the equipotential surface corresponding to V= 30V will be small insulating gap

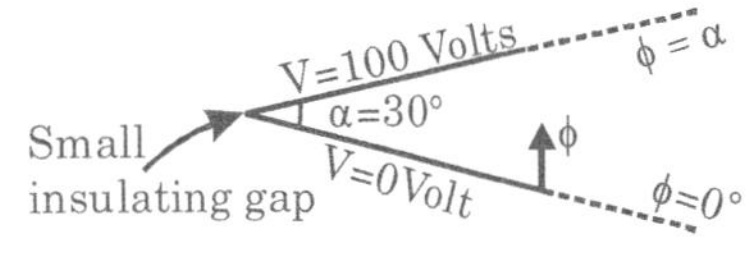

(*a*) $\phi = 15°$
(*b*) $\phi = 12°$
(*c*) $\phi = 9°$
(*d*) $\phi = 6°$ **[ES]**

34. The line integral of the vector potential A around the boundary of a surface S represents

(*a*) flux through in the surface S
(*b*) flux density in the surface S
(*c*) magnetic density
(*d*) current density **[GATE]**

35. A capacitor is made up of two concentric spherical shell. The radii of the inner and outer shells are R_1 and R_2 respectively and $\in$ is the permittivity of the medium between the shells. The capacitance of the capacitor is given by

(a) $\frac{1}{4\pi\in}\left(\frac{1}{R_1}-\frac{1}{R_2}\right)$

(b) $\frac{1}{4\pi\in}\left(\frac{1}{R_1}+\frac{1}{R_2}\right)$

(c) $4\pi\in\frac{R_1R_2}{R_1+R_2}$

(d) $4\pi\in\frac{R_1R_2}{R_1-R_2}$ **[BHEL]**

36. The infinite parallel metal plates are charged with equal surface charge density of the same polarity. The electric field in the gap between the plates is

(a) the same as that produced by one plate

(b) double of the field produced by one plate

(c) dependent on coordinates of field point

(d) zero **[BHEL]**

37. For an ideal multi-conductor transmission line with propagation along the *z*-axis, the electric and magnetic fields between and on the conductors of the line are characterised by

(a) $E_x=0, E_y=0, E_x\neq 0$

(b) $E_x=0, H_z=0$

(c) $E_x\neq 0, E_y\neq 0, E_z=0$

(d) $E_x\neq 0, H_y\neq 0, E_z>0$ **[BHEL]**

38. Consider the following statements:

In electrostatic, the equipotential surface is defined as the surface where

1. Electric field indensity is normal everywhere.
2. Electric field intensity is tangential everywhere.
3. No work is done in moving a charge over it.
4. No charge is present.

Of these statements:

(a) 1 alone, is correct

(b) 3 and 4 are correct

(c) 1 and 3 are correct

(d) 2 and 4 are correct **[BHEL]**

39. The characteristic impedance of an air dielectric co-axial line is 77Ω. If the air is replaced by a dielectric with $\in_r$ = 2.28, then characteristic impedance will be

(a) 33.8Ω (b) 51Ω

(c) 116Ω (d) 175Ω **[BHEL]**

40. At a certain frequency *f*, uniform plane wave is found to have established a wavelength λ in a good conductor. If the source frequency is doubled, then the wavelength would change to

(a) $\frac{\lambda}{\sqrt{2}}$ (b) $\sqrt{2}\lambda$

(c) 2λ (d) 4λ **[ES]**

41. Match List-I with List-II and select the correct answer using the codes given below the lists:

List-I	List-II
A. $\nabla\times\vec{H}=\vec{J}+\frac{\partial\vec{D}}{\partial t}$	1. Faraday's law
B. $\nabla\times\vec{E}=\frac{\partial\vec{D}}{\partial t}$	2. Gass' law
C. $\nabla.\vec{D}=\rho$	3. Ampere's law

Codes:

	A	B	C
(a)	3	1	2
(b)	2	4	3
(c)	3	2	1
(d)	4	2	3

[ES]

42. Kirchhoff s current law for direct currents is implicit in the expression

(a) $\nabla.\vec{D}=\rho$

(b) $\int\vec{J}.\vec{n}\,ds=0$

(c) $\nabla.\vec{B}=0$

(d) $\nabla.\vec{H}=\vec{J}\frac{\partial\vec{D}}{\partial t}$ **[ES]**

43. Two loss-loss transmission lines 1 and 2 shown in the given figure have each a characteristic impedance Z. Line 1 is charged to a voltage of V and line 2 to a voltage of 2V. Then both of them are discharged through resistances Z as shown, if line 1 gets fully discharged in time T, then line 2 will get fully discharged in time

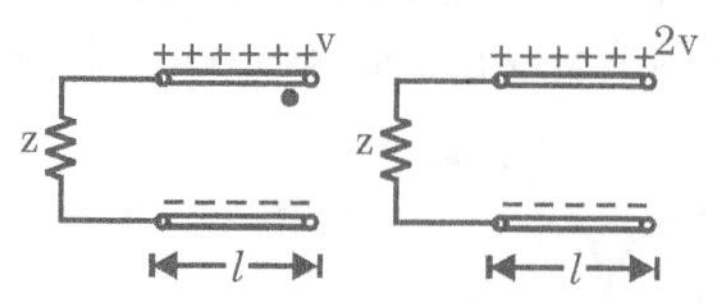

(a) T/4

(b) T/2

(c) T

(d) 2T **[ES]**

44. A parallel plate air capacitor has plates of 1500 cm^2 separated by 5 mm. If a layer of dialectic 2 mm thick and relative permittivity 3 is now introduced between the plates, then the new separation in mm between the plates, so that the capacitance value is unchanged, will be

(*a*) 7.00 (*b*) 6.33

(*c*) 5.67 (*d*) 5.00 **[ES]**

45. The ratio of conduction current density of the displacement current is (symbols have the usual meaning)

(*a*) $\frac{j\sigma}{\omega e}$ (*b*) $\frac{\sigma}{j\omega e}$

(*c*) $\frac{\sigma\omega}{je}$ (*d*) $\frac{\sigma\omega}{j\omega}$ **[ES]**

46. Two circular rings carry equal and opposite charge as shown in the given figure. The dipole moment of the charge distribution is

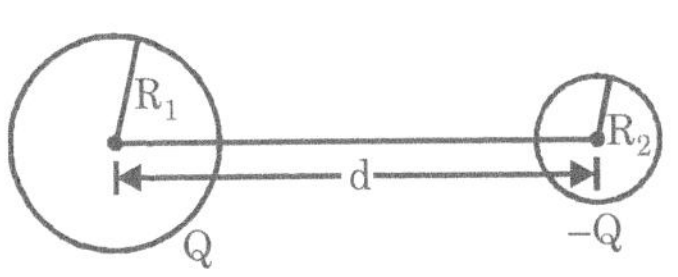

(*a*) $Q\frac{R_2}{R_1}d$ (*b*) $Q\frac{R_1}{R_2}d$

(*c*) $Q\frac{R_1^2 - R_2^2}{d}$ (*d*) Qd **[ES]**

47. An electric dipole (– Q, Q) revolves with a uniform angular velocity 'ω' in a circle about its centre 'O'. When the dipole is in the position shown in the given figure, the direction of the displacement current density at 'O' is

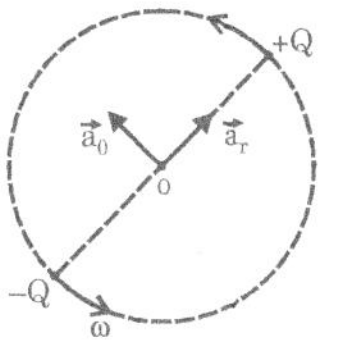

(*a*) $\vec{a}_r$

(*b*) $-\vec{a}_r$

(*c*) $-\vec{a}_\theta$

(*d*) $\vec{a}_\theta$ **[IAS]**

48. A charge distribution is shown in the given figure: Which of the following expressions for the displacement flux density $\overline{D}$ due to the given charge distribution are correct?

1. $\overline{D} = \rho_0 a\bar{i}_x$ for $x < -a$

2. $\overline{D} = -\rho_0(x+a)\bar{i}_x$ for $0 < x < a$

3. $\overline{D} = +\rho_0(x-a)\bar{i}_x$ for $0 < x < a$

4. $\overline{D} = +\rho_0 a\bar{i}_x$ for $x > a$

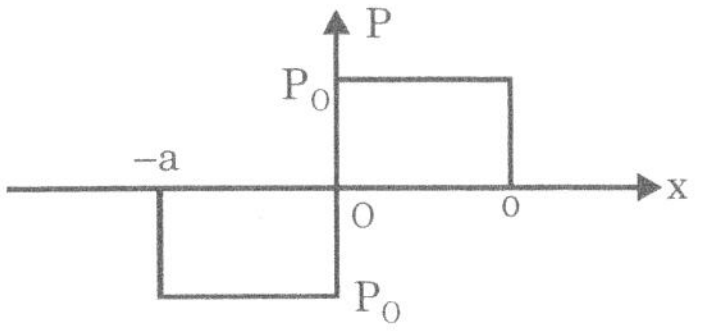

Select the correct answer using the codes given below:

Codes:

(*a*) 1 and 4

(*b*) 2 and 3

(*c*) 1, 2 and 3

(*d*) 2, 3 and 4 **[ES]**

49. Two point charges Q and –Q are located on the surface of a sphere of radius R as shown in the figure. The average electric field in the volume of the sphere is

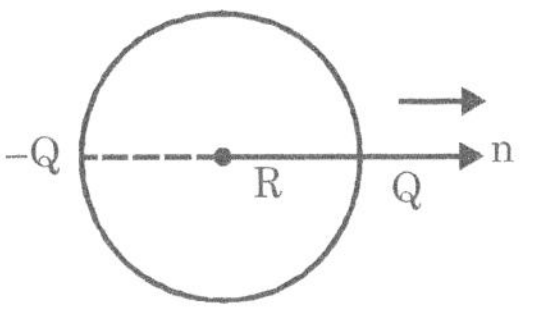

(*a*) inifinite (*b*) $\frac{2Q}{4\pi \epsilon_0 R^2}(-\vec{n})$

(*c*) $\frac{2Q}{4\pi \epsilon_0 R^2}(+\vec{n})$ (*d*) zero **[ES]**

50. If the $\vec{E}$ field in a plane electromagnetic wave travelling along the *z*-axis is given by

$\vec{E} = (\vec{a}_x + \vec{a}_y)f(\omega t - \beta z) + (\vec{a}_x + \vec{a}_y)f(\omega t + \beta z)$

The H field associated with the wave is (z_0 is the characteristic impedance)

(*a*) $(\vec{a}_x + \vec{a}_y)f(\omega t - \beta z)/z_0 + (\vec{a}_x + \vec{a}_y)f(\omega t + \beta z)/z_0$

(*b*) $(\vec{a}_x - \vec{a}_y)f(\omega t - \beta z)/z_0 + (\vec{a}_x + \vec{a}_y)f(\omega t + \beta z)/z_0$

(*c*) $(\vec{a}_y - \vec{a}_x)f(\omega t - \beta z)/z_0 + (-\vec{a}_y - \vec{a}_x)f(\omega t + \beta z)/z$

(*d*) $(\vec{a}_x - \vec{a}_y)f(\omega t - \beta z)/z_0 + (\vec{a}_y + \vec{a}_x)f(\omega t + \beta z)/z$

[IAS]

51. A dielectric material in an electric field gets polarized. If the polarization is non-uniform, then there would be accumulation of

(*a*) bound charge only on the surface

(*b*) bound charge only within the material

(*c*) bound charge within the material as well as on the surface

(*d*) free charge on the surface of the dielectric

[IAS]

52. Consider the following statements:

If vector $\bar{r} = \hat{x}x + \hat{y}y + \hat{z}z$ then

1. $\bar{r}.\nabla\times\bar{r} = 0$
2. $\nabla\times\bar{r} = 3$
3. $\nabla(\bar{r}.\bar{r}) = 2\bar{r}$
4. $\nabla(\nabla\times\bar{r}) = 0$

Among these statements:

(*a*) 1 and 4 are correct
(*b*) 2, 3 and 4 are correct
(*c*) 1, 2 and 3 are correct
(*d*) 1, 2, 3 and 4 are correct **[IAS]**

53. Match List-I (Phenomena) with List-II (Reasons) and select the correct answer using the codes given below the lists:

List-I

A. Force on a unit north pole due to current.
B. Flux density-intensity relation in a magnetic material.
C. Force of attraction between a current carrying conductor.
D. Identically zero magnetic field outside a coaxial cable.

List-II

1. Magnetic Polarisation.
2. Maximisation of magnetic lines of force.
3. Work done along a closed path is equal to total current flow.
4. Force between magnetic elements.

Codes:

	A	B	C	D
(*a*)	3	1	2	4
(*b*)	4	1	2	3
(*c*)	3	2	1	4
(*d*)	4	2	1	3

[ES]

54. Two rectangular loops and an infinite wire, all carrying a current 'I' are shown in the given figure.

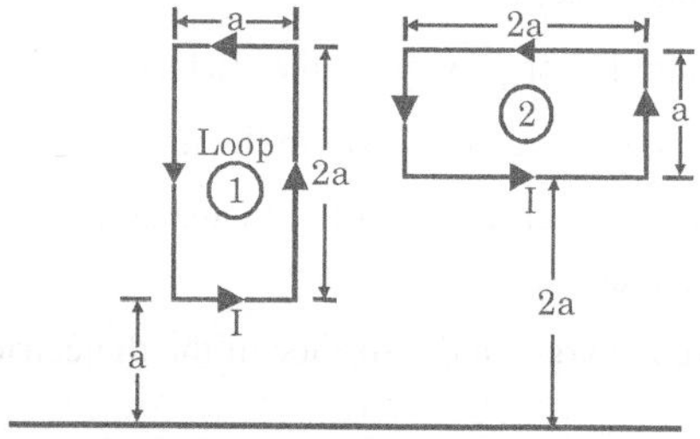

If the force on loop 1 due to the infinite wire were 'F', then the force on loop 2 due to the infinite wire would be

(*a*) $\frac{F}{3}$ (*b*) $\frac{F}{2}$

(*c*) F (*d*) $\frac{3}{2}F$ **[ES]**

55. An electromagnetic field is radiated from

(*a*) a stationary point charge
(*b*) a capacitor with a DC voltage
(*c*) a conductor carrying a DC current
(*d*) an oscillating dipole **[NTPC]**

56. Consider the components of $\vec{E}$ and $\vec{H}$ listed below:

1. E_x 2. E_y
3. E_z 4. H_y
5. H_z

For a plane wave travelling in *x*-direction and linearly polarizes in *y*-direction

(*a*) 1, 2 and 5 exist
(*b*) 1, 3 and 4 exist
(*c*) 3 and 4 exist
(*d*) 2 and 5 exist **[IAS]**

57. The magnetic field intensity $\vec{H}$ in A/m at the point (0, 0, 2) due to a current sheet $\vec{K} = 8\vec{a}_x$ A m lying in the $z = 0$ plane is

(*a*) $2\vec{a}_y$ (*b*) $-8\vec{a}_y$

(*c*) $-4\vec{a}_y$ (*d*) $-\vec{a}_y$ **[IAS]**

58. If a conductor of cross-section A and carrying a current lay is oriented along the *y*-axis in a magnetic field $\vec{B} = B_o\vec{a}_x + B_o\vec{a}_y$ the force density exerted on the conductor is

(*a*) $+\frac{B_oI}{a}\vec{a}_z$ N/m^3 (*b*) $+\frac{B_oI}{a}\vec{a}_x$ N/m^3

(*c*) $+\frac{B_oI}{a}\vec{a}_z$ N/m^3 (*d*) $+\frac{B_oI}{a}\vec{a}_y$ N/m^3 **[ES]**

59. Curl curl $\bar{A}$ is given by

(*a*) $\nabla(\nabla.\bar{A}) - \nabla^2\bar{A}$
(*b*) $\nabla(\nabla.A) - \nabla^2\bar{A}$
(*c*) $\nabla(V.\bar{A}) + \nabla^2\bar{A}$
(*d*) $\nabla(V.A) - \nabla\bar{A}$ **[NTPC]**

60. There concentric spherical shells of radii R_1, R_2, R_3 ($R_1 < R_2 < R_3$) carry charges –1, – 2 and 4 coulombs, respectively. The charge in coulombs on the inner and outer surface respectively, of the outermost shell is
(*a*) 0 and 4
(*b*) 3 and 1
(*c*) – 3 and 7
(*d*) – 2 and 6 **[ES]**

61. An electric dipole AR and a circular loop are shown in the given figure. The plane of the loop is just above the point B. The displacement current through the plane area bounded by the loop and in the direction of I is

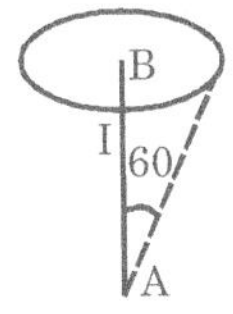

(*a*) 1/4
(*b*) 1/2
(*c*) 1
(*d*) –1/4 **[BHEL]**

62. If v, *w*, *q* for voltage, energy and charge, then v can be expressed as
(*a*) $v = \frac{dq}{dw}$ (*b*) $v = \frac{dw}{dq}$
(*c*) $dv = \frac{dw}{dq}$ (*d*) $dv = \frac{dq}{dw}$ **[GATE]**

63. Vector $\vec{A}$ is expressed as $\vec{A} = \nabla \times \vec{F} + \nabla f$ where F is a vector and f is a scalar function that satisfies $\nabla^2 f + k^2 f = 0$, with $f = 0$ on a closed surface 'S; inside which the differential equation holds. What will be the divergence of $\vec{A}$ integrated over the closed surface? (K^2 is a scalar constant)
(*a*) k^2
(*b*) $-k^2$
(*c*) zero
(*d*) $k^2 f$ **[ES]**

Directions: *The following nine (9) items consist of two statements, one labelled the 'Assertion (A)' and the other labelled the 'Reason (R)'. You are to examine these two statements carefully and decide if the Assertion (A) and the Reason (R) are individually true and if so, whether the Reason is a correct explanation of the Assertion. Select your answers to these items using the codes given below and mark your Answer Sheet accordingly.*

Codes:
(*a*) both A and R are true and R is the correct explanation of A
(*b*) both A and R are true but R is not a correct explanation of A
(*c*) A is true but R is false
(*d*) A is false but R is true

64. **Assertion (A):** For a uniformly charged dielectric solid sphere, the magnitude of radial electric field at a point inside the sphere is proportional to the distance between the centre and the point, while outside the sphere, it is inversely proportional to the square of the distance between the centre and the point.
Reason (R): Field inside the sphere is due to a uniformly charged sphere of radius equal to the distance of the the point from the centre whereas outside the sphere, it is due to the whole charged sphere. **[ES]**

65. **Assertion (A):** A vector $\overline{A} = 2y\,\overline{i}_x - x\,\overline{i}_y$ represents a magnetic field.
Reason (R): $\overline{\nabla}.\overline{A} = 0$ **[IAS]**

66. **Assertion (A):** In a high voltage cable, concentric conducting surfaces are provided in an insulating medium of same permittivity throughout the cross section of the cable to reduce the ratio of maximum to minimum electric stress.
Reason (R): The potential differences between the conducting surfaces redistribute the charge distribution. **[ES]**

67. **Assertion (A):** The electric field intensity ξ is proportional to *x*, the distance from the axis of the conductor of radius, *r*, for $x \geq r$.
Reason (R): Bundled conductors increase the line capacitance from that of a single conductor of the same cross-sectional area. **[ES]**

68. **Assertion (A):** In general, circulation of the magnetic field intensity is zero.
Reason (R): Vector source of circulation is the current density. **[ES]**

69. **Assertion (A):** Surface charge density is considered a physical possibility for either dielectrics or perfect conductors.
Reason (R): Surface current density is assumed only in conjunction with perfect conductors. **[IAS]**

70. Two parallel conducting plates carrying charges + Q and – Q form a capacitor as shown in the figure. A dielectric slab is partly introduced in the gap and released.

Assertion (A): The force acting on the slab will tend to push it outward.

Reason (R): According to Thomson's theorem, in a charge distribution in equilibrium, the energy in the electrostatic field is always minimum. **[ES]**

71. **Assertion (A):** The emf induced in the loop is maximum during the interval the loop is fully inside the region of $\overline{B}$.

Reason (R): Induced emf $= -\frac{\partial}{\partial t}$ (flux linkage). **[ES]**

72. **Assertion (A):** A dipole placed in a uniform field experiences a torque tending to align the dipole axis with the field.

Reason (R): A dipole placed in a uniform field does not experience a translational force. **[ES]**

73. A magnetic field $\vec{B} = (\vec{a}_x + 2\vec{a}_y - 4\vec{a}_z)$ exists at a point. If a test charge moving with a velocity $\vec{V} = V_0(3\vec{a}_x - \vec{a}_y + 2\vec{a}_z)$ experiences no force at a certain point, the electric field at that point will be

(*a*) $\vec{E} = -V_0(3\vec{a}_x - 2\vec{a}_y - 4\vec{a}_z)$

(*b*) $\vec{E} = -V_0(14\vec{a}_y + 7\vec{a}_z)$

(*c*) $\vec{E} = -\frac{V_0}{\mu}(14\vec{a}_y + 7\vec{a}_z)$

(*d*) $\vec{E} = +\frac{V_0}{\mu}(14\vec{a}_y + 7\vec{a}_z)$ **[PGCIL]**

74. What is the work done by a constant force $F = (3\vec{x} + 4\vec{y})$ N in moving an object over a straight line path defined by means of the vector $R = (-\vec{x} + 6\vec{y} + 3\vec{z})m$?

($\vec{x}$, $\vec{y}$ and $3\vec{z}$ are unit vectors parallel to *x*, *y* and z-axis)

(*a*) zero (*b*) 127

(*c*) 15J (*d*) 21J **[IOCL]**

75. The term(s) responsible for attenuation in a source-free wave equation

$$\nabla^2\vec{E} - \mu\mu_0 \in\in_0 \frac{\partial\vec{E}}{\partial^2 t} - \mu\mu_0\sigma\frac{\partial\vec{E}}{\partial t} = 0$$

is/are given by

(*a*) $\mu\mu_0\sigma\frac{\partial\vec{E}}{\partial t}$

(*b*) $\mu\mu_0 \in\in_0 \frac{\partial^2\vec{E}}{\partial^2 t}$

(*c*) $\nabla^2\vec{E}$ and $\mu\mu_0\sigma\frac{\partial^2\vec{E}}{\partial t}$

(*d*) $\mu\mu_0\sigma\frac{\partial\vec{E}}{\partial t}$ and $\mu\mu_0 \in\in_0 \frac{\partial^2\vec{E}}{\partial^2 t}$ **[IOCL]**

76. A parallel plate capacitor with air as dielectric can withstand a voltage V. If the capacitor is half filled with a liquid of dielectric constant 2, then the capacitor will withstand

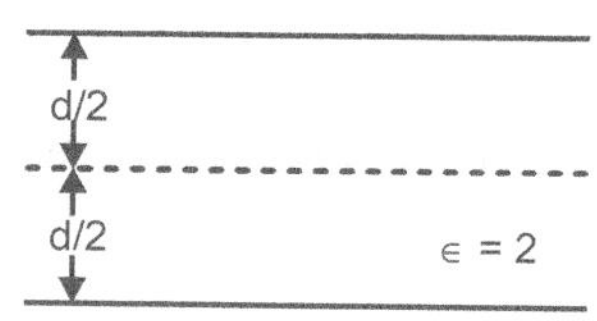

(*a*) 2V (*b*) $\frac{4}{3}$V

(*c*) $\frac{3}{4}$V (*d*) $\frac{1}{2}$V **[IOCL]**

77. Which one of the following statements is correct?

If the Fermi level lies midway between the conduction and valance bands, then the semiconductor is

(*a*) intrinsic (*b*) extrinsic

(*c*) *p*-type (*d*) *n*-type **[IOCL]**

78. Which one of the following statements is correct?

A wave guide can be considered to be analogous to a

(*a*) low pass filter

(*b*) high pass filter

(*c*) band pass filter

(*d*) band stop filter **[IOCL]**

79. For time-varying electromagnetic fields with electric and magnetic field given by |E| and |H| respectively the rate of energy flow per unit area has a magnitude given by

(*a*) |E| |H| cos θ

(*b*) |E| |H| sin θ

(*c*) (|E| + |H|) sin θ

(*d*) (|E| – |H|) cos θ **[IAS]**

80. A thin copper tube and a co-axial cable with a thin outer conductor are shown in the given figure. The copper tube carries a current IA and the coaxial cable carries a load of IA. The pressure on the thin copper of the cable are, respectively

(*a*) radially outward and radially outward

(*b*) radially outward and radially inward

(*c*) radially inward and radially inward

(*d*) radially inward and radially outward

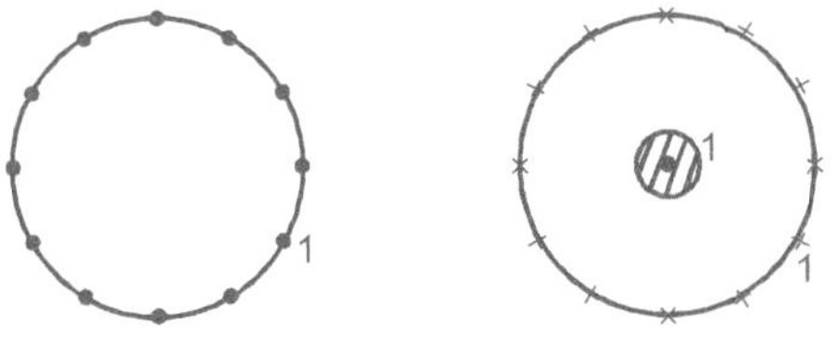

[ES]

81. When a charge is given to a conductor

(*a*) it distributes uniformly all over the surface

(*b*) it distributes uniformly all over the volume

(*c*) it distributes on the surface, inversely proportional to the radius of curvature

(*d*) it stays where it was placed [GATE]

82. Electromagnetic waves may be diffracted

(*a*) by reflection from the ground

(*b*) when encountered with a spherical wave front

(*c*) around the edge of a sharp obstacle

(*d*) while passing through a large slot. [ES]

83. A spherical conductor of radius V with charge '*q*' is placed concentrically inside an uncharged and unearthed spherical conducting shell of inner and outer radii r_1 and r_2 respectively. Taking potential to be zero at infinity, the potential at any point P within the shell ($r_1 < r < r_2$) will be

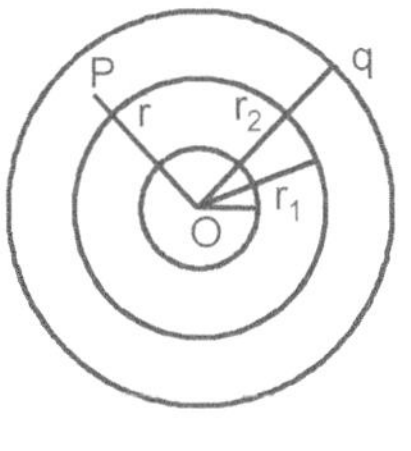

(*a*) $\dfrac{q}{4\pi\varepsilon_0 r}$

(*b*) $\dfrac{q}{4\pi\varepsilon_0 a}$

(*c*) $\dfrac{q}{4\pi\varepsilon_0 r_2}$

(*d*) $\dfrac{q}{4\pi\varepsilon_0 r_1}$ [GATE]

84. The value of electric field at a distance of 1*m* from an infinite line charge of density 1 C/*m* is

(*a*) $\dfrac{1}{2\pi\varepsilon_0}$ (*b*) $2\pi\varepsilon_0$

(*c*) $\dfrac{\varepsilon_0}{2\pi}$ (*d*) $\dfrac{\varepsilon_0}{4\pi}$ [IAS]

85. In the given figure, if the magnetic flux through the rectangular loop '*pqrsp*' due to the infinite line current I is ϕ, the flux linked with double rectangular loop '*pqtuvwrsp*' will be

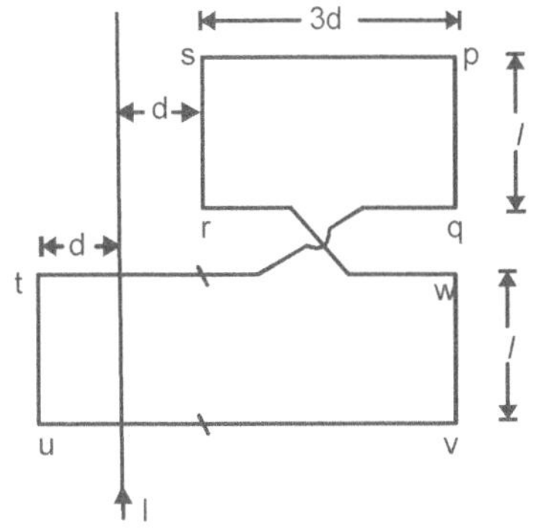

(*a*) 3ϕ (*b*) 2ϕ

(*c*) ϕ (*d*) zero [ES]

86. An electric dipole of moment P is placed in front of a grounded sphere as shown in the figure. The charge induced on the surface of the sphere is

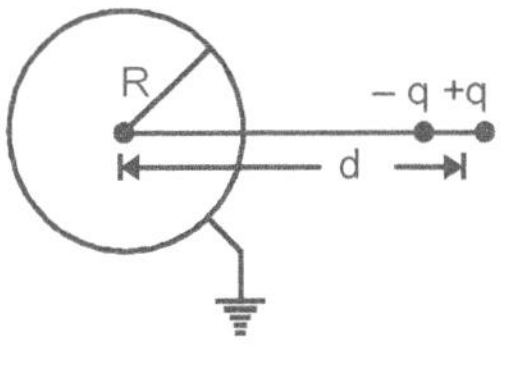

(*a*) zero

(*b*) $\dfrac{PR}{d^2}$

(*c*) $\dfrac{PR^2}{d^3}$

(*d*) $\dfrac{PR^2}{d^4}$ [NTPC]

87. Consider the following statements:

For a uniform place electromagnetic wave

1. The direction of energy flow is the same as the direction of propagation of the wave.
2. Electric and magnetic fields are in time quadrature.
3. Electric and magnetic fields are in space quadrature

Of these statements :

(*a*) 2 alone is correct

(*b*) 1 and 2 are correct

(*c*) 1 and 3 are correct

(*d*) 3 alone is correct [ES]

88. A magnetic field $\overline{B}$ and an electric field $\overline{E}$ are at right angles to each other along the y-axis and x-sxis respectively as shown in the given figure. The path of an electron which starts at the origin will be

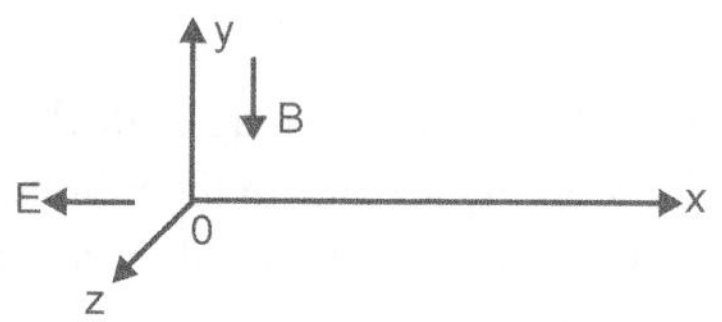

(*a*) straight along the x-axis
(*b*) cycloidal along the x-axis
(*c*) straight along the z-axis
(*d*) cycloidal along the z-axis **[ES]**

89. A loss-less transmission line of finite length is switched on to a voltage source at $t = 0$, as shown in the figure. The ratio of the initial and final values of the current fed by the battery will be

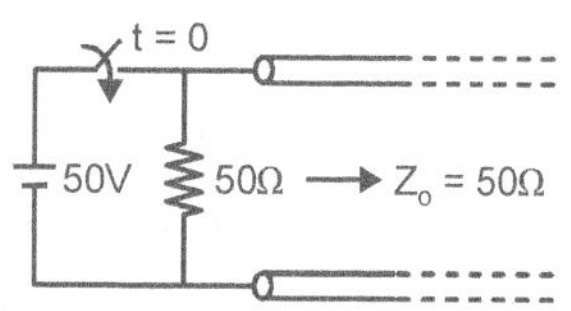

(*a*) 2:1
(*b*) 1:2
(*c*) 1:1
(*d*) 4:1 **[ES]**

90. Inside a hollow conducting sphere
(*a*) electric field is zero
(*b*) electric field is a non-zero constant
(*c*) electric field changes with the magnitude of the charge given to the conductor
(*d*) electric field changes with distance from the centre of the sphere **[GATE]**

91. The electric field intensity of a uniform plane wave in air is

$$\vec{E} = 20e^{-j5z}\vec{a}_x - 25e^{+j5z}\vec{a}_y v/m$$

(*a*) $H_x = +\dfrac{20}{120\pi}\vec{a}_x;\ H_y = -\dfrac{25}{120\pi}\vec{a}_y$

(*b*) $H_x = +\dfrac{25}{120\pi}\vec{a}_x;\ H_y = -\dfrac{20}{120\pi}\vec{a}_y$

(*c*) $H_x = \dfrac{25}{120\pi}\vec{a}_x;\ H_y = -\dfrac{20}{120\pi}\vec{a}_y$

(*d*) $H_x = -\dfrac{20}{120\pi}\vec{a}_x;\ H_y = -\dfrac{25}{120\pi}\vec{a}_y$ **[ES]**

92. For a transmission line terminated by a load, the reflection co-efficient magnitude $|\Gamma|$ and voltage standing wave ratio S are related as
(*a*) $S = 1/(1-|\Gamma|)$
(*b*) $S = 1/(1+|\Gamma|)$
(*c*) $S = (1-|\Gamma|)/(1+|\Gamma|)$
(*c*) $S = (1+|\Gamma|)/(1-|\Gamma|)$ **[ES]**

93. A plane wave propagates in distilled water ($\epsilon_r = 81$). If the peak electric field is 20π V/m, then the peak magnetic field intensity will be
(*a*) 1.50 A/m (*b*) 5.0 A/m
(*c*) 10π A/m (*d*) 20 A/m **[IAS]**

94. Match List-I with List-II and select the correct answer using the codes given below the lists:

List-I	**List-II**
A. $\overline{\nabla}.\overline{D} = \rho$	1. Ampere's Law
B. $\overline{\nabla}.\overline{J} = \dfrac{\partial\rho}{\partial t}$	2. Gauss's Law
C. $\overline{\nabla}\times\overline{H} = \vec{J}_c$	3. Faraday's Law
D. $\overline{\nabla}\times\overline{E} = -\dfrac{\partial\vec{B}}{\partial t}$	4. Continuity equation

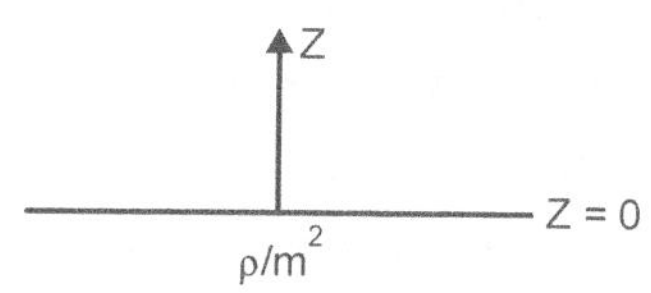

Codes:

	A	B	C	D
(*a*)	4	2	1	3
(*b*)	2	4	1	3
(*c*)	4	2	3	1
(*d*)	2	4	3	1

[ES]

95. An infinite plane surface carries a uniform change density of m² as shown in the figure. Electric field $\vec{E}$ at any point Z is given by

(*a*) $\vec{E} = \dfrac{\rho}{\varepsilon_0}\dfrac{Z}{|Z|}\vec{a}_z$

(*b*) $\vec{E} = \dfrac{\rho}{\varepsilon_0}\dfrac{Z}{|Z|}\vec{a}_z,\quad Z > 0$
$= 0 \quad Z = 0$

(*c*) $\vec{E} = \dfrac{\rho}{\varepsilon_0}\dfrac{Z}{|Z|}\vec{a}_z,\quad |Z| = 0$
$= 0 \quad Z = 0$

(*d*) $\vec{E} = \dfrac{\rho}{\varepsilon_0}\dfrac{Z}{|Z|}\vec{a}_z,\quad Z > 0$
$= -\dfrac{\rho}{\varepsilon_0}\dfrac{Z}{|Z|}\vec{a}_z,\quad Z = 0$
$= 0 \quad Z = 0$ **[ES]**

96. In the case of RC driving-point functions

(*a*) Z_{RC} cannot have a pole at infinity

(*b*) Z_{RC} cannot have a pole at the origin

(*c*) Y_{RC} cannot have a pole at infinity

(*d*) Z_{RC} & Y_{RC} are constant or zero at infinity **[ES]**

97. A straight current-carrying conductor and two conducting loops A and B are shown in the given figure. If the current in the straight wire is decreasing, then the induced currents in the two loops A and B will be

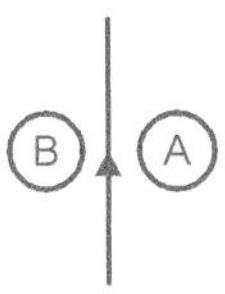

(*a*) clockwise in both A and B

(*b*) anticlockwise in both A and B

(*c*) anticlockwise in A and clockwise in B

(*d*) clockwise in A and anticlockwise in B **[ES]**

98. Consider the following statements associated with a parallel plate capacitor:

1. Capacitance is proportional to area of plates.
2. Capacitance is inversely proportional to distance of separation of plates.
3. The dielectric material is in a state of compression.

Of these statements :

(*a*) 1, 2 and 3 correct (*b*) 1 and 2 are correct

(*c*) 1 and 3 are correct (*d*) 2 and 3 are correct **[ES]**

99.

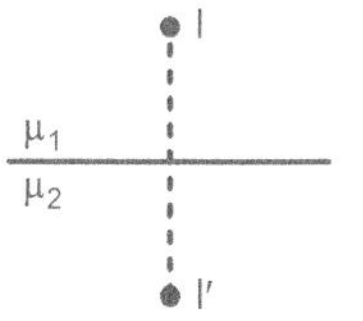

An infinitely long conductor carrying a current I in a region of permeability μ_1 and in the vicinity of another region of permeability μ_2 as shown in the given figure. For calculating the field in the first region, the effect of discontinuity is replaced by an image I′ and with $\mu_1 < \mu_2$. Which one of the following statements is correct in this regard?

(*a*) I and I¢ have the same strengths and direction

(*b*) I and I¢ have opposite direction but the same strength

(*c*) I and I¢ have the same direction but different strengths

(*d*) I and I¢ have opposite direction and different strengths. **[ES]**

100. The image of the dipole of moment P in the grounded sphere shown in the given figure is

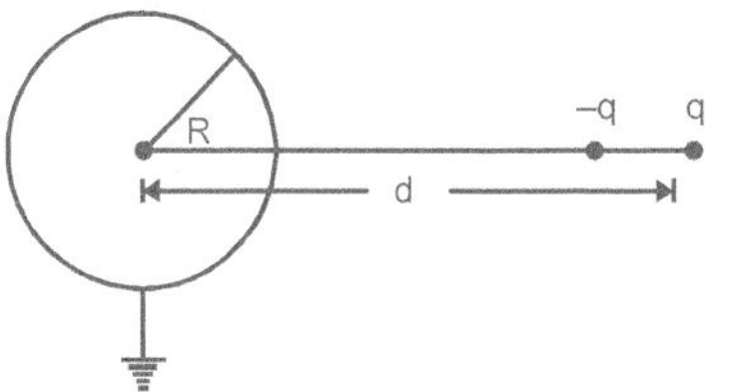

(*a*) a dipole of moment P

(*b*) a dipole of moment $\dfrac{PR}{d}$

(*c*) a dipole of moment $\dfrac{PR^2}{d^2}$

(*d*) not a dipole **[ES]**

ANSWERS

EXERCISE - I

1.(*a*)	2.(*c*)	3.(*d*)	4.(*b*)	5.(*a*)	6.(*a*)	7.(*a*)	8.(*a*)	9.(*a*)	10.(*a*)
11.(*c*)	12.(*c*)	13.(*a*)	14.(*a*)	15.(*d*)	16.(*d*)	17.(*c*)	18.(*b*)	19.(*b*)	20.(*b*)
21.(*a*)	22.(*c*)	23.(*d*)	24.(*c*)	25.(*c*)	26.(*a*)	27.(*d*)	28.(*b*)	29.(*b*)	30.(*c*)
31.(*a*)	32.(*a*)	33.(*c*)	34.(*d*)	35.(*b*)	36.(*c*)	37.(*d*)	38.(*c*)	39.(*d*)	40.(*a*)
41.(*b*)	42.(*d*)	43.(*b*)	44.(*c*)	45.(*a*)	46.(*a*)	47.(*a*)	48.(*c*)	49.(*c*)	50.(*d*)
51.(*b*)	52.(*c*)	53.(*a*)	54.(*c*)	55.(*a*)	56.(*b*)	57.(*d*)	58.(*b*)	59.(*d*)	60.(*b*)
61.(*a*)	62.(*a*)	63.(*a*)	64.(*b*)	65.(*b*)	66.(*d*)	67.(*b*)	68.(*d*)	69.(*d*)	70.(*b*)
71.(*b*)	72.(*d*)	73.(*c*)	74.(*b*)	75.(*a*)	76.(*a*)	77.(*b*)	78.(*c*)	79.(*d*)	80.(*a*)
81.(*d*)	82.(*c*)	83.(*a*)	84.(*a*)	85.(*c*)	86.(*d*)	87.(*d*)	88.(*d*)	89.(*d*)	90.(*c*)
91.(*a*)	92.(*b*)	93.(*b*)	94.(*b*)	95.(*c*)	96.(*d*)	97.(*d*)	98.(*d*)	99.(*c*)	100.(*c*)
101.(*b*)	102.(*a*)	103.(*c*)	104.(*d*)	105.(*b*)	106.(*d*)	107.(*a*)	108.(*c*)	109.(*a*)	110.(*b*)
111.(*a*)	112.(*d*)	113.(*b*)	114.(*b*)	115.(*c*)	116.(*d*)	117.(*d*)	118.(*a*)	119.(*a*)	120.(*c*)
121.(*c*)									

EXERCISE – II

1. (*d*)	**2.** (*b*)	**3.** (*c*)	**4.** (*d*)	**5.** (*a*)	**6.** (*d*)	**7.** (*b*)	**8.** (*c*)	**9.** (*a*)	**10.** (*b*)
11. (*c*)	**12.** (*b*)	**13.** (*a*)	**14.** (*c*)	**15.** (*a*)	**16.** (*d*)	**17.** (*b*)	**18.** (*b*)	**19.** (*b*)	**20.** (*c*)
21. (*c*)	**22.** (*d*)	**23.** (*b*)	**24.** (*b*)	**25.** (*a*)	**26.** (*d*)	**27.** (*c*)	**28.** (*d*)	**29.** (*a*)	**30.** (*b*)
31. (*c*)	**32.** (*b*)	**33.** (*c*)	**34.** (*a*)	**35.** (*d*)	**36.** (*d*)	**37.** (*b*)	**38.** (*c*)	**39.** (*b*)	**40.** (*a*)
41. (*a*)	**42.** (*b*)	**43.** (*c*)	**44.** (*c*)	**45.** (*b*)	**46.** (*d*)	**47.** (*c*)	**48.** (*b*)	**49.** (*b*)	**50.** (*d*)
51. (*c*)	**52.** (*b*)	**53.** (*c*)	**54.** (*b*)	**55.** (*d*)	**56.** (*a*)	**57.** (*c*)	**58.** (*c*)	**59.** (*a*)	**60.** (*c*)
61. (*a*)	**62.** (*b*)	**63.** (*c*)	**64.** (*a*)	**65.** (*a*)	**66.** (*a*)	**67.** (*d*)	**68.** (*d*)	**69.** (*b*)	**70.** (*d*)
71. (*d*)	**72.** (*b*)	**73.** (*b*)	**74.** (*d*)	**75.** (*b*)	**76.** (*c*)	**77.** (*a*)	**78.** (*b*)	**79.** (*b*)	**80.** (*d*)
81. (*a*)	**82.** (*c*)	**83.** (*a*)	**84.** (*a*)	**85.** (*b*)	**86.** (*b*)	**87.** (*c*)	**88.** (*d*)	**89.** (*a*)	**90.** (*a*)
91. (*b*)	**92.** (*d*)	**93.** (*a*)	**94.** (*b*)	**95.** (*a*)	**96.** (*a*)	**97.** (*d*)	**98.** (*a*)	**99.** (*c*)	**100.** (*d*)

EXPLANATIONS

EXERCISE – I

1.

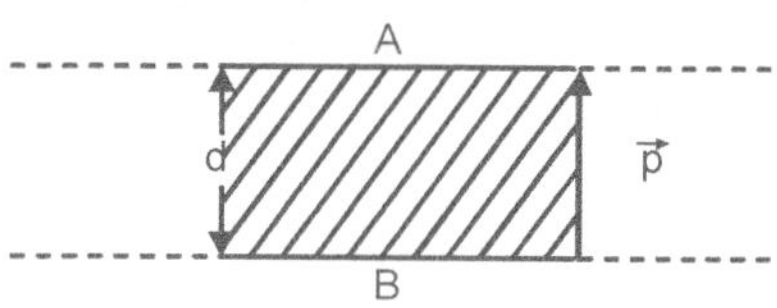

We have, $P = x_e \epsilon_0 E$

$$\therefore \quad E = \frac{P}{x_e \epsilon_0} = \frac{P}{\epsilon_0 (\epsilon_R - 1)}$$

$\therefore$ Potential difference between points A and B,

$$= \frac{P.d}{\epsilon_0 (\epsilon_R - 1)}$$

4.

$$\vec{B} = B_0 \cos \omega \vec{a}_y$$

$$\therefore \quad \frac{dB}{dt} = -B_0 \omega \sin \omega t$$

We have, emf generated

$$= -\frac{d\phi}{dt}$$

$$= -\frac{d}{dt}(\text{Area flux density})$$

$$= -a.b.\, B_0 \omega \sin \omega t$$

$$|\text{emf}| = ab\, B_0 \omega \sin \omega t$$

5.

$$W = -Q \int_{\text{initial}}^{\text{final}} E.dL = \int_{\infty}^{1} \frac{Q}{4\pi r^2 \epsilon_0} dr$$

$$= -\frac{Q}{4\pi \epsilon_0}\left[\frac{1}{r}\right]$$

$$= -\frac{-2}{4\pi \epsilon_0} \times \frac{1}{1} = \frac{-1}{2\pi \epsilon_0}$$

9. For free space, $\eta = 120\pi$

$$\text{Now, } H_y = \frac{E_x}{\eta}$$

$$= \frac{15\cos(6\pi \times 10^8 t - 2\pi z)}{120\pi}$$

$$= \frac{1}{8\pi} \cdot \cos(6\pi \times 10^8 t - 2\pi z)$$

$$\therefore \quad \bar{B} = \mu_o H = \frac{4\pi \times 10^{-7}}{8\pi} \cos(6\pi \times 10^8 t - 2\pi z)$$

$$= 5 \times 10^{-8} \cos(6\pi \times 10^8 t - 2\pi z)\, i_y$$

11.

$$B_1 = 5\vec{a}_x + \vec{a}_z$$

$$\therefore \quad B_{N_1} = 8a_z$$

At boundary: $B_{N_1} = B_{N_2}$

$$\Rightarrow \quad \mu_2 H_{N_2} = B_{N_1}$$

$$\Rightarrow \quad \vec{H}_{N_1} = \frac{8}{6\mu_0} \vec{a}_z \text{ mA/m}$$

$$\vec{H}_{r_1} = \frac{5}{4\mu_0} \vec{a}_x$$

$$\vec{H}_2 = \vec{H}_{t_2} + \vec{H}_{N_2}$$

$$= \frac{5}{4\mu_0} \vec{a}_x + \frac{8}{6\mu_0} \vec{a}_z \text{ mA/m}$$

18. If we consider uniform charge at whole the circle then field at 0 will be zero.

Now we are considering section AB an BCA separately. The field due to section BCA when the AB is uncharged will be opposite and equal to field due to AB (When BCA uncharged).

Hence, field due to BCA only $= -\vec{E}$

20. We have, $\nabla \times E = \frac{\partial B}{\partial t}$

$\Rightarrow$ $E_x . a_x = B_o z\omega \sin \omega t\, a_x$

$\Rightarrow$ $E_x = B_o z\omega \sin \omega t$

24. Direction of flux density produced by all elements of two vertical section of the current shown, at points on PQ is perpendicular to the plane of the paper and outward. This can be checked with $\vec{I} . \vec{B} : \vec{Idi} \times \vec{I}_R$

Then force on an element PQ,

$$d\vec{F} = I\vec{di} \times \vec{B}$$

Keeping in view direction of *di* (which is same as that of I) at points on PQ, direction of force on PQ is along $-\vec{n}$.

25. Velocity of wave in overhead transmission line,

$$v_1 = 3 \times 10^8$$

Velocity in the cable,

$$v_2 = \frac{3 \times 10^8}{\sqrt{4}} = \frac{3}{2} \times 10^8$$

$$\therefore \quad \frac{v_1}{v_2} = 2$$

28. Circular current carrying loop is seen to be lying in the *x-y* plane. To find average direction of magnetic vector potential A at point P_1 one may imagine the loop in two halves. One half which is nearer to point P, effectively carrying current in y-direction and the other half which is relatively further away carrying current-y direction. Since average distances are different, magnetic vector potential caused at P will be different, neralry half dominating. Hence potential will have a direction along *ay*.

29. $\int \vec{B}, \vec{dz}$ from $-\infty$ to $+\infty$ has the same magnitude and sign as *dz* from ∞ to $-\infty$ provided path from ∞ to $-\infty$ is beyond R.

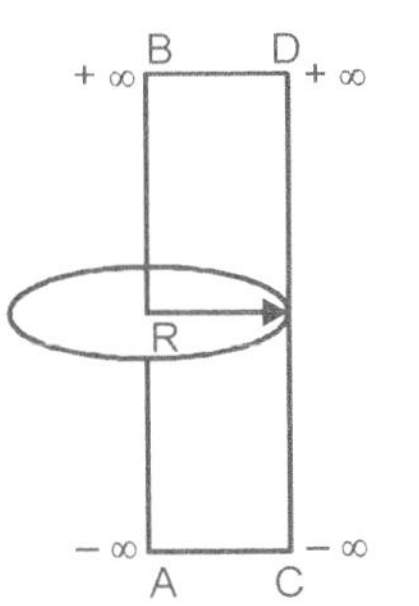

Referring figure $\int_{-\infty}^{\infty} \vec{B} dz$ via AOB is the same as $\int_{-\infty}^{\infty} \vec{B} dz$ via ACDB.

Apply Ampere's circuital law, we have

$$\oint_C B di = \mu o I$$

$$\int_{-\infty}^{\infty} \vec{B}\, di + \int_{\infty}^{-\infty} \vec{B}\, di = z \int_{\infty}^{-\infty} \vec{B}.di$$

$$\therefore \quad \int_{-\infty}^{\infty} \vec{B}.dz = \frac{\mu_0 I}{2} \text{ or } \frac{\mu I}{2}$$

34. $n = \sqrt{\frac{\mu_0 \mu_R}{\epsilon_0 \epsilon_R}} = 120\pi \sqrt{\frac{2}{8}} = 60\pi$

$$H_{oz} = A = \frac{E_{\mu o}}{\eta} = \frac{120\pi}{60\pi} = 2$$

$$B = \omega\sqrt{\mu \in}$$

$$= 10^6 \pi\sqrt{4\pi \times 10^{-7} \times 2 \times 8.85 \times 10^{-12} \times 8}$$

$$= 0.042$$

37. Force per unit area between the plates $= \frac{1}{2} \in E^2$

In first case, $\in_1 = \in_0$

$\therefore$ Force, $F_1 = F = \frac{1}{2} \in E_1^2$

In second case, $\in_2 = \in_0 \in_r$

$$B = \frac{E_1}{\in_r}$$

$$\text{Force per unit area} = \frac{1}{2} \epsilon_0 \epsilon_r \left(\frac{E_1}{\epsilon_r}\right)$$

$$= \frac{1}{2} \epsilon_0 \frac{E_1^2}{\epsilon_r} = \frac{F}{\epsilon_r}$$

38. The plane midway between Q and –Q is the one passing through ABC and perpendicular to the plane of the paper is an equipotential plane. Hence potential at B is the same as that at A or C i.e., IV.

41. R is the resistance per unit length and I is the current flowing

$\therefore$ Power loss occurring in the conductor per unit length $= I^2R$

Thus power is furnished via $\vec{E}$ and $\vec{H}$ fields at the surface of the wire.

Power flow per unit area $= \vec{E} \times \vec{H}$

$\vec{E} \times \vec{H}$ at every point on the surface is radial and directed into the surface.

$$\text{Poynting vector} = \frac{I^2R}{\text{Surface Area}}(-\vec{n})$$

$$= \frac{I^2R}{2\pi r}(-n)$$

where, $\vec{n}$ = unit radial vector directed outward.

42. Force between the wire is given by

$$F = \frac{q^2}{2\pi \in d^2}\left[1 - \frac{1}{(1 - l^2/d^2)}\right]$$

Assuming $l << d$, we get

$$F = \frac{q^2}{2\pi \in d^2}\left[1 - \frac{3l^2}{2d^2}\right]$$

$$\Rightarrow \quad F = \frac{3q^2 l^2}{4\pi \in d^4}$$

$$\therefore \quad \text{Force} \propto \frac{1}{d^4}$$

where, d = distance between the two dipoles

Now when distance between them increase to 2d, then force will become $\frac{1}{16}$ time of the previous value.

47. Mean poynting vector,

$$P = \frac{1}{2}E_m H_m$$

$$= \frac{1}{2}E_m^2\sqrt{\frac{\epsilon_0}{\mu_0}}$$

$$= \frac{1}{2}E_m^2 \frac{\epsilon_0}{\sqrt{\mu_0 \epsilon_0}}$$

$$= \frac{1}{2}E_m^2 \frac{\epsilon_0\sqrt{\mu_0 \epsilon_0}}{\mu_0 \epsilon_0}$$

$$= \frac{1}{2}\frac{E_m^2}{\mu_0 \omega}\left[\text{as } C = \frac{\omega}{\beta} = \frac{1}{\sqrt{\mu_0 \epsilon_0}}\right]$$

$$= \frac{1}{2}\frac{E_m^2}{\mu_0 \omega} = \frac{1}{2}\frac{B\mu_0 C^2 E_m}{\omega}$$

49. For on a charge moving in a magnetic field, $F = q\vec{v} \times \vec{B}$

Since force is perpendicular to both $\vec{V}$ and $\vec{B}$, hence magnitude of velocity $\vec{v}$ is not changed and thus kinetic energy at the entry exist points will be same.

50.

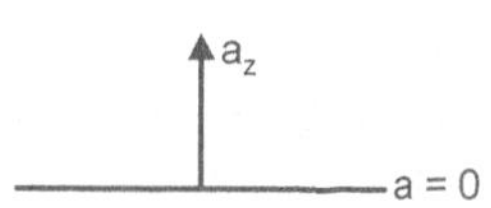

$$\vec{K} = (5\vec{ax} - 6\vec{ay})\text{A/m}$$

$$Ht_1 - Ht_2 = a_{N12} \times K$$

$$\therefore \quad 0 - Ht_2 = a_z(5\vec{ax} - 6\vec{ay})$$

$$\Rightarrow \quad -Ht_2 = +5ay + 6ax$$

$$\therefore \quad BN_1 = BN_2 = 0$$

$$\therefore \quad H_2 = Ht_2 = (6\vec{ax} + 5\vec{ay})$$

51. Electric field C due to B,

$$E_B = 800\underline{|0°}\text{ V/m}$$

Electric field at C due to A

Let AB = AD = CD = 1 m

$$\therefore \quad AC = \sqrt{2}$$

$$E\,\alpha\,\frac{1}{r^2}$$

$$\therefore \quad E_A = \frac{800 \times 1^2}{(\sqrt{2})^2}\underline{|45°}$$

$$= 400\underline{|45°}$$

E at C due to = $800\underline{|90}\ E_D$

∴ Total field at

$$C = E_A + E_B + E_D$$

$$= 400\underline{|45} + 800\underline{|0} + 800\underline{|90}$$

$$= 282.84 + j282.84 + 800 + 800j$$

$$= 1082.84 + 1082.84j$$

$$= \frac{400}{\sqrt{2}}(2\sqrt{2} + 1)\underline{|45°}$$

54. On doubling magnetic field, force on the magnet will be doubled.

Now, time period of oscillation,

$$T\,\alpha\,\frac{1}{\sqrt{F}}$$

$$\therefore \quad T \text{ becomes } \frac{T}{\sqrt{2}}$$

55. Vector magnetic potential,

$$d_A = \frac{\mu_0}{4\pi R}Idl$$

Hence, at P $\quad \int dA = \frac{\mu_0}{4\pi R}\int Idl$

Since there is a closed path for I, hence

$$\oint Idl = 0$$

$$\therefore \quad A \text{ at } P = 0$$

59. Electric field at x, $\vec{E} = \frac{12}{z+2} = 3$ V/cm

61. emf generated in the filament will be given by

$$e = \int_{x=0}^{x=R} B\omega\, dx$$

$$= B\omega\left[\frac{x^2}{2}\right]_0^R = \frac{B\omega R^2}{2}$$

62. Potential due to dipole is given by,

$$v = \frac{Qd\cos\theta}{4\pi \epsilon_0 r^2}$$

Hence to double magnitude of the potential, the distance between the charges should be doubled.

68. When dielectirc constant of the medium between two plates changes, charge on the plates is not affected. Since charge is not affected, the displacement density D also does not change but electric field changes because $d = \epsilon_0 \epsilon_r E$

Thus $$\frac{E_1}{E_2} = \frac{E_r}{1} = \epsilon_r$$

Force per unit area $= \frac{1}{2}\epsilon_0 \epsilon_r E^2$

$$\therefore \quad F_2 = \frac{1}{2}\epsilon_0 \epsilon_r E_2^2$$

$$= \frac{1}{2}\epsilon_0 \epsilon_r \frac{E_1^2}{E_r^2}$$

$$= \frac{1}{2}\frac{\epsilon_0 E_1^2}{\epsilon_r}$$

Thus $$\frac{F_2}{F_1} = \frac{1}{\epsilon_r}$$

$$\therefore \quad \text{Total force} = \frac{P}{\epsilon_r}$$

69. Depth of penetration is given by, $\delta = \frac{1}{\sqrt{\pi f \mu \sigma}}$

$\therefore$ depth of penetration $\alpha \frac{1}{\sqrt{f}}$

When frequency is increased four times, depth of penetration becomes decreased by a factor of two.

70. Characteristic impedance of the medium whith $n = \frac{E}{H}$ is 120 π for free space.

Also phase of H is the reflected component of the wave has to reverse, so that travel of the reflected wave is correctly given by the direction $\vec{E} \times \vec{H}$. $\vec{E}$ and $\vec{H}$ are perpendicular to each other in the incident component or the forward travelling component and backward travelling component of the wave.

$$\therefore \quad \vec{E} \times \vec{H} = \frac{1}{120\pi}[20\cos(\omega t - Bz) - 5\cos(\omega t + Bz)a_y \text{ A/m}]$$

72. For magnetic field, $\vec{\nabla} . \vec{B} = 0$

$$\therefore \quad \left(\frac{\partial}{\partial x}\vec{a}_x + \frac{\partial}{\partial y}\vec{a}_y + \frac{\partial}{\partial z}\vec{a}_z\right) . \left(x\vec{a}_x + y\vec{a}_y + kz\vec{a}_z\right) = 0$$

$$\Rightarrow \quad 1 + 1 + K \times 1 = 0$$

$$\Rightarrow \quad K = -2$$

74. Total charge on spheres 1 and 2 taken together $= -1 + (-2) = -3C$

This charge will induce + 3C on inner surface and –3C on the outer surface of sphere 3.

∴ Total charge on the inner surface of sphere 3 = 3C

∴ Total charge on the outer surface = –3 + 4 = 1C

75. Equivalent currents of the two rings are

$$I_1 = \frac{2\pi a \rho}{2\pi/\omega} = a\rho\omega$$

$$I_2 = \frac{2\pi . 2a\rho}{2\pi/\omega} = 2a\rho\omega$$

Magnetic flux density at 0 shall be inversely proportional to radii and directly proportional to equivalent current

$$\therefore \quad B_1 = \frac{Ka\rho\omega}{a}$$

$$B_2 = \frac{K.2a\rho\omega}{2a}$$

$$\therefore \quad K\rho\omega = K\rho\omega$$

$$\therefore \quad B_1 : B_2 = 1 : 1$$

76. From the direction of magnetic field, we can have the answers. The flux through loop 1st and 3rd loop will cancel out due to opposite direction of field due to two wire.

79. We have, $\vec{E} = 15\sin(\omega t - \pi z)\vec{a}_x + 15\cos(\omega t - \pi z)\vec{a}_y$ V/m

In above field, x and y – components of E have equal magnitudes and also these have a phase difference of 90°, the wave is circularly polarised.

82. Power carried by a wave, $P \propto E^2$ or H^2

$$\therefore \quad P = KE^2$$

$$\Rightarrow \quad E = \sqrt{\frac{P}{K}} = \sqrt{\frac{1}{K}}$$

Reflection coefficient = 1/2

∴ Transmission coefficient

$$= 1 - \frac{1}{3} = \frac{2}{3}$$

∴ Transmitted electric field

$$= \frac{2}{3}E = \frac{2}{3}\sqrt{\frac{1}{K}}$$

Power transmitted or absorbed by the load,

$$P_L = KE^2$$

$$= K.\frac{4}{9}.\frac{1}{K}$$

$$= \frac{4}{9}\omega$$

95. The surface change density ρ equals normal component of electric field $\vec{E}$. Potential difference between the two outer planes is zero.

i.e. $d_2E_2 + 0 - d_1E_1 = 0$

$$\Rightarrow \quad \frac{E_1}{E_2} = \frac{d_2}{d_1}$$

$$\therefore \quad \frac{\rho_1}{\rho_2} = \frac{d_2}{d_1}$$

97. $\in = 2$

$$D = \in_0 E + P$$

$$\therefore \quad \in_0 \in_r E = \in_0 E + P$$

$$\Rightarrow \quad E = \frac{P}{\in_0 (\in_r - 1)} = \frac{P}{\in_0 (2-1)}$$

$$= \frac{P}{\in_0}(-\vec{n})$$

99. Intrinsic impedance,

$$n = \sqrt{\frac{j\omega\mu}{\sigma}} = \sqrt{\frac{\mu\omega}{\sigma}}\,\underline{|45°}$$

101. Displacement current is given by

$$Id = \int_s \frac{\delta D}{\delta t}.ds$$

$$D = \in E$$

$$= 4 \in_0 \times 0.1 t e^{-t}.\, a_x$$

$$D = 0.4 \in_0 t e^{-t}.\, ax$$

$$\frac{\partial D}{\partial t} = 0.4 \in_0 [-t e^{-t} + e^{-t}]$$

At $t = 0$, $\frac{\partial D}{\partial t} = 0.4 \in_0$

$$\therefore \quad Id = 0.4 \in_0 \times 0.1$$

$$= 0.04 \in_0$$

106. At the centre, $v = \frac{Q}{4\pi \in_0 R}$

Now net potential due to all rings

$$v = \frac{Q}{4\pi\in_0}\left(\frac{1}{1} - \frac{1}{2} + \frac{1}{4} - \frac{1}{8} + \frac{1}{10} - - -\right)$$

$$= \frac{Q}{4\pi\in_0}\left[\left(\frac{1}{1} + \frac{1}{4} + \frac{1}{16} + \frac{1}{64} + \ldots\right) - \left(\frac{1}{2} + \frac{1}{8} + \frac{1}{32} + \ldots\right)\right]$$

$$= \frac{Q}{4\pi\in_0}\left[\frac{1}{1-1/4} - \frac{1/2}{1-1/4}\right]$$

$$= \frac{Q}{4\pi\in_0}\left[\frac{4}{3} - \frac{2}{3}\right]$$

$$= \frac{Q}{6\pi\in_0}$$

107. Area of the loop, $A = \pi r^2$

Flux linking the area, $\phi = B \times \pi r^2$

Magnitude of unit induced $= \frac{dQ}{dt}$

$$e(t) = \frac{d}{dt}\{B.\pi r^2\}$$

$$= \{B.\pi 2r\}\left(\frac{dr}{dt}\right)$$

When $B = 0.1$ T, $r = 2m$ and $\frac{dr}{dt} = 2$ m/s

$$\therefore \quad e = 0.1 \times \pi 2 \times 2 \times 2$$

$$= 0.8\pi v$$

110. E field is in perpendicular to the propagation of wave, therefore

$$(4\vec{ax} - 2\vec{ay} + \vec{az}).(4\vec{ax} - 2k\vec{ay} + 4k\vec{az}) = 0$$

$$\Rightarrow \quad 16 + 4K + 4K = 0$$

$$\Rightarrow \quad K = -2$$

111. Magnidude of projection

$$= (6\vec{a}_x + 2\vec{a}_y - 3\vec{a}_z).\frac{(3\vec{ax} - 4\vec{ay})}{\sqrt{3^2 + 4^2}}$$

$$= \frac{18-8}{5} = 2$$

112. $\in_1 E_{N1} = \in_2 E_{N2}$

$$\Rightarrow \quad 2 \times 10\,\vec{a}_z = 5E_{N2}$$

$$\Rightarrow \quad \vec{E}_{N2} = 4\vec{a}_z$$

$$Et_1 = Et_2 = 0$$

$$\therefore \quad E_2 = 4\vec{a}_z$$

Potential difference between $(0, 0, -10^{-3})$ and $(0, 0, 10^{-3})$ and $(0, 0, 10^{-3})$

$$= 10^{-3} \times 10 \times 10^3 + 10^{-3} \times 4 \times 10^3$$

$$= 14\text{V}$$

114. $\vec{B} = x^2\vec{a}_x - x_y\vec{a}_y - kxza_z$

If this represents a magnetic field, then

$$\vec{\nabla}.\vec{B} = 0$$

$$\Rightarrow \quad \left(\frac{\partial}{\partial x}\vec{ax} + \frac{\partial}{\partial y}\vec{ay} + \frac{\partial}{\partial z}\vec{a}_z\right)(x^2\vec{a}_x - xya_y - kxza_z) = 0$$

$$\Rightarrow \quad \frac{\partial}{\partial x}x^2 + \frac{\partial}{\partial y}(-xy) + \frac{\partial}{\partial z}(-kxz) = 0$$

$$\Rightarrow \quad 2x - x - kx = 0$$

$$\Rightarrow \quad 1 - kx = 0$$

$$\Rightarrow \quad k = 1$$

115. $\vec{E} = 0.3\,\vec{ax} + 0.4\,\vec{ay}$

$$\therefore \quad = 0.5\underline{|\tan^{-1}(4/3)}$$

$$= 0.5\underline{|53.13}$$

$$\therefore \quad \cos\theta = \cos 53.13 = 0.6$$

116. When in free space, energy density,

$$W_1 = \frac{1}{2} \times 4 \epsilon_0 E^2$$

Now when $E = a_x E - \vec{a}_y E$

$\therefore$ $|E| = \sqrt{2}E$

$\Rightarrow$ $E = 4E_0$

$$W_2 = \frac{1}{2} \times 4 \epsilon_0 \times 2E^2 = 4\epsilon_0 E^2$$

$\therefore$ $\frac{W_2}{W_1} = 8$

$\Rightarrow$ $W_2 = 8W_1 = 8W$

117. Electric field due to sheet of charge at line charge,

$$E_{sheet} = \frac{P_s}{2 \epsilon_0}(-\vec{a}_n)$$

$\therefore$ Force on the line charge $= \frac{P_s.P_L}{2 \epsilon_0}$

$$= \frac{2 \times 10^{-9} \times 20 \times 10^{-9}}{2 \epsilon_0}$$

EXERCISE - II

3. Energy stored $= \frac{1}{2}CV^2$

$$= \frac{1}{2} \times 40 \times 10^{-6} \times (500)^2$$

$= 5$ Joules

7. $F = BiL$

$\therefore$ $dF = Bi dL$

$\Rightarrow$ $\frac{dF}{dL} = (3a_x + 4a_x) 5a_z = -20a_x + 15a_y$ N/m

9. We have $D = \epsilon_0 E + P$

$P = D - \epsilon_0 E$

$$= D - \frac{\epsilon_0 .D}{\epsilon_0 \epsilon_R}$$

$$= \left(\frac{\epsilon_R - 1}{\epsilon_R}\right) D = \frac{5-1}{5} \times 1$$

$$= \frac{4}{5} = 0.8 \text{ c/m}^2$$

11. Here depth is given by,

$$\delta = \frac{1}{\sqrt{\pi f \mu \sigma}}$$

$\therefore$ $\delta \propto \frac{1}{\sqrt{\sigma}}$

13. Reflection coefficient,

$$\rho = \frac{Z_2 - Z_1}{Z_1 + Z_2}$$

$$= \frac{900 - 600}{600 + 900} = 0.2$$

14. $n = \sqrt{\frac{\mu}{\epsilon}} = \sqrt{\frac{\mu_0 \mu_R}{\epsilon_0 \epsilon_R}}$

$\therefore$ $60\pi = 120\pi \sqrt{\frac{\mu_R}{\epsilon_R}}$

$\Rightarrow$ $0.5 = \sqrt{\frac{\mu_R}{\epsilon_R}}$

$\Rightarrow$ $0.25 = \frac{\mu_R}{\epsilon_R}$

$\therefore$ $\epsilon_R = \frac{1}{0.25} = 4$

15. Poynting vector,

$$P_z = \frac{E_{xo}^2}{n}[\cos^2(\omega t - \beta z)]$$

$\therefore$ $P_z \propto \frac{1}{r^2}$

as field due to inifinite long charge $E \propto \frac{1}{P}$

$$\frac{P_Q}{P_P} = \frac{r_P^2}{r_Q^2} = \frac{R^2}{(R/2)^2} = \frac{4}{1}$$

$\therefore$ $P_Q : P_P = 4 : 1$

18. *Given :* $\epsilon_R = 5$

$D = 2$ c/m^2

We have, $D = \epsilon_0 E + P$

$\therefore$ $P = D - \epsilon_E$

$$= D - \frac{D}{\epsilon_R}$$

$$= \frac{\epsilon_{R-1}}{\epsilon_R}.D$$

$$= \frac{5-1}{5} \times 2 = 1.6 \text{ c/m}^2$$

24. For loss less medium, phase constant β is given by,

$$\beta = \omega\sqrt{\mu\epsilon}$$

$\therefore$ $\beta \propto \sqrt{\epsilon_R}$

28. Ampere's circuital law is, $\int B di = \mu I$

where I is current encolosed by C.

Now if I is taken positive in upward direction, the contour C is to be traversed in anti-clockwise direction, the contour should be traversed in the clockwise direction.

Inspecting the figure for first current, the current direction and contour traversing are consistant there by yielding negative sign and for third current also consistency is not preserved.

Thus $\quad \text{RHS} = \mu_o(I_1 - I_2 - I_3)$

$$= \mu_o(I_1 - I_2 - I_3)$$

$$= \mu_o I$$

$\therefore \quad \left|\int \vec{B}\, d\vec{i}\right| = \mu_o I$

30. $\quad W = F.d$

$$= \frac{1}{2}\frac{Q^2}{C}$$

$\therefore \quad F = \frac{1}{2}.\frac{Q^2}{C.d}$

$$= \frac{1}{2}.\frac{Q^2}{A.\epsilon_0}$$

$\therefore \quad F \propto Q^2$

and $\quad F \propto \frac{1}{A}$

33. Equation of equipotential surface will be given by

$$\phi = \frac{30 \times 30}{100}$$

$$= 9°$$

39. Characteristic impedance is given by,

$$Z_0 = \sqrt{\frac{L}{C}}$$

$$C = \frac{\epsilon A}{d}$$

$\therefore \quad Z_0 = \sqrt{\frac{L.d}{\epsilon A}}$

$\therefore \quad Z_0 \propto \frac{1}{\sqrt{\epsilon}}$

For $\epsilon_r = 2.28$, $\quad Z_0 = \frac{77}{\sqrt{2.28}} = 51\Omega$

40. Wave length, $\lambda = \frac{2\pi}{\beta}$

Where, $\beta = \sqrt{\frac{\omega\mu\sigma}{2}}$ (for good conductors)

$\therefore \quad \lambda = \frac{2\pi\sqrt{2}}{\sqrt{\omega\mu\sigma}}$

This on doubling the frequency λ will becomes $\frac{1}{\sqrt{2}}$ times the original value.

43. The lines are identical and also load resistances are equal, time constants of both are equal. for both of them, the time for the voltage to fall down to some fraction of the original value, will also be equal irrespective of the magnitude of the initial voltage. Therefore, the answer is T.

47. Displacement current density, $I_d = \frac{\partial D}{\partial t}$

48. Displacement flux density at any point outside the slab (directed outward)

$$= \frac{1}{2} \times P \times 1 \times t = \frac{Pt}{2}$$

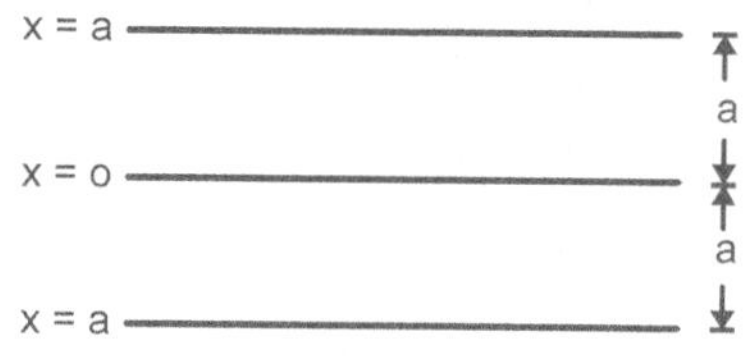

At any point within the slab, at a height say form the lower surface,

$$\text{density} = \frac{1}{2} \times P \times 1 \times h - \frac{1}{2} \times P \times 1 \times (t-h)$$

For points outside the slab

i.e., for $x > a$ and for $-a > x$; $\vec{D} = 0$

For $0 < x < a$,

$$\vec{D} = \left[\frac{1}{2}(-P_0) \times a - \frac{1}{2}(P_0) \times x - \frac{1}{2}(P_0) \times (a-x)\right]$$

$$= P_0(x-a)$$

For $-a < x < 0$

$$\vec{D} = \left[\frac{1}{2}(P_0)a - \frac{1}{2}P_0(-x) + \frac{1}{2}P_0(a+x)\right]$$

$$= -P_0(a+x)$$

49. It is clear that field in the volume of the sphere (assuming that sphere is not a conducting one) is neither ifinite, nor zero and is directed from positive charge to negative charge and without going into the derivation of the value field, by the process of elimination, hence the answer is (*b*).

52. $\quad \vec{r}.(\nabla \times \vec{r}) = \vec{r}\left[\left(\frac{\partial z}{\partial y} - \frac{\partial y}{\partial z}\right)\bar{x} + \left(\frac{\partial x}{\partial z} - \frac{\partial z}{\partial x}\right)\bar{y} + \left(\frac{\partial y}{\partial x} - \frac{\partial x}{\partial y}\right)\bar{z}\right]$

$$= x\left[\left(\frac{\partial z}{\partial y} - \frac{\partial y}{\partial z}\right) + y\left(\frac{\partial x}{\partial z} - \frac{\partial z}{\partial x}\right) + z\left(\frac{\partial y}{\partial x} - \frac{\partial x}{\partial y}\right)\right]$$

Now $\nabla.\vec{r} = \left(\frac{\partial}{\partial x}\bar{x}+\frac{\partial}{\partial y}\bar{y}+\frac{\partial}{\partial z}\bar{z}\right).(\bar{x}x+\bar{y}y+\bar{z}z)$

$$= \frac{\partial x}{\partial x}+\frac{\partial y}{\partial y}+\frac{\partial z}{\partial z}=3$$

Take $\nabla.(\vec{r}\times\vec{r}) = \left(\frac{\partial}{\partial x}\bar{x}+\frac{\partial}{\partial y}\bar{y}+\frac{\partial}{\partial z}\bar{z}\right).(x^2+y^2+z^2)$

$$= 2\overline{xx}+2\overline{yy}+2\overline{zz} = 2\bar{r}$$

$$\nabla(\nabla\times\bar{r}) = \frac{\partial z}{\partial x\partial y}-\frac{\partial y}{\partial z\partial y}+\frac{\partial x}{\partial z\partial y}-\frac{\partial z}{\partial x\partial y}+\frac{\partial y}{\partial x\partial z}-\frac{\partial x}{\partial y\partial y}=0$$

57. $$H = \frac{1}{2}K\times\vec{a}_N$$

$$= \frac{1}{2}(8a_x)\times a_z$$

$$= -4a_r$$

58. $$\vec{F} = \int_{\text{Volume}}(\vec{J}+\vec{B})dv$$

$\therefore$ $$J = \frac{I}{A}\vec{a}_y$$

$\therefore$ $$\text{Force density} = \frac{I}{A}\vec{a}_y\times B_0(\vec{a}_x+\vec{a}_y)$$

$$= \frac{I}{A}B_0(-a_z)\text{N/m}^3$$

61. $$\Psi_A = \frac{qB}{4\pi}\times 2\pi\frac{qB}{2}$$

and $$\Psi_B = \frac{qA}{4\pi}\times 2\pi(1-\cos 60)$$

$$= \frac{qA}{4}$$

Displacement current,

$$I_0 = \frac{d}{dt}(\Psi_A+\Psi_B)$$

$$= \frac{d}{dt}\left(\frac{qB}{2}+\frac{qA}{4}\right)$$

$$= \frac{1}{2}\times I-\frac{1}{4}(-i) = \frac{I}{4}$$

63. $$A = \vec{\nabla}\times\vec{F}+\vec{\nabla}.\vec{F}$$

$$\nabla.A = \vec{\nabla}.\vec{\nabla}\times F+\vec{\nabla}^2.f$$

$$= -K^2 f$$

$\therefore$ $$\vec{\nabla}^2.f+k^2f=0$$

On the surface f is specified to be zero, therefore on the surface $\nabla.A=0$

73. Let charge be q and electric field be E

$\therefore$ $$q\vec{E} = q\vec{v}\times\vec{B}$$

$\Rightarrow$ $$\vec{E} = \vec{v}\times\vec{B}$$

$$= -v_0(3\vec{a}x-\vec{a}y+2\vec{a}z)\times(\vec{a}x+2\vec{a}y-4\vec{a}z)$$

$$= v_0(14\vec{a}y+7\vec{a}z)$$

74. $$\vec{F} = (3\vec{x}+4\vec{y})$$

$$\vec{R} = (-\vec{x}+6\vec{y}+3\vec{z})$$

$\therefore$ Work done $= \vec{F}.\vec{R}$

$$= (3\vec{x}+4\vec{y}).(-\vec{x}+6\vec{y}+3\vec{z})$$

$$= -3+24$$

$$= 21\text{J}$$

76. $$q = CV$$

charge q to be same

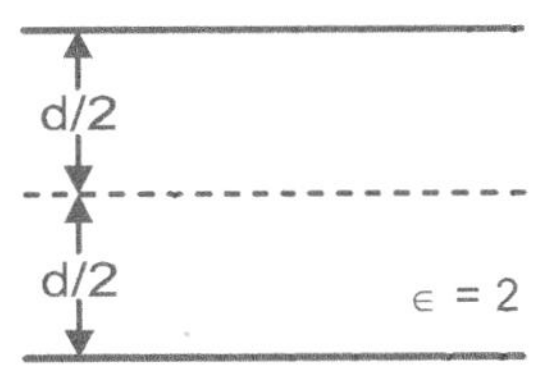

In the second case,

$$\frac{1}{C_{eq}} = \frac{1}{C_1}+\frac{1}{C_2}$$

$$= \frac{d/2}{\epsilon_0\epsilon_R A}+\frac{d/2}{\epsilon_0 A}$$

$$= \frac{d}{\epsilon_0 A}\left[\frac{1}{4}+\frac{1}{2}\right] = \frac{3}{4}\frac{d}{\epsilon_0 A}$$

$\therefore$ $$C_{eq} = \frac{4}{3}.\frac{\epsilon_0 A}{d}$$

For the same charge,

$$C_1V_1 = C_2V_2$$

$\therefore$ $$\frac{\epsilon_0 A}{d}V_1 = \frac{4}{3}.\frac{\epsilon_0 A}{d}V_2$$

$\therefore$ $$V_2 = \frac{3}{4}V_1$$

79. Rate of energy flow,

$$P_z = (E_x\vec{a}x)\times(H_y\vec{a}z)$$

$$= E_xH_y\sin\theta.\vec{a}_z$$

$\therefore$ $$|\vec{P}_z| = |E_x||H_y|\sin\theta$$

87. R is the resistance per unit length and I is the current flowing

$\therefore$ Power loss occurring in the conductor per unit length I^2R.

This power is furnished via $\vec{E}$ and $\vec{H}$ fields at the surface of the wire

$$\vec{P} = \vec{E} \times \vec{H}$$

Since the power is fed from outside through the cylindrical surface of the conductor $\vec{E} \times \vec{H}$ at every point in the surface is radial and directed into the surface.

Pointing vector is

$$\frac{I^2R(-\vec{n})}{\text{Surface area}} = \frac{I^2R}{2\pi r \times l}(-\vec{n})$$

where, $\vec{n}$ = unit radial vector directed outward.

88. Path of the movement can be drawn as shown.

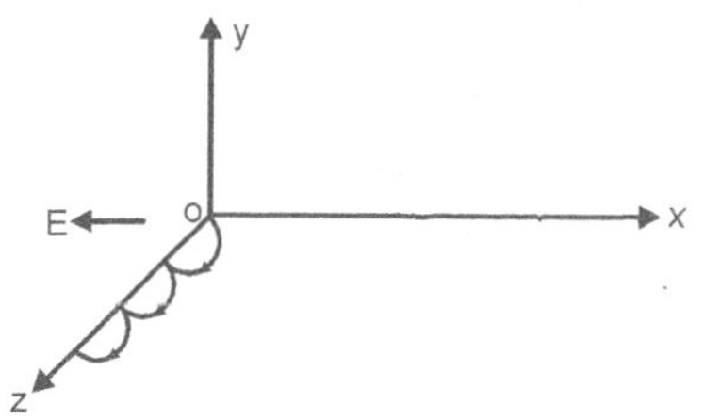

89. A *dc* source is being switched on. Under steady state i.e. at $t = \infty$, battery current would be only, 1A, as the line would then act as open circuit. At the instant of switching, the resistance offered by the line = 50Ω (line is loss less)

$$\therefore \quad \text{Current} = \frac{50}{\frac{50 \times 50}{50 + 50}} = 2\text{A}$$

Thus the derived ratio is 2 : 1

93. We have, $\frac{E}{H} = \sqrt{\frac{\mu}{\epsilon}} = \sqrt{\frac{\mu_r}{\epsilon_r}} \cdot \sqrt{\frac{\mu_0}{\epsilon_0}}$

$$\Rightarrow \quad \frac{20\pi}{H} = 120\pi \times \sqrt{\frac{1}{81}}$$

$$\therefore \quad H = \frac{9 \times 20\pi}{120\pi} = 7.50\ \text{A/m}$$

95. Assuming that the surface is a part of the closed surface $Z > 0$ being the outer region, or the medium in the region $Z < 0$ is

conductor, the electric field $\vec{E}$ exits in the region $Z \geq 0$ only. Application of Gauss's law leads to $\vec{E} = \frac{P}{\epsilon_0} i_z$. The field is independent of the position of the point.

However if the field exists on both the side of the surface, the field is directed normal to and away from the surface on both the sides. Then

$$\vec{E} = \frac{1}{2P} \cdot \frac{i_z}{\epsilon_0};\ Z < 0$$

$$= -\frac{1}{2P} \cdot \frac{i_z}{\epsilon_0};\ Z > 0$$

Combing the two, we get

$$E = \frac{1}{2} \frac{P}{\epsilon_0} \cdot \frac{z}{|z|} i_z$$

100. A dipole consists of two equal and opposite charges separated by a chort distance. The image of charges $+q$ and $-q$ in the grounded sphere will be $\left(\frac{-qR}{d}\right)$ and $\left(\frac{+qR}{(d-l)}\right)$; where l is the distance between $+q$ and $-q$ of the original dipole.

7 CHAPTER POWER SYSTEMS

EXERCISE - I

1. Directional over current relay is used for protection of
   (*a*) long transmission line
   (*b*) large power transformer
   (*c*) ring main distribution line
   (*d*) radial distribution line
2. Buchholz relay is a
   (*a*) oil activated relay
   (*b*) current activated relay
   (*c*) gas activated relay
   (*d*) oil temperature-activated relay
3. The operating characteristics of a reactance relay in the complex impedance plane is a
   (*a*) circle with its origin at the centre of the R-X plane
   (*b*) circle passing through the origin
   (*c*) straight line passing through the origin
   (*d*) straight line parallel to the-X-axis
4. In a static over-current relay, inverse time characteristics are obtained by
   (*a*) a transistor amplifier
   (*b*) an integrating circuit
   (*c*) a transistor switch
   (*d*) a differential circuit
5. By burden of the relay we generally mean
   (*a*) volt-ampere rating of relay
   (*b*) current-rating of relay
   (*c*) voltage rating of relay
   (*d*) watt rating of relay
6. In case electromechanical relay, relay coil is normally designed for
   (*a*) 1 and 5A (*b*) 5 and 10A
   (*c*) 10 and 25A (*d*) 50 and 100A
7. Mho relays have an R-X plane characteristic depicted by
   (*a*) a straight line passing through origin
   (*b*) a straight line parallel to X-axis
   (*c*) a straight line parallel to R-axis
   (*d*) a circle passing through origin
8. Buchholz relay is used for protection in case of a
   (*a*) transformer (*b*) synchronous generator
   (*c*) bus-bar (*d*) induction motor
9. In the case of transmission line protection, over current relay is used
   (*a*) only upto 110 kV
   (*b*) only upto 220 kV
   (*c*) only upto 50 kV
   (*d*) only upto 1100 kV
10. Load flow studies involve solving simultaneous
    (*a*) linear algebraic equations
    (*b*) non-linear algebraic equations
    (*c*) linear differential equations
    (*d*) non-linear differential equations
11. A power system network consists of three elements 0–1,1–2 and 2–0 of per unit impedances 0.2, 0.4 and 0.4 respectively. Its bus impedance matrix is given by

    (*a*) $\begin{matrix} & 1 & 2 \\ 1 & 7.5 & -2.25 \\ 2 & -2.5 & 5.0 \end{matrix}$ (*b*) $\begin{matrix} & 1 & 2 \\ 1 & 0.16 & 0.08 \\ 2 & 0.08 & 0.24 \end{matrix}$

    (*c*) $\begin{matrix} & 1 & 2 \\ 1 & 0.16 & -0.08 \\ 2 & -0.08 & 0.24 \end{matrix}$ (*d*) $\begin{matrix} & 1 & 2 \\ 1 & 0.6 & 0.4 \\ 2 & 0.8 & 0.8 \end{matrix}$

12. When a 50MVA, 11kV, 3-phase generator is subjected to a 3- phase fault, the fault current is $-j5$ *pu* (per unit). When it is subjected to a line-to-line fault, the positive sequence current is $-j4$ *pu*. The positive and negative sequence reactance are respectively
    (*a*) $j0.5$ and $j0.05$ *pu* (*b*) $j0.2$ and $j0.05$ *pu*
    (*c*) $j0.25$ and $j0.25$ *pu* (*d*) $j0.05$ and $j0.05$ *pu*
13. The power generated by two plants are : $P_1 = 50$ MW, $P_2 = 40$MW. If the loss coefficients are, $B_{11} = 0.001$, $B_{22} = 0.0025$ and $B_{12} = -0.0005$, then power loss will be
    (*a*) 5.5 MW (*b*) 6.5 MW
    (*c*) 4.5 MW (*d*) 8.5 MW

14. One current transformer (CT) is mounted over a 3-phase 3-core cable with its sheath and armour removed from the portion covered by the CT. An ammeter placed in the CT secondary would measure

(*a*) the positive sequence current

(*b*) the negative sequence current

(*c*) the zero sequence current

(*d*) three times the zero sequence current

15. The basic circuit of a circulating current system of protection is shown in the figure. To improve the through-fault stability, a stabilizing resistor is connected between the points

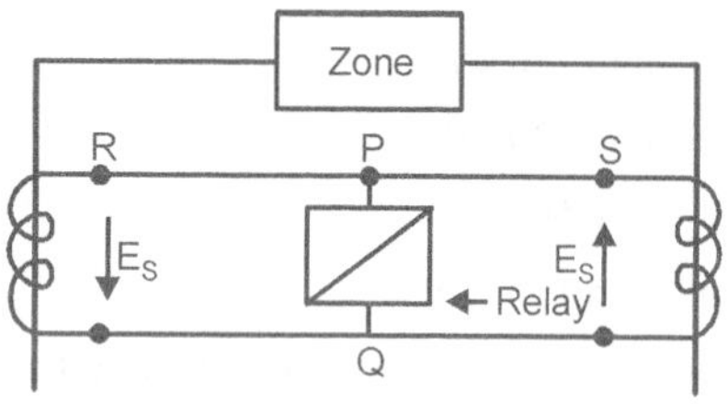

(*a*) R and P in series (*b*) P and S in series

(*c*) P and T in series (*d*) P and Q inparallel

16. The cost function of a 50 MW generator is given by (P_1 is the generator loading)

$F(P) = 225 + 53P + 0.02\,P_1^2$

When 100% loading is applied, the incremental fuel cost (IFC) will be

(*a*) Rs. 55 per MWh (*b*) Rs. 55 per MW

(*c*) Rs. 33 per MWh (*d*) Rs. 33 per MW

17. If the fault current is 3000 A for a relay with a plug setting of 50% and CT ratio of 1000 : 1, the plug setting multiplier would be

(*a*) 1.5 (*b*) 3

(*c*) 4.5 (*d*) 6

18. The critical clearing time of a fault in power systems is related to

(*a*) reactive power limit

(*b*) short-circuit limit

(*c*) steady-state stability limit

(*d*) transient stability limit

19. For a 12-pulse operation of HVDC converters, the most troublesome set of harmonics on the ac side is

(*a*) 23rd and 25th

(*b*) 12th and 24th

(*c*) 11th and 13th

(*d*) 5th and 7th

20. A coal-fired steam power station working at a plant load factor of 80% has one 500 MW generating unit. If the heat content of coal is 2kWh/kg, the overall plant efficiency is 49% and a train load of coal is 2000 metric tons, then the number of trains required daily for the plant is

(*a*) 1 (*b*) 6

(*c*) 10 (*d*) 24

21. The use of fast-acting relays and circuit breakers for clearing a sudden short-circuit on a transmission link between a generator and the receiving end bus improves the transient stability of the machine because the

(*a*) short-circuit current becomes zeros

(*b*) post-fault transfer impedance attains a value higher than that during the fault

(*c*) ordinate of the post-fault power-angle characteristic is higher than that of during-fault characteristic

(*d*) voltage behind the transient reactancy increases to a higher value

22. Consider the following statements:

In case of suspension-type insulators, the string efficiency can be improved by

1. using a longer cross- arm.
2. using a guard ring.
3. grading the insulator discs.
4. reducing the cross-arm length.

(*a*) 1, 2 and 3 are correct

(*b*) 2, 3 and 4 are correct

(*c*) 2 and 4 are correct

(*d*) 1 and 3 are correct

23. A three-phase circuit breaker is rated at 2000 MVA 33kV. Its making current will be

(*a*) 35 kA (*b*) 70 kA

(*c*) 89 kA (*d*) 1671 kA

24. The design of insulation for system above 400 kV, is based upon

(*a*) lightning overvoltage

(*b*) switching surges

(*c*) switching voltage level

(*d*) system load level

25. A 12-bus power system has 3 voltage controlled buses. The dimensions of the Jacobian matrix will be

(*a*) 21×21

(*b*) 21×19

(*c*) 19×19

(*d*) 19×21

26. In Merz Price percentage differential protection of a $\Delta - Y$ transformer, the CT secondaries connection in the primary and secondary windings of the transformer would be in the form of

(*a*) $\Delta - Y$ (*b*) $Y - \Delta$
(*c*) $\Delta - \Delta$ (*d*) $Y - Y$

27. The competence relaying scheme is used for protection of

(*a*) transformer
(*b*) bus-bar
(*c*) synchronous generator
(*d*) transmission line

28. For ground fault, we prefer

(*a*) plain impedance relay
(*b*) direction relay
(*c*) reactance relay
(*d*) over-current relay

29. For the protection of stator winding of a alternator against internal fault involving ground, the relay used, is a

(*a*) biased differential relay
(*b*) directional over-current relay
(*c*) plain impedance relay
(*d*) bucchols relay

30. For the protection of a 3 phase star/delta transformer the CTs for the differential relay should be connected.

(*a*) delta/star
(*b*) delta/delta
(*c*) star/delta
(*d*) star/star

31. Both voltage and current signals are required for

(*a*) a plain over-current relay
(*b*) a differential relay
(*c*) a directional relay
(*d*) a biased differential relay

32. The advantage of neutral earthling is

(*a*) safety of personnel
(*b*) reduction of earth fault current
(*c*) elimination of arcing ground
(*d*) none of the above

33. Shunt capacitors in a substation

(*a*) consume lagging var
(*b*) deliver lagging var
(*c*) consume active power
(*d*) deliver active power

34. A synchronous phase modifier as compared to synchronous motor of the same rating has

(*a*) larger shaft diameter and higher speed
(*b*) smaller shaft diameter and higher speed
(*c*) larger shaft diameter and smaller speed
(*d*) smaller shaft diameter and smaller speed

35. Phase modifier is normally installed increase of

(*a*) short transmission lines
(*b*) medium length lines
(*c*) long lines
(*d*) for any length of lines

36. Power transmission by cable is generally adopted for lines lengths

(*a*) above 100 km (*b*) upto 200 km
(*c*) less than 50 km (*d*) less than 10 km

37. Sheaths are used in cables to

(*a*) provide proper insulation
(*b*) provide mechanical strength
(*c*) prevent ingress of moisture
(*d*) none of these

38. The effect of bonding the cable is

(*a*) to increase the effective resistance and inductance
(*b*) to increase the effective resistance but reduce inductance
(*c*) to reduce the effective resistance and inductance
(*d*) to reduce the effective resistance but increase the inductance

39. The insulating material most commonly used for power cables is

(*a*) PVC (*b*) paper
(*c*) rubber (*d*) none of these

40. Single core cable should have armour mode of

(*a*) magnetic material
(*b*) non-magnetic and non conducting material
(*c*) non-magnetic but conducting material
(*d*) none of these

41. Sulphur hexafluoride cable is insulated by

(*a*) impregnated paper
(*b*) polyvinyl chloride
(*c*) high pressure oil
(*d*) compressed gas

42. The insulation resistance of a cable of length 10km is 1MΩ. Its resistance for 50km length will be

(*a*) 1 MΩ (*b*) 5 MΩ
(*c*) 0.2 MΩ (*d*) none of these

43. The leakage resistance of a 50km long cable is 1 MΩ. For a 100 km long cable it will be

(*a*) 1 MΩ (*b*) 2 MΩ
(*c*) 0.66 MΩ (*d*) none of these

44. The surge impedance of a 50 miles long underground cable is 50 Ω. For a 25 miles length it will be

(*a*) 25 Ω (*b*) 50 Ω
(*c*) 100 Ω (*d*) none of these

45. The capacitance between any two conductor of a 3-core cable with sheath earthed is 3μF. The capacitance per phase will be

(*a*) 1.5 μF (*b*) 6 μF
(*c*) 1 μF (*d*) none of these

46. With bundled conductors the corona inception voltage

(*a*) increases (*b*) decreases
(*c*) remains unaffected (*d*) none of these

47. Bundled conductors are mainly used to

(*a*) increase the shunt capacitance
(*b*) decrease the shunt capacitance
(*c*) increase the series reactance
(*d*) decrease the series reactance

48. Bundled conductors are used to reduce

(*a*) inductance of the line
(*b*) both inductance and capacitance
(*c*) corona loss
(*d*) corona loss and the line inductance

49. The size of conductor on modern EHV lines is obtained based on

(*a*) voltage drop (*b*) current density
(*c*) corona (*d*) skin effect

50. Back to back HVDC is used to

(*a*) increase the transmission capability
(*b*) decrease line losses
(*c*) provide stable interconnection
(*d*) reduce voltage drop

51. The insulation of modern EHV lines is designed based on

(*a*) the lightning voltage
(*b*) corona
(*c*) radio interference
(*d*) switching voltage

52. Shunt compensation in an EHV line is used to

(*a*) improve stability
(*b*) reduce fault level
(*c*) improve the voltage profile
(*d*) substitute for synchronous phase modifier

53. Shunt compensation in an EHV line is restored to

(*a*) improve the stability
(*b*) reduce fault level
(*c*) improve the voltage
(*d*) none of these

54. Main purpose of oil in OCB is to

(*a*) provide insulation
(*b*) provide cooling of contacts
(*c*) quenching arc
(*d*) none of these

55. In a minimum oil circuit breaker the oil is used

(*a*) to act as circuit breaking medium only
(*b*) for circuit braking and providing insulation
(*c*) for providing insulation only
(*d*) none of these

56. The signal most important property that makes SF_6 a very efficient medium for circuit breaking is

(*a*) it is non-toxic and non-inflammable
(*b*) it has high dielectric constant
(*c*) it has high breakdown strength
(*d*) it is highly electronegative in characters

57. For long EHV/UHV transmission lines we normally use

(*a*) air circuit breaker
(*b*) air blast circuit breaker
(*c*) oil circuit breaker
(*d*) isolating switch

58. Arc-chutes are used in

(*a*) oil circuit breakers
(*b*) vacuum circuit breakers
(*c*) SF_6 circuit breakers
(*d*) air blast circuit breakers

59. Compared to the breaking capacity of circuit breaker its making capacity should normally be

(*a*) more (*b*) equal
(*c*) less (*d*) can be anything

60. Isolates used in transmission lines are capable of breaking

(*a*) fault current (*b*) no current
(*c*) charging current (*d*) load current

61. A bus coupler circuit breaker is utilized in a substation for

(*a*) joining the transmission line with station busbar
(*b*) joining main and transfer bus in a substation
(*c*) joining the generator with transfer
(*d*) joining the neutral of generator with earth

62. HRC fuses provide best protection against

(*a*) short-circuit (*b*) lightening

(*c*) sparking (*d*) fire

63. In the protection scheme, relay functions as a

(*a*) switching device

(*b*) sensing device

(*c*) breaking device

(*d*) none of these

64. Relay contacts are normally made up of

(*a*) silver contact (*b*) copper contact

(*c*) platinum contact (*d*) steel contact

65. Under-voltage relays are mostly used for

(*a*) transformer protection

(*b*) bus-bar protection

(*c*) motor protection

(*d*) feeder protection

66. If the positive, negative and zero sequence reactances of an element of a power system are 0.3, 0.3 and 0.8 respectively, then the element would be a

(*a*) synchronous generator

(*b*) synchronous motor

(*c*) static load

(*d*) transmission line

67. For variable heads of near about but less than 30 metres, which type of turbines is used in hydro power stations

(*a*) Pelton (*b*) Kaplan

(*c*) Francis (*d*) None of these

68. Steady-state stability of a power system is improved by

(*a*) reducing fault clearing time

(*b*) using double circuit line instead of single circuit line

(*c*) single pole switching

(*d*) decreasing generator inertia

69. The reflection coefficient of a short-circuited line is

(*a*) – 1 (*b*) +1

(*c*) 0.5 (*d*) zero

70. An overhead line with a surge impedance of 400Ω is connected to a transformer by a short length of cable of surge impedance 100Ω. If a rectangular surge wave of 40 kV travels along the line toward the cable, then the voltage of wave travelling from the junction of the overhead line through the cable towards the transformer would be

(*a*) 16kV (*b*) 32kV

(*c*) 34kV (*d*) 36kV

71. An isolated synchronous generator with transient reactance equal to 0.1 pu on a 100 MVA base. The fault level at the bus is

(*a*) 1000 MVA (*b*) 500 MVA

(*c*) 50 MVA (*d*) 20 MVA

72. An overhead line with surge impedance of 400Ω is terminated through a resistance 'R'. A surge travelling over the line will not suffer any reflection at the junction, if the value of R is

(*a*) 100Ω (*b*) 200Ω

(*c*) 400Ω (*d*) 8000Ω

73. Disruptive corona begins in smooth cylindrical conductor in air at NTP if the electric field intensity at the conductor surface goes up to

(*a*) 21.2 kV (rms)/cm

(*b*) 21.1 kV (peak)/cm

(*c*) 21.1 kV (average)/cm

(*d*) 21.1 kV(rms)/m

74. If a generator of 250 MVA rating has an inertia constant of 6 MJ/MVA, its inertia constant on 100 MVA base is

(*a*) 15 MJ/MVA

(*b*) 10.5 MJ/MVA

(*c*) 6 MJ/MVA

(*d*) 2.4 MJ/MVA

75. When a line-to-ground fault occurs, the current in a faulted phase is 100A. The zero sequence current in this case will be

(*a*) zero (*b*) 33.3A

(*c*) 66.6A (*d*) 100A

76. Galloping in transmission line conductors arises generally due to

(*a*) asymmetrical layers of ice formation

(*b*) vortex phenomenon in light winds

(*c*) heavy weight of the line conductors

(*d*) adoption of horizontal conductor configurations

77. Control rods used in nuclear reactors are made of

(*a*) zirconium (*b*) boron

(*c*) beryllium (*d*) lead

78. In *a* 3-core extra-high voltage cable, a metallic screen around each core-insulation is provided to

(*a*) facilitate heat dissipation

(*b*) give mechanical strength

(*c*) obtain radial electric stress

(*d*) obtain longitudinal electric stress

79. Consider two parallel short transmission lines of impedance Z_A and Z_B respectively Currents I_A and I_B are both lagging and the sending-end voltage is V_S. If the reactance to resistance ratio of both the impedance Z_A and Z_B are equal, then the total current 'I' will

(*a*) lag both I_A and I_B

(*b*) lead both I_A, and I_B

(*c*) lag one of I_A and I_B but lead the other

(*d*) be in phase with both I_A and I_B

80. The per unit impedance of a synchronous machine is 0.242. If the base voltage is increased to 1.1 times, the per unit value will be

(*a*) 0.266 (*b*) 0.242

(*c*) 0.220 (*d*) 0.200

81. Rated breaking capacity (MVA) of a circuit breaker is equal to

(*a*) the product of rated breaking current (kA) and rated voltage (kV)

(*b*) the product of rated symmetrical breaking current (kA) and rated voltage (kV)

(*c*) the product of breaking current (kA) and fault voltage (kV)

(*d*) twice the value of rated current (kA) and rated voltage (kV)

82. Which one of the following grounding methods is used to reduce the tower footing resistance where earth resistance is low?

(*a*) Single driven rod (*b*) Multiple rods

(*c*) Counterpoises (*d*) Plates

83. In the solution of load-flow equation, Newton-Rophson (NR) method is superior to the Gauss-Seidel (GS) method, because the

(*a*) time taken to perform one interation in the NR method is less when compared to the time take in the GS method

(*b*) number of iterations required in the NR method is more when compared to that in the GS method

(*c*) number of iterations required is not independent of the size the system in the NR method

(*d*) convergence characteristics of the NR method are not affected by the selection of slack bus

84. The incremental cost characteristics of the two units in a plant are given by:

$IC_1 = ₹(0.1P_1 + 8.0)$ per MWh

$IC_2 = ₹(0.15\,P_2 + 3.0)$ per MWh

The optimum sharing of load when the total load is 100 MW is

(*a*) $P_1 = 60$MW and $P_2 = 40$ MW

(*b*) $P_1 = 33.3$ MW and $P_2 = 66.7$ MW

(*c*) $P_1 = 40$MW and $P_2 = 60$ MW

(*d*) $P_1 = 66.7$MW and $P_2 = 33.3$ MW

85. If the inductance and capacitance of power system are respectively 1H and 0.01 μF and the instantaneous value of interrupted current is 10A, then the voltage across the breaker contact will be

(*a*) 50kV (*b*) 51kV

(*c*) 60kV (*d*) 100kV

86. An overhead transmission line having surge impedance 'Z_1' is terminated to an underground cable of surge impedance 'Z_2'. The reflection coefficient for the travelling wave at the junction of the line and cable is

(*a*) $\dfrac{Z_1 + Z_2}{Z_1 - Z_2}$ (*b*) $\dfrac{Z^2}{Z_1 + Z_2}$

(*c*) $\dfrac{Z_2 - Z_1}{Z_1 + Z_2}$ (*d*) $\dfrac{Z_1 - Z_2}{Z_1 + Z_2}$

87. The velocity of travelling wave through a cable of relative permittivity 9 is

(*a*) 9×10^8m/s (*b*) 3×10^8m/s

(*c*) 10^8m/s (*d*) 2×10^8m/s

88. Match List-I (Protective schemes) with List-II (Equipment) and select the correct answer using the codes given the lists:

List-I	List-II
A. Mho relays	1. Generation.
B. Inverse time overcurrent relays	2. Transmission line
C. Differential relays	3. Motors.

Codes:

	A	B	C
(*a*)	2	1	3
(*b*)	2	3	1
(*c*)	3	2	1
(*d*)	1	3	2

89. Reversal of power flow in HVDC link by controlling the firing angle simultaneously at both the converter stations results in

(*a*) reversal of voltage polarity of HVDC pole

(*b*) reversal of current in *dc* link

(*c*) reversal of current in *dc* link as well as reversal of voltage polarity of HVDC pole

(*d*) neither reversal of current in *dc* link nor reversal of voltage polarity of HVDC pole

90. A large synchronous generator is feeding power into an infinite bus at slightly lagging power factor. If a total loss of field occurs and the system can supply sufficient reactive power without a large terminal voltage drop, the unit will

(*a*) continue to run as a synchronous generator and no tripping is necessary

(*b*) get short circuited and it should be tripped instantaneously

(*c*) run as an induction generator and it should be tripped after a time delay

(*d*) run as a synchronous motor and it should be tripped after a time delay

91. The negative phase sequence in a three phase synchronous motor exists when the motor is

(*a*) under loaded

(*b*) over loaded

(*c*) supplied without unbalance voltage

(*d*) hot

92. To protect the power transformer (Y – Y with neutral grounded) against fault, what type connection do the CTs have?

(*a*) $\Delta - Y$ connection

(*b*) $\Delta - \Delta$ connection

(*c*) $Y - Y$ connection

(*d*) $Y - \Delta$ connection

93. The most accurate and versatile method of achieving reactive power compensation is by using

(*a*) switched capacitors

(*b*) fixed capacitor with controller reactor

(*c*) saturable reactor with capacitor bank

(*d*) switched capacitor with controlled reactors

94. A hydel power plant of run-off river type should be provided with a pondage so that the

(*a*) firm-capacity of the plant is increased

(*b*) operating head is controlled

(*c*) pressure inside the turbine causing remains constant

(*d*) kinetic energy of the running water is fully utilised

95. If within an untransposed 3-phase circuit of a transmission line, the series impedance of each of the conductors is considered, it is found to contain resistive terms of the form of K $\log_e \left(\frac{d_{12}}{d_{13}}\right)$, K being a constant and d_{12} and d_{13} etc. being spacings between the conductors. These terms represent power transfer from one phase to another. The sum of these terms over the three phases is

(*a*) three times the average

(*b*) $\sqrt{3}$ times the average

(*c*) one-third of the average

(*d*) zero

96. If the current is 3000 A, for an I.D.M.T. relay with a plug setting of 50% and C.T. ratio of 400/ 5, the plug setting multiplier would be

(*a*) 7.5 (*b*) 15.0

(*c*) 18.75 (*d*) 37.5

97. If a circular line of radius R has around its periphery *n* equally spaced points, then the GMD among will be

(*a*) $R(n)^{1/n}$ (*b*) $R(n)^{1/n(n-1)}$

(*c*) $Rn^{1/2}$ (*d*) Rn

98. The incremental generating costs of two generating units are given by

$IC_1 = 0.1\,X + 20$ ₹/MW hr.

$IC_2 = 0.15\,Y + 18$ ₹/MW hr.

where X and Y are power (in MW) generated by the two units. For a total demand of 300 MW, the value (in MW) of X and Y will be respectively for economic load share.

(*a*) 172 and 128

(*b*) 128 and 172

(*c*) 175 and 125

(*d*) 200 and 100

99. In pumped storage scheme, the generator is also used as

(*a*) induction generator or synchronous condenser

(*b*) induction generator or synchronous motor

(*c*) synchronous generator or induction generator

(*d*) synchronous motor or synchronous condenser

100. A transmission line has 0.2 *pu* impedance on a base of 132 kV, 100 MVA. On a base of 220 kV, 50 MVA, it will have a *pu* impedance of

(*a*) $0.2 \times \frac{50}{100} \times \left(\frac{220}{132}\right)^2$

(*b*) $0.2 \times \frac{100}{50} \times \left(\frac{132}{220}\right)^2$

(*c*) $0.2 \times \frac{50}{100} \times \left(\frac{132}{220}\right)^2$

(*d*) $0.2 \times \frac{100}{50} \times \left(\frac{220}{132}\right)^2$

101. The inductance of a three-phase transmission line is 1.2 mH/phase/km. If the spacing of conductors and the radious of the conductors are doubled, then the inductance of the line will be

(*a*) 4.8 mH/phase/km

(*b*) 1.2 mH/phase/km

(*c*) (In 2) × 1.2 mH/phase/km

(*d*) (In 4) × 1.2 mH/phase/km

102. The voltage rating of a multiple shell (petticoat or rainshed) pin-type insulator unit cannot be increased beyond a limiting value by increasing the number of shells, because

(*a*) the internal voltage between shells becomes unequal

(*b*) the leakage-path resistance starts diminishing

(*c*) the disruptive critical voltage for the material of the insulator is reached

(*d*) the puncture voltage of the material of the insulator is reached

103. Consider the following statements:

Distance relays used for protection of transmission lines might operate undesirable due to

1. magnetizing-current in rusk.
2. power swings.
3. series compensating capacitors.
4. switching over voltages.

Of these statements

(*a*) 1 and 2 are correct (*b*) 2 and 3 are correct

(*c*) 1, 2 and 4 are correct (*d*) 1, 3 and 4 are correct

104. A one kilometer long 11kV 3-core three-phase metal sheathed cable gave the following test results:

(*i*) Capacitance between 2 conductors joined to sheath and the third conductor: 0.6 μF

(*ii*) Capacitance between all the three conductors bunched together and sheat: 0.96 μF

The effective capacitance of each core to neutral is

(*a*) 0.12μF (*b*) 0.14μF

(*c*) 0.16μF (*d*) 0.32μF

105. In connections with the arc extinction in circuit breaker, resistance switching is employed where in a resistance is placed in parallel with the poles of the circuit breaker as shown in the figure. This process introduces damping in the LC circuit. For critical damping, the value of '*r*' should be equal to

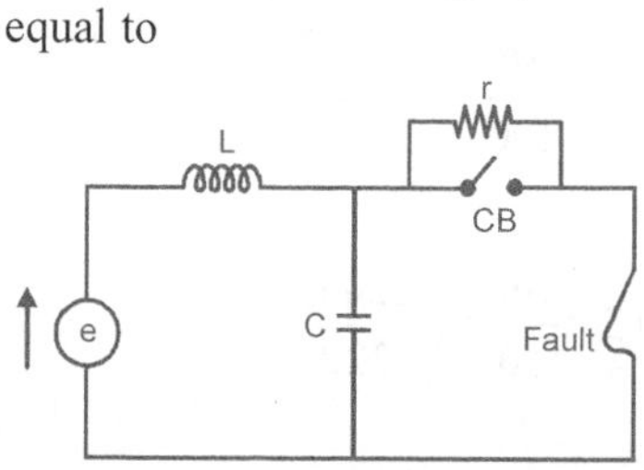

(*a*) $\frac{\sqrt{C}}{L}$

(*b*) $0.5\frac{\sqrt{C}}{L}$

(*c*) $0.5\frac{\sqrt{L}}{C}$

(*d*) $\frac{1}{2\pi}\frac{\sqrt{L}}{C}$

106. The bus admittance matrix (Y_{bus}) of a power system is not

(*a*) symmetric

(*b*) a square matrix

(*c*) a full matrix

(*d*) generally having dominant diagonal elements

107. With a number of generators of MVA capacities S_1, S_2, S_3....S_n and inertia constants H_1,H_2,....H_n respectively connected to the same bus bar in a station, the inertia constant of the equivalent machine on a base of S_n is given by

(*a*) $\sum_{i=1}^{n} H_i S_i$ (*b*) $\sum \frac{S}{S} H$

(*c*) $\sum_{i=1}^{n} \frac{S_i}{S_b} . H_i$ (*d*) $\sum_{i=1}^{n} \frac{S_i}{S_b} . \frac{1}{H_i}$

108. The connections diagram of sequence networks for a particular fault on a power system network is given in the figure. The type of the fault is

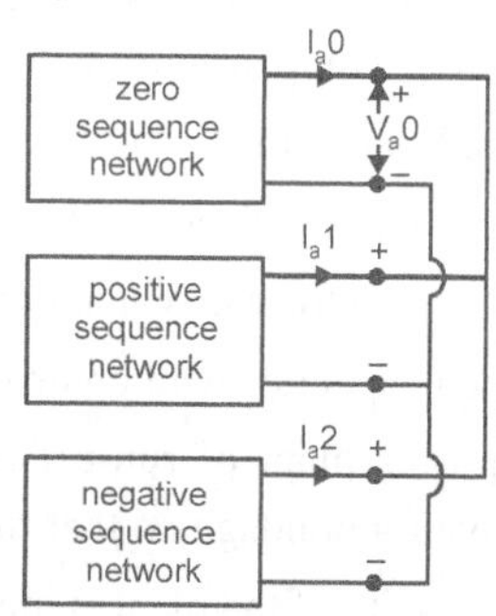

(*a*) single line to ground fault

(*b*) double line to ground fault

(*c*) both (*a*) and (*b*)

(*d*) open circuit

109. Transient stability limit of a power system can be appreciably increased by introducing

(*a*) series inductance

(*b*) shunt inductance

(*c*) series capacitance

(*d*) shunt capacitance

110. Match List-I with List-II and select the correct answer using the codes below the lists:

List-I	List-II
A. Thyrite arrester	1. Tower location.
B. Sag template	2. Cross bonding.
C. Cable sheaths	3. Restriking voltage.
D. Circuit breaker	4. Non-linear resistor.

Codes:

	A	B	C	D
(*a*)	4	1	3	2
(*b*)	4	1	2	3
(*c*)	1	4	3	2
(*d*)	4	3	1	2

111. The inductance of single-phase two-wire power transmission line per kilometer gets doubled when the

(*a*) distance between the wires is doubled

(*b*) distance between the wires is increased four fold

(*c*) distance between the wires is increased as square of original distance

(*d*) radius of the wire is doubled

112. The operation of a nuclear reactor is controlled by controlling the multiplic-ation factor (K) defined as

$$K = \frac{\text{Number of neutors of any one generation}}{\text{Number of neutrons of immediately preceeding generation}}$$

The power-level of the reactor can be increased by

(*a*) raising the value of K above 1 and keeping it at that raised value

(*b*) raising the value of K above 1, but later bringing it back to K = 1

(*c*) lowering the value of K below 1 and keeping it at that lowered value

(*d*) lowering the value of K below 1, but later bringing it back to K = 1

113. The line trap unit employed in carrier current relaying

(*a*) offers high impedance to 50 Hz power frequency signals

(*b*) offers high impedance to carrier frequency signals

(*c*) offers low impedance to carrier frequency signals

(*d*) offers low impedance to carrier frequency signals and high impedance to 50Hz power frequency signals

114. A long-distance overhead transmission line 220 kV rating is to be protected against fault between phases and ground. The fault resistant including that of the ground is found to vary over a wide range. Which one of the following type of relays will give the best performance under the situation indicated above?

(*a*) Overcurrent relay

(*b*) Differential relay with percentage bias

(*c*) Reactance type distance relay

(*d*) Impedance type distance relay

115. Consider the following statements regarding convergence of the Newton-Raphson procedure:

1. It does not converge to a root when the second differential coefficient changes sign.
2. It is preferred when the graph (*x*) is nearly horizontal where it crosses the X-axis
3. It is used to solve algebraic and transcendental equations

Of these statements

(*a*) 1,2 and 3 are correct (*b*) 1 and 2 are correct

(*c*) 2 and 3 are correct (*d*) 1 and 3 are correct

116. In a 400 k V power network, 360 kV is recorded at 400 kV bus. The reactive power observed by a shunt reactor rated for 50 MVAR, 400 kVconnected at the bus is

(*a*) 61.73 MVAR (*b*) 55.56 MVAR

(*c*) 45.0 MVAR (*d*) 40.5 MVAR

117. Which one the following statements is true?

(*a*) Steady-state stability limit is greater than transient stability limit

(*b*) Steady-state stability limit is less than the transient stability limit

(*c*) Steady-state stability limit is less than the transient stability limit

(*d*) No generation can be made regarding the equality or otherwise of the steady-state stability limit and transient stability limit

118. Lighting arresters are used in power systems to protect electrical equipment against

(*a*) direct strokes of lighting

(*b*) overvoltages due to indirect lighting stroke

(*c*) power frequency overvoltages

(*d*) overcurrents due to lighting stroke

119. Corona loss can be reduced by the use of hollow conductors, because

(*a*) the current density is reduced

(*b*) the eddy current in the conductor is eliminated

(*c*) for a given cross section, the radius of the conductor is increased

(*d*) of better ventilation in the conductor

120. The insulation of modern EHV lines is designed based on

(*a*) the lightning

(*b*) corona

(*c*) radio interierence

(*d*) switching voltage

121. A lighting arrestor connected the line and earth in a power system

(*a*) protects the terminal equipment against travelling surges

(*b*) protects the transmission line against direct lighting stroke

(*c*) suppresses high frequency oscillations in the line

(*d*) reflects back the travelling wave approaching it

122. Match List-I (Devices) with List-II (Application) and select the correct answer using the codes given below the lists:

List-I	**List-II**
A. Microprocessor	1. Monitoring surge voltages
B. Breather	2. Digital protection
C. Magnetic links	3. Monitoring surge currents
D. Klydonograph	4. Power transformer

Codes:

	A	B	C	D
(*a*)	4	2	1	3
(*b*)	4	2	3	1
(*c*)	2	4	1	3
(*d*)	2	4	3	1

123. A large *ac* generator supplying power to an infinite bus, has a sudden short-circuit occurring at its terminals. Assuming the prime mover input and the voltage behind the transient reactance to remain constant immediately after the fault, the acceleration of the generator rotors is

(*a*) inversely proportional to the moment of inertia of the machine

(*b*) inversely proportional to the square of the voltage

(*c*) directly proportional to the square of the short circuit current

(*d*) directly proportional to the short-circuit power

124. When a 50MVA, 11kV, 3-phase generator is subjected to a 3-phase fault, the fault current is $-j5$pu (per unit). When it is subject to a line-to-line fault, the positive sequence current is $-j4$pu. The positive and negative sequence reactances are respectively

(*a*) $j0.2$ and $j0.05$ *pu*

(*b*) $j0.2$ and $j0.25$ *pu*

(*c*) $j0.25$ and $j0.25$ *pu*

(*d*) $j0.05$ and $j0.05$ *pu*

125. In the optimum generator scheduling of different power plants, the minimum fuel cost is obtained when

(*a*) only the incremental fuel cost of each plant is the same

(*b*) the penalty factor of each plant is the same

(*c*) the ratio of the incremental fuel cost to the penalty factor each plant is the same

(*d*) the incremental fuel cost of each plant multiplied by its penalty factors is the same

126. Consider the following statements:

A differential relay is used for a 3-phase transformer protection to avoid maloperation due to

1. saturation of current transformer.
2. mismatching of the current ratio for current transformers.
3. difference in connections on both sides of power transformer.
4. current setting multipliers.

Of these statements:

(*a*) 1 and 4 are correct

(*b*) 1 and 2 are correct

(*c*) 1, 2 and 3 are correct

(*d*) 2, 3 and 4 are correct

127. A *dc* reactor is connected in series with each pole of a converter station in order to

(*a*) prevent commutation failures in the inverter

(*b*) supply reactive power to the converter

(*c*) improve system stability

(*d*) increase the power transfer capability

128. In load flow analysis, the load connected at a bus is represented as

(*a*) constant current drawn from the bus

(*b*) constant impedance connected at the bus

(*c*) voltage and frequency dependent source at the bus

(*d*) constant real and reactive drawn from the bus

129. A 3-phase transmission the conductors were arranged in horizontal spacing, with '*d*' as the distance between adjacent conductors. If these conductors are rearranged to form an equilateral triangle with sides equal to '*d*' then the

(*a*) capacitance and the inductance will increase

(*b*) capacitance will increase and the inductance will decrease

(*c*) capacitance and the inductance will remains the same

(*d*) capacitance will decrease and the inductance will increase

130. If the receiving-end voltage and current are numerically equal to the corresponding sending end values, that is $|V_S| = |V_R|$ and $|I_S| = |I_R|$, then such a line is called

(*a*) an infinite line

(*b*) a natural line

(*c*) a tuned line

(*d*) a loss-less line

131. If $|V_S| = |V_R| = 66$ kV for three-phase transmission and reactance is 11 ohms/phase, then the maximum power transmission per phase would be

(*a*) 132 MW

(*b*) 396 MW

(*c*) 66 MW

(*d*) None of these

132. In a 14-bus power system network, there are 5 voltage controlled buses. The size of the Jacobian matrix useful for power flow analysis will be

(*a*) 16×16 (*b*) 23×23

(*c*) 21×21 (*d*) 8×8

133. A thyrite lightning arrester has

(*a*) inverse resistance characteristics

(*b*) a gap

(*c*) efficient earthing

(*d*) a combination of inverse resistance characterised and gap

134. Match the numerical methods given in List-I with List-II and select the correct answer using the codes given below the lists:

Lisl-I	List-II
A. Cholesky method	1. Self-starting method
B. Newton-method	2. Square root method
C. Gauss elimination	3. Quadratic convergence
D. Runge kutta fourth order method	4. Forward reduction & back substitution
	5. Cubic convergence

Codes:

	A	B	C	D
(*a*)	2	1	3	4
(*b*)	4	5	1	3
(*c*)	2	3	4	1
(*d*)	4	5	1	2

135. Match List-I (Coils of a harmonic restrained percentages bias deferential relay used for power transformer protection) with List-II (Currents flowing through the relay coil) and select the correct answer using the codes given below the lists:

List-I	List-II
A. Operational coil	1. Through current.
B. Restraining coil	2. Differential currents.
C. Bias coil	3. Second harmonic current

Codes:

	A	B	C
(*a*)	1	3	2
(*b*)	2	3	1
(*c*)	3	2	1
(*d*)	2	1	3

136. The synchronizing coefficient between two areas 2-area power system is (symbols have usual meanings)

(*a*) $\dfrac{\partial P}{\partial |V|}$ (*b*) $\dfrac{\partial P}{\partial \delta}$

(*c*) $\dfrac{\partial P}{\partial f}$ (*d*) $\dfrac{\partial P}{\partial Q}$

137. A composite transmission and distribution network has lines working at different voltages. Match List-I (Voltage level used) with List-II (Type of relays used for protection) and select the correct answer using the codes given below the lists:

List I	List II
A. 200 kV	1. Directional time lag overcurrent relays.
B. 33 kV	2. Carrier current relays.
C. 11 kV	3. Distance relays.

Codes:

	A	B	C
(*a*)	1	3	2
(*b*)	3	2	1
(*c*)	2	1	3
(*d*)	2	3	1

138. A Bucholz relay is used for

(*a*) protection of a transformer against all internal faults

(*b*) protection of a transformers against external faults

(*c*) protection of transformer against both internal and external faults

(*d*) protection of induction motors

139. Buses for load flow studies are classified as

(*i*) the load bus,

(*ii*) the generator bus and

(*iii*) the slack bus.

The correct combination of the pair of quantities specified having their usual meaning for different buses is

	Load bus	Generator bus	Slack bus
(*a*)	$P, \lvert V \rvert$	P, Q	$P\delta$
(*b*)	P, Q	$P, \lvert V \rvert$	$\lvert V \rvert, \delta$
(*c*)	$\lvert V \rvert, Q$	P, δ	P, Q
(*d*)	P, δ	$Q, \lvert V \rvert$	Q, δ

140. In a power system, each bus or node is associated with four quantities, namely

1. real power.
2. reactive power.
3. bus-voltage magnitude and
4. phase-angle of the bus voltage.

For load-flow solution, among these four, the number of quantities to be specified is

(*a*) any one (*b*) any two

(*c*) any three (*d*) all the four

141. Bus admittance matrix of the network shown in the given figure, for which the marked parameters are per unit impedance, is

(*a*) $\begin{bmatrix} 0.3 & -0.2 \\ -0.2 & 0.2 \end{bmatrix}$ (*b*) $\begin{bmatrix} 0.3 & 0.2 \\ 0.2 & 0.2 \end{bmatrix}$

(*c*) $\begin{bmatrix} 0.3 & -0.2 \\ -0.2 & 0.2 \end{bmatrix}$ (*d*) $\begin{bmatrix} 1.5 & -5 \\ -5 & 5 \end{bmatrix}$

142. Compared with a solid conductor of the same radius, corona appears on a standard conductors at a lower voltage, because standing

(*a*) assists ionisation

(*b*) makes the current flow spirally about the axis of the conductor

(*c*) produces oblique sections to a plane perpendicular to the axis of the conductor

(*d*) produces surfaces of smaller radius

143. The non-uniform distribution of voltage across the units in a string of suspension type insulators is due to

(*a*) unequal self-capacitance of the units

(*b*) non-uniform distance of separation of the units from the tower body

(*c*) the existence of stray capacitance between the metallic junctions of the units and the tower body

(*d*) non-uniform distance between the cross-arm and the units

144. The insulation level of a 400kV EHV overhead transmission line is decided on the basis of

(*a*) lightning overvoltage

(*b*) switching overvoltage

(*c*) corona inception voltage

(*d*) radio and TV interference

145. The power transmission capability of dipolar lines is approximately

(*a*) half that of 3-phase single circuit line

(*b*) the same as that of 3-phase single circuit line

(*c*) twice that of 3-phase single circuit line

(*d*) thrice that of 3-phase single circuit line

146. By using bundled conductors, the critical voltage for formation of corona will

(*a*) increase (*b*) decrease

(*c*) remain the same (*d*) unpredictable

147. The per-unit 'regulation' of a short transm-ission line, having a per-unit resistance voltage drop V_r and a per-unit reactance voltage drop of V_x at rated current and with power-factor cos θ, is given by

(*a*) $(V_r + V_x) \cos\theta$ (*b*) $(V_r + V_x) \sin\theta$

(*c*) $V_r \cos\theta + V_x \sin\theta$ (*d*) $V_r \sin\theta + V_x \cos\theta$

148. The insulation strength of EHV lines is mainly governed by

(*a*) switching overvoltages

(*b*) lightning overvoltages

(*c*) power frequency overvoltage

(*d*) dynamic overvoltages

149. A three-phase transmission line of negligible resistance and capacitance has an inductive reactance of 100 ohms per phase. When the sending-end and receiving-end voltage are maintained at 110 kV, the maximum power that can be transmitted will be

(*a*) 121 MW (*b*) $121\sqrt{3}$ MW

(*c*) 363 MW (*d*) 363 kW

150. If the discharge is 1 m^3/s and the head of water is 1 *m*, then the power generated by the alternator in one hour (assume 100% efficiency of generator and turbine) will be

(*a*) 10 kW (*b*) 73/75 kW

(*c*) 746/75 kW (*d*) 100 kW

151. A string insulator has 4 units. The voltage across the bottom-most unit is 33.33% of the total voltage. Its string efficiency is

(*a*) 25% (*b*) 33.33%
(*c*) 66.67% (*d*) 75%

152. Transmission efficiency of a transmission line increases with the

(*a*) decrease in power factor and voltage
(*b*) increase in power factor and voltage
(*c*) increase in power factor but decrease in voltage
(*d*) increase in voltage but decrease in power factor

153. A voltage control bus is characterised by the specified values of

(*a*) real and reactive powers
(*b*) real power and voltage phase angle
(*c*) real power and voltage magnitude
(*d*) reactive power and voltage magnitude

154. The reactance relay is essentially

(*a*) an over-voltage relay with current restraint
(*b*) an over-current relay with directional restraint
(*c*) a directional relay with voltage restrain
(*d*) a directional relay with current restrai

155. At slack bus, which one of the following combinations of variables is specified

(*a*) $|V|, \delta$ (*b*) P, Q
(*c*) P,|V| (*d*) Q, |V|

(The symbols have their usual meaning)

156. D_s is the GMR of each subconductor of a four subconductor bundle conductor and *d* is the bundle spacing. What is the GMR of the equivalent single conductor?

(*a*) $1.09\sqrt{D_s \times d^3}$ (*b*) $1.09\sqrt{D_s^3 \times d^3}$
(*c*) $1.09\sqrt[4]{D_s^3 \times d^3}$ (*d*) $1.09\sqrt[4]{D_s \times d^3}$

157. The surge impedance of a 100 km long underground cable is 50 ohms. The surge impedance of a 40 km long similar cable would be

(*a*) 20 ohms (*b*) 50 ohms
(*c*) 80 ohms (*d*) 125 ohms

158. The L/C ratio for 132 kV and 400 kV lines are typically 160×10^3 and 62.5×10^3 respectively. The natural 3-phase loading for the two lines will be respectively

(*a*) 108.9 and 2560 MW
(*b*) 44 and 2560 MW
(*c*) 44 and 640 MW
(*d*) 640 and 44 MW

159. Consider the following statements distance relays

1. The effect of an arc may cause the relay to underreach.
2. The effect of an intermediate current source may cause the relay to overreach.
3. The effect of an intermediate load may cause the relay to underreach.
4. The presence of dc offset in the fault current may cause the relay to overreach.

Of these statements:

(*a*) 2 and 3 are correct
(*b*) 1 and 4 are correct
(*c*) 1, 2 and 3 are correct
(*d*) 2, 3 and 4 are correct

160. Which one of the following equations is correct

(*a*) $-AB + CD = -1$
(*b*) $AD + CB = 1$
(*c*) $AB - CD = -1$
(*d*) $-AD + BC = -1$

Where A, B, C and D are generalised circuit constants.

161. Match List-I with List-II and select the correct answer using the codes given below the lists.

List-I

A. Deuterium.
B. Heavy water
C. Uranium – 235
D. Thorium – 232

List-II

1. A fissile fuel of nuclear reactor.
2. A fertile fuel of nuclear reactor.
3. Used as a fuel in fusion reactor.
4. Used as a moderator in nuclear reactor.

Codes:

	A	B	C	D
(*a*)	1	2	3	4
(*b*)	3	4	2	1
(*c*)	3	2	4	1
(*d*)	4	1	2	3

162. For a transmission line with negligible losses, the lagging reactive power (VAR) delivered at the receiving-end, for a given receiving-end voltage, is directly proportional to the

(*a*) square of the line voltage drop
(*b*) line voltage drop
(*c*) line inductive reactance
(*d*) line capacitive reactance

163. Two arrangements of conductors are proposed for a 3-phase transmission line: one with equilateral spacing of 4m and the other, a flat with 4m between the conductors as shown in the given figure

The conductor diameter in each case is 2cm. Assuming that the line is transposed in both cases.

Which one of the following statements would be true?

(C_n = capacitance in F/m line to neutral,

L = inductance in H/m per phase)

(*a*) $C_{n1} = C_{n2}$ and $L_1 > L_2$

(*b*) $C_{n1} > C_{n2}$ and $L_1 < L_2$

(*c*) $C_{n1} < C_{n2}$ and $L_1 > L_2$

(*d*) $C_{n1} > C_{n2}$ and $L_1 = L_2$

164. The economic prospects of nuclear power generation in India to a great extent depends on successful development of

(*a*) fast reactors using enriched uranium as the fuel thus avoiding import of heavy water

(*b*) thermal reactors using natural uranium as the fuel and avoiding import of enriched uranium

(*c*) breeder reactors using thorium as a blanket material and U^{233} as the fuel

(*d*) reactor using plutonium as the fuel, natural uranium as blanket and liquid sodium as the coolant

165. A balanced 3-phase, 3-wire supply feeds balanced star connected resistors. If one of the resistors is disconnected, then the percentage reduction in load will be

(*a*) $33\frac{1}{3}$ (*b*) 50

(*c*) $66\frac{2}{3}$ (*d*) 75

166. When bundle conductors are used in place of single conductors, the effective inductance and capacitance will respectively

(*a*) increase and decrease

(*b*) decrease and increase

(*c*) decrease and remain unaffected

(*d*) remain unaffected and increase

167. For a given power delivered, if the working voltage of a distributor line is increased to *n* times, the cross-sectional area A of the distributor line, would be reduced to

(*a*) $\frac{1}{n}A$ (*b*) $\frac{1}{n^2}A$

(*c*) $\frac{1}{2n^2}A$ (*d*) $\frac{1}{2n}A$

168. The self-inductance of a long cylindrical conductor due to its internal flux linkages is k H/ m. If the diameter of the conductor is doubled, then the self-inductance of the conductor due to its internal flux linkage would be

(*a*) 0.5 KH/m (*b*) KH/m

(*c*) 1.414 KH/m (*d*) 4 K/m

169. Zero sequence currents can flow from a line into a transformer bank if the windings are in

(*a*) grounded star/delta (*b*) delta/star

(*c*) star/grounded star (*d*) delta/delta

170. The voltage across the various discs of a string of suspension insulators having identical discs is different due to

(*a*) surface leakage currents

(*b*) series capacitance

(*c*) shunt capacitances to ground

(*d*) series and shunt capacitances

171. A voltage V is applied to an *ac* circuit resulting in the delivery of a current I, Which of the following expression would yield the true power delivered by the source?

1. Real part of V/I.
2. Real part of VI.
3. I^2 times the real part of $\frac{V}{I}$

Select the correct answer using the codes given below:

Codes:

(*a*) 1 alone (*b*) 1 and 3

(*c*) 2 and 3 (*d*) 3 alone

172. Incremental fuel costs in ₹MWh for a plant consisting of two units are given by

$$\frac{dF_1}{dP_1} = 0.4\,P_1 + 400 \text{ and}$$

$$\frac{dF_2}{dP_2} = 0.48\,P_2 + 320$$

The allocation of loads P_1 and P_2 between the units 1 and 2 respectively for minimum cost of generation for a total load of 900MW is

(*a*) 200 MW and 700 MW

(*b*) 300 MW and 600 MW

(*c*) 400 MW and 500 MW

(*d*) 500 MW and 400 MW

173. The propagation constant of a transmission line is

$0.15 \times 10^{-3} + j1.5 \times 10^{-3}$

The wavelength of the travelling wave is

(*a*) $\frac{1.5\times10^{-3}}{2\pi}$ (*b*) $\frac{2\pi}{1.5\times10^{-3}}$

(*c*) $\frac{1.5\times10^{-3}}{\pi}$ (*d*) $\frac{\pi}{1.5\times10^{-3}}$

174. The identical synchronous machines A and B, running at the same speed, are linked through inductors as shown in the given figure:

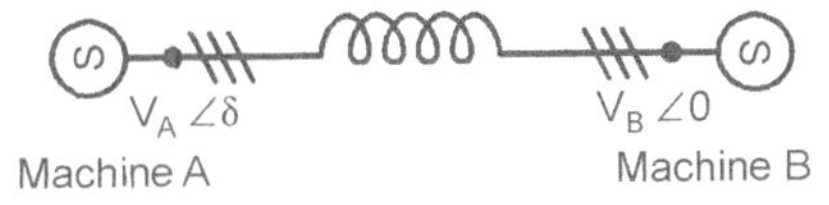

Machine A will supply active and reactive power to machine B when δ is

(*a*) positive and V_A is less than V_B

(*b*) positive and V_A is greater than V_B

(*c*) negative and V_A is less than V_B

(*d*) negative and V_A is greater than V_B

175. Maloperation of differential protection of transformers due to magnetizing inrush current is prevented by

(*a*) setting the current of the relay higher than the maximum value of inrush current

(*b*) keeping the time-setting long enough for the inrush current to fall to a value below the primary operating current of the relay

(*c*) by passing the inrush current from the operating coil of the relay

(*d*) filtering the third harmonic content of the inrush current flowing through the operating coil and passing through the restraining coil

176. "Expanded ACSR" are conductors composed of

(*a*) larger diameter individual strands for a given cross-section of the aluminium strands

(*b*) larger diameter of the central steel strands for a given overall diameter of the conductor

(*c*) larger diameter of the aluminium strands only for a given overall diameter of conductor

(*d*) a filler between the inner steel and the outer aluminium strands to increase the overall diameter of the conductor

177. The function of the earth wire in an extra high voltage line is to

(*a*) prevent earth fault

(*b*) provide a safety measure for any highflyir object

(*c*) provide a shield to the phase conductors from direct lightning stroke

(*d*) provide mechanical strength to the tower

178. An equipment has 0.8 per unit impedance on a base of 66 kV, 100 MVA. Its per unit impedance on a base of 33kV, 100 MVA will be

(*a*) 0.4 (*b*) 0.8

(*c*) 1.6 (*d*) 3.2

179. In a medium or high-head hydroelectric power station, a surge tank is provided to

(*a*) reduce the length of the penstock pipes

(*b*) augment water at the forebay

(*c*) control the pressure variations in the penstock pipes due to sudden load changes

(*d*) control the water flow through the turbine

180. A synchronous motor operates as a synchronous condenser when it is

(*a*) operated at unity power factor

(*b*) under-excited

(*c*) over-excited

(*d*) connected in parallel with condensers

181. In a power station, the cost of generation of power reduces most effectively when

(*a*) diversity factor alone increases

(*b*) both diversity factor and load factor increase

(*c*) load factor alone increases

(*d*) both diversity factor and load factor decrease

182. A bundled conductor in compared to a single conductor line (with same conductor cross-sectional area and same mean distance between conductors) has

	Self GMD	Mutual GMD	Per phase Inductance
(*a*)	Lower	Nearly same	Higher
(*b*)	Higher	Lower	Nearly same
(*c*)	Higher	Nearly same	Lower
(*d*)	Lower	Higher	Higher

183. A suspension type insulators has three units with self-capacitance C and ground capacitance of 0.2 C having a string efficiency of

(*a*) 78% (*b*) 80%

(*c*) 82% (*d*) 84%

184. Arrange the following in the correct sequence in which the flue gas passes through them after coming out of the boiler in a thermal power station:

1. I. D. Fan
2. Air preacher
3. Economiser
4. Electrostatic precipitator

Select the correct answer using the codes given below:

(*a*) 4, 3, 2, 1

(*b*) 3, 2, 1, 4

(*c*) 2, 1, 4, 3

(*d*) 1, 4, 3, 2

185. The convergence characteristic of the Newton- Raphson method for solving a load flow problem is

(*a*) quadratic (*b*) linear

(*c*) geometric (*d*) cubic

186. For reducing tower footing resistance, it is better to employ

(*a*) chemical and counterpoise

(*b*) chemical and ground rods

(*c*) ground rods and counterpoise

(*d*) chemical, ground rods and counterpoise

187. Differential protection of a generator makes use of the principal that, under normal conditions, the current/ currents

(*a*) at the neutral end of a phase winding is zero

(*b*) in each of the phase windings is identical

(*c*) at both ends of the phase winding are equal

(*d*) at the two ends of the phase winding are unequal

188. What is/are the disadvantage(s) of Gauss-Seidel method over Newton's method in load-flow programmes ?

(*a*) More iteration

(*b*) More memory requirement and less accuracy

(*c*) Less iterations but more memory

(*d*) More accuracy and more memory

189. A power system network with a capacity of 100 MVA has a source impedance of 10% at a point. The fault level at that point is

(*a*) 1 0MVA (*b*) 30 MVA

(*c*) 3000 MVA (*d*) 1000 MVA

190. The most appropriate operating speed in rpm of generators used in Thermal, Nuclear and Hydro-power plants would respectively be

(*a*) 3000, 300 and 1500

(*b*) 3000, 3000 and 300

(*c*) 1500, 1500 and 3000

(*d*) 1000, 900 and 750

191. A wave-trap is used at the termination of a HVAC overhead line to a station switchyard to

(*a*) prevent the transformer magnetising harmonics from reaching the overhead line

(*b*) damp the incoming surge waves from the overhead lines

(*c*) attenuate sub-harmonic oscillations

(*d*) provide for carrier communication

192. The power-angle characteristic of salient-pole synchronous machine working with a constant field excitation on an infinite-bus system is shown in the given figure. If the load on the machine is suddenly changed from P_1 to P_2 ($P_2 > P_1$), then the machine will fail to maintain stability if

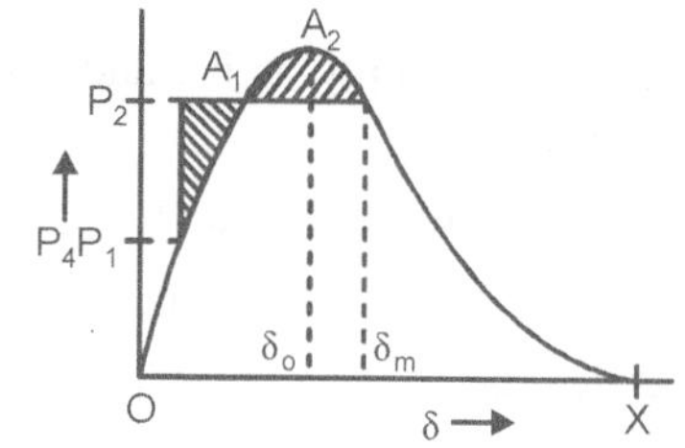

(*a*) shaded are A_1 is grater than shaded are A_2

(*b*) shaded are A_1 is equal to shaded are A_2

(*c*) shaded are A_1 is less than shaded are A_2

(*d*) $\delta_m > \delta_o$

193. An adjustable blade propeller turbine is called a

(*a*) Pelton turbine

(*b*) Kaplan turbine

(*c*) Francis turbine

(*d*) High head turbine

194. The specific speeds of Kaplan, Francis and Pelton turbines are in

(*a*) the increasing order

(*b*) the decreasing order

(*c*) neither increasing nor decreasing

(*d*) none of these

195. The average load factor of thermal power plant is

(*a*) 100% (*b*) 80-95%

(*c*) 50-60% (*d*) 35-45%

196. A super critical boiler is one that operates above the pressure and temperature of following values

(*a*) 100 kg/cm^2 and 540°C

(*b*) 218 kg/cm^2 and 540°C

(*c*) 100 kg/cm^2 and 373°C

(*d*) 218 kg/cm^2 and 373°C

197. The main functions of economiser of a boiler plant is to

(*a*) increase steam production

(*b*) reduce fuel consumption

(*c*) increase steam pressure

(*d*) increase life of the boiler

198. A generating station Which has a high investments and low operating cost is usually operated as a

(*a*) peak load station

(*b*) base load station

(*c*) medium load station

(*d*) none of these

199. Which of the plants is suitable for peak load

(*a*) diesel engine plant

(*b*) steam power plant

(*c*) nuclear power plant

(*d*) hydro power plant

200. A hydro power plant is commonly found in

(*a*) desert area (*b*) hilly area

(*c*) swamps (*d*) grass lands

201. A graphical representation of the discharge and time is known as

(*a*) load curve (*b*) load duration curve

(*c*) monograph (*d*) hydrography

202. In a load duration curve for an integrated power system the uppermost erest representation the energy contributed by

(*a*) base power stations

(*b*) major thermal stations

(*c*) peaking hydro or gas turbine stations

(*d*) non-conventional power stations

203. The most efficient torque-producing actuating structure for induction-type relays is

(*a*) shaded-pole structure

(*b*) watt-hour-meter structure

(*c*) induction-cup structure

(*d*) single-induction-loop structure

204. In the integral control of single-area system, when the system frequency error is reduced to zero.

(*a*) the integral output and the speed changer position attain a constant value

(*b*) the integrator output increases but the speed changer position remains constant

(*c*) the integrator output decreases and the speed changer position also comes down

(*d*) the integrator output decreases but the speed changer position moves up

205. A fault occurring on an end-supplied transmission line is more severe from the point of view of RRRV if it is a

(*a*) long line fault (*b*) short line fault

(*c*) medium line fault (*d*) generator fault

206. The A, B, C constants of a nominal T- circuit are given as

$A = 0.95 \angle 0.5°$ ohm

$B = 85 \angle 70°$ ohm

$C = 0.0005 \angle 90°$ ohm

If the two terminal voltages are held constants at 66 kV, then the steady-state stability limit of the line will be

(*a*) 0.3312 MW (*b*) 3.312 MW

(*c*) 33.12 MW (*d*) 331.2 MW

207. The self GMD of a conductor with three strands each of radius '*r*' and touching each other is

(*a*) $r(0.7788 \times 2 \times 2)^{1/3}$

(*b*) $r(0.7788 \times 2 \times 2 \times 2)$

(*c*) $r(0.7788 \times 2 \times 2)^3$

(*d*) $r(0.7788 \times 2 \times 2)^4$

208. Shunt resistor is connected across the contacts of a circuit-breaker in order to

(*a*) damp out the restriking transient

(*b*) by pass the arc current

(*c*) limit the short-circuit current

(*d*) reduce the damage to the contacts due to arcing

209. Phase comparison relaying is

(*a*) used for providing back-up protection to transmission lines

(*b*) immune to the effects of power surges

(*c*) adversely affected by the currents flowing in the protected line due to mutual induction from another nearly circuit

(*d*) adversely affected by the loss of synchronism between sources of generation beyond the ends of the protected line

210. Which of the following are the advantages of interconnected operation of power systems?

1. Less reserve capacity requirement.
2. More reliability.
3. High power factor.
4. Reduction is short-circuit level.

Select the correct answer using the codes given below:

Codes

(*a*) 1 and 2 (*b*) 2 and 3

(*c*) 3 and 4 (*d*) 1 and 4

211. The inductance per unit length of an overhead line due to internal flux linkages

(*a*) depends on the size of the conductor

(*b*) is independent of the size of conductor and constant

(*c*) depends on the current through the conductor

(*d*) depends on distance between conductors

212. The economizer, a component of steam power plants, is a heat-exchanger utilising the waste heat of

(*a*) bleed-steam to heat the fed water

(*b*) flue-gas to heat the feed water

(*c*) flue-gas to heat the air going into the boiler

(*d*) flue-gas to heat the pulverised coal

213. Shunt compensation in an EHV line is used to
(*a*) improve stability
(*b*) reduce fault level
(*c*) improve the voltage profile
(*d*) substitute for synchronous phase modifier

214. Hollow conductors are used in transmission lines to
(*a*) reduce weight of copper
(*b*) improve stability
(*c*) reduce corona
(*d*) increase power transmission capacity

215. For the transient stability of a power system,
(*a*) $\frac{d\delta}{dt} > 0, \frac{dp}{d\delta} > 0$ (*b*) $\frac{d\delta}{dt} = 0, \frac{dp}{d\delta} > 0$
(*c*) $\frac{d\delta}{dt} < 0, \frac{dp}{d\delta} < 0$ (*d*) $\frac{d\delta}{dt} < 0, \frac{dp}{d\delta} = 0$

216. The time taken for a surge to travel a 600 km long overhead transmission line is
(*a*) 6*s* (*b*) 1*s*
(*c*) 0.02*s* (*d*) 0.002*s*

217. The steady-state stability limit of a synchronous generator can be increased by
(*a*) an increase in its reactance
(*b*) an increase in the excitation of the machine
(*c*) a decrease in the moment of inertia of the machine
(*d*) an incerse in the moment of inertia of the machine

218. In the feed water and steam circuit of a thermal power plant, the condensates flows from the condensate extraction pump to the boiler feed pump and, then to the boiler. The HP and LP heaters are located in this circuit in relation to the boiler feed pump (BFP) as
(*a*) both HP and LP heaters after BFP
(*b*) HP heater before-and, LP heater after BFP
(*c*) both HP and LP heaters before BFP
(*d*) LP heater before- and, HP heater after BFP

219. For equilateral spacing of conductors of an untransposed 3-phase line, we have
(*a*) balanced receiving end voltages and no communication interference
(*b*) unbalanced receiving end voltages and no communication interference
(*c*) balanced receiving end voltages and communication interference
(*d*) unbalanced receiving voltages and communication interference

220. The dielectric stress in a cable is maximum at
(*a*) conductor surface (*b*) bedding
(*c*) lead streath (*d*) armour

221. A single line diagram of a power system is shown in the given figure. The diagonal elements of the Y_{BUS} matrix is given per unit impdences

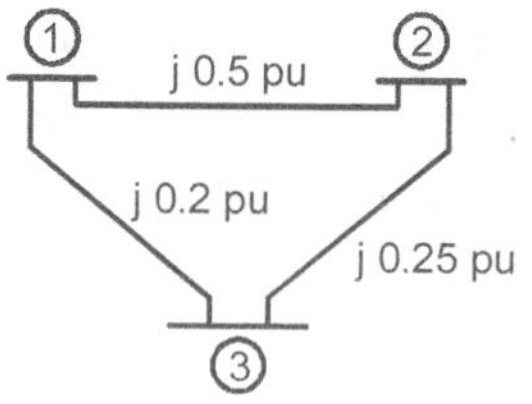

(*a*) $j0.7pu, j0.75pu, j0.45pu$
(*b*) $-j0.7pu, -j0.75pu, -j0.45pu$
(*c*) $-j0.7pu, -j6.0pu, j9.0pu$
(*d*) $j0.7pu, -j6.0pu, j9.0pu$

222. The surge impedance of a 400 km long over head transmission line is 400 ohms. For a 200 km length of the same line, the surge impedance will be
(*a*) 200Ω (*b*) 800Ω
(*c*) 400Ω (*d*) 100Ω

223. A differentiator has transfer function whose
(*a*) phase increases linearly with frequency
(*b*) amplitude remains constant
(*c*) amplitude increase linearly with frequency
(*d*) amplitude decrease linearly with frequency

224. The incremental fuel cost for two generating units are given by
$IC_1 = 25 + 0.2\,P_{G1}$
$IC_2 = 32 + 0.2\,P_{G2}$ where P_{G1} and P_{G2} are real power generated by the units.
The economic allocation for a total load of 250MW, neglecting transmission loss is given by
(*a*) $P_{G1} = 140.25$ MW, $P_{G2} = 109.75$ MW
(*b*) $P_{G1} = P_{G2} = 125$ MW
(*c*) $P_{G1} = 109.75$ MW, $P_{G2} = 140.25$ MW
(*d*) $P_{G1} = 100$ MW, $P_{G2} = 150$ MW

225. Which of the following reason make HRC fuses preferable in power circuits?
1. They have inverse time characteristics.
2. They are consistent in performance,
3. They can be selected for proper discrimination.

Select the correct answer by using the codes given below:
(*a*) 1 and 2 (*b*) 2 and 3
(*c*) 1 and 3 (*d*) 1, 2 and 3

226. The inductance and capacitance of a line are respectively 1.0 H and 0.01μF. If the instantaneous value of interrupted current is 10A, the voltage across the breaker contacts will be

(*a*) 125kV (*b*) 100kV

(*c*) 75kV (*d*) 66kV

227. The impedance per phase of 3-phase transmission line on a base of 100 MVA, 100 kV is 2pu. The value of this impedance on a base of 400 MVA and 400 kV would be

(*a*) 1.5pu (*b*) 1.0 pu

(*c*) 0.5pu (*d*) 0.25pu

228. A 3-phase line having negligible resistance and 22 ohm inductive reactance per phase operates with 110 kV sending-end voltage and 100 kV receiving -end voltage. The maximum power that this line can transmit as

(*a*) $500\sqrt{3}$ MW (*b*) 500 MW

(*c*) 1500 MW (*d*) $\frac{500}{\sqrt{3}}$ MW

229. In load-flow analysis, the load at a bus is represented as

(*a*) a constant current drawn from the bus

(*b*) a constant impedance connected at the bus

(*c*) constant real and reactive powers drawn from the bus

(*d*) a voltage-dependent impedance at the bus

230. The equivalent Thevenin's bus admittance matrix of a two-bus system with identical generators on both buses is

$$\begin{bmatrix} -j30 & +j10 \\ +j10 & -j30 \end{bmatrix}$$

The generator reactance and inter-connecting line reactance will be respectively

(*a*) $j0.05$ and $j0.1$ (*b*) $-j0.05$ and $j0.1$

(*c*) $-j0.05$ and $-j0.1$ (*d*) $j0.1$ and $j0.05$

231. While using air-blast breaker, current chopping is a phenomenon often observed when

(*a*) a long overhead lines is switched off

(*b*) a bank of capacitors is switched off

(*c*) a transformer on no-load is switched off

(*d*) a heavy load is switched off

232. If a short transmission line is delivering to a lagging *pf* load, the sending-end *pf* would be (notation have their usual meaning)

(*a*) $\frac{V_R \cos\phi + IR \sin\phi}{V_s}$ (*b*) $\frac{V_R \cos\phi + IR}{V_s}$

(*c*) $\frac{V_R \sin\phi + IR}{V_s}$ (*d*) $\frac{V_R \sin\phi + IR \cos\phi}{V_s}$

233. Match List-I (Parameter)with List-II (Effects) and select the correct answer using the codes given the Lists:

List-I

A. Percent power lost in transmission.

B. For a given current density, the conductor size.

C. Power handling capacity of a line at a given voltage.

D. Surge impedance of a transmission line.

List-II

1. Decrease with system voltage.
2. Reduces with line length.
3. Remains independent of line length.
4. Increase with line length.

Codes:

	A	B	C	D
(*a*)	1	2	4	3
(*b*)	3	4	2	1
(*c*)	3	2	4	1
(*d*)	1	4	2	3

234. The insulation resistance of a 20 km long underground cable is 8 mega ohm. Other things being the same, the insulation resistance of a 10 km long cable will be

(*a*) 16 mega ohm (*b*) 32 mega ohm

(*c*) 4 mega ohm (*d*) 2 mega ohm

235. A 100 MVA, 11kV, 3-phase, 50 Hz 8-pole synchronous generator has an inertia constant H equal to 4 MJ/MVA. The stored energy in the rotor of the generator at synchronous speed will be

$$H = \frac{E}{G}$$

(*a*) 100MJ (*b*) 400MJ

(*c*) 800MJ (*d*) 12.5MJ

236. The surge impedance loading of a long transmission line can be increased without adversely affecting its stability by connecting

(*a*) lumped inductances in series with the line

(*b*) lumped inductances in shunt with the line

(*c*) lumped capacitances in shunt with the line

(*d*) lumped capacitances in series with the line

237. A transmission line has equal voltages at the two ends, maintained constant by two sources. A third source is to be provided to maintain constant voltage equal to the end voltages at either the midpoint of the line (case 1) or at 75% of the distance from the sending (case 2). Then the maximum power-transfer capabilities of the line in the original case, case 1 and case 2, respectively, will be in the ratio of

(*a*) 1 : 1 : 1 (*b*) 1 : 2 : 1/0.75

(*c*) 1 : 2 : 4 (*d*) 1 : 4 : 16

238. A short length of cable is connected between dead-end tower and sub-station at the end of a transmission line. Which of the following will decrease, when voltage wave is entering from overhead line to cable?

1. Velocity of propagation of voltage wave.
2. Steepness of voltage wave.
3. Magnitude of voltage wave.

Select the correct answer using the codes given below:

Codes:

(*a*) 1 and 2 (*b*) 2 and 3

(*c*) 3 and 1 (*d*) 1, 2 and 3

239. Steady state stability of a power system is the ability of the power system to

(*a*) maintain voltage at the rated voltage level

(*b*) maintain frequency exactly at 50 Hz

(*c*) maintain a spinning reserve margin at all times.

(*d*) maintain synchronism between machine and on external tie lines

240. In an unbalanced 3-phase system, the currents are measured as

$I_a = 0, I_b = 6\angle 60°$ and $I_c = 6\angle -120°$

The corresponding sequence components will be

	I_{a0}	I_{a1}	I_{a2}
(*a*)	zero	$3 - j\sqrt{3}$	$-3 + j\sqrt{3}$
(*b*)	zero	$-3 + j\sqrt{3}$	$3 - j\sqrt{3}$
(*c*)	zero	$-9 + j3\sqrt{3}$	$9 - j3\sqrt{3}$
(*d*)	zero	$9 - j3\sqrt{3}$	$-9 + j3\sqrt{3}$

241. In HVDC transmission system, rectifier firing angle α is kept near

(*a*) 0° (*b*) 15°

(*c*) 30° (*d*) 90°

Directions : *The following items consist of two statements, one labelled as 'Assertion A ' and the other labelled as 'Reason R'. You are to examine these two statements carefully and decide if the Assertion A and the Reason R are individually true and if so, whether the Reason is a correct explanation of the Assertion. Select your answer to these items using the codes given below and mark your answer sheet accordingly.*

Codes:

(*a*) Both A and R true and R is the correct explanation of A

(*b*) Both A and R are true but R is NOT a correct explanation of A

(*c*) A is true but R is false

(*d*) A is false but R is true

242. Assertion (A) : The draft tube in hydro power stations permits a negative of suction head to be established at the runner exit.

Reasons (R) : The turbine can be installed above the tail-race level without loss of head.

Two transformer banks, one connected in Y–Δ and the other in Δ–Y both designed for the same line voltages on the primary and secondary sides, are to be operated in parallel.

243. Assertion (A) : A glazed ceramic insulator has a lower breakdown strength than an unglazed one.

Reasons (R) : Glazing covers pores and cracks and makes the surface non-absorbent.

244. Assertion (A) : An over-excited synchro-nous motor draws a leading current from the mains.

Reasons (R) : A synchronous machine irrespective of its operation as generator or motor has demagnetising armature reaction when over excited.

245. Assertion (A) : To guarantee proper operation of mho type relays under 3-phase faults occuring near the relay location, 'memory action' must be provided.

Reasons (R) : Voltage which is reduced to zero due to short circuit is required for developing pickup torque.

246. Assertion .(A): The input impedance of a lossless transmission line can have two limiting values, zero and infinity.

Reasons (R) : These limiting values will depend upon whether the transmission line is open or shorted at the load end and whether the length of the line is an integral multiple of $\lambda/4$ when λ is the wavelength within the transmission line.

247. Assertion (A) : Residual earth fault relays cannot give satisfactory protection in the case of resonant earthed system.

Reasons (R) : Earth fault currents are very large.

248. Assertion (A) : In the nominal-π model of an overhead transmission line, the value of the sending-end voltage obtained for the normal rated load at the receiving-end at rated voltage, is higher than the actual value.

Reasons (R) : The nominal–π method under emphasize the effect of line capacitance giving a higher value of the sending-end voltage than the actual.

249. **Assertion (A) :** Phase comparison pilot relaying for the protection of a trans-mission line is blocking pilot.

Reasons (R) : The pilot signal is not required from the other end to permit tripping.

250. **Assertion (A) :** In a high-or-medium head hydroelectric power station having a long penstock, a surge-tank is provided near the turbine.

Reasons (R) : A surge-tank stores additional water to be released during the peak-load period.

251. **Assertion (A) :** It is necessary to provide reactive power compensating equipment to generate or absorb reactive volt amps to maintain a voltage profile on transmission system within limits.

Reasons (R) : A synchronous compensator delivers both leading reactive and lagging reactive VARs depending on the excitation.

252. **Assertion (A) :** Shunt capacitors are used in conjunction with lighting arresting for the protection of station equipment against surges.

Reasons (R) : Shunt capacitors reduce the crest of the surges.

253. **Assertion (A) :** In HVDC systems, the *dc* voltage should be as high as possible and the rectifier control angle alpha should be as low as possible.

Reasons (R) : Control of *dc* voltage is exercised by the rectifier and inverter control angles alpha and gama respectively.

254. **Assertion (A) :** The bulk supply into the large urban areas is often carried out by underground cables operating at 132 kV and above. This results in large generation of VARs by the cable and consequent rise in voltage at the receiving end.

Reasons (R) : There VARs may be neutrilised by string series capacitance in the line at appropriate intervals.

255. **Assertion (A) :** Time lag is invariably provided in the majority of protective relaying schemes for the protection of various elements of the electrical power system network.

Reasons (R) : Time lag is provided to permit discrimination between main and back-up protection.

256. **Assertion (A) :** There is no control on tie-line power flow when flat frequency control method is used for controlling the frequency of an interconnected power system.

Reasons (R) : The job is maintaining frequency constant of assigned to all the power stations.

257. **Assertion (A) :** Insulators should have high impulse ratio and lighting arresters should have low impulse ratio.

Reasons (R) : Lighting arrestors are required to divert surge currents to the ground

258. **Assertion (A) :** For the low-flow solution of a large power system Newton-Raphson method is superior to the Gauses Seidel method.

Reasons (R) : In the Gauss-Seidel method, the number of iterations vary between 3 to 5, irrespective of the size of the system.

259. **Assertion (A) :** Mho relay is more suitable for the protection of long lines .

Reasons (R) : Mho relays is stable under power swing conditions.

260. **Assertion (A) :** A relay-type feedback control system is a non-linear system.

Reasons (R) : The signals in some part of a relay-type feedback control system are rectangular waves.

261. **Assertion (A) :** To provide effective protection to the terminal apparatus against lighting, long lead length, to the feeder line and to the ground are recommended for arrester applications.

Reasons (R) : Long lead length offers high impedance path to lightning surge current.

262. **Assertion (A) :** The bus admittance matrix (Y Bus) used in load flow studies is a spare matrix.

Reasons (R) : The elements of Y_{BUS} exhibit the physical interconnections of nodes of the power system.

263. **Assertion (A) :** Reactance relay is more suitable for the protection of short lines.

Reasons (R) : Operation of reactance relay is independent of arc fault resistance.

264. **Assertion (A) :** Mho relay is widely preferred for the protection of long transmission lines.

Reasons (R) : Mho relay is stable under power swing conditions.

EXERCISE - II

1. When a fault occurs on the system a circuit breaker
   (*a*) opens automatically
   (*b*) closes automatically
   (*c*) opens manually
   (*d*) none of these **[NPTC]**
2. The most common type of three phase in unsymmetrical fault is
   (*a*) single line to ground
   (*b*) line to line
   (*c*) double line to ground
   (*d*) three phase **[NTPC]**
3. Which of the following CBs is preferred for EHV applications ?
   (*a*) Air break CB (*b*) Minimum oil CB
   (*c*) Bulk oil CB (*d*) SF_6CB **[NTPC]**
4. A relay performs the function of
   (*a*) fault isolation (*b*) fault detection
   (*c*) fault prevention (*d*) all are correct **[NTPC]**
5. A Mho relay is a
   (*a*) voltage restrained directional relay
   (*b*) voltage controlled over current relay
   (*c*) directional restrained over current relay
   (*d*) directional restrained over voltage relay **[NTPC]**
6. Which is the following statement about corona is incorrect?
   (*a*) corona gives rise to radio interference
   (*b*) corona results in loss of power in transmission
   (*c*) corona discharge can be observed as red lumine science
   (*d*) corona is always accompanied by hissing noise **[NTPC]**
7. In sphere gap, the sphere are made of
   (*a*) aluminium
   (*b*) brass
   (*c*) bronze
   (*d*) any of these **[NTPC]**
8. Surge diverters are
   (*a*) non-linear resistors in series with spark gaps which act as fast switches
   (*b*) are quenching devices
   (*c*) shunt reactors to limit the voltage rise due to Ferranti effect
   (*d*) over voltages of power frequency harmonics **[NTPC]**
9. The breakdown voltage in gases depends on
   (*a*) distance between the electrodes
   (*b*) relative air density
   (*c*) humidity
   (*d*) all of these **[NTPC]**
10. The electrical breakdown strength of insulating material depends on
    (*a*) nature of applied voltage
    (*b*) imperfection in dielectric material
    (*c*) pressure, temperature and humidity
    (*d*) all of these **[NTPC]**
11. The size of the earth wire is determined by
    (*a*) the ampere capacity of the service wires
    (*b*) the atmospheric conditions
    (*c*) the voltage of the service wires
    (*d*) none of these **[NTPC]**
12. The smaller the lagging reactive power drawn by a circuit, its pf will be
    (*a*) better (*b*) poorer
    (*c*) unit (*d*) none of these **[NTPC]**
13. Zero sequence component always flows through
    (*a*) phase wires
    (*b*) neutral wire
    (*c*) earth wire
    (*d*) any of the above **[NTPC]**
14. In order to have lower cost of electrical generation
    (*a*) the load factor and density factor should be high
    (*b*) the load factor should be low but diversity factor high
    (*c*) the load factors should be high but diversity factor low
    (*d*) the load factor and diversity factor should be low **[NTPC]**
15. Liquids are generally used as insulating materials up stresses of about
    (*a*) 100 MV/cm (*b*) 50 MV/cm
    (*c*) 50 kV/cm (*d*) 500 V/cm **[NTPC]**
16. Which of the following techniques is used for the measurements of ac high frequency voltages
    (*a*) peak voltmeter
    (*b*) series resistance micro-ammeter
    (*c*) resistance potential divider
    (*d*) all of these **[NTPC]**

17. A thermal protection switch is able to protect against

(*a*) over load (*b*) over voltage

(*c*) temperature (*d*) short circuit **[NTPC]**

18. Resistance switching is used in

(*a*) bulkoil circuit breakers

(*b*) minimum oil circuit breakers

(*c*) air-blast circuit breakers

(*d*) all types of breakers **[NTPC]**

19. Earthing is necessary to give protection against

(*a*) voltage fluctuation

(*b*) overloading

(*c*) danger of electric shock

(*d*) high temperature of the conductors **[NTPC]**

20. In India least share of power is provided by

(*a*) Thermal power plant

(*b*) Hydro-electric power plants

(*c*) Nuclear power plants

(*d*) Diesel power plants **[BHEL]**

21. Which type of plant has the minimum running cost per Kwh of energy generated

(*a*) Hydro-electric plant

(*b*) Thermal power plant

(*c*) Nuclear power plant

(*d*) Diesel power plant **[BHEL]**

22. What type of insulator will be used if the direction of the transmission line is to be changed?

(*a*) Pin-type (*b*) Suspension type

(*c*) Strain type (*d*) Shakle type **[BHEL]**

23. For a transmission line, the constant A, B, C and D are related as

(*a*) $AD - BC = 1$ (*b*) $BC - AD = 1$

(*c*) $\sqrt{AD - BC} = 0.5$ (*d*) $\sqrt{AD - BC} = 1.2$

[BHEL]

24. Which voltage determines the insulations strength of an EHV line?

(*a*) Switching over voltage

(*b*) Lighting over voltage

(*c*) Power frequency over voltage

(*d*) None of these **[BHEL]**

25. The coefficient of reflection of voltage for a short circuit line is

(*a*) 1.0 (*b*) −1.0

(*c*) 0 (*d*) 2.0 **[BHEL]**

26. Ferranti effect on long overhead lines is experienced when

(*a*) the line is lightly loaded

(*b*) the power factor is unit

(*c*) the power factor is leading

(*d*) corona effect is dominant **[BHEL]**

27. In AC transmission line, difference in phase of voltage at the two ends of the line is due to

(*a*) resistance of line (*b*) system voltage

(*c*) reactance of line (*d*) insulators **[BHEL]**

28. A synchronous machine has higher capacity for

(*a*) leading *p.f*

(*b*) lagging *p.f*

(*c*) it does not depend upon the *pf* of the machine

(*d*) none of these **[BHEL]**

29. More efficient plants are used as

(*a*) base load stations

(*b*) peak load stations

(*c*) both (*a*) and (*b*)

(*d*) none of these **[BHEL]**

30. For a 500 Hz frequency excitation, a 50 km long power line will be modelled as

(*a*) short line

(*b*) medium line

(*c*) lone line

(*d*) data not insufficient for decision **[GATE]**

31. Skin effect does not depend on

(*a*) A nature of material

(*b*) Size of wire

(*c*) Supply frequency

(*d*) Ambient temperature **[BHEL]**

32. An ACSR conductor has specification as 48/7, which of the following statements is correct for the conductor?

(*a*) The conductor has 48 strands

(*b*) The conductors has 55 strands

(*c*) The conductors has 48 aluminium and 7 steel strands

(*d*) The conductor has 48 steel and 7 aluminium strands **[BHEL]**

33. In the underground cable system

(*a*) there are less changes of faults to occur

(*b*) it is difficult to locate fault and repair

(*c*) fault does not occur at all

(*d*) fault occur very rare and difficult to locate and repair **[BHEL]**

34. A Kaplan turbine is

(*a*) inward flow, impulse turbine

(*b*) outward flow, reaction turbine

(*c*) a high head mixed flow turbine

(*d*) low head avail flow turbine **[BHEL]**

35. If a plant has.zero reserve capacity, then plant load factor always

(*a*) equal plant capacity factor

(*b*) is greater than plant capacity factor

(*c*) is less than plant capacity factor

(*d*) none of these **[BHEL]**

36. In India the system adopted for transmission of electric power has

(*a*) three-phase three wire system

(*b*) three-phase four wire system

(*c*) both (*a*) and (*b*)

(*d*) none of these **[BHEL]**

37. The criterion for electron of size of conductor for a feeder is

(*a*) Voltage drop (*b*) Corona loss

(*c*) Temperature rise (*d*) Radio interference **[BHEL]**

38. Which of the following is the protective device against lightening over voltages?

(*a*) Rod gaps (*b*) Surge absorbers

(*c*) Horn gaps (*d*) All of these **[BHEL]**

39. Material generally used for armour of high voltage cables is

(*a*) aluminium (*b*) steel

(*c*) brass (*d*) copper **[BHEL]**

40. The most commonly used generation voltage in India is

(*a*) 11 kV (*b*) 6.6 kV

(*c*) 2.2 kV (*d*) 33 kV **[NTPC]**

41. The insulation resistance of a cable of length 10 km is 1 MΩ. For a length of 100 km of the same cable, the insulation resistance will be

(*a*) 1 MΩ (*b*) 10 MΩ

(*c*) 0.1 MΩ (*d*) 0.01 MΩ **[GATE]**

42. For an unbalanced fault, with paths for zero sequence currents, at the point of fault

(*a*) the negative and zero sequence voltage are minimum

(*b*) the negative and zero sequence voltage are maximum

(*c*) the negative sequence voltage is minimum and zero sequence voltage is maximum

(*d*) the negative sequence voltage is maximum and zero sequence voltage is minimum **[GATE]**

43. Which material is used in controlling chain reaction in a nuclear reactor

(*a*) Thorium (*b*) Heavy water

(*c*) Boron (*d*) Beryllium **[GATE]**

44. If the reference bus is changed in two load flow runs with same system data and power obtained for reference bus taken as specified P and Q in the latter run

(*a*) the system losses will be unchanged but complex bus voltages will change

(*b*) the system losses will change but complex bus voltages remain unchanged

(*c*) the system losses as well as complex bus voltages will change

(*d*) the system losses as well as complex bus voltage will be unchanged **[GATE]**

45. Resistance switching is normally employed in

(*a*) all breakers

(*b*) bulk oil breakers

(*c*) minimum oil breakers

(*d*) air blast circuit breakers **[GATE]**

46. Which one the following is characteristic of an offset MHO relay?

(*a*) 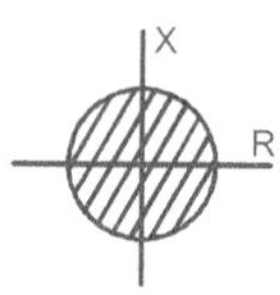

(*b*) 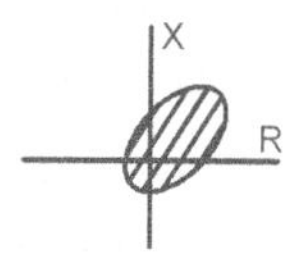

(*c*) 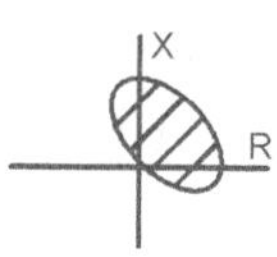

(*d*) 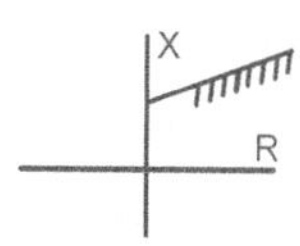

[DRDO]

47. The self-inductance of long cylindrical conductor due to its internal flux linkages is *kH/m*. If the diameter of the conductor is doubled, then the self-inductance of the conductor due to its internal flux linkages would be

(*a*) 0.5 *kH/m* (*b*) *kH/m*

(*c*) 1.414 *kH/m* (*d*) 4 *kH/m* **[NTPC]**

48. Which one of the following is not true regarding HVDC transmission?

(*a*) Corona loss is much more in HVDC transmission

(*b*) The power transmission capability of bipolar line is almost the same as that of single-circuit *ac* line

(*c*) HVDC link can operate between two *ac* system whose frequencies need not be equal

(*d*) There is no distance limitation for HVDC Transmission by underground cable **[NTPC]**

49. The incremental cost characteristic of the two units in a plant are given by

$IC_1 = ₹ (0.1\,P_1 + 8.0)$ per MWh

$IC_2 = ₹ (0.15\,P_2 + 3.0)$ per MWh

The optimum sharing of load when the total load is 100 MW is

(*a*) $P_1 = 60$ MW and $P_2 = 40$ MW

(*b*) $P_1 = 33.3$ MW and $P_2 = 66.7$ MW

(*c*) $P_1 = 40$ MW and $P_2 = 60$ MW

(*d*) $P_1 = 66.7$ MW and $P_2 = 33.3$ MW **[NTPC]**

50. Which of the following circuit breakers take minimum time in installation?

(*a*) Air blast (*b*) Minimum oil

(*c*) Bulk oil (*d*) SF_6 **[NTPC]**

51. A power system is subjected to a fault which makes the zero sequence compound of current equal to zero. The nature of fault is

(*a*) double line to ground fault

(*b*) double line fault

(*c*) line of ground fault

(*d*) three-phase to ground fault **[BHEL]**

52. The incremental generating costs of two generating units are given by

$IC_1 = 0.1\,X + 20$ ₹/MWhr.

$IC_2 = 0.15\,Y + 18$ ₹/MWhr.

Where X and Y are power (in MW) generated by the two units. For a total demand of 300 MW, the values (in MW), of X and Y will be respectively

(*a*) 172 and 128 (*b*) 128 and 172

(*c*) 175 and 125 (*d*) 200 and 100 **[PGCIL]**

53. If a voltage-controlled bus is treated as load bus, then which one of the following limits would be violated?

(*a*) Voltage (*b*) Active power

(*c*) Reactive power (*d*) Phase angle **[PGCIL]**

54. For which of the following reasons is a differential relay biased to avoid maloperation when used for transformer protection?

1. Saturation of CTs.
2. Mismatch to CT ratios.
3. Difference in connection of both sides.
4. Current setting multiplier.

Select the correct answer using the codes given below:

(*a*) 1 and 4 (*b*) 1 and 2

(*c*) 2, 3 and 4 (*d*) 1, 2 and 3 **[PGC1L]**

55. The velocity of travelling wave through a cable of relative permittivity 9 is

(*a*) 9×10^8 m/s

(*b*) 3×10^8 m/s

(*c*) 10^8 m/s

(*d*) 2×10^8 m/s **[BHEL]**

56. A string insulator has 4 units. The voltage across the bottom-most unit is 33.33% of the total voltage. Its string efficiency is

(*a*) 25% (*b*) 33.33%

(*c*) 66.67% (*d*) 75% **[BHEL]**

57. Consider the following statements about distance relays

1. The effect of an arc may cause the relay to underreach.
2. The effect of an intermediate current source may cause relay to overreach.
3. The effect of an intermediate load may cause the relay to under reach.
4. The presence of dc offset in the fault current may cause the relay to overreach.

Of these statements

(*a*) 2 and 3 are correct

(*b*) 1 and 4 are correct

(*c*) 1,2 and 3 are correct

(*d*) 2, 3 and 4 are correct **[BHEL]**

58. A power system network consists of three elements 0–1,1–2 and 2–0 of per unit impedances 0.2, 0.4 and 0.4 respectively. It's bus impedance matrix is given by

(*a*) $\begin{matrix} & 1 & 2 \\ 1 & 7.5 & -2.5 \\ 2 & -2.5 & 5.0 \end{matrix}$ (*b*) $\begin{matrix} & 1 & 2 \\ 1 & 0.16 & -0.08 \\ 2 & -0.08 & 0.24 \end{matrix}$

(*c*) $\begin{matrix} & 1 & 2 \\ 1 & 0.16 & -0.08 \\ 2 & 0.08 & 0.24 \end{matrix}$ (*d*) $\begin{matrix} & 1 & 2 \\ 1 & 0.6 & 0.4 \\ 2 & 0.4 & 0.8 \end{matrix}$ **[BHEL]**

59. A generator is connected to an infinite bus through a double circuit transmission. The fault occuring at the middle of one of the transmission lines is subsequently cleared by opening the circuit breakers at both the ends of the line simult-aneously. The transient stability limit of the system is improved by **[BHEL]**

(*a*) decreasing the excitation of generator

(*b*) decreasing the fault-clearing time

(*c*) increasing the fault-clearing time

(*d*) increasing the transfer reactance between the generator and infinite bus

60. Match List-I (Type of relay) with List-II (For the protection of) and select the correct answer using the codes given below in the lists:

List-I	List-II
A. Buchholtz relay	1. Feeder.
B. Translay relay	2. Transformer
C. Negative sequence relay	3. Generator
D. Directional over current relay	4. Long overhead line

Codes:

	A	B	C	D
(a)	1	2	3	4
(b)	2	1	3	4
(c)	1	2	4	3
(d)	1	4	2	3

[NTPC]

61. A single to ground fault occurs on a 3-phase isolated neutral system with a line to neutral voltage of $V\,kV$. The potentials on the healthy phases rise to a value equal to

(a) $\sqrt{2}$ VkV (b) $\sqrt{3}$ VkV

(c) 3VkV (d) $\frac{1}{\sqrt{3}}$ VkV **[NTPC]**

62. A relay used on short transmission lines is

(a) *mho* relays

(b) *reactance* relay

(c) *impedance* relay

(d) *admittance* relay **[NTPC]**

63. Earthing of electrical equipment is necessary for the protection against

(a) over loading

(b) voltage fluctuation

(c) danger of electric shock

(d) high conductor temperature **[NTPC]**

64. In a string of suspension insulators, the voltages distribution across the different units of a string could be made uniform by the use of a grading ring, because it

(a) forms capacitances with link-pins to carry the changing current from link-pins.

(b) forms capacitances which help to cancel the charging current from link-pins

(c) increases the capacitances of lower insulators units to cause equal voltage drop

(d) decreases the capacitances of upper insulator units to cause equal voltage drop **[NTPC]**

65. The design of insulation for systems above 400 kV is based upon

(a) lightning overvoltage

(b) switching surges

(c) system voltage level

(d) system load level **[NTPC]**

66. If the fault current is 3000 A, for and I.D.M.T relay with a plug setting of 50% and C.T. ratio of 400/5, then plug setting multiplier would be

(a) 7.5 (b) 15.0

(c) 18.75 (d) 37.5 **[NTPC]**

67. The bus admittance matrix (Y_{bus}) of a power system is not

(a) symmetric

(b) a square matrix

(c) a full matrix

(d) generally having dominant diagonal elements **[NTPC]**

68. Match List-I (Parameter) with List-II (Effect) and select the correct answer using the codes given below the lists:

List-I

A. Percent power lost in transmission

B. For a given current density the conductor size

C. Power handling capacity of a line at a given voltage

D. Surge impedance of a transmission line

List-II

1. Decreases with system voltage.
2. Reduces with line length.
3. Remains independent of line length.
4. Increase with line length.

Codes:

	A	B	C	D
(a)	1	2	4	3
(b)	3	4	2	1
(c)	3	2	4	1
(d)	1	4	2	3

[NTPC]

69. The surge impedance of a 100 km long underground cable is 50 ohms. The surge impedance of a 40 km long similar cable would be

(a) 20 ohms (b) 50 ohms

(c) 80 ohms (d) 125 ohms **[NTPC]**

70. Zero sequence currents can flow from a line into a transformer bank if the windings are in

(a) grounded star/delta (b) delta/star

(c) star/grounded/star (d) delta/delta **[NTPC]**

71. The following figure shows the single line diagram of a power system with all reactance marked in per unit (*pu*) on the same base:

The system is on no-load when a 3-phase fault occurs at F on the high voltage side of the transformer T_2 The fault current will be

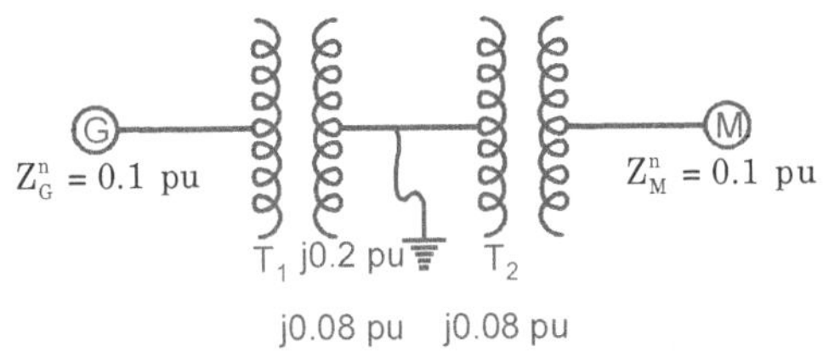

(*a*) $-j0.8187\,pu$ (*b*) $+j0.8187\,pu$
(*c*) $-j8.1871\,pu$ (*d*) $+j8.1871\,pu$ **[NTPC]**

72. In which one of the following models of transmission lines, is the full charging current assumed to flow over half the length of the line only?
(*a*) Equivalent-π
(*b*) Short line
(*c*) Nominal-π
(*d*) Nominal-T **[NTPC]**

73. The following data pertain to two alternators working in parallel and supplying a total load of 80 MW:
Machine 1: 40 MVA with 5% speed regulation
Machine 2: 60 MVA with 5% speed regulation
The load showing between machines 1 and 2 will be

(*a*) $\frac{P_1}{48MW}$ $\frac{P_2}{32MW}$
(*b*) 40 MW, 40 MW
(*c*) 30 MW, 50 MW
(*d*) 32 MW, 48 MW **[PGCIL]**

74. Consider the following statements:
In the case of suspension-type insulators, the string efficiency can be improved by
1. using a longer cross arm
2. using a guard ring
3. grading the insulator discs
4. reducing the cross-arm length
Of these statements
(*a*) 1, 2 and 3 are correct
(*b*) 2, 3 and 4 are correct
(*c*) 2 and 4 are correct
(*d*) 1 and 3 are correct **[PGCIL]**

75. A short transmission line, having its line- impedance angle as theta, is delivering a given power at the receiving-end at a lagging power- factor angle of ϕ. Which one of the following is a set of conditions for which this line will have maximum-and zero-regulation? **[PGCIL]**

	Maximum Regulation	Zero Regulation
(*a*)	$\phi=\theta$	$\phi-\theta=\pi/2$
(*b*)	$\phi=\theta=\pi/2$	$\phi=\theta$
(*c*)	$\phi=\theta$	$\phi+\theta=\pi/2$
(*d*)	$\phi+\theta=\pi/2$	$\phi=\theta$

76. If the fault current is 2kA, the relay setting is 50% and the C. T. ratio is 400/5, then the plug setting multiplier of a relay will be
(*a*) 5 (*b*) 7
(*c*) 8 (*d*) 10 **[PGCIL]**

77. In the solution of load-flow equation, Newton-Raphon (NR) method is superior to the Gauss-Seidal (GS) method, because the
(*a*) time taken to perform one interation in the NR method is less when compared to the time taken in the GS method
(*b*) number of iteration required in the NR method is more when compared to that in the GS method
(*c*) number of iterations required is not independent of the size of the system in the NR method
(*d*) convergence characteristic of the NR method are not affected by the selection of slack bus **[PGCIL]**

78. A line trap in carrier current relaying tuned to carrier frequency presents
(*a*) high impedance to carrier frequency but low impedance to power frequency
(*b*) low impedance to both carrier and power frequency
(*c*) high impedance to both carrier and power frequency
(*d*) low impedance to carrier frequency but high impedance to power frequency **[PGCIL]**

79. For a two-machine system with losses, with the transfer-impedance being resistive, the maximum value of the sending-end power P_{1max} and the maximum receiving-end power P_{2max} will occur at power- angles (δ) in such a manner that
(*a*) both P_{1max} and P_{2max} occur at $\delta < 90°$
(*b*) both P_{1max} and P_{2max} occur at $\delta > 90°$
(*c*) P_{1max} occur at $\delta > 90°$ and P_{2max} at $\delta < 90°$
(*d*) P_{1max} occur at $\delta < 90°$ and P_{2max} at $\delta > 90°$ **[NTPC]**

80. Which of the following are the advantages of interconnected operation of power systems?

1. Less reserve capacity requirement.
2. More reliability.
3. High power factor.
4. Reduction in short-circuit level.

Select the correct answer using the codes given below:

Codes:

(*a*) 1 and 2 (*b*) 2 and 3
(*c*) 3 and 4 (*d*) 1 and 4 **[NTPC]**

81. A hydel power plant of run-off-river type should be provided with a pondage so that the

(*a*) firm-capacity of the plant is increased
(*b*) operating head is controlled
(*c*) pressure inside the turbine casing remains constant
(*d*) kinetic energy of the running water is fully utilised **[NTPC]**

82. The convergence characteristics of the Newton-Rophson method for solving a load flow problem is

(*a*) quadratic (*b*) linear
(*c*) geometric (*d*) cubic **[NTPC]**

83. If, in a short transmission line, resistance and inductance are found to be equal and regulation appears to be zero, then the load will

(*a*) have unity power factor
(*b*) have zero power factor
(*c*) be 0.707 leading
(*d*) be 0.707 lagging **[PGCIL]**

84. Whenever the conductors are dead-ended or there is a change in the direction of transmission line, the insulators used are of the

(*a*) pin type (*b*) suspension type
(*c*) strain type (*d*) shackle type **[PGCIL]**

85. When a line-to-ground fault occurs, the current in a faulted phase is 100A. The zero sequence current in this case will be

(*a*) zero (*b*) 33.3 A
(*c*) 66.6 A (*d*) 100 A **[PGCIL]**

86. Equal area criterion gives the information regarding

(*a*) stability region
(*b*) absolute stability
(*c*) relative stability
(*d*) swing curves **[PGCIL]**

87. Bundled conductors in EHV transmission system provide

(*a*) reduced capacitance
(*b*) increased capacitance
(*c*) increased inductance
(*d*) increased voltage gradient **[PGCIL]**

88. The power generated by two plants are:

P_1 = 50MW, P_2 = 40MW. If the loss coefficient are $B_{11}=0.001$, $B_{22}=0.0025$ and $B_{12}=0.0005$, then the power loss will be

(*a*) 5.5 MW (*b*) 6.5 MW
(*c*) 4.5 MW (*d*) 8.5 MW **[PGCIL]**

89. For a 15-bus power system with 3 voltage controlled bus, the size of Jacobian matrix is

(*a*) 11×11 (*b*) 12×12
(*c*) 24×24 (*d*) 28×28 **[PGCIL]**

90. The power transmission capability of bipolar lines is approximately

(*a*) half that of 3-phase single circuit line
(*b*) the same as that of 3-phase single circuit line
(*c*) twice that of 3-phase single circuit line
(*d*) thrice that of 3-phase single circuit line **[PGCIL]**

91. If $|V_S| = |V_R| = 66$kV for three-phase transmission and reactance is 11 ohms/phase, then the maximum power transmission per phase would be

(*a*) 132 MW (*b*) 396 MW
(*c*) 66 MW (*d*) None of these **[PGCIL]**

92. If the receiving-end voltage and current are numerically equal to the corres-ponding sending-end values, that is, $|V_S| = |V_R| = 66$ kV and $|I_S| = |I_R| = 66$ kV, then such a line is called

(*a*) an infinite line (*b*) a natural line
(*c*) a tuned lines (*d*) a loss-less line **[PGCIL]**

93. The charging current in a transmission line increases due to corona effect because corona increase

(*a*) line current
(*b*) effective line voltage
(*c*) power loss is lines
(*d*) the effective conductor diameter **[PGCIL]**

94. Load frequency control is achieved by properly matching the individual machine's

(*a*) reactive powers
(*b*) generated voltages
(*c*) turbine inputs
(*d*) turbine and generator ratings **[PGCIL]**

95. Impulse ratio of insulators and lighting arrests should be

(*a*) both low

(*b*) high and low respectively

(*c*) low and high respectively

(*d*) both high **[PGCIL]**

96. A voltage of 1000 kV is applied to and overhead line with its receiving end open. If the impedance of the line is 5000 ohms, then the total surge power in the line is

(*a*) 200 MW (*b*) 500 MW

(*c*) 2 MW (*d*) 0.5 MW **[PGCIL]**

97. Corona loss on a particular system at 50Hz is 1kW/km per phase. The corona loss at 60 Hz would be

(*a*) 1 kW/km per phase

(*b*) 0.83 kW/km per phase

(*c*) 1.2 kW/km per phase

(*d*) 1.13 kW/km per phase **[ES]**

98. Steady-state stability of a power system is improved by

(*a*) reducing fault clearing time

(*b*) using double circuit line instead of single circuit line

(*c*) single pole switching

(*d*) decreasing generator inertia **[PGCIL]**

99. In a power system, each bus or node is associated with four quantities, namely

1. Real power.
2. Reactive power.
3. Bus-voltage magnitude and
4. Phase-angle of the bus voltage.

For load-flow solution, among these four, the number of quantities to be specified is

(*a*) any one (*b*) any two

(*c*) any three (*d*) all the four **[PGCIL]**

100. One current transformer (CT) is mounted over a 3-phase 3-core cable with its sheath and armour removed from the portion covered by the CT. An ammeter placed in the CT secondly would measure

(*a*) the positive sequence current

(*b*) the negative sequence current

(*c*) the zero sequence current

(*d*) three times the zero sequence **[PGCIL]**

101. In a DC transmission line

(*a*) it is necessary for the sending end and receiving end to be operated in synchronism

(*b*) the effect of inductive and capacitive reactances are greater than in an AC transmission line of the same rating

(*c*) there are no effects due to inductive and capacitive reactances

(*d*) power transfer capability is limited stability considerations **[PGCIL]**

102. The main criterion for selection the size of a distributor for a radial distribution system is

(*a*) voltage drop (*b*) corona loss

(*c*) temperature rise (*d*) capital cost **[GATE]**

103. In a series resonant circuit, which one of the following will increase on increasing the inductance to four times its value and reducing the capacitance to one-fourth of its value?

(*a*) Resonant frequency

(*b*) Current at resonant frequency

(*c*) Impedance at resonant frequency

(*d*) Selectivity of the circuit **[IAS]**

104. Consider the following statements:

To provide reliable protection for a distributions transformer against over voltages using lightning arresters, it is essential that the

1. Lead resistance is high.
2. Distance between transformer and arrestor is small.
3. Transformer and the arrestor have a common inter-connecting ground.
4. Spark overvoltage of the arrestor is greater than the residual voltage.

Of these statements

(*a*) 1, 3 and 4 are correct

(*b*) 2 and 3 are correct

(*c*) 2, 3 and 4 are correct

(*d*) 1 and 4 are correct **[ES]**

105. The non-uniform distribution of voltage across the units in a string of suspension type insulators is due to

(*a*) unequal self-capacitance of the units

(*b*) non-uniform distance of separation of the units form the lower body

(*c*) the existence of stray capacitance between the metallic junctions of the units and the lower body

(*d*) non-uniform distance between the cross-arm and the units **[BHEL]**

106. The per unit impedance of a synchronous machine is 0.242. If the base voltage is increased by 1.1 times, the per unit value will be

(*a*) 0.266 (*b*) 0.242

(*c*) 0.220 (*d*) 0.200 **[BHEL]**

107. The power angle characteristic of machine- infinite bus system is $P_e = z \sin \delta\, pu$. It is operating at $\delta = 30°$. Which one of the following is the synchronising power coefficient at the operating point?

(*a*) 1.0 (*b*) $\sqrt{3}$

(*c*) 2.0 (*d*) $\frac{1}{\sqrt{3}}$ **[IAS]**

108. The use of fast acting relays and circuit breakers for clearing a sudden short-circuit on a transmissions link between a generator and receiving end bus improves the transient stability of the machine because

(*a*) short-circuit current becomes zero

(*b*) post-fault transfer impedance attains a value higher than that during the fault

(*c*) ordinate of the post-fault power-angle characteristic is higher than that of during-fault characteristic

(*d*) voltage behind the transient reactance increase to a higher value **[BHEL]**

109. In a static over-current relay, how are inverse time characteristics obtained ? **[BHEL]**

(*a*) By using a transistor amplifier

(*b*) By using an integrating circuit

(*c*) By using a transistor switch

(*d*) By using a differe

110. Galloping in transmission line conductors arises generally due to **[BHEL]**

(*a*) asymmetrical layers of ice formation

(*d*) vortex phenomenon in light winds

(*c*) heavy weight of the line conductors

(*d*) adoption of horizontal conductor configurations

111. When bundle conductors are used in place of single conductors, the effective induc-tance and capacitance will respectively

(*a*) increase and decrease

(*b*) decrease and increase

(*c*) decrease and remain unaffected

(*d*) remain unaffected and increase **[PGCIL]**

112. The load carrying capability of a long AC transmission line is **[PGCIL]**

(*a*) always limited by the conductors size

(*b*) limited by stability considerations

(*c*) reduced at low ambient temperatures

(*d*) decreased by the use of bundled conductors of single conductors

113. A transmission line has 0.2 pu impedance on a base of 132 kV, 100 MVA. On a base of 220 kV, 50MVA, it will have a *pu* impedance of

(*a*) $0.2 \times \frac{50}{100} \times \left(\frac{220}{132}\right)^2$

(*b*) $0.2 \times \frac{100}{50} \times \left(\frac{132}{220}\right)^2$

(*c*) $0.2 \times \frac{50}{100} \times \left(\frac{132}{220}\right)^2$

(*d*) $0.2 \times \frac{100}{50} \times \left(\frac{220}{132}\right)^2$ **[NTPC]**

114. Earth wire EHV overhead transmission line is provided to protect the line against

(*a*) lightning surge

(*b*) switching surge

(*c*) excessive fault voltage

(*d*) corona effect **[NTPC]**

115. For a transmission line with negligible losses, the lagging reactive power (VAR) delivered at the receiving-end, for a given receiving-end voltage is directly proportional to the

(*a*) square of the line voltage drop

(*b*) line voltage drop

(*c*) line inductive reactance

(*d*) line capacitive reactance **[BHEL]**

116. Corona losses are minimized, when

(*a*) conductor size is reduced

(*b*) smooth conductor is reduced

(*c*) sharp point are provided in the line hardware

(*d*) current density in conductors is reduced **[BHEL]**

117. The transient stability of the power-system can be effectively improved by

(*a*) excitation control

(*b*) phase shifting transformer

(*c*) single pole switching of circuit breakers

(*d*) increasing the turbine value opening **[GATE]**

118. For a fault at the terminals of a synchronous generator, the fault current is maximum for a

(*a*) 3-phase fault

(*b*) 3-phase to ground fault

(*c*) line-to ground fault

(*d*) line-to-line fault **[GATE]**

119. Reactance relay is normally preferred for protection against

(*a*) earth faults (*b*) phase faults

(*c*) open-circuits faults (*d*) none of these **[GATE]**

120. A three phase over head transmission line has its conductors horizontally spaced with spacing between adjacent conductors equal to '*d*'. If now the conductors of the line are rearranged to form an equilateral triangle of sides equal to '*d*' then

(*a*) average capacitance and inductance will increase

(*b*) average capacitance will decrease and inductance will increase

(*c*) average capacitance will increase and inductance will decrease

(*d*) surge impedance loading of the line increases **[GATE]**

121. A relay most likely to operate unde-sirable on power swings in the

(*a*) over-current relay

(*b*) inverse definite minimum time relay

(*c*) mho relay

(*d*) reactance relay **[IAS]**

122. Consider the following statements:

The calculations performed using short-line approximate model instead of nominal-π model for a medium length transmission line delivering lagging load at a given receiving-end voltage always result in higher

1. sending-end current
2. sending-end power
3. regulation
4. efficiency

Which of these statements are correct?

(*a*) 1 and 2 (*b*) 3 and 4

(*c*) 1, 2 and 3 (*d*) 1, 2 and 4 **[IAS]**

123. The flow-duration curve at a given head of a hydro electric plant is used to determine the

(*a*) total power available at the site

(*b*) total units of energy available

(*c*) load-factor at the plant

(*d*) diversity-factor for the plant **[ES]**

124. The purpose of the boiler feed pump is to

(*a*) pump hot air into the boiler

(*b*) pump pulverised coal into the boiler

(*c*) pump out steam from the boiler

(*d*) pump water into the boiler **[AMIE]**

125. The major function of the condenser is to

(*a*) remove the condensate for boiler feed water

(*b*) condense steam

(*c*) reduce the back pressure so that maximum heat energy can be extracted from steam

(*d*) provide a closed cycle **[AMIE]**

126. Which of the following equipments is installed in steam power plants to reduce air pollution?

(*a*) Desuper heater

(*b*) Air filter

(*c*) Air electrostatic precipitator

(*d*) Stock **[AMIE]**

127. Most of the generator is thermal power plants run at

(*a*) 15000 rpm (*b*) 3000 rpm

(*c*) 1500 rpm (*d*) 1000 rpm **[AMIE]**

128. Most of the steam turbo-alternators are for

(*a*) two poles

(*b*) ten to twenty poles

(*c*) twenty to thirty poles

(*d*) six poles **[ES]**

129. The value of reheat factor for a multistage steam turbine lines in the range of

(*a*) 1.005 to 1.03 (*b*) 1.01 to 1.06

(*c*) 1.02 to 1.1 (*d*) 1.1 to 1.2 **[AMIE]**

130. Pulverised fuel is used for

(*a*) better burning

(*b*) increased calorific value of coal

(*c*) less radiation loss

(*d*) medium size units **[AMIE]**

131. The proximate analysis of coal gives percentage by weight of

(*a*) moisture and volatic matter

(*b*) moisture, volatic matter, fixed carbon and ash

(*c*) carbon, hydrogen, oxygen, nitrozen, sulpher and ash

(*d*) carbon, hydrogen and oxygen **[AMIE]**

132. Heavy water is used in nuclear power plant as

(*a*) fuel (*b*) coolant

(*c*) moderator (*d*) both (*b*) and (*c*) **[AMIE]**

133. The purpose of the moderator in a nuclear power plant is to

(*a*) control the flow of water inlet

(*b*) control the amount of nuclear fuel into the reactor

(*c*) control the nuclear fission or fusion rate by slowing down the neutrons

(*d*) control the steam flow to the turbine **[AMIE]**

134. Graphite is used in nuclear power plant as a

(*a*) fuel (*b*) coolant

(*c*) moderator (*d*) electrode **[AMIE]**

135. The purpose of free-wheeling in a thyristor controlled *ac* to *dc* converter is to

(*a*) reduce the current of its associated thyristor to zero so that commutation can take place

(*b*) share the load current of its associated thyristor

(*c*) conduct the load current when its associated thyristor is turned off

(*d*) maintain voltage across load at constant value **[IAS]**

136. In a HVDC transmission, the equivalent circuit representing the operation of a bridge rectifier is as shown in the given figure:

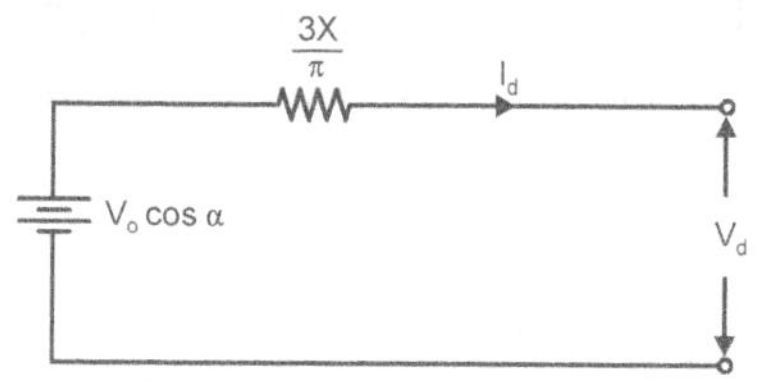

The symbols in the figure have the usual meanings. In this circuit, the voltage drop $\frac{3XI_d}{\pi}$ represents a drop due to

(*a*) ohmic resistance of the thyristors.

(*b*) ohmic resistance of the rectifier transformer

(*c*) commutation in the rectifier

(*d*) leakage reactance of the transformer **[IAS]**

137. In terms of plant powers P_n and P_m loss coefficients B_{mn}, the total transmission loss 'P_L' is

(*a*) $\sum_{m=1}^{N}\sum_{n=1}^{N} B_{mn} P_n$ (*b*) $\sum_{m=1}^{N}\sum_{n=1}^{N} P_m B_{mn}$

(*c*) $\sum_{m=1}^{N}\sum_{n=1}^{N} P_m B_{mn} P_n$ (*d*) $\sum_{m=1}^{N}\sum_{n=1}^{N} 2P_m B_{mn}$ **[IAS]**

138. A star-connected generator supplies power to a star/star connected trans-former as shown in the given figure. The generator voltages are perfectly balanced. As a result,

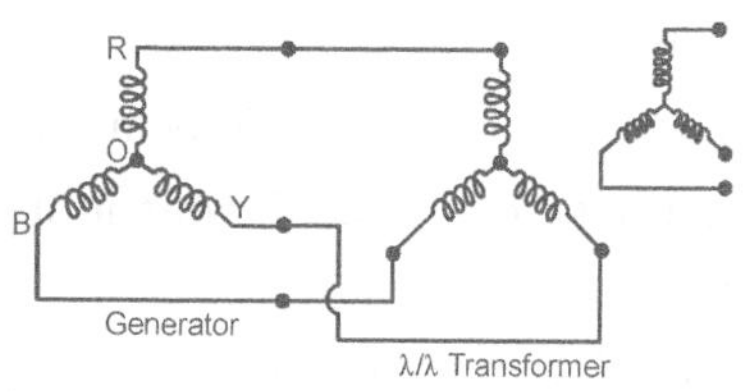

(*a*) third harmonic exciting currents will flow in the primaries of star

(*b*) third harmonic exciting currents cannot flow in the primaries of star

(*c*) the fluxes in the three transformer cores are sinusoidal in time

(*d*) the voltage between 'O'and '*n*' is zero **[IAS]**

139. Given that

Source voltage, $e = E_m \sin(\omega t + a)$, applied at $t = 0$, short-circuit current

$$i = \frac{E_m}{Z}\sin(\omega t + a - \phi) + Ke^{Rt/L}$$

where, $\phi = \tan^{-1}\frac{\omega L}{R}$ and K = constant.

A 3-phase short-circuit occurs at the terminals of an alternator having an effective impedance of $Z = \sqrt{R^2 + \omega^2 L^2}$. If the oscillogram of a phase current shows the presence of an exponentially decaying *dc* component and an *ac* component decaying to its steady-state value, then the *dc* component will be zero when

(*a*) $\alpha = \frac{\pi}{2}$ (*b*) $\alpha = \frac{\pi}{2} + \phi$

(*c*) $\alpha = \frac{\pi}{2} - \phi$ (*d*) $\alpha = \phi$ **[IAS]**

140. In a 3-step distance protection, the reach of the three zones of the relay at the beginning of the first line typically extends upto **[IAS]**

(*a*) 100% of the first line, 50% of the second line and 20% of the third line

(*b*) 80% of the first line, 50% of the second line and 20% of the third line

(*c*) 80% of the first line, 20% of second line and 10% of the third line

(*d*) 50% of first line, 50% of second line and 20% of the third line.

141. A feeder with reactance of 0.2 *pu* has a sending end voltage of 1.2 p.u. If the reactive power supplied at the receiving end of the feeder is 0.3 p.u., the approximate voltage drop in the feeder will be

(*a*) 0.042p.u. (*b*) 0.052p.u.

(*c*) 0.060p.u. (*d*) 0.070p.u. **[IAS]**

142. Ring main distribution is preferred to a radial system because

(*a*) voltage drop in the feeder is less and supply is more reliable

(*b*) voltage-drop in the feeder is less and power factor is high

(*c*) power factor is high and supply is more reliable

(*d*) power factor is high and system is less expensive

[IAS]

143. A single-phase *ac* distributor supplies two single phase loads as shown in the given figure. The voltage drop from A to C is

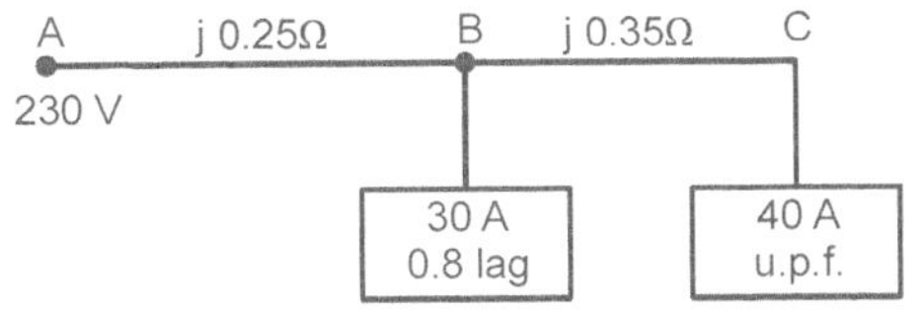

(*a*) $4.5+j30$ V (*b*) $30+j4.5$ V

(*c*) $4.5-j30$ V (*d*) $j30$ V **[ES]**

144. Match List-I with List II and select the correct answer using the codes given below the Lists:

List-I (*Relay used*)

A. Mho relay

B. Negative sequence relay

C. Thermal relay

List-I I (*Equipment*)

1. Transformer
2. Motor
3. Generator
4. Transmission line

Codes:

	A	B	C
(*a*)	2	4	3
(*b*)	1	2	4
(*c*)	3	1	2
(*d*)	4	3	2

[IAS]

145. Match List-I (load) with List-II (relationship) in respect of forward current i_f and the reflected current i_r at the receiving end of the transmission line. Select the correct answer using the codes given below the Lists:

List-I	**List-II**
A. Open-circuit	1. $i_r = 0$
B. Short-circuit	2. $i_r = i_f$
C. $R_L = Zo$	3. Partial reflection
D. $R_L < Zo$	4. $i_r = i_f$

Codes:

	A	B	C	D
(*a*)	2	4	1	3
(*b*)	4	2	3	1
(*c*)	2	4	3	1
(*d*)	4	2	1	3

[IAS]

146. A power system is subjected to a fault which makes the zero sequence component of current equal to zero. The nature of fault is

(*a*) double line to ground fault

(*b*) double line fault

(*c*) line of ground fault

(*d*) three-phase to ground fault **[IAS]**

147. In a thermal power plant, the feed water coming to the economiser is heated using

(*a*) H.P. steam

(*b*) L.P. steam

(*c*) direct heat in the furnace

(*d*) flue gases **[IAS]**

148. In an inverse definite minimum time, electromagnetic type over-current relay, the minimum time feature is achieved because of

(*a*) saturation of the magnetic circuit

(*b*) proper mechanical design

(*c*) appropriate time delay element

(*d*) electromagnetic damping **[IAS]**

149. Earth wire on EHV overhead transmission line is provided to protect the line against

(*a*) lightning surges

(*b*) switching surges

(*c*) excessive fault voltages

(*d*) corona effect **[IAS]**

150. The following sequence currents were recorded in a power system under a fault condition:

$I_{positive} = j1.653$ p.u.

$I_{nagative} = j0.5$ p.u.

$I_{zero} = j1.153$ p.u.

The fault is

(*a*) line to ground

(*b*) three-phase

(*c*) line to line to ground

(*d*) line to line **[IAS]**

151. The following figure shows the single line diagram of a power system with all reactance marked in per unit (pu) on the same base:

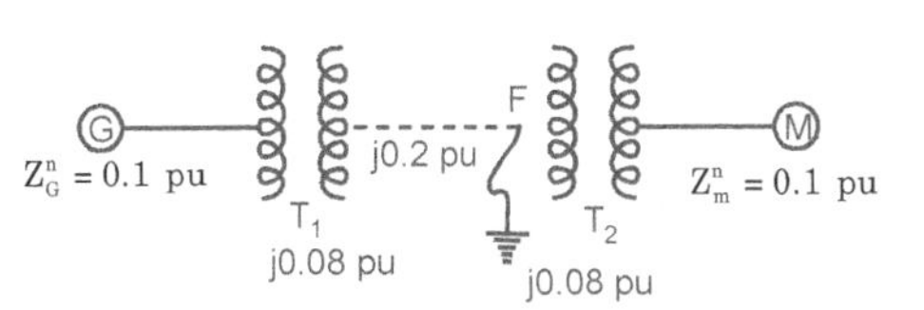

The system is on no-load when a 3-phase fault occurs at 'F' on the high voltage side of the transformer T_2. The fault current will be

(*a*) j–0.8187pu

(*b*) j+0.8187pu

(*c*) j–8.1871pu

(*d*) j+0.1871pu **[ES]**

152. Match List-I (Type of relay) with List-II (For the protection of) and select the correct answer using the codes given below the Lists:

List-I

A. Buchholtz relay.

B. Translay relay.

C. Negative sequence relay.

D. Directional over current relay.

List-II

1. Feeder.
2. Transformer.
3. Generator.
4. Long overhead line.

Codes:

	A	B	C	D
(*a*)	1	2	3	4
(*b*)	2	1	3	4
(*c*)	2	1	4	3
(*d*)	1	2	4	3

[IAS]

153. A loss-less finite generator is connected directly to an infinite bus at rated voltage. Consider the following operating conditions:

A. Generator excitation is maintained constant at a value. Which gives rated voltage at its terminal at rated current.

B. Generator excitation is regulated to maintain rated voltage at its terminal at all loads.

The maximum steady-state power in a A and B will be

	A	B
(*a*)	finite	infinite
(*b*)	infinite	finite
(*c*)	finite	finite
(*d*)	infinite	infinite

[ES]

154. The inertia constant of a 100 MVA, 11 kV water wheal generator is 4. The energy stored in the rotor at the synchronous speed is

(*a*) 400 MJ (*b*) 400 KJ

(*c*) 25 MJ (*d*) 25 KJ **[ES]**

155. A 210 MWgenerator should be cooled by

(*a*) water (*b*) air

(*c*) ice (*d*) hydrogen **[AMIE]**

156. The rotor field of synchronous generators is supplied with

(*a*) dc supply (*b*) ac supply

(*c*) ac or dc supply (*d*) none of these **[AMIE]**

157. When an alternator is delivering a balanced load at unity power factor the phase angle between line voltage and line current is

(*a*) 90° (*b*) 60°

(*c*) 30° (*d*) 0° **[ES]**

158. If two synchronous generators are connected, loss of synchronism will result in

(*a*) stalling of generators

(*b*) wild fluctuations in current

(*c*) wild fluctuations in current and voltage

(*d*) none of these **[AMIE]**

159. The voltage of a bus can be controlled by controlling the

(*a*) phase angle

(*b*) reactive power of the bus

(*c*) active power of the bus

(*d*) phase angle and the reactive power **[AMIE]**

160. The permissible variation of frequency in the power system is

(*a*) ± 1% (*b*) ± 3%

(*c*) ± 5% (*d*) ± 10% **[AMIE]**

161. A gas turbine works on

(*a*) Carnot cycle (*b*) Brayton cycle

(*c*) Dual cycle (*d*) Rankine cycle **[AMIE]**

162. A jet air craft is powered by

(*a*) petrol engine (*b*) diesel engine

(*c*) gas turbine (*d*) solar cells **[AMIE]**

163. In an MHD generator, the conductor is made of

(*a*) alluminium (*b*) copper

(*c*) nichrome (*d*) none of these **[AMIE]**

164. Normaly Z_{BUS} matrix is a

(*a*) Null matrix (*b*) Sparse matrix

(*c*) Full matrix (*d*) Unity matrix

165. Bio-gas plants are suitable for

(*a*) metallurgical industries

(*b*) commercial complexes

(*c*) rural areas

(*d*) coals mines **[AMIE]**

166. A tidal power plant is installed in India near

(*a*) Bay of Bengal (*b*) Vishakhapatnam

(*c*) Goa (*d*) Gulf of Combay [AIME]

167. For under water movement of a subm-arine the power is provided by

(*a*) tidal energy (*b*) MHD generator

(*c*) batteries (*d*) diesel engine **[ES]**

168. A 50Hz 3-phase transmission line of length 100 km has a capacitance of 0.03/πμ F per km. It is represented as a π model. The shunt admittance at each end of the transmission line will be

(*a*) $150 \times 10^{-6} \angle 90°$ mho

(*b*) $100 \times 10^{-6} \angle 90°$ mho

(*c*) $50 \times 10^{-6} \angle 90°$ mho

(*d*) $\frac{10^6}{150} \angle 90°$ mho **[IAS]**

169. For which of the following reasons is a differential relay biased to avoid maloperation when used for transformer protection?

(*a*) Saturation of CT's

(*b*) Mismatch to CT ratios

(*c*) Deference in connection of both sides

(*d*) Current setting multiplier

Select the correct answer using the codes given below:

Codes:

(*a*) 1 and 4 (*b*) 1 and 2

(*c*) 2, 3 and 4 (*d*) 1, 2 and 3 **[IAS]**

170. For an existing *ac* transmission line, the string efficiency is 80%. If *dc* voltage is supplied for the same set up, the string efficiency will be

(*a*) <80% (*b*) 80%

(*c*) 90% (*d*) 100% **[IAS]**

171. The use of high-speed circuit-breakers

(*a*) reduces the short circuit current

(*b*) improves system stability

(*c*) decreases system stability

(*d*) increases the short circuit current **[GATE]**

172. Protection scheme used for detection of loss of excitation of a very large generating unit feeding power into a grid employs

(*a*) undervoltage relay

(*b*) offset mho relay

(*c*) underfrequency relay

(*d*) percentage differential relay **[IAS]**

173. Bundled conductors are employed to improve the

(*a*) appearance of the transmission line

(*b*) mechanical stability of the line

(*c*) current carrying capacity of the line

(*d*) corona performance of the line **[GATE]**

174. Which one of the following matrices reveals the topology of the power system network?

(*a*) Bus incidence matrix

(*b*) Primitive impedance matrix

(*c*) Primitive admittance matrix

(*d*) Bus impedance matrix **[IAS]**

175. The distribution system shown in figure is to be protected by over current system of protection.

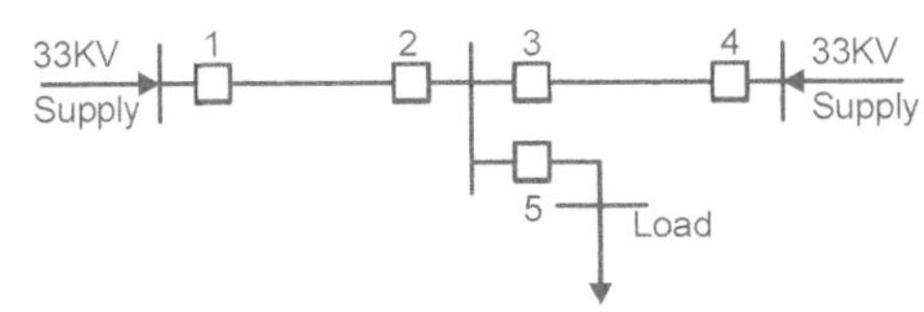

For proper fault discrimination direction over current relays will be required at locations

(*a*) 1 and 4 (*b*) 2 and 3

(*c*) 1, 4 and 5 (*d*) 2, 3 and 5 **[GATE]**

176. The zero sequence reactance (in p.u.) are indicated in the network shown in the given figure. The zero sequence driving-point reactance of node 3 will be

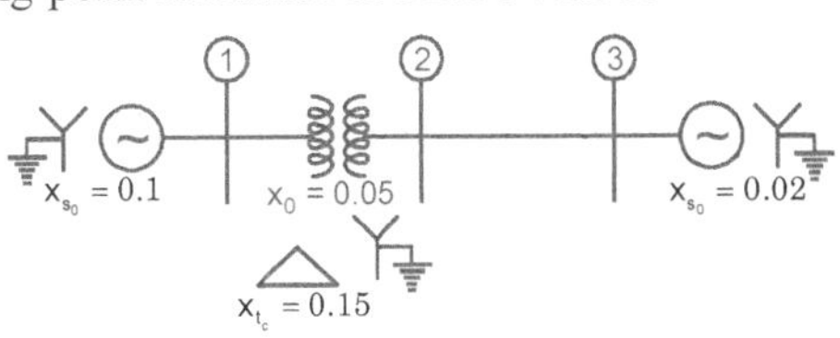

(*a*) 0.05 (*b*) 0.1

(*c*) 0.2 (*d*) 0.3 **[ES]**

177. Which one of the following types of linear couplers is used for extremely reliable high-speed bus protection?

(*a*) Mutual inductors wound on non-magnetic torodial cores

(*b*) Mutual inductors wound on magnetic torodial cores

(*c*) Capacitor voltage transformers

(*d*) Wire-would resistors **[IAS]**

178. A long overhead transmission line is terminated by its characteristic impedance. Under this operating condition, the ratio of the voltage to the current different points along the line will

(*a*) progressively increase from the sending-end to the receiving -end

(*b*) progressively increase from the receiving-end to the sending-end

(c) remain the same at the two ends, but be higher between the two ends being maximum at the centre of the line

(d) remain the same at all points [ES]

179. The coefficient of reflection for current of an open-ended line is

(a) – 1 (b) 1

(c) 0.5 (d) zero [IAS]

180. Bundled conductors in EHV transmission system provide

(a) reduced capacitance

(b) increased capacitance

(c) increased inductance

(d) increased voltage gradient [IES]

181. If power P available from a hydro scheme is given by the formula P = 9.81 QH where Q is the flow rate through the turbine in $\frac{1}{S}$ and H is the head in meteres, then P will be in units of

(a) W (b) HP

(c) kJ/s (d) kWH [ES]

182. To prevent maloperation of differentially connected relay while energising a transformer, the relay restraining coil is biased with

(a) second harmonic current

(b) third harmonic current

(c) fifth harmonic current

(d) seventh harmonic current [ES]

183. Whenever the conductors are dead-ended or there is a change in the direction of transmission line, the insulators used are of the

(a) pin type (b) suspension type

(c) strain type (d) shackle type [ES]

184. A relay has a resistance of 300 ohm. It is switched on to a 110V *dc* supply. If the current reaches 63.2 per cent of its final steady value in 0.002 second then the inductance of the circuit will be

(a) 0.001 H (b) 0.002 H

(c) 0.06 H (d) 0.6 H [IAS]

185. The incremental cost characteristics of two generators delivering 200 MW are as follows

$$\frac{dF_1}{dP_1} = 2.0 + 0.01\ P_1$$

$$\frac{dF_2}{dP_2} = 1.6 + 0.2\ P_2$$

For economic operation, the generations P_1 and P_2 should be

(a) $P_1 = P_2 = 100$ MW

(b) $P_1 = 80$ MW, $P_2 = 120$ MW

(c) $P_1 = 200$ MW, $P_2 = 0$ MW

(d) $P_1 = 120$ MW, $P_2 = 80$ MW [IAS]

186. The annual cost of a generating station can be expressed in the form of Rs (A + B × kW + C × kWh) where A, B, C are constants and kW and kWh represent capacity of the station and energy generated per year respectively. The choice between the "base-load station" and the "peak-load station" basically depends on the fact that

(a) factor B should be less for the peak-load station and factor C should be less for the base-load station

(b) factor C should be less for the peak-load station and factor B should be less for the base load station

(c) the factor B and C should both be less for the base-station

(d) the factor B and C should both be less for the peak-load station. [ES]

187. Nominal-π model is normally used for evaluating the performance of

(a) short line (b) medium line

(c) long line (d) infinite line [IAS]

188. The positive, negative and zero sequence currents of a star connected load for phase R are given as

$IR_1 = (10 - j7)$A

$IR_2 = (5 + j15)$A

$IR_0 = (5 - j8)$A

The neutral current will be

(a) zero (b) $(15 + j24)$A

(c) $(-15 - j24)$A (d) $(15 - j24)$A [IAS]

189. Mutual Geometric Mean Distance (GMD) between '*n*' equally spaced points on a circle of radius '*r*' is equal to

(a) $rn^{\frac{1}{n}}$ (b) $rn^{\frac{1}{n-1}}$

(c) $r(n-1)^{\frac{1}{n}}$ (d) $r(n-1)^{\frac{1}{n-1}}$ [ES]

190. Which of the following acts as a protection against high voltage surges due to lightening and switching?

(a) Breather

(b) Conservator

(c) Horn gaps

(d) Thermal overload relays [ES]

191. Match List-I (Relay) with List-II (Equipment) and select the correct answer using the codes given below the lists:

List-I	**List-II**
A. Mho Relay.	1. Bus bar.
B. Negative Sequence Relay	2. Motor.
C. Thermal Relay	3. Generator.
	4. Transmission line

Codes:

	A	B	C
(*a*)	1	3	2
(*b*)	2	4	1
(*c*)	4	2	3
(*d*)	4	3	2

[IAS]

192. The following lists relate to the relays and their characteristics for 3-zone protection. Match List-I with List-II and select the correct answer using the codes given below the lists:

List-I (*Relays*)

A. Impedance relay

B. Mho relay

C. Reactance relay

List-II (*Relay characteristics*)

1. 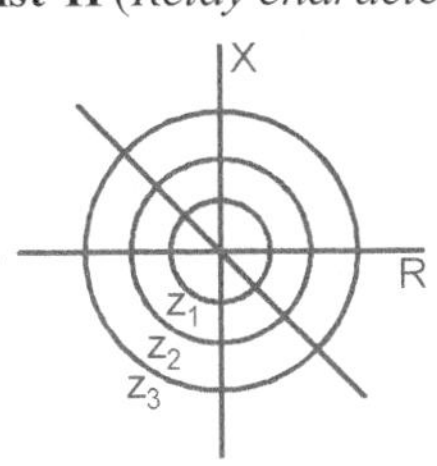

2. 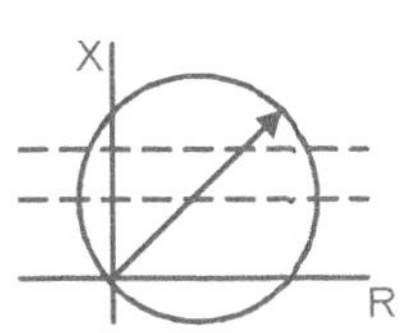

3. 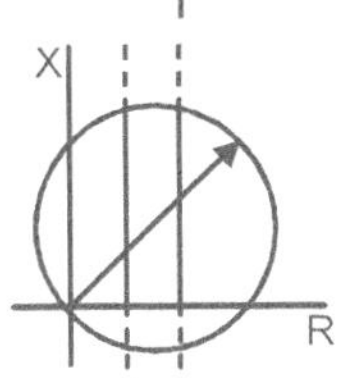

4. 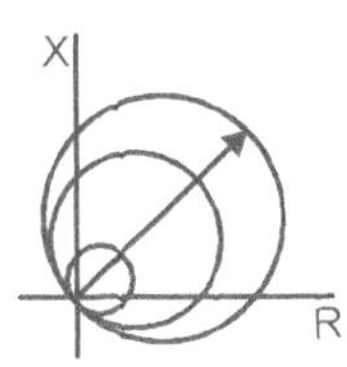

Codes:

	A	B	C
(*a*)	1	2	3
(*b*)	1	4	2
(*c*)	3	2	1
(*d*)	4	3	1

[IAS]

193. A line trap in carrier current relaying tuned to carrier frequency presents

(*a*) high impedance to carrier frequency but low impedance to power frequency

(*b*) low impedance to both carrier and power frequency

(*c*) high impedance to both carrier and power frequency

(*d*) low impedance to carrier frequency but high impedance to power frequency **[ES]**

194. The annual load duration curve of a power supply system may be considered as a straight line from 40MW to 80 MW.

The load factor of the system is

(*a*) 20% (*b*) 40%

(*c*) 60% (*d*) 83.33% **[AMIE]**

195. Diversity factor in a power system is

(*a*) always less than unity

(*b*) normally less than unity

(*c*) always more than unity

(*d*) normally more than unity **[AMIE]**

196. Diversity factor *x* maximum demand is

(*a*) average demand

(*b*) sum of consumer's maximum demands

(*c*) installed capacity

(*d*) generated power **[AMIE]**

197. Plant capacity factor is

(*a*) actual energy produced/maximum possible energy that could have been produced (based on installed capacity)

(*b*) actual energy produced/plant capacity hours for which the plant has been in operation

(*c*) energy generated in a given period/ (maximum demand *x* hours of operation in a given period)

(*d*) none of these **[AMIE]**

198. For stable operation of inter connected power system the passive element that can be used as the inter-connecting elements is

(*a*) vector (*b*) resistor

(*c*) capacitor (*d*) resistor and capacitor **[AMIE]**

199. In an interconnected power system the increase in the field current in any generating units causes

(*a*) increase in active power flow from the unit to rest of the system

(*b*) increase in reactive power from the unit to the rest of system

(*c*) decrease in active power and increase in reactive power flow from the unit

(*d*) increase in both active and reactive power flow from the unit **[AMIE]**

200. To meet the reactive power requirements at the load centres usually

(*a*) shunt capacitors are used

(*b*) series capacitors are used

(*c*) tap changing transformers are used

(*d*) shunt reactors are used **[AMIE]**

201. The area under the load curve represents

(*a*) system voltage
(*b*) current
(*c*) energy consumed
(*d*) maximum demand **[AMIE]**

202. The connected load of a consumer is 2kW fand his maximum demand is 1.5 kW. The demand factor of the consume is

(*a*) 0.75 (*b*) 0.375
(*c*) 1.33 (*d*) none of these **[AMIE]**

203. For the same voltage drop, increasing the voltage of a distributor n times

(*a*) reduces the *x*-section of the conductor by n times
(*b*) increasing the *x*-section of the conductor by *n*-times
(*c*) reduces the *x*-section of the conductor by n^2 times
(*d*) increasing the *x*-section of the conductor by n^2 times **[AMIE]**

204. In a transmission system, the weight of copper used in proportional to

(*a*) E^2 (*b*) E
(*c*) $\frac{1}{E^2}$ (*d*) $\frac{1}{E}$ **[AMIE]**

205. 66 kV is suitable for transmission of power over

(*a*) 30 km (*b*) 66 km
(*c*) 120 km (*d*) 200 km **[AMIE]**

206. ACSR conductors have

(*a*) all conductors made of aluminium
(*b*) outer conductors made of aluminium
(*c*) inner conductors made of aluminium
(*d*) no conductors made of aluminium **[AMIE]**

207. The ratio of puncture voltage to the flashover voltage of a line insulator is

(*a*) equal to 1
(*b*) lower than 1
(*c*) much greater than 1 **[AMIE]**

208. Spinning reserve is

(*a*) the reserve generating capacity which is available for service but not in operation
(*b*) the reserve generating capacity which is connected to the bus and is ready to take load
(*c*) the reserve generating capacity which is in operation but is not in service
(*d*) none of these **[AMIE]**

209. In order to have lower cost of electrical energy generation it is required to have

(*a*) low load factor and low diversity factor
(*b*) low load factor but high diversity factor
(*c*) high load factor but low diversity factor
(*d*) high load factor and high diversity factor **[AMIE]**

210. As per IE rules the maximum allowable variation between declared and actual voltage at consumer's premises should be

(*a*) ± 6% (*b*) ± 8%
(*c*) ± 5% (*d*) none of these **[AMIE]**

211. By increasing the transmission voltage to double of its original value the same power can be despatched keeping the line loss

(*a*) equal to original value
(*b*) half the original value
(*c*) double the original value
(*d*) one-fourth the original value **[AMIE]**

212. The given figure shows electric power input P_e to a lossless synchronous motor as a function of torque angle δ. The load is suddenly increased from P_0 to P_2 and the motor oscillates around δ_2 between δ_0 and δ_m

Consider the following statements derived from the figure regarding the relationship between the motor speed ω and its synchronous speed ω_s at different points in the oscillating cycle:

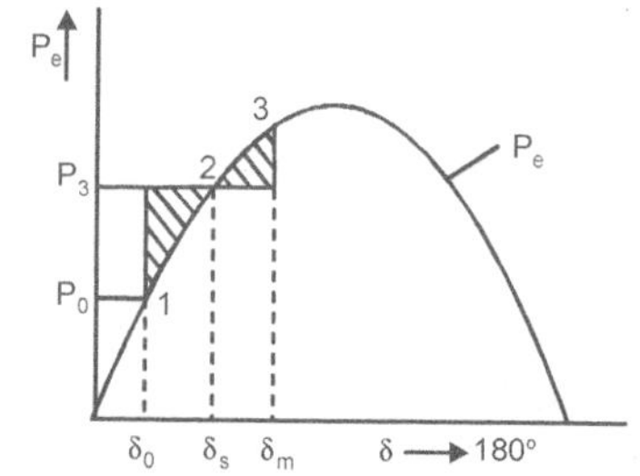

I. At point 1, $\omega = \omega_s$.
II. At point 2, while oscillating from 1 towards 3, $\omega < \omega_s$.
III. At point 3, $\omega < \omega_s$.
IV. At point 2, while oscillating from 3 towards 1, $\omega > \omega_s$.

Which of these statement (*s*) is/are correct?

(*a*) II and IV (*b*) I, II and IV
(*c*) I and III (*d*) III alone **[IAS]**

213. HVDC Transmission is preferred to EHV-AC because

(*a*) HVDC terminal equipment are inexpensive
(*b*) VAR compensation is not required in HVDC systems
(*c*) system stability can be improved
(*d*) Harmonics-problem is avoided **[GATE]**

214. In a string of suspension insulators, the voltage distribution across the different units of a string could be made uniform by the use of a grading ring, because it

(*a*) forms capacitances with link-pins to carry the charging currents from link-pins to earthed structure

(*b*) forms capacitances which help to cancel the charging current from link-pins

(*c*) increase the capacitances of lower insulator units to cause equal voltage drop

(*d*) decrease the capacitances of upper insulator units to cause voltage drop **[IAS]**

215. The reactance of a generator designated "X" is given as 0.25 *pu* based on the generator's name plate rating of 18 kV, 500 MVA. If the base for calculations is changed to 20 kV, 100 MVA, the generator reactance "X" on new base will be

(*a*) 1.025pu (*b*) 0.05pu

(*c*) 0.0405pu (*d*) 0.25pu **[IAS]**

216. In a 400 kV network, 360 kV is recorded at 400 kV bus. The reactive power absorbed by a shunt reactor rated for 50 MVAR, 400 kV connected at the bus is

(*a*) 61.73 MVAR (*b*) 55.56 MVAR

(*c*) 45 MVAR (*d*) 40.5 MVAR **[GATE]**

217. A two-bus system is shown in the given figure. When 1000 MW is transmitted from plant 1 to the load, the transmission loss is 10 MW. The increment fuel costs of the two plants are

$$\frac{dc_1}{dP_1} = 0.02\,P_1 + 16 \text{ and } \frac{dc_2}{dP_2} = 0.04\,P_2 + 20$$

The optimum distribution of total load of 260 MW between the two plants when losses are included but not coordinated is

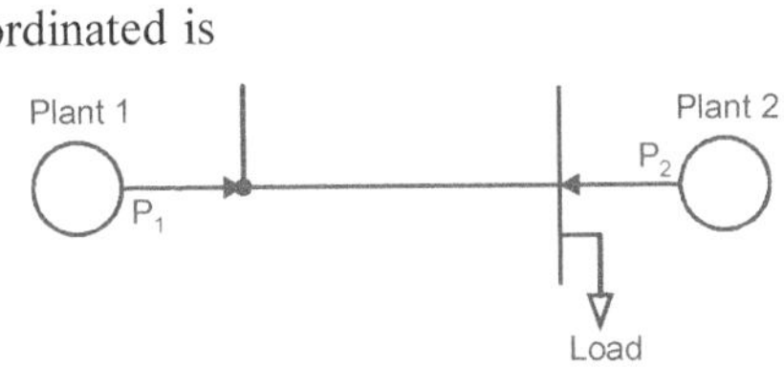

	P_1	P_2
(*a*)	300 MW	50 MW
(*b*)	240 MW	20 MW
(*c*)	130 MW	130 MW
(*d*)	220 MW	60 MW

[IAS]

218. The size of conductor used in power cables depends on

(*a*) operating voltage

(*b*) power factor

(*c*) current to be carried

(*d*) type of insulation used **[IAS]**

219. Match List-I with List-II and select the correct answer using the codes given below the Lists:

List-I (*Type of fuel*)	**List-II** (*Nuclear fuel*)
A. Fertile	1. U-238.
B. Fissile	2. Pu-239.
	3. U-235.
	4. Th-232.

Codes:

	A	B
(*a*)	1, 2	3, 4
(*b*)	2, 3	1, 4
(*c*)	3, 4	1, 2
(*d*)	1, 4	2, 3

[ES]

220. A system is said to be effectively grounded only if R_0/X_1 and X_0/X_1 are respectively (symbols have the usual meanings)

(*a*) ≤ 1 and ≤ 3 (*b*) ≥ 1 and ≥ 3

(*c*) ≤ 2 and ≥ 2 (*d*) ≥ 2 and ≥ 2 **[IAS]**

221. The specific speed of a multiple unit in a hydroelectric station consisting of "*n*" units is equal to the specific speed to single jet unit multiplied by

(*a*) n (*b*) $\frac{1}{n}$

(*c*) $\frac{1}{\sqrt{n}}$ (*d*) $\frac{1}{\sqrt{n}}$ **[IAS]**

222. If the open-and short-circuit input impedances of a loss-less line are respectively Z_1 and Z_2 its characteristic impedance will be

(*a*) $Z_1 + Z_2$ (*b*) $\sqrt{Z_1^2 + Z_2^2}$

(*c*) $\frac{Z_1^2 + Z_2^2}{Z_1 + Z_2}$ (*d*) $\sqrt{Z_1 + Z_2}$ **[ES]**

223. Corona is accompanied by

(*a*) violet visible discharge in darkness

(*b*) hissing sound

(*c*) power loss

(*d*) all of the above [A MIE]

224. To reduce corona effect usually

(*a*) the distance between the conductor is reduced

(*b*) the conductor diameter is reduced

(*c*) bundled conductors are used

(*d*) standard conductors are used **[AMIE]**

225. Corona loss in a transmission line is dependent on

(*a*) diameter of the conductor

(*b*) material of the conductor

(*c*) height of the conductor **[AMIE]**

226. The dielectric strength of air under normal conditions is around

(*a*) 30 kV/cm (*b*) 100 kV/cm

(*c*) 150 kV/cm (*d*) 200 kV/cm **[AMIE]**

227. The inductance of line is minimum when

(*a*) GMD is high

(*b*) GMR is high

(*c*) both GMD and GMR are high

(*d*) GMD is low but GMR is high **[AMIE]**

228. Capacitance of a transmission line

(*a*) increases

(*b*) decreases

(*c*) remains same with increase in its length **[AMIE]**

229. Increase in frequency of transmission line causes

(*a*) nochange in line resistance

(*b*) increase in line resistance

(*c*) decrease in line resistance

(*d*) decrease in line series reactance **[AMIE]**

230. Presence of earth incase of overhead lines

(*a*) increases the capacitance

(*b*) increase the inductance

(*c*) decrease the capacitance

(*d*) decreases the inductance **[AMIE]**

231. Presence of earth incase of overhead lines

(*a*) increases the capacitance

(*b*) increases the inductance

(*c*) decreases the capacitances and increases the inductance

(*d*) does not effect any of the line constants **[AMIE]**

232. Transposition of transposition line is done to

(*a*) reduce line loss

(*b*) reduce skin effect

(*c*) balance line voltage drop

(*d*) reduce corona **[AMIE]**

233. High voltage transmission lines are transposed because then

(*a*) corona losses can be minimized

(*b*) computation of inductance becomes easier

(*c*) voltage drop in the lines can be minimized

(*d*) phase voltage imbalance can be minimized **[AMIE]**

234. Compound with a solid conductor of the same radius corona appears on a stranded conductor at a lower voltage because stranding **[AMIE]**

(*a*) assists ionisation

(*b*) makes the current flow spirally about the axis of the conductor

(*c*) produces oblique sections to a plane perpendicular to the axis of the conductor

(*d*) produces surfaces of smaller radius

235. For a short line if the receiving end voltage is equal to sending end voltage under loaded conditions

(*a*) the sending end power factor is unity

(*b*) the receiving end power factor is unity

(*c*) the sending end power factors is leading

(*d*) the receiving end power factor is leading **[AMIE]**

236. Constant power locus of a transmission once at a particular sending end and receiving end voltage is

(*a*) a straight line (*b*) a circle

(*c*) a parabola (*d*) an ellipse **[AMIE]**

237. For a long distance hv transmission line the receiving end voltage under unloaded conditions is

(*a*) much lower than

(*b*) lower than

(*c*) equal to

(*d*) higher than the sending and voltage **[AMIE]**

238. Ferranti effect on long overhead line is experienced when it is

(*a*) lightly loaded

(*b*) on full load at unity p.f.

(*c*) on full load at 0.8 pf load

(*d*) on any load **[AMIE]**

239. Characteristic impedance of an overhead transmission line is usually in the range of

(*a*) 100 to 200 Ω (*b*) 200 to 300 Ω

(*c*) 0 to 100 Ω (*d*) 400 to 500 Ω **[AMIE]**

240. When the load on a transmission line is equal to the surge impedance loading

(*a*) the receiving end voltage is less than the sending end voltage

(*b*) the sending end voltage is less than the receiving end voltage

(*c*) the receiving end voltage is equal to the sending end voltage

(*d*) none of these **[AMIE]**

241. To increase the transmission capability of a high voltage long line

(*a*) the resistance can be increased

(*b*) the resistance can be decreased

(*c*) the series reactance can be reduced

(*d*) the shunt admittance can be reduced **[AMIE]**

242. Power despatch through a line can be increased by

(*a*) installing series capacitors

(*b*) installing shunt capacitors

(*c*) installing series reactor

(*d*) installing shunt reactor **[AMIE]**

243. When there is interference in on overhead communications line running parallel and in close proximity to an overhead power line, the voltage induced in the communication line in the longitudinal and lateral directions by the power line are due to

(*a*) magnetic induction and electric induction respectively

(*b*) electric induction and magnetic induction respectively

(*c*) both magnetic induction and electric induction

(*d*) magnetic induction only **[ES]**

244. If in a short transmission line, resistance and inductance are found to be equal and regulation appears to be zero, then the load will

(*a*) have unity power factor

(*b*) have zero power factor

(*c*) be 0.707 leading

(*d*) be 0.707 lagging **[ES]**

245. An over-current relay, having a current setting of 12.5% is connected to a supply circuit through a current transformer of ratio 400/5. The pick-up value is

(*a*) 6.25A (*b*) 10A

(*c*) 12.5A (*d*) 15A **[ES]**

246. Which of the following conditions is/are necessary for the parallel operation of two transformers?

(*a*) Percentage impedance should be the same

(*b*) Polarities of both secondaries should be the same

(*c*) Voltage rating should be the same

(*d*) All the conditions mentioned above in (*a*) (*b*) and (*c*) **[ES]**

247. The values of A, B, C and D constants for a transmission line are respectively **[ES]**

(*a*) Z, 0, 1 and 1 (*b*) 0, 1, 1 and 1

(*c*) 1, Z, 0 and 1 (*d*) 1, 1 , Z and 0

248. In terms of power generation and D_{mum} coefficient, the transmission loss for a two-plant system is (Notations have their usual meaning)

(*a*) $P_1^2 B_{11} + 2P_1 P_2 B_{12} + P_2^2 B_{22}$

(*b*) $P_1^2 B_{11} - 2P_1 P_2 B_{12} + P_2^2 B_{22}$

(*c*) $P_2^2 B_{11} + 2P_1 P_2 B_{12} + P_1^2 B_{22}$

(*d*) $P_1^2 B_{11} + P_1 P_2 B_{12} + P_2^2 B_{22}$ **[AMIE]**

249. Which one of the following is not true regarding HVDC transmission?

(*a*) Corona loss is much more in HVDC transmission

(*b*) The power transmission capability of bipolar line is almost the same as that of single-circuit *ac* line

(*c*) HVDC link can operate between two ac systems whose frequencies need not be equal

(*d*) There is no distance limitation for HVDC transmission by underground cable **[IAS]**

250. In the case of a HVDC system, there is

(*a*) charging current but no skin effect

(*b*) no charging current but skin effect

(*c*) neither charging current nor skin effect

(*d*) both charging current and skin effect

Directions : *The following items consist of two statements, one labelled as 'Assertion A ' and the other labelled as 'Reason R'. You are to examine these two statements carefully and decide if the Assertion A and the Reason R are individually true and if so, whether the Reason is a correct explanation of the Assertion. Select your answer to these items using the codes given below and mark your answer sheet accordingly.*

Codes:

(*a*) Both A and R true and R is the correct explanation of A

(*b*) Both A and R are true but R is NOT a correct explanation of A

(*c*) A is true but R is false

(*d*) A is false but R is true

251. Assertion (A) : In the residual type earth compensation used in the distance protection schemes with solidly earthed system, the residual compensation circuit uses auxiliary current transforms.

Reasons (R) : A fraction of the residual current is added to the fault current la for an earth fault on phase '*a*'. This fraction is so choosen that the relay measures the positive sequence impedance Z_1. **[IES]**

252. Assertion (A) : Optimum loading of the power stations interconnected by transmission line is obtained when the incremental fuel cost of each station divided by its penalty factor is the same for all the stations in the system.

Reasons (R) : The total load demand and the transmission losses at any instant must be met by the total generation at that instant and the penalty factor is a functions of the incremental transmission loss. **[ES]**

253. Assertion (A) : Voltage-time insulation coordination curves for the equipment to be protected and the lightning arrester to be employed are used for determining the required insulation levels.

Reasons (R) : Voltage-time insulation coordination curves indicate the time of arrival of the peak of the surge wage at the equipment location. **[ES]**

254. Assertion (A) : The concept of self-GMC is applicable for the calculation of inductance of transmission lines.

Reasons (R) : The radius of a conductor of transmission lines, in the expression for inductance is not the self GMD of the conductor. **[ES]**

255. Assertion (A) : A short length of cable between an overhead line and the substation is effective in reducing the incoming surge voltage.

Reasons (R) : The surge impedance of the cable is quite low when compared with that of the incoming overhead line. **[ES]**

256. Assertion (A) : The conductors are usually transposed in transmission lines.

Reasons (R) : The resistance of each phase has to be equal. **[ES]**

257. Assertion (A) : During the load flow study, a voltage controlled bus is treated as a load bus when the magnitude of the voltage is very much higher or very much lower than the scheduled voltage.

Reasons (R) : A tap changing trans-former is represented as a π network in the load flow model. **[ES]**

258. Assertion (A) : When the primary voltages are the same in both groups, the secondary line voltages of Δ–Y bank lag 60° behind the corresponding secondary voltage of the Y–Δ bank, and parallel operation is not possible.

Reasons (R) : The primary line terminal connections for the Δ–Y bank have not been rotated backward by 240° to make the secondary line voltages of both the groups in phase. **[IES]**

ANSWERS

EXERCISE – I

1. (*c*)	**2.** (*c*)	**3.** (*d*)	**4.** (*b*)	**5.** (*a*)	**6.** (*a*)	**7.** (*d*)	**8.** (*a*)	**9.** (*c*)	**10.** (*b*)
11. (*b*)	**12.** (*b*)	**13.** (*a*)	**14.** (*d*)	**15.** (*c*)	**16.** (*a*)	**17.** (*d*)	**18.** (*d*)	**19.** (*d*)	**20.** (*b*)
21. (*c*)	**22.** (*a*)	**23.** (*c*)	**24.** (*b*)	**25.** (*c*)	**26.** (*b*)	**27.** (*d*)	**28.** (*c*)	**29.** (*a*)	**30.** (*a*)
31. (*c*)	**32.** (*c*)	**33.** (*b*)	**34.** (*b*)	**35.** (*c*)	**36.** (*c*)	**37.** (*c*)	**38.** (*b*)	**39.** (*b*)	**40.** (*c*)
41. (*d*)	**42.** (*c*)	**43.** (*d*)	**44.** (*b*)	**45.** (*b*)	**46.** (*a*)	**47.** (*d*)	**48.** (*d*)	**49.** (*c*)	**50.** (*c*)
51. (*d*)	**52.** (*c*)	**53.** (*a*)	**54.** (*c*)	**55.** (*a*)	**56.** (*d*)	**57.** (*b*)	**58.** (*d*)	**59.** (*a*)	**60.** (*b*)
61. (*b*)	**62.** (*a*)	**63.** (*b*)	**64.** (*a*)	**65.** (*c*)	**66.** (*d*)	**67.** (*b*)	**68.** (*b*)	**69.** (*a*)	**70.** (*a*)
71. (*b*)	**72.** (*c*)	**73.** (*a*)	**74.** (*a*)	**75.** (*b*)	**76.** (*a*)	**77.** (*b*)	**78.** (*c*)	**79.** (*a*)	**80.** (*d*)
81. (*d*)	**82.** (*a*)	**83.** (*b*)	**84.** (*c*)	**85.** (*d*)	**86.** (*c*)	**87.** (*c*)	**88.** (*b*)	**89.** (*b*)	**90.** (*c*)
91. (*c*)	**92.** (*b*)	**93.** (*d*)	**94.** (*d*)	**95.** (*a*)	**96.** (*b*)	**97.** (*a*)	**98.** (*a*)	**99.** (*b*)	**100.** (*c*)
101. (*b*)	**102.** (*a*)	**103.** (*b*)	**104.** (*b*)	**105.** (*c*)	**106.** (*b*)	**107.** (*c*)	**108.** (*d*)	**109.** (*c*)	**110.** (*b*)
111. (*c*)	**112.** (*b*)	**113.** (*b*)	**114.** (*c*)	**115.** (*a*)	**116.** (*d*)	**117.** (*d*)	**118.** (*c*)	**119.** (*c*)	**120.** (*d*)
121. (*a*)	**122.** (*d*)	**123.** (*a*)	**124.** (*a*)	**125.** (*d*)	**126.** (*c*)	**127.** (*c*)	**128.** (*d*)	**129.** (*b*)	**130.** (*b*)
131. (*b*)	**132.** **(c**	**133.** (*d*)	**134.** (*c*)	**135.** (*b*)	**136.** (*b*)	**137.** (*b*)	**138.** (*a*)	**139.** (*b*)	**140.** (*b*)
141. (*d*)	**142.** (*d*)	**143.** (*c*)	**144.** (*b*)	**145.** (*b*)	**146.** (*a*)	**147.** (*c*)	**148.** (*a*)	**149.** (*a*)	**150.** (*c*)

151. (*d*) 152. (*b*) 153. (*c*) 154. (*b*) 155. (*a*) 156. (*d*) 157. (*b*) 158. (*c*) 159. (*) 160. (*d*)

161. (*c*) 162. (*b*) 163. (*b*) 164. (*c*) 165. (*a*) 166. (*b*) 167. (*a*) 168. (*b*) 169. (*a*) 170. (*c*)

171. (*c*) 172. (*c*) 173. (*b*) 174. (*c*) 175. (*c*) 176. (*d*) 177. (*c*) 178. (*d*) 179. (*c*) 180. (*c*)

181. (*c*) 182. (*c*) 183. (*a*) 184. (*b*) 185. (*a*) 186. (*d*) 187. (*c*) 188. (*a*) 189. (*d*) 190. (*d*)

191. (*c*) 192. (*a*) 193. (*b*) 194. (*b*) 195. (*c*) 196. (*b*) 197. (*b*) 198. (*b*) 199. (*a*) 200. (*b*)

201. (*d*) 202. (*c*) 203. (*b*) 204. (*a*) 205. (*d*) 206. (*c*) 207. (*a*) 208. (*a*) 209. (*c*) 210. (*b*)

211. (*b*) 212. (*b*) 213. (*c*) 214. (*c*) 215. (*b*) 216. (*d*) 217. (*b*) 218. (*d*) 219. (*c*) 220. (*c*)

221. (*c*) 222. (*c*) 223. (*c*) 224. (*a*) 225. (*c*) 226. (*b*) 227. (*c*) 228. (*d*) 229. (*c*) 230. (*a*)

231. (*c*) 232. (*b*) 233. (*d*) 234. (*a*) 235. (*b*) 236. (*a*) 237. (*b*) 238. (*d*) 239. (*d*) 240. (*d*)

241. (*c*) 242. (*a*) 243. (*a*) 244. (*c*) 245. (*c*) 246. (*a*) 247. (*c*) 248. (*c*) 249. (*b*) 250. (*c*)

251. (*a*) 252. (*a*) 253. (*c*) 254. (*c*) 255. (*b*) 256. (*b*) 257. (*a*) 258. (*c*) 259. (*a*) 260. (*c*)

261. (*c*) 262. (*a*) 263. (*b*) 264. (*a*)

EXERCISE – II

1. (*a*) 2. (*a*) 3. (*d*) 4. (*b*) 5. (*a*) 6. (*c*) 7. (*d*) 8. (*d*) 9. (*d*) 10. (*d*)

11. (*a*) 12. (*a*) 13. (*b*) 14. (*a*) 15. (*c*) 16. (*a*) 17. (*a*) 18. (*c*) 19. (*c*) 20. (*b*)

21. (*a*) 22. (*b*) 23. (*b*) 24. (*b*) 25. (*b*) 26. (*a*) 27. (*c*) 28. (*a*) 29. (*a*) 30. (*c*)

31. (*b*) 32. (*c*) 33. (*d*) 34. (*d*) 35. (*a*) 36. (*a*) 37. (*c*) 38. (*d*) 39. (*b*) 40. (*a*)

41. (*c*) 42. (*b*) 43. (*b*) 44. (*a*) 45. (*d*) 46. (c) 47. (*b*) 48. (*a*) 49. (*c*) 50. (*d*)

51. (*b*) 52. (*a*) 53. (*c*) 54. (*b*) 55. (*c*) 56. (*d*) 57. (*b*) 58. (*b*) 59. (*b*) 60. (*b*)

61. (*b*) 62. (*b*) 63. (*c*) 64. (*b*) 65. (*b*) 66. (*b*) 67. (*c*) 68. (*b*) 69. (*b*) 70. (*a*)

71. (*c*) 72. (*d*) 73. (*d*) 74. (*a*) 75. (*c*) 76. (*d*) 77. (*d*) 78. (*a*) 79. (*a*) 80. (*a*)

81. (*a*) 82. (*a*) 83. (*c*) 84. (*c*) 85. (*b*) 86. (*b*) 87. (*b*) 88. (*c*) 89. (*d*) 90. (*b*)

91. (*b*) 92. (*a*) 93. (*b*) 94. (*c*) 95. (*b*) 96. (*a*) 97. (*d*) 98. (*b*) 99. (*b*) 100. (*d*)

101. (*c*) 102. (*a*) 103. (*d*) 104. (*b*) 105. (*c*) 106. (*d*) 107. (*b*) 108. (*c*) 109. (*d*) 110. (*a*)

111. (*b*) 112. (*b*) 113. (*c*) 114. (*a*) 115. (*d*) 116. (*b*) 117. (*c*) 118. (*a*) 119. (*c*) 120. (*c*)

121. (*c*) 122. (*a*) 123. (*a*) 124. (*d*) 125. (*c*) 126. (*c*) 127. (*b*) 128. (*a*) 129. (*b*) 130. (*a*)

131. (*b*) 132. (*d*) 133. (*c*) 134. (*c*) 135. (*c*) 136. (*c*) 137. (*c*) 138. (*d*) 139. (*d*) 140. (*c*)

141. (*b*) 142. (*a*) 143. (*c*) 144. (*d*) 145. (*a*) 146. (*b*) 147. (*d*) 148. (*a*) 149. (*a*) 150. (*c*)

151. (*c*) 152. (*b*) 153. (*c*) 154. (*a*) 155. (*d*) 156. (*a*) 157. (*c*) 158. (*a*) 159. (*b*) 160. (*a*)

161. (*b*) 162. (*c*) 163. (*c*) 164. (*b*) 165. (*c*) 166. (*a*) 167. (*c*) 168. (*a*) 169. (*d*) 170. (*d*)

171. (*b*) 172. (*b*) 173. (*d*) 174. (*c*) 175. (*b*) 176. (*c*) 177. (*c*) 178. (*d*) 179. (*a*) 180. (*b*)

181. (*a*) 182. (*a*) 183. (*c*) 184. (*d*) 185. (*d*) 186. (*a*) 187. (*b*) 188. (*a*) 189. (*c*) 190. (*c*)

191. (*d*) 192. (*b*) 193. (*a*) 194. (*c*) 195. (*c*) 196. (*b*) 197. (*a*) 198. (*a*) 199. (*b*) 200. (*a*)

201. (*c*) 202. (*a*) 203. (*a*) 204. (*c*) 205. (*b*) 206. (*b*) 207. (*c*) 208. (*b*) 209. (*d*) 210. (*c*)

211. (*d*) 212. (*b*) 213. (*c*) 214. (*a*) 215. (*c*) 216. (*d*) 217. (*b*) 218. (*c*) 219. (*d*) 220. (*a*)

221. (*d*) 222. (*d*) 223. (*d*) 224. (*c*) 225. (*a*) 226. (*a*) 227. (*d*) 228. (*a*) 229. (*d*) 230. (*a*)

231. (*a*) 232. (*c*) 233. (*d*) 234. (*d*) 235. (*d*) 236. (*d*) 237. (*d*) 238. (a) 239. (*d*) 240. (*c*)

241. (*c*) 242. (*a*) 243. (*d*) 244. (*c*) 245. (*a*) 246. (*d*) 247. (*c*) 248. (*a*) 249. (*a*) 250. (*c*)

251. (*d*) 252. (*a*) 253. (*b*) 254. (*a*) 255. (*a*) 256. (*c*) 257. (*b*) 258. (*a*)

EXPLANATIONS

EXERCISE – I

11. Bus Impedance matrix Z_{Bus} can be formed by using circuit theory knowledge, viz.

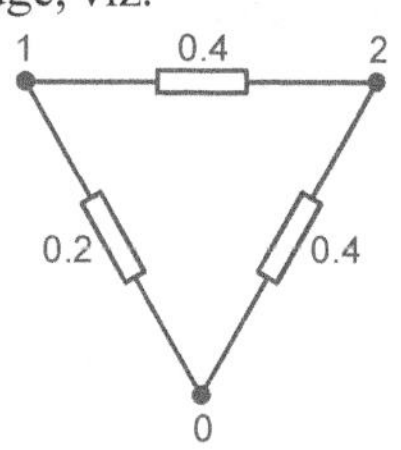

$$E_1 = Z_{11}I_1 + Z_{12}I_2$$

The Z_0 are open circuit impendace

$$E_2 = Z_{21}I_1 + Z_{22}I_2$$

$$Z_{11} = \left.\frac{E_1}{I_1}\right|_{I_2=0} = \frac{0.2 \times 0.8}{0.2+0.4+0.4} = 0.16$$

$$Z_{21} = \left.\frac{E_2}{I_1}\right|_{I_2=0} = \frac{0.4 \times 0.2}{1.0} = 0.08$$

$$Z_{22} = \left.\frac{E_2}{I_2}\right|_{I_2=0} = \frac{0.4 \times 0.6}{1.0} = 0.24$$

Hence bus impedance matrix is given by

$$\begin{bmatrix} 0.16 & 0.08 \\ 0.08 & 0.24 \end{bmatrix}$$

12. For a 3-phase fault

$$I_a = \frac{1}{x_1(pu)}(\text{in } pu)\; -j5 = \frac{1}{x_1(pu)}$$

$\Rightarrow \quad x_1(pu) = j0.2\,pu$

For a line-to-line fault,

$$Ia_1 = \frac{1}{x_1 + x_2}(\text{in } pu)\; -j4 = \frac{1}{j0.2 + x_2(pu)}$$

$\therefore \quad x_2(pu) = j0.05pu$

13. Power loss,

$$P_l = B_{11}P_1^2 + B_{12}P_1P_2 + B_{22}P_2^2$$
$$= 0.0001 \times (50)^2 - 0.0005 \times 50 \times 40 \times 0.0025 \times (40)^2$$
$$= 5.5\text{ MW}$$

16. Incremental fuel cost is given by,

$$\text{IFC} = \frac{dF}{dP_i} = 53 + 0.04\pi$$
$$= 53 + 0.04 \times 50 = 55 \text{ per MWh}$$

17. Pulg setting Multiplier is given by,

$$\text{PSM} = \frac{\text{Fault Current}}{\text{Relay Setting } \times \text{CT ratio}}$$
$$= \frac{3000}{0.50 \times 1000} = 6$$

20. Number of trains required daily

$$= \frac{400000 \times 24}{2000 \times 1000 \times 0.08} = 6$$

23. Rated current of the circuit breaker

$$= \frac{2000}{3 \times \frac{33}{\sqrt{3}}} = 35\text{ KA}$$

$\therefore$ Making current $= 2.55 \times$ rated current

$$= 2.55 \times 35 = 89\text{ KA}$$

42. Insulation resistance of the cable is inversely proportional to the length of the cable.

Hence, insulation resistance for 50 km length $= \frac{1 \times 10}{50}$

$$= 0.2\text{ M}\Omega$$

43. Leakage resistance of cable $\propto \frac{1}{\text{Length}}$

$\therefore$ For 100 km long cable, leakage resistance

$$= \frac{1 \times 50}{100} = 0.5\text{ M}\Omega$$

44. As surge impedance of cable does not depend on its length, hence for 25 miles long, the surge impedance will be same

66. In the given values of reactances, the zero sequence reactance is greater than that of positive and negative sequences. Hence it is fit for transmissions line only.

69. Reflection coefficient for voltage,

$$\eta = \frac{Z_R - Z_0}{Z_R + Z_0} = \frac{0 - Z_0}{0 + Z_0} = -1$$

70. Voltage of the wave

$$= \frac{40 \times 2 \times 100}{500} = 16\,\text{kV}$$

71. Fault level $= \dfrac{1}{X_{total}(pu)} \times 100$

$$= \frac{100}{0.1 + 0.1} = \frac{100}{0.2} = 500\text{ MVA}$$

72. Reflection coefficient is given by

$$n = \frac{Z_2 - Z_1}{Z_1 + Z_2}$$

For no reflection,

$$Z_1 = Z_2$$

74. Inertia constant on 100 MVA base

$$= \frac{6}{10} \times 250 = 15\text{ MJ/MVA}$$

80. $$Z(pu) = \frac{Z(\Omega)\text{MVA}}{(\text{KV})^2}$$

∴ New value of

$$Z(pu) = \frac{0.242 \times (\text{KV})^2}{(1.1\text{KV})^2} = 0.200$$

84. For optimum load sharing,

$$IC_1 = IC_2$$

$\therefore \quad 0.1P_1 + 8 = 0.15\,P_2 + 3$

$\therefore \quad 0.1\,P_1 - 0.15\,P_2 + 5 = 0$

$\Rightarrow \quad P_1 - 1.5\,P_2 + 50 = 0 \qquad ...(i)$

Given: $P_1 + P_2 = 100$ MW $\qquad ...(ii)$

Solving equations (*i*) and (*ii*) we get

$$\cancel{P_1} - 1.5\,P_2 + 50 = 0 \qquad ...(i)$$
$$\cancel{P_1} + P_2 - 100 = 0 \qquad ...(ii)$$
$$- \qquad +$$
$$-2.5\,P_2 + 150 = 0$$

$\Rightarrow \quad P_2 = 60$ MW

$\therefore \quad P_1 = 100 - 60 = 40$ MW

85. $$V = i_a\sqrt{\frac{L}{C}}$$

$$= 10\sqrt{\frac{1}{0.01 \times 10^{-6}}} = 100\,kV$$

87. $$v = \frac{3 \times 10^8}{\sqrt{\epsilon_r}}$$

$$= \frac{3 \times 10^8}{\sqrt{9}} = 10^8\text{ m/s}$$

96. $$\text{P.S.M} = \frac{300}{\frac{400}{5} \times 2.5} = 15$$

98. For economic load sharing

$$IC_1 = IC_2$$

$\therefore \quad 0.1x + 20 = 0.15y + 18$

$\Rightarrow \quad 0.1x - 0.15y + 2 = 0$

$\Rightarrow \quad x - 1.5y + 20 = 0 \qquad ...(i)$

Given: $x + y = 300$

$\Rightarrow \quad x + y - 300 = 0 \qquad ...(ii)$

Solving equations (*i*) and (*ii*), we get

$$-2.5y + 320 = 0$$

$\Rightarrow \quad y = 128$ MW

$\therefore \quad x = 172$MW

101. Inductance of the line,

$$L = 2 \times 10^{-4} \text{ in } \frac{(\text{GMD})\text{ self}}{(\text{GMD})\text{ mutual}}$$

122. Active or Reactive power absorbed by a fixed impedance load is proportional to V^2

$$\text{Required reactive power} = \left(\frac{360}{400}\right)^2 \times 50$$

$$= 40.5\text{ MVAr}$$

131. Given: $|V_s| = |V_R| = 60$kV

$$X = 11\Omega$$

Impedance angle $= g_0$ [Purely Reactive]

$$P_m = \frac{V_s V_r}{\beta} - \frac{AV_r^2}{\beta}\cos(\beta - \beta)$$

$$= \frac{V_s V_r}{\beta} - \frac{AV_r^2}{\beta}\cos(90 - 0)$$

$$= \frac{V_s V_r}{\beta}$$

$$= \frac{66 \times 66}{11}$$

$$= 396\text{ M}\Omega$$

141. $Y_{10} = \frac{1}{0.1} = 10;$

$Y_{12} = \frac{1}{0.2} = 5;$

$Y_{21} = \frac{1}{0.2} = 5;$

$Y_{20} = 0$

$\therefore$ Bus admittance matrix is given by

$$Y_{BUS} = \begin{bmatrix} y_{10} + y_{12} & -y_{12} \\ -y_{21} & y_{20} + y_{21} \end{bmatrix}$$

$$= \begin{bmatrix} 10+5 & -5 \\ -5 & 0+5 \end{bmatrix} = \begin{bmatrix} 15 & -5 \\ -5 & 5 \end{bmatrix}$$

149. $P_{max} = \frac{V_s V_r}{X_s}$

$$= \frac{110 \times 110}{100} = 121 \text{ MW}$$

150. Electirc Power $= \frac{746}{75}$ QHη KW

where, Q = discharge, m³/s

H = water head, in

η = efficiency of turbine and the alternator

For the given values,

$$P = \frac{746}{75} \times 1 \times 1 = \frac{746}{75} \text{ kW}$$

151. String efficiency

$$= \frac{\text{Voltage across the string}}{4 \times \left(\text{voltage across the unit near to conductor}\right)}$$

$$= \frac{V}{4 \times 0.3333V} \times 100 = 75\%$$

158. Natural 3-phase loading or surge impedance loading of a 3-phase line (in MW) $= \frac{V^2}{R_0}$

where v = line voltage in kV, and

R_0 = characteristic resistance of the line

$$\therefore \quad R_0 = \sqrt{\frac{L}{C}}$$

$$\text{Loading} = \frac{132^2}{\sqrt{160 \times 10^3}} = 43.56 \text{ M}\omega \approx 44 \text{ M}\omega$$

For 400-kV line,

$$P_o = \frac{400^2}{\sqrt{62.5 \times 10^3}} = 640 \text{ M}\omega$$

168. $L_{self} = \frac{\gamma_0}{8\pi}$, hence self inductance will be independent of shape and size of conductor. Thus, self inductance will be same.

172. For minimum cost of generation,

$$\frac{dF_1}{dP_1} = \frac{dF_2}{dP_2}$$

$\therefore \quad 0.4P_1 + 80 = 0.48\,P_2$ ...(*i*)

and $\quad P_1 + P_2 = 900$ ...(*ii*)

Solving equations (*i*) and (*ii*), we get

$$P_1 = 400 \text{ MW}$$
$$P_2 = 500 \text{ MW}$$

173. $\alpha + j\beta = (0.15 + j1.5) \times 10^{-3}$

$\therefore \quad \beta = 1.5 \times 10^{-3}$

$$\therefore \text{ Wavelength} = \frac{2\pi}{\beta} = \frac{2\pi}{1.5 \times 10^{-3}}$$

178. $Z_{new(pu)} = \frac{Z_{old(pu)} (KV_{old})^2 \times (MVA_{new})}{(KV_{new})^2 \times (MVA_{old})}$

$$= 0.8 \times \left(\frac{66}{33}\right)^2 \times \left(\frac{100}{100}\right) = 3.2$$

183. $v_{p+1} = v_p + kVp$

$v_2 = v_1 + 0.2v_1 = 1.2v_1$

$v_3 = v_2 + 0.2v_2$

$= 1.2v_1 + 0.2(v_1 + 1.2v_1) = 1.64v_1$

String efficiency

$$= \frac{\text{Voltage across whole string}}{n \times \left(\text{voltage across the unit nearest to conductor}\right)}$$

$$= \frac{v_1 + 1.2v_1 + 1.64v_1}{3 \times 1.64v_1} = 0.78 = 78\%$$

189. Fault level $= \frac{1}{x(pu)} \times 100$

$$= \frac{1}{0.10} \times 100 = 1000 \text{ MVA}$$

206. Maximum power transferred,

$$P_{rm} = \frac{V_s V_r}{\beta} - \frac{AV_r^2}{\beta} \cos(\beta - \alpha)$$

$$= \frac{66 \times 66}{85} - \frac{0.95(66)^2}{85} \cos(70\text{–}0.5)$$
$$= 51.25 - 17.05 = 34 \text{ MW}$$

216. Velocity of surge = 3×10^8

$$\therefore \quad \text{Time taken} = \frac{600 \times 10^3}{3 \times 10^8}$$
$$= 200 \times 10^{-5} = 0.002 \text{ sec.}$$

221. Bus admittance matrix is given by

$$Y_{Bus} = \begin{bmatrix} y_{10} + y_{12} + y_{13} & -y_{12} & -y_{13} \\ -y_{21} & y_{21} + y_{23} & -y_{23} \\ -y_{31} & -y_{32} & y_{31} + y_{32} \end{bmatrix}$$
$$= \begin{bmatrix} -j7 & 2j & 5j \\ 2j & -6j & 4j \\ 5j & 4j & -9j \end{bmatrix}$$

222. Since surge impedance of the line is independent of the length of line, hence for 200km length also it will be 400W

224. For economic alloction of load

$$IC_1 = IC_2$$
$$\therefore \quad 25 + 0.2\,PG_1 = 32 + 0.2\,PG_2$$
$$\Rightarrow \quad PG_1 - PG_2 = 35 \qquad ...(i)$$
$$\text{and} \quad PG_1 - PG_2 = 250 \qquad ...(ii)$$

Solving equations (*i*) and (*ii*), we get

$$2PG_1 = 285$$
$$\therefore \quad PG_1 = 142.5$$
$$\text{and} \quad PG_2 = 107.5$$

226. Voltage across the circuit breaker

$$= 10 \times \sqrt{\frac{1}{0.01 \times 10^{-6}}} = 10\sqrt{10^8}$$
$$= 100000 = 100\text{kV}$$

227. $$Z(pu)_{new} = Z_{pu}(old) \times \left(\frac{kV_{old}}{kV_{new}}\right)^2 \times \left(\frac{MVA_{nes}}{kV_{old}}\right)$$

$$= 2 \times \left(\frac{100}{400}\right)^2 \times \left(\frac{400}{100}\right)$$

$$= 2 \times \frac{1}{16} \times 4 = 0.5\text{pu}$$

228. Maximum power transferred,

$$P_{max} = \frac{V_S V_R}{X}$$
$$= \frac{110 \times 100}{22} = 500 \text{ MW}$$

230. From admittance matrix,

$$y_{11} + y_{12} = -30j$$
$$-y_{12} = j10$$
$$\therefore \quad y_{11} = -20j;$$
$$Z_{11} = 0.05j$$

234. Insulations resistance of the cable $\propto \frac{1}{3}$

For 10km long cable,

$$\text{Insulation resistance} = \frac{8 \times 20}{10}$$
$$= 16 \text{ mega ohm}$$

235. $$\text{Energy stored} = GH$$
$$= 100 \times 4 = 400 \text{ MJ}$$

237. Power transfer in first case,

$$P_1 = \frac{V_S V_R}{X} = P(\text{say})$$

In second case,

$$P_2 = \frac{V_S V_R}{X/2} = 2P$$

In third case,

$$P_3 = \frac{V_S V_R}{0.75\,X} = \frac{P}{0.75}$$

$$\therefore \quad P_1 : P_2 : P_3 = 1 : 2 : \frac{1}{0.75}$$

240. $$\begin{bmatrix} Ia_1 \\ Ia_2 \\ Ia_0 \end{bmatrix} = \frac{1}{3}\begin{bmatrix} 1 & \alpha & \alpha \\ 1 & \alpha^2 & \alpha \\ 1 & 1 & 1 \end{bmatrix}\begin{bmatrix} Ia \\ Ia \\ Ia \end{bmatrix}$$

$$I_{a1} = \frac{1}{3} I_a + \alpha I_b + \alpha^2 I_c$$

$$= \frac{1}{3} \times 0 + e^{j120} \times 6\underline{|60°} + e^{j240}.\,6\underline{|-120}$$

$$= 6\underline{|180} + 6\underline{|-120}$$
$$= -6 - 3 + j5.19$$
$$= -9 + j5.19$$

$$I_{a2} = \frac{1}{3} I_a + \alpha^2 I_b + \alpha I_c$$

$$= \frac{1}{3} \times 0 + e^{j240} \times 6\underline{|60°} + e^{j120} \times 6\underline{|-120}$$
$$= 3 - 5.19j + 6$$
$$I_{a2} = 9 - 5.19J$$

EXERCISE – II

30. $C = f\lambda$

$\therefore \quad 3 \times 10^8 = 500 \times \lambda$

$\Rightarrow \quad \lambda = 6 \times 10^5 = 600\text{km}$

For $f = 50$ Hz, $\lambda = 6000$ km

In 50 Hz lines, line length above 240 km should be considered as ling lines. By analogy for 500 Hz lines, any length about 54 km must be considered as a long line.

41. $R \propto \frac{1}{l}$

For 100 km line,

$$R = \frac{1}{100} \times 10 = 0.1\ \text{M}\Omega$$

49. To have optimum load sharing,

$$IC_1 = IC_2$$

$\therefore \quad 0.1P_1 + 8 = 0.15\,P_2 + 3$

$\Rightarrow \quad 0.1\,P_1 - 0.15\,P_2 + 5 = 0$

Divided by 0.1, we get

$P_1 - 1.5\,P_2 + 50 = 0 \qquad ...(i)$

We have $\quad P_1 + P_2 = 100$ MW

$\Rightarrow \quad P_1 + P_2 - 100 = 0 \qquad ...(ii)$

Subtracting (*i*) from (*ii*), we get

$2.5P_2 - 150 = 0$

$\Rightarrow \quad P_2 = 60\text{MW}$

$\therefore \quad P_1 = 100 - P_2$

$= 100 - 60 = 40$ MW

52. For optimum load sharing,

$$IC_1 = IC_2$$

$\therefore \quad 0.1x + 20 = 0.15y + 18$

$\Rightarrow \quad x - 1.5y + 20 = 0$

and $\quad x + y = 300 \qquad ...(i)$

$\Rightarrow \quad x + y - 300 = 0 \qquad ...(ii)$

Subtracting equation (*i*) from equation (*ii*), we get

$2.5y - 320 = 0$

$\Rightarrow \quad y = 128$ MW

$\therefore \quad x = 300 - 128$

$= 172$ MW

55. Velocity of travelling wave in the cable is given by,

$$v = \frac{3 \times 10^8}{\sqrt{9}} = 10^8\text{m/s}$$

56. String efficiency

$$= \frac{\text{Voltage across the whole string}}{n \times \left(\begin{array}{r}\text{voltage across the unit ajacent}\\ \text{to line conductor}\end{array}\right)}$$

$$= \frac{V}{4 \times 0.3333V} = 0.75 = 75\%$$

58.
$$V_1 = Z_{11}I_1 + Z_{12}I_2$$
$$V_2 = Z_{21}I_2 + Z_{22}I_2$$

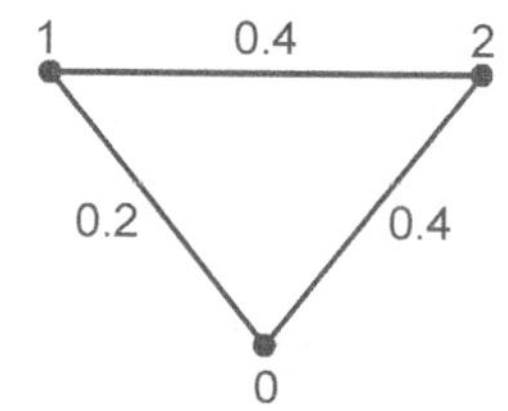

$$Z_{11} = \left.\frac{V_1}{I_1}\right|_{I_2=0} = \frac{0.2 \times 0.8}{0.2 + 0.8} = 0.16$$

$$Z_{22} = \left.\frac{V_2}{I_2}\right|_{I_1=0} = \frac{0.4 \times 0.6}{1.0} = 0.24$$

Let $\quad I_1 = 0 \quad$ and $\quad I_2 = 1$

$$\therefore \quad Z_{12} = \left.\frac{V_2}{I_2}\right|_{I_1=0} = 0.08$$

$$Z_{21} = 0.08$$

66.
$$\text{P.S.M.} = \frac{3000}{\frac{400}{5} \times 2.5} = 15$$

69. Surge impedance of the cable does not depend on the length of the line, hence for 100 km length also the surge impedance will same as that for 40 km.

71. Equivalent reactance seen from the fault point

$$X_{(pu)} = \frac{(j0.3 + j0.08) \times (j0.1 + j0.08)}{j0.1 + j0.2 + j0.08 + j0.08 + j0.1}$$

$= j0.12214$

$$\therefore \quad \text{Fault level current} = \frac{1}{X_{(pu)}}$$

$$= \frac{1}{j0.12214} = -j8.1871$$

76. Plug setting multiplier is given by,

$$\text{P.S.M} = \frac{2000}{\frac{400}{5} \times 2.5} = 10$$

85. For line to ground fault, zero sequence current

$$= \frac{1}{3} \text{ Fault current}$$

$$= \frac{1}{3} \times 100 = 33.3\text{A}$$

88. Power loss is given by,

$$P_2 = B_{11}P_1^2 + 2B_{12}P_1P_2 + B_{22}P_2^2$$

$$= 0.001 \times (50)^2 2(-0.0005) \times 50 \times 40 + 0.025 \times 40^2$$

$$= 4.5 \text{ MW}$$

91. $P_{max} = 396$ MW

96. Surge Power $= \frac{(1000)^2}{5000} = 200$ MW

141. Sending end current $= \frac{0.3}{1.2} = 0.25$

Reactance of the line = 0.2

$\therefore$ Voltage drop in the line $= 0.25 \times 0.2 = 0.05pu$

143. Total voltage drop from A to C

$$= [40 \times j0.35 + [30(0.8 - 0.6j) + 40] \times 0.25j]$$

$$= 4.5 - j30$$

146. $I_{a0} = \frac{1}{3}(I_a + I_b + I_c)$

For double line or line to line fault,

$$I_a = 0$$

$$I_b + I_c = 0$$

$$\therefore \quad I_{a0} = 0$$

150. In the given values

$$I_{positive} = I_{negative} + I_{zero}$$

It is true for LLG fault.

151. Reactance diagram of the system can be drawn as shown.

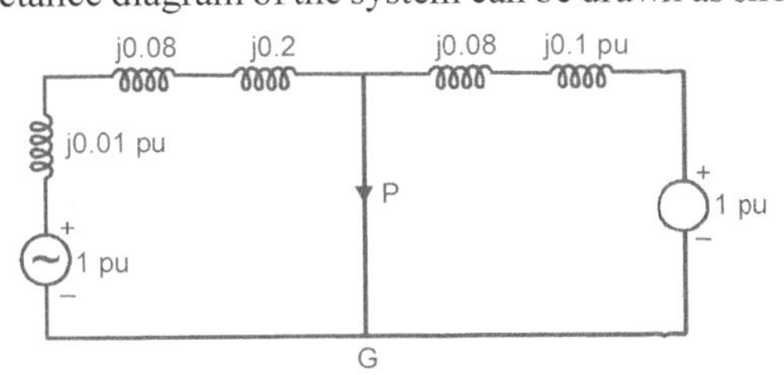

From Thevenin's equivalent circuit

$$V_{Th} = 0.01pu$$

$$Z_{Th} = \frac{j0.38 \times 0.18}{5.26} = j0.122pu$$

$$\therefore \text{ Fault current} = \frac{1.0}{j0.122} = j8.1871pu$$

154. Energy stored = J.H

$$= 4 \times 100 = 400 \text{ MJ}$$

168. Shunt admittance,

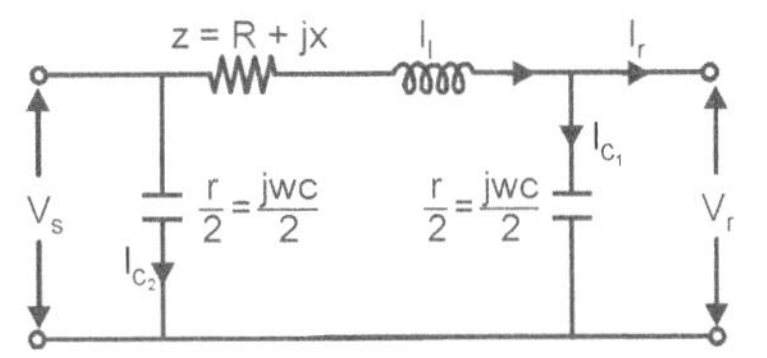

$$\frac{r}{2} = \frac{j\omega c}{z}$$

$$= \frac{j(2\pi \times 50)}{2} \times \frac{0.03}{\pi} \times 10^{-6} \times 100$$

$$= 150 \times 10^{-6} \underline{|90°} \; mho$$

170. When *dc* is applied, then there will be no effect of line to ground capacitance. Hence string efficiency will be 100%.

176. Zero-sequence network for the given circuit is shown below

$\therefore$ Zero-sequence driving point reactance

$$= X_{10} + X_{10} = 0.020pu$$

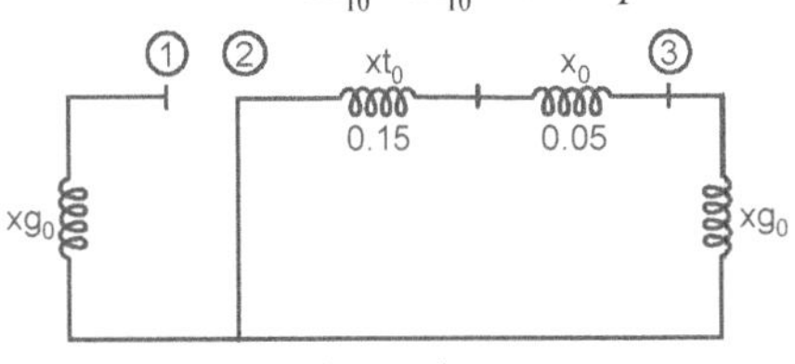

$\therefore$ Zero-sequence driving point reactance

$$= x \text{ to } 0 + x0 = 0.020\pi$$

178. Reflection coefficient,

$$\rho = \frac{Z_2 - Z_1}{Z_2 + Z_1}$$

$$= \frac{1 - Z_1 / Z_2}{1 + Z_1 / Z_2}$$

Since $Z_2 = \infty$, therefor $\rho = 1$

184. $I = I_0(1 - e^{R/L^t})$

Given $I = 0.632 I_0$

$\therefore \quad e^{R/L^t} = 0.368$

$L = 0.6$H

186. A base-load station or plant is to operate round the clock and thus major portion of the energy demanded is produced by base-load station. Therefore relatively "C" for base-load plant should be as small as possible. If "B" too is small, it would be highly economical. Similarly peak-load station or plant is required to supply relatively only a small fraction of the total energy requirement and hence 'C' for such plant may be high in comparison to 'C' for base-load plant. The power capacity requirement may be comparable to base-load capacity and 'B' for '*peak-load*' plant should be comparatively small.

Thus if 'B's are equal 'C' for Base-load station should be smaller and if 'B' s are also unequal, then 'B' for peak load station should be smaller.

188. $$I_R = I_{R0} + I_{R1} + I_{R2}$$
$$= 20A$$
$$I_n = \overrightarrow{I_R} + \overrightarrow{I_R}\underline{|-120°} + I_R\underline{|120°}$$
$$= 20\,[1 + 1\;\underline{|-120°} + \underline{|120°}\,] = 0A$$

202. $$\text{Demand factor} = \frac{1.5}{2} = 0.075$$

215. $$X_{new}(pu) = 0.25 \times \left(\frac{MVA_{new}}{kV_{old}}\right) \times \left(\frac{kV_{old}}{kV_{new}}\right)^2$$
$$= 0.25 \times \left(\frac{100}{500}\right)^2 \times \left(\frac{18}{20}\right) = 0.0405$$

244. Voltage regulation becomes zero for a leading power factor load whose power factor angle is given by

$$\theta = \tan^{-1}\frac{X}{R}$$

where X and R are reactance and resistance of the short line respectively.

Taking reactance (and not inductance as given in the statement) and resistance to be equal

$$\theta = 45°$$

$\therefore$ p.f. of the load = cos 45
$$= 0.707(\text{loading})$$

245. $$\text{Pick up value} = 5 \times 1.25$$
$$= 6.25\,A$$

■■

8 CHAPTER

Electrical Machines

EXERCISE – I

1. The resultant flux density in the airgap of a synchrunous generator is the lowest during^
   (*a*) Open circuit (*b*) Solid short circuit
   (*c*) Full load (*d*) Half load
2. Which one of the following is the primary reason for placing field on rotor in an alternator ?
   (*a*) Small power in field circuit
   (*b*) Insulation of high voltage is made easy on stator than on rotor
   (*c*) Large current in stator
   (*d*) Large power in stator
3. Which one of the following is not a necessary condition to be satisfied for synchronizing an incomming alternator to an already operating alternator ?
   (*a*) Same voltage magnitude
   (*b*) Same frequency
   (*c*) Same prime mover speed
   (*d*) Same phase sequence
4. Match List-I (Motor) with List-II (Application) and select the correct answer using the code given below the lists:

List-I (*Motor*)	**List-II** (*Application*)
A. Hysteresis motor	1. Photocopying machine
B. Shaded pole motor	2. Phonograph turn table
C. Reluctance motor	3. Electric Shover
D. Universal motor	4. Record player

Codes:

	A	B	C	D
(*a*)	4	3	2	1
(*b*)	2	1	4	3
(*c*)	4	1	2	3
(*d*)	2	3	4	1

5. Which one of the following is the correct statements ? Field control of dc shunt motor gives
   (*a*) constant kW drive (*b*) constant torque drive
   (*c*) constant speed drive (*d*) variable load speed drive
6. Dummy coils in dc machines are useful to
   (*a*) Increase the efficiency
   (*b*) Improve the commutation
   (*c*) Reduce the cost of the machine
   (*d*) Maintain mechanical balance of armature
7. Match List-I (Machine components) with List-II (Functions) and select the correct answer using the code given below the lists:

List-I (*Machine components*)
A. Damper winding
B. Interpole winding
C. End ring
D. Conservator

List-II (*Functions*)
1. Used in synchronous motor to provide starting torque.
2. Used in dc generator to neutralize the reactance voltage and cross magneting effect of armature reaction.
3. Used to short circuit rotor bars of cage-type induction motor.
4. Used to protect transformer from damage when oil expands due to rise in temperature.

Codes:

	A	B	C	D
(*a*)	2	1	3	4
(*b*)	1	2	4	3
(*c*)	2	1	4	3
(*d*)	1	2	3	4

8. In a dc machine, for the same values of ϕ, Z and N; which one of the following statements is correct ?
   (*a*) Armature emf is more with wave winding than with lap winding.
   (*b*) Armature emf is less with wave winding than with lap winding.
   (*c*) Armature emf depends on whether the machine is running as a motor or a generator
   (*d*) Armature emf is the same as long as the flux density in the airgap remains the same.

9. In a dc machine, the armature is always on the rotor unlike an ac machine where it could be either on the stator or rotor, because
   (*a*) commutation action would otherwise not possible
   (*b*) armature reaction demagnetising action would otherwise be more
   (*c*) otherwise it would not be possible to place compensating winding in pole shoes
   (*d*) self-starting of dc motor would not be possible otherwise

10. In a 3-phase Induction motor if Pg is the power transferred across the air gap and S is the slip, then ratio of rotor copper loss to the machanical power developed is given by
   (*a*) $\frac{S}{1-S}$
   (*b*) S
   (*c*) $\frac{1}{S}$
   (*d*) $\frac{1-S}{S}$

11. If a saw-cut is made in the iron core of a single phase transformer as shown in the given figure, with secondary terminals open, it will result in

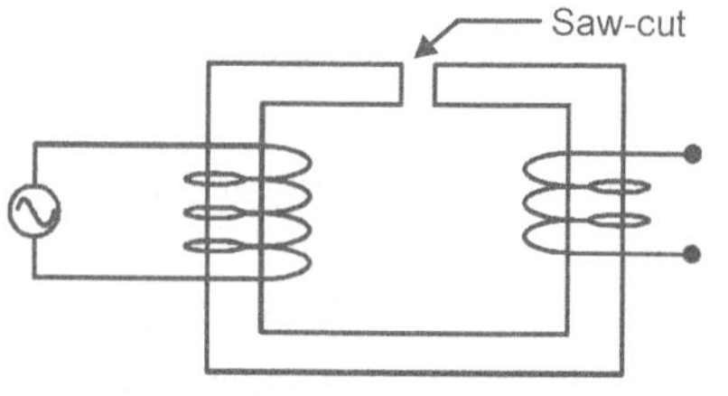

   (*a*) no voltage across secondary terminals due to discontinuity in magnetic circuit
   (*b*) excessively large current drawn by primary terminals due to discontinuity in magnetic circuit
   (*c*) no appreciable change in the operation of the transformer
   (*d*) no transformer action and the primary current will be greatly reduced

12. Match List-I with List-II and select the correct answer using the codes given below the lists.

List-I
A. Injection of voltage into the rotor circuit of a motor
B. Variation of applied voltage of a three-phase induction motor at constant frequency
C. Series parallel control
D. Ward Leonard control of *dc* motors

List-II
1. Steel rolling mills
2. Textile mills
3. Fans and blowers
4. Traction
5. Position controlling a servo system

Codes:

	A	B	C	D
(*a*)	1	2	3	4
(*b*)	4	5	2	1
(*c*)	1	5	2	4
(*d*)	2	3	4	1

13. Match List-I with List-II and select the correct answer using the codes given below the lists:

List-I (*Load pf*)
A. Zero leading
B. Unity
C. Zero lagging
D. Zero with armature terminals suddenly short-circuited

List-II (*Nature of armature flux in an alternator*)
1. Demagnetising.
2. Magnetising.
3. Cross magnetizing.
4. Initially constant, demagnetizing.

Codes:

	A	B	C	D
(*a*)	1	2	4	3
(*b*)	2	3	1	4
(*c*)	4	3	1	2
(*d*)	2	1	3	4

14. Match List-I with List-II and select the correct answer using the codes given below the lists:

List-I	List-II
A. Interpoles	1. Commutation in *dc* machines.
B. Three point starter	2. Lap winding.
C. Dummy coils	3. DC shunt motors
D. Equalising ring	4. Wave winding

Codes:

	A	B	C	D
(*a*)	1	2	3	4
(*b*)	2	3	4	1
(*c*)	1	3	4	2
(*d*)	3	4	1	2

15. A transfer has a core loss of 200 W and a full load copper loss of 800W. The maximum efficiency of the transformer will occur at

(*a*) 0.5 times full load current

(*b*) 0.6 times full load current

(*c*) 0.7 times full load current

(*d*) 0.8 times full load current

16. If the load on a squirrel-cage induction motor, operating on a constant voltage, constant frequency *ac* supply is increased, then

(*a*) power factor angle will increase and slip will decrease

(*b*) power factor angle will decrease and slip will increase

(*c*) both power factor angle and slip will decrease

(*d*) both power factor angle and slip will increase

17. During the slip-test for determining the direct and the quadrature axis synchronous reactances of an alternator, the voltage across the open field circuit terminals is

(*a*) *dc* voltage

(*b*) *ac* voltage of supply frequency

(*c*) *ac* voltage of slip frequency

(*d*) a modulated voltage with an envelop of slip frequency

18. Which single phase induction motor has the lowest speed?

(*a*) Shaded pole (*b*) Universal

(*c*) Hysteresis (*d*) Repulsion

19. For speed higher than say 3000 *rpm*, the machine used is

(*a*) induction motor (*b*) synchronous motor

(*c*) universal motor (*d*) none of these

20. Which of the following motor is preferred for tape recorder ?

(*a*) Shaded pole motor

(*b*) Hysteresis motor

(*c*) Two-value capacitor motor

(*d*) Universal motor

21. Which one of the following statements is not true for an *ac* servo-motor ?

(*a*) Has low inertia rotor

(*b*) Is a single phase motor

(*c*) Has slip torque characteristic as straight line with negative slope

(*d*) Reference voltage is high compared to signal frequency

22. For stability of an *ac* servomotor

(*a*) a negative slop in the torque speed curve is necessary

(*b*) linearized torque-speed curve is essential

(*c*) the ratio of the rotor reactance to rotor resistance should be high

(*d*) none of these

23. A single phase reluctance motor

(*a*) has salient pole rotor structure and runs at subsynchronous speed

(*b*) has salient pole rotor structure and runs at super synchronous speed

(*c*) has salient pole rotor structure and runs at synchronous speed

(*d*) has non-salient pole rotor structure and runs at synchronous speed

24. Reluctance torque is the

(*a*) torque developed in a given synchronous motor as a function of field excitation and the sine of the torque angle

(*b*) torque developed by salient pole synchronous or synchronous induction motors due to variations in air gap produced by armature reaction

(*c*) torque developed by non-salient pole asynchronous motor

(*d*) torque developed by none of these

25. In a single phase hysteresis motor

(*a*) starting torque is caused by both eddy current and hysteresis while the running torque is caused by hysteresis

(*b*) starting as well as running torques are caused by both eddy current and hysteresis

(*c*) starting torque is caused by only eddy current while running torque is caused by only hysteresis

(*d*) starting as well as running torques both are caused by only hysteresis

26. The magnetic circuit of a *dc* shunt generator with negligible armature resistance and brush contact voltage drop remains unsaturated when it delivers full-load current at the rated terminal voltage. The brushes are kept in the interpolar axis and the speed is maintained constant. When the load is switched off, then terminal voltage would

(*a*) remain the same

(*b*) increase

(*c*) decrease

(*d*) becomc zero

27. Under which of the following conditions is a *dc*-motor provided with compensating winding used?

1. Wide range of speed control above normal.
2. Wide range of steady load variation with no speed control.
3. Wide range of rapid variation in load.

Select the correct answer using the codes given below:

(*a*) 1, 2 and 3 (*b*) 2 and 3

(*c*) 1 and 2 (*d*) 1 and 3

28. A *dc* shunt motor having unsaturated magnetic circuit run as at 1000 *rpm* with rated voltage. If the applied voltage is half of the rated voltage, the motor will run at

(*a*) 2000 *rpm* (*b*) 1000 *rpm*

(*c*) 750 *rpm* (*d*) 500 *rpm*

29. The external load characteristics of a *dc* generator at rated speed is assumed to be linear and is as shown in given figure. The load circuit comprises only resistance. When the load current is 50 amperes, the value of the load circuit resistance will be

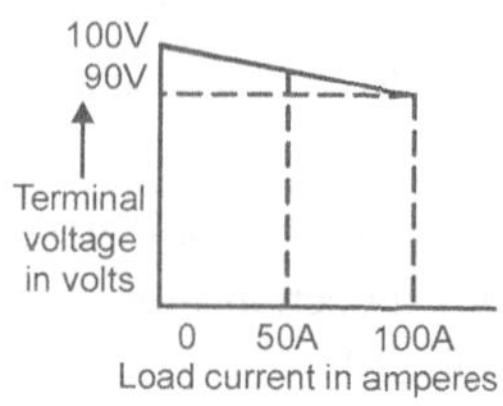

(*a*) 0.9 ohm (*b*) 1.0 ohm

(*c*) 1.9 ohm (*d*) 2.0 ohm

30. Absolute speed of magnetic field in space of a 3-phase, rotor-fed induction motor is

(*a*) synchronous speed N_s

(*b*) rotor speed N_r

(*c*) $(N_s - N_r)$

(*d*) $(N_s + N_r)$

31. It is desirable to elements 5th harmonic voltage from the phase voltage of an alternator. The coils should be short-pitched by an electrical angle of

(*a*) 30° (*b*) 36°

(*c*) 72° (*d*) 18°

32. The armature resistance of 6-pole lap wound *dc* machine is 0.05 Ω. If the armature is rewound using a wave winding, the armature resistance will be

(*a*) 4.45 Ω (*b*) 0.30 Ω

(*c*) 0.15 Ω (*d*) 0.10 Ω

33. For driving a 1000 HP compressor, a synchronous motor is preferred over an induction motor because

1. It will result in lower noise and vibration.
2. It will help attain compressor pressure faster.
3. Of the possibility of operating at better power factor.
4. The starting torque requirement is low

Of these statements

(*a*) 1 and 2 are correct

(*b*) 2 and 3 are correct

(*c*) 1, 2 and 3 are correct

(*d*) 3 and 4 are correct

34. If the supply frequency of a transformer is increased, keeping the supply voltage constant, then

(*a*) both magnetising and core-loss components of current will decrease

(*b*) both magnetising and core components of current will increase

(*c*) magnetizing component of current will, increase, while the core loss component of current will decrease

(*d*) magnetising component of current will decrease, while core loss component of current will increase [ES]

35. Match list-I with List-II and select the correct answer using the codes given below the lists:

List-I (Apparatus)

A. 3-phase induction motor on load

B. Synchronous generator

C. Single-phase transformer on load

D. D.C. series motor

List-II

1.

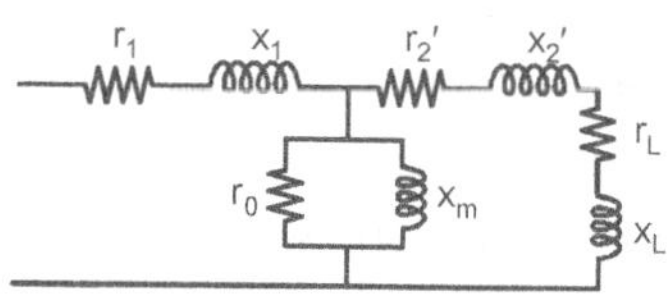

2.

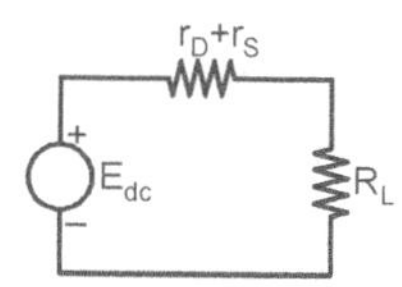

3.

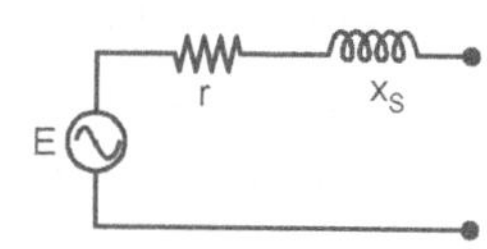

4.

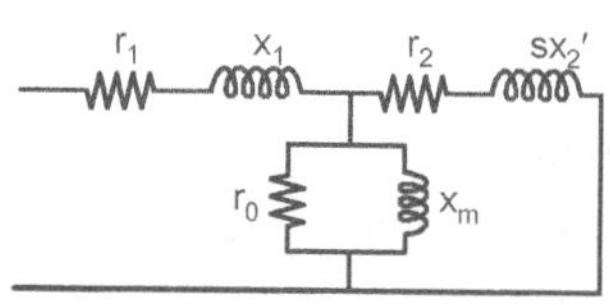

5.

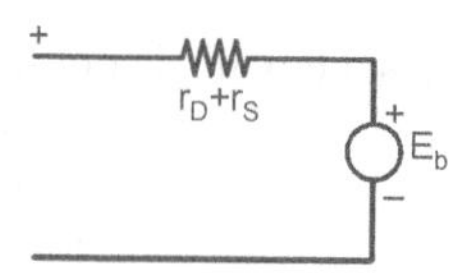

6.

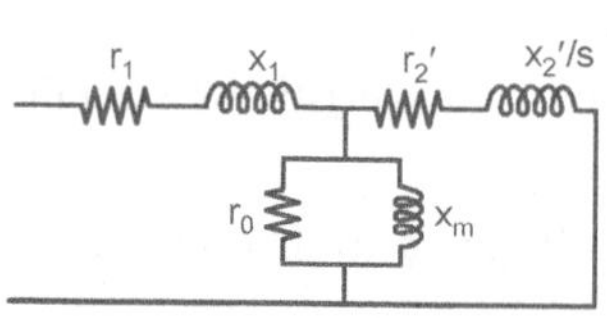

Codes:

	A	B	C	D
(*a*)	6	3	1	2
(*b*)	1	3	6	5
(*c*)	4	3	1	5
(*d*)	6	3	1	5

36. A star-delta transformer can deliver 100 kVA load. If it is reconnected in delta-star, it can deliver a load of

(*a*) 173.2 kVA (*b*) 86.6 kVA

(*c*) 300 kVA (*d*) 100 kVA

37. Consider the following statements regarding the starting of a *dc* shunt motor:

1. External field resistance should be minimum.
2. External field resistance should be maximum.
3. Reduced voltage should be applied to the armature.
4. Increased voltage should be applied to the armature.

Of these statements

(*a*) 1 and 3 are correct (*b*) 1 and 4 are correct

(*c*) 2 and 3 are correct (*d*) 2 and 4 are correct

38. Semiclosed slots or totally closed slots are used in induction motors, essentially to

(*a*) improve starting torque

(*b*) increase pull-out torque

(*c*) increase efficiency

(*d*) reduce magnetising current and improve power factor

39. Which of the following limit the reactive power output of a synchronous generator?

1. Armature current.

2. Field current.

3. Load angle.

4. Prime mover input

Select the correct answer using the codes given below:

(*a*) 1 and 2 (*b*) 2 and 3

(*c*) 3 and 4 (*d*) 1 and 4

40. If a 3-phase induction motor, fed from a rated-voltage and rated-frequency supply is driven by a prime mover in the same direction when operating as a motor, at super synchronous speed, the machine will operate as

(*a*) a reactive power generator

(*b*) a frequency converter

(*c*) an active power generator

(*d*) a rotary converter

41. A 3-phase slip ring induction motor with negligible stator impedance operates on a balanced 3-phase *ac* supply. Consider the following statements in this regard:

The maximum torque developed by the motor

1. Is independent of the value of the rotor circuits resistance.

2. Is directly proportional to the square of the supply voltage.

3. Occurs at a slip whose value is independent of the value of the rotor resistance.

Of these statements

(*a*) 1, 2 and 3 are correct

(*b*) 1 and 2 are correct

(*c*) 2 and 3 are correct

(*d*) 1 and 3 are correct

42. If the field of a *dc* shunt motor gets opened while the motor is running, then the

(*a*) speed of the motor will be reduced

(*b*) motor will attain dangerously high speed

(*c*) armature current will drop

(*d*) armature will oscillate about original speed as the mean speed

43. Match List-I (*dc* Motor) with List-II (characteristics) and select the correct answer using the codes given below the lists:

List-I (*dc Motor*)

A. Cumulative compounded motor

B. Differentially compounded motor

C. Series Motor

D. Shunt Motor

List-II (*characteristics*)

1. Fairly constant speed irrespective of the load
2. It may start in reverse direction.
3. Definite no-load speed.
4. Never started without load.

Codes:

	A	B	C	D
(*a*)	2	3	1	4
(*b*)	3	2	4	1
(*c*)	1	3	2	4
(*d*)	4	1	3	2

44. The residual magnetism of a self-excited *dc* generator is lost. To build up its *emf* again

(*a*) the field winding must be replaced

(*b*) the armature connection must be reversed

(*c*) the field winding connection must be reversed

(*d*) field winding must be excited by loss voltage *dc* supply

45. A universal motor runs at

(*a*) higher speed with *dc* supply and with less sparking

(*b*) higher speed with *ac* supply and with less sparking

(*c*) same speed with both *ac* and *dc* supplies

(*d*) higher speed with *ac* supply but with increased sparking at the brushes

46. A synchronous motor operators at 0.8 *p.f.* lagging. If the field current of the motor is continuously increased
    (*a*) the power factor decreases upto a certain value of field current and thereafter it increases
    (*b*) the armature current increases upto a certain value of field current and thereafter it decreases
    (*c*) the power factor increases upto a certain value of field current and thereafter it decreases
    (*d*) the armature current decreases upto a certain value of field current and thereafter it increases

47. When a balance 3-phase distributed type armature winding is carrying 3-phase balanced currents, the strength of the result rotating magnetic field is
    (*a*) three times the amplitude of each constituent pulsating magnetic field
    (*b*) equal to the amplitude of each constituent pulsating magnetic field
    (*c*) half the amplitude of each constituent pulsating magnetic field
    (*d*) one and a half times the amplitude of each constituent pulsating magnetic field

48. A three phase slip ring induction motor is fed from the rotor side with stator winding short circuited. The frequency of the currents flowing in the short circuited stator is
    (*a*) slip frequency
    (*b*) supply frequency
    (*c*) frequency corresponding to rotor speed
    (*d*) zero

49. Two mechanically coupled alternators delivers power at 50 Hz and 60 Hz respectively. The highest speed of the alternator is
    (*a*) 3600 rpm
    (*b*) 3000 rpm
    (*c*) 600 rpm
    (*d*) 500 rpm

50. In arc welding, in order to obtain steady arc one should use
    (*a*) *dc* series generator
    (*b*) *dc* differentially compounded generator
    (*c*) *dc* shunt generator
    (*d*) *dc* cumulatively compounded generator

51. The effect of leading power factor on the voltage regulation of an alternator is
    (*a*) increasing in nature
    (*b*) decreasing in nature
    (*c*) maintained at constant value
    (*d*) oscillating in nature

52. The regulation obtained by synchronous impedance method is always higher than the actual value because
    (*a*) synchronous reactance is assumed variable while it is not so
    (*b*) effective armature resistance is assumed constant while it is not so
    (*c*) synchronous reactance is assumed constant while it is not so
    (*d*) field current is increased to give short circuit current about thrice of full load current

53. An alternator with higher value of SCR has
    (*a*) poor voltage regulation and lower stability limit
    (*b*) better voltage regulation and higher stability limit
    (*c*) poor voltage regulation and higher stability limit
    (*d*) better voltage regulation and low stability limit

54. When a synchronous generator is designed with lower value of SCR it will
    (*a*) give higher stability limit
    (*b*) have lower sc current
    (*c*) give better voltage regulation
    (*d*) have higher synchronising power

55. Number of slip rings in a 3-phase synchronous motor will be
    (*a*) 0 (*b*) 1
    (*c*) 2 (*d*) 3 or 4

56. The relative speed between the magnetic fields of stator and rotor under steady state operation is zero for
    (*a*) a *dc* machine
    (*b*) an induction machine
    (*c*) a synchronous machine
    (*d*) all the above machines

57. A synchronous motor is supplying a load at unity power factor. If the load on the motor is increased keeping its excitation and terminal voltage constant, the power factor
    (*a*) will remain the same (*b*) will become leading
    (*c*) will become lagging (*d*) none of these

58. A synchronous motor is running clockwise. If the direction of field current is reversed, the motor would
    (*a*) come to stop
    (*b*) run in reversed direction
    (*c*) run as before
    (*d*) runs as before, but at less speed

59. For a 3 phase, 4-pole, 5 Hz synchronous motor the frequency, pole number and the load torque are all the halved. The motor speed will be
    (*a*) 375 rpm (*b*) 75 rpm
    (*c*) 1500 rpm (*d*) 3000 rpm

60. Which motor can conveniently operate at lagging as well as leading power factor ?
(*a*) squirrel cage induction motor
(*b*) wound rotor induction motor
(*c*) synchronous motor
(*d*) *dc* shunt motor

61. A synchronous motor is connected to a constant voltage, constant supply frequency. The motor is initially operated at lagging power factor. When the field current is increased
(*a*) the load angle increases
(*b*) the power input increases
(*c*) the power factor decreases
(*d*) the armature current decreases upto a certain limit and then increases

62. Armature reaction in a synchronous motor at rated voltage and zero power factor (lead) is
(*a*) magnetising (*b*) cross-magnetising
(*c*) both (*a*) and (*b*) (*d*) demagnetising

63. A synchronous motor will deliver maximum power when
(*a*) load angle is equal to internal angle θ
(*b*) input power factor is unity
(*c*) load angle is 45°
(*d*) load angle is 0°

64. Synchronous capacitor is
(*a*) an ordinary static capacitor bank
(*b*) an over-excited synchronous motor driving mechanical load
(*c*) an over excited synchronous motor without mechanical load
(*d*) none of the input

65. In 3-phase synchronous motor, is prevented by
(*a*) dummy coils
(*b*) compensating winding
(*c*) damper winding
(*d*) short pitch winding

66. In a synchronous motor, squirrel cage winding is provided for making motor
(*a*) noise free
(*b*) self starting
(*c*) cheap
(*d*) quick start

67. Damper winding is used in a synchronous motor for
(*a*) power factor improvement
(*b*) efficiency improvement
(*c*) eliminating hunting of the rotor
(*d*) minimising temperature rise

68. A pony motor is basically a
(*a*) small induction motor
(*b*) *dc* series motor
(*c*) *dc* shunt motor
(*d*) double winding *ac*/*dc* motor

69. While starting a synchronous motor its field winding should be
(*a*) kept open
(*b*) short-circuited
(*c*) connected to a *dc* source
(*d*) none of these

70. In case one phase of a 3- phase synchronous motor is short-circuited, the motor will
(*a*) not start
(*b*) run at $\frac{2}{3}$ of synchronous speed
(*c*) run with excessive vibrations
(*d*) take less than the rated load

71. An electric train employing a *dc* series motor is running at a fixed speed. When a sudden slight drop in the mains voltage occurs this would results in a
(*a*) drop in speed and rise in current
(*b*) rise in speed and drop in current
(*c*) rise in speed and rise in current
(*d*) drop in speed with current unaltered

72. A 3-phase 4-pole alternator has 48 stator slots carrying the 3-phase distributed winding. Each coil of the winding is short chorded by one slot pitch. The winding factor is given by
(*a*) $\frac{1}{16}\cot(7.5°)$ (*b*) $\frac{1}{8}\cot(7.5°)$
(*c*) $\frac{1}{16\sin\left(7.5°\right)}$ (*d*) $\cos(7.5°)$

73. A centre zero ammeter connected in the rotor circuit of a 6-pole 50 Hz induction motor makes 30 oscillations in one minute. The rotor speed is
(*a*) 970 (*b*) 990
(*c*) 1010 (*d*) 1030

74. Match List-I with List-II and select the correct answer using the codes given below the lists:

List-I	List-II
A. Stator core	1. Copper.
B. Rotor core	2. Copper alloy.
C. Stator windings	3. Silicon steel.
D. Slip rings	4. Mild steel.
	5. Aluminium.

Codes:

	A	B	C	D
(*a*)	3	5	2	1
(*b*)	4	5	1	2
(*c*)	4	3	1	2
(*d*)	3	4	1	2

75. Cogging and crawling are phenomena associated with

(*a*) cage inductions machines and they are essentially the same

(*b*) squirrel cage induction machines, the former during stating and the latter at a fraction of its rated speed

(*c*) squirrel cage inductions machines, the former at a fraction of its rated speed and the latter during starting

(*d*) wound rotor induction machines and they are reduced by skewing, chording and distributions of winding

76. For purpose of analysis exact equivalent circuit of a transformers is usually replaced by an approximate equivalent circuit. In doing so errors introduced due to winding copper loss and core loss are of differential nature. Due to this, the analysis by approximate equivalent circuits gives fairly satisfactory results. Under the circumstances, which one of the following statements is correct in respect of losses referred to approx, equivalent circuit as compared to exact equivalent circuit?

(*a*) this accounts for somewhat greater primary winding copper loss and less core loss

(*b*) this accounts for somewhat lesser primary winding copper loss and more core loss

(*c*) this accounts for somewhat greater winding copper loss and less core loss

(*d*) this accounts for somewhat lesser secondary winding copper loss and more core loss

77. Match List-I with List-II and select the correct answer using the codes given below the lists:

List-I	List-II
A. Transmitter	1. Delta connected stator
B. Differential transmitter	2. Umbrella shaped rotor
C. Control transformer	3. Damping free rotor
D. Repeater	4. Three-phase rotor
	5. Salient pole rotor

Codes:

	A	B	C	D
(*a*)	5	4	2	1
(*b*)	2	5	4	1
(*c*)	5	4	2	3
(*d*)	3	4	2	1

78. If rated voltage from the power lines is applied to the primary of a single-phase transformer which is operated on load, then

(*a*) both input voltage and current are sinusoidal

(*b*) both input voltage and current are non-sinusoidal

(*c*) input voltage is non sinusoidal and the input current is sinusoidal

(*d*) input voltage is sinusoidal and input current is non-sinusoidal

79. Two 3-phase transformers are to be connected for parallel operation. Which one of the following arrangements is impossible?

(*a*) transformer A: primary Y; secondary Y, Transformer B: primary Δ; secondary Δ

(*b*) transformer A: primary Δ; secondary Y, Transformer B: primary Δ; secondary Δ

(*c*) transformer A: primary Y; secondary Δ, Transformer B: primary Δ; secondary Y

(*d*) transformer A: primary Δ; secondary Δ, Transformer B: primary Δ; secondary Δ

80. In a self-excited induction generator, to keep the frequency of generated voltage constant with the increase in load, the speed of the induction machine should be

(*a*) increased

(*b*) decreased

(*c*) maintained less than the rated synchronous speed

(*d*) maintained more than the rated synchronousspeed

81. Consider the following drive systems:

1. Chopper fed *dc* series motor.
2. Chopper fed *dc* shunt motor.
3. Chopper fed permanent magnet *dc* motor.
4. Brushless *dc* motor drive.
5. Inverter fed 3-phase induction motor.

For the development of a battery operated electric car the possible option are

(*a*) 1, 3, 5 (*b*) 2, 4, 6

(*c*) 1, 5 (*d*) 1, 4, 5

82. The type of *dc* generator used for welding purposes is a

(*a*) series genertor

(*b*) shunt generator

(*c*) cumulatively compounded generator

(*d*) differentially compounded generator

83. In the equivalent circuit of 3-phase induction motor, the mechanical shaft load is represented by the resistance

(*a*) R_2 (*b*) $(R_2/s) + R_2$

(*c*) R_2/s (*d*) $(R_2/s) - R_2$

84. A 220 V *dc* machine has an armature resistance of 1 Ω. If the full load current is 20 A, the difference in the induced voltage when the machine is running as a motor, and as a generator is

(*a*) 20 V (*b*) zero
(*c*) 40 V (*d*) 50 V

85. A separately excited *dc* generator is feeding a *dc* shunt motor. If the load torque on the motor is halved approximately

(*a*) armature current of both motor and generator are halved
(*b*) armature current of motor is halved and that of generator is unaltered
(*c*) armature current of generator is halved and that of motor is unaltered
(*d*) armature currents of both motor and generator are unaltered

86. In case of an armature controlled separately excited *dc* motor drive with closed loop speed control, an inner current loop is useful because it

(*a*) limits the speed of the motor to a safe value
(*b*) helps in improving the drive energy efficiency
(*c*) limits the peak current of the motor to the permissible value
(*d*) reduces the steady speed error

87. An induction motor having full load torque of 60 Nm when delta-connected develops a starting torque of 120 Nm. For the same supply voltage, if the motor is changed to star-connection, the starting torque developed will be

(*a*) 40 Nm (*b*) 60 Nm
(*c*) 90 Nm (*d*) 120 Nm

88. A three phase alternator has negligible stator resistance. A short circuit test is conducted on this alternator. At a particular speed a field current on I_{fl}, is required to drive the rated armature current. If the speed of the alternator is reduced to half, the field current required to maintain rated armature current

(*a*) would be equal to I_{fl}
(*b*) would be equal to 2 I_{fl}
(*c*) would be equal to $I_{fl}/2$
(*d*) cannot be predicted due to insufficient data[GATE]

89. A cumulative compounded long shunt motor is driving a load at rated torque and rated speed. If the series field is shunted by a resistance equal to the resistance of the series field, keeping the torque constant,

(*a*) the armature current increases
(*b*) the motor speed increases
(*c*) the armature current decreases
(*d*) the motor speed decreases

90. In a 3-phase induction machine, motoring, generating and braking operators take place in the range of slip '*s*' given by

(*a*) motoring : $1 > s > 0$
generating : $0 > s > -1$
breaking : $s > 1$
(*b*) motoring : $s > 1$
generating : $1 > s > 0$
breaking : $0 > s > -1$
(*c*) motoring : $s > 1$
generating : $0 > s > -1$
breaking : $1 > s > 0$
(*d*) motoring : $0 > s > -1$
generating : $s > 1$
breaking : $1 > s > 0$

91. A belt-driven cumulative compounded generator is delivering power to the *dc* mains. If the belt snaps, then the machine will run as a

(*a*) cumulative compounded in the same direction
(*b*) differentially compounded motor in the same direction
(*c*) cumulative compounded motor in the opposite direction
(*d*) differentially compounded motor in the opposite direction

92. The main reason of using a hysteresis motor for high quality tape recorders and record players is that

(*a*) its speed is constant (synchronous)
(*b*) it develops extremely steady torque
(*c*) it requires no centrifugal switch
(*d*) its operations is not affected by mechanical vibrations

93. Which type of motor is most suitable for a computer printer drive?

(*a*) Reluctance motor
(*b*) Hysteresis motor
(*c*) Reluctance motor
(*d*) Stepper motor

94. To avoid cogging in a squirrel cage induction motor the following stator slot (Z_1) and rotor slot (Z_2) combination must be avoided

(*a*) $Z_1 = Z_2$ (*b*) $Z_1 - Z_2 = 2P + 1$
(*c*) $Z_1 - Z_2 = 3P + 1$ (*d*) $Z_1 - Z_2 = 3P - 1$

95. Skew of rotor bar eliminates

(*a*) the effect of space harmonics
(*b*) the entire effect of crawling
(*c*) magnetic noise
(*d*) vibration due to unequal force developed on rotor

96. The outer cage of a double squirrel rotor consists of

(*a*) magnese brass (*b*) red copper

(*c*) bronze (*d*) steel

97. In double cage induction motor the inner cage has

(*a*) high resistance and high leakage reactance

(*b*) high resistance and low leakage reactance

(*c*) low resistance and high leakage reactance

(*d*) low resistance and low leakage reactance

98. In case of 3-phase slip ring induction motor as the rotor resistance is increased, the starting torque

(*a*) increases

(*b*) decreases

(*c*) increase upto a certain value of resistance and then remain constant

(*d*) increases upto a certain value of resistance and then decreases

99. Blocked rotor test of an induction motor corresponds in case of a transformer to

(*a*) full load

(*b*) half-full load

(*c*) no load

(*d*) short-circuit operation

100. When the induction motor is stand still the slip is

(*a*) zero

(*b*) 1

(*c*) infinity

(*d*) none of these

101. An increase in number of poles of an induction motor results in

(*a*) decrease in maximum p.f

(*b*) increase in maximum p.f.

(*c*) no change in maximum p.f.

(*d*) cannot be predicted

102. Leakage reactance pre phase of the stator of a polyphase induction motor is 1.0 ohms. The turns per phase of the stator are increased by 10%. The leakage reactance is then equal to

(*a*) $(1.1)^2$

(*b*) $(0.9)^2$

(*c*) $1/(1.1)^2$

(*d*) $1/(0.9)^2$

103. Air gap of a polyphase induction motor is kept small to

(*a*) reduce the possibility of crawling

(*b*) reduce the noise

(*c*) reduce magnetizing current

(*d*) obtain high starting torque

104. In $\frac{V}{f}$ speed control of $3-\phi$ induction motor, with increase in frequency, the maximum torque and slip at which maximum torque occur, would respectively

(*a*) increase and decrease

(*b*) increase and increase

(*c*) decrease and increase

(*d*) decrease and decrease

105. Synchronous speed of an induction motor can be increased by

(*a*) reducing mechanical friction

(*b*) increasing supply voltage

(*c*) increasing number of poles

(*d*) increasing supply frequency

106. In a variable speed induction motor drive $\frac{V}{f}$ is kept constant over a wide range of its frequency variation. The motor operates in the following mode

(*a*) constant power mode

(*b*) constant torque mode

(*c*) variable power and variable torque mode

(*d*) make the power constant

107. In *ac* motor control $\frac{V}{f}$ is kept constant to

(*a*) make maximum use of the magnetic circuit

(*b*) make minimum use of the magnetic circuit

(*c*) maximize the current drawn

(*d*) make the power constant

108. In motor resistance control of induction motor the hardness speed torque characteristic

(*a*) increases (*b*) decreases

(*c*) remain same (*d*) none of these

109. In case of voltage injection method of speed control, the injected emf should be of

(*a*) supply frequency (f)

(*b*) slip frequency (sf)

(*c*) $(1-s)$

(*d*) $(2-s)f$

110. While braking a 3-phase induction motor, its terminal were switched-over to a *dc* source from *ac* supply. The method of braking is

(*a*) regenerative braking (*b*) counter-current braking

(*c*) eddy current braking (*d*) none of these

111. One slip-ring connection of a 3-phase induction motor gets open circuited. The motor will

(*a*) continue to run at the same speed

(*b*) run at a lower speed

(*c*) run at a higher speed

(*d*) stop

112. A 3-phase induction motor runs at super synchronous speed. For self excitation the machine

(*a*) draws real power from the mains

(*b*) draws reactive power from the mains

(*c*) feeds reactive power to the mains

(*d*) generations emf at the expenses of residual magnetism

113. An 8-pole wound rotor induction motor operating at 60 Hz supply is driven at 1800 rpm by a prime mover in the opposite direction of revolving speed. The rotor current frequency is

(*a*) 60 Hz

(*b*) 120 Hz

(*c*) 180 Hz

(*d*) none of these

114. Linear induction motor is used in

(*a*) traction

(*b*) magnetic attraction

(*c*) mechanical workshops

(*d*) textile mills

115. A 3-phase induction motor is fed from a balanced 3-phase supply on the rotor side, short circuiting the stator terminals. The speed of the rotating magnetic fields is

(*a*) zero

(*b*) equal to rotor speed

(*c*) equal to synchronous speed

(*d*) equal to slip speed

116. in particular motor there is a stator phase winding and a short-circuited rotor winding. The motor runs at synchronous speed but does not necessitate any *dc* excitation. The motor is a

(*a*) 3-phase induction motor

(*b*) 3-phase synchronous motor

(*c*) 3-phase reluctance motor

(*d*) hysteresis motor

117. In a double revolving field theory of single phase induction motor, the slip of the forward motor is s then slip of the backward motor is

(*a*) $2s$

(*b*) s

(*c*) $2 - s$

(*d*) $s - 2$

118. Phase splitting can be accomplished in a single phase induction motor

(*a*) only by adding a capacitor in series with the auxiliary winding

(*b*) only by causing the auxiliary winding to have high reactance

(*c*) only by causing the auxiliary winding to have low resistance

(*d*) by any one of the above three methods

119. Centrifugal switch fitted on the rotor will operate when

(*a*) rotor speed reaches its rated condition

(*b*) rotor speed reaches 70% of its rated value

(*c*) rotor speed exceeds synchronous speed

(*d*) rotor speed exceeds 40 percent of its rated value

120. Compared to a resistor split phase motor a capacitor start motor has

(*a*) high starting torque

(*b*) lower starting torque

(*c*) higher running torque

(*d*) lower running torque

121. For successful parallel operation of two single-phase transformers, the most essential conditions is that their

(*a*) percentage impedances are equal

(*b*) polarities are properly connected

(*c*) turns ratios are exactly equal

(*d*) KVA rating are equal

122. The voltage regulation of a transformer at full load and 0.8 power factor leading will be

(*a*) –2.5%

(*b*) zero

(*c*) 1.0%

(*d*) 2.5%

123. A transformer possesses a percentage resistance and percentage reactance of 1% and 4% respectively. Its voltage regulation at power factor 0.8 lagging and 0.8 leading would be

(*a*) 2.4 % and – 0.8 %

(*b*) 3.2 % and – 1.6%

(*c*) 3.2 % and – 3.2 %

(*d*) 4.8 % and – 1.6%

124. A *dc* shunt motor has external resistance of R_a and R_f in the armature and field circuits respectively. Armature current at starting can be reduced by keeping

(*a*) R_f maximum and R_a minimum

(*b*) R_a minimum and R_f maximum

(*c*) R_f minimum and R_a maximum

(*d*) R_a maximum and R_f minimum

125. If the frequency of input voltage of a transformer is increased keeping the magnitude of voltage unchanged, then

(*a*) both hysteresis loss and eddy current loss in the core will increase

(*b*) hysteresis loss will increase but eddy current loss in the core will decrease

(*c*) hysteresis loss will decrease but eddy current loss will increase

(*d*) hysteresis loss will decrease current loss will remain unchanged

126. A 3-phase balanced sinusoidal rated voltage is fed to the star side of a star/delta bank of three single-phase transformers having the same turns ratios. A voltmeter across an open corner of the secondary delta records about 50% of the normal phase voltage. This voltage is due to

(*a*) wrong-polarity connections

(*b*) magnetic non-linearity

(*c*) unequal pu impedances of the transformers

(*d*) defective transformers stamping

127. The main purpose of the conservator in a transformer is to

(*a*) store extra oil to compensate for the loss of oil due to leakage

(*b*) achieve better cooling of the transformer

(*c*) take up the expansion of oil due to heating

(*d*) have the Bucholtz relay fitted

128. A lap wound armature winding, fitted with a commutator an a pair of brushes on it is rotated at a speed N_r in a rotated magnetic field having P poles and rotating at a speed N_r in space, N_f and N_r both being in the same direction, the frequency of the induced voltage across the brushes on the commutator is

(*a*) $\frac{N_f P}{120}$ (*b*) $\frac{(N_f - N_r)P}{120}$

(*c*) $\frac{(N_f + N_r)P}{120}$ (*d*) $\frac{N_r P}{120}$

129. A 2 KVA transformer has iron loss of 150 Watts and full-load copper loss of 250 Watts. The maximum efficiency of the transformer would occur when the total loss is

(*a*) 500 W

(*b*) 400 W

(*c*) 250 W

(*d*) 320 W

130. A starting torque of 80 Am is developed in an induction motor by an auto-transformer starter with a tapping of 30%. If the tapping of auto transformers starter is 60% then the starting torque will be

(*a*) 40 Nm (*b*) 160 Nm

(*c*) 240 Nm (*d*) 320 Nm

131. A pair of similar *dc* shunt generator operate in parallel and supply a common load. It is required to switch off Machine No. 1 and allow machine No.2 to supply the entire load. The following operations are to be used to achieve this:

1. Switch off the main switch of machine No. 1.
2. Reduce the field current of machine No. 1.
3. Increase the field current of machine No 2.
4. Ensure that machine No. 1 just floats

The correct sequence of these operations is

(*a*) 4, 3, 2, 1

(*b*) 2, 3, 4, 1

(*c*) 2, 4, 3, 1

(*d*) 3, 2, 4, 1

132. Consider the following statements about commutating poles which are fitted on most large *dc* shunt motors:

1. The commutating poles are placed in the geometric neutral planes and their number is usually equal to the number of main poles.
2. The winding on the commutating poles is connected in series with the shunt field winding on the main poles.
3. The polarity of the commutating pole must be that of the next main pole further ahead.
4. The commutating poles neutralize the reactance voltage produced in the coil undergoing commutation.

Of these statements

(*a*) 1, 2 and 3 are correct (*b*) 1 and 4 are correct

(*c*) 2, 3 and 4 are correct (*d*) 1, 2 and 4 are correct

133. A 3-phase synchronous motor connected to an infinite bus is operating at half full-load with normal excitation. When the load on the synchronous motor is suddenly increased,

(*a*) its speed will first decrease and then become synchronous

(*b*) its speed will first increase and then become synchronous

(*c*) its speed will fluctuate around synchronous speed and then become synchronous

(*d*) none of these

134. The total electromagnetic torque developed by a motor may be expressed in a general form as

$$T_e = \frac{A.d^2\theta}{dt^2} + B\frac{d\theta}{dt} + C$$

Which of the following are represented by the first, second and the third terms on the right-hand sides of this equation?

1. Accelerating torque.
2. Damping torque.
3. Friction torque.
4. Load torque.

Select the correct combination from the following codes:

Codes:

	First Term $\left(\frac{A.d^2\theta}{dt^2}\right)$	Second Term $\left(B.\frac{d\theta}{dt}\right)$	Third Term (C)
(*a*)	3	2	1 and 4
(*b*)	3	4	1 and 2
(*c*)	1	4	2 and 4
(*d*)	1	2	3 and 4

135. In an induction motor, what is the ratio of rotor copper loss and rotor input

(*a*) 1/S (*b*) S
(*c*) 1 – S (*d*) S/(1 – S)

136. A *dc* shunt generator is running at rated speed developing its rated voltage. Match List-I (condition of operation) with List-II (effect) and select the correct answer using the codes given below the lists:

List-I (*Condition of operation*)

A. The direction of rotation, residual magnetism and field connections are reversed
B. The direction of residual magnetism and field connections are reversed
C. The direction of rotation and connections of the field winding are reversed
D. The direction of rotation and residual magnetism are reversed

List-II (*Effect*)

1. The generator will build up with same polarity
2. The generator will build up with reversed polarity.
3. The generator will not build up.

Codes:

	A	B	C	D
(*a*)	1	3	2	2
(*b*)	3	2	1	2
(*c*)	3	2	2	1
(*d*)	2	3	1	1

137. In the case of an armature controlled *dc* motor, if the equivalent moment of inertial of the motor and load is made small and the armature resistance is reduced, then the motor tends to act like

(*a*) a differentiating circuit
(*b*) an integrating circuit
(*c*) a PID controller
(*d*) a second order critically damped system

138. A *dc* shunt motor is required to drive a constant power load at rated speed while drawing rated armature current. Neglecting saturation and all machine losses, if both the terminal voltage and the field current of the machine are halved, then

(*a*) Small
(*a*) the speed becomes 2 pu (per unit) but armature current remains at 1 pu
(*b*) the speed remains at 1 pu but armature current becomes 2 pu
(*c*) both speed and armature current becomes 2 pu
(*d*) both speed and armature current remain at 1 pu

139. The torque vs armature current characteristics (A, B, C and D) of different types of *dc* motors are shown in given diagram.

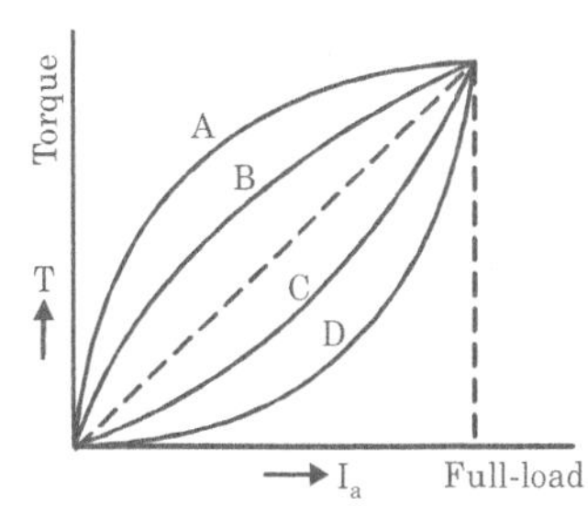

Match the curves A, B, C, and D with the *dc* motors given in List-I and select the correct answer using the codes given below the list:

List-I

A. Shunt
B Cumulative compound
C Series
D Differential compound

Codes:

	A	B	C	D
(*a*)	4	2	1	3
(*b*)	4	1	2	3
(*c*)	3	1	2	4
(*d*)	3	2	1	4

140. The capacitor-start single-phase induction motors develop much larger starting torque in comparison with the "split-phase" motors, because the use of capacitors in the auxiliary winding enables

(*a*) provision of larger number of turns in the auxiliary winding

(*b*) a larger starting current to be drawn from the supply

(*c*) the torque-slip characteristic in general to get a shape to give a large starting torque

(*d*) the current division between the main and auxiliary windings to be conductive for the development of large starting torque

141. The field winding of a 3-phase synchronous motor is short-circuited directly. If a 3-phase balanced voltage is impressed across the stator terminals, then the rotor will

(*a*) rotate at synchronous speed

(*b*) rotate at a speed slightly less than synchronous speed

(*c*) rotated at a speed very near to half the synchronous speed

(*d*) not rotate at all

142. An electric motor with "constant output power" will have a torque-speed characteristic in the form of a

(*a*) straight line through the origin

(*b*) straight line parallel to the speed axis

(*c*) circle about the origin

(*d*) rectangular hyperbola

143. For successful parallel operation of two single phase transformers the essential condition is that their

(*a*) percentage impedances should be equal

(*b*) turns ratios should be exactly equal

(*c*) polarities must be properly connected

(*d*) KVA rating should be equal

144. Which is the common method of cooling a power transformer

(*a*) Air cooling (*b*) Air-blast cooling

(*c*) Oil-cooling (*d*) Natural cooling

145. Good voltage regulation of a transformer means that the difference between its no load voltage and full load voltage is

(*a*) very high (*b*) low

(*c*) very low (*d*) none of these

146. Leakage flux in a transformer depends upon

(*a*) the applied voltage

(*b*) the frequency

(*c*) the load current

(*d*) the mutual flux

147. Two 3-phase *ac* generators are such that one has twice the linear dimensions of the other. The field windings of each are excited to give identical sinusoidal air-gap flux density waveform. Both have same number of stator slots and identical winding patterns. The conductor/slot in big generator is K times that of the smaller one. The value of K to get equal no-load voltage at same frequency is

(*a*) 8 (*b*) 4

(*c*) $\frac{1}{2}$ (*d*) $\frac{1}{4}$

148. A single-phase transformers fed from a rated voltage supply operates at no-load. The core loss of the transformer is assumed to be negligible. The current drawn by the transformer from the supply will have the waveform

(*a*)
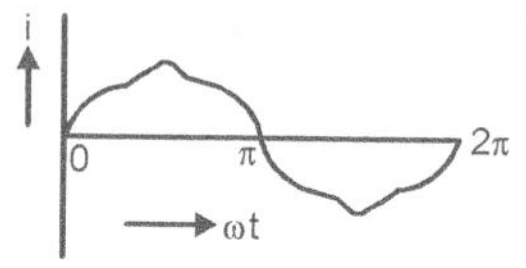

(*b*)
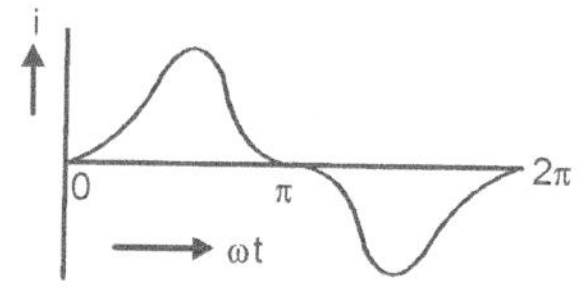

(*c*)
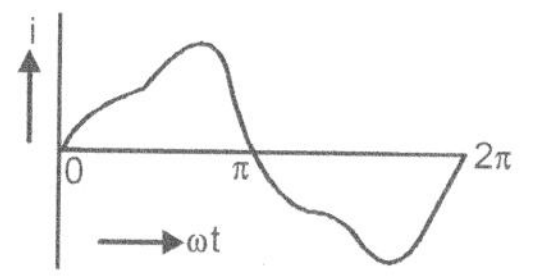

(*d*)
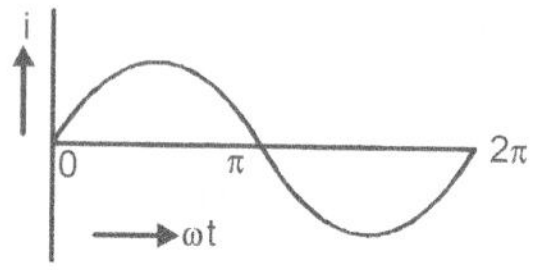

149. The phase angle error in *ac* tachometer generators is entirely removed at any specified slip by introduced a resistor in

(*a*) parallel with the reference winding

(*b*) series with the reference winding

(*c*) series with the output winding

(*d*) parallel with the output winding

150. For computing the value of *d*-axis transient, reactance of a salient pole synchronous machine, the appropriate approximate equivalent circuit is given in where

X_a =Armature leakage reactance

X_{ad} =Factious equivalent to the effect of armature reaction in *d*-axis

X_{fd} =Field winding reactance in *d*-axis

X_{Dd} =Damper winding reactance in *d*-axis

(*a*) 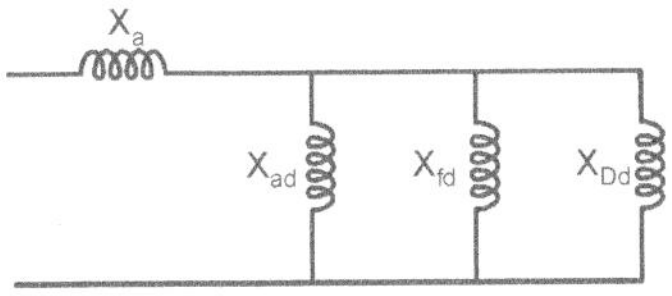

(*b*) 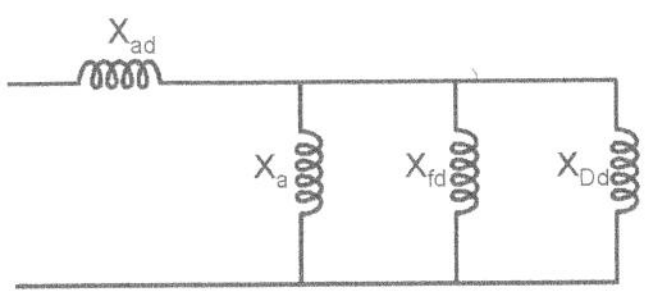

(*c*) 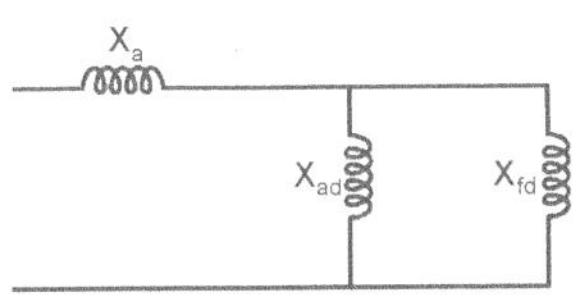

(*d*) 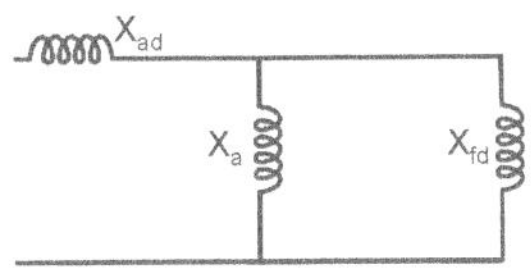

151. A 3-phase synchronous generator, with its armature resistance and the leakage reactance being neglected, is synchronized to an infinite bus and its field excitation is kept constant thereafter. Now the machine is loaded by supplying mechanical input to the shaft so that the load-angle δ reaches value of 60°. Under this conditions, the operating power-factor would be

(*a*) 0.866 leading

(*b*) 0.866 lagging

(*c*) 0.5 leading

(*d*) 0.5 lagging

152. Three single-phase transformers are connected to form a 3-phase transformer bank. The transformers are connected in the following manner:

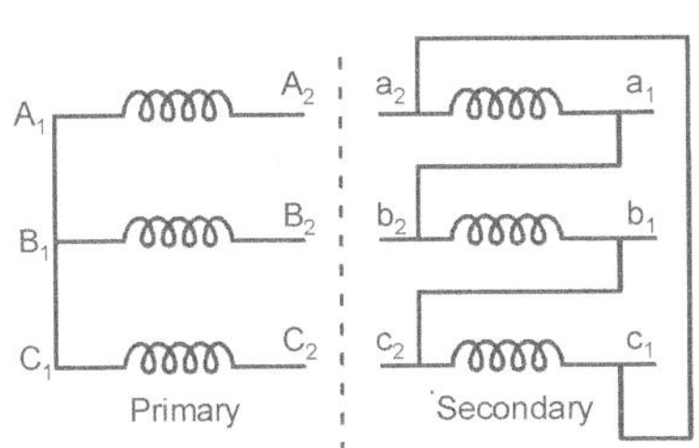

The nomenclature of the above connection for phase sequence ABC is

(*a*) Yd0

(*b*) Yd6

(*c*) Yd11

(*d*) Yd1

153. Consider the following statements:

If a 3-phase squirrel cage induction machine operates at a slip of – 0.05 (i.e minus 0.05), the machine will

1. draw electrical power from the mains.

2. draw mechanical power through the shaft.

3. deliver electrical power to the mains.

Of these statements

(*a*) 1, 2 and 3 are correct

(*b*) 1 and 2 are correct

(*c*) 2 and 3 are correct

(*d*) 1 and 3 are correct

154. The torque speed characteristic of a repulsion motor resembles which of the following *dc* motor characteristic?

(*a*) Separately excited (*b*) Shunt

(*c*) Series (*d*) Compound

155. Which one of the following statements regarding a transformer is correct?

(*a*) In transformers, a laminated core is used to reduce copper losses

(*b*) Core losses in a transformer can be determined by open-circuit test

(*c*) Short-circuit test is conducted on a transformer to determine constant losses

(*d*) In a shall type transformer, the primary and secondary winding are wound on separate limbs [ES]

156. Typical characteristics of a *dc* series motor, operating under loaded condition, are shown in the following figure List-I gives the identification of these characteristics, while List-II gives the description of important load characteristics of *dc* motors working on load.

List-I (*See figure curves labelled*)

A. A

B. B

C. C

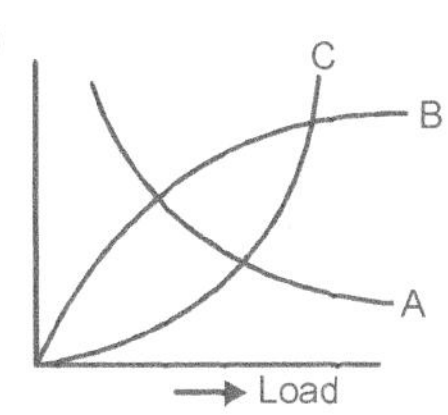

List-II

1. Efficiency Vs. load.
2. Torque Vs. load.
3. Current Vs. load.
4. Speed Vs. load.

Match List-I with List-II and select the correct answer using the codes given below:

Codes:

	A	B	C
(*a*)	4	2	3
(*b*)	2	3	1
(*c*)	3	2	1
(*d*)	4	1	2

157. What is the approximate efficiency of a normal thermal power station ?

(*a*) 30-40% (*b*) 45-55%

(*c*) 20-25% (*d*) 60-70%

158. Consider the principal that the torque in rotating machine is proportional to peak of stator Mmf(F_s) rotor Mmf(F_r) and the angle between them (δ). Following statements relate to F_s, F_r and δ in different machines:

1. δ is fixed in *dc* machine.
2. F_s is fixed in *dc* shunt machine.
3. δ is variable in induction machine.
4. F_r is fixed in synchronous machine.
5. F_r is variable in *dc* machine.
6. F_r is fixed in induction motor.
7. Resultant to F_s and F_r is fixed in synchronous motors.

Of these statements:

(*a*) 1, 2, 3, 4 and 6 are correct

(*b*) 1, 2, 3, 5 and 7 are correct

(*c*) 2, 3, 5, 6 and 7 are correct

(*d*) 1, 2, 4, 5 and 7 are correct

159. Consider the following statements :

A ceiling fan fails to start because

1. Main winding is open-circuited.

2. auxiliary winding is open-circuited.

3. capacitor is short-circuited.

4. supply terminals are reversed.

5. main and auxiliary winding connections are interchanged.

Which of these statements are correct ?

(*a*) 1, 2 and 5 (*b*) 2, 3, 4 and 5

(*c*) 1, 2, and 3 (*d*) 1, 3, and 4

160. The calculate the full-load voltage regulation of synchronous generator at any power factor using mmf method, it is required to have

(*a*) open-circuit and short-circuit characteristics

(*b*) open-circuits and zero power factor characteristics

(*c*) open-circuit characteristics, field current at rated short-circuit and armature resistance

(*d*) short-circuit characteristics, zero power factor characteristics and armature resistance

161. What is the angle between the induced voltage and supply voltage and supply and supply voltage of a synchronous motor under running condition?

(*a*) Zero

(*b*) Greater than zero but ≤ 90°

(*c*) Between 90° and 180°

(*d*) > 180°

162. Two single phase transformers with equal turns have impedance of $(0.5 + j3)$ ohms and $(0.6 + j10)$ ohms with respect to the secondary. If they operate in parallel, how will they share a load of 100 kw at 0.8 p.f. lagging?

(*a*) 50 kw, 50 kw

(*b*) 62 kw, 38 kw

(*c*) 78.2 kw, 21.8 kw

(*d*) 85.5 kw, 14.5 kw

163. Which one of the following figures represents the correct directions of flow of current in the different section of the windings of an auto transformer?

(*a*)

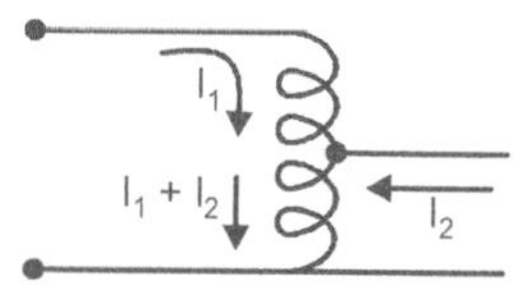

(*b*)

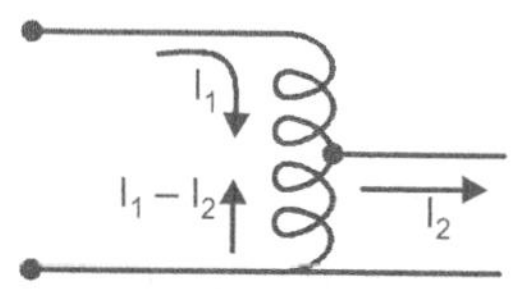

(*c*)

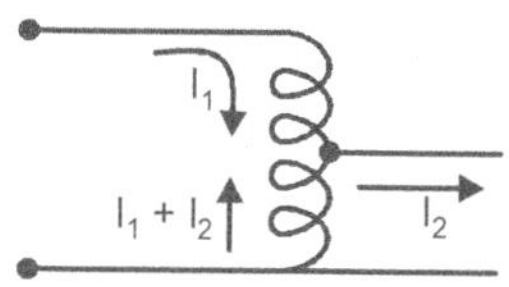

(*d*)

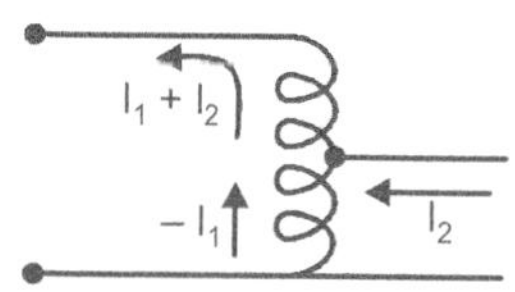

164. Hysteresis loss of a transformer designed at voltage 'V' is P_h. If the transformer is operated from a voltage '*x*V' and frequency '*xf*' the hysteresis loss for a Steinmetz constant of 2 will be

(*a*) x^2P_h (*b*) xP_h

(*c*) P_h (*d*) x^3P_h

165. In the slew range of a stepper motor, it

(*a*) can start, stop and reverse as desired

(*b*) cannot start but can stop and reverse on command

(*c*) cannot start, stop and reverse on command

(*d*) cannot start and synchronize but can reverse on command

166. A single-phase transformer has rating of 15 kVA, 600/120 V. It is recommended as an auto transformer to supply at 720 V primary source. The maximum load it can supply is

(*a*) 90 kVA (*b*) 18 kVA

(*c*) 15 kVA (*d*) 12 kVA

167. A 3-phase induction motor develops a torque 'T' when driven from a balanced 3-phase supply. The terminal voltage and frequency are halved so that the air-gap flux remains the same while the load torque is kept constant. Then the slip

(*a*) is reduced to half

(*b*) remains same as the previous value

(*c*) speed is reduced to half

(*d*) speed remains the same

168. A 4-pole, 3 phase induction motor is running at 4 percent slip at full load. If the speed of the motor is 720 rpm, the supply frequency is

(*a*) $16\frac{2}{3}$ Hz (*b*) 25 Hz

(*c*) 50 Hz (*d*) 60 Hz

169. The relationship between rotor frequency f_2, slip s and the stator supply frequency f_1 is given by

(*a*) $f_1 = sf_2$ (*b*) $f_2 = sf_1$

(*c*) $f_2 = f_1(1-s)$ (*d*) $f_1 = \sqrt{sf}$

170. A 3-phase, 400 V, 4-pole induction motor is fed from a 3-phase, 400 V, 50 Hz supply and runs at 1440 rpm the frequency of rotor emf is

(*a*) 2 Hz (*b*) 50 Hz

(*c*) 48 Hz (*d*) zero Hz

171. The rotor of a 4-polc, 50 Hz, 3-phase slipring induction motor runs in clockwise direction when its stator terminals 1, 2, 3 are connected to supply terminals A, B, C respectively. If 1, 2, 3 are connected to A, C, B respectively of supply terminals and rotor runs in clockwise direction of synchronous speed, then the frequency of induced emf across the open-circuited rotor terminals is

(*a*) 50 Hz (*b*) zero

(*c*) 25 Hz (*d*) 100 Hz

172. For a machine on infinite bus active power can be varies by

(*a*) changing field excitation

(*b*) changing of prime-mover speed

(*c*) both (*a*) and (*b*) above

(*d*) none of these

173. If the prime mover of an alternator supplying load to an infinite bus is suddenly shut down then it will

(*a*) stop

(*b*) continue to run as an alternator

(*c*) continue to run as a synchronous motor in the reverse direction.

(*d*) continue to run as a synchronous motor is the same direction.

174. An induction motor works with

(*a*) dc only (*b*) ac only

(*c*) ac and dc both (*d*) none of these

175. An induction motor has synchronous speed of 1500 rpm. What will be slip when it is running on a speed of 1450 rpm?

(*a*) 3.33 % (*b*) −3.33 %

(*c*) +3 % (*d*) −3 %

176. The full load slip of 960 Hz, 12 pole squirrel cage induction motors is 5%. Its full load speed is

(*a*) 600 rpm (*b*) 570 rpm

(*c*) 500 rpm (*d*) 475 rpm

177. Drop in terminal voltage of an alternator due to armature reaction is countered by

(*a*) damper winding

(*b*) effect of saliency

(*c*) increased prime-mover output

(*d*) automatic voltage regulator

178. Synchronous reactance of alternator represents

(*a*) armature reactance and leakage reactance

(*b*) a reactance operating at synchronous speed

(*c*) field winding reactance

(*d*) a reactance connected in series with a synchronous generator

179. In a synchronous generator delivering lagging power factor load

(*a*) the excitation emf leads terminal voltage by the power angle

(*b*) the excitation emf lags the terminal voltage by the power angle

(*c*) excitation emf leads the terminal voltage by the power factor angle

(*d*) none of these

180. A leading *pf* load on an alternator implies that its voltage regulation shall be

(*a*) positive (*b*) negative

(*c*) zero (*d*) any one of these

181. Squirrel-cage bars placed in the rotor pole faces of an alternator help reducing hunting

(*a*) above synchronous speed only

(*b*) below synchronous speed only

(*c*) above and below synchronous speed both

(*d*) none of these

182. The maximum possible speed at which an alternator can be driven generate 50 Hz and 4000 V is

(*a*) 4000 rpm (*b*) 3600 rpm

(*c*) 3000 rpm (*d*) 1500 rpm

183. Skew of rotor bar eliminates
(*a*) the effect of space harmonics
(*b*) the entire effect of crawling
(*c*) magnetic voice
(*d*) vibration due to unequal force developed on rotor

184. To eliminate rth harmonic from the induced emf in a phase of a synchronous machine the pitch coils must be
(*a*) $\frac{r-1}{r}$ (*b*) $\frac{2r-1}{r}$
(*c*) $\frac{r}{r+1}$ (*d*) $\frac{2r}{r+1}$

185. In which transformer, the tertiary winding is used?
(*a*) star-delta (*b*) delta-delta
(*c*) star-star (*d*) delta-star

186. Two transformers are operating in parallel their load sharing will depend upon
(*a*) cooling methods
(*b*) all the efficiency
(*c*) maximum efficiency
(*d*) per unit impedance

187. Tappings of a transformer are provided
(*a*) at the phase and of *hv* side
(*b*) at the phase end of *hv* side
(*c*) at the neutral end of *hv* side
(*d*) at the middle of *hv* side

188. A 100/5 CT is used in conjuctions with a 5 A ammeter. If the latter read 3.5 A the line current will be
(*a*) 350 A
(*b*) 175 A
(*c*) 120 A
(*d*) 70 A

189. In operating a 400 V transformer of 50 Hz
(*a*) only voltage is reduced in the same proportion as the frequency
(*b*) only kVA rating is reduced in the same proportion as the frequency
(*c*) both voltage and kVA rating are reduced in the same proportions as the frequency
(*d*) none of these

190. A shell type transformer has
(*a*) high eddy current loss
(*b*) reduced magnetic leakage
(*c*) negligible hysteresis loss
(*d*) none of these

191. Distribution transformers have core losses
(*a*) $>$ copper loss
(*b*) $<$ copper losses
(*c*) = copper loss
(*d*) $= \frac{1}{2}$(copper loss)

192. Addition of tubes to the transformer tank improves heat dissipation capacity because of
(*a*) additional cooling surface
(*b*) additional dissipation by radiation only
(*c*) additional dissipation by convection only
(*d*) additional dissipation by radiation and convection both

193. A 3-phase has its primary connected in delta and secondary in star. Secondary to primary turn ratio per phase is 5. For a primary voltage of 400 V, the secondary voltage would be
(*a*) 200 V
(*b*) 80 V
(*c*) 3464 V
(*d*) $80\sqrt{3}$ V

194. In a properly connected Y-Δ transformer, a voltmeter connected in the secondary (opening a none of delta), the voltmeter will read
(*a*) zero voltage
(*b*) line voltage of supply
(*c*) phase voltage in the secondary
(*d*) none of these

195. In the transformer the primary and secondary windings are interlaced for
(*a*) easiness of coil making
(*b*) reduced leakage reactance
(*c*) reduced cost
(*d*) uniform heating

196. Under short circuit condition, the winding of a transformer having 5% impedance will experience a radial force
(*a*) 5 times of full load value
(*b*) 20 times of full load value
(*c*) 25 times of full load value
(*d*) 400 times of full load value

197. Multi-step core is used in a transformer to
(*a*) increase the output
(*b*) decrease the cost of core material
(*c*) decrease the cost of copper
(*d*) increases the efficiency

198. Which is the arrangement of winding in a core type single phase transformer ?

(*a*) Half *lv* inside and half *hv* out side on each core limb

(*b*) *lv* on one core limb and *hv* on the other

(*c*) Sand wiched *lv* and *hv* discs on each core limb

(*d*) Half *hv* inside and half *hv* outside on each core limb

199. The iron loss in a 100 kVA transformer is 1 kw and the full copper losses are 2 kw then maximum efficiency occures at a load of

(*a*) 141.4 kVA (*b*) 70.7 kVA

(*c*) 50 kVA (*d*) 70.7 kW

200. In order to reduce the hysteresis loss

(*a*) core may be laminated

(*b*) silicon steel may be used as the core material

(*c*) core may be constructed with any permanent magnet material such as Alxico

(*d*) core may be impregnated with varnish

201. Grain oriented laminated sheet steel in a transformer reduces

(*a*) copper loss (*b*) eddy current loss

(*c*) hysteresis loss (*d*) none of these

202. Silicon content in iron laminations is kept within 5% as it

(*a*) increases hysteresis loss

(*b*) increases cost

(*c*) makes the material brittle

(*d*) reduces the curie point

203. Consider the following statements:

As a 3-phase induction motor is loaded from no-load to rated load

1. There is an improvement in the power factor.

2. The torque increases almost in proportion to slip.

3. The air-gap flux falls sharply.

(*a*) 1, 2 and 3 are correct

(*b*) 1 and 2 are correct

(*c*) 2 and 3 are correct

(*d*) 1 and 3 are correct

204. A shunt motor is running at its rated speed as rated load. If the field circuit gets suddenly opened

(*a*) torque developed by the motor will be almost zero and the motor will stop without causing any damage to the motor

(*b*) the motor will draw very high armature current but it will continues to drive the load at a reduced speed

(*c*) the motor would tend to race to a high speed while driving the load

(*d*) the motor would draw very high armature current and will stall with consequent damage to the armature

205. Consider the following statements about the utility of delta-connected tertiary windings in star-star connected power transformer:

1. It makes supply available for small loads.
2. It provides low reactance paths for zero sequence currents.
3. It is used to suppress harmonic voltages.
4. It is used to allow flow of earth fault current for operation of protective relays.

Which of these statements are correct?

(*a*) 1 and 2 (*b*) 2, 3 and 4

(*c*) 1, 3 and 4 (*d*) 1, 2, 3 and 4

206. If a 3-phasor induction motor and a 1-phase induction motor of the same kW rating and same winding voltage are compared, then

(*a*) both will be of same size and efficiency

(*b*) the 3-phase induction motor will be bigger in size and less efficient

(*c*) the 3-phase induction motor will be smaller in size and more efficient

(*d*) the 3-phase induction motor will be bigger in size and more efficient

207. In a 3-phase induction motor, the rotor impedance angle at standstill is θ. The load angle δ will be

(*a*) $\pi + \theta$ (*b*) $\pi - \theta$

(*c*) $\frac{\pi}{2} + \theta$ (*d*) $\frac{\pi}{2}$

208. Open-circuit (*oc*) and short-circuit (*sc*) tests on an AC generator are conducted which of the following conditions?

1. *oc* test at nominal flux and *sc* test at nominal current
2. *oc* test at reduced flux and *sc* test at reduced current.
3. *oc* test at zero armature current and *sc* test at reduced flux.
4. *oc* test at zero current and *sc* test at nominal flux.

Select the correct answer using the codes given below:

(*a*) 1 and 3 (*b*) 2 and 3

(*c*) 1 and 4 (*d*) 2 and 4

209. A 2-pole series motor with its two fields coils connected in series run as at a speed of 500 rpm. If the field coils are reconnected in parallel and assuming that the torque is constant and the magnetic circuit is unsaturated, the new speed will be

(*a*) 250 rpm (*b*) $\frac{500}{\sqrt{2}}$ rpm

(*c*) $500\sqrt{2}$ rpm (*d*) 1000 rpm

210. Match List-I (Types of motor) with List-II (Applications) and select the correct answer using the codes given below the Lists:

List-I	**List-II**
(*Types of motor*)	(*Applications*)
A. Shaded pole motor	1. Robotics
B. Capacitor	2. Rockets
C Permanent magnet motor	3. Ceiling fan
D. Stepper motor	4. Tale fan

Codes:

	A	B	C	D
(*a*)	4	3	1	2
(*b*)	3	4	2	1
(*c*)	3	4	1	2
(*d*)	4	3	2	1

211. If the load of a synchronous motor increases while keeping the field excitation constant, then
(*a*) power factor will increase and power angle will decrease
(*b*) power factor will decrease and power angle will increase
(*c*) both power factor and power angle will increase
(*d*) both power factor and power angle will decrease

212. A turbo alternator set feeds power to a 3-phase constant-voltage, constant frequency bus. If the steam supply to the set is cut off, then the set will
(*a*) continue to run at rated speed in the same direction
(*b*) continue to run at a reduced speed in the same direction
(*c*) run at rated speed in the reverse direction
(*d*) come to a stop

213. A transformer does not transform
(*a*) power
(*b*) voltage
(*c*) current
(*d*) impedance

214. Efficiency of a power transformer is around
(*a*) 50% (*b*) 60%
(*c*) 80% (*d*) 95%

215. Transformer action requires a
(*a*) constant magnetic flux
(*b*) increasing magnetic flux
(*c*) alternating magnetic flux
(*d*) alternating electric flux

216. A transformer steps up the voltage by a factor 100. The ratio of current in the primary to that in the secondary is
(*a*) 1 (*b*) 100
(*c*) 0.01 (*d*) 0.1

217. In a transformer operating at constant voltage, if the input frequency increases, the core loss
(*a*) increases
(*b*) decreases
(*c*) remain constant
(*d*) increases as square of the frequency

218. For a particular motor, the cooling time constant is usually
(*a*) smaller than the heating time constant
(*b*) greater than the heating time constant
(*c*) equal to the heating time constant

219. For estimation of power rating of an electric drive, its losses can be considered to be proportional to
(*a*) power
(*b*) $(\text{power})^2$
(*c*) $(\text{power})^3$
(*d*) torque

220. In rotating electrical machines, the insulation temperature limit for class B type is
(*a*) 105° (*b*) 130°
(*c*) 150° (*d*) 180°

221. The basic function of a transformer is to change
(*a*) the level of the voltage
(*b*) the power level
(*c*) the power factor
(*d*) the frequency

222. The frequency of a voltage at the secondary of the transformer is
(*a*) greater than that of primary
(*b*) equal to that of primary
(*c*) less than that of primary
(*d*) none of these

223. Assume that the magnetisation curve, the rated values of voltage, armature current, speed and torque are same for the following motors:
A. series motor
B. shunt motor
C. cumulative compound motor
D. differential compound motor
Match the type of motor with corresponding speed-current characteristics shown in the figure and select the correct answer using the codes given below:

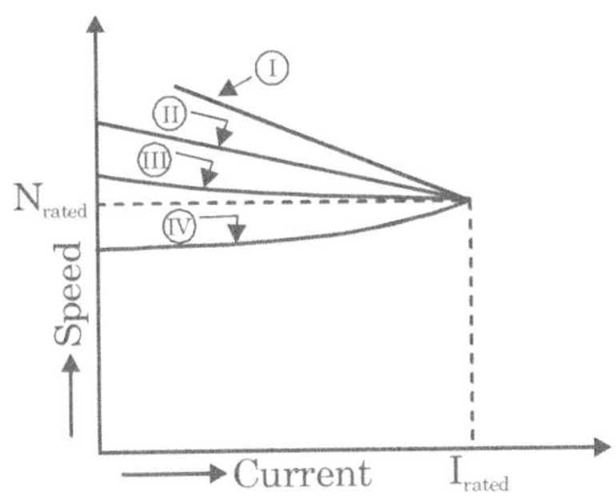

Codes:

	A	B	C	D
(*a*)	II	III	I	IV
(*b*)	I	III	II	IV
(*c*)	I	IV	II	III
(*d*)	II	IV	I	III

224. A dc series motor fed from rated supply voltage is over loaded and its magnetic circuit is saturated. The torque-speed characteristic of this motor will be approximately represented by which curve of Fig. R 1.7 ?

(*a*) curve A (*b*) curve B
(*c*) curve C (*d*) curve D

225. What does the SCR (short circuit ratio) of a synchronous machine field ?

(*a*) $\dfrac{1}{X_s\,(\text{unsaturated})\,\text{pu}}$

(*b*) $\dfrac{1}{X_s\,(\text{unsaturated})\,\text{in ohm}}$

(*c*) $\dfrac{1}{X_s\,(\text{adjusted})\,\text{pu}}$

(*d*) By changing the prime mover speed

226. In a 3-phase power transformer, 5-limbed construction is adopted to

(*a*) suppress the 5th and 7th harmonics
(*b*) suppress the 3rd and its multiple harmonics
(*c*) suppress all the orders of harmonics
(*d*) increase the capacity of the transformer

227. In ideal synchronous motor has no starting torque because the

(*a*) rotor is made up of salient poles
(*b*) relative velocity between the stator and the rotor mmfs is zero
(*c*) relative velocity between stator and rotor mmfs is not zero
(*d*) rotor winding is highly reactive

228. Following a sudden short-circuit at the terminals of a 3-phase unloaded synchronous generator, the initial effect of the pole-face damper windings, is to

(*a*) establish the armature flux through the direct-axis magnetic circuit of the machine
(*b*) allow only partial linkage of the armature flux with the main field winding
(*c*) confine the armature flux to completely link the damper winding
(*d*) repel the armature flux and confine it to the leakage flux path in the air-gap

229. Consider the following statements about single-phase reluctance motor:

1. The starting torque is a function of rotor position.
2. The torque developed at subsynch-ronous speeds varies sinusoidally.
3. Starting is asynchronous but running is synchronous.
4. It pulls synchronism on switching on *dc* excitation.

Of these statements:

(*a*) 1, 2 and 3 are correct
(*b*) 1, 3 and 4 are correct
(*c*) 2, 3 and 4 are correct
(*d*) 1 and 4 are correct

230. A *dc* shunt generator, when driven at its rated speed, is found to be not generating any voltage. Which of the following would account for this?

1. There is no residual magnetism.
2. The connection of the field winding is not proper with respect to the armature terminal.
3. The resistance of the field circuit is greater than the critical field resistance.
4. The load resistance is less than the critical armature resistance.

Codes:

(*a*) 3 and 4 (*b*) 1, 2 and 4
(*c*) 1, 2 and 3 (*d*) 1, 2, 3 and 4

231. Which of the points on the torque-speed curve of induction motor represents operation at a slip greater than 1 ?

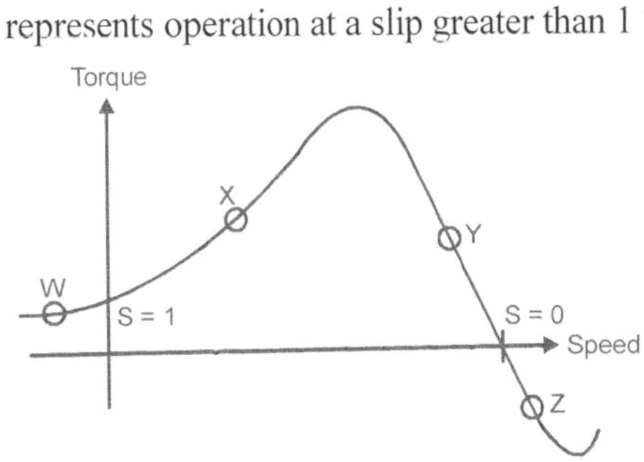

(*a*) W (*b*) X
(*c*) Y (*d*) Z

232. A 6-pole lap-connected *dc* generator with 480 conductors has an armature resistance of 0.06 ohm. If the conductors are reconnected to form a wave winding, other things remaining unchanged, the value of the armature resistance will be

(*a*) 0.01 ohm (*b*) 0.06 ohm
(*c*) 0.36 ohm (*d*) 0.54 ohm

233. Consider the following statements:

Modern alternators are designed for large air gap to have

1. Large inherent regulation
2. Very stable parallel operation.
3. Sinusoidal mmf distribution

Which of these statements are correct?

(*a*) 1, 2 and 3
(*b*) 2, 3 and 4
(*c*) 1, 3 and 4
(*d*) 1, 2 and 4

234. Two generators rated at 200 MW and 400 MW are operating in parallel. Their governor droop characteristic are respectively 4% and 5% from no-load. At no-load, the system frequency is 50 Hz When supplying a load of 600 MW, the system frequency will be

(*a*) 50 Hz (*b*) 49 Hz
(*c*) 516 Hz (*d*) 47.7 Hz

235. Consider the following statements:

The use of Delta-connected tertiary windings in star-star connected power transformers

1. Makes available supply for single-phase loads.
2. Suppresses harmonic voltage.
3. Allows flow of earth fault current for operation of protective devices.
4. Provides low-reactance paths for zero-sequence currents.

Of the statements:

(*a*) 1, 2 and 3 are correct
(*b*) 1, 3 and 4 are correct
(*c*) 2 and 4 are correct
(*d*) 1, 2, 3 and 4 are correct

236. For the system shown in given figure, the phase relation of current 'I' with respect to the voltage V_{AB} is

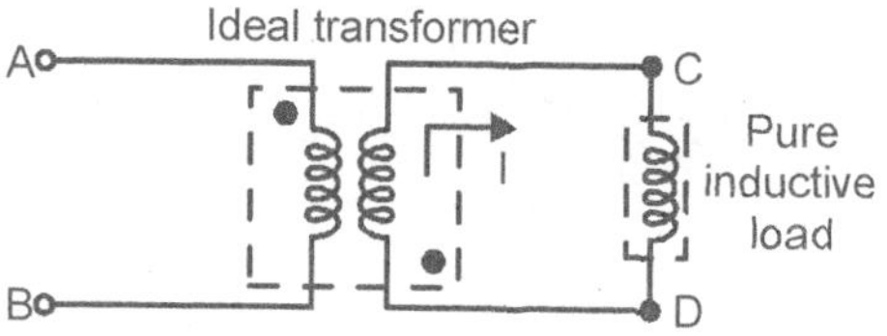

(*a*) zero (*b*) 90° lead
(*c*) 90° lag (*d*) 180°

237. Which of the following will change in a three-phase synchronous motor, as a consequence of excitation variation?

1. Pull-out torque.
2. Torque angle.
3. Magnitude and power factor of stator current
4. Output power.

Select the correct answer using the codes given below:

(*a*) 1 and 3 (*b*) 2 and 4
(*c*) 2, 3 and 4 (*d*) 1, 2 and 3

238. For a rotor-fed 3-phase induction motor, the absolute speed of the rotation magnetic field in space is

(*a*) the synchronous speed
(*b*) the difference between the synchronous speed and the rotor speed
(*c*) the rotor speed
(*d*) zero

239. A 3-phase delta connected squirrel cage induction motor has a starting current of I_Δ and a starting torque of T_Δ at rated voltage. If the starting current and starting torque while the motor is started through star-delta starter and auto-transformer (with 60% voltage) starter alternatively, are I_Y and T_Y and T_{auto} and T_{auto} respectively, then is

$$\frac{I_Y}{I_\Delta}:\frac{I_{auto}}{I_\Delta}:\frac{T_Y}{T_\Delta}:\frac{T_{auto}}{T_\Delta}$$

equal to

(*a*) $\frac{1}{\sqrt{3}}:0.6:\frac{1}{\sqrt{3}}:0.6$ (*b*) $\frac{1}{3}:0.6:\frac{1}{3}:0.36$

(*c*) $\frac{1}{3}:0.36:\frac{1}{\sqrt{3}}:0.6$ (*d*) $\frac{1}{\sqrt{3}}:0.36:\frac{1}{3}:0.36$

240. When a single-phase capacitor start induction motor is running at a steady speed delivering a fixed torque, then

(*a*) peak of forward rotor mmf is equal to the peak of backward rotor mmf
(*b*) peak of forward stator mmf is equal to the peak of backward stator mmf
(*c*) net forward rotating flux is equal to the net backward rotating flux
(*d*) forward flux produced by rotor current is equal to backward flux produced by rotor current

241. A 200 V dc shunt motor delivers and output of 17 kW with an input of 20 kW, The field winding resistance is 50 Ω and armature resistance is 0.04 Ω. Maximum efficiency will be obtained when the total armature copper losses are equal to

(*a*) 2632 W (*b*) 3000 W
(*c*) 3680 W (*d*) 5232 W

242. A 3-phase Y/Δ transformer has per phase turn ratio 'K' and line voltage ratio 'K_1'. If the voltage of D winding lags the voltage of Y winding by an angle ϕ, then K, K_1 and ϕ are related as

(*a*) $K_1 = K$ and $\phi = -30°$

(*b*) $K_1 = \frac{K}{\sqrt{3}}$ and $\phi = -30^0$

(*c*) $K_1 = \sqrt{3}\,K$ and $\phi = -30°$

(*d*) $K_1 = 3K$ and $\phi = -30°$

243. In a word- Leonard drive, if the field current of the *dc* generator is suddenly reduced, the controlled *dc* motor operates

(*a*) still in the motoring mode

(*b*) in dynamic braking mode

(*c*) in counter-current braking mode

(*d*) in regenerative braking mode

244. A *dc* shunt motor running lightly at 1000 *rpm* is operated under plugging with plugging connections left as it is the final speed of the motor will be

(*a*) zero (*b*) 1000 *rpm*

(*c*) – 1000 *rpm* (*d*) – 2000 *rpm*

245. Dynamic braking is very effective if the *dc* motor

(*a*) is series excited

(*b*) is shunt excited

(*c*) is separately excited

(*d*) has cumulative compound excitation

246. For non-reversing *dc* drives it is preferable to use

(*a*) plugging

(*b*) regenerating braking

(*c*) dynamic braking with sperate excitation

(*d*) dynamic braking with self excitation

247. The value of *dc* lead voltage for a single phase semiconductor under current mode is

(*a*) $\frac{V_m}{\pi}(1 + \cos\alpha)$ (*b*) $\frac{V_m}{\pi}(1 - \cos\alpha)$

(*c*) $\frac{V_m}{2\pi}(1 + \cos\alpha)$ (*d*) $\frac{V_m}{2\pi}(1 - \cos\alpha)$

248. Running the machine at no load is inadvisable for

(*a*) *dc* shunt motor (*b*) *dc* series motor

(*c*) induction motor (*d*) synchronous motor

249. The speed of a *dc* motor can be altered

(*a*) only by varying the flux

(*b*) only by varying the armature voltage

(*c*) by varying both the flux and armature voltage

(*d*) none of these

250. A *dc* motor is running with a certain load. The effect of adding an external resistance in the shunt field circuit is

(*a*) to increase the motor speed

(*b*) to reduce the motor speed

(*c*) to reduce the armature current of motor

(*d*) to stop the motor

251. Speed control by variation of field flux results in

(*a*) constant power drive

(*b*) variable power drive

(*c*) constant torque drive

(*d*) variable torque drive

252. A *dc* shunt motor is operating at no load. If the armature is shunted by a resistance, its speed will

(*a*) increase (*b*) decrease

(*c*) remain the same (*d*) be unpredictable

253. A symmetrical 3-phase supply is applied to the primary terminals A_1, B_1 and C_1 of the Y-Y transformer as shown in the given figure. The secondary terminals A_2, B_2 and C_2 give a symmetrical 3-phase voltage. If the terminals H_1 and H_2 of phase I, of the primary are reversed without changing the supply at A_1 B_1, and C_1, then the phase angle between the line voltage V_{A1B2} and V_{C3A2} on the secondary side will be

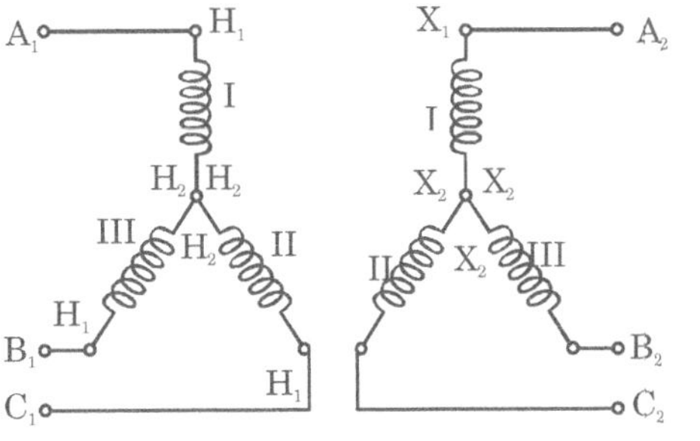

(*a*) 180° (*b*) 120°

(*c*) 60° (*d*) 30°

254. When the supply voltage for an induction motor is reduced by 10 %, the maximum running torque will nearly

(*a*) decrease by 10 % (*b*) decrease by 20 %

(*c*) increase by 10 % (*d*) increase by 20 %

255. Which of the following are the features of a shaded pole motor?

1. Salient-pole stator.
2. Uniform air-gap.
3. Two stator winding one of which is a short-circuited ring.
4. Squirrel cage rotor.

Select the correct answer using the codes given below:

Codes:

(*a*) 1 and 4 (*b*) 2 and 4

(*c*) 1, 3 and 4 (*d*) 1, 2 and 3

256. A synchronous machine connected to infinite bus bars is initially delivering active power to and absorbing lagging reactive power from the bus bars. As increase in the power input to the synchronous machine will cause

(*a*) an increase in both armature current and power factor

(*b*) an increase in armature current and a decrease in power factor

(*c*) a decrease in both armature current and power factor

(*d*) a decrease in armature current and an increase in power factor

257. A 3-phase induction motor requires a starter

(*a*) because there is not back emf induced in the stator winding during starting

(*b*) because the motor does not possess an inherent staring torque

(*c*) because the induced secondary voltage and hence the currents in the windings are abnormally high during starting

(*d*) to increase the starting torque

258. The introduction of interpoles in between the main poles improves the performance of a *dc* machine, because

(*a*) the interpoles produce additional flux to augment the developed torque

(*b*) the flux waveform is improved with reduction in harmonics

(*c*) the inequality of air-gap flux on the top and bottom halves of the armature is removed

(*d*) a counter-emf is induced in the coil undergoing commutation

259. For proper synchronisation of a large synchronous machine to a bus, the frequency of the incoming machine.

(*a*) should be exactly the same as that of the bus

(*b*) should be slightly higher than that of the bus

(*c*) should be slightly lower than that of the bus

(*d*) can be of any value

260. Four speed-torque curves (labelled I, II, III and IV) are shown. Torque in the given figure. That of an *ac* servomotor will be as in the curve labelled

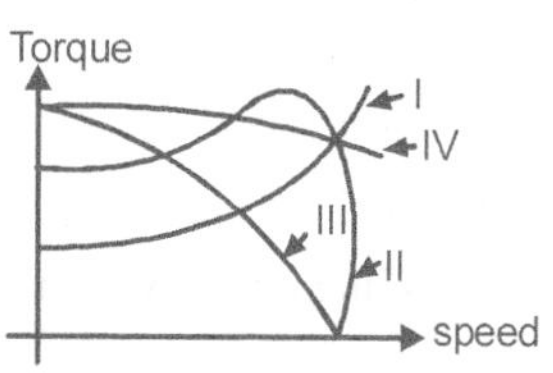

(*a*) I

(*b*) II

(*c*) III

(*d*) IV

261. To have sparkles commutation, the armature reaction effect in a *dc* machine is neutralised by

(*a*) using compensating winding and commutating poles

(*b*) shifting the brush axis from the geometrical neutral axis to the magnetic neutral axis

(*c*) fixing the brush axis in line with the pole axis

(*d*) increasing the field excitation

262. A 3-phase alternator delivers power to a balanced 3-phase load of power factor 0.707 lagging. It is observed that the open-circuit *emf* phasor leads the corresponding terminal voltage phasor by 15°. Neglecting the effect of harmonics, the angle between the axis of the main field *mmf* and the axis of the armature *mmf* will be

(*a*) 30° electrical

(*b*) 60° electrical

(*c*) 90° electrical

(*d*) 150° electrical

263. In a 3-phase induction motor, if the angular speed of the stator field is ω and the actual angular speed of the rotor is w_r and the stator induced *emf* is given by $e_s = N_1 \omega K_{ws} \phi \sin \omega t$, ($N_1$ being the stator turns/phase), the the rotor emf with N_2 as the rotor/turns phase will be K_{ws} = stator winding factor and K_{wr} = rotor winding factor)

(*a*) $e_r = N_2 \omega K w_r \sin \omega t$

(*b*) $e_r = N_2 \omega K w_r \sin \omega_r t$

(*c*) $e_r = N_2(\omega - \omega_r) K_{\omega r} \phi \sin (\omega - \omega_r)t$

(*d*) $e_r = N_2(\omega + \omega_r) K_{\omega r} \phi \sin (\omega - \omega_r)t$

264. A constant voltage generator has

(*a*) minimum efficiency

(*b*) low internal resistance

(*c*) high internal resistance

(*d*) minimum current capacity

265. A transformer designed for operation on 60 Hz supply is worked on 50 Hz supply system without changing its voltage and current rating. When compared with full load efficiency at 60 Hz the transformer efficiency on full load at 50 Hz will

(*a*) increase marginally

(*b*) increase by a factor of 1.2

(*c*) remain unaltered

(*d*) decrease marginally

266. In a loaded *dc* generator, if the brushes are given a shift from the interpolar axis in the direction of rotation, then the commutation will

(*a*) improve with fall of terminal voltage

(*b*) deteriorate with fall of terminal voltage

(*c*) improve with rise in terminal voltage

(*d*) deteriorate with rise in terminal voltage

267. A-3 phase synchronous motor with negligible losses is connected to the supply at rated frequency and constant terminal voltage V. The induced emf of the motor is designated as E. If the motor is now gradually loaded to its rated power, adjusting its excitation to obtain, say 0.8 leading power factor operation, then E-phasor would

(*a*) be less than V- phasor and also lead V-phasor

(*b*) be greater than V-phasor and also lead V-phasor

(*c*) be less than V-phasor and also lag V-phasor

(*d*) be greater than V-phasor and also lag V-phasor

268. The phenomenon of crawling in a-3 phase cage induction motor may be due to

(*a*) unbalanced supply voltage

(*b*) 7th space harmonic of air-gap field

(*c*) 7th time harmonic of voltage wave

(*d*) 5th space harmonic

269. A synchronous motor having a synchronous speed of 1500 *rpm* slows down to 1490 *rpm* due to sudden increase in load on the motor shaft. Assuming a south pole on the stator near the damper bars labelled 1, 2, 3 and 4 as shown in the given figure, the direction of induced force on conductor 3 is

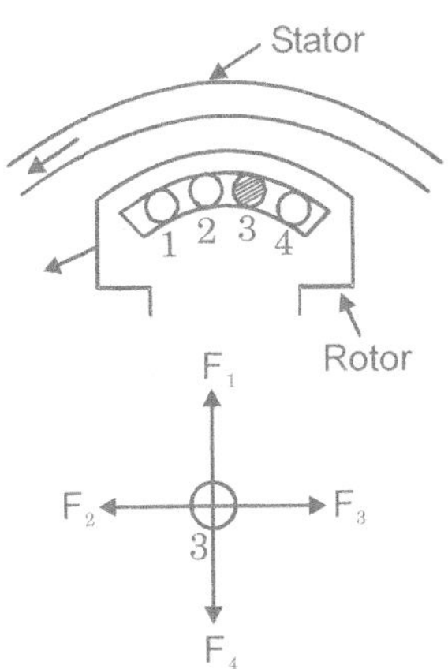

(*a*) F_1 (*b*) F_2

(*c*) F_3 (*d*) F_4

270. If the secondary winding of the ideal transformer of the circuit shown in the given figure has 40 turns, then for maximum power transfer to the 2-ohm resistor, the number of turns required in the primary winding will be

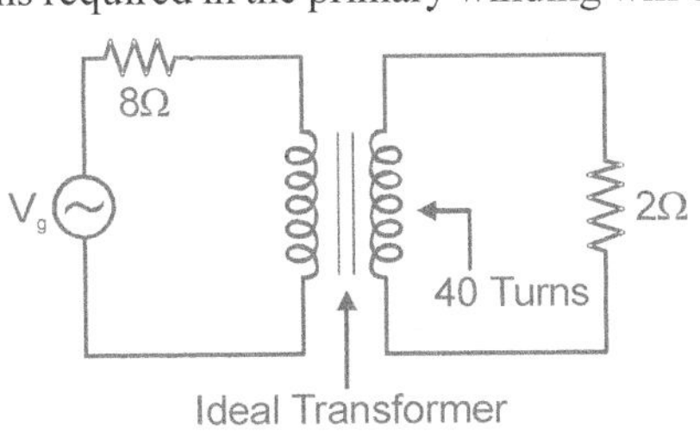

(*a*) 20 (*b*) 40

(*c*) 80 (*d*) 120

271. The sinusoidal distributed balanced three phase windings shown in the given figure are supplied with balanced three-phase currents.

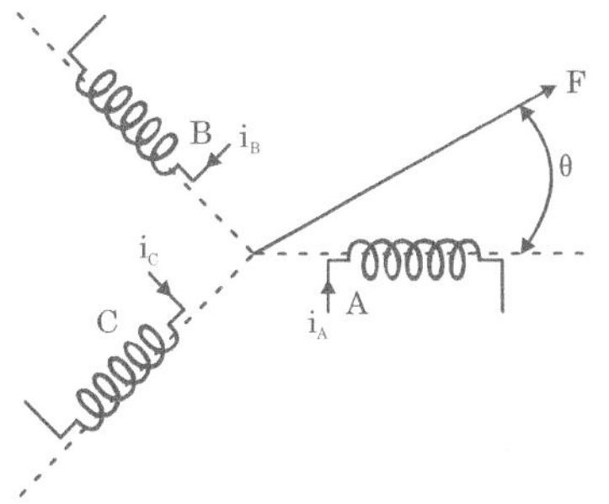

$I_A = I_m \sin\left(\omega t - \frac{\pi}{6}\right)$, $i_B = I_m \sin\left(\omega t - \frac{5\pi}{6}\right)$ and $i_C = I_m$

$\sin\left(\omega - \frac{\pi}{2}\right)$

At the angular instant $wt = \frac{\pi}{2}$, the value of angular position '*q*' of the rotating mmf 'F' with respect to phase A axis will be

(*a*) $-\frac{\pi}{2}$ (*b*) $\frac{\pi}{2}$

(*c*) $\frac{\pi}{6}$ (*d*) $-\frac{\pi}{6}$

272. Two coils 1 and 2 have self and mutual inductances of displacement '*x*' meters as under:

$L_{11} = 2 + 3x, L_{22} = 4 + 6x, L_{12} = 2 - 3x.$

For constant $I_1 = I_A$ and $I_2 = -0.5$ A, the magnitude and direction of the developed electromagnetic force would be

(*a*) 1.5 Newtons acting in the direction to increase *x*

(*b*) 1.5 Newtons acting in the directions to decrease *x*

(*c*) 3.75 Newtons acting in the direction to increase *x*

(*d*) 4.5 Newtons acting in the direction to decrease *x*

273. If a 230 V *dc* series motor is connected to a 230 V *ac* supply,

(*a*) the motor will vibrate violently

(*b*) the motor will run with less efficiency and more sparking

(*c*) the motor will not run

(*d*) the fuse will be blown

274. The rotor output of a 3-phase induction motor is 15 kW and the correspondence slip is 4%. The rotor copper loss will be

(*a*) 600 W (*b*) 625 W

(*c*) 650 W (*d*) 700 W

275. The speed of a *dc* shunt motor can be increased above its normal speed by

(*a*) increasing the fields current

(*b*) decreasing the fields current

(*c*) decreasing the terminal voltage

(*d*) increasing the armature resistance

276. A fly wheel is normally fitted to

(*a*) *dc* series motor driving a constant torque load

(*b*) separately excited *dc* motor driving pulsed torque load

(*c*) cumulatively compounded motor driving torque load

(*d*) differentially compounded motor driving pulsed torque load

277. An insaturated *dc* shunt motor drives a constant torque load at a speed of 1000 *rpm* taking an armature current of 20 A when 230 V is applied to the motor. If the applied voltage is reduced to 115 V the motor runs at

(*a*) 500 *rpm* taking an armature current of 20 A

(*b*) 1000 *rpm* taking an armature current of 20 A

(*c*) 500 *rpm* taking an armature current of 40 A

(*d*) 1000 *rpm* taking an armature current of 40 A

278. A *dc* shunt motor is running at light load, what will happen if the field winding gets opened ?

(*a*) Motor will stop

(*b*) Motor will burn

(*c*) Motor will make noise

(*d*) Motor will pick up high speed

279. A dc shunt motor runs at rated speed. If its field circuit gets-circuited, then soon after this the motor speed would tend to

(*a*) decrease

(*b*) remains unchanged

(*c*) increase

(*d*) fluctuate around its previous speed

280. A *dc* series motor develops a torque of 20 NM at 3 A of load current. If the current is increased to 6 A, then torque developed will be

(*a*) 10 N-m

(*b*) 20 N-m

(*c*) 80 N-m

(*d*) none of these

281. A 10 kVA, 400 V/ 200 Vsingle-phase transformer with a resistance of 3% and a reactance of 6% is supplying a current of 50A to a resistive load. The voltage across the load is

(*a*) 194 V (*b*) 196 V

(*c*) 198 V (*d*) 390 V

282. For a *p*-pole machine, the relation between electrical (θ_e) and mechanical (θ_m) degree is given by

(*a*) $\theta_e = \frac{2}{P}\theta_m$ (*b*) $\theta_e = P.\theta_m$

(*c*) $\theta_e = P.\theta_m$ (*d*) $\theta_e = \frac{P}{2}\theta_m$

283. In a synchronous generator operating at zero pf lagging, the effect of armature reaction is

(*a*) magnetizing

(*b*) demagnetizing

(*c*) cross-magnetizing

(*d*) both magnetizing and cross-magnetizing

284. Which one of the following statements is not correct in respect of synchronous machines ?

(*a*) In salient pole machines, the direct axis synchronous reactance is greater than the quadrature axis synchronous reactance.

(*b*) The damper bars help the motor to self-start

(*c*) Short circuit ratio is the ratio of field current required to produce the rated voltage on open circuit to the rated armature current

(*d*) The V curve of a synchronous motor represents the variation in the armature current with field excitation at a given output power

285. If the dimensions or all the parts of a synchronous generator, and the number of field and armature turns are doubled, then the generated voltage will change by a factor of

(*a*) 1 (*b*) 2

(*c*) 4 (*d*) 8

286. A 3-phase induction motor has full-load slip of 3 per cent at normal voltage. Which one of the following will be the value of slip of the motor if it develops the same torque theoretically while operating at 110 per cent of its normal voltage?

(*a*) 2.48 %

(*b*) 0.248 %

(*c*) 0.483 %

(*d*) 4.83 %

287. A 6-pole, 3-phase, 60 Hz induction motor runs at 1000 rpm developed maximum torque. Rotor resistance per phase is 1.2 ohms. Neglecting stator impedance, then for developing maximum torque at starting the external resistance to be connected in series with each rotor phase will be

(*a*) 7.2 ohms (*b*) 6 ohms

(*c*) 1.44 ohms (*d*) 1.2 ohms

288. If the height to width ratio of the window of core-type transformer is increased, then

(*a*) its leakage reactance and percentage voltage regulation will decrease

(*b*) its leakage reactance and percentage voltage regulation will increase

(*c*) its leakage reactance will be unaffected but total space in the window will increase to accommodate the winding better

(*d*) efficiency will increase and temperature rise will decrease

289. A 4 kVA, 230/115 volt, 400 Hz, single-phase transformer, originally designed for use in aircrafts, its available in a laboratory for obtaining 115 V supply form standard 230 V, 50 Hz mains. If overloading of the device is not permitted, then the net kVA output of the transformer in the laboratory cannot exceed

(*a*) 32 (*b*) 16

(*c*) 4 (*d*) 0.5

290. Match List-I (Parts of electrical machines with List-II) (The approximate nature of the air-gap mmf pattern produced by them) and select the correct answer using the codes given below the Lists:

List-I	**List-II**
A. DC machine, stator	1. Sinusoidal
B. DC machine, rotor	2. Trapezoidal
C. Salient-pole synchronous machine, stator	3. Triangular
D. Squirrel-cage rotor of induction motor	

Codes:

	A	B	C	D
(*a*)	1	2	1	3
(*b*)	3	1	3	1
(*c*)	2	3	1	3
(*d*)	2	3	1	1

291. A 3-phase, 400 V, delta connected alternator has a synchronous impedance of $(0 + j20)$ ohms per phase. If it delivers a balanced load of 12 kVA at zero power factor leading, then the percentage voltage regulation of the alternator is equal to

(*a*) – 50 (*b*) + 50

(*c*) – 150 (*d*) + 150

292. The inductive reactance of a transformer depends on

(*a*) electromotive force

(*b*) magnetomotive force

(*c*) magnetic flux

(*d*) leakage flux

293. A 3-phase induction machine operators on 3-phase fixed frequency ac mains at a per unit slip of 1.5. Consider the following statements regarding the operating condition of the machine

1. It draws, electrical power from the mains.
2. It draws mechanical power through the shaft.
3. It deliver electrical power to the mains.
4. It delivers mechanical power through the shaft.

Of these statements:

(*a*) 1 and 2 are correct (*b*) 1 and 4 are correct

(*c*) 2 and 3 are correct (*d*) 3 and 4 are correct

294. In a normal 3-phase slip-ring induction motor, the developed starting torque will

(*a*) increase with the increase in the value of the rotor resistance

(*b*) decrease with the increase in the value of the rotor resistance

(*c*) not depend upon the value of the rotor resistance

(*d*) increase with increase in the value of the rotor resistance upto a certain value and then decrease with further increase in rotor resistance.

295. Consider the following statements:

The effects of slot harmonies are effectively reduced by

1. skewing the slots.
2. fractional slot winding.
3. short-statements.
4. distributed winding.

Of these statements:

(*a*) 1 and 2 are correct

(*b*) 2, 3 and 4 are correct

(*c*) 1, 3 and 4 are correct

(*d*) 1 and 3 are correct

296. A single-phase induction regulator is a constant input transformer to obtain smooth variation of the output voltage by varying the

(*a*) ratio of turns between primary and secondary windings

(*b*) frequency

(*c*) flux density in the core

(*d*) angle between the magnetic axes of the primary and secondary windings

297. In the phasor-diagram of a round-rotor synchronous generator, the voltage equation is $\overline{E}_f = \overline{V}_t + \overline{I}_a (r + jX_s)$ where E_f = excitation voltage, V_t = terminal voltage, I_a = armature current at lagging power factor, r_a = armature resistance; X_s = synchronous reactance. While remaining synchronized to the bus-bars (infinite bus), if the power input from the prime mover is gradually decreased and finally stopped, it will result in

(*a*) reversal of I_s and E_f lagging V_t

(*b*) reversal of I_s but E_f leading V_t

(*c*) the sign of I_s remaining unchanged, but E_f leading V_t

(*d*) the sign of I_s remaining unchanged, but E_f lagging V_t

298. A dc shunt generator builds up to a voltage of 220 V at no-load while running at its rated speed. If the speed of the generator is raised by 25% keeping the circuit conditions unaltered, then the voltage to which the machine would build up will

(*a*) not change and remain at 220 V

(*b*) increase to 1.25 times 220 V

(*c*) increase to a value lying between 220 V and 1.25 times 220V

(*d*) increase to a value greater than 1.25 times 220 V

299. The magnetising current component of the no-load current of an induction motor is much larger than a corresponding transformer because of

(*a*) addition friction and winding loss in motor

(*b*) different winding configuration on stator and rotor

(*c*) increased flux requirement

(*d*) an air-gap in the magnetic circuit

300. Match List-I (Ways of compounding series and shunt field winding of a dc compound generator) with List-II (Volt-ampere characteristics.at constant speed) and select the correct answer using the codes given below the Lists:

List-I

A. Over compounded

B. Level compounded

C. Differentially compounded

D. Under compounded

List-II

1. 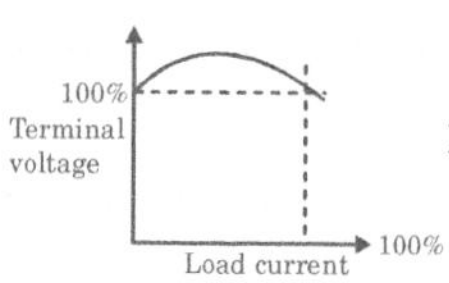

2. 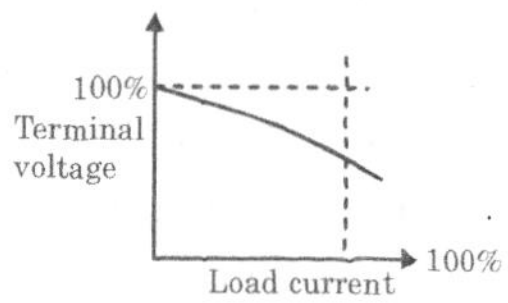

3. 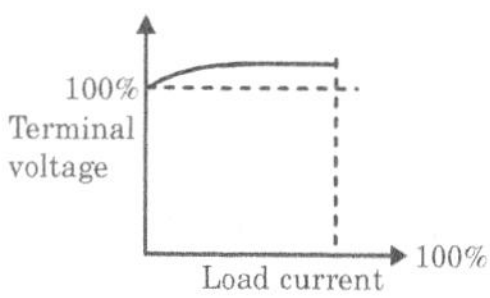

4. 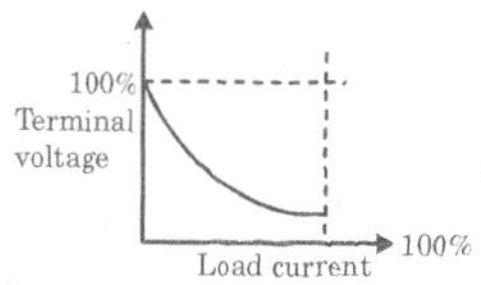

Codes:

	A	B	C	D
(*a*)	3	1	4	2
(*b*)	3	1	2	4
(*c*)	1	3	2	4
(*d*)	1	3	4	2

301. A 3-phase 4-pole slip-ring induction motor is supplied with appropriate voltage at 50 Hz to its stator and voltage at 10 Hz to its rotor through slip-rings. The phase sequence of both the voltages is same. The motor will produce steady torque at a speed of

(*a*) 1500 *rpm* (*b*) 1200 *rpm*

(*c*) 300 *rpm* (*d*) zero

302. The rotor of a 4-pole 3-phase cage induction motors is replaced by a 3- phase, 4-pole wound rotor. When fed with normal supply, the machine will

(*a*) not run

(*b*) run at very low speed

(*c*) run at slightly lower than normal speed

(*d*) run at slightly higher than normal speed

303. A 200V *dc* machine has an armature resistance of 0.5. If the full load armature current is 30A, the induced emf when the machine run

(*i*) as generator, and

(*ii*) as motor will be respectively

(*a*) 230 V, 170 V (*b*) 225 V, 175 V

(*c*) 185 V, 215 V (*d*) 215 V, 185 V

304. If a current-carrying coil is placed in a uniform magnetic field with its plane perpendicular to the direction of magnetic induction then

(*a*) the net force and the torque on the coil both are zero

(*b*) the net force is zero but torque is finite

(*c*) the net force is finite but torque is zero

(*d*) the net force and torque both are finite

305. The direction of rotation of a *dc* shunt motor can be reversed by interchanging

(*a*) the supply terminals

(*b*) the field terminals only

(*c*) the armature terminals only

(*d*) either field or armature terminals

306. When the supply terminals of a *dc* shunt motor are interchanged

(*a*) the motor will stop

(*b*) the motor will run at its normal speed in the same direction as before

(*c*) the direction of rotation will reverse

(*d*) the motor will run much faster in the same direction

307. The speed of a *dc* motor is

(*a*) directly proportional to back emf and flux

(*b*) directly proportional to its back emf and inversely proportional to flux

(*c*) inversely proportional to both back emf and flux

(*d*) directly proportional to the and univers-ally proportional to back emf

308. If the load current and flux of a *dc* motor are held constant and voltage applied across its armature is increased by 15%, the speed of the motor will

(*a*) increase by 5%

(*b*) reduce by 5%

(*c*) remain unaltered

(*d*) depends on other factors

309. A series motor is working drawing a load current of 1 A from the lines. If now the load is reduced such that the current drawn is halved, the speed of the machine (neglecting the saturation and armature resistance would be)

(*a*) unchanged

(*b*) reduced by 50 %

(*c*) reduced by 100 %

(*d*) increased by 100 %

310. A cumulatively compounded dc generator is supplying 20 A at 200 V Now if the series field winding is short circuited, the terminal voltage

(*a*) will remain unaltered at 200 V

(*b*) will rise to 220 V

(*c*) will shoot upto a very high value

(*d*) will become less than 2000 V

311. A *dc* cumulatively compounded generator was operating satisfactorily and Supplying power to an infinite bus when the mechanical power supply from the prime mover failed. The machine will then run as a

(*a*) differentially compounded motor with speed reversed

(*b*) differentially compounded motor with the direction of speed the same as before

(*c*) cumulatively compounded motor with the same direction of speed as before

(*d*) cumulatively compounded motor with speed reversed

EXERCISE – II

1. Two transformers are to be operated in parallel such that they share load in proportion to their KVA ratings. The rating of the first transformer is 500 kVA and its pu leakage impedance is 0.05 pu. If the rating of second transformer is 200 kVA, then its pu leakage impedance is

(*a*) 0.20

(*b*) 0.10

(*c*) 0.05

(*d*) 0.025 **[GATE]**

2. A three-phase cage induction is started by direct-on-line (DOL) switching at the rated voltage. If the starting current drawn is 6 times the full load current and the full load slip is 4%, then ratio of the starting developed torque to the full load torque is approximately equal to

(*a*) 0.24 (*b*) 1.44

(*c*) 2.40 (*d*) 6.00 **[GATE]**

3. A 10 MVA, 11 KV, 3-phase star connected alternator has $X_s = 1.1$ *pu* machine rating. Its X_s is ohm is

(*a*) 26.61 Ω

(*b*) 12 Ω

(*c*) 6.655 Ω

(*d*) 13.31 Ω **[BHEL]**

4. A torque can develop due to interaction between two coil currents if these are

(*a*) along the same axis

(*b*) in quadrature

(*c*) 180° away from each other

(*d*) 45° away from each other **[NTPC]**

5. For a 3-phase transformer, out of the positive, negative and zero sequence impedance, the minimum impedance is

(*a*) zero-sequence impedance

(*b*) positive sequence impedance

(*c*) negative-sequence impedance

(*d*) both positive and negative sequence impedances **[NTPC]**

6. A three-phase induction motor connected in star takes 5 A, when connected in delta, it will take

(*a*) $5/\sqrt{3}$ A (*b*) $5\sqrt{3}$ A

(*c*) 15 A (*d*) 5 A **[NTPC]**

7. The torque of a rotor in an induction motor under running condition is maximum

(*a*) at unity value of the slip

(*b*) at zero value of slip

(*c*) at the value of slip which makes the rotor reactance half of the rotor resistance

(*d*) at the value of slip which makes the rotor reactance per phase equal to its resistance per phase **[ES]**

8. If a synchronous motor is switched on the 3 phase supply with its rotor winding short circuited it will

(*a*) start

(*b*) not start

(*c*) start and continue to run as induction motor

(*d*) start and continue to run as synchronous motor **[BHEL]**

9. In a capacitor start single phase induction motor, the capacitor is connected
   (*a*) in series with the main winding
   (*b*) in parallel with the auxiliary winding
   (*c*) in series with the auxiliary winding
   (*d*) none of these **[PGCIL]**
10. If the rotor of a 3-phase induction motor is assumed purely resistive, then electromagnetic torque is
   (*a*) minimum with load angle of 90°
   (*b*) minimum with load angle of 0°
   (*c*) optimum with load angle of 90°
   (*d*) optimum with load angle of 0° **[PGCIL]**
11. A 4 pole generator with 16 coils has a two layer lap winding. The pole pitch is
   (*a*) 32 (*b*) 16
   (*c*) 8 (*d*) 4 **[GATE]**
12. In a swin burne test on a *dc* machine
   (*a*) all the losses can be determined individually
   (*b*) copper losses cannot be determined individually
   (*c*) mechanical and iron losses can not be determined individually
   (*d*) none of these **[NTPC]**
13. The loss due to eddy currents in a transformer is
   (*a*) directly proportional to the current
   (*b*) indirectly proportional to the current
   (*c*) indirectly proportional to the resistance of the winding
   (*d*) directly proportional to the resistance of the winding **[BHEL]**
14. The speed of the rotating magnetic field in an induction motor is known as the
   (*a*) slip speed
   (*b*) effective speed
   (*c*) shaft speed
   (*d*) synchronous speed **[PGCIL]**
15. A 3-phase, 400-V, 50 Hz 4 pole induction motor is fed from a 3-phase, 400-V supply and runs of 1425 *rpm*. The frequency of the rotor emf is
   (*a*) 2.5 Hz (*b*) 50 Hz
   (*c*) 48 Hz (*d*) Zero **[PGCIL]**
16. In a transformer the phase difference between the primary and secondary voltage of a transformer is nearly
   (*a*) 0 deg (*b*) 90 deg
   (*c*) 120 deg (*d*) 180 deg
17. In a hysteresis motor, the rotor cylinder is made up of
   (*a*) magnetically soft chrome steel
   (*b*) magnetically hard cobalt steel
   (*c*) steel laminations
   (*d*) solid steel **[BHEL]**
18. An alternator with frequency f_1 is to be synchronized with an infinite bus of frequency f. For proper synchronization
   (*a*) $f_1 = f$
   (*b*) $f_1 < f$
   (*c*) $f_1 > f$
   (*d*) either (*b*) or (*c*) **[ES]**
19. A Schrage motor is a
   (*a*) constant power motor with shunt speed-torque characteristics
   (*b*) constant-torque motor with shunt speed-torque characteristics
   (*c*) constant-power motor with series speed torque characteristics
   (*d*) constant torque motor with series speed-torque characteristics **[NHPC]**
20. In a polyphase IM, the effect of skewing rotor slot is to
   (*a*) decrease both starting and maximum torques
   (*b*) increase both starting and maximum torques
   (*c*) decrease both rotor resistance and noise
   (*d*) increase rotor resistance and decrease noise **[NHPC]**
21. The function of oil in a transformer is
   (*a*) to provided insulation and cooling
   (*b*) to provide protection against lightning
   (*c*) to provide protection against circuit
   (*d*) to provide lubrication **[GATE]**
22. Match List-I with List-II and select the correct answer using the codes given below the List:

   List-I (Driving system)
   A. Slip ring induction motor drive with *dc* dynamic breaking arrangement
   B. Variable voltage and frequency synchronous motor drive
   C. DC shunt motor drive with plugging arrangement
   D. DC series motor drive with propulsion dynamic and regenerating breaking

   List-II (Application)
   1. Textile mill's looms
   2. Mine winder
   3. Hoist motion
   4. Ship propulsion

   Codes:

	A	B	C	D
(*a*)	1	2	3	4
(*b*)	4	3	2	1
(*c*)	1	3	2	4
(*d*)	2	3	4	1

[ES]

23. The torque-slip characteristic of a polyphase induction motor becomes almost linear at small values of slip, because in this range of slips

(*a*) the effective rotor circuit resistance is very large compared to the rotor reactance

(*b*) the rotor resistance is equal to the stator resistance

(*c*) the rotor resistance is equal to the rotor reactance

(*d*) the rotor reactance is equal to the stator reactance **[ES]**

24. A 4-pole dynamo with wave wound armature has 51 slots containing 20 conductors in each slot. The induced emf is 357 volts and the speed is 8500 rpm. The flux per pole will be

(*a*) 3.5 mWb (*b*) 1.2 mWb

(*c*) 14 mWb (*d*) 21 mWb **[GATE]**

25. The low voltage winding of a 400/230 V, 1-Phase 50 Hz transformer is to be connected to a 25 Hz supply. In order to keep the magnetization current at the same level as that for normal 50 Hz the supply voltage should be

(*a*) 230 V (*b*) 460 V

(*c*) 115 V (*d*) 65 V **[GATE]**

26. When the supply voltage to an induction motor is reduced by 10%, the maximum torque will decrease by approximately

(*a*) 5% (*b*) 10%

(*c*) 20% (*d*) 40% **[ES]**

27. Capacitance Voltage Transformer (CVT) used to

(*a*) improve the power factor of the transmission

(*b*) reduce losses in a transmission line

(*c*) connect instruments on the LT side

(*d*) reduce the incidence of overvoltage surge on transmission lines **[IAS]**

28. A 6-pole, 50 Hz single-phase induction motor runs at a speed of 900 rpm. The frequency/frequencies of current in the cage rotor will be

(*a*) 5 Hz

(*b*) 5 Hz, 55 Hz

(*c*) 5 Hz, 95 Hz

(*d*) 55 Hz, 95 Hz **[ES]**

29. Consider the following statements:

1. A grid-connected induction generator always supplies leadin g reactive VARs to the bus.
2. An overexcited synchronous motor draws current at lagging power factors.
3. An under excited synchronous generator connected to an infinite bus works at a leading power factor.
4. The load angle of a synchronous machine is the angle between the excitation voltage and the gap voltage.

Of these statements

(*a*) 1 and 2 are correct

(*b*) 2 and 3 are correct

(*c*) 1,3 and 4 are correct

(*d*) 1, 2, 3 and 4 are correct **[ES]**

30. Two similar synchronous generators are working in parallel to supply a common load demand with identical of excitations and steam supplies to their movers. Now, if the steam supply to the prime mover of the generators is increased compared to the other, with field excitation kept unchanged, then

(*a*) its active power component will remain the same but the reactive power will increase

(*b*) its active power will decrease while the reactive power will increase

(*c*) both active and reactive components of power will increase

(*d*) its active power contribution will increase but reactive power contributions of both will remain unchanged **[ES]**

31. Match List-I with List-II and select the correct answer using the codes given below the Lists:

List-I

A. Self-excitation of *dc* generations

B. Speed control of 3-phase induction motor by varying rotor resistance

C. Speed control of a dc motor by field weakening

D. Variation of applied voltage to an induction motor with varying stator frequency

List-II

1. Efficient and smooth speed control over a wide range
2. Constant power applications
3. Associated with variation of load resistances
4. Constant torque application
5. Dangerously high speeds at no-load

Codes:

	A	B	C	D
(*a*)	3	4	5	1
(*b*)	3	4	2	1
(*c*)	3	2	4	5
(*d*)	4	3	2	5

[ES]

32. Consider the following statements:

If the terminal voltage of a *dc* shunt motor is halved with load torque varying as the square of speed,then

1. Speed is halved.
2. Speed remains unaltered
3. Armature current is doubled
4. Armature current remains unaltered

Of these statements:

(*a*) 1 and 3 are correct (*b*) 1 and 4 are correct

(*c*) 2 and 3 are correct (*d*) 2 and 4 are correct **[ES]**

33. A reduction of the voltage applied to a 3-phase induction motor

(*a*) increases the breakdown torque as well as the starting torque

(*b*) increases the breakdown torque but decreases the starting torque

(*c*) decreases the breakdown torque but the starting torque remain unchanged

(*d*) decreases both breakdown and starting torques **[IAS]**

34. Skew is used in induction motors in order to reduce

(*a*) time harmonics (*b*) space harmonics

(*c*) slot harmonics (*d*) reverse rotating fields **[GATE]**

35. With the increase in supply frequency of a transformer, there is

(*a*) decrease in rating (*b*) increase in rating

(*c*) no change in rating (*d*) none of these **[NTPC]**

36. The path of the magnetic flux in a transformer should have

(*a*) low reluctance

(*b*) low resistance

(*c*) high reluctance

(*d*) high resistance **[NTPC]**

37. The object of using compensating winding in *dc* machines is to neutralize the

(*a*) armature reaction in the inter polar zone

(*b*) armature reaction in the commutating zone

(*c*) armature reaction under the pole faces

(*d*) cross-magnetizing armature reaction **[PGCIL]**

38. Two alternators working in parallel supply power to a load the excitation of one of them is increased.

This will result in a change of their

A. reactive powers

B. active powers

C. terminal voltage

D. volt-amperes

(*a*) A, B, C (*b*) A, C, D

(*c*) B, C, D (*d*) A, D **[PGCIL]**

39. A 3-phase induction motor runs at 980 rpm at no load. If its squirrel-cage rotor is replaced by solid iron cylinder, then its no load speed would be

(*a*) zero (*b*) 900 rpm

(*c*) below 980 rpm (*d*) 490 rpm [IOCL]

40. If a separately-excited *dc* generator is with its brushes displaced from neutral axis, then its

(*a*) terminal voltage V_t decrease but armature current I_a increases

(*b*) V_t increases, I_a decreases

(*c*) both V_t and I_a increase

(*d*) both V_t and I_a decrease **[ES]**

41. An induction motor crawls at lower speed. This is the

(*a*) skewing

(*b*) harmonic induction torques

(*c*) high secondary reactances

(*d*) low secondary reactance **[NHPC]**

42. The stator winding of synchro transmitter and receiver are excited by

(*a*) the differential field flux

(*b*) the AC magnetic field of the rotor

(*c*) a DC source

(*d*) compensating winding **[NHPC]**

43. The mmf of an interpole in *dc* machines is proportional to

(*a*) armature current (*b*) field current

(*c*) both (*a*) and (*b*) (*d*) none of these **[NHPC]**

44. While a 2-phase *ac* servomotor is in operation, if the voltage across the control field winding becomes zero, then the motor has a tendency to run as a single-phase induction motor. To prevent this

(*a*) rotor having high mass moment of internal is used for such a motor

(*b*) drag cup type of light rotor with high rotor resistance is preferred

(*c*) a low rotor resistance is used

(*d*) the number of turns in the control field winding used is less than the main reference winding **[ES]**

45. If a induction machine is run at above synchronous speed it acts as

(*a*) a synchronous motor (*b*) an induction generator

(*c*) an induction motor (*d*) none of these **[GATE]**

46. The core flux of a practical transformer with a resistance load

(*a*) is strictly constant with load changes

(*b*) increases linearly with load

(*c*) increase as the square root of the load

(*d*) decreases with increased load **[ES]**

47. In a transformer, if the iron losses and copper losses are 40.5 kW and 50 kW respectively, then at what fraction of load will the efficiency be maximum?

(*a*) 0.8 (*b*) 0.57

(*c*) 0.70 (*d*) 0.9 **[ES]**

48. During a disturbance on a synchronous machine, the rotor swing A to B before finally settling down to a steady state at point C on the power angle curve. The speed of the machine during oscillational is synchronous at point (s)

(*a*) A and B (*b*) A and C
(*c*) B and C (*d*) only at C **[GATE]**

49. When a short-circuit test is conducted on a single-phase transformer, 30% of the rated voltage is required to allow full-load current. The short-circuit power factor is found to be 0.2. The percentage regulation at UPF is

(*a*) 30 (*b*) 29.5
(*c*) 15 (*d*) 6 **[IAS]**

50. In an induction motor, if the air gap is increased

(*a*) speed will reduce
(*b*) efficiency will improve
(*c*) power factor will be lowered
(*d*) breakdown torque will reduce **[GATE]**

51. Which one of the following capacitor-start split-phase induction motors will have the largest value of capacitance?

(*a*) $\frac{1}{2}$ H.P., 3450 rpm
(*b*) $\frac{1}{4}$ H.P., 1725 rpm
(*c*) $\frac{1}{2}$ H.P., 1140rpm
(*d*) $\frac{1}{4}$ H.P., 1140 rpm **[ES]**

52. During hunting of synchronous motor

(*a*) negative phase sequence currents are general
(*b*) harmonic are developed in the armature circuit
(*c*) damper bar develops torque
(*d*) field excitation increase **[GATE]**

53. The energy stored in the magnetic field of a solenoid 30 cm long and 3 cm diameter wound with 1000 turns of wire carrying a current of 10 A is

(*a*) 0.015 joule (*b*) 0.15 joule
(*c*) 0.5 joule (*d*) 1.15 joule **[GATE]**

54. Auto-transformer is used in transmission and distributions

(*a*) when operator is not available
(*b*) when iron losses are to be reduced
(*c*) when efficiency consideration can be ignored
(*d*) when the transformation ratio is small **[GATE]**

55. Two transformers of identical voltages but of current capacities are operating in parallel. For satisfactory load sharing

(*a*) impedance must be equal
(*b*) per-unit impedances must be equal
(*c*) per-unit impedances and $\frac{X}{R}$ ratios must be equal
(*d*) impedance and $\frac{X}{R}$ ratios must be equal **[GATE]**

56. A differentially compounded *dc* motor with interpoles and with brushes on the neutral axis is to be driven as a generator in the same direction with the same polarity of the terminal voltage. It will then

(*a*) be a cumulatively compounded generator but the interpole coil connectors are to be reversed
(*b*) be a cumulatively compounded generator without reversing the interpole coil connections
(*c*) be a differentially compounded generator but the interpole coil connections
(*d*) be a differentially compounded generator but the interpole coil connections are to be reversed **[GATE]**

57. An inductions motor is fed from a balanced three-phase supply at rated voltage and frequency through a bank of three single phase transformers connected in delta-delta. One unit of the bank develops fault and is removed. Then,

(*a*) single phasing will occur and the machine fails to start
(*b*) single phasing will not occur but the motor terminal voltages will become unbalanced and the machine can be loaded to the extent at 57.7% of its rating
(*c*) the machine can be loaded to the extent of 57.7% of its rating with balanced supply as its terminals.
(*d*) the machine can be loaded to the extent of $66\frac{2}{3}$% with balanced supply at its terminals. **[GATE]**

58. Which one of the following can be obtained by the equivalent circuit of an electrical machine?

(*a*) temperature rise in the cores
(*b*) complete performance characteristics of the machine
(*c*) type of protection used in the machine
(*d*) design parameters of the windings **[IAS]**

59. In case of a split phase motor, the phase shift between currents in the two windings is around

(*a*) 30 degrees (*b*) 70 degrees
(*c*) 90 degrees (*d*) 120 degrees **[GATE]**

60. Unbalanced supply voltage given to a 3-phase, delta-connected induction motor will cause
(*a*) zero sequence currents
(*b*) less heating of the motor
(*c*) negative sequence component current
(*d*) all the above **[ES]**

61. The purpose of using interpoles in *dc* machines is to nullify the
(*a*) demagnetizing effect of armature mmf
(*b*) cross-magnetizing effect of armature mmf
(*c*) reactance voltage
(*d*) both cross magnetizing mmf and reactance voltage

62. When the speed becomes more than the synchronous speed during hunting, the damper bars develop
(*a*) synchronous motor torque
(*b*) single-phase induction motor torque
(*c*) induction motor torque
(*d*) induction generator torque **[NHPC]**

63. Which of the following motors is used for rolling mills
(*a*) *dc* shunt motors
(*b*) *dc* cumulative compound motor
(*c*) *dc* series motor
(*d*) *dc* differential compound motor **[NHPC]**

64. The main purpose of using core in a transformer is to
(*a*) decrease iron losses
(*b*) prevent eddy current loss
(*c*) eliminate magnetic hysteresis
(*d*) decrease reluctance of the common magnetic flux path **[NHPC]**

65. Armature winding is one in which working
(*a*) flux is produced by the field current
(*b*) flux is produced by the working current
(*c*) emf is produced by the working flux
(*d*) emf is produced by the leakage flux **[NHPC]**

66. An induction motor in which the stator has the squirrel cage winding is
(*a*) 3-phase squirrel cage motor
(*b*) 1-phase induction motor
(*c*) 1-phase ceiling fan
(*d*) 1-phase table fan **[ES]**

67. Changing the electrical connectors to reverse direction of rotation of a motor running at full speed is called
(*a*) slugging (*b*) plugging
(*c*) dynamic braking (*d*) brush shifting **[NHPC]**

68. The flux responsible for the development of mechanical forces in transformers is
(*a*) primary leakage flux (*b*) total flux
(*c*) mutual flux (*d*) leakage flux **[PGCIL]**

69. The magnetizing inrush current in the transformers is rich in
(*a*) third harmonics
(*b*) fifth harmonics
(*c*) seventh harmonics
(*d*) second harmonics **[PGCIL]**

70. A cylindrical rotor synchronous motor is switched on to the supply with its field windings shorted on themselves. It will
(*a*) not start
(*b*) start but not run at synchronous speed
(*c*) start as an induction motor and then run as a synchronous motor
(*d*) start and run as s synchronous motor **[GATE]**

71. In Fig the characteristic that corresponds to the variation of synchronous reactance of a synchronous motor with field current is

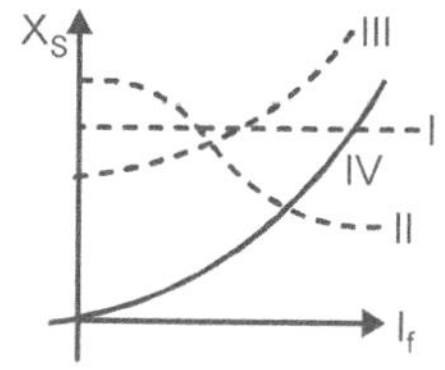

(*a*) curve I
(*b*) curve II
(*c*) curve III
(*d*) curve IV **[GATE]**

72. Which of the following class is mainly used in electrical machines for insulation ?
(*a*) class Y
(*b*) class A
(*c*) class F
(*d*) class C **[IOCL]**

73. The directions of rotation of a *dc* shunt motor can be reversed by interchanging
(*a*) the supply terminal
(*b*) the field terminals only
(*c*) the armature terminals only
(*d*) either field or armature terminals **[IOCL]**

74. The division of active amongst alternators running in parallel depends upon
(*a*) speed-load characteristics of prime movers
(*b*) volt-ampere characteristics of alternators
(*c*) excitations emfs of alternators
(*d*) all of these **[IOCL]**

75. In a 3-phase induction motor, a deep bar rotor is used to obtain

(*a*) more operating efficiency

(*b*) more starting torque

(*c*) more pull-out torque

(*d*) reduced rotor core loss **[NHPC]**

76. Scott-connected transformers can convert

(*a*) 3 to 2 phases (*b*) 2 to 3 phases

(*c*) both (*a*) and (*b*) (*d*) 2 to 4 phase **[IOCL]**

77. A synchronous generator has higher power handling capacity of operating at

(*a*) a leading p.f.

(*b*) a lagging p.f.

(*c*) it does not depend upon the p.f. of the generator

(*d*) it depends upon the load p.f. as generator has no p.f. of its own **[ES]**

78. Most suitable motor for tape recorder will be

(*a*) *ac* series motor

(*b*) reluctance motor

(*c*) hysteresis motor

(*d*) 1-phase induction motor **[ES]**

79. If the supply voltage decreases by 4%, the torque in 3-phase 1 M would be decrease by

(*a*) 4% (*b*) 10%

(*c*) 7.84% (*d*) 8% **[NHPC]**

80. Synchronous machine with high value of SCR has

(*a*) bigger size

(*b*) smaller size

(*c*) machine size independent of SCR

(*d*) none of these **[NHPC]**

81. For a full pitch winding, the generated voltages in both coil sides are

(*a*) exactly in phase

(*b*) in quadrature

(*c*) exactly 180° out of phase

(*d*) approximately 180° out of phase **[PGCIL]**

82. In a coal mine, Ward-Leonard method of speed control as shown in the given figure is used to lift coal from the pit through a crane and bucket arrangement. After lifting the coal to desired level and emptying at the dumping ground, the bucket has to be lowered into the pit. The action to be taken at this stage is to reverse the

(*a*) direction of current in F_2 and slowly increased it

(*b*) current in F_3 and slowly decrease it

(*c*) supply terminal of M_1

(*d*) direction of current in F_1 and slowly increase it **[ES]**

83. At 50 Hz operation, a single-phase transformer has hysteresis loss of 200 W and eddy current loss of 100 W. Its core loss at 60 Hz operation will be

(*a*) 432 W (*b*) 408 W

(*c*) 384 W (*d*) 360 W **[ES]**

84. A 220/440 V, 50 Hz operation, a single-phase transformer operators on 220 V, 40 Hz supply with secondary winding. Then

(*a*) the eddy current loss and hysteresis loss of the transformer decrease

(*b*) the eddy current loss and hysteresis loss of the transformer increase

(*c*) thc hysteresis loss of the transformer increases while eddy current loss remains the same

(*d*) the hysteresis loss remains the same whereas eddy current loss decreases **[GATE]**

85. A synchronous motor on load draws a current at a leading power factor angle ϕ. If the internal power factor angle-which is phase angle between excitation e.m.f. and current in time phasor diagram - is Ψ then the air gap excitation m.m.f. lags the armature m.m.f. by

(*a*) Ψ (*b*) $\frac{\pi}{2}+\Psi$

(*c*) $\frac{\pi}{2}-\Psi$ (*d*) $\Psi+\phi$ **[GATE]**

86. Transformer zero voltage regulation occurs at

(*a*) unity power factor

(*b*) leading p.f.

(*c*) lagging p.f.

(*d*) zero p.f. leading **[BHEL]**

87. Electromagnetic torque in rotating machine is present when

(*a*) air gap is uniform

(*b*) stator winding alone carries current

(*c*) rotor winding alone carries current

(*d*) both stator and rotor winding carry current **[BHEL]**

88. Transformer at no-load behaves like

(*a*) a resistor, pf = 1.0

(*b*) an inductive reactor, pf = 0.2 lagging

(*c*) a capacitive reactor, pf = 0.2 leading

(*d*) an inductive reactor, pf = 0.8 lagging **[BHEL]**

89. In an ideal transformer, the no-load primary current I_o

(*a*) lags behind V_1 by slightly less than 90°

(*b*) is in phase with V_1

(*c*) lead V_1 by 90°

(*d*) lags V_1 by an angle lying between 0° and 90° **[BEL]**

90. The short circuit test on a transformer is conducted to obtain

(*a*) core losses at any load

(*b*) ohmic losses at any load

(*c*) hysteresis losses

(*d*) none of these **[BEL]**

91. Distributions transformers are designed to have maximum efficiency nearly

(*a*) at full load (*b*) at 50% of F.L.

(*c*) at no load (*d*) none of these **[BEL]**

92. DC series motor should never be switched on at no load, because

(*a*) the field current is initially zero

(*b*) the motor does not pickup

(*c*) the speed becomes dangerously high

(*d*) it will take too long to accelerate **[NTPC]**

93. The direction of ratio of a *dc* shunt motor is reversed by

(*a*) reversing armature connection

(*b*) interchanging the armature and field connection

(*c*) adding resistance to the field circuit

(*d*) reversing supply connections **[ES]**

94. In *dc* machines, the air gap length under the poles is not kept uniform so as to

(*a*) obtain a sinusoidal main-field flux

(*b*) obtain better cooling

(*c*) minimise the effect of armature mmf on main field

(*d*) obtain sinusoidal armature mmf wave **[BHEL]**

95. No load an full-load *pf* 5 of a 3-phase induction motors are respectively of the order of

(*a*) 0.2, 0.85 (*b*) 0.5, 0.8

(*c*) 0.7, 0.9 (*d*) 0.3, 0.95 **[IOCL]**

96. A 3-phase, 400 V, 50 Hz 4 pole induction motor is fed from a 3-phase 400 V supply and runs at 1425 rpm. The frequency of the rotor emf is

(*a*) 2.5 Hz (*b*) 50 Hz

(*c*) 48 Hz (*d*) zero **[ES]**

97. In a transformer the exciting current has two component magnetizing constant and core loss component. Neglecting leakage impedance drop,

(*a*) both of them lag the impressed voltage by 90°

(*b*) both of them are in phase with impressed voltage

(*c*) the former lags the impressed voltage by 90° while the latter is in phase with impressed voltage

(*d*) the former is in phase with the impressed voltage while the latter lags the impressed voltage by 90° **[IOCL]**

98. The object of using compensating winding in *dc* machines is to neutralize the

(*a*) armature reaction in the interpolar zone

(*b*) armature reaction in the cumulating zone

(*c*) armature reaction under the pole faces

(*d*) cross-magnetizing armature reaction **[IOCL]**

99. In an induction motor, the rotor input is 600 W and slip is 4% the rotor copper loss is

(*a*) 700 W (*b*) 0.6 W

(*c*) 650 W (*d*) 24 W **[IOCL]**

100. The efficiency of a *dc* generator is maximum when **[ES]**

(*a*) magnetic loss is equal to mechanical loss

(*b*) field ohmic loss is equal to the constant loss

(*c*) stray load loss is equal to the armature circuit

(*d*) armature circuit loss is equal to the sum of rotational loss and field circuit loss

101. Induction motors with open slots have **[ES]**

(*a*) better p.f.

(*b*) improved efficiency

(*c*) more breakdown torque

(*d*) less starting torque

102. For satisfactory performance of a 3-phase 420 V, 60 Hz induction motor, the supply voltage at 50 Hz should be equal to

(*a*) 420 V (*b*) 380 V

(*c*) 350 V (*d*) 400 V **[ES]**

103. The direction of rotation of a 3-phase induction motors is reversed by

(*a*) interchanging the connection of any two phases

(*b*) interchanging the connection of all the three phase

(*c*) rewinding the stator

(*d*) adding a capacitor in any phase **[IOCL]**

104. Open winding are used for **[ES]**

(*a*) *dc* machines

(*b*) *ac* machines

(*c*) *ac* commutator machines

(*d*) both *dc* and *ac* commutator machines

105. The main advantage of distributing the winding in slots is to

(*a*) add mechanical strength to the winding

(*b*) reduce the amount of copper required

(*c*) reduce the harmonics in the generated emf

(*d*) reduce the size of the machine **[IOCL]**

106. Which of the following lubricants is the shaft of the motor lubricated ? **[ES]**

(*a*) Graphite (*b*) Grease

(*c*) Silicon oil (*d*) Mineral oil

107. Primary or secondary winding of a 3-phase transformer in delta helps to suppress
(*a*) third harmonic current
(*b*) third harmonic voltage
(*c*) third harmonic flux
(*d*) all of these **[PGCIL]**

108. The maximum torque that a synchronous motor can deliver is proportional to
(*a*) $\frac{1}{V^2}$ (*b*) $\frac{1}{V}$
(*c*) 1 (*d*) V^2 **[BHEL]**

109. Typical value of SCR for modern turbo alternator is
(*a*) 1 (*b*) 1.2
(*c*) 0.5 (*d*) 1.5 **[ES]**

110. In a synchronous machine connected to infinite bus, if rotor speed departs from synchronous speed, then
(*a*) damping power comes into play
(*b*) synchronizing power comes into play
(*c*) both (*a*) and (*b*) above
(*d*) none of these **[PGCIL]**

111. A 3-phase stator and rotor winding of a slip-ring induction motor is fed by 3-phase ac supply at a frequency *f*. If the rotor speed is N_r and synchronous speed N_s then the absolute speed of resultant flux in space is
(*a*) N_s irrespective of rotor speed N_r
(*b*) $(N_s + N_r)$ if the flux rotors in the same
(*c*) $(N_s - N_r)$ if the flux rotates in the same direction as the rotor
(*d*) N_s irrespective of N_s **[IAS]**

112. When a synchronous motor is running at synchronous speed, the damper winding produces
(*a*) damping torque
(*b*) eddy current torque
(*c*) torques aiding the developing torque
(*d*) no torque **[ES]**

113. A bank of three identical single-phase 250 kVA, 11 kV/230 V transformers is used to provide 400 V low tension supply from a 11 kV, 3-phase substation. The effective kVA rating of the bank will be
(*a*) 250 (*b*) $250\sqrt{3}$
(*c*) 500 (*d*) 750 **[IAS]**

114. Consider the following statements
The steady-state speed of a 3-phase cage induction motor delivering a load torque depends upon
1. supply voltage.
2. supply frequency.
3. load torque-speed characteristic

Of these statements:
(*a*) 1, 2 and 3 are correct
(*b*) 1 and 2 are correct
(*c*) 2 and 3 are correct
(*d*) 1 and 3 are correct **[IAS]**

115. Which of the following connection of transformer will give the highest secondary voltage?
(*a*) delta primary, delta secondary
(*b*) delta primary, star secondary
(*c*) star primary, star secondary
(*d*) star primary, delta secondary **[ES]**

116. If a 400 V, 50 Hz, star connected, 3 phase squirrel cage induction motor is operated from a 400 V, 75 Hz supply, the torque that the motor can now provide while drawing rated current from the supply
(*a*) reduces
(*b*) increases
(*c*) remains the same
(*d*) increases or reduces depending upon the rotor resistance **[ES]**

117. A 3-phase alternator with high rotational inertia and a negligible armature resistance was delivering power to an isolated load when the armature terminals got shot circuited. After a while, the mechanical power input to the shaft of the machine is terminal at item t_0. The short-circuited current I_s circulating in the armature will vary till the machine comes to rest at time t_s. Which one of the curves shown in the following figure best reflects this variation?

(*a*)
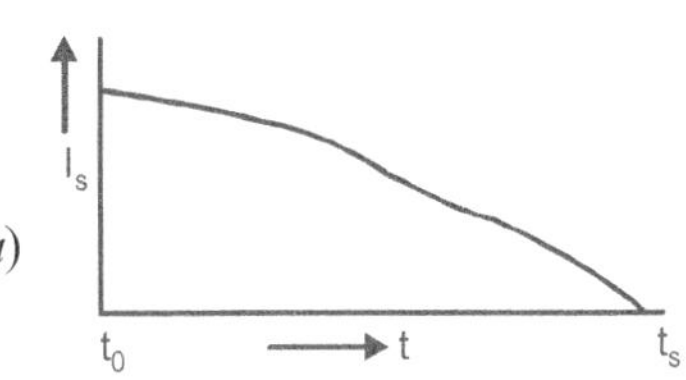

(*b*)
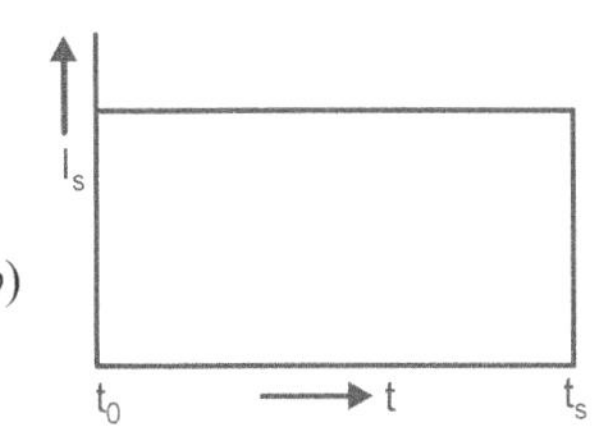

(*c*)
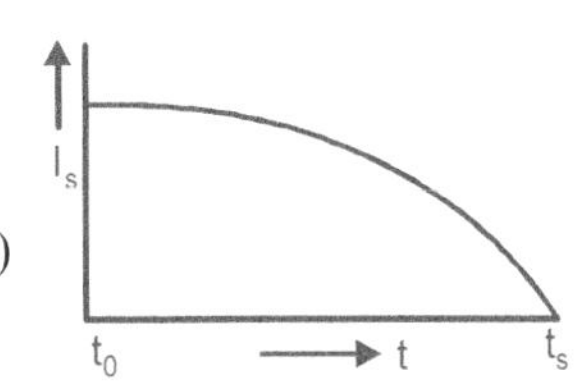

(*d*)
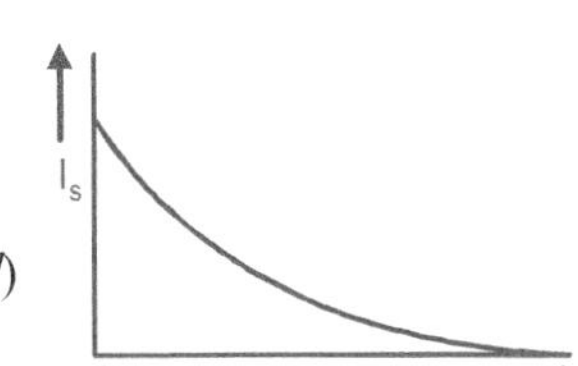

[ES]

118. The voltage across the open-circuited field terminals of a synchronous machine under slip test is
(*a*) *dc*
(*b*) *ac* of slip frequency
(*c*) a modulated supply frequency *ac* voltage with slip frequency envelope
(*d*) *ac* of supply frequency **[IAS]**

119. In a *dc* machine, the armature mmf is
(*a*) rectangular and directed along the interpolar axis
(*b*) triangular and directed along the inter-polar axis
(*c*) triangular and directed along the brushes axis
(*d*) rectangular and directed along the brush axis **[ES]**

120. If we increase field current of a dc generator its emf output
(*a*) increases indefinitely
(*b*) increase till the winding burns
(*c*) increases till the magnetic saturation takes place
(*d*) first increases and then start reducing **[AMIE]**

121. The voltage at the terminals of a dc series generator running at rated rpm and no load will be
(*a*) more than the rated voltage
(*b*) full rated voltage
(*c*) a very small voltage
(*d*) zero **[SCETE. WB]**

122. In a 4-pole, 25 kW, 200 V wave would dc shunt generator the current in each parallel path will be
(*a*) 62.5 A
(*b*) 125 A
(*c*) 31.25 A
(*d*) 250 A **[BTEMP]**

123. A *dc* generator beyond critical resistance will generated
(*a*) maximum power
(*b*) maximum voltage
(*c*) maximum current
(*d*) no voltage **[BTEMP]**

124. If the field circuit resistance of a dc shunt generator exceeds its critical value, then generator
(*a*) fails to build up
(*b*) builds up very high voltage
(*c*) exceeds its current capacity
(*d*) produces power beyond its rating **[AMIE]**

125. For a septic open circuit voltage of a dc generator, the short circuit current will be maximum when it is
(*a*) separately excited
(*b*) shunt connected
(*c*) cumulatively compounded
(*d*) differentially compounded **[ES]**

126. The voltmeter connected across a generator reads voltage same at no-load and at full load (rated) the generator is of the type
(*a*) shunt generator
(*b*) series generator
(*c*) level compounded
(*d*) short-shunt compounded **[BTEMP]**

127. The magnitude of mechanical force in a dc motor is dependent on
(*a*) length of the conductor
(*b*) cross sectional area of the conductor
(*c*) material of the conductor
(*d*) none of these **[AMIE]**

128. The direction of armature current in a dc motor is
(*a*) the same as of the generated emf
(*b*) opposite to that of the generated emf
(*c*) not dependent upon the direction of the armature emf
(*d*) none of the about **[AMIE]**

129. Consider the following statements regarding an electrical machine having cylindrical stator and salient-pole rotor:
1. Reluctance torque is produced when rotor winding is excited.
2. Reluctance torque is produced when stator winding is excited.
3. When both stator and rotor windings are excited, electromagnetic torque is produced
4. When both stator and rotor windings are excited, electromagnetic as well as reluctance torques are produced.

Which of these statements are correct? **[ES]**
(*a*) 2 and 4
(*b*) 1 and 4
(*c*) 1 and 3
(*d*) 2 and 3

130. If the field winding of an unloaded dc shunt motor gets opened while running, it will **[ES]**
(*a*) stop
(*b*) run with reduced speed
(*c*) run with increased speed
(*d*) reach a dangerously high speed

131. A separately excited dc machine, having an armature resistance of 2 ohms was working on a 220 V supply and drawing 10 A armature current from the source when the supply voltage suddenly dropped to 200 V. Assuming that the field circuit source voltage remained unaffected, how will the armature current of the machine react to the change?

(*a*) It will initially rise to 11 A and then settle down to 10 A

(*b*) It will fall momentarily to 9.09 A and then slowly attain 10 A

(*c*) It will reduce to zero first and then settle back to 10 A

(*d*) It will remains unaffected by the change and continue to be 10 A **[IAS]**

132. Which one of the following figures best represents the variation in armature current and speed with time during the starting of a dc shunt motor?

(*a*) 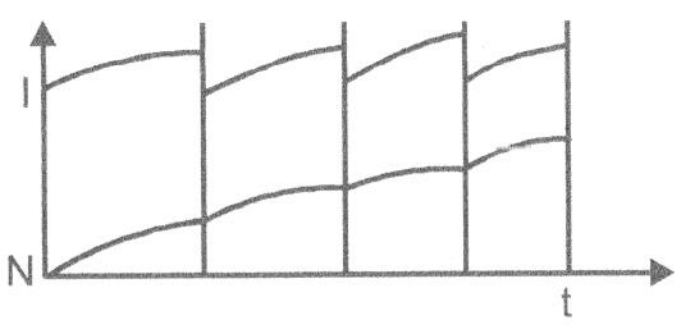

(*b*) 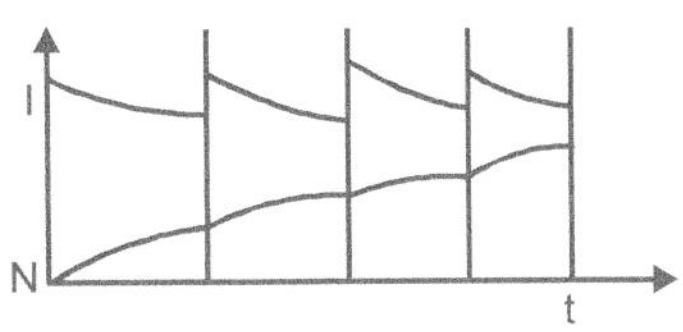

(*c*) 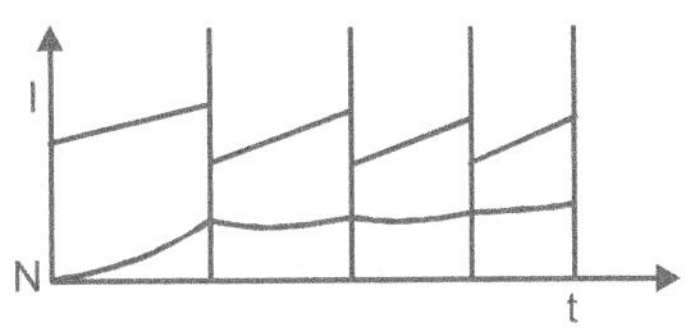

(*d*) 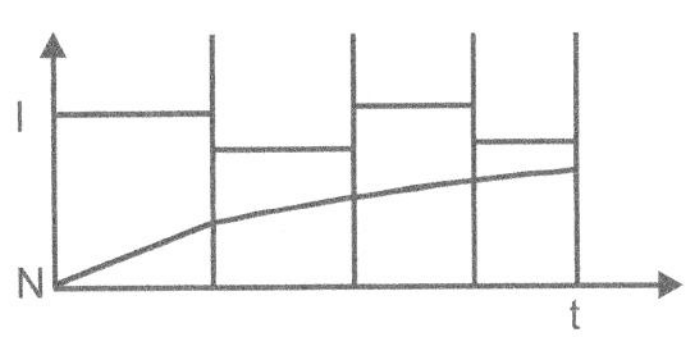

[ES]

133. In an autotransformer of voltage ratio $\frac{V_1}{V_2}$ and $V_1 > V_2$ the fraction of power transferred inductively is **[ES]**

(*a*) $\frac{V_1}{V_2 + V_2}$

(*b*) $\frac{V_2}{V_1}$

(*c*) $\frac{V_1 - V_2}{V_1 + V_2}$

(*d*) $\frac{V_1 - V_2}{V_1}$ **[ES]**

134. The series winding of a cumulatively compounded dc motor is short-circuited while driving a load at rated torque. This results in

(*a*) reduction in both the armature current and the motor speed

(*b*) increase in the armature current and reduction in the motor speed

(*c*) increase in both the armature current as well as the motor speed

(*d*) reduction in the armature current and increase in the motor speed **[IAS]**

135. The given figure shows the equivalent circuit of a 3-phase induction motor with deep-bar rotor. The sequence of the magnitudes of the parameters of resistances and reactances in the ascending order is

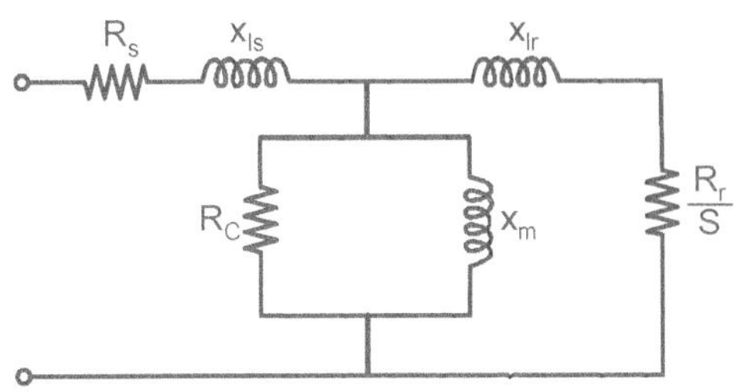

(*a*) $R_s, R_r, x_{ls}, x_{lr}, R_C, x_m$

(*b*) $x_{ls}, x_{lr}, R_r, R_s, x_m, R_C$

(*c*) $R_r, R_s, x_{ls}, x_{lr}, R_C, x_m$

(*d*) $R_s, R_r, x_{ls}, x_{lr}, x_m, R_C$ **[ES]**

136. Which one of the following statements for a dc machine which is provided with interpole winding (IW) as well as compensating winding (CW) is correct? **[ES]**

(*a*) Both IW and CW are connected in series with the armature winding

(*b*) Both IW and CW are connected in parallel with the armature winding

(*c*) IW is connected in series but CW is connected parallel with the armature winding

(*d*) IW is connected in parallel but CW is connected in series with the armature winding

137. At a certain speed and flux, the voltage generated by a dc generator is 230 volts. If the speed is increased by 20 % and the flux is simultaneously reduced by 10 %, then the voltage will be

(*a*) increased by 10 % (*b*) reduced by 20 %

(*c*) increased by 8 % (*d*) decreased by 8 % **[AMIE]**

138. The field coils of a dynamo have an inductance of 8 H and have a resistance of 60 Ω and are connected to 120 v dc source. The steady state current in the field coiled is

(*a*) 15 A (*b*) 2 A

(*c*) 0 A (*d*) none of these **[AMIE]**

139. Approximate physical size of the 50 kω, 1800 rpm dynamo as a function of a 25 kω, 900 rpm dynamo will be

(*a*) 1 (*b*) 1/2

(*c*) 1/4 (*d*) none of these **[AMIE]**

140. The armature mmf wave in a dc machine is

(*a*) sinusoidal (*b*) triangular

(*c*) rectangular (*d*) square **[BTEUP]**

141. Air gap at the pole tips of a *dc* machine is kept more than that at the centre of the pole mainly to reduce

(*a*) reactance voltage

(*b*) effect of armature reaction

(*c*) losses of armature core

(*d*) noise of the machine **[AMIE]**

142. Interpoles are provided in *dc* machine to

(*a*) neutralize the cross-magnetizing component armature reaction

(*b*) neutralize the demagnetizing component of armature reaction

(*c*) reduce iron loss

(*d*) reduce copper losses **[BTEMP]**

143. The function of interpole in *dc* machine is to

(*a*) reduce field winding heating

(*b*) improve commutation

(*c*) compensate for air-gap variation

(*d*) reduce losses **[WBSEB]**

144. Compensating winding in a *dc* machine

(*a*) counteracts armature mmf in the interpolar zone

(*b*) prevents large speed drop

(*c*) shunts most of the armature current

(*d*) none of these **[AMIE]**

145. The armature of a *dc* machine has a resistance of 0.1 Ω and is connected to a 230 V supply. What will be the emf of generator giving 80 A ?

(*a*) 8V (*b*) 22V

(*c*) 238V (*d*) 230V **[AMIE]**

146. With of a carbon brush should be equal to

(*a*) less than the width of one commutator segments

(*b*) the width of 1 to 2 commutator segments

(*c*) the width of 2 to 3 commutator segments

(*d*) the width of more than 3 commutator segment **[AMIE]**

147. The purpose of providing dummy coils in a wave would dc machine armature is to

(*a*) enhance armature

(*b*) amplify voltage

(*c*) mechanical balance for the rotor

(*d*) reduce eddy current loss **[AMIE]**

148. The armature resistance of 96-pole lap wound dc machine is 0.05 Ω. If the armature is rewound using a wave winding, then the armature resistance will be **[ES]**

(*a*) 0.45 (*b*) 0.30

(*c*) 0.15 (*d*) 0.10

149. A frog-leg winding is used in a dc machine to eliminate

(*a*) the use of equalizers in lap winding

(*b*) the use of dummy coils in armature

(*c*) the use of compensation winding

(*d*) reactance voltage produced due to commutation **[AMIE]**

150. A pole lap wound dc machine had an armature current I_a. The conductor current is the armature winding is

(*a*) I_a (*b*) I_a/p

(*c*) PI_a (*d*) none of these **[AMIE]**

151. A lap wound dc machine has 400 conductors and 8 poles. The voltage induced per conductor is 2V. The machine generates a voltage of

(*a*) 100 V (*b*) 200 V

(*c*) 400 V (*d*) 800 V **[AMIE]**

152. The number of parallel paths for a 4-pole duplex lap winding will be

(*a*) 2 (*b*) 4

(*c*) 6 (*d*) 8 **[AMIE]**

153. Minimum number of coils for 400 V, 6 pole lap connected dc machine for a maximum voltage of 15 V between adjacent commutator segment should be

(*a*) 80 (*b*) 160

(*c*) 200 (*d*) 400 **[AMIE]**

154. Two alternators each having 4% sped regulation are working in parallel. Alternator '1' is rated for 12 MW and alternator '2' is rated for 8 MW. When the total load is 10 MW, the loads shared by alternators 1 and 2 would be respectively

(*a*) 4 MW and 6 MW

(*b*) 6 MW and 4 MW

(*c*) 5 MW and 5 MW

(*d*) 10 MW and zero **[ES]**

155. The combined frequency regulation of machines in area of capacity 1500 MW and operating at a nominal frequency of 60 kHz is 0. 1 pu on its own base capacity. The regulation in Hz/MWwill be

(*a*) $\frac{0.1}{1500}$ (*b*) $\frac{60}{1500}$

(*c*) $\frac{6}{1500}$ (*d*) $\frac{60}{150}$ **[ES]**

156. The following lists relate to speed control of dc motors. Match List-I with List-II, and select the correct answer using the codes given below the lists :

List-I

A. Armature voltage control

B. Field current control

C. Use of diverter resistance

D. Rheostatic control

List-II

1. Speeds above base speed
2. Speeds below base speed
3. Poor motor efficiency
4. Speed control of series motor

Codes:

	A	B	C	D
(*a*)	1	2	3	4
(*b*)	2	1	4	3
(*c*)	4	1	3	2
(*d*)	2	3	1	4

[IAS]

157. How can the reactive power delivered by a synchronous generator be controlled ?

(*a*) By changing the prime mover input

(*b*) By changing the excitation

(*c*) By changing the direction of rotation

(*d*) By changing the prime mover speed **[ES]**

158. A single phase transformer has a maximum effeciency of 30 % at full load and unity power factor. Efficiency at half load at the same power factor is **[ES]**

(*a*) 86.7 %

(*b*) 88.26 %

(*c*) 88.9 %

(*d*) 87.8 %

159. The mmf produced by the current of 3-phase induction. motor

(*a*) rotates at the speed of rotor in the air gap

(*b*) is standstill with respect to stator mmf

(*c*) rotates at slip speed with respect to stator mmf

(*d*) rotates at synchronous speed with respect to rotor **[ES]**

160. A dc cumulatively compounded motor delivers rated load torque at rated speed. If the series field is short-circuited, then the armature current and speed will

(*a*) both decrease

(*b*) both increase

(*c*) increase and decrease respectively

(*d*) decrease and increase respectively **[IAS]**

161. The phenomenon of crawling is usually not formed in slip-ring because these motor have

(*a*) no space harmonics

(*b*) high resistance rotor windings

(*c*) no zig-zag leakage

(*d*) comparatively large air-gap **[IAS]**

162. A dc shunt motor is run under the following conditions:

1. Normal voltage with no additional resistance in armature or field circuits.
2. 50 % normal voltage with no additional resistancy in armature or field circuits.
3. Normal voltage with a small external resistance in the armature circuit of a value equal to the armature resistance.
4. Normal speed of the motor would increase in the sequence

(*a*) 1, 2, 3, 4 (*b*) 1, 4, 3, 2

(*c*) 2, 1, 4, 3 (*d*) 2, 3, 1, 4 **[IAS]**

163. The 3-limb, 3-phase delta-star connected transformer are supplied from the same source. One of the transformers is of Dy 1 and the other is of Dy 11 connection. The phase difference between the corresponding phase voltage of the second arics would bc

(*a*) 0° (*b*) 30°

(*c*) 60° (*d*) 120° **[ES]**

164. An unlocked normal field-excited synchronous motor with its negligible losses synchronized to an infinite bus having voltage v_f. Its field excitation is increased till the rated armature current 'I_a' is reached. The magnitude of the induced emf 'E_f' is measured at this excitation. The value of the leakage reactance is determined by

(*a*) $\dfrac{E_r}{l_a}$ (*b*) $\dfrac{V_t}{l_a}$

(*c*) $\dfrac{E_r + V_1}{2l_a}$ (*d*) $\dfrac{E_r - V_1}{l_a}$ **[IAS]**

165. The low-voltage winding of a core-type transformer is subdivided into two equal halves, each of the half the original width of the single winding with the high-voltage winding in between (instead of having the usual construction of low- voltage winding adjacent to the core and surrounded by the high-voltage winding). Such an interlacing of coils would make the combined primary and secondary leakage reactance (in terms of the primary) nearly

(*a*) twice (*b*) equal

(*c*) half (*d*) one-fourth **[ES]**

166. A-3 phase induction motor draws 1000 kVA at a pf of 0.8 lag. A synchronous condenser is connected in parallel to draw an additional 750 kVA at 0.6pf lead. The *pf* of the total load supplied by the main is

(*a*) unity
(*b*) 0.707 lag
(*c*) 0.707 lead
(*d*) zero **[IAS]**

167. Low voltage starting of induction motors is prescribed because

(*a*) high starting current at full voltage would damage the winding
(*b*) it facilities smooth acceleration
(*c*) it increases the life of the motor
(*d*) it is mandated by electric supply authorities to prevent voltage dips at other consumer premises. **[IAS]**

168. The winding of a Q kVA, $\frac{V_1}{V_2}$ volts, three phase D*d* core type transformer are reconnected to work as a single-phase transformer. The power rating and the maximum voltage ratio of the new configuration will be respectively

(*a*) $\frac{Q}{3}$ kVA and $\frac{V_1}{V_2}$ volts
(*b*) 3 Q kVA and $\frac{3V_1}{3V_2}$ volts
(*c*) 2 Q kVA and $\frac{\sqrt{3}V_1}{\sqrt{3}V_2}$ volts
(*d*) $\frac{2V_1}{2V_2}$ volts **[ES]**

169. A dc over compound generator is supplying power to an infinite bus. If the prime mover is accidentally cut off, the dc machine will

(*a*) Small
(*a*) stop running
(*b*) run as cumulatively compounded motor in reverse direction
(*c*) run as differentially compounded motor in reverse direction
(*d*) ran as differentially compounded motor in same direction **[IAS]**

170. A polyphase slip-ring induction machine has the stator and rotor winding wound for

(*a*) the same number of phase but different number of pairs of poles
(*b*) the same number of pairs of poles but different number of phases
(*c*) the same number of phases with number of pairs of poles either equal or different
(*d*) the same number of pairs of poles with number of phases either equal or different **[IAS]**

171. Two alternators A and B are sharing an inductive load equally. If the excitation of alternator A is increased

(*a*) alternator B will deliver more current and alternator A will deliver less current
(*b*) alternator B will deliver less current and alternator A will deliver more current
(*c*) both will continue to share load squally
(*d*) both will deliver more current **[ES]**

172. While plugging a 3-phase induction motor, if one supply terminal gets disconnected-then the motor will

(*a*) continue to run in the same direction
(*b*) stop
(*c*) start rotating in the opposite direction at the same speed
(*d*) start rotating in the opposite direction at reduced speed **[IAS]**

173. The duty cycle of a motor is

Load	Duration
20 kW	120 sec
10 kW	120 sec
stopped	280 sec

The continuous rating of motor suitable for above application would be

(*a*) 20 kw (*b*) 16.4 kw
(*c*) 14.14 kw (*d*) 9 kw **[ES]**

174. "Cogging" in induction motor occurs when

(*a*) number of stator teeth-number of rotor teeth = odd number
(*b*) number of stator teeth-number of rotor teeth = even number
(*c*) number of stator teeth-number of rotor teeth = 0
(*d*) number of stator teeth-number of rotor teeth = negative number **[ES]**

175. The brush-axis of a bipolar dc motor is rotated by 90°. The effect of this rotation an the back emf E_b and the torque developed T_d would be such that

(*a*) both E_b and T_d are unchanged
(*b*) E_b is zero, but T_d is unchanged
(*c*) E_b is unchanged, but T_d is zero
(*d*) both E_b and T_d are zero **[ES]**

176. On which of the following factors does hysteresis loss depend ?

1. Flux density
2. Frequency
3. Thickness of lamination
4. Time

Select the correct answer using the code given below :

(*a*) 2 and 3 (*b*) 1 and 2

(*c*) 3 and 4 (*d*) 1 and 4 **[ES]**

177. Motor A has shallow and wide slots. Motor B has deeper and narrow slots. If both are 3 phase 400 V, 50 Hz, 1440 rpm induction motors, it can be concluded that

(*a*) motor A has more starting torque as compared to motor B

(*b*) motor B has more starting torque as compared to motor A

(*c*) motor A has more pull out torque as compared to motor B

(*d*) motor B has more pull out torque as compared to motor A **[ES]**

178. A 3-phase 20-pole, synchronous generator has 180 stator slots with single-layer full pitch coils. There are 6 conductors per slot and all coils per phase are connected in series. The rotor is driven at 300 rpm and the flux per pole (sinusoidally distributed) is 25 mWb. The voltage induced per phase is nearest to

(*a*) 500 V (*b*) 1000 V

(*c*) 1500 V (*d*) 2000 V **[IAS]**

179. Match List-I (motors) with List-II (applications) and select the correct answer using the codes given below the lists:

List-I

A. 1 -phase induction motor

B. Stepper motor

C. Universal motor

D. Brushless dc motor

List-II

1. Fast-acting servo drive.
2. Hand drilling machine.
3. Computer numerical control.
4. Domestic pump.

Codes:

	A	B	C	D
(*a*)	3	4	1	2
(*b*)	3	4	2	1
(*c*)	4	3	1	2
(*d*)	4	3	2	1

[IAS]

180. A single-phase induction motor with only the main winding excited would produce zero torque at

(*a*) synchronous speed only

(*b*) starting and synchronous speed

(*c*) starting only

(*d*) starting and a speed slightly below synchronous speed **[IAS]**

181. A series drawing an armature current of I_s is operating under saturated magnetic conditions, torque developed in the motor is proportional to

(*a*) $\frac{1}{I_a}$ (*b*) $\frac{1}{I_a^2}$

(*c*) I_a^2 (*d*) I_a **[ES]**

182. Two separate induction motors having 6 poles and 4 poles are cascaded 3-phase, 400 V, 60 Hz is the supply (in rpm) of cascaded set will be

(*a*) 720, 1200, 1800 and 2400

(*b*) 600,1000, 1500 and 3000

(*c*) 720,1200, 1800 and 3600

(*d*) 600, 1200,1800 and 2400 **[IAS]**

183. Open-circuit test on a transformer gives

(*a*) hysteresis loss

(*b*) eddy current loss

(*c*) copper loss

(*d*) sum of hysteresis and eddy current loss **[ES]**

184. Open circuit test in a transformer is performed with

(*a*) rated transformer voltage

(*b*) rated transformer current

(*c*) direct current

(*d*) high frequency supply **[AMIE]**

185. Group-I list different applications and Group-II list the motors for these applications. Match the application with the most suitable motor and choose the right combination among the choices given thereafter

Group-I

A. Food

B. Cassette tape recorder

C. Domestic water pump

D. Escalator

Group-II

1. Permanent magnet dc motor.
2. Single-phase induction motor.
3. Universal motor.
4. Three-phase induction motor.
5. DC series motor
6. Stepper motor

Codes: [ES]

	A	B	C	D
(*a*)	3	6	4	5
(*b*)	1	3	2	4
(*c*)	3	1	2	4
(*d*)	3	2	1	4

186. A single phase induction motor with only the main winding excited would exhibit the following response at synchronous speed [ES]

(*a*) rotor current is zero

(*b*) rotor current is non-zero and is at slip frequency

(*c*) forward and backward rotating fields are equal

(*d*) forward rotating field is more than the backward rotating field

187. In a transformer

(*a*) both open circuit (OC) and short circuit (SC) tests are conducted on *lv* side

(*b*) OC test is conducted on *lv* side and the SC test on *lv* side

(*c*) OC test is conducted on *lv* side and SC test on *lv* side

(*d*) both OC and SC tests are conducted on *lv* side [BTEUP]

188. For a transformer to be tested at full load conditions but consuming only losses from the mains, we do

(*a*) load test

(*b*) open-circuit and short-circuit tests

(*c*) back to back test

(*d*) none of these [AMIE]

189. Efficiency of a transformer is maximum if

(*a*) transformer is unloaded

(*b*) copper losses are equal to iron losses

(*c*) eddy current losses are equal to hysteresis losses

(*d*) it is maximally loaded [AMIE]

190. A transformer has at full-load iron-loss of 900 watts and copper loss of 1600 watts. At what percentages of the load the transformer will have maximum efficiency [ES]

(*a*) 133%

(*b*) 125%

(*c*) 75%

(*d*) 66.6% [SCEIEWB]

191. For continuously running rolling mills with intermittent loading, the most suitable dc drive is a

(*a*) dc series motor

(*b*) dc shunt motor

(*c*) dc differentially compounded motor

(*d*) dc cumulatively compounded motor coupled to a heavy flywheel [ES]

192. The approximate value of efficiency of three phase induction motor running at a slip '*s*' is given by

(*a*) $\frac{1}{1+s}$ (*b*) $\frac{s}{1+s}$

(*c*) $\frac{1-s}{1+s}$ (*d*) $\frac{s}{1-s}$ [IAS]

193. The reluctance power of a salient-pole synchronous motor does not depend upon the field excitation but on the degree of saliency. Expression for reluctance power is (the terms have their usual meaning) [ES]

(*a*) $\frac{V^2}{2} \times \frac{(X_d - X_q)}{X_d X_q} \sin 2\delta$

(*b*) $2V^2 \frac{(X_d - X_q)}{X_d X_q} \sin \delta$

(*c*) $2V^2 \frac{(X_d - X_q)}{X_d X_q} \sin 2\delta$

(*d*) $\frac{V^2}{2} \times \frac{(X_d - X_q)}{X_d X_q} \sin \delta$

194. Given figure shows an electromechanical energy conversion device. The instantaneous value of armature induced emf, bus bar voltage and armature current are respectively *e*, *v* and *i*. The instantaneous values of external torque, armature developed torque and angular velocity of the shaft are respectively T_e T_a and ω_r. If the armature is connected to supply bus, the mode of operation will be

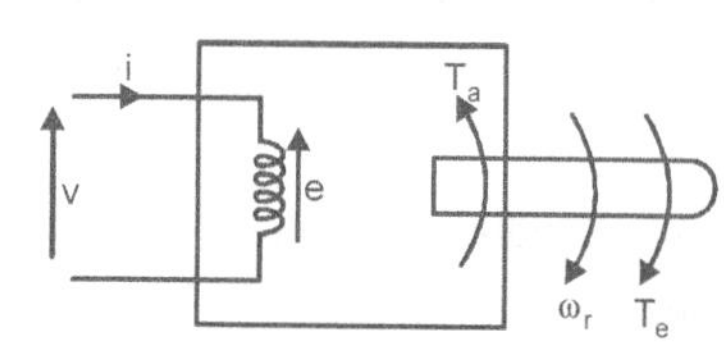

(*a*) both motoring and generating

(*b*) motoring

(*c*) generating

(*d*) braking [IAS]

195. The rotor of a three-phase induction machine is replaced by a rotor with a commutator type winding having no external connection between the brushes. The stator is supplied from a balanced three-phase source and the rotor is driven at twice the synchronous speed in the direction opposite to the stator mmf. The frequency of the emf induced between consecutive brushes will be (*f* is the supply frequency) [ES]

(*a*) zero (*b*) *f*

(*c*) 2*f* (*d*) 3*f*

196. Given that for the slip test on a three-phase synchronous generator

E_1 = impressed maximum voltage per phase

E_2 = impressed minimum voltage per phase

I_1 = maximum current per phase

I_2 = minimum current per phase.

The direct axis synchronous reactance 'X_d' and quadrature axis synchronous reactance 'X_q' will be respectively

(*a*) $\frac{E_1}{I_2}$ and $\frac{E_2}{I_1}$ (*b*) $\frac{E_1}{I_1}$ and $\frac{E_2}{I_1}$

(*c*) $\frac{\sqrt{3}}{2}\frac{E_1}{I_2}$ and $\frac{\sqrt{3}}{2}\frac{E_2}{I_1}$ (*d*) $\frac{\sqrt{3}}{2}\frac{E_1}{I_1}$ and $\frac{\sqrt{3}}{2}\frac{E_2}{I_2}$

[IAS]

197. For a given load, the speed of shaded-pole single-phase induction motor fluctuates slightly because **[ES]**

(*a*) the speed of the magnetic field is not constant

(*b*) the magnitude of the magnetic field is not constant

(*c*) the motor operates at higher value of slips

(*d*) the motor is used with fluctuation load

198. In which one of the following sets of 3-phase transformer, connections with zero sequence current be present in the transformer windings?

(*a*) Primary in star, neutral grounded; secondary in star, neutral not grounded

(*b*) Primary in star, neutral grounded; secondary in delta

(*c*) Primary in star, neutral not grounded; secondary in delta

(*d*) Primary in star, neutral grounded; secondary in delta

[ES]

199. A generator is connected to an infinite bus through a double circuit transmission. The fault occurring at the middle of one of the transmission lines is subsequently cleared by opening the circuit breakers at both the ends of the line simultaneously. The transient stability limit of the system is improved by

(*a*) decreasing the excitation of generator

(*b*) decreasing the fault-clearing time

(*c*) increasing the fault-clearing time

(*d*) increasing the transfer reactance between the generator and infinite bus **[IAS]**

200. In a shaded pole motor, shading coils are used to

(*a*) reduce windage losses

(*b*) reduce friction losses

(*c*) produce rotating magnetic field

(*d*) to protect against sparking **[ES]**

201. A synchronous motor is operating on no-load at unity power factor. If the field current is increased, the power factor will become

(*a*) leading and the current will decrease

(*b*) lagging and the current will increase

(*c*) lagging and the current will decrease

(*d*) leading and the current will increase **[ES]**

202. If the supply frequency in a transformer is doubled, then

(*a*) hysteresis loss also doubles

(*b*) eddy current loss also doubles

(*c*) iron losses double

(*d*) copper losses double **[AMIE]**

203. Core loss in a transformer is usually **[ES]**

(*a*) 1 % to 3 % (*b*) 10 % to 15 %

(*c*) 25 % to 30 % (*d*) 40 % to 50 %

204. Which of the following is minimized by laminating the core of the transformer ?

(*a*) Hysteresis loss

(*b*) Eddy current loss

(*c*) Heat loss

(*d*) Mechanical vibration **[SCETE WB]**

205. The copper loss of a certain transformer at half-fully load is measured as 400 watts. The copper loss at full load will be

(*a*) 800 watts (*b*) 200 watts

(*c*) 400 watts (*d*) 1600 watts **[AMIE]**

206. An open-circuit test on a transformer is performed basically to determine

(*a*) copper loss (*b*) iron loss

(*c*) regulation (*d*) leakage impedance

[AMIE]

207. When a transformer is operating on no load, the primary applied voltage is approximately balanced by

(*a*) primary induced emf

(*b*) secondary induced emf

(*c*) terminal voltage across the secondary

(*d*) voltage drop across the resistance **[ES]**

208. In the transformer circuit mode, the core loss is represented as a

(*a*) series resistance

(*b*) series inductance

(*c*) shunt resistance

(*d*) shunt inductance **[AMIE]**

209. A transformer can have zero regulation at

(*a*) zero pf (*b*) unity pf

(*c*) lagging pf (*d*) leading pf **[SCE TE WB]**

210. In a transformer the hysteresis and eddy current losses depend upon

I. load current

II. maximum flux density in the core

III. type of lamination

IV. supply frequency

The correct statements are

(*a*) I and III (*b*) II and III

(*c*) I and IV (*d*) II and IV **[BTEMP]**

211. The efficiency of two identical transformer under load conditions can be determined by **[ES]**

(*a*) back to back test (*b*) open circuit test

(*c*) short circuit test (*d*) any of these

212. If the field circuit of a shunt motor, equipped with 3-point starter, gets interrupted accidentally while running normally, then the

(*a*) starter arm will fly back to the off position

(*b*) speed of the motor will rise dangerously

(*c*) armature winding will draw heavy current

(*d*) motor will continue to run normally **[IAS]**

213. The fundamental and the second-harmonic component of the field-flux-waves (having their respective positive-maximum at the same spatial position) induce the maximum emf of E1 and E2 in each side of a full-pitch coil respectively when the coil is rotated at certain speed. At the instant when the sides of the coil are cutting the maximum flux, the resultant voltage across the coil terminals will be

(*a*) $E_1 + E_2$ (*b*) $2(E_1 + E_2)$

(*c*) $2\sqrt{E_1^2 + E_2^2}$ (*d*) $2E_1$ **[IAS]**

214. The torque speed characteristic of two phase induction motor is largely affected by **[ES]**

(*a*) voltage

(*b*) speed

(*c*) X/R ratio

(*d*) supply voltage frequency

215. A 500 kVA transformer has constant losses of 500 W and copper losses at full load are 2000 W. Then at what load is the efficiency maximum ?

(*a*) 250 kVA (*b*) 500 kVA

(*c*) 1000 kVA (*d*) 125 kVA **[ES]**

216. Consider the following statements about the operations of a three-phase induction motor running at full load:

If one of the supply line fuse blows off, the motor will

1. stall
2. continue to run with increased slip
3. continue to run with normal supply current
4. continue to run with excessive supply current

Of these statements:

(*a*) 1 alone is correct

(*b*) 3 alone is correct

(*c*) 2 and 3 are correct

(*d*) 2 and 4 are correct **[IAS]**

217. What does the use of higher flux density value in a transformer design load to ?

(*a*) Increase in weight per kVA

(*b*) Decrease in weight per kVA

(*c*) Reduced iron losses

(*d*) Reduced copper losses **[ES]**

218. The losses on a transformer are **[ES]**

I. Copper losses

II. Eddy current losses

III. Hysteresis losses

The constant power loss of a transformer loss is given

(*a*) I only (*b*) I and II only

(*c*) II and III only (*d*) I, II and III

219. The leakage flux in a transformer depends upon the

(*a*) applied voltage (*b*) frequency

(*c*) load current (*d*) mutual flux **[IAS]**

220. Which one of the following is the type of single phase induction motor having the highest power factor at full load ?

(*a*) Shaded-pole type (*b*) Split-phase type

(*c*) Capacitor-start type (*d*) Capacitor-run type **[ES]**

221. An auto-transformer having a transformation ratio of 0.8 supplies a load of 10 kW. The power transferred inductivity from the primary to the secondary is

(*a*) 10 kW (*b*) 8 kW

(*c*) 2 kW (*d*) zero **[ES]**

222. 1 kVA, 230 V, 50 Hz single phase transformer has an eddy currents loss of 30 watts. The eddy current loss when the transformer is excited by a dc source of same voltage will be

(*a*) 30 watts (*b*) more than 30 W

(*c*) less than 30 watts (*d*) zero watts **[AMIE]**

223. A transformer has hysteresis loss of 30 W, at 240 V, 60 Hz. The hysteresis loss at 200 V, 50 Hz will be

(*a*) 28 W (*b*) 25 W

(*c*) 30 W (*d*) 36 W **[ES]**

224. A single phase transformer has specification as 250 kVa, 11000 V/ 415 V, 50 Hz. What are the approximate value of primary and secondary currents ?

(*a*) Primary = 22.7 A, Secondary = 602 A

(*b*) Primary = 602 A, Secondary = 22.7 A

(*c*) Primary = 301 A, Secondary = 22.7 A

(*d*) Primary = 11.4 A, Secondary = 301 A **[AMIE]**

225. The no load current in a transformer lags the applied voltage by

(*a*) 90° (*b*) 0°

(*c*) 75° (*d*) about 115°C **[BTEUP]**

226. In a transformer, the exciting current will be in phase quadrature with the impressed voltage provided

(*a*) only the leakage impedance drop is ignored

(*b*) only the core loss is ignored

(*c*) both the leakage impedance drop and the core loss are ignored

(*d*) only the no load copper loss is ignored **[AMIE]**

227. The flux in transformer core

(*a*) increase with load

(*b*) decrease with load

(*c*) remains constant irrespective of load

(*d*) none of these **[AMIE]**

228. When a transformer is operating on no load, the primary applied voltage is approximately balanced by

(*a*) primary induced emf

(*b*) secondary induced emf

(*c*) terminal voltage across the secondary

(*d*) voltage drop across the resistance and reactance **[AMIE]**

229. In the transformer circuit mode, the core loss is represented as a

(*a*) series resistance (*b*) series inductance

(*c*) shunt resistance (*d*) shunt inductance **[AMIE]**

230. The voltage regulation of a large transformer is mainly influence by

(*a*) no-load current and load power factor

(*b*) winding resistances and load power factor

(*c*) leakage fluxes and load power factor

(*d*) winding resistances and core loss **[ES]**

231. In the statement, "In a star/delta transformer, there exists a phase difference of 'X' between the corresponding phases of primary and secondary 'Y', 'X' and 'Y' stand respectively for

(*a*) 60° and voltages (*b*) 60° and currents **[ES]**

(*c*) 30° and voltages (*d*) 30° and currents

232. Consider the following statements:

To control the speed of a dc shunt motor above the base speed over a reasonably wide range, the motor must

1. have compensating winding.
2. have interpole winding.
3. be started using a 3-point starter.
4. be started using a 4-point starter.

(*a*) 1, 2 and 4 are correct

(*b*) 1, 2 and 3 are correct

(*c*) 1 and 3 are correct

(*d*) 2 and 4 are correct **[IAS]**

233. Consider the following type of drives:

1. Slip energy recover schemes with converter cascade in the rotor circuit.
2. Stator voltage control using ac voltage controller.
3. Constant V/*f* control (V = applied voltage, *f* = frequency).
4. Variable voltage and variable frequency schemes with air-gap flux constant.

Those suitable for blowers and centrifugal pumps ($T \alpha N^2$) include

(*a*) 1, 3 and 4 (*b*) 1, 2 and 3

(*c*) 1, 2 and 4 (*d*) 2, 3 and 4 **[IAS]**

234. If the rotor power factor a 3-phase induction motor is 0.866, the spiral displacement between the stator magnetic field and the rotor magnetic field will be **[ES]**

(*a*) 30° (*b*) 90°

(*c*) 120° (*d*) 150°

235. Consider the following statements about a three-phase synchronous generator synchronized to an infinite bus when its mechanical input is increased gradually with field current held constant:

1. The power factor of the current supplied becomes more lagging.
2. The power factor of the current supplied improves.
3. The power factor remains unity.
4. The load angle is increased.

Of these statements:

(*a*) 1 alone is correct (*b*) 2 alone is correct

(*c*) 2 and 4 are correct (*d*) 3 and 4 are correct **[IAS]**

236. A bipolar *dc* machine with interpoles has a main pole flux of ϕ per pole and an inter-pole flux of ϕ_1 per pole. The yoke of the machine is divided into four quadrants by the main-pole axis and the commutation axis. The flux-distribution in the quadrants will be

(*a*) $\frac{1}{2}(\phi + \phi_1)$ in all the four quadrants

(*b*) $\frac{1}{2}(\phi - \phi_1)$ in all the four quadrants

(*c*) $\frac{1}{2}(\phi + \phi_1)$ in two diametrically opposite quadrants and $\frac{1}{2}(\phi + \phi_1)$ in the remaining two quadrants

(*d*) $\frac{1}{2}(\phi + \phi_1)$ in two adjacent quadrants, and $\frac{1}{2}(\phi - \phi_1)$ in the remaining two quadrants **[IAS]**

237. Consider the following statements:

Transient stability of a synchronous generator feeding power to an infinite bus through a transmission line can be increased by

1. increasing the steam supply-to the turbine driving the generator during fault clearing.
2. connecting resistors at the generator terminals during fault conditions.
3. employing a faster excitation system.
4. quickly throwing off the generator load.

Which of these statements are correct?

(*a*) 2 and 3 (*b*) 3 and 4
(*c*) 1 and 2 (*d*) 2 and 4 **[IAS]**

238. Two single phase 100 kVA transformers, each different leakage, impedance are connected in parallel. When a load of 150 kVA at 0.8 p.f. lagging is applied

(*a*) both transformers will operate at power factor more than 0.8 lagging
(*b*) both transformers will operate at p.f. more than 0.8 lagging
(*c*) one of the transformers will operate at p.f. more than 0.8 lagging and other will operate at p.f. less than 0.8 lagging
(*d*) both transformers will operate at identical power factors. **[ES]**

239. A synchronous generator with a synchronous reactance of 1.3 pu is connected to an infinit bus whose voltage is 1 pu, through an equivalent reactance of 0.2 pu. For maximum output of 1.2 pu, the alternator emf must be

(*a*) 1.5 *pu* (*b*) 1.56 *pu*
(*c*) 1.8 *pu* (*d*) 2.5 *pu* **[IAS]**

240. In *dc* machine, the field system has to be provided on stator unlike synchronous machines wherein it could be on any member, because **[ES]**

(*a*) it reduces field structure iron losses
(*b*) it gives more uniform air-gap flux distribution
(*c*) commutator action is not possible otherwise
(*d*) dc machines are comparatively low rating machines

241. Pole of dc machine are often laminated to

(*a*) reduce pulsation loss
(*b*) reduce armature reaction
(*c*) reduce iron weight
(*d*) dissipate more heat **[AMIE]**

242. The armature of a *dc* machine is laminated to reduce

(*a*) eddy current loss
(*b*) hysteresis loss
(*c*) copper loss
(*d*) friction and windage losses **[AMIE]**

243. In *dc* machine, the armature winding are placed on the rotor because of the necessity for

(*a*) electromechanical energy conversion
(*b*) generation of voltage
(*c*) commutation
(*d*) development of torque **[AMIE]**

244. The main advantage of using fractional pitch winding is to reduce

(*a*) amount of copper in the winding
(*b*) size-of the machine
(*c*) harmonics in the emf
(*d*) cost of the machine **[BTEIIP]**

245. For a P-pole machine, the relation between electrical and mechanical degrees is given by

(*a*) $\theta_{elect} = \frac{P}{Z}\ \theta$ mech

(*b*) $\theta_{elect} = \frac{-Z}{P}\ \theta$ mech

(*c*) $\theta_{elect} = \frac{Z}{P}\ \theta$ mech

(*d*) none of these **[BTEUP]**

246. The direction of emf generated in a *dc* generator can be found out by considering

(*a*) Flemings' right hand rule
(*b*) Flemings' left hand rule
(*c*) Lenz's law
(*d*) Kirchoff's law **[AMIE]**

247. Laminated yoke in dc motor can reduce

(*a*) speed regulation
(*b*) iron loss
(*c*) temperature rise
(*d*) sparking on load **[AMIE]**

248. Pole shoe of a dc machine is laminated for the purpose of

(*a*) decreasing hysteresis loss
(*b*) decreasing eddy current loss
(*c*) decreasing both hysteresis and eddy current loss
(*d*) manufacturing ease **[AMIE]**

249. An alternator of 400 kW is driven by a prime mover of speed regulation 4% and another alternator of 200 kW is driven by a prime mover of speed regulation 3%. The total they can take, without overloading either of the machine is

(*a*) 600 kW (*b*) 500 kW
(*c*) 450 kW (*d*) 400 kW **[IAS]**

250. Two transformers X and Y with identical ratings and dimension have 0.8 mm and 1.2 mm thick laminations respectively. If R_m and X_m are the magnetizing branch parameteres in the equivalent circuit, then

(*a*) R_m values in both are likely to be equal, but X_m of X is likely to be higher than X_m of Y in

(*b*) X_m values in both are likely to be equal, but R_m of X is likely to be higher than R_m of Y

(*c*) X_m values in both are likely to be equal, but R_m of X is likely to be lower than R_m of Y

(*d*) R_m values in both are likely to be equal, but X_m of X is likely to be lower than X_m of Y **[IAS]**

251. If the load on an induction motor is increased from no load to full load, its slip and the power factor will, respectively

(*a*) decrease, decrease (*b*) decrease,increase
(*c*) increase, decrease (*d*) increase, increase **[ES]**

252. In a synchronous machine, the electrical circuits are armature circuit, field circuit and damper circuit and X_d is *d*-axis synchronous reactance; X_d is *d*-axis transient synchronous reactance and X_d is *d*-axis sub-transient synchronous reactance. If there is no damper winding, then **[ES]**

(*a*) $X'_d = X''_d$ (*b*) $X''_d = X_d$
(*c*) $X'_d = X_d$ (*d*) $X_d = X'_d = X''_d$

253. A cumulative compounded dc motor runs at 100 rpm at no-load. On full load, the flux increases by 10%, whereas the full load drop in the combined resistance of the armature and series field is 5%. Neglecting magnetic saturation, the full load speed will be nearly **[ES]**

(*a*) 863 rpm (*b*) 909 rpm
(*c*) 1000 rpm (*d*) 1050 rpm

254. As compared to the use of a single series *dc* motor for electric traction for a given starting time *t* the series-parallel control using two similar motors with time *t*/2 for each series and parallel operation would give a saving starting energy of **[ES]**

(*a*) 100 % (*b*) 50 %
(*c*) 25 % (*d*) zero

255. Consider the following reasons cited for providing damper bars on the pole faces of a synchronous motor:

1. Starting the motor as a squirrel cage induction motor.
2. To reduce the tendency of oscillation of the rotor with load changes.
3. To provide additional induction motor torque besides the main synchronous motor torque.
4. To reduce the effects of slot harmonics causing noise and vibrations.

Of these reasons: **[ES]**

(*a*) 1 alone is valid (*b*) 1 and 2 are valid
(*c*) 1, 2 and 3 are valid (*d*) 1, 2, 3 and 4 are valid

256. The equivalent resistance and equivalent reactance of a transformer are respectively 0.5 Ω and 0.8 Ω. The voltage regulation will be zero when the load power factor is

(*a*) $\frac{-0.8}{\sqrt{0.89}}$ lagging (*b*) $\frac{0.5}{\sqrt{0.89}}$ lagging

(*c*) $\frac{-0.5}{\sqrt{0.89}}$ lagging (*d*) $\frac{0.8}{\sqrt{0.89}}$ lagging **[IAS]**

257. A *dc* shunt wound motor funds application in **[ES]**

(*a*) electric trains (*b*) tape recorders
(*c*) blowers (*d*) steel rolling mills

258. Following controls are considered for dc motors

I. Control of flux
II. Armature resistance control.
III. Supply voltage control.

Which of the above controls plays significant role in the speed control of dc motors?

(*a*) I only (*b*) II and III only
(*c*) I and III only (*d*) I, II and III **[ES]**

259. A 3-phase induction motor is used as an adjustable speed drive from zero to 2.0 per unit speed using a variable frequency invertor. In the speed range of 1.0 to 2.0 per unit, in order to ensure satisfactory operation, which one of the following sets of quantities to be maintained approximately constant? **[ES]**

(*a*) voltage and power (*b*) flux and torque
(*c*) voltage and torque (*d*) flux and power

260. Plugging of *dc* motors is normally done by

(*a*) connecting a resistance across the armature
(*b*) reversing simultaneously the armature and field polarity
(*c*) reversing the field polarity
(*d*) reversing the armature polarity **[IAS]**

261. An induction generator

(*a*) take reactive power from the mains and supplies active power to the mains
(*b*) takes active power from the mains and gives reactive power to the mains
(*c*) supplies both active and reactive powers to the mains
(*d*) consumes both active and inactive powers from the mains **[IAS]**

262. Zero sequence impedance of a 3-phase induction motor is due to

(*a*) stator circuit 3rd time harmonic
(*b*) stator circuit 3rd space harmonic
(*c*) rotor circuit 3rd time harmonic
(*d*) rotor circuit 3rd space harmonic **[ES]**

263. Consider the following statements:

A dc shunt motor starter ensure that

1. armature current is under limit during.
2. field flux is maintained at the maximum.
3. acceleration time is controlled.
4. starting torque is always more than the load torque.

Of these statements:

(*a*) 1, 2, 3 and 5 are correct

(*b*) 1, 2, 3 and 4 are correct

(*c*) 1 and 4 are correct

(*d*) 1, 3 and 4 are correct **[IA S]**

264. In a dc generator operating on load with its brushes on the geometrical neutral axis (GNA), the magnetic neutral axis (MNA) is shifted in the direction of rotation **[ES]**

Now, if the brushes are given a lead of 90° (electrical), then the MNA will

(*a*) shift forward by 90°

(*b*) shift forward by more than 90°

(*c*) shift forward by less than 90°

(*d*) coincide with the GNA

265. Reluctance torque in rotating machine is present, when

(*a*) air gap is not uniform

(*b*) reluctance seen by rotor mmf varies

(*c*) reluctance seen by stator mmf varies

(*d*) reluctances seen by the working mmf varies **[BTEUP]**

266. Rotational losses in electrical machines consist of

(*a*) friction and windage losses

(*b*) stator core, friction and windage losses

(*c*) rotor core, friction and windage losses

(*d*) stray load losses and friction and windage losses **[BTEUP]**

267. The commutator in a dc machine can convert

(*a*) *ac* to *dc*

(*b*) *dc* to *ac*

(*c*) both (*a*) and (*b*)

(*d*) none of these **[BTEUP]**

268. A 4 pole *dc* generator is running at 1500rpm. The frequency of the current in the armature is

(*a*) 25 Hz (*b*) 50 Hz

(*c*) Zero Hz (*d*) 100 Hz **[BTEUP]**

269. The dc generator works on the principle of

(*a*) fleming's left hand rule

(*b*) fleming's right hand rule

(*c*) lenz's law

(*d*) none of these **[AMIE]**

270. A circular metallic disc is placed in a vertical magnetic field of constant induction in the downward direction. If the disc is rotated in horizontal plane, the emf induced will be

(*a*) zero

(*b*) constant independent of disc size

(*c*) increasing in the outward direction

(*d*) decreasing radially in the outward direction **[AMIE]**

271. Emf induced in each conductor of a dc machine is

(*a*) an alternating emf **[ES]**

(*b*) a direct emf

(*c*) a pulsating emf

(*d*) emf of random waveshape

272. Which device illustrate the principle of electromechanical conversion

(*a*) amplifier

(*b*) battery charger

(*c*) transformer

(*d*) electric bell **[BTEMP]**

273. Match List-I (Type of winding) with List-II (Feature) and select the correct answer using the codes given below the Lists: **[ES]**

List-I

A. Fractional slot windings

B. Chorded and distributed windings

C. Damper windings

D. Single-phase windings

List-II

1. Only two-third of the pole pitch is occupied by the winding
2. Effects of slot harmonics are reduced or even eliminated.
3. The effects of phase belt harmonics are reduced
4. Additional asynchronous torque capability in projected pole motors.

Codes:

	A	B	C	D
(*a*)	2	3	4	1
(*b*)	2	3	1	4
(*c*)	3	2	1	4
(*d*)	3	2	4	1

274. While conducting a "slip" test for determination of direct and quadrature-axis synchronous reactance X_d and X_q of salient pole synchronous machine, the rotor of the machine is run with a slip '*s*' and stator supply frequency '*f*'. The frequency of

1. voltage induced across envelope field terminals.
2. envelope of the armature terminal voltage.
3. envelope of the armature current, and
4. armature current.

will be respectively [ES]

(*a*) *sf*, *sf*, *sf* and *f*

(*b*) *sf*, *f*, *sf* and *f*

(*c*) *f*, *sf*, *f* and *sf*

(*d*) f, $(1-s)f$, $(2-s)f$ and sf

275. An alternator is delivering power to a balanced load at unit power factor. The phase angle between the line voltage and the line current is [ES]

(*a*) 90° (*b*) 60°

(*c*) 30° (*d*) 0°

276. A 3-phase wound rotor induction motor, when started with load connected to its shaft, was found to start but settle down at about half synchronous speed. If the rotor winding as well as the stator winding were star connected, then the cause of the malfunctioning could be attributed to

(*a*) one of the stator phase windings being short-circuited

(*b*) one of the supply fuses being blown

(*c*) one of the rotor phases being open circuited

(*d*) two of the rotor phases being open circuited [ES]

277. In the cases of a converter-inverter speed control arrangement for an induction motor operating with $\frac{v}{f}$ constant and with negligible stator impedance,

(*a*) the maximum torque is independent of frequency

(*b*) the maximum torque is proportional to frequency

(*c*) the slip at maximum torque is proportional to frequency [ES]

(*d*) the starting torque is proportional to frequency

278. In machine tool drive application, the speed of a separately excited dc required to be controlled both below and above the rated speed of the motor. Which one of the following method is best suited for this purpose? [ES]

(V_a – armature voltage, I_f = field current and I_f rated = current at rated speed).

(*a*) $I_f = I_f$ rated V_a variable

(*b*) V_a = fixed, I_f variable

(*c*) V_a and I_f variable, with $I_f \leq I_f$ rated

(*d*) V_a and I_f variable, with $I_f \geq I_f$ rated

279. In a transformer fed from a fundamental frequency voltage source, the source of harmonics is the

(*a*) overload

(*b*) poor insulation

(*c*) iron loss

(*d*) saturation of core [ES]

280. A 3-phase induction motor when started picks up speed but runs stably at about half the normal speed. This is because of

(*a*) unbalance in the supply voltages

(*b*) non-sinusoidal nature of the supply voltage

(*c*) stator circuit asymmetry

(*d*) rotor circuit asymmetry [ES]

281. Which motor should not be used for centrifugal pumps ?

(*a*) Shunt

(*b*) Series

(*c*) Cumulatively compound

(*d*) Kifferently compound [ES]

282. If the applied voltage of a certain transformer is increased by 50% and the frequency is reduced to 50% (assuming that the magnetic circuit remains unsaturated), then maximum core flux density will

(*a*) change to three times the original value

(*b*) change to 1.5 times the original value

(*c*) change to 0.5 times as the original value

(*d*) remain the same as the original value [ES]

283. A 3-phase alternator is connected to D*d* transformer. The hysteresis and eddy current losses of the transformer arerespectively 300 W. If the speed of the alternator is reduced by 10%, then the hysteresis and eddy current losses of the transformer will be respectively

(*a*) 228 W and 262.44 W

(*b*) 243 W and 324 W

(*c*) 243 W and 360 W

(*d*) 270 W and 400 W [ES]

284. Which of the following statements regarding skewing of motor bars in a squirrel-cage induction motor are correct?

1. It prevents cogging.
2. It produces more uniform torque.
3. It increases starting torque.
4. It reduces motor 'hum' during its operation.

Select the correct answer using the codes given below.

Codes:

(*a*) 2, 3 and 4 (*b*) 1, 2 and 3

(*c*) 1, 3 and 4 (*d*) 1, 2 and 4 [ES]

285. For a given developed power, a synchronous motor operating from a constant voltage and constant frequency supply, will draw the minimum and maximum armature currents I_{min} and I_{max} respectively, corresponding to

(*a*) I_{max} at unity pf, but I_{min} at zero pf

(*b*) I_{min} at unity pf, but I_{max} at zero pf

(*c*) both I_{min} and I_{max} at unity pf

(*d*) both I_{min} and I_{max} at zero pf [ES]

286. In a shaded pole squirrel cage induction motor the flux in the shaded part always
(*a*) lead the flux in the unshaded pole segment
(*b*) is in phase with the flux in the unshaded pole segment
(*c*) lags the flux in the unshaded pole segment
(*d*) none of these **[AMIE]**

287. The phase sequence of a three-phase alternator will reverse if
(*a*) the field current is reversed keeping the direction of rotation same
(*b*) the field current remains the same but the direction of rotation is reversed
(*c*) the field current is reversed and the number of pole is doubled
(*d*) the number of poles is doubled without reversing the field current

288. The voltage rating for a ceiling motor will be in the range of
(*a*) 50 to 250 W (*b*) 250 to 500 W
(*c*) 50 to 150 W (*d*) 10 to 20 W **[AMIE]**

289. A dc shunt motor does not operate on ac due to
(*a*) low resistance of armature circuit
(*b*) high resistance of field circuit
(*c*) high inductance of field circuit
(*d*) reversal of polarity **[AMIE]**

290. What will happen if a single phase supply is connected to an ordinary dc series motor?
(*a*) Not run, at all
(*b*) Run, but not satisfactorily
(*c*) Armature burnt out
(*d*) Field winding burnt out **[BTEMP]**

291. AC series motors are built with as few field terms as possible to reduce
(*a*) flux
(*b*) eddy current loss
(*c*) reactance
(*d*) speed **[BTEMP]**

292. Speed of the universal motor is
(*a*) dependent on frequency of supply
(*b*) proportional to frequency of supply
(*c*) independent of frequency of supply
(*d*) none of these **[AMIE]**

293. In a single-phase repulsion motor power factor is
(*a*) always leading
(*b*) high at low speed
(*c*) high at high speed
(*d*) always unity **[AMIE]**

294. In a repulsion motor by shifting the brush position
(*a*) speed and direction of rotation can be controlled
(*b*) only speed can be controlled
(*c*) speed, torque and direction of rotation can be controlled
(*d*) only torque can be controlled by varying the speed at constant output power **[AMIE]**

295. To ensure that two-phase induction type servomotor does not run when the control phase voltage is zero, the ratio of
(*a*) stator leakage reactance to rotor phase resistance should be high
(*b*) rotor leakage reactance to rotor phase resistance should be less than one
(*c*) the sum of stator and rotor resistances to the sum of the stator and rotor leakage reactances should be greater than one
(*d*) torque to inertia should be very high **[IAS]**

296. Short-circuit ratio of a synchronous machine is the ratio of the
(*a*) open-circuit voltage to the short-circuit current at a given excitation
(*b*) short-circuit current to the open-circuit voltage at a given excitation
(*c*) excitation current required for the rated voltage on open-circuit to the excitation current required for the rated armature current under-short-circuit
(*d*) excitation current required for the rated armature current under short-circuit to the excitation current required for rated voltage on open-circuit **[IAS]**

297. Consider the following statements in connection with the operating characteristics of a dc series motor:
1. It is a constant torque-variable speed motor.
2. Its torque is inversely proportional to speed.
3. Its starting torque is high.
4. The power drawn from the supply by the motor remains almost constant at different load torques.

Of these statements:
(*a*) 1, 3 and 4 are true
(*b*) 1, 2 and 3 are true
(*c*) 2, 3 and 4 are true
(*d*) 2 and 4 are true **[IAS]**

298. A 240 Vdc series motor takes 40 A when giving its rated output at 1500 rpm. Its resistance is 0.3 ohms. The value of resistance which must be added to obtain rated torque at 1000 rpm is,
(*a*) 6 ohms (*b*) 5.7 ohms **[ES]**
(*c*) 2.2 ohms (*d*) 1.9 ohms

299. A transformer has a resistance of 2% and reactance of 4%. Its regulation at 0.8 power factor lagging a leading respectively are

(*a*) 4 per cent and – 0.8 per cent

(*b*) – 4 per cent and 0.8 per cent

(*c*) – 0.8 per cent and 4 per cent

(*d*) 0.8 per cent and – 4 per cent **[IAS]**

300. Two-3 phase transformers are connected to the same supply of balanced 3-phase voltages on the primary side. The first transformer is connected in star-delta and the second in delta-star on the primary and secondary sides respectively. The phase difference between the line-to-line voltages of the corresponding phases on the secondary sides of the two transformers is

(*a*) 0° (*b*) 30°

(*c*) 60° (*d*) 90° **[IAS]**

301. The stator of a 415, 50 Hz, 6-pole, 970 rpm, 3-phase sliping induction motor is fed with normal voltage. The rotor with slip rings open-circuited is rotated at 500 rpm along the normal direction of its rotation. The frequency of the voltage across slip rings will be

(*a*) 1.5 Hz

(*b*) 25 Hz

(*c*) 26.5 Hz

(*d*) 75 Hz **[IAS]**

302. List-I contains the types of fractional kilowatt motors and List-II their torque-speed characteristics. Match List-I with List-II and select the correct answer using the codes given below the lists:

List-I (Types of motor)

A Hysteresis motor

B. Capacitor-start split phase motor

C. Reluctance motor

List-II (Torque speed characteristic)

1. 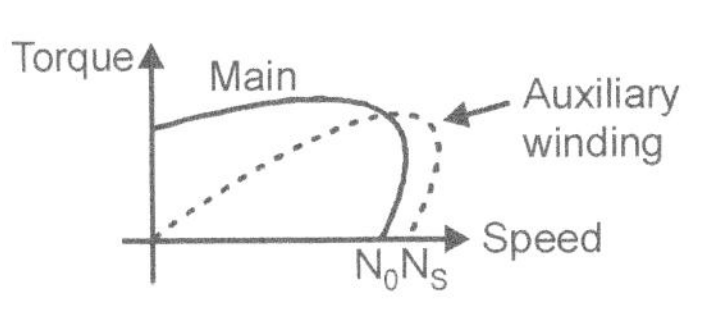

2. 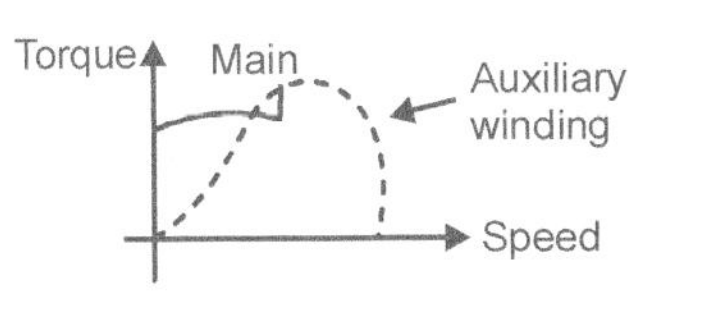

3. 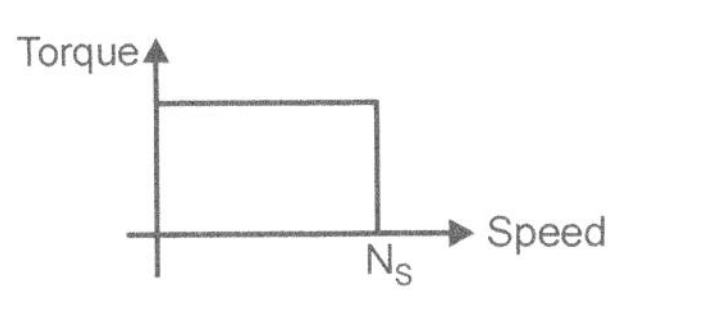

Codes:

	A	B	C
(*a*)	3	1	2
(*b*)	2	3	2
(*c*)	2	1	3
(*d*)	1	3	2

[ES]

303. A synchronous machine working in the motor mode, is fed from an infinite bus and is delivering half full-load. If an increase infield current causes an increase in the armature current, then the motor will

(*a*) deliver reactive power and active power to the bus

(*b*) absorb reactive power and active power from the bus

(*c*) deliver reactive power to the bus and absorb active power from the bus

(*d*) none of these **[IAS]**

304. The single-phase induction motors is started with

(*a*) capacitance connection

(*b*) inductance connection

(*c*) resistance connection

(*d*) shaded pole arrangement **[IAS]**

305. In a 3-phase induction motor, if stator impedance is neglected, then the slip at maximum torque is equal to

(*a*) $\frac{x_2}{r_1}$

(*b*) $\frac{x_2}{r_2}$

(*c*) $\sqrt{\frac{x_2}{r_2}}$

(*d*) $\sqrt{\frac{x_1}{r_1}}$ **[BTEUP]**

306. In a 3-phase induction motor if the leakage reactance is reduced by using open slots

(*a*) starting torque and starting current will decrease but power factor will increase

(*b*) starting torque and starting current both will increase but power will decrease

(*c*) pull out torque will decrease

(*d*) starting current will increase but starting torque will decrease **[AMIE]**

307. In a 3-phase induction motor the mechanical power developed in terms of air gap power P_g is

(*a*) $(s-1)P_g$ (*b*) $\frac{P_g}{1-s}$

(*c*) $\frac{1}{s}$ (*d*) $(1-s)P_g$ **[BTEUP]**

308. The stator referred resistance in the equivalent circuit of an induction motor, representing mechanical output is

(*a*) $\frac{r_1^1}{s}$ (*b*) $r_2^1\left(\frac{1}{s}-1\right)$

(*c*) $\frac{r_2}{s}$ (*d*) $r_1^2\left(\frac{1}{s}-1\right)$ **[AMIE]**

309. The power factor of a lightly loaded induction motor is likely to be maximum when it will operate at **[ES]**

(*a*) full load

(*b*) no-load

(*c*) maximum slip

(*d*) maximum torque

310. The power factor of a lightly loaded induction motor is quite low because

(*a*) the current drawn is largely a magnetising current due to air gap

(*b*) of the current due to air-gap

(*c*) the current drawn is largely a magnetising component due to laminated core

(*d*) the current drawn is largely an energy component due to laminated core **[BTEMP]**

311. The power factor of a squirrel cage induction motor is

(*a*) low at light loads only

(*b*) low at heavy loads only

(*c*) low at light and heavy load both

(*d*) low at rated load only **[AMIE]**

312. The power factor of a squirrel cage induction motor is

(*a*) low at light loads only

(*b*) low at heavy loads only

(*c*) low at light and heavy loads both

(*d*) low at rated load only **[AMIE]**

313. Match List-I with List-II and select the correct answer using the codes given below the lists:

List-I

A. Voltage drop test

B. Hopkinson's test

C. Swineburne's test

D. Retardation test

List-II

1. Efficiency.
2. Separation of iron and friction losses.
3. Open and short circuited armature coils.
4. Temperature rise.

Codes:

	A	B	C	D
(*a*)	3	1	4	2
(*b*)	2	4	1	3
(*c*)	3	4	1	2
(*d*)	2	1	4	3

[ES]

314. In a synchronous generator, a divided winding rotor is preferable to a conventional winding rotor because of

(*a*) higher efficiency

(*b*) increased steady-state stability limit

(*c*) higher short-circuit ratio

(*d*) better damping **[ES]**

315. A 6-pole, 50 Hz would rotor induction motor when supplied at the rated voltage and frequency with slip ring open-circuited, developed a voltage of 100 V between any two rings. If the rotor is driven by and external means at 1000 rpm opposite to the direction of stator field, the frequency of voltage across slip rings will be **[ES]**

(*a*) zero (*b*) 50 Hz

(*c*) 100 Hz (*d*) 200 Hz

316. The interpoles in the dc machines have a tapering shape in order to

(*a*) reduce the overall weight

(*b*) reduce the saturation in the interpole

(*c*) economise on the material required for interpoles and their winding

(*d*) increase the acceleration of commutation **[ES]**

317. In a 3-phase Δ/Y transformer shown in the figure, the phase displacement of secondary line voltag with corresponding primary line voltage will be

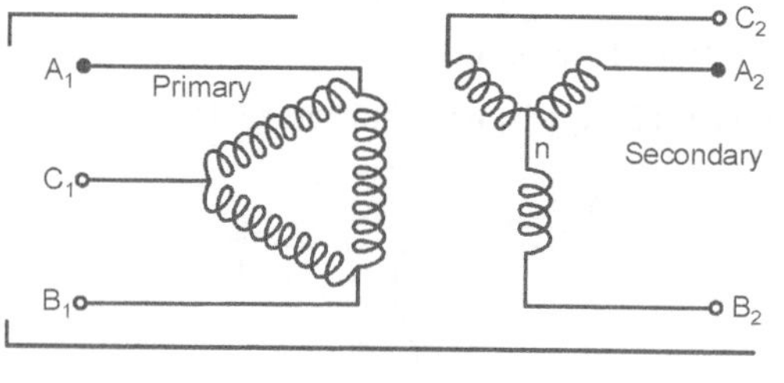

(*a*) zero (*b*) 30° lag

(*c*) 30° lead (*d*) 180° **[ES]**

318. In hand-tool applications, which one of the following single-phase motors is used ?

(*a*) Shaded-pole motor

(*b*) Capacitor start motor

(*c*) Capacitor run motor

(*d*) AC series motor **[ES]**

319. Consider the following statements regarding horse power shaded-pole motor **[ES]**

1. Its directions of rotation is from unshaded to shaded portion of the poles.
2. Its direction of rotation is from shaded to unshaded portion of the poles,
3. It can remain stalled for short period without any harm.
4. It is very poor power factor.

Of these statements:

(*a*) 1, 3 and 4 are correct
(*b*) 2, 3 and 4 are correct
(*c*) 2 and 4 are correct
(*d*) 1 and 3 are correct

320. Consider the following statements regarding speed control of induction motors by means of external resistors:

1. Reduction in speed is a accompanied by reduced efficiency.
2. With a large resistance in the rotor circuit, the speed would very considerably with variation in torque.
3. The method is very complicated.

The disadvantages of such a method of speed control would include

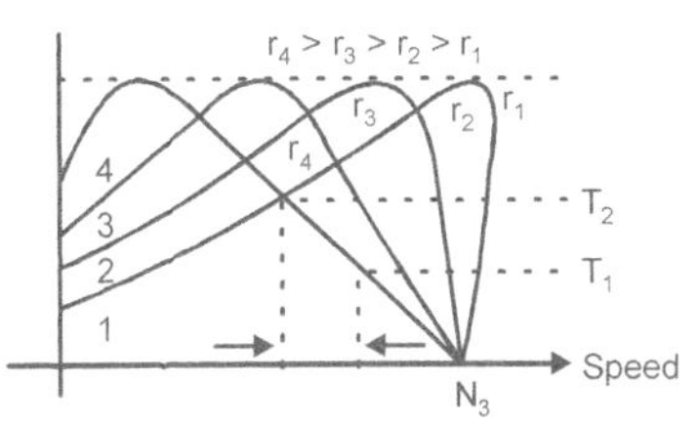

(*a*) 1 and 2 (*b*) 2 and 3
(*c*) 1 and 3 (*d*) 1, 2 and 3 **[ES]**

321. Match List-I (Power fector) with List-II (Armature reaction of an alternator) and select the correct answer using the codes given below the lists:

List-I	**List-II**
A. UPf	1. Fully demagnetizing.
B. ZPF	2. Fully magnetising.
C. ZPF leading	3. Cross-magnetising.
D. Intermediate lagging	4. Partly cross-magnetising and partly demagnetising

Codes:

	A	B	C	D
(*a*)	3	1	4	2
(*b*)	3	1	2	4
(*c*)	1	3	2	4
(*d*)	1	3	4	2

[ES]

322. In a dc machine, the demagnetising effect of armature reaction is due to

(*a*) component of armature mmf along field axis
(*b*) non-sinusoidal nature of armature mmf
(*c*) magnetic saturation in half of the field pole
(*d*) uneven air gap length **[IAS]**

323. A separately excited dc motor is required to be controlled from a 3-phase source for operation in the first quadrant only. The most preferred converter would be

(*a*) fully controlled converter
(*b*) fully controlled converter with free-wheeling diode
(*c*) half-controlled converter
(*d*) sequence control of two series connected fully controlled converter **[ES]**

324. Consider the following statements regarding the design of distributed armature winding in a 3-phase alternator:

(*a*) It reduce the phase belt harmonics
(*b*) It increases utilisation of armature iron and copper
(*c*) It increases rigidity and mechanical strength of the winding
(*d*) It reduces copper in the overhang of the winding **[ES]**

325. If two induction motors A and B are identical except that the air-gap of motor 'A' is 50 % greater than that of motor 'B', then

(*a*) the no-load power factor of A will be better than that of B
(*b*) the no-load power factor of A will be poorer than that of B
(*c*) the core losses of A will be more than those of B
(*d*) the operating flux of A will be smaller than that of B **[ES]**

326. On direct on line (DOL) starting, dc motor is found to rotate in the direction opposite to that for which it was designed. The motor is a

(*a*) series motor
(*b*) shunt motor
(*c*) cumulatively compounded motor
(*d*) differential compounded motor **[IAS]**

327. Which one of the following method would give a higher than actual of regulation of an alternator?

(*a*) ZPF method
(*b*) MMF method
(*c*) EMF method
(*d*) ASA method

328. In a 3-phase squirrel cage induction motor, skewing of rotor slots reduces

(*a*) parasitic torque and noise but increase pullout torque
(*b*) parasitic torque and noise but increases starting torque
(*c*) noise but increase pullout torque and parasitic torque
(*d*) parasitic torque, noise, pullout torque and starting torque **[ES]**

329. A 40 kVA transformer has a core loss of 400 W. The proportion of full-load at maximum efficiency is

(*a*) 50%
(*b*) 62.3%
(*c*) 70.7%
(*d*) 100% **[ES]**

330. Consider the following statements:

Permanent magnet dc motors used in cassette tape recorders have

1. magnets on stator and armature on the rotor.
2. magnets on the rotor and armature on the stator.
3. electronic commutation and no brushes.
4. mechanical commutation and no brushes.
5. automatic speed governors.

Of these statements :

(*a*) 1, 3 and 5 are correct
(*b*) 1, 4 and 5 are correct
(*c*) 2, 3 and 5 are correct
(*d*) 1, 4 and 5 are correct **[ES]**

331. An induction motor and synchronous motor are connected to a common feed line. To operate the feeder line at unity pf, the synchronous motor should be

(*a*) under-excited
(*b*) over-excited
(*c*) normally excited
(*d*) disconnected from the common terminals **[ES]**

332. In a synchronous machine, damper winding are used to

(*a*) help in starting as a motor
(*b*) run it as an induction motor
(*c*) help in starting as a motor and to reduce hunting
(*d*) increase efficiency **[ES]**

333. Load frequency control is achieved by properly matching the individual machine's

(*a*) reactive powers
(*b*) generated voltages
(*c*) turbine inputs
(*d*) turbine and generator rating **[ES]**

334. A 3-phase slip-ring induction motor is wound for 4 poles on stator and 6 poles on rotor. When 3-phase balanced voltage source at 50 Hz is applied to the motor, it will run at

(*a*) 1500 rpm (*b*) 1000 rpm
(*c*) 750 rpm (*d*) zero speed **[ES]**

335. A 3-phase slip-ring induction motor has chopper controlled resistance in the rotor circuit. Its power factor and efficiency can be characterised respectively as

(*a*) good and poor (*b*) poor and good
(*c*) good and good (*d*) poor and poor **[ES]**

336. Match List-I with List-II and select the correct answer using the codes given below the lists : **[ES]**

List-I (Flux due to)

A. Currents in damper winding
B. Currents in filed winding
C. Armature current with unity internal pf
D. Armature currents with zero pf

List-II

1. Direct axis synchronous reactance.
2. Quadrature axis synchronous reactance.
3. Transient reactance.
4. Sub-transient reactance.

Codes:

	A	B	C	D
(*a*)	4	3	1	2
(*b*)	3	4	2	1
(*c*)	3	4	1	2
(*d*)	4	3	2	1

337. A 3-phase induction machine draws active power 'P' and reactive power 'Q' from the grid. If it is operated as a generator, P and Q will respectively be

(*a*) positive and negative
(*b*) negative and negative
(*c*) positive and positive
(*d*) negative and positive **[ES]**

338. Consider the following statements:

1. Interpole winding are connected in series with armature winding.
2. Polarity of interpole must be the same as that of the main pole in advance
3. Distortion of the main field under the pole shoes is not affected by the use of interpoles.

Which of these statements regarding interpole used in dc motors is/are correct?

(*a*) 1 and 2 (*b*) 2 and 3
(*c*) 1 and 3 (*d*) 1 alone **[ES]**

339. A transformer is supplying a unity factor load. The power factor at primary terminals will be

(*a*) about 0.95 lagging (*b*) about 0.95 leading

(*c*) about 0.96 lagging (*d*) unity **[BHEL]**

340. When two identical transformer are available, then the best method to find their efficiency under load condition is

(*a*) short-circuit test (*b*) back to back test

(*c*) open-circuit test (*d*) phase sequence test **[BHEL]**

341. In a 4-pole, 50 Hz induction machine, the slip rings of the machine are open-circuited. The frequency of the voltage across slip rings in 75Hz. The rotor speed is

(*a*) 750 rpm (*b*) 1500 rpm

(*c*) 2250 rpm (*d*) 3750 rpm **[NTPC]**

342. A10 kW, 400 V, 3-phase induction motor is started with direct-on line starter. The motor current and pf at starting will respectively be

(*a*) 120 A, 0.2 (*b*) 20 A, 0.2

(*c*) 120 A, 0.6 (*d*) 20 A, 0.6 **[IOCL]**

343. If dc supply is given to a transformer, it will

(*a*) work without any problem

(*b*) not work

(*c*) give lower voltage than the rated voltage on the secondary side

(*d*) burn the winding **[IOCL]**

344. Hysteresis loss in a transformer depends upon the

(*a*) applied voltage

(*b*) type of core material

(*c*) number of lamination

(*d*) reactance of the winding **[IOCL]**

345. In any transformer, the voltage per turn in primary and secondary remains

(*a*) always same

(*b*) always in ratio of K

(*c*) always different

(*d*) sometimes same **[NHPC]**

346. The reactance per phase as compared to the resistance per phase of an induction motor is

(*a*) quite high (*b*) slightly large

(*c*) almost same (*d*) very small **[NHPC]**

347. An increase in the value of air gap flux density is an induction motor **[ES]**

(*a*) increase iron-loss

(*b*) increase efficiency

(*c*) decrease efficiency

(*d*) both (*a*) and (*c*) above

348. Scott connections are used for

(*a*) single phase to three phase transformation

(*b*) three phase to single phase transformation

(*c*) three phase to two phase transformation

(*d*) any of these **[ES]**

349. While performing the open-circuit and short-circuit tests on a transformers to determine its parameters the status of the low voltage (L.V) and high voltage (H. V) winding will be such that **[ES]**

(*a*) in O.C., L.V. is open and in S.C., H.V. is shorted

(*b*) in O.C., H.V. is open and in S.C., L.V. is shorted

(*c*) in O.C., L.V. is open and in S.C., L.V. is shorted

(*d*) in O.C., H.V. is open and in S.C., H.V. is shorted **[ES]**

350. Match List-I (type of single-phase motor) with List-II (type of appliance) and select the correct answer using the codes given below the List:

List-I	List-II
A. Permanent magnet type	1. Rocket.
B. Shaded pole	2. Refrigerator compressor.
C. Universal	3. Sewing machine.
D. Capacitor start, capacitor run induction motor	4. Photocopying machine.

Codes:

	A	B	C	D
(*a*)	1	2	3	4
(*b*)	1	2	4	3
(*c*)	4	1	3	2
(*d*)	1	4	3	2

[ES]

351. The armature current of a dc motor fed from a thyristor power converter contain ripple. This ripple in the armature current affects.

(*a*) commutation capability of the motor

(*b*) overloaded capacity of the motor

(*c*) torque capability of the motor

(*d*) controllability of the speed of the motor **[ES]**

352. Which of the following devices can be used as a phase advancer? **[ES]**

(*a*) 3 phase induction motor squirrel cage type

(*b*) 3 phase induction motor-slip ring type

(*c*) synchronous motor working at leading power factor

(*d*) synchronous motor working at lagging power factor

353. The speed of a universal motor can be controlled by
(*a*) introducing a variable resistance in series with the motor
(*b*) tapping the field at various points
(*c*) centrifugal mechanism
(*d*) all of these steps **[ES]**

354. Which of the following tests can be conducted on other than shunt machines ?
(*a*) Swinburne's test (*b*) Retardation test
(*c*) Field's test (*d*) Back to back test **[ES]**

355. The winding for an alternator are
I. 36 slots, four poles, span 1 to 3
II. 72 slots, six poles, span 1 to 10
III. 96 slots, six poles, span 1 to 12
The winding having pitch factors of more than 0.9 are
(*a*) I and II only (*b*) II and III only
(*c*) I and III only (*d*) I, II and III **[ES]**

356. An inverted V-curve of synchronous motor shows the variation of
(*a*) field current and dc excitation at constant load
(*b*) supply voltage and field current at constant excitation
(*c*) power factor and supply voltage during hunting
(*d*) supply voltage and excitation current at constant load **[ES]**

357. When starting a differentially connected compound motor, it is desirable to short-circuit the series field winding to
(*a*) avoid very high starting time
(*b*) avoid excessive starting speed
(*c*) prevent the motor from starting in the reverse direction
(*d*) avoid heavy in-rush of current **[IAS]**

358. A single-phase alternator has a synchronous reactance of 2 ohm and negligible resistance. If it supplies 10 A to apurely capacitive load at 200 V, then the generated emf will be
(*a*) 240 volts
(*b*) 220 volts
(*c*) 200 volts
(*d*) 180 volts **[IAS]**

359. For a uniformly distributed winding of an alternator with a phase spread of β degree, the distribution factor is

(*a*) $\frac{\sin\beta}{\beta}$ (*b*) $\frac{2\sin\beta/2}{\beta}$

(*c*) $\frac{\sin\beta/2}{\beta}$ (*d*) $\frac{\sin\beta/2}{2\beta}$ **[IAS]**

360. In synchronous alternator, which of the following coils will have emf closer to sine wave form?
(*a*) Concentrated winding in full pitch coils
(*b*) Concentrated winding in short pitch coils
(*c*) Distributed winding in full pitch coils
(*d*) Distributed winding in short pitch coils **[ES]**

361. The core in electrical machines are generally made of laminations with a view to reduce
(*a*) the eddy current loss
(*b*) the hysteresis loss
(*c*) the copper loss
(*d*) eddy current, hysteresis and copper losses **[ES]**

362. Match List-I with List-II and select the correct answer by using the Codes given below the lists:

List-I	List-II
A. Copper losses	1. Windage losses.
B. Iron losses	2. Hysteresis loss.
C. Mechanical losses	3. I^2R loss in interpole.
	4. Commutator friction loss.
	5. I^2R loss in series field
	6. Bearing friction loss.
	7. Eddy current loss.
	8. I^2R loss in shunt field.

Codes:

	A	B	C
(*a*)	1, 2, 3	4, 5	6, 7, 8
(*b*)	2, 3	1, 4, 5	6, 7, 8
(*c*)	8	1, 2, 3	4, 5, 6, 7
(*d*)	3, 5, 6	2, 7	1, 4, 6

[ES]

363. Breakdown torque of 3-phase induction motor of negligible stator impedance is
(*a*) directly proportional to the rotor resistance
(*b*) inversely proportional to the rotor resistance
(*c*) directly proportional to the reactance
(*d*) inversely proportional to the rotor leakage reactance **[ES]**

364. Consider the following statements regarding the equivalent circuit parameters of an induction motor:
1. Leakage reactance is independent on supply current.
2. Magnetising reactance is dependent on the air-gap flux.
3. Core loss is dependent on the input voltage and frequency
4. Rotor resistance is dependent on the speed.

Of these statements:

(*a*) 2 and 3 are correct

(*b*) 1, 3 and 4 are correct

(*c*) 1, 2, 3 and 4 are correct

(*d*) 1, 2 and 4 are correct **[IAS]**

365. If the applied voltage to a *dc* machine is 230 V, Then the back emf, for maximum power develop is

(*a*) 115 V (*b*) 200 V

(*c*) 230 V (*d*) 460 V **[ES]**

366. Which one of the following methods gives more accurate result for determination of voltage regulation of an alternator

(*a*) mmf method

(*b*) Synchronous impedance method

(*c*) Potier triangle method

(*d*) American Institution Standard method **[ES]**

367. Consider the following statements : **[ES]**

1. In a *dc* series generator, full-load voltage is more than the no load voltage.
2. In a separately excited *dc* generator, full-load voltage is more that the no-load voltage.
3. In a *dc* shunt generator, full-load voltage is less than the no-load voltage.

Of these statements

(*a*) 1, 2 and 3 are correct

(*b*) 1 and 2 are correct

(*c*) 2 and 3 are correct

(*d*) 1 and 3 are correct

368. Starting torque can be obtained in the case of a single phase induction motor with identical main and auxiliary winding by connecting **[ES]**

(*a*) a capacitor across the mains

(*b*) a capacitor in series with the machine

(*c*) a capacitor in series with the auxiliary winding

(*d*) the main and the auxiliary winding in series

369. Crawling of an induction motor is due to

(*a*) space harmonics in the stator flux

(*b*) the number of teeth on stator being equal to the number of teeth on the rotor

(*c*) the fluctuation in load torque

(*d*) unequal distribution of winding **[IAS]**

370. Consider the following statements regarding speed control of dc motors:

1. Word-Leonard method is suitable for constant torque drives.
2. Field control method facilitates speed control below 'base speed'.
3. 'Armature resistance control' method is more efficient when compared to Word Leonard method.
4. Field control method is suitable for constant horse power drives.

Of these statements:

(*a*) 1, 2 and 3 are correct

(*b*) 1, 3 and 4 are correct

(*c*) 2, 3 and 4 are correct

(*d*) 1 and 4 are correct **[ES]**

371. A star/delta starter of a 3-phase induction motor is equivalent to an autotransformer starter hanging a tapping of

(*a*) $\frac{\sqrt{3}}{2}$ (*b*) $\frac{1}{2}$

(*c*) $\frac{1}{\sqrt{3}}$ (*d*) $\sqrt{3}$ **[IAS]**

372. Which one of the following motors running under the specified conditions will have a leading power factor?

(*a*) slip ring induction motor at no-load

(*b*) underexcited synchronous motor

(*c*) overexcited synchronous motor

(*d*) double-cage induction motor at load **[IAS]**

373. A *dc* shunt motor runs at 500 *rpm* at 220 V. A resistance of 4.5 Ω. is added in series with the armature for speed control. The armature resistance is 0.5 Ω. The current to stall the motor will be

(*a*) 44 Amp (*b*) 50 Amp

(*c*) 44.4 Amp (*d*) 60 Amp **[ES]**

374. Can a 50 Hz transformer be used as 25 Hz, if the input voltage is maintained constant at the rated value corresponding to 50 Hz

(*a*) yes, since the voltage is constant, current levels will not change

(*b*) no, flux will be doubled which will drive the core to excessive saturation

(*c*) no, owing to decreased reactance of transformer, input current will be doubled at load

(*d*) yes, at constant voltage, insulation will not be overstressed **[ES]**

375. T is the load torque of a *dc* series motor having linear magnetization and negligible armature resistance speed of the motor is

(*a*) inversely proportional to $\sqrt{T}$

(*b*) directly proportional to $\sqrt{T}$

(*c*) inversely proportional to T^2

(*d*) directly proportional to T^2 **[AMIE]**

376. A *dc* series motor must not be started on no load because it would

(*a*) not develop starting torque

(*b*) foil to start without load

(*c*) run at dangerously high speed

(*d*) draw more current without producing sufficient torque **[AMIE]**

377. In case of a *dc* series motor is possible to have finite no load speed if a resistance is connected across its

(*a*) field terminals

(*b*) armature terminals

(*c*) field and armature together

(*d*) it is always very high at no load **[AMIE]**

378. A *dc* series motor is running at rated speed without any additional resistance in series. If and additional resistance is placed in series, the speed of motor

(*a*) increases

(*b*) decreases

(*c*) remain unchanged

(*d*) oscillate around the rated speed **[AMIE]**

379. A *dc* series motor is used for and over hauling load. It can work stably if

(*a*) the armature is shunted by a resistor

(*b*) the field winding is reversed

(*c*) a resistor is put in series with the machine

(*d*) a diverter is put across the field **[AMIE]**

380. A *dc* shunt motor drives a load at rated speed and supply voltage. If both the load and voltage are halved, the speed of the motor will be

(*a*) almost doubled

(*b*) almost halved

(*c*) rated speed

(*d*) slightly less than rated speed **[AMIE]**

381. Consider the following statements.

A cumulative compounded *dc* motor runs at a speed N, *rpm* drawing an armature current of I amperes. If keeping the load torque constant, the series field winding is short circuited then the **[ES]**

1. speed of the motor will increase
2. armature current will increase
3. armature current will decrease
4. speed of the motor will decrease

Of these statements

(*a*) 3 and 4 are correct

(*b*) 1 and 4 are correct

(*c*) 2 and 4 are correct

(*d*) 1 and 2 are correct

382. In a *dc* shunt generator working on load, the brushes are moved forward in the direction of rotation, as a result of this, commutation will

(*a*) improve but terminal voltage will fall

(*b*) worsen and terminal voltage will fall

(*c*) improve and terminal voltage will rise

(*d*) worsen and terminal voltage will rise **[ES]**

383. A salient pole synchronous motor is running with normal excitation. If the excitation is reduce to zero, **[ES]**

(*a*) it becomes an induction motor

(*b*) it becomes a reluctance motor

(*c*) it remains a synchronous motor

(*d*) the motor would stop

384. Two identical loss-less series motors connected in series across a *dc* supply voltage, run at speed *s* of N_1 and N_2. The ratio of their output powers will be

(*a*) $N_1^2 : N_2^2$ (*b*) 1 : 1

(*c*) $N_1 : N_2$ (*d*) $N_2 : N_1$ **[IAS]**

385. Consider the following statements:

The *emf* induced in the rotor of a 3-phase induction motor is proportional to the

1. relative velocity between stator field and rotor conductors.
2. voltage applied to the stator.
3. slip.

Of these statements:

(*a*) 1, 2 and 3 are correct

(*b*) 1 and 2 are correct

(*c*) 2 and 3 are correct

(*d*) 1 and 3 are correct **[IAS]**

386. The compensating winding in a *dc* machine **[ES]**

(*a*) is located in armature slots for compensation of the armature reaction

(*b*) is located on commutating poles for improving the commutation

(*c*) is located on pole shoes for avoiding the flashover at the commutator surface

(*d*) is located on pole shoes to avoid the sparking at the brushes

387. The stator of a 6-pole, 3-phase induction motor is fed from a 3-phase, 50 Hz supply which contains a pronounced fifth time harmonic. The speed of the fifth space harmonic field produced by the fifth time harmonic in the stator supply will be

(*a*) 200 *rpm* (*b*) 1500 *rpm*

(*c*) 1000 *rpm* (*d*) 5000 *rpm* **[ES]**

388. For an ideal transformer the winding should have

(*a*) maximum resistance on primary side and least resistance on secondary side **[ES]**

(*b*) least resistance on primary side and maximum resistance on secondary side

(*c*) equal resistance on primary and secondary side

(*d*) no ohmic resistance on either side

389. Short-circuit test is performed on a transformer with a certain impressed voltage at rated frequency. If the short-circuited test is now performed with the same magnitude impressed voltage, but at a frequency higher than the rated frequency, then

(*a*) the magnitude of current and power factor will both increase

(*b*) the magnitude of current will decrease but the power factor will increase

(*c*) the magnitude of current will increase but the power factor will decrease

(*d*) the magnitude of current as well as the power factor will decrease **[ES]**

390. The general expression for the instantaneous value of the torque of dynamo-electric, machines is given by:

$$T = \frac{1}{2}(i_1)^2 \frac{dL_1}{d\theta} + \frac{1}{2}(i_2)^2 + \frac{dL_2}{d\theta} + i_1 i_2 \frac{dM}{d\theta}$$

where i_1 and i_2 are rotor and stator currents respectively L_1 and L_2 are rotor winding's self inductances

M is the mutual inductance between stator and rotor windings, and θ is the angular displacement of the rotor.

(*a*) the average value of $(i_1)^2$ and $(i_2)^2$ each is zero

(*b*) the first and the second terms cancel each other

(*c*) $\frac{dL_1}{d\theta}$ and $\frac{dL_2}{d\theta}$ have opposite signs and $\frac{\frac{dL_1}{d\theta}}{\frac{dL_2}{d\theta}}$

(*d*) L_1 and L_2 do not vary with respect to θ. **[ES]**

391. Two series motors are mechanically coupled. One machine is run as motor and the other as generator. The iron and friction losses of the machines will be identical when

(*a*) their speeds are identical

(*b*) their speeds and excitations are identical

(*c*) their speeds are equal and back *emfs* are half the supply voltage

(*d*) their rating and armature sizes are identical **[ES]**

392. Match List-I (Devices) with List-II (Application) and select the correct answer using the codes given below the lists:

List-I	List-II
A. Microprocessor	1. Monitoring surge voltages
B. Breather	2. Digital protection
C. Magnetic links	3. Monitoring surge currents
D. Klydonograph	4. Power transfer

Codes:

	A	B	C	D
(*a*)	4	2	1	3
(*b*)	4	2	3	1
(*c*)	2	4	1	3
(*d*)	2	4	3	1

[ES]

393. The chart given below indicates four different condition under which an induction motor may be called upon to operate, with a constant shaft load

	Applied voltage	Rotor Frequency	Resistance
1.	Normal	Normal	Normal
2.	80%	Normal	Normal
3.	80%	80%	Normal
4.	Normal	Normal	80%

The operating conditions which lead to increasing value of running slips will be in the sequence

(*a*) 1, 4, 3, 2 (*b*) 1, 4, 2, 3

(*c*) 4, 1, 2, 3 (*d*) 4, 1, 3, 2 **[IAS]**

Direction Q. (394-420): *The following items consist of two statements, one labelled the 'Assertion A' and the other labelled the 'Reason R'. You are to examine these two statements carefully and decide explanation of the Assertion. Select your answer to these items using the codes given below and mark your answer to these items using the codes given below and mark your answer sheet accordingly:*

Codes:

(*a*) Both A and R are true and R is the correct explanation of A

(*b*) Both A and R are true but R is not a correct explanation of A

(*c*) A is true but R is false

(*d*) A is false but R is true

394. Assertion (A): A differentially compounded *dc* motor has the possibility of starting in the reversed.

Reasons (R): Owing to higher inductance the series field, the resultant main flux reverse during starting. **[ES]**

395. Assertion (A): The poles and zeros of a reactance function alternate on the s-plane.

Reasons (R): The poles of reactance functions as a function of frequency always has a positive slope. **[ES]**

396. Assertion (A): Multistate *dc* amplifiers are invariably based on two-transistor differential amplifiers.

Reasons (R): Two-transistor differential amplifiers have high gain and high CMMR. [ES]

397. Assertion (A): The iron losses of a transformer increase if the supply frequency is increased with applied voltage remaining constant.

Reasons (R): The flux density in a core is inversely proportional to the frequency of the exciting current. [ES]

398. Assertion (A): The presence of damper winding in the alternator increase the initial value of short circuit current when a fault occurs.

Reasons (R): The damper winding is closely coupled to the field winding and hence does not allow the field flux linkages to decay. [IAS]

399. Assertion (A): The pole shoes of a normal *dc* machine are laminated.

Reasons (R): The flux in the poles is unidirectional. [ES]

400. Assertion (A): DC shunt motors are usually flitted with interpoles.

Reasons (R): Interpoles increases the starting torque of *dc* shunt motors. [IAS]

401. Assertion (A): When a synchronous motor is operating on an infinite bus at a leading power factor an increase in the field current causes an increase in the armature current

Reasons (R): At leading power factors, the excitation voltage of the motor is more than its terminal voltage. The increase in field current causes an increase in the excitation voltage of the motor and thereby in increase in the voltage drop across the synchronous reactance of the motor. [ES]

402. Assertion (A): The core losses of a *dc* shunt machine increases as the machine is loaded.

Reasons (R): The flux of armature current reacts with the main field flux resulting in a destroyed flux distribution in the air-gap of a *dc* machine. [ES]

403. Assertion (A): Squirrel cage induction motors, when connected to supply mains sometimes exhibit a tendency to run at a speed is which nearly one seventh of the synchronous speed.

Reasons (R): Seventh harmonic is present in the applied voltage. [ES]

404. Assertion (A): The core losses of a *dc* shunt machine increases as the machine is loaded.

Reasons (R): The flux of armature currents reacts with the main field flux resulting in a distorted flux distribution in the air-gap of a *dc* machine. [IAS]

405. Assertion (A): The starting torque developed in a single-phase reluctance motor is highly dependent upon the rotor position.

Reasons (R): Rotor saliently offers difference reluctances to the rotating field at different position of the stator. [ES]

406. Assertion (A): In the experimental determination of synchronous impedance of an ac generator, it is possible to take one set of data only from short-circuit test, and obtain there form the short-circuit characteristics of the machine as a straight line passing through the origin of the plot.

Reasons (R): During the short-circuit test, the magnetic circuit of an alternator, remains unsaturated even when the field excitation is high. [IAS]

407. Assertion (A): Solid state regulators are finding favour in speed control of domestic fans, over conventional resistance and inductance type regulators.

Reasons (R): Solid state regulators are compact, relatively less expensive, energy efficient, more reliable and afford noise-free operation. [IAS]

408. Assertion (A): In a salient-pole synchronous motor, the angular displacement of rotor per unit change of power is less than that for a similar cylindrical rotor machine.

Reasons (R): A reluctance torque is developed in the salient-pole machine due to the effect of direct-axis flux on the damper winding. [ES]

409. Assertion (A): Under normal operating conditions, major portion of iron losses of an induction motor occurs in the stator.

Reasons (R): The frequency of the rotor currents under normal operating condition is very small. [IAS]

410. Assertion (A): In a cylindrical rotor synchronous machine, the armature reaction flux Φ_a is in phase with the armature current I_a but in a salient-pole machine Φ_a is not in phase with I_a

Reasons (R): The reluctances along the direct axis and the quadrature axis being different, the ratios of the mmfs and the respective fluxes in the two cases are not equal [ES]

411. Assertion (A) : A self-controlled synchronous motor fed from a CSI, need not have damper windings.

Reasons (R) : Damper windings are responsible for subtransient reactance in synchronous motors. [IAS]

412. Assertion (A): The winding factor of slot harmonics is equal to that of the fundamental.

Reasons (R): Skewing of the rotor slots of an induction motor reduces the effects of slot harmonics. [ES]

413. **Assertion (A):** In a salient pole 3- phase alternator, direct axis synchronous reactance is more than the quadrature axis synchronous reactance.

Reasons (R): The length of air-gap along the direct axis of the salient pole 3-phase alternator is less than that along quadrature axis. [IAS]

414. **Assertion (A):** The no-load current of star/star connected bank of three single phase transformers contains all harmonics except the triplen harmonics.

Reasons (R): Saturation of the core is responsible for the harmonics in the no-load current of a single phase transformer. [IA S]

415. **Assertion (A):** A-3phase induction motor operating on a 3-phase voltage controller for speed control cannot be used for constant torque loads.

Reasons (R): The voltage waveform of the controller is distorted. [IAS]

416. **Assertion (A):** In a 3-phase induction motor, the stator is to be wound for 4 poles but by mistake, the rotor is wound for 2 poles only. When the stator is fed from a 50 Hz, 3-phase supply with rotor winding shorted, the motor does not start up. When, however the wound rotor is replaced by a squirrel-cage rotor, the machine operates without any problem.

Reasons (R): Rotor of an induction motor should not have any poles. [IAS]

417. A 40 MW water turbine is coupled to a 50 MVA 3-phase ac generator rated for 0.8 lagging power factor operation. The losses of the generator are negligable. It is required to deliver a load of 30 MW at 0.5 power factor leading by the set.

Assertion (A): The set can not deliver the load.

Reasons (R): The trubine will be overloaded. [IAS]

418. **Assertion (A):** In a level compounded dc generator, the terminal voltage at fault loss is equal to the terminal voltage at no-load.

Reasons (R): Under load conditions, the armature reaction increases the flux per poles and thus helps maintain a constant terminal voltage. [IAS]

419. **Assertion (A):** DC series motors are know to be capable of operating on ac supply as well. With a very minor modification, dc shunt motors can work on ac supply as well.

Reasons (R): The direction of torque developed in a rotating machine remains unaltered if the directions of conductor current and air-gap flux are reversed simultaneously. [IAS]

420. **Assertion (A):** The rotor resistance of a deep bar type cage induction motor is higher than the running resistance.

Reasons (R): The extra resistance present in the rotor circuit at starting is cut out by a centrifugal switch during normal running. [IAS]

ANSWERS

EXERCISE – I

1. (b)	**2. (b)**	**3. (c)**	**4. (c)**	**5. (a)**	6. (b)	7. (d)	8. (a)	9. (d)	10. (a)
11. (a)	12. (b)	13. (b)	14. (c)	15. (a)	16. (b)	17. (c)	18. (a)	19. (c)	20. (b)
21. (d)	22. (a)	23. (c)	24. (b)	25. (a)	26. (b)	27. (b)	28. (b)	29. (c)	30. (c)
31. (b)	32. (a)	33. (d)	34. (a)	35. (c)	36. (d)	37. (a)	38. (d)	39. (b)	40. (c)
41. (b)	42. (b)	43. (b)	44. (d)	45. (a)	46. (c)(d)	47. (d)	48. (a)	49. (c)	50. (c)
51. (b)	52. (c)	53. (b)	54. (b)	55. (c)	56. (c)	57. (c)	58. (c)	59. (c)	60. (c)
61. (d)	62. (d)	63. (a)	64. (c)	65. (c)	66. (b)	67. (c)	68. (a)	69. (b)	70. (d)
71. (d)	72. (b)	73. (b)	74. (d)	75. (b)	76. (b)	77. (a)	78. (d)	79. (b)	80. (a)
81. (a)	82. (d)	83. (d)	84. (c)	85. (a)	86. (c)	87. (a)	88. (d)	89. (a,b)	90. (a)
91. (b)	92. (b)	93. (d)	94. (a)	95. (a)	96. (a)	97. (c)	98. (d)	99. (d)	100. (b)
101. (a)	102. (a)	103. (c)	104. (a)	105. (d)	106. (b)	107. (a)	108. (b)	109. (b)	110. (d)
111. (b)	112. (b)	113. (c)	114. (a)	115. (d)	116. (c)	117. (c)	118. (d)	119. (b)	120. (a)
121. (b)	122. (c)	123. (b)	124. (d)	125. (d)	126. (c)	127. (c)	128. (b)	129. (c)	130. (d)
131. (b)	132. (b)	133. (c)	134. (d)	135. (b)	136. (a)	137. (d)	138. (b)	139. (b)	140. (d)
141. (b)	142. (d)	143. (c)	144. (c)	145. (c)	146. (c)	147. (c)	148. (b)	149. (c)	150. (c)
151. (a)	152. (d)	153. (c)	154. (c)	155. (b)	156. (d)	157. (a)	158. (b)	159. (c)	160. (b)
161. (b)	162. (c)	163. (b)	164. (b)	165. (c)	166. (a)	167. (b)	168. (b)	169. (b)	170. (a)
171. (d)	172. (d)	173. (d)	174. (b)	175. (a)	176. (b)	177. (d)	178. (a)	179. (a)	180. (d)
181. (c)	182. (c)	183. (a)	184. (a)	185. (c)	186. (d)	187. (d)	188. (d)	189. (c)	190. (b)
191. (b)	192. (a)	193. (c)	194. (b)	195. (b)	196. (d)	197. (d)	198. (a)	199. (b)	200. (b)
201. (b)	202. (c)	203. (b)	204. (d)	205. (d)	206. (c)	207. (c)	208. (a)	209. (d)	210. (d)
211. (c)	212. (a)	213. (a)	214. (d)	215. (c)	216. (b)	217. (b)	218. (b)	219. (b)	220. (b)

221. (*a*) **222.** (*b*) **223.** (*a*) **224.** (*b*) **225.** (*c*) **226.** (*a*) **227.** (*b*) **228.** (*c*) **229.** (*a*) **230.** (*c*)
231. (*a*) **232.** (*d*) **233.** (*b*) **234.** (*d*) **235.** (*d*) **236.** (*c*) **237.** (*b*) **238.** (*a*) **239.** (*d*) **240.** (*c*)
241. (*a*) **242.** (*b*) **243.** (*d*) **244.** (*c*) **245.** (*c*) **246.** (*c*) **247.** (*a*) **248.** (*b*) **249.** (*c*) **250.** (*a*)
251. (*a*) **252.** (*c*) **253.** (*d*) **254.** (*b*) **255.** (*c*) **256.** (*c*) **257.** (*c*) **258.** (*b*) **259.** (*b*) **260.** (*c*)
261. (*b*) **262.** (*b*) **263.** (*a*) **264.** (*b*) **265.** (*d*) **266.** (*a*) **267.** (*a*) **268.** (*b*) **269.** (*a*) **270.** (*c*)
271. (*c*) **272.** (*b*) **273.** (*b*) **274.** (*b*) **275.** (*b*) **276.** (*c*) **277.** (*d*) **278.** (*d*) **279.** (*c*) **280.** (*c*)
281. (*a*) **282.** (*d*) **283.** (*b*) **284.** (*d*) **285.** (*c*) **286.** (*c*) **287.** (*b*) **288.** (*a*) **289.** (*d*) **290.** (*d*)
291. (*c*) **292.** (*d*) **293.** (*a*) **294.** (*a*) **295.** (*a*) **296.** (*d*) **297.** (*d*) **298.** (*c*) **299.** (*d*) **300.** (*a*)
301. (*c*) **302.** (*b*) **303.** (*d*) **304.** (*a*) **305.** (*d*) **306.** (*b*) **307.** (*b*) **308.** (*a*) **309.** (*d*) **310.** (*c*)
311. (*b*)

EXERCISE–II

1. (*b*) **2.** (*b*) **3.** (*d*) **4.** (*b*) **5.** (*a*) **6.** (*b*) **7.** (*d*) **8.** (*c*) **9.** (*c*) **10.** (*d*)
11. (*d*) **12.** (*b*) **13.** (*a*) **14.** (*d*) **15.** (*a*) **16.** (*d*) **17.** (*b*) **18.** (*c*) **19.** (*b*) **20.** (*a*)
21. (*a*) **22.** (*a*) **23.** (*a*) **24.** (*b*) **25.** (*c*) **26.** (*c*) **27.** (*a*) **28.** (*c*) **29.** (*c*) **30.** (*d*)
31. (*b*) **32.** (*c*) **33.** (*d*) **34.** (*c*) **35.** (*b*) **36.** (*a*) **37.** (*c*) **38.** (*b*) **39.** (*c*) **40.** (*d*)
41. (*b*) **42.** (*b*) **43.** (*a*) **44.** (*b*) **45.** (*b*) **46.** (*a*) **47.** (*d*) **48.** (*d*) **49.** (*d*) **50.** (*c*)
51. (*d*) **52.** (*c*) **53.** (*b*) **54.** (*d*) **55.** (*c*) **56.** (*b*) **57.** (*c*) **58.** (*b*) **59.** (*a*) **60.** (*c*)
61. (*d*) **62.** (*d*) **63.** (*b*) **64.** (*d*) **65.** (*c*) **66.** (*c*) **67.** (*b*) **68.** (*d*) **69.** (*d*) **70.** (*b*)
71. (*b*) **72.** (*b*) **73.** (*d*) **74.** (*a*) **75.** (*b*) **76.** (*c*) **77.** (*b*) **78.** (*c*) **79.** (*c*) **80.** (*a*)
81. (*a*) **82.** (*a*) **83.** (*c*) **84.** (*a*) **85.** (*b*) **86.** (*b*) **87.** (*d*) **88.** (*b*) **89.** (*a*) **90.** (*b*)
91. (*b*) **92.** (*c*) **93.** (*a*) **94.** (*c*) **95.** (*a*) **96.** (*a*) **97.** (*c*) **98.** (*d*) **99.** (*d*) **100.** (*d*)
101. (*c*) **102.** (*c*) **103.** (*a*) **104.** (*b*) **105.** (*c*) **106.** (*b*) **107.** (*b*) **108.** (*d*) **109.** (*c*) **110.** (*c*)
111. (*a*) **112.** (*d*) **113.** (*d*) **114.** (*a*) **115.** (*b*) **116.** (*a*) **117.** (*d*) **118.** (*c*) **119.** (*c*) **120.** (*c*)
121. (*c*) **122.** (*a*) **123.** (*d*) **124.** (*a*) **125.** (*c*) **126.** (*c*) **127.** (*a*) **128.** (*b*) **129.** (*d*) **130.** (*c*)
131. (*b*) **132.** (*b*) **133.** (*d*) **134.** (*c*) **135.** (*d*) **136.** (*a*) **137.** (*c*) **138.** (*b*) **139.** (*a*) **140.** (*b*)
141. (*b*) **142.** (*a*) **143.** (*b*) **144.** (*d*) **145.** (*c*) **146.** (*c*) **147.** (*c*) **148.** (*a*) **149.** (*a*) **150.** (*b*)
151. (*a*) **152.** (*d*) **153.** (*b*) **154.** (*b*) **155.** (*c*) **156.** (*b*) **157.** (*d*) **158.** (*c*) **159.** (*b*) **160.** (*b*)
161. (*c*) **162.** (*b*) **163.** (*c*) **164.** (*d*) **165.** (*d*) **166.** (*a*) **167.** (*a*) **168.** (*d*) **169.** (*c*) **170.** (*d*)
171. (*b*) **172.** (*a*) **173.** (*c*) **174.** (*c*) **175.** (*d*) **176.** (*b*) **177.** (*b*) **178.** (*b*) **179.** (*c*) **180.** (*b*)
181. (*d*) **182.** (*c*) **183.** (*d*) **184.** (*a*) **185.** (*b*) **186.** (*d*) **187.** (*c*) **188.** (*c*) **189.** (*b*) **190.** (*c*)
191. (*d*) **192.** (*c*) **193.** (*a*) **194.** (*c*) **195.** (*c*) **196.** (*c*) **197.** (*b*) **198.** (*b*) **199.** (*a*) **200.** (*c*)
201. (*d*) **202.** (*a*) **203.** (*a*) **204.** (*b*) **205.** (*d*) **206.** (*b*) **207.** (*a*) **208.** (*c*) **209.** (*d*) **210.** (*d*)
211. (*a*) **212.** (*a*) **213.** (*a*) **214.** (*c*) **215.** (*a*) **216.** (*d*) **217.** (*b*) **218.** (*c*) **219.** (*c*) **220.** (*d*)
221. (*c*) **222.** (*d*) **223.** (*b*) **224.** (*a*) **225.** (*c*) **226.** (*c*) **227.** (*c*) **228.** (*a*) **229.** (*c*) **230.** (*c*)
231. (*c*) **232.** (*a*) **233.** (*b*) **234.** (*c*) **235.** (*a*) **236.** (*b*) **237.** (*d*) **238.** (*c*) **239.** (*a*) **240.** (*d*)
241. (*a*) **242.** (*a*) **243.** (*c*) **244.** (*c*) **245.** (*d*) **246.** (*a*) **247.** (*b*) **248.** (*b*) **249.** (*a*) **250.** (*d*)
251. (*b*) **252.** (*a*) **253.** (*c*) **254.** (*c*) **255.** (*b*) **256.** (*d*) **257.** (*d*) **258.** (*d*) **259.** (*a*) **260.** (*d*)
261. (*b*) **262.** (*b*) **263.** (*d*) **264.** (*d*) **265.** (*d*) **266.** (*b*) **267.** (*c*) **268.** (*b*) **269.** (*b*) **270.** (*a*)
271. (*a*) **272.** (*d*) **273.** (*a*) **274.** (*a*) **275.** (*c*) **276.** (*a*) **277.** (*a*) **278.** (*c*) **279.** (*d*) **280.** (*b*)
281. (*b*) **282.** (*a*) **283.** (*d*) **284.** (*c*) **285.** (*a*) **286.** (*c*) **287.** (*b*) **288.** (*c*) **289.** (*c*) **290.** (*b*)
291. (*c*) **292.** (*c*) **293.** (*c*) **294.** (*c*) **295.** (*b*) **296.** (*c*) **297.** (*c*) **298.** (*d*) **299.** (*a*) **300.** (*c*)
301. (*b*) **302.** (*a*) **303.** (*d*) **304.** (*a*) **305.** (*b*) **306.** (*b*) **307.** (*d*) **308.** (*b*) **309.** (*a*) **310.** (*a*)
311. (*a*) **312.** (*a*) **313.** (*c*) **314.** (*b*) **315.** (*c*) **316.** (*d*) **317.** (*c*) **318.** (*a*) **319.** (*a*) **320.** (*a*)
321. (*b*) **322.** (*c*) **323.** (*c*) **324.** (*c*) **325.** (*a*) **326.** (*d*) **327.** (*c*) **328.** (*d*) **329.** (*c*) **330.** (*c*)
331. (*b*) **332.** (*c*) **333.** (*c*) **334.** (*d*) **335.** (*c*) **336.** (*d*) **337.** (*d*) **338.** (*b*) **339.** (*a*) **340.** (*b*)
341. (*d*) **342.** (*a*) **343.** (*d*) **344.** (*b*) **345.** (*a*) **346.** (*a*) **347.** (*d*) **348.** (*a*) **349.** (*b*) **350.** (*d*)
351. (*a*) **352.** (*c*) **353.** (*b*) **354.** (*c*) **355.** (*a*) **356.** (*a*) **357.** (*b*) **358.** (*d*) **359.** (*b*) **360.** (*d*)
361. (*a*) **362.** (*d*) **363.** (*d*) **364.** (*b*) **365.** (*a*) **366.** (*c*) **367.** (*d*) **368.** (*c*) **369.** (*a*) **370.** (*d*)
371. (*c*) **372.** (*c*) **373.** (*a*) **374.** (*b*) **375.** (*a*) **376.** (*c*) **377.** (*b*) **378.** (*b*) **379.** (*a*) **380.** (*d*)
381. (*b*) **382.** (*a*) **383.** (*b*) **384.** (*d*) **385.** (*c*) **386.** (*c*) **387.** (*a*) **388.** (*d*) **389.** (*d*) **390.** (*d*)
391. (*b*) **392.** (*d*) **393.** (*c*) **394.** (*c*) **395.** (*a*) **396.** (*a*) **397.** (*c*) **398.** (*b*) **399.** (*d*) **400.** (*c*)
401. (*a*) **402.** (*a*) **403.** (*c*) **404.** (*a*) **405.** (*b*) **406.** (*a*) **407.** (*a*) **408.** (*c*) **409.** (*a*) **410.** (*d*)
411. (*b*) **412.** (*d*) **413.** (*a*) **414.** (*b*) **415.** (*d*) **416.** (*c*) **417.** (*c*) **418.** (*a*) **419.** (*c*) **420.** (*d*)

EXPLANATIONS

EXERCISE – I

15. For maximum efficiency,

$$I_2 = I_{fl}\sqrt{\frac{P_c}{\text{Full load ohmic loss}}}$$

$$= I_{fl}\sqrt{\frac{200}{800}}$$

$$= I_{fl} \times 0.5$$

$$= 0.5\, I_{fl}$$

28. Since, $N \propto \frac{N}{\phi}$

$\therefore$ $V_1 = \frac{V}{2}, \phi = \frac{\phi}{2}$

$\therefore$ $N_1 = N$

29. Since, load characteristics is linear, hence when the load current is 50A, then

$$\text{terminal voltage} = 100 - \left(\frac{100-90}{2}\right)$$

$$= 95$$

$\therefore$ Load circuit resistance

$$= \frac{\text{Terminal Voltage}}{\text{load current}}$$

$$= \frac{95}{50} = 1.9\,\Omega$$

32. For lap connected winding, number of parallel path, $a = P = 6$

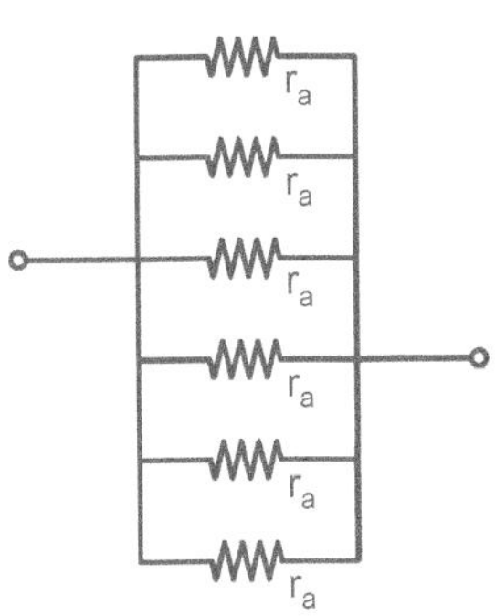

Given : total armature resistance

$$= 0.05\,\Omega$$

$\therefore$ Resistance of one conductor,

$$r_a = 0.05 \times 6$$

$$= 0.30\,\Omega$$

Now, armature is rewound in wave connection, hence $a = 2$

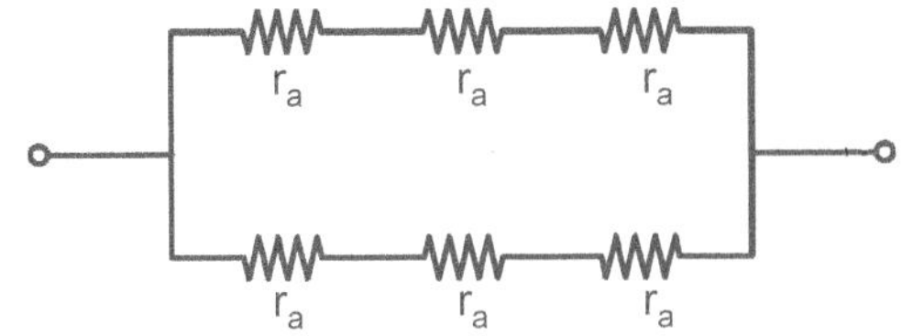

$\therefore$ Total armature resistance

$$= \frac{0.30 + 0.30 + 0.30}{2} = 0.45\,\Omega$$

36. Since kVA delivered by transformer depends on the rating of its winding, hence in both connection delivered kVA will be same.

49. Since, both alternators have same speed, hence for differential frequencies they must have different number of poles

$$f = \frac{p.n}{120}$$

$$\Rightarrow \quad n = \frac{120 f}{p}$$

For 1st, $n_1 = \frac{120 \times 50}{p}$

Put 'p' = 2, 4, 6, 8, 10, we have
n = 3000, 1500, 1000, 750 and 600 rpm

For 2nd, $n_2 = \frac{120 \times 60}{p}$

Put 'p' = 2, 4, 6, 8, 10, 12 we have
n=3600, 1800, 1200, 900, 720 and 600 rpm

Hence they may have common maximum speed of 600 rpm.

53. Higher value of SCR means less value of synchronous reactance as $\text{SCR} = \frac{1}{X_s(pu)}$

Hence low synchronous reactance results on low voltage variation with load and therefore a better voltage regulation low value of X_s also results in higher synchroning power as

$$P_s = \frac{E_f . V_t}{X_s} \cos\delta \,.\, \Delta\delta,$$

hence higher stability limit.

61. Synchronous motor phasor diagram (for lagging p.f.) can be drawn as follows :

Now when E_f is increased then for the constant supply voltage, it is clear that armature current or $I_a X_s$ will first decrease to some value and then will increase.

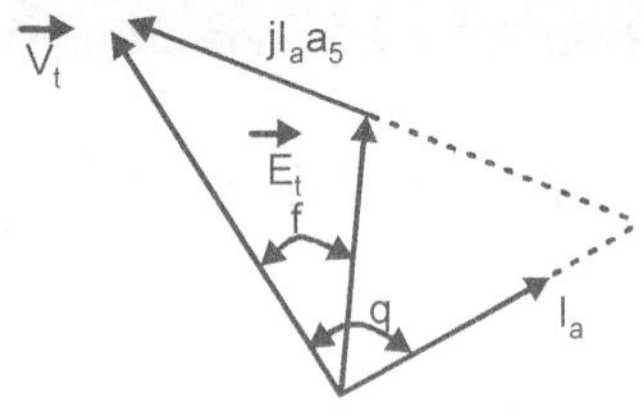

72. Slot per pole per phase, $q = \dfrac{48}{4 \times 3} = 4$

Slot angle $\quad V = \dfrac{2 \times 360}{48} = 15°$

Distribution factor,

$$K_d = \frac{\sin \frac{qr}{2}}{q \sin \frac{r}{2}}$$

$$= \frac{\sin \frac{4 \times 15}{2}}{4 \sin \frac{15}{2}}$$

$$= \frac{1}{8 \sin(7.5)}$$

Pitch factor, $\quad K_p = \cos \dfrac{r}{2}$

$$= \cos \frac{15}{2} = \cos (7.5)$$

$\therefore$ Winding factor, $k_w = k_p \, . \, k_d$

$$= \frac{1}{8} \cot (7.5)$$

73. Synchronous speed,

$$n_s = \frac{120 \times 50}{6}$$

$$= 1000 \text{ rpm}$$

We have, $\quad$ rator frequency, $f_2 = sf$

$$\therefore \quad \frac{30}{60} = s \times 50$$

$$\Rightarrow \quad \text{Slip}, s = \frac{0.5}{50} = 0.01$$

$\therefore \quad$ Rotor speed $= N_s (1 - 5)$

$$= 1000 (1 - 0.01)$$

$$= 990 \text{ rpm}$$

80. Induction generator are employed commonly for generating high frequency *dc* voltages of the order of 1 kHz. Main advantage of induction alternator is that both armature and field windings are on the stator, and rotor with no winding on it, is a laminated structure with open slot. The absence of rotor winding and no sliding contacts on the rotor allows the speed to be raised to maximum possible safe limit for obtaing more output for given machine dimensions.

82. Differentially compounded generator is used for welding purposes because its performance is unaffected by short circuit, as its series flux and shunt flux are opposing each other.

84. When working as a motor,

$$V_t = E_f + I_a r_a$$

$$\therefore \quad E_f = 220 - 20 \times 1$$

$$= 220 \text{ V}$$

When working as a generator

$$E_{f2} = V_t + I_a r_a$$

$$= 220 - 20 \times 1$$

$$= 200 \text{ V}$$

$$E_{f2} - E_{f1} = 240 - 200$$

$$= 40 \text{ V}$$

85. In dc machine, $T_e = K_a \phi I_a$

As torque is halved, and since field flux of shunt motor is constant, hence current I_a should be halved. And therefore generator (armature) current is also halved.

87. In the induction motor, torque generated to proportional to the supply voltage

Also, $\quad Te_{st} \propto V^2$

When motor reconnected in ψ, then its winding will have phase voltage of

$$V_{ph} = \frac{V_L}{\sqrt{3}}$$

But in delta connection, $V_{ph} = V_L$

Hence in star connection,

$$\frac{(Te_{st})_2}{(Te_{st})_1} = \frac{(Vp_{h2})^2}{(Vp_{h1})^2}$$

$$\Rightarrow \quad (Te_{st})_2 = (Te_{st})_1 \times \frac{(V_L/\sqrt{3})^2}{V_L^2}$$

$$= (Te_{st})_1 \times \frac{1}{3}$$

$$= 120 \times \frac{1}{3}$$

$$= 40 \text{ N-m}$$

$$\therefore \quad (Te_{st})_2 = 40 \text{ N-m}$$

88. If speed of the alternator is reduced to half or in other words to maintain rated open circuit voltage, then field current required is two times the previous value

But impedance,

$$Z_s = \frac{\text{Open circuit voltage } E_1}{\text{Short circuit current } I_1}$$

varies with field current and so unless we know the value of Z_s. For various field current it is impossible to decide the field current required to maintain rated current at short-circuited condition with the half reduced speed.

110. When braking in a 3-phase induction motor is done by throwing its terminal to dc source from ac sources, this method of breaking is called dynamic braking.

113. Synchronous speed,

$$N_s = \frac{120 \times 60}{8} = 900$$

Since, rotor is driven in opposite direction at 1800 rpm, therefore

Speed of rotor mmf with respect to stator mmf = 1800 + 900 = 2700

$$\therefore \quad f_2 = \frac{2700 \times 8}{120}$$

$$= 180 \text{ Hz}$$

122. Voltage regulation lagging power factor is given by,

$$V.R. = \epsilon_r \cos\theta_2 + \epsilon_r \sin\theta_2$$

Since, transformers generally have very low, hence ϵ_r can be neglected

$$\therefore \quad V.R. \cong \epsilon_r \sin\theta_2$$

Now, under leading power factor

$$V.R. = \epsilon_r \cos\theta_2 - \epsilon_r \sin\theta_2$$

$$= \epsilon_x \sin\theta_2$$

Hence voltage regulation for leading p.f. will be – 2.5%.

123. At 0.8 lagging p.f.,

$$VR = 0.01 \times 0.8 + 0.04 \times 0.6$$

$$= 0.032 = 3.2\%$$

At 0.8 p.f. leading

$$VR = 0.01 \times 0.8 - 0.04 \times 0.6$$

$$= -0.016 = -1.6\%$$

124. To reduce, the armature current R_a should be maximum as

$$V_t = E_f + I_a \times R_a$$

$$\Rightarrow \quad I_a = \frac{V_t - E_f}{R_a}$$

And to have greater starting torque [$T_e \propto \phi I_a$], the field flux or field current should be maximum. To have maximum field current R_f should be minimum.

125. Hysteresis loss, $P_h = Khf^{1-x} V^x$

Here, $x = 1.6$

and Eddy current loss, $P_e = K_e V^2$

Hence, when the frequency of the input voltage of a transformer is increased, then hystersis loss will decrease and eddy current loss remains unchanged as the supply voltage is unchanged.

129. The maximum efficiency will occur when

fixed iron loss = variable copper loss

$\Rightarrow$ Variable copper loss (also) should be 150 W.

$\therefore$ Total loss = 150 + 150 = 300 W

130. With auto transformer, $Te_{st} \propto x^2$

$$\Rightarrow \quad Te_{st} = kx^2$$

$$\therefore \quad \frac{(Te_{st})_2}{(Te_{st})_1} = \frac{x_2^2}{x_1^2}$$

$$= \frac{(0.60)^2}{(0.30)^2} = 4$$

$$\therefore \quad (Te_{st})_2 = 4 \times 80$$

$$= 320 \text{ N-m}$$

142. Phase emf, $E_{ph} = \sqrt{2}$

Phase emf is affected by

$$K_w = K_p . K_d$$

$$= \frac{\sin\frac{\sigma}{2}}{q\sin\frac{\sigma}{2q}}$$

148. For a sinusoidal applied voltage, flux is a sine wave and the magnetizing current due to saturation is peaky containing a pronounced third harmonic current.

151. The phase diagram under given condition can be drawn as follows:

Here, $\delta = 60°, \theta = 0°$

p.f. = cos 30 = 0.866

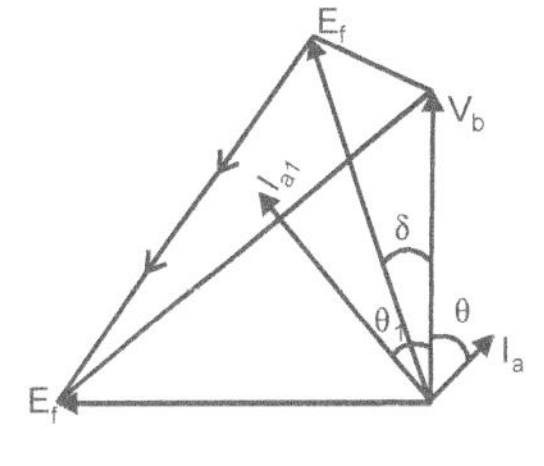

θ_1 = new pf angle

θ = old pf angle

162. Load shared by first transformer,

$$S_a = S \times \frac{\vec{Z}_{eb}}{\vec{Z}_{ea} + \vec{Z}_{eb}}$$

$$= 100 \times \frac{10.01}{10.01 + 3.04}$$

$$= 76.7 \text{ kW}$$

164. $$P_h = Kh . f^{1-y} V^y$$

$$\frac{P_{h2}}{P_{h1}} = \left(\frac{f_2}{f_1}\right)^{1-2}\left(\frac{V_2}{V_1}\right)^2$$

$$= \left(\frac{xf}{f}\right)^{-1}\left(\frac{xv}{v}\right)^2$$

$$= x^{-1} . x^2$$

$$= x$$

$\therefore$ $$P_{h2} = xP_{h1} = xP_h$$

166. Transformer connection to supply 720 V can be drawn as shown :

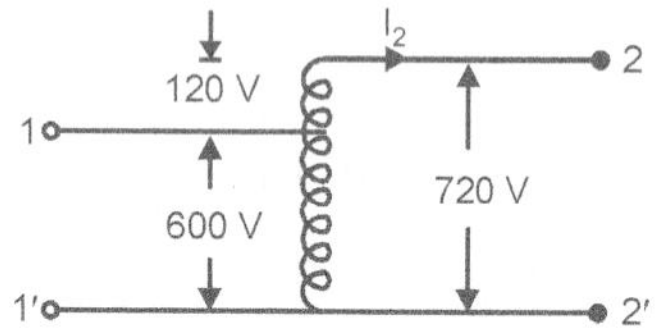

$$\text{Rated current of } hv \text{ winding} = \frac{15000}{600}$$

$$= 25 \text{ A}$$

$$\text{Rated current of } lv \text{ winding} = \frac{15000}{120}$$

$$= 125 \text{ A}$$

$\therefore$ Rating of the autotransformer

$$= 125 \times 720$$

$$= 9000 \text{ VA}$$

$$= 90 \text{ kVA}$$

167. emf induced in one phase of the rotor distributed winding can be written as

$$E = \sqrt{2}\,\pi\, fs\, k\omega\, N_{phr}\,\phi \qquad ...(i)$$

And electromagnetic torque generated is given by

$$T_e = \frac{\pi}{8} P_2\, \phi\, F_2 \sin(90 + \theta_2) \qquad ...(ii)$$

For a constant load torque, ϕ flux has to be constant from equation (*i*) as voltage applied and frequency both are halved, hence slip "*s*" will be constant, hence speed will remain same as the previous value.

168. $$s = 0.04$$

We have, $$N_r = N_s(1 - s)$$

$\therefore$ $$720 = N_s(1 - 0.04)$$

$\Rightarrow$ $$N_s = 750$$

Now, $$f = \frac{p.n}{120}$$

$$= \frac{4 \times 750}{120}$$

$$= 25 \text{Hz}$$

170. $$N_s = \frac{120 \times 50}{4}$$

$$= 1500 \text{ rpm}$$

$$N_r = 1440 \text{ rpm}$$

$\therefore$ $$\text{Slip, } s = \frac{1500 - 1440}{1500}$$

$$= 0.04$$

$\therefore$ Frequency of rotor emf

$$= 50 \times 0.04$$

$$= 2 \text{ Hz}$$

171. When phase sequence is reversed, then stator mmf will rotate in opposite direction

But rotor is moving in same direction, hence

$$\text{slip} = \frac{n_s - (-n_s)}{n_s} = 2$$

Hence, frequency of induced emf across the open circuited rotor terminal

$$= sf = 2 \times 50 = 100 \text{ Hz}$$

175. $$\text{Slip, } s = \frac{1500 - 1440}{1500} \times 100 = 3.33\%$$

176. Synchronous speed, $$N_s = \frac{120 \times 60}{12}$$

$$= 600 \text{ rpm}$$

Slip, $$s = 0.05$$

$\therefore$ $$N_r = N_s(1 - s)$$

$$= 600(1 - 0.05)$$

$$= 570 \text{ rpm}$$

182. We have, $$f = \frac{P.N}{120}$$

$\therefore$ $$N = \frac{120 \times f}{P}$$

For N to be maximum, number of poles P should be minimun

$\therefore$ $$P = 2$$

$\therefore$ $$N_{max} = \frac{120 \times 50}{2}$$

$$= 3000 \text{ rpm}$$

188. $$\text{Turn ratio} = \frac{100}{5} = \frac{I_p}{I_s}$$

$\therefore$ Secondary current, $$I_s = 3.5 \text{ A}$$

$\therefore$ Line current, $$I_p = \frac{100}{5} \times 3.5$$

$$= 70 \text{ A}$$

189. Emf per turn induced in the transformer is given by,

$$\frac{E}{N} = \sqrt{2}\,\pi f \phi_{max}$$

As frequency applied is reduced to 50 Hz, then for maximum flux to be constant voltage applied should also be reduced in the same proportion

Since voltage applied is reduced, hence the kVA output or kVA rating will also be reduced.

193. Ratio of phase voltage $= \frac{V_2P}{V_1P} = 5$

In primary side (Delta connected)

$$V_{phase} = V_{line} = 400 \text{ V}$$

$\therefore$ Secondary phase voltage

$$= 400 \times 5 = 2000 \text{ V}$$

$\therefore$ Secondary line voltage

$$= 2000 \times \sqrt{3}$$

$$= 3464 \text{ V}$$

199. At maximum efficiency,

Variable copper loss = Fixed iron loss

$$= 1 \text{ kW}$$

kVA output at maximum efficiency

$$= (\text{kVA})_{fl} \sqrt{\frac{P_c}{P_{ohmic}}}$$

$$= 100\sqrt{\frac{1}{2}} = 70.7 \text{ kVA}$$

209. When series fields are reconnected in parallel, then effective resistance of the field coil will be reduced hence field flux gets reduced and therefore for the same torque, speed will be increased.

217. Eddy current loss is independent of frequency for constant supply voltage and hysteresis loss will decrease on increasing the supply frequency as

$$P_h = khf^{1-s}\, V^x$$

Here x is equal to 1.6

232. When lap connected, number of parallel path is equal to pole

Hence, in lap connection,

$$\frac{r_a}{6} = 0.06$$

$\Rightarrow$ $r_a = 0.06 \times 6 = 0.36$

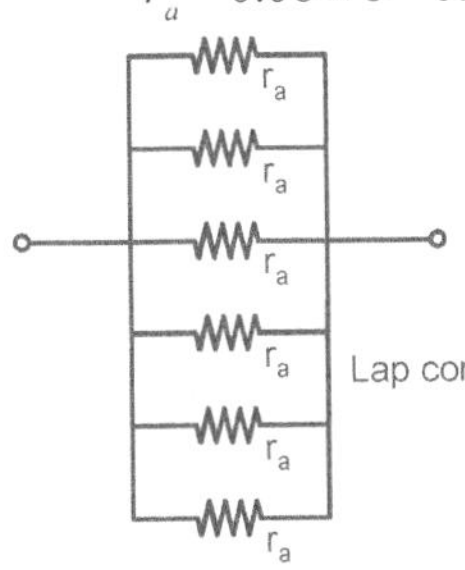

Lap connection

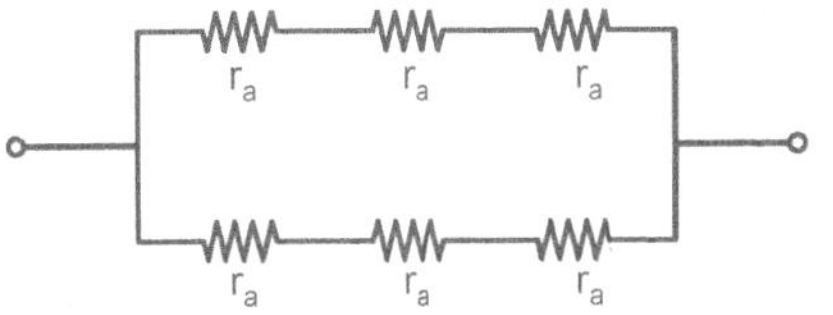

Wave connection

When wave connected, parallel paths = 2

$\therefore$ Total armature resistance $= \frac{0.36 \times 3}{2}$

$$= 0.54\ \Omega$$

234. For generator 1; G_1 : 200 MW,

$$F_1 = 50 - \frac{4 \times 50}{100} \times \frac{P_1}{200}$$

$$= 50 - \frac{P_1}{200}$$

For generator 2; G_1 : 400 MW,

$$F_2 = 50 - \frac{5 \times 5}{100} \times \frac{P_2}{400}$$

$$= 50 - \frac{P_2}{400}$$

$\Rightarrow$ $F_1 = F_2$

$$P_1 + P_2 = 600$$

and $\frac{P_1}{P_2} = 1.6$

$\therefore$ $P_1 = \frac{600}{2.6}$ MW

$\therefore$ $f_1 = 50 - \frac{600}{2.6 \times 100}$

$$= 47.7 \text{ Hz}$$

241. Field winding resistance,

$$r_f = 50\ \Omega$$

Armature winding resistance,

$$r_a = 0.04\ \Omega$$

$$I_a = \frac{20000}{200}$$

$$= 100 \text{ A}$$

$\therefore$ Armature circuit loss $= (100)^2 \times 0.04$

$$= 10000 \times 0.04 = 400 \text{ W}$$

Field circuit loss $= \frac{(200)^2}{50} = 800$

$\therefore$ Total loss $= 20000 - 17000$

$$= 3000 \text{ W}$$

$\therefore$ Fixed lossess $= 3000 - 400$

$= 2600$ W

At maximum efficiency, variable ohmic losses in the armature copper losses be equal to fixed loss.

$\therefore$ Total armature copper loss $= 2600$ W

254. Since, maximum torque $\propto V^2$

Hence, when supply voltage is decreased by 10%, then maximum torque will be decreased by 20%.

257. When induction motor is started without starter, it draws about 5 times the full-load current at starting.

With such large starting current motor can be damaged due to overheating so it requires a starter.

262. The phasor diagram can be drawn as follows.

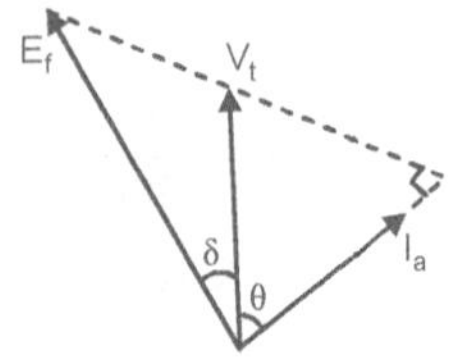

Angle between *ef* and ϕ

$= \delta + \theta$

$= 15° + \cos^{-1}(0.707) = 60°$

265. When lower frequency is reduced to 50 Hz with same value of voltage and current, then loss will remain same but hysteresis loss ($P_h = Khf^{1-x} V^x$) will be increased slightly; hence transformer efficiency will decreased marginally.

266. If brushes are given a backward shift in a generator or toward shift in a motor, the flux per pole is increased and as a result of it, the generator terminal voltage increases and the motor speed decreases.

And further brush shift is carried out to improve the commutation-forward in a generator and backward in a motor.

270. For maximum power transfer to 2Ω resistor, its value should be equal to (equivalent) resistance seen towards transformer from its terminal.

Let N be the number of turns in the primary of the transformer. Then

$$2 = 8\left(\frac{40}{N}\right)^2$$

$$\Rightarrow \quad \left(\frac{N}{40}\right)^2 = 4$$

$$\Rightarrow \quad N = 80$$

274. Power output $= 15$ kW

$= (1 - s)P_g$

$$\therefore \quad P_g = \frac{15000}{1 - 0.04}$$

$= 15625$

Now, copper loss, $P_c = s \,.\, P_g$

$= 0.04 \times 15625$

$= 625$ W

280. In series motor, $T_e = C.\ I_a^2$

When armature current will double, the torque developed will become 4 times.

$$\therefore \quad T_e = 20 \times \left(\frac{6}{3}\right)^2$$

$= 20 \times 4$

$= 80$ N-m

281. Voltage across the load,

$$V_2 = \frac{200}{(1 + 0.03)}$$

$= 194$

286. Motor is operating at a very low value of slip (3%), hence rotor speed is very near to synchronous speed,

Torque equation can be written as

$$T_e = s\frac{mV_e^2}{w_s} \cdot \frac{1}{r^2}$$

Since voltage is reached to 110 percent of its normal value, hence for the same torque

$$0.03 \times V^2 = s \times (1.10\ V)^2$$

$$\Rightarrow \quad s = 0.0248$$

$$\therefore \quad \%s = 248\%$$

287.

$$N_s = \frac{120 \times 60}{6}$$

$= 1200$

$$\therefore \quad \text{Slip, } s = \frac{1200 - 1000}{1200}$$

$= 01.667$

Since, at this slip, motor is generating maximum torque, therefore

at maximum torque, $\frac{r_2}{s} = x_1$

$$\therefore \quad x_1 = \frac{1.2}{0.1667}$$

$$\Rightarrow \quad x_1 = 7.2\ \Omega$$

For maximum starting torque, $s = 1$

$$\therefore \quad \frac{r_1^2}{1} = x_1$$

$$\Rightarrow \quad r_2' = 7.2\ \Omega$$

$\therefore$ External resistance to be added

$= 7.2 - 1.2$

$= 6.0\ \Omega$

289. In transformer, emf induced in the primary is given by

$$E_1 = \sqrt{2}\,\pi f N_1 \phi_{max}$$

We have to keep in mind to use transformer at lower value of frequency, that the maximum flux value should not be exceeded.

For maximum flux to be same,

$$\frac{E_1}{f_1} = \frac{E_2}{f_2}$$

$$\therefore \quad E_2 = \left(\frac{f_2}{f_1}\right) E_1$$

$$= \frac{50}{400} \times 230$$

$$= 28.75$$

$$\therefore \text{ New kVA rating} = \frac{28.75}{230} \times 4.00 = 0.5 \text{ kVA}$$

291. Percentage regulation $= \dfrac{E_f - V_t}{V_t} \times 100$

$$= \frac{jX_s . I_a}{V_t} \times 100$$

$$= \frac{(j20)(30j)}{400} \times 100$$

$$= -150$$

301. Synchronous speed of stator mmf

$$= \frac{120 \times 50}{4} = 1500$$

Synchronous speed of rotor mmf

$$= \frac{120 \times 10}{4} = 300$$

Torque will develop at two speeds and these speed will be (1500 + 300) and (1500 – 300) rpm

Hence, correct answer is 1200 rpm.

303. When machine is running as a generator, then

$$E_{f1} = V_t + I_a r_a$$

$$= 200 + 30 \times 0.5$$

$$= 215 \text{ V}$$

When machine is running as motor, then

$$E_{f2} = V_t - I_a r_a$$

$$= 200 - 30 \times 0.5$$

$$= 185 \text{ V}$$

EXERCISE – II

15. Synchronous speed, $n_s = \dfrac{120 \times 50}{4} = 1500$ rpm

Rotor speed = 1425 rpm

$$\therefore \quad s = \frac{1500 - 1425}{1500}$$

$$= 0.05$$

$$\therefore \quad \text{Frequency of rotor emf} = 0.05 \times 50$$

$$= 2.5 \text{ Hz}$$

24. emf generated, $E = \phi Z n \dfrac{P}{A}$

$$\therefore \; 357 = \phi \times 51 \times 20 \times \frac{8500}{60} \times \frac{4}{2}$$

$$\Rightarrow 357 = 17 \times 8500 \times 2 \times \phi$$

$$\Rightarrow \quad \phi = 1.2 \text{ mWb}$$

25. For transformer, emf induced,

$$E = \sqrt{2}\,\pi f N_1 \phi_{max}$$

Now when frequency is halved to 25 Hz, then to have constant magnetisation mmf or maximum these density supply voltage should be halved.

47. At maximum efficiency kVA

$$= (\text{kVA})_{fl} \sqrt{\frac{P_c}{P_{ohmic}}}$$

$$= (\text{kVA})_{fl} \times \sqrt{\frac{40.5}{50}}$$

$$= 0.9\,(\text{kVA})_{fl}$$

49. $$Z_{eH} = \frac{V_{sc}}{I_{sc}} = \frac{0.30}{1.0} = 0.30$$

$$\therefore \quad r_{e_2} = z_{e_2} \cos\theta_2$$

$$= 0.30 \times 0.2 = 0.06$$

Percentage regulation

$$= (\epsilon_r \cos\theta_2 + \epsilon_r \cos\theta_2) \times 100$$

At U.P.E., $\sin\theta_2 = 0$

$$\therefore \quad \text{V.R.} = (0.06 \times 1 + 0)\,100$$

$$= 0.06 \times 100 = 6\%$$

53. Inductance, $L = \dfrac{N^2 \mu_0 A}{l}$

$$= \frac{(1000)^2 \times 4\pi \times 10^{-7} \times \dfrac{\pi \times 9}{4} \times 10^{-4}}{0.3}$$

$$= \frac{9\pi^2 \times 10^{-5}}{0.3}$$

$\therefore$ Energy stored $= \frac{1}{2}LI^2$

$$= \frac{1}{2} \times \frac{9\pi^2 \times 10^{-5} \times 10^2}{0.3}$$

$$= 0.148\text{ J} \approx 0.15\text{ J}$$

54. The use of Auto-transformer is economical only when turn ratio or transmission ratio is equal to unity to have lesser core material, hence core loss will be reduced. Therefore an autotransformer has higher efficiency than a two winding transformer of the same output.

56. Circuit diagram of a differentially compounded motor can be drawn as follows.

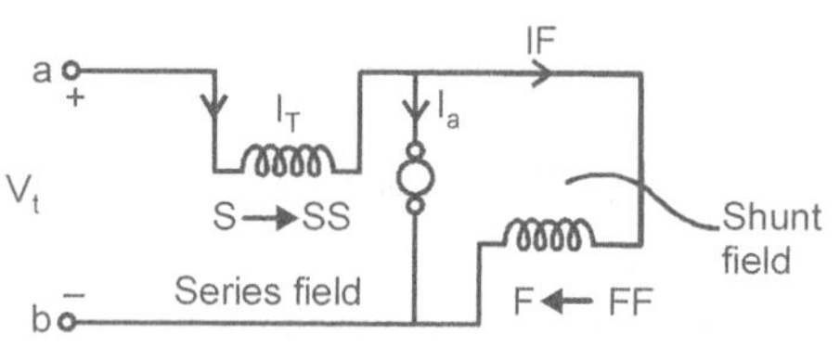

When same dc machine is driven as generator, then current directions can be shown as follows.

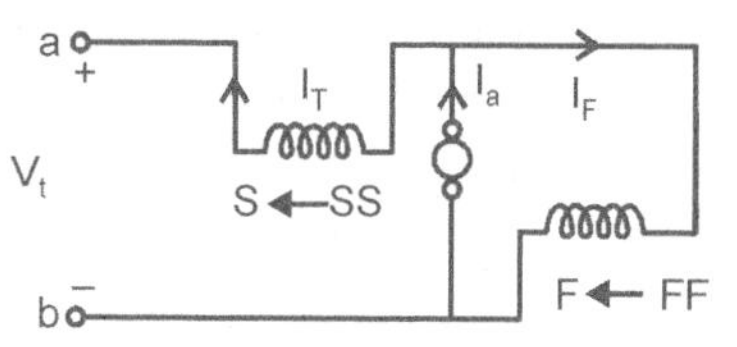

Hence in the case of generator, the series field and shunt fields are adding to each other. Thus, machine now work as a commulatively compounded generator without reversing the interpole coil connection.

57. VA rating of the bank of three transformers in delta

$$= \sqrt{3}\,V_l I_l$$

$$= \sqrt{3}\,V_{ph}\left(\sqrt{3}\,V_{ph}\right)$$

$$= 3V_{ph} \cdot I_{ph}$$

Secondary line voltages from a balanced three-phase system is shown in the phasor diagram.

$\therefore$ VA rating of the open-delta connection

$$= \sqrt{3}\,V_l I_l = \sqrt{3}\,V_{ph} I_{ph}$$

$$\therefore \frac{\text{open-delta VA or rating}}{\text{closed-delta VA or KVA rating}}$$

$$= \frac{\sqrt{3}\,V_{ph} I_{ph}}{3V_{ph} I_{ph}} = 0.577$$

$\therefore$ Motor will have maximum loading of 57.7%

71. Synchronous reactance decrease after some value of failed current because of saturation occur.

79. Torque in the induction motor is proportional to the square of supply voltage

$$T_{el} = kV^2$$

Hence, when supply voltage is decreased by 4%, then torque will be

$$T_{e2} = k(0.96V)^2$$
$$= k\,0.9216\,V^2$$

$\therefore$ Decrease in torque

$$= \frac{kV^2(1 - 0.9216)}{Ky^2} \times 100$$

$$Ky^2 = 7.84\%$$

83. $P_e = KeV^2$

$\therefore$ $P_{e2} = P_{el} = 100$ W

[as voltage is not changed]

$$P_h = kh\,f^{1-x}\,V^x$$

$$\therefore \frac{P_{h2}}{P_{h1}} = \left(\frac{f_2}{f_1}\right)^{1-1.6} = 0.8963$$

$\Rightarrow$ $P_{h2} = 200 \times 0.8963 = 179.27$ W

$\therefore$ Total core loss in second case

$$= 100 + 179.27$$
$$= 279.27\text{ W}$$

84. Eddy current loss, $P_e = KeV^2$

Hysteresis loss, $Y^2 = Khf^{1-x}\,Y^2$

Since secondary voltage is decreased to 220 V, hence eddy current loss will decrease

Now, $$\frac{P_{h2}}{P_{h1}} = \left(\frac{f_2}{f_1}\right)^{1-0.6}\left(\frac{V_2}{V_1}\right)^{1-6}$$

$$= \left(\frac{40}{50}\right)^{-0.6}\left(\frac{220}{440}\right)^{1.6}$$

$$= 1.1432 \times 0.3298$$
$$= 0.377$$

$\therefore$ $P_{h2} = 0.377\,P_{h1}$

Hence hysteresis loss will also decrease.

85. Synchronous motor phasor diagram can be drawn as shown.

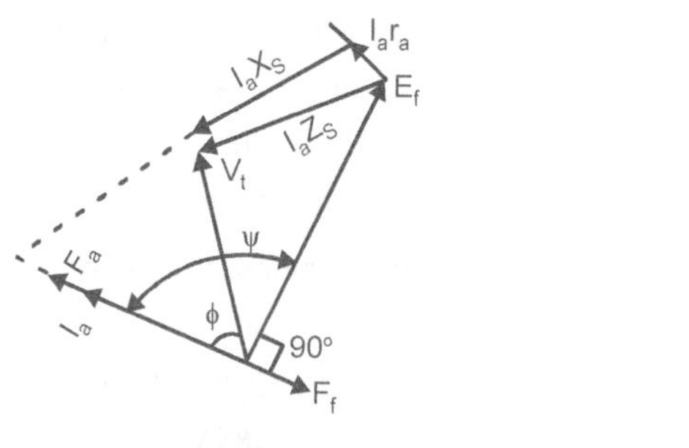

From phasor diagram, angle between I_a and E_f is $\psi + \phi$.

89. *The no load phasor diagram can be drawn as follows :*

No load current consist of I_c and $I\phi$ is core loss and magnetizing components respectively. The angle between V_1' and exciting current I_c is slightly less than 90°.

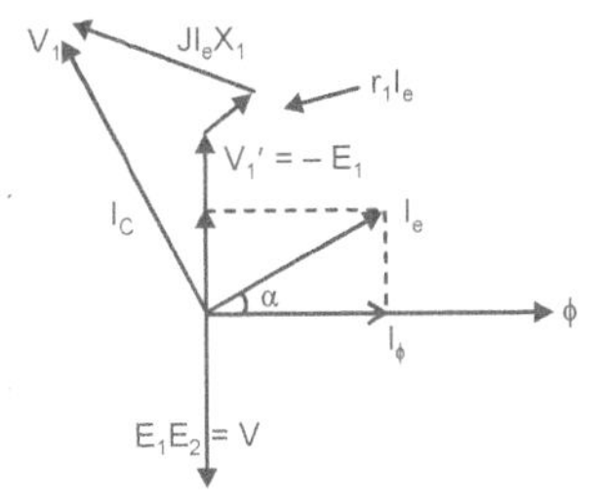

91. Load on a distribution transformer varies over a wide range during 24-hour of a day. Since primary of the distribution transformers are always energised and therefore core loss takes place continuously. Hence distribution transformers are designed to have very low value of core loss. But for reduced core loss the maximum efficiency may occur at about one half of its rated kVA. As for maximum efficiency variable copper loss should be equal to fixed core loss.

92. DC series has torque-speed characteristics as follows:

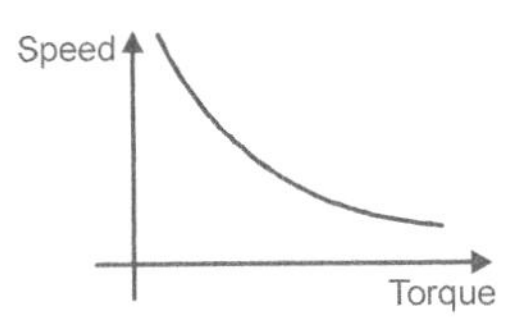

In the starting when speed of motor is zero, then it offers very high starting torque, hence if started at no load then its speed becomes dangerously high.

96. Synchronous speed,

$$n_s = \frac{120 \times 50}{4}$$

$$= 1500 \text{ rpm}$$

Rotor speed, $n_r = 1425$ rpm

$$\therefore \quad \text{Slip, } s = \frac{1500 - 1425}{1500}$$

$$= 0.05$$

∴ Frequency of the rotor emf

$$= sf$$

$$= 0.05 \times 50$$

$$= 2.5 \text{ Hz}$$

99. $P_g = 600$ W

Slip, $s = 0.04$

∴ Rotor copper loss $= s \cdot P_g$

$$= 0.04 \times 600$$

$$= 24 \text{ W}$$

101. Open slots in an induction motor has the effect of increasing air-gap length as a result of which more magnetizing current is required and the operating power factor of the motor is worsened. But an induction motor with open slot has less leakage reactance and therefore, more starting torque.

109. Typical values of SCR are 0.5 to 0.8 for turbo-generators, 1 to 1.4 for hydro-generator and 0.4 for synchronous condenser.

122. Rated current $= \frac{25000}{200} = 125$ A

Since dc machine is wave connected, hence number of parallel path = 2

∴ Current in each parallel path

$$= \frac{125}{2}$$

$$= 62.5 \text{ A}$$

137. Generated emf in a dc machine is given by,

$$E_a = k_a \phi \omega_m$$

$$Eq_1 = k_a \phi \omega_m$$

$$Eq_2 = k_{ax}(0.90\phi) \times (1.2\omega_m)$$

$$= 1.08\, k_a \phi \omega_m$$

$$\therefore \quad \frac{E_{a2} - E_{a1}}{E_{a1}} = \frac{1.08 - 1}{1} \times 100 = 8\%$$

138. Steady state field current $= \frac{120}{60} = 2$A

145.

$$E_a = V_t + I_a r_a$$

$$= 230 + 80 \times 0.1$$

$$= 238 \text{ V}$$

150. For lap-connected winding, number of parallel path is equal to the number of poles.

Hence, current in the winding $= \frac{I_a}{P}$

151. Number of conductor in each parallel path

$$= \frac{400}{8} = 50$$

∴ Generated emf = 50 × 2 = 100 V

153. Number of conductor in each parallel path

$$= \frac{400}{15} = 26.66$$

∴ Total coils = 26.66 × 6 = 160

154. Load shared by alternator first,

$$G_1 = \frac{10}{20} \times 12 = 6 \text{ Mw}$$

Load shared by alternator 2nd,

$$G_1 = \frac{10}{20} \times 8$$

$$= 4 \text{ Mw}$$

163. The connection 0 to 1 signifies that, secondary phase voltage lages the corresponding primary phase voltage by 30°, and connection D_{y11} signifies that the secondary phase voltage loads the corresponding primary phase voltage by 30°. Thus phase difference between corresponding phase voltage of the two secondaries would be 60°.

166. Reactive power drawn by motor,

$$Q_m = 1000 \sin(\cos^{-1} 0.8)$$
$$= -600 \text{ kVAr (lagging)}$$

Reactive power drawn by condensor,

$$Q_c = +750 \sin(\cos^{-1} 0.6)$$
$$= -600 \text{ kVAr (leading)}$$

∴ Total reactive power,

$$Q = Q_m + Q_c$$
$$= 0 \text{ kVAr}$$

∴ $$\text{p.f.} = \cos\left(\tan^{-1}\frac{Q}{P}\right) = 1$$

168. Voltage, $V = \frac{V_1}{2} + \frac{V_1}{2} + V_1$

$$= 2V_1 \text{ (flux density will remains same)}$$

Current carrying capacity of conductor will remain same, therefore New rating of transfer $= 2V_pI_p$

178. Number of total conductors

$$= 6 \times 180$$
$$= 1080$$

Frequency, $$f = \frac{P.n}{120}$$

$$= \frac{300 \times 20}{120}$$
$$= 50 \text{ Hz}$$

Number of turns $= \frac{1080}{2} = 540$

Number of turns (series) (phase)

$$= \frac{540}{3} = 180$$

Slot angle, $$r = \frac{180 \times P}{s}$$

$$= \frac{180 \times 20}{180}$$
$$= 20$$

∴ Slot/pole/phase, $m = \frac{180}{3 \times 20} = 3$

⇒ Breadth factor,

$$K_b = \frac{\sin\frac{mr}{2}}{m \sin\frac{r}{2}}$$

$$= \frac{\sin\frac{3 \times 20}{2}}{3 \sin 10}$$

$$= \frac{\sin 30}{3 \sin 10} = 0.95$$

∴ $$E_p = 4.44\, k_b f N_{ph} \phi$$
$$= 4.44 \times 0.95 \times 50 \times 180 \times 25 \times 10^{-3}$$
$$= 1000 \text{ V}$$

181. Torque developed in dc motors, $T = K\phi I_a$

For series motor, $\phi \propto CI_a$,

But saturation constant at equal to saturated value, hence

$$T \propto I_a$$

182. Synchronous speed,

$$N_{s1} = \frac{120 \times 60}{6}$$
$$= 1200 \text{ rpm}$$

$$N_{s2} = \frac{120 \times 60}{4}$$
$$= 1800 \text{ rpm}$$

$$N_{s3} = \frac{120 \times 60}{P_1 + P_2}$$
$$= \frac{120 \times 60}{6 + 4}$$
$$= 720 \text{ rpm}$$

$$N_{s4} = \frac{120 \times 60}{P_1 - P_2}$$
$$= \frac{120 \times 60}{6 - 4}$$
$$= 3600 \text{ rpm}$$

190. kVA output at maximum efficiency $(\text{kVA})f_l \sqrt{\frac{900}{1600}}$

$$= (\text{kVA})f_l \times \frac{3}{4}$$
$$= 0.75\, (\text{kVA})f_l$$

202. Eddy current loss is given by, $P_e = Kef^2B_m^2$

Hysteresis loss is given by, $P_h = KhfB_m^x$

For a constant value of maximum flux density, hysteresis loss is proportional to the frequency, hence on doubling supply frequency, the hysteresis loss also gezts double.

205. At full load, copper loss $= 400 \times (2)^2 = 1600$ watts

221. $$\frac{\text{Transformed power}}{\text{Input power}} = 1 - K$$
$$= 1 - 0.8$$
$$= 0.2$$
$$\therefore \quad \text{Transformed power} = 0.2 \times 10$$
$$= 2.0 \text{ kW}$$

223. $$P_h = Khf^{1-x} V^x; x = 1.6$$
$$\frac{P_{h2}}{P_{h1}} = \left(\frac{f_2}{f_1}\right)^{-0.6} \left(\frac{v_2}{v_1}\right)^{1.6}$$
$$= \left(\frac{50}{60}\right)^{-0.6} \left(\frac{200}{240}\right)^{1.6}$$
$$= 1.1156 \times 0.747$$
$$= 0.833$$
$$\therefore \quad P_{h2} = 30 \times 0.833$$
$$= 25 \text{ W}$$

239. $$P_m = \frac{E_f V_t}{x} \sin \delta$$
$$P_{max} = \frac{E_f V_t}{x}$$
$$\therefore \quad E_f = \frac{x P_{max}}{V_t}$$
$$= \frac{(0.2 + 1.3) \times 1.2}{1.2}$$
$$= 1.5 \text{ pu}$$

249. For alternator 1 of 200 kW,
$$0.03 = \frac{OA - Ob}{Ob}$$

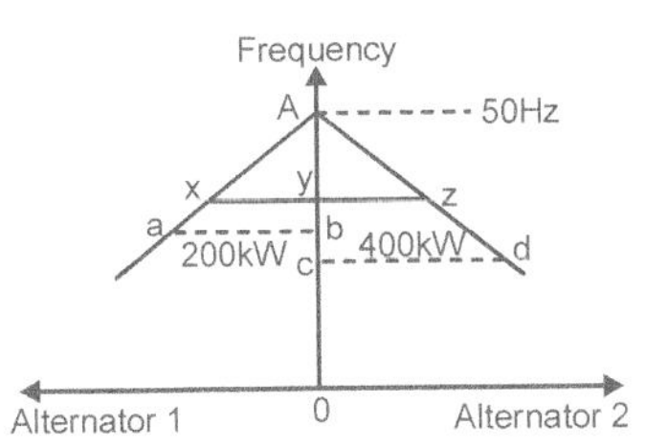

$$\therefore \quad \frac{50 - Ob}{Ob} = 0.03$$
$$\Rightarrow \quad Ob = 48.54 \text{ Hz}$$
For alternator of 400 kW,
$$0.04 = \frac{OA - Oc}{Oc}$$
$$= \frac{50 - Oc}{Oc}$$
$$\Rightarrow \quad OC = \frac{50}{1.04} = 48.077 \text{ Hz}$$
For alternator 1 not be overloaded,
$$xy = 200$$
$$= \frac{200}{50 - Ob}(50 - f)$$
where f is operating frequency
$$f = Ob = 48.544 \text{ Hz}$$
Then $$yz = \frac{400}{50 - 48.077}(50 - f)$$
$$= \frac{400 \times 1.45}{1.923}$$
$$= 302 \text{ kW}$$
$$\text{Total load} = xy + zy$$
$$= 200 + 302$$
$$= 502 \text{ kW}$$

256. For zero voltage regulation,
$$\phi = \tan^{-1}\left(\frac{R}{X}\right) \text{leading}$$
$$= \tan^{-1} \frac{5}{8}$$
$$\Rightarrow \quad \cos \phi = \cos\left(\tan^{-1} \frac{5}{8}\right)$$
$$= \frac{0.8}{\sqrt{0.89}} \text{ leading}$$

268. $$f = \frac{4 \times 1500}{120} = 50 \text{ Hz}$$

282. $$V = 4.44 f B_m A i N$$
$$\therefore \quad \frac{V_2}{V_1} = \frac{f_2 Bm_2}{f_1 Bm_1}$$
$$\Rightarrow \quad Bm_2 = \frac{V_2}{V_1} \times \frac{f_2}{f_1} \times Bm_1$$
$$= 1.5 \times \frac{1}{0.5} \times Bm_1$$
$$= 3Bm_1$$

283. Since voltage is constant, hence eddy current loss will not change with frequency.

299. At 0.8 pf lagging,

$$\text{V.R.} = \in_x \cos\theta_2 + \in_r \sin\theta_2$$
$$= 0.02 \times 0.8 + 0.4 \times 0.6 = 0.04$$

$\therefore$ $\text{V.R.} = 4\%$

At 0.8 pf leading,

$$\text{V.R.} = \in_x \cos\theta_2 - \in_r \sin\theta_2$$
$$= 0.02 \times 0.8 - 0.04 \times 0.6 = -0.08$$

301. Synchronous speed,

$$N_s = \frac{120 \times 50}{6} = 1000 \text{ rpm}$$

Slip, $s = \dfrac{1000 - 500}{1000} = 0.5$

Frequency of the voltage across the slip ring

$$= sf = 0.5 \times 50 = 25 \text{ Hz}$$

329. kVA output at maximum efficiency

$$(\text{kVA})_{o/p} = (\text{kVA})_{fl}\sqrt{\frac{P_c}{P_{oh}}}$$
$$= 40 \times \sqrt{\frac{400}{800}} = 28.284 \text{ kVA}$$

$$\Rightarrow \sqrt{\frac{(\text{kVA})_{o/p}}{(\text{kVA})_{fl}}} = 0.707$$

334. Motor will not run at all. Because for production of electormagnetic torque, the number of rotor poles should be equal to number of stator poles.

341. Synchronous speed $= \dfrac{120 \times 50}{4} = 1500$ rpm

Speed of rotor mmf $= \dfrac{120 \times 75}{4} = 2250$ rpm

$$\text{Slip} = \frac{75}{50} = 1.5$$

Hence rotor is moving in opposite direction to the stator mmf at the speed of 1500 + 2250 = 3750 rpm

358. Generated emf is given by,

$$E_f = \vec{V}_t + \vec{I}_a \times \vec{s}$$
$$= 200 + (10j)(2j)$$
$$= 180 \text{ volts}$$

373. The motor is to be stalled i.e., motor speed is reduced to zero by adding more load on the shaft.

Under this condition, $E_b = 0$

$$\therefore \quad I_a = \frac{220}{4.5 + 0.5} = 44 \text{ A}$$

9 CHAPTER Communication Systems

EXERCISE – I

1. An FM signal with a deviation δ is passed through a mixer, and has its frequency reduced fivefold. The deviation in the output of the mixer is

(*a*) 5δ (*b*) integer minute
(*c*) δ/5 (*d*) δ

2. Which one of the following system offers the best trade off between bandwidth and S/N ratio?

(*a*) PAM (*b*) PDM
(*c*) PPM (*d*) PCM

3. The signals contaminated with large noise are demodulated by

(*a*) envelope detector
(*b*) Synchronous detector
(*c*) envelope detector followed by low-filter
(*d*) envelope detector followed by high pass-filter

4. In FM, modulation index m_f is equal to (Δf is the frequency deviation and f_m is the modulating signal frequency)

(*a*) $\Delta f . f_m$ (*b*) $\frac{f_m}{\Delta f}$
(*c*) $\frac{\Delta f}{f_m}$ (*d*) $\frac{1}{\Delta f . f_m}$

5. High frequency waves follow

(*a*) the line of sight direction
(*b*) the ground wave propagation
(*c*) the ionosphere propagation
(*d*) the curvilinear propagation

6. The figure-of-merit of an earth station in satellite communication system is specified in terms of

(*a*) noise temperature
(*b*) antenna gain
(*c*) gain-to-noise temperature ratio
(*d*) sensitivity of the receiver

7. The fidelity of a receiver relates to variation of the output with the modulating frequency when the output load is

(*a*) resistive
(*b*) inductive
(*c*) capacitive
(*d*) a combination of all the above three

8. The frequency modulated (FM) radio frequency range is nearly

(*a*) 2500-300 MHz (*b*) 150-200 MHz
(*c*) 90-105 MHz (*d*) 30-70 MHz

9. The two fundamental physical limitations faced by an engineer in designing a communication system would include

(*a*) bandwidth and interference
(*b*) interference and distortion
(*c*) distortion and noise
(*d*) noise and bandwidth

10. In the spectrum of a FM wave

(*a*) the carrier frequency cannot disappear
(*b*) the carrier frequency disappears when on the modulation index is large
(*c*) the amplitude of any sideband depends on the modulation index
(*d*) the total number of sidebands depends on the modulation index

11. The radiation pattern of a parabolic antenna is

(*a*) omni-directional
(*b*) a figure of eight
(*c*) highly directional
(*d*) multi-directional

12. A carrier is simultaneously modulated by two sine waves with modulation indices of 0.4 and 0.3. The resultant modulation index will be

(*a*) 1.0 (*b*) 0.7
(*c*) 0.5 (*d*) 0.35

13. The capacity of a channel is given by the

(*a*) number of digits used in coding
(*b*) volume of information it can take
(*c*) maximum rate of information transmitted
(*d*) bandwidth required for information

14. The most useful modulation technique for high fidelity audio broadcasting on radio in current practice is

(*a*) amplitude modulation
(*b*) frequency modulation
(*c*) pulse amplitude modulation
(*d*) pulse code modulation

15. In respect of the block diagram shown in the figure, the input power is 1 mW. The output power P_n will be

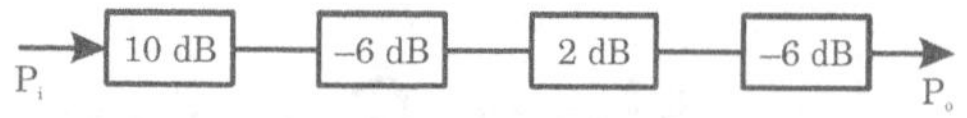

(*a*) 2 mW (*b*) 1 mW
(*c*) 0.5 mW (*d*) zero

16. The rms thermal noise currents of three resistor are I_1, I_2 and I_3 respectively. If these resistors are connected in parallel, the total equivalent noise current 'I' will be

(*a*) $I_1 + I_2 + I_3$ (*b*) $(I_1 I_2 I_3)^{1/3}$
(*c*) $\sqrt{I_1^2 + I_2^2 + I_3^2}$ (*d*) $\sqrt{I_1} + \sqrt{I_2} + \sqrt{I_3}$

17. In a DSB-SC system with 100% modulation, the power saving is

(*a*) 50% (*b*) 66%
(*c*) 75% (*d*) 100%

18. Match List-I with List-II and select the correct answer using the codes given below the lists:

List-I (Systems)

A. Satellite Communication.
B. Line Carrier Telephony.
C. Monophonic Broadcasting.
D. Stereophonic Broadcasting.

List-II (Modulation)

1. SSB.
2. AM.
3. SQPSK/OQPSK.
4. FM.
5. PM.

Codes:

	A	B	C	D
(*a*)	3	4	1	2
(*b*)	1	3	5	2
(*c*)	2	5	1	3
(*d*)	3	1	2	4

19. Which of the following pulse communication system is inherently immune to noise?

(*a*) PPM (*b*) PCM
(*c*) PWM (*d*) PAM

20. Consider the following statements:

In comparison to PCM, a DM system suffers from the disadvantages of

1. Slope overload distortion.
2. Larger bandwidth.
3. More complex circuitry.
4. Inferior performance.

Which of these statements is/are correct?

(*a*) 1 alone (*b*) 1 and 2
(*c*) 2 and 3 (*d*) 4 alone

21. Satellite tracking stations are located in remote area in order to minimise the effect of

(*a*) solar noise (*b*) man-made noise
(*c*) cosmic noise (*d*) thermal noise

22. In the transmission line circuit shown in the given figure, aλ/4 transformer is connected at BB at a distance of d_{min} from the load where the first voltage minimum occurs. What should be the value of R_p the characteristic impedance of the transformer section in order that the input impedance at AA is 50 + *j*0 ohms?

(*a*) 50 ohm (*b*) 75 ohm
(*c*) 25 ohm (*d*) 100 ohm

23. If the distance '*d*' between the radar and the target is reduced to '*d*/2' then the received power becomes 'N' times the power received earlier. The value of 'N' is

(*a*) 2 (*b*) $\sqrt{8}$
(*c*) 4 (*d*) 16

24. Which of the following statement about receivers is false?

(*a*) Double conversion is used to improve image rejection
(*b*) Double conversion is used to improve selectivity
(*c*) Variable sensitivity is used to improve selectivity
(*d*) Variable sensitivity is used to eliminate selective fading

25. In a TV receiver, if the picture is rolling, this may be due to

(*a*) problem in vertical synchronization
(*b*) problem in horizontal synchronization
(*c*) no vertical deflection
(*d*) no horizontal deflection

26. An ergodic process in communication is present if many random signals have

(*a*) identical time averages
(*b*) identical ensembles average
(*c*) identical time and ensemble averages
(*d*) none of these

27. Consider the following statements: The pilot carrier in SSB is used for

1. Provide better noise immunity.
2. Frequency stability.
3. Low power consumption.

Of these statements

(*a*) 1 alone is correct
(*b*) 2 alone is correct
(*c*) 2 and 3 are correct
(*d*) 1, 2 and 3 are correct

28. The core diameter of a step index optical fibre ranges from

(*a*) 300 μm to 500 μm (*b*) 300 mm to 500 mm

(*c*) 30 nm to 40 nm (*d*) 2 μm to 200 μm

29. In India, the maximum frequency deviation in commercial FM broadcasting signals is about

(*a*) 5 kHz (*b*) 15 kHz

(*c*) 75 kHz (*d*) 200 kHz

30. An air-filled rectangular wave-guide have inside dimensions 7 × 3.5 cm. The ratio of cut-off frequency in TE_{01} mode to that in TE_{10} mode is

(*a*) 0.5 (*b*) 0.707

(*c*) 1.0 (*d*) 2.0

31. A loss-less transmission line of length $\lambda/8$ is short circuited at one end. The impedance measured at the other end is jZ at a frequency of 13.5 MHz. If the frequency is raised to 27 MHz, the impedance measured would be

(*a*) $j2Z$ (*b*) $-jZ$

(*c*) zero (*d*) infinity

32. In a certain telemetry system, the measured values are converted to digital form. The digital values can then be transmitted via FSK (binary or quaternary) or BPSK or QPSK systems. Out of these, the best noise immunity can be obtained with

(*a*) binary FSK (*b*) quaternary FSK

(*c*) BPSK (*d*) QPSK

33. The mean square value of the short noise current

(*a*) varies inversely as average current

(*b*) is independent of average current

(*c*) varies as average current

(*d*) varies directly as average current

34. The capacity of a channel is

(*a*) the maximum rate of information transmission

(*b*) the bandwidth required for transmission

(*c*) volumes of information it can take

(*d*) the number of digits used in coding

35. Wave-guides are not used for frequencies

(*a*) below 1 GHz

(*b*) below 10 GHz

(*c*) below 100 GHz

(*d*) below 500 MHz

36. If the antenna diameter in a radar system is increased by a factor of 4, the maximum range will be increased by a factor of

(*a*) $\sqrt{2}$ (*b*) 2

(*c*) 4 (*d*) 8

37. Consider the following features regarding satellite communication:

1. Height of satellite.
2. Downlink frequency.
3. Gain of transmitting antenna.
4. Gain of receiving antenna.

The free space transmission loss depends on

(*a*) 1 and 3 (*b*) 1 and 2

(*c*) 2 and 3 (*d*) 3 and 4

38. A FUNCTION subprogram is invoked in the main program by using

(*a*) a GOTO statement

(*b*) a logical IF statement

(*c*) a CALL statement

(*d*) none of these

39. A frequency-modulated voltage wave is given as

$$V = 10 \sin (2\pi \times 10^8 t + 5 \sin 400 \pi t).$$

The carrier frequency and maximum deviation of the signal are respectively

(*a*) 10 MHz and 10 kHz

(*b*) 100 MHz and 1 kHz

(*c*) 10 MHz and 1 kHz

(*d*) 100 MHz and 10 kHz

40. What is the automatic frequency control voltage of the FM transmitter VCO ?

(*a*) *dc* voltage

(*b*) *sine wave* voltage

(*c*) *square wave* voltage

(*d*) *ramp* voltage

41. In a 100% amplitude modulated signal, if the total transmitted power is P, then the carrier power will be

(*a*) $\frac{2}{3}P$ (*b*) $\frac{1}{2}P$

(*c*) $\frac{1}{3}P$ (*d*) $\frac{1}{4}P$

42. An antenna has radiation resistance of 30 ohm, equivalent resistance of 10 ohm for all other losses and directive gain of 4.0. The power gain of the antenna will be

(*a*) 4.0 (*b*) 3.0

(*c*) 2.0 (*d*) 1.33

43. Two signal have same frequency spectrum but different phase function their power-density, spectrums will be

(*a*) differing in phase

(*b*) differing by e^{wot}

(*c*) differing by w_0

(*d*) identical

44. In microwave communication, the antenna of one repeater must be accurately aligned with the next repeater in the link, because

(*a*) there are at least four carriers with 600-2700 channels per carrier

(*b*) a repeater must amplify the signal received

(*c*) the antenna beam widths are less than 2°

(*d*) the antennas are small in size

45. The ISDN basic rate interface specified as per CCITT standard is

(*a*) one B channel (64 kbps) and one D channel (16 kbps)

(*b*) two B channels (64 kbps)

(*c*) two B channels (64 kbps) and one D channel (16 kbps)

(*d*) two B channels (64 kbps) and one D channel (64 kbps)

46. An A3J transmitter operating at 16 MHz has a frequency stability of 1 part per million. If its transmission is reproduced by a receiver whose frequency stability is 8 parts per million, the maximum frequency error that the output of this receiver could have in reproducing this transmission will be

(*a*) 20 Hz (*b*) 48 Hz

(*c*) 72 Hz (*d*) 144 Hz

47. Consider the following statements regarding digital communication systems:

1. Pulse code modulation (PCM) is used.
2. Quantisation noise occurs in PCM.
3. In PCM, companding is used to protect small signals from quantisation distortion.
4. The Shannon-Hartley law describes signalling rates.

Of these statements:

(*a*) 2, 3 and 4 are correct (*b*) 1, 3 and 4 are correct

(*c*) 1, 2 and 4 are correct (*d*) 1, 2 and 3 are correct

48. Consider the following statements:

Equipment is available to enable carrying more than one voice channel on

1. A pair of wires.
2. A coaxial cable.
3. A radio link.
4. An optic fibre.

Of these statements:

(*a*) 1, 2 and 3 are correct

(*b*) 2, 3 and 4 are correct

(*c*) 1 and 4 are correct

(*d*) 1, 2, 3 and 4 are correct

49. The monostable multivibrator is used for generating PWM because it is a

(*a*) integrator

(*b*) voltage to time converter

(*c*) voltage to frequency converter

(*d*) pulse generator

50. Consider the following statements regarding noise in pulse communication systems:

1. Those forms of pulse modulation which transmit constant amplitude signals can be
2. For PWM and PPM, improved signal-to-noise ratio requires higher bandwidth.
3. In PCM, the quantising noise goes up if the number of standard levels is increased.

Of these statements:

(*a*) 1 and 2 are correct

(*b*) 1 and 3 are correct

(*c*) 2 and 3 are correct

(*d*) 1, 2 and 3 are correct

51. In a continuous data system

(*a*) data may be a continuous function of time at all points in the system

(*b*) data is necessarily a continuous function of time at all points in the system

(*c*) data is continuous at the input and output parts of the system but not necessarily during intermediate processing of the data

(*d*) Only the reference signal is a continuous function of time

52. An AM transmitter radiates 4 kW with an unmodulated carrier wave and 5 kW when it is amplitude modulated. If the same carrier were frequency modulated, then the radiated powar will be

(*a*) 4 kW (*b*) 5 kW

(*c*) 4.5 kW (*d*) 8 kW

53. Entropy essentially is a measure of

(*a*) rate of information

(*b*) average information

(*c*) disorder in information

(*d*) portability of information

54. If the number of bits in a PCM system is increased from n to $n + 1$, then signal-to-quantization noise ratio will increase by a factor

(*a*) $\frac{(n+1)}{n}$ (*b*) $\frac{(n+1)}{n^2}$

(*c*) 2 (*d*) 4

55. Serration pulses in composite video waveform serve the purpose of

(*a*) equalising the change in the integration before the start of the vertical retrace

(*b*) helping vertical synchronisation

(*c*) helping horizontal synchronisation

(*d*) preventing ghost picture

56. A $\lambda/2$ dipole antenna radiates 20 watts at 200 MHz having a gain of 1.64. A receiving $\lambda/2$ dipole antenna is 50 km away from the transmitting antenna. The antennas are positioned for optimum transmission and reception and the effective length of the receiving antenna is λ/π, λ being the wavelength of the transmitted wave. The open circuit voltage in the receiving antenna will be

(*a*) 800 μV (*b*) 400 μV

(*c*) 200 μV (*d*) 100 μV

57. In the double conversion AM receiver, the adjacent channel selectivity is determined mainly by

(*a*) the persecutor and the RF stages

(*b*) the higher frequency IF stages

(*c*) the lower frequency IF stages

(*d*) both the higher and lower frequency IF stages

58. In the narrow band FM system, the bandwidth is (f_m is the modulating frequency)

(*a*) f_m (*b*) $2f_m$

(*c*) $3f_m$ (*d*) $3f_m$

59. The refractive index of the ionosphere (i.e. an ionised layer)

(*a*) is always smaller than 1.0

(*b*) is always equal to 1.0

(*c*) is always greater than 1.0

(*d*) is small than 1.0 during the day and greater than 1.0 during the night

60. For an energy signal $g(t)$ and its time-shifted version $g(t-\tau)$, the .

(*a*) amplitude spectra are identical and their phase spectra are also identical

(*b*) amplitude spectra are identical but their phase spectra are different

(*c*) amplitude spectra are different but their phase spectra are identical

(*d*) amplitude spectra are different and their phase spectra are also different

61. Keyed.' AGC is AGC that

(*a*) works only with Morse code

(*b*) is achieved when keyed by the transmitting signal

(*c*) is used in TV receivers so that transmitted pictures brightness shoes not effect the AGC

(*d*) is used in colour TV receivers so that the transmitted colour has no effect on the AGC

62. Consider the following statements about SSB (Single Side Band) transmission:

1. It requires less power.
2. Transmitter and receiver needed are simple.
3. There is no interference.
4. Bandwidth required is less.

Of these statements:

(*a*) 1, 2 and 3 are correct

(*b*) 1, 2 and 4 are correct

(*c*) 1 and 4 are correct

(*d*) 3 and 4 are correct

63. Consider the following applications:

1. Signal adder
2. Signal multiplier
3. AM demodulator
4. Frequency multiplier
5. Pulse modulator

A balanced modulator is useful as

(*a*) 1, 2, 4 and 5 (*b*) 1, 3, and 5

(*c*) 2, 3, 4 and 5 (*d*) 2, 3 and 4

64. A practical local oscillator in the superheterodyne type of communication receiver is

(*a*) an R-C phase shift oscillator

(*b*) a Colpitts oscillator

(*c*) a Hartley oscillator

(*d*) a phase-shift type of harmonic oscillator that employs L-C resonator

65. A cubical cavity resonator measures 3 cm on each side. The lowest resonant frequency is nearly

(*a*) 6 GHz (*b*) 7 GHz

(*c*) 8 GHz (*d*) 9 GHz

66. The radio altimeter on board an aircraft is basically a

(*a*) doppler radar

(*b*) MTI radar

(*c*) continuous wave radar

(*d*) frequency modulated CW radar

67. Which one of the following systems gives the highest figure-of-merit (a measure of the noise performance)?

(*a*) WBFM (*b*) NBFM

(*c*) AM (*d*) SSB

68. In T. V. transmission, picture signal is amplitude modulated and sound signal is frequency modulated. This is done because

(a) it is not possible to frequency modulate the picture signal
(b) bandwidth requirement is minimised
(c) sound signal is more susceptible to noise than picture signal
(d) synchronisation of picture frames becomes easier

69. In radar engineering, in order to make use narrow beam antennas for achieving accuracy in target direction indication, it is preferable use frequencies in the range of

(a) lower VHF (b) upper VHF
(c) lower UHF (d) upper UHF

70. A Hi-Fi amplifier consists of the following stages:

1. Audio amplifier.
2. Power amplifier.
3. Driver.
4. Pre-driver.
5. Pre-amplifier.

The correct sequence in which these stages appear from the input end to the output end in the amplifier, is

(a) 5, 2, 4, 3, 1 (b) 2, 4, 3, 5, 1
(c) 4, 3, 5, 1, 2 (d) 5, 1, 4, 3, 2

71. The video detector used in TV receiver is a

(a) diode detector
(b) ratio detector
(c) phase detector
(d) balanced demodulator

72. The voltage reflection coefficient for a wave travelling over a transmission line terminated in an impedance equal to the surge impedance of the line is

(a) zero (b) 1
(c) -1 (d) ∞

73. Audio frequency range used in telephone is

(a) 20-20000 Hz (b) 30-30000 kHz
(c) 300-3000 Hz (d) 30-300 MHz

74. A 600 Hz signal is sampled at a rate of 1000 samples per second. The frequency components of sampled signal will consist of

(a) 400 Hz and 600 Hz
(b) 400 Hz, 600 Hz and 1200 Hz
(c) 600 Hz, 800 Hz, 1000 Hz and 1200 Hz
(d) 600 Hz, 1000 Hz, 1200 Hz and 1400 Hz

75. Compounding is used

(a) to overcome quantising noise in PCM
(b) in PWM receivers to reduce impulse noise
(c) to protect small signals in PC from quantising noise
(d) none of these

76. Frequency shifting keying (FSK) system is used presently in communication of data through satellite. The technique uses

(a) some frequency signal for transmission of ONE, and no signal for transmission of ZERO
(b) constant frequency signal for ZERO and no signal for ONE
(c) one set of frequencies for ONE, and another set of frequencies of ZERO
(d) pulses are transmitted through satellite and frequency etc. has no relevance

77. A signal of bandwidth 4 kHz is recorded in a tape recorder at normal speed. If for transmission purposes, the recorder is replayed at 4 times the normal speed, then the transmission bandwidth will be

(a) 16 kHz (b) 8 kHz
(c) 4 kHz (d) 1 kHz

78. The frequency deviation of an FM signal is increased from 25 kHz to 50 kHz. For the same sinusoidal modulating signal of 10 kHz, the improvement in the demodulated signal to noise power ratio will be

(a) 2.0 (b) 2.5
(c) 4.0 (d) 5.0

79. The open and short-circuit impedances of a transmission line at 1.5 kHz are $900 \angle -40°$ ohms and $400 \angle -10°$ ohms respectively. Its characteristic impedance in ohms, will be

(a) $1300 \angle -50°$ (b) $500 \angle -30°$
(c) $600 \angle -25°$ (d) $50 \angle -20°$

80. Quantisation noise arises only in

(a) amplitude modulation circuits
(b) frequency modulation circuits
(c) phase modulation circuits
(d) pulse code modulation circuits

81. An FM signal with a modulation index m_f is passed through a frequency doubler. The wave in the output of the doubler will have a modulation index of

(a) $m_f/2$ (b) m_f
(c) $2\,m_f$ (d) $4\,m_f$

82. A 3A modulation is sometimes used to

(a) reduce the bandwidth required for transmission
(b) reduce the power that must be transmitted
(c) simplify the frequency stability problem in reception
(d) allow the receiver to have a frequency synthesizer

83. The round-trip distance for a satellite relay may be as great as

(*a*) 5000 km (*b*) 50000 km
(*c*) 100000 km (*d*) 500000 km

84. The most commonly used system of modulation for telegraphy is

(*a*) single tone modulation
(*b*) on-off keying
(*c*) frequency shift keying
(*d*) pulse code modulation

85. Which of the following has the least wavelength?

(*a*) X-rays (*b*) Ultra-violet
(*c*) Infra-red (*d*) UHF

86. For a periodic function, the spectral density and the autocorrelation functions form

(*a*) Fourier transform pair
(*b*) Laplace transform pair
(*c*) Kilbert's transform pair
(*d*) Z-transform pair

87. The capacity of a channel is the

(*a*) number of digits used in coding
(*b*) bandwidth required for information
(*c*) volume of information it can take
(*d*) maximum rate of information transmission

88. Delay distortion in transmission lines exists when

(*a*) different frequency components have different attenuation
(*b*) alpha is a constant
(*c*) beta is proportional to frequency
(*d*) beta is a constant

89. The absorption of radiowaves by the atmosphere depends on

(*a*) polarisation of the waves
(*b*) polarisation of the atmosphere
(*c*) distance from the transmitter
(*d*) frequency

90. In PC for 132 standard quantising levels, the maximum error will be

(*a*) $\frac{1}{66}$ of the total amplitude range
(*b*) $\frac{1}{132}$ of the total amplitude range
(*c*) $\frac{1}{264}$ of the total amplitude range
(*d*) $\frac{1}{528}$ of the total amplitude range

91. A transmission line is distortionless if

(*a*) LC=RG (*b*) LR=CG
(*c*) L=GCR (*d*) LG=CR

92. RMS noise voltage at the input of a TV RF amplifier at the ambient temperature of 27°C is 30 μV. The noise voltage at 90°C will be

(*a*) zero (*b*) 30 μV
(*c*) 33 μV (*d*) 36.3 μV

93. Serrations are used in the composite video waveform of a television system to

(*a*) facilitate interlacing
(*b*) aid generation of vertical sync pulse
(*c*) aid vertical synchronisation
(*d*) aid horizontal synchronisation

94. Quantising noise is produced in

(*a*) FDM
(*b*) PCM
(*c*) all modulation systems
(*d*) all pulse modulation systems

95. A particular FM radio station is having frequency deviation equal to 75 kHz. Taking the maximum audio frequency as 15 kHz, the transmission bandwidth is approximately equal to

(*a*) 90 kHz (*b*) 180 kHz
(*c*) 270 kHz (*d*) 360 kHz

96. With a real-time constraint, the transmission bandwidth needed for a digital signal with *r* symbols per second is equal to or greater than

(*a*) $1/r$ (*b*) $r/2$
(*c*) r (*d*) $2r$

97. Flat top sampling of baseband signals leads to

(*a*) aliasing
(*b*) aperture effect
(*c*) under-sampling
(*d*) impossible reconstruction

98. In FM reception, amplitude disturbances due to static change

(*a*) reduce the signal frequencies
(*b*) increase the signal frequencies
(*c*) disturb the signal frequencies
(*d*) do not affect the signal frequencies

99. The autocorrelation function of a rectangular pulse of duration T is

(*a*) another rectangular pulse of duration T
(*b*) another rectangular pulse of duration 2T
(*c*) a triangular pulse of duration T
(*d*) a triangular pulse of duration 2T

100. Match List-I (RM/OSI layer) with List-II (Function) and select the correct answer using the codes given below the lists:

List-I	List-II
A. Network layer	1. Control of packets.
B. Transport layer	2. Data compression.
C. Data link layer	3. Breaks input bit into frames.
D. Presentation layer	4. Breaks data into smaller units.

Codes:

	A	B	C	D
(*a*)	1	4	3	2
(*b*)	4	1	3	2
(*c*)	1	4	2	3
(*d*)	4	1	2	3

101. Which of the following is not an oscillator ?

(*a*) COHO (*b*) STALO
(*c*) MOPA (*d*) MTI

102. Match List-I with List-II and select the correct answer using the codes given below the lists: (B is the bandwidth of the modulating signal)

List-I	List-II
A. NBRM	1. B.
B. SSB	2. B to 2B.
C. PCM	3. 2B.
D. VSB	4. >2B.
	5. >>2B.

Codes:

	A	B	C	D
(*a*)	4	2	5	1
(*b*)	3	1	5	2
(*c*)	3	1	4	2
(*d*)	4	1	5	2

103. Entropy is basically a measure of

(*a*) rate of information
(*b*) average information
(*c*) probability of information
(*d*) channel capacity for transmission information

104. Match List I (Transducer) With List II (Input/Output variables) and select the correct answer using the codes given below the Lists :

List I	List II
A. Electrodynamic generator	1. Gas pressure to resistance change
B. Venturimeter	2. Force to displacement
C. Pirani gauge	3. Motion to voltage
D. Spring balance	4. Flow rate to pressure

Codes:

	A	B	C	D
(*a*)	3	4	1	2
(*b*)	2	1	4	3
(*c*)	2	4	1	3
(*d*)	3	4	2	1

105. The phase shift method for generation of SSB signals is most suitable for

(*a*) data signals
(*b*) weak modulating signals
(*c*) larger bandwidth modulating signals
(*d*) smaller bandwidth modulating signals

106. Doppler effect is used in

(*a*) pulsed radar
(*b*) CW radar
(*c*) MTI (moving target indicator)
(*d*) both CW radar and MTI

107. As far as high frequency propagation is concerned, which of the following layer is of least importance?

(*a*) D-layer (*b*) E-layer
(*c*) F_1 layer (*d*) F_2 layer

108. The IF stage of a receiver employs

(*a*) capacitive coupling
(*b*) impedance coupling
(*c*) double-tuned transformer coupling
(*d*) single-tuned transformer coupling

109. Smith's charts are used to obtain the impedance relations along a

(*a*) lossless transmission line for different load conditions
(*b*) lossy transmission line for different load conditions
(*c*) lossy transmission line terminated at a load equal to line characteristic impedance
(*d*) lossless transmission line terminated at a load equal to line characteristic impedance

110. Under identical conditions FM and PM are indistinguishable for a single modulating frequency. Now if the modulating frequency is increased

(*a*) PM modulation index will remain constant, whereas FM modulation index will increase
(*b*) PM modulation index will remain constant, whereas FM modulation index will decrease
(*c*) PM modulation index will decrease and FM modulation index will increase
(*d*) both PM as well as FM modulation indices will increase

111. The FM modulation index is given by

(a) $\dfrac{\text{maximum frequency}}{\text{minimum frequency}}$

(b) $\dfrac{\text{minimum frequency}}{\text{maximum frequency}}$

(c) $\dfrac{\text{modulating frequency}}{\text{maximum frequency deviation}}$

(d) $\dfrac{\text{maximum frequency deviation}}{\text{modulating frequency}}$

112. Consider the following expression:

$$v = f_1(x - rt) + f_2(x + rt)$$

where f_1 and f_2 represent two travelling waves on a transmission line. In this case

(a) both waves travel in the positive direction of x

(b) both waves travel in the negative direction of x

(c) wave f_2 travels in the positive direction of x but wave f_1 travels in the negative direction of x

(d) wave f_1 travels in the positive direction of x but wave f_2 travels in the negative direction of x

113. Which of the following are the advantages of FM broadcasting over AM broadcasting ?

1. Better S/N ratio
2. Not subject to signal fading
3. Power efficiency is superior
4. Demodulation is simpler

Select the correct answer from the code given below:

(a) 1 and 2 (b) 1, 2 and 4

(c) 2, 3 and 4 (d) 1 and 3

114. In a pulse radar system, the received echo energy is inversely proportional to (R is the range measured)

(a) $(R)^{1/4}$ (b) R^2

(c) R^3 (d) R^4

115. The technique of polarisation diversity is used

(a) in satellite communication

(b) in troposcatter communication

(c) to combat fading in microwave communication

(d) to increase repeater spacing in microwave link

116. In a narrow band FM system, the highest modulating frequency is f_m. The bandwidth of the system will be

(a) $6f_m$ (b) f_m

(c) $2f_m$ (d) $10f_m$

117. Consider the following statements about Loran.

1. It is also known as hyperbolic system of navigation.
2. Standard Loran operates in the frequency range of 1.8 to 2.0 MHz.
3. In Loran, the synchronism between the master and the slave stations is achieved with the help of coaxial links.
4. The range of Loran can be enhanced to about 3500 km over the sea by using a frequency range of 90 to 110 kHz

Of these statements:

(a) 1,2 and 3 are correct

(b) 1, 3 and 4 are correct

(c) 1, 2 and 4 are correct

(d) 2, 3 and 4 are correct

118. The RMS thermal noise voltages of three resistors individually are E_1, E_2 and E_3. If these resistors are connected in series, then total noise voltage is given by

(a) $E_1 + E_2 + E_3$

(b) $\left(\sqrt{E_1} + \sqrt{E_2} + \sqrt{E_3}\right)^2$

(c) $\left(\sqrt{E_1^2 + E_2^2 + E_3^2}\right)$

(d) $(E_1E_2E_3)^{1/3}$

119. Match List-I (Data Comm. schemes) with List-II (Applications) and select the correct answer using the codes given below the lists:

List-I

A. Parallel synchronous

B Parallel asynchronous

C. Serial synchronous

D. Serial asynchronous

List-II

1. Computer to computer.
2. Computer to printer.
3. Buses within computer.
4. Telephone central office inter-communication.
5. Short distance data communication.

Codes:

	A	B	C	D
(a)	1	2	3	5
(b)	5	3	2	4
(c)	5	3	4	2
(d)	1	2	5	3

120. Which one of the following pairs is not correctly matched?

(a) Amplitude demodulation circuit: Diode detector

(b) Frequency demodulation circuit: Ratio detector

(c) Phase demodulation circuit: Coherent detector

(d) Pulse demodulation circuit: TDM

121. The resolution of an 8-bit optical encoder is

(*a*) 0.7° (*b*) 1.4°
(*c*) 2.8° (*d*) 3.6°

122. The directivity of an isotropic antenna is

(*a*) zero (*b*) less than unity
(*c*) unity (*d*) infinity

123. If the aperture of the radar antenna is doubled, then the radar range increase by a factor of

(*a*) $\sqrt{2}$ (*b*) 2
(*c*) 4 (*d*) 16

124. Consider the following statements:

The total number of modes entering an optical fibre depends on

1. Core radius.
2. Wavelength of the optical source.
3. Refractive index profile.

Of these statements:

(*a*) 1 and 2 are correct
(*b*) 2 and 3 are correct
(*c*) 1 and 3 are correct
(*d*) 1, 2 and 3 are correct

125. A video monitor uses 600 lines per frame and 40 frames per second. Each scan line has a maximum of 1000 pixels. To display on the monitor, the highest frequency of the signal required is

(*a*) 24 MHz (*b*) 12 MHz
(*c*) 6 MHz (*d*) 4 MHz

126. When electromagnetic waves are propagated in a waveguide, they

(*a*) travel along the border walls of the wave guide
(*b*) are reflected from the walls but do not travel along them
(*c*) travel through the dielectric without touching the walls
(*d*) travel along all the four walls

127. If on doubling the modulating frequency, the modulation index is halved, and the modulating voltage remains constant, then the system is

(*a*) amplitude modulated
(*b*) frequency modulated
(*c*) phase modulated
(*d*) pulse modulated

128. Match List-I with List-II and select the correct answer by using the codes given below the lists:

List-I	List-II
A. RF amplifier.	1. Amplifies received carrier and sidebands.
B. Loudspeaker.	2. Gives acoustic output.
C. Demodulator	3. Has IF input and AF output.
D. IF amplifier.	4. Fixed tuned to intermediate frequency.

Codes:

	A	B	C	D
(*a*)	1	2	3	4
(*b*)	2	1	3	4
(*c*)	3	4	1	2
(*d*)	1	4	3	2

129. The bandwidth of a PCM system having two quantization levels is B. If the quantization level is increased to 8, the resultant bandwidth will be

(*a*) B (*b*) 2B
(*c*) 3B (*d*) 4B

130. The rms noise voltage of an amplifier due to an input resistor is 20 microvolts in a bandwidth of 2 MHz. If the operating bandwidth is doubled, the noise voltage will become (rounded to nearest whole number).

(*a*) 40 microvolts
(*b*) 35 microvolts
(*c*) 30 microvolts
(*d*) 20 microvolts

131. Control of data flow within networks to ensure maximum efficiency and minimum interference between users, is achieved

(*a*) through the isolation of users
(*b*) through magic-T networks
(*c*) by address coding
(*d*) by using system protocol

132. Van de Graff generator is used

(*a*) for generating microwave frequencies
(*b*) in radar system for bearings
(*c*) as a device for producing high speed neutrons
(*d*) for generating high de voltages

133. A channel has a SNR of 7 and BW 4 kHz. Keeping the channel capacity (BPS) same, the BW is reduced to accommodate more channels. Then SNR

(*a*) decreases
(*b*) increases
(*c*) remains same
(*d*) is marginally affected

134. A PCM system has a range of ± 4 volts and is to be encoded with 3 bits. The step size is
(a) 0 (b) +2V
(c) – 2V (d) 1V

135. The main advantage of TDM over FDM is
(a) because TDM requires more bandwidth than FDM
(b) TDM is less complicated in hardware than FDM
(c) because it has a better SNR
(d) because it does not require filter

EXERCISE - II

1. Vestigial sideband modulation is normally used for
(a) HF point to point communications
(b) manual broad casting
(c) TV broad casting
(d) stereo broad casting [NPTC]

2. When the modulation index to an AM wave is doubled the antenna current is also doubled the AM system being used is
(a) single-sideband full carrier
(b) vestigial sideband
(c) single-sideband, suppressed carrier
(d) double sideband, full carrier [NTPC]

3. Topermit the selection of 1 out of 16 equiprobable events the number of bits required is
(a) 2 (b) $\log_{10}16$
(c) 8 (d) 4 [NTPC]

4. The most common modulation system used for telegraphy is
(a) frequency-shift keying
(b) two-tone modulation
(c) pulse-code modulation
(d) single-tone modulation [NTPC]

5. The Hartely law states that
(a) the maximum rate of information transmission depends on the channel bandwidth
(b) the maximum rate of information transmission depends on the depth of modulation
(c) redundancy is essential
(d) only binary codes may be used [NTPC]

6. In a broadcast superheterodyne receiver the
(a) local oscillator opcrate below the signal frequency
(b) mixer input must be tuned to the signal frequency
(c) local oscillator frequency is normally double the IF
(d) RF amplifier normally double the IF [NTPC]

7. The local oscillator of a broadcast receiver is tuned to a frequency higher than the incoming frequency
(a) to help the image frequency rejection
(b) to permit easier tracking
(c) because otherwise an intermediate frequency could not be produced
(d) to allow adequate frequency coverage without switching [NTPC]

8. The difference between phase and frequency modulation
(a) is purely theoretical because they are the same in practice
(b) is too great to make the two systems compatible
(c) lies in the poorer audio response of phase modulation
(d) lies in the different definition of the modulation index [NTPC]

9. A band-limited signal has no spectral components above the frequency of 100 kHz. The signal can be uniquely determined by its value at uniform intervals of duration less than
(a) 2 μ sec (b) 5 μ sec
(c) 10 μ sec (d) 3 μ sec [NTPC]

10. Which of the following operations are carried out by INSAT?
(a) Telecasting
(b) Parabolic antenna
(c) Chicken-mesh antenna
(d) All of these [BHEL]

11. For satellite communication, the frequency should be
(a) less than the critical frequency of ionosphere
(b) equal to the critical frequency of ionosphere
(c) more than the critical frequency of ionosphere
(d) none of these [BHEL]

12. If the diameter of the parabolic reflectors of a microwave antenna is doubled, its gain will increase by
(a) 0 dB
(b) 2 dB
(c) 4 dB
(d) 6 dB [BHEL]

13. To convert narrow band FM to wide band FM modulation is increased by
(a) frequency multiplication
(b) frequency division
(c) frequency translation
(d) none of these [BHEL]

14. The function of an amplitude limiter in an FM receiver is
   (a) to eliminate any change in amplitude of received FM signal
   (b) to reduce the amplitude of the signal to suit IF amplifier
   (c) to amplify low frequency signals
   (d) none of these [BHEL]

15. The modulation index of a phase-modulated wave is
   (a) proportional to modulating frequency
   (b) proportional to reciprocal of modulating frequency
   (c) same as in frequency modulating
   (d) none of these [NTPC]

16. In FM spectrum, the modulation index increases the number of significant sidebands
   (a) increases
   (b) remains same
   (c) decreases
   (d) none of these [BHEL]

17. The modulation index of an FM wave is changed from 0 to 4. The transmitted power is
   (a) unchanged (b) halved
   (c) increased by 50 % (d) double [NTPC]

18. The intermediate frequency of a superheat receiver is 450 kHz. If it is tuned to 1200 kHz, the image frequency will be
   (a) 750 kHz
   (b) 900 kHz
   (c) 1650 kHz
   (d) 2100 kHz [NTPC]

19. The effective refractive index of the ionosphere for radio waves is
   (a) more than unity
   (b) less than unity
   (c) equal to unity
   (d) either more or less than unity depending on the electron concentration [NTPC]

20. The frequency deviation of an FMl wave with modulation index of 7 and a practical band width of 160 KHz is
   (a) 10 KHz
   (b) 35 KHz
   (c) 70 KHz
   (d) 140 KHz [BHEL]

21. Which of the following requirements is necessary for communication ?
   (a) High transmitter power
   (b) Large band width
   (c) Kigher channel capacity
   (d) None of these [HAL]

22. The frequency range used in frequency modulator is
   (a) 30 MHz to 300 MHz
   (b) 88 MHz to 108 MHz
   (c) 3 MHz to 30 MHz
   (d) 1 MHz to 3 MHz [BHEL]

23. The amount of information in a continuous signal is
   (a) zero (b) 1
   (c) infinite (d) none of these [BHEL]

24. The function of a pre-emphasis circuit in a communication system is to boost
   (a) the higher audio frequencies
   (b) the lower audio frequencies
   (c) the whole audio band
   (d) the modulated wave [BHEL]

25. Maximum voltage will be induced in a loop antenna of
   (a) it is placed parallel to the in-coming wave
   (b) it is placed at right angle to the in-coming wave
   (c) it is placed at 45° to the in-coming wave
   (d) its width is greater than $\lambda/2$

26. Which of the following statements is not correct for an TEM wave ?
   (a) Electromagnetic fields are entirely transversed
   (b) The velocity of the wave is greater than the velocity of light
   (c) The cut-off frequency is zero
   (d) Characteristic impedance is equal to 37 ohms [HAL]

27. When a wave travelling in air enters into a waveguide,
   (a) the phase velocity will increase
   (b) the group velocity will increase
   (c) the phase velocity will decrease
   (d) none of these [HAL]

28. Before transmitting to the satellite the wideband satellite signal is amplified by
   (a) klystron
   (b) traveling wave tube
   (c) tunnel-diode amplifier
   (d) wave guide [HAL]

29. Which of the following is the most reliable measurement for comparing amplifier noise characteristics ?

(*a*) Signal to noise ratio

(*b*) Noise-factor

(*c*) Shot noise

(*d*) Thermal agitation noise **[HAL]**

30. What is the ratio of modulating power to total power at 100 percent modulation ?

(*a*) 1 : 3 (*b*) 1 : 2

(*c*) 2 : 3 (*d*) none of these **[HAL]**

31. The modulation index of an AM wave is changed from 0 to 1. The transmitted power is

(*a*) unchanged

(*b*) halved

(*c*) doubled

(*d*) increased by 50 percent **[HAL]**

32. If the carrier of a 100 percent modulated AM wave is suppressed, then percentage power saving will be

(*a*) 50 (*b*) 150

(*c*) 100 (*d*) 66.66 **[BEL]**

33. The image frequency of a superheterodyne receiver

(*a*) is created within the receiver itself

(*b*) is due to insufficient adjacent channel rejection

(*c*) is not rejected by the IF tuned circuits

(*d*) is independent of the frequency to which the receiver is tuned **[BEL]**

34. A modulator is a system to

(*a*) separate two frequencies

(*b*) impress the information on to a radio frequency carrier

(*c*) extract information from the carrier

(*d*) amplify the audio frequency signal **[BEL]**

35. In AM transmission, the frequency which is not transmitted is

(*a*) upper side frequency

(*b*) lower side frequency

(*c*) audio frequency

(*d*) carrier frequency **[BHEL]**

36. In an AM wave with 100 % modulation the carrier is suppressed. The percentage of power saving will be

(*a*) 100 %

(*b*) 50 %

(*c*) 25 %

(*d*) 66.7 % **[BHEL]**

37. A Klystron operates on the principle of

(*a*) pulse modulation

(*b*) amplitude modulation

(*c*) frequency modulation

(*d*) none of these **[BHEL]**

38. A carrier is simultaneously modulated by sine waves with modulation indices of 30% and 40% respectively. The over all modulation index will be

(*a*) 50%

(*b*) 70%

(*c*) 100%

(*d*) indefinite as modulation by two waves simultaneously is not possible **[BHEL]**

39. In an FM receiver, the channel bandwidth is around

(*a*) 10 kHz (*b*) 20 kHz

(*c*) 75 kHz (*d*) 200 kHz **[HAL]**

40. In a radio receiver, which of the following stages does not need alignment

(*a*) TRF stage

(*b*) IF stage

(*c*) antenna input stage

(*d*) audio stage **[HAL]**

41. Most of the radio broad casting stations are in which frequency band

(*a*) MF (*b*) HF

(*c*) VHF (*d*) UHF **[BEL]**

42. In a superheterodyne receiver, the IF stage has better selectivity than RF stage because

(*a*) frequency has been increased

(*b*) frequency has been lowered

(*c*) constant passband is possible

(*d*) high L to C radio of the circuit **[BEL]**

43. Pulse-width modulation may be generated

(*a*) by differentiating pulse-position modulation

(*b*) with a monostable multivibrator

(*c*) by integrating the signal

(*d*) with a free-running multivibrator **[BHEL]**

44. The biggest disadvantages of PCM is

(*a*) its inability to handle analog signals

(*b*) the high error rate which its quantizing noise introduces

(*c*) its incompatibility with TDM

(*d*) the large bandwidth that are required for it **[BEL]**

45. Quantising noise can be reduced by increasing

(*a*) bandwidth

(*b*) sampling rate

(*c*) number of standard quantum levels

(*d*) all of these **[BEL]**

46. In FM, the noise can be further decreased by

(*a*) decreasing deviation

(*b*) increasing deviation

(*c*) keeping deviation constant

(*d*) none of these **[HAL]**

47. The function of buffer amplifier in transmitters is to provide

(*a*) impedance matching

(*b*) frequency stability of the oscillator

(*c*) amplification of RF signal

(*d*) none of these **[HAL]**

48. During heterodyne process in a receiver, the modulate of the signal

(*a*) remain unaffected

(*b*) decreases

(*c*) increases

(*d*) is eliminated **[NTPC]**

49. For standard 300 to 3400 Hz voice channel, the standard sampling rate is

(*a*) 3000 samples/sec.

(*b*) 4000 samples/sec.

(*c*) 8000 samples/sec.

(*d*) none of these **[BHEL]**

50. In a multi-channel transmission if the sampling pulse width is made smaller, amount of information that can be transmitted would

(*a*) increase (*b*) decrease

(*c*) remain constant (*d*) none of these **[HAL]**

51. Which of the following is correct ?

(*a*) A hybrid transformer is used to isolate the two wire from 4 wire network

(*b*) A hybrid transformer connects a two wire circuit to a four wire connection at repeater stations

(*c*) A hybrid transformer is connected to amplify the signal to make up for the attenuation of the line

(*d*) A hybrid transformer is used to load the line to reduce attenuation **[BHEL]**

52. Consider the following statements regarding transmission lines transmitting high frequency (HF) signals:

HF transmission lines

1. have appreciable ohmic and dielectric losses.
2. have small physical length.
3. are loss-free lines.

Of these statements:

(*a*) 1 and 2 are correct

(*b*) 2 and 3 are correct

(*c*) 3 alone is correct

(*d*) 1, 2 and 3 are correct **[IAS]**

53. Match List-I with List-II and select the correct answer using the codes given below the lists:

List-I	**List-II**
A. AM radio broadcast	1. 2000 to 12000 MHz
B. FM radio broadcast	2. 80 to 150 MHz
C. microwave radio	3. 300 Hz to 4 kHz
D. short wave radio and ratio telephony	4. 500 kHz to 3 MHz
	5. 3 to 30 MHz.

Codes:

	A	B	C	D
(*a*)	3	4	1	5
(*b*)	4	2	1	5
(*c*)	2	4	5	3
(*d*)	4	2	5	3

[IAS]

54. In respect of a rectangular wave-guide of dimension *a* x *b*, match List-I (Mode) with List-II (Cut-off wavelength) and select the correct answer using the codes given below the lists:

List-I	**List-II**
A. $TE_{0,1}$	1. $2a$
B. $TE_{1,0}$	2. $2b$
C. $TE_{2,0}$	3. $\frac{2ab}{\sqrt{a^2+b^2}}$
D. $TE_{1,1}$	4. a
	5. b

Codes:

	A	B	C	D
(*a*)	2	1	3	4
(*b*)	2	1	4	3
(*c*)	4	5	3	1
(*d*)	4	5	1	3

[IAS]

55. A single noise source has rms noise voltages E_1, E_2, E_3 in three different parts of the frequency spectrum. Their combined effect when connected in series is

(*a*) $E_1 + E_2 + E_3$

(*b*) $E_1^2 + E_2^2 + E_3^2$

(*c*) $\sqrt{E_1^2 + E_2^2 + E_3^2}$

(*d*) $\frac{E_1}{\sqrt{2}} + \frac{E_2}{\sqrt{2}} + \frac{E_3}{\sqrt{2}}$ **[ES]**

56. Match List-I with List-II and select the correct answer using the codes given below the lists:

List-I	List-II
A. Collector modulation	1. FM generation
B. Phase shift method	2. DSB generation
C. Balanced modulator	3. AM generation
D. Amplitude limiter	4. SSB generation

Codes:

	A	B	C	D
(*a*)	3	4	1	2
(*b*)	4	3	1	2
(*c*)	3	4	2	1
(*d*)	4	3	2	1

[IAS]

57. Consider the following statements:

Odd line 2 : 1, interlacing in television helps in

1. reducing flicker.
2. simplifying vertical motion circuitry
3. horizontal synchronisation
4. separation of synchronising pules

Of these statements:

(*a*) 1 and 2 are correct

(*b*) 2 and 3 are correct

(*c*) 1 and 4 are correct

(*d*) 2 and 4 are correct **[IAS]**

58. In an FM system, the deviation is 4 kHz when the modulating voltage is 3 V and the modulating frequency is 400 Hz. If the modulating frequency increases to 1000 Hz and its amplitude to 5 V, then the modulation index will

(*a*) remain the same

(*b*) increase

(*c*) decrease

(*d*) be indeterminate **[IAS]**

59. A wave is propagated in a parallel-plane waveguide. The frequency is 6 GHz and plane separation is 3 cm. For the dominant mode match List-I (Quantities) with List-II (Numerical values) and select the correct answer using the codes given below the lists:

List-I	List-II
A. Cut-off wave-length (cm)	1. 5
B. Wave-length in wave guide (cm)	2. 6
C. Phase velocity (megameter/sec.)	3. 9.05
D. Group velocity (megameter/sec.)	4. 166
	5. 543

Codes:

	A	B	C	D
(*a*)	1	2	5	4
(*b*)	2	1	4	3
(*c*)	2	1	5	4
(*d*)	5	4	2	3

[IAS]

60. Consider the following statements:

Delta Modulation has some significant advantage over PCM as

1. it provides greater channel capacity for a given bit-rate.
2. it does not inherently require synchronization.
3. it is more tolerant to system noise.

Of these statements:

(*a*) 1, 2 and 3 are correct

(*b*) 1 and 2 are correct

(*c*) 2 and 3 are correct

(*d*) 1 and 3 are correct **[IAS]**

61. In a Mesh topology of communication network, if there are N stations, the total number of possible full duplex links is

(*a*) $N-1$ (*b*) $N(N-1)$

(*c*) $\frac{N(N-1)}{2}$ (*d*) $\frac{(N-1)}{2}$ **[IAS]**

62. Consider the following statements:

The advantages of KU band for satellite communications are that

1. heavy rainfall does not attenuate the signal.
2. frequencies are not shared with terrestrial microwave radio system.
3. it has a wide footprint.
4. if has smaller antenna.

Of these statements:

(*a*) 3 and 4 are correct

(*b*) 1 and 4 are correct

(*c*) 2 and 4 are correct

(*d*) 1, 3 and 4 are correct **[IAS]**

63. For nearly distortion less received signal in pulse -modulation it is required that the speed should be

(*a*) less than the signal frequency

(*b*) less than twice the signal frequency

(*c*) more than twice the signal frequency

(*d*) none of these

64. Modems are used for

(*a*) modulating digital signals

(*b*) converting analog to digitals and vice versa

(*c*) either of these

(*d*) none of these

65. Which of the following circuits need to be checked to rectify a colour TV receiver showing no colour, monochrome picture and normal sound?

1. Colour killer.
2. 3.58 sub-carrier oscillator.
3. Chrome amplifier.
4. Brightness control.

Select the correct answer using the codes given below:

Codes:

(*a*) 1 and 2 (*b*) 1, 2 and 3

(c) 1, 3 and 4 (*d*) 2, 3 and 4 **[ES]**

66. The original spectrum of a message contains 100 Hz, 200Hz frequency components. It is amplitude modulated by a carrier of 0.9KHz. Which frequency components are contained in the amplitude modulated signal spectrum ?

(*a*) 900, 1000 and 1100 Hz

(*b*) 700, 800 and 900 Hz

(*c*) 700, 800, 900, 1000 and 1100 Hz

(*d*) 100, 200 and 900 Hz **[ES]**

67. In a perfect electronic must system, the bandwidth necessary for high fidelity should be

(*a*) 3 kHz (*b*) 10 kHz

(*c*) 20 kHz (*d*) 40 kHz

68. In radar sets, the range resolution depends upon pulse

(*a*) average power

(*b*) rise time

(*c*) width

(*d*) repetition frequency **[ES]**

69. In television system, VSB, modulation is preferred to SSB in order to

(*a*) save on bandwidth

(*b*) enable demodulation using an envelope detector

(*c*) simplify generation

(*d*) save on signal power **[ES]**

70. In telephone system, the dial tone has a frequency of

(*a*) 1 Hz (*b*) 33 Hz

(*c*) 133 Hz (*d*) 33 KHz

71. Skip distance is independent of

(*a*) frequency

(*b*) state of ionisation

(*c*) transmitting power

(*d*) type of wave propagation **[ES]**

72. A loss-less long transmission line charged to a voltage V and a capacitor C charged to a voltage V/2 are shown in the given figure.

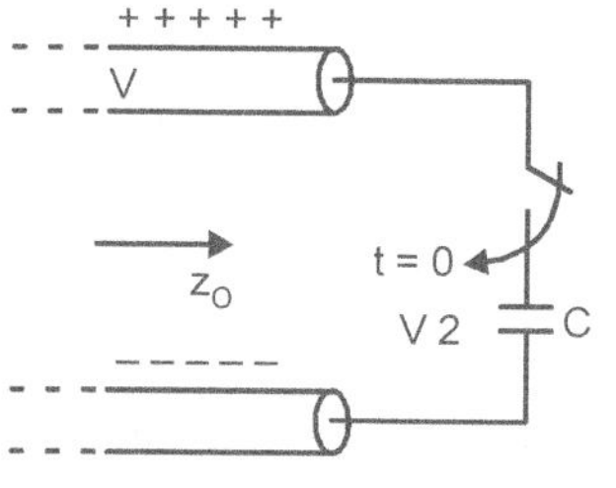

If the switch is closed at $t = 0$, the voltage $v(t)$ across the capacitor for $t \geq 0$ is given by

(*a*) $v(t) = \frac{V}{2}(1 - e^{-t/czo})$

(*b*) $v(t) = \frac{V}{2}(2 - e^{-t/czo})$

(*c*) $v(t) = (2 - e^{-t/czo})$

(*d*) $v(t) = V, t > o = \frac{3v}{4}, t = 0$ **[ES]**

73. In a rectangular wave guide, what is the ratio of wave impedance for TM mode to wave impedance for TE mode (ie Z_{TM}/Z_{TE}) ?

(*a*) 1 (*b*) $\frac{1}{\left[1-(f_c/f)^2\right]}$

(*c*) $[1 - (f_c/f)^2]$ (*d*) $[1 - (f_c/f)]$ **[IAS]**

74. The directive gain of an antenna is 30 dB. If the antenna radiates 7.5 kW, the power density at a distance of 40 km will be

(*a*) $\dfrac{7.5\times10^3}{40\times10^3}\times30\ \text{W/m}^2$

(*b*) $\dfrac{7.5\times10^3}{4\pi\left(40\times10^3\right)^2}\ \text{W/m}^2$

(*c*) $\dfrac{7.5\times10^3\times1000}{4\pi\left(40\times10^3\right)^2}\ \text{W/m}^2$

(*d*) $\dfrac{7.5\times10^3}{\left(40\times10^3\right)^2}\times30\ \text{W/m}^2$ **[ES]**

75. In amplitude modulation, the modulation envelope has a peak value which is double the unmodulated carrier value. What is the value of the modulation index?

(*a*) 25% (*b*) 50%

(*c*) 75% (*d*) 100% **[ES]**

76. A broadcast radio receiver with IF = 455 kHz is tuned to 1500 kHz. The image frequency will be

(*a*) 1045 kHz (*b*) 1500 kHz

(*c*) 1955 kHz (*d*) 2410 kHz **[ES]**

77. If the bandwidth of 1st channel is 'w_i' and the total number of channels accommodated in a time interval of T is 'N', then the minimum bandwidth 'B_{min}' for TDM is equal to

(*a*) $N\sum_{i=1}^{N} w_i$ (*b*) $\dfrac{N}{\sum_{i=1}^{N} w_i}$

(*c*) $\sum_{i=1}^{N} w_i$ (*d*) $\dfrac{1}{\sum_{i=1}^{N} w_i}$ **[ES]**

78. Consider the following statements:

The fundamental characteristic of an FM wave is that δf (maximum deviation of instantaeous frequency from carrier frequency) is.

1. Proportional to the peak amplitude of the information signal.
2. Independent of its frequency.
3. Proportional to the RMS amplitude of the information signal.

Of these statements

(*a*) 1 alone is correct

(*b*) 2 and 3 correct

(*c*) 3 alone is correct

(*d*) 1 and 2 are correct **[ES]**

79. In a binary PCM system, if the number of bits per sample were increased from n to $(n + 1)$, then the signal to quantisation noise ratio will increase by

(*a*) a factor of $\dfrac{(n+1)}{n}$

(*b*) a factor of $\dfrac{(n+1)^2}{n^2}$

(c) $6n$ dB

(*d*) a factor which is independent of n **[ES]**

80. A carrier signal of frequency f_c is modulated by a signal

$2A\cos 2\pi f_1 t - A\sin 2\pi f_2 t.$

The spectrum of the modulated signal will include which of the following frequency components?

1. $f_c+f_1+f_2$
2. $f_c-f_1-f_2$
3. $f_c+2f_1-f_2$
4. $f_c-2f_1+f_2$

Codes:

(*a*) 1 and 2

(*b*) 3 and 4

(*c*) 3 alone

(*d*) 1, 2, 3 and 4 **[ES]**

81. Message switching in computer communication

(*a*) is a store and forward system

(*b*) requires a dedicated path between transmitted and receiver

(*c*) is used only for long messages

(*d*) does not require a buffer at transmitted **[ES]**

82. The noise figure of an amplifier is 6 dB. If the input S/N ratio is 38 dB, then the output S/N ratio will be

(*a*) 44 dB (*b*) 40 dB

(*c*) 36 dB (*d*) 32 dB **[ES]**

83. In a delta modulation scheme, the step width is 1.5 ms. The maximum slope that the staircase can track is

(*a*) 50 V/s (*b*) 55 V/s

(*c*) 60 V/s (*d*) 65 V/s **[ES]**

84. In order to obtain monochrome display in a colour TV monitor, the green, red and blue colours are respectively added in the proportion

(*a*) 0.59, 0.20 and 0.21
(*b*) 0.60, 0.25 and 0.15
(*c*) 0.59, 0.30 and 0.11
(*d*) 0.60, 0.10 and 0.30 **[ES]**

85. Without any filtering, a broadcast station at 1600 kHz is heard together with another station at 2500 kHz on a superheterodyne receiver. The IF employed would have been

(*a*) 900 kHz (*b*) 450 kHz
(*c*) 300 kHz (*d*) 225 kHz **[ES]**

86. Two signals have the same frequency spectrum but different phase functions. Their power density spectrum will

(*a*) differ in phase (*b*) differ by $e^{\omega t o}$
(*c*) differ by ω_o (*d*) be identical **[ES]**

87. The output V_R of the ratio detector and the output V_F of similar Forester-Seeley discriminator are related as

(*a*) $V_F = V_R$
(*b*) $V_F > V_R$
(*c*) $V_F = 0.51\ V_R$
(*d*) $V_F = 2.51\ V_R$ **[ES]**

88. A 10 MHz carrier of peak value 10 V is amplitude modulated by a 10 kHz signal of amplitude 6 V. The amplitude of each side-band frequency is

(*a*) 3V (*b*) 4V
(*c*) 5V (*d*) 6V **[IAS]**

89. The Bootstrap and Miller Integrator type time base generators, in the ideal case, are characterized by amplifier gains of

(*a*) – and + 1 respectively
(*b*) + 1 and – ∞ respectively
(*c*) + 1
(*d*) – ∞ **[IAS]**

90. Equalising pulses in TV are placed during the

(*a*) vertical blanking period
(*b*) horizontal blanking period
(*c*) serration
(*d*) horizontal retrace **[ES]**

91. Single loop ionospheric transmission systems can have a range up to

(*a*) 50 km (*b*) 200 km
(*c*) 500 km (*d*) 1000 km **[ES]**

92. To couple a coaxial line to a parallel wire line, it is best to use a

(*a*) slotted line
(*b*) balun
(*c*) directional coupler
(*d*) quarter-wave transformer **[ES]**

93. DB, DW, and DD directives are used to place data in particular location or to simply allocate space without preassigning anything to space. The DW and DD directions are used to generate

(*a*) offsets
(*b*) full address of variables
(*c*) full address of labels
(*d*) offsets of full address of labels and variables **[ES]**

94. Consider the following statements:

The maximum range of radar can be increased by

1. Increasing the peak transmitted power.
2. Increasing the gain of the receiver.
3. Increasing the diameter of the antenna.
4. Reducing the wavelength used.

Of these statements:

(*a*) 1, 3 and 4 are correct
(*b*) 1, 2, 3 and 4 are correct
(*c*) 2 and 4 are correct
(*d*) 1 and 3 are correct **[ES]**

95. The most useful approach to radar system for monitoring the speed of moving vehicles is:

(*a*) Pulsed radar
(*b*) Monopulse
(*c*) Doppler radar
(*d*) Auto-tracking radar **[ES]**

96. At high frequencies, which of the following types of noise becomes most important?

(*a*) Elicker Noise
(*b*) Shot Noise
(*c*) Thermal Noise
(*d*) Transit time Noise **[ES]**

97. In a line-of-sight system, the receiving and transmitting antenna heights are respectively 10 m and 100 m above the surface of the earth. The range of the system will be approximately

(*a*) 20 km

(*b*) 30 km

(*c*) 40 km

(*d*) 50 km **[ES]**

98. In case of communication involving micro-wave. communication system, repeaters are used approximately at a distance of 20 to 25 km. These repeaters are used to

(*a*) filter out the noise at each repeater place

(*b*) amplify the signal

(*c*) multiplex the signal

(*d*) keep the signal at the line of sight and to provide direction **[ES]**

99. Which one of the following pairs is NOT correctly matched?

(*a*) DSB-SC modulation : Balanced modulator

(*b*) PCM : Pre-emphasis

(*c*) FM : Reactance modulator

(*d*) SSB modulation : Weaver's method **[ES]**

100. A Fortran line is segmented columnwise for specific purposes. Match List-I (purpose) with List-II (column numbers used) and select the correct answer using the codes given below the Lists:

List-I	List-II
A. Statement number	1. 6
B. Arithmetic statement	2. 1 to 80
C. Continuation mark	3. 7 to 72
D. Data	4. 1 to 5

Codes:

	A	B	C	D
(*a*)	4	3	2	1
(*b*)	4	3	1	2
(*c*)	3	4	2	1
(*d*)	3	4	1	2

[ES]

Direction (Q. 101-120) : *The following items consist of two statement, one labelled the "Assertion A" and the other labelled the Reason R. You are to examine these two statements carefully and decide if the Assertion A and the Reason R are individually true and if so, whether the reason is a correct explanation of the Assertion. Select your answers to these items using the codes given below and mark your answer sheet accordingly:*

Codes:

(*a*) Both A and R are true and R is the correct explanation of A

(*b*) Both A and R are true but R is NOT a correct explanation of A

(*c*) A is true but R is false

(*d*) A is false but R is true

101. Assertion (A): The degree of modulation of ratio-frequency voltage present at the input terminals of the diode is less than the degree of modulation of the original radio frequency signal in AM detector.

Reason (R): It depends primarily on the ratio of the source impedance of the radio frequency voltage to the diode load impedance. **[ES]**

102. Assertion (A): PSK is inferior to FSK.

Reason (R): PSK requires less bandwidth than FSK. **[ES]**

103. Assertion (A): A 3-phase thyristorised full converter feeding a purely resistive load with delayed firing angel of thristors will draw reactive power from the ac supply.

Reason (R): The ac supply current at delayed firing angle will be non-sinusoidal and its fundamental component would large the supply voltage. **[ES]**

104. Assertion (A): In a high frequency transm-ission line, when a voltage is placed across the line at the sending end, the voltage and current in the line change with distance.

Reason (R): A transmission line consists of continuous conductors with a cross-sectional configuration that is constant throughout their lengths. **[IAS]**

105. Assertion (A): For continuous communi-cation purposes, satellites are normally placed in geostationary orbit.

Reason (R) : In satellite communication, uplink and downlink frequencies are different. **[ES]**

106. Assertion (A): FM systems employ pre-emphasis and deemphasis.

Reason (R) : Preemphasis and deemphasis in FM provide better protection to lower amplitude modulating signals. **[ES]**

107. **Assertion (A):** The different components of a composite video signal are needed consecutively and not simultaneously at the TV receivers.

Reason (R): The different components of the composite video signal are sent to the receiver on a frequency division basis. **[ES]**

108. **Assertion (A):** In fibre optic communi-cation, the 1.3 μm, window is preferred over the μm window for long distance communication.

Reason (R): At 1.3 μm, not only the attenuation of silica fibre is lower than that at 0.85 μm, the pulse broadening at the long wavelength window is also minimum. **[ES]**

109. **Assertion (A):** A communication satellite is a microwave link repeater.

Reason (R): A communication satellite receives the signal beamed up by an earth station. It amplifies this signal and retransmits it to ground station. **[IAS]**

110. **Assertion (A):** Consider an SSB signal where the modulating signal is a speech signal.

Reason (R): The envelope of an SSB is constant. **[ES]**

111. **Assertion (A):** A low voltage standing wave ratio (VSWR) is the goal in a transmission line.

Reason (R): The higher the VSWR, the greater is the mismatch on the line. **[IAS]**

112. **Assertion (A):** For transmitting audio frequency signals, antennas of several hundred kilometers length would be required.

Reason (R): For efficient radiation of electromagnetic energy to occur from an antenna, the wavelength of the radiated signal must be comparable with the physical dimensions of the antenna. **[ES]**

113. **Assertion (A):** Adaptive delta modulation circuit has less distortion.

Reason (R): Quantisation noise is required. **[IAS]**

114. **Assertion (A):** In a double stub matching arrangement, the spacing between the two fixed stubs is never made half wavelength.

Reason (R): A spacing of $\lambda/2$ between the stubs effectively puts the stubs in parallel and thus adds to more flexibility in adjustment. **[ES]**

115. **Assertion (A):** Commercial PCM generally use a 6-bit code.

Reason (R): This is a compromise between quantisation noise and channel bandwidth requirements. **[IAS]**

116. **Assertion (A):** During inverter operation of a phase-controlled thyristor power converter, the upper value of firing angle range is limited to a value of less than 180°

Reason (R): A thyristor requires a definite amount of time to regain forward blocking capability after its current has become zero. **[ES]**

117. **Assertion (A):** A threshold effect is observed in square law detectors.

Reason (R): Square law detector is nonlinear. **[ES]**

118. **Assertion (A):** Square law detectors, though simple, are not satisfactory for the detection of amplitude modulated signals.

Reason (R): Square law detection leads to large second harmonic distortion, which is proportional to the modulation index of the sinusoidal signal. **[IAS]**

119. **Assertion (A):** Frequency modulation systems are move widely used than amplitude modulation systems.

Reason (R): Frequency modulation systems have less noise than amplitude modulation system. **[IAS]**

120. **Assertion (A):** Low frequencies are more suitable than high frequencies for underwater, communication and for communication with underwater objects.

Reason (R): Electric and magnetic fields in the underwater objects are out of phase by 45°. **[ES]**

ANSWERS

EXERCISE - I

1. (*d*)	2. (*d*)	3. (*b*)	4. (*c*)	5. (*c*)	6. (*c*)	7. (*a*)	8. (*c*)	9. (*d*)	10. (*c*)
11. (*c*)	12. (*c*)	13. (*c*)	14. (*b*)	15. (*b*)	16. (*c*)	17. (*b*)	18. (*d*)	19. (*b*)	20. (*a*)
21. (*b*)	22. (*c*)	23. (*d*)	24. (*d*)	25. (*a*)	26. (*c*)	27. (*c*)	28. (*b*)	29. (*c*)	30. (*d*)
31. (*d*)	32. (*b*)	33. (*d*)	34. (*a*)	35. (*a*)	36. (*c*)	37. (*b*)	38. (*b*)	39. (*b*)	40. (*a*)
41. (*a*)	42. (*b*)	43. (*d*)	44. (*b*)	45. (*c*)	46. (*d*)	47. (*d*)	48. (*b*)	49. (*b*)	50. (*a*)
51. (*b*)	52. (*a*)	53. (*b*)	54. (*a*)	55. (*d*)	56. (*c*)	57. (*a*)	58. (*b*)	59. (*d*)	60. (*b*)
61. (*b*)	62. (*c*)	63. (*b*)	64. (*d*)	65. (*c*)	66. (*d*)	67. (*d*)	68. (*b*)	69. (*a*)	70. (*d*)
71. (*a*)	72. (*a*)	73. (*c*)	74. (*a*)	75. (*c*)	76. (*c*)	77. (*a*)	78. (*a*)	79. (*d*)	80. (*d*)
81. (*c*)	82. (*c*)	83. (*b*)	84. (*c*)	85. (*a*)	86. (*a*)	87. (*d*)	88. (*c*)	89. (*d*)	90. (*c*)
91. (*d*)	92. (*c*)	93. (*a*)	94. (*b*)	95. (*b*)	96. (*d*)	97. (*b*)	98. (*d*)	99. (*d*)	100. (*c*)
101. (*d*)	102. (*b*)	103. (*b*)	104. (*a*)	105. (*d*)	106. (*d*)	107. (*a*)	108. (*c*)	109. (*c*)	110. (*b*)
111. (*d*)	112. (*d*)	113. (*a*)	114. (*a*)	115. (*c*)	116. (*c*)	117. (*d*)	118. (*a*)	119. (*a*)	120. (*c*)
121. (*b*)	122. (*c*)	123. (*d*)	124. (*c*)	125. (*a*)	126. (*b*)	127. (*b*)	128. (*a*)	129. (*c*)	130. (*c*)
131. (*c*)	132. (*b*)	133. (*b*)	134. (*d*)	135. (*c*)					

EXERCISE - II

1.(*c*)	2.(*c*)	3.(*d*)	4.(*a*)	5.(*a*)	6.(*b*)	7.(*d*)	8.(*d*)	9.(*b*)	10.(*b*)
11.(*c*)	12.(*d*)	13.(*a*)	14.(*a*)	15.(*a*)	16.(*a*)	17.(*a*)	18.(*d*)	19.(*b*)	20.(*c*)
21.(*b*)	22.(*b*)	23.(*c*)	24.(*a*)	25.(*a*)	26.(*b*)	27.(*a*)	28.(*b*)	29.(*b*)	30.(*a*)
31.(*d*)	32.(*d*)	33.(*c*)	34.(*b*)	35.(*c*)	36.(*d*)	37.(*d*)	38.(*a*)	39.(*d*)	40.(*d*)
41.(*a*)	42.(*c*)	43.(*b*)	44.(*d*)	45.(*c*)	46.(*b*)	47.(*c*)	48.(*a*)	49.(*c*)	50.(*a*)
51.(*b*)	52.(*a*)	53.(*d*)	54.(*b*)	55.(*c*)	56.(*c*)	57.(*c*)	58.(*c*)	59.(*c*)	60.(*b*)
61.(*b*)	62.(*c*)	63.(*c*)	64.(*b*)	65.(*c*)	66.(*c*)	67.(*c*)	68.(*d*)	69.(*a*)	70.(*b*)
71.(*c*)	72.(*b*)	73.(*c*)	74.(*b*)	75.(*d*)	76.(*d*)	77.(*c*)	78.(*a*)	79.(*a*)	80.(*b*)
81.(*a*)	82.(*d*)	83.(*a*)	84.(*c*)	85.(*b*)	86.(*d*)	87.(*b*)	88.(*a*)	89.(*b*)	90.(*a*)
91.(*d*)	92.(*b*)	93.(*d*)	94.(*a*)	95.(*c*)	96.(*d*)	97.(*d*)	98.(*b*)	99.(*b*)	100.(*b*)
101.(*a*)	102.(*c*)	103.(*b*)	104.(*a*)	105.(*b*)	106.(*a*)	107.(*a*)	108.(*c*)	109.(*b*)	110.(*b*)
111.(*a*)	112.(*a*)	113.(*a*)	114.(*c*)	115.(*d*)	116.(*b*)	117.(*d*)	118.(*a*)	119.(*a*)	120.(*d*)

EXPLANATIONS

EXERCISE – I

4. Modulation index, $m_f = \dfrac{\Delta f}{f_m}$

12. Resultant modulation index,

$$m_t = \sqrt{m_1^2 + m_2^2}$$
$$= \sqrt{0.4^2 + 0.3^2}$$
$$= 0.5$$

15. Total power gain is given by,

$$\frac{P_o}{P_i} = 10 - 6 + 2 - 6 = 0 \text{ dB}$$

$\therefore$ anti log 0 = 1

$\therefore$ $P_o = P_i = 1$ mW

16. Total equivalent noise current,

$$I = \sqrt{I_1^2 + I_2^2 + I_3^2}$$

17.

$$P_t = P_c\left(1 + \frac{m^2}{2}\right)$$
$$= P_c\left(1 + \frac{1^2}{2}\right)$$
$$= P_c\left(\frac{3}{2}\right) = 1.5\,P_c$$
$$P_{SB} = P_c \times \frac{m^2}{2}$$
$$= P_c \times \frac{1}{2}$$
$$= \frac{P_c}{2}$$

$\therefore$ Power saving $= \dfrac{1.5 - 0.5}{1.5} \times 100 = 66.67\%$

22. For a $\lambda/4$ long line, $R_S R_R = R_o^2$

d_{min} is $\lambda/4$ for the characteristic impedance $R_o = 50\,\Omega$

Therefore, resistance at R lead,

Then $\dfrac{\lambda}{4}$ long line with characteristic resistance,

$$R_T = \frac{50}{200} = 12.5\,\Omega$$

Then, $R_o \times 12.5 = R_T^2$

$\therefore$ $R_T = \sqrt{50 \times 1.25} = 25\,\Omega$

23. Power received at radar, $r^1 \propto \dfrac{1}{\gamma^4}$

$$\therefore \quad N = \frac{d^2}{(d/2)^2} = 16$$

30. Cut-off wave length for $TE_{m,n}$ mode,

$$\lambda_o = \frac{2}{\sqrt{\left(\frac{m}{a}\right)^2 + \left(\frac{n}{b}\right)^2}}$$

$\therefore$ Cutt-off frequency, $f_o \propto \dfrac{1}{\lambda_0}$

$$\therefore \quad \frac{f_{01}}{f_{10}} = \sqrt{\frac{\left(\frac{1}{a}\right)^2}{\left(\frac{1}{b}\right)^2}} = \frac{b}{a}$$
$$= \frac{3.5}{7} = 0.5$$

33. Current in FET or BJT under dc conditions is constant at every instant. However circuit from the emitter to the collector consists of a stream of individual electrons or holes, and it is only the time averaged flow measured as constant current. The fluctuation observed in the number of covers called *short noise* given as

$$I_n^2 = 2q\,\text{IdcB}$$

where, $q = 1.6 \times 10^{-19}$ C

and B = bandwidth

36. Maximum range $R_{max} = kd$,

where, k is constant and d is diameter

Then $\dfrac{R_{max\,2}}{R_{max\,1}} = \dfrac{d_2}{d_1} = 4$

39.

$$\omega_c = 2\pi \times 10^8$$
$$\therefore \quad 2\pi f_c = 2\pi \times 10^8$$
$$\Rightarrow \quad f_c = 1 \times 10^8 \text{ Hz} = 100 \text{ MHz}$$
$$\omega_m = 400\,\pi$$
$$\therefore \quad 2\pi f_m = 400\,\pi$$
$$\Rightarrow \quad f_m = 200$$

Now, $\dfrac{\delta}{f_m} = m$

$$\therefore \quad \delta = 5 \times 200$$
$$= 1000 \text{ Hz}$$
$$= 1\text{kHz}$$

41. $$P_t = P_c\left(1+\frac{m^2}{2}\right)$$
$$= P_c\left(1+\frac{1}{2}\right) = P_c \times \frac{3}{2}$$
$$\therefore \quad P_c = \frac{3}{2}P$$

42. Directive gain for an antenna,

$$D(\theta, \phi) = \frac{4\pi r(\theta,\phi)}{P_{rad}}$$

where, $\cup(\theta, \phi)$ is radiation intensity and P_{rad} is the radiated power.

and, power gain, $$G(\theta, \phi) = \frac{4\pi r(\theta,\phi)}{P_{rad}+P_L}$$
$$= D(\theta, \phi) \times \frac{P_{rad}}{P_{rad}+P_L}$$
$$= 4 \times \frac{I^2.30}{I^2.30 + I^2.10} = 3$$

52. Since, in frequency modulated wave, power will be same as that of unmodulated wave, hence for the FM wave, radiated power = 4kW

$$P_t = P_c\left(1+\frac{m^2}{2}\right)$$

Hence for 100% modulation,

$$P_t = P_c\left(1+\frac{1}{2}\right) = \frac{3}{2}P_c$$

Modulating power = S.B. power

$$P_{S.B.} = \frac{m^2}{2}$$
$$P_{S.B.} = \frac{1}{2}P_c$$
$$\therefore \quad \frac{P_{S.B.}}{P_t} = \frac{\frac{1}{2}P_c}{3/2P_c} = \frac{1}{3}$$

57. RF section is a tunable filter and an amplifier that picks up the desired station.

79. $$Z_o = \sqrt{Z_{open}.Z_{so}}$$
$$= \sqrt{900\lfloor -40 \times 400\lfloor -10}$$
$$= \sqrt{36000\lfloor -50} = 600\lfloor -25^\circ$$

92. Mean square value of thermal noise voltage V_{TN} is given by,

$$E\left[V_{TN}^2\right] = 4KTR\Delta f \text{ volts}$$

where, T is absolute temperature and K is Boltsman constant

$$\therefore \quad E\left[V_{TN}^2\right] \propto T$$
$$\Rightarrow \quad E\left[V_{TN}^2\right] = \frac{T_2}{T_1}.E\left[V_{1TN}^2\right]$$
$$\therefore \quad \frac{V_{2TN}}{V_{1TN}} = \sqrt{\frac{T_2}{T_1}}$$
$$\Rightarrow \quad V_{2TN} = \sqrt{\frac{363}{300}} \times 30\mu V = 33\ \mu V$$

95. $$\text{Band Width} = 2(\Delta f + f_m)$$
$$= 2(75+15) = 180\text{ kHz}$$

116. In narrow band FM system,

$$m = \frac{8}{f_m} = 1$$
$$\Rightarrow \quad \delta = f_m$$
$$\Rightarrow \quad \text{bandwidth} = 2f_m$$

121. $$\text{Resolution} = \frac{360}{2^8} = 1.4^\circ$$

123. $$\text{Radar range, } R_{max} = \left[\frac{P_t A_P^2 \lambda^2 S}{(4\pi)^3 P_{min}}\right]$$

130. rms value of noise voltage over a band width B is expressed as

$$V_n^2 = 4KTRB$$

where, K = boltzmann constant; J/K°

T = resistor temperature, and

R = resistor

$$\Rightarrow \quad V_n \propto \sqrt{B}$$
$$\therefore \quad \frac{V_{n2}}{V_{n1}} = \sqrt{\frac{B_2}{B_1}}$$
$$\Rightarrow \quad V_{n2} = 20 \times \sqrt{2} = 30\ \mu V$$

EXERCISE – II

3. Number of bits required = $\log_2 16 = 4$

12. We know $$A = 6\left(\frac{D}{\lambda}\right)^2$$
$$\therefore \quad \frac{A_2}{A_1} = \frac{2^2}{1} = 4$$

In decibel, $$\frac{A_1}{A_2} = 10\log 4 = 6\text{ dB}$$

18. Image frequency, $$f_{si} = 1200 + 2 \times 450$$
$$= 2100\text{ kHz}$$

20. $m_f = \frac{\delta}{f_m} = \frac{\Delta f}{f_m}$

And band width $= 2(\Delta f + f_m)$

$\therefore$ $160 = 2(\Delta f + f_m)$

$\Rightarrow$ $\Delta f + f_m = 80$ ...(*i*)

Now $7 = \frac{\Delta f}{f_m}$

$\Rightarrow$ $\Delta f = 7 f_m$

Putting this in equation (*i*), we get

$7 f_m + f_m = 80$

$\Rightarrow$ $f_m = 10\text{ kHz}$

$\therefore$ Deviation, $\Delta f = 7 \times 10 = 70\text{ kHz}$

25. Input impedance, $Z_s = \frac{Z_0^2}{Z_L}$

Since, Z_L for short circuited line will be 0,

hence $Z_s = \infty$

31. $P_t = P_c\left(1 + \frac{m^2}{2}\right)$

For $m = 0$; $P_{t_1} = P_c(1 + 0) = P_c$

For $m = 1$; $P_{t_2} = P_c\left(1 + \frac{1}{2}\right) = \frac{3}{2}P_c$

$\therefore$ $\frac{P_{t_2} - P_{t_1}}{P_{t_1}} = \frac{\frac{3}{2} - 1}{1} = 0.5 = 50\%$

32. $P_t = P_c\left(1 + \frac{m^2}{2}\right)$

$= P_c\left(1 + \frac{1}{2}\right) = \frac{3}{2}P_c$

$\therefore$ % Power saving $= \frac{3/2 - 1/2}{3/2} \times 100 = \frac{2}{3} \times 100$

$= 66.66\%$

38. $m_t = \sqrt{m_1^2 + m_2^2} = \sqrt{0.3^2 + 0.4^2}$

$= 0.5 = 50\%$

59. Cut off wave length, $\lambda_o = \frac{2a}{m}$

When, $m = 1$

$\lambda_o = 2 \times 3 = 6\text{ cm}$

Wave-length in the wave guide,

$\lambda = \frac{-v}{f}$

$= \frac{3 \times 10^8}{6 \times 10^9} = 5\text{ cm}$

$\vartheta_g = v_c\sqrt{1 - (\lambda/\lambda_o)^2}$

$= 3 \times 10^8\sqrt{1 - (5/6)^2}$

$= 1.66 \times 10^8\text{ m/s}$

$V_P = \frac{v_c}{\sqrt{1 - (5/6)^2}}$

$= \frac{3 \times 10^8}{0.553} = 5.43 \times 10^8$

60. Power density at a distance,

$R = \frac{W}{4\pi R^2} = \frac{7.5 \times 10^3}{4\pi(40 \times 10^3)^2}\text{ W/m}^2$

76. $f_s = 1500\text{ kHz}$

I.F. $= 455\text{ kHz}$

$\therefore$ Image frequency $= 1500 + 2 \times 455 = 2410\text{ kHz}$

82. Noise figure, $F = \frac{\text{Input S/N}}{\text{Output S/N}}$

$\therefore$ F(dB) = Input S/N (dB) = 32

83. Maximum slope $= \frac{75}{1.5} = 50\text{ V/s}$

85. $IF = \frac{2500 - 1600}{2}$

$= \frac{900}{2} = 450\text{ kHz}$

88. Modulation index,

$m = \frac{V_m}{V_c} = \frac{6}{10} = 0.6$

A.W. wave is represented as,

$v = V_c \sin \omega_c t + \frac{mV_c}{2}\cos(\omega_c - \omega_m)t$

$- \frac{mV_c}{2}\cos(\omega_c + \omega_n)t$

$\therefore$ Frequency of sideband $= \frac{mV_c}{2}$

$= \frac{0.6 \times 10}{2} = 3\text{V}$

97. Radius of the earth = 6370 km

$AB = \sqrt{(OB + OA).(OB - OA)}$

$= \sqrt{(6370.1 + 6370)0.1} = 80\sqrt{0.2}$

Similarly, $AC = \sqrt{(6370.1 + 6370)0.1} = 80\sqrt{0.2}$

$\therefore$ Range, BC = BA + AC $\approx$ 50 km

10
CHAPTER

Analog Circuits

EXERCISE - I

1. The dynamic drain resistance of MOSFET is of the order of

(*a*) 10 kΩ (*b*) 500 kΩ
(*c*) 5 kΩ (*d*) 100 kΩ

2. With both juction reverse biased the transistor operates in

(*a*) active region (*b*) cut off region
(*c*) saturation region (*d*) inverted region

3. An amplifier has a gain of 1000 ± 10. Negative feedback is provided such that the gain variation is within 0.1%. Then the amount of feedback β1 is

(*a*) $\frac{1}{10}$ (*b*) $\frac{1}{9}$
(*c*) $\frac{9}{100}$ (*d*) $\frac{9}{1000}$

4. Monolithic IC clip has thickness of about

(*a*) 1 mil (*b*) 5 mils
(*c*) 50 mils (*d*) 200 mils

5. The transconductance g_m of a JFET is of the order of

(*a*) 1 mS (*b*) 1 S
(*c*) 100 S (*d*) 1000 S

6. Most of the small signal transistor are

(*a*) *npn* silicon transistor is plastic package
(*b*) *pnp* silicon transistor in plastic package
(*c*) *npn* germanium transistor in metallic case
(*d*) *pnp* germanium transistor in metalic case

7. FET has offset voltage of about

(*a*) 0.2 volt
(*b*) 0.6 volt
(*c*) 1.1 volt
(*d*) zero

8. If negative feedback is introduced in an amplifire in a shunt current configuration, then the input resistance will

(*a*) increase
(*b*) remain the same
(*c*) decrease
(*d*) increase, decrease or remain the same depending upon input

9. The current in the meter (M) in the op- amp circuit shown is

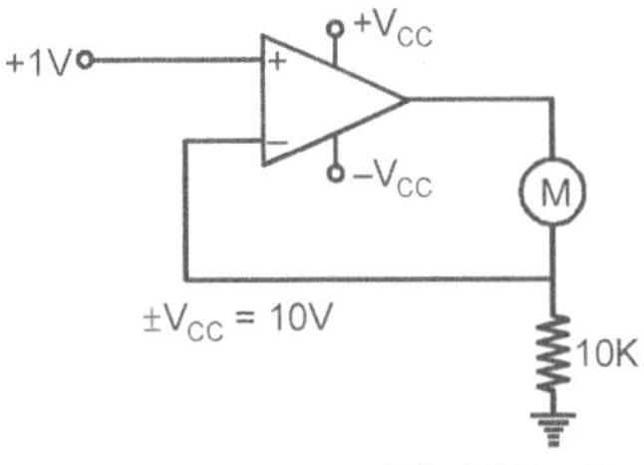

(*a*) 1 mA (*b*) 100 μA
(*c*) 2 mA (*d*) 200 μA

10. Which of the following are the main advantages of class-B push pull power amplifier (using BJTs)

1. Even harmonics tend to cancel out at the output
2. More power output per transister
3. Conversion efficiency can be as high as 78%
4. Absence of cross-over distortion.

Select the correct answer using the code given below

(*a*) 1, 2 and 3 (*b*) 1, 2 and 4
(*c*) 1, 3 and 4 (*d*) 2, 3 and 4

11. An op-amp, having a slew rate of 62.8 V/μsec, is connected in a voltage follower configuration. If the maximum amplitude of the input sinousoid is 10 V, then the minimum frequency at which the slew rate limited distortion would set in at the output is

(*a*) 1.0 MHz (*b*) 6.28 MHz
(*c*) 10.0 MHz (*d*) 62.8 MHz

12. A power amplifier uses transistor of maximum power dissipation rating of 10 watts. Match the operation classes of List-I with possible output power rating of List-II and select the correct answer using the codes given below the lists:

List-I (Operation)	**List-II**
A. Single-ended class	1. 10 W
B. Push-pull class A	2. 20 W
C. Single-ended class B	3. 30 W
D. Push-pull class B	4. 37 W

Codes:

	A	**B**	**C**	**D**
(*a*)	1	1	4	4
(*b*)	2	2	3	3
(*c*)	3	3	4	4
(*d*)	1	2	3	4

13. Schottky barrier diode can be used as
(a) low noise amplifier
(b) variable capacitance device
(c) power supply rectifier
(d) low level detector

14. In a *p-n* junction, to make the depletion region extend predominently into *p*-region, the concentration of impurities in the *p*-region must be
(a) Much less than the concentration of impurities in *n*-region
(b) Much higher than the concentration of impurities in *n*-region
(c) Equal to the concentration of impurities in *n*-region
(d) Zero

15. Which one of the following statements is correct? A photo diode works on the principle of
(a) photo - voltaic effect
(b) photo - conductive effect
(c) photo - electric effect
(d) photo - thermal effect

16. A medium bandwidth high-gain multistage transistor amplifier can be economically realised, if the coupling technique used between differnt stages is
(a) DC (b) RC
(c) transformer (d) LC

17. The following table provides a comparison of the performances of common base (CB), common emitter (CE) and common collector (CC) configuration of a transistor:

Parameter	CB	CE	CC
1. Input impedance	low	medium	high
2. Current gain	high	low	high
3. Voltage gain	>1	> 1	< 1

In this table, the performances of CB, CE and CC have been correctly indicated in respect of parameters
(a) 1, 2 and 3 (b) 1 and 2
(c) 2 and 3 (d) 1 and 3

18. A three-OP-AMP based instrumentation amplifier is shown in the given figure:

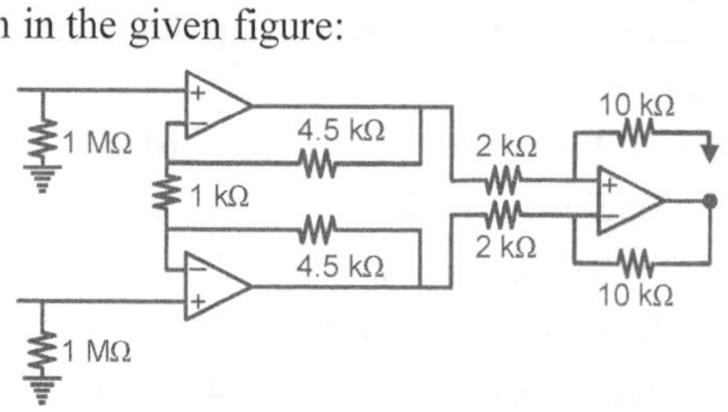

Consider the following statements regarding the amplifier: (Assume that the OP-AMPs are ideal and resistors with the same nominal value are matched)
1. The common mode input impedance is 2 MW.
2. The differential mode gain is 50.
3. The differential mode input impedance is 2 MW
4. The common mode gain is zero.

Of these statements :
(a) 1, 2, and 4 are correct
(b) 1 and 3 are correct
(c) 1 and 4 are correct
(d) 2 and 4 are correct

19. When the reverse voltage across a *p-n* junction diode is gradual decreases, the depletion region inside the diode
(a) does not change in width
(b) initially increase upto a certain width and the decreases
(c) continously increases in width
(d) continously decreases in width

20. Match List-I (Circuit) with List-II (Type of connections) and select the correct answer using the codes given below the Lists:

List-I

A. 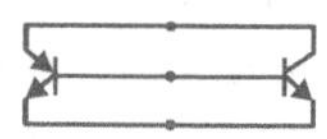

B. 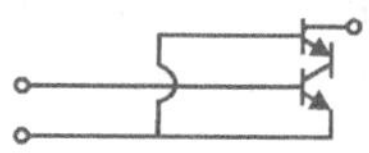

C. 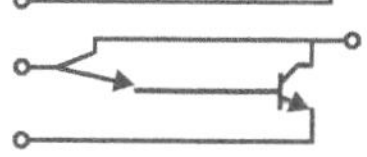

D. 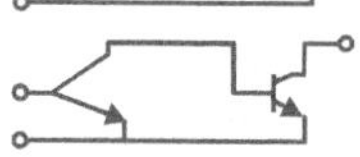

List-II
1. Darlington.
2. Cascade.
3. Parallel.
4. Traction

Codes:

	A	B	C	D
(a)	4	3	1	2
(b)	3	4	2	1
(c)	3	4	1	2
(d)	4	3	2	1

21. Match List-I (Device) with List-II (Property) and select the correct answer using the codes given below the lists:

List-I	List-II
A. Planar	1. Low-cost resistors. monolithic BJT
B. Thick Film	2. Resistors suitable for high frequency use
C. Thin Film	3. Isolator
D. Optocoupler	4. High frequency transistor

Codes:

	A	B	C	D
(a)	2	3	1	4
(b)	4	1	2	3
(c)	4	1	3	2
(d)	2	3	4	1

22. Cascading two amplifiers will result in

(*a*) reduction in overall gain and increase in the overall bandwidth

(*b*) reduction of overall gain and reduction in the overall bandwidth

(*c*) increase in overall gain and increase in the overall bandwidth

(*d*) increase in overall gain and reduction in the overall bandwidth

23. In a properly biased transistor with $\beta = 50$, the voltage across the collector resistance of vlaue 2K is 2,V. The base current in CE and CB configuration would be respectively

(*a*) 0.02 mA, 0.02 mA (*b*) 0.01 mA, 0.02 mA

(*c*) 0.05 mA, 0.01 mA (*d*) 0.01 mA, 0.01 mA

24. Out of the four devices mentioned below, the fasted switching device is

(*a*) JFET (*b*) BJT

(*c*) MOSFET (*d*) Triode

25. In a transistor current I_{CBO} flow in

(*a*) base and emitter leads

(*b*) collector and emitter leads

(*c*) base and collector leads

(*d*) emitter, base and collector leads

26. The output time period of a transistorised monsostable multivibrator using base resistor R_b and coupling capacitor C_b for the output transistor is given by

(*a*) R_bC_b (*b*) $0.69\,R_bC_b$

(*c*) $2R_bC_b$ (*d*) $1.38\,R_bC_b$

27. An npn transistor operated $I_c = 2.5$ *mA* has a collector load resistor of 1 kilo ohms and its emitter is grounded. The voltage gain of this state is about

(*a*) 25 (*b*) 100

(*c*) 250 (*d*) 1000

28. If properly biased, JFET will act as a

(*a*) current controlled current source

(*b*) voltage controlled voltage source

(*c*) voltage controlled current source

(*d*) current controlled voltage source

29. Which one of the following can be achieved by using the Reset terminals in a tunner chip?

(*a*) Keyed oscillations

(*b*) Delayed monostable action

(*c*) Square wave generation

(*d*) Pulse generation

30. Two *p-n* junction diodes are connected back to back to make a transistor. Which one of the following is correct ?

(*a*) The current gain of such a transistor will be high.

(*b*) The current gain of such a transistor will be moderate

(*c*) It cannot be used as a transistor due to large base width

(*d*) It can be used only for pnp transistor

31. Clamper circuit are also known as

(*a*) a restores

(*b*) dc restores

(*c*) voltages to frequency converters

(*d*) sweep circuits

32. An amplifier has a voltage gain of 100. To reduce distortion, 10% negative feedback is employed. The gain of the amplifier with feedback is

(*a*) 1.01 (*b*) 9.09

(*c*) 90.9 (*d*) 101

33. The main functions of the presentation layer of OSI model include

(*a*) encryption and authentification

(*b*) authentification and compression

(*c*) compression and coding

(*d*) encryption and compression

34.

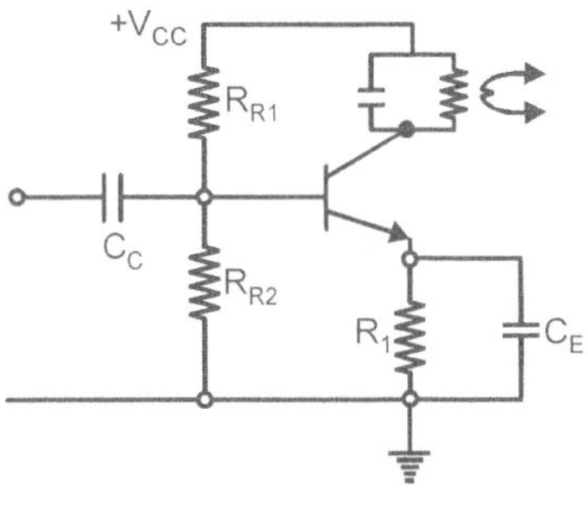

The amplifier circuit as shown in the above figure represents

(*a*) feedback amplifier (*b*) narrow band amplifier

(*c*) high input (*d*) RC-coupled amplifier

35. In Schottky TTL, a Schottky diode is used for

(*a*) forming the gate

(*b*) connecting the resistor

(*c*) clamping of the base collector junction

(*d*) none of these

36. The differential mode voltage gain of the amplifier $= \left(\dfrac{v_0}{v_{i1} - v_{i2}}\right)$ given in the figure is

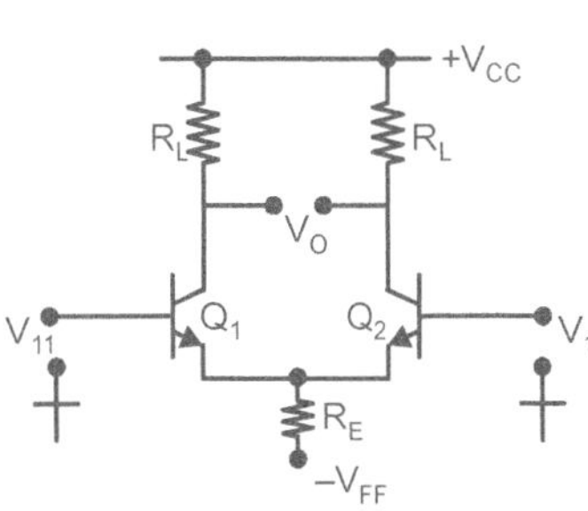

(*a*) $h_{fe}R_L/h_{ie}$ (*b*) R_L/R_E

(*c*) R_L/R_E (*d*) $h_{fe}R_L/R_E$

37. The most preferred high power output stage hi-fi A. F. amplifier is of the
(*a*) Class B push-type
(*b*) Class B single ended type
(*c*) Class A type
(*d*) Class C type

38.

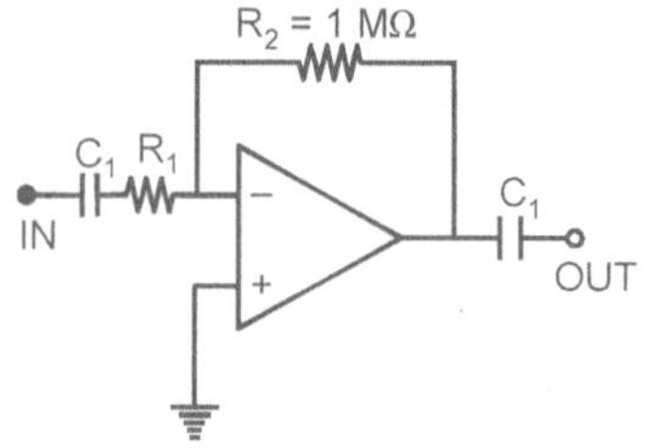

The op-amp as shown in the above figure, has a gain of −100. The value of R_1 is
(*a*) 10 KΩ (*b*) 10.1 KΩ
(*c*) 20 KΩ (*d*) 100 MΩ

39. The function of the base in the transistor is analogues to
(*a*) plate in the triode (*b*) cathode in the triode
(*c*) grid in the triode (*d*) heater in the triode

40. In a transistor, if electrons flow into the emitter
(*a*) holes flow out of emitter
(*b*) electorns flow into the collector
(*c*) electrons flow out of collector
(*d*) holes flow out of collector

41. In a symbol of transistor, the sign of arrow shows.
(*a*) emitter (*b*) base
(*c*) collector (*d*) base and collector both

42. For transistor operation
(*a*) emitter-base diode is always forward biased and the collector-base diode is always reverse-biased.
(*b*) emitter-base diode is reverse-biased and collector-base diode is forward biased.
(*c*) both the emitter-base diode an the collector base diode are forward biased
(*d*) both the emitter-base diode and the collector-base diode are reverse-biased

43. Transistor is
(*a*) current controlled current device
(*b*) current controlled voltage device
(*c*) voltage controlled current device
(*d*) voltage controlled voltage device

44. A common collector amplifier has the highest
(*a*) voltage gain (*b*) current gain
(*c*) power gain (*d*) output impedance

45. The output characteristic curve of a transistor is much more like those of
(*a*) a diode (*b*) a tetrode
(*c*) a triode (*d*) a pentode

46. The input resistance of a common emitter transistor is of the order of
(*a*) 1 MΩ (*b*) 11 KΩ
(*c*) 0.01 Ω (*d*) 0.001 MΩ

47. The gain of a voltage follower is
(*a*) greater than 1 (*b*) exactly equal to one
(*c*) slightly less than 1 (*d*) zero

48. To get maximum undistorted output from a CE amplifier with V_{cc} = 10v, V_{CE}(G) should be of the order of
(*a*) 0.1 V (*b*) 5V
(*c*) 10V (*d*) $10\sqrt{2}$ V

49. When a transistor is used as a switch its operation is confined in
(*a*) cut-off region (*b*) saturation region
(*c*) active region (*d*) (*a*) and (*b*) both

50. The operation of a JFET involves
(*a*) flow of majority carriers
(*b*) flow of minority carriers
(*c*) flow of both carriers
(*d*) none of these

51. The input guage current of a FET is of the order of
(*a*) few micro-amperes (*b*) negligible
(*c*) few millianpers (*d*) few amperes

52. Field effect transistor has
(*a*) large input impedance (*b*) large output impedance
(*c*) large power gain (*d*) small voltage gain

53. A transistor is said to be in a quiscent stage, when
(*a*) emitter junction bias is just equal to collector junction bias
(*b*) no currents are flowing
(*c*) no signal is applied to the input
(*d*) it is unbiased

54. Two identical stages of amplifiers are cascade in R-C coupling. If 10 dB is the mid band voltage gain of each stage, then the overall gain of cascade configuration will be
(*a*) 40 dB (*b*) 20 dB
(*c*) 100 dB (*d*) $20 \log_{10} 20$ dB

55. In an R-C coupled amplifier, the voltage gain
(*a*) remains almost constant over a range of frequency
(*b*) always increase with frequency
(*c*) always decreases with frequency
(*d*) None of these

56. Overall low frequency response of an amplifier of *n* cascades
(*a*) $f_n = \dfrac{f}{\sqrt{2^{1/n} - 1}}$ (*b*) $f_n = f\sqrt{2^{1/n} - 1}$
(*c*) $f_n = f$ (*d*) none of these

57. For an extremely high fidelity amplifier where the magnitude of power output is not a major consideration the amplifier should be operated in

(*a*) class A (*b*) class B
(*c*) class AB (*d*) class C

58. To get maximum undistorted output from a CE amplifier with V_{cc} = 10v, V_{ce}(G) should be of the order of

(*a*) 0.1 V (*b*) 5V
(*c*) 10V (*d*) $\frac{10}{\sqrt{2}}$ V

59. An emitter in a bipolar junction transistor is doped much more heavily than the base as it increases the

(*a*) Emitter efficiency (*b*) Base transport factor
(*c*) Forward current gain (*d*) All the three given above

60. A voltage source having an open-circuit voltage of 100 V an internal resistance of 50 ohm is equivalent to current source of

(*a*) 2 A is parallel with 50 ohm
(*b*) 2 A with 50 ohm is series
(*c*) 0.5 A in parallel with 50 ohm
(*d*) 2 A is parallel with 100 ohm

61. For high speed operation which of the following paraments may have to minimised?

1. Junction capacitance.
2. Dynamic forward resistance.
3. Depletion layer thickness.
4. Charge stores.

Choose the correct set of paraments using the codes giving below:

(*a*) 1 and 2 (*b*) 1, 2 and 3
(*c*) 1 and 4 (*d*) 2, 3 and 4

62. In the circuit shown in the figure, the expression for I_0/I_i is

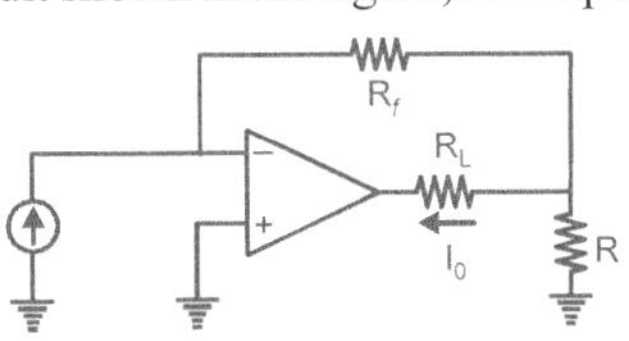

(*a*) $\frac{R_L}{R_L + R}$ (*b*) $\frac{R_f}{R_L}$
(*c*) $1 + \frac{R_f}{R_L}$ (*d*) $1 - \frac{R_f}{R_L}$

63. What is the thermal runway in a bipolar junction transistor biased in the active region due to ?

(*a*) Heating of the transistor emitter region
(*b*) Changes in 'B' which increases with temperature.
(*c*) Base emitter voltage V_{BE} which decrease with rise in temperature.
(*d*) Increase in reverse collector-base saturation current due to rise in internal device temperature.

64. Which one of the following statements is true of phase-shift type and When-bridge type RC oscillators?

(*a*) Both use positive feedback
(*b*) The former uses positivies feedback only whereas the latter used both positive and negative feedback
(*c*) The former uses both positive and negative feedback, whereas the latter uses positive feedback only
(*d*) Both use negative feedback

65. The circuit given in the figure represents a block commonly used in linear ICs. This is basically a

(*a*) current amplifier
(*b*) constant current source
(*c*) constant voltage source
(*d*) voltage amplifier

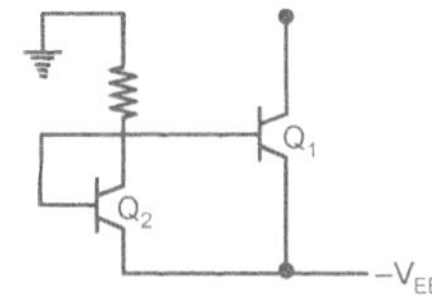

66. A voltage source $V_{AB} = 4 \sin \omega t$ is applied to the terminals A and B of the circuit shown is the given figure. The diodes are assumed to be ideal. The impedance offered by the circuit across the terminals A and B is

(*a*) 5 kΩ
(*b*) 10 kΩ
(*c*) 15 kΩ
(*d*) 20 kΩ

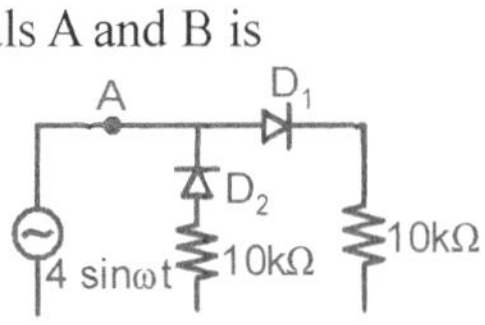

67. In the circuit shown in the given figure, the transistor is in the active mode. V_{BE} = 0.7 volt. The best approximation without knowing h_{fe} for V_c as marked in figure is

(*a*) 1.0 V (*b*) 3.3 V
(*c*) 4.3 V (*d*) 4.7 V

68. Which one of the following oscillators is used for generation of high frequencies?

(*a*) R.C. phase shift (*b*) Wien bridge
(c) L.C. Oscillator (*d*) Blocking oscillator

69. Consider the following four common types of transistors:

1. Point Contact Transistor.
2. Bipolar Junction Transistor.
3. MOS Field Effect Transistor
4. Junction Field Effect Transistor

The correct arrangment of these transistors in the increasing order of input impedance is

(*a*) 1, 2, 4, 3 (*b*) 1, 2, 3, 4
(*c*) 2, 1, 3, 4 (*d*) 2, 1, 4, 3

70. The overall voltage gain in small signal transistor CE amplifier is given by

(*a*) $A_v = \frac{h_{fe}R_L}{h_{je} + R_s}$ (*b*) $A_v = \frac{-h_{fe}R_L}{h_{je} + R_s}$

(*c*) $A_v = \frac{h_{fe}R_L}{h_{je} + R_s/R_L}$ (*d*) $A_v = \frac{-h_{fe}R_L}{h_{je} + R_s/R_L}$

71. An R-C coupled transistor amplifier has a lower 3-dB frequency of 100 Hz due to coupling capacitor. The emitter bypass capacitor is of such a value that the lower 3-dB frequency due to emitter degeneration alone is also 100 Hz. The lower 3-dB frequency of the state will be

(*a*) 64 Hz (*b*) 100 Hz

(*c*) 156 Hz (*d*) 244 HZ

72. A 1 *m*A/ 100 *m*V meter of resistance R*m* = 100 Ω is connected to an ideal OPAMP as shown in the given figure. The circuit across terminals *x*, *x* will now function as a

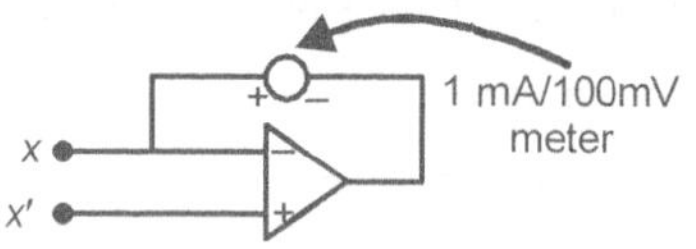

(*a*) 1 mA ammeter of zero resistance

(*b*) 1 mA ammeter with 100Ω zero resistance

(*c*) 100 mV voltmeter with zero resi-stance

(*d*) 100 mV voltmeter with 100Ω resi-stance

73. Match List-I with List-II and select the correct answer using the codes given below the Lists:

List-I	**List-II**
A. Class A amplifire	1. Hi Fidelity
B. Class B amplifier	2. Tuned amplifier
C. Class C amplifier	3. Power amplifier
D. Class AB amplifier	4. Low distortion power amplifier

Codes:

	A	B	C	D
(*a*)	4	3	2	1
(*b*)	1	2	3	4
(*c*)	4	2	3	1
(*d*)	1	3	2	4

74. The gain of an "Op-amp" is nearly

(*a*) 10^3 (*b*) 10^4

(*c*) 10^5 (*d*) 10^6

75. The input impedance of the operational amplifiers is

(*a*) infinite

(*b*) zero

(*c*) very high but not infinite

(*d*) very small

76. A differential amplifier essentially consists of

(*a*) two input and two output terminals

(*b*) only resistors and transistors

(*c*) two transistors

(*d*) two CE amplifier having their amitters directly coupled to each other.

77. Power supply ratio (PSR) of an op-amp should be

(*a*) ideally unity (*b*) as large as possible

(*c*) as small as possible (*d*) none of these

78. The decimal equivalent of binary 10011 is

(*a*) 7 (*b*) 21

(*c*) 17 (*d*) 19

79. Multiplication of two binary number 011 and 110 is

(*a*) 10010 (*b*) 11001

(*c*) 11100 (*d*) 01110

80. Two's complement of a binary number 1010 is

(*a*) 0101 (*b*) 000

(*c*) 0110 (*d*) 1001

81. Which is the octal form of the decimal number 29 ?

(*a*) 29 (*b*) 35

(*c*) 10 (*d*) 13

82. Match List-I with List-II and select the correct answer using the codes given below the lists

List-I (*Device*)	**List-II** (*Biasing mode*)
A. LED	1. Forward bias
B. Zenar	2. Reverse bias
C. Varactor	
D. SCR	

Codes:

	A	B	C	D
(*a*)	1	1	2	2
(*b*)	1	2	1	2
(*c*)	1	2	2	1
(*d*)	2	2	1	1

83. Match List-I (Logic circuit) with List-II (Property) and select the correct answer using the codes given below the lists

List-I	**List-II**
A. DTL	1. High voltage supply.
B. TTL	2. High speed of operation.
C. *n*MOS	3. High packing density.
D. CMOS	4. High noise margin

Codes:

	A	B	C	D
(*a*)	1	3	2	4
(*b*)	2	1	3	4
(*c*)	2	3	4	1
(*d*)	1	2	3	4

84. The bandwidth of an RF tuned amplifier is dependent on

(*a*) Q-factor of the turned output circuit

(*b*) Q-factor of the turned input circuit

(*c*) Quiescent operation point

(*d*) Q-factor of the output and input circuites as well as the quiescent operating point

85. Consider the following statements:
In a CMOS inverter.
1. one transistor is N-channel and the other is P-channel.
2. both the transistor are enhancement type.
3. one transistor is enhancement type and the other is depletion type.
4. both the transistor are N-channel

Of these statements
(*a*) 1 and 2 are correct (*b*) 2 and 4 are correct
(*c*) 1 and 3 are correct (*d*) 3 and 4 are correct

86. In the case of a BJT amplifier, bias stability is achieved by
(*a*) keeping the base current constant
(*b*) changing the base current in order to keep the I_c and V_{CE} constant
(*c*) keeping the temperature constant
(*d*) keeping the temperature in the base current constant

87. Which of the following can be considered to be the advantages of hybrid ICs monolithic ICs?
1. Greater design flexibility.
2. Lower cost.
3. Improved performance, hybrid ICs can combine the best performance features of all ICs : MOS, bipolar, thin film and thick film.
4. Heat removal from hybrid circuit is easier.

Select the correct answer using the codes given below:
Codes:
(*a*) 1, 2 and 3 (*b*) 1, 3 and 4
(*c*) 1, 2 and 4 (*d*) 2, 3 and 4

88. The circuit schematic of a source follower having $A_v = \frac{V_o}{V_i}$ is shown in the figure. The input resistance of this circuit is given by

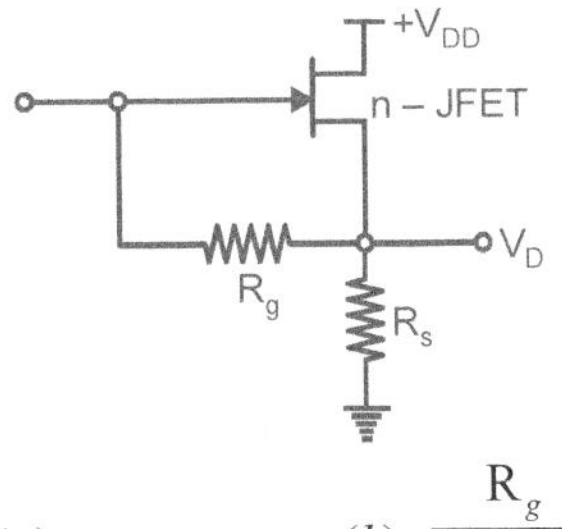

(*a*) $R_g(1+A_v)$ (*b*) $\frac{R_g}{(1+A_v)}$
(*c*) $R_g(1-A_v)$ (*d*) $\frac{R_g}{(1-A_v)}$

89. Match List-I with List-II and select the correct answer using the codes given below the Lists:
List-I (Type of oscillator)
A. Wien Bridge
B. Phase shift
C. Hartly
D. Crystal

List-II (Properties)
1. Stable amplitude and frequency.
2. 1:10 frequency coverage in a single range common
3. Minimum three RC sections required
4. 1:3 frequency coverage in a single range common

Codes:

	A	B	C	D
(*a*)	3	2	4	1
(*b*)	2	3	4	1
(*c*)	3	2	1	4
(*d*)	2	3	1	4

90. If in intrinsic germanium at 300 K (27°C) the charge concentration and mobilities of free electrons and holes are 2.5×10^{13} per cc, $3.8 \times 10^3 cm^2$/V-sec and $1.8 \times 10^3 cm^2$/V-sec respectively, then its resistivity at 300 K is
(*a*) 230×10^3 Ω-cm (*b*) 100 Ω-cm
(*c*) 44.6 Ω-cm (*d*) 22.3 Ω-cm

91. Which of the following classes of amplifier has maximum distortion?
(*a*) Class A (*b*) Class B
(*c*) Class B (*d*) Class C

92. Heat sink is used in the power amplifier circuit
(*a*) to increase the output power
(*b*) to reduce the heat loss in a transistor
(*c*) to increase the voltage gain of the power amplifier
(*d*) to increase the collector dissipation rating of the transistor

93. Negative feedback in amplifier
(*a*) increase its gain (*b*) increases its stability
(*c*) increase its distortion (*d*) decreases its distortion

94. One of the effects of negative feedback in amplifiers is the
(*a*) increase the noise
(*b*) increase in harmonic distortion
(*c*) decrease in bandwidth
(*d*) decrease in harmonic distortion

95. An amplifier has gain without feedback as 10. To make it oscillate β must be
(*a*) 0.2 (*b*) 0.1
(*c*) 0.5 (*d*) 0.8

96. A tuned amplifier amplifies
(*a*) a wide range of frequencies
(*b*) audio frequencies only
(*c*) video frequencies only
(*d*) a narrow band of frequencies

97. For a parallel L-C circuit the resonance occurs when
(*a*) $X_L = \frac{1}{x_c}$ (*b*) $X_L = Q X_c$
(*c*) $X_L = X_c$ (*d*) $X_c = 10 X_L$

98. Thyristor is nothing but a
(*a*) controlled switch
(*b*) controlled transistor
(*c*) amplifier with large current rating
(*d*) amplifier with higher gain

99. Thyristor is a
(*a*) *p-n-p-n* device (*b*) *p-p-p-p* device
(*c*) *n-n-n-n* device (*d*) none of these

100. A SCR is
(*a*) three-layer three terminal device
(*b*) three-layer four-terminal device
(*c*) four-layer three terminal device
(*d*) four-layer four-terminal device

101. A thyristor is a
(*a*) current controlled device
(*b*) bilateral device
(*c*) voltage controlled device

102. The co-amp circuit shown in the given figure is a

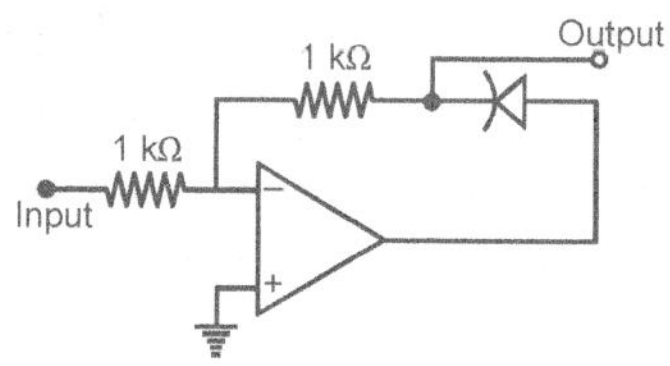

(*a*) precision half-wave rectifier
(*b*) precision full-wave rectifier
(*c*) peak detector
(*d*) level shifter

103. A special purpose diode which uses metal like gold, silver or platinum on one side of the junction, *n*-type doped silicon on another side and has almost no charge storage in the junction, is a
(*a*) schottky diode (*b*) tunnel diode
(*c*) varactor diode (*d*) zener diode

104. The gain of a negative feedback amplifier is 40 *d*B. If the attenuation of the feedback paths is 50 *d*B, then the gain of the amplifier without feedback is
(*a*) 78.92 (*b*) 146.32
(*c*) 215.51 (*d*) 317.23

105. A differential amplifier is basically a/an
(*a*) DC amplifier (*b*) AC amplifier
(*c*) chopper amplifier (*d*) mixer amplifier

106. In LED, light is emitted because
(*a*) recombination of charge carries takes place
(*b*) diode gets heated up
(*c*) light falling on the diode gets amplified
(*d*) light gets reflected due to lens action

107. Which of the following are true for *h*-parameters of transistors ?
1. They are real numbers at audio frequencies
2. They are easy to measure
3. They vary widely with temperature

Select the correct answer using the code given below:
(*a*) 1 and 2 only (*b*) 2 and 3 only
(*c*) 1 and 3 only (*d*) 1, 2 and 3

108. Match List-I (Device) with List-II (Application) and select the correct answer using the code given below the lists :

List-I	List-II
A. *p-n* junction	1. Micro wave generator
B. tunnel diode	2. Low frequency rectifier
C. JFET	3. High frequency rectifier
D. Schottk barrier diode	4. Voltage variable resistor

Codes:

	A	B	C	D
(*a*)	2	3	4	1
(*b*)	4	1	2	3
(*c*)	2	1	4	3
(*d*)	4	3	2	1

109. The low frequency gain of the low pass filter shown in the given figure is

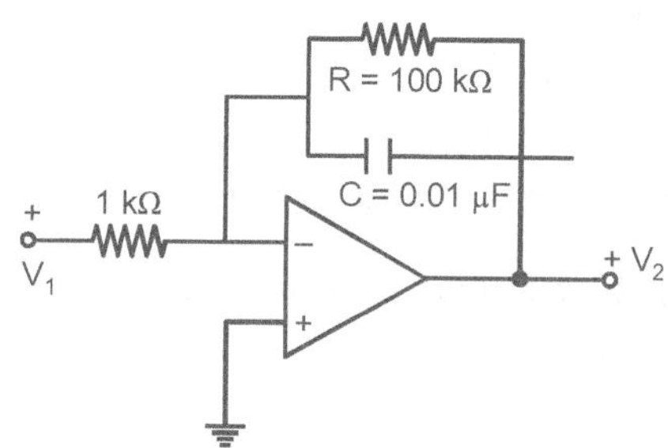

(*a*) 10 dB (*b*) 20 dB
(*c*) 30 dB (*d*) 40 dB

110. The waveform of the output voltage 'V_o' for the circuit shown in the given figure (RC >> 1) is a
(*a*) sinusoidal wave (*b*) squre wave
(*c*) series of spikes (*d*) triangular wave

111. A differential amplifier essentially consists of
(*a*) two input and two output terminals
(*b*) only resistors and transistors
(*c*) two transistors
(*d*) two CE amplifiers having their emitters directly coupled to each other

112. The decimal equivalent of binary 10011 is
(*a*) 7 (*b*) 21
(*c*) 17 (*d*) 19

113. A unit positive step is applied at the input of the circuit shown in the figure. After 20 seconds, the output V_0 will be

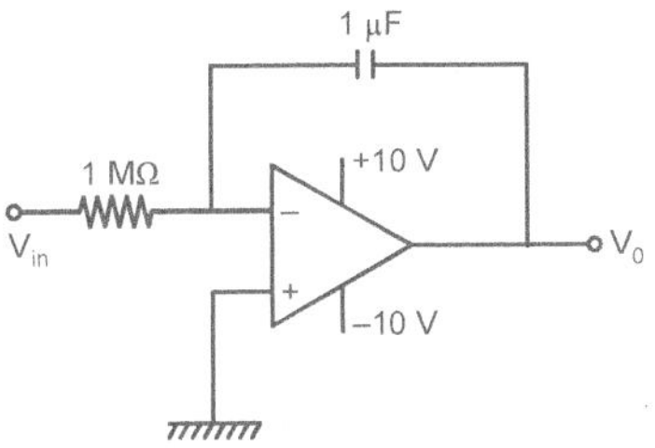

(*a*) +20 V (*b*) +10 V
(*c*) – 10 V (*d*) –20 V

114. The output voltage of the circuit shown in the given figure is

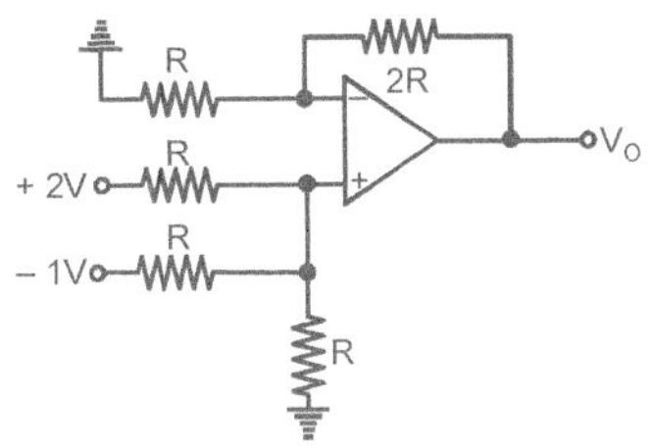

(*a*) +1.0 V (*b*) 1.5 V
(*c*) 2.0 V (*d*) 2.5 V

115. An op-amp circuit is shown in the given figure. The current 'I' is

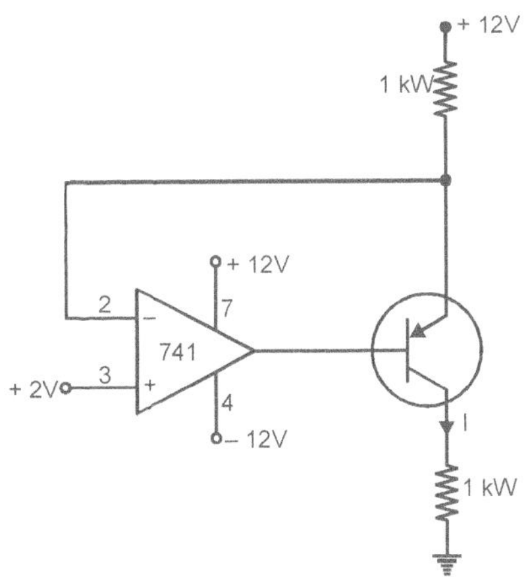

(*a*) zero (*b*) 6 mA
(*c*) 10 mA (*d*) 12 mA

116. Match List-I (Parameters) with List-II (Units) and select the correct answer using the codes given below the Lists:

List-I	**List-II**
A. h_{11}	1. Dimensionless.
B. h_{12}	2. Ohms.
C. h_{22}	3. Siemens.

Codes:

	A	B	C
(*a*)	1	2	3
(*b*)	1	3	2
(*c*)	2	1	3
(*d*)	3	2	1

117. A 10 Mhz square wave clocks a 5-bit ripple counter. The frequency of the third-flip-flop's output would be

(*a*) 2 MHz (*b*) 1.25 MHz
(*c*) 50 MHZ (*d*) 615 MHz

118. Match List- I (Semiconductor parameters) with List-II (Physical processes) and select the correct answer using the codes given below the lists:

List-I

A. Impurity concentration
B. Carrier mobility
C. Carrier life time
D. Intrinsic carrier concentration

List-II

1. Recombination.
2. Band to band transition
3. Scatterring
4. Ion implantation

Codes:

	A	B	C	D
(*a*)	3	4	2	1
(*b*)	4	3	2	1
(*c*)	3	4	1	2
(*d*)	4	3	1	2

119. The senitivity of a photodiode depends upon

(*a*) light intensity and depletion region width
(*b*) depletion region width and excess carrier life time
(*c*) excess carrier life time and forward bias current
(*d*) forward bias current and light intensity

120. The transfer gain for the circuit shown in the figure is given by

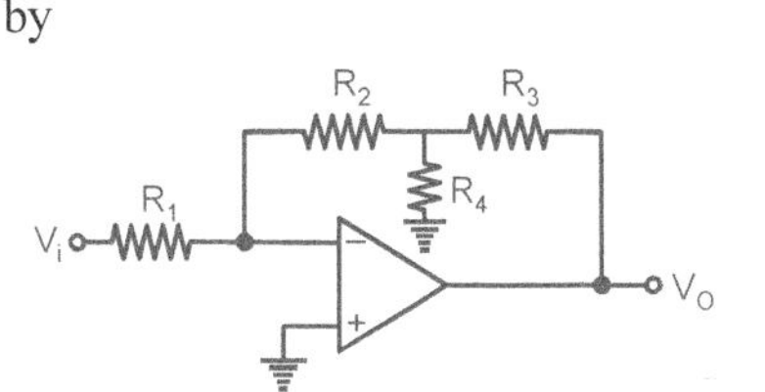

(*a*) $-\left[\dfrac{R_2+R_3+\dfrac{R_2R_4}{R_4}}{R_1}\right]$

(*b*) $-\left[\dfrac{\dfrac{R_3R_4}{R_3+R_4}+R_2}{R_1}\right]$

(*c*) $-\left[\dfrac{\dfrac{R_2R_4}{R_2+R_4}+R_3}{R_1}\right]$

(*d*) $-\left[\dfrac{R_2+R_3}{R_1}\right]$

121. In the *pnp* transistor circuit shown in the given figure, the transistor is in saturation with values of $V_{be} = 0.7$ V, $V_{ce(sat)} = 0.3$ V and $\beta = 20$. The value of I_c will be

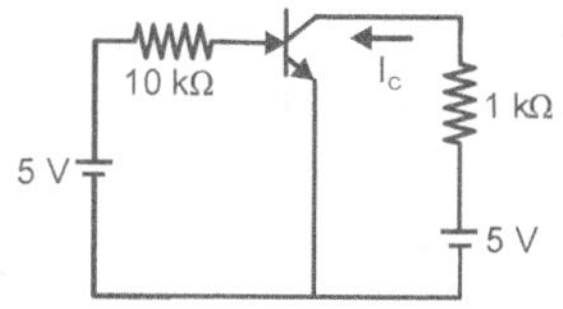

(*a*) 4.7 mA (*b*) 5.3 mA
(*c*) 8.6 mA (*d*) 10 mA

122. Match List-I (Circuits) with List-II (Application) and select the correct answer using the codes given below the lists:

List-I

A. Astable multivibrator
B. Schmitt trigger
C. Bistable multivibrator

List-II

1. Counter.
2. Sweep generator.
3. Voltage to frequency converter.
4. Comparator

Codes:

	A	B	C
(*a*)	3	4	1
(*b*)	3	2	1
(*c*)	1	4	3
(*d*)	1	2	3

123. For the circuit of figure below with an ideal operational amplifier, the maximum phase shift of the output V_{out} with reference to the input V_{in} is

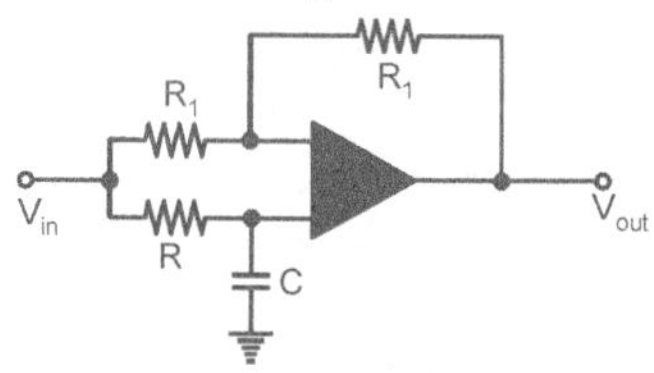

(*a*) 0° (*b*) −90°
(*c*) +90° (*d*) ± 180°

124. A class B push-pull transistor amplifier is operated under optimum conditions. $V_{cc} = 30$ V, Audio Power = 4.5 watts. With the usual simplifying assumptions

(*a*) the DC power drawn from V_{cc} is 9 watts
(*b*) the power dissipated by each transitor is 2.25 watts
(*c*) the reflected resistance in the primary of the transformer is 200 ohms
(*d*) the peak ac collector current is 0.3 amp

125. A transistor has $h_{fe} = 100$, $h_{ie} = 5.2$ k and $r_{bb} = 0$. At room temperature $V_T = 26$ *m*V, the collector current $|I_c|$ will be

(*a*) 10 *m*A (*b*) 5 *m*A
(*c*) 1 *m*A (*d*) 0.5 *m*A

126. An amplifier with an initial open-loop gain of 400 is used an a negative feedback amplifier. The feedback factor is 0.05. If the gain of the amplifier changes by 10% due to temperature, then the closed-loop gain will change approximately by

(*a*) 0.05 % (*b*) 0.1 %
(*c*) 0.5 % (*d*) 1 %

127. In the circuit shown below if the output is taken from point E instead of node C, what will be the result ?

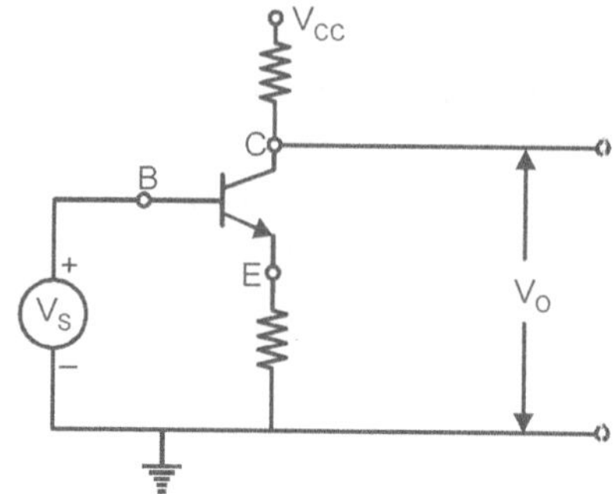

(*a*) An increase in the output impedance
(*b*) A reduction in the output impedance
(*c*) An increase in the input impedance
(*d*) A reduction in the input impedance

128. An n-channel JFET, having a pinch-off voltage (V_p) of −5V, shows a transconductance (g_m) of 1 *m*A/V when the applied gate-to-source voltage (V_{GS}) is – 3V. Its maximum transconductance (in *m*A/V) is

(*a*) 1.5 (*b*) 2.0
(*c*) 2.5 (*d*) 3.0

129. Assume ideal diode and transformer in the circuit shown in the given figure. If $v_t = 40$ Vrms, then the dc current through the load will be

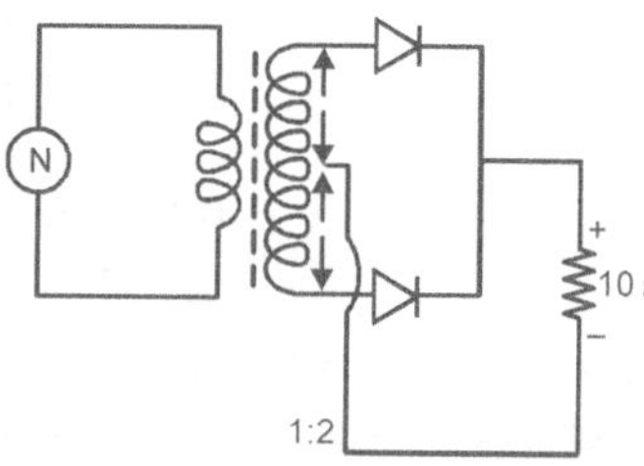

(*a*) 5.760 A (*b*) 2.880 A
(*c*) 1.831 A (*d*) 0.915 A

130. IC74121 TTL one IC is used without connecting any external capacitor between C_{ext} and R_{ext}/C_{ext} pin. The R_{int} pin is connected to V_{cc}. On triggering, one would expect

(*a*) no output pulse
(*b*) an output pulse of duration of the order of seconds
(*c*) an output pulse of variable duration
(*d*) an output pulse of duration of the order of 30-35 nanoseconds

131. The circuit shown in the figure below represents

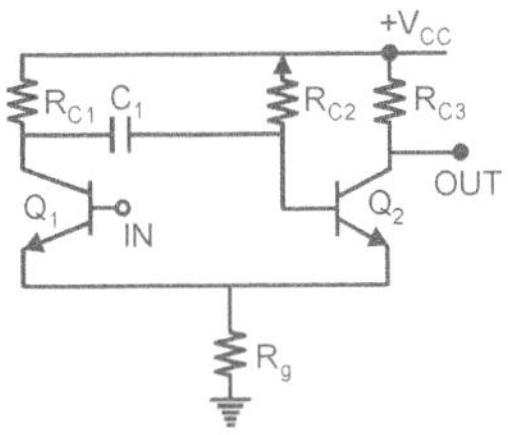

(*a*) a Schmitt trigger
(*b*) an astable multivibrator
(*c*) a monostable multivibrator
(*d*) a bistable multivibrator

132. The open-loop gain of an operational amplifier is 10^5. an input signal of 1 mV is applied to the inverting input with the noninverting connected to the ground. The supply voltage are ± 10 V. The output of the amplifier will be

(*a*) + 100 V (*b*) – 100 V
(*c*) + 10 V (approximately) (*d*) – 10 V (approximately)

133. For any amplifier configuration using a BJT, the input impedance R_i is

(*a*) $h_i(1 + h_{fe})$ (*b*) $\dfrac{h_i}{1 + h_r A_v}$

(*c*) $\dfrac{h_i}{1 - h_r A_v}$ (*d*) $\dfrac{h_i}{1 + h_{fe}}$

134. When a chip with more than two gates is assembled into an IC package, then
(*a*) the system releability is worst
(*b*) the system becomes costlier
(*c*) the total cost remains unaltered
(*d*) the package cost is shared between more gates resulting in a lower cost per gate

135. Extremely low power dissipation and low cost per gate can be achieved in which of the following IC?
(*a*) ECL (*b*) CMOS
(*c*) TTL (*d*) MOS

136. The dynamic drain resistance of a JFET is of the order of
(*a*) 1 kΩ (*b*) 10 kΩ
(*c*) 500 MΩ (*d*) 100 MΩ

137. Ripple factor of a full-wave rectifier is
(*a*) 0.24 (*b*) 0.96
(*c*) 0.48 (*d*) 0.72

138. The input resistance of a JFET is of the order
(*a*) 1 kΩ (*b*) 10 kΩ
(*c*) 10 MΩ (*d*) 100 MΩ

139. Use of IC's in measuring instruments
(*a*) does not involve linear types
(*b*) is not suitable
(*c*) requires development of special circuits
(*d*) utilizes their low power consumption and low ope rating cost

140. An intergrated electronic component is
(*a*) One circuit element of IC
(*b*) a complete electronic function consisting of several IC's formed and interconnected on a single chip of silicon with circuit complexity above 10 gates
(*c*) a discrete device assembly
(*d*) an electronic component included in an electronic circuit assembled on a PCB

141. Avalanche photodiodes are preferred over PIN diodes in optical communication systems because of
(*a*) Speed of operation
(*b*) higher sensitivity
(*c*) larger bandwidth
(*d*) larger power handling capacity

142. For the OP-AMP cascade shown in the given figure, $\dfrac{V_2}{V_s}$ is

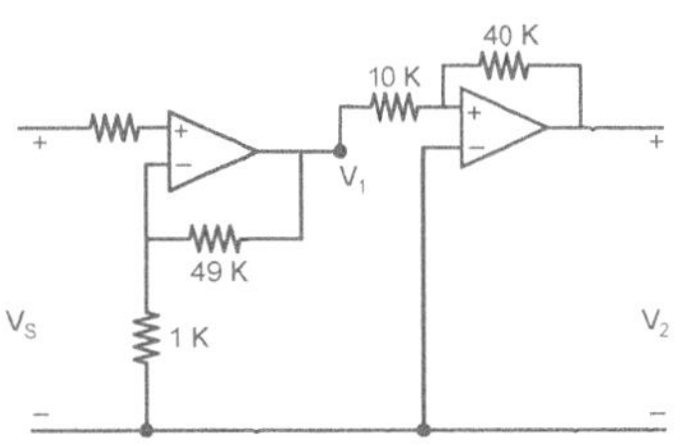

(*a*) –200 (*b*) 196
(c) 200 (*d*) 250

143. 10 mV input to an amplifier gives 5 V output without feedback. With negative feedback,160 mV input is required to give the same output. What is the feedback factor (B) used
(*a*) 3 % (*b*) 5 %
(*c*) 10 % (*d*) 16 %

144. The *h*-parameter equivalent circuit is the most common representation of the junction transistor because
(*a*) it is usable over the entire frequency band of interest to the transistor
(*b*) the parameters values correspond to input impedance, output admittance and current gain for the most common applications of the transistor
(*c*) it represents the admittance parameters of the transistors
(*d*) it is most useful in the ON and OFF states of the transistor

145. In the OPAMP based Wien bridge RC-Osci!lator shown in the given figure, R_2 is invariably a resistors

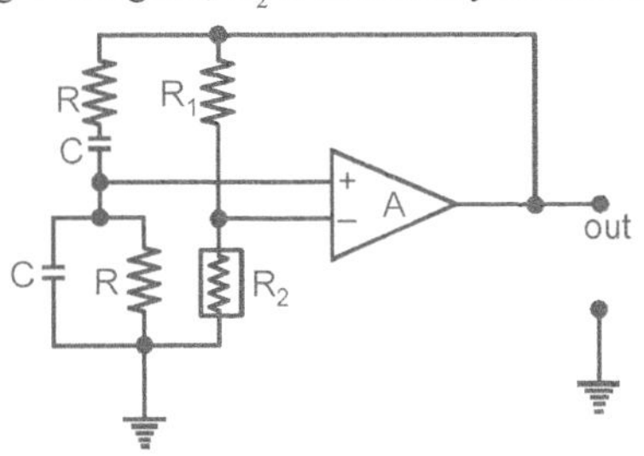

(*a*) negative temperature coefficient
(*b*) positive voltage dependence
(*c*) positive temperature
(*d*) negative voltage dependence

146. An OPAMP has as slew rate of 5 V/ms. The largest sinewave output voltage possible at a frequency of 1 MHz is

(*a*) 10 π volts (*b*) 5 volts
(*c*) $\frac{5}{\pi}$ volts (*d*) $\frac{5}{2\pi}$ volts

147. A feedback amplifier has a closed-loop gain of – 200. It should not vary more than 50%, despite 25% variation in amplifier gain A without feedback. The value of A is

(*a*) 800 (*b*) – 800
(*c*) 1000 (*d*) – 1000

148. What is the hysteris voltages for the Schmitt trigger circuit shown in the given figure, if $V_{sat} = \pm 10$ V?

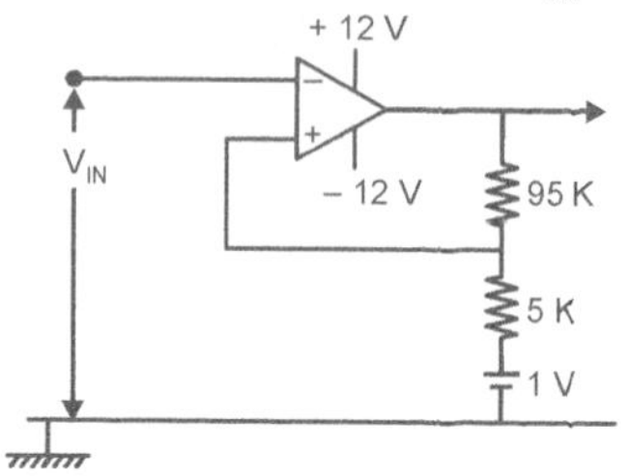

(*a*) 1.90 volts (*b*) 1.54 volts
(*c*) 1.00 volts (*d*) 0.90 volts

149. A Zener diode has a Zener resistance of 5Ω. It the current through the Zener diode changes from 10 mA to 20 mA, the change of voltage across the Zener diode will be

(*a*) 0.05 V (*b*) 0.075 V
(*c*) 0.1 V (*d*) 0.5 V

150. Match List I (Instruction) with List II (Operation) for Intel 8085 and select the correct answer using the codes given below the lists :

List-I	**List-II**
A. PCHL	1. Exchange the top of the stack with the contents of HL pair
B. SPHL	2. Exchange the contents of HL with those of DE pair
C. XTHL	3. Transfer the contents of HL to the stack pointer
D. XCHG	4. Transfer the contents of HL to the programme counter

Codes:

	A	B	C	D
(*a*)	4	3	1	2
(*b*)	3	4	1	2
(*c*)	3	4	2	1
(*d*)	4	3	1	2

151. The circuit connections of OPAMP given in figures (*i*) and (*ii*) represent

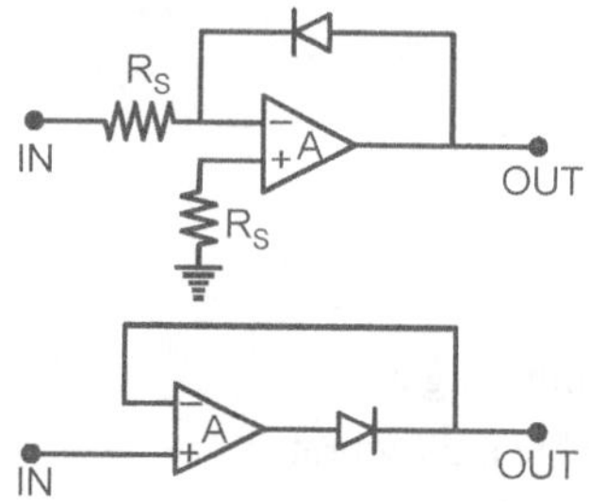

(*a*) logarithmic amplifiers for both figure (*i*) and (*ii*)
(*b*) detectors for both figures (*i*) and (*ii*)
(*c*) detector for figure (*i*) and logarithmic amplifier for figure (*ii*)
(*d*) logarithmic amplifier for figure (*i*) and detector for figure (*ii*)

152. The noise margin of a digital IC is the

(*a*) maximum frequency of extraneous voltage that does not cause a gate to change its state
(*b*) maximum extraneous voltage that does not cause a gate to change its state
(*c*) thermal noise voltage which causes a gate to change its state
(*d*) minimum frequency of extraneous voltage that causes a gate to change its state

153. A junction transistor operating at room temperature with $I_c = 2$mA, where $k_t = 25$ mV has $\beta = 100$.
The values of the parameters g_m in mhos and r_π in ohms will be respectively

(*a*) 0.04 and 2500 (*b*) 0.08 and 1250
(*c*) 0.5 and 800 (*d*) 0.08 and 5000

154. The circuit shown in the figure has Zener regulated *dc* power supply. Assuming that Zener diode is ideal, the MINIMUM value of R_L down to which the output voltage would remain constant is

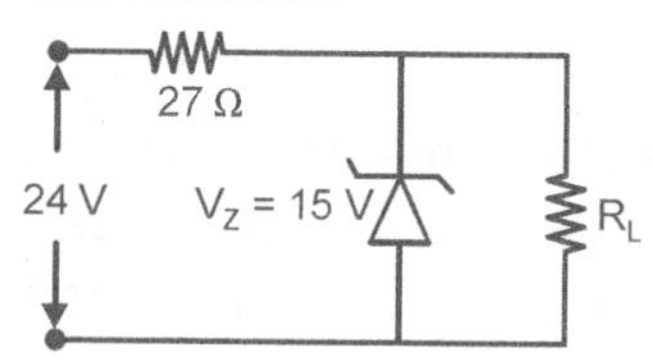

(*a*) 15 ohms (*b*) 24 ohms
(*c*) 27 ohms (*d*) 45 ohms

155. A non-converting OP-AMP summer is shown in the figure. The output voltage V_0 is

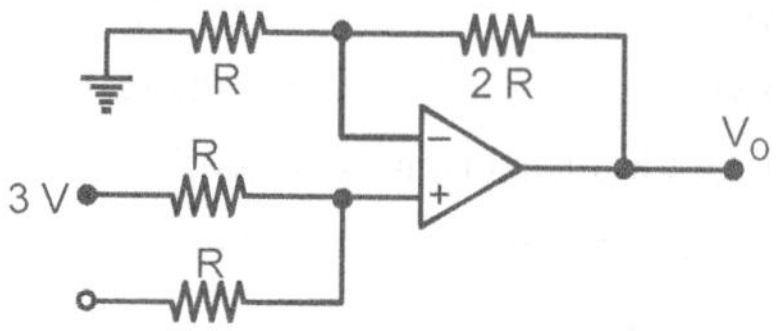

(*a*) sin 100*t* (*b*) $\frac{3}{2}$ sin 100*t*
(*c*) 2 sin g 100*t* (*d*) 2 sin 100*t*

156. An amplifier with mid-band gain $|A| = 500$ has negative feedback $|\beta| = \frac{1}{100}$. If the upper cut-off without feedback were at 60 kHz, then with feedback it would become

(*a*) 10 kHz (*b*) 12 kHz
(*c*) 300 kHz (*d*) 360 kHz

157. Match List-I (Symbols) with List-II (Devices) and select the correct answer using the codes given below the lists:

List-I

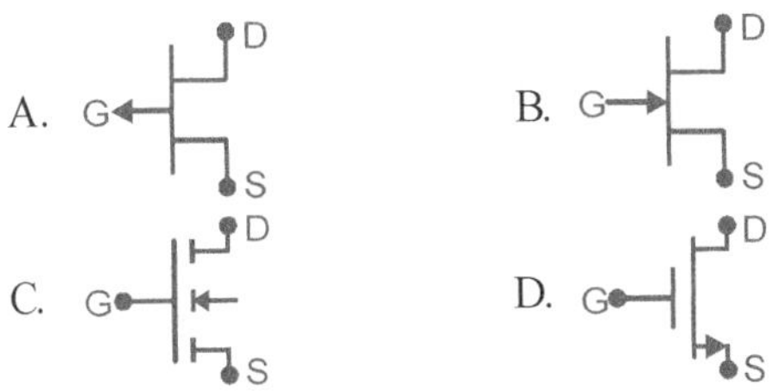

List-II

1. Depletion mode MOSFET
2. *p*-channel JFET
3. *n*-channel JFET
4. Enhancement mode MOSFET

Codes :

	A	B	C	D
(*a*)	2	1	4	3
(*b*)	4	3	2	1
(*c*)	2	3	4	1
(*d*)	4	1	2	3

158. Open-collector gates requires

(*a*) external output pull-up resistors
(*b*) external capacitor to ground
(*c*) a buffer at the
(*d*) higher supply voltage

159. Voltage at V_1 and V_2 of the arrangement shown in the figure will be respectively

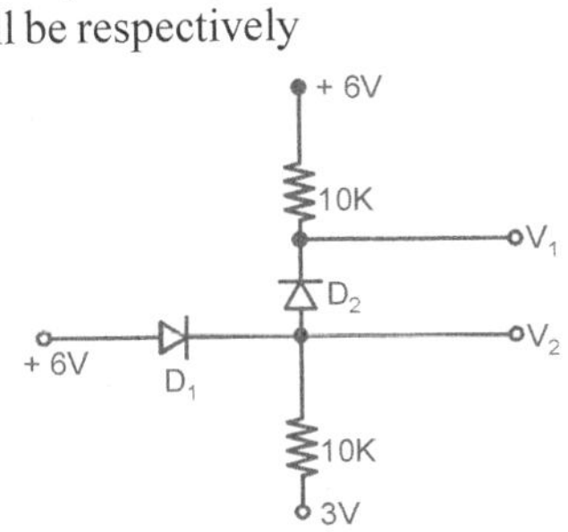

(*a*) 6 V and 5.4 V (*b*) 5.4 V and 6 V
(*c*) 3 V and 5.4 V (*d*) 6 V and 3 V

160. The primary switching delay in an over-driven dipolar junction transistor is contributed by which one of the following characteristic of the transistor?

(*a*) Storage time (*b*) Fall time
(*c*) Rise time (*d*) Charging time

161. Which one of the following is the advantage of base modulation over collector modulation of a transistor Class C amplifier?

(*a*) Requires lower modulation power
(*b*) High power output per transistor
(*c*) Better efficiency
(*d*) Better linearily

162. If an input signal with non-zero dc component is applied to a low pass RC network, the dc component in the output will be

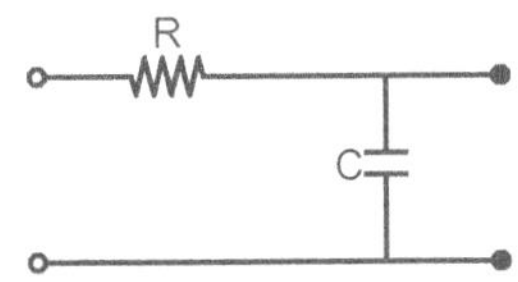

(*a*) the same as that in the input
(*b*) less than that in the input
(*c*) more than that in the input
(*d*) zero

163. Two intially identical sample A and B of pure germanium are doped with doners to concentrations of 1×10^{20} m^{-3} and 3×10^{20} m^{-3} respectively. If the hole concentration in A is 9×10^{12} m^{-3}, then the hole concentration in B at the same temperature will be

(*a*) 3×10^{12} m^{-3} (*b*) 7×10^{12} m^{-3}
(*c*) 11×10^{12} m^{-3} (*d*) 27×10^{12} m^{-3}

164. The voltage-ratio transfer function of an active filter is given by $\frac{V_{2(S)}}{V_{1(S)}} = \frac{(s^2 + \delta)}{(s^2 + \alpha s + \delta)}$

The circuit in question is a

(*a*) low-pass filter (*b*) high-pass filter
(*c*) band-pass filter (*d*) band-reject filter

165. If $\alpha = 0.98$, $I_{co} = 6$ mA and $I_b = 100$ mA for a transistor, then the value of I_c will be

(*a*) 2.3 mA (*b*) 3.1 mA
(*c*) 4.6 mA (*d*) 5.2 mA

166. Consider the following statements regarding RC phase-shift oscillator:

1. The amplifier gain is positive.
2. The amplifier gain is negative.
3. The phase shift introduced by the feedback network is 180°.
4. The phase shift introduced by the feedback network is 360°.

(*a*) 1 and 3 are correct
(*b*) 2 and 3 are corrent
(c) 2 and 4 are correct
(*d*) 1 and 4 are correct

167. A difference amplifier is shown in the figure. Transistor Q_1 and Q_2 have identical parameters. Assuming that V_{BE} = 0.7V and β = 200 for each transistor and given that V_{s1} = V_{s2} = 0, the value of the collector current I_c will be

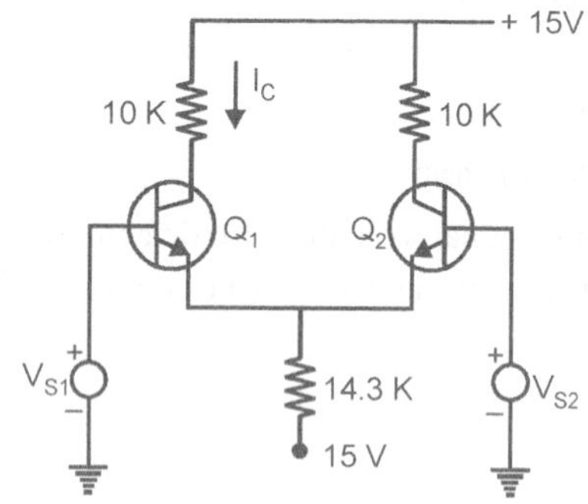

(*a*) 2 mA (*b*) 1 mA

(*c*) 0.5 mA (*d*) 2.5 μA

168. The given shows a silicon transistor connected as a common emitter amplifier. The quiescent collector voltage of the circuit is approximately

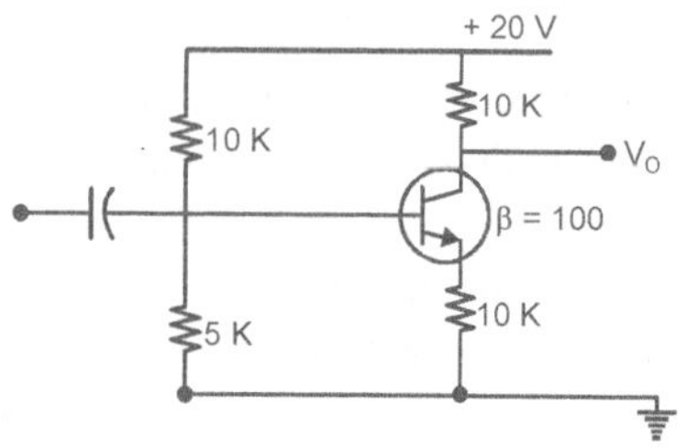

(*a*) 20/3 V (*b*) 10V

(*c*) 14V (*d*) 20V

169. In a Zener diode,

(*a*) only the P region is heavily doped

(*b*) only the N region is heavily doped

(*c*) both P and N regions are heavily doped

(*d*) both P and N regions are lightly doped

170. Large Scale Integration

(*a*) does not require costly process facilities

(*b*) can not be applied to MOS system

(*c*) refers only to integrated electronic silicon

(*d*) refers to technology used to fabricate IC's with complexity above 100 gates

171. A *pnpn* device having no gate is called

(*a*) UJT

(*b*) Triac

(*c*) Shockly diode

(*d*) SCR

172. FET

(*a*) has a very high input impedance

(*b*) depends on minority carrier flow

(*c*) uses a forward biased junction

(*d*) uses a high concentration junction

173. The most important factor governing the cost of IC components is

(*a*) the shape of the component

(*b*) area occupied by the component

(*c*) number of electrode connections

(*d*) location of the component on the slice

174. Which of the following digital IC family can give maximum fan out?

(*a*) ECL (*b*) PMOS

(*c*) CMOS (*d*) HTL

175. Heat sink removes heat from a power transistor mainly by

(*a*) radiation (*b*) conduction

(*c*) nature convection (*d*) forced convection

176. In an *npn* transistor, the function of emitter is

(*a*) to inject holes into base

(*b*) inject electrons into the base

(*c*) to inject electrons into the collector

(*d*) to inject holes into the collector

177. The load current I_L in the given circuit is

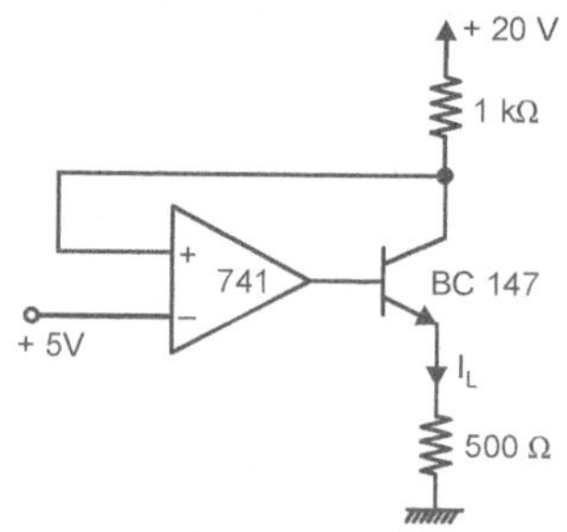

(*a*) 10 *mA* (*b*) $\frac{40}{3}$ *mA*

(*c*) 15 *mA* (*d*) 20 *mA*

178. In the CE configuration of a transistor amplifier, there is

(*a*) no phase change between input and output voltages

(*b*) 180° phase change between input and output currents

(*c*) no phase change between input and output currents and no phase change between input and output voltage

(*d*) no phase change between input and output currents and 180° phase change between input and output voltages

179. An emitter coupled differential amplifier behaves as an excellent double-ended clipper if the magnitude of the input differential voltages exceeds

(*a*) 8 V_T (*b*) 4 V_T

(*c*) 2 V_T (*d*) V_T

180. A tank circuit of R F. amplifier has a coil of Q factor of 100. It is tunned to a frequency of 1 MHz. The bandwidth of the circuit is

(*a*) 1 kHz (*b*) 10 kHz

(*c*) 100 kHz (*d*) 1000 kHz

EXERCISE - II

1. The maximum power rating of a transistor is not exceeded if
   (*a*) Q-point lies in the area bounded by the two axis of output characteristic and the constant power dissipation line
   (*b*) the load line lies in the area bounded by the two axis and the constant power dissipation line
   (*c*) the load line interesects the constant dissipation line
   (*d*) none of these [NTPC]
2. The leakage current in a NPN transistor is due to
   (*a*) the flow of holes from the base to the emitter
   (*b*) the flows of electron from the collector to the base
   (*c*) the flow of holes from collector to the base
   (*d*) the flow of minority carriers from collector to emitter [NTPC]
3. The most heavily doped region in a transistor is
   (*a*) base
   (*b*) collector
   (*c*) emitter
   (*d*) both emitter and collector [NTPC]
4. In a transistor, if electrons flow into the emitter
   (*a*) holes flow out of the emitter
   (*b*) electrons flow into the collector
   (*c*) electrons flow out to the collector
   (*d*) all of these [NTPC]
5. Silicon devices are prepared at high temperature operations as comapred to germanium because
   (*a*) silicon is more thermally stable
   (*b*) silicon can dissipate more power
   (*c*) reverse saturation current is less in case of silicon
   (*d*) all of these [BHEL]
6. The negative feedback is more effective if the loop gain is
   (*a*) small (*b*) large
   (*c*) unity (*d*) zero [BHEL]
7. In order to convert RC-coupled amplifier to a wide band amplifier, it is necessary
   (*a*) to compensate for low frequency drop off
   (*b*) to compensate for high frequency drop-off.
   (*c*) to compensate harmonic phase shifts that are not proportional to the frequency of the harmonics
   (*d*) all of these [NTPC]
8. In its equivalent circuit, a transistor is represented by
   (*a*) two voltages sources and two resistance [NTPC]
   (*b*) two current sources and two resistance
   (*c*) one current source and two resitances
   (*d*) one voltage source, one current source and two resistance
9. In a CE configuration transistor amplifier, there is [NTPC]
   (*a*) no phase change between i/p and o/p voltage
   (*b*) 180° phase change between input and output currents
   (*c*) no phase change between input and output currents
   (*d*) none of these
10. The current stability of a common collector amplifier can be improved by [NTPC]
    (*a*) decreasing emitter and base resistance
    (*b*) increasing emitter and base resistance
    (*c*) decreasing emitter resistance and increasing base resistances
    (*d*) increasing resistance and decreasing base resistance
11. The ON- resistance r_d, ON of an FET is the ratio
    (*a*) V_{DS}/I_D at the origin [NTPC]
    (*b*) V_{DS}/I_D in the saturation region
    (*c*) $\Delta V_{DS}/\Delta I_D$ in the saturation region
    (*d*) VG_S/I_D at the origin
12. In voltage reference diodes [NTPC]
    (*a*) the breakdown mechanism is due to Zener effect
    (*b*) for low voltages and heavily doped material the breakdown is due to avalanche multiplication
    (*c*) for a limiting reverse voltage which can be used as reference, the reverse current increases depending on the circuit connected
    (*d*) the current flow at the reference voltage is due to minority carrier injection
13. TTL has which of the following advantage over CMOS ?
    (*a*) Lower P_D [NTPC]
    (*b*) Use of transistors alone as circiut elements
    (*c*) Greater suitability for LSI
    (*d*) Simpler fabrication process
14. In a transistor, current I_{CBO} [NTPC]
    (*a*) increase with increase of temperature
    (*b*) decrease with increse of temperature
    (*c*) is normally greater for Si transistor than Ge transistor
    (*d*) mainly depends on the emitter base juctions bias
15. The charge carries in a *p*-channel FET are [NTPC]
    (*a*) electrons alone
    (*b*) holes alone
    (*c*) both electrons and holes
    (*d*) may be either electrons or holes
16. An RC circuit is used as a differentiate to produce impluse voltage. If the value of C is 0.1 μf, then value of R for an input pulse width of 1 *ms* is [BHEL]
    (*a*) 1 MΩ (*b*) 10 K
    (*c*) 1 K (*d*) 100 Ω

17. In IC electronic system [BEL]
(a) cost of printed circit is negligible
(b) it is most economical to use types with the optimum complexity so as to get the low cost per gate
(c) the total system is same irrespective of type used
(d) it is desirable to use simple gate type mounted as printed circuit board

18. IC's are economical for use in logic systems because
(a) of the quanity of repeat circuits used
(b) binary circuits are possible only with IC's
(c) the diffusion process is simpler
(d) privated circuit board are inexpensive [BEL]

19. V_{PO} is the pinch off voltage for $V_{GS} = 0$ in an FET. When the gage is reverse biased by V_{GS} the pinch off voltage
(a) will be less than V_{PO}
(b) will be more than V_{PO}
(c) will be the same as V_{PO}
(d) does not depend on V_{GS} [BEL]

20. Which of the following transistor is affected by static electricity? [BEL]
(a) npn transistor (b) IFET
(c) UJT (d) MOSFET

21. Power transistors are invariable provided with [IOCL]
(a) heat sink (b) metallic casing
(c) soldered connections (d) fan for heat removal

22. The JFET can operate in [IOCL]
(a) depletion mode only
(b) enhancement mode only
(c) either depletion or enhancement mode at a time
(d) both depletion and enhancement modes simultaneously

23. In the audio frequency range, as the frequency increases, the noise factor [IOCL]
(a) decreases to a minimum and then increases
(b) increases to a maximum and then decreases
(c) increases continuously
(d) decrease continuously

24. In an ideal balanced differential amplifier, the common-mode gain is [NHPC]
(a) double of that of single-ended amplifier
(b) half of that of single-ended differential amplifier
(c) very high
(d) zero

25. The gain of an amplifier will be more affected by change in transistor parameter in case of [NHPC]
(a) positive feedback (b) negative feedback
(c) zero feedback (d) none of these

26. The highest voltage gain can be obtained from a
(a) CC (b) CE [NTPC]
(c) CB (d) both CE and CB

27. In a MOSFET if gate voltage V_{gs} is increased then
(a) mobile charge carrier density decreases
(b) channel conductivity increases
(c) channel widens
(d) channel conductivity decreases [NTPC]

28. In a feedback amplifier, feedback improves the signal to noise ratio for noise signal
(a) internally generated in the amplifier
(b) present with the amplifier input
(c) both internally generated and present with the input
(d) present at the output [BHEL]

29. When a signal is applied to a narrow band amplifier, then distortion produced in the output is due to
(a) use of transformer coupling
(b) loss of the low frequency components
(c) loss of the high frequency components
(d) phase distortion in the amplifier [BHEL]

30. The 3*dB* cut off frequency of atransistor amplifier is the frequency at which
(a) the current gain increases by 3*dB*
(b) the current gain becomes 3*dB*
(c) the current gain becomes 0.707 of its midrange value
(d) the power gain decreases by 3*dB* [BHEL]

31. When a P-N junction is forward biased
(a) the width of the deplection layer is decreased
(b) the width of deplection layer is increased
(c) the width of the depletion layer remains the same
(d) none of these [BEL]

32. Low frequency response of an RC coupled amplifier can be improved by
(a) increasing the coupling capacitors only
(b) increasing the by-pass capacitors only
(c) increasing the by-pass capacitor as well as coupling capacitor
(d) decreasing the by-pass capacitor [BEL]

33. An op-amp has a differential gain of 10^3 and a CMRR of 100. The output voltage of the op-amp with inputs of 120 μV and 80 μV will be [BEL]
(a) 26 *m*V (b) 41 *m*V
(c) 100 *m*V (d) 200 *m*V

34. In JFET, if the gate voltage V_{gs} is made more negative, then
(a) channel conductivity increases
(b) deplication region decreases
(c) channel conductivity decreases
(d) channel current increases [BEL]

35. An amplifier has a power gain of 50. This gain in dB will be

(*a*) 17 dB (*b*) 31 dB

(*c*) 34 dB (*d*) 68 dB **[BEL]**

36. In a JFET, beyonds the pinch off voltage, as the drain voltage increases; the drain current

(*a*) remains almost constant

(*b*) decreases

(*c*) increases

(*d*) may increase or decrease **[BEL]**

37. The band width of a difference amplifier is comparable to that of a

(*a*) single-stage CE amplifier

(*b*) double-stage CE amplifier

(*c*) cascade amplifier

(*d*) darlington amplifier **[IOCL]**

38. If the reverse-.bias voltage of a PN function is increased within limits, the reverse saturation current will **[IOCL]**

(*a*) decrease (*b*) increase

(*c*) remain unaffected (*d*) none of these

39. A voltage series feedback has **[IOCL]**

(*a*) high input impedance (*b*) low input impedance

(*c*) high output impedance (*d*) high voltage gain

40. EPROM memory content can be erased by exposing it to

(*a*) UV light (*b*) intense heat radiation

(*c*) IR rays (*d*) microwaves **[IOCL]**

41. A wien bridge oscillator is shown in the figure which of the following statements are true, if *f* is the frequency of oscillation?

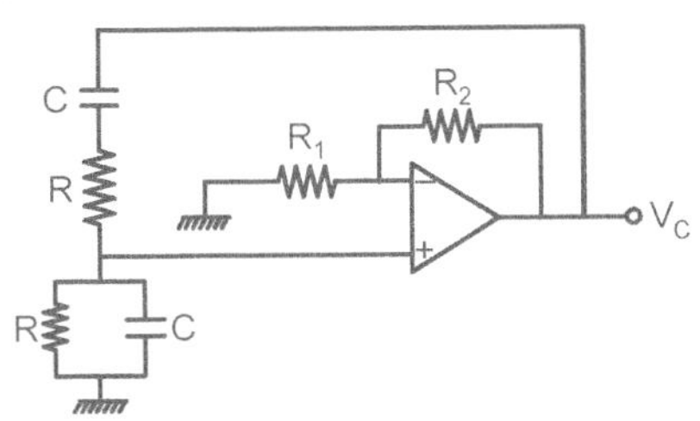

(*a*) For R = 1 K, C = $\frac{1}{2\pi}$ μF, *f* = 1 kHz

(*b*) For R = 3 K, C = $\frac{1}{18\pi}$ μF, *f* = 3 kHz

(*c*) The gain of the op-amp stage should be less than two for proper operation

(*d*) The gain of the op-amp stage should be three for proper operation **[GATE]**

42. Thyristor are not made using germanium becasue

(*a*) it is a stable semiconductors

(*b*) group *v* elements cannot be dopped with it

(*c*) its current handling capacity is less than silicon

(*d*) it has high leakage current **[BIE]**

43. The common emitter forward current gain of the transistor shown is B_F = 100

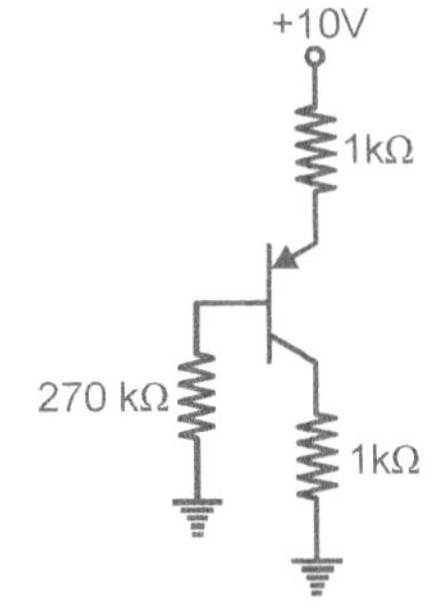

The transistor is operating in **[GATE]**

(*a*) Saturation region (*b*) Cutoff region

(*c*) Reverse active region (*d*) Forward active region

44. The three- terminal linear voltage regulator is connected to a 10 Ω load resistor as shown in the figure.

If V_{in} is 10 V, what is the power dissipated in the transistor?

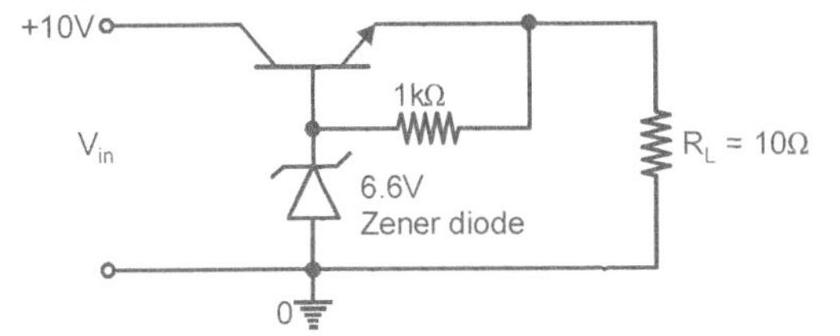

(*a*) 0.6 W (*b*) 2.4 W

(*c*) 4.2 W (*d*) 5.4 W **[GATE]**

45. What are the states of the three ideal diodes of the current shown in the figure below?

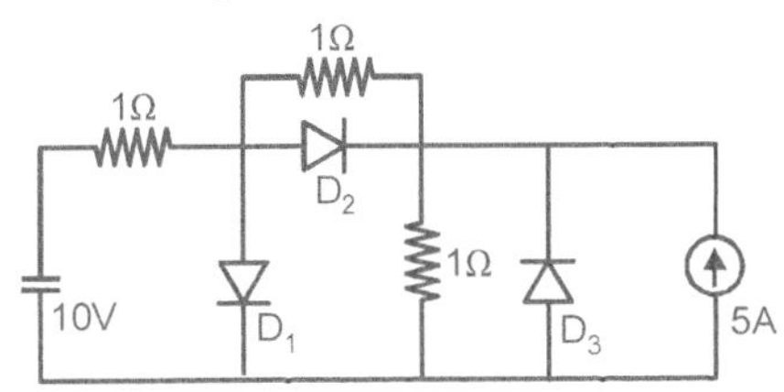

(*a*) D_1 ON, D_2 OFF, D_3 OFF

(*b*) D_1 OFF, D_2 ON, D_3 OFF

(*c*) D_1 ON, D_2 OFF, D_3 ON

(*d*) D_1 OFF, D_2 ON, D_3 ON **[GATE]**

46. The parameters of the circuit shown in the figure are R_i = 1 MΩ, R_o = 10 Ω, A = 10^6 V/V. If V_i = 1 μV, then output voltage, input impedance and output impedance respectively are

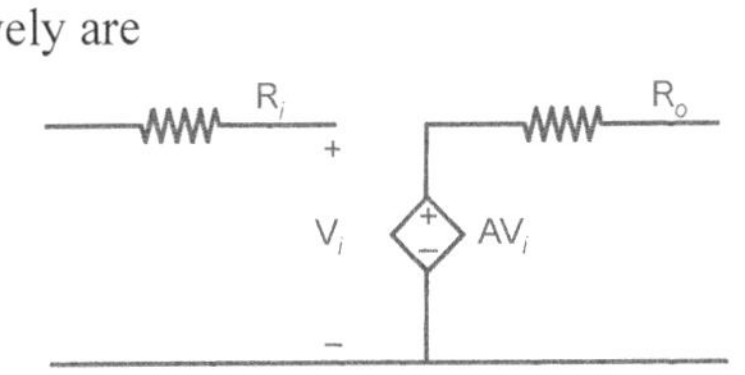

(*a*) 1V, ∞, 10 Ω

(*b*) 1V, 0, 10 Ω

(*c*) 1V, 0, ∞

(*d*) 10 V, ∞, 10 Ω **[GATE]**

47. Assume that D_1 and D_2 in figure are ideal diodes. The value of current I is

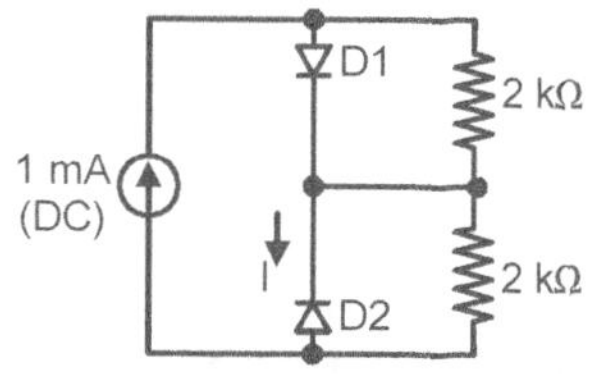

(*a*) 0 mA (*b*) 0.5 mA
(*c*) 1 mA (*d*) 2 mA **[GATE]**

48. Assuming the operational amplifier to be ideal, the gain V_{out}/V_{in} for the circuit shown in figure below is

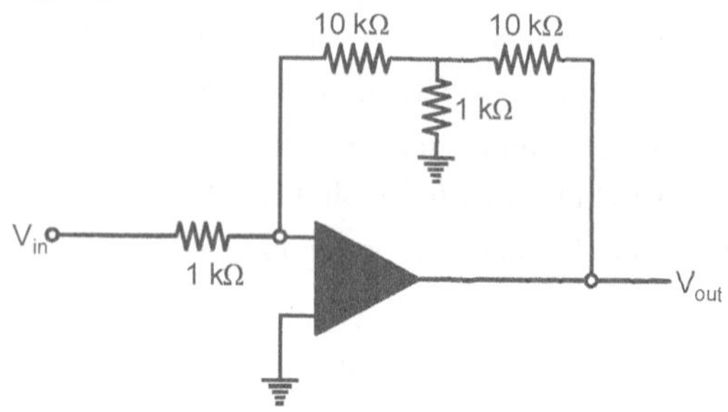

(*a*) –1 (*b*) – 20
(*c*) – 100 (*d*) – 120 **[GATE]**

49. The common emitter amplifier shown in the figure is biased using a 1 mA ideal current source. The approximate base current value is

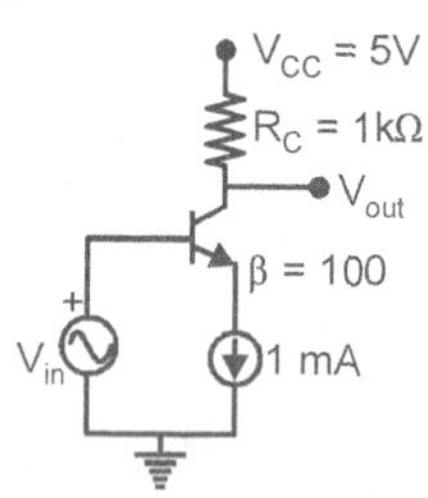

(*a*) 0 μA (*b*) 10 μA
(*c*) 100 μA (*d*) 1000 μA **[GATE]**

50. The advantage of using a function diode as temperature stabilising element in transistor circuit is that **[BHEL]**
(*a*) its resistance temperature coefficient is negative
(*b*) it is made of the same material as that of the transistor
(*c*) its reverse current is low
(*d*) its resistance changes with change in temperature

51. The theoretical maximum efficiency of a full wave rectifier circuit is
(*a*) the same as that of a half wave circuit
(*b*) double as that of a half wave circuit
(*c*) definitely more but less than double as that of a half wave circuit
(*d*) none of these **[IOCL]**

52. For an amplifier, coupling method which gives the highest gain is **[NTPC]**
(*a*) capacitance coupling (*b*) resistance coupling
(*c*) impedance coupling (*d*) transformer coupling

53. High power efficiency of the push-pull amplifier is due to the fact that **[NTPC]**
(*a*) Each transistor conducts on different cycles of the input
(*b*) transistor are placed in CE configuration
(*c*) there is no collector current with no input
(*d*) low forward biasing voltage is required

54. Which of the following transistor are symmetrical in the sence that emmitter and collector or source and drain terminals can be interchanged? **[NHPC]**
(*a*) JFET (*b*) MOSFET
(*c*) NPN transistor (*d*) PNP transistor

55. The gain of an amplifier will be more affected by change in transistor parameters in case of **[NHPC]**
(*a*) positive feedback (*b*) negative feedback
(*c*) zero feedback (*d*) None of these

56. The current due to electron flow in conductron band is
(*a*) less than the hole current in valence band
(*b*) equal to the hole current in valence band
(*c*) greater than the hole current in valence band
(*d*) None of these **[NHPC]**

57. Fermi energy is the amount of energy which **[NHPC]**
(*a*) a valence electron can have at room temperature
(*b*) must be given to an electron to move it to conductron band
(*c*) must be given to a hole to more it to balance band
(*d*) a hole can have at room temperature

58. The charge carriers in an *n*-channel FET are **[IOCL]**
(*a*) electrons alone
(*b*) holes alone
(*c*) both electrons and holes
(*d*) may be either electrons or holes

59. In IC's npn construction is preferred to *pnp* construction because **[BEL]**
(*a*) *npn* construction is cheaper
(*b*) *npn* construction permits higher packing of elements
(*c*) *p*-type base is preferred
(*d*) *n*-type base is preferred to reduce diffusion constant

60. The saturation drain current I_{DS} in an FET equals **[NHPC]**

(*a*) $I_{DSS}\left(1-\frac{V_{GS}}{V_P}\right)$ (*b*) $\frac{V_{2(S)}}{V_{1(S)}}=\frac{\left(s^2+\delta\right)}{\left(s^2+\alpha s+\delta\right)}$

(*c*) $I_{DSS}\sqrt{1-\left(\frac{V_{GS}}{V_D}\right)}$ (*d*) $I^2_{DSS}\left(1-\frac{V_{GS}}{V_P}\right)$

61. The photo-resist process is used **[NTPC]**

(*a*) during high temperature diffusion

(*b*) to prevent photo response

(*c*) to control the etching of SiO_2 from selected regions on a silicon slice

(*d*) to photograph the silicon slice

62. Under normal operating voltages, the reverse current in a silicon diode is about

(*a*) 100 μA (*b*) 1 μA

(*c*) 50 μA (*d*) 200 μA **[NHPC]**

63. In a centre-tap circuit, if voltage between one end of secondary winding and centre-tap is 300 V peak, P.I.V. is

(*a*) 300 V (*b*) 150 V

(*c*) 600 V (*d*) 900 V **[NHPC]**

64. *n*-channel FET's are superior to *p*-channel FET's because

(*a*) they have lower switching time

(*b*) they have lower pinch off voltage

(*c*) they have higher input impedance

(*d*) mobility of charge carrier electron in *n*-channel FET is greater than the mobility of charge carrier hole in *p*-channel FET **[NTPC]**

65. The operation of a JFET involves

(*a*) flow of minority carriers alone

(*b*) flow of majority carriers alone

(*c*) flow of both minority and majority carriers

(*d*) use of a magnetic field **[BHEL]**

66. Monolithic IC design is based on **[NTPC]**

(*a*) extensive use of transistor and diodes

(*b*) extensive use of RC coupling

(*c*) using high value resistors and capacitors

(*d*) making the area of circuit elements as large as possible

67. In a full-wave rectifier, of a.c. supply frequency is 50 Hz, then a.c. supply in the output is **[PGCIL]**

(*a*) 100 Hz (*b*) 25 Hz

(*c*) 50 Hz (*d*) 75 Hz

68. Which of the following device has the highest input impedance ? **[PGCIL]**

(*a*) CE BJT (*b*) CC BJT

(*c*) JFET (*d*) MOSFET

69. In a Wien Bridge oscillator, a FET is included in the feedback loop. **[NHPC]**

(*a*) This is to obtain highly precision frequency

(*b*) This is to stabilize the amplifier loop gain

(*c*) This is to make it a high frequency oscillator

(*d*) This is to operate the amplifier in class C condition.

70. In regard to class A single ended transformer coupled amplifier the following statements are made.

1. The d.c. current flowing through primary cause considerable dissipation of power
2. The output a.c. power obtainable is the same as in the series fed directly coupled load for a given transistor dissipation capability.
3. In the presence of a signal, the transistor dissipation is less than in the case of direct coupled load
4. The drain on the power supply is ideally one-half of what it is in direct coupled amplifier

Of these four statements, the true ones are

(*a*) 1 and 2 (*b*) 3 and 4

(*c*) 2 and 3 (*d*) 2 and 4 **[ONGC]**

71. In an *npn* transistor operating in the active region the main current crossing the collector junction from base side is **[ONGC]**

(*a*) hole drift current

(*b*) hole diffusion current

(*c*) electron diffusion current

(*d*) electron drift current

72. The transconductance g_m of an FET in saturation region equals **[ONGC]**

(*a*) $-\frac{2I_{DSS}}{V_P}\left(1-\frac{V_{GS}}{V_P}\right)$ (*b*) $-\frac{2I_{DSS}}{V_P}\left(1-\frac{V_{GS}}{V_P}\right)^2$

(*c*) $-\frac{2I_{DSS}}{V_P}\left(1-\frac{V_{GS}}{V_P}\right)^{1/2}$ (*d*) $-\frac{1}{V_P}(I_{DSS}.I_{DS})^{1/2}$

73. A transistor with emitter junction forward biased and collector junction reverse biased is said to operate in **[ONGC]**

(*a*) active region (*b*) saturation region

(*c*) cutoff region (*d*) inverted region

74. The most popularly used IC's are **[ONGC]**

(*a*) ECL and CTL (*b*) RTL and RCTL

(*c*) TTL and DTL (*d*) simple gate types

75. In the differentiating circuit given in the figure, the function of R_1 is to **[ES]**

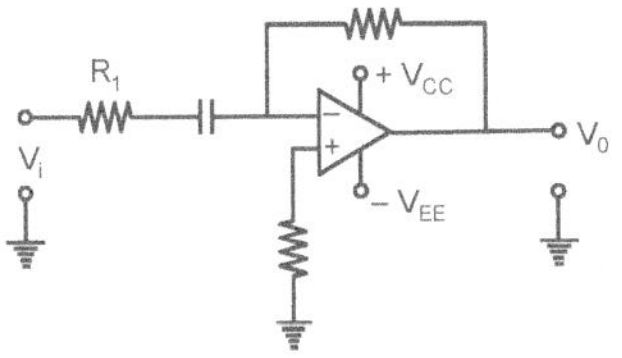

(*a*) enable circuit to a approach ideal differentiator

(*b*) maintain high input impedance

(*c*) eliminate high frequency noise spikes

(*d*) prevent oscillations at high frequencies

76. Consider the following statements:

As compared to Si single crystal grown by the CZ method, the Si single crystal grown by the FZ method should have

1. lesser oxygen contamination.
2. uniform doping throught length of the crystal.
3. more substitutional carbon atoms in the crystal.

Which of these statement pair is correct?

(*a*) 1, 2 and 3 (*b*) 1 and 3

(*c*) 2 and 3 (*d*) 1 and 2 **[IAS]**

77. The CE amplifier is shown in the given figure. The voltage gain of the amplifier is

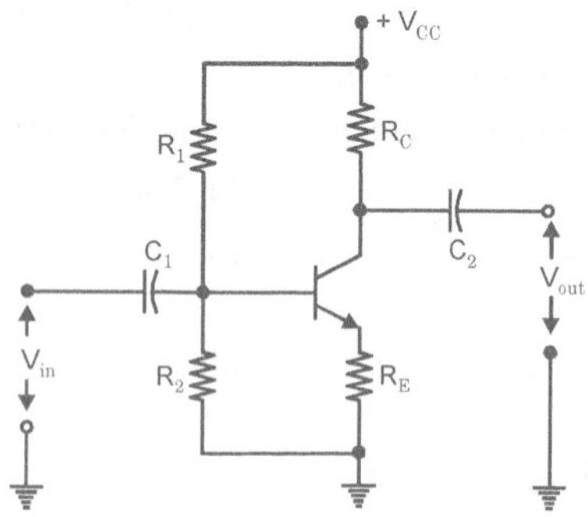

(*a*) $-\frac{R_C}{R_E}$ (*b*) $-\frac{R_E}{R_C}$

(*c*) $-\frac{R_C h_{ie}}{R_1R_2/(R_1+R_2)}$ (*d*) always < 1 **[IAS]**

78. A push-pull amplifier suppresses

(*a*) only the second harmonic component

(*b*) only the odd harmonic components

(*c*) only the even harmonic components

(*d*) both odd and even harmonic components **[IAS]**

79. For a JFET, g_m = 1 mA/V and

$C_{gs} + C_{gd} + C_{gss}$ = 5 pF. The symbols have their usual meaning.

The f_T of the JFET is nearly equal to **[IAS]**

(*a*) 1 MHz (*b*) 5 MHz

(*c*) 30 MHz (*d*) 190 MHz

80. The circuit shown in the figure below uses an ideal Op-Amp working with + 5V and – 5V power supplies. The output voltage V_o is equal to

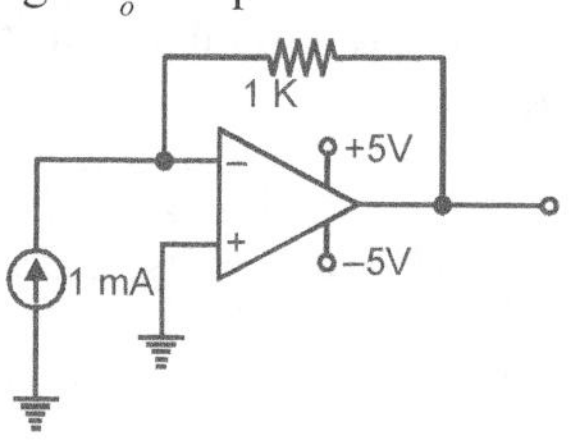

(*a*) +5 V (*b*) +1 V

(*c*) –1 V (*d*) –5 V **[IAS]**

81. The voltage gain of a device is 6dB. This means that the ratio of the amplitudes of the input to output will be approximately

(*a*) 6 (*b*) 1/6

(*c*) 2 (*d*) 1/2 **[IAS]**

82. Microprocessor chip came to the market in **[IAS]**

(*a*) early sixties (*b*) early seventies

(*c*) middle sixties (*d*) late seventies

83. The main drawback of a JFET is its **[NTPC]**

(*a*) high input impedance (*b*) low input impedance

(*c*) higher noise (*d*) lower gain

84. The input gate current of FET is **[NTPC]**

(*a*) a few amperes (*b*) a few milli-amperes

(*c*) a few micro-amperes (*d*) negligibly small

85. In a *p*-type semiconductor **[BHEL]**

(*a*) the hole concentration is less than the electron concentration

(*b*) electrons are the majority carriers

(*c*) at room temperature the hole concentration equals the acceptor concentration

(*d*) electron density does not change with temperature

86. Pinch of voltage V_P for an FET is the drain voltage at which **[BHEL]**

(*a*) significant drain current starts flowing

(*b*) drain current becomes zero

(*c*) all free charges get removed from the channel

(*d*) avalanche break down takes place

87. In switching a junction diode from forward to reverse voltage bias, the recovery of the current to the saturation value is delayed. During this time

(*a*) the current is in the same direction as the forward current

(*b*) the excess minority carriers have drop to zero

(*c*) the excess minority carriers have disappeared and the minority carrier density reaches zero value at the junction

(*d*) the voltage across the diode is in the direction of forward bias **[BEL]**

88. The magnitude of the threshold voltage V_T for enhancement MOSFET is of the order of **[BHEL]**

(*a*) 4 volts (*b*) 10 volts

(*c*) 40 volts (*d*) 100 volts

89. When the gate to source voltage V_{GS} of a *p*-channel JFET is made more positive, the drain current

(*a*) increases

(*b*) decreases

(*c*) remains constant

(*d*) may increase or decrease **[BEL]**

90. Tunnel diodes

(*a*) are mainly operated with reverse bias

(*b*) have a negative resistance region with reverse bias

(*c*) has very high impurity concentration thus increasing the barrier width

(*d*) electrons have a finite probability of penetrating of the potential barrier **[BEL]**

91. The circuit shown in the given figure is best described as

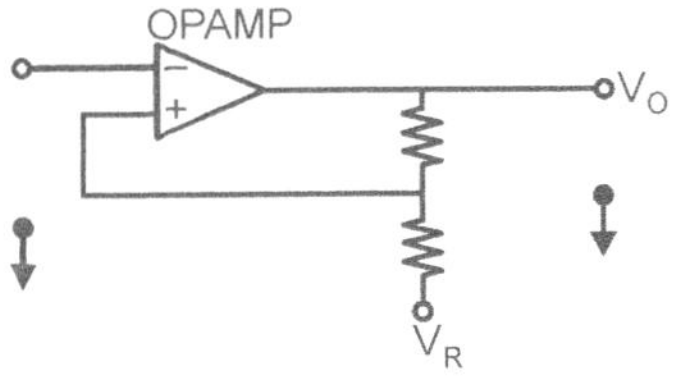

(*a*) voltage follower

(*b*) monostable multivibrator

(*c*) Schmitt trigger

(*d*) high input impedance amplifier **[IAS]**

92. At the junction of a *pn* diode. **[BHEL]**

(*a*) the depletion region has an electrostatic potential which makes *p* side positive with respect to the *n* side

(*b*) the space charge region exists predominantly in the heavily doped side of the junction

(*c*) holes are prevented from diffusing from *n* side to the *p* side

(*d*) the electric field at the junction is of such value as to cause the diffusive tendency to be counterbalanced by the drift tendency of electrons and holes

93. Inter electrode capacitances in an FET are of

(*a*) 1 pF (*b*) 100 pF

(*c*) 0.1 μF (*d*) 1 μF **[BEL]**

94. Main advantage of IC technology is possibility of

(*a*) securing high stability at low cost **[NTPC]**

(*b*) using high values of capacitors

(*c*) fabricating low tolerance resistors

(*d*) reparing individual circuit elements

95. Input resistance of a BiFET op-amp is **[NTPC]**

(*a*) 1 mΩ (*b*) 10^{12} Ω

(*c*) 10^{15} Ω (*d*) 1 kΩ

96. In monolithic IC, isolation may be easily obtained by

(*a*) using a layer of photo resist

(*b*) scribing

(*c*) using reverse biased *pn*-junction

(*d*) using SiO_2 layer **[BEL]**

97. In a *pnp* transistor biased in the active region, the dominant carrier flow is **[NTPC]**

(*a*) excess holes drifting in the base region

(*b*) minority carriers electrons injected from base to emitter

(*c*) minority carriers diffusing in the emitter region

(*d*) due to concentration gradient of holes, holes diffusing in the base region

98. The *p-n-p* transistor circuit as shown in the above figure is in saturation. Given that V_{BE} = 0.7 V and V_{CEsat} = 0.3 V and β_{min} = 20, the value of I_C will be

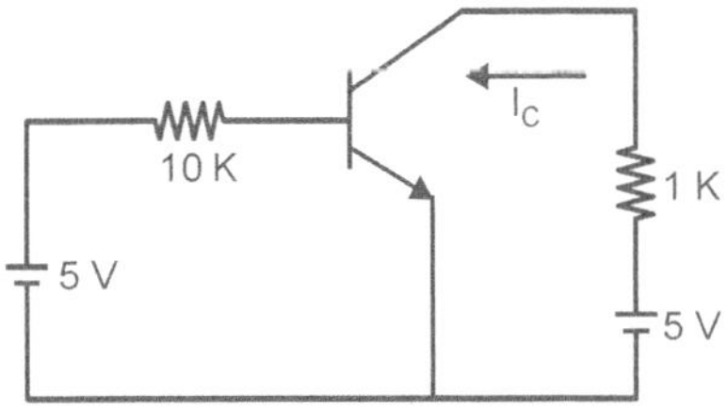

(*a*) 4.7 mA (*b*) 5.3 mA

(*c*) 8.6 mA (*d*) 10 mA **[IAS]**

99. A source follower using an FET usually has a voltage gain which is

(*a*) greater than + 100

(*b*) slightly less than unity but positive

(*c*) exactly unity but negative

(*d*) about – 10 **[ES]**

100. The gain and input resistance of an amplifier in a boot-strap sweep circuit are respectively **[IAS]**

(*a*) unity and zero (*b*) infinity and infinity

(*c*) unity and infinity (*d*) infinity and zero

101. In the circuit shown in the given figure, the type of feedback used is

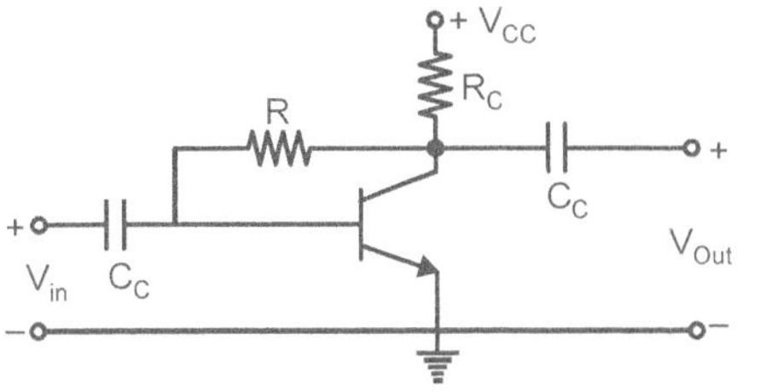

(*a*) voltage series feedback

(*b*) voltage shunt feedback

(*c*) current series feedback

(*d*) current shunt feedback **[IAS]**

102. Consider the following semiconductor diodes :

1. Germanium diode. 2. Sillicon diode.
3. Tunnel diode. 4. Schottky diode.

The correct increasing order of forward voltage drop of these diodes is

(*a*) 1, 2, 3, 4 (*b*) 3, 4, 1, 2

(*c*) 3, 1, 4, 2 (*d*) 1, 3, 2, 4 **[IAS]**

103. Circuit of a feedback amplifier having series type of feedback is shown in the given figure:

The β of feedback network is determined by

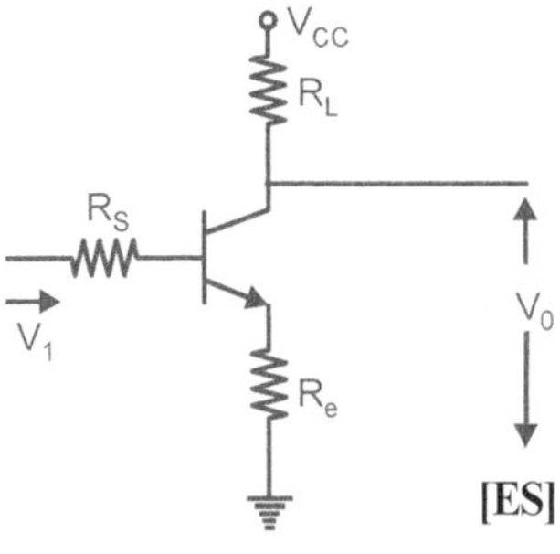

(*a*) $-\frac{R_e}{R_L}$

(*b*) $-R_e$

(*c*) $\frac{r_m}{g_m}$

(*d*) $-\left(\frac{R_e}{R_L}\right)^2$ **[ES]**

104. The g_m of the MOS transistor is 4 mA/V and the common base current gain of the bipolar transistor is 0.99. The transconductance of the composite transistor as shown in the given figure will be

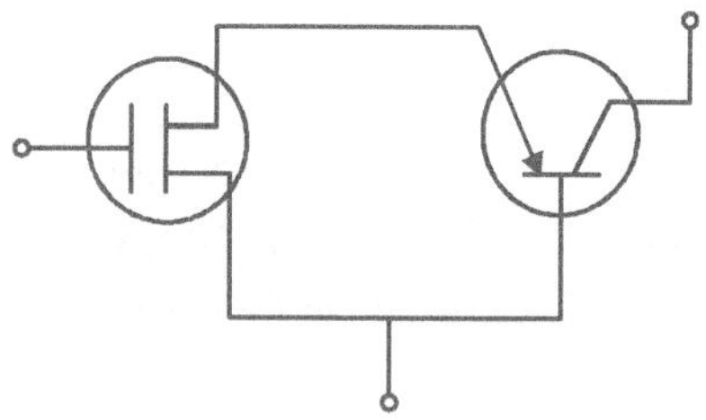

(*a*) 39.6 mA/V (*b*) 3.96 mA/V
(*c*) 1.98 mA/V (*d*) 0.99 mA/V **[IAS]**

105. For a *p*-type semiconductor, Fermi level lies
(*a*) In the middle of the energy gap
(*b*) Close to the bottom of the conduction band
(*c*) Close to the top of the conduction band
(*d*) Close to the top of the valence band **[IAS]**

106. Match **List-I** with **List-II** and select the correct answer using the codes given below the Lists

List-I	**List-II**
A. Class A	1. Less than 180°
B. Class B	2. 180°
C. Class AB	3. Greater than 180°
D. Class C	4. 360°
	5. Usually 50° to 60°

Codes:

	A	B	C	D
(*a*)	2	3	4	5
(*b*)	4	5	2	3
(*c*)	4	2	3	1
(*d*)	1	2	3	4

[IAS]

107. Match **List-I** with **List-II** and select the correct answer using the codes given below this lists:

List-I	**List-II**
A. RC phase shift	1. R.F Oscillator. oscillator
B. UJT oscillator	2. Higher frequency stabillity.
C. L-C oscillator	3. A.F. oscillator.
D. Crystal oscillator	4. Relaxation oscillator.

Codes:

	A	B	C	D
(*a*)	3	4	2	1
(*b*)	2	1	3	4
(*c*)	2	3	1	4
(*d*)	3	4	1	2

[ES]

108. An analogue electronic circuit that measures rms value of the input voltage by averaging the square of the instantaneous voltage level, responds slowly to changes in the input signal due to **[ONGC]**
(*a*) the "square" function built into the circuit
(*b*) the "square-root" function built into the circuit
(*c*) the averaging function built into the circuit
(*d*) none of these

109. As the temperature is increased, the voltage across a diode carrying a constant current **[ONGC]**
(*a*) increases
(*b*) decreases
(*c*) remains constant
(*d*) may increase or decrease depending upon the doping levels in the junction.

110. CMOS has the following advantage over PMOS/NMOS
(*a*) simpler fabrication process **[ONGC]**
(*b*) lower P_D
(*c*) lower input capacitance
(*d*) greater suitability for LSI

111. If the PN junction is heavily doped, the breakdown voltage will be **[ONGC]**
(*a*) more (*b*) neither more nor less
(*c*) more or less (*d*) decreasing

112. A typical medium gain IC Op-Amp **[ONGC]**
(*a*) is suitable for use at low frequency only
(*b*) has only one input terminal
(*c*) has an open loop gain of about 100
(*d*) has an open loop gain of about 2500

113. In an intrinsic semiconductor.
(*a*) with increasing temperature the density of electron hole pairs increases
(*b*) in intrinsic germanium the resistivity increases 6% per degree rise in temperature
(*c*) at low temperature, will above 0 K, electrons move into the valence band
(*d*) holes along with electrons carry current in the conduction band **[BHEL]**

114. A field-effect transistor (FET)
(*a*) has three *pn* junctions
(*b*) uses a forward biased junction
(*c*) depends on the variation of a magnetic field for its operation
(*d*) depends on the variation of a reverse voltage its operation **[ONGC]**

115. In a JFET, the amplification factor μ, transconductance g_m and the dynamic drain resistance are related is

(*a*) $\mu = \frac{g_m}{r_d}$ (*b*) $\mu = \frac{r_d}{g_m}$
(*c*) $\mu = g_m r_d$ (*d*) $\mu = g_m r_d^2$ **[BHEL]**

116. In switching a junction diode forward to reverse bias, the transition interval is the period from the time

(*a*) the diode voltage begins to reverse to the time when the diode has nominally recovered

(*b*) when the diode is forward biased to the time when it is reverse biased

(*c*) of forward bias to the time diode voltage has attained zero value

(*d*) none of these **[NTPC]**

117. Which of the following can be considered to be the advantages of FET amplifiers as compared to BJT amplifiers? **[ONGC]**

1. Higher input impedance
2. Good bias stability
3. Higher gain-bandwidth product
4. Lower noise figure.

Select the correct answer using the codes given below:

(*a*) 1, 2 and 3 (*b*) 1, 2 and 4

(*c*) 2, 3 and 4 (*d*) 1, 3 and 4

118. Consider the following statements: **[IAS]**

1. BJT needs bias stabilization because its I_{CO} increases with increase in V_{BE} and temperature.
2. Thermal run away may be avoided by choosing $V_{CE} < \frac{1}{2} V_{CC}$
3. FET and NJT both need bias stabilization
4. By increasing RE, bias stabilization improves.

Of these statements:

(*a*) 1 and 2 are correct (*b*) 3 and 4 are correct

(*c*) 2 and 3 are correct (*d*) 2 and 4 are correct

119. Regarding CC amplifier consider the statements 1 to 4.

1. It performs a resistance trans-formation from low to high resistance.
2. Its current gain is close to unity.
3. Its voltage gain is close to unity.
4. The frequency range is higher than that of a CE stage:

Of these the only true statements are:

(*a*) 1, 2 and 4 (*b*) 1 and 3 **[IAS]**

(*c*) 2 and 4 (*d*) 3 and 4

120. A single diode operates as a **[IAS]**

(*a*) Full-wave rectifier (*b*) Half-wave rectifier

(c) Bridge rectifier (*d*) Non-rectifier

121. A diffused resistor in an IC **[PGCIL]**

(*a*) can be of any value

(*b*) is formed at the same time as one region of BJT

(*c*) is formed after the transistor diffusion

(*d*) can be only of *n*-type

122. In order to form a structure containing both *pnp* and *npn* transistors, monolithic IC requires **[PGCIL]**

(*a*) four layers (*b*) five layers

(*c*) three layers (*d*) six layers

123. When the gate-to-source voltage V_{GS} of an *n*-channel JET is made and more negative, the drain current

(*a*) increases

(*b*) decreases

(*c*) remains unchanged

(*d*) may increase or decrease **[PGCIL]**

124. A transformer coupled amplifier would give **[ES]**

(*a*) maximum voltage gain (*b*) impedance matching

(*c*) maximum current gain (*d*) larger band width

125. FET voltage divider bias circuit is shown in the given figure. If $I_D = 4$ mA, then V_{GS} and V_{DS} will be respectively

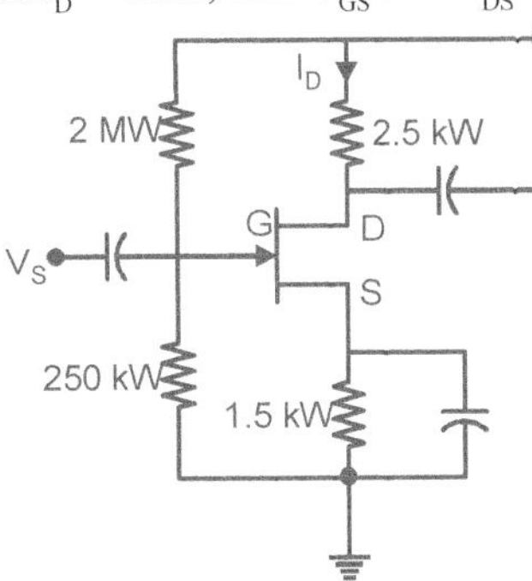

(*a*) –3.78 V and 4 V (*b*) 4 V and –3.78 V **[ES]**

(*c*) –3.78 V and – 4 V (*d*) – 4 V and –3.78 V

126. Consider the following statements in connection with transistor amplifiers in class B and class AB mode

1. In the class B for low signal voltages at the base, the collector current is zero
2. In the class B, there is cross-over distortion
3. In the class AB with no signal the dc power drawn is zero

Which of the statements given above are correct?

(*a*) 1, 2 and 3 (*b*) 1 and 2 only

(*c*) 2 and 3 only (*d*) 1 and 3 only **[IAS]**

127. In the circuit shown in the given figure, the output voltage will be

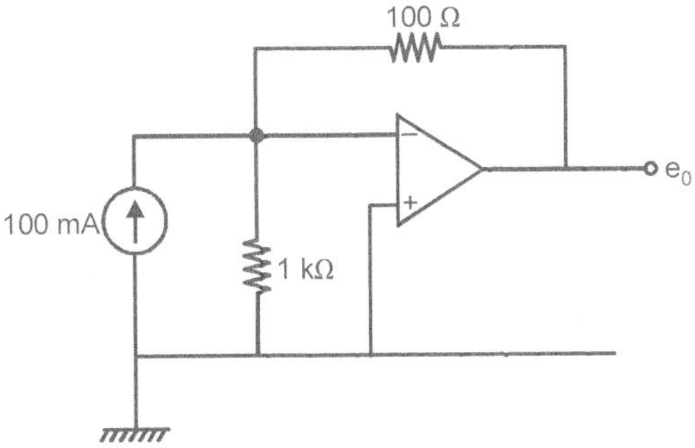

(*a*) 9V (*b*) 10V

(*c*) 11V (*d*) 12V **[ES]**

128. The overall bandwidth of two identical voltage amplifiers connected in cascade will

(*a*) remain the same as that of a single stage

(*b*) be worse than that of a single stage

(*c*) be better than that of single stage

(*d*) be better if stage gain is low and worse if stage gain is high **[ES]**

129. In modern electronic multimeters a FET or MOSFET is preferred over BJT because

(*a*) its input resistance is high

(*b*) its input resistance is high and does not vary with the chagne of range

(*c*) its input resistance is low

(*d*) it is cheaper **[ES]**

130. The Zener diode shown in the circuit has a reverse breakdown voltage of 10 volts. The power dissipation in R_S would be

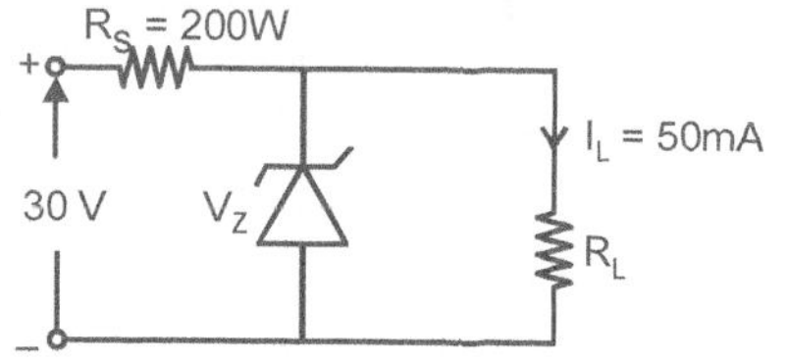

(*a*) 0.5 W (*b*) 1 W

(*c*) 15 W (*d*) 2 W **[ES]**

131. Consider the following processes in the fabrication of I_{CS} **[ES]**

1. Substrate
2. Oxide layer
3. Photo resist and photolithography
4. Etching

The correct sequence of these processes is

(*a*) 1, 2, 3, 4 (*b*) 1, 2, 4, 3

(*c*) 2, 1, 3, 4 (*d*) 2, 1, 4, 3

132. The type of power amplifier which exhibits crossover distortion in its output is **[ES]**

(*a*) Class A (*b*) Class B

(*c*) Class AB (*d*) Class C

133. Match List-I (amplifier configuration) with List-II (application) and select the correct answer using the codes given below the lists:

List-I	List-II
A. CE amplifier	1. Low bandwidth high input impedance amplifier
B. CB amplifier	2. Audio frequency amplifier
C. JFET amplifier	3. Radio frequency amplifier microwave amplifier
D. CC amplifier	4. Microwave amplfier
	5. Buffer amplifier

Codes:

	A	B	C	D
(*a*)	2	1	5	4
(*b*)	2	3	1	5
(*c*)	4	3	1	5
(*d*)	5	4	3	2

[ES]

134. Consider the following statements: **[ES]**

Negative feedback in amplifiers resutls in

1. Reduced voltage gain.
2. Reduced band-width.
3. Increased signal to noise ratio.
4. Reduced distortion.

(*a*) 1 and 2 are correct.

(*b*) 1, 3 and 4 are correct.

(*c*) 2, 3 and 4 are correct.

(*d*) 1 and 4 are correct.

135. Match List-I (Circuit Symbol) with List-II (Device) and select the correct answer using the code given below the lists :

List-I	List-II
A. 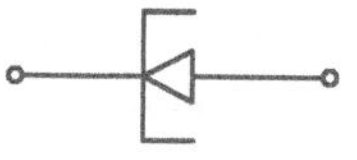	1. N-channel FET
B. 	2. Varactor
C. 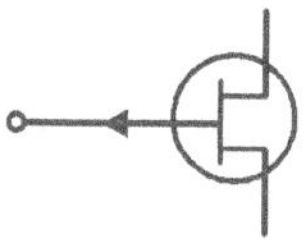	3. Tunnel diode
D. 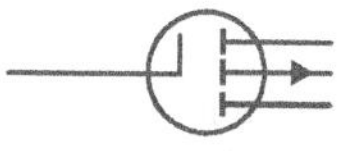	4. P-channel MOSFET

Codes:

	A	B	C	D
(*a*)	3	2	1	4
(*b*)	1	4	3	2
(c)	3	4	1	2
(*d*)	1	2	3	4

[ES]

136. If the unity-gain bandwidth of the operational amplifier shown in the given figure is 4 MHz, then the bandwidth of the amplifier will be

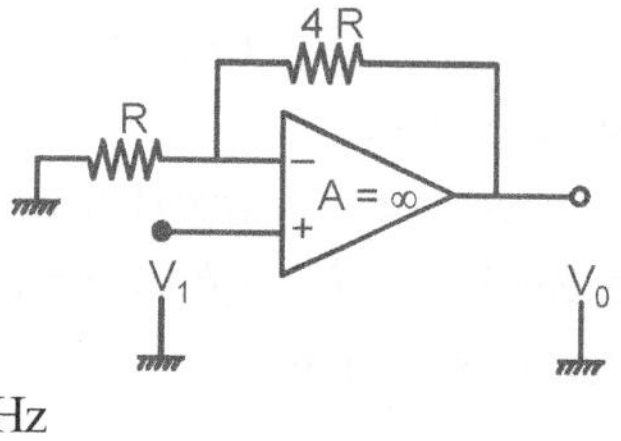

(*a*) 0.1 MHz

(*b*) 0.8 MHz

(*c*) 2 MHz

(*d*) 4 MHz **[IAS]**

137. A non-inverting Op-Amp summer is shown in the given figure the output voltage V_o is

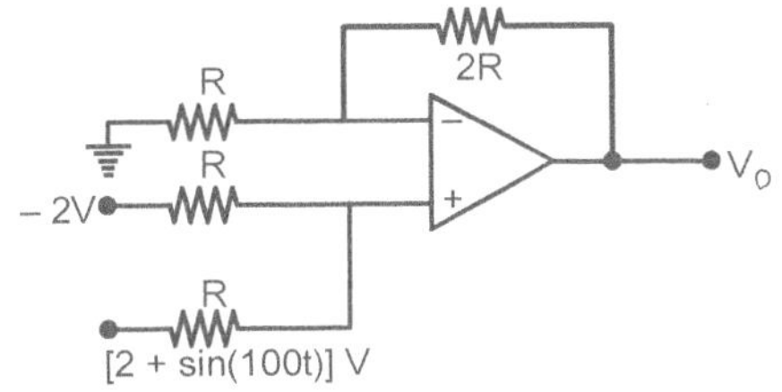

(*a*) (3/2) sin (100*t*) (*b*) 3 sin (100*t*)

(*c*) 2 sin (100*t*) (*d*) sin (100*t*) **[GATE]**

138. Match List-I with List-II and select the correct answer using the codes given below the lists:

List-I

A. Wien bridge oscillator

B. Voltage shunt feedback amplifier

C. Crystal oscillator

D. Current shunt feedback amplifier

List-II

1. Low output impedance
2. RF frequency range
3. Audio frequency range
4. High input impedance
5. High output impedance

Codes:

	A	B	C	D
(*a*)	2	1	3	5
(*b*)	3	4	2	1
(*c*)	5	4	2	1
(*d*)	3	1	2	5

[ES]

139. The main advantage in using a three OP-AMP instrumentation amplifier over a single OP-AMP differential amplifier lies in

(*a*) higher values of CMRR

(*b*) lower noise figure

(*c*) elimination of the need for accurate matching of resistors

(*d*) simplicity of gain adjustment **[ES]**

140. An op-amp has a common mode gain of 0.01 and a differential mode gain of 10^5. Its common mode rejection ratio would be

(*a*) 10^{-7} (*b*) 10^{-3}

(*c*) 10^3 (*d*) 10^7 **[IES**

141. Cascade amplifiers are used in

(*a*) video amplifiers

(*b*) voltage amplifiers

(*c*) power amplifier

(*d*) turned amplifier design **[ES]**

142. The Barkhausen criterion for sustained oscillation is given by

(*a*) $A\beta = 1$ (*b*) $|A\beta| \geq 1$

(*c*) $|A\beta| < 1$ (*d*) $\angle A\beta = 180°$ **[ES]**

143. The circuit shown in the figure given below represents a feedback amplifier. For this circuit, the current and voltage gains would respectively be

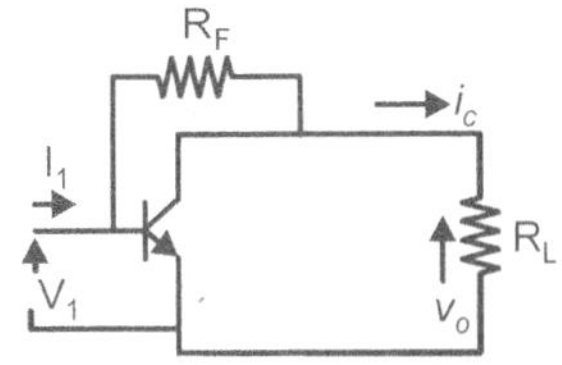

(*a*) $h_{fe}, \frac{h_{fe}.R_L}{h_{ie}}$ (*b*) $\frac{R_F}{R_L}, \frac{h_{fe}.R_L}{h_{ie}}$

(*c*) $h_{fe}, \frac{R_L}{R_F}$ (*d*) $\frac{R_F}{R_L}, \frac{R_L}{R_F}$ **[ES]**

144. The maximum theoretical efficiency of a Class B push-pull transistor amplifier is approximately

(*a*) 25% (*b*) 50%

(*c*) 70.7% (*d*) 78.6% **[ES]**

145. If the differential and common mode gains of a differential amplifier are 50 and 0.2 respectively, then the CMRR will be

(*a*) 10 (*b*) 49.8

(*c*) 50.2 (*d*) 250 **[ES]**

146. If the differential and common mode gains of a differential amplifier are 50 and 0.2 respectively, then the CMRR will be **[ES]**

(*a*) barrier voltage (*b*) breakdown voltage

(*c*) pick-off voltage (*d*) pinch-off voltage

147. In modern electronic multimeters a FET or MOSFET is preferred over BJT because

(*a*) its input resistance is hgih

(*b*) its input resistance is high and does not vary with the change of range

(*c*) its input resistance is low

(*d*) it is cheaper **[ES]**

148. Consider the following features regarding an amplifier:

1. Voltage gain being less than one.
2. High input impedance.
3. High output impedance.
4. High current gain.

Which of these are the characteristic features of an emitter follower amplifier?

(*a*) 1, 2 and 4

(*b*) 1, 2 and 3

(*c*) 1 and 3

(*d*) 2, 3 and 4 **[ES]**

149. The best approximation for V_c in the circuit shown in the figure will be (assume β to be high)

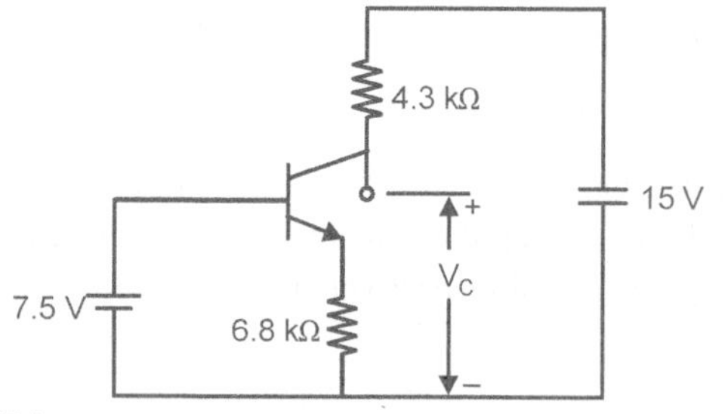

(*a*) 4 V (*b*) 6.8 V
(*c*) 8.7 V (*d*) 10.7 V **[ES]**

150. Consider the following statements: A tunnel diode is

1. Made of Ge or GaAs.
2. An abrupt junction with both sides heavily doped.
3. A hyper abrupt junction with both sides heavily doped.
4. A majority carrier device.

Which of these statements are correct?

(*a*) 1 and 2 (*b*) 3 and 4
(*c*) 1, 3 and 4 (*d*) 1, 2 and 4 **[ES]**

151. The upper cut-off frequencies f_{21} and f_{22} of the two stages of a cascaded amplifier are respectively 5MHz and 3.3 MHz. The equivalent upper cut-off frequency of the cascaded amplifier would be

(*a*) 4.16 MHz (*b*) 3.33 MHz
(*c*) 2 MHz (*d*) 5 MHz **[ES]**

152. In an amplifier, the increase in gain is 12 dB if the frequency is doubled. If the frequency is increased by 10 times, then the increase in gain will be

(*a*) 2.4 dB (*b*) 20 dB
(*c*) 40 dB (*d*) 60 dB **[ES]**

153. An amplifier circuit has an overall current gain of –100 and an input resistance of 10 kΩ with a load resistance of 1 kΩ. The overall voltage gain of the amplifier is

(*a*) 5 dB (*b*) 10 dB
(*c*) 20 dB (*d*) 40 dB **[ES]**

154. The figure of merit of a logic family is given by the product of

(*a*) gain and bandwidth
(*b*) propagation delay time and power dissipation
(*c*) fan out and propagation delay time
(*d*) noise margin and power dissipation **[IAS]**

155. The frequency of an LC circuit of a tuned base turned emitter type of coupled oscillator having resonant frequency of crystal as ω, is

(*a*) slightly less than ω
(*b*) equal to ω
(*c*) slightly more than ω
(*d*) at least 2 ω **[IAS]**

156. To design an RC phase-shift oscillator using a JFET, the absolute minimum amplification factor of the JFET that will be required is

(*a*) 3 (*b*) 17
(*c*) 23 (*d*) 29 **[IAS]**

157. In an operational amplifier the bandwidth (B) in terms of rise time (t_r) is given by

(*a*) $\frac{0.35}{t_r}$ (*b*) $\frac{1}{0.35 t_r}$
(*c*) $0.35\, t_r$ (*d*) $\frac{t_r}{0.35}$ **[IAS]**

158. The slew rate of an operational amplifier is 0.5 v/microsec. The maximum frequency of a sinusoidal input of 2 V rms that can be handled without excessive distortion is

(*a*) 3 kHz (*b*) 30 kHz
(*c*) 300 kHz (*d*) MHz **[IAS]**

Directions: *The following items consist of two statements, one labelled the ' Assertion A' and the other labelled the 'Reason R'. You are to examine these two statements carefully and decide if the Assertion A and the Reason R are individually true and if so, whether the Reason is a correct explanation of the Assertion. Select your answers to these items using the codes given below and mark your answer sheet accordingly.*

Codes:

(*a*) Both A and R are true and R is the correct explanation of A
(*b*) Both A and R are true but R is not a correct explanation of A
(*c*) A is true but R is false
(*d*) A is false but R is true

159. **Assertion (A):** FET is a unipolar device. **[ES]**

Reason (R): In an FET, current flows in only one direction

160. **Assertion (A):** Square wave testing of multistage R-C coupled amplifier is an important tool.

Reason (R): Design of coupling networks is done on a frequency response basis. **[ES]**

161. **Assertion (A):** Common-collector amplifier is generally used as a Buffer amplifier.

Reason (R): The voltage gain of the common-collector amplifier is nearly equal to unity. **[ES]**

162. **Assertion (A):** As the temperature of a transistor increases, the base spreading resistance *r' bb* in the hybrid model at high frequencies decreases.

Reason (R): The mobility of the carriers decreases as temperature increases. **[ES]**

163. **Assertion (A):** Power supply regulation is important with push-pull Class A amplifiers but not with single-ended Class A amplifiers.

Reason (R): Total collector current in a push-pull Class A amplifier remains practically constant. **[IAS]**

164. **Assertion (A):** It is easy to parallel power MOSFETs for increased current handlingcapability.

Reason (R): The ON state resistance of a power MOSFET has a positive temperature coefficient. **[ES]**

165. **Assertion (A):** Direct coupled Class A power Amplifier is not preferred.

Reason (R): The collector circuit efficiency of direct coupled Class A power amplifier is not 25%. **[ES]**

166. **Assertion (A):** A BJT amplifier needs very good bias stabilization circuit as compared to that of an FET amplifier circuit.

Reason (R): The input *p-n* junction of FET amplifier is reverse biased and that of the BJT amplifier is forward biased. **[IAS]**

167. **Assertion (A):** Design of *dc* amplifiers is difficult and involves a complex procedure. **[ES]**

Reason (R): Drift problem is associated with dc amplifiers.

168. **Assertion (A):** High power RF amplifiers are operated as class C amplifiers.

Reason (R): The flywheel effect of the output circuit tuned coil and capacitor restores the sine wave shape to the output signal. **[ES]**

169. **Assertion (A):** The op-amp is known for its very high voltage gain and broad bandwidth.

Reason (R): The op-amp has high R_{in} and CMMR. **[ES]**

170. **Assertion (A):** For a periodic input voltage whose amplitude is large enough to pass through the trip points, a Schmitt trigger produces a rectangular wave-form at the output.

Reason (R): A Schmitt trigger is a comparator with positive feedback and hence prevents noise triggering. **[IAS]**

171. **Assertion (A):** The emitter logic reduces the transistor rise time and hence results in logic gates which switch very fast.

Reason (R): In emitter-coupled logic, the transistors are maintained in unsaturatedcondition. **[ES]**

172. **Assertion(A):** The multistage inter-mediate frequency amplifier of a super heterodyne radio receiver is always staggered tuned.

Reason (R): A very sharp tuned high gain amplifier is obtained maximising the gain and increasing the fidelity of the receiver **[IAS]**

173. **Assertion (A):** In a class B push-pull power amplifier, there is no power loss under no signal condition. **[IAS]**

Reason (R): The two transistors are biased at cut-off.

174. **Assertion (A):** Some compromise is necessary while selecting the value of commutating capacitors of a bistable multivibrator.

Reason (R): Smaller commutating capacitors yield shorter transition time and longer setting time. **[IAS]**

175. **Assertion (A):** Self-biasing method of biasing is generally used for a CE amplifier.

Reason (R): The design of self-biasing circuit is very easy as compared to other biasing circuits. **[IAS]**

176. **Assertion(A):** A monostable multi-vibrator can be used as a counter.

Reason (R): The pulse width of the non-stable multivibrator is a function of its input dc voltage. [IAS]

177. **Assertion (A):** FET is a unipolar device.

Reason (R): The current carriers in a FET are either boles or electrons. **[ES]**

178. **Assertion (A):** The individual active and passive components in IC should be electrically isolated from one another before metallisation.

Reason (R): The epitaxial layer in which these components are fabricated can provide a common leakage path. **[IAS]**

179. **Assertion (A):** An integrator is preferred to a differentiator in an analog computer. **[IAS]**

Reason (R): Integrator is faster than differentiator.

180. **Assertion (A):** To avoid saturation in a transistor switch, a diode is connected between the base and collector terminal of the transistor.

Reason (R): The forward biased diode must be switched off-before the transistor turn-off can commence. **[ES]**

181. **Assertion (A):** Operational amplifiers invariably posses high common mode rejection.

Reason (R): Operational amplifiers are designed to have excellent matching of their input transistors. **[ES]**

182. **Assertion (A):** The gain of an RC coupled CE amplifier falls off at low frequencies. **[ES]**

Reason (R): The low frequency response depends on the transistor junction capacitance.

183. **Assertion (A):** In a bipolar junction transistor, the emitter region width (in comparison to base region width) is immaterial in the design so long as the base doping is kept lower than the emitter doping.

Reason (R): The emitter injection efficiency is also dependent on both emitter and base region width and their relative dopings. **[ES]**

184. **Assertion (A) :** Three-phase trans-formers of different vector groups should not be operated in parallel.

Reason (R): Transformers belong to different vector groups will not have the same phase sequence on the secondary side. **[IAS]**

185. **Assertion (A) :** A MOSFET amplifier offers very high input impedance irrespective of its circuit configuration.

Reason (R): MOSEFT has extremely low gate current. **[ES]**

186. **Assertion (A):** ECL gate has less propagation delay as compared to TTL and CMOS gates.

Reason (R): ECL gate consumes less power. **[IAS]**

187. **Assertion (A):** TTL and CMOS cannot be normally used together.

Reason (R): TTL operates on a (± 5 0.25)V regulated supply voltage and some mA, while the CMOS operates on unregulated supply voltage of + 3 V + 15 V and some μA **[ES]**

188. **Assertion (A):** A bipolar junction transistor (BJT) has collector-base junction forward biased and emitter-base junction as reverse biased. **[ES]**

Reason (R): BJT acts as a current amplification device.

ANSWERS

EXERCISE - I

1. (a)	**2.** (b)	**3.** (d)	**4.** (b)	**5.** (a)	**6.** (a)	**7.** (d)	**8.** (c)	**9.** (b)	**10.** (d)
11. (a)	**12.** (c)	**13.** (d)	**14.** (a)	**15.** (b)	**16.** (c)	**17.** (b)	**18.** (d)	**19.** (d)	**20.** (c)
21. (c)	**22.** (d)	**23.** (a)	**24.** (c)	**25.** (c)	**26.** (b)	**27.** (b)	**28.** (d)	**29.** (d)	**30.** (c)
31. (d)	**32.** (b)	**33.** (d)	**34.** (d)	**35.** (c)	**36.** (a)	**37.** (d)	**38.** (a)	**39.** (c)	**40.** (c)
41. (a)	**42.** (a)	**43.** (a)	**44.** (b)	**45.** (c)	**46.** (b)	**47.** (c)	**48.** (b)	**49.** (d)	**50.** (a)
51. (b)	**52.** (a)	**53.** (c)	**54.** (b)	**55.** (a)	**56.** (a)	**57.** (a)	**58.** (b)	**59.** (d)	**60.** (a)
61. (c)	**62.** (c)	**63.** (b)	**64.** (b)	**65.** (b)	**66.** (b)	**67.** (c)	**68.** (a)	**69.** (d)	**70.** (b)
71. (b)	**72.** (b)	**73.** (d)	**74.** (c)	**75.** (c)	**76.** (a)	**77.** (b)	**78.** (d)	**79.** (a)	**80.** (c)
81. (b)	**82.** (c)	**83.** (d)	**84.** (a)	**85.** (a)	**86.** (b)	**87.** (c)	**88.** (c)	**89.** (b)	**90.** (c)
91. (d)	**92.** (d)	**93.** (b)	**94.** (d)	**95.** (b)	**96.** (d)	**97.** (c)	**98.** (a)	**99.** (a)	**100.** (c)
101. (a)	**102.** (b)	**103.** (a)	**104.** (b)	**105.** (a)	**106.** (a)	**107.** (b)	**108.** (d)	**109.** (d)	**110.** (d)
111. (a)	**112.** (d)	**113.** (b)	**114.** (b)	**115.** (c)	**116.** (c)	**117.** (b)	**118.** (c)	**119.** (a)	**120.** (a)
121. (a)	**122.** (a)	**123.** (d)	**124.** (b)	**125.** (d)	**126.** (c)	**127.** (c)	**128.** (c)	**129.** (c)	**130.** (c)
131. (c)	**132.** (d)	**133.** (b)	**134.** (d)	**135.** (b)	**136.** (c)	**137.** (c)	**138.** (d)	**139.** (d)	**140.** (b)
141. (b)	**142.** (a)	**143.** (a)	**144.** (a)	**145.** (c)	**146.** (d)	**147.** (d)	**148.** (c)	**149.** (a)	**150.** (a)
151. (d)	**152.** (a)	**153.** (b)	**154.** (d)	**155.** (b)	**156.** (d)	**157.** (c)	**158.** (a)	**159.** (d)	**160.** (a)
161. (a)	**162.** (a)	**163.** (a)	**164.** (d)	**165.** (d)	**166.** (b)	**167.** (c)	**168.** (d)	**169.** (c)	**170.** (d)
171. (c)	**172.** (a)	**173.** (b)	**174.** (c)	**175.** (c)	**176.** (b)	**177.** (c)	**178.** (b)	**179.** (b)	**180.** (b)

EXERCISE - II

1. (b)	**2.** (c)	**3.** (c)	**4.** (c)	**5.** (c)	**6.** (b)	**7.** (d)	**8.** (d)	**9.** (c)	**10.** (d)
11. (a)	**12.** (c)	**13.** (d)	**14.** (a)	**15.** (b)	**16.** (c)	**17.** (b)	**18.** (a)	**19.** (a)	**20.** (d)
21. (a)	**22.** (a)	**23.** (d)	**24.** (d)	**25.** (a)	**26.** (b)	**27.** (b)	**28.** (a)	**29.** (c)	**30.** (c)
31. (a)	**32.** (c)	**33.** (b)	**34.** (c)	**35.** (a)	**36.** (a)	**37.** (c)	**38.** (c)	**39.** (a)	**40.** (a)
41. (a),(b)	**42.** (d)	**43.** (c)	**44.** (b)	**45.** (a)	**46.** (a)	**47.** (a)	**48.** (d)	**49.** (b)	**50.** (b)
51. (b)	**52.** (d)	**53.** (c)	**54.** (b)	**55.** (a)	**56.** (c)	**57.** (b)	**58.** (a)	**59.** (d)	**60.** (b)
61. (c)	**62.** (b)	**63.** (c)	**64.** (d)	**65.** (b)	**66.** (a)	**67.** (a)	**68.** (b)	**69.** (b)	**70.** (d)
71. (c)	**72.** (a)	**73.** (a)	**74.** (c)	**75.** (b)	**76.** (a)	**77.** (a)	**78.** (c)	**79.** (c)	**80.** (c)

81. (*d*)	**82.** (*b*)	**83.** (*d*)	**84.** (*d*)	**85.** (*c*)	**86.** (*c*)	**87.** (*c*)	**88.** (*a*)	**89.** (*b*)	**90.** (*d*)
91. (*c*)	**92.** (*d*)	**93.** (*a*)	**94.** (*a*)	**95.** (*b*)	**96.** (*c*)	**97.** (*d*)	**98.** (*a*)	**99.** (*a*)	**100.** (*c*)
101. (*b*)	**102.** (*c*)	**103.** (*a*)	**104.** (*b*)	**105.** (*d*)	**106.** (*c*)	**107.** (*d*)	**108.** (*c*)	**109.** (*b*)	**110.** (*b*)
111. (*a*)	**112.** (*d*)	**113.** (*a*)	**114.** (*d*)	**115.** (*d*)	**116.** (*a*)	**117.** (*b*)	**118.** (*a*)	**119.** (*d*)	**120.** (*b*)
121. (*b*)	**122.** (*b*)	**123.** (*b*)	**124.** (*b*)	**125.** (*a*)	**126.** (*a*)	**127.** (*b*)	**128.** (*b*)	**129.** (*b*)	**130.** (*d*)
131. (*a*)	**132.** (*b*)	**133.** (*b*)	**134.** (*b*)	**135.** (*a*)	**136.** (*b*)	**137.** (*b*)	**138.** (*d*)	**139.** (*c*)	**140.** (*d*)
141. (*a*)	**142.** (*b*)	**143.** (*a*)	**144.** (*d*)	**145.** (*d*)	**146.** (*d*)	**147.** (*b*)	**148.** (*a*)	**149.** (*d*)	**150.** (*a*)
151. (*c*)	**152.** (*c*)	**153.** (*c*)	**154.** (*a*)	**155.** (*a*)	**156.** (*d*)	**157.** (*a*)	**158.** (*b*)	**159.** (*b*)	**160.** (*b*)
161. (*b*)	**162.** (*c*)	**163.** (*a*)	**164.** (*b*)	**165.** (*b*)	**166.** (*c*)	**167.** (*a*)	**168.** (*a*)	**169.** (*b*)	**170.** (*b*)
171. (*a*)	**172.** (*b*)	**173.** (*a*)	**174.** (*b*)	**175.** (*a*)	**176.** (*a*)	**177.** (*a*)	**178.** (*b*)	**179.** (*a*)	**180.** (*a*)
181. (*a*)	**182.** (*c*)	**183.** (*d*)	**184.** (*c*)	**185.** (*d*)	**186.** (*b*)	**187.** (*c*)	**188.** (*d*)		

EXPLANATIONS

EXERCISE - I

23. $$I_c = \frac{2}{2}\text{mA} = 1\text{ mA}$$

Now for both CE and CB configurations

$$\beta = \frac{I_c}{I_b}$$

$$\therefore \quad I_b = \frac{I_c}{\beta} = \frac{1}{50} = 0.02\text{ mA}$$

32. Gain of the amplifier with feedback

$$A_f = \frac{A}{1+\beta A} = \frac{100}{1+0.10\times 100} = \frac{100}{11} = 9.09$$

38. $$\text{Gain} = \frac{1\times 10^6}{R_1} = -100$$

$$\therefore \quad R_1 = 1000\Omega = 10\text{k}\Omega$$

60. Equivalent current $= \frac{100}{50} = 2\text{A}$

Equivalent resistance = 50 ohms

62. Equivalent circuit is shown as follows :

$$R_0 \approx 0$$
$$R_i = \infty$$

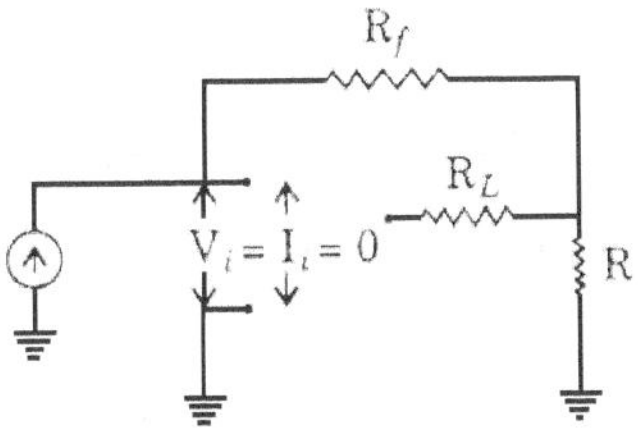

$$R_0 = \frac{R\times R_f}{R\times R_f},\ V_i = \frac{I_i R_f}{1-A_v}$$

$$V_0 = -\frac{R.R_f}{R+R_f},\ V_i = \frac{I_i R_f}{1-A_v}$$

$$A_V = \frac{V_0}{V_i} = \frac{R.R_f I_o}{R+R_f}.\frac{A_v}{R_f I_i}$$

$$\therefore \quad \frac{I_o}{I_i} = \frac{R+R_f}{R} = 1 + \frac{R_f}{R}$$

66. In the positive half cycle, diode D_1 conducts and impedance offered will be 10 kΩ.

In the negative half cycle, diode D_2 coducts and impedance offered will be 10 kΩ.

Hence equivalent impedance offered in one full such as 10 kΩ.

67. $$I_c \approx I_E = \frac{4-0.7}{3.3} = 1\text{ mA}$$

$$V_c = 8 - 3.75\times 1 = 4.3\text{V}$$

71. If frequencies are relating far a part due to by pass capacitor and coupling capacitor, then highest cut-off frequency will determine the lower cut-off frequency of the entire system.

Here, $$f_{LE} = \frac{1}{2\pi R_e C_E} = \frac{1}{2\pi(R_0 + C_1)C_c} = 100\text{ Hz}$$

90. $$\sigma = q(n_n\mu_n + n_p\mu_p)$$
$$= 1.6\times 10^{-19}\times 2.5\times 10^{13}(3.8\times 10^3 + 1.8\times 10^3)$$
$$= 22.4\times 10^{-3}$$

$\therefore$ Resistivity, $$\rho = \frac{1}{\sigma} = \frac{1}{22.4\times 10^{-3}} = \frac{1000}{22.4}$$
$$= 44.6428\ \Omega.\text{cm}$$

102. In positive cycle diode is open, hence

$$V_0 = V_i$$

In negative cycle conducts, then

$$\frac{V_i - 0}{1} = \frac{0 - V_0}{1}$$

$$\Rightarrow \quad V_0 = V_i$$

Hence, current balances as a full-wave rectifier

104. Given: $A_f = 40\,dB$

$\therefore \quad A_f = \text{antilog}_{10}\left(\frac{40}{20}\right) = 100$

Annotation, $\beta = \text{antilog}_{10}\left(\frac{50}{20}\right) \approx 316.227$

We have, feedback gain,

$$\beta' = \frac{1}{316.27}$$

Also, $$A_f = \frac{A}{1 + \frac{1}{316.227} \times A}$$

$\therefore \quad A = 146.32$

109. Gain of the system is given by

$$A = \left|\frac{V_2(s)}{V_1(s)}\right| = \frac{R}{(RCs+1)} = \left|\frac{100}{(j10^{-3w}+1)}\right|$$

$\therefore$ |A| dB = 20 log 100 at low frequencies

115. We can draw equivalent circuit using virtual ground concept as

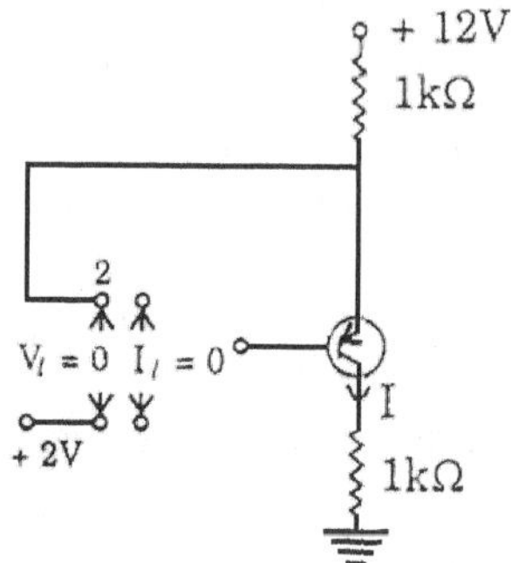

Voltage at point P = Voltage at point 2
= Voltage at point 3 = 2V

$\therefore$ Current in $1k\Omega = \frac{12-2}{1k\Omega} = 10$ mA

As collector current ≈ Emitter current

$\therefore \quad I = 10$ mA

117. Frequency of output at third flip-flop

$$= \frac{1}{2^3} = \frac{1}{8} = 1.25 \text{ MHz}$$

120. $$\frac{V_0 - V_c}{3} = \frac{V_c}{R_4} + \frac{V_c}{R_2}$$

(A and B are at same potential using virtual ground concept)

$$\therefore \quad V_c = \frac{V_0}{R_3} \cdot \frac{R_2R_3R_4}{R_2R_3 + R_3R_4 + R_4 \times R_2}$$

$$= V_0 \cdot \frac{R_2R_4}{R_2R_3 + R_3R_4 + R_4R_2}$$

and $$\frac{V_c}{R_2} + \frac{V_i}{R_1} = 0$$

$$\therefore \quad V_0 \frac{R_4}{R_2R_3 + R_3R_4 + R_4R_2} = \frac{V_i}{R_1}$$

$$\therefore \quad \frac{V_0}{V_i} = \frac{1}{R_1}\left(\frac{R_2R_3 + R_3R_4 + R_4R_2}{R_4}\right)$$

$$= \frac{1}{R_1}\left(R_2 + R_3 + \frac{R_2R_3}{R_4}\right)$$

125. $$h_{ie} = \frac{h_{fe}V_T}{|I_c|} = \frac{100 \times 26 \times 10^{-3}}{|I_c|} = 5.2K$$

$\Rightarrow \quad |I_c| = 0.5$ mA

126. $$A_{f1} = \frac{A}{1+\beta A} = \frac{400}{1+0.05 \times 400} = 19.04$$

and $$A_{f2} = \frac{360}{1+0.05 \times 360} = 18.94$$

Change in closed-loop gain

$$= \frac{A_{f_1} - A_{f_2}}{A_{f_1}} \times 100 = 0.5\%$$

129. $$\frac{V_i}{2V_2} = \frac{I_2}{I_i} = \frac{N_1}{N_2} = \frac{1}{2}$$

$\Rightarrow \quad V_2 = V_i$

$$I_{dc} = \frac{V_{2m}}{\pi R_L} \cdot \frac{2V_{im}}{\pi R_L} \quad \text{...}[V_i = 40V_{rms},$$

$$\Rightarrow \quad V_m = 40\sqrt{2}\,] = \frac{2 \times 40\sqrt{2}}{\pi \times 20} = 1.831\text{A}$$

142. $$\frac{V_2}{V_5} = \left(1 + \frac{49}{1}\right)\left(-\frac{40}{10}\right)$$

$$= 50(-4) = -200$$

146. $V_o = V_m \sin \omega t$

$$\left(\frac{dV_o}{dt}\right)_{max} = |V_m.\omega.\cos \omega t| = V_m \omega$$

$$5 \times 10^6 = V_m \times 2\pi \times 10^6$$

$$\therefore \quad V_m = \frac{5}{2\pi} \text{ volts}$$

149. zoner resistance $\gamma = \frac{dV_z}{dI_z}$

$\Rightarrow \quad dV_z = 10 \times 10^{-3} \times 5 = 0.05$ V

153. $$g_m = \frac{I_c}{K_t} = \frac{2}{25} = 0.08$$

$$\gamma_\pi = \frac{\beta}{g_m} = \frac{100}{0.8} = 1250$$

154. To have constant output voltage, voltages across zoner should be higher than zoner breakdown voltage, so that zoner conducts in reverse direction.

$$\therefore \quad \frac{R_L \times 24}{27 + R_L} \geq V_Z$$

$$\Rightarrow \quad \frac{R_L \times 24}{27 + R_L} \geq 15$$

$\Rightarrow \quad R_L \geq 45$ ohms.

155. Potential of terminals B,

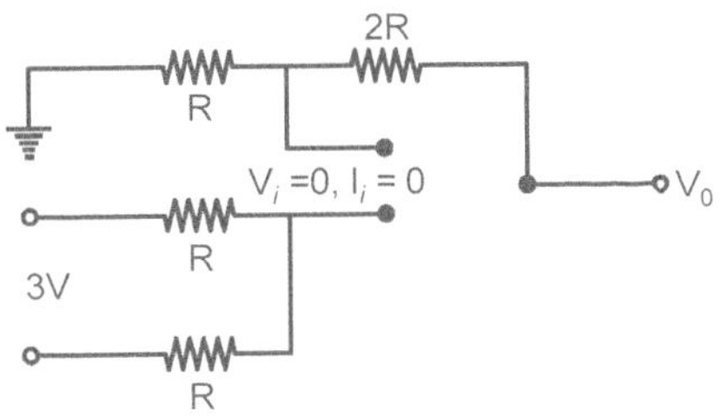

$$V_B = 3 + \sin 100t - \frac{6 + \sin 100t}{2R} \times R$$

$$= 3 + \sin 100t - 3 - \frac{\sin 100t}{2} = \frac{\sin 100t}{2}$$

Since input impedance of op-amp is very large, the input current is negligibly small and points A and B can be considered to be at the same potential

$$V_A = A_B = \frac{\sin 100t}{2}$$

Then, $$\frac{V_o - \frac{\sin 100t}{2}}{2R} = \frac{\frac{\sin 100t}{2}}{R}$$

$\therefore$ $$V_o = \frac{3}{2} \sin 100t$$

156. With feedback, upper cut-off frequency shall become (1 +|A||B|) times original cut-off frequency

$$\therefore\ f_{cf} = \left(1 + \frac{500}{100}\right) \times 60 = 360 \text{ kHz}$$

159. Diode D_1 will conduct only, D_2 will be in off position

Hence $V_1 = 6V$

and voltage V_2 coil equal to the voltage across 10kΩ, hence

$V_2 = 3V$

163. $n_e \cdot n_h = n_i^2$

For sample A :

$$n_e \times n_h = 1 \times 10^{20} \times 9 \times 10^{12} = 9 \times 10^{32}$$

$\therefore$ $n_i^2 = 9 \times 10^{32}$

For sample B :

$$n_e = 3 \times 10^{20}$$

$$n_h = \frac{n_i^2}{n} = \frac{9 \times 10^{32}}{3 \times 10^{20}} = 3 \times 10^{12}$$

164. $$\frac{V_2(s)}{V_1(s)} = \frac{s^2 + \delta}{s^2 + \alpha s + \delta}$$

$$\lim_{s \to 0} \frac{V_2(s)}{V_1(s)} = \lim_{s \to 0} \frac{s^2 + \delta}{s^2 + \alpha . s + \delta} = 1$$

$$\lim_{s \to \infty} \frac{V_2(s)}{V_1(s)} = \lim_{s \to \infty} \frac{s^2 + \delta}{s^2 + \alpha s + \delta} = 1$$

Since in initial and final frequency range the gain is unity, and there should be some drop in gain at mid-frequencies, hence it should be band reject filter.

165. $$I_c = \frac{I_{co}}{1 - \alpha} + \frac{\alpha}{1 - \alpha} I_B = \frac{6}{1 - 0.98} + \frac{0.98}{1 - 0.98} \times 100$$

$= 5.2$ mA

167. For base-emitter loop,

$$15 - 0.7 = 2I_E \times 14.3$$

$\Rightarrow$ $I_E = 0.5$ mA

$\therefore$ $I_c \approx I_E = 0.5$ mA

168. $I_E = 0.596$ mA

$$(V_{CE}) = 20 - 10I_C - 10I_E$$

$$= 20 - 20 \times 0.596 = 10 \text{ V}$$

177. Using virtual ground concept equivalent circuit can be drawn as

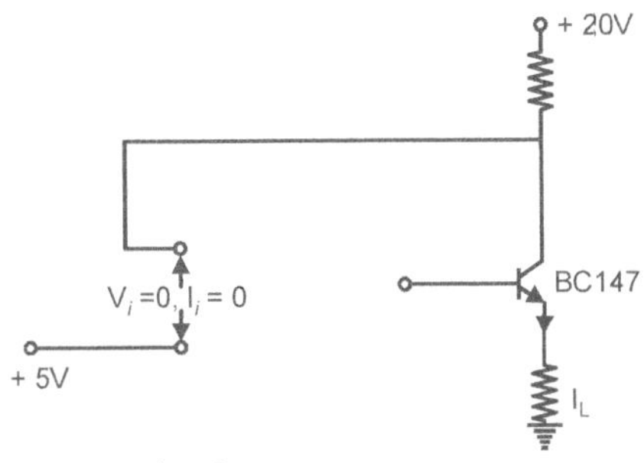

$\therefore$ Voltage at point C

= voltage at point A

= voltage at point B = 5V

$\therefore$ Current, $I = \frac{20 - 5}{1} = 15$ mA

No current flow though A and B, hence all current I flow throw 500 Ω.

$\therefore$ $I_L = I = 15$ mA

180. Band width $= \frac{f_o}{Q} = \frac{1 \times 10^6}{100}$

$= 10000$ Hz $= 10$ kHz

EXERCISE - II

2. Without feedback gain = 1000

Now, variation in gain without feedback

$$= \frac{10}{1000} \times 100 = 1\%$$

Now the feed back gain, variation is to be in limit of 0.1%

$$0.1 = \frac{1}{1 + \beta \times 1000}$$

$\Rightarrow$ $0.1 + 100\beta = 1$

$\Rightarrow$ $$\beta = \frac{9}{1000}$$

33. $$V_o = A_d V_d \left(1 + \frac{1}{CMRR} \cdot \frac{V_c}{V_d}\right)$$

$$= 10^3 \times (120 - 80) \times 10^{-6} \left[1 + \frac{1}{100} \times \frac{(120 + 80)/2}{120 - 80}\right]$$

$$= 40 \times 10^{-3} [1 + 0.025] = 41 \times 10^{-3} \text{ V} = 41 \text{ mV}$$

35. $P(dB) = 10 \log \frac{P_2}{P_1} = 10 \log 50 = 17$ dB

41. For a Wien Bridge oscillator,

$$f = \frac{1}{2\pi RC}$$

(*a*) $R = 1\ k\Omega, C = \frac{1}{2\pi}\mu F$

$$\therefore \quad f = \frac{1}{2\pi\times1\times10^3\times\frac{1}{2\pi}\times10^{-6}} = 10^3\ Hz$$

(*b*) $R = 3\ k\Omega, C = \frac{1}{18\pi}\mu F$

$$\therefore \quad f = \frac{1}{2\pi\times1\times10^3\times\frac{1}{18\pi}\times10^{-6}} = 3\ kHz$$

The gain should be 3 for proper operation

77. Voltage gain is given by,

$$A_v = \frac{h_{fe}R_c}{h_{ie}+(1+h_{fe})R_e}$$

Assuming $h_{fe} >> 1$ and $R_e >> h_{ie}$, we get

$$A_v = -\frac{R_c}{R_e}$$

79. Frequency, $f_T = \frac{g_m}{2\pi C} = \frac{1\times10^{-3}}{2\times3.14\times5\times10^{-12}}$

$= 30\ MHz$

81.
$$A_v(dB) = 20\log\frac{V_2}{V_1}$$

$$\therefore \quad 6 = 20\log\frac{V_2}{V_1}$$

$$\Rightarrow \quad \frac{V_2}{V_1} = 1.99$$

$$\Rightarrow \quad \frac{V_1}{V_2} = 0.5 = \frac{1}{2}$$

98. $I_C = \frac{5-V_{BE}}{1} = \frac{5.0-0.7}{1} = 4.7\ mA$

125. $V_G = \frac{20\times0.25}{2+0.25} = \frac{20}{9}$

$V_{GS} = V_G - I_o R_s = \frac{20}{9} - 4\times1.5 = -3.78\ V$

$V_{DS} = V_{DD} - I_D(R_s + R_D)$

$= 20 - 4\times(2.5+1.5) = 4V$

127. $V_o = -\left[\frac{100}{1000}\times1000\times0.10\right] = -10\ V$

130. Voltage across $R_s = 30 - 10 = 20V$

$\therefore$ Power loss in $R_s = \frac{20^2}{200} = 2W$

136. B.W. × A = 4

Here, $A = \left|\frac{V_0}{V_1}\right| = \left(1+\frac{4R}{R}\right) = 5$

$\therefore$ B.W. × 5 = 4

$\Rightarrow$ B.W. $= \frac{4}{5} = 0.8$ MHz

137. $V_o = [-2+2+\sin(100t)]\times\left(1+\frac{2R}{R}\right)$

$= \sin(100t)\times3 = 3\sin(100t)$

140. $CMRR = \frac{A_d}{A_c} = \frac{10^5}{0.01} = 10^7$

143. Current gain, $A_i = h_{fe}$ for CE stage

Voltage gain, $A_v = \frac{A_I Z_L}{h_{ie}} = \frac{h_{fe}R_L}{h_{ie}}$

145. $CMRR = \frac{A_d}{A_c} = \frac{50}{0.2} = 250$

149. For base emitter loop,

$7.5 - 0.3 = I_E\times6.8$

$\Rightarrow \quad I_E = 1.0588$

$\therefore \quad I_C >> I_E$

$\therefore \quad V_C = 10.45$

151. Equivalent upper cut-off frequency of cascaded amplifier will be lower than the cut-off frequencies of individual amplifiers, hence equivalent cut-off frequency will be 2 MHz.

152. Slope in dB/decade $= \frac{12}{\log_{10}2}$

Hence when frequency is changed to 10 times, number of decade $= \log_{10}10 = 1$

$\therefore$ Increase in gain $= \frac{12}{\log_{10}2}\times1 = 40$ dB

153. Voltage gain $= -\left(\frac{-100\times1}{10}\right) = 10$

$\therefore \quad A_v(dB) = 20\log_{10} = 20$ dB

157. In operational amplifier, product of B.W. and rise time is constant

B.W. × $t_r = 0.35$

$\therefore$ B.W. $= \frac{0.35}{t_r}$

158. $f \le \frac{\text{Slew Rate}}{2\pi V_m}$

$\Rightarrow \quad f \ge \frac{0.5}{10^{-6}\times2\times3.14\times2\times\sqrt{2}}$

$\Rightarrow \quad f \le 30\ kHz$

11 CHAPTER

Digital Circuits

EXERCISE - I

1. Match List-I with List-II and select the correct answer using the orders given below the lists

List-I	**List-II**
A. Monitor program	1. Used to indicate memory location
B. Assembler	2. A combination of letters, symbols and numerals
C. Mnemonic	3. A program that translates symbolic instructions into binary equivalent
D. Program	4. An operating system counter

Codes :

	A	B	C	D
(*a*)	4	3	2	1
(*b*)	4	3	1	2
(*c*)	3	4	1	2
(*d*)	3	4	2	1

2. Function $g = \overline{(A\overline{B}).(\overline{A}B)}$ is

(*a*) $A + B$ (*b*) AB

(*c*) $\overline{A} + \overline{B}$ (*d*) $\overline{A}B + A\overline{B}$

3. What is the total number of memory locations and input - output devices that can be addressed with a process or having 16-bit address bus, using memory mapped I/03.

(*a*) 64 K memory locations and 256 I/O devices

(*b*) 256 I/O devices and 65279 memory locations

(*c*) 64 K memory locations and no I/O devices

(*d*) 64 K memory locations or input - output devices

4. Noise margin of a digital IC is

(*a*) the minimum extraneous voltage that causes a gate to change its state

(*b*) the maximum extraneous voltage that causes a gate to change its state

(*c*) the thermal noise voltage which causes a gate to change its state

(*d*) None of these.

5. Which of the following storages has greatest capacity with less cost per bit ?

(*a*) Coincidence-current memory

(*b*) Magnetic tape

(*c*) Magnetic drum

(*d*) Semi-conductor memory

6. Which of the following statement is not correct ?

(*a*) $X + \overline{X}Y = X$ (*b*) $X(\overline{X} + Y) = XY$

(*c*) $XY + X\overline{Y} = X$ (*d*) $ZX + ZXY = ZX$

7. Match List-I with List-II select the correct answer using the codes given below the lists :

List-I (*Boolean statements*)

A. $A \oplus B = 0$

B. $\overline{A + B} = 0$

C. $\overline{A}.B = 0$

D. $A \oplus B = 1$

List-II (*Conditions under which statements is true*)

1. $A \neq B$

2. $A = B$

3. $A = 1$ and $B = 1$

4. $A = 1$ and $B = 0$

Codes :

	A	B	C	D
(*a*)	1	2	4	3
(*b*)	2	3	4	1
(*c*)	1	3	2	4
(*d*)	2	4	1	3

8. As compared to MOS memories, bipolar memories have

(*a*) slower success time but are cheaper

(*b*) slower access time and are costly

(*c*) faster access time and are cheaper

(*d*) faster access time and are costly

9. The balance equation $Z = A\overline{B} + \overline{A}.B$ can be realised using only

(*a*) AND gates (*b*) OR gates

(*c*) NAND gates (*d*) NOT gates

10. For the logic circuit shown in the figure below, what is the output y equal to ?

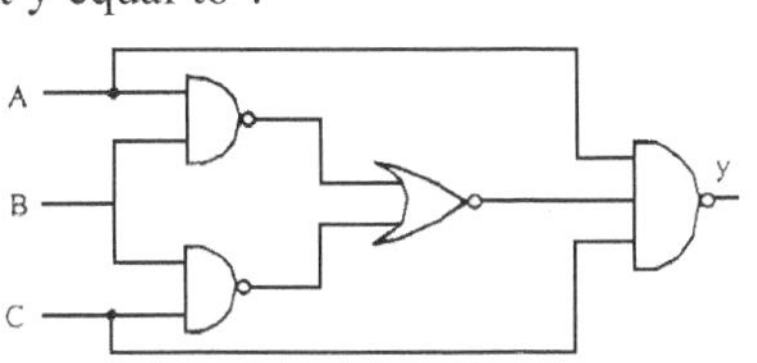

(*a*) $\overline{ABC}$ (*b*) $\overline{AB} + \overline{BC}$

(*c*) $\overline{AB} + \overline{BC} + \overline{A} + \overline{C}$ (*d*) All of these

11. The translator program that converts source code in high level language into machine code line by line is called

(*a*) Assembler (*b*) Compiler

(c) Loader (*d*) Interpreter

12. Suppose one wishes to be certain that, after execution of the statement

if $a > b$, then $x = a$; the value of and 'a' will be the same. Of the following which is the weakest (least restrictive) condition that must necessarily hold before execution of the statement?

(*a*) $(x = a)$ OR $(a > b)$ (*b*) $(a > b)$ AND $(x = a)$

(*c*) $a > b$ (*d*) $x = a$

13. The advantage of immediate addressing is

(*a*) to decode the operand easily

(*b*) to execute the instruction immediately

(*c*) not to fetch operand from memory

(*d*) to allow both positive and negative operands

14. A unit pulse input is given to a single-pole stable digital filter. Its output response will be

(*a*) constant pulses

(*b*) exponentially decaying pulses

(*c*) impulse

(*d*) exponentially rising pulses

15. EPROM memory content can be erased by exposing it to

(*a*) UV light (*b*) Intense heat radiation

(*c*) IR rays (*d*) Microwaves

16. The advantage of magnetic drum storage is

(*a*) high access time

(*b*) large size

(*c*) only read out is possible

(*d*) low access time

17. Which one of the following is a valid arithmetic expression ?

(*a*) X ± Y * SIN (X) (*b*) X + Y (*SIN (X))

(*c*) X − Y (SIN (X) (*d*) X − Y* SIN (X)

18. Exclusive OR followed by an inverter has the property

(*a*) Output high for input word of even parity

(*b*) Output high for input word of odd parity

(*c*) Output is 1's complement of input word

(*d*) Output is 2's complement of input word

19. Which one of the following is the correct bit position for various flags in the flag resister of an 8085 microprocessor?

(*a*)

S_1	Z	AC	*p*	CY

(*b*)

S	Z	AC	*p*	CY

(*c*)

Z	S	AC	*p*	CY

(*d*)

CY	Z	AC	*p*	S

20. The digital multiplexer is basically a combinational logic circuit to perform the operation

(*a*) AND − AND (*b*) OR − OR

(*c*) AND − OR (*d*) OR − AND

21. In 8085 microprocessor, H, L register pair is used for storing

(*a*) address of memory

(*b*) data

(*c*) address of next instruction

(*d*) current instruction

22. The number of memory chips of size 1 K × 4 bits required to build a memory bank of size 16 K × 8 bits is

(*a*) 64 (*b*) 32

(*c*) 16 (*d*) 8

23. While ______ is the fastest unsaturated logic gate______has the excellent noise immunity?

(*a*) ECL; TTL (*b*) TTL; ECL

(*c*) ECL; HTL (*d*) RTL; DTL

24. Semiconductor ROM's are sometimes preferred to semi conductor RAM's because

(*a*) ROM's are cheaper than RAM's

(*b*) ROMs are faster

(*c*) ROMs do not require power supply for their operation

(*d*) program stored in ROMs cannot be altered either by the power failure or by the user

25. Assuming that only the X and Y logic inputs are available and their complements $\overline{X}$ and $\overline{Y}$ are not available, what is the minimum number of two-input *NAND* gates required to implement $X \oplus Y$

(*a*) 2 (*b*) 3

(*c*) 4 (*d*) 5

26. In an 8085 microprocessor, the accumulator contents get altered by the instruction

(*a*) ANI 5C (*b*) CMP C

(*c*) CPI 1A (*d*) ORA A

27. For the product-of-sum expression $L = (X + Y)(\overline{X} + \overline{Z})$, the equivalent sum-of-products expression will be

(*a*) $XY + \overline{X}\,\overline{Z}$ (*b*) $\overline{X}\,\overline{Y} + XZ$

(*c*) $\overline{X}Y + X\overline{Z}$ (*d*) $X\overline{Y} + \overline{X}Z$

28. What are the output bits S (sum) and C (carry) of a Half Adder having input A = 1 and B = 1 ?

	S	**C**
(*a*)	1	1
(*b*)	1	0
(*c*)	0	1
(*d*)	0	0

29. Which one of the following difference equations will satisfy the JK flip-flop Truth Table?

(*a*) $Q_{n+1} = J_n\bar{Q}_n + \bar{K}_n Q_n$

(*b*) $Q_{n+1} = \bar{J}_n \bar{Q}_n + \bar{K}_n Q_n$

(*c*) $Q_{n+1} = J_n Q_n + K_n Q_n$

(*d*) $Q_{n+1} = \bar{J}_n \bar{Q}_n + \bar{K}_n \bar{Q}_n$

30. Consider the following expressions:

1. $Y = f(A, B, C, D) = \Sigma(1, 2, 4, 7, 8, 11, 13, 14)$
2. $Y = f(A, B, C, D) = \Sigma(3, 5, 7, 10, 11, 12, 13, 14)$
3. $Y = f(A, B, C, D) = \pi(0, 3, 5, 6, 9, 10, 12, 15)$
4. $Y = f(A, B, C, D) = \pi(0, 1, 2, 4, 6, 8, 9, 15)$

Which of these expressions are equivalent to the expression, $Y = A \oplus B \oplus C \oplus D$?

(*a*) 2 and 3 (*b*) 1 and 4
(*c*) 2 and 4 (*d*) 1 and 3

31. Which all elements either above or below main diagonal) of a square matrix are zero, then matrix is said to be triangular. The save memory taken by zero elements a lower triangular matrix $A(n \times n)$ is stored as one dimensional array, B(m) where *m* is the size of array. The value *of m* should be equal to

(*a*) $n(n+1)/2$ (*b*) $n(n-1)/2$
(*c*) $(n^2-1)/2$ (*d*) $(n^2+1)/2$

32. Match List-I with List-II and select the correct answer using the codes given below the lists:

List-I (*Binary*)	**List-II** (*Decimal*)
A. 10101010	1. 128
B. LL110000	2. 240
C. 1000 1000	3. 170
D. 10000000	4. 136

Codes :

	A	B	C	D
(*a*)	3	2	4	1
(*b*)	2	3	1	4
(*c*)	2	4	1	3
(*d*)	3	1	2	4

33. A bulb is controlled by two switches A and B. The bulb glows only when either switch A or switch *b* is ON, its Boolean expression is

(*a*) $A\bar{B} + AB$ (*b*) $\overline{AB} + AB$
(*c*) $\bar{A}B + \overline{AB}$ () $A\bar{B} + \bar{A}B$

34. The 4 bit ripple counter which is composed of flip-flops with a propagation delay of 25 ns, would have the maximum counting speed of

(*a*) 1 MHz (*b*) 4 MHz
(*c*) 10 MHz (*d*) 100 MHz

35. A computer system has cache memory of 64 KB, RAM of 32 KB and RAM of 512 KB.
What is the maximum capacity of the memory available to a programmer?

(*a*) 608 KB (*b*) 576 KB
(*c*) 544 KB (*d*) 512 KB

36. If $\bar{x} \times \bar{y} = 0$, then

(*a*) $\bar{x}y + \bar{y}x + xz = x\bar{y} + yz$

(*b*) $\bar{x}\,\bar{y}\,\bar{z} + xyz = xy\bar{z} + \bar{x}\,\bar{y}\,\bar{z}$

(*c*) $\bar{x}y + \bar{y}x = xy + \bar{x}\,\bar{y}$

(*d*) $\bar{x}yz = 1$

37. The FORTRAN expression corresponding to M^{N^L}, without ambiguity, is

(*a*) M** N** L (*b*) M** (N** L)
(*c*) (M**N)**L (*d*) (N** L)** M

38. Given $(125)_R = (203)_5$, the value of radix R will be

(*a*) 16 (*b*) 10
(*c*) 8 (*d*) 6

39. In which one of the following modes of operation, shift register ceases to work as a shift register?

(*a*) SISO (*b*) SIPO
(*c*) PISO (*d*) PIPO

40. Cycle time of a Random Access Memory is the time required between successive

(*a*) accesses to memory (*b*) clocks
(*c*) delete operations (*d*) Read/Write operations

41. The decimal equivalent of the hexadecimal number $(BAD)_{16}$ is

(*a*) 111013 (*b*) 5929
(*c*) 3416 (*d*) 2989

42. In microprocessor based system, DMA facility is required to increases the speed of data transfer between

(*a*) microprocessor and I/O devices
(*b*) microprocessor and memory
(*c*) memory and HO devices
(*d*) microprocessor, I/O devices and memory

43. In a language in which operations are associated right to left instead of left to right [*i.e., a + b + c = a +* (*b + c*)], value of the expression $7 - [16/(3+1) * 2] - 4$ is

(*a*) 7 (*b*) 1
(*c*) –1 (*d*) –5

44. If $(211)_x = (152)_8$ then the value of base '*x*' is

(*a*) 3 (*b*) 5
(*c*) 7 (*d*) 9

45. The function $F = AB\bar{C} + ABC + \bar{A}BC + \bar{A}B\bar{C}$; can be reduced to which one of the following ?

(*a*) F = A (*b*) F = AB
(*c*) F = ABC (*d*) F = B

46. For memory chips of 16 × 4 size have their address buses connected together. The resulting memory system will be of size

(*a*) 64 × 4 (*b*) 16 × 16
(*c*) 32 × 8 (*d*) 256 × 1

47. The number of flip flops required in a decade counter is

(*a*) 2 (*b*) 3
(*c*) 4 (*d*) 10

48. In a ripple counter using edge-triggered J-K flip-flops, the pulse input is applied to

(*a*) clock input of all flip-flops
(*b*) clock input of one flip-flop
(*c*) J and K inputs of one flip-flop
(*d*) K and K inputs of all flip-flops

49. AD/A converterhas 5 V full-scale output voltage and an accuracy of ± 0.2%. The maximum error for any output voltage

(*a*) will be 5 m V
(*b*) will be 10 mV
(*c*) will be 20 mV
(*d*) can be of any value depending on input

50. The disadvantage of a counter- type A/D converter as compared to simultaneous A/D converter is that

(*a*) the resolution is low
(*b*) longer conversion time is required
(*c*) the circuit is more complex
(*d*) its stability is low

51. For emitter-complied logic, the switching speed is very high because

(*a*) negative logic is used
(*b*) the transistors are not saturated when conducting
(*c*) emitter-coupled transistor are used
(*d*) multi-emitter transistors are used

52. A 4-bit module-16 ripple counter uses J-K Flip-Flop. If the propagation delay of each Flip-Flop is 50-nanoseconds the maximum clock frequency that can be used is equal to

(*a*) 20 MHz (*b*) 10 MHz
(*c*) 8 MHz (*d*) 5 MHz

53. Consider the following statements in respect to ECL gate

1. Its switching speed is high.
2. It provides OR and NOR logic operations.
3. Its power dissipation is small as compared to other logic gates.
4. Its logic levels are compatible with other logic family gates.

Which of these statements are correct ?

(*a*) 1 and 2 (*b*) 1, 2 and 3
(*c*) 1, 2 and 4 (*d*) 3 and 4

54. The tree shown in the given figure relates to a data structure. The nodes l, 2 and 3 are respectively.

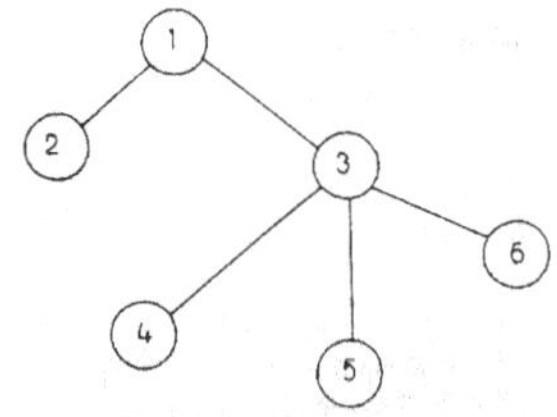

(*a*) parent, root, children (*b*) children, parent, leaves
(*c*) root, children, parent (*d*) root-leaves, parent

55. A half adder has

(*a*) one input and one output
(*b*) two inputs and one output
(*c*) two inputs and two outputs
(*d*) three inputs and two outputs

56. What is the function of control unit in the CPU ?

(*a*) Decoding of program instruction
(*b*) Transfer of data to primary storage
(*c*) Performing logical operations
(*d*) Storage of program instructions.

57. The number of Boolean functions that can be generated by *n* variables is equal to

(*a*) 2^n (*b*) 2^{2^n}
(*c*) 2^{n-1} (*d*) -2^n

58. Consider the flow chart given in the figure: The flow chart is for

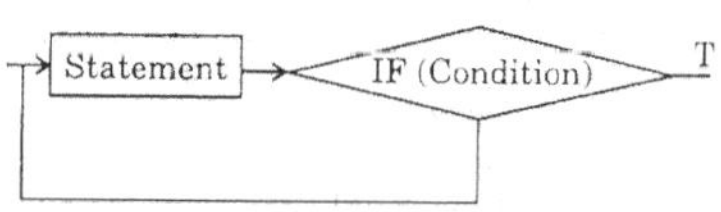

(*a*) IF flow chart is for (*b*) while statements
(*c*) repeat until statement (*d*) Do statement

59. The switching circuit given in the figure can be expressed in binary logic notation as

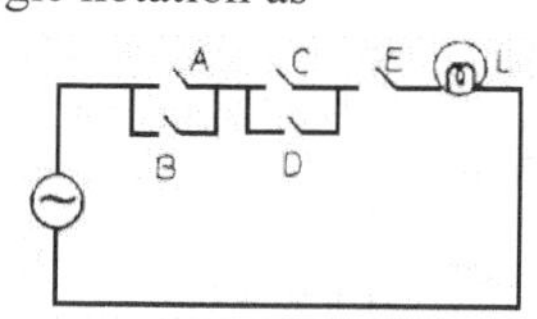

(*a*) L=(A+B)(C+D)E (*b*) L=AB+CD+E
(*c*) L=E+(A+B)(C+D) (*d*) L=(AB+CD)E

60. Which one of the following statements is correct for an 8-bit successive approximation A/D converter?

(*a*) The number of clock cycle required for conversion of any analog signal is 4
(*b*) The number of clock cycle required for conversion of any analog signal is 8
(*c*) The conversion time depends on the analog input signal frequency
(*d*) The clock cycles required for conversion of any analog signal is dependent on signal amplitude

61. Match List-I with List-II and select the correct answer using the codes given below the lists:

List-I	**List-II**
(Boolean identity)	(Boolean expression)
A. $Y\cdot(Y+Z)$	1. $(X+Z)(X+Z)$.
B. $Y+\bar{X}\cdot Z$	2. (Y)
C. $Y\oplus Z$	3. $(Y+Z)$
D. $X+Y\cdot Z$	4. $(X+Z)\cdot(\bar{X}+\bar{Z})$

Codes :

	A	B	C	D
(*a*)	1	2	4	3
(*b*)	2	3	1	4
(*c*)	2	3	4	1
(*d*)	3	2	1	4

62. Which one of the following FORMAT statements is not correctly given?

(*a*) 7890 has format 14

(*b*) 78.90 has format F 5.2

(*c*) 7890 E – 02 has format H 5

(*d*) 7.890 E + 01 has format 9.3[ES]

63. A 4-input multiplexer can be used to implement

(*a*) four combinational functions of 2-variables each

(*b*) two combinational functions of 4-variables each

(*c*) one combinational functions of 4-variables

(*d*) one combinational function of 3-variables

64. If the input to the digital circuit of the figure below, consisting of a cascade of 20 XOR gate is X, then what is the output Y?

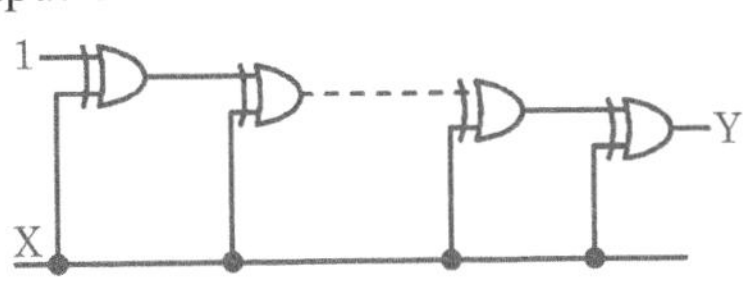

(*a*) 0 (*b*) 1

(*c*) X (*d*) $\bar{X}$

65. For the counter circuit shown in the given figure, the stabilized counting sequence will be

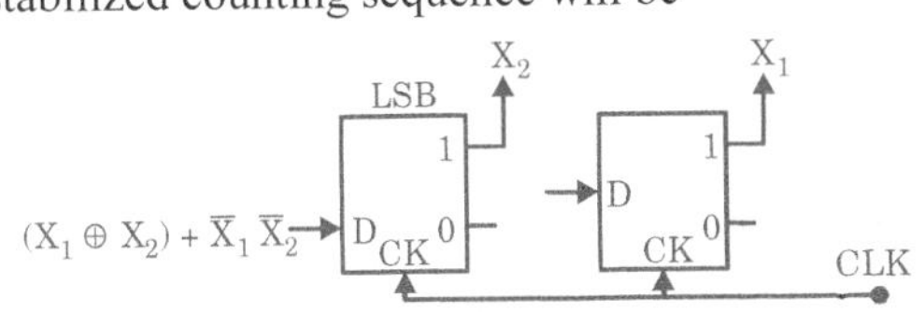

(*a*) 1, 2 repeat (*b*) 1, 2, 3 repeat

(*c*) 3, 2, 1 repeat (*d*) 2, 3, 1 repeat

66. Which of the following statements is not correct?

(*a*) $X+\bar{X}Y=X$

(*b*) $X+(\bar{X}+Y)=XY$

(*c*) $X+X+\bar{Y}=XY$

(*d*) $ZX+Z\bar{X}Y=ZX+ZY$

67. For a JK flip flop, Q_n is output at time step t_n. Which of the following Boolean expression represents Q_{n+1} ?

(*a*) $J_n\bar{Q}_n+\bar{K}_nQ_n$ (*b*) $J_nQ_n+K_n\bar{Q}_n$

(*c*) $\bar{J}_n Q_n+K_n\bar{Q}_n$ (*d*) $J_nQ_n+\bar{K}_n\bar{Q}_n$

68. For reading a 50 × 50 matrix, the valid DIMENSION statement is

(*a*) A (50, 45) (*b*) A (45, 50)

(*c*) A (60, 60) (*d*) None of these

69. Which one of the following is a non-valid BCD code?

(*a*) 0111 1001 (*b*) 0101 1011

(*c*) 01001000 (*d*) 0100 1001

70. Expression $\omega=\dfrac{a^b-c^d}{b-d}$ can be written in FORTRAN language as

(*a*) $\omega=\dfrac{(A**B-C**D)}{(B-D)}$ (*b*) $\omega=\dfrac{(A*B-C*D)}{(B-D)}$

(*c*) $\omega=\dfrac{(A**B+C**D)}{(B+D)}$ (*d*) $\omega=\dfrac{(A*B+C*D)}{(B+D)}$

71. Which of the following storages has greatest capacity with less cost per bit ?

(*a*) Coincidence-current memory

(*b*) Magnetic tape

(*c*) Magnetic drum

(*d*) Semi-conductor memory

72. If $M(ABC)=AB+AC+BC$, then $(\bar{x}_1\bar{x}_2x_3)\ M(\bar{x}_1x_2\bar{x}_3)\ M(x_1\bar{x}_2\bar{x}_3)$ will be

(*a*) $x_1\oplus x_2\oplus x_3$

(*b*) $\bar{x}_1\oplus\bar{x}_2\oplus\bar{x}_3$

(*c*) $\bar{x}_1\bar{x}_2x_3\oplus\bar{x}_1x_2\bar{x}_3\oplus x_1\bar{x}_2\bar{x}_3$

(*d*) $\bar{x}_1\bar{x}_2\oplus\bar{x}_1\bar{x}_3\oplus\bar{x}_2\bar{x}_3$

73. Which of the following are of the combinational logic types?

1. Read Only Memory
2. Random Access Memory
3. Shift Register Memory

Select the correct answer using the codes given below:

Codes :

(*a*) 1 and 2

(*b*) 2 and 3

(*c*) 1 and 3

(*d*) 1, 2 and 3

74. If it is desired to have a 64 × 8 memory and if the memories available are of 16 × 4 size, then the number of memories required will be

(*a*) 8 (*b*) 6

(*c*) 4 (*d*) 2

75. Four logical expressions are given below:

1. $\bar{A}.\bar{B}.\bar{C}.\bar{D}.\bar{E}.\bar{F}.\bar{G}.\bar{H}$
2. $\overline{AB}\cdot\overline{CD}\cdot\overline{EF}\cdot\overline{GH}$
3. $\bar{A}+\bar{B}+\bar{C}+\bar{D}+\bar{E}+\bar{F}+\bar{G}+\bar{H}$
4. $(\bar{A}+\bar{B})(\bar{C}+\bar{D})(\bar{E}+\bar{F})(\bar{G}+\bar{H})$

Two of these expressions are equal. These are

(*a*) 1 and 2 (*b*) 3 and 4
(*c*) 1 and 3 (*d*) 2 and 4

76. Most of the linear *ICs* are based on the two-transistor differential amplifier because of its

(*a*) input voltage dependent linear transfer characteristic
(*b*) high voltage gain
(*c*) high input resistance
(*d*) high CMRR

77. Which one of the following is an example of non-volatile memory?

(*a*) Static RAM (*b*) Dynamic RAM
(*c*) ROM (*d*) Cache memory

78. The output X of the circuit shown in the figure will be

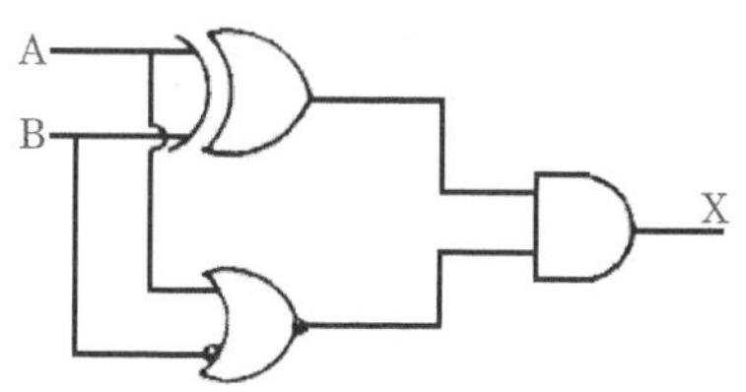

(*a*) AB (*b*) $\bar{A}$ B
(*c*) A $\bar{B}$ (*d*) $\bar{A}\ \bar{B}$

79. The logical expression

A. GE. B AND P. LE. R + 3.0 is equivalent to

(*a*) ((A.GE.B). AND P) LE (R + 3.0)
(*b*) A.GE. (B.AND.P). LER + 3.0
(*c*) (A.GE.B.).AND. (P.LE.(R + 3.0)
(*d*) A.GE. ((B.AND.P). LE. (R + 3.0)

80. Data buffering means

(*a*) a buffer stock of data already stored in a computer unit
(*b*) an electronic circuit which is used to maintain the level of data
(*c*) the data are first collected in an external unit and then processed into the computer
(*d*) application of buffer (unit gain amplifier) at input interface to the computer

81. The 'sum' output in half-adder can be realised by using a single two input gate which should be a

(*a*) exclusive OR gate
(*b*) NOR gate
(*c*) AND gate
(*d*) OR gate

82. The expression $(X+Y)(X+\bar{Y})(\bar{X}+Y)$ is equivalent to

(*a*) $\bar{X}\,\bar{Y}$ (*b*) $\bar{X}Y$
(*c*) $X\bar{Y}$ (*d*) XY

83. After execution of following statements

X = 5
A = 7
IF (A. GT. 5) THEN
IF (A. GT. 10) THEN X = 10
ELSE X = 15

The value assigned to X will be

(*a*) 5 (*b*) 7
(*c*) 10 (*d*) 15

84. The address bus of Inter 8085 is 16 bit wide and hence the memory which can be accessed by this address bus is

(*a*) 2 *k* bytes (*b*) 4 *k* bytes
(*c*) 16 *k* bytes (*d*) 64 *k* bytes

85. Consider the following logic circuit consisting of 2-4 decoder.

for each of decoder $f_0 = 1$ when $i_1 = 0$, $i_2 = 0$
$f_1 = 1$ when $i_1 = 1$; $i_2 = 0$
and so on

Which one of the following gives value of $g(xyz)$?

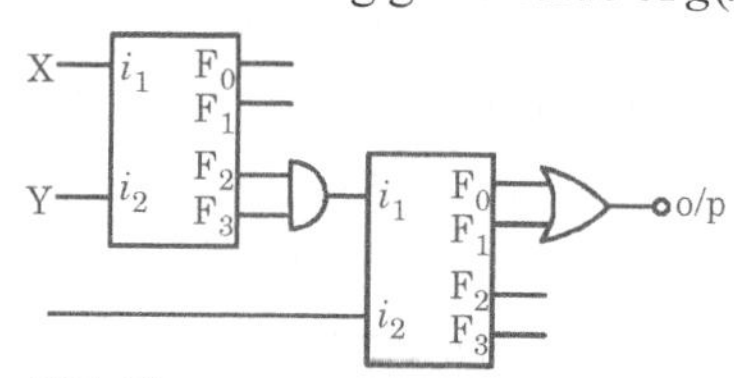

(*a*) $\overline{xy} + \bar{x}\,\bar{y}\cdot\bar{z}$ (*b*) $x \times y \times z$
(*c*) $\left[\overline{xy+\bar{x}\,\bar{y}}\right]\cdot z$ (*d*) 1

86. The maximum time required to perform division using 8-bit binary operand with a clock of 500 MHz is

(*a*) 5.2 μS (*b*) 26 μS
(*c*) 52 μS (*d*) 104 μS

87. One of the measures of speed of the computer system is their MIPS (Million Instruction Per Second) rating. The instruction used for estimating MIPS rating is integer addition. A computer manufacture announces two models of computer system X and Y with following specifications:

Model X : 16 bit, clock frequency 33 MHz, Integer addition time of 4 m/c cycle.

Model Y : 16 bit, clock frequency 66 MHz, Integer addition time 6 m/c cycle.

You may assume m/c cycle to be of the same no. of states for both models. Which one of the following statements is correct?

(*a*) Model Y has MIPS rating doublé that of Model X
(*b*) Model X has MIPS rating one and half times that of model Y
(*c*) Model Y has MIPS rating 4/3 times that of Model X
(*d*) Models X and Y have same MIPS ratings

88. If a mod-6 counter is constructed using 3-flip-flops, the counter will skip

(*a*) 4 counts (*b*) 3 counts
(*c*) 2 counts (*d*) none of the counts

89. Fastest memory cell is

(*a*) Bubble memory
(*b*) Semiconductor RAM
(*c*) Semiconductor memory
(*d*) Superconductor memory

90. The ESC instruction of 8086 may have two formats. In one of the formats, no memory operand is used. Under this format, the number of external op-codes (for the co-processor) which can be specified is-

(*a*) 64 (*b*) 128
(*c*) 256 (*d*) 512

91. The logic function, $f = \overline{A+B} + AB$ is the same as

(*a*) $f=(x+y)\left(\overline{x}+\overline{y}\right)$ (*b*) $f=\overline{\left(\overline{x}+\overline{y}\right)+\left(x+y\right)}$
(*c*) $f=\overline{\left(x.y\right).\left(\overline{x}.\overline{y}\right)}$ (*d*) none of these

92. For the diode matrix shown in the figure the output Y will be

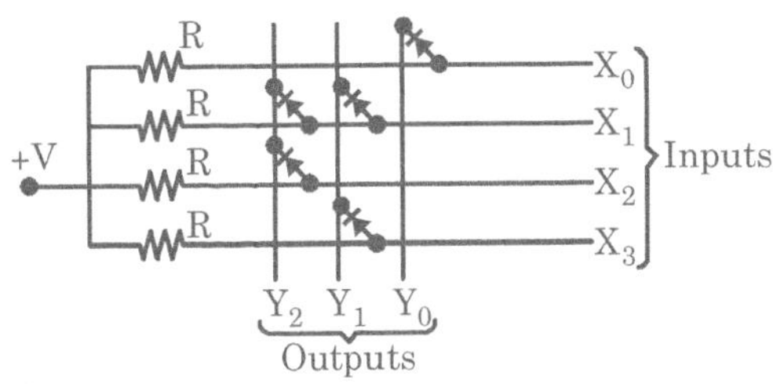

(*a*) X_0X_2 (*b*) X_1X_3
(*c*) X_1+X_3 (*d*) X_2

93. In the circuit shown in the figure, Q = 0 initially. When clock pulses are applied, the subsequent states of 'Q' will be

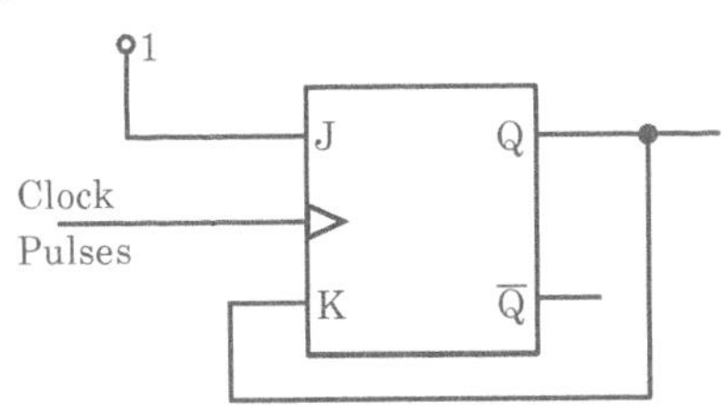

(*a*) 1, 0, 1, 0,........ (*b*) 0, 0, 0, 0,........
(*c*) 1, 1, 1, 1,......... (*d*) 0, 1, 0, 1.....

94. An RRC instruction in 8085 microprocessor instruction set will affect

(*a*) CY, S, Z flag (*b*) CY flag
(*c*) S flag (*d*) Z flag

95. The reduced form of the Boolean expression

$A[B+C(\overline{AB+AC})]$ is

(*a*) $\overline{A}B$ (*b*) $A\overline{B}$
(*c*) AB (*d*) $AB+B\overline{C}$

96. Programmable ROM has a decoder at the input and

(*a*) both these blocks being fully programmable
(*b*) only the former block being programmable
(*c*) only the latter block being programmable
(*d*) both these blocks being partially programmable

97. The number of Flip-flops required to build a mod-19 counter is

(*a*) 4 (*b*) 5
(*c*) 6 (*d*) 7

98. The TRAP is one of the interrupts available in INTEL 8085. Which one of the following statement is true for TRAP ?

(*a*) It is level triggered.
(*b*) It is negative edge triggered.
(*c*) It is positive edge triggered.
(*d*) It is both positive edge triggered and level triggered.

99. The number of bits needed to address 4 K memory is

(*a*) 6 (*b*) 8
(*c*) 12 (*d*) 16

100. Access in magnetic drum memory is

(*a*) completely random
(*b*) sequential and cycle
(*c*) a cyclic sequential
(*d*) partly random and partly cyclic sequential.

101. For a 4- input NOR gate, when only two inputs are to be used , the best option for the unused inputs is to

(*a*) connect them to the ground
(*b*) connect them to V_{cc}
(*c*) keep them open
(*d*) connect them to the used inputs

102. The output of a logic gate is "1" when all its inputs are at logic "O". The gate is either

(*a*) a NAND or an EX-OR gate
(*b*) a NOR or an -OR gate
(*c*) an AND or an EX-NOR gate
(*d*) a NOR or an EX-NOR gate

103. After an operation code has been fetched, which one of the following is the next process that takes place?

(*a*) Reading the operands
(*b*) Displaying the result
(*c*) Decoding the opcode
(*d*) Clearing the accumulator

104. The output of an EX-OR gate with A and B as inputs will be

(*a*) $AB+\overline{AB}$ (*b*) $(A+B)\left(\overline{A+B}\right)$
(*c*) $(A+B)\overline{AB}$ (*d*) $\overline{A+B}+AB$

105. What are the values respectively of R_1 and R_2 in the expression
$(235)R_1 = (565)_{10} = (865)R_2$
(*a*) 8, 16 (*b*) 16, 8
(*c*) 6, 16 (*d*) 12, 8

106. The range of numbers that can be represented in two's complement mode with four binary digits is
(*a*) – 15 to + 15 (*b*) – 8 to + 8
(*c*) – 8 to 7 (*d*) – 7 to + 7

107. A 555 timer can be used as
(*a*) an astable multivibrator only
(*b*) a monostable multivibrator only
(*c*) a frequency divider only
(*d*) an astable multivibrator or a monostable multivibrator or a frequency divider

108. Among the following four, the slowest ADC (analog-to-digital converter) is
(*a*) parallel-comparator (i.e., flash) type
(*b*) successive approximation type
(*c*) integrating type
(*d*) counting type

109. The fastest mode of data transfer from CPU to memory in a microprocessor is
(*a*) memory mapped I/O (*b*) I/O mapped I/O
(*c*) interrupt driven I/O (*d*) DMA

110. For the digital circuit shown in the given figure, the output $Q_3\,Q_2\,Q_1\,Q_0$ = 0001 initially. After a clock pulse appears, the output $Q_3\,Q_2\,Q_1\,Q_0$ will be

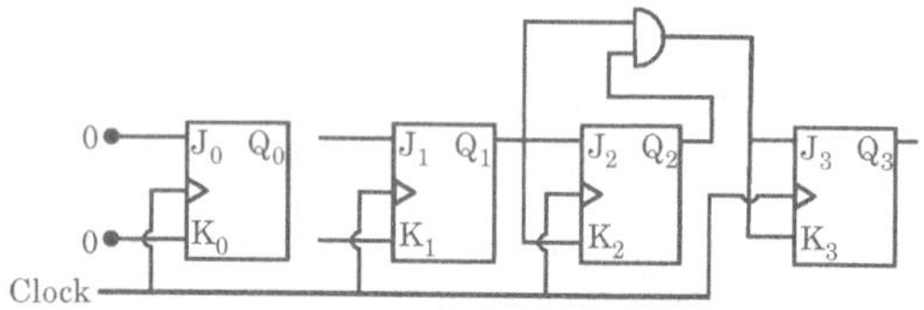

(*a*) 0001 (*b*) 0011
(*c*) 0100 (*d*) 1100

111. The highest priority in 8085 microprocessor system is
(*a*) RST 7.5 (*b*) RST 6.5
(*c*) INTR (*d*) TRAP

112. The binary representation of hexadecimal 'C3' is
(*a*) 1111 (*b*) 110011
(*c*) 111100 (*d*) 11000011

113. Which one of the following sets of status signals is used in an 8085 microprocessor to identify uniquely the seven machine cycle?
(*a*) $IO/\overline{M}$, $\overline{RD}$, $\overline{WR}$
(*b*) S_1, S_0, $\overline{INTA}$
(*c*) $IO/\overline{M}$, S_1, S_0
(*d*) S_2, S_1, S_0

114. In the case of a 16-bit processor, a single instruction is enough to process a function. For processing the same function,
(*a*) more than one 8-bit processors will be required to work in sequence
(*b*) more than one 8-bit processors will be required to work in parallel
(*c*) a long sequence of instructions will be required for a 8-bit processor
(*d*) the same instruction will do for a 8-bit processor also

115. When a program is being executed in an 8085 microprocessor, its Program Counter contains
(*a*) the number of instructions in the current program that have already been executed
(*b*) the total number of instructions in the program being currently executed
(*c*) the memory address of the instruction that is being currently executed
(*d*) the memory address of the instruction that is to be executed next

116. Positive logic in a logic circuit is one in which
(*a*) logic 0 and 1 are represented by 0 and positive voltage respectively
(*b*) logic 0 and 1 are represented by then by negative and positive voltages respectively
(*c*) logic 0 voltage level is higher than logic 1 voltage level
(*d*) logic 0 voltage level is lower than logic 1 voltage level

117. Match List-I which List-II and select the correct answer by using the Codes given below the lists:

List-I	List-II
Hexadecimal	**Octal**
A. 68	1. 150
B. 8C	2. 214
C. 4F	3. 117
D. 4D	4. 135

Codes:

	A	B	C	D
(*a*)	3	1	4	2
(*b*)	1	2	3	4
(*c*)	3	2	4	1
(*d*)	4	1	2	3

118. The number of flip-flops required in a decade counter is
(*a*) 2 (*b*) 3
(*c*) 4 (*d*) 10

119. It is desired to clean up ragged looking pulses that have been distorted during transmission from one place to another. Which of the following device will be appropriate?
(*a*) Multiplexer (*b*) D/A converter
(*c*) JK flip-flop (*d*) Schmitt trigger

120. 3 × 8 decoder with two enable inputs is to be used to address 8 blocks of memory. What will be the size of each memory block when addressed from a sixteen bit bus with two MSBs used to enable the decoder?

(*a*) 2 K (*b*) 4 K

(*c*) 16 K (*d*) 64 K

121. Match **List-I** with **List-II** and select the correct answer by using the codes given below the lists:

List-I (Octal)	List-II (Binary)
A. 75	1. 010110.
B. 65	2. 110101.
C. 37	3. 111101.
D. 26	4. 011111.

Codes:

	A	B	C	D
(*a*)	3	1	4	2
(*b*)	2	1	3	4
(*c*)	3	2	4	1
(*d*)	4	2	3	1

122. In a line printer used to print the output of a computer, the paper is required to be advanced by a line spacing as soon as printing of the line is over. Which one of the following motors is best suited for this application?

(*a*) DC motor (*b*) Synchronous motor

(*c*) Induction motor (*d*) Stepper motor

123. The most economical manner in which a high -speed decade counter can be realised is by using

(*a*) synchronous binary counter

(*b*) ripple binary counter

(*c*) ring counter

(*d*) Johnson counter

124. It is required to construct a 2^n to –1 multiplexer by using 2 to 2 multiplexers only. How many of 2 - to – 1 multiplexers are needed?

(*a*) n (*b*) 2^n

(*c*) 2^{n-1} (*d*) 2^{n+1}

125. There are four Boolean variables x_1, x_2, x_3 and x_4. The following functions are defined on sets of them:

$f(x_1, x_2, x_3) = \Sigma(3, 4, 5)$

$g(x_4, x_3, x_2) = \Sigma(1, 6, 7)$

$h(x_4, x_3, x_2, x_1) = fg$

Then $h(x_4, x_3, x_2, x_1)$ is

(*a*) zero (*b*) $\Sigma(3, 12, 13)$

(*c*) $\Sigma(3, 4, 5, 1, 6, 7)$ (*d*) $\Sigma(3, 12, 15)$

126. Accumulation is a device which

(*a*) stores a number

(*b*) adds two number

(*c*) adds and stores previously stored number to another number

(*d*) None of these.

127. Number of comparators required to build a 5-bit Analog to Digital Converted (ADC) is

(*a*) 5 (*b*) 11

(*c*) 21 (*d*) 31

128. An 8-bit microprocessor uses 2's complement representation. How will the decimal number-41 appear?

(*a*) 1101 1010 DOE (*b*) 1101 1001 DEA

(*c*) 1101 0000 DOH (*d*) 1101 1111 DAH

129. According to De Morgan's second theorem

(*a*) A NAND gate is always compli-mentary to an, AND gate.

(*b*) A AND gate is equivalent to a bubbled NAND gate.

(*c*) A NAND gate is equivalent to a bubbled AND gate.

(*d*) A NAND gate is equivalent to a bubbled OR gate.

130. Consider the following functions of an 8086 microprocessor:

1. Temporary storage of data
2. Storing offset of a memory address in DS.
3. String instructions.
4. JNLE instructions.
5. JCXZ instructions.

Which of these functions require the use of the SI and DI registers?

(*a*) 1, 2 and 3 (*b*) 1, 2 and 4

(*c*) 2, 3 and 5 (*d*) 3, 4 and 5

131. The ASCII code is a

(*a*) 5-bits code (*b*) 7-bits code

(*c*) 9-bits code (*d*) 11-bits code

132. The voltage levels for a negative logic system

(*a*) must necessarily be negative

(*b*) could be negative or positive

(*c*) must necessarily be positive

(*d*) must necessarily be either zero or 5 V.

133. Which one of the following is the truth table of the given logic circuit?

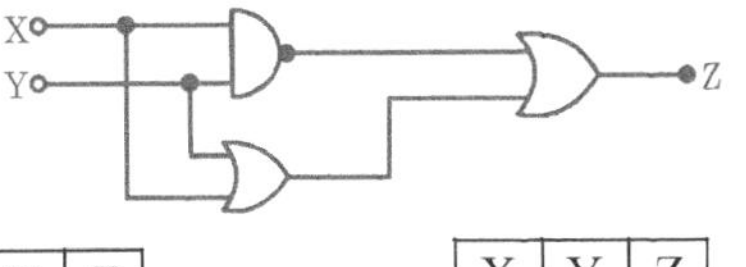

(*a*)

X	Y	Z
0	0	1
0	1	1
1	0	1
1	1	1

(*b*)

X	Y	Z
0	0	0
0	1	1
1	0	1
1	1	1

(*c*)

X	Y	Z
0	0	1
0	1	0
1	0	0
1	1	0

(*d*)

X	Y	Z
0	0	0
0	1	1
1	0	1
1	1	0

134. In a 16-bit microprocessor, words are stored in two consecutive memory locations. The entire word can be read in one operation provided the first

(*a*) word is even

(*b*) word is odd

(*c*) memory location is odd

(*d*) memory address is even.

135. 7483 is a

(*a*) TTL binary adder (*b*) TIL clock

(*c*) 8-bit adder-subtractor (*d*) none of these

136. The states of a RS flip-flop are given in the following table:

States	R	S	Q_n	Q_{n+1}
1	0	0	1	1
2	0	1	0	1
3	1	0	1	0

The mode of operation of states 1, 2 and 3 are respectively

(*a*) indeterminate, set and reset

(*b*) prohibited, set and hold

(*c*) set, hold and reset

(*d*) hold, set and reset.

137. PROM is

(*a*) Permanent read only memory

(*b*) Polarised read only memory

(*c*) Positive read memory

(*d*) Programmable read only memory.

138. Match List-I (type of instruction) with List-II (instruction) and select the correct answer using the codes given below the Lists:

List-I	**List-II**
A. One byte instruction	1. MOV A, data.
B. Two bytes instruction	2. MOV A, B.
C. Three bytes instruction	3. MOV A, M.
D. Register indirect addressing	4. JMP addr.

Codes :

	A	B	C	D
(*a*)	1	2	4	3
(*b*)	2	1	4	3
(*c*)	2	1	3	4
(*d*)	1	2	3	4

139. The larger the RAM of a computer, the faster is its speed, since it eliminates

(*a*) need for ROM

(*b*) need for external memory

(*c*) frequency disk I/O s

(*d*) need for a data-wide path.

140. An 8085 microprocessor based system drives a multiplexed 5-digit, /-segment display. The digits are refreshed at rate of 500 Hz. The ON time for each digit (assumed equal) is

(*a*) 4×10^{-3} S (*b*) 0.4×10^{-3} S

(c) 10^{-2} S (*d*) 2.5×10^{-2} S

141. Control bus is used for transmitting and receiving control signals between

(*a*) processor and key board

(*b*) processor and various devices

(*c*) processor and memory

(*d*) input devices and memory.

142. A subroutine can return

(*a*) only one value (*b*) only two values

(*c*) only three values (*d*) any number of values.

143. The simplified form of a logic function $Y = \overline{(A\overline{B}).(\overline{A}B)}$ is

(*a*) $A+B$ (*b*) AB

(*c*) $\overline{A}+\overline{B}$ (*d*) $\overline{A}B + A\overline{B}$

144. Match List-I with List-II and select the correct answer using the codes given below the lists:

List-I (FORTRAN statements)

A. GO TO statement

B. Arithmetic IF statement

C. Logic IF statement

D. Computed GO TO statement

List-II

1. Three way branching.
2. Unconditional transfer of control.
3. Two way branching.
4. Multi-way branching.

Code :

	A	B	C	D
(*a*)	1	2	3	4
(*b*)	2	4	1	3
(*c*)	2	1	3	4
(*d*)	4	3	1	2

145. An overflow occurs in integer binary arithmetic when

(*a*) a carry comes from the most significant

(*b*) the sign of the first operand is different from the sign of the result

(*c*) exclusive NOR of carry from the most significant bit and carry from the next most significant bit is '1'

(*d*) exclusive OR of carry from the most significant bit and carry from the next most significant bit is '1'.

146. Which one of the following statements is true?

(*a*) Logical errors can be detected during the compilation of a program.

(*b*) Syntax errors are generally detected during the execution of a program.

(*c*) Syntax errors arise from the improper construction of program statement.

(*d*) Logical errors are generally detected during the linking of a program.

147. By placing an inverter between both input of an S-R flip-flop, the resulting flip-flop becomes

(*a*) J-K flip-flop
(*b*) D-flip-flop
(*c*) T-flip-flop
(*d*) Master slaver slave JK flip-flop.

148. Which one of the following statements is true of SP register in 8086 microprocessor ?

(*a*) It can be used for reading instruction from memory.
(*b*) The result of arithmetic operation always goes to SP.
(*c*) It can be used for reversing the string.
(*d*) The control command first goes to SP.

149. For emitter-coupled logic, the switching speed is very high because

(*a*) negative logic is used
(*b*) the transistors are not saturated when conducting
(*c*) emitter-coupled transistors are used
(*d*) multi-emitter transistors are used.

150. For the value of $x = 2, y = 3, z = 4$, the statements

$X = Y + Z, Y = X - Y, Z = Y * Z$

will assign the final computed value to Z as

(*a*) 4 (*b*) 12
(*c*) 15 (*d*) 16

EXERCISE - II

1. The reduced form of the Boolean expression

$A[B + C(\overline{AB + AC})]$ is **[BHEL]**

(*a*) $\bar{A}B$ (*b*) $A\bar{B}$
(c) AB (*d*) $AB + B\bar{C}$

2. Cached and interleaved memories are ways of speeding up memory access between CPUs and slower RAM. Which memory models are best suited (improves performance the most) for which programs?

1. Cache memory is best suited for small loops.
2. Interleaved memory is best suited for small loops.
3. Interleaved memory is best suited for large sequential code.
4. Cached memory is best suited for large sequential code.

(*a*) 1 and 2 are true (*b*) 1 and 3 are true
(*c*) 4 and 2 are true (*d*) 4 and 3 are true **[BEL]**

3. Each instruction in an assembly program has the following fields:

1. Label field 2. Mnemonic field
3. Operand field 4. Comment field

The correct sequence/order of these field is:

(*a*) 1, 2, 3, 4 (*b*) 1, 2, 4, 3
(c) 2, 1, 3, 4 (*d*) 2, 1, 4, 3 **[BEL]**

4. The interface chip used for data transmission between 8086 and a 16-bit ADC is

(*a*) 8259 (*b*) 8255
(*c*) 8253 (*d*) 8251 **[BEL]**

5. Consider the following Assembly Language program:

```
MVI A,    30 H
ACI,      30 H
XRA,      A
POP,      H
```

After the execution of the above program, the contents of the Accumulator will be **[BEL]**

(*a*) 30 H (*b*) 60 H
(*c*) 00 H (*d*) contents of Stack

6. DB, DW and DD directives are used to place data in particular locations or to simply allocate space without preassigning anything to space. The DW and DD directions are used to generate **[BEL]**

(*a*) offsets
(*b*) full address of variables
(*c*) full address of labels
(*d*) offsets or full address of labels and variables [BEL]

7. Match List-I (Request) with List-II (Device) and select the correct answer using the codes given below the Lists:

List-I	**List-II**
(*a*) Interrupt	1. I, O device
(*b*) Wait	2. DMA controller
(*c*) Hold	3. Memory

Code :

	A	B	C
(*a*)	1	2	3
(*b*)	1	3	2
(*c*)	2	1	3
(*d*)	2	3	1

[BHEL]

8. The reference bit is used for the purpose of **[BHEL]**

(*a*) Implementing *LRU* page replacement algorithm
(*b*) Implementing *NRU* (Not Recently used) algorithm
(*c*) To check if the page table entry is in cache memory
(*d*) Check if the page has been written into recently

9. Stack pointer in the 8085 microprocessor is a
   (*a*) 16 bit register that point to stack memory locations
   (*b*) 16 bit accumulator
   (*c*) memory location in the stack
   (*d*) flag register used for the stack **[BHEL]**

10. Consider the following set in instructions:
   STC
   CMC
   MOV A, B
   RAL
   MOV B, A
   This set of instructions
   (*a*) doubles the number in Register by *B*
   (*b*) divides the number in Register by 2
   (*c*) multiples B by A
   (*d*) adds A and B **[BHEL]**

11. The *TRAP* is one of the interrupts available in *INTEL* 8085. Which one of the following statements is true of *TRAP*?
   (*a*) It is a level triggered. **[NTPC]**
   (*b*) It is negative edge triggered.
   (*c*) It is positive edge triggered.
   (*d*) It is both positive edge triggered and level triggered.

12. If we use 3 bits in the instruction word to indicate if an index register is to be used and if necessary, which one is to be used, then the number of index registers to be used in the machine will be
   (*a*) 3 (*b*) 6
   (*c*) 7 (*d*) 8 **[NTPC]**

13. Which of the following are required for a multimedia PC ?
   1. CD-ROM drive, speaker and sound card.
   2. Modem and network card.
   3. Hardware needed to display videos and animation.
   4. Software needed to display videos and animation.
   Select the correct answer using the codes given below:
   Codes:
   (*a*) 1, 2 and 3 (*b*) 1, 2, 3 and 4
   (*c*) 1, 2 and 4 (*d*) 1, 3 and 4 **[NTPC]**

14. A personal computer has typically **[NTPC]**
   (*a*) 5 to 10 kilobytes of main memory
   (*b*) 10 to 100 kilobytes of main memory
   (*c*) 100 to 256 kilobytes of main memory
   (*d*) 256 kilobytes to 1 megabyte of main memory

15. In a 8085 microprocessor, the following sequence of instructions is executed:
   STC
   CMC
   MOVE A, B
   RAL
   MOVE B, A
   After the last instruction, the output will
   (*a*) rotate the contents of the accumulator and store it in B
   (*b*) get the contents of B register into accumulator and rotate it to left by one bit
   (*c*) double contents of *B* register
   (*d*) manipulate carry in A and B **[NTPC]**

16. In 8085 microprocessor system, the direct addressing instruction is
   (*a*) MOV A, B (*b*) MOV B, OAH
   (*c*) MOV C, M (*d*) STA addr **[NTPC]**

17. In 8085 microprocessor, the value of the most significant bit of the result following the execution of any arithmetic or Boolean instruction is stored in the
   (*a*) carry status flag
   (*b*) auxiliary carry status flag
   (*c*) sign status flag
   (*d*) zero status flag **[BEL]**

18. The highest priority in 8085 microprocessor system is
   (*a*) RST 7.5 (*b*) RST 6.5
   (*c*) INTR (*d*) TRAP **[BEL]**

19. How many address lines are needed to address each memory location in a 2048×4 memory chip?
   (*a*) 10 (*b*) 11
   (*c*) 8 (*d*) 12 **[IOCL]**

20. An advantage of memory interlacing is that
   (*a*) a larger memory is obtained
   (*b*) effective speed of the memory is increased
   (*c*) the cost of the memory is reduced
   (*d*) a non-volatile memory is obtained **[IOCL]**

21. If an original MIB microcomputer operates at 5 MHz with an B bit bus and a newer version operates at 20 MHz with a 32 bit bus, compute (approximately) the maximum speed-up possible
   (*a*) 2 (*b*) 4
   (*c*) 6 (*d*) 16 **[IOCL]**

22. Consider the following statements:
   1. A total of about one million bytes can be directly addressed by the 8086 microprocessor.
   2. The 8086 has thirteen 16-bit registers.
   3. The 8086 has eight flags.
   4. Compared to 8086 the 80286 provides a higher degree of memory protection.
   Of these statements:
   (*a*) 2, 3 and 4 are correct
   (*b*) 1, 3 and 4 are correct
   (*c*) 1, 2 and 4 are correct
   (*d*) 1, 2 and 3 are correct **[BHEL]**

23. In an 8085 microprocessor, after the execution of XRA A instruction
(*a*) the carry flag is set [GATE]
(*b*) the accumulator contains FF_H
(*c*) the zero flag is set
(*d*) the accumulator contents are shifted to left by one bit

24. A Direct Memory Access (DMA) transfer implies
(*a*) Direct transfer of data between memory and accumulator
(*b*) Direct transfer of data between memory and HO devices without the use of microprocessor
(*c*) Transfer of data exclusively within microprocessor registers
(*d*) A last transfer of data between microprocessor and HO devices. [NTPC]

25. A *n*-bit A/D converter is required to convert an analog input in the range 0-5 Vto an accuracy of 10 mV. The value of *n* should be
(*a*) 16 (*b*) 10
(*c*) 9 (*d*) 8 [NHPC]

26. D Flip-Flop can be made from a J-K Flip-Flop by making
(*a*) J = K (*b*) J = K = 1
(*c*) J = 0, K = 1 (*d*) $J = \overline{K}$ [BHEL]

27. The computer program which converts statements written in high level language to object code is known as
(*a*) assembier
(*b*) operating system
(*c*) object-oriented software
(*d*) none of these [GATE]

28. A microprocessor or is ALU [NHPC]
(*a*) and control unit on a single chip
(*b*) and memory on a single chip
(*c*) register unit and I/O device on a single chip
(*d*) register unit and control unit on a single chip

29. The output of *a* JK Flip-flop toggles when
(*a*) J = 1, K = 0 (*b*) J = 0, K = 1
(*c*) J = 1, K = 1 (*d*) J = 0, K = 0 [NHPC]

30. In Intel 8085 A microprocessor ALE signal is made high to
(*a*) Enable the data bus to be used as low order address bus.
(*b*) To latch data D_0-D_7 from data bus
(*c*) To disable data bus
(*d*) To achieve all the functions listed above [BEL]

31. How many and which types of machines cycles are needed to execute *PUSH PSW* by an Intel 0885 *A* microprocessor?
(*a*) 2, Fetch and Memory write
(*b*) 3, Fetch and 2 Memory write
(*c*) 3, Fetch and 3 Memory write
(*d*) 3, Fetch, Memory read and Memory write [BEL]

32. Which one of the following is not a vectored interrupt ?
(*a*) TRAP (*b*) INTR
(*c*) RST 3 (*d*) RST 7.5 [NTPC]

33. After an arithmetic operation, the Flag register of a 8085 microprocessor has the following look:

D_7	D_6	D_5	D_4	D_3	D_2	D_1	D_0
1	0	×	1	×	0	×	1

The arithmetic operation has resulted in [NTPC]
(*a*) A carry and an odd parity has resulted in
(*b*) Zero and the auxiliary carry Flag being set
(*c*) A number with even parity and 1 as the MSB
(*d*) A number with odd parity and 0 as the MSB.

34. Floppy disc is a [NTPC]
(*a*) flexible storage (*b*) permanent storage
(*c*) ROM (*d*) Last storage

35. Dynamic memory cells are constructed using [NTPC]
(*a*) transistors (*b*) Flip-flop
(*c*) MOSFET (*d*) FET*s*

36. The program counter in a 8085 microprocessor is a 16-bit register because
(*a*) it counts 16 bits at a time
(*b*) here are 16 address at a time
(*c*) it facilitates the user storing 16-bit data temporarily
(*d*) it has to fetch two 8-bit data at a time [BHEL]

37. Which of the following 8085 assembly language instructions does not affect the content of the accumulator?
(*a*) CMA (*b*) CMPM
(*c*) DAA (*d*) A [IOCL]

38. A ring counter is same as [IOCL]
(*a*) up-down counter (*b*) parallel counter
(*c*) shift register (*d*) None of these

39. Which of the following logics exhibit relatively highest power dissipation per gate?
(*a*) TTL (*b*) ECL
(*c*) CMOS (*d*) LSI [IOCL]

40. Which List-I (Interrupts) with List-II (corresponding characteristics) and select the correct answer using the codes given below the lists:

List-I	**List-II**
A. TRAP	1. Level triggered
B. INTR	2. Non maskable
C. RST 7.5	3. For increasing the number of interrupts
D. RST 6.5	4. Positive edge triggered

Codes : [BEL]

	A	B	C	D
(*a*)	2	4	3	1
(*b*)	1	4	3	2
(*c*)	1	3	4	2
(*d*)	2	3	4	1

41. Choose the correct statement [BEL]
(*a*) Hardware is merely the preparation of printed circuit board for the connection of micro processor mP
(*b*) Software is to provide the required IC chips for μP
(*c*) Hardware is to develop programmes and to feed in to the computer
(*d*) Software is the impermanent in format conal structure

42. According toBoolean algebra
$(1 + A + B + C)$ is equal to
(*a*) A+B+C (*b*) ABC
(*c*) 1+ABC (*d*) 1 [HAL]

43. The ECL can be used to switch frequencies as high as
(*a*) 1 MHz (b) 100 MHz
(*c*) 500 MHz (*d*) 10 MHz [HAL]

44. Which one of the following statement for Intel 8085 is correct? [BEL]
(*a*) Program counter (PC) specifies the address of the instruction last
(*b*) PC specifies the address of the instruction being executed
(*c*) PC specifies the address of the instruction to be executed
(*d*) PC specifies the number of instructions executed so for

45. Which one of the following statements is correct?
(*a*) ROM is a Read/Write memory
(*b*) PC points to the last instruction that was executed
(*c*) Stock works on the principle of LIFO
(*d*) All instruction affect the flags [BEL]

46. The number 149 in octal code will be
(*a*) 154 (*b*) 178
(*c*) 254 (*d*) 225 [BHEL]

47. If $(327)_9 = (x)_5$, then the value of x given by [BEL]
(*a*) 327 (*b*) 268
(*c*) 2033 (*d*) 3302

48. Which one of the following logic families is the fastest ?
(*a*) I^2L (*b*) ECL
(*c*) TTL (*d*) CMOS [BHEL]

49. After completing the execution, microprocessor returns to
(*a*) Halt state (*b*) Fetch state
(*c*) Execute state (*d*) Interrupt state [BHEL]

50. A DADH instruction is same as [NHPC]
(*a*) Shifting each bit one position to the left
(*b*) Shifting each bit one position to the right
(*c*) Shifting each bit one position to the left with a zero inserted in LSB position
(*d*) Shifting each bit one position to the right with a zero inserted in LSB position

51. A program counter is a storage register for [NHPC]
(*a*) location of data in memory
(*b*) location of instruction in memory
(*c*) binary code for the operation to be performed
(*d*) address of the next instruction to be executed

52. A block of successive memory locations that is accessible on a last-in, first-out basic is called
(*a*) accumulator (*b*) register
(*c*) pointer (*d*) stack [NHPC]

53. The propagation delay of a digital IC is the difference between
(*a*) initiating clock pulse and triggering of gates
(*b*) the application of input and its presence at the output
(*c*) the triggering of a Flip-flops and change in its state
(*d*) None of these [NHPC]

54. Static RAM is preferred over dynamic RAM when the requirement is of
(*a*) slow speed of operation
(*b*) larger storage capacity
(*c*) lower access time
(*d*) lower power consumption. [IAS]

55. In a microprocessor, the address of the next instruction to be executed, is stored in
(*a*) stack pointer
(*b*) address latch
(*c*) program counter register
(*d*) general purpose [GATE]

56. In the 2's complement, Adder-Subtractor [NTPC]
(*a*) sign magnitude number represent negative numbers
(*b*) the sign magnitude number have '1' as leading bit
(*c*) the sign magnitude numbers have '0' as leading bit
(*d*) 2's complement represents positive numbers

57. Given two numbers A and B in sign magnitude representation in an eighth bit form A = 00011110; B = 10011100, the corresponding decimal number respectively are
(*a*) 30 and 156 (*b*) 14 and – 12
(*c*) 30 and – 28 (*d*) – 28 and 30 [NHPC]

58. Given the decimal number – 19, an eight bit 2's complement representation is given by
(*a*) 11101110 (*b*) 11101101 [NTPC]
(*c*) 11101100 (*d*) none of these

59. In one of the code for transfer of numbers, the code for the succeeding number differ from that of the number only in the change of a single digit. The code under consideration is
(*a*) ASCII (*b*) Excess 3 Code [NTPC]
(*c*) Gray Code (*d*) None of these

60. The Gray Code for number 6 is
(*a*) 1100 (*b*) 1001
(*c*) 0101 (*d*) 0110 **[NHPC]**

61. In the binary number 110.101, the fractional part has the value
(*a*) 0.125 (*b*) 0.625 **[NTPC]**
(*c*) 0.875 (*d*) 0.5

62. A four bit number is given as 0110. Its 2's complement is
(*a*) 1001 (*b*) 1000
(*c*) 1010 (*d*) none of these **[BEL]**

63. Indicate which of the following logic gate can be used to realise all possible combinational logic functions ?
(*a*) OR gates only (*b*) NAND gates only
(*c*) EX-OR gates only (*d*) NOR gates only **[BEL]**

64. The 'excess 3 code' for number 3 is
(*a*) 0110 (*b*) 0011
(*c*) 0001 (*d*) 1001 **[BEL]**

65. In the 2's complement Adder-Subtractor given the numbers

X = 00011000 and Y = 11110000.

They stand respectively for
(*a*) 24 and 240 (*b*) 18 and 15
(*c*) 6 and – 15 (*d*) 24 and – 16 **[IOCL]**

66. The Gray Code for number 2 is
(*a*) 0010 (*b*) 0011
(*c*) 1000 (*d*) 1001 **[NTPC]**

67. Which one of the following is not a vectored interrupt?
(*a*) TRAP (*b*) INTR **[NTPC]**
(*c*) RST 7.5 (*d*) RST 3

68. Given the binary 00000111, its 2's complement is obtained and expressed in hexadecimal form. The result is
(*a*) F8 (*b*) F9 **[NHPC]**
(*c*) – 79 (*d*) none of these

69. By taking 2's complement again of the 2's complement of a binary, one gets
(*a*) the 1's complement
(*b*) the 2's complement
(*c*) the original number
(*d*) the sign magnitude form of the number **[ES]**

70. A four bit number is given as 1001. Its 1's complement is
(*a*) 1001 (*b*) 1110
(*c*) 0110 (*d*) 0111 **[NHPC]**

71. The ASCII Code is for information interchange by a binary code for
(*a*) numbers only
(*b*) alphabets only
(*c*) alpha numerics and other common symbols
(*d*) none of these **[NHPC]**

72. In 8085 microprocessor, the value of the most significant bit of the result following the execution of any arithmetic or Boolean instruction is stored in the
(*a*) carry status flag
(*b*) auxiliary carry status flag
(*c*) sign status flag
(*d*) zero status flag **[NHPC]**

73. An Intel 8085 processor is executing the program given below.

```
        MVI/A, 10H
        MVI/B, 10H
BACK :  NOP
        ADD/B
        RLC
        JNC BACK
        HLT
```

The number of times that the operation NOP will be executed is equal to
(*a*) 1 (*b*) 2 **[NHPC]**
(*c*) 3 (*d*) 4

74. Match List-I (Type of instruction) with List-II (Instruction) and select the correct answer using the codes given below the Lists:

List-I	**List-II**
A. One byte instruction	1. MOV A, data.
B. Two byte instruction	2. MOV A, B
C. Three byte instruction	3. MOV A, M.
D. Register indirect addressing	4. JMP addr.

Codes : **[NHPC]**

	A	B	C	D
(*a*)	1	2	4	3
(*b*)	2	1	4	3
(*c*)	2	1	3	4
(*d*)	1	2	3	4

75. In a 16-bit microprocessor, words are stored in two consecutive memory locations. The entire word can be read in one operations provides the first
(*a*) word is even
(*b*) word is odd
(*c*) memory location is odd
(*d*) memory address is even **[NHPC]**

76. In a parallel processing system with processors, each processor makes a request at most every 2 microseconds and each request can ask for a maximum of 64 bits. These requests are made over a high speed shared but with a maximum throughput of 100 megabits. Assume that with each request, 40 bits are required for transmission of the data-address and protocol. Approximately, what is the maximum value of *k* for which the system can operate without any delays at peak load?
(*a*) 1 (*b*) 2 **[NHPC]**
(*c*) 3 (*d*) 4

77. In a multiprocessor configuration, two co-processors are connected to the host 8086 processor. The two co-processor instruction sets

(*a*) must be the same

(*b*) may be overlap

(*c*) must be disjoint

(*d*) must be the same as that of the host **[NHPC]**

78. An interrupt in which the external device supplies its address as well as the interrupt request, is known

(*a*) vectored interrupt

(*b*) maskable interrupt

(*c*) polled interrupt

(*d*) non-maskable interrupt **[NHPC]**

79. Consider the following statements:

I. A 33 MHz 486 has a higher MIPS rating than a 33 MHz 386.

II. A 33 MHz 486 has a higher MFLOPS rating than a 33 MHz 386.

Of these statements

(*a*) both I and II are true

(*b*) both I and II are false

(*c*) I is true but II is false

(*d*) I is false but II is true **[NHPC]**

80. Number of bits needed to address 4 K memory is

(*a*) 6 (*b*) 8

(*c*) 12 (*d*) 16 **[BEL]**

81. Block search is used to locate a record in an order file consisting of N records. The optimum size of the block is

(*a*) $\sqrt{N}$ (*b*) $3\sqrt{N}$

(*c*) $\frac{N}{2}$ (*d*) $\frac{N}{3}$ **[BEL]**

82. If "T" is the duration of the linear sweep and 'τ' is its time constant, then the slope error 'e_s' is given by

(*a*) $e_s = \frac{\tau}{T}$ (*b*) $e_s = \frac{T}{\tau}$

(*c*) $e_s = \frac{8\tau}{T}$ (*d*) $e_s = \frac{8T}{\tau}$ **[ES]**

83. A central processing unit of a computer contains

(*a*) control and arithmetic logic unit

(*b*) control and RAM

(*c*) control, clock, RAM and ROM

(*d*) control, ALU, clock, RAM and ROM **[BEL]**

84. Following two function are to be realized by using a 3 L : 8 L demultiplexer and additional appropriate type of gates that would result in minimum chip count.

f_1(A, B, C) = ΠM (1, 3, 5, 7)

f_2(A, B, C) = Σm (0, 2, 4, 5, 6, 7)

Which one of the following gates would fulfil this requirement?

(*a*) One 4-input NAND + one 8-input NAND

(*b*) Three 2-input ANDs

(*c*) Seven 2-input NANDs

(*d*) One 4-input AND + one 8-input NAND **[ES]**

85. The race-around conditions is NOT presented in

(*a*) RS flip-flop

(*b*) JK flip-flop

(*c*) D flip-flop

(*d*) master-slave JK flip-flop. **[IAS]**

86. In computer terminology 1M Byte memory means

(*a*) 1000000 bytes (*b*) 1000024 bytes **[ES]**

(*c*) 1024000 bytes (*d*) 1048576 bytes

87. If negative logic is used, the diode gate shown in the given figure will represent

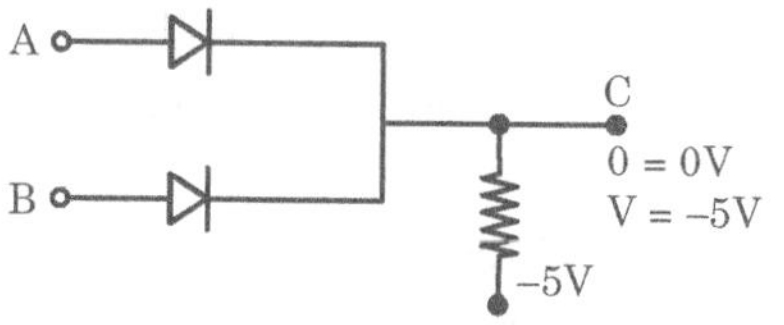

(*a*) OR gate (*b*) AVD gate

(c) NOR gate (*d*) NAND gate **[ES]**

88. In a time base generator, a constant current source (I) is used to drive a capacitance (C). The output voltage across the capacitance as a function of time (t) is given by

(*a*) $\frac{I}{C}t$ (*b*) $\frac{C}{I}t$

(*c*) ICt (*d*) $\frac{t}{IC}$ **[IAS]**

89. Consider the following statements:

A function subprogram is used because the

1. value of only one variable is to be computed
2. values of more than one variable are to be computed.
3. subprogram consists of more than one statement.
4. name of the variable is the same as the name of the subprogram.

Of these statements

(*a*) 1 and 3 are correct

(*b*) 2, 3 and 4 are correct

(*c*) 1, 3 and 4 are correct

(*d*) 2 and 4 are correct **[ES]**

90. A differentiator is rarely used in analog computers, because it

(*a*) reduces the gain

(*b*) decreases the output of the amplifier

(*c*) amplifier noise, drift and other unwanted disturbances

(*d*) may cause oscillation. **[IAS]**

91. Two Boolean variables FLAG I and FLAG 2 are assigned values FALSE and TRUE respectively. Which one of the following expressions is valid and has a TRUE value?

(*a*) FLAG I . AND . FLAG 2
(*b*) FLAG I < FLAG 2
(*c*) FLAG I OR FLAG 2
(*d*) FLAG I > FLAG 2. **[ES]**

92. Match List-I with List-II and select the correct answer using the codes given below the lists

List-I	List-II
A. Immediate addressing	1. LDA 30 SC
B. Implied addressing	2. MOV A, B
C. Register addressing	3. LXI H, 2050
D. Direct addressing	4. RRC

Codes :

	A	B	C	D
(*a*)	3	2	1	4
(*b*)	3	4	2	1
(*c*)	3	2	4	1
(*d*)	2	3	1	4

[IAS]

93. Match List-I and List- II containing typical computer instructions, and select the correct answer using the codes given below the list:

List-I

A. [ADD] [X]

B. [ADD] [X] [Y] [Z]

List-II

1. Two-address instruction.
2. Three-address instruction.
3. Four-address instruction.
4. One-address instruction.

Codes:

	A	B
(*a*)	1	2
(*b*)	4	1
(*c*)	4	2
(*d*)	2	3

[ES]

94. The circuit given in the figure is to be used to implement the function Z = *f*(A, B) = $\bar{A}$ + B. What value should be select for I and J?

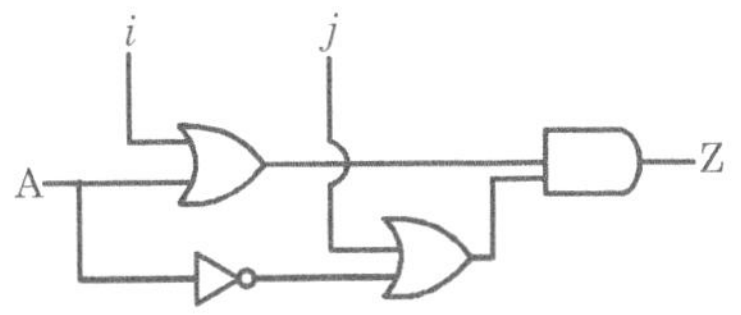

(*a*) I = O, J = B (*b*) I = I, J = I
(*c*) I = B, J = I (*d*) I = $\bar{B}$, J = O **[ES]**

95. The Boolean expression for the output Y in the logic circuit is

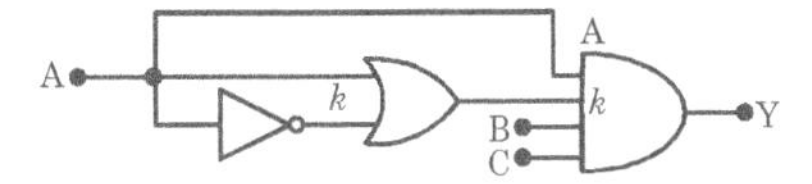

(*a*) $A\bar{B}C$ (*b*) ABC
(*c*) $\bar{A}BC$ (*d*) $\overline{ABC}$ **[ES]**

96. Three devices A, B and C have to be connected to a 8085 microprocessor. Device A has highest priority and device C has the lowest priority. In this context which of the following is correct assignment of interrupt? **[GATE]**

(*a*) A uses TRAP, B uses RST 5.5 and C uses RST 6.5
(*b*) A uses RST 7.5, B uses RST 6.5 and C uses RST 5.5
(*c*) A uses RST 5.5, B uses RST 6.5 and C use RST 7.5
(*d*) A uses RST 5.5, B uses RST 6.5 and C uses TRAP

97. The contents of the control world register of INTEL 8253 (Programmable Interval Timer) is given below:

D7	D6	D5	D4	D3	D2	D1	D0
0	1	1	0	0	0	0	0

This control word sets **[ES]**

(*a*) Counter 0 to Mode 0 (*b*) Counter 1 to Mode 0
(*c*) Counter 2 to Mode 0 (*d*) Counter 0 to Mode 1

98. In 8085 microprocessor system, the direct addressing instruction is

(*a*) MOV A, B (*b*) MOV B,OAH
(*c*) MOV C, M (*d*) SAT addr **[ES]**

99. There is a sequential circuit which has "N" number of states. If states S_i and S_j can be distinguished by an input sequence of length n,

(*a*) $n \geq N$ (*b*) $n \geq N - 1$
(*c*) $n \leq N$ (*d*) $n \leq N - 1$ **[IAS]**

100. If the number of bits in input and output codes is 4 and 8 respectively for a ROM, then the memory of this chip is equal to

(*a*) 12 bits (*b*) 32 bits
(*c*) 128 bits (*d*) 256 bits **[ES]**

101. The number of flip-flop required to build a mod-19 counter is

(*a*) 4 (*b*) 5
(*c*) 6 (*d*) 7 **[ES]**

102. For the JK flip-flop shown in the given figure. J = $\bar{Q}$ and K = 1. If it is initially cleared and then clocked for 6 pulses, the sequence at the Q-output will be

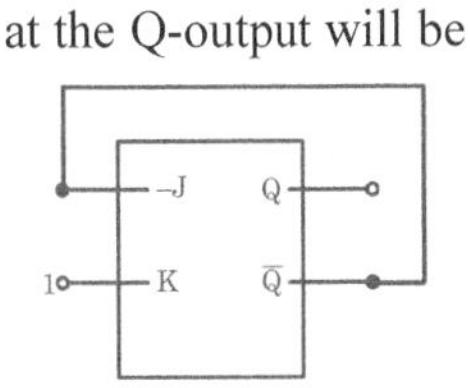

(*a*). 010101 (*b*) 010000
(*c*) 011001 (*d*) 010010 **[IAS]**

103. A sequence of instruction is executed as follows : begin ecode the instruction fetched:

if operand in memory then fetch operand; execute instruction;

This sequence of instruction has performed

(*a*) fetch cycle

(*b*) both fetch and execute cycles

(*c*) execute and decode cycles

(*d*) execute cycle **[ES]**

104. In the combinational logic circuit shown in the given figure, the input square wave is phase-inverted at the output when the control input is 1. It is passed on as it is when the control input is 0. The logic circuit is a 2-input

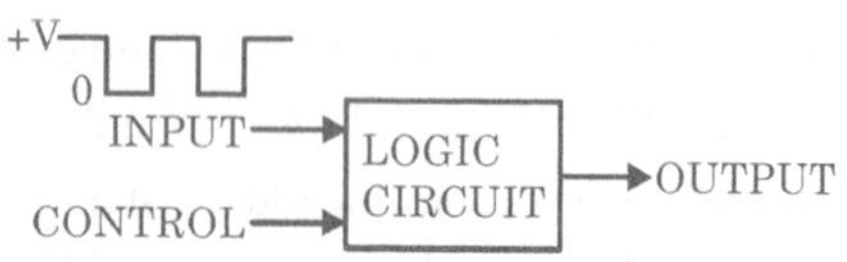

(*a*) NAND gate (*b*) NOR gate **[IAS]**

(*c*) Exclusive OR gate (*d*) Exclusive NOR gate.

105. Consider the following assembly language code for a simple computer having one register only: **[BEL]**

LOAD X

ADD Y

STORE T1 STORE T2

MULTZ MULT Tl

ADD Tl STORE R

Which one of the following arithmetic expressions does it evaluate?

(*a*) $R : Z^2(1 + X + Y)^2$

(*b*) $R : Z^2(X + Y)^2 + (X + Y)$

(*c*) $R : Z(X + Y)^2 + (X + Y)^2$

(*d*) $R : Z(X + Y)^2 + (X + Y)$[ES]

106. In case of dynamic memory **[ES]**

(*a*) contents tend to decay over a period of time

(*b*) contents are retained without distortion

(*c*) power consumption is low

(*d*) the speed is low as compared to static memory.

107. The INTEL 8080 microprocessors has a set of 91 instructions. The op-code to implement this instruction set shouldbe of at least

(*a*) 6 bits (*b*) 4 bits

(*c*) 7 bits (*d*) 8 bits. **[IAS]**

108. Given that N is the actual number of instructions executed and CPI is the average number of cycles per instruction, then

(*a*) both CISE (Complex instruction set computer) and R1SE (Reduced instruction set computer) processors aim to reduce N at the expense of CPI

(*b*) CISE processors aim to reduce N at the expense of CPI whereas RISC processor aim to reduce CPI at the expense of N

(*c*) RISC processors aim to reduce N at the expense of CPI whereas CAST processors aim to reduce CPI at the expense of N

(*d*) both CISC and RISC processors ai m to reduce CPI at the expense of N **[IAS]**

109. The binary equivalent of hexadecimal number 4F2D is

(*a*) 0101111100101100 (*b*) 0100111100101100

(*c*) 0100111000101101 (*d*) 0100111100101101 **[ES]**

110. For $x \geq 0, y \geq 0$, A (x, y) is defined as under:

$A(0, y) = y + 1$

$A(x + 1, 0) =$ $A(x, 1)$, and

$A(x + 1, y + 1) = A(x, A(x + 1, y))$

Then, for all positive integer values of y, A $(1, y)$ is

(*a*) 2 (*b*) $y + 1$

(*c*) $y + 2$ (*d*) $2y + 3$ **[ES]**

111. After completing the execution, microprocessor returns to

(*a*) Halt state (*b*) Fetch state **[ES]**

(*c*) Execute state (*d*) Interrupt state

112. Which of the following hexagonal sum is equivalent to hexadecimal A 8/7?

(*a*) 2 CH + 4 FH (*b*) 5 EH + 1 AH

(*c*) 3 BH + 6 DH (*d*) 5 AH + 2 CH **[ES]**

113. A full adder circuit may be constructed by using

(*a*) two 2-input AND gates and two 3-input OR gates

(*b*) two half adder and a 2-input AND gate **[IES]**

(*c*) two half adder and a 2-input OR gate

(*d*) two 2-input AND gates and two3-input XOR gales

114.

AB \ CD	00	01	11	10
00	1	1	0	1
01	0	1	0	0
11	0	0	0	0
10	1	1	0	1

The Karnaugh map of a Boolean function is shown in the above figure. The simplified function is **[IES]**

(*a*) AB + CD + BD

(*b*) A′D′ + B′C + A′C

(*c*) (A′+ B′) (C + D′) (B′+ D)

(*d*) (A + B′) (C + D′)(B + D′)[IAS-2000]

115. Volatile storage is a device in which data in which data are lost **[IES]**

(*a*) with time

(*b*) when power is removed

(*c*) when programma is over

(*d*) None of these.

116. The decimal equivalent of hexadecimal number 2A OF is

(*a*) 17670 (*b*) 17607
(*c*) 17067 (*d*) 10767 **[ES]**

117. Which one of following signal combinations will generate $\overline{\text{MEMR}}$ signal?

(*a*) $\text{IO}/\overline{\text{M}} \wedge \overline{\text{RD}}$ (*b*) $\text{IO}/\overline{\text{M}} \vee \overline{\text{RD}}$

(*c*) $\overline{\text{IO}/\overline{\overline{\text{M}}}} \vee \overline{\text{RD}}$ (*d*) $\overline{\text{IO}/\overline{\text{M}}} \wedge \overline{\text{RD}}$ **[ES]**

118. The circuit schematic shown in the following figure represents a 4-bit its Karnaugh Map is a as shown. The output of the circuit is given by **[IES]**

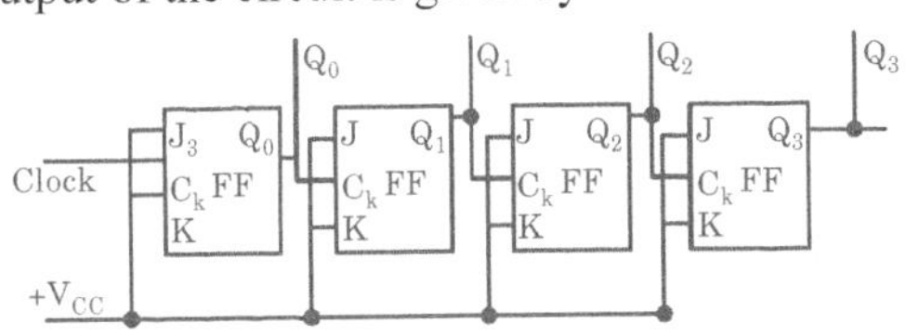

(*a*) static shift register (*b*) dynamic shift register
(*c*) asynchronous counter (*d*) synchronous counter

119. A combinational circuit has input A, B and C and its Karnoaugh Map is a as shown. The output of the circuit is given by

C \ AB	00	00	11	10
0		1		1
1	1		1	

(*a*) $\left(\overline{\text{A}}\text{B}+\text{A}\overline{\text{B}}\right)\text{C}$ (*b*) $\left(\overline{\text{A}}\text{B}+\text{A}\overline{\text{B}}\right)\overline{\text{C}}$

(*c*) $\overline{\text{ABC}}$ (*d*) $\text{A} \oplus \text{B} \oplus \text{C}$ **[ES]**

120. Theoutput Q_n of a J-K flip-flop is zero. Itchanges to 1when a clock pulse is applied. The input J_n, and K_n are respectively

(*a*) 1 and X (*b*) 0 and X
(*c*) X and 0 (*d*) X and 1 **[ES]**

121. To realise the given truth table from the circuit shown in the figure, the input to J in terms of A and B would have to be

TRUTH TABLE

A	B	Q_{n+1}	J
0	0	Q_n	0
0	1	1	1
1	0	Q_n	0
1	1	0	X

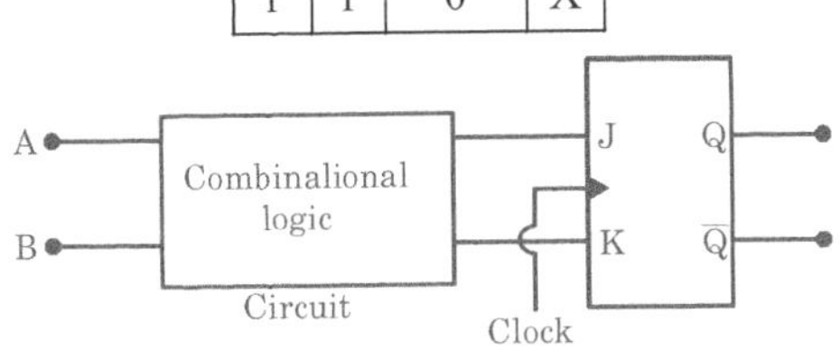

(*a*) $\overline{\text{AB}}$ (*b*) $\overline{\text{A}}$
(c) $\overline{\text{B}}$ (*d*) AB **[ES]**

122. A, B, C and D are input bits and Y is the output bit in the XOR gate circuit of the figure below. Which of the following statements about the sum S of A, B, C, D and Y is correct ?

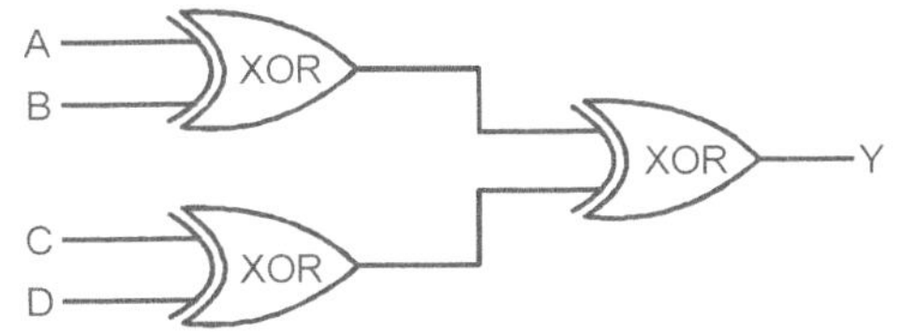

(*a*) S is always either zero or odd **[GATE]**
(*b*) S is always either zero or even
(*c*) S =1 only if the sum of A, B, C and D is even
(*d*) S = 1 only if the sum of A, B, C and D is odd

123. 2,764 is 65, 536-bit EPROM organised as 8,192 words of 8 bits each. It has

(*a*) 1 address lines and 10 data lines
(*b*) 12 address lines and 10 data
(*c*) 13 address lines and 8 data lines
(*d*) 15 address lines and 12 data lines **[ES]**

124. All to the following are non-volatile memories EXCEPT

(*a*) ROMs (*b*) Semiconductor RAM
(*c*) PROMs (*d*) EPROMs **[ES]**

125. Direct-memory-access channel faclitates data to move into and out of the system

(*a*) without subroutine
(*b*) with equal time delay
(*c*) without programme intervention
(*d*) on first come first serve basis **[ES]**

126. A semiconductor Read-only-Memory basically is

(*a*) a sequential circuit with flip-flop **[GATE]**
(*b*) a sequential circuit with flip-flop and gates
(*c*) a set of flip-flop memory elements
(*d*) a combinational logic circuit

127. Scratch-pad memory is

(*a*) first-in first-out memory
(*b*) last-in first-out memory
(*c*) local permanent memory
(*d*) local temporary memory **[ES]**

128. Given an algebraic or transcendental equation, exact solution of which is NOT obtainable mathematically, the computer

(*a*) can be used to solve the equation using numerical methods
(*b*) can solve the equation and provide an expression for the roots
(*c*) can solve the equation follows some fixed pattern
(*d*) cannot solve the equation. **[ES]**

129. In a half-adder having two inputs A and B and two output, (S and C are the Sum and Carry output bits respectively), the Boolean expressions for S and C in terms of A and B is

(*a*) $S = \overline{A}B + A.\overline{B}; C = A.B$

(*b*) $S = AB + \overline{A}.B; C = A + B$

(*c*) $S = \overline{A}.\overline{B} + AB; C = A + \overline{B}$

(*d*) $S = \overline{A} + A.\overline{B}; C = \overline{A} + B$ **[ES]**

130. Which of the following statement is true?

(*a*) ICs area always linear

(*b*) Digital circuits are linear circuits

(*c*) AND gate is a logic circuit whose output is equal to its highest input

(*d*) In a four-input AND circuit, all input must be UP for the output to be UP. **[ES]**

131. The total number of inhibit drives required for a 16 K, 24-bit / word memory with a 3-D organisation is

(*a*) 4

(*b*) 8

(*c*) 16

(*d*) 24 **[ES]**

132. A memory system has a total of 8 memory chips, each with 12 address lines and 4 data lines. The total size of memory system is

(*a*) 6 K bytes

(*b*) 32 K bytes **[GATE]**

(*c*) 48 K bytes

(*d*) 64 K bytes

133. A digital circuit which compares two numbers A_3, A_2, A_1, A_0, $B_3, B_2 B_1, B_0$ is shown in figure below. To get output Y = 0, choose one pair of correct input numbers

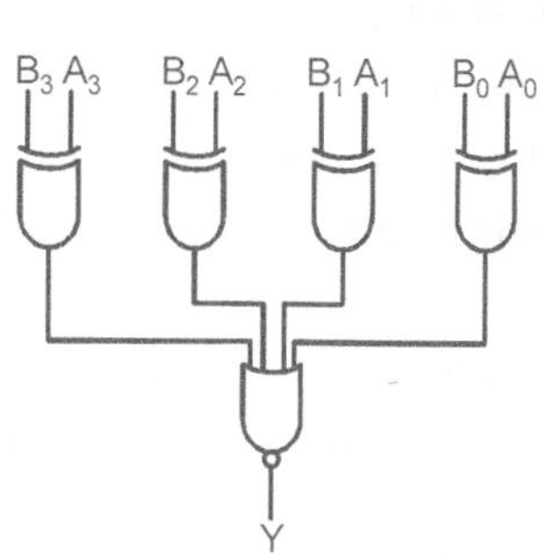

(*a*) 1010,1010

(*b*) 0101,0101

(*c*) 0010,0010

(*d*) 0010,1011 **[GATE]**

134. What is the 2's complement of the word 0011 0101 1001 1100 ?

(*a*) 1100 1010 0110 0100

(*b*) 1100 1010 0110 0011

(*c*) 1100 1010 0110 0101

(*d*) 0011 0101 1001 1101 **[ES]**

135. Match List-I (name of Logic gate) with List-II (Propagation delay) and select the correct answer using the codes given below the lists

List-I	**List-II**
A. DTL	1. 8 ns
B. TTL	2. 10 ns
C. ECL	3. 25 ns
D. CMOS	4. 1 ns

Code : **[GATE]**

	A	B	C	D
(*a*)	3	2	1	4
(*b*)	4	3	2	1
(*c*)	3	2	4	1
(*d*)	2	1	4	3

136. Match List-I with List-II and select the correct answer using the codes given below the Lists:

List-I (*function/circuit*)

A. D-flip – flop

B. T-flip – flop

C. Exclusive OR

D. Half adder

List-II (*circuit realisation*)

1. 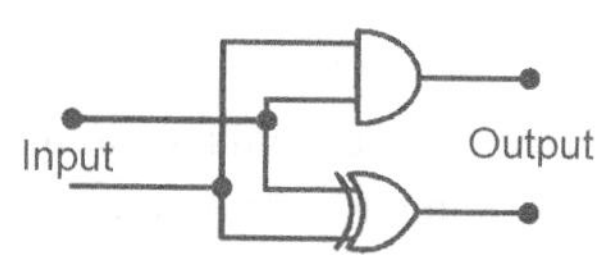

2. 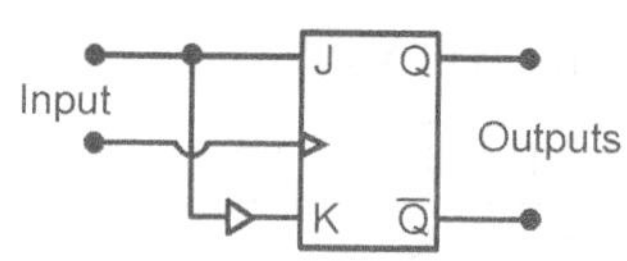

3. 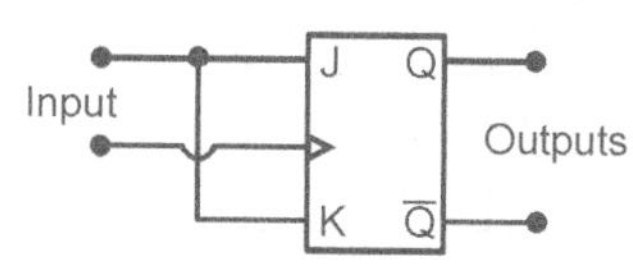

4. 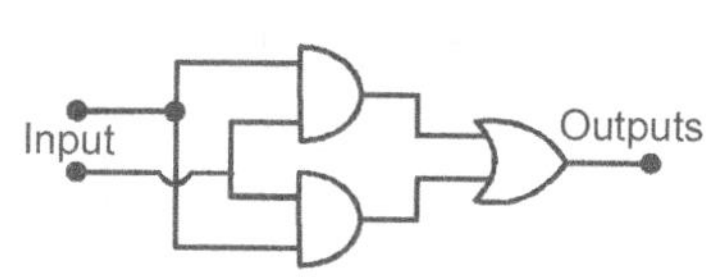

Codes:

	A	B	C	D
(*a*)	2	3	1	4
(*b*)	2	3	4	1
(*c*)	3	2	4	1
(*d*)	3	2	1	4

[ES]

137. An 8-input multiplexer is to be built with a tree network of 2-input multiplexers. The number of 2-input multiplexers required is

(*a*) 4

(*b*) 8

(*c*) 6

(*d*) 7 **[IAS]**

138. The minimized expression for the K-map shown in the given figure is **[BEL]**

CD \ AB	00	01	11	10
00	1	0	0	1
01	1	1	1	1
11	0	0	0	0
10	1	0	0	1

(*a*) $\overline{A}\,\overline{B}C\overline{D} + A\overline{B}C\overline{D} + \overline{A}\,\overline{B}\,\overline{C} + A\overline{B}\,\overline{C} + B\overline{C}D$

(*b*) $\overline{C}D + \overline{A}\,\overline{B}\,\overline{C} + A\overline{B}\,\overline{C} + \overline{A}\,\overline{B}C\overline{D} + A\overline{B}C\overline{D}$

(*c*) $\overline{B}C\overline{D} + \overline{B}\,\overline{C}\,\overline{D} + \overline{C}D$

(*d*) $\overline{B}\,\overline{D} + \overline{C}D$

Directions: *The following items consist of two statements, one labelled the 'Assertion A' and the other labelled the 'Reason R', You are to examine these two statements carefully and decide if the Assertion A and the Reason R are individually true and if so, whether the Reason is a correct explanation of the Assertion. Select your answers to these items using the codes given below and mark your answer sheet accordingly,*

Codes:

(*a*) Both A and R are true and R is the correct explanation of A

(*b*) Both A and R are true but R is not a correct explanation of A

(*c*) A is true but R is false

(*d*) A is false but R is true.

139. Assertion(A): Machine language program is written in hexadecimal.

Reason (R): Microprocessor can under-stand hexadecimal number system. **[ES]**

140. Assertion (A): Read only memory (ROM) is a random access memory.

Reason (R): Time taken to access any location of ROM is the same. **[IES]**

141. Assertion (A): A digital multiplexer can also be used to implement combinational logic functions.

Reason (R): In a combinational circuit, the current output depends on the previous outputs also.

142. Assertion (A): Among RAM and ROM is used as firmware memory.

Reason (R): ROM can be accessed for alteration. **[ES]**

143. Assertion (A): The pair of terms algorithm and program are analogous to the pair information and data respectively.

Reason (R): Algorithm and information refer to an abstract concept, program and data refer to their concrete representation. **[ES]**

144. Assertion (A): Asynchronous sequential circuits are difficult to design.

Reason (R) : Asynchronous sequential circuits use external clock.

145. Assertion (A): A PC controlled robot works in a polluted environment. **[ES]**

Reason (R): PC gets virus in the polluted environment.

146. Assertion (A): Hardware stack registers speed up the operation of a computer. **[ES]**

Reason (R): Hardware stack registers reduce the number of memory references required.

147. Assertion (A): MOSFET RAM are usually slower than bipolar RAMs.

Reason (R): MOSFETRAMs in com-parison with bipolar RAMs can be fabricated in much larger capacity in a single IC and consume much less power. **[IAS]**

148. Assertion (A): The auto increment and auto decrement modes are seldom found in modern processors.

Reason (R): Using two instructions instead of one provides flexibility in implementing a pipeline. **[IAS]**

ANSWERS

EXERCISE - I

1. (*a*)	2. (*d*)	3. (*d*)	4. (*b*)	5. (*c*)	6. (*a*)	7. (*c*)	8. (*d*)	9. (*c*)	10. (*d*)
11. (*d*)	12. (*a*)	13. (*b*)	14. (*d*)	15. (*a*)	16. (*a*)	17. (*a*)	18. (*a*)	19. (*b*)	20. (*c*)
21. (*a*)	22. (*b*)	23. (*c*)	24. (*d*)	25. (*d*)	26. (*d*)	27. (*d*)	28. (*c*)	29. (*a*)	30. (*b*)
31. (*a*)	32. (*a*)	33. (*d*)	34. (*c*)	35. (*a*)	36. (*a*)	37. (*b*)	38. (*d*)	39. (*d*)	40. (*a*)
41. (*d*)	42. (*c*)	43. (*d*)	44. (*c*)	45. (*d*)	46. (*b*)	47. (*c*)	48. (*b*)	49. (*d*)	50. (*b*)
51. (*b*)	52. (*a*)	53. (*c*)	54. (*c*)	55. (*c*)	56. (*a*)	57. (*b*)	58. (*c*)	59. (*a*)	60. (*b*)
61. (*c*)	62. (*b*)	63. (*c*)	64. (*b*)	65. (*b*)	66. (*d*)	67. (*a*)	68. (*d*)	69. (*b*)	70. (*a*)
71. (*c*)	72. (*a*)	73. (*a*)	74. (*a*)	75. (*d*)	76. (*d*)	77. (*c*)	78. (*b*)	79. (*c*)	80. (*c*)
81. (*a*)	82. (*d*)	83. (*b*)	84. (*d*)	85. (*d*)	86. (*b*)	87. (*c*)	88. (*c*)	89. (*c*)	90. (*d*)
91. (*b*)	92. (*c*)	93. (*a*)	94. (*b*)	95. (*c*)	96. (*a*)	97. (*b*)	98. (*d*)	99. (*c*)	100. (*d*)
101. (*a*)	102. (*d*)	103. (*c*)	104. (*c*)	105. (*b*)	106. (*a*)	107. (*d*)	108. (*c*)	109. (*d*)	110. (*b*)
111. (*d*)	112. (*d*)	113. (*c*)	114. (*c*)	115. (*d*)	116. (*d*)	117. (*c*)	118. (*c*)	119. (*d*)	120. (*d*)
121. (*b*)	122. (*d*)	123. (*a*)	124. (*c*)	125. (*d*)	126. (*c*)	127. (*a*)	128. (*c*)	129. (*d*)	130. (*d*)
131. (*d*)	132. (*b*)	133. (*a*)	134. (*d*)	135. (*a*)	136. (*d*)	137. (*d*)	138. (*c*)	139. (*c*)	140. (*b*)
141. (*b*)	142. (*d*)	143. (*c*)	144. (*c*)	145. (*c*)	146. (*c*)	147. (*b*)	148. (*a*)	149. (*b*)	150. (*d*)

EXERCISE - II

1. (*c*)	2. (*b*)	3. (*a*)	4. (*b*)	5. (*c*)	6. (*c*)	7. (*c*)	8. (*b*)	9. (*a*)	10. (*a*)
11. (*d*)	12. (*a*)	13. (*d*)	14. (*d*)	15. (*c*)	16. (*d*)	17. (*c*)	18. (*d*)	19. (*a*)	20. (*b*)
21. (*d*)	22. (*a*)	23. (*c*)	24. (*b*)	25. (*c*)	26. (*d*)	27. (*d*)	28. (*d*)	29. (*c*)	30. (*a*)
31. (*b*)	32. (*c*)	33. (*a*)	34. (*a*)	35. (*c*)	36. (*b*)	37. (*b*)	38. (*c*)	39. (*b*)	40. (*d*)
41. (*d*)	42. (*d*)	43. (*c*)	44. (*c*)	45. (*c*)	46. (*d*)	47. (*c*)	48. (*b*)	49. (*b*)	50. (*c*)
51. (*d*)	52. (*d*)	53. (*d*)	54. (*a*)	55. (*c*)	56. (*c*)	57. (*c*)	58. (*b*)	59. (*c*)	60. (*c*)
61. (*b*)	62. (*c*)	63. (*b,d*)	64. (*a*)	65. (*d*)	66. (*b*)	67. (*d*)	68. (*b*)	69. (*c*)	70. (*c*)
71. (*c*)	72. (*c*)	73. (*c*)	74. (*d*)	75. (*d*)	76. (*d*)	77. (*d*)	78. (*a*)	79. (*a*)	80. (*c*)
81. (*a*)	82. (*b*)	83. (*a*)	84. (*a*)	85. (*d*)	86. (*d*)	87. (*b*)	88. (*a*)	89. (*a*)	90. (*c*)
91. (*c*)	92. (*b*)	93. (*c*)	94. (*b*)	95. (*b*)	96. (*b*)	97. (*b*)	98. (*d*)	99. (*d*)	100. (*b*)
101. (*d*)	102. (*a*)	103. (*b*)	104. (*c*)	105. (*c*)	106. (*a*)	107. (*c*)	108. (*b*)	109. (*a*)	110. (*c*)
111. (*b*)	112. (*c*)	113. (*c*)	114. (*b*)	115. (*a*)	116. (*d*)	117. (*b*)	118. (*c*)	119. (*d*)	120. (*a*)
121. (*c*)	122. (*d*)	123. (*c*)	124. (*b*)	125. (*c*)	126. (*d*)	127. (*d*)	128. (*a*)	129. (*a*)	130. (*b*)
131. (*d*)	132. (*b*)	133. (*c*)	134. (*a*)	135. (*c*)	136. (*b*)	137. (*a*)	138. (*d*)	139. (*c*)	140. (*a*)
141. (*c*)	142. (*a*)	143. (*a*)	144. (*a*)	145. (*c*)	146. (*b*)	147. (*b*)	148. (*d*)		

EXPLANATIONS

EXERCISE - I

2. $\overline{\left(A\overline{B}\right)+\left(A+\overline{B}\right)} = \overline{\left(AB\right)}+\left(\overline{\overline{A}B}\right) = A\overline{B}+\overline{A}B$

3. In memory mapped I/O scheme, there is only one address space. Address space is defined as the set of all possible addresses that a microprocessor can generate. Some addresses are assigned to memories and some address to I/O devices. An I/O device is also treated as a memory location and one address is assigned to it.

In the given problem, address bus is to 16 bit.

∴ Number of total locations to be address

$= 2^{16} = 65536 = 64\text{ K}$

23. Main elements of a bipolar IC are resistors, diodes and transistors. There are two types of operations in bipolar ICs.

(*i*) Saturated

(*ii*) Non-saturated.

In saturated logic, the transistor in the ICs are driven to saturation, whereas in the case of non-saturated logic, transistors are not driven into saturation of the non-saturated bipolar logic families Emitter-coupled logic (ECL) is fastest one.

27. KV map can be drawn as follows :

Z \ XY (F)	00	01	11	10
0	0			
1	0		0	0

From KV map, $f = \overline{X}\overline{Y} + XZ$

29. Truth table of J-K flip-flop can be drawn as shown below :

J_n	K_n	Output (Q_{n+1})
0	0	Q_n
1	0	1
0	1	0
1	1	$\overline{Q}_n$

$$\therefore \quad Q_{n+1} = J\overline{Q}_n + \overline{K}_n Q_n$$

33. Truth table can be constructed as follows.

A	B	Output
0	0	0
0	1	1
1	0	1
1	1	0

∴ Expression for output $f = \overline{A}B + A\overline{B}$

34. We have, $\dfrac{1}{f} \geq Nt_d + T_s$

$\Rightarrow \quad \dfrac{1}{f} \geq Nt_d$

$\Rightarrow \quad \dfrac{1}{f} \geq 100 \times 10^{-9}$

$\therefore \quad f \leq 10\text{ MHz}$

35. Total memory capacity $= 64 + 32 + 512$

$= 608\text{ kB}$

38. $(125)_R = (203)_5$

$= 3 \times 5^0 + 0 \times 2 \times 5^2$

$= 3 + 50 = 53$

Hence, R should be equal to 6

39. Register in which data entered or out in serial form of called *shift register*. In the fourth option both input and output are in parallel form. So it can not be shift register hence ruled out.

41. Decimal equivalent

$= 0 \times 16^0 + A \times 16 + B \times 16^2$

$= 13 + 10 \times 16 + 11 \times 16^2 = 2989$

43. $7 - [16/(3+1) * 2] - 4$

$= 7 - [16/8] - 4$

$= 7 - 2 - 4 = 1$

44. $(211)_x = (152)_g$

$= 1 \times 8^2 + 5 \times 8 + 2 \times 8^0 = 106$

Hence x is equal to 7

46. Number of address line in each 16×4 chip is such that

$2^n \geq 16$

$\therefore \quad n = 4$

Since all the chips have their address buses connected together, hence resulting memory litigation will be same but their word size will be $4 \times 4 = 16$

Hence size of resulting memory system $= 16 \times 16$

47. Number of flip - flop required is such that

$2^n \geq 10$

∴ Minimum number of flip- flops = 4

49. Accuracy of a D/A converter is a measure of the difference between actual output voltage and expected output voltage. It is specified as a percentage of full scale or maximum output voltage.

Since full scale output voltage = 5V, and

accuracy = ± 2%

∴ Maximum error for any output voltage

$$= \frac{5 \times 0.2}{100}$$

$$= \frac{1}{100}\text{ V} = 10\text{ mV}$$

52. In ripple counter, clock is applied only on the 1st flip-flop and hence maximum clock frequency can be limited only by 1st flip flop

∴ Maximum clock frequency

$$= \frac{1}{50 \times 10^{-9}} = 20 \text{ MHz}$$

59. A and B combined as (A + B)

C and D combined as (C + D)

From switching circuit, the logic function can be written as

L = (A + B) (C + D)E

61.
$$\begin{aligned} Y.(Y+Z) &= Y.Y + YZ \\ &= Y + YZ \\ &= Y.(1+Z) \\ &= Y \end{aligned}$$

Take $x \oplus Z = X\overline{Z} + \overline{X}Z$

Now take $(x+z)\left(\overline{X}+\overline{Z}\right)$

$$\begin{aligned} &= X\overline{X} + \overline{Z}\overline{X} + \overline{Z}X + \overline{Z}.Z \\ &= Z\overline{X} + \overline{Z}X \\ &= x \oplus Z \end{aligned}$$

Hence correct answer is (*c*).

69. Take 011 1001 = 79

01011011 = 5 B

Since second number has second digit greater than I hence it is not a valid BCD code.

74. Number of total memory chip required

$$= \frac{64 \times 8}{16 \times 4}$$

78. $x = \left(\overline{A}B + A\overline{B}\right)\left(\overline{A + \overline{B}}\right)$

$$\begin{aligned} &= \left(\overline{A}B + A\overline{B}\right)\left(\overline{A} + B\right) \\ &= \left(\overline{A}B.\overline{A}B + A\overline{B}.\overline{A}B\right) \\ &= \overline{A}B + 0 \\ &= \overline{A}B \end{aligned}$$

82. $(X+Y)\left(X+\overline{Y}\right)\left(\overline{X}+Y\right)$

$$\begin{aligned} &= \left(X.X + Y.Y + X\overline{Y} + Y\overline{Y}\right)\left(\overline{X}+Y\right) \\ &= X\overline{X} + YX\overline{X} + X\overline{Y}\overline{X} + XY + YXY + X\overline{Y}Y \\ &= XY + YXY = XY(1+Y) \\ &= XY \end{aligned}$$

84. Addressable memory = 2^{16}

$= 65536 \approx 64$ k bytes.

85.
$$\begin{aligned} g(x,y,z) &= F_0 + F_1 = \overline{i_2}t + Zi_2 \\ &= i_2 = F_2 = F_3 = \overline{x}y.zy = 0 \end{aligned}$$

$$\therefore \quad g(x,y,z) = \overline{Z} + Z = 1$$

87. $\text{MODELX : MIPS} = \frac{33}{4}$

$\text{MODELY : MIPS} = \frac{66}{6}$

$$\therefore \quad \frac{(\text{MIPS})_Y}{(\text{MIPS})_X} = \frac{66}{6} \times \frac{4}{33} = \frac{4}{3}$$

93. Truth table of JK flip-flop is as shown below.

J	K	Q_+
0	0	Q
0	1	0
1	0	1
1	1	$\overline{Q}$

From the given figure J is always 1

Initially, J = 1, K = Q = 0

∴ After first pulse, AQ = K = 1

After second pulse, Q = K = 0

After third pulse, Q = K = 1

97. Let number of rcquired flip-flop be N.

For mod-19 counter,

$$2^N \geq 19$$

$$\therefore \quad N = 5$$

99. Let numbers of bits needed be N

Hence, to address 4 K memory, we should have

$$2^N \geq 4000$$

∴ Minimum number of bits = 12

104. Output of Ex-OR gate is given by

$$\begin{aligned} \overline{A}B + A\overline{B} &= A\overline{A} + \overline{A}B + A\overline{B} + B\overline{B} \\ &= \overline{A}(A+B) + \overline{B}(A+B) \\ &= (A+B)\left(\overline{A}+\overline{B}\right) \\ &= (A+B)\left(\overline{A.B}\right) \end{aligned}$$

110. When a clock pulse arrives,

$J_0 = 0$, So $Q_{n+1} = Q_n$

$K_0 = 0$, or Q_n will remain 1

$J_1 = K = 1$; so Q will become $\overline{Q}$ or Q = 1

120. From 16 bit bus two MSBs are used to enable the decoder hence total address = 2^{14} = 16384 ≈ 16K

133.
$$\begin{aligned} Z &= \overline{xy} + (x+y) \\ &= \overline{x} + \overline{y} + x + y \\ &= 1 + 1 = 1 \end{aligned}$$

∴ Output is always equal to "1" independent of value of *x* and *y*.

150.
$$\begin{aligned} X &= Y + X = 3 + 4 = 7 \\ Y &= X - Y = 7 - 3 = 4 \end{aligned}$$

$$\therefore \quad Z = Y * Z = 4 \times 4 = 16$$

EXERCISE - II

1. $A\left[B+C\left(\overline{AB+AC}\right)\right]$

$= A\left[B+C\left(\overline{AB}.\overline{AC}\right)\right]$

$= A\left[B+C\left(\overline{A}+\overline{B}\right).\left(A+C\right)\right]$

$= A\left[B+C\left(\overline{A}.\overline{A}+\overline{BA}+\overline{AC}+\overline{BC}\right)\right]$

$= AB+AC\overline{A}+AC\ \overline{BA}+AC\ \overline{AC}+AC\ \overline{BC}$

9. Stack pointer SP, holds the address of the stack top. Stack is sequence of memory locations defined by the programmer. The stack is used to save the content of a register during execution of a program. The last memory location of the occupied portion of the stack is called *stack top*. Stack pointer is 16 bit register.

11. TRAP is edge as well as level triggered interrupt. This means that TRAP should go high and stay high until it is acknowledged.

TRAP is non-maskable interrupt and it has the highest priority overall interrupt.

19. Number of memory locations = 2048

$\therefore$ Number of minimum address line ("n" *s.t*) $2n \geq 2048$

Hence minimum number of address line = 11

23. After execution of XRA A instruction, content of register A is EXCLUSIVE Ored with the content of the accumulator. The result is placed in the accumulator. All status flags are affected.

Since, $A\overline{A}+\overline{A}A=0$

hence result is zero. So, zero flag is set.

25. $$\text{Accuracy} = \frac{10\times10^{-3}}{5} = 2\times10^{-3}$$

It should be equal to $\frac{1}{2^n}$.

Hence n should be 9, so that $\frac{1}{2^9}=\frac{1}{512}$

$\approx 2\times10^{-3}$

27. High level language is converted into object codes statements wise by interpreter. It does not translate the entire program at a time. It takes up one statement of a high level language program at a time, translate it and then execute it. Then after it takes up the next statement of the high-level language program and repeat the process.

46.

		Remainder
8	149	
8	18	5(LSD)
8	2	2
	2	2(MSD)

Octal code will be 225.

47. $(327)_9 = 7\times9^0+2\times9+3\times9^2$

$= 7+18+243 = 268$

Now

		Remainder
5	268	
5	53	3(LSD)
5	10	3
5	2	0
	0	2(MSD)

$\therefore\ x = 2033$

57. First number $= 0+1\times2+1\times2^2+1\times2^3+1\times2^4$

$= 30$ (+ve value)

Second number $= 0+0+1\times2^2+1\times2^3+1\times2^4$

$= 28$ (–ve value)

58. + 19 can be write as 00010011

One's complement is 11101101

Two's complement is 11101101

61. Value of fractional part $= \frac{1}{2}+0+\left(\frac{1}{2}\right)^2$

$= 0.5+0.125 = 625$

62. One's complement = 1001

$\therefore$ Tow's complement = 1010

67. –N in two's complement form is 1011

$\therefore$ (N + 1) in Two's complement form is 1011 + 1 = 1010

68. 00000111

It's one's complement = 11111000

$\therefore$ Two's complement = 11111000 + 1

= 11111001

= F9

70. Number = 1001

It's one's complement = 0110

86. 2^{20} = 1048576 bytes

88. $q = CV$

$\therefore\quad V = \frac{1}{C}q = \frac{1}{C}\int idt$

Since current I is constant, therefore

$V = \frac{1}{C}.I\int dt = \frac{I}{C}t$

94. $Z = (A+I)(\overline{A}+J)$

$= A\overline{A}+AJ+\overline{A}I+IJ$

$= \overline{A}I+AJ+IJ$

Clearly, I should be 1 since A is required

So $Z = \overline{A}+AJ+J$

$J = B, Z = \overline{A}+AB+B$

$\Rightarrow\quad Z = \overline{A}+B(1+A)$

$= A+B$

Hence, I = 1, J = B

97. Control word Register for 8253 is

D7	D6	D5	D4	D3	D2	D1	D0
SC_1	SC_0	RL_1	RL_1	M_2	M_1	M_0	BCD

SC_1 & SC_0 are to select counter SC_1 & $SC_0 = 01$ for counter 1 $M_2M_1M_0$ are to select mode $M_2M_1M_0 = 0\,0\,0 \Rightarrow$ Mode 0

100. 4 × means 4 address space of 8 bit size each.

Hence total number of bits available

$= 4 \times 8 = 32$

102. Truth Table of JK flip-flop is

J	K	Q_+
0	0	Q
0	1	0
1	0	1
1	1	$\overline{Q}$

Initially $\left.\begin{matrix} J = \overline{Q} = 1 \\ K = 1 \end{matrix}\right\} \Rightarrow Q = \overline{Q} = 1$

After first pulse $Q_+ = \overline{Q} = 1$

Now $J = \overline{Q} = 0$ Hence $Q_+ = 0$

$K = 1$

$\therefore$ $J = \overline{Q} = 1$ Hence $Q_+ = 1$

and $K = 1$

Hence the sequence will be 010101.

104. The given logic function can be only implemeneted using Ex-OR gate. This can be explained as follows:

For Ex-OR gate output, $X = A\overline{B} + \overline{A}B$

Suppose A is control signal and B is input

When control input A = 1,

$$X = 1.\overline{B} + 0.B = \overline{B}$$

When A = 0, $\quad X = 0 + 1.B = B$

110.

$$\begin{aligned} A(1,y) &= A(0, A, (a, y-1)) \\ &= A(1, y-z) + 1 \\ &= (0, A(1, y-1) + 1 \\ &= A(1, y-z) + z \end{aligned}$$

This will repeat y times and will yield

$$\begin{aligned} A(1,y) &= A(1,0) + y \\ &= A(0,1) + y \\ &= 1 + 1 + y \\ &= y + z \end{aligned}$$

119. Output $= \overline{AB}C + \overline{A}B\overline{C} + ABC + A\overline{BC}$

$= \overline{C}\left(\overline{AB} + A\overline{B}\right) + C\left(AC + \overline{AB}\right)$

$= A \oplus B \oplus C$

120. Truth Table of JK flip-flop can be drawn as shown below:

J	K	Q_+
0	0	Q
1	0	1
0	1	0
1	1	$\overline{Q}$

Hence Q changes from 0 to 1 when J = 1 & K could be any value.

121.

A	B	Q_{n+1}	J
0	0	Q_n	0
0	1	1	1
1	0	Q_n	0
1	1	0	X

$\therefore$ $J = \overline{A}B + AB = B\left(\overline{A} + A\right) = B$

123. Let address lines be N.

$\therefore \quad 2^N \geq 8192$

$\Rightarrow \quad N \geq 13$

Since each word is of 8 bit and there should be 8 data line

A	B	S	C
0	0	0	0
0	1	1	0
1	0	1	0
1	1	0	1

$\therefore \quad S = \overline{A}B + A\overline{B}$

$C = AB$

12 CHAPTER Signals and Systems

EXERCISE – I

1. The Fourier transform X(*f*) of the periodic delta functions

$$x(t) = \sum_{k=-\infty}^{\infty} \delta(t - kT) \text{ is}$$

(*a*) $T \sum_{k=-\infty}^{\infty} \delta(f - kT)$

(*b*) $T \sum_{k=-\infty}^{\infty} \delta\left(f - \frac{k}{T}\right)$

(*c*) $\frac{1}{T} \sum_{k=-\infty}^{\infty} \delta\left(f - \frac{k}{T}\right)$

(*d*) $\frac{1}{T} \sum_{k=-\infty}^{\infty} \delta(f - kT)$

2. The colour TV picture signal is a

(*a*) single-channel, one-dimensional signal

(*b*) single-channel, three-dimensional signal

(*c*) three-channel, one-dimensional signal

(*d*) three-channel, three-dimensional signal

3. Match List I with List II and select the correct answer using codes given below the lists:

List I (*Functions*)	**List I** (*Fourier Transforms*)
A. $\exp(-\alpha t)\, u(t), \alpha > 0$	1. $\frac{1}{(\alpha + j2\pi f)^2}$
B. $\exp(-\alpha \lvert t \rvert), \alpha > 0$	2. $\frac{1}{\alpha + j2\pi f}$
C. $t \exp(-\alpha t)\, u(t), \alpha > 0$	3. $\left(f - \frac{\alpha}{t_0}\right)$
D. $\exp(j2\pi\alpha t/t_0)$	4. $\frac{2\alpha}{\alpha^2 + (2\pi f)^2}$

Codes :

	A	B	C	D
(*a*)	3	1	4	2
(*b*)	2	4	1	3
(*c*)	3	4	1	2
(*d*)	2	1	4	3

4. Match List I with List II and select the correct answer using codes given below the lists :

List I

(*Type of signal*)

A. Real and even symmetric

B. Real and odd symmetric

C. Imaginary and even symmetric

D. Imaginary and odd symmetric

List II

(*Property of Fourier transform*)

1. Imaginary and even symmetric

2. Real and even symmetric

3. Real odd symmetric

4. Real odd symmetric

Codes :

	A	B	C	D
(*a*)	1	4	2	3
(*b*)	2	4	1	3
(*c*)	1	3	2	4
(*d*)	2	3	1	4

5. Consider the following statements:

1. Fourier transform is special case of Laplace transform.
2. Region of convergence need not be specified for Fourier transform.
3. Laplace transform is not unique unless the region of convergence is specified.
4. Laplace transform is a special case of Fourier transform.

Which of these statements are correct?

(*a*) 2 and 4 (*b*) 4 and 1

(*c*) 4, 3 and 2 (*d*) 1, 2 and 3

6. The response of a linear, time-invariant, discrete-time system to a unit step input $u(n)$ is the unit impulse $\delta(n)$. The system response to a ramp input $n\,u(n)$ would be

(*a*) $u(n)$ (*b*) $u(n-1)$

(*c*) $n\,\delta(n)$ (*d*) $\sum_{k=0}^{\infty} k\delta(n-k)$

7. If $X(z) = \dfrac{z + z^{-3}}{z + z^{-1}}$ then $x(n)$ series has

(a) alternate 0s (b) alternate 1s

(c) alternate 2s (d) alternate – 1s

8. For a Z-transform

$$X(z) = \frac{z\left(2z - \frac{5}{6}\right)}{\left(z - \frac{1}{2}\right)\left(z - \frac{1}{3}\right)}$$

Match List-I with List-II and select the correct answer using codes given below the lists:

List I

(*The sequences*)

A. $\left[\left(\frac{1}{2}\right)^n + \left(\frac{1}{3}\right)^n\right] u(n)$

B. $\left(\frac{1}{2}\right)^n u(n) - \left(\frac{1}{3}\right)^n$

C. $-\left(\frac{1}{2}\right)^n u(-n-1) + \left(\frac{1}{3}\right)^n u(n)$

D. $-\left[\left(\frac{1}{2}\right)^n + \left(\frac{1}{3}\right)^n\right] u(-n-1)$

List II

(*The region of convergence*)

1. $\frac{1}{3} < |z| < \frac{1}{2}$

2. $|z| < \frac{1}{3}\ u(-n-1)$

3. $|z| < \frac{1}{3}$ and $|z| > \frac{1}{2}$

4. $|z| > \frac{1}{2}$

Codes :

	A	B	C	D
(a)	4	2	1	3
(b)	1	3	4	2
(c)	4	3	1	2
(d)	1	2	4	3

9. A system can be represented in the form of state equation as

$$\underline{S}(n+1) = A\,\underline{S}(n) + Bx(n)$$
$$y(n) = C\,\underline{S}(n) + Dx(n)$$

where A, B, C and D are matrices, $\underline{S}(n)$ is the state vector, $x(n)$ is the input and $y(n)$ is the output.

The transfer function of the system $H(z) = \dfrac{Y(z)}{X(z)}$ is given by

(a) $A(ZI - B)^{-1} C + D$ (b) $B(ZI - C)^{-1} D + A$

(c) $C(ZI - A)^{-1} B + D$ (d) $A(ZI - A)^{-1} C + B$

10. Match List I with List II and select the correct answer using codes given below the lists:

List I

(*Fourier series and Fourier transforms*)

A. Fourier series

B. Fourier transform

C. Discrete time Fourier

D. Discrete Fourier transform

List II

(*Their properties*)

1. Discrete, periodic

2. Continuous, periodic

3. Discrete, aperiodic

4. Continuous, aperiodic

Codes :

	A	B	C	D
(a)	3	4	2	1
(b)	1	2	4	3
(c)	3	2	4	1
(d)	1	4	2	3

11. The units of the spectrum obtained by Fourier transforming the covariance function of a stationary stochastic process is

(a) power per Hertz (b) energy per Hertz

(c) power per second (d) energy per second

12. Two independent signals X and Y are known to be Gaussian with mean values x_0 and y_0 and variances σ_x^2 and σ_y^2. A signal Z = X – Y is obtained from them. The mean z_0, variance σ_z^2 and p.d.f. $p(z)$ of the signal Z are given by

(a) $x_0 \sim y_0$, $\sigma_x^2 \sim \sigma_y^2$, Gaussian

(b) $x_0 + y_0$, $\sigma_x^2 + \sigma_y^2$, Rayleigh

(c) $y_0 - x_0$, $\sigma_y^2 - \sigma_x^2$, uniform

(d) $x_0 - y_0$, $\sigma_x^2 + \sigma_y^2$, Gaussian

13. If the cumulative distribution function is $F_x(x)$, then the probability density function $f_x(x)$ is given as

(*a*) $\int F_x(x)dx$ (*b*) $\frac{d}{dx}F_x(x)$

(*c*) $\int F_x(-x)dx$ (*d*) $\frac{d}{dx}F_x(-x)$

14. The Fourier transform of $e^{-\pi t^2}$ is $e^{-\pi f^2}$; then the Fourier transform of $e^{-\alpha t^2}$ is

(*a*) $\left(\frac{1}{\alpha}\right)\cdot e^{-\alpha f^2}$ (*b*) $\sqrt{\frac{\pi}{\alpha}}e^{-\frac{\pi^2 f^2}{\alpha}}$

(*c*) $\frac{1}{\sqrt{\pi\alpha}}e^{-\alpha\pi^2 f^2}$ (*d*) $\sqrt{\pi\alpha}\,e^{-\frac{f^2}{\pi^2\alpha}}$

Directions: *The following three (3) items consist of two statements: one labelled the 'Assertion A' and the other labelled the 'Reason R'. You are to examine these two statements carefully and decide if the Assertion A and the Reason R are individually true and if so, whether the Reason is a correct explanation of the Assertion. Select your answers to these items using the codes given below and mark your Answer Sheet accordingly.*

Codes:

(a) Both A and R are true and R is the correct explanation of A

(*b*) Both A and R true but R is NOT a correct explanation of A

(c) A is true but R is false

(*d*) A is false but R is true

15. Assertion (A) :

The stability of the system is assured if the Region of Convergence (ROC) includes the unit circle in the Z-plane.

Reason (R) :

For a casual stable system all the poles should be outside the unit circle in the Z-plane.

16. Assertion (A) :

The signals $a^n\, u(n)$ and $a^n\, u(-n-1)$ have the same Z-transform, $\frac{Z}{Z-a}$.

Reason (R) :

The Region of Convergence (ROC) for $a^n\, u(n)$ is $|Z| > |a|$, whereas the ROC for $a^n\, u(-n-1)$ is $|Z| < |a|$.

17. Assertion (A) :

For a rational transfer function H(*z*) to be casual, stable and casually invertible, both the zeros and the poles should lie within the unit circle in the Z-plane.

Reason (R) :

For a rational system, ROC is bounded by poles.

18. If $X(s) = \frac{1}{(1+s)}$ is the given network function, then which one of the following describes the correct Bode plot?

(*a*) A straight line with a slope of –20 dB/decade passing through the point 0 dB and 1 rad/s

(*b*) A straight line with a slope of 20 dB/decade passing through the point 0 dB and 1 rad/s

(*c*) Constant at 0 dB for $\omega < 1$ and a slope of + 20 dB/decade for $\omega > 1$

(*d*) Constant at 0 dB for $\omega < 1$ and a slope of – 20 dB/decade for $\omega > 1$

19. $y[n] = \sum_{k=-\infty}^{n} x[k]$

Which one of the following systems is inverse of the system given above?

(*a*) $x[n]=y[n]-y[n-1]$ (*b*) $x[n]=y[n]$

(*c*) $x[n]=y[n+4]$ (*d*) $x[n]=ny[n]$

20. What is the phase angle of the composite sinusoidal signal resulting from the addition of $v_1(n) = \sin[5\pi n]$ and $v_2(n) = \sqrt{2}\cos[5\pi n]$?

(a) $+5\pi$ (*b*) -5π

(c) $+\pi$ (*d*) $-\frac{\pi}{3}$

21. Match List - I with List - II and select the correct answer using the codes given below the lists:

List I

(*Equation Connecting x(n) y(n) of four discrete time system*)

A. $y(n+2)+y(n+1)+y(n)=2x(n+1)+x(n)$

B. $n^2y^2(n)+y(n)=x^2(n)$

C. $y(n+1)+ny(n)=4nx(n)$

D. $y(n+1)y(n)=4x(n)$

List II

(*System Input and Output Category*)

1. Linear, timevariable dynamic
2. Linear, time-invarient dynamic
3. Non-linear, time- variable, dynamic
4. Non-linear, time-invariant dynamic
5. Non-linear, time-variable memoryless

Codes :

	A	B	C	D
(*a*)	3	5	2	1
(*b*)	3	2	5	4
(*c*)	2	3	5	1
(*d*)	2	5	1	4

22. To which one of the following difference equations, the impulse response

$h(n) = \delta(n+2) - \delta(n-2)$ corresponds?

(*a*) $y(n+2) = x(n) - x(n-2)$

(*b*) $y(n-2) = x(n) - x(n-4)$

(*c*) $y(n) = x(n+2) + x(n-2)$

(*d*) $y(n) = -x(n+2) + x(n-2)$

23. Laplace transforms of $f(t)$ and $g(t)$ are $F(s)$ and $G(s)$, respectively. Which one of the following expressions gives the inverse Laplace transform of $F(s)\,G(s)$?

(*a*) $f(t)\,g(t)$ (*b*) $\dfrac{f(t)}{g(t)}$

(*c*) $f(t) + g(t)$ (*d*) $f(t) * g(t)$

24. The impulse response of a linear time-invariant system is a rectangular pulse of duration T. It is excited by an input of a pulse of duration T. What is the filter output waveform?

(*a*) Rectangular pulse of duration T

(*b*) Rectangular pulse of duration 2 T

(*c*) Triangular pulse of duration T

(*d*) Triangular pulse of duration 2 T

25. Consider the following statements:

In a ferroelectric material

1. all domains are lined up in the direction of applied field giving rise to saturation
2. if the field is reduced to zero, many domains remain aligned
3. the remnant polarization can be eliminated only if the material is heated above Curie temperature

Which of the statements given above are correct?

(*a*) 1, 2 and 3

(*b*) 2 and 3

(*c*) 1 and 3

(*d*) 1 and 2

26. The v-i characteristic of an element is shown in the below figure. The element is

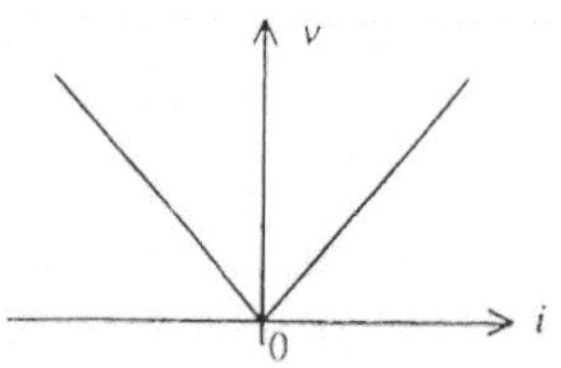

(a) non-linear, active, non-bilateral

(*b*) linear, active, non-bilateral

(c) non-linear, passive, non-bilateral

(*d*) non-linear, active, bilateral

27. The sum of two or more arbitrary sinusoids is

(a) always periodic

(*b*) periodic under certain conditions

(c) never periodic

(*d*) periodic only if all the sinusoids are identical in frequency and phase

28. The average value of the full-wave rectified sine wave with period π, and a peak value of V_m is

(a) $0.707\,V_m$ (*b*) $0.500\,V_m$

(c) $0.637\,V_m$ (*d*) $0.318\,V_m$

29. Match List I with List II and select the correct answer using the codes given below the Lists :

List I	List II
A. Free and Forced response	**1.** Discrete time system
B. Z-transforms	**2.** Dirichlet conditions
C. Probability	**3.** Non-homo-geneous theory differential equation
D. Fourier series	**4.** Random processes

Code :

	A	B	C	D
(*a*)	1	3	2	4
(*b*)	3	1	2	4
(*c*)	1	3	4	2
(*d*)	3	1	4	2

30. The relationship between the input $x(t)$ and the output $y(t)$ of a system is

$$\frac{d^2y}{dt^2} = x(t-2)u(t-2) + \frac{d^2x}{dt^2}$$

The transfer function of the system is

(a) $1 + \dfrac{s^2}{e^{2s}}$ (*b*) $1 + \dfrac{e^{-2s}}{s^2}$

(c) $1 + \dfrac{e^{2s}}{s^2}$ (*d*) $1 + \dfrac{s^2}{e^{-2s}}$

31. Which one of the following is NOT a correct statement about the state-space model of a physical system ?

(a) State-space model can be obtained only for a linear system.

(*b*) Eigenvalues of the system represent the roots of the characteristic equation.

(c) $\dot{X} = AX + BU$ represents linear state-space model of a physical system.

(*d*) $X(t)$ represents the state vector of the system.

32. Match List I with List II and select the correct answer using the codes given below the lists :

List I (*Nature of Periodic*)

A. Impulse train

B. Full-wave rectified sine function

C. $\sin\frac{2\pi t}{6}.\cos\frac{4\pi t}{6}$

D.

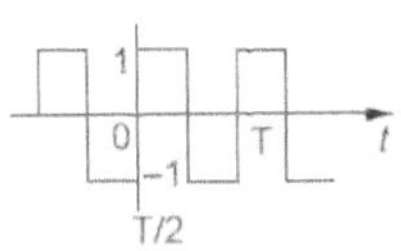

List II

(*Properties of Spectrum Function*)

1. Only even harmonics are present

2. Impulse train with strength 1/T

3. $\alpha_3 = \frac{1}{2j}$; $\alpha_{-3} = -\frac{1}{2j}$;

$\alpha_1 = -\frac{1}{2j}$; $\alpha_{-1} = \frac{1}{2j}$

4. Only odd harmonics are present.

5. Both even and odd harmonics ar present.

Codes :

	A	B	C	D
(*a*)	5	2	3	4
(*b*)	2	1	4	3
(*c*)	5	2	4	3
(*d*)	2	1	3	4

33. The covariance function of a band limited white noise is a/an

(a) dirac delta function

(*b*) exponentially decreasing function

(c) sinc function

(*d*) sinc^2 function

34. The inverse Fourier transform of δ (f) is

(a) *u*(*t*) (*b*) 1

(c) δ (t) (*d*) $e^{j2\pi t}$

35. If $\left(\frac{27s+97}{s^2+33s}\right)$ is the Laplace transform of f(t), then $f(0^+)$ is

(a) zero (*b*) $\frac{97}{33}$

(c) 27 (*d*) infinity

36. The Laplace transform of the waveform shown in the figure below is

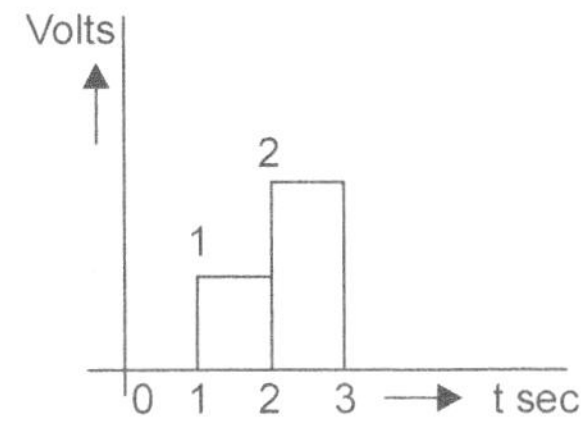

(a) $\frac{1}{s}\left[e^{s}+e^{2s}+2e^{3s}\right]$

(*b*) $\frac{1}{s}\left[e^{s}+e^{2s}+2e^{-3s}\right]$

(c) $\frac{1}{s}\left[e^{-s}+e^{-2s}+2e^{-3s}\right]$

(*d*) $\frac{1}{s}\left[e^{-s}+e^{-2s}-2e^{-3s}\right]$

37. Match List I with List II (and select the correct answer using the codes given below the Lists :

	List I (***F***(*s*))		**List II** (*f*(*t*))
A.	$\frac{10}{s(s+10)}$	**1.**	$10\,\delta(t)$
B.	$\frac{10}{s(s^2+10)}$	**2.**	$(e^{-10t}\cos 10t).u(t)$
C.	$\frac{(s+10)}{(s+10)^2+100}$	**3.**	$(\sin 10t).u(t)$
D.	10	**4.**	$(1-e^{-10t}).u(t)$

Codes :

	A	B	C	D
(*a*)	3	4	1	2
(*b*)	4	3	1	2
(*c*)	3	4	2	1
(*d*)	4	3	2	1

38. The Laplace transform of sin ωt is

(a) $\frac{s}{s^2+\omega^2}$

(*b*) $\frac{\omega^2}{s^2+\omega^2}$

(*c*) $\frac{s^2}{s^2+\omega^2}$

(*d*) $\frac{\omega}{s^2+\omega^2}$

39. Given : $L\,f(t) = F(s) = \int_0^{\infty} f(t)e^{-st}dt$

Which of the following expression are correct ?

1. $L[f(t-a)u(t-a)] = F(s)e^{-sa}$

2. $L\,f(t) = \frac{-dF(s)}{ds}$

3. $L\,(t-a)f(t) = asF(s)$

4. $L\,\frac{dF(t)}{dt} = sF(s) - f(o_+)$

Select the correct answer using the codes given below :

(a) 1, 2 and 3 (b) 1, 2 and 4

(c) 2, 3 and 4 (d) 1, 3 and 4

40. Two rectangular waveforms of duration T_1 and T_2 seconds are convolved. What is the shape of the resulting waveform?

(a) Triangular

(b) Rectangular

(c) Trapezoidal

(d) Semicircular

41. A discrete LTI system is non-causal if its impulse response is

(a) $a^n u(n-2)$ (b) $a^{n-2}u(n)$

(c) $a^{n+2}u(n)$ (d) $a^n u(n+2)$

Direction: *The following 2 (two) items consist of two statements, one labelled as the 'Assertion (A)' and the other as 'Reason (R)'. You are to examine these two statements carefully and select the answers to these items using the codes given below :*

Codes :

(a) Both A and R are individually true and R is the correct explanation of A.

(b) Both A and R are individually true but R is NOT the correct explanation of A

(c) A is true but R is false.

(d) A is false but R is true.

42. **Assertion (A) :**

A periodic function satisfying Dirichlet conditions can be expanded into Fourier series.

Reason (R) :

A periodic function can be reconstructed from $\frac{a_0}{2} + \sum_{n=1} a_n \cos n\omega_0 t + \sum_{n=1} b_n \sin n\omega_0 t$ for very large n, excluding infinity.

43. **Assertion (A) :**

In the exponential Fourier representation of a real-valued periodic function f(t) of frequency f_0, the coefficients of the terms $e^{j2\pi n_0 t}$ and $e^{-j2\pi n f_0 t}$ are negative of each other.

Reason (R) :

The discrete magnitude spectrum of f(t) is even and the phase spectrum is odd.

44. The range of values of *a* and *b* for which the linear time invariant system with impulse response

$$h(n) = \begin{cases} a^n, & n \ge 0 \\ b^n, & n < 0 \end{cases}$$

will be stable is

(a) $|a| > 1, |b| > 1$

(b) $|a| < 1, |b| < 1$

(c) $|a| < 1, |b| > 1$

(d) $|a| > 1, |b| < 1$

45. Which one of the systems described by the following input-output relations is time invariant?

(a) $y(n) = n\,x(n)$

(b) $y(n) = x(n) - x(n-1)$

(c) $y(n) = x(-n)$

(d) $y(n) = x(n) \cos 2p f_o n$

46. Match List I with List II and select the correct answer using codes given below the lists :

List I (*Input-output relation*)	**List II** (*Property of the system*)
A. $y(n) = x(n)$	1. Nonlinear, non-casual
B. $y(n) = x(n^2)$	2. Linear, non-casual
C. $y(n) = x^2(-n)$	3. Linear, casual
D. $y(n) = x^2(n)$	4. Nonlinear, casual

Codes :

	A	B	C	D
(a)	1	4	3	2
(b)	3	2	1	4
(c)	1	2	3	4
(d)	3	4	1	2

47. Consider the following statements related to Fourier series of a periodic waveform:

1. It expresses the given periodic waveform as a combination of dc component, sine and cosine waveforms of different harmonic frequencies.
2. The amplitude spectrum is discrete.
3. The evaluation of Fourier coefficients gets simplified if waveform symmetries are used.
4. The amplitude spectrum is continuous.

Which of the above statements are correct?

(a) 1, 2 and 4 (b) 2, 3 and 4

(c) 1, 3 and 4 (d) 1, 2 and 3

48. What is the Laplace transform of the waveform shown below?

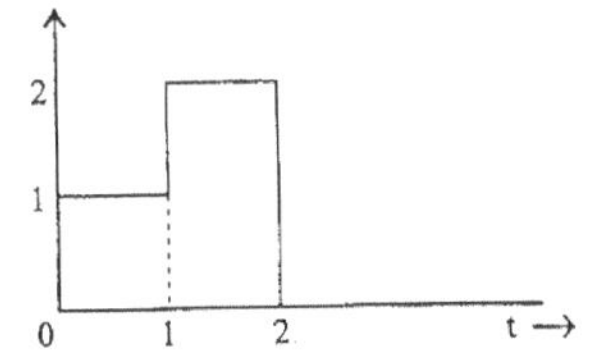

(a) $F(s)=\frac{1}{s}+\frac{1}{s}e^{-s}-\frac{2}{s}e^{-2s}$

(b) $F(s)=\frac{1}{s}-\frac{1}{s}e^{-s}+\frac{2}{s}e^{-2s}$

(c) $F(s)=\frac{1}{s}+\frac{1}{s}e^{s}+\frac{2}{s}e^{2s}$

(d) $F(s)=\frac{1}{s}-\frac{1}{s}e^{-2s}-\frac{2}{s}e^{-s}$

49. For the signal shown below :

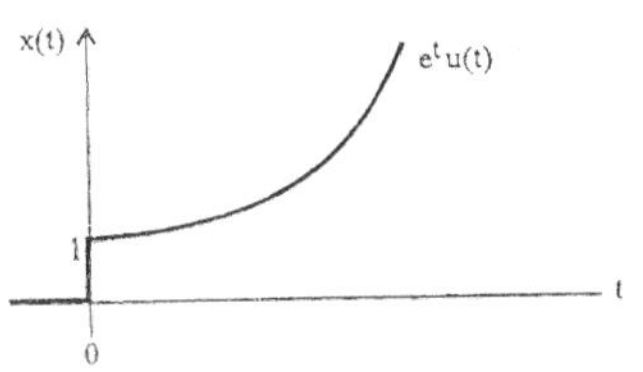

(a) Only Fourier tranform exists

(b) Only Laplace transform exists

(c) Both Laplace and Fourier transforms exist

(d) Neither Laplace transform nor Fourier transform exists

50. Let $x[n]=a^n u[n]$; $h[n]=b^n u[n]$

What is the expression for $y[n]$, for a discrete-time system?

(a) $\sum_{k=-\infty}^{\infty} a^k u[k]\, b^{n-k} u[n-k]$

(b) $\sum_{k=-\infty}^{\infty} a^n u[k]\, b^{n-k} u[n-k]$

(c) $\sum_{k=-\infty}^{\infty} a^k u[n-k]\, b^{n} u[k]$

(d) $\sum_{k=-\infty}^{\infty} a^{n-k} u[k]\, b^{n-k} u[n-k]$

51. The output $y[n]$ of a discrete time LTI system is related to the input x[n] as given below:

$$y[n]=\sum_{k=0}^{\infty} x[k]$$

Which one of the following correctly relates the z-transforms of the input and output, denoted by X(z) and Y(z), respectively ?

(a) $Y(z)=(1-z^{-1})X(z)$

(b) $Y(z)=z^{-1}X(z)$

(c) $Y(z)=\frac{X(z)}{1-z^{-1}}$

(d) $Y(z)=\frac{d\,X(z)}{dz}$

52. A signal represented by

$x(t)=5\cos 400\pi t$

is sampled at a rate 300 sample/s. The resulting samples are passed through an ideal low pass filter of cut-off frequency 150 Hz. Which of the following will be contained in the output of the LPF ?

(a) 100 Hz

(b) 100 Hz, 150 Hz

(c) 50 Hz 100 Hz

(d) 50 Hz, 100 Hz, 150 Hz

53. Match List I (Discrete Time Signal) with List II (Transform) and select the correct answer using the code given below the Lists :

List I (*Discrete Time Signal*)	**List II** (*Transform*)
A. Unit step function	1. 1
B. Unit impulse function	2. $\frac{z-\cos\omega T}{z^2-2z\cos\omega T+1}$
C. sin ωt, $t=0, T, 2T, \ldots\ldots$	3. $\frac{z}{z-1}$
D. cos ωt, t = 0, T, 2T	4. $\frac{z\sin\omega T}{z^2-2z\cos\omega T+1}$

Codes :

	A	B	C	D
(a)	2	4	1	3
(b)	3	1	4	2
(c)	2	1	4	3
(d)	3	4	1	2

54. Which one of the following is the inverse z-transform of

$X(z) = \frac{z}{(z-2)(z-3)}\ |z|<2$?

(*a*) $\left[2^n - 3^n\right] u(-n-1)$

(*b*) $\left[3^n - 2^n\right] u(-n-1)$

(c) $\left[2^n - 3^n\right] u(n+1)$

(*d*) $\left[2^n - 3^n\right] u(n)$

55. The auto-correlation function $R_x(\tau)$ of a random process has the property that $R_x(0)$ is equal to

(*a*) square of the mean value of the process

(*b*) mean squared value of the process

(c) smallest value of $R_x(t)$

(*d*) $\frac{1}{2}\ [R_x(t) + R_v(-t)]$

56. The graph shown below represents a waveform obtained by convolving two rectangular waveforms of duration

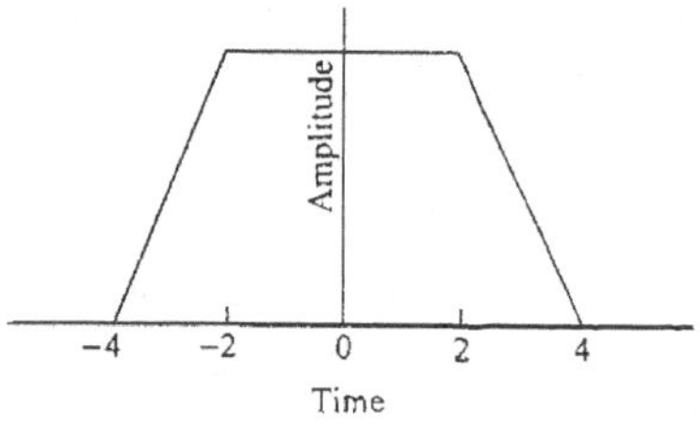

(a) four units each

(*b*) four and two units respectively

(c) six and three units respectively

(*d*) six and two units respectively

57. The discrete time Fourier transform (DTFT) of x[n] = 2(3)n u[−n] is equal to

(*a*) $\frac{2}{1-e^{j\Omega/3}}$ (*b*) $\frac{2}{1+e^{j\Omega/3}}$

(c) $2\left(\frac{1+e^{j\Omega/3}}{1-e^{j\Omega/3}}\right)$ (*d*) $2\left(\frac{1-e^{j\Omega/3}}{1+e^{j\Omega/3}}\right)$

Directions : *The following two (2) items consist of two statements, one labelled as the 'Assertion (A)' and the other as 'Reason (R)'. You are to examine these two statements carefully and select the answers to these items using the codes given below :*

Codes :

(*a*) Both A and R are individually true and R is the correct explanation of A

(*b*) Both A and R are individually true but R is not the correct explanation of A

(c) A is true but R is false

(*d*) A is false but R is true

58. **Assertion (A) :** Even if two different approaches are used to solve Laplace equation, we will end up in the same solution.

Reason (R) : It is mainly because in the process of solution, parameters relating to potentials and their spatial rates of change may assume different values.

59. **Assertion (A) :** A microstrip line can not support pure TEM mode of propagation.

Reason (R) : A microstrip line suffers from various forms of losses.

60. What is the Nyquist rate for the signal

$x(t) = \cos 2000\ \pi t + 3 \sin 6000\ \pi t$?

(*a*) 2 kHz

(*b*) 4 kHz

(*c*) 12 kHz

(*d*) 6 kHz

61. If response of a system to an input does not depend on the future values of the input, then the system is

(*a*) aperiodic

(*b*) causal

(*c*) anticipatory

(*d*) discrete

62. A second-order system is critically damped when the roots of its characteristic equation are

(*a*) negative, real and unequal

(*b*) complex conjugates

(*c*) negative, real and equal

(*d*) positive, real and equal

Directions: *The following, item consist of two statements, one labelled as the 'Assertion (A)' and the other as 'Reason (R)'. You are to examine these two statements carefully and select the answers to these items using the codes given below:*

Codes:

(*a*) Both A and R are individually true, and R is the correct explanation of A

(*b*) Both A and R are individually true but R is not the correct explanation of A

(*c*) A is true but R is false

(*d*) A is false but R is true

63. Assertion (A) :

A periodic function satisfying Dirichlet's conditions can be expanded into a Fourier series.

Reason (R) :

A Fourier series is a summation of weighted sine and cosine waves of the fundamental frequency and its harmonics.

64. Consider the following systems:

1. $y[k] = x[k] + a_1 x[k-1] - b_1 y[k-1] - b_2 y[k-2]$
2. $y[k] = x[k] + a_1 x[k-1] + a_2 x[k-2]$
3. $y[k] = x[k+1] + a_1 x[k] + a_2 x[k-1]$
4. $y[k] = a_1 x[k] + a_2 x[k+1] - b_1 y[k-2]$

Which of the systems given above represent recursive discrete systems?

(*a*) 1 and 4 (*b*) 1 and 2

(*c*) 1, 2 and 3 (*d*) 2, 3 and 4

65. A linear constant parameter continuous-time system has the following A matrix in its state representation X = AX + Bu:

$$A = \begin{bmatrix} 0 & \alpha \\ \alpha & 0 \end{bmatrix}$$

Which one of the following is the natural response of the system following the application of any input?

(*a*) $Me^{\alpha t} + Ne^{-\alpha t}$ (*b*) $(M + Nt)e^{\alpha t}$

(*c*) $(M + Nt)e^{-\alpha t}$ (*d*) $M \cos \alpha t + N \sin \alpha t$

66. For half-wave (odd) symmetry, with T_0 = period of $x(t)$, which one of the following is correct?

(*a*) $x(t \pm T_0/2) = -x(t)$

(*b*) $x(t \pm T_0/2) = x(t)$

(*c*) $x(t \pm T_0) = -x(t)$

(*d*) $x(t \pm T_0) = x(t)$

67. A square wave is defined by

$$x(t) = \begin{cases} A, & 0 < t < T_0/2 \\ -A, & T_0/2 < t < T_0 \end{cases}$$

It is periodically extended outside this interval. What is the general coefficient a_n in the Fourier expansion of this wave?

(*a*) 0 (*b*) $\dfrac{2A(1-\cos n\pi)}{n\pi}$

(*c*) $\dfrac{2A(1+\cos n\pi)}{n\pi}$ (*d*) $\dfrac{2A(1-\cos n\pi)}{[(n+1)\pi]}$

68. What is the inverse Laplace transform of $\dfrac{e^{-as}}{s}$?

(*a*) e^{-at} (*b*) $u(t-a)$

(*c*) $\delta(t-a)$ (*d*) $(t-a)\,u(t-a)$

69. Match List - I with List - II pertaining to Fourier Representation Periodicity Properties and select the corrrect answer using the codes given below the lists:

List I (*Time Domain Property*)	List II (*Frequency Domain Property*)
A. Continuous	1. Periodic
B. Discrete	2. Continuous
C. Periodic	3. Non-periodic
D. Non-periodic	4. Discrete

Codes :

	A	B	C	D
(*a*)	3	4	1	2
(*b*)	2	4	1	3
(*c*)	2	1	4	3
(*d*)	3	1	4	2

70. Given : $x_1(t) = e^{k_1 t}\, u(t)$ and $x_2(t) = e^{-k_2 t}\, u(t)$.

Which one of the following gives their convolution?

(*a*) $\dfrac{\left[e^{k_1 t} - e^{-k_2 t}\right]}{[k_1 + k_2]}$ (*b*) $\dfrac{\left[e^{k_1 t} - e^{-k_2 t}\right]}{[k_2 - k_1]}$

(*c*) $\dfrac{\left[e^{k_1 t} + e^{-k_2 t}\right]}{[k_1 + k_2]}$ (*d*) $\dfrac{\left[e^{k_1 t} + e^{-k_2 t}\right]}{[k_2 - k_1]}$

71. Which one of the following must be satisfied if a signal is to be periodic for $-\infty < t < \infty$?

(*a*) $x(t + T_0) = x(t)$

(*b*) $x(t + T_0) = \dfrac{dx(t)}{dt}$

(*c*) $x(t + T_0) = \int_t^{T_0} x(t)\,dt$

(*d*) $x(t + T_0) + x(t) + kT_0$

72. What is the output of the system with $h[n] = \left(\dfrac{1}{2}\right)^n u(n)$ in response to the input $x[n] = 3 + \cos\left(\pi n + \dfrac{\pi}{3}\right)$?

(*a*) $y[n] = 3 + \dfrac{1}{3}\left[\cos\left(\pi n + \dfrac{\pi}{3}\right)\right]$

(*b*) $y[n] = 3 + \dfrac{2}{3}\left[\cos\left(\pi n + \dfrac{\pi}{3}\right)\right]$

(*c*) $y[n] = 1 + \dfrac{2}{3}\left[\sin\left(\pi n + \dfrac{\pi}{3}\right)\right]$

(*d*) $y[n] = 6 + \dfrac{2}{3}\left[\cos\left(\pi n + \dfrac{\pi}{3}\right)\right]$

73. A discrete time system has impulse response

$h(n) = a^n u(n+2), |a| < 1.$

The system is

(*a*) stable, casual and memory less

(*b*) unstable, casual and has memory

(*c*) stable, non-casual and has memory

(*d*) unstable, non-casual and memory less

74. What is the number of roots of the polynomial $F(z) = 4z^3 - 8z^2 - z + 2$,

lying outside the unit circle?

(*a*) 0

(*b*) 1

(*c*) 2

(*d*) 3

75. Which one of the following gives the cross-correlation ($Y_{xy}(k)$) of two finite length sequences $x(n) = \{1, 3, 1, 2\}$ and $y(n) = \{1, 2, 1, 3\}$?

(*a*) {3, 10, 8, 14, 7, 5, 2}

(*b*) {2, 10, 7, 14, 6, 6, 3}

(*c*) {3, 9, 8, 14, 7, 5, 2}

(*d*) {2, 10, 6, 14, 7, 5, 3}

76. Which one of the following gives the average value or expectation of the function g(X) of the random variable X? (Given *f*(X) is the probability density function)

(*a*) $E[g(X)] = \int_{-\infty}^{\infty} g(X)\, dX$

(*b*) $E[g(X)] = \int_{-\infty}^{\infty} g(X) f(X)\, dX$

(*c*) $E[g(X)] = \int_{-\infty}^{\infty} g^*(X)\, dX$

(*d*) $E[g(X)] = \int_{-\infty}^{\infty} \left[\frac{g(X)}{f(X)}\right] dX$

77. Consider the waveform shown in the figure below.

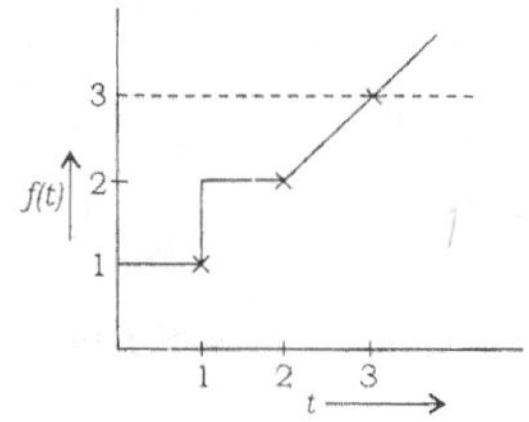

Which one of the following gives the correct description of the waveform

(*a*) $u(t) + u(t-1)$

(*b*) $u(t) + (t-1)\,u(t-1)$

(*c*) $u(t) + u(t-1) + (t-2)\,u(t-2)$

(*d*) $u(t) + (t-2)\,u(t-2)$

EXERCISE – II

1. Consider two signals

$x_1(t) = e^{j20t}$

and $x_2(t) = e^{(-2+j)t}$.

Which one of the following statements is correct ?

(*a*) Both $x_1(t)$ and $x_2(t)$ are periodic

(*b*) $x_1(t)$ is periodic but $x_2(t)$ is not periodic

(*c*) $x_2(t)$ is periodic but $x_1(t)$ is not periodic

(*d*) Neither $x_1(t)$ nor $x_2(t)$ is periodic **[IES]**

2. Match List-I with List-II and select the correct answer using the code given below the lists:

List I (*CT Function*)	**List II** (*CT Fourier Transform*)
A. $e^{-t}u(t)$	1. $\frac{2}{1+\omega^2}$
B. $x(t) = \begin{cases} 1, \lvert t\rvert \le 1 \\ 0, \lvert t\rvert > 1 \end{cases}$	2. $j\omega\, X(j\omega)$
C. $\frac{dx(t)}{dt}$	3. $\frac{1}{1+j\omega}$
D. $\frac{2}{1+t^2}$	4. $\frac{2\sin\omega}{\omega}$

Codes :

	A	B	C	D
(*a*)	1	4	2	3
(*b*)	3	2	4	1
(*c*)	1	2	4	3
(*d*)	3	4	2	1

[IES]

3. The system characterized by the equation

$y(t) = ax(t) + b$

is

(*a*) linear for any value of *b*

(*b*) linear if $b > 0$

(*c*) linear if $b < 0$

(*d*) non-linear **[IES]**

4. What is the Laplace transform of

$x(t) = -e^{2t} u(t) * (tu(t))$?

(*a*) $\frac{-1}{s^2(s+2)}$ (*b*) $\frac{-1}{s^2(s-2)}$

(*c*) $\frac{1}{s^2(s-2)}$ (*d*) $\frac{-1}{s(s-2)}$ **[IES]**

5. What is the inverse z transform of X (z)?

(a) $\frac{1}{2\pi j}\oint X(z)z^{n-1}dz$

(b) $\frac{1}{2\pi j}\oint X(z)z^{n+1}dz$

(c) $\frac{1}{2\pi j}\oint X(z)z^{1-n}dz$

(d) $2\pi j\oint X(z)z^{-(n+1)}dz$ **[IES]**

6. The impulse response of a system is $h(n) = a^n\, u(n)$. What is the condition for the system to be BIBO stable?

(a) a is real and positive

(b) a is real and negative

(c) $|a| > 1$

(d) $|a| < 1$ **[IES]**

7. The region of convergence of z transform $x\,|n|$ consists of the values of z for which $x\,|n|r^{-n}$ is

(a) absolutely integrable

(b) absolutely summable

(c) unity

(d) < 1 **[IES]**

8. Match List-I with List-II and select the correct answer using the code given below the lists:

List I (*Type*)	**List II** (*Application*)
A. Speech signal	1. The received signal of a radar system monitoring variation in prevalent weather condition
B. Non-stationary signal	2. One dimensional signal where amplitude varies with time
C. Ram signal	3. Signals of coupled system of nonlinear difference equation
D. Chaotic signal	4. Ensemble of unpredica-table wave-forms

Codes :

	A	B	C	D
(a)	2	1	4	3
(b)	4	3	2	1
(c)	2	3	4	1
(d)	4	1	2	3

[IES]

9. For a random variable x having the probability density function (PDF) as shown in the figure below, what are the values of the mean and the variance, respectively?

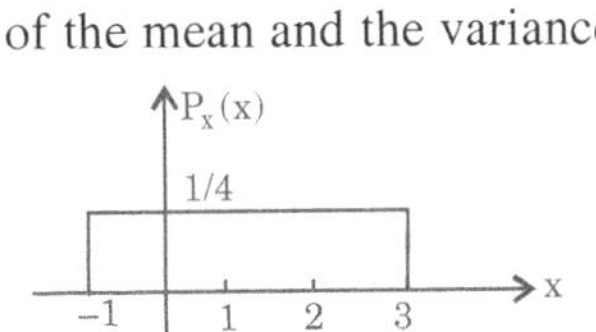

(a) $\frac{1}{2}$ and $\frac{2}{3}$ (b) 1 and $\frac{4}{3}$

(c) 1 and $\frac{2}{3}$ (d) 2 and $\frac{4}{3}$ **[IES]**

Directions: *The next items consist of two statements, one labelled as the 'Assertion (A)' and the other as 'Reason (R)'. You are to examine these two statements carefully and select the answers to these item using the codes given below:*

Codes :

(a) Both A and R are individually true and R is the correct explanation of A

(b) Both A and R are individually true but R is **not** the correct explanation of A

(c) A is true but R is false

(d) A is false but R is true

10. Assertion (A) :

The impulse response is only a function of the terms in natural response.

Reason (R) :

The differentiation and differencing operations eliminate the constant terms associated with the particular solution in the step response and change only the constants associated with exponential terms in the natural reponse.

[IES]

11. $x[n]$ is defined as

$$x[n] = \begin{cases} 0, \text{for } n < -2 \text{ or } n > 4 \\ 1, \text{ otherwise} \end{cases}$$

Determine the value of n for which $x[\,-n-2]$ is guaranteed to be zero.

(a) $n < 1$ and $n > 7$

(b) $n < -4$ and $n > 2$

(c) $n < -6$ and $n > 0$

(d) $n < -2$ and $n > 4$ **[IES]**

12. Match List-I with List-II and select the correct answer using the code given below the lists:

List I (*Application of Signals*)	**List II** (*Definition*)
A. Reconstruction	1. Sampling rate is chosen significa-ntly greater than the Nyquist rate
B. Over sampling	2. A mixture of continuous and discrete time signals
C. Interpolation	3. To convert the discrete time sequence back to a continuous time signal and then resample
D. Decimation	4. Assign values between samples and signals

Codes :

	A	B	C	D
(a)	3	4	1	2
(b)	2	1	4	3
(c)	3	1	4	2
(d)	2	4	1	3

[IES]

13. The continuous time system described by

$$y(t) = x(t)^2$$

is

(*a*) causal, linear and time-varying
(*b*) causal, non-linear and time-varying
(*c*) non-causal, non-linear and time-invariant
(*d*) non-causal and time-variant **[IES]**

14. The discrete- time signal $[n]$ is given as

$$x[n] = \begin{cases} 1, & n = 1, 2 \\ -1, & n = -1, -2 \\ 0, & n = 0 \text{ and } |n| > 2 \end{cases}$$

Which one of the following is the time-shifted signal $y[n] = x[n+3]$?

(*a*) $y[n] = \begin{cases} 1, & n = -1, -2 \\ -1, & n = -4, -5 \\ 0, & n = -3n < -5 \text{ and } n > -1 \end{cases}$

(*b*) $y[n] = \begin{cases} 0, & n = -1, -2 \\ -1, & n = -4, -5 \\ 1, & n = -3n < -5 \text{ and } n > -1 \end{cases}$

(*c*) $y[n] = \begin{cases} 1, & n = -1, -2 \\ 0, & n = -4, -5 \\ -1, & n = -3n < -5 \text{ and } n > -1 \end{cases}$

(*d*) $y[n] = \begin{cases} -1, & n = 1, 2 \\ 1, & n = 4, 5 \\ 0, & n = 3n > 5 \text{ and } n < 1 \end{cases}$

[IES]

15. 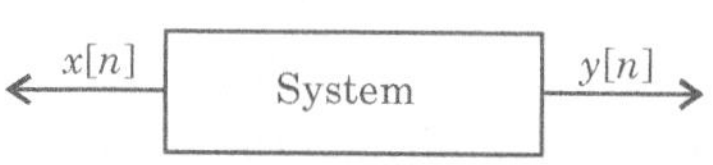

For the system shown,

$x[n] = k\delta[n]$,

and $y[n]$ is related to $x[n]$ as

$$y[n] - \frac{1}{2}y[n-1] = x[n]$$

What is $y[n]$ equal to?

(*a*) k
(*b*) $(1/2)^n k$
(*c*) nk
(*d*) 2^n **[IES]**

16. For continuous-time systems state variable representation is

$$\frac{d}{dt}q(t) = A\,q(t) + bx(t)$$

What is the corresponding represen-tation for discrete-time system?

(*a*) $\frac{d}{dt}q[n] = A\,q[n] + bx[n]$
(*b*) $q[n+1] = A\,q[n] + bx[n]$
(*c*) $q[n] = A\,q[n-1] + bx[n]$
(*d*) $\frac{d}{dn}[q(qn+1)] = A\,q[n-1] + bx[n-1]$ **[IES]**

17. If the Fourier transform of $x(t)$ is $\frac{2}{\omega}\sin(\pi\omega)$, then what is the Fourier transform of $e^{j5t}x(t)$?

(*a*) $\frac{2}{\omega-5}\sin(\pi\omega)$
(*b*) $\frac{2}{\omega}\sin\{\pi(\omega-5)\}$
(*c*) $\frac{2}{\omega+5}\sin\{\pi(\omega+5)\}$
(*d*) $\frac{2}{\omega-5}\sin\{\pi(\omega-5)\}$ **[IES]**

18. $F(s) = \frac{8s+10}{(s+1)(s-2)^3}$ is equal to

(*a*) $\frac{2}{s+1} + \frac{4}{(s+2)^3} - \frac{4}{(s+2)^2} - \frac{2}{s+2}$
(*b*) $\frac{2}{s+1} + \frac{6}{(s+2)^3} - \frac{2}{(s+2)^2} - \frac{2}{s+2}$
(*c*) $\frac{2}{s+1} + \frac{6}{(s+2)^3} - \frac{2}{(s+2)^2} + \frac{2}{s+2}$
(*d*) $\frac{4}{s+1} + \frac{6}{(s+2)^3} - \frac{2}{(s+2)^2} - \frac{4}{s+2}$ **[IES]**

19. For the function $x(t)$, $x(s)$ is given by

$$x(s) = e^{-s}\left[\frac{-2}{s(s+2)}\right]$$

Then, what are the initial and final values of $x(t)$, respectively ?

(*a*) 0 and 1 (*b*) 0 and – 1
(*c*) 1 and 1 (*d*) – 1 and 0 **[IES]**

20. The Laplace transform $X(s)$ of a function $x(t)$ is

$$X(s) = \left(\frac{1-e^{-sT}}{s}\right)$$

What is the wave shape of $x(t)$?

(*a*) 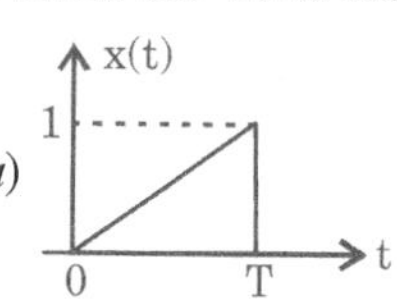

(*b*) 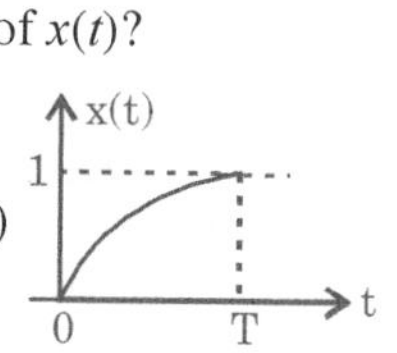

(*c*) 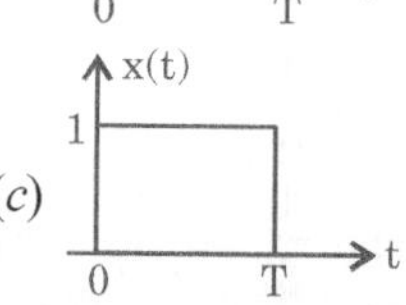

(*d*) 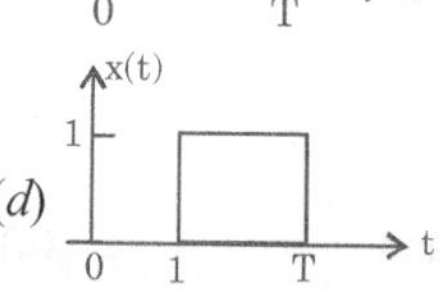

[IES]

21. What is the inverse Fourier transform of $u(\omega)$?

(a) $\frac{1}{2}\delta(t)+\frac{1}{\pi t}$ (b) $\frac{1}{2}\delta(t)$

(c) $2\,\delta(t)+\frac{1}{\pi t}$ (d) $2\,\delta(t)+sgn\,(t)$ **[IES]**

22. A signal $v(n)$ is defined by

$$v[n]=\begin{cases}1; & n=1\\ -1; & n=-1\\ 0; & n=0 \text{ and } |n|>1\end{cases}$$

What is the value of the composite signal defined as $v[n] + v[-n]$?

(a) 0 for all integer values of n

(b) 2 for all integer values of n

(c) 1 for all integer values of n

(d) –1 for all integer values of n **[IES]**

23. The lengths of two discrete time sequences $x_1(n)$ and $x_2(n)$ are 5 and 7, respectively. What is the maximum length of a sequence $x_1(n) * x_2(n)$?

(a) 5 (b) 6

(c) 7 (d) 11 **[IES]**

24. In the graph shown below, which one of the following expresses v(t) ?

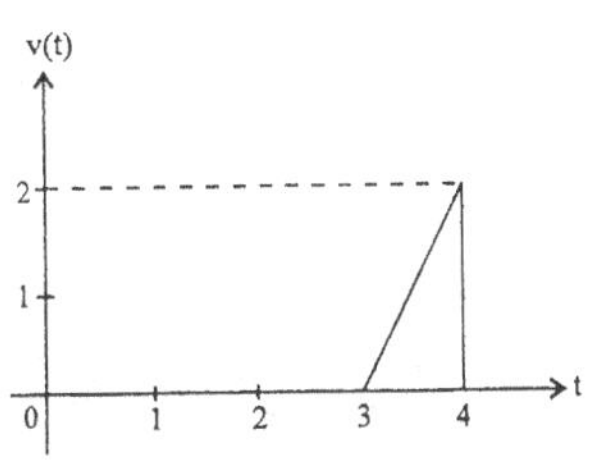

(a) $(2t+6)\,[u(t-3)+2u(t-4)]$

(b) $(-2t-6)\,[u(t-3)+u(t-4)]$

(c) $(-2t+6)\,[u(t-3)+u(t-4)]$

(d) $(2t-6)\,[u(t-3)-u(t-4)]$ **[IES]**

25. The governing differential equations connecting the output $y(t)$ and the input $x(t)$ of four continuous time systems are given in the List I and List II respectively. Match List I with List II and select the correct answer using the code given below the Lists :

List I (*Equation*)	**List II** (*System Category*)
A. $2t\frac{dy}{dt}+4y=2tx$	1. Linear, time invariant and dynamic
B. $y\frac{dy}{dt}+4y=2x$	2. Non-linear, time invariant and dynamic
C. $4\frac{d^2y}{dt^2}+2\frac{dy}{dt}+y=3\frac{dx}{dt}$	3. Linear, time-variable and dynamic
D. $\left(\frac{dy}{dt}\right)^2+2ty=4\frac{dx}{dt}$	4. Non-linear, time variable and dynamic
	5. Non-linear, time invariant & non dynamic

Codes :

	A	B	C	D
(a)	3	2	1	4
(b)	4	1	5	3
(c)	3	1	5	4
(d)	4	2	1	3

[IES]

26. Consider the following statements about linear time-invariant (LTI) continuous time systems :

1. The output signal in an LTI system with known input and known impulse response can always be determined.
2. A causal LTI system is always stable.
3. A stable LTI system has an impulse response, h(t) which has a finite value when integrated over whole of the time axis, i.e. $\int_{-\infty}^{+\infty} h(\lambda)d\lambda$ is finite.

Which of the statements given above are correct?

(a) 1 and 3 (b) 1 and 2

(c) 2 and 3 (d) 1, 2 and 3 **[IES]**

27. The unit sample response of a discrete system is 1 $\frac{1}{2}$ $\frac{1}{4}$ 0 0 0 ..... For and input sequence 1 0 1 0 0 0 ......, what is the output sequence ?

(a) 1 $\frac{1}{2}$ $\frac{1}{4}$ $\frac{1}{2}$ $\frac{1}{4}$ 0 0 ....

(b) 1 0 $\frac{1}{4}$ 0 0 .....

(c) 2 $\frac{1}{2}$ $\frac{5}{4}$ 0 0 ....

(d) 1 $\frac{1}{2}$ $\frac{5}{4}$ $\frac{1}{2}$ $\frac{1}{4}$ 0 0 ..... **[IES]**

28. Which one of the followig represents the phase response of the function

$$H(s)=\frac{s^2+\omega_0^2}{s^2+(\omega_0/Q)s+\omega^2}\ ?$$

(a)
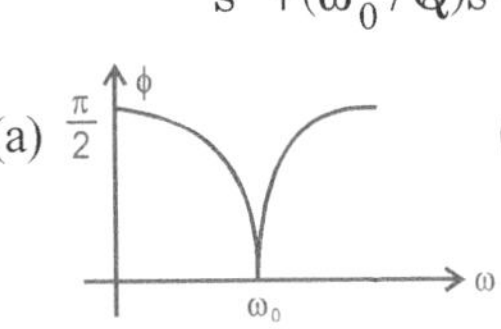

(b)
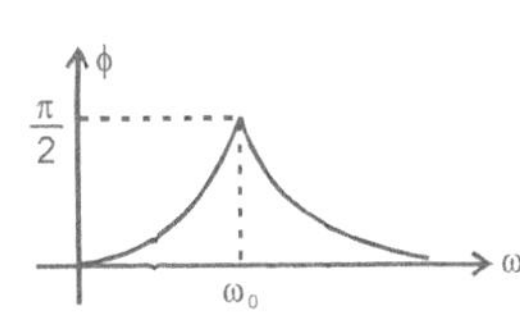

(c)
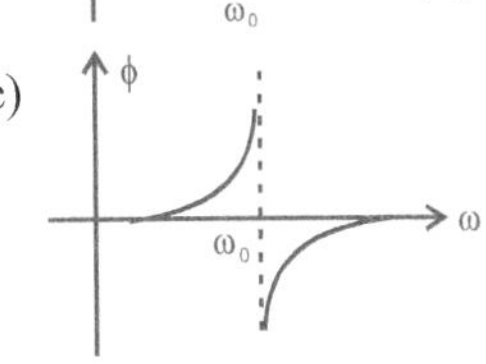

(d)
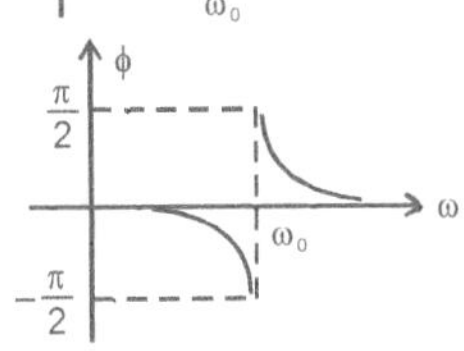

[IES]

29. What does the transfer function of a system describe for the system ?

(*a*) Only zero-input response

(*b*) Only zero-state response

(c) Both zero-input and zero-state responses

(*d*) Neither zero-input response not zero-state response

[IES]

30. Match List I with List II and select the correct answer using the code given below the Lists:

List I (*Time Function*)	**List II** (*Fourier Spectrum/ Fourier Transform*)
A. Periodic Function	1. Continuous spectrum at all frequencies
B. Aperiodic Function	2. $\delta(\omega)$
C. Unit Impulse $\delta(t)$	3. Line discrete spectrum
D. sin ωt	4. 1

Codes :

	A	B	C	D
(*a*)	4	2	3	1
(*b*)	3	1	4	2
(*c*)	4	1	3	2
(*d*)	3	2	4	1

[IES]

31. Which one of the following is the correct relation ?

(*a*) $F(at) \leftrightarrow aF(\omega/a)$

(*b*) $F(at) \leftrightarrow aF(a\omega)$

(*c*) $F(t/a) \leftrightarrow aF(\omega/a)$

(*d*) $F(at) \leftrightarrow (1/a)F(\omega/a)$ **[IES]**

32. The Fourier transform of a function is equal to its two-sided Laplace transform evaluated

(*a*) on the real axis of the s-plane

(*b*) on a line parallel to the real axis of the s-plane

(*c*) on the imaginary axis of the s-plane

(*d*) on a line parallel to the imaginary axis of the s-plane

[IES]

33. If the Fourier transform of $f(t)$ is $f(j\omega)$, then what is the Fourier transform of $f(-t)$?

(*a*) $F(j\omega)$

(*b*) $F(-j\omega)$

(*c*) $-F(j\omega)$

(*d*) Complex conjugate of $F(j\omega)$ **[IES]**

34. If $f(t)$ is an even function, then what is its Fourier transform $F(j\omega)$?

(*a*) $\int_0^{\infty} f(t)\cos(2\omega t)dt$ (*b*) $2\int_0^{\infty} f(t)\cos(\omega t)dt$

(*c*) $2\int_0^{\infty} f(t)\sin(\omega t)dt$ (*d*) $\int_0^{\infty} f(t)\sin(2\omega t)dt$

[IES]

35. What is the laplace transform of cos $\omega_0 t$?

(*a*) $\frac{\omega_0}{s^2+\omega_0^2}$

(*b*) $\frac{s}{s^2+\omega_0^2}$

(*c*) $\frac{\omega_0}{(s+\omega_0)^2}$

(*d*) $\frac{s}{(s+\omega_0)^2}$ **[IES]**

36. Match List I with List II and select the correct answer using the code given below the lists :

List I

(*Function in time domain f(t)*)

A. $\sin \omega_0 t\, u(t-t_0)$

B. $\sin \omega_0(t-t_0)\, u(t-t_0)$

C. $\sin \omega_0(t-t_0)\, u(t)$

D. $\sin \omega_0 t\, u(t)$

List II

(*Corresponding Laplace transform f(s)*)

1. $\frac{\omega_0}{s^2+\omega_0^2}$

2. $\left\{\frac{\omega_0}{s^2+\omega_0^2}\right\}e^{-t_0 s}$

3. $\frac{e^{-t_0 s}}{\sqrt{s^2+\omega_2^0}}\sin\left(\omega_0 t_0+\tan^{-1}\frac{\omega_0}{s}\right)$

4. $\frac{1}{\sqrt{s^2+\omega_2^0}}\sin\left(\omega_0 t_0-\tan^{-1}\frac{\omega_0}{s}\right)$

Codes :

	A	B	C	D
(*a*)	3	1	4	2
(*b*)	4	2	3	1
(*c*)	3	2	4	1
(*d*)	4	1	3	2

[IES]

37. Which one of the following is the impulse response of the system whose step response is given as

$C(t) = 0.5(1-e^{-2t})u(t)$?

(*a*) $e^{-2t}u(t)$ (*b*) $0.5\delta(t)+e^{-2t}u(t)$

(*c*) $0.5\delta(t)-0.5e^{-2t}u(t)$ (*d*) $0.5e^{-2t}u(t)$

[IES]

38. The convolution of $f(t)$ with itself is given to be $\int_0^t F(\tau)d\tau$. Then what is $f(t)$?

(*a*) The unit ramp function

(*b*) Equal to 1

(*c*) The unit step function

(*d*) The unit impulse function **[IES]**

39. If $X(z)$ is $\dfrac{1}{1-z^{-1}}$ with $|z| > 1$, then what is the corresponding $x(n)$?

(*a*) e^{-n} (*b*) e^{n}

(*c*) $u(n)$ (*d*) $\delta(n)$ **[IES]**

40. What is the spectral density of white noise ?

(*a*) A constant

(*b*) $\delta(\omega)$

(*c*) $[\delta(\omega)]^2$

(*d*) A step function in ω **[IES]**

41. A random variable X is defined by the double exponential distribution

$$P_x(x) = ae^{-b|x|}, -\infty < x < \infty$$

where a and b are +ve constants.

What is the relation between a and b so that $p_x(x)$ is a probability denstiy function ?

(*a*) $a = b/2$ (*b*) $b = a/2$

(*c*) $a = b$ (*d*) $a = 1/b$ **[IES]**

42. If a linear time invariant system is excited by a true random signal like white noise, the output of the linear system will have which of the following properties ?

(*a*) Output will be a white noise

(*b*) Output will be periodic

(*c*) Output will not be random

(*d*) Output will be correlated or coloured noise **[IES]**

43. Which of the following is/are not a property/properties of a power spectral density functions $S_x(\omega)$?

(*a*) $S_x(\omega)$ is a real function of ω

(*b*) $S_x(\omega)$ is an even function of ω

(*c*) $S_x(\omega)$ is a non-positive function of ω i.e. $S_x(\omega) \le 0$ for all ω

(*d*) All of these **[IES]**

44. The output of a linear system for any input can be computed in which one of the following ways ?

(*a*) Only by summation of impulse responses by convolution integral

(*b*) Only by summation of step responses by superposition integral

(*c*) Neither (*a*) nor (*b*)

(*d*) Using either (*a*) or (*b*) **[IES]**

45. Energy of a power signal is

(*a*) finite (*b*) zero

(*c*) infinite (*d*) between 1 and 2 **[IES]**

46. Algebraic expression for Z transform of $x[n]$ is X(z). What is the algebraic expression for Z transform of $e^{j\omega_o n}X[n]$?

(*a*) $X(z - z_0)$ (*b*) $X(e^{-j\omega_o} z)$

(*c*) $X(e^{j\omega_o} z)$ (*d*) $X(z)\, e^{j\omega_o z}$ **[IES]**

47. Which one of the following rules determines the mapping of s-plane to z-plane ?

(*a*) Right half of the s-plane maps into outside of the unit circle in z-plane

(*b*) Left half of the s-plane maps into inside of the unit circle

(*c*) Imaginary axis in s-plane maps into the circumference of the unit circle

(*d*) All of these **[IES]**

48. A system can be completely described by a transfer function if it is

(*a*) non-linear and continuous

(*b*) linear and time-varying

(*c*) non-linear and time-invariant

(*d*) linear and time-invariant **[IES]**

49. What is the Z-transform of the signal $x[n] = \alpha^n u(n)$?

(*a*) $X(z) = \dfrac{1}{z-1}$

(*b*) $X(z) = \dfrac{1}{1-z}$

(*c*) $X(z) = \dfrac{z}{z-\alpha}$

(*d*) $X(z) = \dfrac{1}{z-\alpha}$ **[IES]**

50. The response of a linear, time-invariant system to a unit step is

$s(t) = (1 - e^{-t/RC})\, u(t)$,

where $u(t)$ is the unit step.

What is the impulse response of this system ?

(*a*) $e^{-t/RC}$

(*b*) $e^{-t/RC}\, u(t)$

(*c*) $1/RC\, \{e^{-t/RC}\, u(t)\}$

(*d*) $\delta(t)$ **[IES]**

51. What is the output as $t \to \infty$ for a system that has a transfer function $G(s) = \dfrac{2}{s^2 - s - 2}$; when subjected to a step input ?

(*a*) –1 (*b*) 1

(*c*) 2 (*d*) Unbounded **[IES]**

52. The outputs of two systems S_1 and S_2 for the same input $x[n] = e^{jxn}$ are 1 and $(-1)^n$, respectively. Which one of the following statements is correct ?

(*a*) Both S_1 and S_2 are linear time invariant (LTI) systems

(*b*) S_1 is LTI but S_2 is not LTI

(*c*) S_1 is not LTI but S_2 is LTI

(*d*) Neither S_1 nor S_2 is LTI **[IES]**

53. A discrete-time signal x[n] has Fourier transform $X(e^{j\omega})$. Match List-I with List-II and select the correct answer using the code given below the lists :

List - I (*Signal*)	List - II (*Fourier Transform*)
A. $x[-n]$	1. $X^*(e^{-j\omega})$
B. $nx[n]$	2. $X(e^{-j\omega})$
C. $x^*[n]$	3. $e^{-j\omega}X(e^{j\omega})$
D. $x[n-1]$	4. $j\frac{d}{d\omega}X(e^{j\omega})$

Codes :

	A	B	C	D
(*a*)	1	3	2	4
(*b*)	2	4	1	3
(*c*)	1	4	2	3
(*d*)	2	3	1	4

[IES]

54. What is the laplace transform of a delayed unit impulse function $\delta(t-1)$?

(*a*) 1

(*b*) zero

(*c*) exp $(-s)$

(*d*) s **[IES]**

55. A real signal x(t) has Fourier transform X(f). Which one of the following is correct?

(*a*) Magnitude of X(*f*) has even symmetry while phase of X(*f*) has odd symmetry

(*b*) Magnitude of X(*f*) has odd symmetry while phase of X(*f*) has even symmetry

(*c*) Both magnitude and phase of X(*f*) have even symmetry

(*d*) Both magnitude and phase of X(*f*) have odd symmetry

[IES]

56. Consider the circuit shown below :

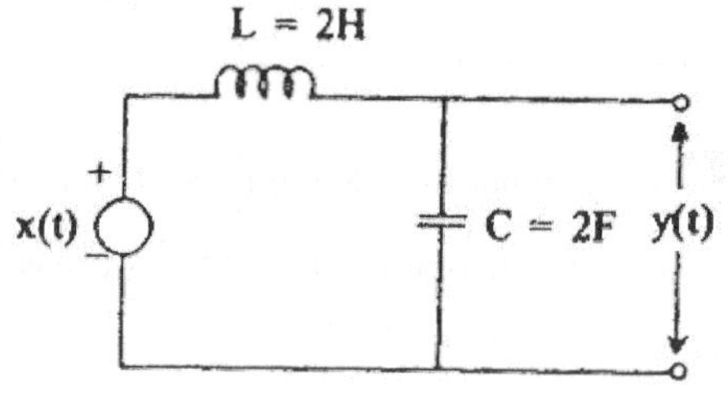

$x(t)$: Input voltage

$y(t)$: Output voltage

What is the natural response of this system ?

(*a*) A sinusoid with constant amplitude

(*b*) A growing sinusoid

(*c*) Zero

(*d*) A decaying sinusoid

[IES]

57. The relation between input $x(t)$ and output $y(t)$ of a continuous-time system is given by

$$\frac{dy(t)}{dt}+3y(t)=x(t).$$

What is the forced response of the system when $x(t) = k$ (a constant) ?

(*a*) k (*b*) $k/3$

(*c*) $3k$ (*d*) 0 **[IES]**

58. The discrete time signal x(n) is defined by

$$x(n)=\begin{cases}1 & n=1\\ -1 & n=-1\\ 0 & n=0 \quad \text{and}[n]>1\end{cases}$$

Which one of the following is the composite signal $y(n) = x(n) + x(-n)$ for all integer values of n ?

(*a*) 0 (*b*) 2

(*c*) ∞ (*d*) $-\infty$ **[IES]**

59. Which one of the following is the mathematical representation for the average power of the signal $x(t)$?

(*a*) $\frac{1}{T}\int_0^T x(t)dt$

(*b*) $\frac{1}{T}\int_0^T x^2(t)dt$

(*c*) $\frac{1}{T}\int_{-T/2}^{T/2} x(t)dt$

(*d*) $\underset{T\to\infty}{\text{Lt}}\frac{1}{T}\int_{-T/2}^{T/2} x^2(t)dt$ **[IES]**

60. Which one of the following systems described by the following input-output relations is non-linear ?

(*a*) $y(n) = n\,x(2n)$ (*b*) $y(n) = x(n^2)$

(*c*) $y(n) = n^2x(n)$ (*d*) $y(n) = x^2(n)$ **[IES]**

61. The mathematical model of the signal shown below is

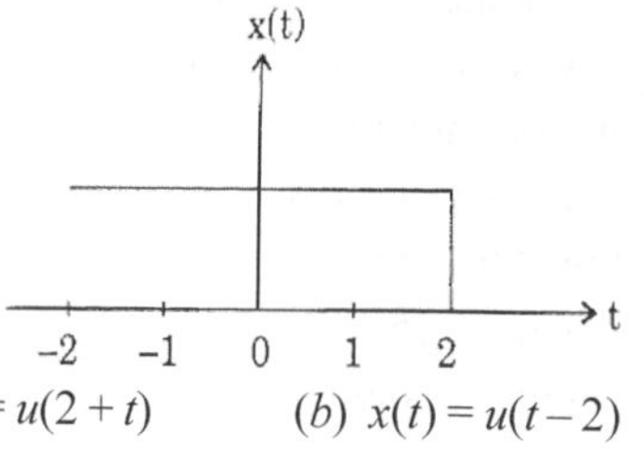

(*a*) $x(t) = u(2+t)$ (*b*) $x(t) = u(t-2)$

(*c*) $x(t) = u(2-t)$ (*d*) $x(t) = u(t-1)$ **[IES]**

62. Number of state variables of discrete time system, described by

$$y[n]-\frac{3}{4}y[n-1]+\frac{1}{8}y[n-2]=x[n]$$

is

(*a*) 2 (*b*) 3

(*c*) 4 (*d*) 1 **[IES]**

63. Unit step response of the system described by the equation $y(n) + y(n-1) = x(n)$ is

(*a*) $\frac{Z^2}{(Z+1)(Z-1)}$

(*b*) $\frac{Z}{(Z+1)(Z-1)}$

(*c*) $\frac{Z+1}{Z-1}$

(*d*) $\frac{Z(Z-1)}{(Z+1)}$ **[IES]**

64. The transfer function for the state representation of continuous time LTI system:

$$\dot{q}(t) = Aq(t) + bx(t)$$

$$y(t) = cq(t) + dx(t)$$

is given by

(*a*) $c(sI - A)^{-1} b + d$

(*b*) $b(sI - A)^{-1} c + d$

(*c*) $c(sI - A)^{-1} d + b$

(*d*) $d(sI - A)^{-1} b + c$ **[IES]**

65. The frequency response $H(\Omega)$ of a system for impulse sequence response

$$h[n] = \delta[n] + \delta[n-1]$$

is

(*a*) $H(\Omega) = 2\cos\left(\frac{\Omega}{2}\right) \angle -\frac{\Omega}{2}$

(*b*) $H(\Omega) = \cos\Omega \angle -\Omega$

(*c*) $H(\Omega) = 2\cos\Omega \angle -\frac{\Omega}{2}$

(*d*) $H(\Omega) = 2 \angle -\frac{\Omega}{2}$ **[IES]**

66. The Fourier transform of unit step sequence is

(*a*) $\pi\delta(\Omega)$

(*b*) $\frac{1}{1-e^{-j\Omega}}$

(*c*) $\pi\delta(\Omega) + \frac{1}{1-e^{-j\Omega}}$

(*d*) $1 - e^{-j\Omega}$ **[IES]**

Directions : *Each of the next two (2) items consists of two statements, one labelled as the 'Assertion (A)' and other as Reason (R)'. You are to examine these two statements carefully and select the answers to these items using the codes given below:*

Codes:

(*a*) Both A and R are individually true and R is the correct explanation of A

(*b*) Both A and R are individually true but R is ***not*** the correct explanation of A

(*c*) A is true but R is false

(*d*) A is false but R is true **[IES]**

67. Assertion (A) :

A linear system gives a bounded output if the input is bounded.

Reason (R) :

The roots of the characteristic equation have all negative real parts and the response due to initial conditions decay to zero as time t tends to infinity. **[IES]**

68. Assertion (A) :

The system described by

$$y^2(t) + 2y(t) = x^2(t) + x(t) + c$$

is a linear and static system.

Reason (R) :

The dynamic system is characterized by differential equation. **[IES]**

69. If random process $X(t)$ and $Y(t)$ are orthogonal, then

(*a*) $S_{XY}(f) = 0$ (*b*) $S_{XY}(f) = S_X(f) = S_Y(f)$

(*c*) $R_{XY}(\tau) = h(\tau)$ (*d*) $H(f) = 0$ **[IES]**

70. A first order circuit is excited with a dc source. The current $i(t)$ through any element of the circuit can be written as (i_f and i_i are final and initial values, respectively, of the current)

(*a*) $i_i - (i_i - i_f)\, e^{-\frac{t}{T}}$ (*b*) $i_f - (i_f - i_i)\, e^{-\frac{t}{T}}$

(*c*) $i_i - (i_f - i_i)\, e^{-\frac{t}{T}}$ (*d*) $i_f - (i_i - i_f)\, e^{-\frac{t}{T}}$ **[IES]**

71. If output of a linear system for step input is $t^2 e^{-2t}$, then the transfer function is

(*a*) $\frac{s}{(s+1)^2}$ (*b*) $\frac{2s}{(s+1)^3}$

(*c*) $\frac{s}{s^2(s+1)}$ (*d*) $\frac{1}{(s+1)^3}$ **[IES]**

72. Consider the following statements regarding the use of Laplace transforms and Fourier transforms in circuit analysis:

1. Both make the solution of circuit problems simple and easy.
2. Both are applicable for the study of circuit behaviour for $t - \alpha$ to α.
3. Both convert differential equations to algebraic equations.
4. Both can be used for transient and steady state analysis.

Which of the above statements are correct?

(*a*) 1, 2, 3 and 4 (*b*) 2, 3 and 4 only

(*c*) 1, 2 and 4 only (*d*) 1, 3 and 4 only **[IES]**

73. If response of LTI continuous time system to unit step input is $\left(\frac{1}{2} - \frac{1}{2}e^{-2t}\right)$, then impulse response of the system is

(*a*) $\left(\frac{1}{2} - \frac{1}{2}e^{-2t}\right)$ (*b*) (e^{-2t})

(*c*) $(1 - e^{-2t})$ (*d*) constant **[IES]**

74. A signal $x_1(t)$ and $x_2(t)$ constitute the real and imaginary parts respectively of a complex valued signal $x(t)$. What form of waveform does $x(t)$ possess ?

(*a*) Real symmetric

(*b*) Complex symmetric

(*c*) Asymmetric

(*d*) Conjugate symmetric **[IES]**

75. A function of one or more variables which conveys information on the nature of physical phenomenon is called

(*a*) noise (*b*) interference

(*c*) system (*d*) signal **[IES]**

76. The output $y(t)$ of a continuous-time system S for the input $x(t)$ is given by

$$y(t) = \int_{-\infty}^{t} x(\lambda)\, d\lambda$$

Which one of the following is correct ?

(*a*) S is linear and time-invariant

(*b*) S is linear and time-varying

(*c*) S is non-linear and time-invariant

(*d*) S is non-linear and time-varying **[IES]**

77. What is the period of the sinusoidal signal $x(n) = 5 \cos [0.2\,\pi n]$?

(*a*) 10 (*b*) 5

(*c*) 1 (*d*) 0 **[IES]**

78. Transfer function of a certain system is

$$\frac{Y(s)}{U(s)} = \frac{1}{s^4 + 5s^3 + 8s^2 + 6s + 3}$$

Which one of the following will be the A, B matrix pair of state variable representation of this system ?

(*a*) $\begin{bmatrix} 0 & 1 & 0 & 0 \\ 0 & 0 & 1 & 0 \\ 0 & 0 & 0 & 1 \\ -3 & -6 & -8 & -5 \end{bmatrix} \begin{bmatrix} 0 \\ 0 \\ 0 \\ 1 \end{bmatrix}$

(*b*) $\begin{bmatrix} 0 & 1 & 0 & 0 \\ 0 & 0 & 1 & 0 \\ 0 & 0 & 0 & 1 \\ -5 & -8 & -6 & -3 \end{bmatrix} \begin{bmatrix} 0 \\ 0 \\ 0 \\ 1 \end{bmatrix}$

(*c*) $\begin{bmatrix} 1 & 0 & 0 & 0 \\ 0 & 1 & 0 & 0 \\ 0 & 0 & 1 & 0 \\ -3 & -6 & -8 & -5 \end{bmatrix} \begin{bmatrix} 1 \\ 0 \\ 0 \\ 0 \end{bmatrix}$

(*d*) $\begin{bmatrix} 0 & 1 & 0 & 0 \\ 0 & 0 & 1 & 0 \\ 0 & 0 & 0 & 1 \\ -3 & -6 & -8 & -5 \end{bmatrix} \begin{bmatrix} 1 \\ 0 \\ 0 \\ 0 \end{bmatrix}$ **[IES]**

79. Separation property of state-transition matrix is

(*a*) $\phi(t - t_0) = \phi(t)\,\phi(t_0)$

(*b*) $\phi(t - t_0) = \phi^{-1}(t)\,\phi^{-1}(t_0)$

(*c*) $\phi(t - t_0) = \phi(t)\,\phi^{-1}(t_0)$

(*d*) $\phi(t - t_0) = \phi^{-1}(t)\,\phi^{-1}(t_0)$ **[IES]**

80. When $y(t) \xrightarrow{FT} Y(j\varpi)$;

$x(t) \xrightarrow{FT} X(j\varpi)$;

$h(t) \xrightarrow{FT} H(j\varpi)$

what is $Y(j\varpi)$?

(*a*) $\dfrac{X(j\varpi)}{H(j\varpi)}$ (*b*) $X(j\varpi)H(j\varpi)$

(*c*) $X(j\varpi) + H(j\varpi)$ (*d*) $X(j\varpi) - H(j\varpi)$ **[IES]**

81. For a series R-L-C circuit, the character-istic equation is given as

$$s^2 + \frac{R}{L}s + \frac{1}{LC} = 0$$

If $\dfrac{R}{2L}$ is denoted by α and $\dfrac{1}{\sqrt{LC}}$ by β, then under the condition of $\beta^2 > \alpha^2$, the system will be

(*a*) critically damped

(*b*) under damped

(*c*) undamped

(*d*) over damped **[IES]**

Directions : *Each of the next three (3) items consists of two statements, one labelled as the 'Assertion (A)' and the other as 'Reason (R)'. You are to examine these two statements carefully and select the answers to these items using the codes given below :*

Codes :

(a) Both A and R are individually true and R is the correct explanation of A

(b) Both A and R are individually true but R is ***not*** the correct explanation of A

(c) A is true but R is false

(d) A is false but R is true

82. **Assertion (A) :**

There are no convergence issues with discrete-time Fourier series in general.

Reason (R) :

A discrete-time signal is always obtained by sampling a continuous-time signal. **[IES]**

83. **Assertion (A) :**

The z-transform of output of an ideal samples is given by

$$Z[f(t)] = K_0 + \frac{K_1}{z} + \frac{K_2}{z^2} + \ldots + \frac{K_n}{z^n}$$

Reason (R) :

The relationship is the result of application of $z = e^{-sT}$, where T stands for the time gap between the samples. **[IES]**

84. **Assertion (A) :**

Sinusoidal response in an a.c. circuit can be represented in rectangual form as $v = V\cos\theta + j\,V\sin\theta$, exactly in a way similar to vectors which we come across for force analysis in mechanics.

Reason (R) :

Phasors are nothing but vectors. **[IES]**

85. Let $x(n)$ be a real-valued sequence that is a sample sequence of a wide-sense stationary discrete-time random process. The power density function of this signal is

(*a*) real, odd and non-negative
(*b*) real, even and non-negative
(*c*) purely imaginary, even and negative
(*d*) purely imaginary, odd and negative **[IES]**

86. A system defined by

$$y[n] = \sum_{k=-\infty}^{n} x[k]$$

is an example of

(*a*) invertible system
(*b*) memoryless system
(*c*) non-invertible system
(*d*) averaging system **[IES]**

87. Which one of the following represents the linear mathematical model of the physical system shown in the figure below ?

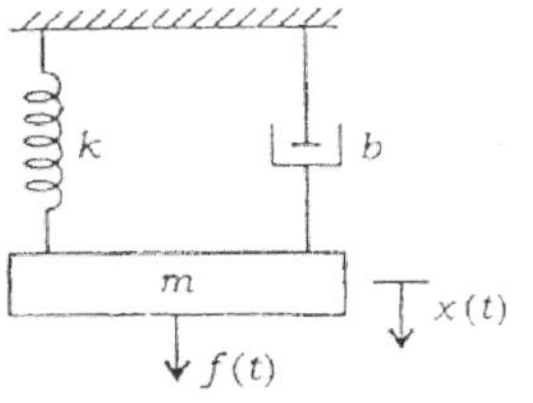

(*a*) $m\frac{d^2x(t)}{dt^2} + b\frac{dx(t)}{dt} + kx(t) = f(t)$

(*b*) $m\frac{d^2x(t)}{dt^2} + b\frac{dx(t)}{dt} + kx(t) = 0$

(*c*) $m\frac{d^2x(t)}{dt^2} + b\frac{dx(t)}{dt} + kx(t) + f(t) = 0$

(*d*) $m\frac{d^2x(t)}{dt^2} + b\frac{dx(t)}{dt} - kx(t) - f(t) = 0$ **[IES]**

88. If the state space equation of a system is

$$x = \begin{bmatrix} -3 & 1 \\ 0 & -2 \end{bmatrix} x$$

and initial state value $x(0) = [10 - 10]^T$, the steady-state value is $[A\ B]^T$, what are the respective values of A and B?

(*a*) 0, 0 (*b*) −3, −2
(*c*) 0, 1 (*d*) ∞, ∞ **[IES]**

89. Which one of the following functions is a periodic one ?

(*a*) $\sin(10\pi t) + \sin(20\pi t)$
(*b*) $\sin(10t) + \sin(20\pi t)$
(*c*) $\sin(10\pi t) + \sin(20t)$
(*d*) $\sin(10t) + \sin(25t)$ **[IES]**

90. What is the average power for periodic non-sinusoidal voltages and currents ?

(*a*) The average power of the fundamental component alone
(*b*) The sum of the average powers of the harmonics excluding the fundamental
(*c*) The sum of the average powers of the sinusoidal components including the fundamental
(*d*) The sum of the root mean square power of the sinusoidal components including the fundamental **[IES]**

91. If a random process X(t) is ergodic, then statistical averages

(*a*) and time averages are different
(*b*) and time averages are same
(*c*) are greater than time averages
(*d*) are smaller than time averages **[IES]**

92. Z and laplace transform are related by

(*a*) $s = \ln z$ (*b*) $s = \frac{\ln z}{T}$
(*c*) $s = z$ (*d*) $s = \frac{T}{\ln z}$ **[IES]**

93. Convolution of two sequences $X_1[n]$ and $X_2[n]$ is represented as

(*a*) $X_1(z) * X_2(z)$ (*b*) $X_1(z).X_2(z)$
(*c*) $X_1(z) + X_2(z)$ (*d*) $X_1(z) / X_2(z)$ **[IES]**

94. Decimation is the process of

(*a*) retaining sequence values of $X_p[n]$ other than zeroes
(*b*) retaining all sequence values of $X_p[n]$
(*c*) dividing the sequence value by 10
(*d*) multiplying the sequence value by 10 **[IES]**

95. Consider the following statements:

1. The Laplace transform of the unit impulse function is $s \times$ Laplace transform of the unit ramp function.
2. The impulse function is a time derivative of the ramp function.
3. The Laplace transform of the unit impulse function is $s \times$ Laplace transform of the unit step function.
4. The impulse function is a time derivative of the unit step function.

Which of the above statements are correct?

(*a*) 1 and 2 only
(*b*) 3 and 4 only
(*c*) 2 and 3 only
(*d*) 1, 2, 3 and 4 **[IES]**

96. Consider the following statements:

Fourier series of any periodic function $X(t)$ can be obtained if

1. $\int_0^1 [x(t)]dt < \infty$
2. Finite number of discontinuous exist within finite time interval t.

Which of the above statements is/are correct?

(*a*) 1 only (*b*) 2 only
(*c*) Both 1 and 2 (*d*) Neither 1 nor 2 **[IES]**

97. Frequency scaling [relationship between discrete time frequency (Ω) and continuous time frequency (ω)] is defined as

(*a*) $\omega = 2\Omega$ (*b*) $\omega = 2\,T_S/\Omega$
(*c*) $\Omega = 2\,\omega/T_S$ (*d*) $\Omega = \omega\,T_S$ **[IES]**

98. If $X(\omega) = \delta(\omega - \omega_0)$, then $X(t)$ is

(*a*) $e^{-j\omega_0 t}$ (*b*) $\delta(t)$
(*c*) $\frac{1}{2\pi}e^{j\omega_0 t}$ (*d*) 1 **[IES]**

99. For distortionless transmission through LTI system phase of $H(\omega)$ is

(*a*) constant
(*b*) one
(*c*) zero
(*d*) linearly dependent on ω **[IES]**

100. Consider the following statements:

1. The coefficients in the polynomials $p(s)$ and $q(s)$ must be real and positive.
2. Poles and zeroes of $z(s)$ must be conjugate if imaginary or complex.

Which of these statements are associated with the driving point function $Z(s) = \frac{p(s)}{q(s)}$?

(*a*) Both 1 and 2
(*b*) 1 only
(*c*) 2 only
(*d*) Neither 1 nor 2 **[IES]**

ANSWERS

EXERCISE – I

1. (*c*)	2. (*b*)	3. (*b*)	4. (*b*)	5. (*d*)	6. (*d*)	7. (*a*)	8. (*c*)	9. (*c*)	10. (*a*)
11. (*b*)	12. (*d*)	13. (*b*)	14. (*b*)	15. (*c*)	16. (*b*)	17. (*b*)	18. (*d*)	19. (*a*)	20. (*d*)
21. (*d*)	22. (*b*)	23. (*d*)	24. (*d*)	25. (*a*)	26. (*d*)	27. (*d*)	28. (*d*)	29. (*d*)	30. (*b*)
31. (*a*)	32. (*b*)	33. (*a*)	34. (*d*)	35. (*c*)	36. (*d*)	37. (*d*)	38. (*d*)	39. (*b*)	40. (*c*)
41. (*d*)	42. (*c*)	43. (*a*)	44. (*c*)	45. (*b*)	46. (*b*)	47. (*d*)	48. (*a*)	49. (*b*)	50. (*a*)
51. (*c*)	52. (*a*)	53. (*b*)	54. (*a*)	55. (*b*)	56. (*d*)	57. (*a*)	58. (*a*)	59. (*d*)	60. (*d*)
61. (*b*)	62. (*c*)	63. (b)	64. (*a*)	65. (*a*)	66. (b)	67. (*a*)	68. (*b*)	69. (*b*)	70. (*a*)
71. (*a*)	72. (*)	73. (*a*)	74. (*c*)	75. (*)	76. (*b*)	77. (*c*)			

EXERCISE – II

1. (*b*)	2. (*d*)	3. (*d*)	4. (*b*)	5. (*a*)	6. (*d*)	7. (*b*)	8. (*c*)	9. (*b*)	10. (*b*)
11. (*c*)	12. (*c*)	13. (*d*)	14. (*a*)	15. (*b*)	16. (*b*)	17. (*d*)	18. (*b*)	19. (*b*)	20. (*c*)
21. (*a*)	22. (*a*)	23. (*d*)	24. (*a*)	25. (*d*)	26. (*a*)	27. (*d*)	28. (*d*)	29. (*b*)	30. (*b*)
31. (*d*)	32. (*c*)	33. (*b*)	34. (*b*)	35. (*b*)	36. (*c*)	37. (*a*)	38. (*c*)	39. (*c*)	40. (*a*)
41. (*a*)	42. (*c*)	43. (*c*)	44. (*d*)	45. (*c*)	46. (*b*)	47. (*d*)	48. (*d*)	49. (*c*)	50. (*c*)
51. (*d*)	52. (*d*)	53. (*b*)	54. (*c*)	55. (*a*)	56. (*a*)	57. (*b*)	58. (*a*)	59. (*d*)	60. (*d*)
61. (*c*)	62. (*a*)	63. (*a*)	64. (*a*)	65. (*a*)	66. (*c*)	67. (*a*)	68. (*c*)	69. (*a*)	70. (*b*)
71. (*b*)	72. (*b*)	73. (*b*)	74. (*d*)	75. (*d*)	76. (*a*)	77. (*a*)	78. (*b*)	79. (*c*)	80. (*b*)
81. (*b*)	82. (*d*)	83. (*c*)	84. (*a*)	85. (b)	86. (*a*)	87. (*a*)	88. (*a*)	89. (*a*)	90. (*d*)
91. (*b*)	92. (*b*)	93. (*b*)	94. (*a*)	95. (*b*)	96. (*c*)	97. (*d*)	98. (*c*)	99. (*d*)	100. (*a*)

EXPLANATIONS

EXERCISE – I

18. Constant at 0dB for $\omega < 1$ and a slope of. – 20 dB/decade for $\omega > 1$.

19.
$$y[n] = \sum_{k=-\infty}^{n} x[k]$$
$$y[n] = \sum_{k=-\infty}^{n-1} x[k] + x[n]$$
$$y[n] = y[n-1] + x[n]$$
$$\therefore \quad x[n] = y[n] - y[n-1]$$

28.
$$V_{ave} = \frac{1}{2\pi}\int_0^{\pi} V_m \sin\omega t\, dt$$
$$= \frac{V_m}{2\pi}[-\cos\omega t]_0^{\pi}$$
$$= \frac{V_m}{\pi}$$
$$= 0.318\, V_m$$

30. Given : $\dfrac{d^2y}{dt^2} = x(t-2)\,u(t-2) + \dfrac{d^2x}{dt^2}$

Taking laplace transform
$$s^2 Y(s) = (e^{-2s} + s^2)\, X(s)$$
$$\therefore \quad T(s) = \frac{Y(s)}{X(s)} = 1 + \frac{e^{-2s}}{s^2}$$

35. We know
$$f(0^+) = \lim_{s\to\infty} sF(s)$$
$$= \lim_{s\to\infty} s\left(\frac{27s+97}{s^2+33s}\right)$$
$$= \lim_{s\to\infty}\left(\frac{27+\frac{97}{s}}{1+\frac{33}{s}}\right)$$
$$= 27$$

36. $v(t) = u(t-1) + u(t-2) - 2u(t-3)$

Taking laplace transform
$$V(s) = \frac{e^{-s}}{s} + \frac{e^{-2s}}{s} - \frac{2e^{-3s}}{s}$$
$$= \frac{1}{s}\left[e^{-s} + e^{-2s} - 2e^{-3s}\right]$$

38.
$$v(t) = \sin\omega t$$
$$V(s) = \int_0^{\infty} v(t)\, e^{-st}\, dt$$
$$= \int_0^{\infty}\frac{1}{2}\left[e^{\omega t} - \frac{1}{2}e^{-\omega t}\right]e^{-st}\,dt$$
$$= \frac{1}{2}\int_0^{\infty}[e^{(\omega - s)t} - e^{(\omega + s)t}]\,dt$$
$$= \frac{\omega}{s^2+\omega^2}$$

39.
$$x(f) = \int_{-\infty}^{\infty} e^{-\alpha t^2}.e^{-j2\pi ft}.dt$$

Put $t^2\alpha = t'^2\pi$
$$= \sqrt{\frac{\pi}{\alpha}}\int_{-\infty}^{\infty} e^{-\pi t'^2}.e^{-j2\pi f.t'\frac{\pi}{\infty}}.dt$$
$$= \sqrt{\frac{\pi}{\alpha}}\cdot e^{-\pi^2\frac{f^2}{\alpha}}$$

48. $f(t) = \left[u(t) + u(t-1) - 2u(t-2)\right]$
$$F(s) = \frac{1}{s} + \frac{e^{-s}}{s} - \frac{2}{s}e^{-2s}$$

49. The Fourier transform does not exist because the signal is unstable.
$$\int_0^{\infty}\left(e^t\, v(t)\right) dt = \infty$$

50.
$$Y[n] = \sum_{k=-\infty}^{\infty} x[k]h[n-k]$$
$$= \sum_{k=-\infty}^{\infty} a^k\, u[k] b^{n-k}\, u[n-k]$$

51.
$$Y(n) - Y(n-1) = X(n)$$
$$(1 - z^{-1})Y(z) = X(z)$$
$$\Rightarrow \quad Y(z) = \frac{X(z)}{1-z^{-1}}$$

52.
$$y(t) = x(t) \times \cos 2\pi.\ 300\, t$$
$$= 5\cos 400\,\pi t \times \cos 600\,\pi t$$
$$= \frac{5}{2}[\cos 2.500\,\pi t + \cos 2.100\,\pi t]$$

L. P. F. of cut off frequency = 150 Hz

$$y(t) = \frac{5}{2}[\cos 2.100\,\pi t] = 100 \text{ Hz}$$

53. $x(n) = u(n)$

$$x(z) = \frac{1}{1-z^{-1}} = \frac{z}{z-1}$$

54. Given : $X(z) = \dfrac{z}{(z-2)(z-3)}$

$$= \frac{A}{(z-2)} + \frac{B}{(z-3)}$$

Solving, we get

$$A = \left.\frac{z}{z-3}\right|_{z=2} = \frac{2}{-1} = -2 \;;$$

and $$B = \left.\frac{z}{z-2}\right|_{z=3} = 3$$

$$\Rightarrow \qquad X(z) = -\frac{2}{(z-2)} + \frac{3}{(z-3)}$$

$$x(n) = [2^n - 3^n]\,u(-n-1)$$

55. $R_x(0) = \varepsilon\,[y^2(t)]$

56. Length of given trapezoidal = 8

60. $x(t) = \cos 2000\,\pi t + 3\sin 6000\,\pi t$.

$x(t)$ is band-limited with $f_M = 3000$ Hz.

Hence, Nyquist rate is 6000 Hz or 6 kHz.

67. $a_n = \dfrac{2}{T_0}\displaystyle\int_0^{T_0} x(t)\,.\cos n\omega t.\,dt$

$$= \frac{2}{T_0}\int_0^{T_{0/2}} A\cos n\omega t\,dt + \frac{2}{T_0}\int_{T_{0/2}}^{T_0} (-A)\,.\cos n\omega t.\,dt$$

$$= \frac{2}{T_0}\left[\left.\frac{A\sin n\omega t}{n\omega}\right|_0^{T_{0/2}}\right] - \frac{2}{T_0}\left[\left.\frac{A\sin n\omega t}{n\omega}\right|_{T_{0/2}}^{T_0}\right]$$

$$= \frac{2A}{nT_0\omega}[\sin n\pi - \sin 2n\pi + \sin n\pi]$$

$$= \frac{2A}{n\pi}\sin n\pi\,(1-\cos n\pi)$$

$$= 0$$

70. $x_1(t)+x_2(t) = \displaystyle\int_{-\infty}^{\infty} x_1(t-\tau)\,x_2(\tau).\,d\tau$

$$= \int_{-\infty}^{\infty} e^{k_1(t-\tau)}\,u(t-\tau)\,.\,e^{-k_2 t}\,.\,u(\tau)\,.\,d\tau$$

$$= \int_0^t e^{k_1(t-\tau)}\,.\,e^{-k_2\tau}\,.\,d\tau$$

$$= e^{k_1 t}\int_0^t e^{-(k_1+k_2)\tau}\,.\,d\tau$$

$$= e^{k_1 t}\left[-\left.\frac{e^{-(k_1+k_2)\tau}}{k_1+k_2}\right|_0^t\right]$$

$$= \frac{e^{k_1 t} - e^{-k_2 t}}{k_1+k_2}$$

77. $f(t) = u(t) - u(t-1) + 2\,[u(t-1) - u(t-2)] + tu(t-2)$

$$f(t) = u(t) + u(t-1) + (t-2)\,u(t-2)$$

EXERCISE – II

4. $X(s) = L\left[-e^{-2t}u(t)\right]\,.\,L[tu(t)]$

Using $L[x_1(t) * X_2(t)] = X_1(s)\,.\,X_2(s)$, we get

$$X(s) = \frac{-1}{(s-2)}\cdot\frac{1}{s^2}$$

$$= -\frac{1}{s^2(s-2)}$$

9. $$P_x(x) = \begin{cases}\dfrac{1}{4}, & -1 \le x \le 3\\ 0, & \text{otherwise}\end{cases}$$

$$E(x) = \mu_x = \int_{-\infty}^{\infty} xP_x(x).\,dx$$

$$= \int_{-1}^{3}\frac{1}{4}.x.\,dx = 1$$

$$E[x^2] = \int_{-\infty}^{\infty} x^2.P_x(x).\,dx$$

$$= \int_{-1}^{3}\frac{x^2}{4}.\,dx$$

$$= \frac{7}{3}$$

Then, variance $\sigma_x^2 = E[x^2] - (E[x])^2$

$$= \frac{7}{3} - 1 = \frac{4}{3}$$

22. $$v[n] = \begin{cases} 1 \;; & n = 1 \\ -1 \;; & n = -1 \\ 0 \;; & n = 0 \text{ and } |n| > 1 \end{cases}$$

$$v[-n] = \begin{cases} 1 & n = -1 \\ -1 & n = 1 \\ 0 & n = 0 \text{ and } |n| > 1 \end{cases}$$

$v[n] + v[-n] = \{0$, for all values of $n\}$

23. Here, $l = 5,\ k = 7$

$\therefore$ Maximum length of a sequence

$$= 5 + 7 - 1 = 11$$

24. $$\frac{2t\,dy}{dx} + 4y = 2\,tx$$

Hence time variant (because t is present, linear and dynamics.

$\frac{y\,dy}{dt} + 4y = 2x$, non linear, time invariant.

25.

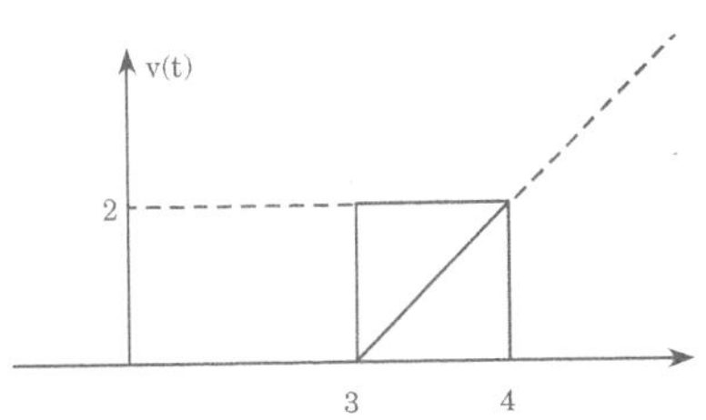

$$v(t) = (2t-6) \times \left[u(t-3) - u(t-4)\right]$$

27. $$h[n] = \delta[n] + \frac{1}{2}\delta[n-1] + \frac{1}{4}\delta[n-2]$$

$$x[n] = \delta[n] + \delta[n-2]$$

$$y[n] = x[n] \times h[n]$$

$$= \left[\delta[n] + \delta[n-2]\right]$$

$$\times\left[\delta[n] + \frac{1}{2}\delta[n-1] + \frac{1}{4}\delta[n-2]\right]$$

$$= \delta[n] + \frac{1}{2}\delta[n-1] + \frac{5}{4}\delta[n-2]$$

$$+\frac{1}{2}\delta[n-3] + \frac{1}{4}\delta[n-4]$$

28. $$H(j\omega) = \frac{-\omega^2 + \omega_0^2}{-\omega^2 + \left(\frac{\omega_0}{Q}\right)(j\omega) + \omega_0^2}$$

$$H(j\omega) = -\tan^{-1}\left(\frac{\omega_0}{Q\left(\omega_0^2 - \omega^2\right)}\right)$$

37. $$C(s) = 0.5\left(\frac{1}{s} - \frac{1}{s+2}\right)$$

$$C(s) = \frac{1}{s(s+2)}$$

$$H(s) = \frac{C(s)}{R(s)} = \frac{1}{s+2}.$$

Impulse response,

$$h(t) = L^{-1}H(s) = e^{-2t} \times u(t).$$

40. $$S_{xx}(\omega) = \int_{-\infty}^{\infty} R_{XX}(\tau).\bar{e}^{j\omega\tau}.d\tau$$

$$= \frac{\eta}{2} \text{ (constant)}$$

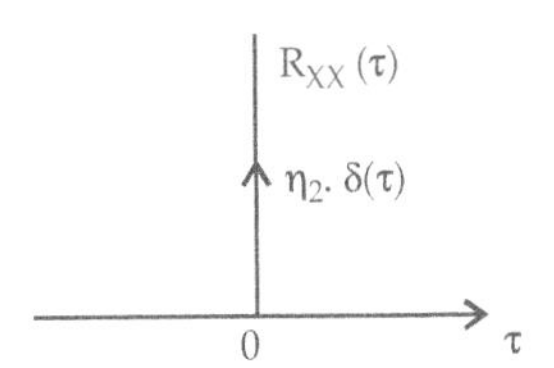

41. $$\int_{-\infty}^{\infty} p_x(x).dx = 1$$

$$\Rightarrow \int_{-\infty}^{\infty} a.e^{-b|x|}.d_x = 1$$

$$\Rightarrow \int_{-\infty}^{0} ae^{+bx}.dx + \int_{0}^{\infty} a\,e^{-bx}.dx = l$$

$$\Rightarrow \frac{a}{b} + \frac{a}{b} = 1$$

$$\Rightarrow \frac{2a}{b} = 1 = a = \frac{b}{2}$$

49. $$X(z) = \sum_{n=-\infty}^{\infty} \alpha^n U(n) z^{-n}$$

$$= \sum_{n=0}^{\infty} \alpha^n z^{-n}$$

$= 1 + \alpha z^{-1} + \alpha^2 z^{-2} + ...$

$= \dfrac{1}{1 - \alpha z^{-1}}$

$= \dfrac{z}{z - \alpha}$

50. $s(t) = (1 - e^{-t/RC}).\, u(t)\, S(s)$

$= \dfrac{1}{s} - \dfrac{1}{s + \dfrac{1}{RC}}$

$= \dfrac{1}{RC} \cdot \dfrac{1}{s(s + \dfrac{1}{RC})}.$

$\therefore \quad H(s) = \dfrac{S(s)}{R(s)}$

$= \dfrac{1}{RC} \dfrac{1}{\left(s + \dfrac{1}{RC}\right)} \delta(t)$

$= L^{-1}\, \delta(t) \left\{ \dfrac{1}{RC} \cdot \dfrac{1}{\left(s + \dfrac{1}{RC}\right)} \right\}$

$= \dfrac{1}{RC} \cdot e^{-\frac{t}{RC}} . u(t)$

51. $c(t) = G(s).R(s)$

$= \dfrac{2}{s^2 - s - 2} \cdot \dfrac{1}{s}$

$c(\infty) = \underset{s \to 0}{lt}\; G(s)$

$= \underset{s \to 0}{lt} \dfrac{2}{(s^2 - s - 2)s}$

$= \infty.$

57. $\dfrac{dy(t)}{dt} + 3y(t) = x(t)$

Taking Laplace tranform of both sides, we have

$sY(s) + 3\, Y(s) = X(s)$

$\Rightarrow \quad Y(s) = \dfrac{X(s)}{s+3} = \dfrac{k}{s(s+3)}$

$\Rightarrow \quad y(t) = \dfrac{k}{3}\left[1 - e^{-\frac{t}{3}}\right]$

$\therefore$ Forced response $= \dfrac{k}{3}.$

61.

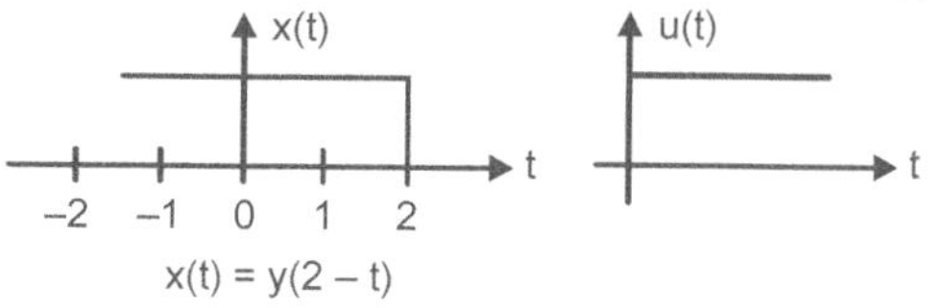

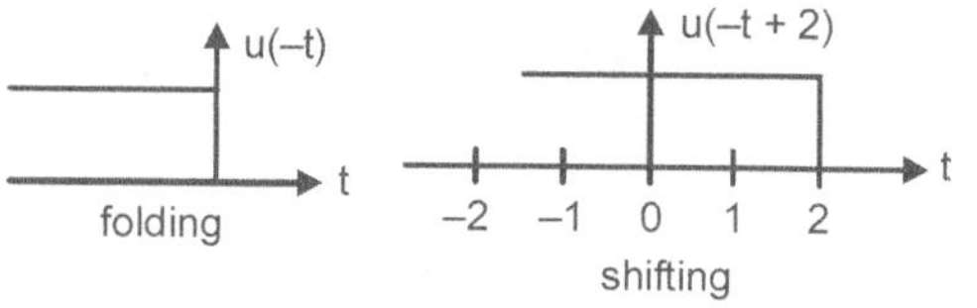

62. $y[n] - \dfrac{3}{4} y[n-1] + \dfrac{1}{8} y[n-2] = x[n]$

Above system is 2nd order, so '2' state variable are required.

63. $y[n] + y[n-1] = x[n]$

Taking z transfer

$y(z)(1 + z^{-1}) = x(z)$

$\Rightarrow \quad y(z) = \dfrac{z}{(z+1)} x(z)$

$x(n) = u(n)$

and $\quad x(z) = \dfrac{1}{1 - z^{-1}} = \dfrac{z}{z-1}$

$y(z) = \dfrac{z^2}{(z+1)(z-1)}$

64. $\dot{q}(t) = A\, q(t) + bx(t)$

$y(t) = c\, q(t) + dx(t)$

Taking Laplace transform with initial condition zero

$sq(s) - Aq(s) = bx(s)$

$q(s) = [sI - A]^{-1}\, b\, x(s)$

$y(s) = cq(s) + d\, x(s)$

$T(s) = \dfrac{y(s)}{x(s)}$

$= c[sI - A]^{-1} b + d$

= Transfer function

65. $h[n] = \delta[n] + \delta[n-1]$

Taking Z-transform

$H[z] = 1 + e^{-Z}$

$H(j\Omega) = 1 + e^{-j\Omega}$

$= e^{-j\Omega/2}\left[e^{j\Omega/2} + e^{-j\Omega/2}\right]$

$= 2\cos\left(\dfrac{\Omega}{2}\right) \angle -\dfrac{\Omega}{2}$

66. Fourier transfer of unit step sequence is

$$\pi\delta(\Omega)+\frac{1}{1-e^{-j\Omega}}$$

69. If random process $x(t)$ and $y(t)$ are orthogonal, then

$$S_{XY}(f)=0$$

70. $i=i_f-(i_f-i_i)\,e^{-t/T}$

71. Given : $\quad f(t)=t^2 e^{-2t}$

$$\therefore \quad y(s)=\frac{2}{(s+2)^3}$$

$$y(s)=\frac{1}{s}$$

Transfer function,

$$T(s)=\frac{y(s)}{x(s)}=\frac{2s}{(s+2)^3}$$

This answer is not available in given options.

72. *Laplace and Fourier transform*

(1) Both make the solution of circuit problem simple and easy.

(2) Both are applicable for the study of circuit behaviour for $t-\alpha$ to α.

(3) Both convert differential equations to algebraic equation.

(4) Both can be used for transient and steady state analysis.

All above statements are correct.

73. $$\delta(t)=\frac{d}{dt}\{u(t)\}$$

Impulse response (unit step response)

$$=\frac{d}{dt}\left\{\frac{1}{2}-\frac{1}{2}e^{-2t}\right\}$$

$$=0-\frac{1}{2}(-2)e^{-2t}$$

$$=e^{-2t}$$

77. $$0.2\pi=\omega=\frac{2\pi}{T}$$

$$\Rightarrow \quad T=10$$

78. Taking Inverse Laplace to transfer function,

$$y^4+5y^3+8y^2+6y^1+3y=u$$

Putting $\quad x_1=y$

$$x_2=\dot{y}=y^1$$

$$x_3=\ddot{y}=y^2$$

$$x_4=\dddot{y}=y^3$$

We have, $\quad \dot{x}_4=-x_1-6x_2-8x_3-5x_4+u$

Then the state equations are

$$\begin{bmatrix}\dot{x}_1\\ \dot{x}_2\\ \dot{x}_3\\ \dot{x}_4\end{bmatrix}=\begin{bmatrix}0&1&0&0\\0&0&1&0\\0&0&0&1\\-1&-6&-8&-5\end{bmatrix}\begin{bmatrix}x_1\\x_2\\x_3\\x_4\end{bmatrix}+\begin{bmatrix}0\\0\\0\\1\end{bmatrix}u$$

91. A random process is called *ergodic process* if the time-average are equal to ensemble average, ensemble averages, the average is taken over the ensemble of waveforms, keeping the time constant or fixed.

If a random process $x(t)$ is ergodic, then statistical averages and time averages are same.

92. $$Z=e^{sT}$$

$$\Rightarrow \quad ln\,Z=sT$$

$$\Rightarrow \quad \frac{lnZ}{T}=s$$

93. $$y(n)=X_1[n]*X_2[n]$$

Taking Z transfrom both size

$$y(z)=X_1(z)\cdot X_2(z).$$

94. Decimation is the process of retaining sequence values of $X_p[n]$ other than zeros.

95. $\delta(t)\Rightarrow L\{\delta(t)\}=1$

Laplace transform of unit steps is

$$L\{u(t)\}=\frac{1}{s}$$

$$s\times L\{u(t)\}=s\times\frac{1}{s}=1=L\{\delta(t)\}$$

$$\delta(t)=\frac{d}{dt}\{u(t)\}$$

96. Fourier series of any periodic function $x(t)$ can be obtained if it follow the Drichlet's condition.

1. $x(t)$ must be integrable in time period 'T'.
2. $x(t)$ must have finite number of discontinuity in time interval 'T'.

Given :

(1) $\int_0^1 |x(t)|\,dt<\infty$

(2) Finite number of discontinuous exist within finite time interval 't'.

Here $x(t)$ follow the Drichlet's condition.

Both (1) & (2) statements are correct.

97. Frequency scaling [relationship between discrete time frequency (Ω) and continuous time frequency (ω)] is defined as $\Omega = \omega T_S$.

98. $$X(\omega) = \delta(\omega - \omega_0)$$

$$\therefore \quad X(t) = \frac{1}{2\pi}\int_{-\pi}^{\pi} \delta(\omega - \omega_0)e^{j\omega t}\,d\omega$$

$$= \frac{1}{2\pi}e^{j\omega_o t}$$

99. For distortionless transmission through LTI system, phase of $H(\omega)$ is linearly dependent on ω.

100. $Z(s) = \dfrac{p(s)}{q(s)}$ for this

(1) The coefficients in the polynomials $p(s)$ and $q(s)$ must be real and positive.

(2) Poles and zeros of $Z(s)$ must be conjugate if imaginary or complex.

13

CHAPTER

Microprocessor

EXERCISE – I

1. Match List I with List II and select the correct answer using the coded given below the lists :

	List I		List II
A.	Immediate addressing	1.	LDA 30 SC
B.	Implied addressing	2.	MOV A, B
C.	Register addressing	3.	LXI H, 2050
D.	Direct addressing	4.	RRC

Codes :

	A	B	C	D
(*a*)	3	4	2	1
(*b*)	2	1	3	4
(*c*)	3	1	2	4
(*d*)	2	4	3	1

2. Consider the following statements:

The 8085 microprocessor will not enter into bus idle machine cycle whenever

1. INTR interrupt is recognised.
2. RST x.5 instruction is recognised.
3. DAD rp instruction is executed.

Which of the statements given above is/are correct?

(*a*) 1 only

(*b*) 2 only

(*c*) 1 and 2

(*d*) 2 and 3

3. The expression computed by the above program and stored in AX is

```
push    a       pop     bx
push    b       add     ax, bx
push    c
POP     AX
POP     BX
SUB     AX, BX
```

(*a*) a + b – c

(*b*) c + b – a

(*c*) c – b + a

(*d*) c – b – a

4. Match List I with List II and select the correct answer using the codes given below the Lists :

List I (*8085 Register*)	List II (*8086 Register*)
A. A	**1.** CH
B. H	**2.** AL
C. L	**3.** BL
D. B	**4.** BH

Codes :

	A	B	C	D
(*a*)	4	2	3	1
(*b*)	2	4	1	3
(*c*)	4	2	1	3
(*d*)	2	4	3	1

5. Consider the Motorola 68008, 68010, 68012 and 68020 microprocessors. In systems that are severely constrained by the space available for the printed circuit board, it is better to use

(*a*) 68008 (*b*) 68010

(*c*) 68012 (*d*) 68020

6. Consider the following features in an 8085 microprocessor system with memory mapped I/O :

1. I/O devices have 16-bit addresses

2. I/O devices are accessed using IN and OUT instructions

3. There can be maximum of 256 input devices and 256 output devices

4. Arithmetic and logic operations can be directly performed with the I/O data

Select the correct answer using the codes given below :

Codes :

(*a*) 1, 2 and 4 (*b*) 1, 3 and 4

(*c*) 2 and 3 (*d*) 1 and 4

7. In 8085 microprocessor the value of the most significant bit of the result following the execution of any arithmetic or Boolean instruction is stored in

(*a*) carry status flag

(*b*) auxiliary carry status flag

(*c*) sign status flag

(*d*) zero status flag

8. In a microprocessor when a CPU is interrupted, it
   (*a*) stops execution of instructions
   (*b*) acknowledges interrupt and branches of subroutine
   (*c*) acknowledges interrupt and continues
   (*d*) acknowledges interrupt and waits for the next instruction from the interrupting device
9. Which one of the following is not a characteristic of RISC processor design?
   (*a*) One instruction per cycle
   (*b*) Register-to-register operations only
   (*c*) Simple address modes
   (*d*) Register-to-memory operations only
10. The following micro operations are part of interrupt cycle of a control unit:
    1. MAR ← save-address
       PC ← routine-address
    2. MBR ← (PC)
    3. Memory ← (MBR)

    Which one of the following is the correct order of their occurrence?
    (*a*) 1, 2 and 3 (*b*) 2, 3 and 1
    (*c*) 2, 1 and 3 (*d*) 3, 1 and 2
11. Which of the following is not a characteristic of transparent DMA mode of I/O operation ?
    (*a*) The external logic steals cycles from the CPU
    (*b*) The normal rate of execution is slowed down
    (*c*) Only one word can be transferred at a time
    (*d*) Data is transferred to/from memory directly
12. A typical cell, for a dynamic RAM can be implemented by using how many MOS transistors?
    (*a*) Six (*b*) Five
    (*c*) One (*d*) Two
13. The contents of accumulator in an 8085 microprocessor are altered after the execution of the instruction?
    (*a*) CMPC (*b*) CP13A
    (*c*) ANISC (*d*) ORAA
14. In an 8085 microprocessor after the execution of XRAA instruction
    (*a*) the carry flag is set
    (*b*) the accumulator contains FF_H
    (*c*) the content of accumulator is shifted by one
    (*d*) the zero flag is set
15. A microprocessor has 24 address lines and 32 data lines. If it uses 10 bits of op code, the size of its Memory Buffer Register is
    (*a*) 22 bits (*b*) 24 bits
    (*c*) 32 bits (*d*) 14 bits
16. In the 8086 instruction ADD, DX, [BX] [CI], the addressing mode of source operand is
    (*a*) Register
    (*b*) Register Indirect
    (*c*) Based Indexed
    (*d*) Direct
17. A microprocessor based system can perform many different functions, because
    (*a*) its operation is controlled by software
    (*b*) it is digital system
    (*c*) it uses a RAM
    (*d*) it can be controlled by input and output devices
18. The interface chip used for data transmission between 8085 and a 16 bit ADC is
    (*a*) 8251
    (*b*) 8253
    (*c*) 8255
    (*d*) 8259
19. Following question consists of two statements, one labelled as the 'Assertion (A)' and the other as 'Reason (R)'. You are to examine these two statements carefully and select the answers to these items using the codes given below:

    Codes:
    (*a*) Both A and R are individually true and R is the correct explanation of A
    (*b*) Both A and R are individually true but R is **not** the correct explanation of A
    (*c*) A is true but R is false
    (*d*) A is false but R is true

    Assertion (A) **:**
    The main difference between a micropro-cessor and a microcon-troller is that the former does not have any on-chip main memory whereas latter has.

    Reason (R) :
    A microprocessor does not need memory to run programs.
20. Which of the following are included in the architecture of a computer ?
    1. Addressing modes, CPU
    2. Instruction set, data formats
    3. Secondary memory, operating system

    Select the correct answer from the codes given below :
    (*a*) 1 and 2
    (*b*) 1 and 3
    (*c*) 2 and 3
    (*d*) 3 only

21. A 3 × 8 decoder with two enable inputs is to be used to address 8 blocks of memory. What will be the size of each memory block when addressed from a sixteen bit bus with two MSBs used to enable the decoder ?

(*a*) 2K (*b*) 4K
(*c*) 16K (*d*) 64K

22. Match List I with List II and select the correct answer using the code given below the Lists :

List I (*Type of Memory*)	List II (*Used As*)
A. DRAM	1. Cache memory
B. SRAM	2. Main memory
C. Parallel Access Registers	3. BIOS memory
D. ROM	4. CPU Registers

Codes :

	A	B	C	D
(*a*)	1	4	2	3
(*b*)	3	4	2	1
(*c*)	1	2	4	3
(*d*)	3	2	4	1

23. Which of the following are the memory performance parameters ?

1. Access time and Latency
2. Block size and Block access time
3. Cycle time and Bandwidth

Select the correct answer from the codes given below :

(*a*) 1 only
(*b*) 1 and 2 only
(*c*) 2 and 3 only
(*d*) 1, 2 and 3

24. What is the address space of 8086 CPU ?

(*a*) One Megabyte
(*b*) 256 Kilobyte
(*c*) 1 K megabytes
(*d*) 64 Kilobytes

25. Which of the following counters can be used to divide the clock frequency of a microprocessor by 5 ?

(*a*) 3 bit counter (*b*) 5 bit counter
(*c*) mod 3 counter (*d*) mod 5 counter

26. Identification of highest priority interrupt can be achieved in a minimum time by which of the following schemes ?

(*a*) Hardwired polling
(*b*) Priority encoder circuit
(*c*) Software polling
(*d*) Subdivision of interrupt register into a number of sub-registers which are checked parallely

27. Which one of the following is used as the interface chip for data transmission between 8086 and a 16-bit ADC ?

(*a*) 8259 (*b*) 8255
(*c*) 8253 (*d*) 8251

28. Which one of the following addressing modes is used in the instruction PUSH B ?

(*a*) Direct (*b*) Register
(*c*) Register indirect (*d*) Immediate

29. Following items consists of two statements : one labelled as the 'Assertion (A)' and the other as 'Reason (R)'. You are to examine these two statements carefully and select the answers to these items using the code given below :

Codes :

(*a*) Both A and R are individually true and R is the correct explanation of A
(*b*) Both A and R are individually true but R is not the correct explanation of A
(*c*) A is true but R is false
(*d*) A is false but R is true

Assertion (A) :

DMA is faster than either Interrupt initiated I/O or Polling based I/O for very large data transfers.

Reason (R) :

DMA takes control of the system buses and needs no processor intervention during the data transfer.

30. The following micro-operations are part of interrupt cycle of a control unit :

(*a*) MAR ← save-address
(*b*) PC ← routine address
(*c*) MBR ← (PC)
(*d*) Memory ← (MBR)

31. A memory system of size 16 k bytes is required to be designed using memory chips which have 12 address lines and 4 data lines each. What is the number of such chips required to design the memory system ?

(*a*) 2 (b) 4
(*c*) 8 (*d*) 16

32. Consider the following statements :

1. In 8085 microprocessor, the pairs of general purpose registers for 16-bit operation are A-B, C-D and H-L.
2. In 8085 microprocessor, the data and the address multiplexed bus is known as ($AD_7 - AD_0$).
3. In 8085 microprocessor, ALE is used to operate multiplexed data address bus.

Which of the statements given above are correct ?

(*a*) 1, 2 and 3 (b) 2 and 3 only
(*c*) 1 and 3 only (*d*) 1 and 2 only

33. Match List I with List II and select the correct answer using the code given below the Lists :

List I (*Instruction of 8085*)	**List II** (*Function of Instruction*)
A. RST 6.5	1. Highest priority of interrupts
B. INTR	2. Vector interrupt
C. TRAP	3. Interrupt request
D. HLDA	4. Hold acknowledge

Code :

	A	B	C	D
(*a*)	1	2	3	4
(*b*)	1	2	3	4
(*c*)	2	3	1	4
(*d*)	2	3	4	1

34. Which of the following are features in an 8085 microprocessor system with memory mapped I/O ?

1. I/O devices have 16-bit addresses.
2. I/O devices are accessed using IN and OUT instructions.
3. There can be maximum of 256 input devices and 256 output devices.

Select the correct answer using the code given below :

Code :

(*a*) 1, 2 and 3
(*b*) 1 and 3 only
(*c*) 2 and 3 only
(*d*) 1 and 2 only

35. Match List I with List II and select the correct answer using the code given below the Lists :

List I	**List II**
A. 8255A	1. Programmable Peripheral Interface
B. 8279	2. Programmable Interval Timer
C. 8254/8253	3. DMA Controller
D. 8237/8257	4. Programmable Keyboard/Display Interface

Codes :

	A	B	C	D
(*a*)	2	4	1	3
(*b*)	1	3	2	4
(*c*)	2	3	1	4
(*d*)	1	4	2	3

36. Consider the following statements :

The advantage of cycle stealing in DMA is that

1. it increases the maximum I/O transfer rate
2. it reduces the interference by the DMA controller in CPU's memory access
3. it is beneficially employed for I/O devices with shorter bursts of data transfer

Which of the statements given above are correct ?

(*a*) 1 and 2 only
(*b*) 1 and 3 only
(*c*) 2 and 3 only
(*d*) 1, 2 and 3 only

37. Following items consit of two statements: one labelled as the 'Assertion (A)' and the other as 'Reason (R)'. You are to examine these two statements carefully and select the answers to these items using the codes given below:

Codes :

(*a*) Both A and R are individually true and R is the correct explanation of A
(*b*) Both A and R are individually true but R is not the correct explanation of A
(*c*) A is true but R is false
(*d*) A is false but R is true

Assertion (A) :

The development of a microprocessor based product requires the design of program and the hardware.

Reason (R) :

The design effort for an electronic product follows the same basic steps used in the development of software.

38. Which of the following instructions of an 8086 microprocessor uses the contents of a CX register as a counter?

1. LOCK
2. LOOP
3. ROTATE

Slect the correct answer using the code given below :

(*a*) Only 1 and 2
(*b*) Only 1 and 3
(*c*) Only 2 and 3
(*d*) 1, 2 and 3

39. For which of the following devices, is DMA the most suitable?

(*a*) Keyboard
(*b*) Mouse
(*c*) Joy stick
(*d*) Hard disk

40. Which one of the following pairs is not correctly matched?
(*a*) Horizontal micro-instruction : CISC-instructions
(*b*) Multiplier control : 2's complement
(*c*) Multiplier control unit : Encoding by function
(*d*) Vertical micro-instruction : RISC-instructions

41. Which one of the following pairs of 8086 microprocessor's interrupt is not correctly matched?
(*a*) INT 0 : Divide by zero
(*b*) INT 1 : Single step
(*c*) INT 2 : Break point
(*d*) INT 4 : Overflow

42. Which one of the following register of 8085 microprocessor is not a part of programming model?
(*a*) Instruction register
(*b*) Memory address register
(*c*) Status register
(*d*) Temporary data register

43. The following sequence of instructions is executed by an 8085 microprocessor :
1000 LXI SP, 27 FF
1003 CALL 1006
1006 POP H
What are the contents of the steak pointer (SP) and the HL register pair after completion of execution of these instructions ?
(*a*) SP = 27 FF, HL = 1003
(*b*) SP = 27 FD, HL = 1003
(*c*) SP = 27 FF, HL = 1006
(*d*) SP = 27 FD, HL = 1006

44. What are the number of machine cycles n, and the types of machine cycles carried out for PUSH B ?
(*a*) $n = 2$, fetch and memory write
(*b*) $n = 3$, fetch and 2 memory write
(*c*) $n = 3$, fetch, memory write and read
(*d*) $n = 3$, fetch and 3 memory read

45. Match List I (2 pins of 8086) with List II (Status) and select the correct answer using the code given below the lists :

	List I (*2 Pins of 8086*)		**List II** (*Status*)
	$\overline{BHE}$	A_0	What is read/written ?
A.	0	0	1. 1 byte from/to odd address
B.	0	1	2. 1 byte from/to even address
C.	1	0	3. 1 16-bit word
D.	1	1	4. NOP

Code :

	A	B	C	D
(*a*)	4	2	1	3
(*b*)	3	1	2	4
(*c*)	4	1	2	3
(*d*)	3	2	1	4

46. Consider the following statements :
The SIM instruction outputs the contents of accumulator to define
1. interrupt mask bit.
2. interrupt pending bit.
3. serial input data line.
4. serial output data line.

Which of the statements given above are correct?
(*a*) 1 and 2 (*b*) 2 and 3
(*c*) 3 and 4 (*d*) 1 and 4

47. An Intel 8085 processor is executing the program given below:

```
        MVI A, 10H
        MVI B, 10H
BACK:   NOP
        ADD B
        RLC
        JNC BACK
        HLT
```

The number of times that the operation NOP will be executed is
(*a*) 1 (*b*) 2
(*c*) 3 (*d*) 4

48. Which one of the following is not a vectored interrupt?
(*a*) TRAP
(*b*) INTR
(*c*) RST7.5
(*d*) RST3

49. The greatest negative number which can be stored in computer that has 8-bit word length and uses 2's complement arithmetic is
(*a*) –256 (*b*) –255
(*c*) –128 (d) –127

50. In a microcomputer, WAIT states are used to
(*a*) make the processor wait during a DMA operation
(*b*) make the processor wait during a power interrupt processing
(*c*) make the processor wait during a power shutdown
(*d*) interface slow peripherals to the processor

51. Which of the following statement(s) is correct?

(*a*) PROM contains a programmable AND array and a fixed OR array.

(*b*) PLA contains a fixed AND array and a programmable OR array

(*c*) PROM contains a fixed AND array and programmable OR array

(*d*) None of these

52. Following sequence of instructions are executed by an 8085 microprocessor :

```
1000    LXI SP, 27 FF
1003    CALL 1006
1006    POP H
```

Contents of the stack pointer (SP) and the HL register pair on completion of execution of these instructions are

(*a*) SP = 27 FF, HL = 1003

(*b*) SP = 27 FD, HL = 1003

(*c*) SP = 27 FF, HL = 1006

(*d*) SP = 27 FD, HL = 1006

53. In a 8085 microprocessor system with memory mapped I / O,

(*a*) I / O devices have 8-bit addresses

(*b*) I / O devices are accessed using IN and OUT instructions

(*c*) there can be a maximum of 256 input devices and 256 output devices

(*d*) arithmetic and logic operations can be directly performed with the I / O data.

54. A computer system has a 4K word cache organised in blockset associative manner with 4 blocks per set, 64 words per block. The number of bits in the SET and WORD fields of the main memory address formula is

(*a*) 15, 4

(*b*) 6, 4

(*c*) 7, 2

(*d*) 4, 6

55. In the program below, number of times the FIRST and SECOND JNZ instructions cause the control to be transfered to LOOP are respectively

```
            MVI H, 02H
            MVI L, 05H
LOOP      : DCR L
FIRST     : JNZ LOOP
            DLR H
SECOND    : JNZ LOOP
```

(*a*) 5 and 2 (*b*) 4 and 1

(*c*) 259 and 1 (*d*) 260 and 1

56. A dynamic RAM call which hold 5V has to be refreshed every 20 m secs, so that stored voltage does not fall below 0.5V. It call has a constant discharge current of 0.1 pA, then storage capacitance of the call is

(*a*) 4×10^{-6} (*b*) 4×10^{-9}

(*c*) 4×10^{-12} (*d*) 4×10^{-15}

57. An I/O processor controls the flow of information between

(*a*) Cache memory and I/O devices

(*b*) Main memory and I/O devices

(*c*) Two I/O devices

(*d*) Cache and main memory

58. On receiving an interrupt from an I/O device, the CPU

(*a*) halts for a predetermined time

(*b*) branches off to the interrupt service routine after completion of the current instruction

(*c*) branches off to the interrupt service routine immediately

(*d*) hands over control of address bus and data bus to the interrupting device

59. In microprocessor based systems DMA facility is required to increase the speed of data transfer between the

(*a*) microprocessor and the I/O devices

(*b*) microprocessor and the memory

(*c*) memory and the I/O devices

(*d*) memory and the reliability system

60. When compared with an RS232C serial port, the USB (Universal Serial Bus)

(*a*) supports a lower range of peripherals

(*b*) supports a faster transfer rate

(*c*) does not support, 'Hot plug-ability'

(*d*) controller in PC can not detect the presence or absence of USB devices

61. If the CALL instruction of 8085 in the main program is conditional then RETURN instruction in the subroutine can be

(*a*) conditional

(*b*) conditional or unconditional

(*c*) can be determined by LDA instruction

(*d*) unconditional

62. The 8085 has two registers known as primary data pointers. These are registers

(*a*) B and C

(*b*) D and E

(*c*) H and L

(*d*) C and D

63. READY signal in 8085 is useful when the CPU communicates with

(*a*) a slow peripheral device

(*b*) a fast peripheral device

(*c*) a DMA controller chip

(*d*) a PPI chip

64. In the following 8085 program how many times (decimal) is the DCRC executed?`

```
LOOP:    MVIC, 78 H
         DCRC
         JNZ LOOP
         HLT
```

(*a*) 119 (*b*) 120

(*c*) 78 (*d*) 77

65. A single instruction to clear the lower four bits of the accumulator in 8085 assembly language is

(*a*) XRI 0FH

(*b*) ANI F0H

(*c*) XRI F0H

(*d*) ANI 0FH

66. Which of the following instructions copies a byte of data into the accumulator from the memory address given in the instruction?

(*a*) LDA address

(*b*) LDAX B

(*c*) LHLD address

(*d*) STA address

67. An instruction used to set the carry flag in a computer can be classified as

(*a*) data transfer

(*b*) arithmetic

(*c*) logical

(*d*) program control

68. In an 8085 μP system, the RST instruction will cause an interrupt

(*a*) only if an interrupt service routine is not being executed

(*b*) only if a bit in the interrupt mask is made 0

(*c*) only if interrupts have been enabled by an EI instruction

(*d*) none of these

69. An I/O processor control the flow of information between

(*a*) cache memory and I/O devices

(*b*) main memory and I/O devices

(*c*) two I/O devices

(*d*) cache and main memories

70. MOS/LSI Digital system are becoming popular due to

(*a*) high packing density and low cost

(*b*) high packing density, small power consumption

(*c*) small size, low cost and small power consumption

(*d*) high packing density, low cost and small power consumption

71. If instruction RST is written in a program, then problem will jump to location

(*a*) 0020 H (*b*) 0024 H

(*c*) 0028 H (*d*) 002 CH

72. In a generic microprocessor, instruction cycle time is

(*a*) shorter than machine cycle time

(*b*) larger than machine cycle time

(*c*) exactly double the machine cycle time

(*d*) exactly the same as the machine cycle time

73. The data bus in 8080A/8085 microprocessor is a group of

(*a*) eight bidirectional lines that are used to transfer 8 bits between the microprocessor and its I/O and memory

(*b*) eight lines used to transfer data among the registers

(*c*) eight unidirectional lines that are used for I/O devices

(*d*) sixteen bidirectional lines that are used for data transfer between the microprocessor and memory

74. The TRAP is one of the interrupts available in INTEL 8085. The TRAP is

(*a*) level triggered

(*b*) negative edge triggered

(*c*) positive edge triggered

(*d*) both positive edge triggered and level triggered

75. Block search is used to locate a record in an ordered file consisting of N records. The optimum size of the block is

(*a*) $\sqrt{N}$ (*b*) $3\sqrt{N}$

(*c*) $\frac{N}{2}$ (*d*) $\frac{N}{3}$

76. If we use 3 bits in the instruction word to indicate if an index register is to be used and if necessary, which one is to be used, then the number of index registers to be used in the machine will be

(*a*) 3 (b) 6

(*c*) 7 (d) 8

77. In a multiprocessor configuration, two co-processors are connected to the host 8086 processor. The two co-processor instruction sets

(*a*) must be the same

(*b*) may overlap

(*c*) must be disjoint

(*d*) must be the same as that of the host

78. An advantage of memory interlacing is that

(*a*) a larger memory is obtained

(*b*) effective speed of the memory is increased

(*c*) the cost of the memory is reduced

(*d*) a non-volatile memory is obtained

79. Match List-I (Request) with List-II (Device) and select the correct answer using the codes given below the Lists :

List-I	List-II
A. Interrupt	1. I/O device
B. Wait	2. DMA controller
C. Hold	3. Memory

Codes:

	A	B	C
(*a*)	1	2	3
(*b*)	1	3	2
(*c*)	2	1	3
(*d*)	2	3	1

80. Consider the following statements:

1. A 33 MHz 486 has a higher MIPS rating than a 33 MHz 386.
2. A 33 MHz 486 has a higher MFLOPS rating than a 33 MHz 386.

Of these statements

(*a*) both I and II are true

(*b*) both I and II are false

(*c*) I is true but II is false

(*d*) I is false but II is true

81. The Intel Pentium processor address buts consists of two sets of signals :

29 address lines (A31 : A3)

and B Byte Enable (BE7# : BEO#) lines.

If the address on address lines (A31 : A3) is 0000 0108 (H) and all Byte Enable lines (BE7 # ; BE0#) are assorted, then processor's intention is to address

(*a*) one byte at address 0000 0108 (H)

(*b*) eight byte at locations 0000 0101 to 0000 0108 (H)

(*c*) eight bytes at locations 0000 0108 (H) to 0000 010F (H)

(*d*) one byte at location 0000 010F (H)

82. A personal computer has typically

(*a*) 5 to 10 kilobytes of main memory

(*b*) 10 to 100 kilobytes of main memory

(*c*) 100 to 256 kilobytes of main memory

(*d*) 256 kilobytes to 1 megabyte of main memory.

83. Match List-I (Type of instruction) with List-II (Instruction) and select the correct answer using the codes given below the Lists :

List-I	List-II
A. One byte instruction	1. MOV A, data
B. Two byte instruction	2. MOV A, B
C. Three byte instruction	3. MOV A, M
D. Register indirect addressing	4. JMP addr.

Codes :

	A	B	C	D
(*a*)	1	2	4	3
(*b*)	2	1	4	3
(*c*)	2	1	3	4
(*d*)	1	2	3	4

84. Consider the following set of instructions :

```
STC
CMC
MOV A,B
RAL
MOV B,A
```

This set of instructions

(*a*) doubles the number in Register by B

(*b*) divides the number in Register by 2

(*c*) multiples B by A

(*d*) adds A and B

85. The larger the RAM of a computer, the faster is its speed, since it eliminates

(*a*) need for ROM

(*b*) need for external memory

(*c*) frequent disk I/Os

(*d*) need for a data-wide path

86. DB, DW and DD directives are used to place data in particular locations or to simply allocate space without preassigning anything to space. The DW and DD directions are used to generate

(*a*) offsets

(*b*) full address of variables

(*c*) full address of labels

(*d*) offsets or full address of labels and variables

87. In a 8085 microprocessor, the following sequence of instructions is executed :

```
STC
CMC
MOVE A,B
RAL
MOVE B,A
```

After the last instruction, the output will

(*a*) rotate the contents of the accumulator and store it in B

(*b*) get the contents of B register into accumulator and rotate it to left by one bit

(*c*) double contents of B register

(*d*) manipulate carry in A and B

88. The highest priority in 8085 microprocessor system is

(*a*) RST 7.5 (*b*) RST 6.5

(*c*) INTR (*d*) TRAP

89. In 8085 microprocessor system, the direct addressing instruction is

(*a*) MOV A, B (*b*) MOVB, OAH

(*c*) MOV C, M (*d*) STA addr

90. Each instruction in an assembly program has the following fields:

1. Label field
2. Mnemonic field
3. Operand field
4. Comment field

The correct sequence/order of these fields is :

(*a*) 1, 2, 3, 4 (*b*) 1, 2, 4, 3

(*c*) 2, 1, 3, 4 (*d*) 2, 1, 4, 3

91. An interrupt in which the external device supplies its address as well as the interrupt request, is called

(*a*) vectored interrupt

(*b*) maskable interrupt

(*c*) polled interrupt

(*d*) non-maskable interrupt

92. In 8085 microprocessor, the value of the most significant bit of the result following the execution of any arithmetic or Boolean instruction is stored in the

(*a*) carry status flag

(*b*) auxiliary carry status flag

(*c*) sign status flag

(*d*) zero status flag

93. Consider the following statements :

1. A total of about one million bytes can be directly addressed by the 8086 microprocessor.
2. The 8086 has thirteen 16-bit registers.
3. The 8086 has eight flags.
4. Compared to 8086 the 80286 provides a higher degree of memory protection

Of these statements

(*a*) 2, 3 and 4 are correct

(*b*) 1, 3 and 4 are correct

(*c*) 1, 2 and 4 are correct

(*d*) 1, 2 and 3 are correct

94. Following program is written for an 8085 microprocessor to add two bytes located at memory addresses 1FFE and 1FFF

```
LXI H, 1FFF
MOV     B, M
INR     L
MOV     A,M
ADD     B
INR     L
MOV     M, A
XOR     A
```

On completion of the execution of the program, the result of addition is found

(*a*) in the register A

(*b*) at the memory address 1000

(*c*) at the memory address 1F00

(*d*) at the memory address 2000

95. A memory system has a total of 8 memory chips, each with 12 address lines and 8 data lines. Total size of the memory system is

(*a*) 6 kbytes

(*b*) 32 kbytes

(*c*) 48 kbytes

(*d*) 64 kbytes

96. When a program is being executed in an 8085 microprocessor, its Program Counter contains

(*a*) number of instructions in the current program that have already been executed

(*b*) total number of instructions in the program being executed

(*c*) memory address of the instruction that is being currently executed

(*d*) memory address of the instruction that is to be executed next

EXERCISE – II

1. Which of the following actions detect locations, and remove mistakes from a programme routine?
   (*a*) Erase (*b*) Debug
   (*c*) Diagnose (*d*) Emulate **[GATE]**
2. Direct–Memory Access Channel (DMA) facilitates data to move into and out of the system
   (*a*) on first-come first-serve basis
   (*b*) with equal time delay
   (*c*) without sub – routine
   (*d*) without programme intervention **[NTPC]**
3. Every processor must necessarily have
   (*a*) data bus
   (*b*) data bus and address bus
   (*c*) control bus
   (*d*) data bus, a control bus and an address bus **[GATE]**
4. A microprocessor with a 12–bit address bus will be able to access
   (*a*) 1 kilobyte of mcmory
   (*b*) 4 kilobytes of memory
   (*c*) 8 kilobytes of memory
   (*d*) 0.4 kilobytes of memory **[IOCL]**
5. Which of t he following flag conditions are not available in 8085 processor?
   (*a*) Zero flag
   (*b*) Parity flag
   (*c*) Overflow flag
   (*d*) Auxiliary carry flag **[NTPC]**
6. The frequency of the driving network connected between pin 1 and 2 of a 8085 chip must be
   (*a*) twice of the desired clock frequency
   (*b*) equal to the desired clock frequency
   (*c*) four times the desired clock frequency
   (*d*) none of these **[BHEL]**
7. READY signal in 8085 is useful when the CPU communicates with
   (*a*) slow peripheral device
   (*b*) fast peripheral device
   (*c*) DMA controller chip
   (*d*) PPI chip **[GATE]**
8. A DAD H instruction is same as
   (*a*) shifting each bit one position to the left
   (*b*) shifting each bit one position to the right
   (*c*) shifting each bit one position to the left with a zero inserted in 1st position
   (*d*) shifting each bit one position to the right with a zero inserted in 1st position
9. Return from a subroutine is affected by
   (*a*) a jump instruction
   (*b*) an RST instruction
   (*c*) a RET instruction
   (*d*) a hardware interrupt signal **[GATE]**
10. PSW stands for
    (*a*) accumulator contents
    (*b*) flag byte
    (*c*) accumulator and the flag byte
    (*d*) accumulator and temporary register byte **[NTPC]**
11. During a DMA transfer, the processor (check the incorrect statement)
    (*a*) continues its normal operations
    (*b*) suspends its normal operations
    (*c*) needs to initiate read (write) command
    (*d*) needs to check if the input/output device is ready for data transfer **[NTPC]**
12. In 8085, interrupts excepts TRAP are disabled (check the incorrect statement) by
    (*a*) a DI instruction
    (*b*) a system reset
    (*c*) acknowledgement of a previous interrupt
    (*d*) none of these **[BHEL]**
13. A USART chip provides
    (*a*) half duplex operation
    (*b*) full duplex operation
    (*c*) duplex operation
    (*d*) full duplex operation but cannot work in asynchronous mode **[BHEL]**
14. A high on RESET OUT line signifies that
    (*a*) all the registers of the CPU are being reset
    (*b*) all the registers and counters are being reset
    (*c*) all the registers and counters are being reset and in addition this signal can be used to reset external support chips
    (*d*) processing can begin when this signal goes high **[NTPC]**
15. Normally a microprocessor cycles between
    (*a*) Fetch and Halt states
    (*b*) Fetch and Interrupt states
    (*c*) Fetch and Execute states
    (*d*) Halt and Execute states **[NTPC]**
16. After completing the execution, microprocessor return to
    (*a*) Halt state (*b*) Fetch state
    (*c*) Execute state (*d*) Interrupt state **[ES]**

17. The chip select access time for reading ROM contents is
   (*a*) the delay between application of proper chip select signal and the stable output address
   (*b*) the delay between the previous valid output data and the next change in address
   (*c*) the time for which the output data remains valid when the device is no longer selected
   (*d*) maximum time for which the valid address can be changed **[NTPC]**

18. The delect to output float time is
   (*a*) the delay between application of proper chip select signal and the stable output data
   (*b*) the delay between the previous valid output data and the next change in address
   (*c*) the time for which the output data remains valid when the device is no longer selected
   (*d*) maximum time for which the valid address can be changed **[GATE]**

19. A microprocessor, on arrival of RESET signal returns, from HALT state to
   (*a*) execute (*b*) fetch
   (*c*) interrupt (*d*) none of these **[ES]**

20. The decoding circuit has been used to generate active low chip. Signal for a microprocessor peripheral (address line are designated as A 0 to A7 for I/O address)

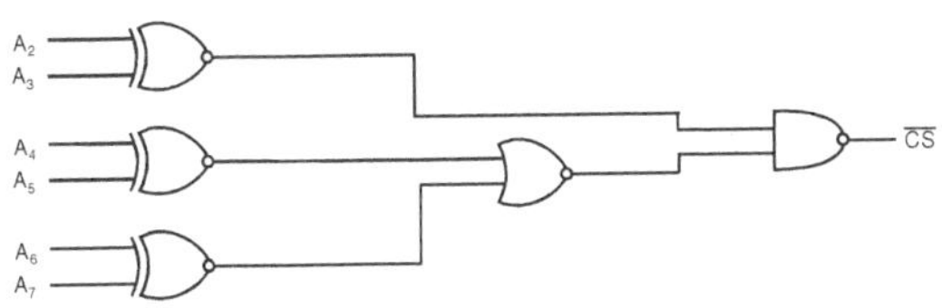

   (*a*) 60H to 63 H
   (*b*) A 4 to A 7 H
   (*c*) 71H to 73 H
   (*d*) 70 H to 73 H **[NTPC]**

21. When a CPU is interrupted, it
   (*a*) stops execution of instructions
   (*b*) acknowledges interrupt and branches to a subroutine
   (*c*) acknowledges interrupt and continues
   (*d*) acknowledges interrupt and waits for the next instruction from the interrupting device **[NTPC]**

22. An 'Assembler' for a microprocessor is used for
   (*a*) assembly of processors in a production line
   (*b*) creation of new programmes using different modules
   (*c*) translation of a program from assembly language to machine language
   (*d*) translation of a higher level language into English text **[IOCL]**

23. A single instruction to clear the lower four bits of the accumulator in 8085 assembly language is
   (*a*) XRI OFH (*b*) ANI FOH
   (*c*) XRI FOH (*d*) ANI OFH **[NTPC]**

24. Which of the following statements is true?
   (*a*) ROM is a Read/ Write memory
   (*b*) PC points to the last instruction that was executed
   (*c*) Stack works on the principle of LIFO
   (*d*) All instructions affect the flags **[IOCL]**

25. In a virtual memory system, the address space specified by the address lines of the CCPU must be .... than the physical memory size and .... than the secondary storage size.
   (*a*) smaller, smaller (*b*) smaller, larger
   (*c*) larger, smaller (*d*) larger, larger **[NTPC]**

26. An example of a spooled device is
   (*a*) line printer used to print the output of a number of jobs
   (*b*) terminal used to enter input data to a running program
   (*c*) secondary storage device in a virtual memory system
   (*d*) graphic display device **[GATE]**

27. A ROM is used to store the table for multiplication of two 8-bit unsigned integers. The size of ROM required is
   (*a*) 256×16 (*b*) $16K \times 8$
   (*c*) $4K \times 16$ (*d*) $64K \times 16$ **[NTPC]**

28. Number of machine cycles required for RET instruction in 8085 microprocessor is
   (*a*) 1 (*b*) 2
   (*c*) 3 (*d*) 5 **[IOCL]**

29. A 100 K byte memory is managed using variable partitions but no compaction. It currently has two partitions of sizes 200 K bytes and 260 K bytes respectively. The smallest allocation request in K bytes that could be denied is for
   (*a*) 151 (*b*) 181
   (*c*) 231 (*d*) 541 **[NTPC]**

30. Four jobs to be executed on a single processor system arrive at time 0+ in the order A, B, D, D. Their burst CPU time requirements are 4, 1, 8, 1, time units respectively. The completion time of a under round robin scheduling with time slice of one time unit is
   (*a*) 10 (*b*) 4
   (*c*) 8 (*d*) 9 **[NTPC]**

31. The advantage of using segment registers in 8086 and above microprocessors are that they
   (*a*) allow the memory capacity to be 1 megabyte even though the addresses associated with the individual instructions are only 16 bits wide

(*b*) allow the instruction, data or stack portion of a program to be more than 64K bytes long by using more than one code, data or stack segment

(*c*) facilitate the use of separate memory area for a program, its data and the stack

(*d*) all of these **[GATE]**

32. The set of commands which give directions to the assembler during the assembly process but are not translated into machine instructions are called

(*a*) mnemonics (*b*) directives

(*c*) identifiers (*d*) operands **[IOCL]**

33. Intel 80286 microprocessor was designed to serve as the CPU in a multitasking operating systems as it has

(*a*) 8 MHz operating frequency

(*b*) real address mode to support lower microprocessors of the same family

(*c*) protection and virtual memory capability

(*d*) signals to interface with co-processor **[BHEL]**

34. Intel 80486 offers high performance than 80386 because

(*a*) no chip memory management is present on 80486

(*b*) multitasking support is provided by 80486 and not by 80386

(*c*) 80386 has no chip cache memory available

(*d*) all of these **[NTPC]**

35. RISC is characterised by

(*a*) hardwire control design with no micro codes

(*b*) fixed instruction format

(*c*) executing most of the instructions in a signal clock cycle

(*d*) all of these **[GATE]**

36. The theoretical dividing line between Reduced Instruction Set Computing (RISC) chips and Complex Instruction Set Computing (CISC) chips is

(*a*) number of pins in the chip

(*b*) number of address and data lines

(*c*) instruction execution rate to be one instruction per clock cycle

(*d*) none of these **[BHEL]**

37. The microprocessor contains ROM chip which contains

(*a*) control function

(*b*) arithmetic functions

(*c*) instructions to execute data

(*d*) memory functions **[NTPC]**

38. The process of causing an unplanned branching operation to occur, usually initiated by external system is called

(*a*) debugging (*b*) masking

(*c*) interrupt (*d*) iteration **[IOCL]**

39. An instruction register is storage for

(*a*) location of data in memory

(*b*) location of instruction in memory

(*c*) binary code for the operation to be performed

(*d*) address of the next instruction to be executed **[ES]**

40. A program counter is a storage register for

(*a*) location of data in memory

(*b*) location of instruction in memory

(*c*) binary code for the operation to be performed

(*d*) address of the next instruction to be executed **[ES]**

41. Stack pointer is a register which comes into use

(*a*) whenever a data is read from the memory

(*b*) whenever a data is written into the memory

(*c*) whenever the output variable is sent out of the CPU

(*d*) whenever an interrupt or high priority call comes from external devices **[NTPC]**

42. In microprocessor architecture, flag indicates

(*a*) number of microprocessor

(*b*) name of the manufacturer

(*c*) internal status of the CPU

(*d*) bit-size of the microprocessor **[NTPC]**

43. A microprocessor is a minimum combination of

(*a*) μP and clock

(*b*) μP, clock and ROMs

(*c*) μP, clock, RAMs and ROMs

(*d*) μP, clock, RAM, ROM, PIA and ACIA **[IOCL]**

44. ACK indicates reception of

(*a*) correct data

(*b*) incorrect data

(*c*) insufficient data

(*d*) sufficient data **[IOCL]**

45. Accumulation is a device which

(*a*) stores a number

(*b*) adds two numbers

(*c*) adds and stores previously stored number to another number

(*d*) none of these **[BHEL]**

46. Address usually is

(*a*) alphabetical (*b*) numerical

(*c*) alpha-numerical (*d*) none of these **[BHEL]**

47. The basic elements of a microprocessor are

(*a*) ALU, memory

(*b*) ALU, control unit

(*c*) ALU, memory, I/O device

(*d*) ALU, control unit, memory **[NTPC]**

48. Flag is a character for
(*a*) identification of a word
(*b*) occurance of some condition
(*c*) marking a tagging
(*d*) all of these **[NTPC]**

49. Mnemonic symbols are used
(*a*) to denote address
(*b*) to employ hamming code
(*c*) to denote error
(*d*) to assist human memory **[BHEL]**

50. What is the state of the signal at X?

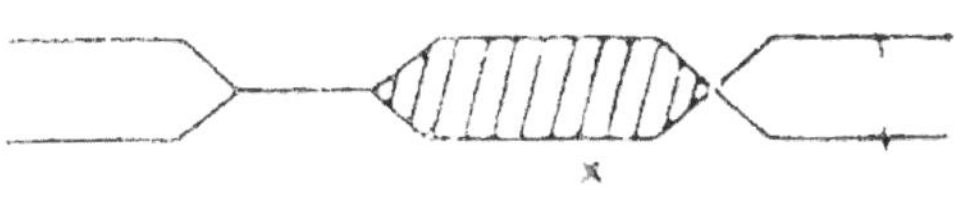

(*a*) Don't know (*b*) Floating
(*c*) Changing their states (*d*) All high **[GATE]**

51. An interrupt in which the external device supplies its address as well as the interrupt request is called
(*a*) vectored interrupt
(*b*) maskable interrupt
(*c*) non-maskable interrupt
(*d*) designated interrupt **[IOCL]**

52. When used with I/O devices, the term intelligent implies
(*a*) a colour output capability
(*b*) speech processing capability
(*c*) high speed printing capability
(*d*) features to support offline and online tasks **[NTPC]**

53. A microprocessor is also referred to as
(*a*) the chip that does some calculations for the computer
(*b*) the computer on a chip
(*c*) the chip that is responsible for data transfer
(*d*) none of these **[BHEL]**

54. A microprocessor is called an n-bit microprocessor depending on
(*a*) register's length
(*b*) size of internal data bus
(*c*) size of external data bus
(*d*) none of these **[NTPC]**

55. Two operands can be checked for equality using
(*a*) OR-operation (*b*) AND-operation
(*c*) X-OR operation (*d*) none of these **[ES]**

56. Setting contents of a register to zero can be efficiently done by
(*a*) Mov immediate instruction using zero as immediate data
(*b*) AND immediate instruction using zero as immediate data
(*c*) XORing register with itself
(*d*) none of these **[NTPC]**

57. Stack memory is used
(*a*) to provide additional memory to the base memory
(*b*) to save return addresses of a subroutine
(*c*) to save the status of the microprocessor
(*d*) none of these **[GATE]**

58. 8085 microprocessor is a
(*a*) zero address microprocessor
(*b*) one address microprocessor
(*c*) two address microprocessor
(*d*) none of these **[BHEL]**

59. SHIFT LEFT instruction causes all bits shifted from one position to the left with rightmost bit set to zero. The effect is to
(*a*) multiply by 2
(*b*) divide by 2
(*c*) SET the most significant bit
(*d*) none of these **[IOCL]**

60. The selection of a microprocessor for an application is done keeping in mind
(*a*) speed compatibility of microprocessor with peripherals
(*b*) the time critical behaviour of the application
(*c*) the size of program required to implement certain functions
(*d*) all of these **[NTPC]**

61. Which of the following signal is used when a microprocessor wants to address the memory?
(*a*) $\frac{I_O}{\overline{M}}$ (*b*) Status signals
(*c*) ALE (d) HOLD and HLDA **[ES]**

62. Which of the following signal is used when a peripheral device request the microprocess to have a DMA operation?
(*a*) $\frac{I_O}{\overline{M}}$ (*b*) READY
(*c*) HOLD and HLDA (*d*) $\overline{RD}$ and $\overline{WR}$ **[NTPC]**

63. Program Counter is used to
(*a*) store address of the next instruction to be executed
(*b*) store temporary data to be used in arithmetic operations
(*c*) store the status of the microprocessor
(*d*) none of these **[GATE]**

64. The synchronization between microprocessor and memory is done by
(*a*) ALE signal
(*b*) HOLD signal
(*c*) READY signal
(*d*) none of these **[NTPC]**

65. Intel 8080A and 8085A differ in
   (*a*) number of address lines
   (*b*) number of data signal
   (*c*) instruction set
   (*d*) number of interrupt [GATE]

66. In 8085 Program Status Word consists of
   (*a*) accumulator contents
   (*b*) flags
   (*c*) both accumulator and flag
   (*d*) status bits [NTPC]

67. Intel 8080A and Motorola 6800 microprocessor differ in
   (*a*) number of address lines
   (*b*) number of data lines
   (*c*) instruction set
   (*d*) instruction set and operating frequency [ES]

68. As compared to 16-bit microprocessor, 8 bit microprocessor are limited in
   (*a*) speed
   (*b*) directly addressable memory
   (*c*) data handling capability
   (*d*) all of these [BHEL]

69. WAIT states are used
   (*a*) to give slow devices additional time to put out valid data
   (*b*) to insert a deliberate delay
   (*c*) during I/O operation
   (*d*) none of these [BHEL]

70. Intel 8085A and 8086A differ in
   (*a*) number of address lines
   (*b*) number of data lines
   (*c*) operating frequency
   (*d*) all of these [NTPC]

71. The register which contains the instruction to be executed is called
   (*a*) instruction register
   (*b*) index register
   (*c*) memory address register
   (*d*) memory data register [BHEL]

72. The register which keeps track of the execution of a program and which contains the memory address of the instruction currently being executed is called
   (*a*) index register
   (*b*) memory address register
   (*c*) program counter
   (*d*) instruction register [GATE]

73. The register which holds address of the location to or from which data are to be transferred is called
   (*a*) index register
   (*b*) instruction register
   (*c*) memory address register
   (*d*) memory data register [NTPC]

74. The register which contains the data to be written into or read out of the addressed location is called
   (*a*) index register
   (*b*) memory address register
   (*c*) memory data register
   (*d*) program counter [NTPC]

75. The process of fetching and executing instructions one at a time in the order of increasing addresses is known as
   (*a*) instruction execution
   (*b*) straight line sequencing
   (*c*) instruction fetching
   (*d*) random sequencing [GATE]

76. What is the minimum number of operations required in a microprocessor with 8 data pins to read a 32-bit word?
   (*a*) 1 (*b*) 2
   (*c*) 4 (*d*) 8 [BHEL]

77. The flow and timing of data to and from the microprocessor is regulated by
   (*a*) control pins (*b*) address pins
   (*c*) data pins (*d*) power pins [NTPC]

78. A microprocessor with 12 address lines is capable of addressing
   (*a*) 1024 locations (*b*) 2028 locations
   (*c*) 4096 locations (*d*) 64 K locations [IOCL]

79. The stack pointer in the 8085 microprocessor is a
   (*a*) 16 bit register which points to stack memory locations
   (*b*) 16 bit accumulator
   (*c*) memory location in the stack
   (*d*) flag register used for the stack [IOCL]

80. How many address lines are needed to address each memory location in a 2048 × 4 memory chip?
   (*a*) 10 (*b*) 11
   (*c*) 12 (*d*) 18 [NTPC]

81. Which of the following enables peripherals to pass a signal down the bus to the next device on the bus during polling of the device?
   (*a*) DMA (*b*) interrupt vectoring
   (*c*) daisy chain (*d*) cycle stealing [NTPC]

82. An opcode
   (*a*) translates a mnemonic (*b*) instructs the CPU
   (*c*) stores data (*d*) all of these [BHEL]

83. A triple bus architecture based processor has
(*a*) 3 control buses
(*b*) 3 address buses
(*c*) 3 data buses
(*d*) 2 data buses and a control bus **[NTPC]**

84. Which of the following is not an 8-bit microprocessor?
(*a*) INTEL 8085 (*b*) MOTOROLA 6800
(*c*) ZILOG 280 (*d*) FAIR CHILD 9440 **[ES]**

85. The register whose contents may be added to or subtracted from the operand address prior to or during the execution of an instruction is called
(*a*) index register (*b*) control register
(*c*) address register (*d*) none of these **[GATE]**

86. Which of the following cycle is required to fetch and execute information?
(*a*) Clock cycle (*b*) Tri cycle
(*c*) Introduction cycle (*d*) Memory cycle **[NTPC]**

87. The three buses associated with three-bus system are I/O bus, memory bus and
(*a*) address bus
(*b*) unibus
(*c*) direct memory access bus
(*d*) data bus **[BHEL]**

88. The bus which is used to transfer data from main memory to peripheral device is
(*a*) data bus (*b*) input bus
(*c*) DMA bus (*d*) output bus **[NTPC]**

89. Which of the following statements is not applicable to serial transmission of data?
(*a*) One bit at a time
(*b*) Faster method of transmission
(*c*) LSB transferred first
(*d*) Only one wire used **[IOCL]**

90. An interrupt which can be temporarily ignored by the counter is called
(*a*) vectored interrupt
(*b*) non-maskable interrupt
(*c*) maskable interrupt
(*d*) low priority interrupt **[NTPC]**

91. In microprocessor
(*a*) program is stored in memory and data is stored in registers
(*b*) program is stored in registers and data is stored in memory
(*c*) both program and data are stored in memory
(*d*) both program and data are stored in registers **[ES]**

92. A microprocessor contains
(*a*) most of the control and arithmetic logic functions of a computer
(*b*) most of the RAM
(*c*) most of the ROM
(*d*) peripheral drivers **[NTPC]**

93. A PC in a micro-computer
(*a*) counts the number of instructions executed in a run
(*b*) counts the number of programs run after starting
(*c*) counts the points to the next executable instruction
(*d*) points to the present instruction being executed **[ES]**

94. In a 8-bit microcomputer having 8K bytes of RAM memory, the length of SP will be
(*a*) 5 (*b*) 8
(*c*) 11 (*d*) 13 **[IOCL]**

95. An instruction cycle is made up of
(*a*) one or more execute cycles
(*b*) one or more fetch cycles
(*c*) one opcode and one execute cycle
(*d*) none of these **[IOCL]**

96. The number of minimum clock cycles in a machine cycle for 8085 are
(*a*) 1 (*b*) 2
(*c*) 3 (*d*) 5 **[NTPC]**

97. In a 8-bit microprocessor, the fetch cycle required to fetch a 8 byte instruction will be
(*a*) 1
(*b*) 2
(*c*) 3
(*d*) depends on computer design **[BHEL]**

98. The maximum integer which can be stored on an 8-bit accumulator is
(*a*) 2 (*b*) 200
(*c*) 224 (*d*) 255 **[BHEL]**

99. The address bus of inlet 8085 is 16 bit wide and hence the memory which can be accessed by this address bus is
(*a*) 112 KB (*b*) 4 KB
(*c*) 16 KB (*d*) 64 KB **[NTPC]**

100. Microprocessor can be
(*a*) non-programmable
(*b*) macro-programmable
(*c*) micro-programmable
(*d*) all of these **[ES]**

ANSWERS

EXERCISE – I

1. (*a*)	**2.** (*c*)	**3.** (*c*)	**4.** (*d*)	**5.** (*d*)	**6.** (*b*)	**7.** (**c**)	**8.** (*d*)	**9.** (*c*)	**10.** (*a*)
11. (*a*)	**12.** (*a*)	**13.** (*a*)	**14.** (*d*)	**15.** (*d*)	**16.** (*c*)	**17.** (*a*)	**18.** (*c*)	**19.** (*c*)	**20.** (*a*)
21. (*a*)	**22.** (*c*)	**23.** (*d*)	**24.** (*a*)	**25.** (*d*)	**26.** (*a*)	**27.** (*b*)	**28.** (*c*)	**29.** (*a*)	**30.** (*a*)
31. (*b*)	**32.** (*b*)	**33.** (*c*)	**34.** (*b*)	**35.** (*d*)	**36.** (*c*)	**37.** (*c*)	**38.** (*c*)	**39.** (*d*)	**40.** (*c*)
41. (*c*)	**42.** (*c*)	**43.** (*a*)	**44.** (*b*)	**45.** (*d*)	**46.** (*d*)	**47.** (*c*)	**48.** (*d*)	**49.** (*c*)	**50.** (*d*)
51. (*c*)	**52.** (*d*)	**53.** (*d*)	**54.** (*d*)	**55.** (*c*)	**56.** (*d*)	**57.** (*c*)	**58.** (*b*)	**59.** (*c*)	**60.** (*b*)
61. (*b*)	**62.** (*c*)	**63.** (*a*)	**64.** (*a*)	**65.** (*b*)	**66.** (*a*)	**67.** (*b*)	**68.** (*c*)	**69.** (*b*)	**70.** (*d*)
71. (*c*)	**72.** (*d*)	**73.** (*d*)	**74.** (*d*)	**75.** (*a*)	**76.** (*a*)	**77.** (*d*)	**78.** (*b*)	**79.** (*c*)	**80.** (*a*)
81. (*a*)	**82.** (*d*)	**83.** (*d*)	**84.** (*b*)	**85.** (*c*)	**86.** (*c*)	**87.** (*c*)	**88.** (*d*)	**89.** (*d*)	**90.** (*a*)
91. (*c*)	**92.** (*c*)	**93.** (*b*)	**94.** (*d*)	**95.** (*b*)	**96.** (*d*)				

EXERCISE – II

1. (*b*)	**2.** (*d*)	**3.** (*d*)	**4.** (*a*)	**5.** (*c*)	**6.** (*a*)	**7.** (*a*)	**8.** (*c*)	**9.** (*c*)	**10.** (*c*)
11. (*a*)	**12.** (*d*)	**13.** (*b*)	**14.** (*c*)	**15.** (*c*)	**16.** (*b*)	**17.** (*a*)	**18.** (*c*)	**19.** (*b*)	**20.** (*a*)
21. (*b*)	**22.** (*c*)	**23.** (*b*)	**24.** (*c*)	**25.** (*b*)	**26.** (*a*)	**27.** (*d*)	**28.** (*c*)	**29.** (*b*)	**30.** (*d*)
31. (*d*)	**32.** (*b*)	**33.** (*c*)	**34.** (*c*)	**35.** (*d*)	**36.** (*c*)	**37.** (*c*)	**38.** (*c*)	**39.** (*c*)	**40.** (*d*)
41. (*d*)	**42.** (*c*)	**43.** (*d*)	**44.** (*a*)	**45.** (*c*)	**46.** (*b*)	**47.** (*b*)	**48.** (*d*)	**49.** (*d*)	**50.** (*b*)
51. (*a*)	**52.** (*d*)	**53.** (*b*)	**54.** (*b*)	**55.** (*c*)	**56.** (*c*)	**57.** (*b*)	**58.** (*b*)	**59.** (*a*)	**60.** (*d*)
61. (*c*)	**62.** (*c*)	**63.** (*a*)	**64.** (*c*)	**65.** (*d*)	**66.** (*c*)	**67.** (*d*)	**68.** (*d*)	**69.** (*a*)	**70.** (*d*)
71. (*b*)	**72.** (*c*)	**73.** (*c*)	**74.** (*c*)	**75.** (*b*)	**76.** (*c*)	**77.** (*a*)	**78.** (*c*)	**79.** (*a*)	**80.** (*b*)
81. (*c*)	**82.** (*b*)	**83.** (*c*)	**84.** (*d*)	**85.** (*a*)	**86.** (*c*)	**87.** (*c*)	**88.** (*c*)	**89.** (*b*)	**90.** (*c*)
91. (*a*)	**92.** (*a*)	**93.** (*c*)	**94.** (*d*)	**95.** (*a*)	**96.** (*c*)	**97.** (*d*)	**98.** (*d*)	**99.** (*d*)	**100.** (*d*)

EXPLANATIONS

EXERCISE – I

3.

AX = c

BX = b

SUB AX BX = $c - b$

BX = a

ADD, AX, BX = 0101

c
b
a

4.

8085 – 8086 Register

A – AL

H – BH

L – BL

B – CH

57. An I/O processor controls the flow of information between two I/O devices.

DMA controls flow between memory and I/O devices.

58. On receiving an interrupt from an I/O device, the CPU branches off to the interrupt service routine after completion of the current instruction.

59. In microprocessor based systems DMA facility is required to increase the speed of data transfer between memory and the I/O devices.

61. RETURN instruction in subroutine can be conditional or unconditional.

62. H & L registers known as *primary data* pointers in 8085 µp.

63. READY signal is 8085 is useful when the CPU communicates with a slow peripheral device.

64. ANI ⇒ Logical AND is operated between accumulater and data.

From ANI F0H

Let accumulater

has → x x x x x x x x

byte 1 1 1 1 0 0 0 0

Accumulater x x x x 0 0 0 0

output